VERTEBRATE ICHNOLOGY

VERTEBRATE ICHNOLOGY

TRACKS AND TRAILS, CONSUMPTION, DIGGING AND REPRODUCTION, GEOCONSERVATION

Edited by

Spencer G. Lucas

New Mexico Museum of Natural History, Albuquerque, NM, United States

Adrian P. Hunt

Flying Heritage and Combat Armor Museum, Everett, WA, United States

Hendrik Klein

Independent Researcher, Saurierwelt Paläontologisches Museum, Neumarkt, Germany

ELSEVIER

Elsevier
Radarweg 29, PO Box 211, 1000 AE Amsterdam, Netherlands
125 London Wall, London EC2Y 5AS, United Kingdom
50 Hampshire Street, 5th Floor, Cambridge, MA 02139, United States

ISBN: 978-0-443-38351-9

For information on all Elsevier publications visit our website at
https://www.elsevier.com/books-and-journals

Publisher: Candice Janco
Acquisitions Editor: Peter Llewellyn
Editorial Project Manager: Joshua Mearns
Production Project Manager: Kumar Anbazhagan
Cover Designer: Christian Bilbow

Typeset by TNQ Tech

Contents

6. Vertebrate dentalites

Adrian P. Hunt and Spencer G. Lucas

7. Regurgitalites

Adrian P. Hunt and Spencer G. Lucas

8. Consumulites

Adrian P. Hunt and Spencer G. Lucas

9. Vertebrate coprolites

Adrian P. Hunt and Spencer G. Lucas

10. Gastroliths

Spencer G. Lucas and Adrian P. Hunt

11. Pabulites, digestilites, micturalites and other trace fossil evidence of predation and consumption

Adrian P. Hunt and Spencer G. Lucas

III

The ichnology of digging and reproduction

12. Vertebrate burrows

Ricardo N. Melchor and M. Cristina Cardonatto

13. The ichnology of reproduction in fossil tetrapods

Lida Xing, Qiyan Chen, Donghao Wang and Jens N. Lallensack

IV

The ichnology of geoconservation

14. Vertebrate ichnosites as geological heritage: A global synthesis of the best practices in geoconservation

Carlos Neto de Carvalho, Adrian P. Hunt, Luis Alcalá, Tobias Fischer, Charles Helm, Kevin Page, Andrea Baucon, Evelyn Kustatscher, Girolamo Lo Russo, Luca Arena, Karen Moreno and Lida Xing

Contributors

Luis Alcalá Parque de las Ciencias, Granada, Spain

Luca Arena DISTAV, Università di Genova, Genova, Italy

Andrea Baucon DISTAV, Università di Genova, Genova, Italy

Zain Belaústegui Departament de Dinàmica de la Terra i de L'Oceà, Facultat de Ciències de la Terra, Institut de Recerca de la Biodiversitat, Universitat de Barcelona, Barcelona, Spain

Aaron B. Camens College of Science and Engineering, Flinders University, Adelaide, SA, Australia

M. Cristina Cardonatto Departamento de Geología, Universidad Nacional de La Pampa, Santa Rosa, La Pampa, Argentina

Qiyan Chen School of the Earth Sciences and Resources, China University of Geosciences (Beijing), Beijing, China

Paula Dentzien-Dias Departamento de Paleontologia e Estratigrafia, Geosciences Institute, Federal University of Rio Grande Do Sul, Rio Grande Do Sul, Brazil

Silverio Feola Instituto de Ciencias de la Tierra y Ambientales de La Pampa, Universidad Nacional de La Pampa and Consejo Nacional de Investigaciones Científicas y Técnicas, Santa Rosa, La Pampa, Argentina

Tobias Fischer Natur- und UNESCO Geopark TERRA.vita, Osnabrück, Germany

Donald Goldstein Department of Earth Sciences, University of Connecticut, Storrs, CT, United States

Charles Helm African Centre for Coastal Palaeoscience, Nelson Mandela University, Gqeberha, South Africa

Adrian P. Hunt Flying Heritage and Combat Armor Museum, Everett, WA, United States

Hendrik Klein Saurierwelt Paläontologisches Museum, Neumarkt, Germany

Evelyn Kustatscher Museum of Nature South Tyrol, Bolzano, Italy

Jens N. Lallensack Departamento de Ciência da Computação, Universidade Federal de Minas Gerais, Belo Horizonte, Brazil

Weronika Łaska Institute of Evolutionary Biology, Faculty of Biology, Biological and Chemical Research Centre, University of Warsaw, Warsaw, Poland

Girolamo Lo Russo Museo di Storia Naturale di Piacenza, Piacenza, Italy

Spencer G. Lucas New Mexico Museum of Natural History, Albuquerque, NM, United States

Ricardo N. Melchor Departamento de Geología, Universidad Nacional de La Pampa, Santa Rosa, La Pampa, Argentina; Instituto de Ciencias de la Tierra y Ambientales de La Pampa, Universidad Nacional de La Pampa and Consejo Nacional de Investigaciones Científicas y Técnicas, Santa Rosa, La Pampa, Argentina

Nicholas J. Minter School of the Environment and Life Sciences, University of Portsmouth, Portsmouth, United Kingdom

Karen Moreno Universidad Austral de Chile, Facultad de Ciencias, Instituto de Ciencias de la Tierra, Valdivia, Chile

Fernando Muñiz Department of Crystallography, Mineralogy and Agricultural Chemistry, Faculty of Chemistry, University of Seville, Seville, Spain

Carlos Neto de Carvalho Geology Office of the Municipality of Idanha-a-Nova, Naturtejo UNESCO Global Geopark, Idanha-a-Nova, Portugal; Earth Surface Processes Group (RG2) – Instituto D. Luiz, Faculdade de Ciências da Universidade de Lisboa, Lisbon, Portugal

Kevin Page Geodiversity & Heritage, Thornedges, Longbarn, Sandford, Crediton, Devon, United Kingdom; Camborne School of Mines (University of Exeter), Penryn Campus, Penryn, United Kingdom

Donghao Wang School of the Earth Sciences and Resources, China University of Geosciences (Beijing), Beijing, China

Lida Xing Frontiers Science Center for Deep-time Digital Earth, China University of Geosciences (Beijing), Beijing, China; School of the Earth Sciences and Resources, China University of Geosciences (Beijing), Beijing, China; Dinosaur Odyssey Science Museum, Guangxi, China; Department of Natural History, Tyrolean State Museums, Hall in Tirol, Austria

PART I

The ichnology of locomotion

CHAPTER

1

The Quaternary record of vertebrate tracks and trackways

Carlos Neto de Carvalho[1], *Charles Helm*[2], *Ricardo N. Melchor*[3], *Silverio Feola*[3], *Donald Goldstein*[4] and *Aaron B. Camens*[5]

[1]Geology Office of the Municipality of Idanha-a-Nova, Naturtejo UNESCO Global Geopark, Idanha-a-Nova, Portugal; [2]African Centre for Coastal Palaeoscience, Nelson Mandela University, Gqeberha, South Africa; [3]Instituto de Ciencias de la Tierra y Ambientales de La Pampa, Universidad Nacional de La Pampa and Consejo Nacional de Investigaciones Científicas y Técnicas, Santa Rosa, La Pampa, Argentina; [4]Department of Earth Sciences, University of Connecticut, Storrs, CT, United States; [5]College of Science and Engineering, Flinders University, Adelaide, SA, Australia

1. Introduction

The Quaternary Period or the system of deposits laid down during it represents a time interval with major consequences for the evolution of life. It spans from 2.58 million years ago to the present, according to the International Commission on Stratigraphy (ICS). As of 2023, the Quaternary is divided into two epochs: the Pleistocene (2.58 Ma to 11.7 ka) and the Holocene (11.7 ka to today). A third epoch, the Anthropocene, has recently been proposed, but it is not officially recognized by the ICS. The Quaternary is marked by numerous extinctions related to climate changes and the expansion of hominins (with implications for the exploitation of habitats as much as resources) during the last glacial period. Although the Quaternary geological record is preserved in greater detail than that for earlier periods, due to its young age the ichnological record only benefits from tectonics and especially diagenesis in exceptional cases: the habitats where vertebrate traces can be found are mostly related to coastal (marine, lake) and volcanic-related sedimentary environments, fluvial terraces and caves. In these environments, active erosional agents are needed to expose track-bearing surfaces (except in caves, which

Vertebrate Ichnology
https://doi.org/10.1016/B978-0-443-38351-9.00004-9

require the opposite), and exposure is commonly brief before the same agents erase the ichnological data. For these reasons, the ichnology of the Quaternary represents a challenge.

The study of the Quaternary vertebrate tracks can offer an insight into group demographics, behavioral ecology, palaeogeography and human-animal interactions (Wiseman et al., 2022). These are especially important for species or groups of animals that are now extinct. Lockley and Harris (2010, p. 36–37) discussed a fundamental ichnological principle: "The similarity between extant and extinct species decreases as temporal distance from the present increases – older organisms are generally less like extant ones than are younger ones. Therefore, the problems of determining trackmaker affinity increase with the increasing age of the fossil footprints." The Pleistocene is very close to the geologically most recent end of the time range for fossil tetrapod traces, and this allows the implementation of experimental ichnological. Neoichnology, the study of freshly made traces nowadays, involves, ideally, watching a trackmaker creating tracks in natural or artificial substrates of variable composition and under different ecological constraints and then examining the tracks (e.g., Milàn and Bromley, 2006). Alternatively, observations based on extant trackmakers can be consulted, and a trackmaker can be assigned to a species, genus, or family with varying degrees of confidence. The experience of professional trackers has only recently been applied to support ichnological studies, with interesting results (e.g., Pastoors et al., 2015; Neto de Carvalho et al., 2020c).

In contrast, there is little such capacity in paleoichnology. Not only can the trackmaker not be observed making the tracks, but it might, in fact, not be known from the body fossil record, and tracks might be the only indication of its existence. The Pleistocene, and especially the Late Pleistocene, can be viewed as a "gray area" in this regard. Many of the trackmaker taxa are extant, in forms that do not differ greatly from their Pleistocene ancestors. Modern tracking manuals are therefore valuable in identifying Pleistocene tracks and traces, something which cannot be said even for Pliocene or Miocene tracks. This approach is still valid for Middle Pleistocene deposits, and that is the main reason why most of the works in the Quaternary, especially in Africa and Oceania and dealing with hominins, are directed to identify a trackmaker without following an ichnotaxonomical approach based on morphological characters of the footprints. The exceptions are where ichnologists are involved in or coordinate the studies. These factors can lead to Pleistocene ichnology being regarded as "easy," compared to the analysis of tracks and traces from older epochs and periods. Indeed, the body fossil record and the general similarity of tracks to those of extant species are useful. However, they should not be regarded as shortcuts to trackmaker identification. A cautious, disciplined approach is required, whereby tracks and traces are first considered on their own merits and described according to their morphology, supported by a large number of examples in good preservational condition, without jumping to conclusions about trackmaker identity. The possibility, albeit remote, which the tracks were made by a species that is hitherto unknown in both the trace and body fossil records should be considered. The consequences of errors are serious, and a cautious approach, whereby tracks are analyzed first on their morphology, followed by attempts to assign them to a trackmaker if feasible, is required. While Quaternary ichnology may present challenges, it may deliver many unanticipated benefits and surprises, as we summarize in this chapter.

The focus of the present work is to present a comprehensive assessment of the ichnological studies of the Quaternary formations worldwide, with special emphasis on the ichnotaxonomy and possible producers. Although many papers have been published, especially in regional journals, since the 1870s, we have made our review supported on the seminal paper of McDonald et al. (2007) and counted on reviews made recently for the Cenozoic of continental regions or countries (Hunt and Lucas, 2007, for North America; Pillola et al., 2020, for Italy; Neto de Carvalho et al., 2016, for Portugal; Bell, 2020, for Great Britain; Rodriguez de la Rosa et al., 2004, for Mexico; Leonardi, 1994, for Latin America, Matsukawa and Shibata, 2015, for Japan), environmental contexts (Houck et al., 2009, for volcanogenic environments), or even animal groups (birds: Lockley and Harris, 2010; Lockley et al., 2021a; artiodactyls: Neto de Carvalho et al., 2022; canids: Melchor et al., 2018; camelids: Lucas and Hunt, 2007; proboscideans: Altamura and Serangeli, 2023).

2. What we have learned from the ichnology of Quaternary deposits

Consideration of the Quaternary ichnosites from Africa reveals that only in a few cases was formal ichnotaxonomy used. In a few cases, this involved the erection of new ichnogenera and ichnospecies. It appears that, in Africa, most authors do not embrace including formal ichnotaxonomic discussion in their research papers. Description of non-hominin vertebrate tracks is rare, except for South Africa, and mostly used to identify the ecological community or potential prey available for humans.

A complete and fully indexed compilation of tetrapod tracks from Latin America (originally finished in 1989) was presented by Leonardi (1994). This author presented precise and concise information on locality (maps) by country, track-bearing unit, coordinates, basin, stratigraphic information, age, paleoecological information, and ichnotaxonomy (when available). However, most of the ichnosites present in South America were found and described in the last 3 decades, especially in Argentina, where the work of the late Silvia Aramayo, Teresa Manera de Bianco, and colleagues in the latest Pleistocene beach deposits from Pehuen Co Paleoichnological Site has resulted in the description of several ichnotaxa. With 17 ichnogenera and 20 ichnospecies assigned to mammals and birds described in Pehuen Co (Aramayo et al., 2015; Feola 2022), this is the most diverse Quaternary ichnosite in the world. The Geological, Paleontological, and Archaeological Provincial Reserve Pehuen Co - Monte Hermoso is, for that reason, on the Tentative List of World Heritage Sites from UNESCO. In Argentina, the work has been more systematic, whereas, in other countries of South America, ichnological works in Quaternary deposits are in their infancy or still lacking.

Identification of Pleistocene vertebrate traces in North America dates back to at least 1882, with numerous reports from Carson City, Nevada (H.W. Harkness, Joseph LeConte, and Charles Gibbs) followed quickly in 1883 by E.D. Cope and with George Davidson's descriptions of mammoth prints. Some of the early descriptions from this site claimed possible hominin footprints, but this was dismissed by LeConte (1882) and Marsh (1883), who attributed the tracks correctly to a ground sloth. Stock (1925, 1936) elaborated on the footprints of xenarthrans, reporting that the prints at Carson City correspond in size to the sloth *Mylodon harlani* (Stock, 1925: p. 197). Fossil tracks from the Tularosa Basin (New Mexico) were reported as early as 1932 (Lucas et al., 2002) and investigated in 1981 by geologists from the New Mexico Bureau of Geology and Mineral Resources (Santucci et al., 2014). Since Donald Curry's report of Pleistocene bird tracks at Death Valley, California (Curry, 1939), numerous works have extended this research (Reynolds, 1999, 2006) and have discussed Pleistocene fossils in the nearby Anza-Borrego Desert State Park, California (Remeika, 2006).

Cave deposits yielded big cat footprints, starting with Simpson's paper on the tracks of *Panthera onca augusta* in Craighead Caverns, Tennessee (Simpson, 1941). Tracks found in Tennessee cross the boundary between the Pleistocene and Holocene. Abundant tracks of jaguars (*Panthera onca*), trapped in Jaguar Cave, Tennessee, are preserved in several passageways. The animals died before finding an exit (Robbins et al., 1981, p. 377; Watson et al., 2005, p. 25). Other cat tracks were reported from caves subsequently by Oesch (1969), Robbins et al. (1981), and Graham et al. (1996). Sinkhole deposits in South Dakota that contained mammoth tracks from the trapped animals were reported by Laury (1980), and bird tracks were also later documented (Falkingham et al., 2010).

Found in 1992, the long trail at the Brennan Mastodon Site in the state of Michigan was announced by Fisher (1994). Based on the scale of the prints, they were assigned to the more locally common mastodons instead of mammoths (Fisher, 2018). Weight estimates were made using tracks and model scaling of these animals (Hubbard et al., 2000), with applications to other proboscidean tracksites.

In 1999, an assemblage of Pleistocene tracks was discovered on a newly exposed surface in a Reservoir in Alberta (Wally's Beach), Canada (Hills and McNeil, 1999; McNeil et al., 1999a,b), which yielded several new ichnospecies (McNeil et al., 2007). The demographic study based on track measurements supported the hypothesis of a declining proboscidean population during the Late Pleistocene (McNeil et al., 2005).

Holocene localities for non-hominid trace fossils in North America are uncommon. The most famous site is, of course, Acahualinca, in Nicaragua, associated with hominin trackways. The associated non-hominin vertebrate tracks have been referred to and sometimes figured (Crawford, 1891; Williams, 1952; Lockley et al., 2008; Lucas et al., 2023). Lockley et al. (2008) were able to confirm the tracks of opossum, deer, and a bird track assigned to the ichnogenus *Anatipeda*. Lucas et al. (2023) added bison tracks from a nearby site and suggested that they indicated a later survival for bison in Central America than previously realized.

The utility of fossil tracks goes beyond the identification of the tracks and track makers. The extent of the deposits in the Tularosa Valley, N.M., USA, and Wally's Beach, Alberta, Canada, have allowed for complex studies of human inter- and intraspecific interactions (Bustos et al., 2018; Urban et al., 2018, 2019; Bennett et al., 2020), sediment deformation and biomechanics (Bennett et al., 2019), and population dynamics (McNeil et al., 2005). Vincelette (2021) and Renders and Vincelette (2023) used the drawings of McNeil et al. (2007) to determine the gait of the Late Pleistocene horse *Equus lambei* and to suggest that their speed indicated that they might be evading predators. McNeil et al. (2004) suggested that the Wally's Beach site is a multi-generic trample ground, interpreted as evidence of interactions between diverse animals in a short time frame, perhaps as a mixed herd. The extent of the trample grounds and the lithology of the sinkhole at Hot Springs, South Dakota, USA, allowed Laury (1980) to determine that this was a death trap and that the mammoth prints represented multiple animals struggling to escape the steep-sided pit. Fisher (2018) discussed the matriarchal family structure that could be discerned by the sequence of tracks at the Brennan *Mastodon* Site in Southern Michigan. The solitary male moved across, followed later by a mixed-age group of females (inferred from print size) and a calf. In addition to the analysis of fossil tracks, comparisons with related living species have been used to understand the activities of fossil trace makers (Benner et al., 2008). Neoichnological studies of Benner et al. (2008) helped not only find a possible match for a trace fossil, *Broomichnium flirii*, made by a sculpin (*Cottus* sp.) but also to determine the hunting activity that the fish was doing at the time.

There is considerable disparity in the utilization of ichnotaxonomic principles in North America. Clearly more ichnotaxonomic work is needed, particularly on bird and mammal tracks (Hunt and Lucas, 2007). Many works do not refer to the tracks by their ichnogenera. They are referred to instead as tracks of the suspected track maker: artiodactyl, bear, buffalo, horse, camel, mammoth, mastodon, sloth, carnivores (wolf, tiger, and cats), and birds.

The countries that make up Oceania are extremely varied in their geological origins. The Australian continent (including Australia, Papua New Guinea, and Western New Guinea [Indonesia]) houses some of the oldest rocks on the planet and has been geologically stable for tens if not hundreds of millions of years. In contrast, the recently recognized Zealandia (Mortimer et al., 2017) is largely submerged, and the few emergent parts (New Caledonia and New Zealand) are made up of much younger rocks. The other 23 countries that makeup Oceania consist of volcanic sea mounts and coral atolls and are largely restricted to Pleistocene- and Holocene-aged rocks. It is thus unsurprising that the vertebrate ichnological record of Oceania is dominated by trace fossils from Australia and New Zealand (e.g., Lockley et al., 2007, 2008; McDonald et al., 2007).

Houck et al. (2009) noted that the majority of tetrapod footprints preserved in pyroclastic sediments occur in volcanic arcs around the Pacific rim, but these appear to be largely restricted to the northern fringe, with very little representation in Oceania. The rapid formation of "beachrock" through aragonite or calcite cementation of beach sands (Fosberg and Carroll, 1965; Welch, 1999) could also conceivably represent a key mechanism for the preservation of vertebrate traces on the islands of Oceania, but no trace fossils have been reported to date.

The Quaternary tetrapod trace fossil record of Oceania is also unusual in that no authors have yet erected ichnotaxa from known traces nor attempted to assign these traces to existing ichnotaxa (see McDonald et al., 2007; Lockley and Harris, 2010; Lockley et al., 2021a and references therein). Although no tetrapod traces have been published from the many other Pacific islands making up Oceania, the coastal aeolianites and beachrock present on several of these islands seem like a logical place to begin the search. In the Marshall Islands, Fosberg and Carroll (1965) record beachrock and beach ridges that appear similar to the Australian aeolianites in which tracks are preserved. Both beachrock and aeolianite have also been recorded on Norfolk Island, the latter dating to ca. 22 ka (Rich et al., 1983, Pee et al., 2003). Aeolianites are also present along the southern coast of the Isle of Pines in New Caledonia (Balouet, 1991), and beachrock has been recorded on Niue Island (Wheeler and Aharon, 1997).

The Quaternary tetrapod trace fossil record in Asia is apparently restricted to Japan, South Korea in Jeju Island, a poorly known trampled surface in China that is related to the prosperous period of the Silk Road during the Tang dynasty, and one tracksite with human footprints recently described in Saudi Arabia (Stewart et al., 2020). Despite the huge coastline and the existence of so many lakes and large estuaries of important rivers, apparently, no attempts were made to record vertebrate trace fossils in Quaternary deposits. The exception is Japan, where Matsukawa and Shibata (2015) comprehensively showed the extensive work that has been done to document mammal and bird tracksites for almost one century, especially with regard to the track records of Pleistocene megafauna. Jeju Island was declared a World Heritage Site in 2007 and a UNESCO Global Geopark, mainly for the volcanic features including the Songaksan Tuff with tracks.

Finally, the ichnological studies of Quaternary successions in Europe are mostly concentrated in Great Britain, the Iberian Peninsula, Germany, and Italy. Bell et al. (2000) and Bell (2020) summarized the great number of mainly Holocene sites that have been found in intertidal mudflats under constant erosion. Some of them, such as Formby, Severn Estuary, and Goldcliff, became very important for the application of ichnological methods and approaches to archaeological problems (Allen, 1997; Bell, 2020), especially to measure the tempo of the major decline in biodiversity by human pressure on coastal habitats (Burns et al., 2022). Iberian coastal deposits, Italian volcaniclastic deposits, and German lake shores have been very important complements to the knowledge of the European trace fossil record with snapshots of daily life, hominin ecology, and food procurement, in very short windows of opportunity for track preservation, in sites such as the classic Schöningen or the "Devil's Trails," and the newly found Matalascañas Trampled Surface (Panarello et al., 2017; Altamura and Serangeli, 2023; Neto de Carvalho et al., 2023). Nowadays, all these three sites are interpreted for the general public, showcasing the contributions of ichnology. Finally, bioturbation and bioerosion in caves have long been documented together with archaeological studies but only very recently gained importance with more systematic descriptions, including formal ichnotaxonomic treatment (Diedrich, 2011b), in caves from France, Italy, and Romania (see examples in Pastoors and Lenssen-Erz, 2021).

A "gray area" where ichnology and tracking approaches are starting to be applied is in the interpretation of engraved rock art. Previously considered exclusive to archaeology, as it is not the result of direct interactions between the depicted track's producer and the rock panel but an interpretation based on the knowledge of the "artist," some attempts were made to provide analyses of this type of rock art in Namibia and Australia, based on experienced tracker's observations working together with scientists with ichnological background (Ouzman et al., 2002; Lenssen-Erz et al., 2023). This is just one more example where ichnology applied to archaeological problems (Ichnoarcheology sensu Baucon et al., 2008; see chapter on Hominin trackways) can unravel and support exciting ideas, as this type of rock art can be found in other parts of the world and was produced in different time periods.

3. Description of relevant ichnosites and ichnotaxa by continent

3.1 Africa

3.1.1 Ethiopia

Altamura et al. (2017) described well-preserved evidence of hippo activity in volcanic sediments dated to 700 ka at Gombore II-2, Melka Kunture, in the Upper Awash Valley. Multiple hippo tracks attributed to *Hippopotamus* cf. *amphibius* include very deep printed examples. It is claimed that the shape of the legs and feet of the hippos are thus shown. Tracks converged into a deep hippo trail. Detailed descriptions of hippo tracks are provided, but there is no formal ichnotaxonomy. Altamura and Mussi (2017) and Altamura et al. (2018, 2020) also described hundreds of tracks of hominins, bovids of various sizes, small gazelles, suids, equids, possible hyenas, and birds from the same area. Analysis of the volcanic and fluvio-lacustrine sequence (1.2–0.85 Ma) at the Gombore II open-air museum site identified 17 track-bearing beds in total.

3.1.2 *Kenya*

The second report of early hominin tracks from Africa (the first being Laetoli) came from Koobi Fora (Turkana Basin) in Kenya. Koobi Fora was initially dated to 1.6–1.5 Ma, but subsequently, new dates of 1.43 Ma were obtained by Bennett et al. (2014a). The Koobi Fora Formation succession shows fluvial and marginal lacustrine environments, where tracks were noted first in profile, then layers were cleared to reveal them in bedding plane view, including 20 different stratigraphic track-bearing layers (Laporte and Behrensmeyer, 1980; Behrensmeyer and Laporte, 1981): hippopotamus (large tracks were attributed to *Hippopotamus gorgops*, small tracks were attributed to a juvenile of this species or to *Hippopotamus aethiopicus*), bovid tracks and a hominin trackway. In total, 89 vertebrate tracks were identified. Adding to those, Bennett et al. (2014a) identified 240 tracks of five types attributed to "punting" or bottom-walking hippopotami (potentially including pygmy species or juveniles), 70 m south of the hominin and other tracks excavated by Behrensmeyer and Laporte (1981). Track morphology showed four digits with nails; manus and pes were inferred to be similar. This was the first recorded description of fossilized mammal swim tracks.

A large tridactyl avian track resembles that of a bustard or crane and a small avian track resembled that of a shorebird (wader) were recorded, although formal ichnotaxonomy was not applied. Feibel (1987) described in the upper part of the KBS Member, dating from 1.88 ± 0.02 to 1.64 ± 0.03 Ma, unornamented, shallow, dish-shaped structures preserved in concave epirelief on sandstones. These structures were interpreted to result from the nest-excavating activities of cichlid fish. The ichnogenus *Piscichnus* Feibel (1987) has since been applied to other circular-to-ovate depressions and dish-shaped unconformities that were a result of biogenic activity in aquatic environments (see chapter on Vertebrate Burrows in this volume).

Another important site in northwestern Kenya, especially for the study of hominin behavior, is Ileret, dated to 1.53–1.51 Ma (Roach et al., 2016). Data from 481 tracks include 97 hominin footprints attributed to *Homo erectus*. Fine preservation was described as related to muddy, cohesive substrates. A deltaic setting near a lakeshore bordered by open grasslands was inferred. The remaining tracks were mostly artiodactyl (bovid and suid tracks). Avian tracks were reported, including those of Anatidae, Ciconiidae, Gruidae, small shorebirds (waders), and "distinctive *Pelecanus* tracks." Very large avian tracks were tentatively ascribed to the extinct giant marabou stork (*Leptoptilos falconeri*). Other tracks reported were attributed to white rhinoceros (*Ceratotherium simum*), elephant (*Elephas or Loxodonta*), *Hippopotamus,* and *Crocodylus* (Bennett et al., 2014a). No formal ichnotaxonomy was applied. Roach et al. (2018) again mentioned but did not describe in detail a great number and diversity of non-hominin tracks: Bovidae/Suidae (325) Aves (29), Hippopotamidae (13), Reptilia (4), Carnivora (1), Cercopithecidae (1), Equidae (2), Proboscidea (1), Rhinocerotidae (2), and Tubulidentata (1).

In the Late Pleistocene deposits of Lake Bogoria, in the Kenya Rift Valley, Scott et al. (2008, 2012) described tracks on the deltaic Sandai Plain, including those of bovids, suids (warthog?), and birds (including flamingo). Tracks were inferred to have been made on exposed, moist lake-marginal mudflats. Well-preserved bovid tracks were thought to represent buffalo, oryx, greater kudu, and hartebeest. Flamingo mound nests and intense bioturbation due to flamingo activity producing a discrete ichnofabric were described.

3.1.3 *Namibia*

Holocene track sites have been reported in Namibia since Wilmer (1893) described the first tracks of elephant and rhinoceros. In a lagoon deposit at Walvis Bay, Kinahan (1996, 2014) summarized the presence of very well-preserved tracks of humans, zebra, ostrich, and *Ardea* sp., with special emphasis on livestock and domestic dogs from the last few centuries, and indicators of late precolonial contact and trade on the Namib Desert coast. Kinahan et al. (1991) and Smith et al. (1993) also described elephant tracks in the lower Kuiseb River. Finally, Bennett et al. (2010, 2014b) described Walvis Bay Holocene sites on dried floodplains dated to ~1.7–0.5 ka. It is mentioned that tracks in the region include those of domestic cattle, sheep/goats, buffalo, giraffes, hyenas, elephants, and a variety of birds, with indications of movement in large herds. Bennett et al. (2010a,b) simply noted the presence of hyena tracks, whereas Morse et al. (2013) included a small image (her Fig. 1.2L) of two track pairs attributed to a hyena. At least one of the probable manus tracks appears larger than the pes print behind it. No further descriptions were provided, and no formal ichnotaxonomy was presented.

Namibia is rich in hunter-gatherer rock art from the Later Stone Age. A recent research project (Lenssen-Erz et al., 2023) entailed indigenous tracking experts from the Kalahari analyzing engraved animal tracks and human footprints in a rock art region in central Western Namibia, the Doro! nawas mountains. The experts were able to define the species, sex, age group, and exact limb of the specific animal or human depicted in more than 90% of the 513 engravings they analyzed. The animal track engravings encompass 39 species, including herbivores, felines, other predators, birds, and primates.

3.1.4 *South Africa*

Presently, in South Africa, we find the most active ichnological research in Africa, with comprehensive descriptions of tracks and formal ichnotaxonomy of new vertebrate behaviors (Fig. 1.1). A summary of the understanding of Cape south coast vertebrate ichnology is found in Helm et al. (2020). Helm's studies document 326 Pleistocene vertebrate ichnosites in aeolianites and cemented foreshore deposits. The ichnotaxonomical treatment of the tracks can be found in Table 1.1. Sections on tracks of reptiles, avians, golden moles, hominins, lions, unidentified carnivores, hyenas, elephants, equids, rhinoceroses, hippopotamuses, giraffes, and bovids are included here. The first record of Pleistocene vertebrate tracks in southern Africa and the first discovery of an open hominin tracksite anywhere were described by Mountain (1966). Hominin, carnivore, and avian tracks in hyporelief were found on an aeolianite surface. The possibility of brown hyena (*Hyaena brunnea*) and crane/oystercatcher/dikkop trackmakers is mentioned, but no evidence is provided to support this. The site was subsequently dated to 126.6 ± 8.4 ka (Jacobs and Roberts, 2009). Tankard (1976) described a large-carnivoran trackway on an aeolianite surface at Langebaan on the west coast, attributed

FIGURE 1.1 Ichnotaxa from South Africa. (A) The holotype of *Natatorichnus subarenosa*; hypichnia, oblique view; scale bar = 10 cm. (B) The holotype of *Natatorichnus sulcatus*; epichnia, orthogonal view; scale is in cm. (C) The holotype of *Australochelichnus agulhasii*; epichnia, tangential view; trackway width = 9.5 cm. (D) The holotype of *Marinerichnus latus*; epichnia, oblique view; scale bar = 10 cm. (E) The holotype of *Anguichnus linearis*, crossed by a *Syncerus antiquus* trackway; epichnia, oblique view; hammer length = 33 cm.

"probably to *Hyaena brunnea*" (the brown hyena or strandwolf). Roberts (2008) was unable to locate it. The tracks and trackway do not appear to exhibit the distinctive features of hyena tracks such as forefoot tracks being substantially larger than hindfoot ones (Tankard, 1976: Fig. 21). The attribution of the trackway to *H. brunnea* may therefore be controversial.

Roberts and Berger's (1997) main focus was three "hominid" footprints that were dated at ~117 ka from the Langebaan site, although it is mentioned that "hyena and small antelope footprints are similarly preserved on different foresets." Roberts (2008) made the definitive account of the Nahoon and Langebaan "hominin tracksites." Nearby hyena tracks are again mentioned for the Langebaan site, and "tetrapod tracks", possible felid tracks, and avian trackways for the Nahoon site are mentioned without more detailed attribution. The size difference provided for front- and hindfoot track lengths for the purported Langebaan brown hyena tracks is less substantial than it should be to provide confident identification. In Still Bay, Roberts et al.'s (2008) main focus was on the tracks of the African elephant (*Loxodonta africana*). In addition, it is stated that the aeolianites "contain a rich archive of fossil

TABLE 1.1 The vertebrate ichnotaxa formally described in the Quaternary formations of Africa.

Ichnotaxon	Diagnosis/description	Producer	Formation	Age	Locality	Source
Anguichnus linearis	A large rectilinear, shallow, slightly sinuous surface trace expressed as a shallow furrow up to 9 cm wide. A thin central groove occurs sporadically in or near the midline of the furrow. This differs from other undulatory traces attributed to vertebrates and invertebrates in width of trace, markedly greater wavelength (w), and lesser amplitude (a) of undulations if they are present, i.e., generally very low a/w values.	Puff adder (*Bitis arietans*)	Waenhuiskrans	Late Pleistocene	Walker Bay Nature Reserve, Western Cape Province, South Africa	Helm et al. (2023d)
Australochelichnus agulhasii	Continuous subsymmetrical trails ~10 cm wide with median ridge averaging 35%–40% of trail width. Median ridge separating posteriorly angled rhomb-shaped flipper or paddle traces slightly wider than long, where length is measured parallel to trail axis. Left and right paddle traces show regular alternating pattern, with anterior margins on each side offset about half a paddle trace length relative to opposite side. In cross section, median ridge appears in convex epirelief, whereas lateral paddle traces are subhorizontal to slightly concave epireliefs. Anterior margins of paddle traces are crescent to semicircular in shape and anteriorly convex, with posterior margins marked by identical anteriorly convex traces of previously registered paddle marks, which repeat to create a highly regular series of traces on each side. Median margins of paddle traces are angled at about 110 degrees to the direction of travel marked by trackway midline, and curving postero-laterally to create outer trackway margin.	Hatchling Loggerhead turtle (*Caretta*)	Waenhuiskrans/ Klein Brak	Middle to Late Pleistocene	East of Still Bay, Western Cape Province, South Africa	Lockley et al. (2019)
Marinerichnus latus	A continuous symmetrical trail 14–15 cm wide, with median ridge about 30% as wide as whole trail. Median ridge separating posteriorly angled arcuate paddle traces, longer than wide, where length is measured parallel to trail axis. Left and right paddle traces show symmetrical arrangement about the median ridge/trackway midline. In cross section, median ridge appears in convex epirelief, whereas lateral paddle traces are concave epireliefs, up to 15 mm deep. Anterior margins of paddle traces are arcuate anteriorly and antero-laterally with posterior margins marked by identical anteriorly convex traces of previously registered paddle marks. Median margins of paddle traces are angled at about 130 degrees to the direction of travel marked by trackway midline and curve postero-laterally to create outer trackway margin.	Hatchling Leatherback turtle (*Dermochelys*)	Waenhuiskrans/ Klein Brak	Middle to Late Pleistocene	East of Still Bay, Western Cape Province, South Africa	Lockley et al. (2019)

Megapecoripeda isp.	A single track with asymmetrical shape exhibiting a large even-toed ungulate track pattern. Poor preservation with minimal detail, and the marginal ridge is very weakly developed. Medial digit shifted distally and slightly diverging digit impressions. Maximum length 7.2 cm; maximum width 8.0 cm.	Medium to large-bodied member of Artiodactyla (c.f., Bovidae)	Tuff I^F, upper Bed I	Middle Pleistocene	Olduvai Gorge, Ngorongoro Conservation Area, Tanzania	Plint and Magill (2021)
Musaltipes isp.	Trackway is symmetrical about the mid-line, with tracks occurring as pairs indicating a hopping or bounding gait (rather than an alternating gait pattern). The maximum straddle is ~4.0 cm. Track dimensions: ~1.5 cm long, and ~1.0 cm wide. Individual tracks exhibit an oval configuration.	Cape gerbil (*Gerbilliscus afra*)	Waenhuiskrans	Middle to Late Pleistocene	East of Still Bay, Western Cape Province, South Africa	Lockley et al. (2022)
Natatorichnus subarenosa	Sand-filled, subhorizontal, sublinear flattened, convex-up transmitted trail trace, ~10 cm wide. Traces are generally upwardly and anteriorly convex into a low roof or ridge. They are transversely divided into short *spreite* expressed as low amplitude, *en échelon*, anteriorly convex bulges, which tend to be arranged sequentially in alternating left-right patterns.	*Eremitalpa* golden mole	Waenhuiskrans	Middle to Late Pleistocene	East of Still Bay, Western Cape Province, South Africa	Lockley et al. (2021b)
Natatorichnus sulcatus	Flattened, sand filled, subhorizontal, linear to meandering, transmitted trail or shallow burrow trace, up to ~10 cm wide. Traces exhibit ridge or roof in mid-line collapsed into a central furrow or sulcus. They are transversely divided into short *spreite* expressed as low amplitude, *en échelon*, anteriorly convex bulges, which tend to be arranged sequentially in alternating left-right patterns. The degree of collapse of the trail roof, and hence the width of the collapsed section, may vary.	*Eremitalpa* golden mole	Nahoon	Middle to Late Pleistocene	Woody Cape, Addo Elephant National Park, Eastern Cape Province, South Africa	Lockley et al. (2021b)
Pecoripeda isp.	Bovid track pattern with interdigital sulcus and the lateral digit impression larger than the medial digit impression. Indentation in tuff corresponds to the sharp edge of the underside of the keratinous hoof-wall. Well preserved. Maximum length 6.5 cm; maximum width 5.4 cm.	medium to large-bodied member of Artiodactyla (cf. Bovidae)	Tuff I^F, upper Bed I	Middle Pleistocene	Olduvai Gorge, Ngorongoro Conservation Area, Tanzania	Plint and Magill (2021)
Piscichnus brownii	A shallow, circular, dish-shaped structure oriented concave upward in a horizontal plane.	Cichlid fish nesting behavior	Upper part of the KBS Member, Koobi Fora Formation	Early Pleistocene 1.88 ± 0.02 to 1.64 ± 0.03 Ma	Northeastern shore of Lake Turkana, Kenya	Feibel (1987)

mammalian trackways including carnivore, proboscidean (elephant), antelope, equid, rodent as well as reptilian (Testudinidae)." A date range of 140 ± 8 ka − 91 ± 5 ka was established through OSL/AAR. Helm et al. (2019a) focused on Roberts Rock and Megafauna Rock. The former, described in Roberts et al. (2008), split in two, exposing numerous bovid tracks, before slumping into the sea. The latter contained several elephant tracks, a long-horned buffalo track (*Syncerus antiquus*), a possible rhinoceros track, and a possible composite giant Cape horse (*Equus capensis*) track.

The extensive coastal outcrops of the Cape south coast of South Africa became the focus of extensive ichnological studies in recent years. A summary of tracksites identified within the Goukamma Nature Reserve, including tracks of avian, elephant, artiodactyl, equid, and carnivore affinity, was made by Helm et al. (2018c). Elephant tracks were attributed to *Loxodonta africana*. Large equid tracks were attributed to the extinct giant Cape horse (*Equus capensis*). Based on size, large felid tracks were compared with the extinct Cape African lion (*Panthera leo melanochaitus*). Large artiodactyl tracks exhibited a length-to-width ratio of <1, a morphology not evident in any extant bovid trackmaker from this region. Neoichnological study compared relative track dimensions of different types of cattle with different horn lengths, supporting the attribution of the fossil tracks to the extinct long-horned buffalo (*Syncerus antiquus*). Subsequent OSL dating shows the Goukamma sites to be in the 71−75 ka range. Tracks attributed to *Loxodonta africana* were described at 35 sites by Helm et al. (2022c), including two trunk drag impressions in a serpentine pattern on either side of an elephant trackway (the first such report). The concept of subsurface seismic communication between elephants leaving a trace fossil signature was discussed by Helm et al. (2023b). Helm et al. (2021) suggested that elephant (*Loxodonta africana*) tracks were precursors to some potholes on the Cape South Coast paleobeaches. The main site was subsequently dated to 110 ± 7 ka; another site is from MIS 11 (~400 ka).

Twenty-six equid tracksites were recorded by Helm et al. (2023a) from the Middle and Late Pleistocene (~161−43 ka) aeolianites, of which 13 could be attributed to the giant Cape horse. One comprised a trackway with 12 tracks. Eight sites with smaller tracks could be attributed to the quagga (*Equus quagga quagga*).

Helm et al. (2018a) is the first report of giraffe tracks on an aeolianite truncation surface subsequently dated through OSL to 109 ± 9 ka. Identification was made on the basis of extant giraffe tracks being uniquely large and long within Bovidae, relatively narrow compared with the largest bovids, and with a distinctive shape of the interdigital sulcus in well-preserved tracks. The first hominin tracksite from South Africa's Cape South Coast, subsequently dated through OSL to between 89 ± 6 ka and 76 ± 5 ka, was found by Helm et al. (2018b) at Brenton-on-Sea. Nearby tracks (not described further) include those of elephants, bovids, large carnivores, and equids.

A summary of tracksites identified within the Robberg Nature Reserve, including tracks of elephants, bovids, equids, and rhinoceros affinity, as well as golden mole burrow traces, can be found in Helm et al. (2019b). Elephant tracks were attributed to *Loxodonta africana*, the largest, as much as 70 cm long (probably transmitted tracks), together with the ones from SW Portugal (Neto de Carvalho et al., 2020c) and ground sloth pes prints (*Neomegatherichnum pehuencoensis*) reaching 1.05 m-long, but 0.7−0.8 m on average (Feola, 2022), are the largest tracks to be recorded since the Mesozoic. Small equid tracks were attributed to the quagga (*Equus quagga quagga*), the extinct subspecies of the plains zebra. The rhinoceros tracks were attributed to the black rhinoceros (*Ceratotherium simum*); distinguishing features include a relatively small outer toe, which is situated relatively far from the middle toe. The first detailed description of "sandswimming" golden mole traces was made by Lockley et al. (2021b), assigned to an *Eremitalpa*-like species (Family Chrysochloridae). In effect, these are "hybrids" between trackways and burrows. A new ichnogenus was erected, *Natatorichnus*, with two ichnospecies, *N. subarenosa* and *N. sulcatus*, depending on whether the roof wall of the structure stayed intact or collapsed (Fig. 1.1A and B). Six infaunal golden mole (Family Chrysochloridae) burrow trace sites were also described (see also *Natatorichnus* in the chapter on Vertebrate Burrows). A trackway and burrows of a small rodent-sized bounding mammal (attributed to the Cape gerbil, *Gerbilliscus afra*) were reported by Lockley et al. (2022). Such small-mammal trackways are rare in the fossil record, occurring at only three Cenozoic sites.

The first report on seal traces was made by Helm et al. (2022b). Both in-motion and resting traces were described and dated to ~75 ka (MIS 5a).

A summary of 29 avian ichnosites, with a focus on six sites with large tracks, was presented by Helm et al. (2020b). These include well-preserved flamingo tracks (larger than those of extant greater flamingo) and unusual flamingo feeding traces described for the first time. One site shows probable large crane tracks, whereas the record of large tracks that do not correspond to any extant southern African avian may represent a possibly extinct species. Helm et al. (2017) also reported about 50 anisodactyl avian tracks, including two long trackways made visible from a distance through preferential adherence of biofilm to hyporelief tracks. These were attributed to a probable spurfowl (Phasianidae) or guineafowl (Numididae) trackmaker. Recently, 12 more newly identified avian ichnosites were described by Helm et al. (2023c), including long avian trackways.

Large reptile tracks and swim traces attributed to the Nile crocodile (*Crocodylus niloticus*) and to a monitor lizard (probably the water monitor, *Varanus niloticus*) were presented by Helm et al. (2020c). They are the first described reptile swim traces in Africa. More substantive evidence of large reptile (probably *Crocodylus niloticus*) swim traces was detailed by Helm and Lockley (2021). Interclaw distances were used to estimate crocodile lengths of 1.15–1.7 m. The first report on tortoise tracks and traces, representing large to very large trackmakers and dated to the Middle and Late Pleistocene (MIS 6–MIS 4), was made by Helm et al. (2022a). They were attributed either to a large form of the extant leopard tortoise (*Stigmochelys pardalis*) or to an extinct species. On the other hand, the first report of hatchling sea turtle trackways was made by Lockley et al. (2019). They were produced during the short interval when the trackmakers headed for the sea. One assemblage of approximately parallel trackways indicates smaller loggerhead turtle hatchlings with alternating gaits and contrasts with a wider trackway with a symmetrical gait pattern, indicating a leatherback turtle hatchling. Paleoenvironmental implications include a warm phase in the Late Pleistocene (possibly MIS 5e sea level high-stand) as sea turtles require warm temperatures for nesting, and their current distribution range ends much further to the northeast. Trackways of inferred loggerheads were assigned to *Australochelichnus agulhasii* igen. et isp. nov. (Fig. 1.1C), and the inferred leatherback trackway to *Marinerichnus latus* igen. et isp. nov. (Lockley et al., 2019, Fig. 1.1D).

The first report on a fossil snake trail, attributed to a puff adder (*Bitis arietans*) engaged in rectilinear motion was just presented by Helm et al. (2023d). A new ichnogenus and ichnospecies—*Anguichnus linearis*—were erected by these authors (Fig. 1.1E).

3.1.5 *Tanzania*

The first tracks reported from Olduvai Gorge in Tanzania, famed for its Plio-Pleistocene hominin body fossils, were reported by Hay (1976). The "BPT"—Bird Print Tuff—in lower Bed II apparently contains abundant shorebird tracks. Ashley and Liutkus (2002) described large, trough-shaped structures thought to be hippo trails in the Lowermost Bed II. Later, they were confirmed by Plint and McGill (2021). Tracks are in a 1.8 Ma volcanic tuff. Three tracks are present, two of which were assigned to the ichnogenus *Pecoripeda* (one with "exceptional level of detail"), and one to *Megapecoripeda*. Track morphology suggests track makers were medium to large-bodied members of Artiodactyla, probably Bovidae.

Finally, more than 400 hominin tracks, plus tracks of a group of zebra and four bovids, were reported by Liutkus-Pierce et al. (2016) from volcaniclastic deposits at Engare Sero on the shore of Lake Natron, dated to between 19.1 and 5.76 ka.

3.2 South America

3.2.1 *Argentina*

Aramayo and Manera de Bianco (1987a,b) are complementary contributions that documented the discovery of the famous Pehuen Co paleoichnological site (Late Pleistocene, Buenos Aires Province), and provided the first ichnotaxonomical treatment of the fossil tracks (synthesis and updates of this work can be found in Aramayo, 2004; Aramayo and Manera de Bianco, 2009). The stratigraphy and age of the site are also described, radiocarbon dated to 12 $\pm$ 0.11 ka (Aramayo and Manera de Bianco, 1996). The paleoichnological record shows more than 100 trackways and hundreds of isolated tracks of different species spread across 5 km of coastal outcrops. There are ichnofossils made by mammals of large- and medium size, most of them extinct, as well as different birds that wandered along the pond shorelines (Bayón et al., 2011). Eleven new ichnogenera and ichnospecies were diagnosed, and four bird morphotypes were left in open nomenclature. The proposed ichnotaxa can be found in Table 1.2. Later, Aramayo et al. (2015) presented an ichnotaxonomical review of the Pehuen Co paleoichnological site and proposed a new ichnotaxa on the basis of new discoveries. The described tracks include 12 mammal and five bird ichnogenera, and 13 and two ichnospecies, respectively (Table 1.2), making this the most diverse ichnosite known for the Quaternary (see, e.g., Figs. 1.2 and 1.3). The more common ichnotaxa are *Neomegatherichnum pehuencohensis* (Fig. 1.2A and B), *Lamaichnum guanicoe* (Fig. 1.2K) and *Phoenicopterichnum rector* (Fig. 1.3C–E). A trackway located at Pehuen Co, assigned to *Megatherium*, the largest known ground sloth, was analyzed to infer the locomotory behavior that explains the inferred bipedal pattern of the track proposed by Casamiquela (1983), based on skeletal and track measures (Blanco and Czerwonogora, 2003). Additional studies have suggested an alternative hypothesis for the locomotion of ground sloths that explains the apparently bipedal trackways due to overprinting of the pes on the manus (Feola et al., 2018; Feola, 2022), as suggested by McDonald (2007). Other quadrupedal trackways are those of mylodontid ground sloths (*Mylodontidichnum rosalensis*, Fig. 1.2D), proboscideans (*Proboscipeda australis*, Fig. 1.2E), macraucheniids (*Eumacrauchenichnus patachonicus*, Fig. 1.2F), equids (*Hippipeda* isp., Fig. 1.2I), and artiodactyls (*Pecoripeda commune*, Fig. 1.2L).

TABLE 1.2 The ichnotaxa described in Quaternary formations of South America.

Ichnotaxa	Diagnosis/description	Producer	Formation	Age	Locality	Source
Acugnaichnus dorregoensis	Footprints belonging to a bipedal mammal, plantigrade, of large size. Footprints similar to *Megatherichnum oportoi* and *Falsatorichnum calceocannabius*, from which differ—besides in size and geological age—by the morphology of the region corresponding to the metatarsal V, with the shape of a caudal hook or spur. Very short pace; average 35 cm.	Scelidotheriinae (Xenarthra, Mylodontidae)	Unnamed unit	Plio-Pleistocene	Monte Hermoso, Buenos Aires, Argentina	Casamiquela (1983)
Acugnaichnus dorregoensis	Large footprints, with noticeable heteropody, with manus (strongly compressed, with length/width ratio less than 1) comparatively smaller than pes (anteroposteriorly expanded, with length/width ratio ranging from 1.30 to 1.75). Round-shaped forefoot imprints, having a long postero-medially directed wedge shaped claw-trace. Hindfoot impressions exhibiting slight anterior-widening and pronounced pedal-arching (with curvilinear longitudinal axis), being ovoidal-ellipsoidal (bilaterally convex) to crescent-shaped (with concave internal margin and convex external border), depending on if they are poorly or well preserved respectively; subtriangular antero-internal notch; falciform antero-external prolongation delimited forwardly by a deep antero-lateral gap and rearwardly by a lesser penetrating posterolateral indentation forming a caudal hook. Rim extended along the anterior and lateral margins of the ichnite.	*Scelidotherium leptocephalum* and *Catonyx tarijensis* (Xenarthra, Mylodontidae)	Lower part of the Inferior section of the Luján Formation	Upper Pleistocene	Laguna del Monte, Buenos Aires Province, Argentina	Casamiquela (1983), Toledo and Arregui (2022)

(*Continued*)

TABLE 1.2 The ichnotaxa described in Quaternary formations of South America.—cont'd

Ichnotaxa	Diagnosis/description	Producer	Formation	Age	Locality	Source
Aramayoichnus rheae (Fig. 1.3A and B)	Bipedal trackway of high pace angulation with mesaxonic tridactyl footprints of moderate size (footprint length usually greater than 10 cm), footprint length/width ratio less than 2, footprints with small positive rotation from the midline. Footprints with three thick digits directed forward, showing clear phalangeal pad impressions and claw marks. At least half of the length of digit III impression projecting forward of the imaginary line formed by the tip of lateral digits, digit IV impression longer than that of digit II. Divarication of digits II–IV is commonly smaller than 80 degrees. Digits proximally separated from or joined to a deep, circular metatarsal–phalangeal pad impression. Interdigital web trace absent. Differs from *Anchisauripus* by the presence of a metatarsal–phalangeal pad impression, greater digit divarication and a lower footprint length/width ratio.	Rheiform birds	Agua Blanca Sequence	Late Pleistocene	Pehuen Co, Buenos Aires Province, Argentina	Aramayo et al. (2015)
Aramayoichnus isp.	No description provided, only figured (Fig. 1.2B)	Rheiform bird	Luján Formation	Late Pleistocene	Miramar, Buenos Aires Province, Argentina	Agnolín et al. (2019)
Canipeda gracilis (Fig. 1.2G–H)	*Canipeda* with manus larger than pes imprints, footprint length between 35 and 65 mm, footprint length/width ratio between 1.0 and 1.3, divarication of outer digits (II–V) from 30 to 70°, divarication of inner digits (III-IV) from 10 to 20°, metapodial pad impression triangular in the manus and subrounded to elliptical in the pes. *Canipeda longigriffa* has a larger footprint length/width ratio (1.6–1.9) and a smaller divarication of outer digits (II–V less than 30 degrees). *Canipeda sanguinolenta* is larger (footprint length 75–85 mm) and exhibits arcuate metapodial pads with strongly bilobed rear end.	Medium-sized fox (body mass about 7 kg), similar to the largest representatives of the genus *Lycalopex*, including the extant *L. gymnocercus* and *L. culpaeus*, and probably, the extinct *Lycalopex* cf. *L. ensenadensis*.	Agua Blanca Sequence	Late Pleistocene	Pehuen Co, Buenos Aires Province, Argentina	Vyalov (1965), Melchor et al. (2018)

Charadriipeda isp. (Fig. 1.3F)	Footprint (average) length: 44 mm, width 46 mm. Divarication of digits II-III: 55 degrees, III-IV: 55, II-IV: 110 degrees.Tridactyl webbed avian footprints of medium to large size.	Short-legged birds. (Aves, Anseriformes)	Agua Blanca Sequence	Late Pleistocene	Pehuen Co, Buenos Aires Province, Argentina	Aramayo et al. (2015)
Dolichotichnus marae	Semiplantigrade footprints of moderate size. Manus footprints of subcircular shape, with four divergent digits (II–V) of similar size ending in sharp claw marks, and a kidney-shaped padded palm. Pes footprints longer than the manus imprint, circular palm pad, and three parallel digits with claws, central digit (III) longer than the lateral ones (II–IV) and having a deep claw mark. The pes oversteps the manus and is rotated inward with respect to the midline.	Cursorial rodent relative of the extant *Dolichotis* (Caviidae) or mara.	Agua Blanca Sequence	Late Pleistocene	Pehuen Co, Buenos Aires Province, Argentina	Aramayo et al. (2015)
Eumacraucheni-chnus patachonicus (Fig. 1.2F)	Quadrupedal trackway composed of large equidimensional tridactyl footprints with manus and pes of similar size and shape. Pace length/footprint length ratio about 5, high pace angulation (125–150 degrees). Manus and pes footprints subcircular in outline, anterior edge with three broad undulations corresponding to short and stout digits, digit III is wider and longer, smooth sole/palm surface. The posterior part of the footprints shows a characteristic medial concavity of the footprint edge.	The most likely producer is *Macrauchenia patachonica* (Litopterna, Macraucheniidae)	Agua Blanca Sequence	Late Pleistocene	Pehuen Co, Buenos Aires Province, Argentina	Aramayo and Manera de Bianco (1987a), Aramayo et al. (2015)
Eumacrauchenichnus patachonicus?	Tridactyl fossil footprints with an average length of 196 mm, width of 208 mm and depth of up to 50 mm.	*Macrauchenia patachonica*	Santa Clara Formation	Late Pleistocene	Mar Chiquita, Buenos Aires, Argentina	Medina and Tassara (2023)

(*Continued*)

TABLE 1.2 The ichnotaxa described in Quaternary formations of South America.—cont'd

Ichnotaxa	Diagnosis/description	Producer	Formation	Age	Locality	Source
Felipeda miramarensis (Fig. 1.2J)	Large felipedid footprints diagnosable on the basis of the following unique combination of characters: manus prints notably larger than pes prints, four subequal unclawed digits arranged in symmetric arc, in front of a broad, bilaterally symmetrical and kidney-shaped interdigital pad (formed by the fusion of three aligned longitudinal lobes), and manus print with third digit more anteriorly positioned than the fourth. The manus and pes imprints are nearly as transversely wide as anteroposteriorly long.	*Smilodon populator* (Felidae)	Luján Formation	Late Pleistocene	Miramar, Buenos Aires Province, Argentina	Agnolín et al. (2019)
Glyptodontichnus pehuencoensis	Large, plantigrade padded footprints with pentadactyl, subcircular to fan-shaped pes footprint somewhat larger than the subelliptical, tridactyl manus footprint. Pes imprint with short, wide and blunt digits; divarication of lateral digits (I–II and IV–V) is wider (40–50 degrees) than in the central digits (5–10 degrees). Manus footprint with almost parallel, wide digits (II to IV) ending in a short tip directed forward; digit IV is slightly divergent from digits II and III (angle between III and IV: 20 degrees), clear subcircular metapodial palm imprint.	A species of the genus *Glyptodon* Xenartha, Cingulata, Glyptodontia).	Agua Blanca Sequence	Late Pleistocene	Pehuen Co, Buenos Aires Province, Argentina	Aramayo et al. (2015)
Gruipeda isp. (Fig. 1.3G–H)	Tetradactyl, anisodactyl tracks that are as long (including hallux) as wide. Two trackways of 33 and 24 footprints were recorded. A few tracks are tridactyl, most of them (90%) are tetradactyl. Digits II to IV are large and directed forward, digit I (hallux) is directed backward. Trackway: length 6.70 m (33 footprints), width 0.14 m. Footprint (average) length (hallux included) 100 mm, width 100 mm. Digits length (average): I: 26 mm, II and IV: 30 mm, III: 50 mm. Divarication of digits: I– II: 120 degrees, II-IV. 140 degrees (Aramayo et al., 2015).	Tentatively compared with those of the extant "chaja" *Chauna torquata* (Anseriformes, Anhimidae)	Agua Blanca Sequence	Late Pleistocene	Pehuen Co, Buenos Aires Province, Argentina	Aramayo et al. (2015)

Hippipeda isp. (Fig. 1.2I)	Footprints with a hemi-ellipsoidal shape, the anterior part is remarkably deep and shows a marginal rim. At the back, a frog is easily distinguished. The trackway has two footprints clearly preserved, and two undertracks. The trackway length is 1.40 m, and the width is 0.35 m. One of the well-preserved footprints is a manus (length 150 mm, width 120 mm) and the other one is a pes (length 130 mm, width 100 mm). The manus footprint is slightly larger and more elongated than the pes footprint. Length/width ratio is: 1.25 for the manus and 1.30 for the pes.	The most likely producers are representatives of *Equus* or *Hippidion* (Perissodactyla, Equidae)	Agua Blanca Sequence	Late Pleistocene	Pehuen Co, Buenos Aires Province, Argentina	Aramayo et al. (2015)
Hippipeda isp.	Deep tracks of a single-toed equid, which are ovoid shaped. The largest footprint measuring 7.0 cm in footprint length and 6.0 cm in footprint wide. The sole is well-pronounced and concaved downward. Both the tracks are bordered anteriorly by displacement rims and micro faults.	Comparable in size with *Hippidion*	Unnamed unit	Late Pleistocene	Tamanduá de Cima, Pernambuco, Brasil	Oliveira et al. (2019)
Hippipeda isp.	Forefoot and rearfoot imprints exhibit the same morphostructural pattern, only distinguishable through their dimensions and proportions (slight heteropody). In this regard, the manus (length: 80−150 mm; width: 70−120 mm; length/width ratio: 1.14−1.25) is comparatively larger and proportionally longer than the pes (length: 60−130 mm; width: 50−100 mm; length/width ratio: 1.20−1.30). The hoof wall print constitutes a deep perimetral flange, of epicycloidal to U-shaped outline, with curvilinear and rectilinear lateral walls (quarters), respectively; a uniformly convex frontal wall (toe); and a nearly constant thickness (8−10 mm) that determines the	*Equus* or *Hippidion* (Peryssodactyla, Equidae)	Unnamed unit	Late Pleistocene	Laguna del Monte, Argentina	Oliva and Arregui (2018)

(*Continued*)

TABLE 1.2 The ichnotaxa described in Quaternary formations of South America.—cont'd

Ichnotaxa	Diagnosis/description	Producer	Formation	Age	Locality	Source
	circumference of the track. On each side, this structure defines a V-shaped postero-external angle (heel buttress) and then bends and points forward and inwardly (antero-medially) in a wedge-shaped bar mark separated from its opposite by the frog apex.					
cf. *Hippipeda* isp.	Footprints of hemi-ellipsoidal outline, with convex central bottom and concave lateral and marginal outline. Measurements: footprint length 100–140 mm, footprint width 90–100 mm, depth 50 mm.	*Hippidion* or *Equus (Amerhippus)*	Santa Clara Formation	Late Pleistocene	Santa Clara del Mar, Argentina	Tassara et al. (2005)
Ichnhippus cotaposi	Imprints belonging to a medium-sized mammal, quadruped and functionally monodactyl (with a hoof or similar structure). Pace angulation near straight. Digitigrade. Subequal palm and plantar surface, leaving irregular ovoidal impressions, deeper in the anterior part and characterized by a paired wing-shaped structure, placed in the anterior half. Tail imprints absent. Measurements: stride length 1 m, pace angulation 160 degrees	*Hippidion* (Equidae)	Unnamed sequence	Plio-Pleistocene	Quebrada de La Chimba, Antofagasta, Chile	Casamiquela and Chong-Díaz (1975)
Iribarnichnum megamericanum	Impressions corresponding to a bipedal, plantigrade mammal, enormous in size, larger than that inferred for *Megatherichnum oportoi*, whose footprints are distinguished by their larger size and an apparently less elongated silhouette (in addition to a geologically different age).	*Scelidotherium*	?	Plio-Pleistocene	Monte Hermoso, Buenos Aires, Argentina	Casamiquela (1983)

Lamaichnum guanicoe (Fig. 1.2K)	*Lamaichnum* with footprint length ranging from 40 to 120 mm.	*Lama guanicoe* (Artiodactyla, Tylopoda, Camelidae)	Agua Blanca Sequence	Late Pleistocene	Pehuen Co, Buenos Aires Province, Argentina.	Aramayo and Manera de Bianco (1987b), Aramayo et al. (2015)
Lamaichnum guanicoe	Average footprint length 126 mm, average width 107 mm.	*Lama guanicoe*	Mar Chiquita Formation	Holocene	Mar Chiquita, Buenos Aires, Argentina	Medina and Tassara (2022)
Lamaichnum guanicoe	Imprints produced are almost bisymmetrical, typically longer than wide (length/width ratio higher than 1), with length ranging from 65 to 120 mm and width varying from 45 to 90 mm. Forefoot and hindfoot tracks are nearly identical (strong homopody). These footprints consist of two acuminate, subparallel, digital impressions, which are invariably undivided and uniformly concave, indicating the presence of smooth, antero-posteriorly elongated, toe cushions.	Camelids	Unnamed unit	Late Pleistocene	Laguna del Monte, Argentina	Oliva and Arregui (2018)
Lamaichnum guanicoe	Description. The footprints were made by a quadrupedal didactyle animal, in which digits III and IV are printed into the substrate. These imprints present an interclavular gap or anterior cleft in the front and a posterior cleft along with an interdigital septum. The posterior cleft might not be preserved or might be less pronounced. Seven of the observed ichnites have evidence of a displacement rim in the anterior part, which is useful for the identification of the direction in which the animal was moving. Measurements: the maximum anteroposterior length recorded range from 12.4 to 7.6 cm, maximum lateromedial width range from 13.5 to 6.6 cm.	Members of the Tribe Lamini (Camelidae)	Unnamend unit	Late Pleistocene.	Punta Pelluco, Chile	Campos-Medina et al. (2022)

(*Continued*)

TABLE 1.2 The ichnotaxa described in Quaternary formations of South America.—cont'd

Ichnotaxa	Diagnosis/description	Producer	Formation	Age	Locality	Source
Lamaichnum tulipensis	Quadrupedal trackway composed of medium- to large-sized didactyl and digitigrade footprints with each digit represented by a single pad, manus and pes of roughly similar shape, posterior end biconvex and anterior pointed. Manus is slightly larger and deeper imprinted than the pes. Manus footprints consist of two elongated concavities with a pointed anterior end facing outwards, the central ridge between both concavities is short and widens anteriorly. Pes footprints consist of two elongated depressions with parallel anterior ends. The size of the footprints is greater than those observed in *Lamaichnum guanicoe.*	Camelid of large size	Agua Blanca Sequence	Late Pleistocene	Pehuen Co, Buenos Aires Province, Argentina	Aramayo and Manera de Bianco (1987b), Aramayo et al. (2015)
Lamaichnum isp.	Subdigitigrade, didactyl footprints with pointed anterior ends and rounded posterior ends, measuring ~5.5 cm in footprint length with each digit ~1.8 cm wide.	Relative of *Lama guanicoe*	Unnamed unit	Late Pleistocene	Tamanduá de Cima, Pernambuco, Brasil	Oliveira et al. (2019)
Mustelidichnum enigmaticus (ichnogenus in need of revision)	Footprints attributed to a small to medium-sized plantigrade mammal, manus and pes indistinguishable, with short digits and apparently of equal length; palm/plantar surface of subquadrangular outline.	Family Mustelidae?	Agua Blanca Sequence	Late Pleistocene	Pehuen Co, Buenos Aires Province, Argentina	Aramayo and Manera de Bianco (1987b)
Mylodontidichnum rosalensis (Fig. 1.2D)	Quadrupedal trackway composed of medium to large-sized, plantigrade footprints; pes subelliptical, manus subcircular. Pes footprint larger than 0.5 m (average footprint length/width ratio greater than 2), typically with an arched	Mylodontids (Xenarthra, Tardigrada, Mylodontidae)	Agua Blanca Sequence	Late Pleistocene	Pehuen Co, Buenos Aires Province, Argentina	Aramayo and Manera de Bianco (1987a), Aramayo et al. (2015)

	outline and narrower posterior part, pes footprint axis parallel to the trackway midline or slightly inwardly directed, moderate pace angulation. Pes footprint with a marginal ridge in the lateral and anterior margin, up to three broad and blunt digit imprints, related to three short, diverging and laterally oriented claw marks. Manus footprint rarely preserved, subcircular, slightly longer than wide, manus footprint length about half of pes footprint length, manus footprint imprinted anterior and/or lateral to the pes footprint.					
cf. *Mylodontidichnum rosalensis*	Trackway composed by two pes prints and three manus prints. Pes prints of subelliptical outline with a marginal ridge in the anterior and lateral side. Manus prints are subelliptical and lack a marginal ridge. Measurements. Pes print length 0.40–0.45 m, width 0.30–0.33 m. Manus print length 0.28–0.33 m, width 0.24–0.26 m. Preserved length of trackway 2.75 m, trackway breadth 0.66 m, pace length 0.72 m, pace angulation 166 degrees.	*Scelidotherium*, *Glossotherium* or *Lestodon*	Santa Clara Formation	Late Pleistocene	Santa Clara del Mar, Argentina	Tassara et al. (2005)
Neomegatherichnum pehuencoensis (Fig. 1.2A and B)	Large plantigrade footprints preserved in concave epirelief. Manus footprints circular to subcircular, commonly longer than wide, footprint length about 40% of the associated pes, placed anterior-medial to the pes footprint, large inwardly directed claw mark. Pes footprints subelliptical, up to 1 m long, wider anteriorly, long axis parallel to the trackway midline, most display a ridge located adjacent to the anterior and lateral margins of the footprint, occasional large	Megatheriids (Xenarthra, Tardigrada, Megatheriidae)	Agua Blanca Sequence	Late Pleistocene	Pehuen Co, Buenos Aires Province, Argentina	Aramayo and Manera de Bianco (1987a), Aramayo et al. (2015)

(Continued)

TABLE 1.2 The ichnotaxa described in Quaternary formations of South America.—cont'd

Ichnotaxa	Diagnosis/description	Producer	Formation	Age	Locality	Source
	claw mark at the medial-anterior corner of the footprint that is directed inward or backward. Pace angulation in a straight trackway is 85°–115 degrees.					
Pecoripeda commune (Fig. 1.2L)	Quadrupedal trackway composed of small- to medium-sized (50–120 mm long), hoofed didactyl footprints, digit imprints with flat or slightly concave surface; separations between digits are commonly absent or poorly developed. The anterior border is acutely marked by both hooves; lateral sides are straight, while the posterior border is slightly convex.	Several families of Pecora, including Cervidae	Agua Blanca Sequence	Late Pleistocene	Pehuen Co, Buenos Aires Province, Argentina	Aramayo and Manera de Bianco (1987b), Aramayo et al. (2015)
Pecoripeda commune	Average footprint length 90 mm, average width 75 mm.	Cervidae footprints	Mar Chiquita Formation	Holocene	Mar Chiquita, Buenos Aires, Argentina	Medina and Tassara (2022)
Pecoripeda isp.	A pair of digit imprints resembling a small cervid, measuring 2.3 cm in footprint length and 3.3 cm in footprint width. The didactyl footprint shows pointed anterior outline and rounded posterior ends.	Cervidae footprints	Unnamed unit	Late Pleistocene	Tamanduá de Cima, Pernambuco, Brasil	Oliveira et al. (2019)
cf. *Odocoileinichnum commune* (=cf. *Pecoripeda commune)*	Pair of parallel furrows with straight margin, flat or slightly concave bottom, anterior end acuminate, rear end narrower and slightly convex, lacking interdigital sulcus. Measurements: Length 60–90 mm, width 38–70 mm.	Cervidae	Santa Clara Formation	Late Pleistocene	Santa Clara del Mar, Argentina	Aramayo and Manera de Bianco (1987b), Tassara et al. (2005)

Phoenicopterichnum rector (Fig. 1.3C–E)	Large tridactyl palmate bird footprints with a fan-shaped outline and a length/width ratio less than 1. Imprint of digit III projects anteriorly from the digital web impression, in an amount of approximately a quarter of the footprint length. In the anterior margin, two clear curved lines join the lateral digit imprints with that of digit III. Posterior margin of footprint slightly convex.	Relatives of flamingoes (Phoenicopteridae).	Agua Blanca Sequence	Late Pleistocene	Pehuen Co, Buenos Aires Province, Argentina	Aramayo and Manera de Bianco (1987b), Melchor et al. (2012), Aramayo et al. (2015)
Porcellusignus isp. *(lapsus calami)*	No description provided, only figured (Fig. 1.2B)		Luján Formation	Late Pleistocene	Miramar, Buenos Aires Province, Argentina	Agnolín et al. (2019)
Proboscipeda australis (Fig. 1.2E)	Large circular to slightly oval footprints (length: 0.23–0.27 m, width: 0.23–0.30 m), manus similar to the pes and slightly larger, oval footprints can be wider than long or longer than wide. Commonly 3 to 5 digit imprints in the anterior margin of the sole or palm, sometimes appearing as slight undulations on the anterior margin with a roughly rectangular shape, with their long axis tangential to the margin of the footprint. Sole/palm surface irregular, cracked and wrinkled.	*Notiomastodon* (Proboscidea, Gomphotheriidae)	Agua Blanca Sequence	Late Pleistocene	Pehuen Co, Buenos Aires Province, Argentina	Aramayo and Manera de Bianco (1987a), Aramayo et al. (2015)
cf. *Proboscipeda australis*	Large circular to slightly oval footprints (length: 0.23–0.27 m, width: 0.23–0.30 m), manus similar to the pes and slightly larger, oval footprints can be wider than long or longer than wide. Commonly 3 to 5 digit imprints in the anterior margin of the sole or palm, sometimes appearing as slight undulations on the anterior margin with a roughly rectangular shape, with their long axis tangential to the margin of the footprint. Sole/palm surface irregular, cracked and wrinkled.	*Notiomastodon* (Proboscidea, Gomphotheriidae)	?	Late Pleistocene	Punta Pelluco, Chile	Aramayo and Manera de Bianco (1987a), Campos-Medina et al. (2022)

(Continued)

TABLE 1.2 The ichnotaxa described in Quaternary formations of South America.—cont'd

Ichnotaxa	Diagnosis/description	Producer	Formation	Age	Locality	Source
Proboscipeda australis	Large and deeply imprinted footprints of roughly circular/ovoidal outline, which are bilaterally symmetrical, almost isodiametrical and usually surrounded by a noticeable annular ridge (marginal rim). Anatomical differences between forefoot and hindfoot impressions are nonexistent. Nevertheless, the manus (length: 240–290 mm; width: 250–300 mm), typically wider than long (length/width ratio: 0.80–0.96), is invariably larger than the pes (length: 230–270 mm; width: 230–250 mm), isodiametric and/or longer than wide (length/width ratio: 1.00–1.10).	*Notiomastodon* (Proboscidea, Gomphotheriidae)	Unnamed unit	Late Pleistocene	Laguna del Monte, Buenos Aires Province, Argentina	Oliva and Arregui (2018)
Felipeda (=Pumaeichnum) biancoi (ichnogenus in need of revision)	Footprints belonging to a digitigrade mammal, of medium size; each little imprint presents the impression of a subtrapezoidal to subtriangular palmar/plantar pad, and the impression of four digital pads. The morphology and dimensions of the footprints match those of the living puma, *Felis* (*Puma*) *concolor*.	Medium sized felid	Agua Blanca Sequence	Late Pleistocene	Pehuen Co, Buenos Aires Province, Argentina	Aramayo and Manera de Bianco (1987b)
cf. *Roepichnus* isp.	A trackway composed of four tridactyl footprints with interdigital web imprint. Footprint length 145 mm, footprint width 120 mm. Digit III is longer than the lateral ones and extend beyond the web imprint. Lateral digits are shorter and are curved inward in the distal end, total divarication (II-IV) 80°.	Anseriform bird	Agua Blanca Sequence	Late Pleistocene	Pehuen Co, Buenos Aires Province, Argentina	Feola et al. (2023)

Ursichnus sudamericanus	Quadrupedal trackway with plantigrade, pentadactyl footprints showing marked heteropody. The manus footprint (11.5–15 cm long, 10–19 cm wide) with oval digit imprints showing parallel orientation and inward rotation with respect to the midline. Palm impression well defined and separated from the digital pad impressions by a ridge, clear claw marks, manus heel impression absent. Pes footprint narrower than manus footprint (20–23 cm long, 10.5–13 cm wide) with a broad sole separated from the heel impression by a notch, digit imprints not connected with the sole, digit III imprint longer.	Carnivore of the Ursidae family (subfamily Tremarctinae)	Agua Blanca Sequence	Late Pleistocene	Pehuen Co, Buenos Aires Province, Argentina	Aramayo et al. (2015)

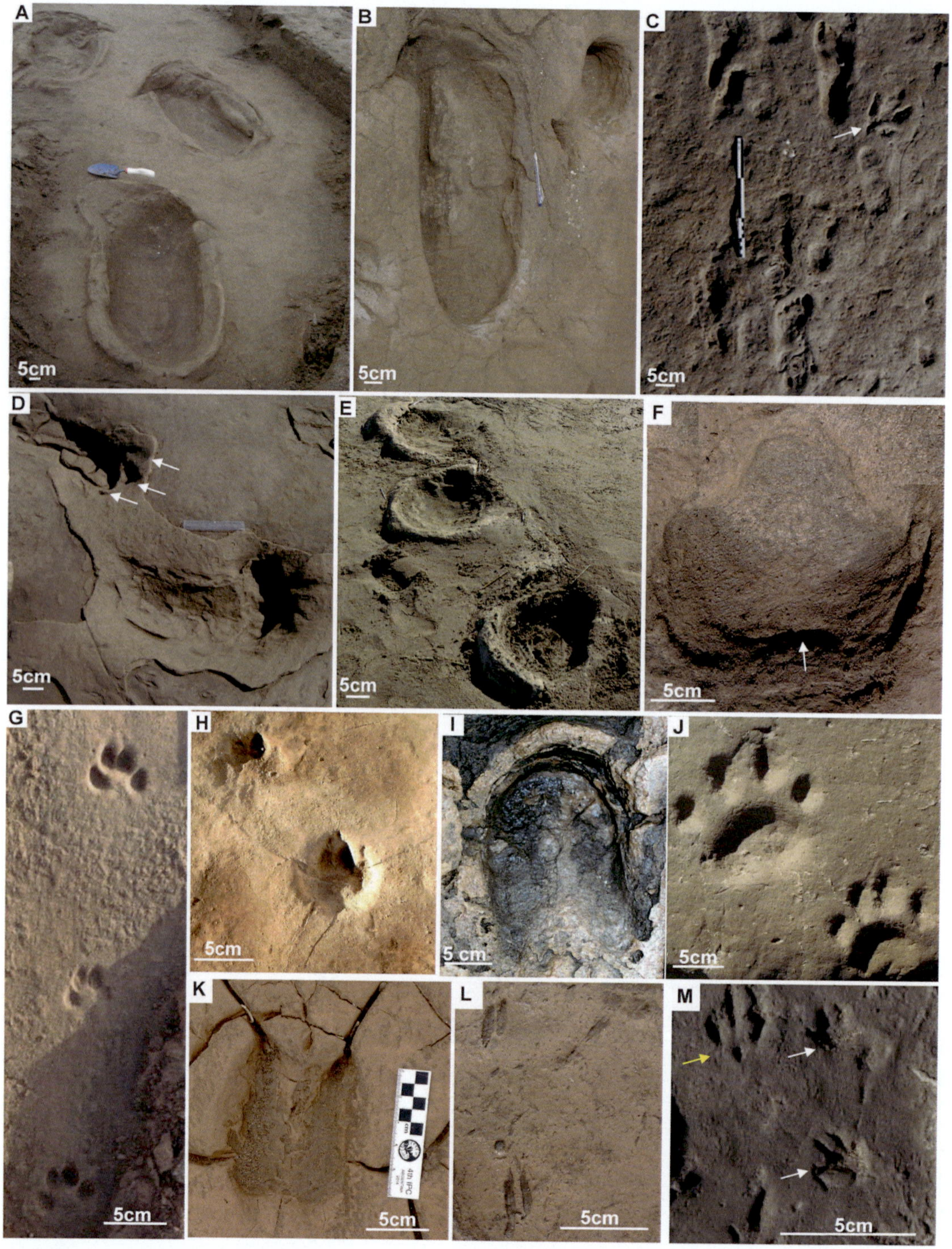

FIGURE 1.2 Mammal ichnotaxa defined in Quaternary deposits from Argentina. (A) Megatheriid tracks from Pehuen Co (Buenos Aires) corresponding to the neotype of *Neomegatherichnum pehuencoensis* (from Aramayo et al., 2015).

When describing Neogene fossil tracks from northwest Argentina, an emended diagnosis for the bird ichnotaxon proposed by Pehuen Co, *Phoenicopterichnum* (Fig. 1.3C–E), was proposed by Melchor et al. (2012). This emendation is based on the type ichnospecies from Pehuen Co. Associated shorebird ichnotaxa include *Charadriipeda* isp. (Fig. 1.3F) and *Gruipeda* isp. (Fig. 1.3G and H); and those of cursorial birds were also identified (Fig. 1.3A and B). Feola et al. (2023b) also introduced the possible occurrence of the bird ichnogenus *Roepichnus* in Pehuen Co. A revision of the ichnology of canid footprints and application to the findings of the Pehuen Co is provided by Melchor et al. (2018). *Pehuencoichnum gracilis* Aramayo and Manera de Bianco (1987b) was synonymized under *Canipeda gracilis* (Vyalov, 1965), and an emended diagnosis for *C. gracilis* (Fig. 1.2G and H) was proposed. The most likely producer of *C. gracilis* in Pehuen Co was a medium-sized fox similar to the extant representatives of *Lycalopex*.

The analysis of tracks and trackways corresponding to gomphotheriids, equids, and camelids, from the Pehuen Co and Laguna del Monte tracksites (both in Buenos Aires Province), allowed Oliva and Arregui (2018) to recognize several supposed pathological signs in the fossil tracks that reflect some of the main pedal affections and alterations experienced by living members of the group. These purported pathological modifications have no ichnotaxonomical implications (Bertling et al., 2006). *Proboscipeda australis* (Aramayo and Manera de Bianco, 1987a) was found to present occasional fissures in its hoofprints, interpreted as fractures on the hornwall (vertical) and cracking of the sole (horizontal). The authors proposed to revise diagnoses of some of the ichnogenera studied (including *Lamaichnum*) to consider pathologies. However, this is not a recommended procedure as pathologies are not accepted ichnotaxobases as they do not represent morphological traits.

Feola et al. (2023a) made a detailed paleoenvironmental assessment and obtained new ages of the Pehuen Co. The track-bearing levels belong to the middle member of the Agua Blanca Sequence. The lower part represents dense to dilute alluvial deposits that were pedogenized. The middle part contains alternating sandy sheet-flood and lacustrine mudflat deposits with sand patches of eolian origin. Most of the fossil tracks with better preservation quality were recorded in the mudflats of a shallow lake sourced by unconfined fluvial

(B) Oblique view of pes (*left*) and manus prints (*right*) of *N. pehuencoensis* from Pehuen Co. (C) Several human footprints and isolated rheid tracks (arrowed) (Bayón et al., 2011) observed at Monte Hermoso (Buenos Aires, Argentina). (D) Mylodontid tracks corresponding to the neotype of *Mylodontidichnum rosalensis* from Pehuen Co (Aramayo et al., 2015) showing three digits with claw marks (arrowed). (E) Oblique view of three proboscidean tracks found at Pehuen Co and assigned to *Proboscipeda australis* by Feola (2022). (F) Tridactyl track attributed to *Eumacrauchenichnus patachonicus* from Pehuen Co. Note prominent central digit and posterior notch (arrow). (G) Canid tracks from Pehuen Co corresponding to the holotype trackway of *Pehuencoichnum gracilis* (Aramayo and Manera de Bianco, 1987b) later combined under *Canipeda gracilis* by Melchor et al. (2018). (H) Right manus-pes pair of *Canipeda gracilis* (Melchor et al., 2018) studied from two slabs collected at Pehuen Co. (I) Isolated equid monodactyl track from Pehuen Co assigned to *Hippipeda* isp. (Aramayo et al., 2015). (J) Holotype of *Felipeda miramarensis* from Agnolin et al. (2019) composed of tetradactyl manus (*left*) and pes (*right*) from Miramar (Buenos Aires, Argentina). (K) Isolated camelid didactyl track from Pehuen Co assigned to *Lamaichnum guanicoe* by Feola (2022). (L) Couple of cervid didactyl imprints assigned to *Pecoripeda commune* from Pehuen Co. (M) Small mammalian tracks compared to *Porcellusignum* isp. (*White arrows*) and rheid tracks (*yellow arrows*) coming from Miramar and assigned by Agnolin et al. (2019). *(A–D, I, L) Courtesy of Teresa Manera de Bianco; (J and M) Courtesy of Federico Agnolin.*

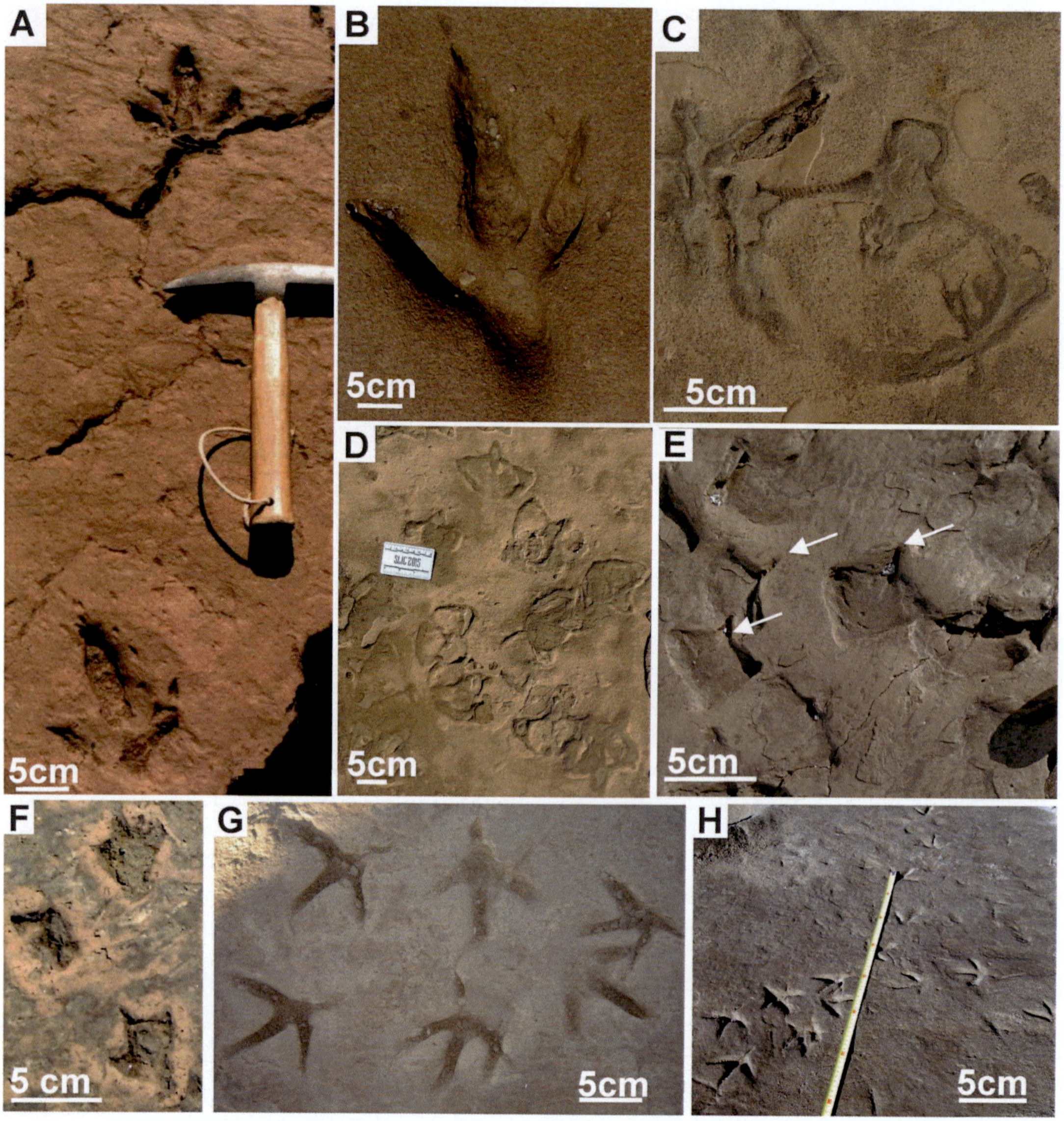

FIGURE 1.3 Bird ichnotaxa described in Quaternary deposits from Argentina (continued). (A) Two consecutive rheid tridactyl avian tracks from a bipedal trackway of *Aramayoichnus rheae* (Aramayo et al., 2015), Pehuen Co. (B) Tridactyl track with clear claw imprints from the type trackway corresponding to *Aramayoichnus rheae* observed at Pehuen Co (Aramayo et al., 2015). (C) Isolated tridactyl webbed track from Pehuen Co assigned to *Phoenicopterichnum rector*. (D) Several tracks of a tridactyl webbed avian attributed to *Phoenicopterichnum rector* found at Pehuen Co (Feola 2022). (E) Moderately preserved avian tracks from Pehuen Co, showing a fan-shaped outline and assigned by Feola (2022) to *Phoenicopterichnum rector*. The arrows point to digit III of some footprints. (F) Ttridactyl webbed avian track of medium to large size from a trackway of *Charadriipeda* isp. (Aramayo et al., 2015), Pehuen Co. (G) Tetradactyl anisodactyl tracks compared with *Gruipeda* isp. (Aramayo et al., 2015), Pehuen Co. (H) Oblique photo of several tetradactyl tracks compared with *Gruipeda* isp. *(C–E) Courtesy of Teresa Manera de Bianco.*

currents. The percentage of eolian deposits increases up section. Locally, a sandy plain of fluvial origin is developed at the top of the section. The new OSL ages, along with a previous radiocarbon date, indicate that the sedimentation of the track-bearing levels would have extended approximately between 28.5 and 12 ka.

In addition to numerous bones, some tracks assigned to middle-sized ground sloths and *Macrauchenia* were found in Playa del Barco, to the west of the Pehuen Co ichnosite (Bayón et al., 2011). Both of them are conspicuous members of the faunistic association of the Late Pleistocene, dated by radiocarbon method to 16.44 ± 0.32 ka.

Fossil tracks were found in lacustrine deposits of the Santa Clara Formation (Mar Chiquita, Buenos Aires Province). Ichnotaxa described by Tassara et al. (2005) include cf. *Mylodontidichnum rosalensis* (both trackways and trampled grounds), cf. *Lamaichnum guanicoe*, cf. *Hippipeda* isp., and cf. *Odocoileinichnum commune* (=*Pecoripeda commune*). Also, scratch traces associated with cf. *M. rosalensis* were described. The presence of *Eumacrauchenichnus patachonicus* in the Santa Clara Formation was briefly documented by Medina and Tassara (2023). These tracks have a minimum age of 21 ± 0.7 ka (overlying tuff level).

The ichnospecies *Felipeda miramarensis*, from the Punta Hermengo site, Buenos Aires Province, was described by Agnolín et al. (2019). This ichnospecies is assigned to machairodontine felids and particularly to *Smilodon populator* (Fig. 1.2J). This site also contains tracks compared with *Porcellusignum* and *Aramayoichnus* (Fig. 1.2M).

Quintana et al. (1998, 2001) provided a brief account of the finding of fossil tracks in a Late Pleistocene–Holocene archaeological site (rocky shelter) in the Tandil hills (Buenos Aires Province). The tracks were attributed to large mammals (14 cm in diameter, 10 cm deep), but the poor preservation precluded a definite ichnotaxonomic assignment.

In October 1832, Charles Darwin found the first fossil remains recovered from the Lower Pliocene layers of the Farola Monte Hermoso fossil locality, stratigraphically named the Monte Hermoso Formation, on the coast of Buenos Aires Province (Bayón et al., 2011). Farther east, Late Pleistocene deposits from the Monte Hermoso site (near the homonymous city), overlying those from the Monte Hermoso Formation, include two new ichnotaxa assigned to ground sloths that were proposed by Casamiquela (1983). The ichnotaxa are *Acugnaichnus dorregoensis* and *Iribarichnum megamericanum*. However, both were erected on the basis of poorly preserved material. Politis and Bayón (1995) found some 420 human footprints at Monte Hermoso 1 site, associated with those of other mammals and birds (Fig. 1.2C), in a former lagoon between the dunes near the ocean surrounded by the typical humid interdune vegetation. There are 18 AMS radiocarbon dates from the site and two conventional ones, which indicate that the marsh lasted between 7920 and 6600 years 14C BP (Bayón et al., 2011).

The ichnotaxon *Acugnaichnum dorregoensis* was compared with the prints made by scelidotheriines (*Scelidotherium* sp.) on the basis of two isolated and poorly preserved tracks from the Late Pleistocene (?) of the Laguna del Monte site, Buenos Aires Province (Toledo and Arregui, 2022). A detailed anatomical comparison of scelidotheriine distal limbs with some fossil tracks is provided. The anatomical correlation builds upon the original suggestion of Casamiquela (1983) that the producers of *Acugnaichnus dorregoensis* were scelidotheres, on the basis of what is interpreted as the impression of the metatarsal V.

The record of trampled surfaces with numerous artiodactyl fossil tracks in the Holocene Mar Chiquita Formation, Buenos Aires Province, was preliminarily described by Medina and Tassara (2022). The tracks were compared with *Lamaichnum guanicoe* and *Pecoripeda commune*.

Alonso (1987, 2012) described a facies zonation of bird tracks from the (modern) Laguna de los Pozuelos of Jujuy Province, northwest Argentina. Flamingo-like tracks from the Pleistocene Blanca Lila Formation are considered a potential fossil analog (no detailed ichnotaxonomical treatment is provided). Laguna de los Pozuelos is considered a modern analog of continental borate deposits where bird track types are arranged in three zones: (1) terrestrial to emergent mudflat (sand and silt), dominated by tetradactyl bird tracks with marked phalangeal pads (mainly Passeriformes, occasionally Rheiformes); (2) submerged mudflat to shallow littoral (silt and clay), characterized by tridactyl bird tracks, including small palmate tracks (assigned to Charadriiformes and Anseriformes); and (3) a deeper littoral zone (clay and evaporite muds), characterized by large palmate bird tracks (comparable with those of Phoenicopteriformes and Anseriformes). These three zones cover a water depth range from subaerial to 50 cm deep.

Finally, Leonardi (1994) mentioned the presence of bird tracks in Pleistocene terraces at Sijes II site, Salta Province.

3.2.2 Brazil

Latest Pleistocene fluvial deposits of Tamanduá de Cima, Pernambuco, contain four morphotypes of mammal tracks: *Lamaichnum* isp., *Pecoripeda* isp., *Hippipeda* isp, and those of a large tridactyl mammal with poor preservation, probably produced by *Toxodon* (Oliveira et al., 2019). The latter is the first track that was attributed to *Toxodon* in South America. The sequence was dated between 19 and 15 ka.

Canid tooth traces, as the first bioerosional structures for the Quaternary of South America, were described by Araújo-Júnior et al. (2017) in Late Pleistocene-early Holocene age mammal-bone remains recovered from Lajedo da Escada caves, state of Rio Grande do Norte, northeastern Brazil. The ichnospecies *Machichnus bohemicus* and *Machichnus fatimae* isp. nov. were recognized. *Machichnus fatimae* and *M. bohemicus* observed in *Glyptotherium* sp. are indicative of predation by an adult individual of the large-sized canid *Protocyon troglodytes* outside the cave. *Machichnus bohemicus* recognized in *Eremotherium laurillardi* and *Smilodon populator* was generated by either a juvenile individual of *P. troglodytes* or an adult of the small-sized canid *Cerdocyon thous*, in a scavenging context, outside the cave.

3.2.3 Chile

The oldest record of lamines (Camelidae: Lamini) and proboscideans (Proboscidea: Gomphotheriidae) in the southwestern margin of the Andes mountain range was described recently by Campos-Medina et al. (2022) in the Late Pleistocene at the Pelluco fossil forest sanctuary, Llanquihue Province, southern Chile. Tracks were assigned to *Lamaichnum guanicoe* and *Proboscipeda* cf. *P. australis*. The tracks described as *L. guanicoe* have poor-to-moderate preservation quality. The sequence containing the tracks was dated between 54.3 and 44.3 ka (see Campos-Medina et al., 2022).

In the city area of Antofagasta, near Quebrada la Chimba, Plio-Pleistocene deposits in an outcrop of a few hundred meters showed numerous trackways made by various individuals of the same species. The best are 12 tracks organized in two trackways and first identified as *Ichnhippus cotaposi* (Casamiquela and Chong Díaz, 1975), an ichnogenus that is probably a junior synonym of *Hippipeda* (Vyalov, 1966). The subcircular tracks are preserved in sandstone and display a morphology similar to modern horse tracks. They occur in a Plio-Pleistocene shallow marine sequence with poor age control, which probably represents a beach facies.

3.3 North America

3.3.1 *Bahamas*

The review of the last interglacial (Marine Isotope Stage 5e) stratigraphic record from the Boiling Hole exposure in northern Eleuthera Island revealed the occurrence of two vertically stacked upward shallowing sequences of oolitic coastal deposits showing beach facies (Kindler et al., 2008). One paleosurface shows a 2 m-long trackway made of 19 tracks and compared to *Charadriipeda*. This is the first reported occurrence of vertebrate trace fossils from the Bahama Archipelago. Two consecutive avian tracks from the Grotto Beach Formation (Pleistocene) of San Salvador Island (Bahamas) were described by Martin and Whitten (2015), occurring in cross-bedded oolitic-bioclastic grainstone. The tracks are in beach facies located below a paleosol dated from Marine Isotope Substage 5e (ca. 120 ka). They are compared with those of modern gulls. No ichnotaxonomic assignment was proposed.

The tracks of birds have also been found in Holocene deposits on Caicos and San Salvador in the Bahamas (White et al., 2012).

3.3.2 *Bermuda*

During the last interglacial (ca. 100 ka), there was the deposition of carbonate sands during storms along southern Bermuda. A trackway of the semipalmate webbing shorebird willet *Catoptrophorus* was described from Charles Island, in the foreset of the lee side of a beach ridge, probably indicating sheltering (Hearty and Olson, 2011).

3.3.3 *Canada*

Wally's Beach is a Late Pleistocene site in southern Alberta at the St. Mary Reservoir, which yields abundant tracks (McNeil et al., 1999a,b, 2004, 2005, 2007; Hills and McNeil, 1999; Milot and Burns, 2006). Numerous tracks and trackways of woolly mammoth (*Mammuthus primigenius*), Pleistocene Western camel (*Camelops hesternus*), Mexican half-ass (*Equus conversidens*), ancient bison (*Bison b. antiquus*), and caribou (*Rangifer tarandus*) have been found (McNeil et al., 2007, Fig. 4). Multiple trackways demonstrate social behavior in mammoths, camelids, and equids. Fossil mammoth (*Mammuthus primigenius)* tracks were found in eolian sediment dated 11,300–11,000 years BP. By comparing the size distribution of 51 of these tracks with those of modern African elephants (*Loxodonta africana*), the age distribution of this mammoth population was determined by McNeil et al. (2005). Containing far fewer juveniles than would be expected for an expanding or stable population, these tracks provided the first evidence that a living mammoth population, coexisting with human inhabitants, was in decline.

Gibbard and Dreimanis (1978) described proboscidean tracks from Pleistocene lake sediments in southwestern Ontario.

3.3.4 *Mexico*

The Pie de Vaca Formation yields a diverse ichnofauna of Pleistocene age preserved in lacustrine strata in the town of Tepexi de Rodríguez, south-central Puebla (Applegate et al., 1984; Cabral Perdomo and Applegate, 1991a,b, 2000; Rodríguez de la Rosa et al., 2004). The main track site contains nine camelid trackways, three trackways of large felids, and trackways of small artiodactyl, small ungulates, and birds. Other smaller track sites in the area yield tracks of proboscideans, birds (Phoenicopteriformes and Charadriiformes), camelids, felids, small ungulates, and artiodactyls (Cabral Perdomo, 1995, 2000). Cabral-Perdomo (2000, 2013) made a brief account of the ichnofossil localities of the Pie de Vaca Formation. A total of 283 tracks and 14 trackways were identified that are linked with mammals and birds. The camelid tracks were assigned to *Lamaichnum guanicoe* and were linked to members of the genus *Hemiauchenia*, which is the most common camelid of the late Cenozoic of Mexico (Cabral-Perdomo et al., 2018). The tracks of felids were only referred to as Felipedidae due to the poor morphological preservation and were probably produced by a medium to large-sized felid such as a machairodontid or pantherine. Two-toed artiodactyl tracks from the Plio-Pleistocene Pie de Vaca Formation were used to erect the new ichnotaxon *Tayassuichnum felixarangutii* on the basis of five tracks composing a single trackway, with moderate to poor preservation quality (Bravo Cuevas et al., 2019). Their size and morphology are comparable with those produced by extant peccaries, and the ichnotaxon is considered comparable with *Entelodontipus* Casanovas-Cladellas and Santafé-Llopis, 1982 (see chapter in this volume on Paleogene/Neogene vertebrate tracks). A member of *Platygonus*, the most common peccary for the late Cenozoic of North America, is proposed as the potential track maker.

Relatively poorly preserved footprints interpreted as those of humans, canids, felids, cloven-hoofed mammals, and a bird are preserved on desiccation-cracked layers of tephra near the top of the Xalnene Ash in a quarry about 1 km west of Cerro Toluquilla, near Puebla, Mexico. The Xalnene Ash is interpreted to have been deposited from base surges and subaqueous density flows along a lake shoreline (Gonzalez et al., 2006). Although the age of the tracksite, as well as the human footprints, are disputed, it can be confidently attributed to the Pleistocene.

Dávalos-Álvarez et al. (2007) described the Cenozoic stratigraphy of the Tehuacán Valley, Oaxaca, Mexico. They included a brief description and illustration of supposed felid and camelid footprints. The felid footprints were assigned to *Canipeda* isp. by Melchor et al. (2018) because of the overall morphology and presence of claw traces that are not present in felid footprints. The camelid footprints are roughly comparable with *Lamaichnum*.

Human and poorly preserved ungulate tracks occur in tufa at Cuatro Ciènegas (Coahuila) (Rodríguez de la Rosa et al., 2004; Gonzalez et al., 2006; Gonzalez et al., 2007). Gonzalez et al. (2007) mentioned the presence of poorly preserved cervid tracks and a possible single duck track. Cuatro Ciènegas site was dated to ca. 7.24 ka (González et al., 2009; Felstead et al., 2014).

Large tracks of probably jaguar or cougar were first identified by Dugès (1897) in Plio-Pleistocene deposits at San Juan de los Lagos, State of Jalisco. However, the exact age of the track-bearing beds is uncertain, but most likely Late Pleistocene (Lucas, 2008). This author briefly described and illustrated the *Avipeda* and *Felipeda* tracks. Rodríguez de la Rosa and Guzmán Gutiérrez (2012) provide the first comprehensive description and ichnotaxonomy of this large tracksite formed in lacustrine deposits. Four avian track morphotypes were mentioned, although no ichnotaxonomic assignment was made. In addition, *Lamaichnum macropodum*, the new ichnotaxon *Mitsupes dugesii*, and proboscidean tracks were also described. *Mitsupes dugesii* was attributed to a feline producer within the Machairodontinae. *Mitsupes* is considered a junior synonym of Felipeda (see chapter on Paleogene/Neogene vertebrate tracks in this volume).

3.3.5 Nicaragua

On the southern shore of Lake Managua, at the El Cauce, located in the western area of the current City of Managua, hundreds of footprints were revealed in a quarry in 1874. Nowadays, it is the Acahualinca Footprint Museum (Lucas et al., 2023). They were preserved in consolidated volcanic mud covered by volcanic ash, dated to 2–2.2 ka (Schmincke et al., 2010; Lucas et al., 2023). A total of 322 human footprints, together with rarer (33) nonhuman tracks first noted by Crawford (1891), including deer (Odocoileinae), a dog, opossum, a large peccary, lizards, and birds (Brown, 1947; Williams, 1952; Bryan, 1973; Lockley et al., 2008). Bison and tapir tracks have been documented and illustrated from the nearby and stratigraphically equivalent El Recreo Site (Williams, 1952). The casts of some of the bison tracks are apparently all that is left from this site (Lucas et al., 2023). The bison tracks are important as they document the much later survival of bison in Central America than previously thought (Alvarado et al., 2008; Lucas and Alvarado, 2017). Lockley et al. (2008) reviewed the nonhuman tracks at Acahualinca and were able to document only a single bird track the size of the Canadian goose assigned to the ichnogenus *Anatipeda* and trackways of opossum and deer. Tracks were also found near San Rafael del Sur, on the coast, 64 km SE of Managua.

3.3.6 USA

Hunt and Lucas (2007) provided the first comprehensive review of Cenozoic vertebrate tracksites in North America. The ichnotaxa formally erected can be found in Table 1.3. Here, we include the Quaternary sites described by these authors and update the list of sites by state with new information. The highest latitude tracksites are found in southwestern and central Alaska. Lea (1996) described deformational structures interpreted as vertebrate hoof- and foot-tracks within Upper Pleistocene aeolian sand sheet deposits in the stabilized Kantishna sand sea and in the Nushagak lowland. Artiodactyl and proboscidean tracks, probably from those of woolly mammoth, are interpreted in cross-sections. Preservation of roots and rhizoliths within the eolian deposits suggests that vegetated sand sheets may have formed a locally important grazing habitat for large herbivores during at least part of the last glaciation.

Tracks and trackways have occasionally been uncovered in the course of mining the erosional remnants of a zeolite bed in the Bowie Zeolite Mines of the San Simon Valley, State of Arizona (Eyde, 1978, 1982; Thrasher, 2007). Tracks represent small and large camelids, a proboscidean, a carnivore, and, from a nearby quarry, aquatic birds, including the swan *Cygnus* (*C.*) sp. The tracks are Plio-Pleistocene (late Blancan) in age (Thrasher, 2007). Gardner (1940) reported Pleistocene tracks that could represent an artiodactyl, ursid, or sloth in northwestern Arizona. Elephant-foot-sized depressions on a buried occupation surface at the Murray Springs archaeological site in Arizona were interpreted as mammoth tracks (Haynes, 1973; Haynes and Huckell, 2007), found in a paleosurface dated around 13 ka. Besides making tracks in saturated surficial sediments, mammoths had also scraped the ground in places within a dry or sluggish stream bed, possibly in search of water (Haynes, 1991). Tracks of equids and bovids were also found.

A mammalian ichnofauna is present in the lacustrine marginal sediments in the Borrego Badlands of Anza-Borrego Desert State Park, California (Bucheim et al., 1999; Remeika, 2006). The tracksites range in age from Middle-to-Late Pleistocene (Irvingtonian ca. 850 ka to Rancholabrean; Remeika, 2006). The San Felipe Ichnofauna was introduced in Remeika (2006), including the large llamoid (Camelidae) hoofprints *Megalamaichnum albus* and the

TABLE 1.3 The ichnotaxa described from Quaternary formations of North America.

Ichnotaxa	Diagnosis/description	Producer	Formation	Age	Locality	Source
Avipeda isp.	Tridactyl tracks in which digit III is 20–23 mm long, digit III is ~35% longer than the lateral digits and total digit divarication is ~100-120°.	–	–	Likely Late Pleistocene	San Juan de los Lagos, Jalisco State, Mexico	Lucas (2008)
cf. *Avipeda*	Nonwebbed tridactyl track. Length varies from 3 to 5.5 cm (average 4.2 cm), and width from 4.5 to 7 cm (average 6 cm). Digit III usually is the most deeply impressed (up to 0.8 cm). The digits are relatively broad and their tips are U to V shaped and without claw impressions. Tracks with three digit prints mostly have a relatively pronounced heel region where the three digits merge together.	It is suggested that the trackmaker belonged to the Order Charadriiformes and was quite possibly one of the following: American oystercatcher (*Haematopus palliates*), greater yellowlegs (*Tringa melanoleuca*), blacknecked stilt (*Himantopus mexicanus*), or stilt sandpiper (*Calidris himantopus*).	The oolitic-peloidal Unit II	Late Pleistocene (MIS 5e)	Boiling Hole cove, Eleuthera Island, Bahamas	Kindler et al. (2008)
Bijugopeda anterofossa (*Bijugopeda* is a *nomen dubium* based on an undiagnostic and incomplete(?) camel footprint, according to Lucas and Hunt, 2007). But see a different opinion in the chapter on Paleogene/Neogene vertebrate tracks in this volume.	Large-sized tracks. Imprints of manus and pes of similar size and shape. Width of impression equals, or more typically, exceeds length, giving an overall laterally oval appearance. No claws visible along rounded anterior edge. Individual digit impressions relatively thick, leaving a narrow medial trough and small medial pocket. Anterior of digital impressions narrower than posterior, resulting in the widest part of the medial pocket being offset anteriorly. Medial trough open both anteriorly and posteriorly. Axes of impressions are convergent anteriorly and posteriorly, anterior interdigital angle is obtuse, generally greater than 100°.	Bison, likely *Bison antiquus*	Pleistocene glacial till, sand and gravel unconformably overlying the St. Mary River Formation	Late Pleistocene	Wally's Beach, Alberta, Canada	McNeil et al. (2007)

Bijugopeda bisymmetrica (*Bijugopeda* is a nomen dubium or junior synonym of *Lamaichnum* based on an undiagnostic and incomplete(?) camel footprint, according to Lucas and Hunt, 2007). But see a different opinion in chapter on Paleogene/neogene vertebrate tracks	Large sized tracks. Width of impression approximately equals, or occasionally exceeds, length, giving an overall circular appearance. Individual digit impressions of uniform thickness, leaving a wide medial trough and large medial pocket. Digital impressions of constant thickness, making the directionality difficult to determine. Medial trough open both anteriorly and posteriorly. Axes of impressions are convergent anteriorly and posteriorly, anterior interdigital angle is acute to right-angled, generally being 90° or slightly less.	Caribou, likely *Rangifer tarandus*	Pleistocene glacial till, sand and gravel unconformably overlying the St. Mary River Formation	Late Pleistocene	Wally's Beach, Alberta, Canada	McNeil et al. (2007)
Broomichnium flirii (Fig. 1.4A)	*Broomichnium* with a long straight median imprint opposite the paired imprints. Outer and inner imprints may show striations, which correspond to small projections in the imprint. Commonly occurs in a linear series of several individual imprints, but may also occur in isolation. Note: possibly represents a fish launching off the substrate.	Small freshwater sculpin, *Cottus* sp.	Sediments of Pleistocene Lake Hitchcock	Late Pleistocene (17.9 −12.6 cal ka)	Connecticut, Massachusetts, Vermont and New Hampshire, USA	Benner et al. (2008)
Camelopichnum (considered *nomem nudum*: no photo or description available)	Large camel track, measuring 27 cm by 24 cm	*Camelops hesternus and/or C. huerfanensis*	Borrego badlands	Middle Pleistocene	Anza-Borrego Desert State Park, California, USA	Remeika (2006)
Canipeda isp. (originally described as felid footprint)	Large metapodial pad, digital pads with a distal pointed end, probably suggesting the presence of claws	Unknown canid	"Sedimentos lacustres Villa Alegría"	Pliocene-Pleistocene	Oaxaca, Mexico	Dávalos-Álvarez et al. (2007), Melchor et al. (2018)

(*Continued*)

TABLE 1.3 The ichnotaxa described from Quaternary formations of North America.—cont'd

Ichnotaxa	Diagnosis/description	Producer	Formation	Age	Locality	Source
Chelipus isp. (junior synonym of *Canipeda* according to Sarjeant et al., 2002)	Paw prints in the *Chelipus* trackway are identical to, but larger than, those made by modem coyotes.	*Canis latrans*	Borrego badlands	Middle Pleistocene	Anza-Borrego Desert State Park, California, USA	Remeika (2006)
Felipeda isp. (synonymized with *Mitsupes dugesii*, Rodríguez de la Rosa and Guzmán Gutierrez (2012))	Two cat tracks are 103 × 117 (width x length) and 84 × 113, respectively. They have four oval pads that form a semicircle in front of the metacarpal/metatarsal pad, and no claw tip imprints.		–	Likely Late Pleistocene	San Juan de los Lagos, Jalisco State, Mexico	Lucas (2008)
Hippipeda cardstoni (Fig. 1.4D)	Moderate sized impressions of the *Hippipeda* type. Manus and pes prints similar in size and shape. Hoof walls well-defined, relatively broad, and of constant thickness around the circumference of the track. Anterior section of hooves semicircular with parallel or slightly expanding walls extending posterior, so that the width is approximately equal to the length of the tracks. Posterior bars slightly oblique and continuous, creating isolated sole imprint in the center of the impression. Shallow frog exists at the posterior.	*Equus conversidens* (*E. lambei* according to Renders and Vincelette, 2023) (*E. scotti* in Fossil Lake)	Pleistocene glacial till, sand and gravel unconformably overlying the St. Mary River Formation; Marble Bluff biotite tuff	Late Pleistocene	Wally's Beach, near Cardston, Alberta, Canada; Fossil Lake, eastern Oregon, USA	McNeil et al. (2007), Vincelette (2021), Renders and Vincelette (2023), Retallack et al. (2018)
Hippipeda downsi	Ovoid to rectangular in shape, distinguished by a single central digital pad impression. There is no evidence or impressions of lateral digits (side toes). The digital pad (sole) is bordered anteriorly by a strongly curved digit and hoof wall. The sole is well pronounced, concave downward to the bar (the support structure of the hoof wall	*Dinohippus* sp. or *Equus simplicidens*	Borrego badlands	Middle Pleistocene	Anza-Borrego Desert State Park, California, USA	Remeika (1999, 2006)

	that extends from the heel and ends at the point of the frog), defining the posterior ends of the pad. The pad surrounds a nondescript V-shaped frog (a rubbery cushionlike structure in the rear of the underside of the hoot) that is either subtly expressed or absent. The overall shape of the manus (forefoot) print is, in most examples, wider than long. Overall length of adult footprints is 6.2 cm; overall print breadth is 12.0 cm.					
Lamaichnum borregoensis (considered junior synonym of *Lamaichnum guanicoe* by Lucas and Hunt, 2007)	Small- to moderate-sized tracks that average 12 cm in length, showing digital cushions (pads), and have an oval to rounded-rectangular shape, with two distinct weight-bearing hooves (pads plus toenail) in each foot of both manus and pes The medial (toward the center) and lateral (toward the outside) hooves are nearly identical. The axes of these hooves are parallel or splayed outward to the front, separated by a linear interdigital (between the toes) crest line. The claw-like toes point forward. Each hoof is broadest at the heel, which is round to sharply parabolic in shape with apexes (points) rounded in the manus and pointed to round in the pes.	*Hemiauchenia* sp.	Borrego badlands	Middle Pleistocene	Anza-Borrego Desert State Park, California, USA	Remeika (1999, 2006)
Lamaichnum macropodum	Footprints of suboval outline with two clearly defined hooves, separated by a V-shaped interdigital space, with a slightly concave depression in the heel. Footprint length 160 mm, footprint width 117. One of the digits is slightly narrower.	Large camelid	San Juan de Los Lagos site	Plio-Pleistocene	Jalisco, México	Sarjeant and Reynolds (1999), Rodríguez de la Rosa and Guzmán Gutiérrez (2012)

(*Continued*)

TABLE 1.3 The ichnotaxa described from Quaternary formations of North America.—cont'd

Ichnotaxa	Diagnosis/description	Producer	Formation	Age	Locality	Source
Lamaichnum sarjeanti (Fig. 1.4C)	A large ichnospecies of *Lamaichnum*. Prints heart-shaped in outline. Claws are absent resulting in blunt rounded termination of the distal digital cushion. A narrow and deep interclavular gap penetrates almost half of the length of the impression, widening toward the center to form a small medial pocket; proximal gap is shallow and forms a wide U-shape. Axis of impressions roughly symmetrically curved with the greatest divergence located two-thirds toward the posterior, interdigital angle approximately 40°, but changing according to the substrate. Digits roughly equal in length.	*Camelops* (*Camelops hesternus* in Fossil Lake)	Pleistocene glacial till, sand and gravel unconformably overlying the St. Mary River Formation; Marble Bluff biotite tuff	Late Pleistocene	Wally's Beach, Alberta, Canada; Fossil Lake, eastern Oregon, USA.	McNeil et al. (2007), Retallack et al. (2018)
cf. *Lamaichnum* isp. (originally described as felid footprint)	Suboval to semicircular footprints with two claw marks separated anteriorly.	Unknown camelid	"Sedimentos lacustres Villa Alegría"	Pliocene-Pleistocene	Oaxaca, Mexico	Dávalos-Álvarez et al. (2007)
Megalamaichnum albus (considered a junior synonym of *Lamaichnum macropodum* by Lucas and Hunt (2007))	Moderate- to large-sized (overall length 18.5 cm by greatest breadth 14.5 cm) llamoid tracks. The medial and lateral hooves, two per foot, are mirror images in outline separated by a narrow linear interdigital crest line. Hoofprints are widest near the heel, and taper forward to an apex at the toe; the rounded heel forms a broad parabolic curve.	*Blancocamelus* or possibly *Camelops minidokae*, or *Paleolama* sp.	Borrego badlands	Middle Pleistocene	Anza-Borrego Desert State Park, California, USA	Remeika (2001, 2006)
Mammuthichnum isp. (*nomem nudum* according to Lucas et al. (2007))		*Mammuthus columbi*	Borrego badlands	Middle Pleistocene	Anza-Borrego Desert State Park, California, USA	Remeika (2006)

Mitsupes dugesii; Combined under *Felipeda dugesii* by Abbassi et al. (chapter Paleogene/Neogene vertebrate footprints)	Felid-like footprints of medium size (9–11 cm long, 9–10 cm wide), with an asymmetrical metapodial pad with slightly subtriangular to strongly reniform shape. Manus deeply imprinted in contrast to the shallower impressions of the pes, indicating a heavy weight on the front of the body.	Machairodontinae (Homotherinii)	San Juan de Los Lagos site	Plio-Pleistocene	Jalisco, México	Rodríguez de la Rosa and Guzmán Gutiérrez (2012)
Pecoripeda enigmii (*nomen nudum*)	A medium-sized species of *Pecoripeda*. Digital impressions of manus are longer than wide. Outer margins of pads are slightly convex, with the pad margins curving in more strongly at anterior termination. Inner margins of pads are gently concave. Toes form blunt points that are directed forward, slightly converging with an interdigital angle of approximately 15°. Pads are typically splayed so that the widest part of track is toward the anterior. Individual pads may or may not be in contact at the posterior of the impression. Dew claw impressions are common. The pes impression is shorter, giving an overall trapezoidal shape to the pes, with the width nearly equal to the length. The outer pad margins are more strongly convex, and the inner margins more strongly concave, giving the overall shape of each digit impression as a recurved wedge. Hoof terminations are sharp points and face forward with no, or only slight interdigital convergence. Pads are splayed with the widest part of the overall impression at approximately the mid-point. Individual pads are usually not in contact with the posterior of the impression. Dew claws are rarer in pes than manus.	Unknown species of artiodactyl	Pleistocene glacial till, sand and gravel unconformably overlying the St. Mary River Formation	Late Pleistocene	Wally's Beach, Alberta, Canada	McNeil et al. (2007)

(*Continued*)

TABLE 1.3 The ichnotaxa described from Quaternary formations of North America.—cont'd

Ichnotaxa	Diagnosis/description	Producer	Formation	Age	Locality	Source
Piscichnus isp.	Bowl-shape trace fossils, ovoid to tubular in plan view, composed of a pebbly, fine-grained sandstone core surrounded by an adjacent zone containing crudely imbricated pebbles. The cross-sectional width and height of the columnar structures is between 4.4–15.6 cm and 26.1–54.5 cm, respectively. Inclined or recumbent occurrences typically contain one side that dips at a shallow angle (30°), whereas the opposing side is steeply inclined to near vertical (60°-85°).	Walrus feeding excavations	Older terrace deposits of the Columbia River	Middle Pleistocene	North Cove, Willapa Bay, Washington, USA	Gingras et al. (2007)
Piscichnus isp.	Most visible in cross-section as a dish-shaped unconformity. The deformation caused by the original unconformity is propagated upward, in some cases 30 cm, via draping by later varves. In some *Piscichnus* there are multiple unconformities within one vertical sequence, possibly indicating biological reactivation of the site. In plan view, excavations show *Piscichnus* to be between 10 and 12 cm in diameter and approximately 2 cm deep.	Freshwater fish	Varved clay and silt to fine sand	Late Pleistocene (17.9 −12.6 cal ka)	Glacial Lake Hitchcock and glacial lakes in the Connecticut Valley and Glacial Lake Marrimack	Brenner et al. (2009)
Proboscipeda panfamilia (Fig. 1.4B)	Large sized impressions of the *Proboscipeda* type. Only the pes prints typically visible, as the manus impressions are overprinted by the pes in typical locomotion. Impressions are circular to suboval in outline, with the width being the greatest dimension. Three to five	*Mammuthus primigenius*	Pleistocene glacial till, sand and gravel unconformably overlying the St. Mary River Formation	Late Pleistocene	Wally's Beach, Alberta, Canada	McNeil et al. (2007)

	short wide toe impressions visible on the anterior portion of the impression. In cross section the mechanics of the footstep cycle can be seen with indications of the deep initial heel strike, a central raised rollover section and a final deep toe pushoff at the anterior.					
Proboscipeda panfamilia	As in modern elephants the wide prints were identified as manus, and the narrower ones as pes. Not all prints were as clear with indications of nails: some were broadly circular with concentric bedding, and may be undertracks impressed on moister ground. A particularly unusual feature of this trackway is the differential depth, with the right manus and pes imprints much deeper than the left pes, and left manus intermediate depth, in the middle of the trackway. Another unusual feature is that the tracks are closely spaced in a sinuous path, unlike the near-linear arcs and well-spaced large spherical tracks of elephants elsewhere	*Mammuthus columbi*	Marble Bluff biotite tuff; gypsiferous sediments	Late Pleistocene	Fossil Lake, Oregon, USA; White Sands National Park, New Mexico, USA	McNeil et al. (2007), Retallack et al. (2018), Bennett et al. (2020)
Tayassuichnum felixarangutii	The specimens are preserved in concave epirelief. The number and arrangement of the footprints are indicative of a quadrupedal individual. The footprints have a rounded shape, with two hoof imprints that correspond to digits III and IV. The manus imprints are somewhat larger than the pes imprints. Footprint length of manus/pes is about 5 cm. Hoof imprints are longer than wide. Toes are splayed. Tips of the hooves directed forwards. Dew claws imprints absent.	Peccary footprints (Artiodactyla, Tayassuidae)	Pie de Vaca Formation	Plio-Pleistocene	Pie de Vaca, México	Bravo Cuevas et al. (2019)

(*Continued*)

TABLE 1.3 The ichnotaxa described from Quaternary formations of North America.—cont'd

Ichnotaxa	Diagnosis/description	Producer	Formation	Age	Locality	Source
Undichna britannica	Occur as two out of-phase waves in the median and occasionally discontinuous-to-continuous waves toward the outer margin of the trail. Wavelength of the caudal fin trail (highest amplitude wave) ranges from 3.5 to 30 cm and amplitude ranges from 1 to 9.5 cm. Using relationships of fish length to body waves, fish making U. britannica could range from 3.2 to 72.2 cm	Coldwater fish	Varved clay and silt to fine sand	Late Pleistocene (17.9 −12.6 cal ka)	Glacial Lake Hitchcock and glacial lakes in the Connecticut Valley and Glacial Lake Marrimack	Benner et al. (2009), Higgs (1988)
Undichna bina	Consists of two adjacent, in-phase waves, separated by some distance. The low amplitude of these waves indicates a more anterior appendage creating the trail. They range in wavelength from 2.5 to 7 cm and have low amplitudes from 0.25 to 0.33 cm.	Coldwater fish	Varved clay and silt to fine sand	Late Pleistocene (17.9 −12.6 cal ka)	Glacial Lake Hitchcock and glacial lakes in the Connecticut Valley and Glacial Lake Marrimack	Anderson (1976), Benner et al. (2009)
Undichna simplicitas	A more complex trace fossil preserving traces of in-phase waveforms at the outer margins or interior of the trail in addition to one or more median, higher amplitude out of phase waves. Included in the ichnospecies are trails that preserve discontinuous outer in-phase waves. Caudal amplitudes range from 1.5 to 3.8 cm and wavelengths of 6.2 −9.7 cm. Total length estimates for fish making this trail range from 7 to 42.2 cm.	Coldwater fish	Varved clay and silt to fine sand	Late Pleistocene (17.9 −12.6 cal ka)	Glacial Lake Hitchcock and glacial lakes in the Connecticut Valley and Glacial Lake Marrimack	Anderson (1976), Benner et al. (2009)

Undichna unisulca	It is characterized by a single sinusoidal wave. The epichnial groove (or hypichnial ridge) may be more or less broad depending on substrate conditions at the time the trace was made and the mode of locomotion exhibited by the fish. Some of the largest and smallest trails are *U. unisulca* with wavelengths of up to 20 cm and amplitudes up to 6.5 cm, and wavelengths down to 2 cm and amplitudes as low as 0.8 cm. Total length estimates for fish making this trail range from 3.2 to 72.2 cm.	Coldwater fish	Varved clay and silt to fine sand	Late Pleistocene (17.9 −12.6 cal ka)	Glacial Lake Hitchcock and glacial lakes in the Connecticut Valley and Glacial Lake Marrimack	Gibert et al. (1999), Benner et al. (2009)

smaller *Lamaichnum borregoensis*, the latter attributed to *Hemiauchenia* sp. *Camelopichnum* was named for the prints produced by *Camelops hesternus* and/or *C. huerfanensis*. *Hippipeda downsi*, erected by Remeika (1999), is related to *Dinohippus* sp. or *Equus simplicidens*. A *Chelipus* trackway shows paw prints comparable to but larger than those of the coyote *Canis latrans*. Middle Pleistocene *Mammuthichnum* was attributed to *Mammuthus columbi*. Sarjeant et al. (2002) assigned the ichnospecies *Chelipus therates* to *Canipeda therates*, which was considered a junior synonym of *Canipeda gracilis* by Melchor et al. (2018). Remeika (1999) named two camelid ichnospecies from the Blancan Fish Creek Canyon ichnofauna in the Anza-Borrego Desert State Park. One of them, *Lamaichnum borregoensis*, has a diagnosis that does not distinguish it from *L. guanicoe*. Indeed, the holotype of *L. borregoensis* (Remeika, 1999) is 118 mm long and is morphologically well within the range of variation of Barstow Formation specimens of *L. guanicoe*. Therefore, *L. borregoensis* is considered a junior subjective synonym of *L. guanicoe* (Lucas and Hunt, 2007). *Megalamaichnum* was also considered by these authors to be a junior synonym of *Lamaichnum tulipensis* Aramayo and Manera de Bianco (1987b), and *M. albus* a junior synonym of *Lamaichnum macropodum* (Sarjeant and Reynolds, 1999). *Mammuthichnum* was coined by Remeika (1999) without a diagnosis of the ichnogenus or designation of a type ichnospecies and therefore is considered a *nomen nudum*. Finally, *Camelopichnum* was named by Remeika (2006) without presenting any image, therefore it must be considered a *nomen nudum*.

Reynolds (1999) reported Pleistocene (late Irvingtonian) mammal tracks from Lake Tecopa sediments near Shoshone in southern Death Valley, California. This ichnofauna includes equid tracks (Reynolds, 2006). Rector (1979, 1999) and Rector et al. (1983, 1984) described human and animal tracks (raccoon, coyote, ungulates) from the Quaternary of Oro Grande, Mojave Desert. The age of the deposits probably ranges from the latest Pleistocene to the Holocene.

At Chickamauga/Chattanooga National Military Park, two caves (Kitty City, 27 Spider) preserve large felid prints and claw marks on the walls, which are presumed to represent attempts by cats to escape from pit traps (Santucci et al., 2006).

Concerning other bioerosional evidence, possible Pleistocene rub sites have been found at Sonoma Coast State Park in northern California (Parkman, 2002, 2009). These sites are interpreted as places where large mammals scratched themselves against blueschist boulders and sea stacks, smoothing and polishing some parts of the rock surfaces. The polished areas are often located much higher off the ground than can be reached by modern terrestrial mammals, such as cattle or horses (Haynes, 2012). The smoothing affected high points on the rock surfaces rather than entire surfaces, indicating it was a mechanical process of contact abrasion rather than a chemical process such as weathering or dissolution. Peterson (2003) and Parkman (2009) indicated other possible rub sites in New Mexico, Nevada, Minnesota, and Wisconsin.

In 2008, Benner et al. discovered locomotion trace fossils in sediments of glacial Lake Hitchcock in Connecticut, Massachusetts, New Hampshire, and Vermont. Using these traces and extensive neoichnological experiments, they were able to reinterpret the ichnogenus *Broomichnium* (Kuhn, 1958) to include a new ichnospecies, *Broomichnium flirii*, as the locomotion trace of a freshwater fish (sculpin) (Fig. 1.4A). Later work (Benner et al., 2009) extended the number of fish species likely present in glacial lakes Hitchcock and Merrimack based on multiple ichnospecies of *Undichna* Anderson (1976). These include *Undichna bina*, *U. unisulca*, *U. simplicitas*, and *U. britannica*. Abundant invertebrate ichnofossils, including *Cochlichnus*, indicated that the glacial lakes supported a "thriving ecosystem."

In southeastern Michigan, Fisher (1994) reported Late Pleistocene mastodon trackways in pond-margin sediments. The Brennan Mastodon Site preserved the skeletal remains of a

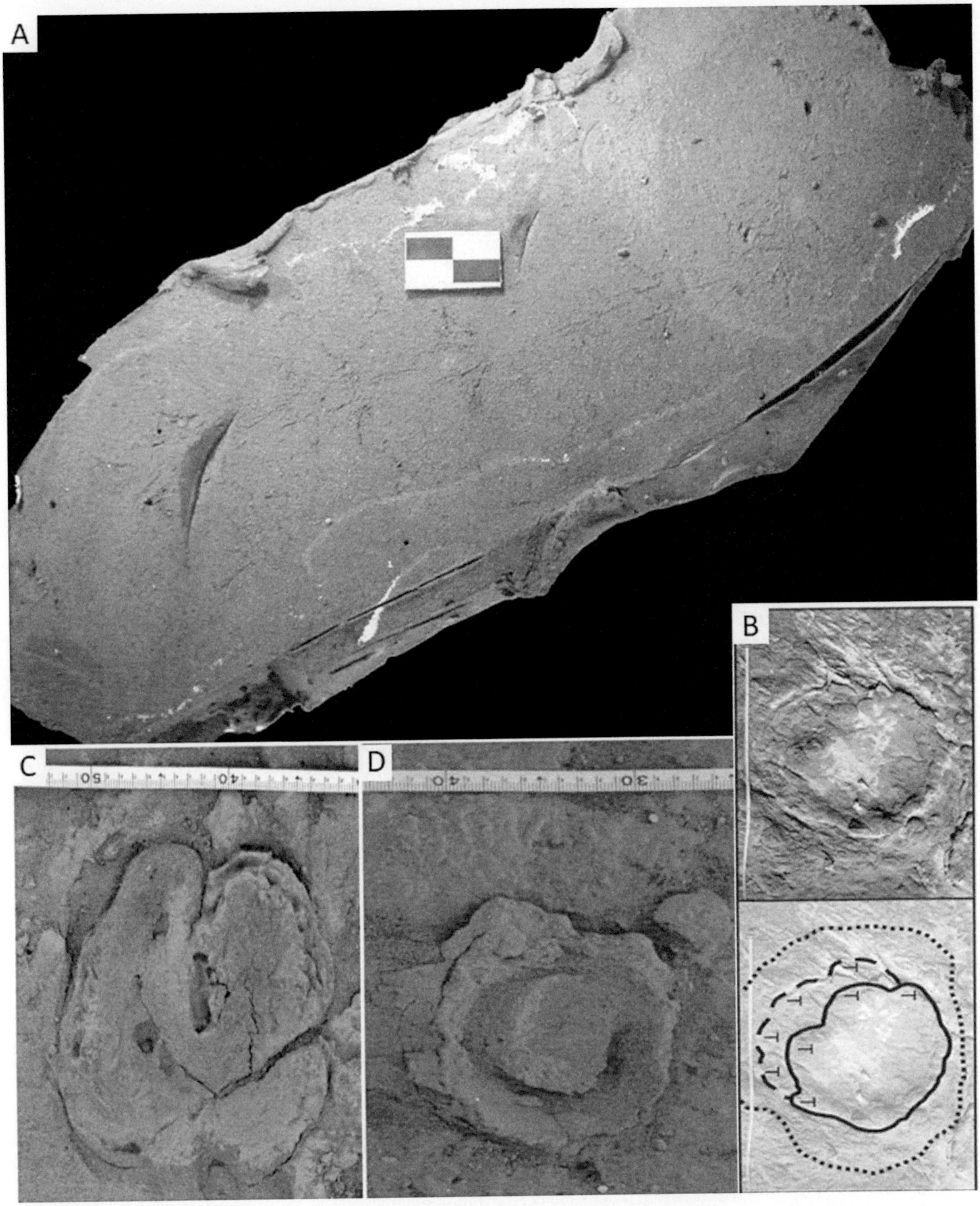

FIGURE 1.4 Vertebrate ichnotaxa described in the Quaternary of North America. (A) Holotype of *Broomichnium flirii* Benner et al. (2008). Pleistocene, Connecticut River Valley, USA. Photo of the original specimen on clay. This specimen no longer exists but a plaster cast is preserved in the Museum of Comparative Zoology, Harvard. (B) Photo and interpretative diagram of *Proboscipeda panfamilia* McNeil et al. (2007). Pleistocene, Alberta, Canada including the overlapping pes (solid line) and manus prints (dashed line in contact with the pes imprint), with the indication of the toe prints (T) and the outline of the pressure pad indicated by the dotted line (Fig. 2 from McNeil et al., 2007). (C) *Lamaichnum sarjeanti* McNeil et al. (2007). Pleistocene, Alberta, Canada. Holotype from Wally's Beach (Fig. 9 from McNeil et al., 2007). (D) *Hippipeda cardstoni* McNeil et al. (2007). Pleistocene, Alberta, Canada. Track photo from Wally's Beach (Fig. 14 from McNeil et al., 2007). *(A) Photo: J. Benner.*

mastodon, but also trackways. Tracks occur as deformations made by animals walking through shallow water of the lake margin. These trackways were interpreted as mastodon, rather than mammoth because the stride length relative to the size of individual prints indicated a proboscidean with a longer torso than was characteristic of mammoths (Fisher, 2018). The first paleobiological use of these tracks was to estimate body mass based on the displacement of sediment produced by individual prints, fore and hind (Hubbard et al., 2000). According to Fisher, they record the passage of a solitary adult male, followed by a matriarchal herd consisting of one fully adult female, one or two other females slightly younger and smaller, and a calf (Fisher, 2018).

In Greene County, Missouri, the Riverbluff Cave Site yields thousands of trackways and claw marks of the Pleistocene age (Forir et al., 2007). In this preliminary account, three types of large traces were mentioned to be produced by a large ursid, possibly *Arctodus simus* (claw marks), a large felid (claw marks), and a peccary (tracks). In Perry County, Oesch (1969) noted the presence of "hundreds of footprints of a large feline" in Berome Moore cave (Graham et al., 1996). The trackway can be traced for over 1.6 km. Oesch (1969) suggested that the trackmaker may have been *Panthera onca* rather than the contemporary *Puma concolor* or *Smilodon.*

In the Nebraska Sand Hills, Loope (1986) recognized tracks in cross-section in large aeolian bedforms of Holocene age. He interpreted these tracks as being produced by bison.

In the 1880s, prisoners from the Nevada State Prison at Carson City uncovered a Pleistocene lacustrine margin that preserved a very large tracksite (LeConte, 1882; Gibbes, 1882; Harkness, 1882; Marsh, 1883; Marche, 1986; Lockley and Hunt, 1995). Tracks are very diverse and represent birds, sloths, cervids, proboscideans, equids, felids, and canids (Lockley and Hunt, 1995). The tracks were first described by LeConte (1882, 1883) and ascribed to giant humans. Marsh (1883) provided the first interpretation that the tracks were made by a sloth, specifically a mylodont, and this interpretation has been followed by subsequent workers (see McDonald, 2007).

In southern New Mexico, fossil tracks from the Late Pleistocene of the Otero Formation on the White Sands Missile Range (Lucas et al., 2002, and references therein; Morgan et al., 2002) were made by proboscideans (mammoth) and camelids. The Tularosa Basin corresponds to sediments deposited along the margin of a shallow saline lake (Lake Otero). Mammoth tracks mostly head east-west and suggest that mammoths traveled to and along the shorelines of Lake Otero (Lucas et al., 2007; Hunt and Lucas, 2007). Beginning in 2011, systematic field documentation of vertebrate tracks of this megatracksite was initiated at White Sands National Monument. Hundreds of proboscidean tracks and extensive trackways were discovered and photogrammetrically documented on the southern shore of Lake Lucero in the monument. Additionally, camelid tracks, a few unidentified artiodactyl tracks, and large and small carnivore tracks were also photogrammetrically documented during field inventories (Santucci et al., 2014). Tracks and trackways of Proboscidea (mammoth/mastodon), Folivora (ground sloth), Carnivora (canid and felid), and Cetartiodactyla (bovid and camelid) are all known from this site, along with those of humans (Bustos et al., 2018; Bennett et al., 2021). Bennett et al. (2020) described very long human trackways intersected by giant ground sloths and mammoth trackways. The giant ground sloth tracks show behavior consistent with human predator awareness, while mammoth tracks show no such apparent concern. In fact, Bustos et al. (2018) found contemporaneous ground sloth and human trackways, in which sloth trackways show evidence of evasion and defensive behavior when associated with human tracks. Behavioral inferences from these trackways indicate prey selection and suggest that humans were harassing, stalking, and/or hunting the giant ground sloth. Radiocarbon ages of terrestrial pollen, together with OSL ages of sediments, point to an age interval between ca. 23 and ca. 21 ka for the track levels (Pigati et al., 2023). However, there has been strong criticism of these dates (e.g., Lepper, 2023; see also the chapter on hominin trackways in this volume).

Blackwater Locality Number One in New Mexico—the original Clovis archaeological site, after which Clovis stone spear points were named—contains spring conduits and wells attributed to human digging, at least one that is Late Pleistocene in age. The well at the site is a fairly narrow shaft or "circular pit." It may be possible that a mammoth dug it using its trunk to grasp bundles of sediments and throw them aside (Haynes, 2012).

Gregarious behavior of Columbian mammoths (*Mammuthus columbi*) was revealed in the Pleistocene locality of Fossil Lake, Oregon (Retallack et al., 2018). A tracksite composed of 117 tracks includes one 20-m-long adult trail, partial trackways of three additional adults, and a yearling and a baby all heading generally west. The matriarchal group moved through landscapes of sagebrush and grassland ca. 43 ka. The authors interpreted a limping female attended by concerned juveniles. Single tracks attributed to *Lamaichnum sarjeanti* and *Hippipeda cardstoni* were mentioned and are the first figured (Retallack et al., 2018). In the Oregon Caves National Monument, a single bear print and 20 bear scratch marks are preserved in cave sediments (Santucci et al., 2001, 2006). Packard and Allison (1980a,b; Packard et al., 1981) described very large bear tracks from the Pleistocene of Lake County. These tracks are so large that they could only have been produced by *Arctodus simus*, the giant short-faced bear.

In South Dakota, a large Late Pleistocene sinkhole in Hot Springs preserves mammoth skeletons together with isolated proboscidean tracks, and the uppermost levels are heavily bioturbated by large tetrapods (presumably *Mammuthus* sp.) (Laury, 1980; Agenbroad, 1984). Falkingham et al. (2010) described didactyl and tridactyl tracks of small birds from the same location feeding or drinking at the water's edge.

Jaguar Cave in north–central Tennessee is named for both jaguar fossils (dated between 35 and 10 ka) and presumed jaguar felid tracks (Robbins et al., 1981; Watson et al., 2005; Willey et al., 2005).

Simpson (1941) described jaguar tracks from Late Pleistocene deposits in Craighead Caverns, Monroe County, and included actualistic studies of a living jaguar. Wright et al. (1982) described Pleistocene footprints from Centerville Cave, Hickman County. The ichnofauna includes jaguar, bobcat, and otter tracks.

Santucci et al. (1998) noted the presence of camel and bird tracks in the Quaternary Coalpits Lake deposits of Zion National Park, Utah. These specimens have not been formally described.

At quarry walls in Middle Pleistocene marginal-marine coarse-grained deposits located in Willapa Bay, Washington, Gingras et al. (2007) described the bowl-shaped *Piscichnus* intersecting bivalve equilibrichnia *Siphonichnus*. They were interpreted by these authors as feeding excavations using hydraulic jetting by walrus herds (see also detailed treatment of the ichnogenus in the chapter on Vertebrate burrows in this volume).

3.4 Oceania

3.4.1 Australia

Lockley and Harris (2010) noted that no avian ichnotaxa had been described from Australia, nor had any Australian traces been ascribed to existing avian ichnotaxa. Camens and Worthy (2019) subsequently assigned Pliocene traces to species of *Anatipeda*, *Phoenicopterichnum*, and *Koreanaornis*, but, to date, despite numerous traces having been described, no attempt has been made to insert any Australian vertebrate traces (avian or otherwise) from the Quaternary into the existing ichnotaxonomic framework.

Despite the lack of an ichnotaxonomic framework for tetrapod tracks from Australia, fossil traces have been described from a number of sites, most commonly occurring in the abundant aeolianites present along much of the southern Australian coast. In Western Australia, the majority of these sediments are included in the Tamala Limestone (Playford et al., 1976; Bell, 2007b) while in South Australia and Victoria, they largely fall within the Bridgewater Formation (Boutakoff, 1963).

The first fossil tracks described from Australia created some controversy as they came from deep within the Pleistocene sediments of the Bridgewater Formation and were ascribed to humans (Officer, 1893; Klaatch, 1906; Douglas, 2004). The prints were in fact most likely hopping traces of a kangaroo, but the block housing the prints (originally at the Warrnambool Museum) has been lost. In association with the purported human prints, small bird (digits ~60 mm long), emu, wombat, kangaroo, and dingo (most likely thylacine, Fig. 1.5B) prints were also reported (Officer, 1893). Prints from the same location belonging to the giant, flightless bird *Genyornis newtoni* were figured by Klaatch (1906) and additional *G. newtoni* prints were described in more detail by Vickers-Rich and Gill (1976) and Vickers-Rich (1979). Additional isolated traces from ca. 110 ka eolian sediments have been attributed to emus (Playford, 1988, p.25), the giant extinct kangaroo *Protemnodon* (Hand, 1989; Belperio and Fotheringham, 1990, Fig. 1.5A) and an unidentified bird (Bell and De Merlo, 1969). A subsequent study by A.B.C. after the relocation of the Bell and De Merlo slab suggests it is a landing trace from a large raptor. More recently, a relatively diverse ichnofauna dated to 135.4 ± 5.9 ka has been described from the Bridgewater Formation on Kangaroo Island, South Australia, and it includes traces made by a range of birds, reptiles, and mammals (Camens et al., 2018, Fig. 1.5C).

Vertebrate trace fossils are much rarer in inland Australia, and only a handful of pre-human sites are known. Tracks of *Diprotodon optatum*, the largest marsupial ever to have lived, are known from Lake Callabonna in South Australia (Tedford, 1973, 1984; Wells and Tedford, 1995, Fig. 1.5D). On the basis of these trackways, Camens and Wells (2009) calculated a traveling speed of 6.3 km/hour for *D. optatum*. In western Victoria, the tracks of a range of megafaunal animals, including *Diprotodon*, *Protemnodon*, and *Thylacoleo carnifex*, are preserved in reworked volcanic sediments (ca. 110–60 ka). The site also includes skeletal remains, making it one of a handful of Quaternary sites globally that preserve contemporaneous trace and body fossils (Carey et al., 2011; Camens and Carey, 2013).

A small lake near Clare Bay on the west coast (5470 ± 90 BP, Gara and Cane 1988; Hand, 1989; Belperio and Fotheringham, 1990) referred to by the Traditional Owners as "Nyeerina jinna," preserves the tracks of humans, kangaroos, wallabies, wombats, emus and possibly dingoes, but these tracks have not been described in the literature (Hand, 1989; Bates, 2010, p. 123; West Kimberley Place Report, 2012, p. 139).

Late Pleistocene (19.3 ± 1.3 ka–19.8 ± 1.2 ka) human footprints have been described from Willandra Lakes in New South Wales (Webb et al., 2006; Webb, 2007; Westaway, 2010; Westaway et al., 2013), together with marsupial and bird prints (Westaway, 2010; Westaway et al., 2013), but the latter remain undescribed.

Animal tracks are commonly represented in the rock art of Indigenous Australians, and the possibility that some of these depictions represent the tracks of extinct megafaunal species has long been considered. A depiction of a large, 5-toed foot from Pimba in South Australia was suggested to belong to the giant marsupial herbivore *Diprotodon optatum* (Basedow, 1914), and several depictions of bird tracks have been allocated to *Genyornis newtoni* (Hall et al., 1951; Tindale, 1951; Ouzman et al., 2002). The antiquity of human occupation in Australia (up to 65 ka; Clarkson et al., 2017) means that there is a distinct possibility that the tracks of extinct megafauna

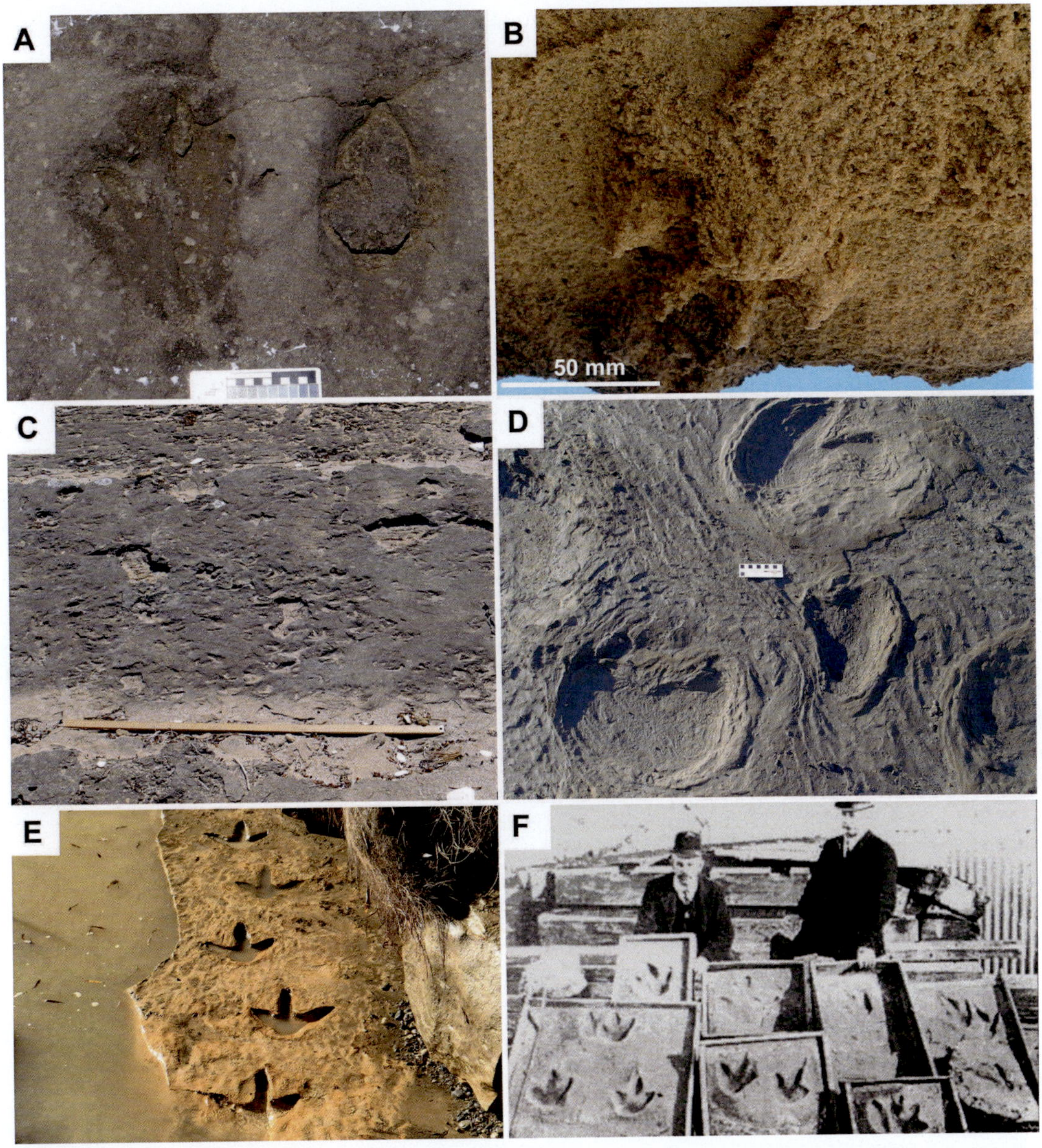

FIGURE 1.5 Examples of Quaternary tetrapod footprints in Australia and New Zealand. (A) *Protemnodon* footprints from Clare Bay, Australia; (B) thylacine footprint from Western Australia; (C) bioturbated pavement from Kangaroo Island, south Australia; (D) *Diprotodon optatum* trackway from Victoria, Australia; (E) moa trackway from Otago, New Zealand (Ian Griffin, Otago museum); (F) A selection of moa footprint casts from various New Zealand locations (Lockley et al., 2007).

could be depicted in rock art, but such depictions should be interpreted carefully due to difficulties in determining the age of the art and the source of inspiration.

On the walls of Tight Entrance Cave in Western Australia are preserved thousands of claw marks. They have been attributed to *Thylacoleo carnifex*, the largest marsupial predator ever to evolve in Australia, and are the only definitive evidence of its ability to climb (Arman and Prideaux, 2016). It is also possible that some of the smaller scratches could be attributed to the Tasmanian Devil (*Sarcophilus harrisii*).

3.4.2 *New Zealand*

New Zealand is unique among the world's continental land masses in having a modern fauna that evolved in a near-complete lack of terrestrial mammals (for exceptions, see Worthy et al., 2006; Hand et al., 2018). Birds have evolved to fill many of the niches that are filled by mammals on other continents, especially those relating to large herbivores. It is thus unsurprising that the Quaternary track record of New Zealand is dominated by avian traces (and moas in particular), with mammals only being represented after relatively recent (ca. 700–750 BP, Wilmshurst et al., 2011) human colonization.

Moa tracks (Figs. 1.5E and F) occur primarily in estuarine and dune sediments around the coast of the North Island of New Zealand, but, in recent times, prints have also been recorded from South Island (see later). Several studies have previously summarized the moa track record of New Zealand, including Oliver (1949), Anderson (1989), Worthy and Holdaway (2002), and Lockley et al. (2007) and Hunt and Lucas (2024).

The earliest records of moa tracks, found near the confluence of the Tūranganui and Taruheru River and Waikanae Creek at Poverty Bay (Gisborne), were published in 1872 (Gillies, 1872; Williams, 1872), and prints from this site have been mentioned in several subsequent studies (Cockburn-Hood, 1875; Plant, 1877; Owen, 1879, p. 451–453; Voy, 1880; Lydekker, 1891, p. 350; Hill, 1895, 1914; Benham, 1913). These tracks had clearly been known for some time prior to European discovery by the local Māori iwi, who believed that they represented the footprints of the famous Māori ancestor Rongokako (Hill, 1914). Hood (1874) and Owen (1879) referred the prints to species of *Dinornis*, and Lydekker (1891) indicated that, while some could be referred to *Dinornis*, others may have been made by species of *Emeus* or *Anomalopteryx* (Worthy and Holdaway (2002), note that Owen's *Dinornis dromioides* = *Anomalopteryx didiformis*).

There are numerous other sites around North Island where moa tracks have subsequently been recorded including Manawatu River (Hill, 1895, 1914; Wilson, 1913); Napier (Hill, 1914); Tangimoana (Oliver, 1949); New Plymouth (Stidolph, 1946); Marton (Millener, 1981; Worthy and Holdaway, 2002; Lockley et al., 2007); Lake Taupo (Worthy and Holdaway, 2002; Lockley et al., 2007); Ruamahanga River (Collen and Vella, 1984); Henderson Bay, Bay of Plenty, Matata and Whakatane (Ohiwa Harbor) (Gregory et al., 2006; Lockley at el., 2007); Muriwai (Hayward, 2016); and Mosquito Beach near Kaipara South Head (Hayward, 2022).

In May 2019, the first moa tracks discovered on the South Island were recorded from the bed of the Kyeburn River in Maniototo. The latter occurrence has recently been estimated to 1.0 +/− 0.5 Ma and the ichnotaxon *Tapuwaemoa manunutahi* was erected in association with these prints (Thomas et al., 2025). It is believed the tracks were made by a moa most similar to *Emeus crassus*. It is currently unclear as to whether the sediments they are imprinted in are of late Pliocene or Early Pleistocene age (Fleury et al., 2023). A second occurrence of moa prints from South Island was reported in August 2022 from near Pareora, South Canterbury. Although they are yet to be published, photogrammetry images are available via Sketchfab (Howe, 2022).

Hunt and Lucas (2024) were the first workers to place the moa tracks of New Zealand into an ichnological framework with the erection of ichnofamily Moapidae.

They also erected the ichnotaxa *Turanganuipus worthyi, Moapus tennysoni, Dinornipus oweni, Gisbornepus angustus, Aotearoapus lockleyorum and Tutaenuipus woodsi*. These ichnotaxa and *Tapuwaemoa manunutahi* (Thomas et al., 2025) are the only ones to have been erected in association with any Quaternary vertebrate tracks from Oceania.

In addition to tracks, an unusual trace fossil attributed to moas has been identified in New Zealand (Fig. 1.5C). Horn (1989) suggested that game trails of compacted earth (some of which contain gizzard stones) could be recognized in various parts of the country where humans were not known to have lived and where livestock were not present. If these paths can be demonstrated unequivocally to have been formed by the frequent passage of moa, they may represent an ichnofabric rarely represented in other parts of the world.

Alexander (1983) is one of the very few authors to have delved into the locomotory capabilities of moa as indicated by trace fossils; in general, most ichnological investigations have been limited to descriptions of moa tracks. Based on measurements taken from published depictions of moa prints (Williams, 1872; Owen, 1879; Hill, 1895; Wilson, 1913), Alexander (1983) predicted movement speeds of 0.8–2 m/second for the trace makers.

Besides moa, the only pre-human avian trace fossil reported from New Zealand is a kiwi-like track described by Hutton (1899) from Manaroa. It is a clawed, tridactyl print 98 mm long and 112 mm wide with digit impressions more gracile than are seen in moa. More recent calculations from a new discovered trackway resulted in an estimated movement speed of 0.47 m/second (Fleury et al., 2023). The exact provenance is unknown, but Hutton suggested that it came from the Waitemata Group sandstones, and if this were the case, then it is of Early Miocene age.

A site on Motutapu Island, near Auckland, preserves the footprints of humans and dogs in ash from the eruption of the nearby Rangitoto Volcano (Nichol, 1981, 1982; Gregory, 2005) and is dated to around 600 BP (Lindsay et al., 2011). The site extends across half a hectare and also includes the remains of garden beds contemporaneous with the tracks, but despite its cultural significance, it has not received further attention since initial excavations in the 1980s.

3.5 Asia

3.5.1 China

In our present knowledge, the only ichnosite from the Quaternary of China is Dingdingmen in the city of Luoyang, Henan Province. The Dingding Gate was the south-central and main gate of the outer city wall of Luoyang, the eastern capital city during the Sui and Tang dynasties. According to historical sources, the gate was built in 605 AD and was destroyed near the end of the Northern Song Dynasty. Therefore the archaeological remains of the Dingding Gate site span four periods: the early middle Tang, the late middle Tang, the period between the Tang and Northern Song dynasties (918–960 AD), and the Northern Song period (Tang Luoyang City Archaeological Team, the Institute of Archaeology, CASS and Luoyang Municipal Archaeological Team, 2004). In the 2007 excavation, significant evidence relating to the Silk Road were found in front of the gate, including rutted wheel tracks, human footprints, and camel hoof prints. After 3D scanning the whole surface, it was reburied in order to protect the trampled surface. A cast was made, which was then placed on top of the original surface (Fig. 1.6A). The artificial soil used for casting has now settled and is the same in appearance as the original (Su et al., 2020). The heritage value of the Dingding Gate site rests

FIGURE 1.6 Vertebrate ichnosites and ichnotaxa from Asia. (A) The cast of the trampled surface found in front of Dingding Gate, now exhibited as part of the World Heritage Site of Luoyang Silk Road, Henan province, China. (B) *Cervipes rex* single hoofprint from the lower Pleistocene Oyamada formation of the Kazusa Group, Hino City, Tokyo.

on the claim that it was the imperial gate of the Sui and Tang dynasties with close material connections to the Silk Road. For this reason, the reconstitution of the Dingding Gate was made in 2009 (Museum of the Site of the Dingding Gate) to protect the exposed ruins and to mark the importance of the site. Special attention was given to the abundance of camel hoofprints, which are direct evidence of the prosperity of commercial exchanges with the western regions at that time. Together with historical records, the Dingding Gate was recognized as the eastern origin of the Silk Road from the seventh to the tenth centuries (Su et al., 2020). Luoyang Silk Road Dingding Gate was subsequently included as one of the sites of the 2014 UNESCO World Heritage Silk Roads: The Routes Network of Chang'an-Tian Shan Corridor. The Western Han (202 BC–AD 8) village of Sanyangzhuang archaeosite include field systems so well preserved due to burial by a flood of the Yellow River, that the ploughing furrows and the prints of human feet and cattle hooves were preserved. The cattle hoofprints are figured in Qin et al. (2019) but no descriptions are provided.

3.5.2 *Cyprus*

The only tracksite so far found in Cyprus occurs in the aeolianites of Paphos (SW coast), along vertical exposures in the Apostolou Pavlou Avenue (Milàn et al., 2015). Some additional tracks were found in the archaeological site of Tombs of the Kings. Tracks occur in vertical exposures (i.e., cross-section) in densities as much as 10 tracks/m^2, with 50–150 mm of exposed size, a few reaching up to 230 mm. The larger ones were considered to be produced by dwarf hippopotami and elephants. The age of the aeolianite was attributed by Milàn et al. (2015) to be Late Pleistocene to Early Holocene without geochronological control. Nearby aeolianites have been dated recently to the Late Pleistocene (late MIS 5a-early MIS3) (Tsakalos et al., 2018; Polidorou and Evelpidou, 2021).

3.5.3 *Japan*

Since the first description of mammal tracks in the Plio-Pleistocene of Japan by Saito (1928), the Japanese trace fossil record became the most important resource for understanding the Cenozoic, especially the Pleistocene vertebrate ichnofaunas in Asia (see Table 1.4). Hundreds of tracksites have been described and geochronologically well constrained, which Matsukawa and Shibata (2015), in their very comprehensive work, divided into 14 areas located in the islands of Honshu and Kyushu.

The *Kitakami Lowland in Hanamaki* is the northernmost site with vertebrate tracks in Japan and the first to be described (Saito, 1928). Pleistocene vertebrate tracks were found in two

YCM-GP3001 housed in the Yokosuka City Museum, Kanagawa Prefecture. (C) *Elaphuruformipes tamaensis* (replica) from the lower Pleistocene Oyamada formation of the Kazusa Group, Hino City, Tokyo. YCM-GP3003 housed in the Yokosuka City Museum, Kanagawa Prefecture. (D) cf. *Proboscipeda* attributed, based on age, to *Stegodon aurorae* from the lower Pleistocene Oyamada formation of the Kazusa Group, Hino City, Tokyo. Replica is housed in the Yokosuka City Museum, Kanagawa Prefecture, YCM-GP1996. (E) General view of a proboscidean ichnosite attributed to *Stegodon aurorae* from the lower Pleistocene Oyamada formation, Hino City, Tokyo. Replica is housed in the Yokosuka City Museum, Kanagawa Prefecture, YCM-GP1995. (F) cf. *Proboscipeda* attributed, based on the age, to *Palaeoloxodon naumanni* from the upper-Pleistocene Umibata sand and mud member of the Nojiri-ko Formation, Shinano town, Nagano Prefecture. Replica is housed in the Nojiri-ko museum, Shinano town, and the registered number is Z9-25.

TABLE 1.4 The ichnotaxa described in Quaternary formations of Japan.

Ichnotaxa	Diagnosis/description	Producer	Formation	Age	Locality	Source
Cervipus rex	Large, cloven-hoofed track, with two tear-drop-shaped impressions, narrowing gently at one third of length from the posterior of hoof to a sharply pointed tip. Hooves convergent posteriorly, divergent anteriorly. Dew claw impressions circular if impressed. Trackway is narrow, subequally long and outwardly convex. Lengths of track are about 10 cm and widths about 9 cm. The tip of the tracks is pointed.	*Cervus (Nipponicervus) praenipponicus*	Oyamada Formation	Early Pleistocene	River bank of Tamagawa River, Sakae-cho, Hino City, Tokyo, Japan	Matsukawa et al. (2007), Matsukawa and Shibata (2015)
Cervipus rex	No description	*Cervus*	Iwasakishiden, Tsuchisawa and Momooka formations; Okui Formation; Kobuku Formation	Early Pleistocene	Kitakami Lowland, Tomi, Tondabayashi	Matsukawa et al. (2007), Matsukawa and Shibata (2015)
cf. *Cervipus rex*	No description	*Cervus*	Akashi Formation	Early Pleistocene	Akashi coast	Matsukawa et al. (2007), Matsukawa and Shibata (2015)
Elaphuruformipes tamaensis (considered a *nomen nudum* in the Chapter on Paleogene/Neogene vertebrate tracks)	Cloven-hoofed tracks, consisting of pair of elongated kidney-shaped impressions. Blunt tips of both hooves. Wide space between two hooves. Functionally didactyl, quadruped narrow trackway with subparallel, subequally long and outwardly convex impressions. Length of tracks (5–13 cm) is longer than width (4 –12 cm).	*Elaphurus*	Oyamada Formation	Early Pleistocene	River bank of Tamagawa River, Sakae-cho, Hino City, Tokyo, Japan	Matsukawa et al. (2007)

Elaphuruformipes tamaensis Matsukawa et al. (2007)	No description	*Elaphurus*	Iwasakishiden, Tsuchisawa and Momooka formations; Okui Formation; Kobuku Formation	Early Pleistocene	Kitakami Lowland, Tomi, Tondabayashi	Matsukawa and Shibata (2015)
cf. *Elaphuruformipes tamaensis* Matsukawa et al. (2007)	No description	*Elaphurus*	Akashi Formation	Early Pleistocene	Akashi coast	Matsukawa and Shibata (2015)
cf. *Proboscipeda*	Fore foot tracks are subcircular and hind tracks are elliptical. Hind foot track is a little longer than the fore foot track. Narrow gauge trackways	*Stegomastodon Stegodon aurorae*	Terada, Hanno and Oyamada formations; Iwasakishiden, Tsuchisawa and Momooka formations; Okui Formation; Kobuku Formation	Early Pleistocene	Kitaasakawa River at Narahara; Hachioji, Tamagawa River; Akishima and Hino; Kitakami Lowland; Tomi	Matsukawa et al. (2007)
cf. *Proboscipeda*	Proboscidean-type, wide-gauge trackways	*Palaeoloxodon neumanni*	Nojiri-ho Formation	Late Pleistocene	Nojiri-ho	Matsukawa et al. (2007)

tracksites in the Iwasakishinden Formation (1.9 ± 0.0 Ma to 1.2 ± 0.2 Ma), two tracksites in the Tsuchisawa Formation (1.8 ± 0.3 Ma to 1.4 ± 0.2 Ma), and three tracksites in the Momooka Formation in Isawa (1.0 ± 0.3 Ma to 0.53 ± 0.22 Ma) (Oishi and Yoshida, 1998; Oishi et al., 2002). A total of 365 tracks were identified in these floodplain deposits along the Kitakami River and tributaries, mainly artiodactyl tracks, including *Elaphuruformipes tamaensis* and *Cervipus rex* (Matsukawa and Kato, 2011; see below), proboscidean tracks cf. *Proboscipeda*, and carnivore and bird tracks (Kinoshita and Todori, 1991, 1993; Research Group for fossil footprints of Isawa-gawa, 1993; Matsukawa and Kato, 2011).

In West Tokyo, Matsukawa et al. (2004, 2007), after Baba et al. (2000) and the Research Group for Fossil Footprints from the Tamagawa riverbed (2001, 2002), described the predominance of tracks of proboscideans and even-toed ungulates in the Early Pleistocene of the lower part of the Kazusa Group (Terada Formation—2.3 ± 0.1 Ma to 2.1 ± 0.1 Ma; Hanno Fm.—about 2.1. to 1.6 Ma; and Oyamada Formation—about 1.3 Ma). The only ichnotaxa formalized in Asia are from the Oyamada Formation and correspond to *Cervipus rex* (Fig. 1.6B), compared with a *Cervus (Nipponicervus)-like* producer and the larger *Elaphuruformipes tamaensis* (Fig. 1.6C), attributed to a large *Elaphurus*-like tracemaker (Matsukawa et al., 2007). *Elaphuruformipes* is considered a junior synonym of *Entelodontipus* (see the chapter on Paleogene/Neogene vertebrate tracks in this volume).

The proboscidean tracks were assigned to cf. *Proboscipeda* and produced by *Stegodon*; most likely *S. protoaurorae* in the Terada Formation and *S. aurorae* in the upper formations (Fig. 1.6D and E), due to the stratigraphic range of the two species. The presence of two parallel trackways composed of large and small proboscidean tracks suggests a female adult together with a juvenile walking along a shoreline (Matsukawa and Shibata, 2015) in a succession dominated by lowland, swamp environments. Bird tracks were also found and compared to Ciconiidae (Mukaiyama and Fukushima, 2008). Matsukawa et al. (2007) found the same ichnoassemblage in the Hanno Formation (Akishima) and Terada Formation (Hachiouji) and compared them (see later) with similar ichnofaunas from the Kobiwako Group (ca. 2 Ma) and the Omma Formation (Saikawa) (0.9—0.8 Ma).

Many proboscidean tracks and rare artiodactyl tracks were described in the upper Okui Formation (1.3 ± 0.1 Ma) along the riverbed of the Chikuma River, close to the city of Tomi (Nagamori et al., 2005; Matsukawa and Shibata, 2015). Proboscidean tracks were attributed to *S. aurorae* and suggest gregarious behavior, together with the presence of *Cervipus rex* and *Elaphuruformipes tamaensis* in a succession developed in a fluvial system of an intermountain basin at an altitude of about 1000 m a.s.l (Sato and Matsukawa, 2014).

Many proboscidean tracks and trackways have been reported from 24 tracksites in the Kobiwako Group, of which 15 tracksites occur in the Early Pleistocene Gamo Formation (Yasugawa) with a date of 2.27 ± 0.44 Ma, and two tracksites in the Middle Pleistocene Katata Formation (Katata) and the equivalent Zeze Formation, dated between 0.87 ± 0.07—0.37 ± 0.38 Ma (Kamei et al., 1989; Okamura, 1993; Okamura et al., 1993, 1995). The Gamo Formation includes proboscidean tracks produced by *S. aurorae*, 16 tracksites of cloven-hoofed tracks with sharp toe impressions compared with *Cervus (Sika) nippon*, and round toe impressions related to *Elaphurus davidianus* (Takahashi and Okamura, 2004).

One tracksite was found, including seven stratigraphic beds of the upper part of the Omma Formation (<0.899 ± 0.025 Ma) at Ishikawa Prefecture (Okamura et al., 2004). In back marsh deposits, 100 proboscidean tracks were reported, with >48 artiodactyl tracks with round toe impressions attributed to *Elaphuroides shikamai*, and only 2–3 tracks with sharp toe impressions. Bird tracks attributed to the Family Ardeidae were also found.

In Tondabayashi, 60 stratigraphic levels of the Kokubu Formation (1.01 ± 0.24 Ma) yielded proboscidean, artiodactyl, and bird tracks (see Matsukawa and Shibata, 2015). The proboscidean tracks may have been produced by *S. aurorae* and show gregarious behavior in female-dominated herds, while the artiodactyl tracks include the presence of *Elaphuruformipes tamaensis* and *Cervipus rex*. The bird tracks may be related to the Family Ciconiidae?

Finally, on the island of Kyushu, Okamura (2001) described vertebrate tracks in Maruishibana Formation (Himeshima).

Besides the Katata/Zeze formation tracksites, the other Middle Pleistocene tracksite was found on the Akashi coast, in the upper part of the Osaka Group. Morimoto and Tsuda (1937) reported artiodactyl tracks that can no longer be found and attributed them to *Cervus* and *Elaphurus*. Matsukawa and Shibata (2015) made an attempt to compare them with *Cervipus rex* and *Elaphuruformipes tamaensis*.

The only Late Pleistocene tracksites (2) were found in the lacustrine deposits of the Nojiri-ko Formation (42,550 ± 530 to 17,460 ± 340 BP) at Nagano Prefecture (Footprint fossils research group for Nojiri-ko Excavation, 1992). Thirty-five proboscidean tracks comprising three trackways were attributed to *Palaeoloxodon naumanni* (Fig. 1.6F), and two artiodactyl short trackways were compared to *Cervipus rex* (Matsukawa et al., 2007; Fossil mammal Research group for Nojiri-ko excavation, 2008).

The ichnoassemblages dominated by proboscidean and artiodactyl tracks show a strong recurrence in the Pleistocene fluvial-dominated deposits from Japan. Early Pleistocene tracksites prevail, with many fewer dated from the Middle and Late Pleistocene. It is clear the difficulties to identify producers for this time span simply from track morphology and size. *Elaphuruformipes tamaensis* and *Cervipus rex*, the only ichnotaxa formally introduced in Japan, occur throughout the entire Pleistocene record despite the diversity of the deer body fossil records in Japan. For the proboscideans, Matsukawa and Shibata (2015) noticed that pes length changes through time and discussed the relationship between pes length and proboscidean body size. Trackways attributed to *Stegodon* species are narrower-gauge when compared to the wide-gauge ones attributed to *Palaeoloxodon naumanni*. Such differences between narrow- and wide-gauge trackways could have genus-level significance when it comes to identifying the most likely producer. About 40,000 horse hoofprints have been discovered at the Shiroi site, Central Japan. The prints were covered with pumice layer dated to 6th century AD. Inoue and Sakaguchi (1997) estimated the withers height of the horse by measuring the hoofprints and by analysing them. They concluded that the Shiroi horse was the same medium-type as the modern Kiso horse which has about from 125 cm to 135 cm withers height.

3.5.4 Republic of Korea

Phreatomagmatic eruptions of Mt. Songak in Jeju Island produced pyroclastic surge deposits known as the Songaksan tuff. Reworked tephra was deposited in low-lying coastal areas to form the Hamori Formation. According to Kim and Kim (2006), the geologic age of the track-bearing strata is thought to be 15,161 ± 70 yr BP. However, a combination of radiocarbon, U-series, and OSL dating methods has shown that the footprint horizons should be coeval with a volcanic eruption dated ca. 3.75 ka (Sohn et al., 2015). Coastal outcrops of the

Songaksan tuff and the Hamori Formation contain tracks of hominins, artiodactyls, proboscideans, birds, and carnivores (Kim and Kim, 2004; Kim et al., 2004). Kim et al. (2009) presented an account of the track-bearing strata, including 505 hominins, >1000 artiodactyl (deer), eight proboscidean, five carnivores, 85 birds (of eight different types), and two fish tracks and trackways. No ichnotaxonomical studies were included. Two of the track-bearing levels contain several large circular or irregularly shaped tracks attributed to woolly mammoths (*Mammuthus primigenius*). The presence of these elephant tracks would be indicative of Pleistocene age, at least for these stratigraphic levels.

3.5.5 Saudi Arabia

A track-bearing unit between 121 ± 11 and 112 ± 10 ka in age was recently described in the Alathar paleolake deposit in the Nefud desert by (Stewart et al., 2020). These authors focused on the hominin footprints among the 376 tracks recorded but also identified very abundant elephant and camel tracks indicating herds (i.e., adults and juveniles), giant buffalo possibly *Syncerus*, wild ass, and a medium-sized bovid, including some body fossils. The paleolake surface is heavily trampled, which most likely reflects a dry season during which herbivores congregated around diminishing water supplies. No formal ichnotaxonomy has been developed.

3.6 Europe

3.6.1 Austria

Some isolated tracks of a large bovid and most probably horse were described by Ebner and Gräf (1978) from a river terrace at Mutzenbaal near Graz, dating from the Late Pleistocene to Early Holocene. Also, cave-bed bear depressions are well known in Austrian caves (Ehrenberg, 1967). Fish tracks and trails (cf. *Undichna* and cf. *Broomichnium*, according to Benner et al., 2009) were described in Würm age varves of the Austrian Tyrol (Fliri et al., 1970, 1971). Fliri et al. (1971) noted that the large *U. britannica* could have been made by *Esox linnaeus* (pike) of over 1 m in length, as there previously had been the recovery of a few partial specimens of *Esox* body fossils.

3.6.2 Denmark

The Lodbjerg dune system on the North Sea Coast of NW Jylland includes soil horizons containing thousands of tracks and trackways of cattle, first described by Broholn (1946). A fen deposit dated to 1265 ± 145 cal BC shows tracks of cattle, sheep or goat, red deer, horse, and dog or wolf (Milàn et al., 2006, 2007). In SE Jylland, the late Glacial Lille Slotseng kettle-hole basin dated to 13.6 ka cal BP provided a single reindeer (*Rangifer tarandus*) track in a drill core (Noe-Nygaard et al., 2007).

3.6.3 France

Tracks of cave bears have been reported, although not in detail, from several caves in France, including Grotte Bijoux (Kurten, 1976), the Grotte de Bruniquel (Rouzaud et al., 1995), and the Chauvet Cave (Garcia, 2005). In the Chauvet Cave, dated to ca. 28 ka, hibernation nests were found (Chauvet et al., 1996), as well as in the Rouffignac Cave (Plassard, 1999). Clottes (2003) determined cave bear occupation in Chauvet Cave through beds and tracks, besides one wolf/dog and ibex prints. The "canid footprints" described by García (2005) in Chauvet Cave were allocated to *Canipeda* isp. by Melchor et al. (2018). Casteret (1948) described terminal Pleistocene hyena tracks and coprolites from a den in the Aldène Cave, along with bear tracks and hominin tracks. The photographs of the well-preserved hyaenid tracks (Casteret, 342 1948, p. 410) show closely packed digit impressions in a 'jigsaw pattern'.

Open-air tracksites in France include L'Ouziere in Jersey, Channel Islands (Patton, 1993). Besides, Bertran and Fouéré (2020) found at Soulac-sur-Mer, Médoc, cattle footprints in peat and dune deposits disturbed by cattle trampling, as they examined a sequence of estuarine and aeolian formations ranging from millennia to centuries ago.

3.6.4 Germany

At Mertloch (Neuwied Basin, Rhineland-Palatinate), well-preserved imprints attributed to Allerød (ca. 12.9 ka cal BP) animal species were documented from a short-lived surface within a Middle Laacher See Tephra ash layer (Baales, 2002). Syneruptive rain produced a muddy ash surface that preserved the animal tracks over a surface of several thousands of square meters. The animal tracks can be attributed predominantly to female capercaillie (*Tetrao urogallus*), with a trackway 70 m-long made by a brown bear (*Ursus arctos*), a red deer (*Cervus elaphus*) with a calf, and several horses (*Equus* sp.) (Baales and Berg, 1997; Baales, 2002). The season of the eruption was supported by foal tracks found at Mertloch, which represent neonate animals (Baales and von Berg, 1999).

The site of Schöningen, in Germany (Lower Saxony), dated to ca. 320–300 ka, brought to light the largest assemblage of fossilized straight-tusked elephant (*Palaeoloxodon antiquus*) tracks ever found at a Paleolithic excavation (Altamura and Serangeli, 2023). Herds of elephants (juveniles and adults) and several other species of herbivores (including a large bovid and cervids) congregated along the muddy shores of a paleolake (Altamura et al., 2023). This is the northernmost occurrence of tracks known for *Palaeoloxodon antiquus*.

Bottrop-Welheim, located in the Emscher River valley in the Ruhr district, is one of the most well-known tracksites in Germany (Frechen, 1995). A 150 m^2 tracksite related to a waterhole environment with 600 tracks referred to reindeer (16 trackways), wolf (synonymized under *Canipeda* by Melchor et al., 2018), a large bovine (aurochs or bison), horse, and cave lion were described by Koenigswald et al. (1995, 1996). *Anser* sp. is the only bird represented in the track record. A trackway of *Pantheraichnus bottropensis* of the Late Pleistocene lion *Panthera leo spelaea* has been documented from Bottrop (Diedrich, 2011a; Antón and Salesa, 2012). *Pantheraichnus* is considered a junior synonym of *Felipeda*, and *P. bottropensis* is a synonym of *Felipeda biancoi* (Abbassi et al., this book, chapter Paleogene/Neogene vertebrate tracks).

In the Eemian (MIS 5e) of Neumark-Nord in Saxony Anhalt, tracks of Equidae, Bovidae, and Cervidae were recorded in muddy substrate (Laurat et al., 2007).

Finally, cave-bed bear depressions are well known in German caves (Rabeder and Hofreiter, 2004).

3.6.5 Gibraltar

Catalan Bay, on the eastern side of "the Rock", shows a large cliff front and rampant dune system reaching 300 m high and dated from the Middle-to-Late Pleistocene. In old quarry fronts from the bottom of the aeolianite, Muñiz et al. (2019) described tracks, mostly in sections, of straight-tusked elephants (including a large depression interpreted as a resting trace), red deer, caprines, possible aurochs, and a large felid (*Panthera pardus?*), together with a single print of a Neanderthal child. Since then, new ichnological data have been documented by Fernando Muñiz and colleagues from the famous caves of Gorham's and Vanguard (not yet described), almost completely filled with aeolianite, which is contributing to the understanding of the paleoecology of this World Heritage Site.

3.6.6 *Greece*

The purported latest record of proboscidean activity known in Europe on the basis of tracks was found on the SW coast of the island of Rhodes by Milàn et al. (2007a,b). The Kattavia aeolianite recorded poorly preserved tracks from the Late Pleistocene to the Early Holocene age. Seventy-nine tracks found in sections were divided into four size classes. Only four tracks were found exposed on bedding planes. The three smallest size classes were probably produced by artiodactyls (including camels), while the largest class was produced by proboscideans (Milàn et al., 2007a,b). The age of the aeolianite was considered speculative due to the lack of geochronological constraints.

Also in Rhodes, Nielsen et al. (2020) recently described bowl-shaped feeding traces in the Cape Arkhangelos Formation in lower shoreface facies under the occasional influence of storms, dated to the Pleistocene. These traces may suggest ray foraging behavior and were assigned to the ichnospecies *Piscichnus waitemata* (see under *Piscichnus* in the chapter on Vertebrate Burrows in this volume). Foreshore deposits dating from ca. 45 BCE to 430 AD on the east coast of Rhodes show two different-sized undertracks attributed to artiodactyls (cervids or pigs and cattle) and humans traveling along the shore (Bromley et al., 2009).

Other islands with bioturbated aeolianites were indicated by Milàn et al. (2015): in the Phalasama area of southern Crete (scarce small tracks); a large number of tracks occur on the island of Zakynthos where the size suggests Quaternary elephants or humans; other possible tracks may occur on Elafonisos island.

3.6.7 *Hungary*

The Vértesszőlős quarry is a Paleolithic site located in northwestern Hungary. The site is dated from the Middle Pleistocene (ca. 310 ka). A short distance from where the Samu (*H. heidelbergensis*) remains were found is an exposed surface of calcareous mudstone, preserving numerous fossil tracks made by a range of mammals and birds (Tanaka et al., 2021). The track site (site III) is now protected as an open-air museum as part of the Vértesszőlős Hungarian National Museum. Site III extends for approximately 40 m^2, and the record consists of 125 tracks, 106 of which were identified as produced by mammals (e.g., bison, bear, rhinoceros, and stag) and some others attributed to avian trackmakers (Kretzoi and Dobosi, 1990).

3.6.8 *Ireland*

Very recently, the first vertebrate tracksite was found in the Holocene of West Ireland. Westley et al. (2023) identified an organic-rich clay exposed in the intertidal zone of a beach near Streedagh (Co. Sligo). The tracks were dated between 5.8 and 4 ka. There were 47 tracks showing three morphotypes that were attributed to different artiodactyls.

3.6.9 *Italy*

Pillola et al. (2020) offered a review of the Pleistocene non-hominid vertebrate ichnofossil record of Italy (see also Table 1.5). A few large tracks, possibly left by *Mammuthus meridionalis*, have been documented in the aeolian-reworked alluvial sand sheet deposits of Upper

TABLE 1.5 The ichnotaxa described in Quaternary formations of Europe.

Ichnotaxa	Diagnosis/description	Producer	Formation	Age	Locality	Source
Avipeda isp.		Small passerine	Lahar	Late Pleistocene (ca. 25 ka cal BP	Tor Vergata-Via Gaudenzio, Rome, Italy	Cerilli et al. (2020)
Bifidipes aeolis (nomen nudum: see Abbassi et al., this book, chapter Paleogene/Neogene vertebrate tracks).	*Bifidipes* associated with considerable disturbance of the surrounding substrate including microfolding and microfaulting in eolian sediments. Details vary with consistency and attitude of substrate surface. Pes may coincide with or fall behind manus; stride irregular. Trackways commonly follow irregular course.	*Myotragus balearicus*	Coastal aeolianites	Middle-Late Pleistocene	Southeastern Mallorca Island, between Cap Blanc and Cap des Moro; Southwestern Mneorca Island; Northern Cabrera Island (Cala Santa Maria)	Fornós et al. (2002)
Bifidipes aeoli (lapsus cal.)	Individual bilobate tracks that can be observed in both hyporelief and vertical section. They present widths between 10 and 16 cm and a maximum depth of 5 cm. The base is divided by a ridge or a groove depending on the type of preservation.	*Cervus elaphus*	Catalan Bay Sands	Middle-Late Pleistocene	Eastern Gibraltar	Fornós et al. (2002), Muñiz et al. (2019)
Bifidipes cf. *aeoli (lapsus cal.)*	Tracks preserved as convex hyporeliefs that form a trackway with two clear (cloven-hoofed) impressions. Their width ranges from 4–5 cm.	Caprinae	Catalan Bay Sands	Middle-Late Pleistocene	Eastern Gibraltar	Fornós et al. (2002), Muñiz et al. (2019)
Bifidipes aff. *B. aeolis*	The tracks corresponding to internal/nearly internal vertical sections showing distinctly the insertion within the sediment of the forefoot (manus) and hindfoot (pes), moderately superimposed on each other. Sometimes a hoof-like distal termination seems to be present.	*Nesogoral*, and, maybe, *Asoletragus gentryi*	Capo Mannu dune complex	Early Pleistocene	Western Sardinia, Italy	Fornós et al. (2002), Fanelli et al. (2007), Pillola et al. (2020)

(Continued)

TABLE 1.5 The ichnotaxa described in Quaternary formations of Europe.—cont'd

Ichnotaxa	Diagnosis/description	Producer	Formation	Age	Locality	Source
Bifidipes isp.	Symmetrical didactyl prints, isolated or organized in small trackways, normally preserved in concave epirelief and more rarely observed in convex hyporelief and in sections Individually, the footprints show a rounded (undertrack) to elliptical shape, are deep (more than 5 cm) and presented as direct shallow impressions or undertracks, being bordered by pressure pads and radial fractures. Sometimes small circular pits produced by the dew claws are present. Size differences between front and hind limbs were not identified.	Red deer *Cervus elaphus*	Malhão aeolianite; Murração aeolianite	Middle Pleistocene (187 ± 11 ka)	SW Portugal	Neto de Carvalho et al. (2003, 2016), Neto de Carvalho (2009, 2011).
Bifidipes isp.		*Praemegaceros caziotti*		Late Pleistocene	Northern and western coasts of Sardinia	Fanelli et al. (2007), Pillola et al. (2020)
Bovinichnus uripeda	Tetradactyl, paraxonic, and subsymmetrical tracks that are as wide as long; central (III–IV) digit impressions are subparallel, with broad distal end pointing forward, and a narrow, linear interdigital space; internal surfaces of hoofprints flat to slightly concave, external surfaces slightly to markedly convex; hoofprints widest near the heel, but tapering only to minor degree between the heel and the apex; rounded-to-rectangular posterior dewclaw (digits II and V) prints that are much smaller than central digits, printed next to or close to the heel; manus and pes of closely similar form and size.	*Bos primigenius*	Matalascañas Trampled Surface and Cape Trafalgar aeolianite (including Trafalgar Trampled Surface)	Middle Pleistocene (151 ± 11 ka) to Late Pleistocene (107 ± 2 ka to 78 ± 8 ka)	El Asperillo Cliff, Matalacañas (Huelva Province), and Cape Trafalgar, (Cadiz Province), Andalusia –Spain	Neto de Carvalho et al. (2022)

Bovipeda isp. (*nomen nudum;* Vyalov, 1966) proposed Bovipedidae (ichnofamily) but there is no ichnogenus with that name)	Double impression of bovine feet, which due to the spatial relationship should refer to the left front hoof on which the left rear hoof would overlap. The impressions of the abaxial and axial left posterior third phalanges have a length of 120 mm and a width of 40 mm; the impressions of the left front hoof are not measurable, as they are strongly disturbed by the impression above. The walls of the impression are strongly sloping and rounded at the external margin. The total length and width of the track is 170 mm	*Bos primigenius*	Lahar	Late Pleistocene (ca. 25 ka cal BP	Casetta Mistici and Tor Vergata-Via Gaudenzio, Rome, Italy	Cerilli et al. (2020)
Buboichnus vicentinus	Moderate-sized tetradactyl zygodactyl tracks with well-developed hallux impression; digital pads recognizable and with claw marks attached or independent. Digits II and III point anteriorly with digits I and IV pointing posteriorly with variable divarication angle, but showing mesaxonic symmetry.	Eagle-owl *Bubo bubo*	Telheiro aeolianite	Late Pleistocene ca. 82 ka	Quebrada stream at the Telheiro beach, Vila do Bispo, Portugal	Neto de Carvalho et al. (2023a)
Canipeda isp.	The footprint is preserved as concave epirelief. It is digitigrade and symmetrical with four well-developed toe pads and relative claw marks, and its central pad is not clearly visible.	*Cynotherium*		Middle-Late Pleistocene	Porto Paglia, Sardinia, Italy	Zoboli and Pillola (2018), Pillola et al. (2020)
Canipeda isp. (previously described by Neto de Carvalho, 2009, 2011; Neto de Carvalho et al., 2016 as *Bestiopoda* isp.)	Quadrupedal trackways with shallow symmetric digitigrade footprints, made of 5 pad impressions leaving a central high between them by sand expulsion under pressure. Elliptical shaped tracks 5 cm long.	*Vulpes vulpes*	Malhão aeolianite	Middle Pleistocene (187 ± 11 ka)	Angra da Vaca, Odemira, Portugal	Neto de Carvalho (2009, 2011), Neto de Carvalho et al. (2016)

(*Continued*)

TABLE 1.5 The ichnotaxa described in Quaternary formations of Europe.—cont'd

Ichnotaxa	Diagnosis/description	Producer	Formation	Age	Locality	Source
Canipeda isp.	Semidigitigrade, tetradactyl, paraxonic, tracks longer (90 mm) than wide (65 mm). Each track is composed of an anterior arc of similar sized and elliptical digital pad impressions, separated from a rounded, triangular metapodial pad impression. Clear triangular claw impressions occur separated from the digital pads, indicating canid-like footprints.	*Canis lupus*	Matalascañas Trampled Surface	Middle Pleistocene (151 ± 11 ka)	El Asperillo Cliff, Matalacañas (Huelva Province), Andalusia –Spain	Neto de Carvalho et al. (2020c)
Cervipeda/Bifidipes isp.	Large hoof prints with an average size range of 120–135 mm in length and 75 mm broad, and a few trackways with measured stride lengths between 1000–1320 mm. Tracks are elliptical in outline. Impressions of third and fourth toes are well-developed mirror images with inner hoof surfaces sometimes in median or posterior contact, linear or concave at the front. In some cases, they are separated by a continuous interdigital space, usually broadest at the front. They can be widest near the heel or in the central part, tapering to an apex of angular or sharply parabolic shape, directed forward, and convergent, parallel or divergent according to the pace of the moving animal.	*Cervus elaphus*	Matalascañas Trampled Surface	Middle Pleistocene (151 ± 11 ka)	El Asperillo Cliff, Matalacañas (Huelva Province), Andalusia –Spain	Neto de Carvalho et al. (2020c)

Cervipeda isp.	The track is 100 mm long and 70 mm wide at the back. The impressions of the soles of the two third phalanges are both 30 mm wide and have the axial side parallel, furthermore they are separated by a space (fillet) of constant width of 10 mm from the anterior apex to the posterior side.	*Cervus elaphus*	Lahar	Late Pleistocene (ca. 25 ka cal BP	Casetta Mistici, Rome, Italy	Cerilli et al. (2020)
Charadriipeda isp.	Tridactyl, anisodactyl and mesaxonic tracks, of moderate size and slender digits, commonly showing claw marks. These are formed by a longer digit III (30 −55 mm), with divarication angles with toes II and IV of between 30° and 42°. Digit II measures 15 −30 mm and digit IV, between 18 and 38 mm. These two digits are linear or bend upwards. There is no impression of the hallux. The flat areas are related to interdigital webbing.	Shorebirds	Malhão aeolianite and Telheiro aeolianite	Middle-to-Late Pleistocene (OSL dating of 187 ± 11 ka and ca. 82 ka, respectively)	Pessegueiro islet, Sines, and Quebrada stream at Telheiro beach, Vila do Bispo, Portugal	Figueiredo et al. (2018), Neto de Carvalho et al. (2016, 2023a)
Corvidichnus odemirensis	Bipedal trackway of high pace angulation with mesaxonic tetradactyl, anisodactyl tracks of moderate size with small positive rotation with the midline. Tracks with three comparable-sized digits directed forward showing clear phalangeal pad impressions, and a long hallux pointing backward, but slightly offset from the middle line; divarication angles between digits II-IV are small (i.e., ~70°); track length/width ratio less than 2.	Western jackdew	Malhão aeolianite	Middle Pleistocene (OSL dating of 187 ± 11 ka	Aivados Sul Beach, Odemira, Portugal	Neto de Carvalho et al. (2023a)

(Continued)

TABLE 1.5 The ichnotaxa described in Quaternary formations of Europe.—cont'd

Ichnotaxa	Diagnosis/description	Producer	Formation	Age	Locality	Source
Felipeda lynxi	Each footprint, digitigrade-made and circular-shaped, about 5.7 cm long, shows four asymmetrical toe impressions (with leading toe), without claw prints. These toe impressions are distributed in front of the heel pad with bilobed leading edge.	Iberian lynx *Lynx pardinus*	Malhão aeolianite	Middle Pleistocene (OSL dating of 187 $\pm$ 11 ka)	Pessegueiro islet, Sines, Portugal	Panin and Avram (1962), Neto de Carvalho (2014), Neto de Carvalho et al. (2016)
Gruipeda aff. *maxima*	Tetradactyl anisodactyl tracks longer than wide, of relatively large size, 10 cm long by 7 cm wide. They exhibit four digits, II and IV directed forward and diverging at an angle of about 65°, digit I directed backwards. Digits II and IV are similar in length, with sides almost parallel for most of their length and tapering abruptly to a pointed tip. Digit I is short, spurlike and tapering; its axis is offset at a low angle to the right from that of digit III. The impressions of digit II and IV are connected proximally; digit I is separate.	Rallidae	Malhão aeolianite	Middle Pleistocene (OSL dating of 187 $\pm$ 11 ka)	Aivados Sul Beach, Odemira, Portugal	Panin and Avram (1962), Neto de Carvalho et al. (2023a)
Hippipeda isp.	Tracks that are monodactyl, mesaxonic, subcircular, with the front part markedly convex that tends to expand laterally where the margins have a subparallel trend, are very similar to each other in shape and size and it is difficult to distinguish the manus from pes impressions. Sometimes the conceivable presence of the V-shaped frog at the rear margin of the footprint (tough barely discernible between the two posterior wall bars.	*Equus* sp.	LS7 pyroclastic unit	Middle Pleistocene	Foresta Ichnosite, Central Italy	Pillola et al. (2020)

Hippipeda isp.	The external edge of the wall may be very clear, the internal one is less delineated, while an internal concentric line can be seen. The frog is quite well recognizable, even if not particularly marked. The external width of the footprint is approximately 130 mm; the frog is 75 mm long and 80 mm wide at the base.	*Equus ferus*	Lahar	Late Pleistocene (ca. 25 ka cal BP	Casetta Mistici, Rome, Italy	Cerilli et al. (2020)
Leporidichnites malhaoi	The fore prints are circular, 4 cm in size, sometimes organized in groups of 3 forming the angles of an equilateral triangle. The hind prints are elongated, up to 12 cm long and 5 cm wide, occurring in pairs and juxtaposed. Toe impressions are not preserved. Sediment expulsion rims provide the direction of motion.	*Oryctolagus cuniculus*	Malhão aeolianite	Middle Pleistocene (187 ± 11 ka)	Aivados Sul, Pessegueiro islet and Malhão, Odemira, Portugal	Neto de Carvalho (2009), Neto de Carvalho (2009, 2011), Neto de Carvalho et al. (2016)
Pantheraichnus bottropensis Diedrich, 2001a (=*Felipeda biancoi* after Abbassi et al., Chapter Paleogene/Neogene Vertebrate tracks in this volume))	Large felid footprints, each with four round-oval digit pad II-V imprints in front of the large pad. In the best preserved imprints this large pad is oval shaped in the manus and more kidney-shaped or rectangular in the pes. The distances between the digit pad impressions are larger in the manus than in the pes imprints, in ideal cases. The manus is much longer than its width (142/136 mm), whereas the pes imprints are similar in length and width (143/143 mm).	*Panthera leo spelaea*	Periglacial Emscher River valley flooding plain deposit 35,000–42,000 BP	Late Pleistocene	Bottrop-Welheim (Welheimer Mark), North Rhine-Westphalia, north-west Germany	Diedrich (2011a)

(*Continued*)

TABLE 1.5 The ichnotaxa described in Quaternary formations of Europe.—cont'd

Ichnotaxa	Diagnosis/description	Producer	Formation	Age	Locality	Source
Pecoripeda (Cervipeda) isp.	Four large impressions of the left and right medial digit III and lateral digit IV hoof, and, on the back, smaller impressions of the dewclaws, the vestigial hooves for digits II and V. The large concave hyporeliefs of the two hooves, show the typical elongated, subcuneiform shape with a convex, parabolic external and slightly concave internal contour, an angled distal and wide caudal part. The dewclaw impressions are considerably smaller, nearly fused each other, and subquadrangular in shape.	Medium-sized artiodactyl such as roe deer	LS7 pyroclastic unit	Middle Pleistocene	Foresta Ichnosite, Central Italy	Pillola et al. (2020)
Piscichnus waitemata	A bowl-shaped or subcylindrical structure oriented vertical to bedding, with a steep, smooth margin. The structure can be associated with a concave-up, dish-shape structure at the top.	Ray feeding behavior	Cape Arkhangelos Formation	Pleistocene	Southern Kallithea	Gregory (1991), Nielsen et al. (2020)
Presbyorniformipes isp.	Moderate-to- relatively large-size web-footed tracks. They are longer than wide, with two dimensional classes, of 50 and 85 mm long and 44 and 75 mm wide, and asymmetrical outer digits of different size curving inwards. Toes are 10 mm wide and all clearly show nodes at phalangeal joints, terminating in short claws. Angles of divergence of the lateral digits (II and IV) exhibit spans of 62° and 74°. Hallux is usually preserved as a short point	Anserinae	Matalascañas Trampled Surface	Middle Pleistocene (151 ± 11 ka)	El Asperillo Cliff, Matalacañas (Huelva Province), Andalusia –Spain	Neto de Carvalho et al. (2020c)

	depression 10 mm in diameter, and rarely, with a comma or spur-like shape turned in, reaching 15 mm in the smaller form, and connected to the other toes. The larger tracks show a slightly variable pace length of 135–175 mm along the seven tracks and stride lengths of 250–290 mm, the smaller tracks a variable pace length up to 154 mm and a stride length of 250 mm. Both show a typical in-turned gait, with an angle of 130° and 120°, respectively. The morphotype B occurs as a set of three asymmetrical tracks showing wide phalangeal pads (17 mm for the digit III) with clearly-defined sharp claws and interdigital membrane. The tracks are as long as they are wide (60 mm) with a short divarication angle of 72° between digits II and IV.					
Proboscipeda panfamilia	Few tracks are well preserved and show some anatomical details (i.e., toe/nail impressions), whereas other potential footprints cannot be firmly validated because they are located in an area that has undergone strong anthropic alteration and currently is subject to degradation, due to the action of surface runoff water. The tracks, nearly rounded to oval in shape, are preserved as convex hyporeliefs, bordered by more or less inflated and prominent pressure pads.	Probably *Palaeoloxodon antiquus*	LS7 pyroclastic unit	Middle Pleistocene	Foresta Ichnosite, Central Italy	McNeil et al. (2007), Palombo et al. (2018), Pillola et al. (2020)

(*Continued*)

TABLE 1.5 The ichnotaxa described in Quaternary formations of Europe.—cont'd

Ichnotaxa	Diagnosis/description	Producer	Formation	Age	Locality	Source
Proboscipeda panfamilia	Very large circular tracks; anteriorly-directed, short and blunt toe marks when present. Occasionally, large and flat sole surfaces show evidence of ornamentation. Pes is oval to subrounded and show four nail pads, and manus is circular. Trackways show the typical elephantine gait.	*Palaeoloxodon antiquus*	Matalascañas Trampled Surface; Malhão aeolianite	Middle Pleistocene (151 ± 11 ka and 187 ± 11 ka)	El Asperillo Cliff, Matalacañas (Huelva Province), Andalusia −Spain; Malhão, Odemira, Portugal	McNeil et al. (2007), Neto de Carvalho (2009, 2011), Neto de Carvalho et al. (2016, 2020a,b, 2021)
Proboscipeda panfamilia	Isolated manus/pes couples preserved as convex hyporeliefs or concave epireliefs. Furthermore, proboscidean tracks are clearly visible in transverse sections in the field. The nature of the sediment did not allow the impression and preservation of detailed morphological features, and free digit impressions are not preserved in all specimens.	*Mammuthus lamarmorai*		Middle Pleistocene	Funtana Morimenta (Gonnesa), Sardinia, Italy	McNeil et al. (2007), Pillola and Zoboli (2017), Pillola et al. (2020)
Proboscipeda isp.	Very large tracks observable only in vertical section, with a bowl shape and a symmetric-asymmetric profile. Their depth varies between 12 and 25 cm and their width between 22 and 34 cm. In some tracks the deepest part can be slightly displaced toward the front and, in some cases, a marginal ridge is observed in the upper part.	*Palaeoloxodon antiquus*	Catalan Bay Sands	Middle-Late Pleistocene	Eastern Gibraltar	Muñiz et al. (2019)

Suidichnus galani	Tetradactyl, paraxonic, and sub-symmetrical tracks that are longer than wide, with a general trapezoidal shape; digits III and IV symmetric, with rounded-to-squarish outline and sharp or blunt toe tip traces, wider in the back of the hoofprint, with narrow interdigital ridge but splayed in the anterior part of the fore print; lateral dewclaws II and V impressions conical to comma-shaped, smaller than the main toes III and IV, longer than wide, and projected laterally from behind the main toes, with a wider angle in the fore print than in the hind print.	*Sus scrofa*	Matalascañas Trampled Surface	Middle Pleistocene (151 ± 11 ka)	El Asperillo Cliff, Matalacañas (Huelva Province), Andalusia –Spain	Neto de Carvalho et al. (2020b)
Ursichnus isp.	The best-preserved track, a sub-oval, roughly kidney-shaped large depression, would correspond to the manus. Five small oval depressions along its convex edge, i.e., in front of the pad, may correspond to digit impressions. Some pits in front of the 'digit impression' would correspond to claw marks, considering that the claw marks of bear sometimes may be slightly marked if the footprints are left on soft surfaces.	*Ursus*	LS7 pyroclastic unit	Middle Pleistocene	Foresta Ichnosite, Central Italy	Pillola et al. (2020)

(*Continued*)

TABLE 1.5 The ichnotaxa described in Quaternary formations of Europe.—cont'd

Ichnotaxa	Diagnosis/description	Producer	Formation	Age	Locality	Source
Ursichnus isp.	Badly preserved tracks and undertracks with five claw prints and the inward rotation of the claw prints with respect to the midline are typical of a bear walking gait, also indicating the direction of movement Average track length is 15,3 cm and average track breadth is 18,3 cm; the strong variation of track length, length/width ratio of 0,89, and the rare impression of the heel is possibly related with an animal walking upslope.	Ursidae	Medo Tojeiro/ Almograve aeolianite	Middle Pleistocene	Cavalo beach, Odemira, Portugal	Neto de Carvalho and Belo (2022)

Valdarno "Sabbie del Palazzetto," pre-Olduvai Early Pleistocene (Ghinassi et al., 2013). The Middle Pleistocene (ca. 600 ka) Atella site in the northern Basilicata includes large tracks found in sections that were attributed to *Palaeoloxodon antiquus* by Fabiano and Zucchelli (2003) due to their size and the presence of osteological remains of this species. Without details of pad morphology, if other possible producers can be discarded, then these could be the first proboscidean tracks described in the Pleistocene of Europe and possibly the oldest of *P. antiquus*. The Middle Pleistocene Casal de' Pazzi site in the eastern periphery of Rome shows fluvial silt deposits with the record of five trackways related to passeriform or columbiform birds and a single track in the section that is probably attributable to *Palaeoloxodon antiquus* (Altamura and Gioia, 2020).

The ichnosite known as "Ciampate del Diavolo" (the "Devil's Trails") is located on the northeastern slope of the Roccamonfina volcano (Campania region, southern Italy). The track level corresponds to an ignimbrite radiometrically dated at 349 ± 3 ka, i.e., during the glacial phase of MIS 10. This surface reveals footprints of hominins together with tracks of an indeterminate medium-sized artiodactyl, a horse (*Equus* sp.), a bear (*Ursus* sp.), and an elephant, probably a young *Palaeoloxodon antiquus* (Panarello et al., 2017; Palombo et al., 2018; Pillola et al., 2020).

The Sardinian ichnofossil record is related to coastal and aeolian contexts. During the Pleistocene glacial phases, the western continental shelf of Sardinia emerged due to global sea level fall, resulting in the extension of dune fields exposed to the NW winds (Pillola et al., 2020). This scenario led to the formation of a wide range of sedimentary environments on the northern and western coasts of Sardinia, which allowed the preservation of the tracks left by large mammals, such as the bovid tracks from the Early Pleistocene of Capo Mannu (Fanelli et al., 2007, 2008; Kotsakis et al., 2008; Pillola and Zoboli, 2017; Zoboli and Pillola, 2018). Almost 1000 tracks and trackways recognized in six tracksite areas and 50 stratigraphic beds, preserved in both bedding planes and vertical sections, were attributed to *Praemegaceros cazioti*, an endemic cervid (Fanelli et al., 2007). The authors concluded that frequentation of coastal habitats by this species could be related to the ingestion of salt. Pillola and Zoboli (2017) described isolated *Proboscipeda panfamilia* manus-pes couples and a cross-section from the Middle-Late Pleistocene aeolianites of Gonnesa (SW Sardinia). The track size and the occurrence of body fossils in the same deposits indicate that the trackmaker was likely the Sardinian dwarf mammoth. In Porto Puglia, one isolated canid track (*Canipeda* isp.), likely referable to *Cynotherium*, together with cervid and proboscidean tracks, was described by Zoboli and Pillola (2018).

Several mammal tracks have also been recently found in the Late Pleistocene (ca. 25 ka) of Casetta Mistici and San Gaudenzio—Via della Sorbona (Rome) on the surface of lahar sediments, arising from the crater of the Albano Lake (Cerilli et al., 2020). A predominance of *Equus ferus* trackways, as well as those attributed to aurochs and red deer, and few tracks of a small passerine were described following an outdated ichnotaxonomy based on the studies of Vyalov (1965, 1966).

Numerous tracksites have been found and documented in cave environments in Italy. The Covolo Fortificato di Trene site is located in the middle of the great Po-Venetian plain in Northern Italy. This cave shows remains dated to the early Epigravettian, including traces of *Ursus spelaeus* (Allegranzi and Barolomei, 1956). Also, at the Gola di Frasassì in Genga (Ancona), which includes the Grotta Grande del Fiume, Bocchini and Coltorri (1978) recorded the presence of claw marks and tracks of *Ursus spelaeus*. The Bàsura Cave (Toirano, Liguria) contains ichnofabrics generated by humans, bears, and canids and dated to ca. 14 ka (Avanzini et al., 2018). Tracks, claw scratches, hibernation, or gestation nests are referred to as both juvenile and adult *Ursus arctos* and are documented by Citton et al. (2017) and Avanzini et al. (2021). The canid tracks are older than or contemporary with the human footprints in close association, which could shed new light on dog domestication in the Upper Paleolithic (Romano et al., 2019).

In the Holocene, mammal tracks have been found on the uppermost levels of the Pomici di Avellino eruption (Somma-Vesuvius Volcano, 3945 $\pm$ 10 BP), at Croce del Papa, and at the Bronze Age village of Afragola-Badagnano (Vesuvius Volcano) (see Mastrolorenzo et al., 2006 and Di Vito et al., 2019, and references therein).

3.6.10 *Netherlands*

At the Neolithic site of Kilhorn, cattle hoof prints mainly defined a path across a channel and also radiated from a freshwater well (Bakels and Zeiler, 2005).

3.6.11 *Poland*

The Quaternary inland dunes of the European Sand Belt in the eastern part of the belt in Poland provided several localities with hoofprints of various ungulate mammals, both wild and domestic, as well as footprints of humans (Koster et al., 1993; Hsieh et al., 2022; Uchman et al., 2024). Six footprint-bearing horizons from a total of eight were radiocarbon-dated using charcoals, found either in their nearest underlying paleosol (buried soil) layers or the horizon itself, to constrain their maximum age. The oldest dated sites, Laszczyny and Unin, have paleosols dated to as early as the fifth to sixth century AD, while those at Mnin and Oddziały have dates in the range of the 18th–19th centuries (Hsieh et al., 2022). Trample grounds and horizons with high densities of tracks are associated with other signs of intensifying usage of dune habitats. Some of the smallest ungulate hoofprints excavated from many of these sites, about 3 to 5 cm long, potentially belong to roe deer, *Capreolus capreolus*, if they were made by wild animals, or to domestic ovicaprids (sheep or goats). Somewhat larger or wider tracks may be attributed to wild boar, *Sus scrofa*. Larger than these (up to about 10 cm) but elongate and similar in shape to those of the smaller roe are large deer-like tracks (possibly *Cervus elaphus*). Large hoofprints mostly ranging from around 10 to 15 cm, likely produced by domestic cattle and horses, are seen, often at sites associated with human footprints, for instance at Oddziały, or Ochotnik.

3.6.12 *Portugal*

Since the first findings of cloven-hoofed prints in the Pessegueiro islet in 2001, several sites have been described and treated ichnotaxonomically in the aeolianites from the Middle-to-Late Pleistocene (ca. 200 ka to 78 ka, dated by OSL) cropping out on the SW coast of Portugal, between Pessegueiro islet and Telheiro beach. Abundant tracks and trackways morphologically compared to red deer occur together with those of straight-tusked elephants (Fig. 1.7B; sometimes showing gregarious behavior), the Iberian lynx *Lynx pardinus*, the fox *Vulpes vulpes*, brown bear, and hare/rabbit (Fig. 1.7C; Neto de Carvalho et al., 2003; Neto de Carvalho, 2009, 2011, 2014; Neto de Carvalho et al., 2020a; Neto de Carvalho and Belo, 2022). Recently, the first avian ichnotaxa from the Pleistocene of Europe were formally described

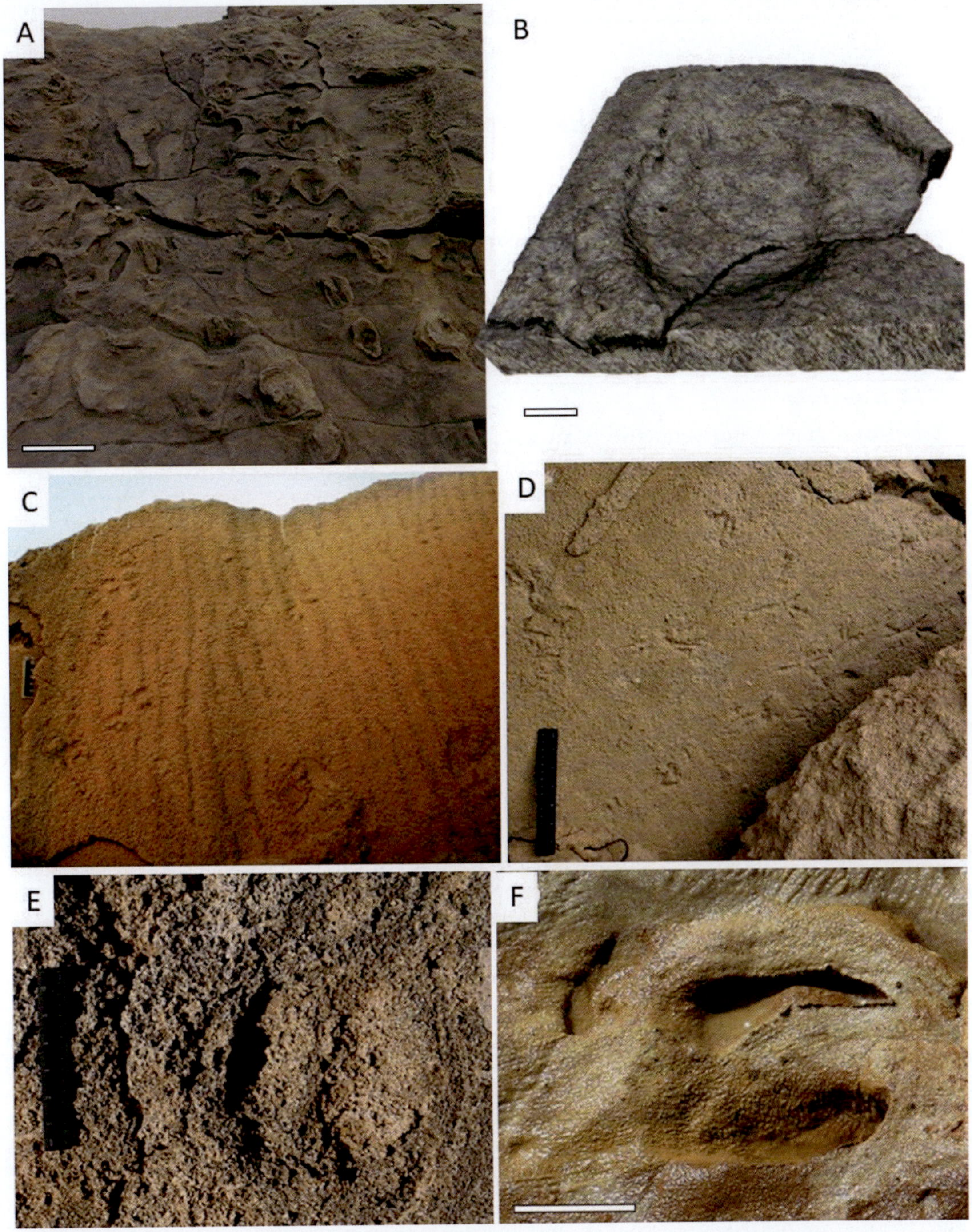

FIGURE 1.7 Ichnotaxa from Portugal and Spain. (A) *Bifidipes aeolis* (*nom. nud.*) trackway across slope in carbonate aeolianite foreset surface from the Late Pleistocene of s'Estret des Temps (SSE island of Mallorca); epichnia, oblique view; scale bar is 10 cm (courtesy of Luis Cáceres). (B) *Proboscipeda panfamilia* from the Middle Pleistocene of Malhão aeolianite, SW Portugal. Manus-pes couplet with clear toe prints; epichnia, 3D model in real colors; scale bar is 10 cm (courtesy of João belo). (C) *Leporidichnites malhaoi* trackway showing hopping gait from the Middle Pleistocene of the Pessegueiro islet aeolianite; hypichnia in a rippled interdune surface together with cervid underprints. (D) Holotype of *Corvidichnus odemirensis* from the Middle Pleistocene of the Aivados aeolianite, SW Portugal; epichnia; almost orthogonal view. (E) *Bovinichnus uripeda* in the type section, Late Pleistocene sandy backshore facies from Cape Trafalgar (SW Spain); epichnia, orthogonal view. F—*Suidichnus galani* from the Middle Pleistocene Matalascañas Trampled surface (SW Spain); epichnia, oblique view; scale bar is 5 cm.

(Neto de Carvalho et al., 2023a) and included locomotion of the western jackdaw (*Corvidichnus odemirensis*; Fig. 1.7D), besides the locomotion and predation/feeding behaviors of a large eagle owl (*Buboichnus vicentinus*). The last trace fossil may correspond to the first evidence of a raptorial bird-prey interaction in the fossil record. Typical shorebird tracks and trackways attributed to gulls (Laridae) (Figueiredo et al., 2018), curlews, and others tentatively compared with Rallidae, such as the Eurasian coot, are also discussed.

The Algar do Vale da Pena is a cave situated in the Natural Park of Serras de Aire e Candeeiros (Leiria, Portugal). The Late Pleistocene trace fossils attributed to *Ursus arctos* include at least 189 tracks interpreted as bear claw marks, from juvenile to adult sizes (Estraviz-López and Mateus, 2019). These were the first bear scratches found in Portuguese caves.

3.6.13 Romania

Cave bear tracks and scratches from the Late Pleistocene (early to middle Weichselian) were first described in the Urşilor Cave in the western Carpathians, NW Romania, by Terzea (1978). More recently, Diedrich (2011b) assigned the tracks of cave bears to *Ursichnus europaeus*, and more than 140 hibernation beds identified in this cave allowed the erection of the ichnogenus and ichnospecies *Ursalveolus carpathicus*. These nests were also found in the Oase Cave, SW Romania (Quilés et al., 2006). Tens of thousands of bear claw scratches are found on the slopes and top of a clay plateau.

3.6.14 Spain

The first vertebrate tracks described in Pleistocene deposits of the Iberian Peninsula are those from the Ses Piquetes quarry in Santanyí, Island of Mallorca (Fornós and Pons-Moya, 1982). This was the first record of *Myotragus balearicus* tracks, an endemic goat. Thousands of similar tracks (Fig. 1.7A) were described by Fornós et al. (2002) between Cap Salines and Cap des Moro, on the south coast of Mallorca (see also Bromley, 2001), dated to the Late Pleistocene (MIS 5b/a, four and 3; Fornós et al., 2009). They were described as *Bifidipes aeolis* based mostly on sedimentological criteria, and for this reason, this ichnospecies is not considered valid (Abbassi et al., chapter Paleogene/Neogene vertebrate tracks). *Bifidipes aeolis* is also found on other islands from the Balearic archipelago, including the northern coast of the island of Cabrera (Fornós and Sábat, 1998). On the SW coast of the island of Menorca, *Myotragus* is commonly associated with tracks of the dormouse *Hypnomys* (Rodentia, Myoxidae) (Quintana, 1993).

In SW Spain, recently, a coastal trampled paleosurface was uncovered by storms, close to the town of Matalascañas, Province of Huelva. Thousands of tracks and some trackways have been documented since then, and every time the storms and tides are conjugated to remove almost 2 m of beach sand to expose parts of what was called the "Matalascañas Trampled Surface" (Neto de Carvalho et al., 2020c). Along with a few hominin trackways confirmed as belonging to Neandertals by geochronology and lithic industry (Neto de Carvalho et al., 2023b), potential game prints were found, including those of red deer, aurochs, wild boar (Fig. 1.7F), straight-tusked elephants, and three types of waterbirds, including geese and waders, besides wolf (Neto de Carvalho et al., 2020a,b,c, 2021, 2022). Revision of the valid artiodactyl ichnogenera by Neto de Carvalho et al. (2022) and the track length/width ratio can be very useful to distinguish between the presence of aurochs and steppe bison.

Particularly important is the presence of hundreds of tracks attributed to straight-tusked elephants, with the predominance of well-preserved tracks of newborn calves and juveniles (Fig. 1.7E), indicating a snapshot in time of the occurrence of matriarchal herds in a favorable reproduction habitat (Neto de Carvalho et al., 2021).

Straight-tusked elephant and aurochs tracks were also found and/or described in two other Pleistocene sites of SW Spain, including Cape Trafalgar and Chipiona beach (Neto de Carvalho et al., 2022; Goméz Gutiérrez et al., 2023). Panera et al. (2014) described in the Manzanares Valley complex terrace of Butarque with at least four tracks matching the size and features of elephant tracks found together with *Palaeoloxodon antiquus* bones and Neanderthal lithic tools related to the scavenging of the elephant (Silva et al., 2012). The Arriaga IIa site in the Manzanares Valley has minimum thermoluminescence date of 134–133 ka. A circular pit was located below the elephant remains and was suggested to be a hole excavated by the straight-tusked elephant to get water from the dry riverbed (Rus and Santonja, 2011).

The Cape Trafalgar tracksite described by Neto de Carvalho et al. (2022) includes the earliest tracks attributed to aurochs (Late Pleistocene; Fig. 1.7E), besides roe deer/caprine tracks, rare large proboscidean tracks, and hominin footprints. The "Trafalgar Trampled Surface" might represent the arrival of a "beach–loving" aurochs herd of bulls for enjoying digestion while controlling possible predators in an open landscape.

One of the early papers on mammal tracks in aeolianite deposits from the Iberian Peninsula is from Flor (1989). He described deer tracks, almost all in vertical sections, in the small aeolianite on Górliz beach in Vizcaya (N Spain). They may represent different species, sexual dimorphism, or different ontogenetic stages (adults and juveniles). Although considered Late Pleistocene in age, the aeolianite was later radiocarbon dated at 6020–5710 $\pm$ 50 BP (Cearreta et al., 1990). The more rounded didactyl tracks are comparable to those of reindeer, a species commonly found in northern Spain during the Pleniglacial.

3.6.15 *Sweden*

Hundreds of traces of unshod horses dominate the area around a Swedish Bronze Age building in coastal *fjord* sediments of Ullunda site dated to 3.4–3.2 ka is accompanied by tracks of cattle/elk (Price 1995). They provide the earliest evidence for the horse in Sweden and demonstrate the involvement of horses in everyday activities (Bell and Walker, 2005). Another footprint was identified at Karstorp (Alexanderson and Bernhardson, 2019), and considered to likely be from a reindeer.

3.6.16 *Switzerland*

At Sion–Anciens Arsenaux archeosite, Valais, goat and domestic cattle hoofprints were found together with the oldest known plough marks in Europe, dating from the beginning of the 5th millennium BCE. They bear witness to the use of animal traction quite soon after the establishment of an agro-pastoral economy in the Alpine region (van Willigen et al., 2024).

3.6.17 *United Kingdom*

Many ichnological sites, especially from the Holocene, have been documented in Great Britain. In total, Bell et al. (2000) identified 93 sites, of which 39 present animal tracks. This list of sites was revised by Bell (2020). Swanscombe in eastern London is an internationally famous Middle Pleistocene locality in the Lower Thames Valley. It is dated to within Marine

Isotope Stage (MIS) 11 (424–374 ka) and represents a trampled horizon where only isolated tracks were identified (but not described in detail). The tracks represent a deer, a horse, a large bovid, most probably rhinos, and a straight-tusked elephant (Sutcliffe, 1985; Davis, 1996).

Tracks have been found across the intertidal zone of the Severn Estuary (Aldhouse-Green et al., 1992; Aldhouse-Green and Housley, 1993; Allen, 1997; Allen et al., 2003; Scales, 2007; Bell, 2013). Multiple species are represented by the tracks, including red deer, aurochs, sporadic horses, wolf or dog, and humans. This Flandrian tidal mudflat formed over the last 9-8 ka and was originally studied by Allen (1997) in a seminal paper where field and laboratory experiments on track-making and sediment deformation concluded that the moisture content of the sediment at the time was crucial to the preservational conditions of the tracks. Over thousands of years, the sediments recorded the transition from hunting to domesticated cattle and sheep/goats. Past patterns of animal husbandry may be illuminated by data on age and sex (Allen et al., 2003). Tracks of juvenile bovids and ovicaprids found around Bronze Age buildings at Redwick, Severn Estuary, show this site to have been occupied in the Spring and Summer (Bell, 2001). Barr and Bell (2016) compared the tracks of primitive rare breed sheep and cattle with prehistoric examples to establish the age structure of the herds and test the hypothesis of seasonal estuarine salt marsh utilization.

The intertidal mid-Holocene mudflats of Formby Point, on the Sefton Coast at the mouth of the Mersey estuary in northwest England, have been a great source of ichnological information. The Formby track-bearing beds form one of the world's largest known concentrations of Holocene vertebrate tracks. Aurochs, red deer, roe deer, wild boar, canid, unshod ponies, and predominantly feeding trackways of oystercatchers (*Haematopus ostralegus*), crane (including circular trackways related to courting, and parenting behavior) and other wading birds tracks have been regularly studied together with a large number of hominin footprints in four main sectors along of a 4 km stretch of coast, since the works of Tooley reported ungulate tracks in the 70s (Cowell et al., 1993; Roberts et al., 1996; Huddart et al., 1999; Bell et al., 2000; Roberts, 2004, 2009; Roberts and Worsley, 2008; Wiseman et al., 2022). Bioturbation was developed between >9 and 1.2 ka (Roberts, 2009, 2014; Burns et al., 2022). In addition, a dune-edge peat stratum dated to ca. 3 ka contains hoofprints of domestic oxen (Roberts, 2009; Burns, 2021). This stratigraphic record, with 31 track beds, 401 large animals, and 192 human individuals determined from tracks, provides fair evidence for the Holocene decline in large mammal biodiversity (Burns et al., 2022). Humans formed part of a rich Mesolithic intertidal ecosystem with aurochs, red deer, roe deer, wild boar, beaver, and lynx. With the advent of agriculture in Great Britain (after ca. 5.5 ka), human footprints dominate the Neolithic and later beds, alongside a significant fall in large mammal species richness. The decline in large mammals' diversity could be the result of habitat shrinkage following sea level rise and the development of agriculture, as well as hunting pressure from a growing human population. In this way, Burns et al. (2022) set out a new way to measure long-term change in biodiversity via multilayer monitoring of dated track beds.

An eroding intertidal peat bed at Low Hauxley, Northumberland, provided animal tracks of two general types: cloven-hoofed and large, more circular-shaped prints (Eadie and Waddington, 2013). The cloven hoof prints may represent several different types of animals (cervids, wild boar, bovid) as there is variation in shape, size, and form. This site was dated to the late Mesolithic period in the final centuries of the sixth millennium cal BP. At Cresswell, another tracksite was identified (Waddington, 2014).

Goldcliff is the most important ichnological area in Wales. Four excavated Mesolithic settlement areas are associated with extensive areas of tracks, including those of red and roe deers, the occasional aurochs, and many birds (Scales, 2002, 2004, 2006; Bell, 2007a). At the stratigraphic level, there are six sites with buildings and trackways interpreted as seasonal settlements of pastoralists who grazed animals on the saltmarsh during the Spring and Summer periods of lush vegetation growth and reduced tidal range. Rectangular buildings are partly encircled by depressions covered by animal tracks, mostly of cattle, with some sheep/goats and a few pigs (Bell, 2013). A latter-dated site at Goldcliff also shows tracks of cattle (Bell et al., 2000). At Lydstep, Pembrokeshire and Manorbier (Murphy et al., 2014), Redwish Rhyl, North Wales, Kenfig, Borth and Port Eynon, Gower, late Mesolithic tracks of hominins and deer were found (Bell, 2007a,b; Bennett et al., 2010a,b; Bell, 2020).

Other tracksites in England include the Shaugh Moor Bronze Age site in Dartmoor, with many tracks of cattle and sheep but occasionally horse and badger as well (Smith et al., 1981). At Dovey Estuary, tracks of hominins, cattle, sheep/goats, and possibly bear on the peat surface date between ca. 2–1 ka cal BC (Bell, 2020). In the East Anglia Fengate, tracks of animals, including cattle and people, were found on the edge of a paleochannel (Pryor, 1998). Must Farm and Westward Ho! in Devon are other tracksites indicated in Bell (2020).

Table 1.5 presents the ichnotaxa erected and described in Europe.

Acknowledgments

We would like to thank Luis Caceres (University of Huelva), Masaki Matsukawa (Tokyo Gakugei University), and Kenishiro Shibata (Yokosuka City Museum) for the provided photos of *Bifidipes aeolis* from Mallorca Island and from Japan, respectively. Also to Jacob Benner (University of Tennessee) and Spencer Lucas (New Mexico Museum of Natural History and Science) for providing photos of *Broomichnium flirii* from Connecticut Valley, and McNeil et al. (2007) from Alberta. Teresa Manera de Bianco and Federico Agnolin (Museo Argentino de Ciencias Naturales "Bernardino Rivadavia") kindly provided with the photos from Argentinean sites. João Belo (flyGIS) provided the 3D model of *Proboscipeda panfamilia* from SW Portugal. Participation in RNM was funded by project PICT2019-114 from the Agencia Nacional de Promoción Científica y Tecnológica of Argentina and project G22 from Universidad Nacional de La Pampa, both from Argentina. The editor, Spencer Lucas, is greatly appreciated for his careful work with the manuscript.

References

Agenbroad, L.D., 1984. Hot Springs, South Dakota: entrapment and taphonomy of columbian mammoth. In: Martin, P.S., Klein, R. (Eds.), Quaternary Extinctions: A Prehistoric Revolution. University of Arizona Press, Tucson, pp. 113–127.

Agnolín, F.L., Chimento, N.R., Campo, D.H., Magnussen, M., Boh, D., De Cianni, F., 2019. Large carnivore footprints from the late Pleistocene of Argentina. Ichnos 26, 119–126. https://doi.org/10.1080/10420940.2018.1479962.

Aldhouse-Green, S.H.R., Housley, R.A., 1993. The Uskmouth mattock: a radiocarbon date. Archaeol. Cambrensis 117, 340.

Aldhouse-Green, S.H.R., Whittle, A., Allen, J.R.L., Caseldine, A.E., Culver, S.J., Day, M.H., Lunqvist, J., Upton, D., 1992. Prehistoric human footprints from the Severn Estuary at Uskmouth and Magor Pill, Gwent, Wales. Archaeol. Cambrensis 141, 14–55.

Alexander, R.M., 1983. On the massive legs of a moa (*Pachyornis elephantopus*, Dinornithes). J. Zool. 201, 363–376.

Alexanderson, H., Bernhardson, M., 2019. Late Glacial and Holocene sand drift in northern Götaland and Värmland, Sweden: sediments and ages. GFF 141 (2), 84–105.

Allegranzi, A., Barolomei, G., 1956. Unghiate di orso speleo nei Colli Berici (Vicenza). Ann. Univ. Ferrara - Sezione 9 Sci. Geol. Mineral. 2, 207–212.

Allen, J.R.L., 1997. Subfossil Mammalian Tracks (Flandrian) in the Severn Estuary, S.W. Britain: mechanics of formation, preservation and distribution. Philos. Transact. R. Soc. B 352 (1352), 481–518.

Allen, J.R.L., Bell, M., Scales, R., 2003. Animal and human footprint-tracks in archaeology: description and significance. Archaeol. Severn Estuary 14, 55–68.

Alonso, R.N., 1987. Valoración icnoavifaunística de ambientes boratíferos. IV Congreso Latinoamericano de Paleontología. Asociación Boliviana de Paleontología, Santa Cruz de La Sierra, Bolivia, pp. 586–597.

Alonso, R.N., 2012. Icnitas de aves en depósitos de boratos y su contribución a la reconstrucción paleoambiental. Rev. Paléobiol. 11, 429–445.

Altamura, F., Bennett, M.R., D'Août, K., Gaudzinski-Windheuser, S., Melis, R.T., Reynolds, S.C., Mussi, M., 2018. Archaeology and ichnology at Gombore II-2, Melka Kunture, Ethiopia: everyday life of a mixed-age hominin group 700,000 years ago. Sci. Rep. 8, 2815. https://doi.org/10.1038/s41598-018-21158-7.

Altamura, F., Bennett, M.R., Marchetti, L., Melis, R.T., Reynolds, S.C., Mussi, M., 2020. Ichnological and archaeological evidence from Gombore II OAM, Melka Kunture, Ethiopia: an integrated approach to reconstruct local environments and biological presences between 1.2 and 0.85 Ma. Quat. Sci. Rev. 244, 106506. https://doi.org/10.1016/j.quascirev.2020.106506.

Altamura, F., Gioia, P., 2020. Casal de' Pazzi (Roma, RM). Notiziario di Preistoria e Protostoria 7 (1), 15–17.

Altamura, F., Lehmann, J., Rodríguez-Álvarez, B., Urban, B., van Kolfschoten, T., Verheijen, I., Conard, N.J., Serangeli, J., 2023. Fossil footprints at the late Lower Paleolithic site of Schöningen (Germany): a new line of research to reconstruct animal and hominin paleoecology. Quat. Sci. Rev. 310, 108094. https://doi.org/10.1016/j.quascirev.2023.108094.

Altamura, F., Mussi, M., 2017. Archeologia e impronte fossili nel sito acheuleano di Gombore II (0,85 Ma), Melka Kunture, Etiopia. Scienze dell'Antichità 23 (1), 21–35.

Altamura, F., Mussi, M., Melis, R.T., 2017. A middle Pleistocene hippo tracksite at Gombore II-2 (Melka Kunture, Upper Awash, Ethiopia). Palaeogeogr. Palaeoclimatol. Palaeoecol. 470, 122–131.

Altamura, F., Serangeli, J., 2023. A tale of many tracks: an overview of fossil proboscidean footprints at Paleolithic sites around the world, with a particular focus on Schöningen, in Germany. J. Mediterr. Earth Sci. 15. https://doi.org/10.13133/2280-6148/18132.

Alvarado, G.E., Lucas, S.G., Gómez, L.D., 2008. Evidencias directas e indirectas sobre la probable coexistencia de bisontes y el ser humano en Centroamérica durante el Holoceno. Rev. Geol. Am. Cent. 39, 253–265.

Anderson, A., 1976. Fish trails from the Early Permian of South Africa. Palaeontology.

Anderson, A., 1989. Prodigious Birds: Moas and Moa-Hunting in Prehistoric New Zealand. Cambridge University Press, Cambridge.

Antón, M., Salesa, M.J., 2012. Walking with lions: the missed potential of Quaternary palaeoichnology. Quat. Sci. Rev. 49, 106–108.

Applegate, S.P., Espinosa-Arrubarrena, L., López-Neri, P., 1984. Las calizas litográficas de la cantera Tlayúa en Tepexi de Rodríguez, Puebla, México. Sociedad Geológica Mexicana, Convención Geológica Nacional 7, México Excursión to Tepexi de Rodríguez, State of Puebla, pp. 1–83.

Aramayo, S.A., 2004. Late Quaternary paleoichnological sites at Southern Atlantic coast of Buenos Aires Province, Argentina. In: Kim, J.Y., Kim, K.-S., Park, S.I., Shin, M.-K. (Eds.), International Symposium on Quaternary Footprints of Hominids and Other Vertebrates. Jeju, Korea, pp. 112–130.

Aramayo, S.A., Manera de Bianco, T., 1987a. Hallazgo de una icnofauna continental (Pleistoceno tardío) en la localidad de Pehuén Có, Provincia de Buenos Aires, Argentina. In: Parte I: Edentata, Litopterna, Proboscidea. 4° Congreso Latinoamericano de Geología, Actas, Santa Cruz de la Sierra, Bolivia, pp. 516–531.

Aramayo, S.A., Manera de Bianco, T., 1987b. Hallazgo de una icnofauna continental (Pleistoceno tardío) en la localidad de Pehuén Co, Provincia de Buenos Aires, Argentina. In: Parte II: Carnivora, Artiodactyla y Aves. 4° Congreso Latinoamericano de Geología, Actas, Santa Cruz de la Sierra, Bolivia, pp. 531–547.

Aramayo, S.A., Manera de Bianco, T., 1996. Edad y nuevos hallazagos de icnitas de mamiferos y Aves en el yacimiento paleoicnologico de Puhuen-Co (Pleistoceno tardio), Provincia de Buenos Aires, Argentina. Asociación Paleontológica Argentina. Publicación Especial 4, 47–57.

Aramayo, S.A., Manera de Bianco, T., 2009. Late Quaternary paleoichnological sites from the southern Atlantic coast of Buenos Aires, Argentina: mammal, bird and hominid evidence. Ichnos 16, 25–32. https://doi.org/10.1080/10420940802470714.

Aramayo, S.A., Manera de Bianco, T., Bastianelli, N.V., Melchor, R.N., 2015. Pehuén Co: updated taxonomic review of a late Pleistocene ichnological site in Argentina. Palaeogeogr. Palaeoclimatol. Palaeoecol. 439, 144–165. https://doi.org/10.1016/j.palaeo.2015.07.006.

Araújo-Júnior, H.I., de Souza Barbosa, F.H., da Silva, L.H.M., 2017. Overlapping paleoichnology, paleoecology and taphonomy: analysis of tooth traces in a Late Pleistocene-early Holocene megafaunal assemblage of Brazil and description of a new ichnotaxon in hard substrate. Palaeogeogr. Palaeoclimatol. Palaeoecol. 468, 122–128. https://doi.org/10.1016/j.palaeo.2016.12.007.

Arman, S.D., Prideaux, G.J., 2016. Behaviour of the Pleistocene marsupial lion deduced from claw marks in a southwestern Australian cave. Sci. Rep. 6 (1), 21372.

Ashley, G.M., Liutkus, C.M., 2002. Tracks, trails and trampling by large vertebrates in a Rift Valley paleo-wetland, Lowermost Bed II, Olduvai Gorge, Tanzania. Ichnos 9 (1–2), 23–32.

Avanzini, M., Romano, M., Citton, O., Salvador, I., Arobba, D., Caramiello, R., Firpo, M., Rellini, I., Negrino, F., Clementi, L., Zunino, M., Giannotti, S., Starnini, E., Conventi, M., 2018. Ichno-archeology of a human Palaeolithic ecosystem: the human and animal footprints in the Grotta della Básura (Toirano, Northern Italy). Alp. Mediterr. Quat. 31, 39–42.

Avanzini, M., Salvador, I., Starnini, E., Arobba, D., Caramiello, R., Romano, M., Citton, O., Rellini, I., Firpo, M., Zunino, M., Negrino, F., 2021. Following the father steps in the bowels of the Earth: the ichnological record from the Bàsura Cave (Upper Palaeolithic, Italy). In: Pastoors, A., Lenssen-Erz, T. (Eds.), Reading Prehistoric Tracks. Springer, Cham, pp. 251–276.

Baales, M., 2002. Auf der Fährte spätglazialer Pferde bei Mertloch (Neuwieder Becken, Mittelrhein, Deutschland). Bonn. Zool. Beitr. 50 (3), 109–133.

Baales, M., Berg, A.V., 1997. Tierfährten in der Allerødzeitlichen Vulkanasche des Laacher ee-Vulkans bei Mertloch, Kreis Mayen-Koblenz. Archaol. Korresp. 27, 1–12.

Baales, M., von Berg, A., 1999. Völlig unerwartet: Tierfährten von Pferden, Braunbären, Rot- und Auerwild in Ablagerungen des allerfrdzeitlichen Laacher See-Vulkans (ca. 12.9 ky cal BP) bei Mertloch (Kr. Mayen-Koblenz, Neuwieder Becken, Rheinland-Pfalz, Deutschland). Tier und Museum 6, 68–74.

Baba, M., Matsukawa, M., Koarai, K., Hayashi, K., Okubo, A., Ito, M., 2000. Let's reanimate ancient animals based on Elephant tracks for possible teaching development. J. Earth Sci. Educ. Japan 53, 269–281.

Bakels, C., Zeiler, J., 2005. Neolithic subsistence. In: Louwe Kooijmans, L.P., van den Broeke, P.W., Fokkens, H., van Gijn, A.L. (Eds.), The Prehistory of the Netherlands Volume 1. Amsterdam University Press, pp. 311–366.

Balouet, J.-C., 1991. The fossil vertebrate record of New Caledonia. In: Vertebrate Palaeontology of Australasia, pp. 1381–1409.

Barr, K., Bell, M., 2016. Neolithic and Bronze Age ungulate footprint-tracks of the Severn Estuary: species, age, identification and the interpretation of husbandry practices. Environ. Archaeol. https://doi.org/10.1080/14614103.2016.1222691.

Basedow, H., 1914. Aboriginal rock carving of great antiquity in South Australia. J. Roy. Anthropol. Inst. G. B. Ireland 44, 195–211.

Bates, D., 2010. The Passing of the Aborigines: A Lifetime Spent Among the Natives of Australia. Lulu. com.

Baucon, A., Privitera, S., Morandi Bonacossi, D., Canci, A., Neto de Carvalho, C., Kyriazi, E., Laborel, J., Laborel-Deguen, F., Morhange, C., Marriner, N., 2008. Principles of Ichnoarchaeology: new frontiers for studying past times. In: Avanzini, M., Petti, F.M. (Eds.), Italian Ichnology. Studi Trentini Science Naturali, Acta Geologica, vol 83, pp. 43–72.

Bayón, C., Manera, T., Politis, G., Aramayo, S., 2011. Following the tracks of the first South Americans. Evol. Educ. Outreach 4, 205–217.

Behrensmeyer, A.K., Laporte, L.F., 1981. Footprints of a Pleistocene hominid in northern Kenya. Nature 289, 167–169.

Bell, K.N., De Merlo, J.A., 1969. Some fossil bird tracks. Victoria Naturalist 86 (5), 134–135.

Bell, M., 2001. Environmental archaeology in the Severn Estuary: progress and prospects. Estuarine archaeology. The Severn and beyond. Archaeol. Sever. Estuary 11, 69–104.

Bell, M., 2007a. Prehistoric coastal communities: the Mesolithic in Western Britain. CBA Research Report 149. York.

Bell, M., 2007b. Banded sediments with footprint-tracks. In: Bell, M. (Ed.), Prehistoric Coastal Communities: The Mesolithic in Western Britain. Council for British Archaeology, York, pp. 48–55.

Bell, M., 2013. The Bronze age in the Severn Estuary. Council for British Archaeology Research Report 172. York.

Bell, M., 2020. Making One's Way in the World: The Footprints and Trackways of Prehistoric People. Oxbow Books, Oxford.

Bell, M., Caseldine, A.E., Neumann, H. (Eds.), 2000. Prehistoric Intertidal Archaeology in the Welsh Severn Estuary, 120. Council of British Archaeology, Research Report, York.

Bell, M., Walker, M.J.C., 2005. Late Quaternary Environmental Change Physical and Human Perspectives. Routledge.

Belperio, A.P., Fotheringham, D.G., 1990. Geological setting of two Quaternary footprint sites, western South Australia. Aust. J. Earth Sci. 37, 37–42.

Benham, W.B., 1913. Notes on footprints of the moa. Transact. New Zealand Instit. 45, 211.

Benner, J.S., Ridge, J.C., Taft, N.K., 2008. Pleistocene freshwater fish (Cottidae) trackways from New England (USA) glacial lakes and a reinterpretation of the ichnogenus *Broomichnium* Kuhn. Palaeogeogr. Palaeoclimatol. Palaeoecol. 260, 375–388.

Benner, J.S., Ridge, J.C., Knecht, R.J., 2009. Timing of post-glacial reinhabitation and ecological development of two New England, USA, drainages based on trace fossil evidence. Palaeogeogr. Palaeoclimatol. Palaeoecol. 272, 212–231.

Bennett, M.R., Bustos, D., Belvedere, M., Martinez, P., Reynolds, S., Urban, T., 2019. Soft sediment deformation below mammoth tracks at White Sands National Monument (New Mexico) with implications for biomechanical inferences from tracks. Palaeogeogr. Palaeoclimatol. Palaeoecol. 527, 25–38.

Bennett, M.R., Bustos, D., Odess, D., Urban, T.M., Lallensack, J.N., Budka, M., Santucci, V.L., Martinez, P., Wiseman, L.A., Reynolds, S.C., 2020. Walking in mud: remarkable Pleistocene human trackways from White Sands National Park (New Mexico). Quat. Sci. Rev. 249, 106610.

Bennett, M.R., Bustos, D., Pigati, J.S., Springer, K.B., Urban, T.M., Holliday, V.T., Reynolds, S.C., Budka, M., Honke, J.S., Hudson, A.M., 2021. Evidence of humans in North America during the last glacial maximum. Science 373, 1528–1531.

Bennett, M.R., Gonzalez, S., Huddart, D., Kirby, J., Toole, E., 2010a. Probable Neolithic footprints preserved in intertidal peat at Kenfig, South Wales (UK). PGA (Proc. Geol. Assoc.) 121, 66–76.

Bennett, M.R., Liutkus, C.M., Thackeray, F., Morse, S.A., McClymont, J., Stratford, D., 2010b. Footprints of the Kuiseb Delta, Namibia. Digging Stick 27 (3), 1–4.

Bennett, M.R., Morse, S.A., Falkingham, P.L., 2014a. Tracks made by swimming hippopotami: an example from Koobi Fora (Turkana Basin, Kenya). Palaeogeogr. Palaeoclimatol. Palaeoecol. 409, 9–23.

Bennett, M.R., Morse, S.A., Liutkus-Pierce, C., Mcclymont, J., Evans, M., Crompton, R.H., Thackeray, F., 2014b. Exceptional preservation of children's footprints from a Holocene footprint site in Namibia. J. Afr. Earth Sci. 97, 331–341.

Bertling, M., Braddy, S.J., Bromley, R.G., Demathieu, G.D., Genise, J.F., Mikuláš, R., Nielsen, J.K., Nielsen, K.S.S., Rindsberg, A.K., Schlirf, M., Uchman, A., 2006. Names for trace fossils: a uniform approach. Lethaia 39, 265–286.

Bertran, P., Fouéré, P., 2020. The Holocene estuarine and coastal dune deposits of Soulac sur-Mer (Médoc peninsula, southwest France). Quaternaire 31 (3), 231–248.

Blanco, R.E., Czerwonogora, A., 2003. The gait of *Megatherium* Cuvier 1796 (Mammalia, Xenarthra, Megatheriidae). Senckenberg. Biol. 83 (1), 61–68.

Bocchini, A., Coltorri, M., 1978. Unghiate ed impronte de *Ursus spelaeus* nella Grotta del Fiume nella Gola di Frasassi (Ancona). Rassegna Speleologia Italiana Memoria XII, 138–141.

Boutakoff, N., 1963. The Geology and Geomorphology of the Portland Area. Department of Mines, Victoria. Memoir 22.

Bravo Cuevas, V.M., Jimenez Hidalgo, E., Cabral Perdomo, M.A., Contreras-López, M., 2019. The first Mexican record of peccary footprints (Artiodactyla, Tayassuidae) from the late Cenozoic of Puebla: ichnotaxonomy and palaeobiological considerations. Hist. Biol. 31 (9), 1135–1144. https://doi:10.1080/08912963.2018.1424844.

Broholn, H.C., 1946. Danmarks Broncealder 3. Copenhagen, pp. 1–278.

Bromley, R.G., 2001. Tetrapod tracks deeply set in unsuitable substrates: Holocene musk oxen in fluid earth (East Greenland) and Pleistocene caprines in aeolian sand (Mallorca). Bull. Geol. Soc. Den. 48 (2), 209–215.

Bromley, R.G., Milàn, J., Uchman, A., Hansen, K.S., 2009. Rheotactic *Macaronichnus*, and human and cattle trackways in Holocene beachrock, Greece: reconstruction of paleoshoreline orientation. Ichnos 16, 103–117.

Brown, R.W., 1947. Fossil plants and human footprints in Nicaragua. J. Paleontol. 21, 38–40.

Bryan, A.L., 1973. New light on ancient Nicaraguan footprints. Archaeology 26, 146–147.

Bucheim, H.P., Leggitt, V.L., Biaggi, R.E., Loewen, M.A., 1999. Palaeoenvironments of avian trackway sites within the "Olla member," Palm Springs Formation, Anza-Borrego Desert State Park. San Bernardino County Museum Association Quarterly 46 (2), 47–52.

Burns, A., 2021. The Mesolithic footprints retained in one bed of the former saltmarshes at Formby Point, Sefton Coast, North West England. In: Pastoors, A., Lenssen-Erz, T. (Eds.), Reading Prehistoric Tracks. Springer, Cham, pp. 295–316.

Burns, A., Woodward, J., Conneller, C., Reimer, P., 2022. Footprint beds record Holocene decline in large mammal diversity on the Irish Sea coast of Britain. Nat. Ecol. Evol. 6, 1553–1563.

Bustos, D., Jakeway, J., Urban, T.M., Holliday, V.T., Fenerty, B., Raichlen, D.A., Budka, M., Reynolds, S.C., Allen, B.D., Love, D.W., Santucci, V.L., Odess, D., Willey, P., Mcdonald, H.G., Bennett, M.R., 2018. Footprints preserve terminal Pleistocene hunt? Human-sloth interactions in North America. Sci. Adv. 4 (4), eaar7621.

Cabral Perdomo, M.A., 1995. Los Icnofósiles de Vertebrados Terrestres del Terciario Tardío del área de Tepexi de Rodríguez, Estado de Puebla. Universidad Nacional Autónoma de México, tesis de licenciatura, México, D.F., p. 107.

Cabral-Perdomo, M.A., 2000. Pleistocene vertebrate tracks and traces from the Pie de Vaca Formation. Guide Book of the Field Trips. Society of Vertebrate Paleontology Annual Meeting 2000, 115–119.

Cabral Perdomo, M.A., 2013. Icnofósiles de vertebrados terrestres del Cenozoico Tardío en el área de "Pie de Vaca", Tepexi de Rodríguez, Estado de Puebla. Palaontol. Mexic. 3, 50–57.

Cabral Perdomo, M.A., Applegate, S.P., 1991a. Preliminary report of large fossil mammals from the Pie de Vaca formation in Tepexi de Rodríguez, Puebla, México: First record of this kind of ichnofossils in México. J. Vertebr. Paleontol. 11 (Suppl. no.3), 19A.

Cabral Perdomo, M.A., Applegate, S.P., 1991b. Las huellas fósiles de camélidos de la localidad "Pie de Vaca" en Tepexi de Rodríguez, Puebla: Un informe preliminar. In: II Congreso Nacional de Paleontología, Sociedad Mexicana de Paleontología, Libro de Resúmenes, vol 95.

Cabral Perdomo, M.A., Bravo-Cuevas, V.M., Pérez-Pérez, A., García-Cabrera, N., 2018. Descripción de las huellas de camélidos y félidos de la localidad Pie de Vaca, Cenozoico Tardío de Puebla, centro de México y algunas consideraciones paleobiológicas. Bol. Soc. Geol. Mex. 70 (2), 397–416. https://doi.org/10.18268/bsgm2018v70n2a9.

Camens, A., Carey, S.P., 2013. Contemporaneous trace and body fossils from a Late Pleistocene lakebed in Victoria, Australia, allow assessment of bias in the fossil record. PLoS One 8 (1), e52957.

Camens, A.B., Carey, S.P., Arnold, L.J., 2018. Vertebrate trace fossils from the late Pleistocene of Kangaroo Island, South Australia. Ichnos 25 (2–3), 232–251. https://doi.org/10.1080/10420940.2017.1337633.

Camens, A.B., Wells, R.T., 2009. Diprotodontid footprints from the Pliocene of Central Australia. J. Vertebr. Paleontol. 29, 863–869.

Camens, A.B., Worthy, T.H., 2019. Pliocene avian footprints from the Lake Eyre Basin, South Australia. J. Vertebr. Paleontol. 39, e1676764.

Campos–Medina, J., Moreno, K., Rojas, J., Gómez, G., García, J.L., Buldrini, K.E., Lüthgens, C., Rodríguez, E., Álvarez, R., Abarzúa, A.M., 2022. The oldest record of the Lamini Tribe and Proboscidea order in the southwestern margin of the Andes Mountain range: late Pleistocene mammalian footprints at the Pelluco fossil forest sanctuary. J. S. Am. Earth Sci. 118, 103940. https://doi.org/10.1016/j.jsames.2022.103940.

Carey, S.P., Camens, A., Cupper, M.L., Grün, R., Hellstrom, J.C., McKnight, S.W., Mclennan, I., Pickering, D.A., Trusler, P., Aubert, M., 2011. A diverse Pleistocene marsupial trackway assemblage from the Victorian Volcanic Plains, Australia. Quat. Sci. Rev. 30 (5–6), 591–610.

Casamiquela, R.M., 1983. Pisadas del Pleistoceno (¿Superior?) del Balneario de Monte Hermoso, Buenos Aires. In: La confirmación del andar bipedal en los megateroideos Cuadernos del Instituto Superior Juan XXIII 4, pp. 5–21.

Casamiquela, R.M., Chong Díaz, G., 1975. Icnitas (Mammalia, Equidae?) en rocas del Plio-Pleistoceno de la costa, provincia de Antofagasta. In: Primer Congreso Argentino de Paleontología y Bioestratigrafía, Actas, vol. 2, pp. 621–632.

Casteret, N., 1948. The footprints of prehistoric man: a vivid new evidence of our ancestors of fifteen thousand years ago. Illus. Lond. News 1 (10), 408–410.

Cearreta, A., Edeso, J.M., Merino, A., Ugalde, T., Ugarte, F.M., 1990. Las dunas litorales de Barrika (costa occidental de Bizkaia). Kobie 19, 77–83.

Cerilli, E., Celant, A., Giordano, G., Carboni, G., 2020. Casetta Mistici e Tor Vergata-San Gaudenzio (Roma): impronte di animali ed umane su una superficie di lahar. In: Anzidei, A.P., Carboni, G. (Eds.), Roma Prima del Mito. Archaeopress Publishing, Oxford, pp. 15–32.

Chauvet, J.-M., Deschamps, E.B., Hillaire, C., 1996. Dawn of Art: The Chauvet Cave. Harry N. Abrams, New York.

Citton, P., Romano, M., Salvador, I., Avanzini, M., 2017. Reviewing the upper Pleistocene human footprints from the 'Sala dei Misteri' in the Grotta della Bàsura (Toirano, northern Italy) cave: An integrated morphometric and morphoclassificatory approach. Quat. Sci. Rev. 169, 50–64.

Clarkson, C., Jacobs, Z., Marwick, B., Fullagar, R., Wallis, L., Smith, M., Roberts, R.G., Hayes, E., Lowe, K., Carah, X., 2017. Human occupation of northern Australia by 65,000 years ago. Nature 547, 306–310.

Clottes, J., 2003. Return to Chauvet Cave. Thames and Hudson, London.

Cockburn-Hood, T., 1875. 3. Remarks upon the footprints of the Dinornis in the Sand Rock at Poverty Bay, New Zealand, and upon its recent extinction. Proc. R. Soc. Edinb. 8, 236–240.

Collen, J., Vella, P., 1984. Hautotara, Te Muna and Ahiaruhe formations, middle to late Pleistocene, Wairarapa, New Zealand. J. Roy. Soc. N. Z. 14, 297–317.

Cowell, R.W., Milles, A., Roberts, G., 1993. Prehistoric footprints on Formby Point beach, Merseyside. In: Middleton, R. (Ed.), North West Wetlands Survey, Annual Report 1993, pp. 43–52.

Crawford, J., 1891. Neolithic man in Nicaragua. Am. Geologist 8, 160–161.

Curry, D., 1939. Tertiary and Pleistocene mammal and bird tracks in Death Valley. Bull. Geol. Soc. Am. 50 (12), 1971–1972.

Dávalos-Álvarez, O.G., Nieto-Samaniego, A.F., Alaniz-Álvarez, S.A., Martínez-Hernández, E., Ramírez-Arriaga, E., 2007. Estratigrafía cenozoica de la región de Tehuacán y su relación con el sector norte de la falla de Oaxaca. Rev. Mex. Ciencias Geol. 24, 197–215.

Davis, P., 1996. The footprint surfaces at the Barnfield Pit, Swanscombe. In: Conway, B., McNabb, J., Ashton, N. (Eds.), Excavations at Barnfield Pit, Swanscombe 1968-72. British Museum Press, London, pp. 169–185.

Diedrich, C.G., 2011a. Late Pleistocene steppe lion *Panthera leo spelaea* (Goldfuss, 1810) footprints and bone records from open air sites in northern Germany – evidence of hyena-lion antagonism and scavenging in Europe. Quat. Sci. Rev. 30, 1883–1906.

Diedrich, C.G., 2011b. An overview of the ichnological and ethological studies in the Cave Bear Den in Urşilor Cave (Western Carpathians, Romania). Ichnos 18, 9–26.

Douglas, J.G., 2004. The Nature of Warrnambool, Warrnambool. Warrnambool Field Naturalists Club Incorporated.

Di Vito, M.A., Talamo, P., de Vita, S., Rucco, I., Zanchetta, G., Cesarano, M., 2019. Dynamics and effects of the Vesuvius Pomici di Avellino Plinian eruption and related phenomena on the Bronze Age landscape of Campania region (Southern Italy). Quat. Int. 499, 231–244.

Dugès, A., 1897. Felis fósil de San Juan de los Lagos. Naturaleza 2 (8), 421–423.

Eadie, G., Waddington, C., 2013. Rescue Recording of an Eroding Inter-tidal Peat Bed at Lower Hauxley, Northumberland (6109). Archaeological Research Service 2013/17, pp. 3–23.

Ebner, F., Graf, W., 1978. Späteiszeitliche Säugetierfährten in Terassensedimenten des Ilztales. Jahresberichtv der Landesmuseum und Joanneum Graz NF 8, 175–180.

Ehrenberg, K., 1967. Der Höhlenbär, sein Vorkommen und seine Beziehungen zur Umwelt. Mitteilungen der Österreichischen Arbeitsgemeinschaft für Urund Frühgeschichte 18, 34–50.

Estraviz-López, D., Mateus, O., 2019. Tracks and multiple skeletons of brown bear (Ursus arctos) in Algar do Vale da Pena, Portugal. Zubia - Monogr. 31, 77–82.

Eyde, T.H., 1978. Bowie Zeolite: an Arizona industrial mineral. State of Arizona Bureau of Geology and Mineral Technology Fieldnotes 8 (4), 1–5.

Eyde, T.H., 1982. Zeolite deposits in the Gila and San Simon Valleys of Arizona and New Mexico. New Mexico Bureau of Mines and Mineral Resources. Circular 182, 65–71.

Fabiano, M., Zucchelli, M., 2003. Impronte di elefante sulle rive del lago pleistocenico di Atella (PZ). Studi Ecol.l Quaternario 25, 9–22.

Falkingham, P.I., Agenbroad, L.D., Thompson, K., Manning, P.I., 2010. Bird tracks at the Hot Springs Mammoth Site, South Dakota, USA. Ichnos 17, 34–39.

Fanelli, F., Palombo, M.R., Pillola, G.L., Ibba, A., 2007. Tracks and trackways of *"Praemegaceros" cazioti* (Depéret, 1897) (Artiodactyla, Cervidae) in Pleistocene coastal deposits from Sardinia (Western Mediterranean, Italy). Bolletino della Soc. Paleontol. Ital. 46 (1), 47–54.

Fanelli, F., Pillola, G.L., Palombo, M.R., Ibba, A., 2008. The Pleistocene cervid tracks of Porto Paglia. In: Fossil Mammalian Biotas of Sardinia. Fieldtrip Guide-Book, Euromam, Sardinia, pp. 72–77.

Felstead, N.J., Gonzalez, S., Huddart, D., Noble, S.R., Hoffmann, D.L., Metcalfe, S.E., Leng, M.J., Albert, B.M., Pike, A., Gonzalez-Gonzalez, A., Concepcion Jiminez-Lopez, J., 2014. Holocene-aged human footprints from Cuatrociénegas Basin, NE Mexico. J. Archaeol. Sci. 42, 250–259.

Feola, S.F., 2022. Icnología y Sedimentología del Yacimiento Paleoicnológico de Pehuen Co (Pleistoceno tardío), provincia de Buenos Aires. Doctoral Thesis. Universidad Nacional del Sur, 420 pp.

Feola, S.F., Manera de Bianco, T., Melchor, R.N., 2018. Bipedal Ground Sloths? the Footprint Evidence from the Pleistocene of Argentina. Fourth Latin-american Symposium on Ichnology, Santa Marta, Colombia.

Feola, S.F., Manera de Bianco, T., Melchor, R.N., 2023a. Fooprints comparable to *Roepichnus* at the Pehuen Co Paleoichnological site (Late Pleistocene), Buenos Aires province, Argentina. In: 4th Virtual Palaeontological Congress. Resúmenes (170).

Feola, S., Manera, T., Tatumi, S., Kinoshita, A., Barbosa, F., Baffa, O., Yee, M., Melchor, R., 2023b. High-resolution sedimentological and stratigraphic framework for the Pehuen Co Paleoichnological site, Late Pleistocene, Buenos Aires, Argentina. J. S. Am. Earth Sci., 104512 https://doi.org/10.1016/j.jsames.2023.104512.

Feibel, C.S., 1987. Fossil fish nests from the Koobi Fora Formation (Plio-Pleistocene) of northern Kenya. J. Paleontol. 61 (1), 130–134.

Figueiredo, S., Neto de Carvalho, C., Belo, J., 2018. Novos dados acerca das pegadas de aves dos eolianitos do Plistocénico Superior da Ilha do Pessegueiro (SO de Portugal): interpretações tafonómicas e taxonómicas resultantes do levantamento 3D. Boletim do Centro Português de Geo-história e Pré-história 1 (1), 47–52.

Fisher, D., 1994. Late Pleistocene proboscidean trackways in pond-margin sediments in south-eastern Michigan. J. Vertebr. Paleontol. 14 (3), 25A.

Fisher, D., 2018. Paleobiology of Pleistocene proboscideans. Annu. Rev. Earth Planet Sci. 46 (1), 229–260.

Fleury, K., et al., 2023. The moa footprints from the Pliocene - early Pleistocene of Kyeburn, Otago, New Zealand. J. R. Soc. N. Z. 54, 620–642. https://doi.org/10.1080/03036758.2023.2264789.

Fliri, F., Bortenschlager, S., Felber, H., Heissel, W., Hilscher, H., Resh, W., 1970. Der Bänderton von Baumkirchen (Inntal, Tirol) eine neue Schlüsselstelle zur Kenntnis der Würm-Vereisung der Alpen. Z. Gletscherkd. Glazialgeol. 6, 5–35.

Fliri, F., Hilscher, H., Markgraf, V., 1971. Weitere Untersuchungen zür Chronologie der alpinen Vereisung (Bänderton von Baumkirchen, Inntal, Nordtirol). Z. Gletscherkd. Glazialgeol. 7, 5–24.

Flor, G., 1989. Estructuras de deformación por pisadas de cérvidos en la duna cementada de Gorliz (Vizcaya, N de España). Boletin de la Sociedad Geologica de España 2 (1–2), 23–29.

Footprint Fossils Research Group for Nojiri-ko Excavation, 1992. Footprint fossils of Naumann's elephant discovered from the upper Pleistocene Nojiri-ko Formation. Earth Sci. 46, 385–404.

Forir, M., Ciampaglio, C., Ryan, N., 2007. Preliminary investigation of the trackways and claw marks within the Riverbluff Cave system, Springfield, Missouri. New Mexico Museum of Natural History and Science Bulletin 42.

Fornós, J.J., Bromley, R.G., Clemmensen, L.B., Rodríguez-Perea, A., 2002. Tracks and trackways of *Myotragus balearicus* Bate (Artiodactyla, Caprinae) in Pleistocene aeolianites from Mallorca (Balearic Islands, Western Mediterranean). Palaeogeogr. Palaeoclimatol. Palaeoecol. 180, 277–313.

Fornós, J.J., Clemmensen, L.B., Gómez-Pujol, L., Murray, A., 2009. Late Pleistocene carbonate aeolianites on Mallorca, Western Mediterranean: a luminescence chronology. Quat. Sci. Rev. 28 (25–26), 2697–2709.

Fornós, J.J., Pons-Moya, J., 1982. Icnitas de *Myotragus balearicus* del yacimiento de Ses Piquetes (Santanyi, Mallorca). Bull. Soc. Hist. Nat. Balears 26, 135–144.

Fornós, J.J., Sábat, F., 1998. Aspectes geológics de l'Arxipélag de Cabrera. In: Fornós, J.J. (Ed.), Aspectes Geológics de les Balears. U.I.B., Palma de Mallorca, pp. 67–93.

Fosberg, F.R., Carroll, D., 1965. Terrestrial sediments and soils of the northern Marshall Islands. Atoll Res. Bull. 113, 1–156.

Fossil Mammal Research Group for Nojiri-ko Excavation, 2008. Footprint fossils of deers from the Nojiri-ko Formation during the 16th Nojiri-ko excavation. Bulletin of the Nojiri-ko Museum 16, 33–41.

Frechen, M., 1995. Lumineszenz-Datierungen der pleistozänen Tierfährten von Bottrop-Welheim. Munchner Geowissenschaftliche Abdhandlungen 27, 63–80.

Gara, T., Cane, S., 1988. Environmental, Anthropological and Archaeological Background to the Nullarbor Plains. ANUTECH Pty. Ltd, Canberra (Unpublished).

Garcia, M.-A., 2005. Ichnologie genérale de la grotte Chauvet. Bull. Soc. Prehist. Fr. 102, 103–108.

Gardner, W.A., 1940. Place of the gods. Nat. Hist. 45, 40–43 and 54–55.

Ghinassi, M., Fidolini, F., Magi, M., Sagri, M., 2013. Depositional environments of the Plio-Pleistocene Upper Valdarno Basin (Tuscany, Italy). Ital. J. Geosci. 132, 33–53.

Gibbard, P.L., Dreimanis, A., 1978. Trace fossils from late Pleistocene glacial lake sediments in southwestern Ontario, Canada. Can. J. Earth Sci. 15, 1967–1976.

Gibbes, C.D., 1882. Prehistoric footprints found in the sandstone quarry of the Nevada State Prison. Proc. Calif. Acad. Sci. 1882, 1–8.

Gibert, J.M., et al., 1999. The fish trace fossil *Undichna* from the Cretaceous of Spain. Palaeontology 42 (3), 409–427.

Gillies, T.B., 1872. On the occurrence of footprints of the moa at Poverty Bay. New Zealand Instit. Proc. Transact. 4, 127–128.

Gingras, M.K., Armitage, I.A., Pemberton, G., Clifton, H.E., 2007. Pleistocene walrus herds in the Olympic Peninsula area: trace–fossil evidence of predation by hydraulic jetting. Palaios 22 (5), 539–545. https://doi.org/10.2110/palo.2005.p05-120r.

Gómez Gutiérrez, P., Muñiz, F., Neto de Carvalho, C., Toscano, A., Chacón, J.A.M., Belaústegui, Z., Galán, J.M., Rodríguez-Vidal, J., Cáceres, L.M., Ruiz, F., Ramírez-Cruzado, S., Belo, J., Cachão, M., Proença Cunha, P., Figueiredo, S., Moreira, N., Baucon, A., Finlayson, G., Finlayson, S., Giles Guzmán, F., Finlayson, C., 2023. Tras el rastro de *Palaeoloxodon antiquus* en el suroeste ibérico: nuevas evidencias icnológicas. In: Fiegl, J.P.-R., Fernández, B., Onrubia, M., Díaz-Acha, Y. (Eds.), Libro de Resumenes XXV Bienal de la Real Sociedad Española de Historia Natural, Málaga, pp. 159–160.

Gonzalez, A.H.G., Lockley, M.G., Rojas, C.S., Espinoza, J.L., Gonzalez, S., 2007. Notes on the re-discovery of a 'lost' hominid footprint site from the Cuatro Ciénegas basin (Coahuila), Mexico. New Mexico Museum of Natural History and Science Bulletin 54, 11–15.

Gonzalez, A.H., Lockley, M., Rojas, C., Espinoza, J.L., Gonzalez, S., 2009. Human tracks from quaternary tufa deposits, Cuatro Cienegas, Coahuila, Mexico. Ichnos 16, 12–24.

Gonzalez, S., Huddart, D., Bennett, M.R., Gonzalez-Huesca, A., 2006. Human footprints in Central Mexico older than 40,000 years. Quat. Sci. Rev. 25, 201–222.

Graham, R.W., Farlow, J.O., Vandike, J.E., 1996. Tracking ice age felids: identification of tracks of *Panthera atrox* from a cave in southern Missouri, U.S.A. In: Stewart, K.M., Seymour, K.M. (Eds.), Palaeoecology and Palaeoenvironments of Late Cenozoic Mammals: Tributes to the Career of C.S. (Rufus) Churcher. University of Toronto Press, Toronto, pp. 331–345.

Gregory, A., 2005. Footprints from Rangitoto's birth join volcanic display. N. Z. Herald, 1st Nov. https://www.nzherald.co.nz/nz/footprints-from-rangitotos-birth-join-volcanic-display/OANTYMSYLK7GIVKJ42BE6EVDCU/.

Gregory, M.R., 1991. New trace fossils from the Miocene of Northland, New Zealand: Rorschachichnus amoeba and Piscichnus waitemata. Ichnos 1, 195–205. https://doi.org/10.1080/10420949109386352.

Gregory, M.R., Hikuroa, D., Phillips, K., Campbell, K.A., Martin, A., 2006. Moa trackways and footprints: Three new discoveries. In: Stewart, B., Wallace, C., Lecointre, J., Reyners, M. (Eds.), Geological Society of New Zealand and New Zealand Geophysical Society Joint Conference, 2006 Massey University, NZ. Geological Society of New Zealand, p. 44.

Hall, F., McGowan, R., Guleksen, G., 1951. Aboriginal rock carvings: a locality near Pimba, SA. Record S. Aust. Mus. 9, 375–380.

Hand, S.J., 1989. Ancient Aborigines make a lasting impression. Aust. Geogr. 014.

Hand, S.J., Beck, R.M., Archer, M., Simmons, N.B., Gunnell, G.F., Scofield, R.P., Tennyson, A.J., De Pietri, V.L., Salisbury, S.W., Worthy, T.H., 2018. A new, large-bodied omnivorous bat (Noctilionoidea: Mystacinidae) reveals lost morphological and ecological diversity since the Miocene in New Zealand. Sci. Rep. 8 (1), 235–245.

Harkness, H.W., 1882. Footprints found at the Carson State Prison. In: Proceedings of the California Academy of Science, pp. 1–7.

Hay, R.L., 1976. Geology of the Olduvai Gorge: A Study of Sedimentation in a Semiarid Basin. University of California Press, Berkeley, 203 pp.

Haynes Jr., C.V., 1973. Exploration of a mammoth-kill site in Arizona. National Geographic Society Research Reports 1966 Projects 125–126.

Haynes Jr., C.V., 1991. Geoarchaeological and paleohydrological evidence for a Clovis-Age drought in North America and its bearing on extinction. Quat. Res. 35 (3), 438–450.

Haynes Jr., C.V., Huckell, B.B. (Eds.), 2007. Murray Springs: A Clovis Site with Multiple Activity Area in the San Pedro Valley, Arizona. Anthropological Papers of the University of Arizona Number 71.

Haynes, G., 2012. Elephants (and extinct relatives) as earth-movers and ecosystem engineers. Geomorphology 157–158, 99–107.

Hayward, B.W., 2016. Moa Footprints in Pleistocene Dune Sand Sequence. Geoscience Society of New Zealand Newsletter, West Auckland.

Hayward, B.W., 2022. Geological setting and interpretation of fossil moa footprints, Mosquito Beach, Kaipara South Head. Geocene. Auckland Geology Club 29, 2–7.

Hearty, P.J., Olson, S.I., 2011. Preservation of trace fossils and molds of terrestrial biota by intense storms in mid-last interglacial (MIS 5c) dunes on Bermuda, with a model for development of hydrological conduits. Palaios 26, 394–405.

Helm, C.W., Carr, A.S., Cawthra, H.C., De Vynck, J.C., Dixon, M.G., Lockley, M.G., Stear, W., Venter, J.A. 2022a. Large Pleistocene tortoise tracks on the Cape south coast of South Africa. Quaternary Research. Published online. https://www.doi.org/10.1017/qua.2022.50.

Helm, C.W., Lockley, M.G., Moolman, L., Cawthra, H.C., De Vynck, J.C., Dixon, M.G., Stear, W., Thesen, G.H.H., 2022c. Morphology of Pleistocene elephant tracks on South Africa's Cape south coast, and probable elephant trunk drag impressions. Quaternary Research 105, 100–114. https://doi.org/10.1017/qua.2021.32.

Helm, C.W., Anderson, R.J., Buckley, L.G., Cawthra, H.C., De Vynck, J.C., 2017. Biofilm assists recognition of avian trackways in Late Pleistocene coastal aeolianites, South Africa. Palaeontol. Afr. 52, 78–84.

Helm, C.W., Carr, A.S., Cawthra, H.C., De Vynck, J.C., Dixon, M.G., Gräbe, P.-J., Thesen, G.H.H., Venter, J.A., 2023a. Tracking the extinct giant Cape zebra (*Equus capensis*) on the Cape south coast of South Africa. Quat. Res. 114, 178–190. https://doi.org/10.1017/qua.2023.1.

Helm, C.W., Carr, A.S., Cawthra, H.C., De Vynck, J.C., Dixon, M.G., Paterson, A., Rust, R., Stear, W., Thesen, G.T.T., Van Berkel, F., Van Tonder, M., 2023b. Elephant seismicity: ichnological and rock art perspectives from South Africa. Proc. Geol. Assoc. https://doi.org/10.1016/j.pgeola.2023.09.006.

Helm, C.W., Carr, A.S., Cawthra, H.C., De Vynck, J.C., Dixon, M., Stear, W., Stuart, C., Stuart, M., Venter, J.A., 2022b. Possible Pleistocene pinniped ichnofossils on South Africa's Cape south coast. J. Coast Res. 38 (4), 735–749. https://doi.org/10.2112/JCOASTRES-D-21-00131.1.

Helm, C.W., Cawthra, H.C., Cowling, R.M., De Vynck, J.C., Marean, C.W., McCrea, R.T., Rust, R., 2018a. Palaeoecology of giraffe tracks in Late Pleistocene aeolianites on the Cape south coast. South Afr. J. Sci. 114 (1/2). https://doi.org/10.17159/sajs.2018/20170266.

Helm, C.W., Cawthra, H.C., Combrink, X., Helm, C.J.Z., Rust, R., Stear, W., Van den Heever, A., 2020a. Pleistocene large reptile tracks and probable swim traces on South Africa's Cape south coast. South Afr. J. Sci. 116 (3/4). https://doi.org/10.17159/sajs.2020/6542. Art. #6542.

Helm, C.W., Cawthra, H.C., Cowling, R.M., De Vynck, J.C., Lockley, M.G., Marean, C.W., Thesen, G.H.H., Venter, J.A., 2020b. Pleistocene vertebrate tracksites on the Cape south coast of South Africa and their potential palaeoecological implications. In: Cleghorn, N., Potts, A.J., Cawthra, H.C. (Eds.), The Palaeo-Agulhas Plain: A Lost World and Extinct Ecosystem. Quaternary Science Reviews, vol 235, 105857. https://doi.org/10.1016/j.quascirev.2019.07.039.

Helm, C.W., Cawthra, H.C., De Vynck, J.C., Dixon, M., Stear, W., 2021. Elephant tracks – a biogenic cause of potholes in Pleistocene South African coastal rocks. J. Coast Res. 37 (1), 59–74. https://doi.org/10.2112/JCOASTRES-D-20-00064.1.

Helm, C.W., Cawthra, H.C., De Vynck, J.C., Lockley, M.G., McCrea, R.T., Venter, J., 2019a. The Pleistocene fauna of the Cape south coast revealed through ichnology at two localities. South Afr. J. Sci. 115 (1/2). https://doi.org/10.17159/sajs.2019/5135. Art. #5135.

Helm, C.W., Cawthra, H.C., Hattingh, R., Hattingh, S., McCrea, R.T., Thesen, G.H.H., 2019b. Pleistocene trace fossils of Robberg Nature Reserve. Palaeontol. Afr. 54, 36–47.

Helm, C.W., Lockley, M.G., 2021. Pleistocene reptile swim traces confirmed from South Africa's Cape south coast. South Afr. J. Sci. 117 (3/4). https://doi.org/10.17159/sajs.2021/8830. Art.#8830.

Helm, C.W., Lockley, M.G., Carr, A.S., Cawthra, H.C., De Vynck, J.C., Dixon, M.G., Gräbe, P.-J., Rust, R., 2023c. New Pleistocene avian ichnosites on South Africa's Cape south coast. Ostrich. https://doi.org/10.2989/00306525.2023.2255385.

Helm, C.W., Lockley, M.G., Cawthra, H.C., De Vynck, J.C., Dixon, M.G., Stear, W., Venter, J.A., 2023d. Pleistocene fossil snake traces on South Africa's Cape south coast. Ichnos. https://doi.org/10.1080/10420940.2023.2204231.
Helm, C.W., Lockley, M.G., Cawthra, H.C., De Vynck, J.C., Helm, C.J.Z., Thesen, G.H.H., 2020c. Large Pleistocene avian tracks on the Cape south coast of South Africa. Ostrich 91 (4), 275–291. https://doi.org/10.2989/00306525.2020.1789772.
Helm, C.W., McCrea, R.T., Cawthra, H.C., Cowling, R.M., Lockley, M.G., Marean, C.W., Thesen, G.H.H., Pigeon, T., Hattingh, S., 2018b. A new Pleistocene hominin tracksite from the Cape south coast, South Africa. Sci. Rep. 8 (1), 3772. https://doi.org/10.1038/s41598-018-22059-5.
Helm, C.W., McCrea, R.T., Cawthra, H.C., Thesen, G.H.H., Mwankunda, J.M., 2018c. Late Pleistocene trace fossils in the Goukamma Nature Reserve, Cape south coast, South Africa. Palaeontologia Africana 52, 89–101.
Higgs, R., 1988. Fish trail in the Upper Carboniferous of South West England. Palaeontology 31, 255–272.
Hill, H., 1895. On the occurrence of moa-footprints in the bed of the Manawatu River, near Palmerston North. Transact. R. New Zealand Instit. 27, 476–477.
Hill, H., 1914. The moa-legendary, historical, and geological: why and when the moa disappeared. Transact. Proc. New Zealand Instit. 46, 330–351.
Hills, L.V., McNeil, P.E., 1999. First reported Canadian proboscidean trackway from the St. Mary reservoir, Alberta, Canada. Abstracts of papers. J. Vertebr. Paleontol. 19 (S3), 1–93, 51A.
Hood, T.H.C., 1874. Dinornis: Footprints and Recent Excavations. Proc. R. Society Edinb. 8, 236–240.
Horn, P.L., 1989. Moa tracks: an unrecognized legacy from an extinct bird. N. Z. J. Ecol. 12, 45–50.
Houck, K.J., Lockley, M.G., Avanzini, M., 2009. A survey of tetrapod tracksites preserved in pyroclastic sediments, with special reference to footprints of hominids, other mammals and birds. Ichnos 16 (1), 76–97.
Howe, P., 2022. Photogrammetry image of a single moa footprint. South Canterbury Museum. https://sketchfab.com/3d-models/pachyornis-moa-footprint-b8ba67ee3b2c441086cd1bb26abd19ce.
Hsieh, S., Ninard, K., Uchman, A., Łaska, W., Łapcik, P., 2022. Mammal tracks record anthropogenic environmental influence in late-Holocene inland dunes of the European Sand Belt, Poland. Holocene 33 (1), 61–80.
Hubbard, W.J., Fisher, D.C., Kardulias, P.N., 2000. Mastodon body weight estimates from footprints and a scale model. Curr. Res. Pleistocene 17, 122–123.
Huddart, D., Roberts, G., Gonzalez, S., 1999. Holocene human and animal footprints and their relationships with coastal environmental change, Formby Point, NW England. Quat. Int. 55, 29–41.
Hunt, A.P., Lucas, S.G., 2007. Cenozoic vertebrate trace fossils of North America: ichnofaunas, ichnofacies and biochronology. In: Lucas, S.G., Spielmann, Lockley, M.G. (Eds.), Cenozoic Vertebrate Tracks and Traces. New Mexico Museum of Natural History and Science Bulletin, vol 42, pp. 17–41.
Hunt, A.P., Lucas, S.G., 2024. An overview of the ichnology of moa tracks and other traces from the late Cenozoic of New Zealand. In: Taylor, L.H., Raynolds, R.G., Lucas, S.G. (Eds.), Vertebrate Paleoichnology: A Tribute to Martin Lockley, 95. New Mexico Museum of Natural History and Science Bulletin, pp. 179–197.
Hutton, F.W., 1899. On the footprint of a kiwi-like bird from Manaroa. Transact. Proc. New Zealand Instit. 31, 486.
Inoue, M., Sakaguchi, H., 1997. Estimating the withers height of the ancient Japanese horse from hoof prints. Anthropozoologica 25, 119–128.
Jacobs, Z., Roberts, D.L., 2009. Last Interglacial Age for aeolian and marine deposits and the Nahoon fossil human footprints, Southeast Coast of South Africa. Quat. Geochronol. 4 (2), 160–169.
Kamei, T., Ishigaki, S., Tamura, M., 1989. Fossil footprints from the Kobiwako Group. Kagaku 59, 320–323.
Kim, J.Y., Kim, K.S., 2004. Hominid and other animals footprints from the Cenozoic Hamori Formation of Jeju Island. In: Proceedings of the 2004 Spring Meeting of the Korean Earth Science Society, pp. 25–31.
Kim, K.S., Kim, J.Y., 2006. Review on the stratigraphy and geological age of the hominid footprints-bearing strata, Jeju Island, Korea. J. Korean Earth Sci. Soc. 27, 236–246.
Kim, J.Y., Kim, K.S., Lee, C.Z., Lim, J.D., 2004. Occurrence of hominid and other vertebrate footprints of Jeju Island, Korea. In: Kim, J.Y., Kim, K.-S., Park, S.I., Shin, M.-K. (Eds.), Proceedings of International Symposium of the Quaternary Footprints of Hominids and Other Vertebrates, Namjejigun, pp. 1–26.
Kim, K.S., Kim, J.Y., Kim, S.H., Lee, C.Z., Lim, J.D., 2009. Preliminary report on hominid and other vertebrate footprints from the Late Quaternary strata of Jeju Island, Korea. Ichnos 16, 1–11.
Kinahan, J., 1996. Human and domestic animal tracks in an archaeological lagoon deposit on the coast of Namibia. S. Afr. Archaeol. Bull. 51, 94–98.

Kinahan, J., 2014. Cattle paths and the choreography of late pre-colonial contact and trade on the Namib Desert coast. S. Afr. Archaeol. Bull. 69, 96–102.

Kinahan, J., Pallet, J., Vogel, J., Ward, J., Lindeque, M., 1991. The occurrence and dating of elephant tracks in the silt deposits of the lower !Kuiseb River, Namibia. CIMBEBASIA 13, 37–43.

Kindler, P., Curran, H.A., Marty, D., Samankassou, E., 2008. Multiple sedimentary sequences, bird tracks, and lagoon beaches in last interglacial oolites, Boiling Hole, north Eleuthera Island, Bahamas. In: Park, L.E., Freile, D. (Eds.), Proceedings of the Symposium on the Geology of the Bahamas, vol 13, pp. 169–181.

Kinoshita, T., Todori, Y., 1991. Fossils from the Upper Motohata Formation – fossil footprints. Bulletin of the Kitakami City Museum 8, 142–153.

Kinoshita, T., Todori, Y., 1993. On the Motohata Formation and *Stegodon* cf. *aurorae* Matsumoto in Isawa River, Iwate Pref. Bulletin of the Kitakami City Museum 9, 123–136.

Klaatch, H., 1906. II. Australia (Victoria, South Australia, Western Australia). Z. Ethnol. 38 (4/5), 776–795.

Koenigswald, W., Walders, M., Sander, M., 1995. Jungpleistozane tierfahrten aus der Emscher-Niederterrasse von Bottrop-Welheim. Munchner Geowissenschaftliche Abdhandlungen 27, 5–50.

Koenigswald, W., Walders, M., Sander, M., 1996. The Upper Pleistocene tracksite Bottrop-Welheim (Germany). Acta Zoologica Cracoviensia 39, 235–244.

Koster, E.A., Castel, I.I.Y., Nap, R.L., 1993. Genesis and sedimentary structures of late Holocene aeolian drift sands in northwest Europe. Geol. Soc. Spec. Publ. 72, 247–267.

Kotsakis, T., Palombo, M.R., Angelone, C., Melis, R.T., Fanelli, F., September 16–21, 2008. Madriola – Capo Mannu. In: Palombo, M.R., Pillola, G.L., Kotsakis, T. (Eds.), Fossil Mammalian Biotas of Sardinia. Fieldtrip Guidebook, Euromam, Sardinia, pp. 58–64.

Kretzoi, M., Dobosi, V., 1990. Vértesszőlős – Man, Site, Culture. Akadémiai Kiadó, Budapest, 554 pp.

Kuhn, O., 1958. Die Fährten der vorzeitlichen Amphibien und Reptilien. Druckhaus Meisenbach & Oeben KG, Bamberg.

Kurten, B., 1976. The Cave Bear Story. Life and Death of a Vanished Animal. Columbia-University Press, New York.

Laporte, L.F., Behrensmeyer, A.K., 1980. Tracks and substrate reworking by terrestrial vertebrates in quaternary sediments of Kenya. J. Sediment Petrol. 50, 1337–1346.

Laurat, T., Jurkénas, D., Brühl, E., 2007. Spuren in Ton – Pleistozäne Tierspuren vom mittelpaläolithischen Fundhorizont Neumark-Nord 2/2 (Gieseltalt, Sachsen-Anhalt). Hallesches Jarhb. Geowiss. 23, 195–198.

Laury, R.L., 1980. Paleoenvironment of a late Quaternary mammoth-bearing sinkhole deposit, Hot Springs, South Dakota. Geol. Soc. Am. Bull. 91, 465–475.

Lea, P.D., 1996. Vertebrate tracks in Pleistocene eolian sandsheet deposits of Alaska. Quat. Res. 45 (2), 226–240. https://doi.org/10.1006/qres.1996.0023.

LeConte, J., 1882. The recent discoveries of fossil footprints in Carson, Nevada. Am. Sci. 16, 921–923.

LeConte, J., 1883. Carson footprints. Nature 18, 101–102.

Lenssen-Erz, T., Pastoors, A., Uthmeier, T., Ciqae, T., Kxunta, U., Thao, T., 2023. Animal tracks and human footprints in prehistoric huntergatherer rock art of the Doro! nawas mountains (Namibia), analysed by present-day indigenous tracking experts. PLoS One 18 (9), e0289560. https://doi.org/10.1371/journal.pone.0289560.

Leonardi, G., 1994. Annotated Atlas of South American Tetrapod Footprints (Devonian to Holocene). Publication of the Companhia de Perquisa de Recursos Minerais, Brasilia, 248 pp.

Lepper, B., 2023. Questioning the age of human and megafauna footprints at White Sands National Park. Mammoth Trumpet 36 (4), 116–118.

Lindsay, J., Leonard, G., Smid, E., Hayward, B., 2011. Age of the Auckland Volcanic Field: a review of existing data. N. Z. J. Geol. Geophys. 54, 379–401.

Liutkus-Pierce, C.M., Zimmer, B.W., Carmichael, S.K., McIntosh, W., Deino, A., Hewitt, S.M., McGinnis, K.J., Hartney, T., Brett, J., Mana, S., Deocampo, D., 2016. Radioisotopic age, formation, and preservation of Late Pleistocene human footprints at Engare Sero, Tanzania. Palaeogeogr. Palaeoclimatol. Palaeoecol. 463, 68–82. https://doi.org/10.1016/j.palaeo.2016.09.019.

Lockley, M.G., Abbassi, N., Helm, C.W., 2021a. Large, unwebbed bird and bird-like footprints from the Mesozoic and Cenozoic: a review of ichnotaxonomy and trackmaker affinity. Lethaia 54, 969–987.

Lockley, M.G., Cawthra, H.C., De Vynck, J.C., Helm, C.W., McCrea, R.T., Nel, R., 2019. New fossil sea turtle trackway morphotypes from the Pleistocene of South Africa highlight role of ichnology in turtle palaeobiology. Quat. Res. 92, 626–640. https://doi.org/10.1017/qua.2019.40.

Lockley, M.G., Gregory, M.R., Gill, B., 2007. The ichnological record of New Zealand's moas: a preliminary summary. New Mexico Museum of Natural History and Science Bulletin 42, 73–78.

Lockley, M., Harris, J., 2010. On the trail of early birds: a review of the fossil footprint record of avian morphological evolution and behavior. In: Ulrich, P.K., Willett, J.H. (Eds.), Trends in Ornithological Research. Fargo, North Dakota. Nova Publishers.

Lockley, M.G., Helm, C.W., Cawthra, H.C., De Vynck, J.C., Perrin, M.R., 2021b. Pleistocene golden mole and 'sand-swimming' trace fossils from the Cape coast of South Africa. Quat. Res. 101, 169–186. https://doi.org/10.1017/qua.2020.97.

Lockley, M.G., Helm, C.W., Cawthra, H.C., De Vynck, J.C., Dixon, M.G., Venter, J.A., 2022. Pleistocene small-mammal and arthropod trackways from the Cape south coast of South Africa. Quat. Res. 1–15 107, 178–192. https://doi.org/10.1017/qua.2021.77.

Lockley, M.G., Hunt, A.P., 1995. Dinosaur Tracks and Other Fossil Footprints of the Western United States. Columbia University Press, New York.

Lockley, M.G., Lucas, S.G., Hunt, A.P., Garcia Vasquez, R., Espinoza, E., 2008. Non-hominid vertebrate tracks from the Holocene of Nicaragua. New Mexico Museum of Natural History and Science Bulletin 44, 393–397.

Loope, D.B., 1986. Recognizing and utilizing vertebrate tracks in cross section: Cenozoic hoofprints from Nebraska. Palaios 1, 141–151.

Lucas, S.G., 2008. Late Cenozoic fossil vertebrate assemblages from Jalisco, Mexico. In: Lucas, S.G., Morgan, G.S., Spielmann, J.A., Prothero, D.R. (Eds.), Neogene Mammals, New Mexico Museum of Natural History & Science Bulletin, vol 44, pp. 51–64.

Lucas, S.G., Allen, B.D., Morgan, G.S., Myers, R.G., Love, D.W., Bustos, D., 2007. Mammoth footprints from the Upper Pleistocene of the Tularosa Basin, Doña Ana County, New Mexico. In: Lucas, S.G., Spielmann, J.A., Lockley, M.G. (Eds.), Cenozoic Vertebrate Tracks and Traces. New Mexico Museum of Natural History and Science Bulletin, vol 42, pp. 149–154.

Lucas, S.G., Alvarado, G.E., 2017. Vertebrate paleontology in Central America: a narrative and analytical history. Geol. Soc. London Special Publications 442, 155–169. https://doi.org/10.1144/SP442.9.

Lucas, S.G., Espinoza, E., Alvarado, G.E., 2023. Acahualinca, Nicaragua, a scientifically significant site of ancient human footprints in the New World. In: Clary, R.M., Pyle, E.J., Andrews, W.M. (Eds.), Geology's Significant Sites and Their Contributions to Geoheritage. Geological Society, London, Special Publications, p. 543. https://doi.org/10.1144/SP543-2022-255.

Lucas, S., Hunt, A., 2007. Ichnotaxonomy of camel footprints. In: Lucas, S.G., Spielmann, J.A., Lockley, M.G. (Eds.), Cenozoic Vertebrate Tracks and Traces. New Mexico Museum of Natural History & Science, Bulletin, vol 42, pp. 155–168.

Lucas, S.G., Morgan, G.S., Hawley, J.W., Love, D.W., Myers, R.G., 2002. Mammal footprints from the Upper Pleistocene of the Tularosa Basin, Doña Ana County, New Mexico. In: New Mexico Geological Society Guidebook, 53rd Field Conference, Geology of White Sands, pp. 285–288.

Lydekker, R., 1891. Catalogue of the Fossil Birds in the British Museum (Natural History), Fleet Street. Taylor and Francis, London.

Marche, J.D.I.I., 1986. Extraordinary petrifications: footprints at Nevada State Prison. Terra 24 (5), 12–18.

Marsh, O.C., 1883. On the supposed human foot-prints recently found in Nevada. Am. J. Sci. 152, 139–140.

Martin, A.J., Whitten, M.J., 2015. First known fossil bird tracks (Pleistocene) on San Salvador Island, Bahamas. Geol. Acta 13 (1), 63–68.

Mastrolorenzo, G., Petrone, P., Pappalardo, L., Sheridan, M.F., 2006. The Avellino 3780-yr-B.P. catastrophe as a worst-case scenario for a future eruption at Vesuvius. Proc. Natl. Acad. Sci. 103, 4366–4370.

Matsukawa, M., Kato, T., 2011. The Plio-Pleistocene terrestrial ecosystem of the Kitakami Lowland, northeast Japan. Bulletin of the Tokyo Gakugei University, Nat. Sci. 63, 53–66.

Matsukawa, M., Shibata, K., 2015. Review of Japanese Cenozoic (Miocene-Modern) vertebrate tracks. Ichnos 22, 261–290. https://doi.org/10.1080/10420940.2015.1064407.

Matsukawa, M., Baba, K., Lockley, M.G., 2007. A mammalian ichnofauna from Plio-Pleistocene terrestrial deposits of west Tokyo, Japan. In: Lucas, S.G., Spielmann, J.A., Lockley, M.G. (Eds.), Cenozoic Vertebrate Tracks and Traces. New Mexico Museum of Natural History & Science, Bulletin, vol 42, pp. 185–199.

Matsukawa, M., Nankinishi, R., Shibata, K., Baba, K., Horoaki, A., 2004. Mammalian footprints from Plio-Pleistocene deposits around Tokyo and their implication for paleoecosystem reconstruction using food-web and energy-flow model. In: Kim, J.Y., Kim, K.-S., Park, S.I., Shin, M.-K. (Eds.), International Symposium on Quaternary Footprints of Hominids and Other Vertebrates. Jeju, Korea, pp. 130–132.

McDonald, H.G., 2007. Biomechanical inferences of locomotion in ground sloths: integrating morphological and track data. New Mexico Museum of Natural History and Science Bulletin 42.

McDonald, H.G., White, R.S., Lockley, M.G., Mustoe, G.E., 2007. An indexed bibliography of Cenozoic vertebrate tracks. In: Lucas, S.G., Spielmann, J.A., Lockley, M.G. (Eds.), Cenozoic Vertebrate Tracks and Traces. New Mexico Museum of Natural History and Science Bulletin, vol 42, pp. 275–302.

McNeil, P.E., Hills, L.V., Kooyman, B., Tolman, S., 1999a. First reported Canadian proboscidean trackway, probably mammoth, from the St. Mary Reservoir, Alberta, Canada. In: McCracken, A.D. (Ed.), Canadian Paleontology Conference Program and Abstracts CPC-IX. University of Calgary, AB, Canada, p. 19.

McNeil, P., Hills, L., Kooyman, N.B., Tolman, S., 1999b. First reported Canadian proboscidean trackway, probably mammoth, from the St. Mary Reservoir, Alberta, Canada. 3rd British Columbia Paleontological Symposium, Victoria 51.

McNeil, P.E., Hills, L.V., Kooyman, B., Tolman, S., 2004. Late Pleistocene geology and fauna of the Wally's Beach Site (Dh Pg 8) Alberta, Canada. In: Kooyman, B., Kelly, J. (Eds.), Archaeology on the Edge: New Perspectives from the Northern Plains. University of Calgary Press, Calgary, pp. 79–94.

McNeil, P.E., Hills, L.V., Kooyman, B., Tolman, S., 2005. Mammoth tracks indicate a declining late Pleistocene population in southwestern Alberta, Canada. Quat. Sci. Rev. 24, 1253–1259.

McNeil, P., Hills, L.V., Tolman, M.S., Kooyman, B., 2007. Significance of latest Pleistocene tracks, trackways, and trample grounds from southern Alberta, Canada. In: Lucas, S.G., Spielmann, J.A., Lockley, M.G. (Eds.), Cenozoic Vertebrate Tracks and Traces. New Mexico Museum of Natural History and Science Bulletin, vol 42, pp. 209–223.

Medina, R., Tassara, D., 2022. Presencia de un pisadero en la desembocadura del arroyo los Cueros, partido de Mar Chiquita (Provincia de Buenos Aires). Actas del XXI Congreso Geológico Argentino Puerto Madryn, Chubut.

Medina, R., Tassara, D., 2023. Regristro de un nuevo icnotaxón de un mamífero del Pleistoceno Tardío en camet Norte (Provincia de Buenos Aires, Argentina). Libro de Resúmenes Reunión de Comunicaciones de la Asociación Paleontológica Argentina, Publicación Electrónica de la APA 23 (R), R100.

Melchor, R.N., Cardonatto, M.C., Visconti, G., 2012. Palaeonvironmental and palaeoecological significance of flamingo-like footprints in shallow-lacustrine rocks: An example from the Oligocene–Miocene Vinchina Formation, Argentina. Palaeogeogr. Palaeoclimatol. Palaeoecol. 315–316, 181–198.

Melchor, R.N., Feola, S.F., Manera de Bianco, T., 2018. Canid paleoichnology: taxonomic review and producers of *Canipeda* from the late Pleistocene of Argentina. Ichnos 26, 85–107. https://doi.org/10.1080/10420940.2018.1447466.

Milàn, J., Bromley, R.G., 2006. True tracks, undertracks and eroded tracks, experimental work with tetrapod tracks in laboratory and field. Palaeogeogr. Palaeoclimatol. Palaeoecol. 231, 253–264.

Milàn, J., Bromley, R.G., Titschack, J., Theodorou, G., 2007a. A diverse vertebrate ichnofauna from the quaternary eolian oolite, Rhodes, Greece. In: Bromley, R.G., Buatois, L.A., Mángano, G., Genise, J.F., Melchor, R.N. (Eds.), Sediment-Organism Interactions: A Multi-Faceted Ichnology. SEPM Special Publication, vol 88, pp. 333–343. https://doi.org/10.2110/pec.07.88.0333.

Milàn, J., Clemmensen, L.B., Buchardt, B., Noe-Nygaard, N., 2006. Tracking the Bronze Age fauna: preliminary investigations of a new Late Holocene tracksite, Lodbjerg dune system, northwest Jylland, Denmark. Hantkeniana 5, 42–45.

Milàn, J., Clemmensen, L.B., Buchardt, B., Noe-Nygaard, N., 2007b. A Late Holocene tracksite in the Lodbjerg dune system, northwest Jylland, Denmark. New Mexico Museum of Natural History and Science Bulletin 42, 241–250.

Milàn, J., Theodorou, G., Loope, D.B., Panayides, I., Clemmesnsen, L.B., Gkioni, M., 2015. Vertebrate tracks in late Pleistocene-early Holocene (?) carbonate aeolianites, Paphos, Cyprus. Ann. Soc. Geol. Pol. 85 (3), 507–514. https://doi.org/10.14241/asgp.2015.012.

Millener, P.R., 1981. The Quaternary avifauna of the North Island, New Zealand. Ph.D. thesis, University of Auckland, New Zealand, p. 897.

Milot, P., Burns, J., 2006. Refined paraffin wax used to preserve 11,000 year-old camelid footprints in lakebed sediments, St. Mary's Reservoir, southern Alberta, Canada. J. Vertebr. Paleontol. 26, 101A.

Morgan, G.S., Lucas, S.G., Hawley, J.W., Love, D.W., Myers, R.G., 2002. Mammal footprints from Pleistocene Lake Otero, Tularosa Basin, White Sands Missile Range, Dona ana county, New Mexico. N. M. Geol. 24, 67.

Morimoto, Y., Tsuda, S., 1937. Discovery of fossil footprints in Akashi City, Hyogo Prefecture. Mag. Nat. Hist. 35 (60), 409–421.

Morse, S.A., Bennett, M.R., Liutkus -Pierce, C., Thackeray, F., McClymont, J., Savage, R., Crompton, R.H., 2013. Holocene footprints in Namibia: the influence of substrate on footprint variability. Am. J. Phys. Anthropol. 151, 265–279.

Mortimer, N., Campbell, H.J., Tulloch, A.J., King, P.R., Stagpool, V.M., Wood, R.A., Rattenbury, M.S., Sutherland, R., Adams, C.J., Collot, J., 2017. Zealandia: Earth's hidden continent. GSA Today 27, 27–35.

Mountain, E.D., 1966. Footprints in calcareous sandstone of Nahoon Point. South Afr. J. Sci. 66, 103–111.

Mukaiyama, T., Fukushima, T., 2008. Fossil footprints of birds discovered from the lower part of Kazusa Group (early Pleistocene) exposed on the river floor in the middle stream of Tama River, Kanto district, Japan. Bull. Assoc. Kanto Quat. Res. 29, 39–51.

Muñiz, F., Cáceres, L.M., Rodríguez Vidal, J., Neto de Carvalho, C., Belo, J., Finlayson, C., Finlayson, G., Finlayson, S., Izquierdo, T., Abad, M., Jimenez-Espejo, F.J., Sugisaki, S., Gómez, P., Ruiz, F., 2019. Following the last Neanderthals: Mammal tracks in Late Pleistocene coastal dunes of Gibraltar (S Iberian Peninsula). Quat. Sci. Rev. 217, 297–309. https://doi.org/10.1016/j.quascirev.2019.01.013.

Murphy, K., Caseldine, A., Barker, K., Fielding, S., Burrow, S., Carlsen, S., 2014. Mesolithic human and animal footprints at Lydstep Haven, Pembrokeshire, 2010: the environmental context. Archaeol. Cambrensis 163, 23–41.

Nagamori, H., Yoshikawa, H., Hatakeyama, K., Terao, M., Tanabe, T., 2005. Proboscidean footprint fossils from the Lower Pleistocene Okai Formation on the river beds of Chikuma River, Saku basin, Nagano Prefecture, central Japan. Bull. Geol. Soc. Japan 73, 441–442.

Neto De Carvalho, C., 2009. Vertebrate tracksites from the Mid-Late Pleistocene eolianites of Portugal: The first record of elephant tracks in Europe. Geol. Q. 53 (4), 407–414.

Neto de Carvalho, C., 2011. Pegadas de vertebrados nos eolianitos do Plistocénico Superior do Sudoeste Alentejano, Portugal. Comunicações Geológicas 98, 99–108.

Neto de Carvalho, C., 2014. Pegadas de Lince-Ibérico (*Lynx pardinus*) no Plistocénico superior da Ilha do Pessegueiro (Portugal). Comunicações Geológicas 101 (Esp. I), 513–516.

Neto de Carvalho 2020c. Tracking the last elephants in Europe during the Würm Pleniglacial: the importance of the Late Pleistocene aeolianite record in SW Iberia. Ichnos 27 (3), 352–360.

Neto de Carvalho, C., Belaústegui, Z., Toscano, A., Muñiz, F., Belo, J., Galán, J.M., Gómez, P., Cáceres, L.M., Rodríguez-Vidal, J., Cunha, P.P., Cachão, M., Ruiz, F., Ramirez-Cruzado, S., Giles-Guzmán, F., Finlayson, G., Finlayson, S., Finlayson, C., 2021. First tracks of newborn straight-tusked elephants (*Palaeoloxodon antiquus*). Sci. Rep. 11, 17311. https://www.nature.com/articles/s41598-021-96754-1.

Neto de Carvalho, C., Belo, J., 2022. Bear-claw deep tracks in the Pleistocene at north of Praia do Cavalo, Odemira (Portugal). Ciências da Terra/Earth Sci. J. 20 (1), 21–25.

Neto de Carvalho, C., Belo, J., Figueiredo, S., Cunha, P.P., Muñiz, F., Belaústegui, Z., Cachão, M., Rodríguez-Vidal, J., Cáceres, L.M., Baucon, A., Murray, A.S., Buylaert, J.-P., Zhang, Y., Ferreira, C., Toscano, A., Gómez, P., Ramírez-Cruzado, S., Finlayson, G., Finlayson, S., Finlayson, C., 2023a. Coastal raptors and raiders: New bird tracks in the Pleistocene of SW Iberian Peninsula. Quat. Sci. Rev. 313, 108185.

Neto de Carvalho, C., Figueiredo, S., Belo, J., 2016. Vertebrate tracks and trackways from the pleistocene eolianites of SW Portugal. Comunicações Geológicas 103 (Esp. I), 101–116.

Neto de Carvalho, C., Figueiredo, S., Muñiz, F., Belo, J., Cunha, P.P., Baucon, A., Cáceres, L.M., Rodríguez-Vidal, J., 2020a. Tracking the last elephants in Europe during the Würm Pleniglacial: the importance of the Late Pleistocene aeolianite record in SW Iberia. Ichnos 27 (3), 352–360. https://doi.org/10.1080/10420940.2020.1744586.

Neto de Carvalho, C., Muñiz, F., Belaústegui, Z., Belo, J., Gómez, P., Toscano, A., Cachão, M., Cunha, P.P., Ruiz, F., Cáceres, L.M., Rodríguez-Vidal, J., Finlayson, G., Finlayson, S., Giles-Guzmán, F., Finlayson, C., 2020b. Paleoecological implications of large-sized wild boar tracks recorded during the Last Interglacial (MIS 5) at Huelva (SW Spain). Palaios 35 (12). https://doi.org/10.2110/palo.2020.058.

Neto de Carvalho, C., Muñiz, F., Cáceres, L.M., Belaústegui, Z., Rodríguez-Vidal, J., Belo, J., Moreira, N., Cachão, M., Cunha, P.P., Figueiredo, S., Galán, J.M., Zhang, Y., Gómez, P., Toscano, A., Ruiz, F., Ramírez-Cruzado, S., Giles-Guzmán, F., Finlayson, G., Finlayson, S., Finlayson, C., 2022. Aurochs roamed along the SW coast of Andalusia (Spain) during Late Pleistocene. Sci. Rep. 12, 9911. https://doi.org/10.1038/s41598-022-14137-6.

Neto de Carvalho, C., Muñiz, F., Cáceres, L.M., Rodríguez-Vidal, J., Medialdea, A., Del Val, M., Cunha, P.P., García, J.M., Giles-Guzmán, F., Carrión, J.S., Belaústegui, Z., Toscano, A., Gómez, P., Galán, J.M., Belo, J., Cachão, M., Ruiz, F., Ramirez-Cruzado, S., Finlayson, G., Finlayson, S., Finlayson, C., 2023b. Neanderthal footprints in the "Matalascañas trampled surface" (SW Spain): new OSL dating and Mousterian lithic industry. Quat. Sci. Rev. 313, 108200.

Neto de Carvalho, C., Muñiz, F., Galán, J.M., Rodríguez-Vidal, J., Gómez, P., Toscano, A., Ruiz, F., Belo, J., Belaústegui, Z., Cáceres, L.M., Finlayson, G., Finlayson, S., Giles-Guzmán, F., Finlayson, C., 2020c. First vertebrate tracks and palaeoenvironment in a MIS-5 context in the Doñana National Park (Huelva, SW Spain). Quat. Sci. Rev. 243. https://doi.org/10.1016/j.quascirev.2020.106508.

Neto de Carvalho, C., Saltão, S., Campos Ramos, J., Cachão, M., 2003. Pegadas de *Cervus elaphus* nos Eolianitos Plistocénicos da Ilha do Pessegueiro (SW Alentejano, Portugal). Ciências da Terra (UNL) nº esp. 5, 36–40.

Nichol, R., 1981. Preliminary report on excavations at the Sunde site, N38/24, Motutapu Island. New Zealand Archaeol. Assoc. Newsletter 24, 237–256.

Nichol, R., 1982. The eruption history of Rangitoto: reappraisal of a small New Zealand myth. J. Roy. Soc. N. Z. 22 (3), 159–180.

Nielsen, J.K., Nara, M., Jacobsen, A.R., 2020. Bowl-shaped structures in a Pleistocene clastic carbonate wedge on the Island of Rhodes, Greece. Ichnos 27 (3), 326–333.

Noe-Nygaard, N., Milàn, J., Hede, M.U., Holm, J., 2007. A reindeer track from a drill core, and lake basin development of the late Glacial Lille Slotseng kettle-hole basin, South-East Jylland, Denmark. Bull. Geol. Soc. Den. 55, 85–95.

Oesch, R.D., 1969. Fossil Felidae and Machairondontidae from two Missouri caves. J. Mammal. 50, 367–368.

Officer, C.G.W., 1893. The discovery of supposed human footprints on aeolian rock at Warrnambool. The Victorian Naturalist 9, 32–39.

Oishi, M., Yoshida, H., 1998. Investigation on the Pleistocene fossil footprints in the lower part of the Isawa River, Kitakami Lowland, northeast Japan. Research Report of the Iwate Prefectural Museum 14, 61–65.

Oishi, M., Terasawa, H., Sakai, M., Fujii, T., 2002. Fossil footprints of Early Pleistocene large mammals from the construction site of the Hanamaki City Museum, Hanamaki City, Iwate Prefecture, northeast Japan. Research Report of the Hanamaki City Museum Project Office 1, 73–94.

Okamura, Y., 1993. Fossil footprints from the Echi River, Shiga Prefecture. Research Reportof the Lake Biwa Museum Project Office 1, 81–95.

Okamura, Y., 2001. Fossil footprints at Ajima Town and nearby areas, Oita Prefecture, Japan. Research Report of the Lake Biwa Museum 18, 169–187.

Okamura, Y., Takahashi, K., Lake Biwa Research Associates, 1993. Fossil foot prints of a bird from the Kobiwako Group, Central Japan. Fossil (Kaseki) 55, 9–15.

Okamura, Y., Takahashi, K., Yamamoto, H., Matsuura, N., Okuwa Fossil Footprint Research Group, 2004. Footprint fossils of *Elaphurus* (Cervinae, Mammalia) from the riverbed of the Saikawa River, Okuwa, Kanazawa, central Japan. J. Fossil Res. 37, 68–75.

Okamura, Y., Tamura, M., Takahashi, K., Lake Biwa Museum Research Associates, 1995. An outline of fossil footprints from the Kobiwako group, Central Japan. Research Report of the Lake Biwa Museum Project Office 3, 135–199.

Oliva, C., Arregui, M., 2018. Mammalian ichnopathology: a case study of holartic ungulates (Gomphotheriidae, equidae, Camelidae) of the late Pleistocene of south America. Ichnotaxomic implications. Bol. Soc. Geol. Mex. 70 (2), 417. https://doi.org/10.18268/bsgm2018v70n2a10.

Oliveira, E.V., Bélo, P.S., Fambrini, G.L., Sial, A.N., Silva, A.K.B., Barreto, A.M.F., 2019. A new late Pleistocene ichnological site with mammal footprints from Brazil. J. S. Am. Earth Sci. 94, 102216. https://doi.org/10.1016/j.jsames.2019.102216.

Oliver, W.R.B., 1949. The moas of New Zealand and Australia. Dom. Mus. Bull. 15.

Ouzman, S., Taçon, P.S., Mulvaney, K., Fullagar, R., 2002. Extraordinary engraved bird track from north Australia: Extinct fauna, dreaming being and/or aesthetic masterpiece? Camb. Archaeol. J. 12, 103–112.

Packard, E.L., Allison, I.S., 1980a. Fossil bear tracks in Lake County, Oregon. Oregon Geol. 42, 1–2.

Owen, R., 1879. Memoir on the ornithichnites, or foot-prints of species of Dinornis. In: Owen, R. (Ed.), Memoirs on the Extinct Wingless Birds of New Zealand; with an Appendix on those of England, Australia, Newfoundland, Mauritius, and Rodriguez (2 vols), pl. CXVI. Van Voorst, London, pp. 451–453.

Packard, E.L., Alison, I.S., 1980b. Correction Oregon Geol. 42, 71–72.

Packard, E.L., Allison, I.S., Cressman, L.S., 1981. Fossil mammal tracks in Lake County, Oregon. Oregon Acad. Sci. Proc. 2, 14.

Palombo, M.R., Panarello, A., Mietto, P., 2018. Did elephants meet humans along the Devil's Path? A preliminary report. Alpin. Mediterr. Quat. 31, 83–87.

Panarello, A., Palombo, M.R., Biddittu, I., Mietto, P., 2017. Fifteen years along the Devil's Trails": new data and perspectives. Alpine Medit. Quat. 30 (2), 137–154.

Panera, J., Rubio-Jara, S., Yravedra, J., Blain, H.-A., Sesé, C., Pérez-González, A., 2014. Manzanares Valley (Madrid, Spain): A good country for Proboscideans and Nenaderthals. Quat. Int. 326–327, 329–343.

Panin, N., Avram, E., 1962. Noe urme de vertebrate in Miocenul Subcarpatilor Ruminesti. Studii si Cercetari de Geologie 7, 455–484.

Parkman, B., 2002. Mammoth rocks: where Pleistocene giants got a good rub? Mammoth Trumpet 18 (1), 4–7 20.

Parkman, B., 2009. Rubbing Rocks, Vernal Pools, and the First Californians: Pursuing the Rancholabrean Hypothesis. http://www.parks.ca.gov/?page_id=23566.

Pastoors, A., Lenssen-Erz, T. (Eds.), 2021. Reading Prehistoric Human Tracks. Springer, Switzerland, 436 pp.

Pastoors, A., Lenssen-Erz, T., Ciqae, T., Kxunta, /U., Thao, T., Bégouën, R., Biesele, M., Clottes, J., 2015. Tracking in Caves. Experience based reading of Pleistocene human footprints in French caves. Camb. Archaeol. J. 25 (3), 551–564. https://doi.org/10.1017/S0959774315000050.

Patton, M.A., 1993. Neolithic and later archaeology. In: Keen, D. (Ed.), Quaternary of Jersey. QRA Field Guide, London, pp. 48–56.

Pee, S.T., Duvall, G., Girault, J.P., 2003. Eolian calcarenite and heavy minerals from the southwest Pacific. Carbonates Evaporites 18, 108–119.

Peterson, K.A., 2003. The rubbing post: a hypothesis for Pleistocene fauna agencies in the formation of anomalous polished rock surfaces in Nevada. 3rd International Mammoth Conference, Program and Abstracts. Government of the Yukon, Palaeontology Program, Occasional Papers in Earth Science 5, 126.

Pigati, J.S., Springer, K.B., Honke, J.S., Wahl, D., Champagne, M.R., Zimmerman, S.R., Gray, H.J., Santucci, V.L., Odess, D., Bustos, D., 2023. Independent age estimates resolve the controversy of ancient human footprints at White Sands. Science 382, 73–75.

Pillola, G.L., Zoboli, D., 2017. Dwarf mammoth footprints from the Pleistocene of Gonnesa (Southwestern Sardinia, Italy). Boll. Soc. Paleontol. Ital. 56, 57–64. https://doi.org/10.4435/BSPI.2017.05.

Pillola, G.L., Palombo, M.R., Panarello, A., Zoboli, D., 2020. The Pleistocene non-hominid vertebrate ichnofossil record of Italy. In: Romano, M., Citton, P. (Eds.), Tetrapod Ichnology in Italy: The State of the Art. J. Medit. Earth Sci., vol 12, pp. 193–212.

Plant, J., 1877. Footprint of "Dinornis", bones of "Dinornis" and "Dodo. Proceedings of the Literary and Philosophical Society of Manchester 16, 181–182.

Plassard, J., 1999. Rouffignac Das Heiligtum der Mammuts. Thorbecke Verlag, Stuttgart.

Playford, P.E., Cockbain, A.E., Low, G.H., 1976. Geology of the Perth Basin Western Australia. Geol. Surv. West. Aust. Bull. 124, 1–311.

Plint, T., McGill, C., 2021. Large mammal tracks in 1.8-million-year-old volcanic ash (Tuff IF, Bed I) at Olduvai Gorge, Tanzania. Ichnos 28 (2), 114–124. https://doi.org/10.1080/10420940.2021.1930540.

Playford, P.E., 1988. Guidebook to the Geology of Rottnest Island, Perth, Geological Society of Australia and Geological Survey of Western Australia.

Polidorou, M., Evelpidou, N., 2021. Geomorphology of the coastal sand dune fields and their association with the palaeolandscape evolution of Akrotiri Peninsula, Lemesos, Cyprus. Geosciences 11 (11), 448. https://doi.org/10.33901/geosciences11110448.

Politis, G.G., Bayón, C., 1995. Early Holocene footprints and sea mammals in the tidal zone of the Argentinian foreshore. Past 20, 5–6.

Price, N., 1995. Houses and horses in the Swedish Bronze Age. Past 20, 10–12.

Pryor, F., 1998. Farmers in Prehistoric Britain. Tempus, Stroud.

Qin, Z., Storozum, M., Zhao, H., et al., 2019. Cereals, soils and iron at Sanyangzhuang: Western Han agricultural production in the Central Plains. Antiquity 93 (369), 685–701.

Quilés, J., Petrea, C., Moldovan, O., Zilhão, J., Rodrigo, R., Rougier, H., Constantin, S., Milota, S., Gherase, M., Sarcina, L., Trinkaus, E., 2006. Cave bears (*Ursus spelaeus*) from the Pestera cu Oase (Banat, Romania): Paleobiology and taphonomy. C. R. Palevol 5, 927–934.

Quintana, C., Martínez, G.A., Osterrieth, M.L., Mazzanti, D.L., 1998. Icnites de mamíferos en un reparo rocoso de Tandilia Oriental (Pleistoceno tardío-holoceno temprano), pcia. de Buenos Aires. Tercera Reunión Argentina de Icnología y Primer Reunión de Icnología del Mercosur. Resúmenes (26).

Quintana, C.A., Martinez, G.A., Osterrieth, M.L., Mazzanti, D.L., 2001. Mammal footprints in a cave on the eastern edge of Tandilia range, late Pleistocene of Buenos Aires Province, Argentina. Curr. Res. Pleistocene 18, 110–111.

Quintana, J., 1993. Descripción de un rato de *Myotragus* e icnitas de *Hyponomys* del yacimento cuaternario de Ses Penyes d'es Perico (Ciutadellad Menorca, Balears). Paleontolgía i Evolucío 26–27, 271–279.

Rabeder, G., Hofreiter, M., 2004. Der neue Stammbaum der Hohlenbären. Hohle 55, 1–19.

Rector, C., 1979. 5,000-year-old footprints on the Mojave River, California, USA. Antiquity 53, 149–150.

Rector, C.H., 1999. Human and animal trackway at Oro Grande. San Bernardino County Museum Assoc. Quarterly 46 (2), 53–55.

Rector, C., Swenson, J.D., Wilke, P.J., 1983. Archaeological Studies at Oro Grande, Mojave Desert, California. San Bernardino County Museum Association, Redlands.

Rector, C., Swenson, J.D., Wilke, P.J., 1984. Archaeological Studies at Oro Grande, Mojave Desert, California. J. Calif. Great Basin Anthropol. 6, 280–282.

Remeika, P., 1999. Identification, stratigraphy, and age of Neogene vertebrate footprints from the Vallecito-Fish Creek Basin, Anza-Borrego Desert State Park, California. In: San Bernadino County Museum Association Quarterly, vol 46, pp. 37–45.

Remeika, P., 2001. The Fish Creek Canyon ichnofauna: A Pliocene (Blancan) vertebrate footprint assemblage from Anza-Borrego Desert State park, California. In: Santucci, V.L., McClelland, L. (Eds.), Proceedings of the 6th Fossil Resource Conference, Geological Resources Division, Technical Report. NPS/NRGRD/GRDTR-01/01, pp. 55–75.

Remeika, P., 2006. Fossil footprints of Anza-Borrego. In: Jefferson, G.T., Lindsay, L. (Eds.), Fossil Treasures of the Anza-Borrego Desert: The Last Seven Million Years. Sunbelt Publications, El Cajon, pp. 311–327.

Renders, E., Vincelette, A., 2023. New studies on the gaits displayed by Miocene, Pliocene and Pleistocene fossil horse trackways. In: Animal Science – Annual Volume 2023. IntechOpen. https://doi.org/10.5772/intechopen.112157.

Research Group for Fossil Footprints from the Tamagawa riverbed, 2001. Report of the fossil footprints of *Stegodon aurorae* from the Plio-Pleistocene Kazusa Group exposed at the riverbed of Tama-gawa in the western part of Kanto Plain, Japan. Education Board in Akishima City, 15 pp.

Research Group for Fossil Footprints from the Tamagawa riverbed, 2002. The fossil footprints from the Kazusa Group at the riverbed of Asa-kawa in the western part of Kanto Plain. Abstract of the Meeting of Chigakudantai (the Association for the Geological Collaboration in Japan) 55, 198–201.

Research Group for Fossil Footprints of Isawa-gawa, 1993. Research report of the Isawa-gawa mammalian ichnoassemblage. Misuzawa City Board of Education and Kanegasaki City Board of Education, Iwate Prefecture, 61 pp.

Retallack, G.J., Martin, J.E., Broz, A.P., Breithaupt, B.H., Matthews, N.A., Walton, D.P., 2018. Late Pleistocene mammoth trackway from Fossil Lake, Oregon. Palaeogeogr. Palaeoclimatol. Palaeocol. 496, 192–204.

Reynolds, R.E., 1999. Pleistocene mammal tracks near Shoshone, southern Death Valley. San Bernardino County Museum Association Quarterly 46 (2), 27–29.

Reynolds, R.E., 2006. Horse hoof prints in the fossil record. In: Reynolds, R.E. (Ed.), Making Tracks across the Southwest, the 2006 Desert Symposium Field Guide and Abstracts from Proceedings. California State University, Desert Studies Consortium and LSA Associates, Inc., Fullerton, pp. 25–28.

Rich, P., Van Tets, G., Orth, K., Meredith, C., Davidson, P., 1983. The prehistory of Norfolk biota; its place in the Southwest Pacific, and comparisons with other island fossil avifaunas. In: Schodde, R., Fullargar, P.J., Hermes, N. (Eds.), A Review of Norfolk Island Birds: Past and Present. Australian National Parks and Wildlife Service, Canberra.

Roach, N.T., Hatala, K.G., Ostrofsky, K.R., Villmoare, B., Reeves, J.S., Du, A., Braun, D.R., Harris, J.W., Behrensmeyer, A.K., Richmond, B.G., 2016. Pleistocene footprints show intensive use of lake margin habitats by *Homo erectus* groups. Sci. Rep. 6, 26374.

Roach, N.T., Du, A., Hatala, K.G., Ostrofsky, K.R., Reeves, J.S., Braun, D.R., Harris, J.W., Behrensmeyer, A.K., Richmond, B.G., 2018. Pleistocene animal communities of a 1.5 million–year–old lake margin grassland and their relationship to *Homo erectus* paleoecology. J. Hum. Evol. 122, 70–83.

Robbins, L.M., Wilson, R.C., Watson, P.J., 1981. Paleontology and archeology of Jaguar Cave, Tennessee. In: Beck, B.F. (Ed.), Eighth International Congress of Speleology Proceedings, pp. 377–380.

Roberts, D.L., 2008. Last Interglacial hominid and associated vertebrate fossil trackways in coastal eolianites, South Africa. Ichnos 15, 190–207.

Roberts, G., 2004. Ephemeral, subfossil mammalian and avian tracks within intertidal-zone, Flandrian sediment exposures at Formby Point, Sefton Coast, North West England. In: Kim, K.-Y., Kim, K.-S., Park, S.-I., Shin, M.-K. (Eds.), Proceedings of the International Symposium on the Quaternary Footprints of Hominids and Other Vertebrates, pp. 54–83.

Roberts, G., 2009. Ephemeral, subfossil mammalian, avian and hominid footprints within Flandrian sediment exposures at Formby Point, Sefton Coast, North West England. Ichnos 16, 33–48. https://doi.org/10.1080/10420940802470730.

Roberts, G., 2014. The Lost World of Formby Point. The Alt Press, Liverpool.

Roberts, G., Gonzales, S., Huddart, D., 1996. Intertidal Holocene footprints and their archaeological significance. Antiquity 70, 647–651.

Roberts, D., Berger, L.R., 1997. Last Interglacial (c. 117 kyr) human footprints from South Africa. South Afr. J. Sci. 93 (8), 349–350.

Roberts, D.L., Bateman, M.D., Murray-Wallace, C.V., Carr, A.S., Holmes, P.J., 2008. Last Interglacial fossil elephant trackways dated by OSL/AAR in coastal aeolianites, Still Bay, South Africa. Palaeogeogr. Palaeoclimatol. Palaeoecol. 257 (3), 261–279.

Rodríguez-de la Rosa, R., Aguillón-Martínez, M., López-Espinoza, J., Eberth, D., 2004. The fossil record of vertebrate tracks in Mexico. Ichnos 11 (1–2), 27–37.

Roberts, G., Worsley, A., 2008. Evidence of human activity in mid-Holocene coastal Palaeoenvironments of Formby, northwest England. In: Lewis, J., Stanistreet, J. (Eds.), Sand and Sea: Sefton's Coastal Heritage. Sefton Council: Bootle. Sefton's Coastal Heritage.

Rodríguez de la Rosa, R.A., Guzmán Gutiérrez, J.J., 2012. Huellas de aves y mamíferos del Neógeno de Jalisco. Palaontol. Mexic. 42, 147–158.

Romano, M., Citton, P., Salvador, I., Arobba, D., Rellini, I., Firpo, M., Negrino, F., Zunino, M., Starnini, E., Avanzini, M., 2019. A multidisciplinary approach to a unique Palaeolithic human ichnological record from Italy (Bàsura Cave). eLife 8, e45204. https://doi.org/10.7554/eLife.45204.

Rouzaud, F., Soulier, M., Lignereux, Y., 1995. La grotte de Bruniquel. Spelunca 60, 27–34.

Rus, I., Santonja, M., 2011. Arriaga IIa. In: Santonja, M. (Ed.), Los Exploradores de los Valles. Comunidad de Madrid, Madrid, pp. 56–57.

Saito, F., 1928. Geology of the region of Hanamaki yielding fossil walnuts, Iwate. J. Geogr. 40, 250–259.

Santucci, V.L., Hunt, A.P., Lockley, M.G., 1998. Fossil vertebrate tracks in National Park Service areas. Dakoterra 5, 107–114.

Santucci, V.L., Kenworthy, J., Kerbo, R., 2001. An Inventory of Paleontological Resources Associated with National Park Caves. National Park Service Geologic Resources Division Technical Report NPS/NRGRD/GRDTR-01/02, 50 pp.

Santucci, V.L., Hunt, A.P., Nyborg, T., Kenworthy, J.P., 2006. Additional fossil vertebrate tracks in National Park Service areas. New Mexico Museum of Natural History and Science Bulletin 34, 152–158.

Santucci, V.L., Tweet, J., Bustos, D., Von Haden, J., Verela, P., 2014. Vertebrate paleontological resources from national park service areas in New Mexico. In: Lucas, S.G., Sullivan, R.M. (Eds.), Fossil Vertebrates of New Mexico. New Mexico Museum of Natural History & Science Bulletin, vol 64, pp. 1–9.

Sarjeant, W.A.S., Reynolds, R.E., 1999. Camelid and horse footprints from the Miocene of California. San Bernardino County Mus. Geol. Paleontol. Q. 46 (2), 3–19.

Sarjeant, W.A.S., Reynolds, R.E., Kissell-Jones, M.M., 2002. Fossil creodont and carnivore footprints from California, Nevada and Wyoming. In: Reynolds, R.E. (Ed.), Between the Basins: Exploring the Western Mojave and Southern Basin and Range Province: Fullerton. California State University, Desert Studies Consortium, pp. 37–50.

Sato, K., Matsukawa, M., 2014. Plio-Pleistocene terrestrial ecosystem of the Komoto Group in Nagano Prefecture, Japan. Bulletin of the Tokyo Gakugei University. Nat. Sci. 66, 95–131.

Scales, R., 2002. Footprint-tracks at Goldcliff East 2002. Archaeol. Severn Estuary 13, 31–35.

Scales, R., 2004. Footprint-tracks at Goldcliff East: the fauna and a simple methodology for field recording. Archaeol. Severn Estuary 14, 35–40.

Scales, R., 2007. Footprint-tracks of human and animals. In: Bell, M. (Ed.), Prehistoric Coastal Communities. CBA Research Report 149. Council .of British Archaeology, pp. 139–157.

Scott, J.J., Renaut, R.W., Owen, R.B., 2008. Preservation and paleoenvironmental significance of a footprinted surface on the Sandai Plain, Lake Bogoria, Kenya Rift Valley. Ichnos 15, 208–231.

Scott, J.J., Renaut, R.W., Owen, R.B., 2012. Impacts of flamingos on saline lake margin and shallow lacustrine sediments in the Kenya Rift Valley. Sediment. Geol. 277–278, 32–51. https://doi.org/10.1016/j.sedgeo.2012.07.007.

Schmincke, H., Kutterolf, S., Wendy Perez, W., Rausch, J., Freundt, A., Strauch, W., 2010. Walking through volcanic mud: the 2,100-year-old Acahualinca footprints (Nicaragua) I: Stratigraphy, lithology, volcanology and age of the Acahualinca section. Int. J. Earth Sci. 99, 279–292.

Silva, P.G., López-Recio, M., Tapias, F., Roquero, E., Morín, J., Rus, I., Carrasco-García, P., Giner-Robles, J.L., Rodríguez-Pascua, M.A., Pérez-López, R., 2012. Stratigraphy of the Arriaga Palaeolithic sites. Implications for the geomorphological evolution recorded by thickened fluvial sequences within the Manzanares River valley (Madrid Neogene Basin, Central Spain). Geomorphology 196, 138–161.

Simpson, G.G., 1941. Discovery of jaguar bones and footprints in a cave in Tennessee. Am. Mus. Novit. 1131, 1–12.

Smith, K., Coppen, J., Wainwright, G.J., Beckett, S.C., 1981. The Shaugh Moor Project: Third report – settlement and environmental investigations. Proc. Prehist. Soc. 47, 205–273.

Smith, R.M.H., Mason, T.R., Ward, J.D., 1993. Flash-flood sediments and ichnofacies of the Late Pleistocene Homeb Silts, Kuiseb River, Namibia. Sediment. Geol. 85, 579–599. https://doi.org/10.1016/0037-0738(93)90103-C.

Sohn, Y.K., Yoon, W.S., Ahn, U.S., Kim, G.B., Lee, J.-H., Ryu, C.K., Jeon, Y.M., Kang, C.H., 2015. Stratigraphy and age of the human footprint-bearing strata in Jeju Island, Korea: controversies and new findings. J. Archaeol. Sci. Rep. 4, 264–275.

Stewart, M., Clark-Wilson, R., Breeze, P.S., Janulis, K., Candy, I., Armitage, S.J., Ryves, D.B., et al., 2020. Human footprints provide snapshot of last interglacial ecology in the Arabian interior. Sci. Adv. 6, eaba8940. https://doi.org/10.1126/sciadv.aba8940.

Stidolph, R.H.D., 1946. no title (describes moa footprints from New Plymouth). Notornis 2, 16.

Stock, C., 1925. Cenozoic gravigrade edentates of western North America, 331. Carnegie Institute of Washington Publication, p. 206 pp.

Stock, C., 1936, July 26–27. Sloth tracks in the Carson Prison. Westways.

Su, X., Sigley, G.G., Song, C., 2020. Relational authenticity and reconstructed heritage space: a balance of heritage preservation, tourism, and urban renewal in Luoyang Silk Road Dingding Gate. Sustainability 12, 5830. https://doi.org/10.3390/su12145830.

Sutcliffe, A.J., 1985. On the Track of Ice Age Mammals. British Museum (Natural History), London.

Takahashi, K., Okamura, Y., 2004. Neogene and Pleistocene footprint fossils of Japan, with special reference to footprints from the Plio-Pleistocene Kobiwako Group, central Japan. In: Kim, J.Y., Kim, K.-S., Park, S.I., Shin, M.-K. (Eds.), International Symposium on Quaternary Footprints of Hominids and Other Vertebrates. Jeju, Korea, pp. 84–93.

Tanaka, I., Markó, A., Hyodo, M., Strickson, C.E., Falkingham, P.l., 2021. A re-analysis of Chibanian Pleistocene tracks from Vértesszőlős, Hungary, employing photogrammetry and 3D analysis. Ann. Soc. Geol. Pol. 91, 75–83.

Tang Luoyang City Archaeological Team, the Institute of Archaeology, CASS and Luoyang Municipal Archaeological Team, 2004. Excavation Report on the Dingding Gate. Chin. Archaeol. 6, 87–94.

Tankard, A.J., 1976. Pleistocene history and coastal morphology of the Ysterfontein–Elands Bay area, Cape Province. Ann. S. Afr. Mus. 69, 73–119.

Tassara, D., Aramayo, S., Osterrieth, M., Scian, R., 2005. Paleoicnología de mamíferos de la Formación Santa Clara (Pleistoceno tardío) en la zona costera del Partido de Mar Chiquita (Provincia de Buenos Aires) Argentina. Actas XVI Congreso Geológico Argentino: La Plata, Argentina 4, 219–226.

Tedford, R.H., 1973. The Diprotodons of Callabonna. Aust. Nat. Hist. 17, 349–354.

Tedford, R.H., 1984. The Diprotodons of Callabonna. In: Archer, M., Clayton, G. (Eds.), Vertebrate Zoogeography and Evolution in Australasia. Hesperian Press, Marrickville, NSW, pp. 999–1002.

Terzea, E., 1978. Depot de remplisage et mammiferes quaternaires de "Peştera Urşilor" de Chişău, departament de Bihor (note preliminaire). Travaux de l'Institut de Speologie Emile Racovitza 17, 139–144.

Thrasher, L., 2007. Fossil tracks and trackways from the Bowie Zeolite Mines, Graham county, Arizona. In: Lucas, S.G., Speilmann, Lockley, M.G. (Eds.), Cenozoic Vertebrate Tracks and Traces. New Mexico Museum of Natural History and Sciences Bulletin, vol 42, pp. 269–273.

Thomas, D.B., Fleury, K., Paterson, M., Hayward, B.W., Erickson, R., 2025. A short trackway of tridactyl fossil footprints discovered in the Kaipara region of the North Island of New Zealand. N. Z. J. Geol. Geophys 1–15. https://doi.org/10.1080/00288306.2025.2472831.

Tindale, N.B., 1951. Comments on supposed representations of giant bird tracks at Pimba. Record S. Aust. Mus. 9 (4), 381–382.

Toledo, N., Arregui, M., 2022. Concurrent evidence from ichnology and anatomy: The scelidotheriine ground sloths (Xenarthra, Folivora) from the Pleistocene of Argentina. Hist. Biol. 35 (2), 284–292. https://doi.org/10.1080/08912963.2022.2035379.

Tsakalos, E.P., Athanassos, C., Tsipas, P., Triantaphyllou, M., Geraga, M., Papatheodorou, G., Fillipako, E., Christodoulakis, J., Kazantzaki, M., 2018. Luminiscence geochronology and paleoenvironmental implications of coastal deposits of southeast Cyprus. Archaeol. Anthropol. Sci. 10 (11), 41–60.

Uchman, A., Hsieh, S., Ninard, K., Łapcik, P., Łaska, W., 2024. Bioturbation in stabilized Quaternary inland dunes of the European Sand Belt in Poland. Palaeogeogr. Palaeoclimatol. Palaeoecol. 640, 112091.

Urban, T.M., Bennett, M.R., Bustos, D., Manning, S.W., Reynolds, S.C., Belvedere, M., Odess, D., Santucci, V.L., 2019. 3-D radar imaging unlocks the untapped behavioral and biomechanical archive of Pleistocene ghost tracks. Sci. Rep. 9, 16470.

Urban, T.M., Bustos, D., Jakeway, J., Manning, S.W., Bennett, M.R., 2018. Use of magnetometry for detecting and documenting multi-species Pleistocene megafauna tracks at White Sands National Monument, New Mexico, U.S.A. Quat. Sci. Rev. 199, 206–213.

van Willigen, S., Ozainne, S., Guélat, M., Haller, A.L.G., Haller, M., 2024. New evidence for prehistoric ploughing in Europe. Humanit. Soc. Sci. Commun. 11, 372. https://doi.org/10.1057/s41599-024-02837-5.

Vickers-Rich, P., 1979. The Dromornithidae, an extinct family of large ground birds endemic to Australia. Bureau of Mineral Resources. Geol. Geophys. Bull. 184, 1–71.

Vickers-Rich, P., Gill, E., 1976. Possible dromornithid footprints from Pleistocene dune sands of southern Victoria, Australia. Emu 76, 221–223.

Vincelette, A., 2021. Determining the gait of Miocene, Pliocene, and Pleistocene horses from fossilized trackways. Fossil Record 24, 151–169.

Voy, C.D., 1880. On the occurrence of footprints of Dinornis at Poverty Bay, New Zealand. Am. Nat. 14, 682–684.

Vyalov, O.S., 1965. Sztratigrafija Neogenovüh Molassez Predkarpatszkovo Progiva. Naukova Dumka, Kiev, 178 pp.

Vyalov, O.S., 1966. Sledy Zhiznedeyatelnosti organizmov i ikh paleontologicheskoe znacheniye. Naukova Dumka (Akademiya Nauk Ukrainskoy SSR Institut Geologii I Geokhimii Goryuchikh Iskopayemykh), Kiev, 219 pp.

Waddington, C., 2014. Rescued from the Sea: An Archaeologist's Tale. Archaeological Research Services, Bakewell.

Watson, P.J., Kennedy, M.C., Willey, P., Robbins, L.M., Wilson, R.C., 2005. Prehistoric footprints in Jaguar Cave, Tennessee. J. Field Archaeol. 30 (1), 25–43.

Webb, S., 2007. Further research of the Willandra Lakes fossil footprint site, southeastern Australia. J. Hum. Evol. 52, 711–715.

Webb, S., Cupper, M.L., Robins, R., 2006. Pleistocene human footprints from the Willandra Lakes, southeastern Australia. J. Hum. Evol. 50 (4), 405–413.

Welch, D., 1999. Fossilised human footprints on the coast of north Western Australia. Artefact J. Archaeol. Anthropol. Soc. Victoria 22, 3–10.

Wells, R.T., Tedford, R.H., 1995. *Sthenurus* (Macropodidae: Marsupialia) from the Pleistocene of Lake Callabonna, South Australia. Bull. Am. Mus. Nat. Hist. 225, 1–111.

West Kimberley Place Report, 2012. Department of Climate Change, Energy, the Environment and Water, Western Australia, pp. 1–276. https://www.dcceew.gov.au/sites/default/files/env/pages/ed0b4e39-41eb-4cee-84f6-049a932c5d46/files/ahc-final-assessment-full.pdf.

Westaway, M., 2010. Footprints of the First Australians. Australas. Sci. 31, 14–18.

Westaway, M.C., Cupper, M.L., Johnston, H., Graham, I., 2013. The Willandra Fossil Trackway: Assessment of ground penetrating radar survey results and additional OSL dating at a unique Australian site. Aust. Archaeol. 76, 84–89.

Westley, K., Cooper, A., Plets, R., 2023. Holocene animal tracks from the intertidal zone in the west of Ireland. Biol. Environ. 123B (1), 25–37.

Wheeler, C., Aharon, P., 1997. Geology and hydrogeology of Niue. Geology and Hydrogeology of Carbonate Islands. In: Developments in Sedimentology, 54. Elsevier, Amsterdam, pp. 537–564.

White, B., Soja, C., Anonymous, 2012. Ichnology of Holocene avian tracks, Caicos and San Salvador Island, Bahamas archipelago. Abstr. Progr. Geol. Soc. Am. 44 (7), 68.

Willey, P., Stollen, J., Crothers, G., Watson, P.J., 2005. Preservation of prehistoric footprints in Jaguar Cave, Tennessee. J. Cave Karst Stud. 67 (1), 61–68.

Williams, H., 1952. Geological observations on the ancient human footprints near Managua. Contribut. Am. Anthropol. Hist. 52, 1–31.

Williams, W.L., 1872. On the occurrence of footprints of a large bird found at Turanganui, Poverty Bay. New Zealand Institute. Proc. Transact. 4, 124–127.

Wilmer, H.C., 1893. The relation of the sand dune formation on the south west coast of Africa to the local wind currents. Transact. South Afr. Philos. Soc. 5, 326–329.

Wilmshurst, J.M., Hunt, T.L., Lipo, C.P., Anderson, A.J., 2011. High-precision radiocarbon dating shows recent and rapid initial human colonization of East Polynesia. Proc. Natl. Acad. Sci. U. S. A. 108 (5), 1815–1820.

Wilson, K., 1913. Footprints of the moa. Transact. New Zealand Instit. 45, 211–212.

Wiseman, A.L.A., Vicari, D., Belvedere, M., de Groote, I., 2022. Neolithic track sites from Formby Point, England: new data and insights. J. Archaeol. Sci. Reports 44, 103546.

Worthy, T.H., Holdaway, R.N., 2002. The Lost World of the Moa: Prehistoric Life of New Zealand. Indiana University Press.

Worthy, T.H., Tennyson, A.J., Archer, M., Musser, A.M., Hand, S.J., Jones, C., Douglas, B.J., McNamara, J.A., Beck, R.M., 2006. Miocene mammal reveals a Mesozoic ghost lineage on insular New Zealand, southwest Pacific. Proc. Natl. Acad. Sci. USA 103 (51), 19419–19423.

Wright, G.R., Crist, B., Riddle, T.L., Corgan, J.X., Tomlinson, G., 1982. Pleistocene footprints from Centerville Cave, Hickman county, Tennessee. J. Tenn. Acad. Sci. 57 (2), 44.

Zoboli, D., Pillola, G.L., 2018. New evidences of mammal tracks from the Pleistocene of Gonnesa area (southwestern Sardinia, Italy). J. Mediterr. Earth Sci. 10, 69–71.

CHAPTER

2

Walk the world and beyond: Hominin footprints

Carlos Neto de Carvalho[1], *Charles Helm*[2], *Donald Goldstein*[3], *Aaron B. Camens*[4], *Zain Belaústegui*[5], *Andrea Baucon*[6] and *Fernando Muñiz*[7]

[1]Geology Office of the Municipality of Idanha-a-Nova, Naturtejo UNESCO Global Geopark, Idanha-a-Nova, Portugal; [2]African Centre for Coastal Palaeoscience, Nelson Mandela University, Gqeberha, South Africa; [3]Department of Earth Sciences, University of Connecticut, Storrs, CT, United States; [4]College of Science and Engineering, Flinders University, Adelaide, SA, Australia; [5]Departament de Dinàmica de la Terra i de L'Oceà, Facultat de Ciències de la Terra, Institut de Recerca de la Biodiversitat, Universitat de Barcelona, Barcelona, Spain; [6]DISTAV, Università di Genova, Genova, Italy; [7]Department of Crystallography, Mineralogy and Agricultural Chemistry, Faculty of Chemistry, University of Seville, Seville, Spain

1. Hominin ichnology and ichnological applications in archeology and anthropology

Footprints are the physical signature of life in dynamic interaction with an evolving ecological space. Reading the tracks has been an existential necessity for the procurement of food resources. Tracking implies reasoning, using induction, deduction, and abduction in order to generate new knowledge (Pastoors and Lenssen-Erz, 2021). For this reason, Liebenberg (1990) claimed tracking as the origin of science. What about tracking our own lineage? Hunter-gatherers track human footprints to understand the location of relatives from the same social group or to evaluate proximity and number of potential enemies. Can we use a similar approach to recognize people and their actions in more distant pasts? Can we track hominin evolutionary paleoecology?

The reevaluation of the hominin track record by Lockley and Roberts (2004), Lockley et al. (2007b), Kim et al. (2008), and Lockley et al. (2008) brought hominin footprints to the realm of ichnology, the science that focuses on organism—substrate interactions. "Hominid Ichnology"

Vertebrate Ichnology
https://doi.org/10.1016/B978-0-443-38351-9.00007-4

was a term introduced by Lockley (1998) and corresponds to the study of traces "made by hominids" (Kim et al., 2008). However, Kim et al. (2008) pointed out that footprints are just a part of "Hominid Ichnology." Besides footprints, "Hominid Ichnology" comprises all the different types of traces left by hominins in natural environments, such as marks left by butchering activities, lithic industry, structures, and even artistic expressions. A great part of hominin-related structures can be considered as traces (e.g., rock art, stone tools, or graves; see Hasiotis et al., 2007; Lockley et al., 2008; Lockley et al., 2016), although the traditional methods of ichnology do not always find application to all of them. In fact, structures created by human technology should be excluded from ichnotaxonomic treatment (Bertling et al., 2006). Agreeing with Lockley et al. (2016), "there is no scientific justification for confining discussion of Hominid Ichnology only to footprints," including also handprints or digit prints we should say. Nevertheless, our approach for this chapter focuses mainly on limb prints and trackways, representing direct or indirect records (such as shoes, wheels, etc). This chapter explores the application of ichnological methods and approaches to the hominin track record, under the broader realm of Ichnoarcheology (Baucon et al., 2008), which is the study of traces, including footprints and coprolites, therefore bioturbation and biodeposition, but also bioerosion structures, found in archeological contexts, or on archeological materials and substrates (Fig. 2.1). Hominin tracksites may occur where skeletal

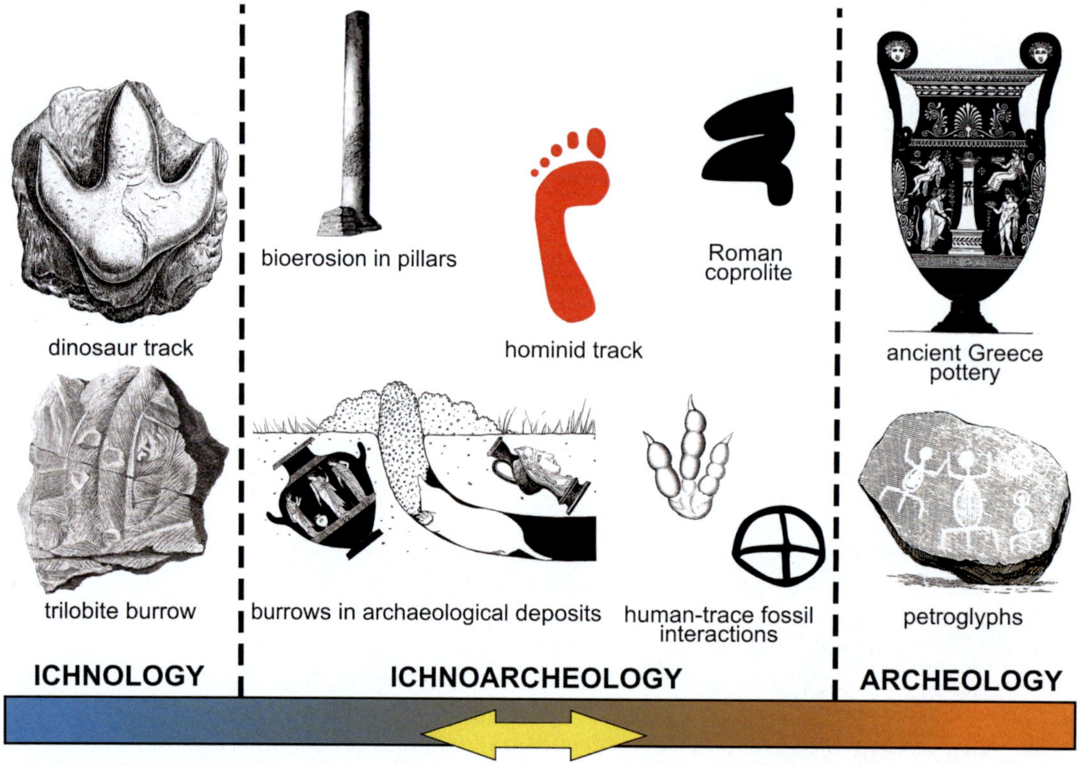

FIGURE 2.1 Ichnoarcheology as the application of ichnological concepts and methods to archeological problems. *Adapted from Baucon et al. (2008).*

remains and stone artifacts are absent, being the sole evidence of hominins present in the environment, sometimes providing evidence of ecological interactions with other animals.

Besides the claims for hominin-like foot morphology in the Miocene Trachilos footprints (Giérlinski et al., 2017), and in the Pliocene Laetoli, *Australopithecus afarensise* and another hominin—attributed footprints (Leakey and Hay, 1979; Suwa, 1984; Meldrum et al., 2011; McNutt et al., 2021), all hominin tracksites currently known are Quaternary in age and show footprints that follow in detail the morphology of a modern human foot. The diagnosis of human footprints, attributed to the ichnogenus *Hominipes*, was established by Kim et al. (2008, p. 131), as "pentadactyl, entaxonic, elongate footprints of a biped. From the anterior to the posterior end prominent features include a large, elongate-oval digit I impression, a transverse ball, and a sub-rounded heel trace. Toes are short and not divergent. Trackway is narrow, with short walking step 2–3 times the size of the footprint length. Footprint axis is parallel to trackway axis or slightly rotated outward." The elongate, pentadactyl footprint shows digit impressions progressively shorter, where digit I is adducted. The heel impression is oval to subrounded. Heel, lateral margin, ball, and digit I are typically better impressed, reflecting the biomechanics of the foot. The medial longitudinal arch between the heel and ball has long been considered to be diagnostic of a modern human foot (Tuttle, 2008; Holowka et al., 2017; Hatala et al., 2023a) and a critical adaptation for bipedal locomotion. Pace angulation is high, typically about 160–170 degrees (Kim et al., 2008). Footprint rotation angle depends on slope angle and substrate conditions. Also, speed and gait may depend on these parameters, as well as behavior. Alexander (1984) combined stride length and estimated leg length (stature) from footprint length to calculate the speed of progression. The ratio between the footprint width and length (footprint index) varies from 0.29 to 0.48. Trackway width is closely related to body width.

Tuttle (2008) proposed criteria for the identification of hominin tracks. Similar principles were addressed by Lockley et al. (2007b) and Kim et al. (2008). Tuttle's (2008) criteria involved the alignment and shape of the hallux, relative digit lengths, and the presence of a prominent medial longitudinal arch. Hominin footprints have the potential to provide information and inferences on the foot size, stature, body mass, gait, speed, and behaviors of both individuals and groups (Kim et al., 2008). According to Bennett and Morse (2014), fossil footprints can provide evidence of: (1) hominin presence at a location in space and time; (2) anatomical and biomechanical information about the producer by comparing modern humans with other hominin species; (3) kinematic information about speed and gait, as well as substrate conditions; (4) ecological information where other vertebrate tracks are present; and (5) behavioral ecology and the interaction of one trackmaker with another or other vertebrates. Footprints do not move and are not formed over a length of time. Hence, they are evidence of a real presence that is limited in time and space, and they provide a sort of snapshot that captures a fleeting moment in a person's and/or animal's long-gone life (Altamura, 2019). The better the preservation of the footprints and the more extensive the trackways are, the greater the potential to infer biomechanical and paleobiological information (Bennett and Reynolds, 2021).

Webb et al. (2006) and, in a more systematic way, Pastoors et al. (2015, 2017, 2021), Lenssen-Erz and Pastoors (2021), and Pastoors and Lenssen-Erz (2021) have opened pathways for including indigenous expertise in the study of ancient human footprints, which has been mainly applied to caves. Except for Willandra Lakes in Australia, such an alternative approach, based on empirical observations and interaction with the environment, accumulated over many generations, has been vital for subsistence, but has not been applied to open-air sites. At these sites, indigenous knowledge could optimally be applied to analyze and interpret ecological, social interaction and behavioral data from footprints. However, it is important to realize the lack of testability and validation of the observations provided through indigenous collaboration. Notwithstanding, the application of indigenous tracking abilities to rock art has yielded some very interesting results (Lenssen-Erz et al., 2023). Pioneering studies in rock art from Namibia, indeed, showed that in hundreds of engravings of animal tracks, indigenous trackers identified not only the animal species but also determined sex and age class and which of the four legs was depicted. Considering that these depictions were produced by hunter-gatherers, reading and interpreting them are best attempted by hunter-gatherers themselves (Lenssen-Erz and Pastoors, 2021). As rightly pointed out by these authors, the application of trackers' knowledge can be highly beneficial if cross-checked with the results of ichnological studies.

A more systematic approach to the gathering, analysis, and interpretations of footprints was outlined by Allen et al. (2003). The geochronological age of a footprint-bearing level, obtained preferably from multiple samples and/or methods, is often crucial in determining its archeological or anthropological significance and enabling a comparison with other sites (Bennett and Morse, 2014). Detailed dating methodologies are now essential tools in the study of hominin tracksites. Although there are obstacles to precisely dating some of the sites, there is a great deal of importance in getting the dates right, in order to help clarify the possible hominin producer (i.e., Avanzini et al., 2021; Neto de Carvalho et al., 2023) or even the series of events in the peopling of the Americas (e.g., Lockley et al., 2008; Lucas, 2017; Moreno et al., 2019; Pigati et al., 2023). Claims of dates older than the oldest hominin remains and artifacts have been controversial. The problem was defined succinctly by Lucas (2017), stating that footprints are sedimentary structures and that the sediment containing the footprints can only provide the "maximum limiting age" for the footprints, which are bracketed by the age of overlying sediment defining the minimum age of the tracks.

Challenges in recording and preserving the footprints have been ongoing (see the chapter on Vertebrate Ichnosites as Geological Heritage chapter in this book) and have resulted in some innovations in accurately recording high-resolution, three-dimensional morphology of the footprints through optical laser scanning, digital photogrammetry, digital scanning, and 3-D radar (Huddart et al., 2008; Bustos et al., 2018; Urban et al., 2019; Pastoors and Lenssen-Erz, 2021; Bennett and Reynolds, 2021; Lucas et al., 2024, Fig. 5 therein). Three-dimensional models of footprints and trackways, and the resulting analysis through a diversified array of software packages available, are now considered standard procedure to any investigation. Open-air tracks are subject to erosion from the effects of water, wind, and desiccation. Arid environments may be favorable to the formation of footprints, but are poor environments for their preservation (Lockley and Rodríguez-de la Rosa, 2009). Cave footprints have been buried in falling blocks, mined, eroded by water on occasion, and have been overprinted by accretional deposits of minerals in the caverns. All types of prints and their associated relics have been vandalized, carelessly crushed by later explorers, or destroyed by construction and mining, both ancient and modern (Graham et al., 1980; Stuart, 1980; Watson, 1983; McNatt, 1996; Simek et al., 1998; Moniz-Nakamura, 2003; Willey et al., 2005). Speaking about Wyandot Cave, Indiana, USA, Horace Hovey (1878) noted that "Indian foot-prints were visible in all parts of the new cave when first explored; and I saw them in 1854, although now they are obliterated." In El Salvador, footprints found in 1955

(Haberland and Groebe, 1957) were removed in some cases as curiosities and destroyed by tractors in others. The scientists were able to recover and study some of the prints after being led to the site by a workman. Franklin (2002, p. 77) stated that footprints in Xanadu Cave, Tennessee, noted in 1981 (Sims et al.), had been destroyed by "heavy modern caving traffic." They were fortunately able to locate more bare footprints in an upper passageway. Even the best efforts of local authorities to protect or cover tracksites are not always successful and may be overcome by natural erosion causes (Lucas et al., 2024).

Garcia (1986) made one of the first attempts to compile the bibliography that existed on human footprints. The published record of fossilized hominin tracks has increased steadily in recent decades. However, most of the ichnosites currently known reveal only a few footprints, and many of them show poor preservation. Comprehensive lists of ichnosites were provided by Lockley et al. (2008 – 63 sites) and Bennett and Morse (2014 – 44 sites), beginning with the Pliocene tracks at Laetoli, dated around 3.7 Ma. Lockley et al. (2016) provided a further summary. Subsequent publications have added substantially to this inventory, including many newly described African sites. Bell (2020) provided a list of 77 sites, with a focus on the United Kingdom, but with several sites not studied yet. Pastoors and Lenssen-Erz (2021) included in a single publication the latest findings at some of the most important tracksites, especially from European caves. Refinements have followed, e.g., Gierlinski et al. (2017) reported purported hominin tracks from the Late Miocene (~6.05 Ma) of Crete. More recently, Hatala et al. (2023b) provided a global summary of hominin ichnology, with special regard to the new ichnosite discoveries between 2016 and 2021 and with an East African focus. Hunt and Lucas (2007) presented an account of Cenozoic vertebrate trace fossils from North America in which there is a compilation of the references on hominin tracks.

We introduce the latest comprehensive list of hominin ichnosites currently known worldwide. Several ichnosites included in this chapter were barely studied, if not only mentioned. Even less common are tracksites that were studied using different approaches and that resulted in more than one published work. Lack of attention to hominin footprints can probably be explained by incorrect perceptions that these are exceptionally rare "curiosities" rather than sources of data that yield insights on a par with skeletal fossils or artifacts. Also, sedimentological conditions for preservation of the footprint record, such as sediment composition, moisture content, grain size, texture, yield strength, cohesiveness, rate of consolidation, and/or the presence and nature of microbial mats (Allen, 1997; Marty et al., 2009), must be considered. After being produced, footprints can be easily destroyed through a variety of taphonomic processes, including the duration of exposure before burial, and trampling (e.g., Smith et al., 2019), but also the present setting and the time of exposure under the prevailing erosion agents. In recent years, however, discoveries of hominin footprints have surged in frequency, shining important new light on anatomy, locomotion, behaviors, and environments from a wide variety of time periods and places (Hatala et al., 2023b). In fact, a single track level affords a degree of association seldom achieved in the skeletal fossil record (Hatala et al., 2022). For this reason, hominin tracksites have been a source of interest and of new inputs and bridges for intertwining disciplines such as anthropology (e.g., morphology, biometrics), paleontology (e.g., ichnology, taphonomy, evolution), archeology (e.g., maker, social dynamics), and paleobiology (ecology, biogeography, behavior). The substantial increase in the number of hominin ichnosites has contributed greatly to raising the awareness of archeologists of the potential to find footprints and other vertebrate tracks in excavations and their usefulness as direct evidence for hominin presence.

A broad-based interpretation of hominin ichnology by Lockley et al. (2016) indicates that there have been at least four major threshold events that starkly punctuate the ichnologic record. The first was bipedalism, the second was art expression and tool making, the third was the spread of hominins to Australia and the New World, and the fourth was the ability to set foot on other celestial bodies, including human-operated machines. This chapter is organized through these major threshold events, i.e., starting by traveling around the world synthesizing what we have learned so far from the hominin track record; then providing an account on the early use of our limbs in interaction with natural and artificial substrates to deliberately create symbolic meanings and artistic expressions or improving limb capabilities through the development of footwear; and finally, demonstrating that aerospace technology development in recent decades is leaving preserved human imprints far beyond our world.

2. Hominin ichnosites

In accordance with Bennett and Morse (2014), hominin tracksites can be grouped by geographic regions (in this chapter through the definition of continents, countries or territories, regions, and sites), sedimentary facies in which they are preserved, and their absolute age and therefore potential species of trackmakers. This provides their archeological and anthropological significance and frames ethological and paleoecological interpretations. The sites summarized in Tables 2.1–2.6 are organized by continent (e.g., Fig. 2.2 for Africa) and within it by numerical age, preferably from the oldest to the most recent; geographical coordinates for easy application to Google Earth are provided whenever public; otherwise, the coordinates are just indicative of the location region. Preservation follows the stratinomic classification of trace fossils by Martinsson (1970) regarding the relative position to the top (epichnia or epirelief), within (endichnia) and base (hypichnia or hyporelief) of a stratum, and the quality of the record is qualitatively ranked as poor (footprint outline without the preservation of details), moderate (identification of ball and hallux, angle between digits II and V, maximum heel or ball depths), or good (discrimination of toes and other morphological details). Numbers of footprints and trackways, as well as the main references for their study, are indicated when possible.

2.1 Africa

2.1.1 Eritrea

2.1.1.1 Aalad-Amo

Announced in 2016, this site was not yet studied at that point. Several fossilized hominin footprints were found in the Danakil desert. The tracked surface, covering several square meters, was brought to light by erosion at the bottom of a dried-out stream bed. About a dozen footprints are preserved on the surface of a sandy-silty layer dating to 0.8 Ma (Altamura, 2019). They may have been made by at least two individuals moving from north to south and are associated with ungulate tracks. Studies are in progress, but taking into account the obtained age, the footprints can be tentatively attributed to *Homo erectus*.

TABLE 2.1 African hominin tracksites.

Site	Region	Country	Geographic coordinates	Age BP and method(s)	Substrate	Footprint preservation	# Footprints trackways	Inferred trackmaker	Main references
Laetoli	Ngorongoro-Lengai UNESCO Global Geopark	Tanzania	~03° 13′ S; ~35° 11′ E	3.66 Ma – 40Ar/39Ar	Volcanic ash deposits (boundary between the Lower- and Upper Footprint Tuffs)	Epichnial good	Site G: 3 trackways, 3 trackmakers, >70 tracks; Site S: 2 larger trackmakers, 14 tracks; Site A: 1 trackway, five tracks, a different species of hominin with a different gait pattern has been inferred.	*Australopithecus afarensis* for Site G and Site S, no species attribution for Site A	Leakey and Hay (1979), Leakey and Harris (1987), Suwa (1984), White and Suwa (1987), Meldrum (2007), Deino (2011), Masao et al. (2016), McNutt et al. (2021)
Ileret	Turkana Basin	Kenya	~04° 19′ N ~36° 16′ E	1.53–1.51 Ma	Marginal lacustrine, interbedded with volcanic tuffs	Epichnial good	Cluster of five sites, with 97 tracks; at least 10 trackways	*Homo erectus*	Bennett et al. (2009), Dingwall et al. (2013), Hatala et al. (2016, 2017), Roach et al. (2016, 2018), Villmoare et al. (2019)
Koobi Fora	Turkana Basin	Kenya	~03° 52′ N; ~36° 15′ E	1.43 ± 0.01 Ma	Shallow subaqueous	Profile and epichnial; poor to moderate	7 tracks, 1 trackway	*Homo erectus*	Laporte and Behrensmeyer (1980), Behrensmeyer and Laporte (1981)
Gombore II open Air museum	Melka Kunture, Upper Awash Valley	Ethiopia	8°42.284′ N 38° 36.098′E	1200–850 Ka; $^{40}Ar/^{39}Ar$	Silt layer in fluvial and lacustrine sediments, interbedded with volcanic tephras	Epichnial, moderate	8 tracks, including those of young children	*Homo erectus*	Altamura et al. (2020)
Aalad-Amo	Danakil Desert	Eritrea	14°50′. 00″ N 39° 51′ 00″ E	800 ka		Epichnial	Footprints attributed to at least two individuals	*Homo erectus*	Altamura (2019)
Gombore II-2	Melka Kunture, Upper Awash Valley	Ethiopia	08°42.284′ N; 38° 36.098′ E	700 ka; K–$Ar^{40}Ar/^{39}Ar$	Silty, preserved by ash-flow surge	Epichnial, poor to moderate	11 tracks, including those of young children	*Homo heidelbergensis*	Altamura et al. (2018)

(*Continued*)

TABLE 2.1 African hominin tracksites.—cont'd

Site	Region	Country	Geographic coordinates	Age BP and method(s)	Substrate	Footprint preservation	# Footprints trackways	Inferred trackmaker	Main references
Nahoon point	East London, Cape east coast	South Africa	~33° 00′ S; ~27° 57′ E	124 ± 4 ka; OSL	Eolianite	Hyporelief, good	3 tracks, 1 trackway, well preserved	*Homo sapiens*	Mountain (1966), Roberts (2008), Jacobs and Roberts (2009)
Kraalbaai, Langebaan	Cape west coast	South Africa	~33° 08′ 48″ S; 18° 01″ 43″ E	~117 ka	Eolianite	Epichnial, poor to moderate	3 tracks, 1 trackway	*Homo sapiens*	Roberts and Berger (1997), Roberts (2008).
Lake Bogoria	Sandai Plain	Kenya	~22° 47′ N; ~36° 03′ E	Late Pleistocene	Exposed, moist lake-marginal mudflats	Epichnial, good	1 track incomplete	*Homo sapiens*	Scott et al. (2008)
Engare Sero	Lake Natron	Tanzania	~02.65° S; ~35.9° E	19.1–5.76 ka; $^{40}Ar/^{39}Ar$, ^{14}C	Volcaniclastic	Epichnial, good	>400 tracks, 23 trackways	*Homo sapiens*	Richmond et al. (2011), Liutkus-Pierce et al. (2016), Hatala et al. (2020)
Brenton-on-Sea	Cape south coast	South Africa	34° 04′ 33″ S; 23° 01′ 49″ E	Between 89 ± 6 ka and 76 ± 5 ka; OSL	Eolianite	Hypichnial and profile, good	40 tracks, at least two trackways	*Homo sapiens*	Helm et al. (2018), Helm et al. (2023a)
Goukamma (west)	Goukamma Nature Reserve, Cape south coast	South Africa	34° 02′ 52″ S; 22° 50′ 35″ E	71 ± 5 ka; OSL	Eolianite	Epichnial, good	three tracks, no trackways; associated ammoglyphs	*Homo sapiens*	Helm et al. (2020, 2023a)
Gericke's point	Garden Route National Park, Cape south coast	South Africa	34° 02′ 06″ S; 22° 45′ 53″ E	153 ± 10 ka; OSL	Eolianite	Hypichnial, poor to moderate	7 tracks, 1 trackway	*Homo sapiens*	Helm et al. (2020, 2023a)
Goukamma (central)	Goukamma Nature Reserve, Cape south coast	South Africa	34° 03′ 00.″ S; 22° 51′ 17″ E	75 ± 5 ka; OSL	Eolianite	Hypichnial and epichnial, poor to moderate	32 tracks, at least two trackways	*Homo sapiens*	Helm et al. (2020, 2023a)
Langebaan Lagoon east shore	Cape west coast	South Africa	33.10985° S, 18.03837° E	Late Pleistocene	Eolianite	Hypichnial, poor to moderate	2 tracks, 1 trackway, poor	*Homo sapiens*	Helm et al. (2022)

Kleinkrantz	Garden Route National Park, Cape south coast	South Africa	34° 00′ 51″ S; 22° 41′ 37″ E	Late Plesitocene	Eolianite	Epichnial, good (shod)	at least two tracks	*Homo sapiens* (shod)	Helm et al. (2023b)
Goukamma (east)	Goukamma Nature Reserve, Cape south coast	South Africa	34° 03′ 00.″ S; 22° 51′ 22″ E	Late Pleistocene	Eolianite	Epichnial, good (shod)	4 tracks, 1 trackway	*Homo sapiens* (shod)	Helm et al. (2023b)
Woody Cape	Addo Elephant National Park, Cape east coast	South Africa	33° 44′ 52″ S; 26° 15′ 23″ E	Late Pleistocene	Eolianite	Epichnial, moderate (shod)	4 tracks, 1 trackway	*Homo sapiens* (shod)	Helm et al. (2023b)
Larache	Northwest coast	Morocco	35° 09′ N; 06° 11′ W	90.3 ± 7.6 ka; OSL	Paleobeach deposit	Epichnial, poor	85 tracks, at least five trackmakers	*Homo sapiens*	Sedrati et al. (2024)
Sebkra El Azrag	Azrag area	Islamic Republic of Mauritania	22° 44′ 7.51″ N, −12° 28′ 16.82″ W	Lower Holocene	Paleolake sediments	Epichnial and internal molds, poor to moderate	118 tracks, 11 trackways	*Homo sapiens*	Mafart (2006)
Walvis Bay	Namib Desert	Namibia	~22° 58′ S; ~14° 29′ E	Last two centuries?	Lagoon saltpan deposit	Epichnial, good	Many tracks, 1 trackway of 5 tracks	*Homo sapiens*	Kinahan (1996)
Kuiseb delta	Walvis Bay, Namib Desert	Namibia	~23° 00′ 43″ S; ~14° 31′ 51″ E	~1.7–0.5 ka	Dried floodplain	Epichnial, good	many tracks; one trackway contained 70 tracks	*Homo sapiens*	Bennett et al. (2010b), Morse et al. (2013)
Kuiseb delta	Walvis Bay, Namib Desert	Namibia	~23° 01′ 30″ S; ~14° 31′ 24″ E	1.5 ka		Epichnial good	very well preserved tracks of children	*Homo sapiens*	Bennett et al. (2014a)

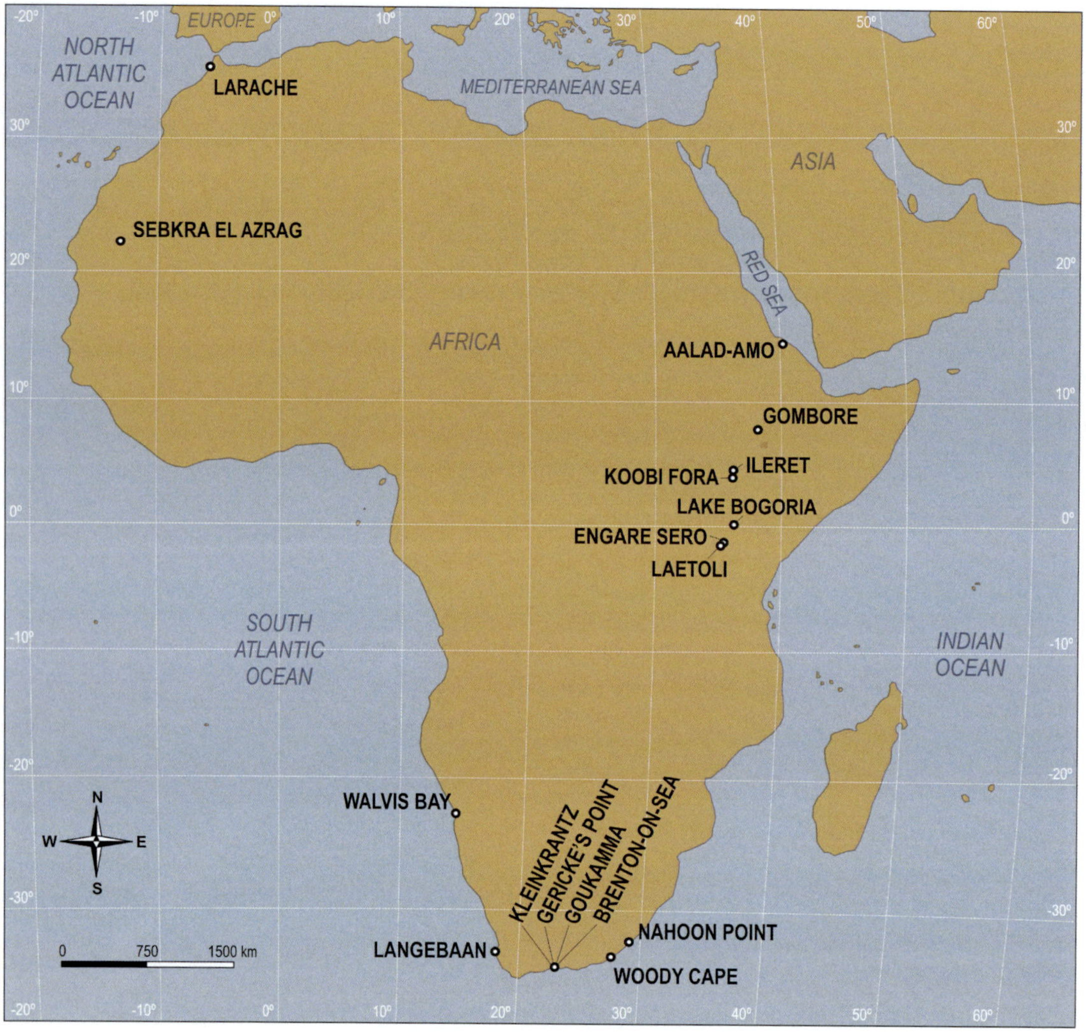

FIGURE 2.2 Map of Africa with location of the hominin ichnosites.

2.1.2 *Ethiopia*

2.1.2.1 Gombore II

The oldest of two Ethiopian ichnosites was reported by Altamura et al. (2020). Analysis of a 1200–850 ka volcanic and fluviolacustrine sequence at the Gombore II Open-Air Museum site involved 17 numbered surfaces. Eight purported in situ hominin footprints, possibly made by adults or subadults, young children, and a child <1 year, were reported from three surfaces. The vertebrate trackmaker assemblage also included hippopotamus, various sizes of bovids, suids, a possible hyena, and birds.

The youngest hominin footprints within the Early/Middle Pleistocene East African cluster, also from Ethiopia, were reported by Altamura et al. (2018). The footprints, from the Gombore II-2 site at Melka Kunture in the Upper Awash Valley, were dated to 700 ka and were attributed to *H. heidelbergensis*. Together with evidence of flint knapping and a butchered hippopotamus carcass, and tracks attributed to gazelles, hippos, equids, suids, and birds, the hominin footprints were inferred to have been part of a trampled silty waterbody margin. Eleven hominin footprints were reported and analyzed, with some attributed to individuals as young as 12 months (Altamura and Mussi, 2017). Mussi et al. (2023) provided a summary of all the excavations at Melka Kunture.

2.1.3 Kenya

2.1.3.1 Ileret

The Ileret site in Kenya, dated to 1.53–1.51 Ma, has become the second-most intensively researched hominin ichnosite in Africa, with no less than seven high-profile published articles. It was first described by Bennett et al. (2009), who reported on 20 hominin footprints, which were attributed to *H. erectus*. It was claimed that these footprints provided the oldest evidence of an essentially modern, human-like foot, with a relatively adducted hallux, a medial longitudinal arch, and medial weight transfer before push-off (Bennett et al., 2009; Holowka et al., 2017). Dingwall et al. (2013) compared the footprints with those of habitually unshod local adults, with a conclusion from the fossil trackways of speeds of between 0.45 and 2.2 m/s suggesting trackmakers that were relatively heavy and tall. Hatala et al. (2016) inferred the presence of cooperative multimale groups from clusters of large footprints (which suggested human-like social behaviors) and also concluded that the footprints provided the oldest evidence for modern, human-like weight transfer, confirming the presence of a longitudinally arched foot.

By 2017, Ileret had expanded from one site (Bennett et al., 2009) to five distinct, excavated sites with 97 hominin footprints, and the intensive use of lake margin habitats by *Homo erectus* groups was inferred (Roach et al., 2016). Hatala et al. (2017) confirmed foot sizes consistent with those of modern humans. Ten trackways were analyzed, leading to estimates of walking speeds of 0.45–1.58 m/s. The results were incorporated into earlier conclusions regarding social behaviors. Roach et al. (2018) used the data on the hominin and other vertebrate tracks to examine hominin interactions with the paleoenvironment, and the use of a marginal lacustrine habitat was interpreted as a possible hominin adaptive strategy. Finally, Villmoare et al. (2019) used the Ileret hominin footprints to investigate sexual dimorphism, concluding that East African *Homo erectus* was more dimorphic than modern *Homo sapiens*.

2.1.3.2 Koobi Fora

Following the first descriptions of the Pliocene tracks at Laetoli in Tanzania, the next East African site to be described, just a year later, was at Koobi Fora in eastern Turkana lake. The first report (Laporte and Behrensmeyer, 1980) addressed the reworking of sedimentary substrates in the Koobi Fora Formation by terrestrial vertebrates, and the resulting trackways and bioturbation in fluvial and lake-margin strata (20 track-bearing layers were noted in a 17-meter stratigraphic section). Discussion of the seven hominin tracks, initially identified in 1978, was brief, along with tracks of hippopotamus, "antelope," and birds. A more detailed report on the hominin footprints appeared the following year (Behrensmeyer and Laporte, 1981). The footprints, from a marginal lacustrine environment, were initially reported as being 1.6—1.5 Ma in age, but a subsequent report by Bennett et al. (2014a) indicated an age of 1.43 Ma. Footprints had initially been noted in profile, whereupon track-bearing layers were cleared to reveal the tracks in epirelief. The trackway of seven hominin footprints, probably registered in a shallow subaqueous substrate, included two partial footprints; hominin footprint length was measured at ~26 cm, and the probable trackmaker was *H. erectus*. Hatala et al. (2024) described in Area 103 the first evidence of two diferent patterns of Pleistocene hominin bipedalism record in the same track-bearing level, dated >1.52 Ma. This reveals habitat coexistence between *H. erectus* and *Paranthropus boisei*.

2.1.3.3 Lake Bogoria

In 2008, a single hominin footprint was described from the deltaic Sandai Plain at Lake Bogoria in Kenya (Scott et al., 2008). It exhibited a prominent hallux trace and heel impression and was identified within an assemblage of bovid, suid, and avian tracks. The tracks, preserved in epirelief, were inferred to have been made on exposed, moist, lake-marginal mudflats. Although an absolute age could not be precisely determined, the track-bearing layer was interpreted as being of Late Pleistocene age.

2.1.4 Mauritania

2.1.4.1 Sebkra El Azrag

Mafart (2006) reported 118 hominin tracks, including 11 trackways, from the sebkra El Azrag site in the Islamic Republic of Mauritania. The longest trackway contained 21 footprints. A pathological gait pattern was inferred in the case of one trackway. The tracksite was dated as early Holocene.

2.1.5 Morocco

2.1.5.1 Larache

A welcome addition to the series of published reports of new hominin tracksite discoveries from Africa came in 2023 from Larache, on the northwest coast of Morocco (Sedrati et al., 2024). Eighty-five hominin footprints, dated through OSL to 90.3 ± 7.6 ka, were reported on a "now-indurated beach surface." The tracks were interpreted as representing at least five individuals of different ages and footprint sizes. Being the oldest tracks attributed to *Homo sapiens* in northern Africa, and being from a paleoenvironment broadly similar to that of the southern African tracks, this discovery might open the door to the identification of further hominin prints from North Africa, just as new hominin ichnosites continue to be discovered in South Africa.

2.1.6 Namibia

2.1.6.1 Walvis Bay

Kinahan (1996) first drew attention to Holocene hominin ichnosites containing very well-preserved footprints on the coast of Namibia. Tracks of livestock and domestic dogs were also reported. The assemblage was thought to date to the last few centuries (Kinahan, 2014). This was followed by further published reports of ichnosites from the region from the Kuiseb Delta, east of Walvis Bay. Bennett et al. (2010b) described well-preserved hominin tracks of various sizes (the smallest measuring less than 10 cm) on dried floodplains, dated to

~1.7–0.5 ka. The longest trackway comprised 70 footprints. The site also contained tracks of hyena, goat/sheep, cattle, buffalo, elephant, giraffe, and birds. Morse et al. (2013) further interpreted the tracks in relation to substrate variability. Bennett et al. (2014a) documented a new site from the region, dated to 1.5 ka, dominated by very well-preserved footprints of children. The footprints of three size classes of hominins were noted, ranging from 11.4 to 20.6 cm, representing $\sim 9 \pm 2$ individuals.

2.1.7 *South Africa*

Globally, there is only one reported hominin tracksite attributed to *Homo sapiens* and older than 40 ka that is not from Africa (Saudi Arabia; see also the discussion about Theopetra cave). Of the remaining 13 unequivocal or probable hominin tracksites that have thus far been reported, the great majority have been identified in aeolianites on the Cape south coast of South Africa (Fig. 2.3). Although some of the South African sites have been identified in sea caves or overhangs, at the time of registration, these would all have been open-air sites and are best considered as such. Since 2018, several open-air hominin ichnosites in aeolianites, most of them in situ, have been reported from the Cape south coast of South Africa. The relatively large grain size that characterizes cemented dunes and beaches implies that track morphology is often inferior to that of tracks registered in finer-grained substrates such as volcanic ash, silt, or mud. Furthermore, surface exposures tend to be relatively small. However, an advantage is that the surfaces, either in situ or on fallen loose blocks and slabs that come to lie on the beach at cliff bases, are naturally exposed and do not require excavation. It is serendipitous that an area containing substantial archeological evidence of the cognitive emergence of *H. sapiens* in the Pleistocene happens to contain one of the finest trace-fossil records in the world from this time period.

2.1.7.1 Nahoon point

The first open-air hominin tracksite ever to be described was from Nahoon Point, near East London, on South Africa's east coast (Mountain, 1966). This was also the first report of Pleistocene vertebrate tracks from southern Africa and comprised a trackway of three hominin tracks together with a carnivore trackway and avian tracks, all in situ and in hyporelief on an aeolianite surface. Soon after discovery, the site suffered a collapse, but the track-bearing slab was recovered and is housed in the East London Museum. One of the hominin tracks, in particular, exhibited well-defined digit impressions (Fig. 2.3A). The tracks were eventually reliably dated through OSL to 124 ± 4 ka (Jacobs and Roberts, 2009).

2.1.7.2 Langebaan

The next site to be described was from an in situ, open-air aeolianite surface on the shores of the Langebaan Lagoon on South Africa's west coast (Roberts and Berger, 1997). A trackway comprising three tracks in epirelief on an in situ aeolianite surface was described, and dated to ~117 ka. Due in part to the risk of modern graffiti, the track-bearing slab was airlifted to the Iziko South Africa Museum in Cape Town (see the chapter on Vertebrate Tracksites as Geoheritage in this book). While the hominin origin of the Langebaan tracks was generally acknowledged, and they have entered the pantheon of hominin tracksites, acceptance was not universal: Bennett and Morse (2014: p. 71–72) noted that "not all

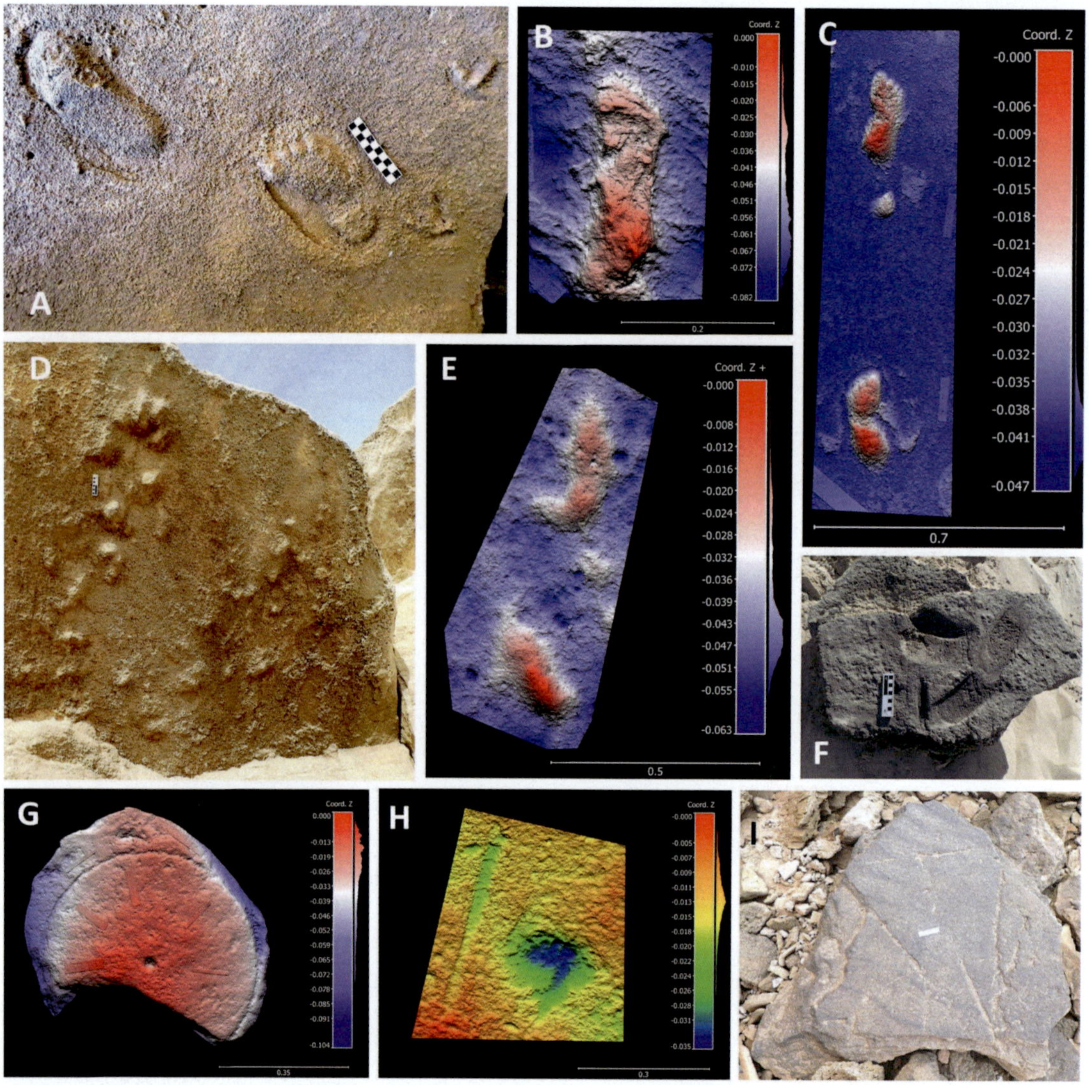

FIGURE 2.3 Hominin ichnosites from the Cape coast of South Africa. (A) Two of the Nahoon point hominin footprints (avian tracks are also evident). (B) 3G photogrammetry image of the oldest known *Homo sapiens* footprint, dated to ~153 ka, from Gericke's point in the Garden Route National Park. (C) 3D photogrammetry image of a short hominin trackway at Brenton-on-Sea. (D) Hominin footprints in the Goukamma nature Reserve. (E) 3D photogrammetry image of probable hominin footprints from Langebaan on the west coast. (F) Possible shod-hominin footprints from Kleinkrantz in the Garden Route National Park. (G) 3D photogrammetry image of a circular putative ammoglyph from the Garden Route National Park. (H) 3D photogrammetry image of a hominin forefoot impression beside a linear groove feature in the Goukamma Nature Reserve. (I) Triangular putative ammoglyph east of Still Bay. All scale bars = 10 cm; vertical and horizontal scales in all photogrammetry images are in meters.

authorities are convinced that they are in fact human tracks." A detailed summary of the Nahoon and Langebaan sites followed (Roberts, 2008).

The first possible hominin footprints on South Africa's west coast to be described since the Langebaan tracks announcement of 1997 followed (Helm et al., 2022). Two footprints with suboptimal morphological detail, evident in hyporelief, were described from an in situ ceiling of a small recess in coastal cliffs, near the town of Langebaan (Fig. 2.3E). This find arguably lends support to the contention that the initially described Langebaan footprints are of hominin origin.

2.1.7.3 Brenton-on-Sea

Brenton-on-Sea was the first of the new tracksites found in South Africa following a systematic study (Helm et al., 2018) and was subsequently dated through OSL to between 89 ± 6 ka and 76 ± 5 ka (Helm et al., 2023a). Forty hominin footprints were identified in hyporelief and in profile on two aeolianite surfaces in a coastal sea cave. Digit impressions were noted in nine footprints, one of which remains the best-preserved hominin footprint from the Cape south coast (Fig. 2.3C). One trackway, registered while moving down a gentle dune slope, exhibited a jogging gait pattern, making it the oldest reported ichnological evidence of a running hominin. Although the site lies in a cave, it is best regarded as an open-air site as the paleoenvironmental setting was an open dune surface.

2.1.7.4 Goukamma Nature Reserve

Helm et al. (2020) reported on three newly identified hominin tracksites in Cape south coast aeolianite deposits. One of these, in the Goukamma Nature Reserve, had already been reported as an ammoglyph site (Helm et al., 2019). Registered in epirelief, it contained three hominin forefoot impressions (on three distinct layers) associated with multiple subparallel groove features and smaller rounded depressions (Fig. 2.3H). It has subsequently been dated through OSL to 71 ± 5 ka (Helm et al., 2023a). Another site was also identified in the Goukamma Nature Reserve. It was registered by a group of hominins of various ages/sizes, moving across and up a dune surface, and contained 32 footprints (Fig. 2.3D). Unusually, not only were the tracks evident in situ on the hyporelief surface, but they also were apparent in epirelief on a fallen slab below. The tracks were dated through OSL, yielding an age of 75 ± 5 ka.

2.1.7.5 Gericke's Point

A new site was situated at Gericke's Point in the Garden Route National Park. It contained at least seven in situ hominin footprints in hyporelief on the ceiling of a small cave, as well as unidentifiable, more eroded footprints. One of the footprints exhibited a vague hallux impression (Fig. 2.3B). Of interest is that OSL dating established that this is the oldest tracksite thus far attributed to *Homo sapiens*, with an age of 153 ± 10 ka.

2.1.8 Tanzania

2.1.8.1 Laetoli

Simply stated, Laetoli in Tanzania includes the most famous hominin tracksites in the world. More than twice the age of the next oldest hominin tracksite in Africa (Ileret, Kenya), the first site to be described from East Africa, and unequivocally exhibiting trackways of bipedal Pliocene hominins, its presence is felt across all subsequent hominin ichnological endeavor on the continent. It was discovered in 1976 and first reported 3 years later (Leakey and Hay, 1979; Day and Wickens, 1980; Leakey and Harris, 1987; Leakey et al., 1987). The site has been dated through $^{40}Ar/^{39}Ar$ to 3.66 Ma (Deino, 2011). The trackways occur in volcanic ash deposits from the boundary between the Lower and Upper Footprint Tuffs. Extensive discussion on trackmaker gait and attribution has ensued, tending toward *Australopithecus afarensis* (e.g., Suwa, 1984; Meldrum, 2007a,b). Initially, three trackways were described at Site G: G-1 for footprints of a smaller trackmaker (estimated height 1.10–1.15 m) and G-2/3(4) for two (or three, according to Musiba et al., 2011) overlapping trackways with larger footprints with an estimated trackmaker height of 1.32–1.52 m (White and Suwa, 1987). In 2015, 14 further footprints were found at "Site S," 150 m from Site G: one trackway of 13 footprints was made by a larger individual (Masao et al., 2016). In 2019, another trackway from Site A, which had fallen into obscurity, was reexcavated. It contained footprints with different morphology and gait pattern, leading to the conclusion that at least two distinct hominin taxa had coexisted at the site (McNutt et al., 2021), with one form more human-like (sites G and S) and the second form less human-like and more ape-like (site A). The distinctions of the Laetoli footprints when compared with modern unshod human footprints (Raichlen et al., 2010) are expressed in the introduction of the new ichnotaxon *Praehominipes laetoliensis* (Meldrum et al., 2011). With the most recent findings increasing the number of nonoverprinted footprints, it is important to review the validity of this ichnotaxon.

2.1.8.2 Engare Sero

First reported in 2011 (Richmond et al., 2011), the terminal Pleistocene Engare Sero site on the shore of Lake Natron in Tanzania eventually yielded more than 400 footprints. The prints were interpreted as having been made by mixed-age and mixed-sex groups and included 23 trackways, some of which provide evidence of a running gait. One assemblage of 17 trackways was attributed to 14 adult females walking together in a similar direction, together with two adult males and one juvenile (Hatala et al., 2020). The track-bearing layer, in volcaniclastic deposits, was dated to 19.1–5.76 ka and described as "the most abundant and best-preserved footprint site of anatomically modern *Homo sapiens* currently known in Africa" (Liutkus-Pierce et al., 2016).

2.2 South America

Very few sites with human footprints have been described in South America (Fig. 2.4), but they are important to understand the expansion of humans in this continent during the Late Pleistocene (Table 2.2).

2.2.1 *Argentina*

2.2.1.1 Pehuén Co paleoichnological site

On the shores of temporary interdune ponds of the Pampean Late Pleistocene, an exceptional record of animal tracks and trackways crops out on platforms along 5000 m of the modern coast (Bayón et al., 2011). The process was repeated quite regularly, each after a short period of time, generating overlapping layers. The paleoichnological record shows more than 100 trackways and hundreds of isolated tracks belonging to 24 species of mammals and birds

FIGURE 2.4 Human tracksites studied in South America.

(Aramayo et al., 2015). A trackway of 13 consecutive footprints, with alternated traces of the right and left foot, shows the walking gait of a single human individual. The length of each footprint ranges between 33 and 35 cm, the maximum width between 13 and 15 cm, the depth between 1 and 2 centimeters, and the interstep distance varying between 54 and 59 cm. None of the footprints have toe prints, which is clearly the result of walking in a highly desiccated substrate, as also indicated by the presence of mud cracks (Fig. 2.5A). The assignment of this trace to humans is based particularly on the angle between the footprints, which is 160 degrees. Besides, two isolated footprints with preserved toe prints were found, although without stratigraphic context (Fig. 2.5B). As shown in Fig. 2.5A, in the same layer, there is a parallel *Megatherium* trackway. In addition, it contains *Macrauchenia*, artiodactyl,

TABLE 2.2 South American human tracksites.

Site	Region	Country	Geographic coordinates	Age and method(s)	Substrate	Footprint preservation	#Footprints trackways	Main references
Pilauco	Osorno	Chile	40° 34′08″S 73°06′12″W	ca. 15.6 ka cal radiocarbon	Swamp-related sandy-peat layer	Epichnial, poor	A single footprint	Moreno et al. (2019)
Monte Verde		Chile	41°30′25″S 73°12′01″W	ca. 14.5 ka BP, radiocarbon, OSL	Sand		3 footprints	Dillehay (1989), Dillehay et al. (2008)
Pehuén co	Bueno Aires	Argentina	39°00′13″S 61°32′30″W	12 ± 0.11 ka radiocarbon	Lagoon mudstone	Epichnial, moderate	One trackway and two isolated footprints	Aramayo (2004, 2009), Aramayo et al. (2015)
Laguna La María	Santa Fé	Argentina	33°58′S 61°34′W	9 ka?	Lagoon fine sediments	Epichnial	3 footprints, only one complete	Lockley et al. (2008)
Monte Hermoso	Buenos Aires	Argentina	38°59′19″S 61°20′54″W	7.125 ± 0.075 −6,0.795 ± 0.12 ka radiocarbon	Interdune pond mudstone	Epichnial, moderate	Hundreds of footprints	Zavala et al. (1992), Bayón and Politis (1996), Quattrocchio et al. (2008), Bayón et al. (2011)

and flamingo trackways that cross the human one in different directions. Nevertheless, the hunting scenario hypothesized by Bayón et al. (2011) is difficult to support, with totally different preservational scenarios for the *Megatherium* and human trackways.

2.2.1.2 Monte Hermoso 1

Politis and Bayón (1995) found on the Argentinean coast of Monte Hermoso, near Buenos Aires, hundreds of children, women, and men's footprints registered in the soft mud, including trackways (Bayón et al., 2011). The assemblages also include human bone remains of two individuals—found in the backshore—scanty faunistic remains and a few wood artifacts. The several overlying layers with footprints indicate that the place was periodically visited for hundreds of years. The footprints found in the western sector belong to adults walking toward a predominantly SE-NW direction, suggesting a transit path (Bayón et al., 2011). The eastern sector is composed mainly of small footprints of children and women wandering along the beach collecting food resources around the pond (Bayón and Politis, 1996).

The Geological, Paleontological, and Archeological Provincial Reserve of Pehuén Co-Monte Hermoso is protected by law and was included in the Tentative List of UNESCO for World Heritage (see the chapter on Vertebrate Ichnosites as Geological Heritage in this book).

FIGURE 2.5 Human ichnosites from the Buenos Aires coast of Argentina (credits: Teresa Manera de Bianco). (A) Human trackway following from close a giant ground sloth trackway; they were produced in very different substrate conditions, which enable questioning of the hunting interpretation. Pehuén Co (Fig. 9 of Bayón et al., 2011). (B) PCO-05–01. One of the best-preserved human footprints in Pehuén Co (left print of an adult). (C) Left footprint of a child from the site of Monte Hermoso.

2.2.1.3 Laguna La María

Near Villa Cañas, Province of Santa Fé, a single complete and two badly preserved footprints were found at the shore of Laguna la María (Lockley et al., 2008). They are dated around 9 ka, but no further details are known.

2.2.2 *Chile*

2.2.2.1 Monte Verde

A campsite buried in the north terrace of Chinchihuapi Creek, which formed around 14.5 ka cal BP, included the remains of a long, tent-like dwelling, hearths, human footprints, edible plants, and wood, reed, bone, and stone artifacts (Dillehay et al., 2015). At Monte Verde, three footprints assigned to a small adult or a large adolescent were reported (Dillehay, 1989). No further study was developed.

2.2.2.2 Pilauco

A single footprint was described in detail at the archeological site of Pilauco, southern Chile, by Moreno et al. (2019). The ichnological analysis of these authors confidently assigns the trace to the ichnospecies *Hominipes modernus*. Some characteristics of the Pilauco print include an elongated distal hallux, lateral digit impressions obliterated by the collapsed sediment, and sediment lumps inside and around the trace (Moreno et al., 2019). Due to its dimensions, general shape, and taphonomic attributes, the footprint of Pilauco corresponds to a right foot impression of an adult human. The trace fossil is associated with megafauna bones, plant material, and unifacial lithic tools. The age of the footprint is constrained by seven ^{14}C dates to c. 15.6 ka cal BP (Moreno et al., 2019).

2.3 North America

Human footprint sites in North America have been found primarily in shoreline and volcanic ash deposits and in caves (Fig. 2.6). By 2012, more than 1000 caves in the southeastern United States had been examined, by one estimate (Simek et al., 2012). With more than 9000 caves in Tennessee and "thousands more in Alabama, Georgia and the upper south," there is much more to learn, and there may be many footprints left to find. All of the tracksites currently known from Central and North America (Table 2.3) are of modern humans, named *Hominipes modernus* by Kim et al. (2008) (designated holotype at Acahualinca, Nicaragua (Lucas et al., 2024), and are dated to the latest Pleistocene or the Holocene (Lockley et al., 2008).

2.3.1 Belize

2.3.1.1 Footprint cave

Also known as Actun Chek (McNatt, 1996: p 91), this cave was discovered by Tom Miller in 1976, during a reconnaissance of the Caves Branch area (Graham et al., 1980). Numerous prints of barefooted humans were discovered "along a stream bank deep within the main passage" of the cave. The cave is noted for mining activity, the presence of digging marks in the red clay, for artwork, clay vessels, and wall sculptures (Graham et al., 1980; McNatt, 1996). The area called "the Ledge" served as a center for the ritual breaking of vessels. The ritual use of caves was a widespread practice in the Mayan civilization (Brady and Stone, 2019). The vessels are dated to 850–1000 AD (Graham et al., 1980). The footprints have been photographed, but they have not been analyzed, molded, or otherwise preserved, and they remain "susceptible to damage from both floods and casual visitors" (McNatt, 1996).

2.3.2 Canada

2.3.2.1 Calvert Island

Duncan McLaren and his colleagues uncovered a series of barefoot human footprints on Calvert Island, British Columbia, Canada (Site EjTa4), impressed in a paleosol 60 cm below the beach surface. Analysis of entrained plant debris confirmed an age for the track horizon of between 13.317 and 12.633 ka BP (McLaren et al., 2018, 2021). This date has been questioned by Lucas (2017), citing evidence that the prints may only be 2000 years old, pointing out that the tested charcoal "may have been introduced into the sediment long before the footprint was

FIGURE 2.6 Human tracksites referenced in North America continent.

impressed." Further testing may be needed to resolve this. The footprints were impressed in a clay soil substrate, just above high tide, and subsequently covered in black sand washed over the tracks during a storm (McLaren et al., 2021). In total, 29 footprints and additional partial footprints were recovered during the field seasons of 2015 and 2016. They ranged in quality from poor impressions to impressions clearly showing the toes and arch. Many had displacement rims similar to footprints left on intertidal surfaces from minor slippage during

TABLE 2.3 Human ichnosites in the North American continent.

Site	Region	Country	Geographic coordinates	Age BP and methods	Substrate	Footprint preservation	# Footprints trackways	Main references
Toluquilla	Puebla	Mexico	18°54′N 98°07′W	~43–38 ka AMS radiocarbon of molluscs, electron spin resonance of mammoth bone, OSL (Gonzalez et al., 2006); (disputed by Renne et al., 2005, 1.3 Ma with argon-argon dating and paleomagnetic data)	Ash	Epichnial, poor preservation Later authors question human origin although reliable trackways are present	269 marks including "human and other animals"	Gonzalez et al. (2006), Huddart et al. (2008), Renne et al. (2005), Morse et al. (2010)
White sands, Tularosa Valley	New Mexico	USA	32°46′46.21″N 106°10′21.65″W	23.4 ± 2.5 ka to 22.6 ± 2.3 ka, radiocarbon >21.5 ± 1.9 ka OSL Disputed (Haynes, 2022; Oviatt et al., 2022; Rachal et al., 2024)	Playa shoreline Gypsum sand	Endichnial Imprints into thin layers	Hundreds. 447 + at locality 3	Bustos et al. (2018), Haynes (2022), Oviatt et al. (2022), Pigati et al. (2023), Rachal et la. (2024)
Calvert island	British Columbia	Canada	51°39′50.81″N 128°05′51.12″W	13.317–12.6 ka Radiocarbon, entrained charcoal.	Beach	Epichnial, barefoot some with arch and toe	29 + partials	McLaren et al. (2018), Lucas (2017)
Cuatrociénegas	Coahuila	Mexico	26°58′N 102°59′W	10.55–7.24 ka	Ash/Tufa	Epichnial, moderate	17	Gonzalez et al. (2007), 2009, Felstead et al. (2014)
Oro Grande	California	USA	34°35′N117°20′W	~5.07 ka Radiocarbon of charcoal lens	Silty Riverine mud	Epichnial, barefoot, deep clear prints	54 4+ track makers	Rector (1979, 1983, 1999), Willey et al. (2009)
Jaguar cave	Tennessee	USA	36°22′N84°53′W	4.695–4.53 ka Radiocarbon of charcoal samples from cane torches	Cave mud	Epichnial Bare feet, Excellent definition	272 + 9 trackmakers	Robbins et al. (1981), Kennedy et al. (1984), Watson et al. (2005), Willey et al. (2005)
Crystal cave	Kentucky	USA	37°12′39.17″N 86°03′16.40″W	4.15–3.67 ka Radiocarbon from torch charcoal	Impressions in mud dust	Epichnial	12	Watson (1969), Crothers et al. (2002), Willey et al. (2008)

Upper and lower Salts cave	Kentucky	USA	37°11′13.24″N 86°06′04.49″W	~3.1–2.2 ka radiocarbon	Impressions in cave floor	Lost, were clear, sandal imprints (1875). bare feet, 1969	15+	Putnam (1875a,b), Watson (1969, 1974), Crothers et al. (2002)
Fisher ridge cave	Kentucky	USA	37°11′N85°58′W	~3.175–2.75 ka Radiocarbon on Charred cane and oak	Mud floor	Slippered imprints	Appx. 18 (two people)	Watson (1983, 1986), Kennedy et al. (1984), Crothers et al. (2002)
Acahualinca	Managua	Nicaragua	12°09′38.70″N 86°17′30.30″W	6.6 to 5.8 ka; 2.12 ± 0.12 ka radiocarbon	Ash	Epichnial, Excellent condition, *Hominipes modernus* Holotype from this site	346 (322 human) in 12 trackways.	Williams (1952), Bice (1979), Lockley et al. (2007a,b), Schmincke et al. (2009, 2010), Lucas (2017)
Third unnamed cave	Tennessee	USA	36°22′N84°53′W (private property)	4.35–2 ka Radiocarbon	Muddy Sediment	Epichnial; barefoot impressions	Appx 6	Fergusen (1983), Simek et al. (1998), Franklin (1999), Crothers et al. (2002)
Swan point	Alaska	USA	64°09′09″N 145°50′32″W	~1.84 ka radiocarbon	Loess mud layer	Epichnial, Shod foot	One	Smith et al. (2019)
La Rama	Usulután	El Salvador	13°20′N 88°26′W	ca. 1.8–1.2 ka radiocarbon (stated as AD 200–800	Sands	Epichnial	Nine in situ	Haberland and Groebe (1957)
Pocket cave pit house 4	Arizona	USA	36°57′N 111°54′W	~1.55–1.45 ka Dendrochronology and archeological evidence	Carbonized on a thin clay floor	Epichnial, bare feet, excellent clarity	Appx, 12 + 2 forefoot At least three sizes	Morris (1959, 1980), Hays-Gilpin et al. (1998), Willey et al. (2009)
Mud glyph cave	Tennessee	USA	36°10′N86°45′W	~1.49 ka Radiocarbon	Mud	Impression in mud wall	1 modified into a representation of a turtle	Faulkner et al. (1984), Faulkner (1986), Watson (1986), Willey et al. (2008)
Footprint cave: (Actun Chek) Caves branch	El Progresso	Belize	17°07′03.07″N 88°53′25.72″W	ca. 1.15 to 1 ka Based on artifact comparison with others from the region (stated as 850–1000 AD)	Cave floor mud near stream	Cave floor barefoot imprints. in main Branch, Photographed	Unspecified (Numerous)	Graham et al. (1980), McNatt (1996)
Xanadu cave #W3	Tennessee	USA	36°22′N84°53′W	~875 Radiocarbon from charcoal	Dried mud floor	Epichnia; barefoot impression	Two located	Sims et al. (1981) Franklin (2002)

(*Continued*)

TABLE 2.3 Human ichnosites in the North American continent.—cont'd

Site	Region	Country	Geographic coordinates	Age BP and methods	Substrate	Footprint preservation	# Footprints trackways	Main references
Naj Tunich Peten (UNESCO heritage site)	Poptún	Guatemala	16°16.32″N 89°15.35W	~733–762 AD Calendar artifact dates (Stuart) 400–900 AD (UNESCO) 2.75 to 0.32 ka BP (Radiocarbon)	Cave Floor Mud	Epichnial; Bare footprints in clay along the wall edges (Silent Wall)	Unspecified	Stuart (1981), Brady and Stone (1986), Lucas (2017)
Sequoia cavern	Alabama	USA	34°40′06.34″N 85°36′19.07″W	~520 ± 50 Radiocarbon on torch charcoal	Mud floor	Barefoot imprints	Seven	Sneed (1984), Watson (1983, 1986), Willey et al. (2005, 2008)
Kau valley	Hawaii	USA	19°22′06.96″N 155°21′53.55″W	~480–200 radiocarbon	Ash lapilli	Epichnial Excellent	1733 (441+ People)	Doerr (1931); Jaggar (1921, 1930), Moniz Nakamura (2003, 2009)
Footprint cave	Virginia	USA	36°59′N 82°05′ W	~430–375 radiocarbon Pine torch charcoal	Mud floor	Epichnial	Appx. 30+	Willey et al. (2005, 2008, 2009), Crothers (1997)
Guaimaca, El Portillo de la Crucita	Francisco Morazán	Honduras	14°24′54.48″N 86°50′15.96″W	Undated	Ash	Epichnial, good	Three footprints	Veliz (1978)
El Pando del Venado, La Cuchilla, La Ocotera	Jalisco	Mexico	21°50′N103°08′W 20°46′N103°27′W 20°32′N104°26′W	Unknown	Tufa/ash	Unspecified	Not stated	Rodriguez-de la Rosa et al. (2004)
Cucurpe	Sonora	Mexico	30°19′N 110°42′W	Untested	Ash and lapilli	Epichnial, poor	Not stated	Houck et al. (2009)
Lon Odell cave	Missouri	USA	37°29′N93°59′W	Unknown (untested)	Mud floor	NA	≥10	Beard (1997), Willey et al. (2008)

emplacement. Footprints range between 15 and 28 cm in length and 6–13 cm width. The authors concluded that the footprints were made by at least three individuals.

2.3.3 *El Salvador*

2.3.3.1 La Rama (Usulután)

Workers constructing a farm road in 1955, near the town of Usulután, El Salvador, discovered "footprints in stone" on the "hacienda La Carrera" (Haberland and Grebe, 1957). They saved some slabs out of curiosity, but destroyed the rest during construction. Haberland and Groebe (1957) were led to the construction site by one of the workers and there, at 1.2 m below the surface, they uncovered nine footprints, belonging to five or more persons, walking parallel. Based on the evidence of fragments of pottery in a coeval layer nearby, the authors proposed an age for the footprints of ca. 1.7–1.1 ka BP.

2.3.4 *Guatemala*

2.3.4.1 Naj Tunich cave

Discovered in 1979 (Brady and Fahsen, 1991; UNESCO, 2012), and first explored in 1980 (Stuart, 1981), Naj Tunich was added to the UNESCO World Heritage tentative list in 2012, due primarily to its spectacular cave art. Footprints have been found in a passage leading to the deepest east branch chamber of the cave. Due to looting, discovered soon after the cave was first explored, the cave was placed under government protection (Brady and Fahsen, 1991; Stuart, 1981). Stuart found a group of glyphs that yielded a precise date of December 18, 738 AD and another "in the same scribe's handwriting" from 3 days earlier, probably the date that the glyphs were created (Stuart, 1981). The style of the glyphs, Stuart wrote, is consistent with vases from the period 600– to 900 AD found elsewhere in the region. Radiocarbon and tree ring dating of numerous objects in the cave by Brady and Fahsen (1991) give a long range of occupation from 2750 to 320 years BP. In addition to the glyphs, evidence of small-scale mining in Naj Tunich suggests that the cave was used as a ceremonial site, perhaps a pilgrimage center (Brady and Rissolo, 2006). Unshod footprints, attributed to pilgrims, have been found in the mud of the approach tunnel leading to the "Silent Well" (Brady and Stone, 1986). No further elaboration about the number, quality, or age of the footprints is given.

2.3.5 *Honduras*

2.3.5.1 Guaimaca

Three barefoot prints were found on the hill called El Portillo de la Crucita, about 5.5 km south of the town of Guaimaca (Veliz, 1978). Three archeologists, Dr. Julia Zúñiga, Mr. Jorge Varela, and Dr. Veliz, found four fragmentary footprints that they did not count in their assessment and three full footprints. Two of the footprints, a right and a left print, were facing south, and one right footprint faced north. These directions may be irrelevant, as the author stated that the rocks may have been moved from their original position. The feet were described as being wider than the modern foot but shorter, attributed to habitual barefooted walking. Dimensions for the prints are: length, 21–24 cm, width across the toes, 13–14 cm, and heel width, 6–8 cm (Veliz, 1978). The prints are preserved as superficial imprints on rhyolite. One print was left in-situ, and the other two prints were deposited in institutions in Honduras, one in the UNAH (Universidad Nacional Autónoma de Honduras) and the other in the IHAH (Instituto Hondureño de Antropología e Historia). A sketch of the print now in IHAH (Veliz, 1978, Fig. 3 therein) exhibits five toes, an arch, and a rounded heel.

Unfortunately, the footprints were not found with any cultural artifacts or charcoal, and no radiometric dating of the rock was completed.

2.3.6 *Mexico*

Human footprints are known from several localities in Mexico, although several are problematic and not described (Rodríguez de la Rosa et al., 2004; Lockley et al., 2007a, 2007b). Rodríguez de la Rosa et al. (2004) and Lucas (2008) indicated three human tracksites in the state of Jalisco ((El Pando (Banco?), del Venado (Venado), La Cuchilla (Cerro?) La Ocotera)), in tuffaceous strata. In addition, Lockley et al. (2007a) and Huddart et al. (2008) mentioned vaguely other tracksites in the states of Yucatan, Guerrero, México (Amanalco de Becerra), Chihuahua, Oaxaca (Santo Tomás de Jalieza), and Sonora (Cucurpe). Aveleyra Arroyo de Anda (1950) mentioned a human tracksite at Rincon de Guadalupe (Mexico State) that was also the subject of an earlier lost manuscript (Lockley, 2007).

2.3.6.1 Toluquilla (Puebla)

The oldest putative tracks from Xalnene Ash near Puebla are purportedly Late Pleistocene in age (Gonzalez et al., 2003). The footprint surface shows multiple "trails" of adults and children, together with clear animal tracks and trackways. From the 269 tracks identified by Huddart et al. (2008), 60% were attributed to humans, with 36% of these being children. The prints are set in a course medium of ash lapilli (Applegate et al., 1984) along a former lake shoreline and are not as distinct as at some other well-known sites (Huddart et al., 2008). The track-bearing level has been variously dated as older than 40 ka BP (Gonzalez et al., 2006; Huddart et al., 2007) or 1.3 Ma (Renne et al., 2005; Gonzalez et al., 2006). Both dates are highly improbable for human footprints in this part of the world, and it is likely that these features are in fact inorganic (Lockley et al., 2007a). In fact, Morse et al. (2010) demonstrated that the supposed "footprints" were the effects of the quarrying activity of the local tuff.

2.3.6.2 Cuatro Cienégas (Coahuila)

Two sets of footprints have been found in the Cuatro Cienégas basin. Two individual prints were excavated in 1961, but were not deposited in a museum for several years, finally ending up in the Museo Regional de la Laguna at Torreon. They were later displayed by the Museo del Desierto, Saltillo (Coahuila) after it opened in 1999 (Gonzalez et al., 2009). The footprint site was not rediscovered until 2006 (Gonzalez et al., 2007, 2009). An additional 15 complete prints were copied for study, and one partial print was extracted. The two museum footprints, ~24 cm long and 10 cm wide, are clear and well preserved. There are several trackways at the footprint site. The trackway designated as "number 1" has prints averaging 27 cm long by 10 cm wide and has an average stride length of 74–81 cm. Footprints in the main trackways are well preserved and show heel, ball, and toe impressions. There are several animal trackways at the site as well, including a large cervid, and possibly a duck (Gonzalez et al., 2007, 2009). The sediments containing the footprints have been dated between 10.55 and 7.24 ka (Felstead et al., 2014), the oldest in Mexico according to the study's authors. The older footprints are in the local museum, and the younger sets are those found in situ. Although the climate was generally arid, the interpreted local conditions of groundwater discharge have created a local oasis, capable of supporting a diverse biota (Felstead et al., 2014).

2.3.6.3 Cucurpe (Sonora)

This small site with several human footprints is figured but not described in detail by Houck et al. (2009: Fig. 10).

2.3.7 *Nicaragua*

2.3.7.1 Acahualinca

One of the most well-known and well-preserved human footprint sites in the Americas is located on the shores of Lake Managua, among a chain of volcanoes that were the source of the volcaniclastic material into which the footprints are impressed (Lockley et al., 2007a). There are three sites of footprints that have been found here, designated El Recreo, documented but now built over (Brown, 1947; Williams, 1952; Lucas et al., 2024), El Cauce, the main site and home of the Huellas de Acahualinca Track Museum, and an unnamed site 4.8 km east at Portalueza, noted by Williams but which has not been further studied. Several excellent papers have been written reviewing this famous site in a worldwide context (Lockley et al., 2008, 2016; Lucas, 2017) and as a focused study (Lockley et al., 2007a, 2009; Lucas et al., 2024; Schmincke et al., 2009, 2010).

The footprints were first found at El Cauce during the quarrying of building stone (Flint, 1884; Johnson, 1884; Lucas et al., 2024) and were reported on scientifically by Earl Flint, who examined the site for the Peabody Museum (Flint, 1884, 1886, 1888). Flint uncovered "six rows of impressions, breaking through a layer of rock." At least five of the footprints were extracted; four were sent to the Peabody Museum, with an additional print being sent directly to Dr. Brinton, of the Academy of Natural Sciences, Philadelphia (Brinton 1887). Flint characterized the rock as "volcanic detritus" from a volcanic eruption. As described later by Crawford (1891), "men, dogs and horses fled to Managua to shelter themselves from the highly heated cinders and ashes ejected from volcano Masaya." Williams (1952) described the footprint-bearing layer as a volcanic mudflow, a lahar, emplaced in a stream channel through which the people walked in an NNW direction. He said that there was no evidence that the footprints were "made in haste" (Williams, 1952, p. 22), a conclusion that Lockley et al. (2009) agreed with. Schmincke et al. (2009, 2010) suggested that the people were moving briskly, over a muddy surface, to try to escape from a major volcanic eruption from the Masaya Caldera. Mapping and paleontological work on the footprints at the site accelerated with Martin Lockley's work in 2006 (Lucas et al., 2024), who, together with collaborators, published detailed descriptions of the prints, maps of the trackways, and defined the new ichnogenus and ichnospecies *Hominipes modernus* from a holotype at this site (Lockley et al., 2007a, 2009; Kim et al., 2008). In all, 322 human footprints were mapped and documented, but there were many more in a trample ground that were not counted in this total.

At first, Flint (1884) refrained from designating a date for the footprints, saying "it is useless to speculate on the lapse of time that has passed since their occurrence," but he later described the tracks as "Pre-Adamite" (Flint, 1886) and was attempting to discern the true age of the footprints based on what he thought were contemporary shells, titling one article "Human footprints in the Eocene" (1888). Brinton (1887) gave some of these shells to his colleague Angelo Heilprin, at the Academy of Natural Sciences, who wrote that he thought that from the condition of the shells and their species were most likely "post-Pliocene." Flint's statement that the layer was Eocene was also refuted by Peet (1889). Modern attempts to date the site began in the 1970s with Bryan reporting the radiocarbon age of soil just below the footprint layer as 5945 ± 145 years BP (Bryan, 1973; Lucas et al., 2024), but gave an estimate of 3 ka BP to account for soil development. Bice (1979) tested material just above the footprint layer, getting a reading of ∼6.5 ka years BP. The two radiocarbon readings bracket the footprint layer, giving an estimated age of between 6.5 ka and 5.8 ka BP (Lucas et al., 2024). Schminke et al. (2009, 2010) correlated the tephras at Acahualinca to the Masaya Triple Layer (see Schmincke et al., 2009, Fig. 11 therein), which is dated at 2120 ± 120 years BP (Lucas 2017; Lucas et al., 2024; Schmincke et al., 2009, p 491). This discrepancy in ages allows for radically different archeological interpretations, either a group of "lake adapted Archaic-period hunter gatherers made them" about 6500 years BP, for which corroborating archeological evidence of artifacts is absent, or a more recent "Late Formative agricultural people" had their lives upended by the eruptions of Masaya, which deposited the triple layer at ∼ 2200 years BP (Lucas et al., 2024). Lucas et al. (2024) rightly argue for the need for more direct evidence to constrain these dates.

A shelter was first built over the trackway in 1942 to protect the footprints. New buildings were constructed in the 1980s, but in 2015, the tracks were flooded, and the museum was closed to the public for planning and for reconstruction of the museum buildings (Lucas et al., 2024). The Acahualinca Tracks Museum was considered one of the most important hominin ichnosites in the world by Kim et al. (2008) because it combines accessibility, a large sample of well-preserved trackways, and reliable geochronological data (see the chapter on Vertebrate Ichnosites as Geological Heritage).

2.3.8 *USA*

2.3.8.1 Sequoia Caverns (Alabama)

Sequoia Caverns, in DeKalb County, was the site of the discovery, in 1981, of unshod human footprints and torch fragments (Sneed, 1984). The prints were photographed later that year and, in 1982, an expedition with Dr. Patty Jo Watson collected torch charcoal for dating and made measurements of the seven prints. Dr. Watson concluded that the prints were made by one individual, probably during a single trip to the cave. Radiocarbon analysis by the Smithsonian Radiocarbon Laboratory determined that the torch fragments date from 520 ± 50 BP (Watson, 1983; Willey et al., 2005). The owners of the cave installed a sign marking the "Barefoot Room," in order to help protect the prints (Sneed, 1984).

2.3.8.2 Swan Point (Alaska)

Swan Point, in the Tanana Valley of central Alaska, is the site of a single footprint associated with a housepit excavated in 2017 and 2018 (Smith et al., 2019). The footprint of an individual wearing a soft soled moccasin or similar shoe is impressed in the sediment of stratum layer V, a yellowish-brown loess interpreted as being an outdoor living surface. Radiocarbon dating of layer V and the coterminous housepit floor yield an average age of 1840 BP. The footprint was preserved as a plaster mold and photographic scans, which were converted to three-dimensional models. Analysis of the print and comparisons with known moccasin prints resulted in an estimated length of the foot as 20.28–20.54 cm and a width at the ball of the foot of 7.41 cm. Based on this information, the authors concluded that the footprint was most likely made by a preteen individual of 8–9 years old.

2.3.8.3 Pocket Cave (Arizona)

Pocket Cave, one of several cave house sites along the Atahonez Canyon, a tributary canyon of the Red Wash in the Prayer Rock District of northeastern Arizona, is the site of numerous archeological finds. Earl Morris noted the presence of these sites in 1928 (Morris, 1959). Many artifacts were found during the Bernheimer expedition for the American Museum of Natural History in 1930, and more in 1931 during an expedition by Earl Morris. An extended discussion of the sandals, pottery, weavings, pictographs, nets, feathers, animal and vegetable remains recovered from the Prayer Rock District can be found in Elizabeth Morris' 1959 thesis. Among the articles noted are several "soot-blackened muddy footprints," found in Pit House 4 in Pocket Cave (Morris, 1959, p 183; Morris, 1980; Hays-Gilpin et al., 1998, see Fig. 3.4 therein; Willey et al., 2009). The footprints represented three or more people, including one child, based on the footprints' relative sizes (Morris 1959, 1980). There are also two handprints, adjacent to one of the radial floor ridges, suggesting to Morris (1959, 1980) that one individual squatted in front of the ridge with their heels resting on the ridge. She extrapolated from this foot position and associated objects, Yucca root (the source of cleansing shampoo), and a coiled basket that might have held water, that the person may have been washing their hair. Morris (1959, 1980) also noted that besides the footprints, "distinctly visible" fingerprints can be found impressed into the roof mud layer of Pit House 4. It has been suggested that distinctive individual footprints, both bare and shod, may have been considered a symbol of prayer (Hays-Gilpin et al., 1998, p. 122).

Dates of occupation for Pocket Cave based on a combination of sandal style descriptions and tree ring chronology yield two phases, an early occupation in the late fifth century, and a reoccupation in the seventh century AD (Hays-Gilpin et al., 1998).

2.3.8.4 Oro Grande (California)

The Oro Grande site in The Mojave Desert, near Victorville, California, is the location of a trackway of barefoot human prints found during excavations in 1977 and 1978 (Rector, 1999). The footprints are dated to 5070 ± 120 Years BP, based on radiocarbon dating of an associated charcoal lens thought to be a hearth, in the layer containing the footprints, 95 cm below the ground surface. A total of 54 human footprints, between 14 and 26 cm in length, are thought to represent up to five individuals (Rector, 1979, 1983, 1999). Three sets were recorded "in stride" (Rector, 1999). One of the prints has six toes, and another was interpreted as being from a pigeon-toed individual. The researchers found associated artifacts including flakes and drills on what is surmised to be a camp on an adjacent sandbar, and the tracks of raccoons, ungulates, and coyotes in the same layer, one of which is overprinted by a human footprint (Rector, 1999). The trackmakers, including three juvenile and two adult individuals, may have foraged a riparian zone with dense vegetation. The 28 best prints were preserved by latex molds and resin casts, and the entire set of tracks was measured, mapped, and photographed onsite (Rector, 1979, 1983).

2.3.8.5 Kau Valley (Hawaii)

Human footprints impressed in ash were discovered in a 13 km^2 area of the Kau Desert on Kilauea in 1919 by Ruy H. Finch (Moniz Nakamura, 2009). According to legend, the unshod tracks (Fig. 2.7A) belonged to a doomed party of Chief Keoua's warriors killed by an

FIGURE 2.7 Some examples of human ichnosites from USA. (A) Well-preserved right forefoot print on the ask layer of Ka'ū desert, Kilauea; scale bar is 5 cm. (B) Typical trampled surface preservation at the Kau valley (A and B courtesy: National parks Service). (C) Digital elevation model of trackways in white sands national monument, new Mexico; scale bar is 220 mm. (D) Detail of C showing a real image of the well-preserved footprints produced by an adult and teen walking one after another (C and D courtesy: National parks Service).

explosive eruption of Kilauea in 1790, between battles with the new Hawaiian King, Kamehameha (Mayor and Sarjeant, 2001). Historically, the civil war in Hawaii began with the death of King Kalaniopuu in 1782 (Doerr, 1931). The old king named his son Kiwalao as his successor, and his nephew Kamahameha as district chief over much of the island of

Hawaii, angering Keoua, the half-brother of the new king. The civil war devolved into small skirmishes between the two claimants until 1790, when Keoua took advantage of Kamahameha's absence (he was fighting in Maui), to overrun the north of Hawaii. Kamehameha returned to the island of Hawaii and pushed Keoua from the north to Hilo, in the south (Doerr, 1931). At this point, Keoua moved his army toward Kau along a route that skirted Kilauea's crater. He split his forces into three columns. Unfortunately, their march was interrupted by the eruption of Kilauea. On the third day of the eruption, on a trail near the crater's rim, the central column of about 80 people was destroyed by "one or more directed blasts" of ash from the eruption (Finch, 1947). Oral tradition describes the loss as "the warriors of two small villages," which was considered a bad omen (Ellis, 1827, p 199). Among the 80 people killed were not just warriors, but family groups that accompanied them as well (Doerr, 1931). The legendary and historical records in fact account for some of the trackways found in the footprint parcel, but the complete picture is more complex (Swanson and Christiansen, 1973; Moniz Nakamura, 2009).

In 1938, Hawaii National Park was expanded to include the footprint site. The ichnosite covers an area of 4284 acres, and footprints have been found even outside of its boundaries. Field surveys of the site in and outside of the park parcel in 1998 and 2000 identified 1733 footprints attributed to at least 441 individuals associated with numerous archeological artifacts including temporary shelters and trails (Moniz Nakamura, 2003, 2009). The footprints are found in multiple layers of pisolitic ash that fell as mud during eruptions of Kilauea and that increase in thickness toward the crater (Jaggar, 1921, 1930, 1934). Many of the footprints are well defined and exhibit several components identified by Meldrum (2004) as hallmarks of the modern human foot: a well-developed medial longitudinal arch, a broad rounded heel with a well-developed heel pad, and a rounded ball medial to the hallux (Fig. 2.9F). Meldrum (2000) suggests that the gap between the hallux and the lateral toes seen in the Hawaiian prints is typical of people who "habitually" walk barefooted.

Jaggar (1921, 1930, 1934) identified two layers of footprints and attributed both to the 1790 eruption, saying "This succession proves conclusively that the print-making was contemporaneous with those explosive eruptions" (Jaggar, 1930). The upper layer was seen as the second, more violent part of the same eruption cycle. The probable cause of death of the party was due to a pyroclastic base surge enveloping the victims (Swanson and Christiansen, 1973; Swanson et al., 2015).

Moniz Nakamura (2009) revisited the question of age and concluded that the archeological and geologic evidence indicates that, besides the group trapped in the 1790 eruption, the footprints found within and outside the borders of the park were most likely accumulated from a continuous use of the corridor trails for at least 200 years before the 1790 eruption (Fig. 2.7B). Her reasoning combines historical patterns, archeological evidence, and the eruption history of Kilauea. The location of the footprints is along a corridor used to traverse the island from ancient time, a natural saddle between the Kilauea caldera and the slopes of Mauna Loa. A modern highway utilizes the same corridor. Less abundant footprints not associated with the movement of the army are found moving in an east-to-west as well as west-to-east direction. Over 500 structures have been found along the route, including temporary shelters dating from up to 200 years before the eruption (Moniz Nakamura, 2003, 2009). Radiocarbon dating shows that the lower, older layer, noted by Jaggar, was deposited approximately 400 BP (Meldrum, 2004). This suggests that the footprint site on Hawaii records the passage of humans along this part of Hawaii for the most part in the course of normal life, in addition to the tragic mass death in 1790.

2.3.8.6 Salts Cave (Kentucky)

Salts Cave is a part of the larger Mammoth Cave National Park. F. W Putnam, of the Peabody Museum, recounting his exploration of Salts Cave (he calls it Salt Cave), notes the numerous fire pits and implements but then goes on to describe "the most important discovery" (Putnam, 1875a: p. 412), about 3 miles from the entrance, in a side chamber in which "the foot of a white man had never stepped before." In a chamber discovered by Putnam's guides, Cutliff and Lee, his party found the "imprints of feet that had been shod with peculiar braided moccasons or sandals" (Putnam, 1875b, p. 319). He described the prints as leaving the impressions of "naked heel and toe and the braided covering to the sole of the foot." In these side chambers, the party collected numerous discarded, braided sandals. Later authors have not referred to these footprints from Upper Salts Cave. Colonel Bennett Young (1910) entered Salts Cave through what was then known as the Hazen entrance in 1894, and noted, besides flint flakes, human and animal bones, and burned cane torches, "footprints of the men who, hundreds of years ago, walked along these majestic avenues." That entrance, currently known as the Pike Chapman entrance, is located about 3 miles from the entrance used by Putnam. It is a collapsed entrance to Salts Cave that had been reopened and collapsed several times, including a reopening by Hazen in 1893. The footprints seen by Young may be the same as those seen by Putnam. Mammoth Cave scientists suspect that the footprints found by Putnam were destroyed during looting in the late 19th and early 20th centuries (Rickard Toomey, personal communication, 2024).

Lower Salts Cave, discovered in 1953, yielded 15 recorded bare footprints in an area called Indian Avenue (Watson, 1969: p. 29) and in the nearby, connected Crystal Cave, Upper Crouchway (now recognized as part of the Unknown Cave, Mammoth Cave system). This was the site of a discovery in 1954 of series of footprints by William Austin and Jack Lehrberger (Watson, 1969, 1974; Crothers et al., 2002, 2007). Based on radiocarbon measurements of torch fragments, the prints in the Upper Crouchway are dated to 3.7 ka BP (Willey et al., 2008). Prints in Salts Cave measured from 19 to 24.5 cm in length. Those in Crystal Cave (Unknown Cave) ranged from 24.5 to 26 cm (Watson, 1974). Numerous discarded, braided sandals were found in Lower Salts Cave. Ages for the associated burned torches and other remnants range between 3.14 and 2.24 ka BP (Watson ed., 1974: p. 235). Salts Cave and Unknown Cave were found to be connected in 1961 (NPS website; Watson, 1969) and then discovered to be connected with Mammoth Cave in 1972 (Bruker and Watson, 1976). Watson (1969) figured one print from each site, side by side on page 63 (plate 14). It is not known whether the footprints in Indian Avenue are still extant. The passage is long and tight, and the prints may have been inadvertently destroyed. There is still the open question of the location of the prints found in a side chamber by Putnam and his guides in 1875, but not referred to after the initial report. Putnam's guides, Cutliff and Lee, were well-known explorers who ranged all over the cave, and the description of a small chamber about 3 miles from the entrance is not specific enough to pinpoint the spot. There are also references to footprints in Mammoth Cave from the early 1800s, but those prints were not recorded and evidently have not survived (G. Crothers, personal communication, 2024). Different people, exploring connected caves though different entrances, can create quite a mystery.

2.3.8.7 Fisher Ridge Cave (Kentucky)

Fisher Ridge Cave is located east of Mammoth Cave National Park. Approximately 18 footprints were found in a muddy passage, testifying to the exploration by two humans wearing moccasins during the Archaic Period (Watson, 1983; Kennedy et al., 1984; Crothers et al., 2002; Willey et al., 2008). Radiocarbon dates were obtained in 1983 from a piece of oak (800 BCE) and cane charcoal from a torch remnant (1225 BCE). There is no evidence of the ancient use of the cave beyond that of exploration.

2.3.8.8 Lon Odell Cave (Missouri)

Lon Odell Cave, in Dade County, southwestern Missouri, was the subject of a story in Beard (1997) recounting an ancient exploration of the cave that Beard was able to recreate using the locations of abandoned torches and occasional footprints. Much of the standard evidence, torch marks on the walls, abandoned torches, and up to 10 barefoot prints were found here, but the recounting is not rigorous, and Beard specifically states that the charcoal from the torches was not tested, so that the age of the footprints is in question.

2.3.8.9 White Sands National Monument (New Mexico)

The western shores of Paleo-lake Otero, at White Sands National Monument in the Tularosa Valley, are the site of a recently discovered suite of trackways and footprints of animals and unshod humans that Bustos et al. (2018) propose represented a hunt of sloths by humans during the terminal Pleistocene. Nonhominin tracks had been reported from the site since 1932 and had been photographed during an expedition in 1981. The earlier footprints of mammoths were described as being epichnial pedestals in convex relief (Lucas et al., 2002). The human footprints described in 2018 appear as negative relief, impressed tracks. The visible prints are result of the compaction and recrystallization of the gypsum sand under the track, and their subsequent enhanced resistance to erosion aided by gypsum cements and microbial mats. They are visible under specific moisture conditions, which enhance the contrast of the tracks to the surrounding sediment (Lucas et al., 2002; Bustos et al., 2018). The site contains more than 100 sloth and human trackways. About 10 of the human trackways at the site were found superimposed on sloth trackways in such a way as to suggest that the human was stalking the sloth, intentionally walking in its tracks. Other indications of a possible hunt include the presence of a series of toe impressions leading toward the sloth trackways in a way to suggest a stealthy approach. The sloth trackways show sharp changes of direction when in association with the human prints, possibly indicating evasive behavior. The trackways cross-cut and show evident interaction (Bustos et al., 2018). Sediments in the adjacent escarpment date between 33 and 10 ka BP. The authors proposed that the best fit for the age of the tracks was sometime between the commencement of the erosion of the lake bed of Lake Otero, 15.56–10 ka, which coincides with Clovis culture dates. They conceded that finding an exact date was problematic because of the reworking of sediment from other outcrops in the region.

Studies reported in 2018–19 introduced methods that extended the ability to visualize the tracks at White Sands and added to the knowledge of species that shared the landscape (Urban et al., 2018). Two trackways of human footprints on the eastern edge of the paleolake, extending for over 800 m, were investigated by surface surveys, partial excavation, and 3-D imaging (Urban et al., 2019), revealing that one trackway, going northward, was made earlier, and the other, going southward, was made later. This determination was made by imaging showing that a concurrent mammoth track deformed and partially obliterated two of the northbound tracks and was superimposed by one of the southbound tracks. Radar data at the site also confirmed the tracks of ground sloths, carnivores, and artiodactyls. Urban et al. (2019) did not propose a date for the tracksite other than Late Pleistocene, based on the "Current N. American dates for an initial human presence along with the dates for mammoth extinction on the continent."

This same series of trackways, White Sands National Park (WHSA) "double trackway," was revisited and further analyzed (Bennett et al., 2020), giving more detailed biometric data for the two human trackmakers. By this time, the known trackway had been extended in length to 1.5 km. The site has a total of over 427 human footprints. The work was done by a combination of hand work, finding the track outlines and brushing away the infilled dust to reveal the prints, and taking photogrammetric digital orthophotos of 90 of the 140 excavated footprints for analysis. The uniform depth of the footprints suggested that they were impressed over a firm subsurface layer. The double trackway is accompanied by the occasional footprints of a child, estimated to be less than 3 years old. These trackways are not continuous, and the impression given was of a child being held and set down briefly. At one location, there are several child and adult trackways in a complex array, indicating a longer stop or rest from carrying the child. The adult footprint length averages about 24.7 cm northbound and 23.3 cm southbound, corresponding to a modern age range of 12–14 years old for those footprints. However, it is noted that female feet reach their maximum length at about age 14, and if the trackmakers were female, they could be adults. The similarities of the trackways may indicate a round trip by the same individual, but the authors do not rule out a different person of the similar stature. The cross-cutting relationship of the mammoth track, seen in the earlier work by Urban et al. (2019), is confirmed in this chapter. In addition, the tracks of a ground sloth rising up on its hind legs appear to be contemporaneous with the human trackways. Bennett et al. (2020) proposed that the human trackmakers were moving in haste, with a child, through a "potentially dangerous landscape with potential carnivores."

Work done on a second eastern basin site, WHSA locality 2 (Bennett et al., 2021), indicated to the authors that the site dated not from the Clovis period, ($\sim$13 ka BP), but from at least 23 ka BP, "with evidence of occupation spanning approximately two millennia." The footprints are found on eight surfaces in total. Most of the footprints are attributed to teenagers and children, with fewer clearly adult footprints visible (Fig. 2.7C). The footprints are often clear, exhibiting heel pads, arches, and toe prints (Fig. 2.7D). Mammoth and canid tracks were found in the layers associated with the human prints. More recently, Bennett et al. (2025) described linear features associated with human footprints, which were interpreted to be drag marks formed by travois pulled by humans, presumably during the transport of resources. Although the appealing claim of the earliest evidence for the use of transport technology, the simple dragging of firewood cannot be discarded. The proposed age of the footprint layers was determined through radiocarbon dating of the entrained seeds of an aquatic plant, *Ruppia cirrhosa*, tested for each layer. They yielded dates of from 22.86 ± 0.32 to 21.13 ± 0.25 ka BP. The authors concluded that humans were present in North America before the ice advances of the Last Glacial Maximum closed off the possible migration corridors.

This dating claim was quickly challenged by Haynes (2022), who contended that the hard water effect (dismissed by Bennett et al., 2021) or the redeposition of robust *Ruppia* seeds from older strata or cross-lake locations could give artificially old dates for the deposit. Haynes proposed that the proper date for the site would be closer to ca. 13 ka BP, during the Clovis Period, at the Pleistocene–Holocene boundary or closer to radiocarbon dates taken earlier, from the western edge of Lake Otero, of about 15.3 ka BP. He did not rule out an earlier date, but stated that there needs to be more work done to determine the correct date. Pigati et al. (2022) replied to Haynes, stating that some of the assumptions by Haynes are based on studies of the western margin of Lake Otero and cannot be applied to the eastern margin, which experienced a different depositional environment. They pointed out that the human footprints are "intercalated with and stratigraphically below tracks of Pleistocene megafauna." The authors also refuted Haynes objection based on the hard-water effect in which aquatic plants incorporate older carbon from the host waters. They questioned whether the seeds themselves could have been the result of redeposition. Oviatt et al. (2022) also cast doubt on the *Ruppia*-based dating. In fact, Rachal et al. (2024) suggested that the seeds could have been sourced from deeper water (hence, subject to the hard water effect), with broken pieces of plant moved about by wave action. Pigati et al. (2023) attempted to answer the question of age using an independent measurement of pollen grains in the sediments. Ages determined by radiocarbon analysis of the pollen ranged from 23.4 ± 2.5 ka to 22.6 ± 2.3 ka, and the OSL age was determined to be $> 21.5 \pm 1.9$ ka, supporting the claim first made by Bennett et al. (2021). Pigati et al. (2023) show that three independent measures converging on the same age range were a robust support for the "conclusion that humans were present in North America during the Last Glacial Maximum."

Fiedel (2022) has suggested that people from the north might have found a corridor to the south. He notes that a wolf clade "apparently followed this route" ending up in Wyoming, at Natural Trap Cave at ca. 26 ka BP. However, the matter is not closed. The most recent paper discussed an evaluation of the radiocarbon dates for the *Ruppia* seeds, calibrated using the modern analog of *Ruppia* seeds in a remnant of the Lake Otero wetland complex in the Tularosa Basin (Rachal et al., 2024). This study found margins of error for the age determination of ~7500 years too old, sufficient to support an estimated age for the trackway layers of 15.5–13.5 ka BP, consistent with the Clovis period and the latest survival of the megafauna in the basin. Rachel et al. took samples from the same locality as Bennett et al. (2021), which they characterize as seed balls that form in rough water conditions and are responsible for spreading the seeds around to hospitable environments. This evidence of transport, they say, negates the claim of Bennett et al. (2021) that the site was in an area materially different from the modern analog, less disturbed, shallower, with fresher water and with consequent lower hard water effects, in which the plants incorporate older carbon from the alkaline water. In a final repartee, Rachal et al. addressed the report on pollen ages (Pigati et al., 2023), countering that unlike the stable shallow shoreline depicted by Bennett et al. (2021) and by Pigati et al. (2023), they see a dynamic, unstable environment in which seeds and pollen grains could be transported to the site, redeposited and mixed, introducing discrepancies in the ^{14}C dating. This basic difference in the interpretation of the sedimentary environment is at the heart of the problem.

2.3.8.10 Jaguar Cave (Tennessee)

Jaguar Cave, Fentress County, Tennessee, formerly Wolf River Cave, contains the largest collection of preserved footprints of any US cave, with about 274 documented hominin footprints in the c. 400-meter passage called Aborigine Avenue (Robbins et al., 1981; Crothers et al., 2002; Watson et al., 2005; Willey et al., 2005). Discovered in 1976, the unshod footprints are impressed in the passageway wherever there was soft mud (see Willey et al., 2008, Fig. 1 for an on-site photograph). The best footprints exhibit clear toe prints and arches. They record the exploration of the cave on one or possibly two occasions by nine individuals (Robbins et al., 1981), sometime between 4.7-4.5 ka BP (Crother et al., 2002; Watson et al., 2005; Willey et al., 2008), the earliest record of ancient cave exploration in the Eastern Woodlands of the United States to date (Crothers et al., 2002). The people formed a mixed group of males, females, and one possible adolescent (Watson et al., 2005). Foot lengths measured from casts of nine representative footprints, made at the site, vary between 21 and 26.2 cm (Robbins, 1978, 1981: Table 1; Watson et al., 2005: Table 4). Where the footprints were not preserved, in drier sections of the cavern, the passage of the explorers was attested by numerous charcoal fragments and torch marks, samples of which were used to determine the date of passage.

2.3.8.11 Mud Glyph Cave (Tennessee)

Mud Glyph Cave, also in Fentress County, is known for hundreds of drawings on the cave walls (glyphs), and a single footprint impressed low on a cave wall, modified to a glyph of a turtle (Faulkner, 1986, 1988; Willey et al., 2008). Before its discovery, in 1979, evidence from the eastern United States indicated that caves were used for practical purposes as domiciles, mines, burials, or exploration (Faulkner et al., 1984; Simek et al., 2012). Mud Glyph Cave, however, was the first found that was clearly used for ritual or ceremonial purposes (Watson, 1986). Since that time, there have been several other caves found in the southeastern USA containing human footprints (Faulkner, 1986), but in Faulkner's words (1988), footprints are "usually obliterated by recent traffic." Although the artwork was emplaced between 1485 and 190 BP (Faulkner et al., 1984; Faulkner, 1986; Willey et al., 2008), during the Late Woodland and Mississippian periods, there is a connection to the descendants of the creators of the artwork. The footprint is found on the right wall of the cave between meter stakes 55 and 56. It is the print of a right foot of "a person negotiating the narrow passage … by pushing his foot up on the wall" (Faulkner et al., 1984). The foot is partially superimposed by a rectangular human mask. The heel has been incorporated into an image of a turtle's shell. It appears that the same artist created both the turtle and the mask at the same time with the same implement (Faulkner et al., 1984, Fig. 4).

2.3.8.12 Third Unnamed Cave (Tennessee)

Third Unnamed Cave, also in Fentress County, contains about six footprints in muddy sediment that appear to have been connected to chert mining (Franklin, 1999, 2002; Willey et al., 2008). They are located in a chamber about 60 m from the main flint mining chamber. A preliminary report was made (Fergusen, 1983) in which the footprints were noted, and it was mentioned that the cave had experienced low modern caving activity and was, therefore, relatively undisturbed. Associated with the footprints were torch marks, digging sticks, flint-knapping debris, and petroglyphs (Franklin, 1999, 2002). The footprints were thought to be from prehistoric indigenous people that had inhabited the site around 3000 years ago, during the Archaic Period. This age estimate would prove to be accurate when torch and charcoal fragments were tested later (Franklin, 1999, 2002), yielding date estimates from 4.35 to 2.01 ka BP (Simek et al., 1998; Crothers et al., 2002; Willey et al., 2005).

2.3.8.13 Xanadu Cave (Tennessee)

Xanadu Cave, also in Fentress County, about 2 km from third Unnamed Cave, was the site of exploration by Mississippian peoples (Franklin, 2002). It is a large cave with about 38 miles of passages. Reports by Sims et al. (1981) noted signs of ancient activity in Xanadu, in the form of footprints and cane charcoal, but they also noted that "heavy caving traffic" had obliterated the prints. Two unshod footprints were found by Franklin in an "upper maze passage" accessed by the Zoroaster entrance. Five charcoal fragments found in the passage were tested for radiocarbon ages, and the median age ranged from 860 to 875 BP (Franklin, 2002, p. 77). Although there were no obvious signs of ancient mining, the footprints and charcoal marks recorded the ancient exploration of the cave.

2.3.8.14 Footprint Cave (Virginia)

Footprint Cave, at Russell County, was noted in Crothers' 1997 inventory for the Virginia Department of Historic Resources (Crothers, 1997). Willey et al. (2005) summarized the findings as 30 footprints impressed in mud. The site was dated using pine torch charcoal from two torches associated with the footprints, to 370 ± 50 and 380 ± 60 cal BP (G. Crothers, personal communication, 2024).

2.4 Oceania

2.4.1 Australia

The first fossil footprints described from Australia created some controversy as they came from deep within the Pleistocene sediments of the Bridgewater Formation and were ascribed to humans (Officer, 1893; Klaatsch, 1906; Douglas, 2004). The prints were, in fact, most likely hopping traces of a kangaroo (Fig. 2.9C), but the block housing the prints (originally at the Warrnambool Museum) has been lost. A handful of sites around Australia contain human trace fossils, but they are largely restricted to Holocene sediments (Fig. 2.8). Late Pleistocene (19.3 ± 1.3 ka−19.8 ± 1.2 ka) human footprints have been described from Willandra Lakes in New South Wales (Webb et al., 2006; Webb, 2007; Westaway, 2010; Westaway et al., 2013). The site, located in the Garnpung-Leaghur interlake region, now includes over 700 prints, from at least 26 individual trackways (Westaway et al., 2013), making it one of the largest human footprint sites in the world (Fig. 2.9A and B). Prints were measured between 13 and 30 cm in length. One footprint was registered by an adult almost 2 meters tall, calculated to be running at 20 km/h. The site is of great cultural significance as it appears to depict the activities of a family group that lived in the area, including people of a range of ages, people walking, people sprinting while hunting, and even a one-legged man moving across the claypan (Webb et al., 2006; Webb, 2007, 2021; Westaway et al., 2013). The site has World Heritage status (Willandra Lakes Region World Heritage Area).

TABLE 2.4 Human tracksites in Oceania.

Site	Region	Country	Geographic coordinates	Age BP and method(s)	Substrate	Footprint preservation	# Footprints Trackways	Main references
Warrnambool	Victoria	Australia	38°23′S 142°29′E	Last interglacial (~130 ka) based on OSL dating of nearby rocks	Eolianite		Two footprints interpreted as human (kangaroo tracks)	Officer (1893), Klaatsch (1906)
Willandra lakes	New South Wales	Australia	33°29′S 143°05′E	19.8 ± 1.2 ka–19.3 ± 1.3 ka OSL	Lake hardpan, gypseous silty clay	Epichnial, good	Over 700 footprints and 26 trackways in 700 m^2 area.	Webb et al. (2006), Webb (2007, 2021), Westaway (2010), Westaway et al. (2013)
Pillie lake	Eyre Peninsula	Australia		<6–5 ka radiocarbon	carbonate			DeDekker et al. (1982)
Clare Bay (Nyeerina jinna)	Eyre Peninsula	Australia	31°52′S 132°28′E	5.47 ± 0.19 ka, radiocarbon	Marls from groundwater seepage in coastal lake	Epichnial, moderate to good	Several footprints and two trackways of adults and children	Belperio and Fotherigham (1990)
Lombardina	Broome	Australia	16°30′S, 122°53′E	1.92 ± 0.04 ka OSL	Beachrock	Epichnial, moderate	Eleven in two trackways, one overprinting the other	Welch (1999)
Motutapu (Sunde)	Auckland	New Zealand	36°45′ 06.01″S 174°54′43. 03″E	552 ± 7 to 504 ± 5 cal C14, K–Ar and thermoluminescence	Rangitoto ash layer	Epichnial, good	Human and dog footprints; tens, possible hundreds, representing at least 8 individuals	Nichol (1981, 1982)
Lake Siwi	Yasur	Vanuatu	19°31′ 06″S 169°26′ 29″E	ca. 1690 AD undated	Seasonal lake clay with desiccation cracks	Epichnial, good	Many prints from multiple individuals, including at least one child.	Firth et al. (2014)
Crab Creek	Broome	Australia	17°58′S 122°17′E	undated (7ka according to Mayor and Sarjeant, 2001)				Welch (1999), Mayor and Sarjeant (2001)

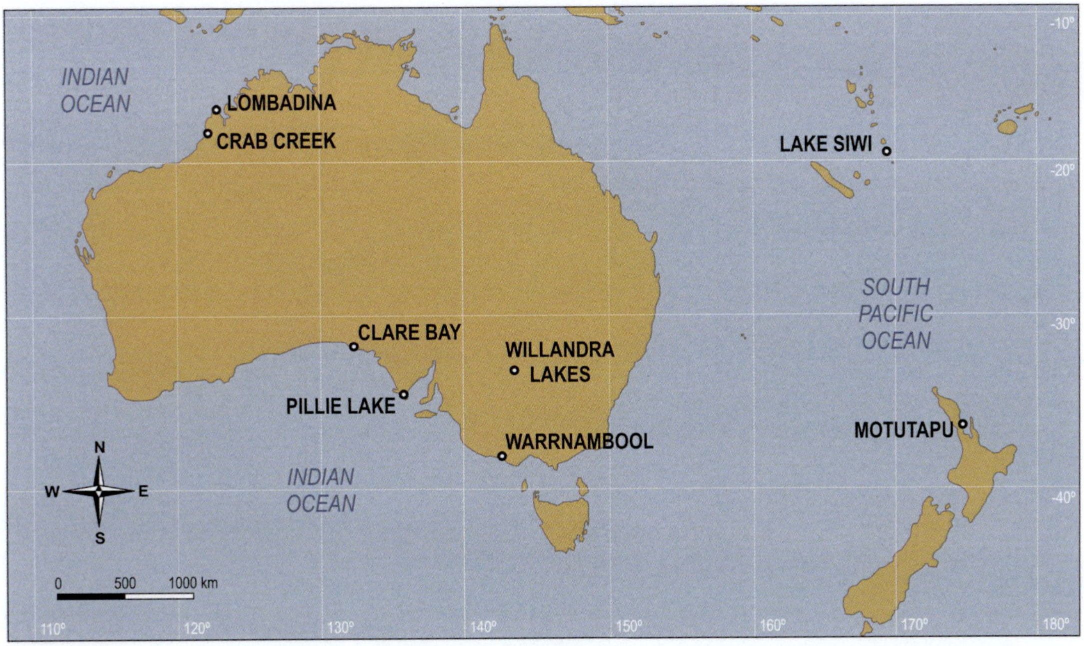

FIGURE 2.8 Map of Oceania with the location of the human-footprint ichnosites (Kau valley site in Hawaii is placed in the North America map for convenience).

In South Australia, human prints are known to occur at Pillie Lake on the southern Eyre Peninsula (5.08 ± 0.14–2.47 ± 0.1 ka, DeDekker et al., 1982) and at a small lake NW of Clare Bay on the west coast (5.47 ± 0.19 ka, e.g., Belperio and Fotheringham, 1990) referred to by the Traditional Owners as "Nyeerina jinna." The latter includes several tracks and trackways of adults and children, with footprint lengths of 20 and 12 cm, respectively. Two sites are known near Broome in Western Australia at Lombadina (1.92 ± 0.04 ka) and Crab Creek (Welch, 1999; Mayor and Sarjeant, 2001) (Fig. 2.4C).

2.4.2 *New Zealand*

2.4.2.1 Motutapu Island

Footprints have the potential to provide the first unequivocal records of human presence on a continent (e.g., Bennett et al., 2021; Pigati et al., 2023), and the only site described from New Zealand has been dated close to the initial timing of human arrival. The Sunde site on Motutapu Island, near Auckland, preserves the footprints of humans and dogs in ash from an eruption of the nearby Rangitoto Volcano (Nichol, 1981, 1982) and is dated to around 600 BP (Lindsay et al., 2011; Fig. 2.9D). The site extends across half a hectare and also includes garden beds, but, despite its cultural significance, it has not received further attention since initial excavations in the 1980s. The site is protected under the Motutapu Archeological and Historic Landscapes.

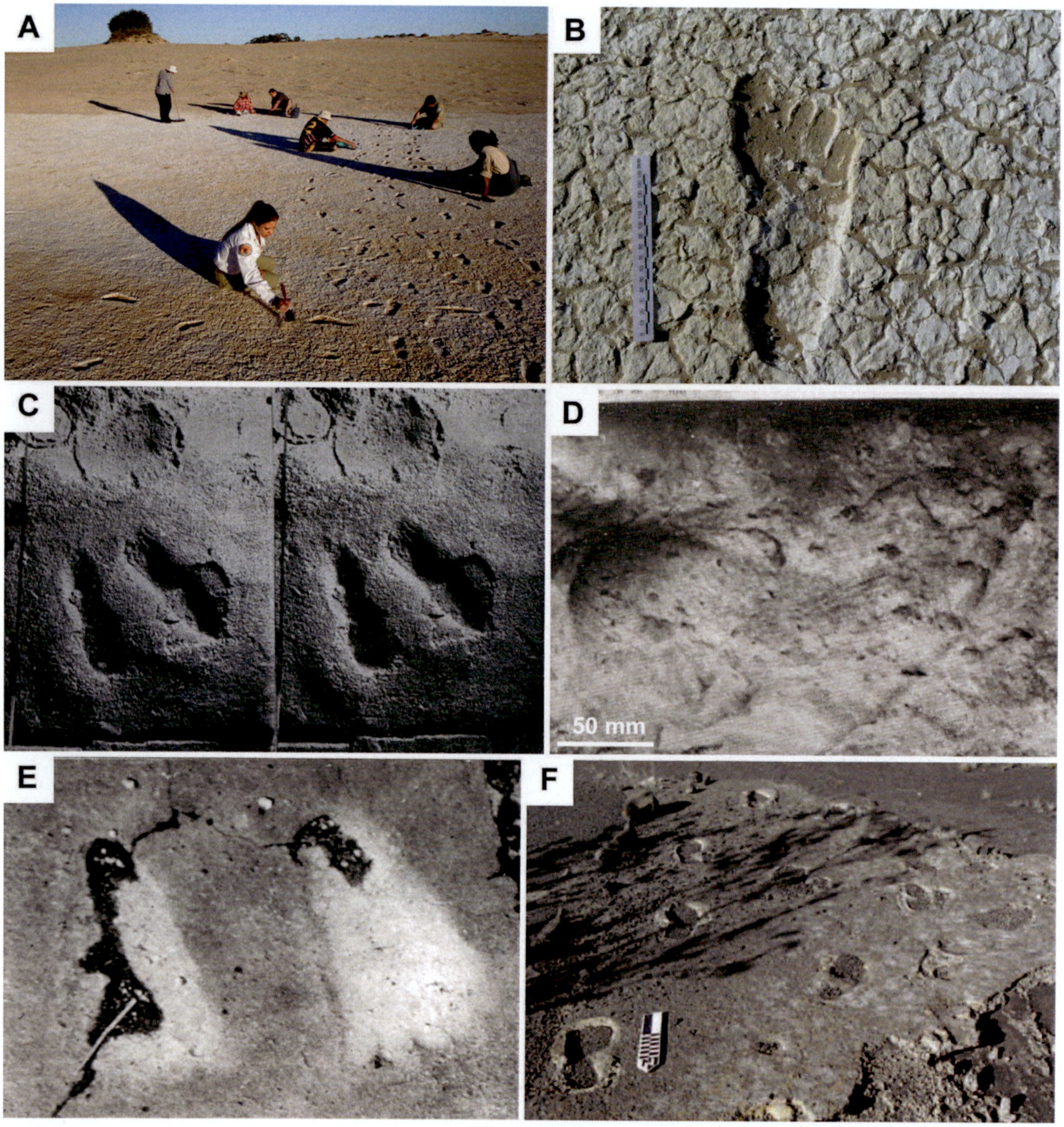

FIGURE 2.9 Examples of hominin tracksites in Oceania. (A) Part of the human trackway surface at Willandra lakes, Australia. (Photo Michael Amendolia); (B) Close-up of a footprint from Willandra lakes; (C) Purported human traces from Warrnambool, Australia, later shown to be from a kangaroo (Klaatsch, 1906); (D) Human footprint from Motutapu, New Zealand (Nicol, 1981); (E) Carved human footprints from Kiribati (Turbott, 1949); (F) Human traces from Hawaii (Nakamura, 2009).

2.4.3 *Kiribati (Gilbert Islands)*

A number of "footprints" have been recorded in the beach rock of Tarawa Atoll in Kiribati (Gilbert Islands). The largest has 12 toes and is nearly 1.5 m long, is attributed to the giant Tabuariki (Turbott, 1949), and is clearly a carving. A number of other prints of large size are also reported by Turbott (1949) from throughout the islands of Kiribati (Fig. 2.9E). While it seems likely that they were all carved by people, none have been figured, and there remains a possibility that real human footprints are preserved in the beachrock there.

2.4.4 *Vanuatu*

There is a single published record of fossil human footprints from Vanuatu (Firth et al., 2014). The tracks seem to record a group of humans, including a child, moving across a lake surface that was preserved by a tephra layer dated ca. 1690 AD. An ichnological study of the site is still missing.

2.5 Asia

2.5.1 *China*

2.5.1.1 Quesang (or Qiusang) river (Tibet)

Hot spring travertine deposits located between 4060 and 4260 m a.m.s.l. on the high Central Tibetan Plateau (Fig. 2.10) were deliberately impressed while still soft for a long period of time, up to 226 ka as claimed by Wang et al. (2023). The five handprints with outspread fingers and 17 footprints found in two panels 1.2 km distant were made by at least six different people, including 2 children, 7 and 12 years old (using modern humans growth curves). Two hands (21.2 cm long) and two feet (26–27 cm long) are larger than the average hand and foot sizes of the present Tibetan males (Zhang et al., 2003). They were considered a deliberate artistic act, the world's most ancient (Zhang et al., 2021; Chen et al., 2022 see below) and the oldest archeological site in the Plateau. Their attribution to Denisovans by these authors is remarkable. However, according to Huisheng et al. (2024), this claim ignores the high susceptibility of open-site calcite travertine to surface retreat by weathering and the likely removal of uranium by rainwater. Also, deposition of detrital thorium can occur in such porous and weathered carbonates. Both processes are known to increase significantly age estimates of reprecipitated carbonate deposits. Tang et al. (2022) have reported the presence of two Tibetan writing characters on the hardened surface of the Quesang travertine. These authors suggested that they should be less than 1300 years old, and they have presented evidence that the symbols were made before the travertine had crystallized.

2.5.1.2 Dingdingmen (Henan)

Luoyang in the Province of Henan was an international metropolis in the Silk Road during the Sui and Tang dynasties. At the Dingding Gate, an area covering over 12,500 m^2 was excavated in 2007 where human footprints together with hundreds of camel tracks and some carriage wheel tracks were found. Because Luoyang is a World Heritage Section of the Silk Road, in order to preserve this direct evidence of the use of the Silk Road from the 7th to 10th century, the Museum of the Site of the Dingding Gate was built over the site (see the chapter on

TABLE 2.5 Hominin ichnosites from Asia.

Site	Region	Country	Geographic coordinates	Age BP and method(s)	Substrate	Footprint preservation	# Footprints Trackways	Inferred trackmaker	Main references
Quesang river	Tibet	China	30°3′ N, 90°57′ E	227–169 and 20–7.4 ka 230Th/234U and TL, but see Huisheng et al. (2024). 16–7 ka (U–Th, 14C and OSL); likely <1.3 ka	Hot spring travertine	Epichnial, well preserved	Five handprints and 17 footprints deliberately impressed	Denisovans? Very likely *H. sapiens*	Zhang and Li (2002), Zhang et al. (2003, 2021), Meyer et al. (2017), Tang et al. (2022), Wang et al. (2023), Huisheng et al. (2024)
Alathar	Nefud Desert	Saudi Arabia	–	121 ± 11 to 112 ± 10 ka OSL	Sandy–silt diatomite	Epichnial, moderately preserved	Seven footprints	*H. sapiens*	Stewart et al. (2020)
Jeju island	–	Republic of Korea	33°13′01.89″ N, 126°17′ 35.83″ E	25–19 ka AMS 14C and Radiocarbon	Volcaniclastic shoreline deposit	Epichnial, poor to moderate	505 footprints, At least nine trackways	*H. sapiens*	Kim and Kim (2004, 2006), Kim et al. (2004, 2009, 2010)
Paphos	Paphos	Cyprus	34°46′29.91″N 32°24′26.16″E	Late Pleistocene	Carbonate aolianites	Cross section, no anatomical details	Several prints with a size relatable to humans	Human?	Milàn et al. (2015)
Barcın Höyük	Anatolia	Turkey	40°18′N, 29°36′E	8.4 ka cal	Wet freshly plastered floor of a building	Epichnial, well preserved	A pair of footprints of an adult male 169.9 cm tall and weighing 71.9 kg	*H. sapiens*	Atamtürk et al. (2018)

Yenikapı	Istanbul	Turkey	40°59′N, 28°56′E	ca. 8 ka	Swamp clay	Epichnial, seemly well preserved	A total of 2080 shod footprints were documented, the smallest measuring 15.9 cm from heel to toe and the largest 28.9 cm;	*H. sapiens*	Kızıltan and Polat (2013)
Demirköprü	Manisa	Turkey	38°39′N, 28°20′E	4.7 ± 0.7 ka Cosmogenic 36Cl and combined U–Pb and (U–Th)/He zircon geochronology	Hydrovolcanic (phreatomagmatic) ash layer	Epichnial, well preserved	200 footprints, at least three trackways, two of them moving in the same direction and the third in opposite direction	*H. sapiens*	Ozansoy (1968, 1969), Valladas et al. (1976)
Qatna		Syria	34°50′06″N 36°51′57″E	ca. 3.8–3.6 ka, radiocarbon	–	Epichnial, good	Trampled surface with shod footprints	*H. sapiens*	Baucon et al. (2008)
Gunma	Honshu Island	Japan	36°33′N 138°52′E	1.6–0.72 ka	Clay	Epichnial	Hundreds?	*H. sapiens*	Harada and Noto (1984), Houck et al. (2009)
Dingdingmen	Henan	China	34°37′53″N 112°28′01′E	1 ka	Soil	Epichnial	Human footprints and camel trackways	*H. sapiens*	Su et al. (2020)

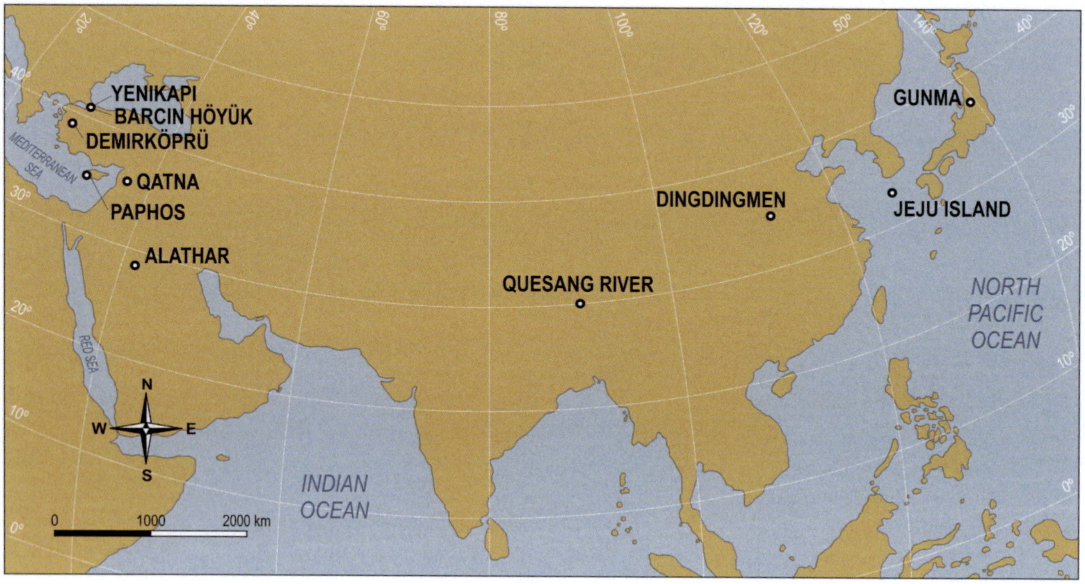

FIGURE 2.10 Map of Asia with the location of the ichnosites.

Vertebrate Ichnosites as Geological Heritage in this book). A cast of artificial soil was placed on top of the reburied trampled ground (Su et al., 2020). No detailed study of the footprints and tracks seems to have been published.

2.5.2 *Cyprus*

2.5.2.1 Paphos

In this island located geographically in western Asia, Milàn et al. (2015) described numerous vertebrate tracks in Late Pleistocene carbonate aeolianites in and around the town of Paphos, on the southwest coast. The sizes of the tracks range from 4 to 23 cm in diameter, with the majority between 5 and 15 cm in diameter. Most of the tracks are found in vertical section and are bowl-shaped, without any morphological details. The possibility of some of them being produced by *H. sapiens*, presented by the authors, is just circumstantial and not supported by morphological evidence.

2.5.3 *Japan*

Many human footprints have been recorded from more than 130 excavations of paddy fields from the northern Honshu and southern Kyushu islands, ranging from 1.9–1.8 ka BP to 1400–1600 AD (Takahashi and Okamura, 2004; Matsukawa and Shibata, 2015). Matsukawa et al. (2007) estimated population levels from the human footprint record and compared them with rice production estimates from an ancient village in Shizuoka Prefecture, as an example. They came to the conclusion that the population almost doubled from about 300–200 BCE to about 1400–1600 AD. No detailed studies of the ichnosites are published, though, except to a certain extent for Gunma.

2.5.3.1 Gunma

Hundreds of human footprints were reported from agricultural sites in Gunma Prefecture by Harada and Noto (1984). They were dated from 1.6 to 0.72 ka BP and occur in several ash layers.

2.5.4 *Republic of Korea*

2.5.4.1 Jeju Island

Animal tracks and human footprints were described in a coastal section on Sagye beach area, SW Jeju Island, by Kim and Kim (2004) and Kim et al. (2004). More than 20 stratigraphic horizons with tracks were recorded in a section 2 m in thickness. In total, 505 hominin footprints were documented in a single horizon, 12–26 cm long and 6–12 cm wide, including nine trackways (Fig. 2.11A). They were produced by at least nine humans, from children to adults. Kim et al. (2010) dated the tracksite to 25–19 ka cal BP. Furthermore, the presence of proboscidean footprints attributable to woolly mammoths in the footprint-bearing strata supports the radiocarbon dating results. However, Sohn et al. (2015) contradicted the obtained ages with a new date of ca. 3.7 ka BP for the basal part of the Songaksan Tuff. The site is now one of the geosites of the Jeju Island UNESCO Global Geopark.

2.5.5 *Saudi Arabia*

2.5.5.1 Alathar (Nefud Desert)

The oldest securely dated evidence for *Homo sapiens* in the Arabian Peninsula corresponds to seven hominin footprints found in a lake deposit of the Nefud desert (Stewart et al., 2020). The tracksite was dated between ca. 121 and 112 ka, corresponding to the Last Interglacial, and therefore, more humid climate conditions compared with that at present. The Alathar paleolake was interpreted by these authors as a stopping place to drink and forage. In fact,

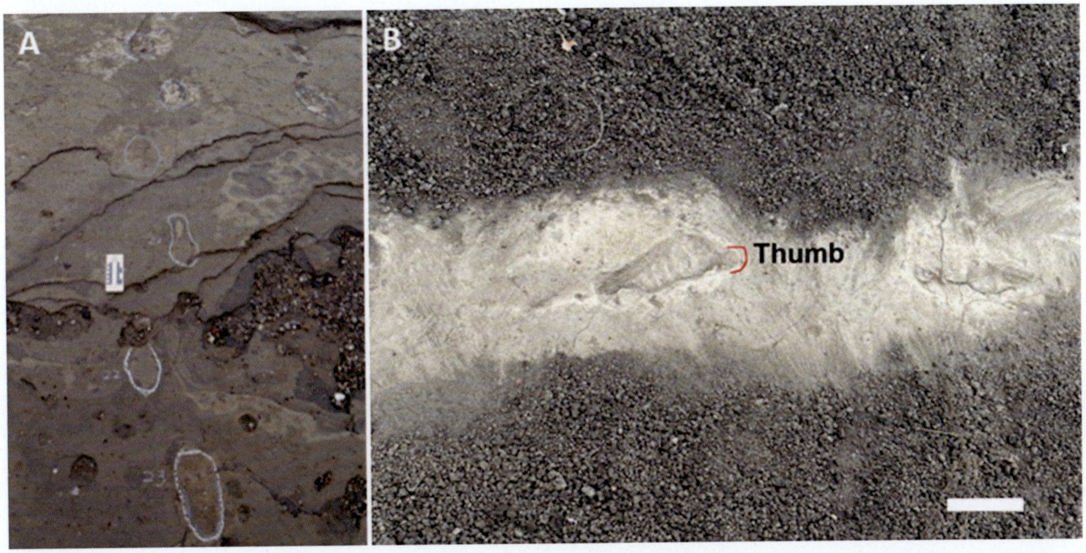

FIGURE 2.11 Examples of human tracksites in Asia. (A) Trackway in Sagye beach, Jeju island. (Credit: Jeju Special Self-Governing Province World Heritage Headquarters). (B) Exposed human trackway from Çakallar site, at the Kula-Salihli UNESCO Global Geopark, Turkey. (Credit: Serdar Aytaç).

what the footprints show is 2–3 individuals walking with similar distances from each other, in the same direction. They may place anatomically modern humans in vivo out of Africa, for the first time during the MIS 5.

2.5.6 *Syria*

2.5.6.1 Qatna

Archeological excavations conducted by an Italo-Syrian expedition at Mishrifeh, ancient Qatna (Central Western Syria), discovered a large pottery workshop on the summit of the site's acropolis. The ceramic manufacturing area reached its maximum extent during the Middle Bronze Age II (c. 3.8–3.6 ka BP) and was abandoned in the following Late Bronze Age I. On the surface of a trodden floor located between two platforms (equipped with kilns and other installations), dozens of human and equine tracks, some of the earliest evidence of the use of horses in Syria, were preserved (Baucon et al., 2008; see the following section on shod footprints). No detailed ichnological study has been performed.

2.5.7 *Turkey*

2.5.7.1 Barcın Höyük

A pair of preserved footprints was described at the site of Barcın Höyük, a Neolithic site located in northwestern Turkey (Atamtürk et al., 2018). They were dated to approximately 6.4 ka cal BC. The measurements and analyses confirm that the footprints are the bare left and right foot of a single individual, likely to be of an adult male 169.9 cm tall and weighing 71.9 kg. They were impressed within a freshly plastered floor indoors. The plaster was made of loam with straw mixed in (see the following section on footprints in human-made substrates).

2.5.7.2 Yenikapı

Excavations at the site of Yenikapı in İstanbul yielded over 2000 footprints, some barefooted and others wearing some sort of footwear (Kızıltan and Polat, 2013). Unfortunately, a dedicated study is yet to be developed.

2.5.7.3 Demirköprü (Çakallar prehistoric fossil human footprints)

In the Kula-Salihli UNESCO Global Geopark region, over 200 fossilized, well-preserved footprints belonging to at least three different individuals were found near Çakallar volcano (Ozansoy, 1968, 1969). Different foot sizes indicating multiple individuals have been identified (Valladas et al., 1976; Yalçınlar, 1987). Pace lengths between 75 and 80 cm indicate a normal walking speed rather than running (Fig. 2.11B). Small prints potentially indicate the use of staffs, and canid tracks were found alongside human footprints (Ozansoy, 1968; Yalçınlar, 1987; Akdeniz, 2011). Initially dated at ca. 250 ka (Ozansoy, 1968), zircon ages for crustal xenoliths from scoria deposits directly overlying the footprints yielded an age of 4.7 $\pm$ 0.7 ka (Ulusoy et al., 2019). The morphology of the footprints coincides with a fully modern foot (Lockley et al., 2016). Approximately 60 footprints were excavated and moved to the General Directorate of Mineral Exploration and Research Natural History Museum in Ankara. The printed surface was covered, until protection measures to preserve the remaining fragile footprints in situ can be developed (Aytaç et al., 2022). About 2 km from the ichnosite, rock art in a sheltered location between gneissic boulders was interpreted to depict the erupting Çakallar cone (Akdeniz, 2011). Three handprints without thumb and forefinger are located above the cone. Because of unequivocal evidence from the footprints that humans witnessed a volcanic eruption, this pictograph is uniquely significant among several prehistoric art pieces interpreted to illustrate an erupting volcano (Valladas et al., 1976; Yalçınlar, 1987; see the following section on Rock Art).

2.6 Europe

As with North America, Europe has a long tradition of studying hominin ichnology. It was started in caves, and only more recently, the number of open-air sites has been increasing, especially with the archeological studies developed on the coasts of the United Kingdom (Fig. 2.12 and Table 2.6). Many Late Paleolithic caves with hominin tracks have been mentioned or illustrated without detailed descriptions (see Lockley et al., 2016). Here we present just the best studied cave examples, which are mostly found in France and Italy.

2.6.1 Denmark

2.6.1.1 Fehmarn Belt Tunnel

Human footprints from the Neolithic period have been known in Denmark since the work of Broholn (1946). They were never studied in detail. Recently, the Museum of Lolland-Falster reported the finding of two short trackways associated with fishing gear and a fish trap in fjord fine deposits, during the construction of a tunnel (Cooper-White, 2014).

2.6.2 France

The study of ancient footprints in France gained momentum between the 1970s and the early 2000s, with the finding of several ichnosites in limestone caves. The oldest record is still the single footprint found in Terra Amata and attributed to *H. heidelbergensis*. The recent finding of hundreds of footprints chronologically attributed to Neanderthals highlighted the importance of the French fossil record for the study of hominin foot evolution, especially in coastal deposits where, apparently, they have been less investigated.

2.6.2.1 Terra Amata

Terra Amata is an open-air Lower Paleolithic site near the delta of the Paillon River, SE France. A single hominin footprint found in coastal deposits at Nice was attributed to *Homo heidelbergensis* (De Lumley, 1966, 1967; De Lumley et al., 2011). Meldrum (2006) suggests that the footprint lacks a modern medial longitudinal arch, like the ones from Roccamonfina in Italy. The cast of this well-preserved footprint can be found in the Museum of Natural History of Nice.

2.6.2.2 Biache-Saint-Vaast

One purported human footprint was found in an ungulate trampled surface in a river terrace, at Biache-Saint-Vaast, in northern France (Tuffreau, 1988). The "footprint" is poorly preserved and trampled without any morphological feature of a human foot. In the same layer, remains of two Neanderthal skulls were found and dated at ca. 236 ka. Although identified as "potential footprint" by Duveau et al. (2019), nothing indicates that in the photo provided in Tuffreau (1988). A cast of the trampled surface is kept at the Arkéos Archeological Museum of Douai, and a detailed analysis of the "footprint" from this cast will be required to include it in the very short list of Neanderthal tracksites.

FIGURE 2.12 Distribution of the main studied hominin ichnosites in Europe.

2.6.2.3 Le Rozel

Le Rozel on the coast of Normandy is a Neanderthal occupation site for processing prey, especially deer and horses. In total, 595 footprints and 8 handprints (Duveau, 2021) dated between ca. 86 and 75 ka (Mercier et al., 2019), associated with archeological materials and abundant faunal remains, represent by far the largest known Neanderthal ichnological assemblage to date (Duveau et al., 2019). However, 338 footprints were rejected because they were affected by taphonomic influences. In fact, before the first publications about this ichnosite, only four sites yielding a total of eight footprints had been assigned to

TABLE 2.6 Hominin ichnosites in Europe.

Site	Region	Country	Geographic coordinates	Age BP and method(s)	Substrate	Footprint preservation	# Footprints Trackways	Inferred trackmaker	Main references
Trachilos	Crete	Greece	35°30′85.7″N 23°37′66′E	6.05 Ma Magneto- and biostratigraphic data with cyclostratigraphic data based on magnetic susceptibility	Marginal marine calcarenites	Poor? Three well-preserved footprints	28 prints including 2 trackways?	Biped hominin versus habitual or nonhabitual biped hominoid?	Gierliński et al. (2017), Meldrum and Sarmiento (2018), Kirscher et al. (2021)
Happisburgh	Norwich	UK	52°49′31.10″N 1°32′06.09″E	0.9–0.85 Ma	Estuarine sand/silt sediments	Epichnial, poor	152 footprints with possible trackways. Only one footprint presents clear toe impressions	*H. antecessor*	Ashton et al. (2014), Ashton (2021)
Terra Amata	Nice	France	43°41′52″ N 7°17′22″E	ca. 380 ka (although Villa (1983), claims an age of ca. 230 ka	Beach sand	Epichnial, moderate	A single footprint	*H. heidelbergensis*	De Lumley (1966, 1967), Meldrum (2006, 2011)
Ciampate del Diavolo	Caserta Province	Italy	41°19′57.07″N 14°01′29.50″E	349 ± 3 ka 40Ar/39Ar	Ignimbrite from Roccamonfina volcano	Epichnial, poor	81 footprints and four trackways	Hominin indet,	Mietto et al. (2003), Avanzini et al. (2008), Panarello et al. (2017a,b, 2018, 2020)
Schöningen	Lower Saxony	Germany	52°07′N 10°59′E	ca. 300 ka	Lacustrine mudstone	Epichnial, poor	3 potential footprints	*H. heidelbergensis*	Altamura et al. (2023)

(*Continued*)

TABLE 2.6 Hominin ichnosites in Europe.—cont'd

Site	Region	Country	Geographic coordinates	Age BP and method(s)	Substrate	Footprint preservation	# Footprints Trackways	Inferred trackmaker	Main references
Matalascañas trampled surface	Huelva Province	Spain	37°01′03.18″N 6°35′11.96″W	295.8 ± 17 ka (Mayoral et al., 2022) 151.1 ± 11 ka (Neto de Carvalho et al., 2023) OSL	Paleosol in a coastal lacustrine-eolianite succession	Epichnial, poor-to-moderate	Several footprints and few short trackways	*H. neanderthalensis*	Mayoral et al. (2021, 2022)
Catalán Bay	–	Gibraltar	36°08′14.07″N 5°20′29.44″W	28.45 ± 3.01 ka OSL (Muñiz et al., 2019) corrected to 140.1 ± 9.3 ka OSL (Finlayson et al., under publ.	Coastal eolianite	Hyporelief, poor	A single footprint	*H. neanderthalensis*	Muñiz et al. (2019)
Theopetra cave	Thessaly	Greece	39°40′56.968″N 21°40′54.876″E	46.3 ± 1.6 ka radiocarbon 150 ± 15 ka and 57 ± 6 TL of burnt flint	Cave mud	Epichnial, poor	4 footprints	Possibly *H. neanderthalensis* although *H. sapiens* is not discarded	Manolis et al. (2000), Facorellis et al. (2001), Valladas et al. (2007), Kyparissi-Apostolika and Manolis (2021)
Cape Trafalgar	Cadiz	Spain	36°10′54.64″N 6°02′05.00″W	99 ± 12 ka OSL	Coastal eolinite	Cross section, poor	Purported single footprint	*H. neanderthalensis*	Neto de Carvalho et al. (2022)
Le Rozel	Normandy	France	49°28′20.92″N 1°50′25.58″W	ca. 86–75 ka OSL	Coastal eolianite sandy mud	Epichnial, poor to moderate	595 footprints, only 12 well-preserved; five trackways with two to three footprints; eight handprints	*H. neanderthalensis*	Duveau et al. (2019), Mercier et al. (2019), Duveau (2021), Duveau et al. (2021)

Vârtop cave	Alba	Romania	46°32′26″N 22°46′50″E	67.8 to 62 ka U-series	Cave moonmilk	Epichnial, only one footprint well preserved	Three footprints	*H. neanderthalensis*	Onac et al. (2005), Onac et al. (2021)
Ciur-Izbuc cave	Transylvania	Romania	46°50′58.43″N 22°23′31.31″E	ca. 36.9–29 ka AMS 14C of bear bones	Cave mud	Epichnial, moderate to good	400 footprints, originally good	*H. sapiens*	Rusu et al. (1969), Rişcuţia and Rişcuţia (1970), Webb et al. (2014), Harvati and Roksandic (2016), Onac et al. (2021)
Chauvet cave	Ardèche	France	44°20′46″ N 4°30′22″E	37–28 ka radiocarbon	Cave mud	Epichnial	20 footprints in one trackway	*H. sapiens*	Chauvet et al. (1996), Harrington (1999), Clottes (2003), Garcia (2005), Quiles et al. (2016)
Cussac cave	Dordogne	France	45°42′N0°51′E	31–28 ka cal radiocarbon	Cave mud	Epichnial, good?	28 footprints, four of them forming a trackway	*H. sapiens*	Ledoux et al. (2017, 2021a,b)
Ariège cave	Occitanie	France	42°58′N1°25′E	30–20 ka	Cave mud	Epichnial	Three trackways	*H. sapiens*	Bahn and Vertut (1988), Garcia (1999, 2001)
Casetta Mistici and Tor Vergata	Rome	Italy	41°50′N 12°37′E	23.734–23.228 ka BC AMS14C	Lahar	Epichnial, poor-to-moderate	Two isolated footprints in Casetta Mistici and one trackway made of 13 footprints, plus two isolated prints	*H. sapiens*	Cerilli et al. (2020)

(Continued)

TABLE 2.6 Hominin ichnosites in Europe.—cont'd

Site	Region	Country	Geographic coordinates	Age BP and method(s)	Substrate	Footprint preservation	# Footprints Trackways	Inferred trackmaker	Main references
Pech-Merle cave	Lot	France	44°30′27″N1°38′39″E	25–15 ka? radiocarbon	Cave clay	Epichnial, poor to good	18 footprints	*H. sapiens*	Duday and Garcia (1983), Pastoors et al. (2015, 2017)
Lascaux cave	Dordogne	France	44°56′36.80″N 1°2′31.60″E	ca. 17 ka radiocarbon	Cave clay	Epichnial	Footprints made by adolescents	*H. sapiens*	Barriere and Sahly (1964)
Tuc d'Audubert cave	Ariège	France	43°01′59″N1°12′13″E	17.2 to 16.5 ka cal radiocarbon	Cave clay	Epichnial, good	Several trackways and isolated footprints	*H. sapiens*	Vallois (1927), Bégouën et al. (2009), Pastoors et al. (2015)
Ojo Guareña cave	Castile and Leon	Spain	43°02′01′N3°39′50″W	15.6 ka and 4.3 ka cal radiocarbon	Cave sand	Epichnial, good	Over 1200 footprints including trackways and isolated footprints	*H. sapiens*	Marcos (2001), Ortega et al. (2021)
Fontanet cave	Ariège	France	42°48′44″N 1°38′44″E	15.4–15.1 ka radiocarbon	Cave clay	Epichnial	Over 80 footprints, handprints and knee prints; 8 trackways	*H. sapiens*	Clottes (1975), Pastoors and Lenssen-Erz (2021)
Grotta della Bàsura	Imperia, La Spezia e Savona Province	Italy	44°08′08.6″N 8°12′00.28″E	14.5 ± 0.42 ka cal radiocarbon	Cave deposits	Epichnial, moderate to good	Over 100 traces including, footprints, handprints, and heel prints	*H. sapiens*	Pales (1960), Citton et al. (2017), Avanzini et al. (2018, 2021), Romano et al. (2019)

Niaux cave	Ariège	France	42°49′17″N1°35′38″E	12.9–12.4 ka radiocarbon	Cave clay	Epichnial, good	About 40 footprints	*H. sapiens*	Cartailhac and Breuil (1907), Malvesin-Fabre and Robert (1953), Clottes and Simonnet (1972), Pales (1976), Garcia et al. (1990), Clottes et al. (1992)
Formby point	Liverpool	UK	53°33′50″N 3°06′00″W	>9 ka to 1.2 ka radiocarbon OSL	Silty-sands	Epichnial, good	145 trackways	*H. sapiens*	Cowell et al. (1993), Roberts et al. (1996), Huddart et al. (1999b), Roberts (2004, 2009, 2014), Burns (2021), Burns et al. (2022), Wiseman et al. (2022)
Aldène cave	Hérault	France	43°21′17″N 2°41′56″E	8.9 ± 0.013 −7.79 ± 0.00 ka radiocarbon	Cave mud	Epichnial	400 footprints, main trackway 30 m long	*H. sapiens*	Casteret (1948), Ambert et al. (2000), Galant et al. (2021)
East Goldcliff	Wales	UK	51°32′N2°53′W	7.5–7.2 ka cal radiocarbon	Well-laminated siltstone	Epichnial, moderate to good	270 footprints	*H. sapiens*	Bell et al. (2000), Scales (2002, 2004, 2007), Bell (2007a,b), Barr and Bell (2016), Barr (2018), Bell (2020)
Hauxley Low	Northern England	UK	55°19′N 1°32′W	7.3–7.2 ka cal; 7.2–7 ka cal radiocarbon	Peat	Epichnial, Apparently poor (interpreted as shod)	90 footprints with some short trackways	*H. sapiens*	Eadie and Waddington (2013), Waddington (2014)

(*Continued*)

TABLE 2.6 Hominin ichnosites in Europe.—cont'd

Site	Region	Country	Geographic coordinates	Age BP and method(s)	Substrate	Footprint preservation	# Footprints Trackways	Inferred trackmaker	Main references
Fehmarn Belt tunnel	Lolland island	Denmark	54°35′N11°17′E	7–4 ka?	Siltstone?	Epichnial, poor to moderate	Two short trackways	*H. sapiens*	Cooper-White (2014)
Uskmouth	Wales	UK	51°32′N2°58′W	6.18 ± 0.08 ka AMS	Well-laminated siltstone	Epichnial, poor to moderate	Three trackways	*H. sapiens*	Aldhouse-Green et al. (1992), Aldhouse-Green and Housley (1993)
Kenfig	Wales	UK	51°30′N 3°43′W	5.11 ± 0.05 −3.81 ± 0.04 ka radiocarbon	Peat	Epichnial		*H. sapiens*	Doyle (2007), Bennet et al. (2010a)
St. Ouen's Bay	Jersey Island	UK	49°13′41″N 2°14′03″W	6–4 ka? undetermined	Peat	Epichnial, moderate	At least one footprint	*H. sapiens*	Patton (1993)
Clairvaux	Jura	France	46°34′N 5°45′E	6–4 ka?	Lacustrine fine sediments			*H. sapiens*	Pétrequin and Pétrequin (1988)
Magor Pill	Wales	UK	51°33′N 2°48′W	5.72 ± 0.08 ka radiocarbon	Mudflat silt	Epichnial, good		*H. sapiens*	Aldhouse-Green et al. (1992)
Zeewijk	–	Netherlands	52°27′N 4°35′E	4.6–4.45 ka cal radiocarbon	–	Epichnial	Eight footprints	*H. sapiens*	Theunissen et al. (2014)
Foissac cave	Aveyron	France	44°30′09″N 2°00′24″E	4.6–4.2 ka cal radiocarbon	Cave mud	Epichnial, poor	Four footprints	*H. sapiens*	Garcia and Duday (1983)
Lydstep Haven	South Wales	UK	51°39′11″N 4°45′28″W	4.345–3.95 ka cal BC radiocarbon	Peat	Epichnial, poorly preserved	Trampled surface with adult and children-sized footprints	*H. sapiens*	Murphy et al. (2014)

Afragola, Nola and Palma Campania	Campania	Italy	40°55′N 14°18′E; 40°55′N 14°31′E; 40°52′N14°33′E	ca. 3.78 ka BP radiocarbon	Surge ash deposit	Epichnial, moderate	Thousands of footprints and several trackways	*H. sapiens*	Mastrolorenzo et al. (2006), Di Vito et al. (2019), Giampaola et al. (2007), Laforgia et al. (2009)
Redwick	Wales	UK	51°32′N 2°50′W	3.3.-2.9 ka cal radiocarbon	Coastal fine sediments	Epichnial	Footprints of adults and children	*H. sapiens*	Bell (2013)
Saint-Jean-le-Thomas		France	48°43′N1°30′W	Neolithic-Bronze age transition undetermined				*H. sapiens*	Billard and Bernard (2016)
Dovey estuary	Wales	UK	52°31′N 4°03′W	3–2 ka cal radiocarbon	Peat	Epichnial		*H. sapiens*	
Aosta	Aosta Valley	Italy	45°44′28″N 7°18′54″E	Iron Age 2.5–2.1 ka stratigraphy	–	Epichnia, moderately preserved?	Several footprints including trackways	*H. sapiens*	Avanzini et al. (2020)
Moregine Pompei	Campania	Italy	40°44′N 14°29′E	79 AD	Ash	Epichnia	Three footprints	*H. sapiens*	Avanzini et al. (2020)
Island of Rhodes	–	Greece	36°04′39.15″N 28°02′17.04″E	45 BC-430 AD AMS	Beachrock deposit	Epichnia, poor	One isolated footprint and a trackway of three prints	*H. sapiens*	Bromley et al. (2009)
Rawthey cave	Cumbria	UK	54°19′N 2° 31′W	815 ± 40–605 ± 40 radiocarbon	Clay	Epichnia	Two footprints	*H. sapiens*	Chamberlain et al. (1998)
Cueva Tempranas	Llanes	Spain	43°25′N 4°51′W	Undated	Clay	Epichnia, poor-to-well(?)	Trampled surface with a small trackway	*H. sapiens*?	Noval Fonseca (2007)
Fenland		UK	52°34′N 0°06′08″E	Undated				*H. sapiens*	Pryor (1998)

Neanderthals. According to Duveau et al. (2019), this ichnoassemblage provides direct evidence for the size and composition of a Neanderthal social group. Footprints were made by a small group comprising different age classes, from one year to adults, with a majority of children. The footprints are distributed by six stratigraphic subunits, considered each of them as a single brief occupation event (Duveau et al., 2019). In total, 88 footprints considered longitudinally complete were analyzed from the densest stratigraphic subunit (D3b-4), which extends over 92 m^2. The footprint lengths range from 11.4 to 28.7 cm, allowing an estimate of trackmaker heights of between 66.9 and 179.9–190.3 cm. A minimum of four individuals, more likely 10–14 individuals, constituted a social group mainly composed of children (51.4% of the total footprints; Duveau, 2021). Despite Duveau et al. (2019) concluding that Le Rozel, together with El Sidrón skeletons in Spain, is the only site providing reliable information of the composition of Neanderthal social groups, Duveau (2021) remarked that is impossible to know whether the D3b-4 track-making group is representative or not of the whole social group.

Only a subset of 12 best-preserved footprints were used in geometric morphometric analysis, leading Duveau et al. (2019) to conclude that the Le Rozel footprints were closer to the footprints of anatomical modern humans, although wider, especially at the midfoot, suggesting a less gracile morphology and a less pronounced plantar vault. The authors did not take into the account the morphological variability resulting from trampling in sand under different levels of moisture. Moreover, the small number of footprints used limited interpretations. Other studies show that the Neanderthal footprint index can be within the variability range of anatomical modern humans (Neto de Carvalho et al., 2023).

The Le Rozel sample was likely derived from a mixed-age Neanderthal group. Ruff et al. (2021) revised body mass estimates from Duveau et al. (2019) by taking into account BMI, suggesting that this group included more adults than initially proposed and highlighting the difficulties in differentiating small adult from adolescent body sizes.

Casts of 65 footprints were made, and 62 original footprints were hardened and directly extracted to be stored in the Direction Régionale des Affaires Culturelles at Caen.

2.6.2.4 Cussac Cave

In the Cussac Cave, Ledoux et al. (2017, 2021a) described 28 footprints, with footprint lengths ranging from 19 to 30 cm, and some handprints; 4 footprints form a trackway. In some cases, the hallux is clear, although no other toe has left an impression. Therefore, these footprints were interpreted as representing shod feet (see the following section on the record of shod humans).

2.6.2.5 Chauvet Cave

Exploration of the Chauvet Cave near Vallon-Pont-d'Arc in southern France has yielded some of the oldest footprints of anatomically modern humans in Europe. A single trackway of a young child with an estimated age of 8 years extends for about 50 m across the cave floor. The footprints lead to the so-called Room of Skulls, where a number of bear skulls have been found (Harrington, 1999). A canid trackway seems to be contemporaneous with the child's trackway. However, a combined statistical and geometric-morphometric approach to separate wolf from some dog prints (Ledoux and Boudadi-Maligne, 2015) casts substantial doubt on the identification of Chauvet canid tracks as belonging to an extremely early example of a domesticated dog. Moreover, other canid tracks of different-sized animals do exist at Chauvet. Tracks of different bear species were also recognized by Garcia (2005). Most interesting is also the presence of Ibex tracks in the deepest section of the cave described by this author, and why Ibex were deep in a cave is so far unknown.

Chauvet Cave is a UNESCO Word Heritage Site and includes some of the oldest and best-preserved animal paintings and hand stencils in the world, dated to the Aurignacian. This cave was never open to the public.

2.6.2.6 Tuc d'Audoubert Cave

Exceptionally well-preserved trackways and isolated footprints are attributed to juvenile humans (Bégouën et al., 2009; Pastoors et al., 2015). Over 300 prints are known in the upper gallery of the cave (Pastoors et al., 2021). They represent a maximum of five visits carried out by two to six individuals according to the interpretation of indigenous trackers. The remarkable distribution of heel imprints at the "Salle des Talons" indicates activities around the extraction of clay and the making of drawings in the cave floor (Bégouën et al., 2009). In total, 183 imprints from five individuals were counted, and it is still an open question as to why no complete footprint is present in the "Salle des Talons."

2.6.2.7 Niaux Cave

One of the earliest detailed descriptions of well-preserved hominin footprints (Cartailhac and Breuil, 1907), the Niaux caves have been described and illustrated in detail by Pales (1976) and apparently include patterned arrangements of footprints that have been interpreted as deliberate activity of children at play (Lockley and Meyer 2000). In total, 38 footprints were identified produced by youngsters 9–12 years in age (Pales, 1976). The indigenous trackers identified a girl (7–13 years) as their sole producer (Pastoors et al., 2015).

2.6.2.8 Aldène Cave

In the Aldène Cave, five sectors show over 400 human footprints, and among them the main trackway extends over about 30 m (Galant et al., 2021). About 20 people visited the cave in a group composed mostly of children, with only a few adults. For these footprints, an age of ca. 8 ka was assigned by Ambert et al. (2000, 2007), and they are associated with hyena tracks and cave bear nests, but the behavior of the trackmakers, in relation to these animals, if any, is unclear. Also, if proved to be hyena tracks, they represent a much earlier occupation of the cave by these animals, still during the Pleistocene, than the human footprint record. In connection with the footprints, there are traces of torches on the walls and remains found on the floor (Galant et al., 2021). This contemporaneity resulted from a single visit. The site is still under study.

2.6.2.9 Ariège Cave

At Ariège Cave, three trackways of children were recorded (Bahn and Vertut, 1988). The footprints extend for about 50 m across the cave floor and may date to between 30 and 20 ka (García 1999, 2001).

2.6.2.10 Fontanet Cave

In Fontanet Cave, footprints suggest that a child may have followed a puppy or fox into a cave, but this inference can hardly be confirmed or denied on the basis of available analysis (Bahn and Vertut, 1988; Lockley et al., 2016). Besides footprints, there are some handprints and a large number of slipping marks. The number of producer individuals are two to six, including children. The group walked together, sometimes fast, slipping and kneeling (Pastoors and Lenssen-Erz, 2021).

2.6.2.11 Lascaux Cave

Tracks have been reported from the famous Lascaux Cave (Barriere and Sahly, 1964). The footprints were made by adolescents. No detailed study was provided, though.

2.6.2.12 Pech-Merle Cave

About 12 footprints were described in Pech-Merle Cave by Duday and Garcia (1983) and Pastoors et al. (2015, 2017). It includes traces that suggest an individual with a walking stick. The indigenous trackers identified five different producers among adults and youngsters (Pastoors et al., 2017).

2.6.2.13 Clairvaux

Human footprints were briefly mentioned at the lacustrine setting of Clairvaux (Pétrequin and Pétrequin, 1988). They are associated with a Neolithic settlement dated at 6–4 ka BP.

2.6.2.14 Saint-Jean-le-Thomas

The size of human footprints encountered at the fish-weir site of Saint-Jean-le-Thomas on the French coast—dating from the Neolithic to Bronze Age transition—shows the presence of children and adults (Billard and Bernard, 2016).

2.6.2.15 Foissac Cave

The cave of Foissac provided a brief description of one footprint belonging to a youngster no more than 10 years in age (Garcia and Duday, 1983).

2.6.3 Germany

2.6.3.1 Schöningen

The ca. 300 ka Paleolithic sites of Schöningen in northern Germany yielded several localities with archeological and paleontological remains representing a rich interglacial lacustrine paleoenvironmental record of the late Middle Pleistocene. Recently, Altamura et al. (2023) presented the first ichnological study, which provides insights on Schöningen's paleoenvironment and a snapshot of the mammals once living in the area. Herds of elephants and other species of herbivores congregated along the muddy shores of a paleolake (see chapter in this volume on Quaternary Vertebrate Tracks and Trackways). In addition, the authors described three potential hominin footprints, probably belonging to a late *Homo heidelbergensis*, which were found at Schöningen 13 II-2 Untere Berme. Two of the three human footprints at Schöningen were attributed to young individuals.

2.6.4 Gibraltar

2.6.4.1 Catalán Bay

"Catalán Bay Sands" is the largest accumulation of Middle-to-Late Pleistocene eolian deposits of Gibraltar, with its highest deposits reaching 300 m a.s.l. in the eastern side of "the Rock." The morphology and structures of the "Catalan Bay Sands" point to their genesis due to the accumulation of cliff-foot and rampant dunes developed against the steep Jurassic limestone slopes of the Rock. These dunes and the studied tracks were therefore generated from MIS 6 to probably MIS 2, between ca. 140 ka (according to new unpublished OSL dating) and after 30 ka. A single hominin footprint showing typical five-toe morphology (with the D1 parallel to the longitudinal axis of the track), and wide forefoot morphology (Fig. 2.13A), was

FIGURE 2.13 Examples of hominin tracksites from the Iberian Peninsula. (A) 3D digital model of the footprint from Catalán Bay, Gibraltar; sole bed and cross-sectional views. Despite the poor preservation, the model enabled to complete the hallux and locate digits 2 and 5, besides calculating the maximum depths at the heel and the hallux in a typical transfer of the foot pressure from touch-down through kick-off phase downslope. (B) Short Neanderthal trackway together with red deer tracks at the Matalascañas Trampled surface, Spain. (C) Detail of footprint 3 in B showing the track of a deer intersecting the human footprint. (D) Right footprint with the mark of the hallux produced by a 1–2 year old child on a very deformable sediment. Scales in cm.

described by Muñiz et al. (2019). The track registers the touch-down through kick-off phase, resulting in a short medial arch. The deeper impression of the heel, poorly preserved D2–D5 prints (Fig. 2.13A) together with the development of an expulsion rim in the toe region, and an abnormally extended D1 are evidence for dragging through the sediment during the forward swing of the foot while descending down the dune. This movement caused displacement on the downslope side of the track as well as some slippage of the plantar surface as shown by the convex outer edge. Toe traces are in this circumstance not clearly visible (Kim et al., 2008). Moreover, the footprint appears to have been made in a soft sandy substrate, which rarely preserves digit casts (Helm et al., 2018). The first toe is slightly abducted, reflecting the forced separation of the toes by sediment both during indenting and withdrawal (Bennett and Morse, 2014). It was considered to be produced by a Neanderthal infant, and new OSL dating of the footprint level places it clearly in the late Middle Pleistocene (MIS 6).

2.6.5 Greece

2.6.5.1 Crete

A remarkable site was found at the Trachilos beach, Kissamos municipality in western Crete. In a marginal marine succession of Messinian age belonging to the Roka Formation, Gierliński et al. (2017) described in an area of less than 4 m^2, more than 50 footprints that according to the authors presented hominin-like characteristics. Individual trackways were difficult to discern due to the density of prints in the trampled surface. However, the majority of footprints and two claimed narrow-gauge trackways showed an orientation toward the SW. Footprints were between 94 and more than 223 mm in length. The best footprints have clear displacement rims and sharply defined edges. Only three, especially well-preserved prints provided morphological information. There is no gap between the large, robust hallux and the other alongside, slender digits. Digits II–V become progressively shorter. A large, rounded forefoot ball impression is also visible. Morphometric analysis provided by Gierliński et al. (2017) shows that the footprints have outlines that are distinct from modern nonhominin primates and resemble those of hominins. Nevertheless, the morphology of the sole print is not hominin-like, being proportionally shorter with a narrow tapering heel, and lack of a longitudinal medial arch. Gierliński et al. (2017) attributed these footprints to a small, habitually bipedal hominin with hominin-like pedal digits and ball combined with an ape-like sole lacking a bulbous heel, which would bring substantial implications for early hominin biogeography, as well as for the development of bipedality. Footprints may be from a bipedal hominin (Gierlinski et al., 2017; Kirscher et al., 2021) or a bipedal hominoid (Crompton, 2017).

Trachilos is remarkable and controversial in the ichnological study of bipedalism because it is 2.39 Ma older than the hominin trackways from Laetoli. Meldrum and Sarmiento (2018) claim the poor preservation of the footprints and suggested that they may have been made at best by a nonhominin primate with an adducted hallux and do not provide evidence of habitual bipedality. A new attention to the tracksite, especially the examination in place of the trackways identified by Gierliński et al. (2017), was performed by one of the authors of this chapter (CNC). During the geoconservation assessment made by the University of Crete in May 2024, the tracksite was uncovered (see the chapter on Vertebrate Ichnosites as Geological Heritage in this book), enabling to analyze the two track-bearing levels. In level B2, bird footprints were removed together with nine footprints that were stolen by a local (B. Fassoulas, pers. comm.). One of the examined "trackways" (Gierlinski et al., 2017, Fig. 7B) is composed of three footprints produced by left feet with different lengths. The other "trackway" was impossible to identify. A "pair" of footprints interpreted by the same authors as a "stationary individual" show lengths of 84 and 104 mm, and were more likely made by two producers with different sizes. Nevertheless, they are bona fide footprints with good preservation, despite being mostly infilled with sediment from the overlying bed. They may have been produced by a small primate, but to prove that the footprints were made by a habitually bipedal, distinct trackways are required. New prospections in the same unit are underway, and the reexcavation of the site in the near future may provide clarity about this unique tracksite.

2.6.5.2 Theopetra Cave

Theopetra Cave at Meteora includes four footprints organized in a "trackway" but showing two different, small producers, estimated from footprint length to be between 3 and 4 years old and 98.5 and 106 cm in height (Manolis et al., 2000; Kyparissi-Apostolika and Manolis, 2021). These footprints, including others that still require study, are associated with Levallois industry in the same layer and were redated by Valladas et al. (2007) to ca. 130 ka. In fact, thermoluminescence dating of burnt Mousterian flint specimens provided an age range for the stratigraphic sequence of the cave of between 150 ± 15 ka and 57 ± 6 ka, although there was evidence of postdepositional mixing. Despite the age and the Mousterian industry, Kyparissi-Apostolika and Manolis (2021) do not dismiss the possibility of the footprints having been produced by *H. sapiens*. Distinguishing between the footprints of anatomically modern humans and the track record of Neanderthals on the basis of footprint morphology is still not reliable (Lockley et al., 2016) despite a few claims to the contrary (Onac et al., 2005; Duveau et al., 2019).

2.6.5.3 Island of Rhodes

Bromley et al. (2009) described three consecutive footprints in sandy beachrock dating from Roman times, on the east coast of the island of Rhodes. There is the preservational possibility that they were shod footprints (see the following section on shod humans).

2.6.6 *Italy*

The ichnological record of hominin traces from Italy is somewhat diversified. In recent years, the development and dissemination of methodologies and technological applications have facilitated the reanalysis of this record, enabling deeper interpretations favored by the integration of external data, both geological and paleontological (Avanzini et al., 2020).

2.6.6.1 Ciampate del Diavolo

The oldest evidence of hominin footprints from Italy is represented by the well-known 'Devil's Trails' ichnosite at Tora e Piccilli ("Foresta" area, Roccamonfina volcano, Caserta province, southern Italy). Hominin trackways were described from the mid Pleistocene Roccamonfina volcano site, dated at 349 ± 3 ka (Mietto et al., 2003; Scaillet et al., 2008; Avanzini et al., 2008; Panarello et al., 2017b, 2020). The hominin and animal tracks are preserved in an area of 2000 m^2. The first ichnological survey in the "Devil's Trails" ichnosite (Avanzini et al., 2008) showed the existence of at least three main trackways, which were imprinted on a zeolite-rich deposit produced by a pyroclastic flow (Mietto et al., 2003). However, a new trackway and two other possible directions of walking have been recently described, raising the number of human trackways at least to four (Panarello et al., 2020). Moreover, some lithic tools have been found within the site and in its close surroundings (Panarello et al., 2017b, 2020). One trackway of 27 footprints shows a zig-zag pattern descending a steep (30–80 degrees) slope formed by the surface of a pyroclastic flow. Another trackway of 19 prints follows a gently curved course and shows slipping traces very likely indicating that the individual descended the slope in a less deliberate manner (Avanzini et al., 2008). The trackways are unusual because the three segments reveal individual descending steep slopes, occasionally slipping and leaving handprints in the volcanic ash substrate. Oblique step lengths of 600 mm and stride lengths of 1.2 m are typical of this site.

Panarello et al. (2018) reanalyzed the handprint by means of high-resolution digital photogrammetry. The handprint was interpreted as fully compatible with those characterizing actual human hands. Based on general morphology and print length and width, the material corresponds well to the variability detected in anthropometric values for both Pleistocene hominins and present-day humans and correlates very well with previous stature estimates (Avanzini et al., 2008; Panarello et al., 2018). Details of foot morphology are not preserved. The footprint E3 (Trackway E) is almost completely preserved and has a total length of 27 cm (Panarello et al., 2020). However, most footprints lack clear toe impressions, even though faint traces of the hallux are recognized. They are about 20 cm long and 10 cm wide, suggesting adult individuals no taller than 1.56 m, walking at an average speed of 1.09 m/s (based on the method of Alexander, 1984). The footprints of Trackway E seem left by a taller individual (Panarello et al., 2020), around 174.2 cm in height.

Panarello et al. (2017a) described additionally the P1-Pathway with a total length of 53.19 m and showing a total width ranging from 1.40 m to a maximum of 3.21 m. The authors recognized on the surface several superimposed human traces, some of which were referable to the footprints forming trackways A and B, followed by a postlithification smoothing of the traces linked by Panarello et al. (2017a) to the passage of both humans and animals and, in historical times, to the quarrying activities and frequent foot traffic across the site. Therefore, the authors considered the P1-Pathway as the oldest known human "pathway" in the world.

2.6.6.2 Casetta Mistici and San Gaudenzio-Tor Vergata

Recently, Cerilli et al. (2020) briefly described two ichnosites in Rome that were detected during construction works. In a lahar deposit dated 23.734–23.228 ka BP, two isolated, poorly preserved adult footprints were found in Casetta Mistici, and a trackway made of 13 poor-to-moderately preserved footprints of an adult walking slowly, plus two isolated prints, were described at the university campus of Tor Vergata.

2.6.6.3 Grotta della Bàsura

The Bàsura Cave is located at Toirano, western Liguria (Northern Italy). Initially found by Chiapella (1952) and described by Pales (1954) and Tongiorgi and Lamboglia (1954), the first ichnological study was performed by Pales (1960) on 13 plaster casts of the best preserved footprints, identifying the putative producers as "Neanderthal-type" humans. Blanc (1960) inferred that some individuals, including juveniles, traveled into the cave using the heels. He inferred evidence of putative prehistoric ritual activity (Blanc, 1960). Such interpretation was also supported by the presence in the room of a stalagmite, defined by the author as an "acephalous sphinx" or a "zoomorphic stalagmite," preserving a surface of great interest for intentional finger fluting-drawing (see the following section on Rock Art) attributable to several individuals. Recent surveys in the inner rooms of the cave recognized 107 human traces represented by both partial and complete footprints, handprints, and several human body traces, including knee traces and others left by portions of the legs, and produced by a minimum number of five individuals, from a 3-year-old child to a fully grown man (Citton et al., 2017; Romano et al., 2019; Avanzini et al., 2021). Avanzini et al. (2018) documented 35 hominin footprints together with abundant traces of bears and wolves (or dogs), which are dated at 14.7–14 ka cal BP (Molleson et al., 1972). The reduced number of individuals indicated by the tracks, and a close association with the human prints, could shed new light on dog domestication in the Upper Paleolithic. For the first time in the global human ichnological record, the evidence of crawling locomotion was reported by Romano et al. (2019).

2.6.6.4 Afragola, Nola, and Palma Campania

Footprints are found in association with deposits created by the 3780 BP eruption of Vesuvius, known as the Avellino Plinian eruption, that fell on Afragola, Nola, and Palma Campania, Bronze Age villages near Naples, Italy (Mastrolorenzo et al., 2006; Di Vito et al., 2019). These devastating ash falls created a remarkable footprint record NNW of Vesuvius, where thousands of footprints are oriented away from the volcano, preserved on a first layer before the massive eruption and the collapse of the eruptive column with consequent pyroclastic flows. The best investigated village from an ichnological point of view is Afragola (Giampaola et al., 2007). However, the collected data and the geo/ichnological evidence are almost the same as for other surrounding sites for a radius of at least 15 km (Avanzini et al., 2020). In these areas, thousands of human and animal footprints have been found on various depositional levels both inside and outside villages. These could "testify to an *en masse* exodus from the devastated zone" (Mastrolorenzo et al., 2006). Footprints recorded in all the stratigraphic horizons, including in flood and lahar deposits overlying the surge bed, are good indicators that "the ongoing exodus occurred during both the ash fall and the post-eruption rainstorms and floods" (Mastrolorenzo et al., 2006: p. 4368). The presence of longitudinal grooves in association with some of the trackways may be indicative of individuals dragging possessions and heavy objects. However, speeds calculated from footprints vary from 2.5 to 5.4 km/h in 89.6% of the sample (Laforgia et al., 2009), which do not seem compatible with the catastrophic living scenario created for the prints.

2.6.6.5 Moregine-Pompei

In Moregine, a small village located about 600 m south of the walls of the ancient city of Pompei, the remains of a majestic Roman villa indicate that it was destroyed by the same eruption that buried Pompei and therefore can be dated back to 79 AD. Here casts of three unshod footprints are exposed. The arrangement of the footprints suggests that they could not belong to the same trackmaker. According to Avanzini et al. (2020), the two complete footprints are about 24.5 cm in length. Both footprints show a very raised and well-defined plantar arch, which demonstrates that they were left by unshod feet. It is not known whether the footprints on display were the only ones to be found, as no scientific publication has been dedicated to this site.

2.6.6.6 Aosta

Archeological excavations in an area occupied by the former gym of the CONI (National Sport Federation) brought to light a series of attendance plans related to agricultural activities (Avanzini et al., 2020). On the last of these levels, a series of footprints of at least two adults and a child are documented, as well as animal tracks. The available data show narrow trackways with aligned step sequences of apparently barefoot individuals, which are still under study.

2.6.7 Netherlands

2.6.7.1 Zeewijk

Bakels and Zeitel (2005) and Theunissen et al. (2014) briefly report the occurrence of human and animal footprints in Neolithic sites from the Netherlands. In Zeewijk, NW Netherlands, Theunissen et al. (2014) identified eight possible footprints in a dwelling site.

2.6.8 Romania

2.6.8.1 Vârtop Cave

The first Neanderthal footprints were described in Vârtop Glacier Cave. Onac et al. (2005) reported three footprints dated between ca. 67.8 and 62 ka (Onac et al., 2021) that were attributed to a single Neanderthal, reflecting the robust anatomy expected for these humans. Two footprints are partial, consisting of either heel or forefoot impressions. A third footprint is longitudinally complete and well preserved, 22 cm long, and 10.6 cm wide. It was removed and preserved at the Museum of the Institute of Speleology in Cluj-Napoca, and since then, the other two footprints were stolen from the cave. Vârtop Glacier Cave has been a national monument since 1957.

2.7.8.2 Ciur-Izbuc Cave

In the Ciur-Izbuc Cave, Webb et al. (2014) revised the record of 400 footprints (Rusu et al., 1969) that were dated between 36.5 and 29 ka cal BP. Six to seven individuals were responsible for the footprints. The footprints extend for 75 m and represent a single visit to the cave. The authors claim these to be the oldest direct traces left by anatomically modern humans in a European cave. Three quarters of these footprints were subsequently damaged or destroyed. Unfortunately, insufficient footprints remain to measure movement variables such as stride length. This was declared a national protected site.

2.6.9 Spain

2.6.9.1 Matalascañas Trampled Surface (Andalusia)

This ichnosite in SW Spain was described for the first time in 2020, and since then, regular studies (summarized in Neto de Carvalho et al., 2023) have been detailing the ecology of a coastal interdune lacustrine system that functioned as a congregation site where different mammals (including humans) and birds converged, presumably for water and food resources, and possibly also for reproduction. Mayoral et al. (2021, 2022) described hominin footprints in the same site, which they named the "Hominin Trampled Surface" (*HTS*). Previously regarded as being older than 106 ± 19 ka, the "Matalascañas Trampled Surface" (*MTS*, or lateral equivalent of the HTS according to Mayoral et al., 2021) was dated at 295.8 ± 17.8 ka by Mayoral et al. (2022), opening a discussion about the most likely hominin producer of the purported footprints. No doubt, hominin footprints with unambiguous anatomical definition can be found in the MTS (Fig. 2.13B–D), showing a rounded heel impression, longitudinal arch developed between the heel and ball impressions, sometimes with toe prints especially the prominent hallux and, most important, with depth distribution in the footprint and sediment displacement rims compatible with moving foot biomechanics, or organized in narrow gauge trackways (Fig. 2.13B). From the footprints described by Mayoral et al. (2021), 31 were considered longitudinally complete, but only very few, if any, are illustrated showing reliable morphological features, and no trackways were described in an area of 564 m^2. It is clear from Fig. 2.1B of Mayoral et al. (2021) that the "HTS" is the present foreshore where the coastal erosion has been severely affecting the MTS every time it was exposed. The strong erosion of sedimentary heterogeneities in the present intertidal zone, such as tracks (including those of hominins), rhizoliths 10–20 cm in diameter, desiccation cracks, and structural fractures, may have been responsible for several apparent "footprints."

The association of Pleistocene hominin footprints and lithic industries is only known through a few occurrences worldwide and can be decisive together with geochronology in determining the hominin responsible for the footprints. This is the case of the MTS, where lithics were found in direct association with hominin footprints (Neto de Carvalho et al., 2023). The lithic assemblage is composed of 126 items. The archeological evidence in the MTS shows technical characteristics that fit within the Mousterian technological variabilities widespread in the Iberian Peninsula and in the region, during the MIS 6/5 until MIS 3. The classical association of Mousterian industry (Mode 3) to Neanderthals in Europe, together with new OSL dating of the MTS at 151 ± 11 ka BP (Neto de Carvalho et al., 2023), allowed for confident attribution of the hominin footprints and trackways. Therefore, MTS represents an extraordinary case where Neanderthal footprints are temporally (no more than hours or a day) and spatially associated with other tracks and trackways of large herbivores, as well as to in situ Mousterian lithics. The overstepping of animal tracks together with slow walking gait determined in the hominin trackways allowed for an inference of possible stalking behavior (Neto de Carvalho et al., 2021), and the presence of lithic tools showing expeditious knapping in a short-range distance from the footprints provided compelling evidence for animal processing in the surroundings (Neto de Carvalho et al., 2023).

2.6.9.2 Cape Trafalgar (Andalusia)

Neto de Carvalho et al. (2022) described a new ichnosite in the Cape Trafalgar area (Cádiz) dated from the Last Interglacial (107 ± 2 ka) and composed of a trampled surface in beach sandstone with aurochs trackways (the oldest known), together with several track-bearing beds in aeolianite with roe deer/caprine, proboscidean, and a possible single human footprint in cross section. In the trampled surface IV, together with large hoofprints attributable to aurochs, occurs the cross section of a footprint produced by a flexible foot. According to the composite morphology of the footprint, it can be interpreted as divided in three parts, recalling the human foot anatomy, posture, and kinematics: (1) the deepest part, corresponding to a rounded heel impression; (2) a convex medial part, with the shallowest impact depth, related to a dorsiflexed midfoot; and (3) a forefoot with deeper impression of the metatarsophalangeal area. The track proportions of the three parts are human-like. The age of the Cape Trafalgar aeolianite is much older than the earliest evidences of modern humans in the Iberian Peninsula, so the purported footprint was attributed to a Neanderthal. The possible hominin track from Cape Trafalgar, fitting most of the criteria defined for hominin tracks in cross section (Helm et al., 2019), together with Late Pleistocene (~90 ka) hominin tracks found in Brenton-on-Sea site on the Cape south coast at South Africa (Helm et al., 2018), represent the first attempts to compare a fossil track in cross section with human foot dynamics. Recently, a trampled surface with poorly preserved hominin footprints and aurochs tracks was identified in a lower bed of the beach succession in Cape Trafalgar. This constitutes an example of citizen science, as the new tracksite was found by an informed tourist (Florian Hipp).

2.6.9.3 Ojo Guareña (Castile and León)

In this cave near Burgos, over 1200 footprints were reported (Ortega et al., 2021), covering an area of 5652 m^2. In the Sala and Galerías de las Huellas, at least 18 trackways were produced by 9–11 individuals, possibly males. The superimposition of the footprints in opposite directions is indicative of round trips, which suggests a single inspection of the cave (Ortega et al., 2021). Dates of 15.6 ka were obtained from carbonized wood. New radiocarbon dating provided an age of only 4.3 ka cal BP (Ortega et al., 2021). As noted by Lockley et al. (2008), a "light patina of carbonate" gives the tracks "a very ancient appearance" (Marcos, 2001: p. 35). The Ojo Guareña cave system was declared a natural monument by the Government of Castile and Leon.

2.6.9.4 Tempranas Cave (Asturias)

This ichnosite located near Posada in the Llanes region of Asturias (Noval Fonseca, 2007) remains undated at the present time. The cave was found during construction works of the Cantabrico highway. Noval Fonseca (2007) presented the Galerías de la Huellas briefly. It is possible to see a trampled surface where a short trackway with some well-preserved footprints is oriented toward the exit. Detailed analyses of the trampled surface and dating are still missing. Finger flutings were also found in some of the cave walls.

2.6.10 United Kingdom

Bell (2020) provided a very comprehensive list of the many archeological sites with human footprints preserved in the United Kingdom. The footprints exposed briefly on the Norfolk coast at Happisburgh provided the oldest known (not counting the controversial site of Trachilos in Greece) hominin tracksite outside of Africa. Ichnosites on the Welsh coast are the most common. At Lydstep (Pembrokeshire), Rhyl (North Wales) and Port Eynon, Gower, late Mesolithic tracks of hominins and red deer were found (Bell, 2007a; Murphy et al., 2014; Philp, 2018). Other South Wales sites, where the footprint tracks are preserved in peat, laminated silts, or at the interface between peats and silts, include Kenfig (Bennett et al., 2010b), Magor Pill (Aldhouse-Green and Housley, 1993), Redwick (Bell, 2013), Manorbier (Murphy et al., 2014), and several sites in the Severn Estuary (Aldhouse-Green et al., 1992; Allen, 1997; Bell et al., 2000; Bell, 2007a, 2013). Footprints of similar age to the Welsh examples have been found on the Lancashire coast (Ribble Estuary; Roberts, 2009). Walney Island, Cumbria, Titschwell, and Norfolk, are among the sites indicated by Roberts (2014). Due to the difficulties in tracking the regional bibliography, all the sites where human footprints were found in United Kingdom, here we present the best studied ones providing relevant bibliography.

2.6.10.1 Happisburgh

The oldest hominin footprints found in the United Kingdom are at Happisburgh, dated at 0.9–0.85 Ma (Ashton et al., 2014). They represent the oldest footprints confidently attributed to hominins outside Africa and some of the earliest evidence for hominin occupation in Great Britain. They were formed on the intertidal mudflat of a river estuary. Of 152 footprints recorded in a 12 m^2 surface before their lost to erosion, 49 were clear enough to indicate the direction of travel (preferably to SW), and 12 were complete enough for detailed analysis, indicating at least five individuals including adults and children, with foot lengths between 160 and 172 mm (Ashton et al., 2014). The family group was interpreted by Ashton (2021) as resident in the area, without taking into consideration the high mobility of these hominin groups in the search for resources.

2.6.10.2 Uskmouth

Four hominin trackways, including three lengthy ones, were described at Uskmouth, Severn River estuary (Allen, 1997). They were produced probably by adult males, including a very large individual 2 m tall (not including the likely deformation of the sandy substrate), and a young person or child (Aldhouse-Green et al., 1992; Aldhouse and Housley, 1993).

2.6.10.3 East Goldcliff

The Goldcliff footprints are dated to as early as 7.5 ka cal BP. The finds are in the intertidal zone of the Severn River estuary where archeological investigations have revealed traces of a succession of settlements and activity areas (Bell, 2007a,b). The most striking evidence is in the form of 270 human footprints that crisscross the ancient mudflats and have been preserved in several beds (Scales, 2002, 2004, 2007; Bell, 2007a,b, 2020; Barr and Bell, 2016;). Associated with the human footprints are tracks of red and roe deers, the occasional aurochs, and many seabirds. Among the human footprints, 21 individuals have been identified. Of these, four were adults (over 14 years old), seven were "subadults" (11–14), two were older children (7–11), while the majority of the footprints (67%) were made by eight young children (3–6-year-old). The best preserved prints occurred in fine-grained sediments, indicating activity during the summer season. The trackways indicate the movements of a group of people across the mudflats. This high proportion of child footprints highlights the activities of group members who are usually invisible in the archeological record and are here seen to be fully engaged with the life of the community from an early age (Bell, 2020).

2.6.10.4 Formby Point

The mid-Holocene coastal environments of Formby Point, northwest England, show abundant animal tracks and human footprints exposed ephemerally within 3 km in the intertidal zone (Huddart et al., 1999a,b). At Formby, there are two stratigraphically separated groups of prints, with a total of 31 track and footprint-bearing beds, dated from >9 ka to 1.2 ka BP (Burns et al., 2022). In total, 401 large animals and 192 humans were recorded from the footprints and 145 trackways (Roberts et al., 1996; Burns, 2021), being one of the world's largest known concentrations of Holocene vertebrate tracks and one of the best studied ichnosites. They represent the persistent return to the mudflats by generations of people.

Additionally, 17 trackways and 61 isolated tracks (181 human footprints in total) were discovered and described recently by Wiseman et al. (2022). They enabled inferences of the gender, stature, and speed of the trackmakers. These footprint lengths show that children frequently ventured out onto the mudflats together with an older youth or young adult/female or without an accompanying person (Burns, 2021). The human trackways show a bimodal direction, with most of the trackmakers traveling in a southwest direction, toward the paleo-coastline, with fewer trackmakers traveling inland in a northeast direction; some trackmakers were either walking side by side or were following one another hours/days later as indicated by parallel trackways, while one trackway shows an individual running, eventually coming to a stop with both feet together on the ground (Wiseman et al., 2022). The trackways suggest slower-paced activities by the women and children such as for shellfish gathering, and a longer stride length for the men with imprints, suggesting running after hunted game (Roberts et al., 1996). In fact, a staggered wide-straddled gait suggests attempting to influence the animal's direction of movement (Roberts, 2009). At the Neolithic site, one arthritic individual and a pregnant woman have been inferred (Burns, 2021). According to Bennett and Morse (2014), the methods to estimate speed and the sex of the trackmaker are not clear. Notwithstanding, assertions around the presence of deformities and pathologies described by Roberts were confirmed by these authors' observations at the site. Burns et al. (2022), using multibed monitoring of dated trampled beds for measuring long-term changes in biodiversity, argued that the post-Mesolithic decline in large mammal activity in the area could be the result of habitat shrinkage following sea-level rise, and the development of agriculture, as well as hunting pressures from a growing human population.

2.6.10.5 Lydstep Haven

In 2010, winter storms stripped sand off parts of the beach at Lydstep Haven, revealing intertidal peats, on the surface of which were impressed human and animal tracks. The human footprints were of adults and children, dated to 4345–3950 cal BC (Murphy et al., 2014). No detailed study was presented, maybe due to the poor preservation of the footprints.

2.6.10.6 Kenfig

Footprints within peats have been recorded at Kenfig (Bennett et al., 2010a). They occur in the lower of two intertidal peat layers. The footprints are thought to be probably Neolithic in age, although there is some uncertainty about their date, probably >8 ka (Doyle, 2007). Bell (2020) provided an age range of 5.7–4.2 cal BP.

2.6.10.7 Hauxley Low

Another large group of Mesolithic human and animal footprints dated at about 7.3 ka cal BP has been reported at Hauxley Low, Northumberland (Eadie and Waddington, 2013). A peat deposit was found in the intertidal area covering 213 m^2 and revealed 90 human prints and 88 animal tracks. Human footprints were of various sizes, although in none of them could individual toes be identified. This, together with their generally smoothed edge appearance, suggests that the humans were wearing footwear of some sort. The human footprints vary in size from 8 cm to an average of around 22 cm.

2.6.10.8 Magor Pill

The coastal area of Magor Pill provided a record of human footprints, including adults and children (Aldhouse-Green and Housley, 1993). Aldhouse-Green et al. (1992) dated the ichnosite of Magor Pill earlier than 5.72 ± 0.08 ka BP.

2.6.10.9 St. Ouen's Bay

On the beach of the St. Ouen's Bay on the island of Jersey, Patton (1993) briefly reports the finding of some footprints attributed to the Neolithic. The 3D model of a poorly preserved left footprint with a deep heel, a poorly defined hallux, and an expulsion rim in the forefoot area can be found at sketchfab.com/3d-models/human-footprint-in-Neolithic-land-surface.

2.6.10.10 Dovey Estuary

Hominin and animal tracks are found on the peat surface dating to between 3 and 2 ka cal BP.

2.6.10.12 Redwick

Bell (2013) reported human footprints, including those of children and adults in a coastal setting dated at 1.3–0.9 cal BC.

2.7.10.12 Fenland

In East Anglia, tracks of cattle and people were found on the edge of a paleochannel (Pryor, 1998).

2.6.10.13 Rawthey Cave

Two clear unshod footprints of children were recorded, which were likely to have been made by different individuals. A number of other depressions, which could also have been the remains of footprints, were noted, but positive identification was not possible (Chamberlain et al., 1998). Radiocarbon dates of 605 ± 40 and 815 ± 40 BP provide a terminus ante quem for the human footprint surface (Hedges et al., 1998).

3. Leaving prints in human-made substrates

The invention of pottery introduced fundamental shifts in human subsistence practices and socio-symbolic behaviors. The earliest pottery, used for cooking, has been thought to have appeared in China, at 20–19 ka BP in Xianrendong Cave (Jiangxi Province), during the Last Glacial Maximum (Wu et al., 2012). The early date shows that pottery was first made and used thousands of years before the emergence of agriculture. Before, the oldest known ceramic artifact was dated as early as ca. 25 ka BP, during the late Paleolithic, Gravettian period. It is a statuette of a woman, named the Venus of Dolní Věstonice, from a small prehistoric settlement near Brno, in Czechia. In this location, hundreds of clay figurines representing Ice Age animals were also uncovered near the remains of a horseshoe-shaped kiln. For 25,000 years, the process of producing pottery and ceramic building materials has been the same, from hand shaping, sometimes with the use of a mold and tools, letting dry under the sun lying down on the ground before being fired. It is during the process of shaping and drying that ceramics become a very favorable substrate on which to record prints of the producer, relatives, workers, as well as domesticated animals and wildlife. Ceramics represent an optimal substrate to preserve traces because of their geotechnical properties (clay-rich sediment with variable moisture), open-air drying (lying on the ground for several days), fast "diagenesis" (pyrometamorphism), resistance, and abundance (providing high preservation potential) (Baucon et al., 2008). Therefore, imprinted ceramic building material especially can give a good insight into the contemporary fauna and environment of the tilery. Notwithstanding, it should be stressed that the ceramic materials provide information about the place of their production and not necessarily the place of their recovery, since pottery and ceramic building material was frequently transported great distances (Baucon et al., 2008).

Paleodermatoglyphics as defined by Bartsocas et al. (1982) is the study of fingerprints, palm-prints or footprints—the so-called papillary terrain or friction skin—on prehistoric objects, mostly ceramics, in order to provide some biological or other characteristics of the people who made the prints. A fragmentary fingerprint on the left side of the Venus of Dolní Věstonice I figurine back was analyzed by Valsik (1951) and (Králík et al., 2002). The structure was identified as a negative of human friction skin based on the *minutiae* (small variations from the strictly parallel direction of the ridges, irregularities of direction, discontinuities and branching), ridge breadth, and other markers. The epidermal ridge breadth of the fingerprints in ceramic material is defined as the distance from the edge of the shadow of the ridge to the edge of the shadow of the adjacent ridge, measured upright/vertically to their direction. Epidermal ridge breadth correlates with the age during the growth period of an individual. These authors estimated the age of the artist, based on the fingerprint ridge breadth, taking into account the clay shrinkage during drying and firing, and by applying the equation of Kamp et al. (1999), as being somewhere between 7 and 15 years old. As rightly mentioned, this estimate is valid if the age/ridge breadth relation in the Paleolithic was similar to the present-day one.

Bricks and tiles were manufactured in vast quantities for centuries throughout the Roman world, which makes ceramic building material the most frequent find in Roman archeological sites. Animal footprints on fragments of Roman bricks and tiles are particularly common and distributed across the whole Empire, from what is now England to the Iberian Peninsula and Italy to Hungary, Romania, Israel, and Turkey (Cram and Fulford, 1979; Gudea, 2004; Sousa et al., 2007; Sammartino and Bisconti, 2010; Bar-Oz and Tepper, 2010; Bennett, 2012; Dobosi, 2016; Oğuş, 2021). Roman ceramic building materials were made using removable mold-boards and allowed to dry laid down on the ground in a brickyard/tileyard for up to 2 weeks, depending on the weather conditions and working season. Yards typically would contain tiles or bricks in various states of hardness. This was the time when most of the unintentional prints occurred on their drying upper surfaces, most frequently in the form of animal and human footprints (Fig. 2.14). After drying to "leather-hardness," they were then fired; this preserved permanently any prints that the brick had acquired in its earlier and softer condition (Brodribb 1987, 125). For this reason, animal footprints occurring on ceramic building materials provide a good complement to determine living animal diversity and ecological conditions of the site where they were produced.

The Roman military bathhouse in the fort-village complex at Vindolanda (Northumberland, England), occupied from about 50 to 415 AD, is considered among the largest collections of animal prints in ceramic building materials (Bennett, 2012). The 111 fragments of ceramics with animal footprints found in Vindolanda include domestic dog (*Canis familiaris*), semiferal cat (*Felis sylvestris*), domestic pig (*Sus scrofa*), sheep (*Ovis ammon*), goat (*Capra hircus*), and cattle (*Bos taurus*) (Higgs, 2001). Over 80% of tracks identified by Higgs were made by dogs. Five-size classes of dogs were determined by Bennett (2012). This is one more piece of evidence supporting the Bennett et al. (2016) conclusion that Vindolanda dogs are less morphologically diverse than modern dogs but more diverse than dogs of the British Neolithic and Iron Age. Some fragments also carry human finger and palm-prints and impressions made by tools, nails, textiles, and bits of wood. The presence of the domestic cat in an urban area like Vindolanda makes more sense, especially where dogs were so common. Wild cats do not tolerate humans and prefer habitats where there is almost no human intervention.

FIGURE 2.14 Example of a footprint in human-made substrates, from the Roman settlement of Fafião (NW Lisbon, Portugal) with the left footprint of a small one-year-old baby (up to 10% shrinkage of fired clay included), showing deep impression of the hallux and no longitudinal medial arch, in comparison with the shallow four digits (no claw impressions) of the right forepaw of a large dog printed on the edge, therefore revealing different moments of substrate consistency. This is clear evidence that children's footprints found in Roman ceramic materials do not necessarily represent labor force (specimen n. F/R/76/109, Archeological Museum of S. Miguel de Odrinhas).

In a detailed study of the Silchester tiles, Cram and Fulford (1979) identified the tracks of 91 individual animals, almost exclusively belonging to domestic animals with one possible crow and a possible fox track. Fifty-five tracks were made by dogs or cats, 22 by small ruminants, 6 by cattle, 7 by birds (mostly chicken), and 1 by a horse. Brigetio (Komárom-Szony, Hungary) was one of the major towns of Roman Pannonia. Tiles were manufactured in the legionary tilery of Brigetio during the second and third centuries AD (Dobosi, 2016). According to this author, 10 dog, 5 cat, 2 sheep/goat, and 1 chicken tracks were identified, all from domestic animals kept around the tilery. It is not surprising that no tracks of wild animals were found since it was a large urban area. Among the human sandal prints, there is a well-preserved, almost complete impression of a hobnailed children's shoe.

Tracks are sometimes the only source of information about the animals living in the area. In the neighborhood of Lousada (Portugal), the granite-derived soil is highly acidic, causing the animal bones to decompose, and making tracks the only way to attest to the presence of pigs, goats/sheep, and cats at the site (Sousa et al., 2007). The presence of tracks of terrestrial wild animals in mostly small rural sites, on the other hand, makes it reasonable to assume that the drying area was not surrounded by a fence and was therefore readily accessible for wildlife. A good example is the Roman tilery of Casa Campacci in Livorno, Italy, where two thirds of the 18 tile fragments with footprints belong to wild animals, including fox (*Vulpes vulpes*), wild cat (*Felis sylvestris*), weasel (*Mustela nivalis*), roe deer (*Capreolus capreolus*), and wild boar (*Sus scrofa*). The footprints of domestic animals at this site were limited to dogs and cats (Sammartino and Bisconti, 2010). In the Roman port settlement of Andriake, a suburb of Myra (Turkey), located on a small river, the tracks found in ceramics are mostly from wild animals, including golden jackal (*Canis aureus*), lynx or caracal, and wild goat (*Capra aegagrus*) (Öz, 2022).

Human-made pavements are another substrate highly prone to preserving the ichnological record. It is widely common to find human and animal tracks and trackways in sidewalks and gardens that were produced when the fresh-made pavement was still drying. A good example of application of this kind of man-made substrate to Ichnology is provided by Avanzini et al. (2011). These authors describe boot tracks preserved in a cement layer attributed to a soldier from the Austro-Hungarian army found in a tunnel of the First World War Valmorbiawerk fort in the Trentino Alto Adige region (Italy).

Archeological excavations conducted at Mishrifeh, ancient Qatna (Central Western Syria), discovered a large pottery workshop on the summit of the site's acropolis. The ceramic manufacturing area was built at the beginning of the second millennium BC, reached its maximum extension during the Middle Bronze Age II (c. 1800–1600 BCE), and was abandoned in the following Late Bronze Age I (c. 1600–1400 BCE). On the surface of a trodden floor located between two platforms, dozens of human and equine footprints were preserved (Baucon et al., 2008). The horse footprints are particularly interesting since they represent one of the earliest archeological pieces of evidence of this animal in Syria. The archeological context suggests that horses were used to transport the pottery manufactured in this area. The morphological analysis of human footprints shows that the producers were either adults or children working in the pottery manufacturing area; workers were not barefooted but wore simple sandals, probably consisting of a sole fastened to the foot by a strap or a fabric band.

4. Human prints and rock art

In popular culture, trace fossils have frequently been interpreted as the manifestation of supernatural entities. The structures known as "Ciampate del Diavolo" ("Devil's footprints", Italy) are an example of a kratophanic interpretation (Baucon et al., 2008): an ancient legend interprets the footprints as produced by the Devil, who has the supernatural power of walking on lava. The "Devil's Footprints" are actually hominin trackways preserved in ash deposits (Mietto et al., 2003), as already presented in this chapter.

Lockley and Meyer (2000) and Kim et al. (2008) noted that rock art is itself a type of hominin ichnology. However, noting that ichnology can bring very little to cave art studies, which fall under the realm of cave art experts, or anthropologists at best, there are exceptions that must be summarized here.

Our Paleolithic ancestors deserve full credit for being the authors of the first vertebrate track field guides (Seilacher, 2007). Handprints and hand stencils have been a common symbol in cave art for dozens of thousands of years. In the Maros Pangkep UNESCO Global Geopark, located at south Sulawesi, Indonesia, most of hundreds of caves contain painted art, including colored hand stencils as motifs, some of them drawn at least 45.5 ka ago; they are considered the earliest cave paintings (Brumm et al., 2021), since the cave art of Maltravieso (Cáceres, Spain), which includes handprints, dated at 65 ka and attributed to

Neanderthals (Hoffmann et al., 2018), has been widely contested (e.g., Aubert et al., 2018; White et al., 2020). Unfortunately, none of these important sites have had a morphological study of handprints. Nevertheless, increasing interest has been paid to the reconstruction of the biological identity of individuals who left handprints on the prehistoric caves and rock shelters around the world. Based on the size and shape of prehistoric handprints, attempts have been made to estimate age (e.g., Gunn, 2006; Bednarik, 2008), body height (Manhire, 1998), or laterality (e.g., Uomini, 2009) of their producers, although recent attempts to estimate sex from handprints depicted in Paleolithic cave art, using morphometric data from recent populations, cannot be sustained (Galeta et al., 2014). There are 231 stenciled hands in the cave of Gargas, located in the French Pyrenees and dating back around 27 ka. What makes this cave unique is that 144 handprints are missing finger joints, segments, or entire fingers, suggesting that fingers may have been mutilated, either in a ceremony, e.g., for atonement or sacrifice, but could also have resulted from frostbite, or just manual dexterity, claiming that they correspond to signs of an "alternate" or "non-primary" sign language, like those still employed by a number of bimodal (speaking and signing) human groups in hunter-gatherer populations, like the Australian first nations or the Plains indigenous people (Rikardo et al., 2021).

Carved, painted, and incised footprints are often considered sacred, and the literature on their symbolism is wide ranging (see references in Atamtürk et al., 2018). Artwork depicting footprints carries symbolic connotations because it is often considered a sign either of an invisible deity or a revered holy person (Bertilsson, 2013), or the recording of a pilgrimage or passage, as the in situ material proof of "been there" (Takács, 2007). A representation of a foot or feet, bare or shod, could be votive or a celebration of a visit to a sacred place. Globally, gods, giants, and heroes (such as Herakles, Dionysos, Asklepios/Serapis, Mithras, Jesus, Buddha, Mohammed, saints and prophets, fairies, Devil, King Arthur, but also Virgin Mary and Isis), all left behind footprints that are well represented in popular religions and explained through legends. This symbolism can be found in different time periods and in different religions, being recurrent from Ancient Egypt to medieval times, and related to Greco-Roman religions, Islamism, Buddhism, or Christianism.

Baucon et al. (2008) considered the potential inclusion of podomorphs (footprint representations on hard rock) and "ichnohierophanies" (traces of religious or spiritual significance) under the subject of ichnoarcheology. A good example can be found in the village of Barreiro de Besteiros (central Portugal), a protected national monument, called the Alagoa rock. More than 110 podomorphs are carved in the rock (Gomes and Monteiro, 1977). The carvings reveal features in common with natural ichnoassociations (in particular size disparities and foot proportionality). Therefore, real feet were used as models for the rock art: the authors outlined their own feet with a pointed tool. Moreover, morphometric analysis revealed individuals of different ages—children, teenagers and adults—all participated in the creative process, bare-footed or wearing simple sandals (Gomes and Monteiro, 1977; Baucon et al., 2008).

We have been using an ichnological approach to study the podomorphs from Oleiros at Naturtejo UNESCO Global Geopark (Central Portugal). This was an important sacred or passage site at the top of the Sesmarias hill attributed to the Bronze Age, whose location possibly was related to an ancient road, retained evident importance when religious systems changed, and marked nowadays by the Christian chapel of Christ the King built in 2014. Two subhorizontal rocks are characterized by 38 likely shod- and unshod-feet podomorphs, sometimes

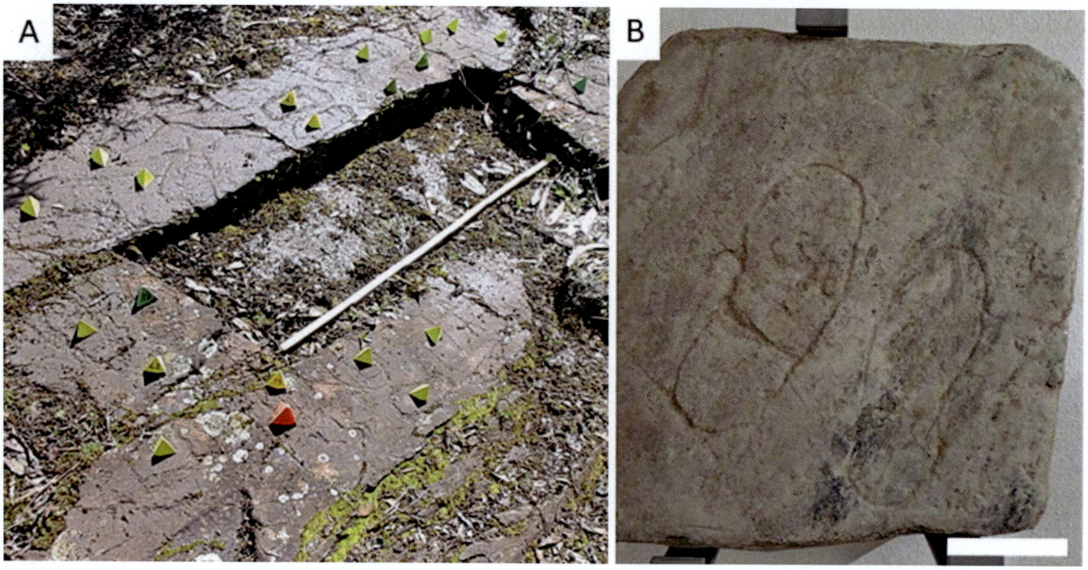

FIGURE 2.15 Podomorph records from different periods found in Portugal. (A) Sesmarias Bronze age rock carvings (Oleiros municipality) in metagreywackes with numerous examples (marked with pyramids) made from shod and unshod feet; ruler is 1 m. (B) Senhora d'Aires example of Roman plantae peda, with broader right, and slender left shod feet placed side by side, carved in local white marble; Viana do Alentejo. Scale bar is 10 cm.

associated with pairs (Fig. 2.15A), engraved using three different techniques and with numerous overlays. According to Caninas et al. (2008), there are very fine contour podomorphs drawn, using narrow grooves or filiform lines. A second way of representation of feet consists of marking the contour with a deeply pitted, wide groove. There is a third way of representing feet, which consists of a deep, coarse, and extensive pitting filling the outline. Age classes derived from foot length indicate the presence in Sesmarias of toddlers, but mostly children and a few adults. About 70% of the feet represented are left ones, the feet most favorable for drawing by the typical dextral handedness. *Plantae pedum* were also very common in the Roman period. The Roman settlement of Senhora de Aires at Viana do Alentejo, southern Portugal, was populated by a Hellenic community where a temple dedicated to Mars (equivalent to the Ares god of Greek mythology) was built. An *ex-voto,* presumably engraved by a couple (presumably from right-man and left-woman feet), was found there (Fig. 2.15B).

Impressed footprints that more clearly fall under the realm of ichnology, on the other hand, may have an equally high ritual standing. The Barcın Höyük pair of unshod footprints are an example of intentional, symbolically charged art, according to Atamtürk et al. (2018). This includes the fact that the pair of footprints were a result of standing and not walking; they were formed by firmly pressing the sole of the foot into the ground and with the distribution of the body weight fully between the heel, metatarsus, and toes. Moreover, the left print was positioned immediately above a buried cow skull, at the entrance of a domiciliary structure. Along with the fact that the footprints were made with bare feet immediately after the floor had been renewed, this suggests that the replastering of the floor atop the bucranium marked a time of symbolic renewal (Atamtürk et al., 2018).

Handprints and footprints made in hot spring travertine were found near the Quesang River, at an altitude of 4260 m a.s.l., on the Tibetan Plateau (Zhang and Li, 2002; Zhang et al., 2003). The U–Th and TL dating of the travertine shows intervals of ca. 226–169 ka and 20–7.4 ka, claiming that geothermal energy resources have attracted hominins, for reasons of both survival and recreation, over a very long period of time (Wang et al., 2023). Thus, the authors claim the intentionality of the imprints, which have been positioned into the available space to create a careful composition, as the oldest "parietal" art (Zhang et al., 2021). This outstanding artistic creation was attributed by the authors to Denisovan-like hominins, meaning they would be not only the first footprints and handprints, but also the first behavioral evidence attributed to Denisovans. However, Huisheng et al. (2024) pointed out the unsuitable application of U–Th dating method to weathered travertines, questioning the obtained ages, at least by this method. Tang et al. (2022) found Tibetan characters associated with one of the panels and also made when the travertine was still soft, reducing the age of printing to <1.3 ka. The hypothesis of artistic behavior and also dating of the hand- and footprints are intriguing but, agreeing with Hatala et al. (2022), they remain open questions worthy of continued investigation.

The concept of "ammoglyphs" was introduced by Helm et al. (2019, 2021), a term that was coined to describe a pattern made by humans in unconsolidated sand, now evident in rock. This would constitute a novel form of paleoart. Several potential sites were identified on South Africa's Cape south coast and ascribed to a possible human origin with varying degrees of confidence. These included circular, triangular, cross-hatched, fan-shaped, radial and linear patterns, and a possible sand-sculpture that would represent the oldest known example of representational art of another species. A cross-hatched pattern, engraved in ochre and dated to 77 ka, had previously been reported from Blombos Cave on the same coastline (Henshilwood et al., 2002). Other potential causative agents (including modern graffiti) were considered and, in most cases, could be excluded. Clearly, ammoglyphs are more compelling when they are associated with hominin footprints or knee impressions on the same surface (Fig. 2.3H). Some of the purported ammoglyphs have been recovered and are accessioned in the Blombos Museum of Archeology in Still Bay. The bed with circular ammoglyph (Fig. 2.3G) has been dated to 136 $\pm$ 8 ka, and the one with triangular pattern (Fig. 2.3I) to 139 $\pm$ 10 ka.

In the Middle Paleolithic of Europe, purported Neanderthal engravings were claimed by Marquet et al. (2023) at La Roche-Cotard cave. They were dated by OSL at 57 $\pm$ 3 ka. The finger-flutings, or digital traces, were produced on Turonian-age, yellow, tuff soft surface layers of the cave walls, in a total of eight panels revealing sets. The majority of the 84 traces were made by fingers laid flat, while a few rare ones appear to have been made by a finger on the side. The compositions seem organized and deliberate. The possibility of these structures being produced during private excavations between 1846 and 1912 was discussed by Marquet et al. (2023). These authors used discrimination of patinas between modern marks and ancient prints. Several other examples of finger-fluting attributed to anatomically modern humans were mentioned in section 2 of this chapter, especially in caves from France. Bednarik (1984, 1986) present a first review of Paleolithic finger-fluting in Europe and Australia. An example of cave art based on finger-fluting in very different longitudes is described by Cooper et al. (2016) on Mona island (Puerto Rico). Here, 10 cave halls exhibit 250 motifs made by the indigenous people before the arrival of the Spanish, with radiocarbon dates of 1272–1387 cal AD and 1420–1458 cal AD.

5. The record of shod humans

The wearing of shoes represents a behavioral choice, since the trackmaker actually does not touch the sediment (Milàn and Bromley, 2009). Morphology of tracks will vary according to the type and shape of footwear and may reflect behavioral adaptation to climate conditions, functionality, and raw materials available. Shod footprints may provide data most relevant to the origin and function of footwear (Ledoux et al., 2021a, 2021b). However, it is important to take into account the preservational quality of the footprints. Footprints poorly preserved due to erosion, or even underprints, can be easily misinterpreted as evidence for shoe wearing.

The global record of sites from which shod-hominin tracks have been considered is sparse. Helm et al. (2023b) provided the first compilation of purported records of shod tracks and consider western Europe and South Africa to be the most suitable areas in which to search for ichnological evidence of shod hominins. The development of criteria for the identification of shod-hominin tracks by Helm et al. (2023b) includes: a general hominin track outline; rounded anterior and posterior ends; evidence of high-quality preservation in a suitable substrate, e.g., crisp track margins; absence of diagnostic features of barefoot hominin tracks such as clear digit impressions; and diagnostic features such as traces of strap attachment points. Therefore, the footwear trace record may provide evidence for technological developments in shoemaking (Trinkaus and Shang, 2008; Milàn and Bromley, 2009).

A report drew attention to the likelihood that Middle Stone Age hominins on South Africa's Cape coast fashioned footwear (Helm et al., 2023b). Evidence from three sites was presented, two from the south coast (one at Klein-krantz in the Garden Route National Park and one in the Goukamma Nature Reserve) and one from the east coast in the Woody Cape coastal portion of the Addo Elephant National Park. Criteria were presented for the identification of shod-hominin tracks, including the presence of crisp track margins that would suggest relatively high-fidelity morphological preservational detail. At the Klein-krantz site, possible strap attachment points were identified within the tracks (Fig. 2.3F). None of the sites were directly dated, but the presence of nearby dated sites at Kleinkrantz and Goukamma suggests a late-Middle- or early-Late Pleistocene age (ca. 148–73 ka). The identification of a Middle Stone Age bone-tool industry at Blombos Cave on the Cape south coast (Henshilwood et al., 2001), and the resulting ability to fashion items of clothing from hides, provides independent support for such findings. However, Helm et al. (2023b) did not consider the evidence conclusive, even though suggestive, of the presence of shod-hominin trackmakers using hard-soled sandals.

Perhaps even older than the South African ichnosites, two of the four hominin tracks described from Theopetra Cave, Greece, in clay sediments dated to 130 ka, were considered to be of a shod hominin (Kyparissi-Apostolika and Manolis, 2021). The footprints were made by young children 3–4 years old (probably *Homo neanderthalensis*). They show a simple contour but are sharply defined, indicating for these authors that the individual was wearing footwear. The poor preservation of the footprints should raise caution regarding the claim for the most ancient evidence of footwear.

Duveau et al. (2021), in describing 257 hominin tracks from Le Rozel in France, dated to ~80 ka and clearly attributed to *Homo neanderthalensis*, suggested that some were made by shod individuals. This was based on the lack of clear digit impressions on relatively complete footprints. The variable preservation shown by these footprints in sandy mud suggests more caution in such interpretations.

Ledoux et al. (2021a, 2021b) reported 28 hominin tracks of Gravettian age (~31–28 ka), recorded in clay substrates in Cussac Cave, France. Seven tracks were complete, but digit impressions were shallow or absent. This contrasted with well-registered finger impressions of other tracks, and areas where adjacent digit impressions of bear tracks were clearly preserved. Evidence of medial longitudinal arches was present.

Clottes (1975) suggested that Magdalenian hominin tracks in Fontanet Cave, France, may have been made by an individual wearing a soft shoe or sock. This appears to be the first published reference to possible ancient shod-hominin footprints. However, this interpretation was questioned by Pastoors et al. (2015). Possible shod footprints were also reported at Jaguar Cave in Tennessee, dated to ~5 ka (Robbins et al., 1981; Bennett and Morse, 2014).

The Yenikapı Neolithic site was found at the point where the Marmaray and Metro stations meet, in the region of Istanbul. In an area of 8 × 20 m, Kızıltan and Polat (2013) excavated a surface with 2080 well-preserved shod prints, dated to ca. 8 ka BP. The ichnosite still requires a proper ichnological study.

The Bronze Age site of ancient Qatna in Syria provided a trampled surface with human and horse footprints dated at 3.8–3.6 ka BP. The trampled surface represents a passageway for personnel operating in the workshop, a surface for drying the pottery, and for cooling and storing the pots after firing and before transport to their final destination. According to Baucon et al. (2008), the morphological preliminary analysis of footprints shows that the producers were either adults or children working in the pottery manufacturing area; workers were not barefooted but wore simple sandals, probably consisting of a sole fastened to the foot by a strap or a fabric band.

In upper foreshore beachrock dating from Roman times, on the east coast of Rhodes Island, a trackway was interpreted as comprising shod human prints (Bromley et al., 2009). A narrow-gait trackway composed of three footprints without evidence for the hallux implied that the producer was wearing shoes.

Finally, an ancient feature resembling a shod human footprint was recently discovered adjacent to a buried prehistoric house pit dating to ~1840 cal yr BP at the Swan Point site in central Alaska (Smith et al., 2019). The footprint was characterized as belonging to a child 9–12 years old wearing soft-sole footwear (probably moccasins) and represents the first of its kind from the subarctic and the first described from Alaska.

6. Expanding the human footprint: Astroichnology

"That's one small step for (a) man, one giant leap for mankind" are the famous words proclaimed by Neil Armstrong when he first stepped onto the Moon (July 20, 1969). The footprints left by Neil Armstrong and Edwin Aldrin on the lunar surface represent a monumental shift in hominin evolution, serving as iconic landmarks of an evolutionary milestone (Kim et al., 2008; Lockley et al., 2016). These footprints, along with the tracks left by lunar vehicles and rovers utilized on Mars, constitute ichnologic evidence of significant recent hominin evolution events. Specifically, they mark the first instance of human beings setting foot on a celestial body other than Earth (Spennemann, 2004; Lockley et al., 2016). These footprints underscore the significance of astronaut tracks within the realm of astroichnology—the application of ichnology in the quest for present and past extraterrestrial life (Baucon et al., 2017, 2020, 2021). Astroichnology, though a relatively nascent field, offers insights into the potential presence of life beyond Earth, using the principles of ichnology. The footprints and tracks left by astronauts and their vehicles on extraterrestrial bodies serve as a critical data source and analog for astroichnologists (Baucon et al., 2017).

A comprehensive review of the ichnological relevance of astronaut tracks on the Moon has been conducted by Díaz-Martínez et al. (2021). This study has further cemented the value of these extraterrestrial footprints in understanding the evolutionary trajectory of hominins, as well as in the broader context of astrobiology and astroichnology. The footprints offer a tangible marker of human exploration beyond our home planet, contributing significantly to our understanding of hominin evolution and the potential for life in the Universe (Baucon et al., 2017, 2020, 2021).

Astronaut tracks have been described by Baucon et al. (2017) as plantigrade, elongate concave epireliefs with transverse ornamentation and no digits (Fig. 2.16A). These specific morphological attributes of the astronaut tracks are direct reflections of human technological advancements. The transverse ornamentation composed of treads and absence of digits are clear indications of the shod nature of these tracks. The unique conditions on the Moon, which are inhospitable to human life without the aid of protective gear, necessitate astronauts to wear spacesuits. This requirement, in turn, impacts the nature of the astronaut tracks, as they are essentially the imprints of the protective boots of the spacesuits. Hence, the ichnological record on the Moon essentially becomes a reflection of human technological behavior (Díaz-Martínez et al., 2021).

It is noteworthy to point out that these tracks provide valuable insights into the specifics of human technology at the time of the lunar missions, offering a unique perspective on the intersection of technology and evolution in the extreme environment of the Moon. As such, astronaut bootprints are much like the shod-hominin tracks that indicate that humans may have worn footwear while traversing harsh dune surfaces during the Middle Stone Age (Helm et al., 2023b).

The bootprints left by astronauts on extraterrestrial surfaces, particularly the Moon, offer invaluable behavioral and biogeographic insights as they serve as evidence of human behavior beyond the Earth and document the initial human exploration of another celestial body. Beyond their behavioral and biogeographic significance, astronaut bootprints also serve as critical sources of geotechnical information, specifically regarding the lunar soil. The imprints left by astronauts have proven instrumental in gathering data about the physical properties of the lunar regolith (Kaydash et al., 2011; Mitchell et al., 1972). Thus, the significance of astronaut bootprints extends beyond their symbolic value, contributing to our scientific understanding of the Moon's surface.

It is noteworthy to mention that astronaut trackways are typically accompanied by a unique set of bioturbational and bioerosional structures, which are reflective of various human behavioral activities associated with space exploration (Baucon et al., 2017). These activities resulted in the creation of a distinctive trace fossil association on the Moon, produced by an array of geological sampling tools used for investigating the lunar regolith. Tools used for lunar exploration encompass a broad spectrum, including scoop-like instruments such as the

FIGURE 2.16 Technotraces on the moon and Mars (bioturbation and bioerosion). (A) Bootprint, Apollo 11 (1969). (B) Astronaut trackway, Apollo 12 (1969). (C) Astronaut trackways and crewed rover trails, Apollo 15 (1971). (D) Landing traces of Surveyor III (1967). (E) Opportunity trails (2010). (F) Curiosity wheel trail (2014). (G) Drilling boring produced by Curiosity (2015). *Source: NASA Image and Video Library.*

Contingency Soil Sampler, spade-like tools such as the Trenching Tool, and rake-like devices, including the Lunar Soil Rake (Allton, 1989). Additionally, cylindrical drill marks observed on the Moon have been created by an assortment of tools. Some of these tools were manually operated, such as core tubes, while others were powered, for instance, the Apollo Lunar Surface Drill (Allton, 1989). Furthermore, it is vital to highlight that similar structures have also been created by unmanned rovers on the planet Mars (Baucon et al., 2017).

Beyond the Earth, the activities of human beings (or at least of their technology) have left different technotraces (sensu Díaz-Martínez et al., 2021) in other planets or even other celestial bodies (e.g., Mars, Venus, the Moon, Titan, or different asteroids and comets). Among them, the most iconic are probably those preserved on the surface of the Moon, some of them already discussed, but which mainly consist of (Fig. 2.16):

a) Bootprints (Fig. 2.16A and B). Twelve north American men astronauts have stepped on the lunar surface (U.S. missions Apollo 11, 12, 14, 15, 16 and 17, from 1969 to 1972) (e.g., Lund, 2018);
b) Trails of the crewed rover Lunar Roving Vehicle (U.S. missions Apollo 15, 16 and 17, from 1971 to 1972; Fig. 2.16C) (e.g., Lund, 2018);
c) Trails of different uncrewed rovers, e.g., the Russian rovers Lunokhod 1 and 2, landed in 1970 and 1973, respectively (Harvey, 2007; Lund, 2018; Gaddis et al., 2023); the Chinese rovers Yutu and Yutu-2, landed in 2013 and 2019, respectively (Gaddis et al., 2023); the Indian rover Pragyan, landed in 2023 (e.g., Gaddis et al., 2023); the Japanese rovers LEV-1 and LEV-2/SORA-Q, landed in 2024;
d) Traces of soft landing of different landers/lunar modules, e.g., the U.S. Surveyor 1 and 3 (Fig. 2.16D) to 7 landed between 1966 and 1968 (Lund, 2018; Zhang et al., 2022; Gaddis et al., 2023); seven Russian landers from the Luna program landed between 1963 and 1976 (Harvey, 2007; Lund, 2018; Gaddis et al., 2023); the Chinese Chang'e 3–5 landed between 2013 and 2020 (Zhang et al., 2022; Gaddis et al., 2023); the Indian Chandrayaan-3 landed on 2023 (Gaddis et al., 2023); the Japanese SLIM landed on 2024 (Gaddis et al., 2023);
e) Traces of take-off of different landers/lunar modules, e.g., the crewed U.S. missions Apollo 11, 12, and 14–17 took off and returned to the Earth (e.g., Lund, 2018);
f) Traces of crashed modules, e.g., the first human-made object to reach the Moon (crashing) was the spacecraft Luna 2 (Soviet Union) in 1959 (Harvey, 2007; Lund, 2018; Gaddis et al., 2023); Ranger 4 was the first U.S. spacecraft to impact the Moon in 1962 (Lund, 2018; Gaddis et al., 2023); Beresheet (Israel) crashed on 2019 (e.g., Gibney, 2019);
g) Drilling borings. See Baucon et al. (2017), Díaz-Martínez et al. (2021) further;
h) Other traces such as traces of transport and placement of scientific equipment or surface sampling marks (see Díaz-Martínez et al., 2021).

With respect to Mars, the main technotraces are the result of uncrewed robotic missions:

a) Rover trails (Geissler et al., 2010; Shayler et al., 2005; Zhang et al., 2022), e.g., rover Sojourner (USA) landed on Mars in 1997; rovers Spirit (USA) and Opportunity (USA) landed in 2004 (Fig. 2.16E); rover Curiosity (USA) landed in 2011 (Fig. 2.16F); rover Perseverance (together with the helicopter Ingenuity) (USA) landed in 2020; rover Zhurong (China) landed in 2021 (Wu et al., 2021);

b) Traces of soft landings, e.g., Viking 1 and Viking 2 (USA) landed in 1976 (Shayler et al., 2005); Mars Pathfinder (USA) landed in 1997 (Shayler et al., 2005); Phoenix (USA) landed in 2008 (Grover III and Cichy, 2011); InSight (USA) landed in 2018 (Banerdt et al., 2020); Tianwen-1 (China) landed in 2021 (Wu et al., 2021);

c) Traces of crashed modules, e.g., Mars 3 (Soviet Union) crashed in 1971 (Shayler et al., 2005);

d) Drilling borings (Fig. 2.16G), e.g., the coring drill system of Curiosity (Abbey et al., 2019).

In relation to Venus, the technotraces mainly consist of landing and drilling traces left by different uncrewed spacecrafts (Venera 4 to 14; Vega 1 and 2) sent by the Soviet Union within the Vera and Venera programs between 1967 and 1985 (e.g., Harvey, 2007). In fact, Venera 3 (crashed in 1966) became the first probe to reach the surface of another planet (e.g., Johnson, 1979).

The space probe Huygens, built and operated by the European Space Agency (ESA), landed on Titan, the largest moon of Saturn, in 2005; so, landing traces were very likely produced on its surface (Lebreton et al., 2017).

Several asteroids have been reached by human technology. For example, the U.S. space probe NEAR Shoemaker landed on the regolith surface of the near-Earth stony asteroid 433 Eros in 2001 (Veverka et al., 2001); in 2020, the spacecraft OSIRIS-REx (USA) impacted and collected samples from 101,955 Bennu (a carbonaceous near-Earth asteroid) producing alterations on its regolith surface (Lauretta et al., 2017); the Japanese spacecraft Hayabusa2 landed on the regolith surface of the near-Earth asteroid 162,173 Ryugu on 2018, collected samples, deployed three rovers (Hibou, Owl, and Mascot) and took off to return to the Earth (Tsuda et al., 2020).

Comets have also been altered by human activity. The U.S. spacecraft Deep Impact collected samples in 2005 from the nucleus of the Jupiter-Family comet Tempel 1 (9P/Tempel) forming an impact crater 150 m in diameter (A'Hearn et al., 2005). The robotic lander Philae (ESA) and the spacecraft Rosetta (ESA) landed on the Jupiter-Family comet 67P/Churyumov-Gerasimenko in 2014 and 2016, respectively (e.g., Ulamec et al., 2016).

In sum, human space exploration results in a recurrent association of traces, which include (1) astronaut bootprints, (2) *Trypanites*-like borings created by core samplers and probe impacting, and (3) bioturbational structures formed by scoop-like tools, rovers, soft landings, and crashed modules. This raises an intriguing question: does this set of ichnites constitute an "astronaut ichnofacies," indicative of extreme environments?

In the Seilacherian sense, this ichnoassociation of traces does not fit the definition of an ichnofacies. This is primarily because it does not recur consistently in space and time, as the conventional concept of ichnofacies necessitates (e.g., Buatois and Mángano, 2011; MacEachern et al., 2007, 2012). Therefore, while these structures provide significant insights into the environmental conditions concurrent with their formation, they diverge from the traditional understanding of Seilacherian ichnofacies. Nevertheless, it is important to underline that the "astronaut ichnoassociation" is dominated by vertebrate ichnites, and as such it may fit within the concept of biotaxonomical ichnofacies. Classic vertebrate ichnofacies are biotaxonomical in nature; hence, they are neither directly comparable nor can they be subsumed within Seilacherian ethoichnofacies, which are categorized based on behavior (Hunt and Lucas, 2016; Lockley, 2007).

The bootprints left by astronauts on the lunar surface hold profound cultural significance as they bear witness to the location where humans first ventured onto another celestial body (Spennemann, 2004). These traces serve as enduring symbols of human achievement, marking a pivotal moment in our evolutionary journey. On the Moon, the lack of atmosphere enhances the preservation potential of astronaut trackways, whereas the emerging trend of space tourism, which includes proposals for lunar heritage tourism, poses a potential threat to the preservation of this invaluable ichnological heritage (Spennemann, 2004). As potential sites for space tourism, astronaut track sites are particularly at risk, thus raising substantial concerns for heritage management. Consequently, there is a pressing need to extend the concept of planetary protection to include ichnological protection (sensu Baucon et al., 2017). This would require a comprehensive approach that considers not only the physical preservation of these sites but also the promotion of their cultural and scientific value. In this way, we can ensure that these symbolic footprints of our space explorations continue to inspire future generations.

7. Conclusions

This chapter explores the application of ichnological methodologies and approaches to the study of the hominin track record, under the realm of ichnoarcheology. An updated review of the hominin track, organized through the major threshold events that definitely influenced the record of hominin evolution (*sensu* Lockley et al., 2016), summarizes 115 ichnosites worldwide. It is an unprecedented compilation resulting from the burst of interest and applications in different areas of knowledge during the past 2 decades. The substantial increase in the number of hominin ichnosites has been largely caused by the raising awareness of archeologists of the potential to find footprints in excavations and their usefulness as direct evidence for human presence.

The African continent reveals some of the oldest hominin footprints, attributed to *A. afarensis*, at the most famous tracksite of Laetoli, almost 3.7 Ma in age. These footprints have been hotly debated regarding locomotion and speed, body size, and behavior, since they preserve early direct evidence of human-like bipedal biomechanics. The most important *H. erectus* tracksites are found in Ethiopia and Kenya, with ages between 1.5 and 0.8 Ma.

The cradle of our own species shows early prints in coastal South Africa. In Europe, the most ancient hominin tracksite is Happisburgh in the United Kingdom, attributed to *H. heidelbergensis*. However, the Trachilos tracksite in Crete is remarkable for exhibiting some hominin-like characteristics in the footprints that were confidently dated to around 6 Ma. This tracksite requires further studies and functional comparisons with contemporaneous fossils of primate feet in order to unlock new perspectives about the evolution of bipedalism. Europe and Western Asia can provide new evidence of the evolutionary paleoecology of Neanderthals, as well as the timing for the expansion of the anatomical modern humans from their African origins. With few exceptions, Eastern Asia remains a huge territory to explore ichnologically, although the discovery of new species, such as *H. floresiensis* in the Philippines, opens the possibility to find new tracksites with important paleobiologial information. Tracksites in the Americas and Oceania have been mostly used as direct evidence for the early colonization and expansion routes by our species.

Pathways for including indigenous expertise in the study of ancient hominin footprints have recently been explored in caves. This complementary approach to ichnology, based on empirical observations and interaction with the environment, accumulated over many generations and has been vital for the subsistence of hunter-gatherers, but has not been widely applied in open-air sites, where indigenous knowledge could optimally be applied to analyze and interpret ecological, social interaction and behavioral data from footprints.

Human-made substrates, such as pavements and ceramics, are also explored in this chapter due to their importance in archeological settlements for the past 25 ka. They provide optimal conditions for the preservation of traces because of their geotechnical properties (fine sediment with variable moisture), open-air drying, fast "diagenesis" in the case of ceramics (pyrometamorphism) and gypsum, lime, and cement plasters (crystallization), resistance, and abundance (providing high preservation potential). Therefore, imprinted ceramic building material and lime plaster indoor and outdoor pavements can give a good insight into daily activities, the contemporary fauna, and environment.

Rock art has long seen limited application of ichnology. The situation has improved with the recent application of knowledge from indigenous trackers, but much potential still can be unlocked by ichnologists working together with rock art experts. Based on the size and shape of prehistoric handprints and stenciled hands, and even dermatoglyphs, attempts have been made to estimate age, body height, or laterality of their producers, although recent attempts to estimate sex from handprints depicted in Paleolithic cave art, using morphometric data from recent populations, cannot be sustained. We also took into consideration the potential inclusion of podomorphs (foot contoured on hard rock), ammoglyphs, finger-flutings, and other deliberate compositions of foot- and handprints in natural and man-made substrates subjected to fast diagenesis, under the realm of ichnoarcheology. If their chronology, symbolic/ritualistic representation, and cognitive significance can be left for other sciences, the study of their morphology to estimate producer's size, age, and even handedness or pathologies, and especially the way they interacted with the substrates to achieve the imagined composition, should not be left behind by ichnologists. One promising avenue to document the passage of humans in caves not covered in this volume or in this chapter is the study of torch marks on cave walls (Watson, 1969; Fosaaen, 2021; Galant et al., 2021; Toomey, personal communication, 2024). Unlike footprints on the ground, which can be easily obliterated, the dark charcoal marks from torches are not as subject to destruction by casual or careless visitors moving along the passageways. The torches were made of various materials including twined bundles of river cane (Watson, 1969). The marks are created when humans trim the ash off of the end of a torch by tapping them against the walls of the cave in order to refresh the flame (Watson, 1969; Toomey, pers comm. 2024). Individual cane marks can be distinctive. They can be understood as biodepositional traces and therefore approachable from Ichnology.

The ichnological record of footwear may extend back as early as the Middle Paleolithic, and may have evolved independently in both Neanderthals and Anatomically Modern Humans. In fact, shod footprints provide data most relevant to the origin and function of footwear, much before the oldest preserved pair of sagebrush sandals. However, it is important to take into account the preservational quality of the footprints. Footprints poorly preserved due to erosion, or even underprints, can be easily misinterpreted as evidence for shoe wearing. Footwear is probably the technological advance most related to hominin ichnology.

The shod prints offer a tangible marker of human behavioral exploration beyond home, the expansion of our species to habitats and regions to which we are not physiologically adapted to. Therefore, bootprints preserved on the Moon and other traces left by vehicles during exploration of other planets, moons, and asteroids, represent our extended phenotype sensu Turner (2000). They contribute significantly to understanding the evolutionary trajectory of hominins, as well as to the broader context of astrobiology, with applications of ichnology in the quest for habitable substrates to present and past extraterrestrial life or astrochnology.

Acknowledgments

The authors would like to thank Dr. George Crothers, William S. Webb Museum of Anthropology, University of Kentucky, and Dr. Rickard Toomey, Mammoth Cave National Park, for the details about the Mammoth Cave System and Footprint Cave (Virginia), which can only be gained by experience. The authors hope that their rendition of their knowledge does justice to their words. Dr. Charalambos Fassoulas (University of Crete) is greatly acknowledged for permitting new observations in the Trachilos tracksite and for fruitful discussions during the geoconservation assessment of the site. Dr. Summer Roper Todd from Hawai'i Volcanoes National Park and Dr. David Bustos from White Sands National Park are greatly acknowledged for providing photos. Dr. Aejin Lee from Jeju Special Self-Governing Province, World Heritage Office, Geopark Team, generously provided the photo of a trackway in Sagye (Jeju Island) ichnosite. Dr. A. Serdar Aytaç, coordinator of Kula Salihli UNESCO Global Geopark, was very helpful in providing information and photo regarding the present situation of the Demirköprü site. The authors thank Dr. Silverio Feola, Instituto de Ciencias de la Tierra y Ambientales de La Pampa, Argentina, for the photos of the Argentinean sites. They also thank the archeologist Dr. Ricardo Campos from the Archeological Museum of Odrinhas for the possibility to study the footprints in ceramics and explanations about the archeological settings. Florian Hipp (Austria) for communicating his finding in Cape Trafalgar. Finally, our best thanks to the Editor Spencer G. Lucas (New Mexico Museum of Natural History & Science) and Matthew R. Bennett (Bournemouth University) for patiently reviewing the English grammar of this long text and insightful comments that greatly improved this chapter.

References

Abbey, W., Anderson, R., Beegle, L., Hurowitz, J., et al., 2019. A look back: the drilling campaign of the Curiosity rover during the Mars Science Laboratory's Prime Mission. Icarus 319, 1–13.

Akdeniz, E., 2011. Some evidence on the first known residents of Katakekaumene (Burned Lands). Mediterr. Archaeol. Archaeom. 11 (1), 69–74.

Aldhouse-Green, S.H.R., Housley, R.A., 1993. The Uskmouth mattock: a radiocarbon date. Archaeol. Cambrensis 117, 340.

Aldhouse-Green, S.H.R., Whittle, A., Allen, J.R.L., Caseldine, A.E., Culver, S.J., Day, M.H., Lunqvist, J., Upton, D., 1992. Prehistoric human footprints from the Severn Estuary at Uskmouth and Magor Pill, Gwent, Wales. Archaeol. Cambrensis 141, 14–55.

Alexander, R.M., 1984. Stride length and speed for adults, children, and fossil hominids. Am. J. Phys. Anthropol. 63, 23–27.

Allen, J.R.L., 1997. Subfossil mammalian tracks (Flandrian) in the Severn Estuary, S.W. Britain: mechanics of formation, preservation and distribution. Philos. Transact. R. Soc. B 352 (1352), 481–518.

Allen, J.R.L., Bell, M., Scales, R., 2003. Animal and human footprint-tracks in archaeology: description and significance. Archaeol. Severn Estuary 14, 55–68.

Allton, J.H., 1989. Catalog of Apollo Lunar Surface Geological Sampling Tools and Containers. National Aeronautics and Space Administration, Lyndon B. Johnson Space Center, Houston, Texa, pp. 97–pp.

Altamura, F., 2019. Ichnology and archaeology in the African record: a complementary approach. In: Lernia, S., Gallinaro, S. (Eds.), Archaeology in Africa, vol. 8. Arid Zone Archaeology Monographs, pp. 127–140.

Altamura, F., Mussi, M., 2017. Archeologia e impronte fossili nel sito acheuleano di Gombore II (0,85 Ma), Melka Kunture, Etiopia. Scienze dell'Antichità 23 (1), 21–35.

Altamura, F., Bennett, M.R., D'Août, K., Gaudzinski-Windheuser, S., Melis, R.T., Reynolds, S.C., Mussi, M., 2018. Archaeology and ichnology at Gombore II-2, Melka Kunture, Ethiopia: everyday life of a mixed-age hominin group 700,000 years ago. Sci. Rep. 8, 2815. https://doi.org/10.1038/s41598-018-21158-7.

Altamura, F., Bennett, M.R., Marchetti, L., Melis, R.T., Reynolds, S.C., Mussi, M., 2020. Ichnological and archaeological evidence from Gombore II OAM, Melka Kunture, Ethiopia: an integrated approach to reconstruct local environments and biological presences between 1.2 and 0.85 Ma. Quat. Sci. Rev. 244, 106506. https://doi.org/10.1016/j.quascirev.2020.106506.

Altamura, F., Lehmann, J., Rodríguez-Álvarez, B., Urban, B., van Kolfschoten, T., Verheijen, I., Conard, N.J., Serangeli, J., 2023. Fossil footprints at the late Lower Paleolithic site of Schöningen (Germany): a new line of research to reconstruct animal and hominin paleoecology. Quat. Sci. Rev. 310, 108094. https://doi.org/10.1016/j.quascirev.2023.108094.

Ambert, P., Colomer, A., Galant, P., 2000. Datations mesolithiques des empreintes humaines de l'etage Cathala de la grotte d'Aldéne (Cesseras, Hérault). Earth Planetary Sci. 331, 67–74.

Ambert, P., Galant, P., Guendon, J.-L., Colomer, A., 2007. Les gravures et les empreintes humaines de la grotte d' Aldène (Cesseras, Hérault) dans leur contexte chronologique et culturel. Bull. Musee Anthropol. Prehist. Monaco 47, 3–36.

Applegate, S.P., Espinosa-Arrubarrena, L., López-Neri, P., 1984. Las calizas litográficas de la cantera Tlayúa en Tepexi de Rodríguez, Puebla, México. Sociedad Geológica Mexicana, Convención Geológica Nacional 7, México Excursión to Tepexi de Rodríguez, State of Puebla, pp. 1–83.

Aramayo, S.A., 2004. Late quaternary paleoichnological sites at Southern Atlantic coast of Buenos Aires Province, Argentina. In: Kim, J.Y., Kim, K.-S., Park, S.I., Shin, M.-K. (Eds.), International Symposium on Quaternary Footprints of Hominids and Other Vertebrates. Jeju, Korea, pp. 112–130.

Aramayo, S.A., Manera de Bianco, T., 2009. Late Quaternary paleoichnological sites from the southern Atlantic coast of Buenos Aires, Argentina: mammal, bird and hominid evidence. Ichnos 16 (1/2), 25–32. https://doi.org/10.1080/10420940802470714.

Aramayo, S.A., Manera de Bianco, T., Bastianelli, N.V., Melchor, R.N., 2015. Pehuén Co: updated taxonomic review of a late Pleistocene ichnological site in Argentina. Palaeogeogr. Palaeoclimatol. Palaeoecol. 439, 144–165. https://doi.org/10.1016/j.palaeo.2015.07.006.

Ashton, N., 2021. The Happisburgh footprints and their connections with the past. In: Pastoors, A., Lenssen-Erz, T. (Eds.), Reading Prehistoric Tracks. Springer, Switzerland, pp. 153–158.

Ashton, N., Lewis, S.G., De Groote, I., Duffy, S.M., Bates, M., Bates, R., Hoare, P., Lewis, M., Parfitt, S.A., Peglar, S., Williams, C., Stringer, C., 2014. Hominin footprints from early Pleistocene deposits at Happisburgh, UK. PLoS One 9, e88329.

Atamtürk, D., Özbal, R., Gerritsen, F., Duyar, I., 2018. Analysis and interpretation of Neolithic period footprints from Barcin Höyük, Turkey. Mediterr. Archaeol. Archaeomet. 18 (1), 163–174.

Avanzini, M., Mietto, P., Panarello, A., De Angelis, M., Rolandi, G., 2008. The Devil's Trails: Middle Pleistocene human footprints preserved in a volcanoclastic deposit of Southern Italy. Ichnos 15 (3/4), 179–189.

Aubert, M., Brumm, A., Huntley, J., 2018. Early dates for "Neanderthal cave art" may be wrong. J. Hum. Evol. 125, 215–217.

Avanzini, M., Bernardi, M., Petti, F.M., 2011. Soldier tracks in a First World War Fort (Valmorbiawerk, Trento, Italy). Ichnos 18, 72–78.

Avanzini, M., Romano, M., Citton, O., Salvador, I., Arobba, D., Caramiello, R., Firpo, M., Rellini, I., Negrino, F., Clementi, L., Zunino, M., Giannotti, S., Starnini, E., Conventi, M., 2018. Ichno-archeology of a human Palaeolithic ecosystem: the human and animal footprints in the Grotta della Bàsura (Toirano, Northern Italy). Alpine Mediterr. Quater. 31, 39–42.

Avanzini, M., Citton, P., Mietto, P., Panarello, A., Raia, P., Romano, M., Salvador, I., 2020. Human footprints from Italy: the state of the art. J. Mediterr. Earth Sci. 12, 213–232.

Avanzini, M., Salvador, I., Starnini, E., Arobba, D., Caramiello, R., Romano, M., Citton, O., Rellini, I., Firpo, M., Zunino, M., Negrino, F., 2021. Following the father steps in the bowels of the Earth: the ichnological record from the Bàsura Cave (Upper Palaeolithic, Italy). In: Pastoors, A., Lenssen-Erz, T. (Eds.), Reading Prehistoric Tracks. Springer, Switzerland, pp. 251–276.

Aveleyra Arroyo de Anda, L., 1950. Prehistoria de Mexico. Editiones Mexicanas, S.A., 167 pp.

Aytaç, A.S., Demir, T., Maddy, D., Bridgland, D.R., 2022. The Kula–Salihli UNESCO Geopark: spectacular records of quaternary volcanism, fluvial and landscape. Proc. Geologists' Assoc. https://doi.org/10.1016/j.pgeola.2022.06.008.

A'Hearn, M.F., Belton, M.J.S., Delamere, W.A., Kissel, J., et al., 2005. Deep impact: excavating comet Tempel 1. Science 310, 258–264.

Bahn, P.G., Vertut, J., 1988. Images of the Ice Age. Facts on file, New York.

Bakels, C., Zeiler, J., 2005. Neolithic subsistence. In: Louwe Kooijmans, L.P., van den Broeke, P.W., Fokkens, H., van Gijn, A.L. (Eds.), The Prehistory of the Netherlands Volume 1. Amsterdam University Press, pp. 311–366.

Banerdt, W.B., Smrekar, S.E., Banfield, D., Giardini, D., et al., 2020. Initial results from the InSight mission on Mars. Nat. Geosci. 13, 183–189.

Barr, K., 2018. Prehistoric Avian, Mammalian and *H. sapiens* Footprint-Tracks From Intertidal Sediments as Evidence of Human Palaeoecology. PhD Thesis. University of Reading.

Bar-Oz, G., Tepper, Y., 2010. Out on the tiles: animal footprints from the Roman site of Kefar 'Othnay (Legio), Israel. Near E. Archaeol. 73 (4), 244–247.

Barr, K., Bell, M., 2016. Neolithic and Bronze Age ungulate footprint-tracks of the Severn Estuary: species, age, identification and the interpretation of husbandry practices. Environ. Archaeol. https://doi.org/10.1080/14614103.2016.1222691.

Barriere, C., Sahly, A., 1964. Les empreintes Humaines de Lascaux. Miscelanea en homenaje a Abate H. Breuil, Barcelona, pp. 173–180.

Bartsocas, C.S., 1982. Paleodermatoglyphics. In: Bartsocas, C.S. (Ed.), Progress in Dermatoglyphic Research. A. Liss, Inc., New York.

Baucon, A., Privitera, S., Morandi Bonacossi, D., Canci, A., Neto de Carvalho, C., Kyriazi, E., Laborel, J., Laborel-Deguen, F., Morhange, C., Marriner, N., 2008. Principles of Ichnoarchaeology: new frontiers for studying past times. In: Avanzini, M., Petti, F.M. (Eds.), Italian Ichnology. Studi Trentini Science Naturali, Acta Geologica, vol 83, pp. 43–72.

Baucon, A., Neto de Carvalho, C., Barbieri, R., Bernardini, F., Cardini, A., Cavalazzi, B., Celani, A., Felletti, F., Ferretti, A., Schoenlaub, H.P., Todaro, A., Tuniz, C., 2017. Organism-substrate interactions and astrobiology: potential, models, methods. Earth Sci. Rev. 171, 141–180.

Baucon, A., Neto Carvalho, C., Felletti, F., Cabella, R., 2020. Ichnofossils, cracks or crystals? A test for biogenicity of stick-like structures from Vera Rubin Ridge, Mars. Geosciences 10. https://doi.org/10.3390/geosciences10020039.

Baucon, A., Neto de Carvalho, C., Briguglio, A., Piazza, M., Felletti, F., 2021. A predictive model for the ichnological suitability of the Jezero crater, Mars: searching for fossilized traces of life-substrate interactions in the 2020 Rover Mission Landing Site. PeerJ 9, e11784. https://doi.org/10.7717/peerj.11784.

Bayón, C., Politis, G., 1996. Estado actual de las investigaciónes en el sitio Monte Hermoso 1 (Provincia de Buenos Aires). Arqueología (Inst.de Ciencias Antropológicas, Facultad de Filosofia y Letras) Universidad de Buenos Aires 6, 83–116.

Bayón, C., Manera, T., Politis, G., Aramayo, S., 2011. Following the tracks of the first South Americans. Evo Edu Outreach 4, 205–217.

Beard, J.B., 1997. Lon Odell Memorial Cave, a Dade County treasure. Missouri Cave and Karst Conservancy Digest 4, 43–49.

Bednarik, R.G., 1984. Die Bedeutung der paläolithischen Fingerlinientradition. Anthropologie 23, 73–79.

Bednarik, R.G., 1986. Parietal finger markings in Europe and Australia. Rock Art Res. 3, 30–61.

Bednarik, R.G., 2008. Children as Pleistocene artists. Rock Art Res. 25, 173–182.

Bégouën, R., Fritz, C., Tosello, G., Clottes, J., Pastoors, A., Faist, F., 2009. Le Sanctuaire secret des bisons. Il y a 14000 ans, dans la caverne du Tuc d'Audoubert. Editions d'art Somogy Association Louis Bégouën, Paris.

Behrensmeyer, A.K., Laporte, L.F., 1981. Footprints of a Pleistocene hominid in northern Kenya. Nature 289, 167–169.

Bell, M., 2007a. Prehistoric coastal communities: the mesolithic in Western Britain. CBA Research Report 149. York.

Bell, M., 2007b. Banded sediments with footprint-tracks. In: Bell, M. (Ed.), Prehistoric Coastal Communities: The Mesolithic in Western Britain. Council for British Archaeology, York, pp. 48–55.

Bell, M., 2013. The Bronze Age in the Severn Estuary. Council for British Archaeology Research Report 172. York.

Bell, M., 2020. Making One's Way in the World: The Footprints and Trackways of Prehistoric People. Oxbow Books, Oxford.

Bell, M., Caseldine, A.E., Neumann, H. (Eds.), 2000. Prehistoric Intertidal Archaeology in the Welsh Severn Estuary, vol 120. Council of British Archaeology, Research Report, York.

Belperio, A.P., Fotheringham, D.G., 1990. Geological setting of two Quaternary footprint sites, western South Australia. Aust. J. Earth Sci. 37, 37–42.

Bennett, D., 2012. Life history information from tracks of domestic dogs (*Canis familiaris*) in ceramic building materials from a Roman bath-house at Vindolanda, Northumberland, England. Archaeofauna 21, 7–36.

Bennett, D., Campbell, G., Timm, R.M., 2016. The dogs of Roman Vindolanda, Part I: morphometric techniques useful in differentiating domestic and wild canids. Archaeofauna 25, 79–106.

Bennett, M.R., Morse, S.A., 2014. Human Footprints: Fossilised Locomotion? Springer, 228 pp.

Bennett, M.R., Reynolds, S.C., 2021. Inferences from footprints: archaeological best practice. In: Pastoors, A., Lenssen-Erz, T. (Eds.), Reading Prehistoric Human Tracks. Springer, Switzerland, pp. 15–39.

Bennett, M.R., Harris, J.W.K., Richmond, B.G., Braun, D.R., Mbua, E., Kiura, P., Olago, D., Kibunjia, M., Omuombo, C., Behrensmeyer, A.K., et al., 2009. Early hominin foot morphology based on 1.5-million-year-old footprints from Ileret, Kenya. Science 323 (5918), 1197–1201.

Bennett, M.R., Gonzalez, S., Huddart, D., Kirby, J., Toole, E., 2010a. Probable Neolithic footprints preserved in inter-tidal peat at Kenfig, South Wales (UK). Proc. Geol. Assoc. 121, 66–76.

Bennett, M.R., Liutkus, C.M., Thackeray, F., Morse, S.A., McClymont, J., Stratford, D., 2010a. Footprints of the Kuiseb Delta, Namibia. The Digging Stick 27 (3), 1–4.

Bennett, M.R., Morse, S.A., Falkingham, P.L., 2014a. Tracks made by swimming hippopotami: an example from Koobi Fora (Turkana Basin, Kenya). Palaeogeogr. Palaeoclimatol. Palaeoecol. 409, 9–23.

Bennett, M.R., Morse, S.A., Liutkus-Pierce, C., Mcclymont, J., Evans, M., Crompton, R.H., Thackeray, F., 2014b. Exceptional preservation of children's footprints from a Holocene footprint site in Namibia. J. Afr. Earth Sci. 97, 331–341.

Bennett, M.R., Bustos, D., Odess, D., Urban, T.M., Lallensack, J.N., Budka, M., Santucci, V.L., Martinez, P., Wiseman, L.A., Reynolds, S.C., 2020. Walking in mud: remarkable Pleistocene human trackways from White Sands National Park (New Mexico). Quat. Sci. Rev. 249, 106610.

Bennett, M.R., Bustos, D., Pigati, J.S., Springer, K.B., Urban, T.M., Holliday, V.T., Reynolds, S.C., Budka, M., Honke, J.S., Hudson, A.M., 2021. Evidence of humans in North America during the last glacial maximum. Science 373, 1528–1531.

Bennett, M.R., Urban, T.M., Bustos, D.F., Reynolds, S.C., Jolie, E.A., Strehau, H.C., Odess, D., Springer, K.B., Pigati, J.S., 2025. The ichnology of White Sands (New Mexico): linear traces and human footprints, evidence of transport technology? Quat. Sci. Adv. 17, 100274.

Bertilsson, U., 2013. Divine footprints. Traces of cosmological archetypes and prehistoric religion on the rock faces. In: Anati, E. (Ed.), Art as Source of History. Edizione del Centro, Capo di Ponte, pp. 163–172.

Bertling, M., Braddy, S.J., Bromley, R.G., Demathieu, G.R., Genise, J., Mikuláš, R., Nielsen, J.K., Nielsen, K.S.S., Rindsberg, A.K., Schlirf, M., Uchman, A., 2006. Names for trace fossils: a uniform approach. Lethaia 39, 265–286. https://doi.org/10.1080/00241160600787890.

Bice, D.C., 1979. Tephra correlation and the age of human footprints near Managua, Nicaragua. Geological Society of America abstracts with programs 11, 388.

Billard, C., Bernard, V., 2016. Pêcheries de Normandie: Archéologie et histoire des pêcheries littorales du département de la Manche. Presses Universitaires de Rennes, collection Archéologie et Culture, Rennes.

Blanc, A.C., 1960. Le palline d'argilla della Grotta della Bàsura. Rivista di Studi Liguri 26, 9–25.

Brady, J., Fahsen, F., 1991. The discovery of a new Maya cave painting site in Guatemala. Explor. J. 69 (2), 52–55.

Brady, J., Rissolo, D., 2006. A reappraisal of ancient Maya cave mining. J. Anthropol. Res. 62 (4), 471–490.

Brady, J., Stone, A., 1986. Naj Tunich: entrance to the Maya underworld. Archaeology 39 (6), 18–25.

Brady, J.E., Stone, A., 2019. Maya caves. In: White, W.B., Culver, D.C., Pipan, T. (Eds.), Encyclopedia of Caves, third ed. Academic Press, London, pp. 686–690.

Brinton, D.G., 1887. On an ancient human footprint from Nicaragua. Proc. Am. Phil. Soc. 24, 437–444.

Brodribb, G., 1987. Roman Brick and Tile. Sutton Publishing, Gloucester.

Broholn, H.C., 1946. Danmarks Broncealder 3. Copenhagen, pp. 1–278.

Bromley, R.G., Milàn, J., Uchman, A., Hansen, K.S., 2009. Rheotactic *Macaronichnus*, and Human and Cattle trackways in Holocene beachrock, Greece: reconstruction of paleoshoreline orientation. Ichnos 16, 103–117.
Brown, R.W., 1947. Fossil plants and human footprints in Nicaragua. J. Paleontol. 21, 38–40.
Bruker, R.W., Watson, R.A., 1976. The Longest Cave. Knopf, New York, 316 pp.
Brumm, A., Oktaviana, A.A., Burhan, B., Hakim, B., Lebe, R., Zhao, J.-X., Sulistyarto, P.H., Ririmasse, M., Adhityatama, S., Sumantri, I., Aubert, M., 2021. Oldest cave art found in Sulawesi. Sci. Adv. 7, eabd4648.
Bryan, A.L., 1973. New light on ancient Nicaraguan footprints. Archaeology 26, 146–147.
Buatois, L.A., Mángano, M.G., 2011. Ichnology: Organism-Substrate Interactions in Space and Time. Cambridge University Press, Cambridge/New York.
Burns, A., 2021. The Mesolithic footprints retained in one bed of the former saltmarshes at Formby Point, Sefton Coast, North West England. In: Pastoors, A., Lenssen-Erz, T. (Eds.), Reading Prehistoric Tracks. Springer, Switzerland, pp. 295–316.
Burns, A., Woodward, J., Conneller, C., Reimer, P., 2022. Footprint beds record Holocene decline in large mammal diversity on the Irish Sea coast of Britain. Nat. Ecol. Evol. 6, 1553–1563.
Bustos, D., Jakeway, J., Urban, T.M., Holliday, V.T., Fenerty, B., Raichlen, D.A., Budka, M., Reynolds, S.C., Allen, B.D., Love, D.W., Santucci, V.L., Odess, D., Willey, P., Mcdonald, H.G., Bennett, M.R., 2018. Footprints preserve terminal Pleistocene hunt? Human-sloth interactions in North America. Sci. Adv. 4 (4), eaar7621.
Caninas, J.C., Sabrosa, A., Henriques, F., Monteiro, J.L., Carvalho, E., Batista, Á., Chambino, M., Henriques, F.R., Monteiro, M., Canha, A., Carvalho, L., Germano, A., 2008. Novos dados para o conhecimento da Pré-História recente do Maciço Central na Beira Interior Sul. Tumuli e gravuras rupestres na Serra Vermelha e na Serra de Alvélos (Oleiros -Castelo Branco). Actas das I Jornadas do Património, Belmonte 1–38.
Cartailhac, E., Breuil, H., 1907. Une seconde campagne aux cavernes ornées de Niaux (Ariège) et Gargas (Haute-Pyrénées). Académie des Inscriptions et Belles-Lettres 213–222.
Casteret, N., 1948. The footprints of prehistoric man: a vivid new evidence of our ancestors of fifteen thousand years ago. Illus. Lond. News 110, 408–409.
Cerilli, E., Celant, A., Giordano, G., Carboni, G., 2020. Casetta Mistici e Tor Vergata-San Gaudenzio (Roma): impronte di animali ed umane su una superficie di lahar. In: Anzidei, A.P., Carboni, G. (Eds.), Roma Prima del Mito 1. Geologia, Ambiente e Fonti Archeologiche. Archaeopress Publishing, Oxford, pp. 15–32.
Chamberlain, A.T., Sellars, W., Murphy, P., Goddard, A., 1998. The archaeology of Rawthey Cave, Sedbergh, Cumbria. Archaeol. North 14, 11–20.
Chauvet, J.-M., Deschamps, E.B., Hillaire, C., 1996. Dawn of Art: The Chauvet Cave. Harry N. Abrams, New York.
Chen, F., Xia, H., Jia, Z., Zhang, D., 2022. Earliest hand- and footprint art indicates that Denisovans may have occupied the interior of the high-altitude Tibetan Plateau since 200 thousand years ago. Sci. China Earth Sci. 65, 769–772. https://doi.org/10.1007/s11430-021-9899-y.
Chiappella, V.G., 1952. Orsi e uomini preistorici nella Grotta della Strega. Rivista del Comune di Genova 29, 22–29.
Citton, P., Romano, M., Salvador, I., Avanzini, M., 2017. Reviewing the upper Pleistocene human footprints from the 'Sala dei Misteri' in the Grotta della Bàsura (Toirano, northern Italy) cave: An integrated morphometric and morphoclassificatory approach. Quat. Sci. Rev. 169, 50–64.
Clottes, J., 1975. Circonscription de Midi-Pyrénées. Gall. Prehist. 18 (2), 613–650.
Clottes, J., 2003. Return to Chauvet Cave. Thames and Hudson, London.
Clottes, J., Simonnet, R., 1972. Le réseau René Clastres de la Caverne de Niaux. Bull. Soc. Prehist. Fr. 69, 293–323.
Clottes, J., Valladas, H., Cachier, H., Arnold, M., 1992. Des dates pour Niaux et Gargas. Bulletin de la Société Préhistorique Française 89 (9), 270–274.
Cooper, J., Samson, A., Nieves, M., Lace, M., Caamano-Dones, J., Kambesis, P., Cartwright, C., del Olmo Frese, L., 2016. The Mona Chronicle: the archaeology of early religious encounter in the New World. Antiquity 90, 1054–1071.
Cooper-White, M., 2014. Human Footprints Discovered in Denmark Date Back 5,000 Years. Retrieved 28 January 2024. www.huffpost.com/entry/human_footprints_denmark_n_6152126.
Cowell, R.W., Milles, A., Roberts, G., 1993. Prehistoric footprints on Formby Point beach, Merseyside. In: Middleton, R. (Ed.), North West Wetlands Survey, Annual Report 1993, pp. 43–52.
Cram, L., Fulford, M., 1979. Silchester tile making - the faunal environment. In: McWhirr, A. (Ed.), Roman Brick and Tile. B.A.R. (International Series), vol 68, pp. 124–226.
Crawford, J., 1891. Neolithic man in Nicaragua. Am. Geologist 8, 160–161.

Crompton, R.H., 2017. Making the case for possible hominin footprints from the Late Miocene (c. 5.7 Ma) of Crete? Proc. Geol. Assoc. 128, 692–693.
Crothers, G.M., 1997. Archaeological site inventory form 40RU0090. Submitted to Virginia Department of Historic Resources, Richmond, Virginia.
Crothers, G.M., Faulkner, C.H., Shimek, J.F., Watson, P.J., Willey, P., 2002. Woodland cave archaeology in eastern north America. In: Anderson, D.G., Mainfort, J.,R.C. (Eds.), The Woodland Southeast. Tuscaloosa. University of Alabama Press, pp. 502–524.
Crothers, G.M., Willey, P., Watson, P.J., 2007. Cave archaeology and the NSS: 1941-2006. J. Cave Karst Stud. 69 (1), 27–34.
Day, M.H., Wickens, E.H., 1980. Laetoli Pliocene hominid footprints and bipedalism. Nature 286, 385–387.
De Lumley, H., 1966. Les fouilles de Terra Amata à Nice. Premiers résultats. Bulletin of the Museum of Anthropology and Prehistory of Monaco 13, 29–51.
De Lumley, H., 1967. Découverte d'habitats de l'Acheuléen ancien dans des dépôts mindéliens sur le site de Terra Amata: Etude sédimentologique. Académie Science Paris 264, 801–804.
De Lumley, M.-A., Lamy, P., Mafart, B., 2011. Une empreinte de pied humain acheuléen dans la dune littorale du site de Terra Amata. Ensemble stratigraphique C1b. In: De Lumley, H. (Ed.), Terra Amata: Nice, Alpes-Maritimes, France 2: Palynologie, anthracologie, faunes, mollusques, écologie et biogéomorphologie, paléoanthropologie, empriente de pied humain, coprolithes. CNRS, Paris, pp. 483–507.
DeDekker, P., Bauld, J., Burne, R.V., 1982. Pillie Lake, Eyre Peninsula, South Australia: modern environment and biota, dolomite sedimentation, and Holocene history. Trans. Roy. Soc. S. Aust. 106, 169–181.
Deino, A.L., 2011. $^{40}Ar/^{39}Ar$ dating of Laetoli, Tanzania. In: Harrison, T. (Ed.), Paleontology and Geology of Laetoli: Human Evolution in Context. Springer, pp. 77–97.
Di Vito, M.A., Talamo, P., de Vita, S., Rucco, I., Zanchetta, G., Cesarano, M., 2019. Dynamics and effects of the Vesuvius Pomici di Avellino Plinian eruption and related phenomena on the Bronze Age landscape of Campania region (Southern Italy). Quat. Int. 499, 231–244.
Díaz-Martínez, I., Cónsole-Gonella, C., Citton, P., de Valais, S., 2021. Half a century after the first bootprint on the lunar surface: the ichnological side of the Moon. Earth Sci. Rev. 212, 103452. https://doi.org/10.1016/j.earscirev.2020.103452.
Dillehay, T.D., 1989. Monte Verde: a late Pleistocene settlement in Chile. In: Volume I: The Paleo-Environment and Site Context. Smithsonian Institution Press, Washington.
Dillehay, T.D., Ramirez, C., Pino, M., Collins, M.B., Rossen, J., Pino-Navarro, J.D., 2008. Monte Verde: seaweed, food, medicine and the peopling of South America. Science 320, 784–786.
Dillehay, T.D., Ocampo, C., Saavedra, J., Sawakuchi, A.O., Vega, R.M., Pino, M., Collins, M.B., Cummings, L.S., Arregui, I., Villagran, X.S., et al., 2015. New archaeological evidence for an early human presence at Monte Verde, Chile. PLoS One 10 (11), e0141923.
Dingwall, H.L., Hatala, K.G., Wunderlich, R.E., Richmond, B.G., 2013. Hominin stature, body mass, and walking speed estimates based on 1.5 million-year-old fossil footprints at Ileret, Kenya. J. Hum. Evol. 64, 556–568.
Dobosi, L., 2016. Animal and human footprints on Roman tiles from Brigetio. DissArch 3 (4), 117–133.
Doerr, J.E., 1931. The footprints of Kau. Nat. Notes 1 (7), 52–56.
Douglas, J.G., 2004. The Nature of Warrnambool, Warrnambool. Warrnambool Field Naturalists Club Incorporated.
Doyle, K.L., 2007. Historian Amazing Find. Glamorgan Gazette.
Duday, H., Garcia, M.A., 1983. Les empreintes de l'Homme préhistorique. La grotte du Pech-Merle à Cabrerets (Lot): une relecture significative des traces de pieds humains. Bull. Soc. Prehist. Fr. 208, 215. https://doi.org/10.3406/bspf.1983.5445.
Duveau, J., 2021. The Le Rozel footprints: snapshots of Neandertal groups in the Late Pleistocene. A combined morphometric and experimental approach. Bulletins et Carniers de la Société d'Anthropologie de Paris 33 (2), 53–64.
Duveau, J., Berillon, G., Verna, C., Laisné, G., Cliquet, D., 2019. The composition of a Neandertal social group revealed by the hominin footprints at Le Rozel (Normandy, France). Proc. Natl. Acad. Sci. U. S. A. 116 (39), 19409–19414. https://doi.org/10.1073/pnas.1901789116.
Duveau, J., Berillon, G., Verna, C., 2021. On the tracks of Neandertals: the ichnological assemblage from le Rozel (Normandy, France). In: Pastoors, A., Lenssen-Erz, T. (Eds.), Reading Prehistoric Tracks. Springer, Switzerland, pp. 183–200.

Eadie, G., Waddington, C., 2013. Rescue Recording of an Eroding Inter-tidal Peat Bed at Lower Hauxley, Northumberland (6109). Archaeological Research Service 2013/17, pp. 3–23.
Ellis, W., 1827. In: Fisher, Son, H., Jackson, P. (Eds.), Narrative of a Tour through Hawaii, or Owhyhee. London, 480 pp.
Facorellis, Y., Kyparissi-Apostolika, N., Maniatis, Y., 2001. The cave of Theopetra, Kalambaka: radiocarbon evidence for 50,000 years of human presence. Radiocarbon 43 (2B), 1029–1048.
Faulkner, C.H. (Ed.), 1986. The Prehistoric Native American Art of Mud Glyph Cave. University of Tennessee Press, Knoxville, 124 pp.
Faulkner, C.H., 1988. A study of seven southeastern glyph caves. North Am. Archaeol. 9 (3), 223–246.
Faulkner, C.H., Deane, B., Earnest, H.H., 1984. A Mississippian Period ritual cave in Tennessee. Am. Antiq. 49, 350–361.
Felstead, N.J., Gonzalez, S., Huddart, D., Noble, S.R., Hoffmann, D.L., Metcalfe, S.E., Leng, M.J., Albert, B.M., Pike, A., Gonzalez-Gonzalez, A., Concepcion Jiminez-Lopez, J., 2014. Holocene-aged human footprints from Cuatrociénegas Basin, NE Mexico. J. Archaeol. Sci. 42, 250–259.
Fergusen, L.G., 1983. An Archaeological investigation of TCS#FE60: a cave in north central Tennessee. The NSS Bulletin Quarterly journal of the National Speleological Society 45, 35.
Fiedel, S.J., 2022. Initial human colonization of the Americas, redux. Radiocarbon 64 (4), 845–897.
Finch, R.H., 1947. Kilauea in 1790 and 1823. The Volcano Letter 496, 1–2.
Firth, C.W., Handley, H.K., Cronin, S.J., Turner, S.P., 2014. The eruptive history and chemical stratigraphy of a post-caldera, steady-state volcano: Yasur, Vanuatu. Bull. Volcanol. 76, 1–23.
Flint, E., 1884. Human footprints in Nicaragua. Am. Antiquarian 6, 112–114.
Flint, E., 1886. Pre-Adamite footprints. Am. Antiquarian 8, 230–233.
Flint, E., 1888. Human footprints in the Eocene. Am. Antiquarian 10, 252–254.
Fosaaen, N., 2021. Unmapped Darkness: The Archaeology of Salts Cave's Early Woodland Miners, [Video]. YouTube. https://www.youtube.com/watch?v=uqjCuGLzgHo.
Franklin, J.D., 1999. The rime of the ancient miners. In: MA Thesis, Department of Anthropology. University of Tennessee, Knoxville, 153 pp. https://trace.tennessee.edu/utk_gradthes.
Franklin, J.D., 2002. The prehistory of Fentress County, Tennessee: an archeological survey. In: PHD Thesis, Department of Anthropology. University of Tennessee, Knoxville. https://trace.tennessee.edu/utk_graddiss/3809.
Gaddis, L.R., Joy, K.H., Bussey, B.J., Carpenter, J.D., Crawford, I.A., Elphic, R.C., Halekas, J.S., Lawrence, S.J., Xiao, L., 2023. Recent exploration of the Moon: science from lunar missions since 2006. In: Neal, C.R., Gaddis, L.R., Jolliff, B.L., Lawrence, S.J., Mackwell, S.J., Shearer, C.K., Valencis, S.N. (Eds.), New Views of the Moon 2, vol 89. Reviews in Mineralogy & Geochemistry, pp. 1–51.
Galant, P., Ambert, P., Colomer, A., 2021. Prehistoric speleological exploration in the Cave of Aldène in Cesseras (Hérault, France): human footprint paths and lighting management. In: Pastoors, A., Lenssen-Erz, T. (Eds.), Reading Prehistoric Human Tracks. Springer International Publishing, Cham, Switzerland, pp. 277–293. https://doi.org/10.1007/978-3-030-60406-6_15.
Galeta, P., Bruzek, J., Lázničková-Galetová, M., 2014. Is sex estimation from handprints in prehistoric cave art reliable? A view from biological and forensic anthropology. J. Archaeol. Sci. 45, 141–149.
Garcia, M.-A., 1986. Pour une bibliographie des emprientes humaines et animales. Publications scientifiques du Muséum national d'Histoire naturelle, Paris.
Garcia, M.-A., 1999. La piste de pas humains de la grotte Chauvet à Vallon-Pont d'Arc. Newsletter on Rock Art 24, 1–4.
Garcia, M.-A., 2001. Les empreintes et les traces humaines et animals. In: Clottes, J. (Ed.), La grotte Chauvet: l'art des origines. Threshold, Paris, pp. 34–43.
Garcia, M.-A., 2005. Ichnologie genérale de la grotte Chauvet. Bull. Soc. Prehist. Fr. 102, 103–108.
Garcia, M.-A., Duday, H., 1983. Grotte de Foissac (Aveyron) A propos d'une découverte récente ou de l'ichnologie comme mode d'approche des structures préhistoriques en grotte. Bull. Soc. Prehist. Fr. 80, 184–187. https://doi.org/10.3406/bspf.1983.5441.
Garcia, M.-A., Duday, H., Courtaud, P., 1990. Les empreintes du Réseau Clastres. Préhist. Ariég. 45, 167–174.
Geissler, P.E., Sullivan, R., Golombek, M., Johnson, J.R., Herkenhoff, K., Bridges, N., Vaughan, A., Maki, J., Parker, T., Bell, J., 2010. Gone with the wind: Eolian erasure of the Mars Rover tracks. J. Geophys. Res. 115, E00F11.

Giampaola, D., Laforgia, E., Amato, L., Boenzi, G., 2007. Linea A.V. - Afragola. Impronte di uomini e animali nel Bronzo Antico. In: Atti della XL Riunione Scientifica dell'Istituto Italiano di Preistoria e Protostoria: Strategie di insediamento fra Lazio e Campania in età preistorica e protostorica, Roma, Napoli, Pompei. Istituto Italiano di Preistoria e Protostoria v. II, pp. 928–931.

Gibney, E., 2019. Israeli spacecraft crash-lands on Moon. Nature 568, 286.

Gierliński, G.D., Niedźwiedzki, G., Lockley, M.G., Athanassiou, A., Fassoulas, C., Dubicka, Z., Boczarowski, A., Bennett, M.R., Ahlberg, P.E., 2017. Possible hominin footprints from the late Miocene (c. 5.7 Ma) of Crete? Proc. Geol. Assoc. 128 (5–6), 697–710.

Gomes, M.V., Monteiro, J.P., 1977. As rochas decoradas da Alagoa, Tondela – Viseu. O Arqueólogo Português 7–9, 145–162.

Gonzalez, S., Huddart, D., Bennett, M.R., Gonzalez-Huesca, A., 2006. Human footprints in Central Mexico older than 40,000 years. Quat. Sci. Rev. 25, 201–222.

Gonzalez, A.H.G., Lockley, M.G., Rojas, C.S., Lopez Espinoza, J., Gonzalez, S., 2007. Notes on the rediscovery of a 'lost' hominid footprint site from the Cuatro Cienegas Basin (Coahuila), Mexico. In: Lucas, S., Spielmann, Lockley, M. (Eds.), Cenozoic Vertebrate Tracks and Traces, vol 42. Museum of Natural History and Science Bulletin, New Mexico, pp. 11–15.

Gonzalez, S., Jimenez-Lopez, J.C., Hedges, R., Huddart, D., Ohman, J.C., Turner, A., Pompa y Padilla, J.A., 2003. Earliest humans in the Americas: new evidence from Mexico. J. Hum. Evol. 44, 379–387.

Gonzalez, A.H., Lockley, M., Rojas, C., Espinoza, J.L., Gonzalez, S., 2009. Human tracks from quaternary tufa deposits, Cuatro Cienegas, Coahuila, Mexico. Ichnos 16, 12–24.

Graham, E., McNatt, L., Gutchen, M.A., 1980. Excavations in Footprint Cave, Caves Branch, Belize. J. Field Archaeol. 7, 153–172.

Grover III, Cichy, 2011. Overview of the Phoenix entry, descent, and landing system architecture. J. Spacecraft Rockets 48 (5), 706–712.

Gudea, A., 2004. Domestic animals at Porolissum: a preliminary study on the basis of animal footprints on bricks. Acta Musei Porolissensis 26, 339–353.

Gunn, R.G., 2006. Hand sizes in rock art: interpreting the measurements of hand stencils and prints. Rock Art Res. 23, 97–112.

Haberland, W., Groebe, W.H., 1957. Pre-historic footprints from El Salvador. Am. Antiq. 22 (3), 282–285.

Harada, T., Noto, T., 1984. Volcanic Calamity season. Bulletin of the Gunma Prefectural Museum of History 5, 1–21.

Harrington, S.P.M., 1999. Human footprints at Chauvet Cave. Archeology 52 (5).

Harvati, K., Roksandic, M., 2016. The human fossil record from Romania. Early Upper Paleolithic European mandible, and Neanderthal admixture. In: Harvati, K., Roksandic, M. (Eds.), Paleoanthropology of the Balkans and Anatolia. Springer, Netherlands, pp. 51–68.

Harvey, B., 2007. Soviet and Russian Lunar Exploration. Springer, Chichester (UK), p. 317.

Hasiotis, S.T., Platt, B.F., Hembree, D.I., Everhart, M., 2007. The trace-fossil record of vertebrates. In: Miller, W.I.I.I. (Ed.), Trace Fossils-Concepts, Problems, Prospects. Elsevier, Amsterdam.

Hatala, K., Roach, N., Ostrofsky, K., Wunderlich, R.E., Dingwall, H.L., Villmoare, B., Green, D.J., Harris, J.W.K., Braun, D.R., Richmond, B.G., 2016. Footprints reveal direct evidence of group behavior and locomotion in *Homo erectus*. Sci. Rep. 6, 28766. https://doi.org/10.1038/srep28766.

Hatala, K.G., Roach, N.T., Behrensmeyer, A.K., 2023a. Fossil footprints and what they mean for hominin paleobiology. Evol. Anthropol. 32 (1), 39–53.

Hatala, K.G., Roach, N.T., Behrensmeyer, A.K., Falkingham, P.L., et al., 2024. Footprint evidence for locomotor diversity and shared habitats among early Pleistocene hominins. Science 286, 1004–1010.

Hatala, K.G., Roach, N.T., Ostrofsky, K.R., Wunderlich, R.E., Dingwall, H.L., Villmoare, B.A., Green, D.J., Braun, D.R., Harris, J.W., Behrensmeyer, A.K., Richmond, B.G., 2017. Hominin track assemblages from Okote Member deposits near Ileret, Kenya, and their implications for understanding fossil hominin paleobiology at 1.5 Ma. J. Hum. Evol. 112, 93–104.

Hatala, K.G., Gatesy, S.M., Falkingham, P.L., 2023b. Arched footprints preserve the motions of fossil hominin feet. Nat. Ecol. Evol. 7, 32–41. https://doi.org/10.1038/s41559-022-01929-2.

Hatala, K.G., Falkingham, P.L., Gatesy, S.M., 2022. Arched footprints reveal bipedal kinematics, not arched foot anatomy, of fossil hominins. American Journal of Physical Anthropology 177 (S73), 80–81.

Hatala, K.G., Harcourt-Smith, W.E., Gordon, A.D., Zimmer, B.W., Richmond, B.G., Pobiner, B.L., Green, D.J., Metallo, A., Rossi, V., Liutkus-Pierce, C.M., 2020. Snapshots of human anatomy, locomotion, and behavior from Late Pleistocene footprints at Engare Sero, Tanzania. Sci. Rep. 10, 7740. https://doi.org/10.1038/s41598-020-64095-0.

Haynes, C.V., 2022. Evidence for humans at White Sands National Park during the last glacial maximum could actually be for Clovis people ~ 13,000 years ago. PaleoAmerica 8 (2), 95–98.

Hays-Gilpin, K., Deegan, A.C., Morris, E.A., 1998. Prehistoric sandals from northeastern Arizona: the Earl H. Morris and Ann Axtell Morris research. University of Arizona Anthropological Papers 62, 150 pp.

Hedges, R.E.M., Pettitt, P.B., Ramsey, C.B., Van Klinken, G.J., 1998. Radiocarbon dates from the Oxford AMS system: archaeometry datelist 26. Archaeometry 40, 437–455. https://doi.org/10.1111/j.1475-4754.1998.tb00848.x.

Helm, C.W., McCrea, R.T., Cawthra, H.C., Cowling, R.M., Lockley, M.G., Marean, C.W., Thesen, G.H.H., Pigeon, T., Hattingh, S., 2018. A new Pleistocene hominin tracksite from the Cape south coast, South Africa. Sci. Rep. 8 (1), 3772. https://doi.org/10.1038/s41598-018-22059-5.

Helm, C.W., Cawthra, H.C., De Vynck, J.C., Helm, C.J.Z., Rust, R., Stear, W., 2019. Patterns in the sand: a Pleistocene hominin signature along the South African coastline? Proc. Geol. Assoc. 130 (6), 719–740. https://doi.org/10.1016/j.pgeola.201b9.08.004.

Helm, C.W., Lockley, M.G., Cawthra, H.C., De Vynck, J.C., Dixon, M.G., Helm, C.J.Z., Thesen, G.H.H., 2020. Newly identified hominin trackways from the Cape south coast of South Africa. South Afr. J. Sci. 116 (9/10). https://doi.org/10.17159/sajs.2020/8156. #8156.

Helm, C.W., Cawthra, H.C., De Vynck, J.C., Helm, C.J.Z., Rust, R., Stear, W., 2021. Large geometric patterns from the Middle Stone Age in aeolianites on the Cape south coast, South Africa. Rock Art Res. 38 (1), 10–22.

Helm, C.W., Cawthra, H.C., De Vynck, J.C., Hattingh, R., Lockley, M.G., 2022. Possible Pleistocene hominin tracks from South Africa's west coast. South Afr. J. Sci. 118 (1/2). https://doi.org/10.17159/sajs.2022/11842. Art. #11842.

Helm, C.W., Carr, A.S., Lockley, M.G., Cawthra, H.C., De Vynck, J.C., Dixon, M.G., Stear, W., 2023a. Dating the Pleistocene hominin ichnosites on South Africa's Cape south coast. Ichnos 30 (1), 49–68. https://doi.org/10.1080/10420940.2023.2204231.

Helm, C.W., Lockley, M.G., Cawthra, H.C., De Vynck, J.C., Dixon, M.G., Rust, R., Stear, W., Zipfel, B., 2023b. Possible shod-hominin tracks on South Africa's Cape coast. Ichnos 30 (2), 79–97. https://doi.org/10.1080/10420940.2023.2249585.

Henshilwood, C.S., d'Errico, F., Marean, C.W., Milo, R.G., Yates, R., 2001. An early bone tool industry from the Middle Stone Age at Blombos Cave, South Africa: implications for the origins of modern human behaviour, symbolism and language. J. Hum. Evol. 41 (6), 631–678. https://doi.org/10.1006/jhev.2001.0515.

Henshilwood, C.S., d'Errico, F., Yates, R., Jacobs, Z., Tribolo, C., Duller, G.A.T., Mercier, N., Sealy, J.C., Valladas, H., Watts, I., Wintle, A.G., 2002. Emergence of modern human behavior: middle Stone Age engravings from South Africa. Science 295 (5558), 1278–1280. https://doi.org/10.1126/science.1067575.

Higgs, W., 2001. Catalogue of animal-imprinted ceramic building materials. In: Birley, A. (Ed.), Vindolanda's Military Bath Houses. The Excavations of 1970 and 2000. The Vindolanda Trust, Bardon Mill, Northumberland, pp. 51–65.

Hoffmann, D.L., Standish, C.D., García-Diez, M., Pettitt, P.B., Milton, J.A., Zilhão, J., et al., 2018. U-Th dating of carbonate crusts reveals neandertal origin of iberian cave art. Science 359, 912–915. https://doi.org/10.1126/science.aap7778.

Holowka, N.B., Neill, M.C., Thompson, N.E., Demes, B., 2017. Chimpanzee and human midfoot motion during bipedal walking and the evolution of the longitudinal arch of the foot. J. Hum. Evol. 104, 23–31. https://doi.org/10.1016/j.jhevol.2016.12.002.

Houck, K.J., Lockley, M.G., Avanzini, M., 2009. A survey of tetrapod tracksites preserved in pyroclastic sediments, with special reference to footprints of hominids, other mammals and birds. Ichnos 16 (1), 76–97.

Hovey, H.C., 1878. Discoveries in western caves. Am.J. Sci. Arts. Ser. 3 (16), 465–471, 91–96.

Huddart, D., Bennett, M.R., Gonzalez, S., Velay, X., 2008. Ash analysis and preservation of Pleistocene human and animal footprints: an example from Toluquilla, Valsequillo Basin (Central Mexico). Ichnos 15 (3), 232–245.

Huddart, D., Gonzalez, S., Robert, G., 1999a. The archaeological record and mid-Holocene palaeoenvironments around Liverpool Bay. J. Quat. Sci. 14, 563–574.

Huddart, D., Roberts, G., Gonzalez, S., 1999b. Holocene human and animal footprints and their relationships with coastal environmental change, Formby Point, NW England. Quat. Int. 55, 29–41.

Huddart, D., Bennett, M.R., Gonzalez, S., Velay, X., 2007. Analysis and preservation of Pleistocene human and animal footprints: an example from Toloquilla, Valsequillo basin (central Mexico). Ichnos (in press).

Huisheng, T., Wangdui, X., Jie, Y., Anni, J., Ge, C., Lanying, S., Bednarik, R.G., 2024. The hand and footprints at Qiusang in Tibet: recommendations for dating rock art by U-Th. J. Archaeol. Sci. 161, 105899.

Hunt, A.P., Lucas, S.G., 2016. The case for archetypal vertebrate ichnofacies. Ichnos 23, 237–247. https://doi.org/10.1080/10420940.2016.1164153.

Hunt, A.P., Lucas, S.G., 2007. Cenozoic vertebrate trace fossils of North America: ichnofaunas, ichnofacies and biochronology. In: Lucas, S.G., Spielmann, Lockley, M.G. (Eds.), Cenozoic Vertebrate Tracks and Traces, vol 42. New Mexico Museum of Natural History and Science Bulletin, pp. 17–41.

Jacobs, Z., Roberts, D.L., 2009. Last interglacial age for aeolian and marine deposits and the Nahoon fossil human footprints, southeast coast of South Africa. Quat. Geochronol. 4, 160–169. https://doi.org/10.1016/j.quageo.2008.09.002.

Jaggar, T., 1921. Fossil human footprints in Kau Desert. Hawaii Volcano Observatory Monthly Bulletin 9, 114–118.

Jaggar, T.A., 1930. The Human footprints in Kilauea's ash bed. The Volcano Letter 273, 1–3.

Jaggar, T.A., 1934. Fossil human footprints in the Kau Desert. Paradise of the Pacific 46 (5), 5.

Johnson, G.H., 1884. Human footprints on sandstone near Managua in Nicaragua. Am. J. Sci. 27, 239–240.

Johnson, N.L., 1979. Handbook of soviet lunar and planetary exploration. NASA STI/Recon Technical Report A 80, 276 pp.

Kamp, K.A., Timmerman, N., Lind, G., Graybill, J., Natowski, I., 1999. Discovering childhood: using fingerprints to find children in the archaeological record. Am. Antiq. 64 (2), 309–315.

Kaydash, V., Shkuratov, Y., Korokhin, V., Videen, G., 2011. Photometric anomalies in the apollo landing sites as seen from the lunar reconnaissance orbiter. Icarus 211, 89–96.

Kennedy, M.C., Hensley-Martin, C., Watson, P.J., 1984. CRF archaeological project, 1983. Cave Research Foundation Annual Report for 1983, 21–23.

Kim, J.Y., Kim, K.S., 2004. Hominid and other animals footprints from the Cenozoic Hamori Formation of Jeju Island. Proceedings of the 2004 Spring Meeting of the Korean Earth Science Society 25–31.

Kim, K.S., Kim, J.Y., 2006. Review on the stratigraphy and geological age of the hominid footprints-bearing strata, Jeju Island, Korea. J. Korean Earth Sci. Soc. 27, 236–246.

Kim, J.Y., Kim, K.S., Lee, C.Z., Lim, J.D., 2004. Occurrence of hominid and other vertebrate footprints of Jeju Island, Korea. In: Kim, J.Y., Kim, K.-S., Park, S.I., Shin, M.-K. (Eds.), Proceedings of International Symposium of the Quaternary Footprints of Hominids and Other Vertebrates, Namjejigun, pp. 1–26.

Kim, J.Y., Kim, K.S., Lockley, M.G., Matthews, N., 2008. Hominid ichnotaxonomy: an exploration of a neglected discipline. Ichnos 15 (3–4), 126–139. https://doi.org/10.1080/10420940802467868.

Kim, K.S., Kim, J.Y., Kim, S.H., Lee, C.Z., Lim, J.D., 2009. Preliminary report on hominid and other vertebrate footprints from the Late Quaternary strata of Jeju Island, Korea. Ichnos 16, 1–11.

Kim, C.B., Kim, J.Y., Kim, K.S., Lim, H.S., 2010. New age constraints for hominin footprints found on Jeju Island, South Korea. J. Archeol. Sci. 37, 3338–3343.

Kinahan, J., 1996. Human and domestic animal tracks in an archaeological lagoon deposit on the coast of Namibia. S. Afr. Archaeol. Bull. 51, 94–98.

Kinahan, J., 2014. Cattle paths and the choreography of late pre-colonial contact and trade on the Namib Desert coast. S. Afr. Archaeol. Bull. 69, 96–102.

Kirscher, U., Atfy, H., Gärtner, A., Dallanave, E., et al., 2021. Age constraints for the Trachilos footprints from Crete. Sci. Rep. 11, 19427.

Kızıltan, Z., Polat, M.A., 2013. The neolithic at Yenikapı, Marmaray Metro project rescue excavations. In: Özdoğan, M., Başgelen, N., Kuniholm, P. (Eds.), The Neolithic in Turkey. New Excavations and New Research, vol. 5. Arkeoloji ve Sanat Yayınları, pp. 113–165.

Klaatsch, H., 1906. II. Australia (Victoria, South Australia, western Australia). Z. Ethnol. 38 (4/5), 776–795.

Králík, M., Novotný, V., Oliva, M., 2002. Fingerprint on the Venus of Dolní Věstonice I. Anthropologie 40 (2), 107–113.

Kyparissi-Apostolika, N., Manolis, S., 2021. Reconsideration of the antiquity of the middle Palaeolithic footprints from Theopetra Cave (Thessaly, Greece). In: Pastoors, A., Lenssen-Erz, T. (Eds.), Reading Prehistoric Human Tracks. Springer International Publishing, Cham, Switzerland, pp. 169–182.

Laforgia, E., Boenzi, G., Amato, L., Di Vito, M.A., Fattore, L., Stanzione, M., Viglio, F., 2009. The Vesuvian "Pomici di Avellino" eruption and Early Bronze Age settlement in the middle Clanis valley. Méditerranée 112, 101–107.
Laporte, L.F., Behrensmeyer, A.K., 1980. Tracks and substrate reworking by terrestrial vertebrates in Quaternary sediments of Kenya. J. Sediment. Petrol. 50, 1337–1346.
Lauretta, D.S., Balram-Knutson, S.S., Beshore, E., Boynton, W.V., et al., 2017. OSIRIS-REx: sample return from asteroid (101955) Bennu. Space Sci. Rev. 212, 925–984.
Leakey, M.D., Harris, J.M. (Eds.), 1987. Laetoli: A Pliocene Site in Northern Tanzania. Clarendon Press, Oxford., p. 561
Leakey, M.D., Hay, R.L., 1979. Pliocene footprints in the Laetoli Beds at Laetoli, northern Tanzania. Nature 278, 317–323.
Leakey, M.D., Robbins, L.M., Tuttle, R.H., 1987. The hominid footprints. In: Leakey, M.D., Harris, J.M. (Eds.), Laetoli: A Pliocene Site in Northern Tanzania. Clarendon Press, Oxford, pp. 490–523.
Ledoux, L., Boudadi-Maligne, M., 2015. The contribution of geometric morphometric analysis to prehistoric ichnology: the example of large canid tracks and their implication for the debate concerning wolf domestication. J. Archaeol. Sci. 61, 25–35. https://doi.org/10.1016/j.jas.2015.04.020.
Ledoux, L., Fourment, N., Maksud, F., Delluc, M., Costamagno, S., Goutas, N., Klaric, L., Laroulandie, V., Salomon, H., Jaubert, J., 2017. Traces of human and animal activity (TrAcs) in Cussac Cave (Le Buisson-de-Cadouin, Dordogne, France): preliminary results and perspectives. Quat. Int. 430, 141–154. https://doi.org/10.1016/j.quaint.2016.06.002.
Lebreton, J.P., Witasse, O., Sollazzo, C., Blancquaert, T., Couzin, P., Schipper, A.M., Jones, J.B., Matson, D.L., Gurvits, L.I., Atkinson, D.H., Kazeminejad, B., Pérez-Ayúcar, M., 2017. An overview of the descent and landing of the Huygens probe on Titan. Nature 438 (8), 758–764.
Ledoux, L., Berillon, G., Fourment, N., Jaubert, J., 2021a. Reproduce to understand: experimental approach based on footprints in Cussac cave (southwestern France). In: Pastoors, A., Lenssen-Erz, T. (Eds.), Reading Prehistoric Tracks. Springer, Switzerland, pp. 67–89.
Ledoux, L., Berillon, G., Fourment, N., Muth, X., Jaubert, J., 2021b. Evidence of the use of soft footwear in the Gravettian cave of Cussac (Dordogne, France). Sci. Rep. 11, 22727.
Lenssen-Erz, T., Pastoors, A., 2021. Epistemic aspects of indigenous knowledge and it implications for the Archaeology of Prehistoric human tracks. In: Pastoors, A., Lenssen-Erz, T. (Eds.), Reading Prehistoric Tracks. Springer, Switzerland, pp. 101–118.
Lenssen-Erz, T., Pastoors, A., Uthmeier, T., Ciqae, T., Kxunta, /U., Thao, T., 2023. Animal tracks and human footprints in prehistoric huntergatherer rock art of the Doro! nawas mountains (Namibia), analysed by present-day indigenous tracking experts. PLoS One 18 (9), e0289560. https://doi.org/10.1371/journal.pone.0289560.
Liebenberg, L.W., 1990. The Art of Tracking: The Origin of Science. David Philip, Claremont.
Lindsay, J.M., Leonard, G.S., Smid, E.R., Hayward, B.W., 2011. Age of the Auckland volcanic field: a review of existing data. N. Z. J. Geol. Geophys. 54 (4), 379–401.
Liutkus-Pierce, C.M., Zimmer, B.W., Carmichael, S.K., McIntosh, W., Deino, A., Hewitt, S.M., McGinnis, K.J., Hartney, T., Brett, J., Mana, S., Deocampo, D., 2016. Radioisotopic age, formation, and preservation of Late Pleistocene human footprints at Engare Sero, Tanzania. Palaeogeogr. Palaeoclimatol. Palaeoecol. 463, 68–82. https://doi.org/10.1016/j.palaeo.2016.09.019.
Lockley, M.G., 1998. The vertebrate tracks record. Nature 396, 429–432.
Lockley, M., 2007. A tale of two Ichnologies: the different goals and potentials of invertebrate and vertebrate (tetrapod) ichnotaxonomy and how they relate to ichnofacies analysis. Ichnos 14, 39–57. https://doi.org/10.1080/10420940601006818.
Lockley, M.G., Roberts, G., 2004. In the footprints of our ancestors: a brief overview of the hominid track record. In: Kim, J.Y., Kim, K.S., Park, S.I., Shin, M.K. (Eds.), Proceedings of International Symposium on the Quaternary Footprints of Hominids and Other Vertebrates, Namjejugun, Korea. pp. 27–43.
Lockley, M.G., Meyer, C.A., 2000. Dinosaur Tracks and Other Fossil Footprints of Europe. Columbia University Press, New York.
Lockley, M.G., Rodríguez-de la Rosa, R., 2009. Preservation of human tracks in arid environments. Ichnos 16, 98–102.
Lockley, M.G., Garcia-Vasquez, R., Espinoza, E., Lucas, S.G., 2007a. Notes on a famous but 'forgotten' human footprint site from the Holocene of Nicaragua. New Mexico Museum of Natural History and Science Bulletin 42, 97–102.

Lockley, M.G., Kim, J.Y., Roberts, G., 2007b. The Ichnos project: a re-evaluation of the hominid track record. New Mexico Museum of Natural History and Science Bulletin 42, 79–89.

Lockley, M.G., Roberts, G., Kim, J.Y., 2008. In the footprints of our ancestors: a brief overview of the hominid track record. Ichnos 15 (3/4), 106–125.

Lockley, M.G., Garcia-Vasquez, R., Espinoza, E., Lucas, S.G., 2009. America's most famous hominid footprints: history and context of tracks from Nicaragua. Ichnos 16, 55–69.

Lockley, M., Meldrum, J., Kim, J.Y., 2016. Major events in hominin evolution. In: Mángano, M., Buatois, L. (Eds.), The Trace-Fossil Record of Major Evolutionary Events. Topics in Geobiology 40. Springer, Dordrecht, pp. 411–448. https://doi.org/10.1007/978-94-017-9597-5_8.

Lucas, S.G., 2008. Late Cenozoic fossil vertebrate assemblages from Jalisco, Mexico. In: Lucas, S.G., Morgan, G.S., Spielmann, J.A., Prothero, D.R. (Eds.), Neogene Mammals. In: Museum of Natural History & Science Bulletin 44, 51–64.

Lucas, S.G., 2017. Human footprints and the peopling of the Americas. Paleoamerica 3 (2), 97–100.

Lucas, S.G., Morgan, G.S., Hawley, J.W., Love, D.W., Myers, R.G., 2002. Mammal footprints from the upper Pleistocene of the Tularosa Basin, Doña Ana County, New Mexico. New Mexico Geological Society Guidebook, 53rd Field Conference. Geology of White Sands 258–288.

Lucas, S.G., Espinoza, E., Alvarado, G.E., 2024. Acahualinca, Nicaragua, a scientifically significant site of ancient human footprints in the New World. In: Clary, R.M., Pyle, E.J., Andrews, W.M. (Eds.), Geology's Significant Sites and Their Contributions to Geoheritage. Geological Society, London, Special Publications, p. 543. https://doi.org/10.1144/SP543-2022-255.

Lund, T., 2018. Early Exploration of the Moon: Ranger to Apollo, Luna to Lunniy Korabl. Springer, Berlin/Heidelberg, Germany, 391 pp.

MacEachern, J.A., Pemberton, S.G., Bann, K.L., Gingras, M., 2007. The ichnofacies paradigm: a fifty-year retrospective. In: Miller III, W. (Ed.), Trace Fossils. Concepts, Problems, Prospects. Elsevier, Amsterdam, pp. 52–77.

MacEachern, J.A., Bann, K.L., Gingras, M.K., Zonneveld, J.-P., Dashtgard, S.E., Pemberton, S.G., 2012. The ichnofacies paradigm. In: Knaust, D., Bromley, R.G. (Eds.), Trace Fossils as Indicators of Sedimentary Environments. Developments in Sedimentology 64. Elsevier, Amsterdam, pp. 103–138.

Mafart, B., 2006. Anthropological study of Holocene human footprints of the sebkra El Azrag (Islamic Republic of Mauritania). Anthropologie 110, 766–787.

Malvesin-Fabre, G.,L.–R., Robert, R., 1953. Empreintes de pieds humains prehistoriques de la Caverne de Niaux (Ariege). Préhistoire et Spéléologie Ariégeoises 7, 37–48.

Manhire, A., 1998. The role of hand prints in the rock art of the south-western Cape. South African Archaeologica Bulletin 53, 98–108.

Manolis, S.K., Aiello, L.C., Henessy, R., Kyparissi-Apostolika, N., 2000. The middle Paleolithic footprints of Theopetra cave (Thessaly, Greece). In: Kyparissi-Apostolika, N. (Ed.), Proceedings of the International Conference Theopetra Cave: 12 Years of Excavation and Research. Greek Ministry of Culture and Institute for Aegean Prehistory, Athens, pp. 87–93.

Marcos, E.L., 2001. El descubrimiento de las Huellas Prehistoricas: Ojo Guareña. Cubía. Boletín del Grupo Espeliológico Edelweiss 3, 34–35.

Marquet, J.-C., Freiesleben, T.H., Thomsen, K.J., Murray, A.S., Calligaro, M., Macaire, J.-J., Robert, E., Lorblanchet, M., Aubry, T., Bayle, G., Bréhéret, J.-G., Camus, H., Chareille, P., Egels, Y., Guillaud, É., Guérin, G., Gautret, P., Liard, M., O'Farrell, M., Peyrouse, J.-B., Thamó-Bozsó, E., Verdin, P., Wojtczak, D., Oberlin, C., Jaubert, J., 2023. The earliest unambiguous Neanderthal engravings on cave walls: La Roche-Cotard, Loire Valley, France. PLoS One 18 (6), e0286568.

Martinsson, A., 1970. Toponomy of trace fossils. In: Crimes, T.P., Harper, J.C. (Eds.), Trace Fossils. Geological Journal, Special Issue, vol. 3, pp. 323–330.

Marty, D., Strasser, A., Meyer, C.A., 2009. Formation and taphonomy of human footprints in microbial mats of present-day tidal-flat environments: implications for the study of fossil footprints. Ichnos 16 (1–2), 127–142. https://doi.org/10.1080/10420940802471027.

Masao, F.T., Ichumbaki, E.B., Cherin, M., Barili, A., Boschian, G., Iurino, D.A., Menconero, S., Moggi-Cecchi, J., Manzi, G., 2016. New footprints from Laetoli (Tanzania) provide evidence for marked body size variation in early hominins. Elife 5, e19568. https://doi.org/10.7554/eLife.19568.

Mastrolorenzo, G., Petrone, P., Pappalardo, L., Sheridan, M.F., 2006. The Avellino 3780-yr-B.P. catastrophe as a worst-case scenario for a future eruption at Vesuvius. Proc. Natl. Acad. Sci. 103, 4366–4370.

Matsukawa, M., Shibata, K., 2015. Review of Japanese Cenozoic (Miocene-Modern) vertebrate tracks. Ichnos 22, 261–290. https://doi.org/10.1080/10420940.2015.1064407.

Matsukawa, M., Ashikawa, N., Suzuki, Y., Lockley, M.G., 2007. Ancient human tracks from Japan: clues to population dynamics and ancient dietary habits. In: Kim, J.Y., Kim, S.K. (Eds.), Proceedings of International Symposium on the Conservation and Application of Hominid Footprints. Jeju Island, Korea, pp. 79–88.

Mayor, A., Sarjeant, W.A., 2001. The folklore of footprints in stone: from classical antiquity to the present. Ichnos 8 (2), 143–163.

Mayoral, E., Díaz-Martínez, I., Duveau, J., Santos, A., Ródríguez-Ramírez, A., Morales, J.A., Morales, L.A., Díaz-Delgado, R., 2021. Tracking late Pleistocene Neandertals on the Iberian coast. Sci. Rep. 11, 4103.

Mayoral, E., Duveau, J., Santos, A., Ródríguez-Ramírez, A., Morales, J.A., Rivera-Silva, J., Gómez-Olivencia, A., Díaz-Martínez, I., 2022. New dating of the Matalascañas footprints provides new evidence of the Middle Pleistocene (MIS 9 8) hominin paleoecology in southern Europe. Sci. Rep. 12, 17505.

McLaren, D., Fedje, D., Dyck, A., Mackie, Q., Gauvreau, A., Cohen, J., 2018. Terminal Pleistocene epoch human footprints from the Pacific coast of Canada. PLoS One 13 (3), e0193522.

McLaren, D., Mackie, Q., Fedje, D., 2021. Experimental re-creation of the depositional context in which Late Pleistocene tracks were found on the Pacific coast of Canada. In: Pastoors, A., Lennsen-Erz, T. (Eds.), Reading Prehistoric Human Tracks. Springer, Cham, Switzerland, pp. 91–100.

McNatt, L., 1996. Cave archaeology of Belize. J. Cave Karst Stud. 58 (2), 81–99.

McNutt, E.J., Hatala, K.G., Miller, C., Adams, J., Casana, J., Deane, A.S., Dominy, N.J., Fabian, K., Fannin, L.D., Gaughan, S., Gill, S.V., Gurtu, J., Gustafson, E., Hill, A.C., Johnson, C., Kallindo, S., Kilham, B., Kilham, P., Kim, E., Liutkus-Pierce, C., Maley, B., Prabhat, A., Reader, J., Rubin, S., Thompson, N.E., Thornburg, R., Williams-Hatala, E.M., Zimmer, B., Musiba, C.M., DeSilva, J.M., 2021. Footprint evidence of early hominin locomotor diversity at Laetoli, Tanzania. Nature 600 (7889), 468–471. https://doi.org/10.1038/s41586-021-04187-7.

Meldrum, D., 2000. Fossilized footprints in the Ka'u Desert, Hawaii. Am. J. Phys. Anthropol. Conf. Proc. 226–227.

Meldrum, D.J., 2004. Fossilized Hawaiian footprints compared with Laetoli hominid footprints. In: Meldrum, D.J., Hilton, C.E. (Eds.), From Biped to Strider: The Emergence of Modern Human Walking, Running, and Resource Transport. Kluwer Academic/Plenum Publishers, New York, pp. 63–83.

Meldrum, D.J., 2006. Sasquatch: legend meets science. A Tom Doherty Associates Book, New York, pp. 1–297.

Meldrum, D.J., 2007a. Ichnotaxonomy of the earliest hominid footprints. New Mexico Museum of Natural History and Science Bulletin 42, 225–231.

Meldrum, D.J., 2007b. Renewed perspective on the Laetoli trackways: the earliest hominid footprints. New Mexico Museum of Natural History and Science Bulletin 42, 233–239.

Meldrum, J., Sarmiento, E., 2018. Comments on possible Miocene hominin footprints. Proc. Geologists' Assoc. 129 (4), 577–580.

Meldrum, D.J., Lockley, M.G., Lucas, S.G., Musiba, C., 2011. Ichnotaxonomy of the Laetoli trackways: the earliest hominin footprints. J. Afr. Earth Sci. 60, 1–12.

Mercier, N., Martin, L., Kreuter, S., Moineau, V., Cliquet, D., 2019. Dating the palaeolithic footprints of "Le Rozel" (Normandy, France). Quat. Geochronol. 49, 271–277.

Meyer, M.C., Aldenderfer, M.S., Wang, Z., Hoffmann, D.L., Dahl, J.A., Degering, D., Haas, W.R., Schlütz, F., 2017. Permanent human occupation of the central Tibetan Plateau in the early Holocene. Science 355, 64–67.

Mietto, P., Avanzini, M., Roland, G., 2003. Human footprints in Pleistocene volcanic ash. Nature 422, 133.

Milàn, J., Bromley, R.G., 2009. Do shod humans leave true tracks? Ichnos 16, 124–126.

Milàn, J., Theodorou, G., Loope, D.B., Panayides, I., Clemmesnsen, L.B., Gkioni, M., 2015. Vertebrate tracks in late Pleistocene-early Holocene (?) carbonate aeolianites, Paphos, Cyprus. Ann. Soc. Geol. Pol. 85 (3), 507–514. https://doi.org/10.14241/asgp.2015.012.

Mitchell, J.K., Houston, W.N., Scott, R.F., Costes, N.C., Carrier III, W.D., Bromwell, L.G., 1972. Mechanical properties of lunar soil: Density, porosity, cohesion, and angle of friction. Proc. Lunar Sci. Conf. 3rd 3235–3253.

Molleson, T.I., Oakley, K.P., Vogel, J.C., 1972. The antiquity of the human footprints of Tanna della Basura. J. Hum. Evol. 1, 467–471.

Moniz-Nakamura, J., 2003. Keonehelelei – the Falling Sands: uncovering the origin of preserved footprints and associated Ka'u Desert Features – a historical overview and archeological survey of the Ka'u Desert. National Park Service Pacific Island Cluster: Publications in Anthropology 2, 237.

Moniz-Nakamura, J.J., 2009. Hominid footprints in recent volcanic ash: new interpretations from Hawaii Volcanoes National Park. Ichnos 16, 118–123.

Moreno, K., Bostelmann, J.E., Macías, C., Navarro-Harris, X., De Pol-Holz, R., Pino, M., 2019. A late Pleistocene human footprint from the Pilauco archaeological site, northern Patagonia, Chile. PLoS One 14, e0213572.

Morris, E.A., 1959. Basketmaker Caves in the Prayer Rock District, Northeastern Arizona. The University of Arizona Proquest Dissertations Publishing, 5803763, 639pp.

Morris, E.A., 1980. Basketmaker Caves in the Prayer Rock District, Northeastern Arizona, vol 35. Anthropological papers of the University of Arizona, 158pp.

Morse, S.A., Bennett, M.R., Gonzalez, S., Huddart, D., 2010. Techniques for verifying human footprints: reappraisal of pre-Clovis footprints in Central Mexico. Quat. Sci. Rev. 29, 2571–2578.

Morse, S.A., Bennett, M.R., Liutkus-Pierce, C., Thackeray, F., McClymont, J., Savage, R., Crompton, R.H., 2013. Holocene footprints in Namibia: the influence of substrate on footprint variability. Am. J. Physical Anthropol. 151 (2), 265–279.

Mountain, E.D., 1966. Footprints in calcareous sandstone of Nahoon Point. South Afr. J. Sci. 66, 103–111.

Muñiz, F., Cáceres, L.M., Rodríguez Vidal, J., Neto de Carvalho, C., Belo, J., Finlayson, C., Finlayson, G., Finlayson, S., Izquierdo, T., Abad, M., Jimenez-Espejo, F.J., Sugisaki, S., Gómez, P., Ruiz, F., 2019. Following the last Neanderthals: mammal tracks in Late Pleistocene coastal dunes of Gibraltar (Siberian Peninsula). Quat. Sci. Rev. 217, 297–309. https://doi.org/10.1016/j.quascirev.2019.01.013.

Murphy, K., Caseldine, A., Barker, K., Fielding, S., Burrow, S., Carlsen, S., 2014. Mesolithic human and animal footprints at Lydstep Haven, Pembrokeshire, 2010: the environmental context. Archaeol. Cambrensis 163, 23–41.

Musiba, C., Matthews, N., Noble, T., Kim, J.Y., Dominguez-Rodrigo, M., 2011. How many individuals left their footprints at Laetoli? Reinterpretation of the trackways based on recently acquired 3D data. J. Vertebr. Paleontol. 31, 164.

Mussi, M., Méndez-Quintas, E., Panera, J., Altamura, F., Di Bianco, L., Briatico, G., Mutri, G., Piarulli, F., Rubio-Jara, S., Giancarlo Ruta, G., Sánchez-Dehesa Galán, S., Serodio Domínguez, A., Melis, R.T., 2023. Melka Kunture (alto Awash, Etiopía) entre 2.000.000 y 5.000 años. Salduie 23 (1), 1–11. https://doi.org/10.26754/ojs_salduie/sald.202318553.

Neto de Carvalho, C., Belaústegui, Z., Toscano, A., Muñiz, F., Belo, J., Galán, J.M., Gómez, P., Cáceres, L.M., Rodríguez-Vidal, J., Cunha, P.P., Cachão, M., Ruiz, F., Ramirez-Cruzado, S., Giles-Guzmán, F., Finlayson, G., Finlayson, S., Finlayson, C., 2021. First tracks of newborn straight-tusked elephants (*Palaeoloxodon antiquus*). Sci. Rep. 11, 17311. https://www.nature.com/articles/s41598-021-96754-1.

Neto de Carvalho, C., Muñiz, F., Cáceres, L.M., Belaústegui, Z., Rodríguez-Vidal, J., Belo, J., Moreira, N., Cachão, M., Cunha, P.P., Figueiredo, S., Galán, J.M., Zhang, Y., Gómez, P., Toscano, A., Ruiz, F., Ramírez-Cruzado, S., Giles-Guzmán, F., Finlayson, G., Finlayson, S., Finlayson, C., 2022. Aurochs roamed along the SW coast of Andalusia (Spain) during Late Pleistocene. Sci. Rep. 12, 9911. https://doi.org/10.1038/s41598-022-14137-6.

Neto de Carvalho, C., Muñiz, F., Cáceres, L.M., Rodríguez-Vidal, J., Medialdea, A., Del Val, M., Cunha, P.P., García, J.M., Giles-Guzmán, F., Carrión, J.S., Belaústegui, Z., Toscano, A., Gómez, P., Galán, J.M., Belo, J., Cachão, M., Ruiz, F., Ramirez-Cruzado, S., Finlayson, G., Finlayson, S., Finlayson, C., 2023. Neanderthal footprints in the "Matalascañas trampled surface" (SW Spain): new OSL dating and Mousterian lithic industry. Quat. Sci. Rev. 313, 108200.

Nichol, R., 1981. Preliminary report on excavations at the Sunde site, N38/24, Motutapu Island. New Zealand Archaeol. Assoc. Newslett. 24, 237–256.

Nichol, R., 1982. Fossilised human footprints in Rangitoto Ash on Motutapu Island. Geol. Soc. New Zealand Newslett. 51, 11–13.

Noval Fonseca, M.A., 2007. Cuevas tempranas (Posada, Llanes): fragil rastro del pasado. In: Excavaciones Arqueologicos en Asturias 1999-2002. Gobierno del Principado de Asturias, Servicio de Publicaciones, Oviedo, pp. 207–213.

Officer, C.G.W., 1893. The discovery of supposed human footprints on aeolian rock at Warrnambool. The Victorian Naturalist 9, 32–39.

Oğuş, B., 2021. Animal footprints on Roman Tiles from Perge and Aizanoi. Adalya 24, 229–248.

Onac, B.P., Viehmann, I., Lundberg, J., Lauritzen, S.-E., Stringer, C., Popitã, V., 2005. U-Th ages constraining the Neanderthal footprint at Vârtop Cave, Romania. Quat. Sci. Rev. 24, 1151–1157.

Onac, B.P., Veres, D.S., Stringer, C., 2021. Hominin footprints in caves from Romanian Carpathians. In: Pastoors, A., Lenssen-Erz, T. (Eds.), Reading Prehistoric Tracks. Springer, Switzerland, pp. 201–210.

Ortega, A.I., Ruiz, F., Martín, M.A., Benito-Calvo, A., Vidal, M., Bermejo, L., Karampaglidis, T., 2021. Prehistoric human tracks in Ojo Guareña Cave System (Burgos, Spain): The Sala and Galerías de las Huellas. In: Pastoors, A.,

Lenssen-Erz, T. (Eds.), Reading Prehistoric Human Tracks. Springer International Publishing, Cham, Switzerland, pp. 317–342. https://doi.org/10.1007/978-3-030-60406-6_17.
Oviatt, C.G., Madsen, D.B., Rhode, D.R., Davis, L.G., 2022. A critical assessment of claims that human footprints in the Lake Otero basin, New Mexico date to the Last Glacial Maximum. Quater. Res. 111, 138–147.
Öz, C., 2022. An ichnoarchaeological perspective: animal footprints on Roman tiles found in Andriake. Tüba-Ked 26, 157–166. https://doi.org/10.22520/tubaked2022.26.008.
Ozansoy, F., 1968. Türkiye Pleistosen fosil insan ayak izleri. Maden Tetkik ve Arama Enstitüsü Dergisi 72, 204–208.
Ozansoy, F., 1969. Pleistocene human footprints in Turkey. Bull. Miner. Res. Explor. Inst. Turk. 72, 146–150.
Pales, L., 1954. Les empreintes de pieds humains de la "Tana della Basura" (Toriano). Revista di Studi Liguri 20, 5–12.
Pales, L., 1960. Les empreintes de pieds humains de la «Grotta della Basura (Toriano). Revue des Études Ligures 26, 25–90.
Pales, L., 1976. Les empreintes de pieds humains dans les caverns. Archives de l' Institut Paléontologie Humaine 36, 1–166.
Panarello, A., Palombo, M.R., Biddittu, I., Mietto, P., 2017a. Fifteen years along the Devil's Trails": new data and perspectives. Alpine Mediterra. Quater. 30 (2), 137–154.
Panarello, A., Santello, L., Farinaro, G., Bennet, M.R., Mietto, P., 2017b. Walking about the oldest human fossil pathway (Roccamonfina volcano, Central Italy)? J. Archaeol. Sci. Report 13, 476–490.
Panarello, A., Mazzardo, L., Mietto, P., 2018. The Devil's touch: a first dataset from what could be the oldest human handprint ever found (Central-Southern Italy). Alpine Mediterr. Quater. 31, 37–47.
Panarello, A., Palombo, M.R., Biddittu, I., Di Vito, M.A., Farinaro, G., Mietto, P., 2020. On the devil tracks: unexpected news from Foresta ichnosite (Roccamonfina volcano, central Italy). J. Quat. Sci. https://doi.org/10.1002/jqs.3186.
Pastoors, A., Lenssen-Erz, T. (Eds.), 2021. Reading Prehistoric Human Tracks. Springer, Cham, Switzerland, p. 436.
Pastoors, A., Lenssen-Erz, T., Ciqae, T., Kxunta, /U., Thao, T., Bégouën, R., Biesele, M., Clottes, J., 2015. Tracking in Caves. Experience based reading of Pleistocene human footprints in French caves. Camb. Archaeol. J. 25 (3), 551–564. https://doi.org/10.1017/S0959774315000050.
Pastoors, A., Lenssen-Erz, T., Breuckmann, B., Ciqae, T., Kxunta, U., Rieke-Zapp, D., Thao, T., 2017. Experience based reading of Pleistocene human footprints in Pech-Merle. Quat. Int. 430, 155–162. https://doi.org/10.1016/j.quaint.2016.02.056.
Pastoors, A., Lenssen-Erz, T., Ciqae, T., Kxunta, /U., Thao, T., Bégouën, R., Uthmeier, T., 2021. Episodes of Magdalenian hunter-gatherers in the upper gallery of Tuc d'Audoubert (Ariège, France). In: Pastoors, A., Lenssen-Erz, T. (Eds.), Reading Prehistoric Tracks. Springer, Switzerland, pp. 211–249.
Patton, M.A., 1993. Neolithic and later archaeology. In: Keen, D. (Ed.), Quaternary of Jersey: Field Guide. Quaternary Research Association, London, pp. 48–56.
Peet, S.D., 1889. The age of the Nicaraguan foot-prints. Am. Antiquarian 11, 120–121.
Pétrequin, A.-M., Pétrequin, P., 1988. Le Néolithique des Lacs: Préhistoire des lacs de Chalain et de Clairvaux (4000–2000 av J.-C.). Editions Errance, Paris.
Philp, R., 2018. Changing Tides: The Archaeological Context of Sea Level Change in Prehistoric South Wales. PhD Thesis. Cardiff University.
Pigati, J.S., Springer, K.B., Holliday, V.T., Bennett, M.R., Bustos, D., Urban, T.M., Reynolds, S.C., Odess, D., 2022. Reply to "evidence for humans at White Sands National Park during the last glacial maximum could actually be for Clovis Peope ~ 13,000 years ago" by vance Haynes. Jr.. PaleoAmerica 8 (2), 99–101.
Pigati, J.S., Springer, K.B., Honke, J.S., Wahl, D., Champagne, M.R., Zimmerman, S.R., Gray, H.J., Santucci, V.L., Odess, D., Bustos, D., 2023. Independent age estimates resolve the controversy of ancient human footprints at White Sands. Science 382, 73–75.
Politis, G.G., Bayón, C., 1995. Early Holocene footprints and sea mammals in the tidal zone of the Argentinian foreshore. Past 20, 5–6.
Pryor, F., 1998. Farmers in prehistoric Britain. Tempus, Stroud.
Putnam, F.W., 1875a. Archaeological explorations in Indiana and Kentucky. Am. Nat. 9, 410–415.
Putnam, F.W., 1875b. Archaeological researches in Kentucky. Proc. Boston Soc. Nat. Hist. 17, 314–332.
Quattrocchio, M., Borromei, A.M., Deschamps, C.M., Grill, S.C., Zavala, C.A., 2008. Landscape evolution and climate changes in the late Pleistocene–Holocene, southern Pampa (Argentina): evidence from palynology, mammals and sedimentology. Quat. Int. 181, 123–138.

Quiles, A., Valladas, H., Bocherens, H., et al., 2016. A high-precision chronological model for the decorated Upper Paleolithic cave of Chauvet-Pont d'Arc, Ardèche, France. Proc. Natl. Acad. Sci. U. S. A. 113, 4670–4675.

Rachal, D.M., Dello-Russo, R., Cuba, M., 2024. The Pleistocene footprints are younger that we thought: correcting the radiocarbon dates of the *Ruppia* seeds, Tularosa Basin, New Mexico. Quater. Res. 1–12. https://doi.org/10.1017/qua.2023.74.

Raichlen, D.A., Gordon, A.D., Harcourt-Smith, W.E.H., Foster, A.D., Haas, W.R.J., 2010. Laetoli footprints preserve earliest direct evidence of human-like bipedal biomechanics. PLoS One 5 (3), e9769.

Rector, C.H., 1979. 5,000-year-old footprints on the Mojave river, California, USA. Antiquity 54, 149–150.

Rector, C.H., 1983. Appendix B. Ancient human and animal trackways. In: Rector, C.H., Swenson, J.D., Wilke, P.J. (Eds.), Archaeological Studies at Oro Grande, Mojave Desert, California. San Bernardino County Museum Association, Redlands, California, 181 pp.

Rector, C.H., 1999. Human and animal trackway at Oro Grande: San Bernardino County Museum Association Quarterly 46 (2), 53–55.

Renne, P.R., Feinberg, G.M., Waters, M.R., Arroyo-Cabrales, J., Ocoa-Castillo, P., Perez-Campa, M., Knight, K.B., 2005. Age of Mexican ash with alleged "footprints". Nature 438, 7–8.

Richmond, B.G., Hatala, K.G., Harcourt-Smith, W.E.H., Rossi, V., Metallo, A., Liutkus, C.M., Pobiner, B.L., Gordon, A.D., Dingwall, H.L., Green, D.J., Olle Moita, G., Brett, J., 2011. Early modern human footprint assemblage from Engare Sero, Tanzania. PaleoAnthropology A31.

Rikardo, R., Etxepare, R., Irurtzun, A., 2021. Gravettian hand stencils as sign language formatives. Phil. Trans. Biol. Sci. 376 (1824). https://doi.org/10.1098/rstb.2020.0205.

Rişcuţia, C., Rişcuţia, I., 1970. Étude anthropologique des emprientes humaines de la grotte "Ciurului-Izbuc". In: Orghidan, T. (Ed.), Livre du centenaire "Émile G. Racovitza. Editura Academiei RSR, Bucharest, pp. 529–533.

Roach, N.T., Hatala, K.G., Ostrofsky, K.R., Villmoare, B., Reeves, J.S., Du, A., Braun, D.R., Harris, J.W., Behrensmeyer, A.K., Richmond, B.G., 2016. Pleistocene footprints show intensive use of lake margin habitats by *Homo erectus* groups. Sci. Rep. 6, 26374.

Roach, N.T., Du, A., Hatala, K.G., Ostrofsky, K.R., Reeves, J.S., Braun, D.R., Harris, J.W., Behrensmeyer, A.K., Richmond, B.G., 2018. Pleistocene animal communities of a 1.5 million–year–old lake margin grassland and their relationship to *Homo erectus* paleoecology. J. Hum. Evol. 122, 70–83.

Robbins, L.M., 1978. The individuality of human footprints. J. Forensic Sci. 23 (4), 778–785.

Robbins, L.M., Wilson, R.C., Watson, P.J., 1981. Paleontology and archeology of Jaguar Cave, Tennessee. Proc. Int. Congr. Speleol. 377–380.

Roberts, G., 2004. Ephemeral, subfossil mammalian and avian tracks within intertidal-zone, Flandrian sediment exposures at Formby point, Sefton Coast, North West England. In: Kim, K.-Y., Kim, K.-S., Park, S.-I., Shin, M.-K. (Eds.), Proceedings of the International Symposium on the Quaternary Footprints of Hominids and Other Vertebrates, pp. 54–83.

Roberts, D.L., 2008. Last Interglacial hominid and associated vertebrate fossil trackways in coastal eolianites, South Africa. Ichnos 15, 190–207.

Roberts, G., 2009. Ephemeral, subfossil mammalian, avian and hominid footprints within Flandrian sediment exposures at Formby Point, Sefton Coast, North West England. Ichnos 16, 33–48. https://doi.org/10.1080/10420940802470730.

Roberts, G., 2014. The Lost World of Formby Point. The Alt Press, Liverpool.

Roberts, D., Berger, L.R., 1997. Last Interglacial (c. 117 kyr) human footprints from South Africa. South Afr. J. Sci. 93 (8), 349–350.

Roberts, G., Gonzalez, S., Huddart, D., 1996. Intertidal Holocene footprints and their archaeological significance. Antiquity 70, 647–651.

Rodríguez de la Rosa, R.A., Arguillón-Martínez, M.C., Lopez-Espinosa, J., Eberth, D.A., 2004. The fossil record of vertebrate tracks in Mexico. Ichnos 11, 27–38.

Romano, M., Citton, P., Salvador, I., Arobba, D., Rellini, I., Firpo, M., Negrino, F., Zunino, M., Starnini, E., Avanzini, M., 2019. A multidisciplinary approach to a unique Palaeolithic human ichnological record from Italy (Bàsura Cave). Life. https://doi.org/10.1101/52940.

Ruff, C.B., Wunderlich, R.E., Hatala, K.G., Tuttle, R.H., Hilton, C.E., D'Août, K., Webb, D.M., Hallgrímsson, B., Musiba, C., Baksh, M., 2021. Body mass estimation from footprint size in hominins. J. Hum. Evol. 156, 102997.

Rusu, T., Viehmann, I., Racovita, G., Crazium, V., 1969. Primele urme de pasi ale omului preistoric din pesteule Românici. Ocrotirea Nat. 13, 191–200.

Sammartino, F., Bisconti, M., 2010. Indagine archeozoologica sulle impronte di mammiferi conservate sui laterizi della fornace romana di Casa Campacci, Livorno. In: Lenzi, A., Leoni, L., Baldacci, C., Brizzi, B., De Santi, C.,

Domenici, V., Feri, E., Lenzi, P., Montesarchio, E., Piombanti, P.P., Santinelli, M. (Eds.), Codice Armanico 2010 – Terzo congresso di scienze naturali Ambiente toscano, Pisa, pp. 17–23.
Scaillet, S., Vita-Scaillet, G., Guillou, H., 2008. Oldest human footprints dated by Ar/Ar. Earth Planet Sci. Lett. 275, 320–325.
Scales, R., 2002. Footprint-tracks at Goldcliff east 2002. Archaeol. Severn Estuary 13, 31–35.
Scales, R., 2004. Footprint-tracks at Goldcliff East: the fauna and a simple methodology for field recording. Archaeol. Severn Estuary 14, 35–40.
Scales, R., 2007. Footprint-tracks of human and animals. In: Bell, M. (Ed.), Prehistoric Coastal Communities. CBA Research Report 149. Council of British Archaeology, pp. 139–157.
Schmincke, H.U., Kutterolf, S., Perez, W., Rausch, J., Freundt, A., Strauch, W., 2009. Walking through volcanic mud: the 2100 year old Acahualinca footprints (Nicaragua) I: stratigraphy, lithology, volcanology and age of the Acahualinca section. Bull. Volcanol. 71, 479–493.
Schmincke, H., Kutterolf, S., Wendy Perez, W., Rausch, J., Freundt, A., Strauch, W., 2010. Walking through volcanic mud: the 2,100-year-old Acahualinca footprints (Nicaragua) II: The Acahualinca people, environmental conditions and motivation. Int. J. Earth Sci. 99, 279–292.
Scott, J.J., Renaut, R.W., Owen, R.B., 2008. Preservation and paleoenvironmental significance of a footprinted surface on the Sandai Plain, Lake Bogoria, Keny Rift Valley. Ichnos 15 (3/4), 209–232.
Sedrati, M., Morales, J.A., Duveau, J., M'rini, A., Mayoral, E., Díaz-Martínez, I., Anthony, E.J., Bulot, G., Sedrati, A., Le Gall, R., Santos, A., Rivera-Silva, J.J., 2024. A Late Pleistocene Hominin footprint site on the North African Coast of Morocco. Sci. Rep. 14, 1962. https://doi.org/10.21203/rs.3.rs-3511552/v1.
Seilacher, A., 2007. Trace Fossil Analysis. Springer, Berlin.
Shayler, D.J., Salmon, A., Shayler, M.D., 2005. Marswalk One. First Steps on a New Planet. Springer, Chichester (UK), 244 pp.
Simek, J.F., Blankenship, S.A., Cressler, A., Douglas, J.C., Wallace. A., Weinand, D., Welborn, H., 2012. The prehistoric cave art and archaeology of Dunbar Cave, Montgomery County, Tennessee. J. Cave Karst Stud. 74 (1), 19–32.
Simek, J.F., Franklin, J.D., Sherwood, S.C., 1998. The context of early southeastern prehistoric art: a report on the archaeology of 3rd Unnamed Cave. Am. Antiq. 63, 663–677.
Sims, J.R., Lewis, C., Clark, M.K., 1981. Xanadu. NSS News 39, 168–175.
Smith, G.M., Parsons, T., Harrod, R.P., Holmes, C.E., Reuther, J.D., Potter, B.A., 2019. A track in the Tanana: forensic analysis of a Late Holocene footprint from central Alaska. J. Archaeol. Sci. Reports 24, 900–912.
Sneed, J.M., 1984. The Sequoyah Caverns story. National Speleological. Society News 42, 178–181.
Sohn, Y.K., Yoon, W.S., Ahn, U.S., Kim, G.B., Lee, J.-H., Ryu, C.K., Jeon, Y.M., Kang, C.H., 2015. Stratigraphy and age of the human footprint-bearing strata in Jeju Island, Korea: controversies and new findings. J. Archaeol. Sci. Reports 4, 264–275.
Sousa, L., Nunes, M., Gonçalves, C., 2007. Tegulae com marcas de oleiro e pegadas de animais no concelho de Lousada. Oppidum 2, 57–74.
Spennemann, D.H.R., 2004. The ethics of treading on Neil Armstrong's footprints. Space Policy 20, 279–290. https://doi.org/10.1016/j.spacepol.2004.08.005.
Stewart, M., Clark-Wilson, R., Breeze, P.S., Janulis, K., Candy, I., Armitage, S.J., Ryves, D.B., et al., 2020. Human footprints provide snapshot of last interglacial ecology in the Arabian interior. Sci. Adv. 6, eaba8940. https://doi.org/10.1126/sciadv.aba8940.
Stuart, G.E., 1981. Maya art treasures discovered in cave. Natl. Geogr. 160 (2), 220–235.
Su, K., Sigley, G.G., Song, C., 2020. Relational authenticity and reconstructed heritage space: a balance of heritage preservation, tourism, and urban renewal in Luoyang, Silk Road, Dingding Gate. Sustainability 12, 5830.
Suwa, G., 1984. Could *Australopithecus afarensis* have made the hominid tracks at Laetoli? J. Physical Anthropol. 963, 224–225.
Swanson, D.A., Christiansen, R.L., 1973. Tragic base surge in 1790 at Kilauea volcano. Geology 1 (2), 83–86.
Swanson, D.A., Weaver, S.J., Houghton, B.F., 2015. Reconstructing the deadly eruptive events of 1790 CE at Kilauea Volcano, Hawaii. GSA Bull. 127 (3–4), 503–515.
Takács, S., 2007. Divine and human feet: records of pilgrims honouring Isis. In: Elsner, J., Rutherford, I. (Eds.), Pilgrimage in Graeco-Roman and Early Christian Antiquity: Seeing the Gods. Oxford University Press, Oxford, pp. 353–369.
Takahashi, K., Okamura, Y., 2004. Neogene and Pleistocene footprint fossils of Japan, with special reference to footprints from the Plio-Pleistocene Kobiwako Group, central Japan. In: Kim, J.Y., Kim, K.-S., Park, S.I., Shin, M.-K.

(Eds.), International Symposium on Quaternary Footprints of Hominids and Other Vertebrates. Jeju, Korea, pp. 84–93.

Tang, H., Xiagewangdui, Lv, H., 2022. Hand and foot prints at Qiusang in Tibet and related issues. J. Hebei Normal Univ. (Philos. Soc. Sci. Ed.) 45 (5), 77–85.

Theunissen, E.M., Brinkkemper, O., Lauwerier, R.C.G.M., Smit, B.I., van der Jagt, I.M.M., 2014. A mosaic of habitation at Zeewijk (The Netherlands). Nederland se Archeologische Rapporten 47, 196.

Tongiorgi, E., Lamboglia, N., 1954. La grotta di Toirano (Grotta della Bàsura o della Strega). In: Istituto internazionale di studi liguri. Genova, p. 47.

Trinkaus, E., Shang, H., 2008. Anatomical evidence for the antiquity of human footwear: Tianyuan and Sanghir. J. Archeol. Sci. 35, 1928–1933.

Tsuda, Y., Saiki, T., Terui, F., Nakazawa, S., Yoshikawa, M., Watanabe, S., Hayabusa2 Project Team, 2020. Hayabusa2 mission status: landing, roving and cratering on asteroid Ryugu. Acta Astronaut. 171, 42–54.

Tuffreau, A., 1988. Les habitats du Paléolithique inferiéur et moyen dans le Nord de la France (Nord, Pas-de-Calais, Somme). Rev. Archeol. Picardie 1, 91–104.

Turbott, I., 1949. The footprints of Tarawa. J. Polyn. Soc. 58, 193–197.

Turner, J.S., 2000. The Extended Organism; the Physiology of Animal-Built Structures. Harvard University Press, 235 pp.

Tuttle, R.H., 2008. Footprint clues in hominid evolution and forensics: lessons and limitations. Ichnos 15 (3/4), 158–165.

Ulamec, S., Fantinati, C., Maibaum, M., Geurts, K., Biele, J., Jansen, S., Küchemann, O., Cozzoni, B., Finke, F., Lommatsch, V., Moussi-Soffys, A., Delmas, C., ÓRourke, L., 2016. RosettaLander - landing and operations on comet 67P/Churyumov–Gerasimenko. Acta Astronaut. 125, 80–91.

Ulusoy, I., Akif Sarıkaya, M., Schmitt, A.K., Şen, E., Danisík, M., Gümüs, E., 2019. Volcanic eruption eye-witnessed and recorded by prehistoric humans. Quat. Sci. Rev. 212, 187–198.

UNESCO World Heritage Convention, 2012. The Caves of Naj Tunich. https://whc.unesco.org/en/tentativelists/5739/. (Accessed 2 January 2024).

Uomini, N.T., 2009. The prehistory of handedness: archaeological data and comparative ethology. J. Hum. Evol. 57 (4), 411–419.

Urban, T.M., Bennett, M.R., Bustos, D., Manning, S.W., Reynolds, S.C., Belvedere, M., Odess, D., Santucci, V.L., 2019. 3-D radar imaging unlocks the untapped behavioral and biomechanical archive of Pleistocene ghost tracks. Sci. Rep. 9, 16470.

Urban, T.M., Bustos, D., Jakeway, J., Manning, S.W., Bennett, M.R., 2018. Use of magnetometry for detecting and documenting multi-species Pleistocene megafauna tracks at White Sands National Monument, New Mexico, U.S.A. Quat. Sci. Rev. 199, 206–213.

Valladas, H., Mercier, N., Froget, L., Joron, J.L., Reyss, J.L., Karkanas, P., Panagopoulou, E., Kyparissi-Apostolika, N., 2007. TL age-estimates for the Middle Paleolithic layers at Theopetra Cave (Greece). Quat. Geochronol. 2, 303–308.

Valladas, H., Pons-Branchu, E., Dumoulin, J.P., Quiles, A., Sanchidrían, J.L., Medina-Tekkaya, I., 1976. Insanlara ait fosil ayak izleri. Yeryuvarı ve. Insan (1/2), 8–10.

Vallois, H.V., 1927. Étude des empreintes de pieds humains de Tuc d'Audoubert, de Cabrerets, de Ganties. Institut International d'Anthropologie, III session, Amsterdam, pp. 328–338.

Valsik, J.A., 1951. The interpretation of the fingerprints from Dolni Vestonice. Zpravy anthropologické. Spolecnosti 4 (5–6), 94–95.

Veliz, V., 1978. Huellas Humanas en Guaimaca, Honduras. Instituto Hondureno de Antropologia e Histoiria, Tegucigalpa 2 (4), 263–269.

Veverka, J., Farquhar, B., Robinson, M., Thomas, P., et al., 2001. The landing of the NEAR-Shoemaker spacecraft on asteroid 433 Eros. Nature 413 (27), 390–393.

Villa, P., 1983. *Terra Amata* and the Middle Pleistocene Archaeological Record of Southern France. Publications in Anthropology. University of California, 304 pp.

Villmoare, B., Hatala, K.G., Jungers, W., 2019. Sexual dimorphism in *Homo erectus* inferred from 1.5 Ma footprints near Ileret, Kenya. Sci. Rep. 9, 7687.

Waddington, C., 2014. Rescued from the Sea: an archaeologist's tale. Archaeological Research Services, Bakewell.

Wang, L., Zhang, H., Zhang, D.D., Cheng, H., Zhang, S., Li, T., Zhang, Y., Wang, X., Wu, Z., Wang, Y., Chen, F., 2023. New evidence of prehistoric human activity on the central Tibetan plateau during the early to middle Holocene. Holocene 33 (10), 1196–1206. https://doi.org/10.1177/09596836231183060.

Watson, P.J., 1969. The prehistory of Salts Cave, Kentucky. Illinois State Museum Reports of Investigations 16, 1–86.

Watson, P.J. (Ed.), 1974. Archeology of the Mammoth Cave Area. Academic Press, New York, 255 pp.

Watson, P.J., 1983. CRF Archaeological Project and Shellmound Archaeological Project – 1982. Cave research Foundation Annual Report, pp. 13–15.

Watson, P.J., 1986. Prehistoric cavers in the Eastern Woodlands. In: Faulkner, C.H. (Ed.), The Prehistoric Native American Art of Mud Glyph Cave. University of Tennessee Press, Knoxville, pp. 109–116.

Watson, P.J., Kennedy, M.C., Willey, P., Robbins, L., Wilson, R.C., 2005. Prehistoric footprints in Jaguar Cave, Tennessee. J. Field Archaeol. 30, 25–43.

Webb, S., 2007. Further research of the Willandra Lakes fossil footprint site, southeastern Australia. J. Hum. Evol. 52, 711–715.

Webb, S., 2021. An echo from a footprint: a step too far. In: Pastoors, A., Lenssen-Erz, T. (Eds.), Reading Prehistoric Human Tracks: Methods & Material. Springer, pp. 397–412.

Webb, S., Cupper, M.L., Robins, R., 2006. Pleistocene human footprints from the Willandra Lakes, southeastern Australia. J. Hum. Evol. 50 (4), 405–413.

Webb, D., Robu, M., Moldovan, O., et al., 2014. Ancient human footprints in Ciur-Izbuc Cave, Romania. Am. J. Phys. Anthropol. 155, 128–135.

Welch, D., 1999. Fossilised human footprints on the coast of north Western Australia. Artefact J. Archaeol. Anthropol. Soc. Victoria 22, 3–10.

Westaway, M., 2010. Footprints of the First Australians. Australas. Sci. 31, 14–18.

Westaway, M.C., Cupper, M.L., Johnston, H., Graham, I., 2013. The Willandra Fossil Trackway: Assessment of ground penetrating radar survey results and additional OSL dating at a unique Australian site. Aust. Archaeol. 76, 84–89.

White, T.D., Suwa, G., 1987. Hominid footprints at Laetoli; facts and interpretations. Am. J. Phys. Anthropol. 72, 485–514.

White, R., Bosinski, G., Bourrillon, R., Clottes, J., Conkey, M.W., et al., 2020. Still no archaeological evidence that Neanderthals created Iberian cave art. J. Hum. Evol. 144, 102640.

Willey, P., Stolen, J., Crothers, G., Watson, P.J., 2005. Preservation of prehistoric footprints in Jaguar Cave, Tennessee. J. Cave Karst Stud. 67, 61–68.

Willey, P., Watson, P.J., Crothers, G., Stolen, J., 2008. Holocene human footprints in North America. Ichnos 15, 1–6.

Willey, P., Watson, P.J., Crothers, G., Stolen, J., 2009. Holocene human footprints in North America. Ichnos 16, 70–75.

Williams, H., 1952. Geological observations on the ancient human footprints near Managua. Contribut. Am. Anthropol. Hist. 52, 1–31.

Wiseman, A.L.A., Vicari, D., Belvedere, M., de Groote, I., 2022. Neolithic track sites from Formby Point, England: new data and insights. J. Archaeol. Sci.: Reports 44, 103546.

Wu, X., Zhang, C., Goldberg, P., Cohen, D., Pan, Y., Arpin, T., Bar-Yosef, O., 2012. Early Pottery at 20,000 Years Ago in Xianrendong Cave, China. Science 336, 1696–1700. https://doi.org/10.1126/science.1218643.

Wu, X., Liu, Y., Zhang, C., Wu, Y., Zhang, F., Du, J., Liu, Z., Xing, Y., Xu, R., He, Z., Lin, Y., Zou, Y., 2021. Geological characteristics of China's Tianwen-1 landing site at Utopia Planitia, Mars. Icarus 370, 114657.

Yalçınlar, I., 1987. Gediz Vadisinde Prehistoric Insanların Ayak Izleri, vol. 2. Istanbul ÜniversitesiCografya Dergisi, pp. 71–80.

Young, B.H., 1910. The Prehistoric Men of Kentucky. John P. Morton & Company, Louisville, Ky, 329 pp.

Zavala, C., Grill, S., Martínez, D., Ortiz, H., González, R., 1992. Análisis paleoambiental de depósitos cuaternarios. Sitio Paleoicnológico Monte Hermoso I, Provincia de Buenos Aires. La Plata: Actas de las Terceras Jornadas Geológicas Bonaerenses 31–37.

Zhang, D.D., Li, S.H., 2002. Optical dating of Tibetan human hand and footprints: An implication for the palaeoenvironment of the last glaciation of the Tibetan Plateau. Geophys. Res. Lett. 29, 161–163.

Zhang, D.D., Li, S.H., He, Q.E., Li, B.S., 2003. Human Settlement of the last glaciation on the Tibetan plateau. Curr. Sci. 84, 701–704.

Zhang, D.D., Bennett, M.R., Cheng, H., et al., 2021. Earliest parietal art: Hominin hand and foot traces from the middle Pleistocene of Tibet. Sci. Bull. 66, 2506–2515.

Zhang, X., Zhang, G., Xie, H., Gao, M., Wen, Y., 2022. A review of sampling exploration and devices for extraterrestrial celestial bodies. Space Sci. Rev. 218, 59.

CHAPTER

3

Evidence from fossil tetrapod tracks of gregarious behavior, ichnopathologies, and locomotion

Adrian P. Hunt[1], *Spencer G. Lucas*[2] *and Hendrik Klein*[3]

[1]Flying Heritage and Combat Armor Museum, Everett, WA, United States; [2]New Mexico Museum of Natural History, Albuquerque, NM, United States; [3]Saurierwelt Paläontologisches Museum, Neumarkt, Germany

1. Introduction

Vertebrate tracks are traces left by living animals, so they provide abundant evidence of behavior, locomotion, and paleoecology (e.g., Enriquez et al., 2022). In this volume, Xing et al. (2025) review traces related to reproduction, and Hunt and Lucas (2021) review ichnological evidence of predation and feeding. Here we briefly discuss the significance of tracks for providing evidence for gregariousness and other behavior, determining speed, assessing gait, and recognizing pathologies. The majority of the work on these topics relates to dinosaurs for several reasons: (1) volume of literature, although there are arguably a similar number of papers on pre-Jurassic tracks; (2) the taxophile effect *sensu* Hunt et al. (2018) given elevated interest in dinosaurs relative to other vertebrates; and (3) the fact that nonavian dinosaurs were very successful organisms but have limited modern analogs, so their paleobiology and behavior are poorly understood and of broad interest.

2. Social behavior

2.1 Introduction

Hitchcock (1836) first noted that four theropod trackways of *Eubrontes* (*Ornithichnites*) *giganteus*, at what is now Dinosaur Footprints Reservation in Massachusetts, were subparallel. He argued that these Early Jurassic tracks indicated group behavior in the trackmakers.

Vertebrate Ichnology
https://doi.org/10.1016/B978-0-443-38351-9.00006-2

Subsequently, he reported another 17 localities of *Eubrontes* (*Ornithichnites giganteus*) that indicated gregarious behavior in what he considered to be giant birds (Hitchcock, 1848; Getty et al., 2012). In an influential paper, Ostrom (1972) identified 22 trackways of *Eubrontes giganteus* from the Dinosaur Footprints Reservation, of which 19 were parallel. Combining this evidence with similar orientations of ornithopod (Albritton, 1942) and sauropod (Bird, 1941, 1944) trackways from the Early Cretaceous of Texas (Figs. 3.1 and 3.2), Ostrom (1972) concluded that there

FIGURE 3.1 Putative gregarious behavior from a transgressive site in the Glen Rose Formation (Lower Cretaceous) at the Davenport Ranch, Banderas, Texas, USA. Note multiple sauropod trackways of *Brontopodus* moving from the bottom to the top of the image. These consist of a large U-shaped pes print in the rear and a smaller lunate manus print in front. A trackway of the theropod *Irenesauripus* is perpendicular to the sauropod tracks and moving from right to left. Asset # AN1384967, American Museum of Natural History Library.

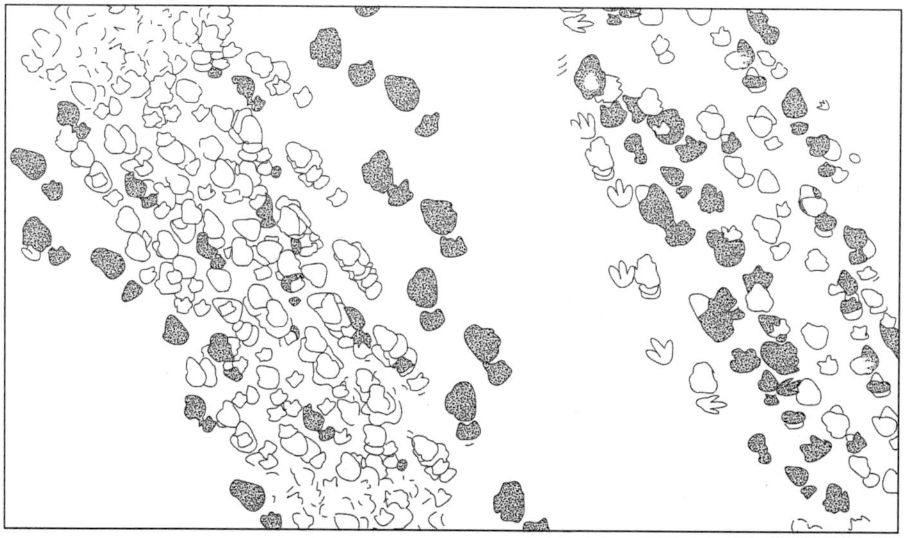

FIGURE 3.2 Map of a portion of the Davenport Ranch transgressive tracksite from the Early Cretaceous Glen Rose Formation of Texas, USA. Note the parallel trackways of the sauropod track *Brontopodus* isp. with different size classes and a limited number of tridactyl theropod tracks (*Irenesauripus* isp.).

was strong evidence for gregarious behavior in several lineages of nonavian dinosaurs. This formed an important part of the new canon of the Dinosaur Renaissance, which interpreted dinosaurs not as giant lizards but as "advanced" endotherms (e.g., Bakker, 1975, 1986).

2.2 Ichnological recognition of gregarious behavior

Gregariousness is the grouping of animals of the same species due to social behavior that can be temporary (e.g., feeding, mating) or permanent. The best evidence of gregarious behavior in fossil tetrapods comes from the ichnological record. There are two types of tracksites related to gregarious behavior (q.v., Morse et al., 2010): (1) sites where several individuals moved across an area in subparallel alignment, which we term transgressive sites (trackway sites of Morse et al., 2010: Figs. 3.1 and 3.2); and (2) sites where tracks of multiple individuals are concentrated at a location, but they are not necessarily aligned in a common direction, which we term (after Morse et al., 2010) as congregation sites. A small number of sites (e.g., Clayton Lake State Park: Hunt and Lucas, 1998b; Lucas and Dalman, 2016; Lucas et al., 2024b: Fig. 3.3) may fall in both categories. The vast majority of the literature on gregarious behavior inferred from trackways relates to transgressive sites (Table 3.1).

There are several factors to consider when evaluating a transgressive site for evidence of gregarious behavior (e.g., García-Ortiz and Pérez-Lorente., 2014):

FIGURE 3.3 Orthophoto from drone of the Clayton Lake State Park congregation tracksite from the Early Cretaceous of New Mexico, USA showing contours (from Lucas et al., 2024b). Tracks occur across the contact of the Lower Cretaceous (upper Albian) Mesa Rica and Pajarito formations. The vast majority of the tracks are of bipedal ornithopods, most of which can be assigned to *Caririchnium* isp.

TABLE 3.1 Putative gregarious transgressive tracksites.

Age	Location	Biotaxa	Ichnotaxa	No. of trackways	Age class	References
Early Permian	New Mexico, USA	Pelycosaur	*Dimetropus*	2	Adult	Hunt et al. (1993b, Fig. 7), MacDonald (1994, p. 129), Lockley and Hunt (1995a, Fig. 2.20)
Early Permian	Arizona, USA		"*Chelichnus*"	Several, up to 12 (on foreset)	Adult	Lockley and Hunt (1995a, Fig. 2.11)
Early Permian	Colorado, USA		"*Chelichnus*"	Several, up to 3 (on foreset)	Adult	Lockley and Hunt (1995a, Figs. 2.10 and 2.13)
Early Permian	Utah, USA		"*Chelichnus*"	2 (on foreset)	Adult	Lockley et al. (1998, Fig. 4)
Early Triassic	Poland	Archosaur	*Protochirotherium hauboldi*	5	Adult	Klein and Niedźwiedzki (2012)
Early Triassic	Germany	Dicynodont	*Dicynodontipus*	3	Adult	Haubold (1984)
Middle Triassic	Arizona, USA	Dicynodont	*Therapsipus cumminsi*	2	Adult	Hunt et al. (1993a)
Middle Triassic	Arizona, USA	Archosaur	*Isochirotherium marshalli*	2	Adult	Klein et al. (2024)
Middle Triassic	Arizona, USA	Archosaur	*Synaptichnium diabloense*	3	Adult	Peabody (1948)
Middle Triassic	France	Archosaur	*Isochirotherium felenci*	4	Adult	Courel and Demathieu (1976)
Middle Triassic	Germany	Archosaur	*Chirotherium barthii*	3	Adult	Rühle v. Lilienstern (1939)
Late Triassic	Bolivia	Suchia	*Brachychirotherium*	several	Adult	Apesteguía et al. (2021)
Late Triassic	Switzerland	Theropod		5	Adult	Furrer (1993)
Late Triassic	Colorado, USA	Sauropod	*Eosauropus cimarronensis*	5, 3	Adult	Lockley et al. (2001)
Late Triassic	Utah, USA	Sauropod	*Eosauropus cimarronensis*	2	Adult	Lockley et al. (1992a)
Late Triassic	New Mexico, USA	Prosauropod	*Evazoum*	6	Adult	Cotton et al. (1988, Fig. 4), Lucas et al. (2010)
Late Triassic	New Mexico, USA	Sauropodomorph	*Eosauropus cimarronensis*	4, 4, 4, 4	Adult	Conrad et al. (1987), Lockley and Hunt (1995a, Fig. 3.16)

TABLE 3.1 Putative gregarious transgressive tracksites.—cont'd

Age	Location	Biotaxa	Ichnotaxa	No. of trackways	Age class	References
Early Jurassic	Connecticut, USA	Theropod	*Eubrontes giganteus*	Several, up to 22	Adult	Hitchcock (1836, 1848), Ostrom (1972)
Early Jurassic	China	Small theropod		many	Adult	Shikama (1942)
Early Jurassic	Utah, USA	Small ornithischian	*Anomoepus*	3	Adult	Lockley et al. (1998b), Lockley and Gierlinski (2006)
Early Jurassic	Utah, USA	Theropod	*?Eubrontes*	5	Adult	Lockley et al. (2006b), Milner et al. (2011)
Early Jurassic	Utah, USA	Prosauropod	*Otozoum*	9	Adult	Lockley (2011)
Early Jurassic	Colorado, USA	Prosauropod	*Otozoum*	5	Adult	Lockley et al. (2004, Fig. 19C)
Early Jurassic	Nevada, USA	Tritylodontid	*Brasilichnium*	12 (on foreset)	Adult and juvenile	Rowland and Mercadante (2014)
Early Jurassic	Utah, USA	Tritylodontid	*Brasilichnium*	Several, up to 4 (on foreset)	Adult	Lockley and Hunt (1995a, Fig. 4.21)
Early Jurassic	France	Theropod	*Eubrontes*	4	Adult	de Lapparent and Montenat (1967)
Early Jurassic	Italy	Sauropod		3	Adult	Avanzini and Leonardi (1993)
Early Jurassic	Poland	Sauropod	*Parabrontopus*	4	Juvenile	Gierlinski and Pienkowski (1999)
Early Jurassic	Hungary	Theropod	*Komlosaurus*	6	Adult	Ösi et al. (2011, Fig. 3).
Middle Jurassic	Morocco	Sauropod		6	Adult	Lallensack et al. (2019)
Middle Jurassic	England	Sauropod, possibly two ichnotaxa		About 40	Adult	Day et al. (2002)
Middle Jurassic	China	Thyreophoran	*cf. Stegopodus*	2	Adult	Xing et al. (2019)
Middle Jurassic	Pakistan	Sauropod		3	Adult	Malkani (2007)
Late Jurassic	Colorado, USA	Sauropod	*Parabrontopodus*	5	Juvenile	Lockley (1991b, 1995), Lockley et al. (1986)

(Continued)

TABLE 3.1 Putative gregarious transgressive tracksites.—cont'd

Age	Location	Biotaxa	Ichnotaxa	No. of trackways	Age class	References
Late Jurassic	Colorado, USA	Sauropod	*Parabrontopodus*	6	Adult	Lockley et al. (1986), Lockley and Schumacher (2021), Schumacher and Lockley (2022)
Late Jurassic	Utah, USA	Sauropod	*Brontopodus*	3	Adult	Barnes and Lockley (1994), Foster and Lockley (2006, 2024)
Late Jurassic	Utah, USA	Small ornithopod	*Dinehichnus*	16	Adult	Lockley et al. (1998a, Fig. 1)
Late Jurassic	Portugal	Small ornithopod	*Dinehichnus*	3	Adult	Lockley et al. (1998a, Fig. 4)
Late Jurassic	Portugal	Theropod	*"Eutynichnium lusitanicum"*	4	Adult	Gomes (1915–1916)
Late Jurassic	Portugal	Theropod	*Megalosauripus*	6	Adult	Santos et al. (1995), Lockley and Meyer (2000, Fig. 7.11)
Late Jurassic	Portugal	Sauropod		7	Subadult	Lockley et al. (1994a), Lockley and Meyer (2000, Fig. 7.14)
Late Jurassic	Spain	Ornithopod		several	Adult	García-Ramos et al. (2002: p. 176; García-Ramos et al., 2004: p. 103; 2006a,b: p. 146–147), Piñuela et al. (2002, Piñuela et al. (2016)
Late Jurassic/ Early Cretaceous	Spain	Bird or small theropod	*Archaeornithipus*	dozens	Adult	Fuentes Vidarte (1996)
Early Cretaceous	South Korea	Large ornithopods		Many localities – up to 16	Adult	Lee et al. (2000), Lockley et al. (2006d, 2012), Kim et al. (2009)
Early Cretaceous	South Korea	Sauropod		3	Adult	Lockley et al. (2012, Fig. 5)
Early Cretaceous	Japan	Small ? ornithopod	*Toyamasauripus*	33	Adult	Matsukawa et al. (1997)
Early Cretaceous	Canada	Large ornithopods		11	Adult	Currie (1983)
Early Cretaceous	Canada	Ankylosaurs	*Tetrapodosaurus borealis*	2 groups, several animals in each	Adult	McCrea (2000, 2001, 2003), McCrea et al. (2014a)

TABLE 3.1 Putative gregarious transgressive tracksites.—cont'd

Age	Location	Biotaxa	Ichnotaxa	No. of trackways	Age class	References
Early Cretaceous	Brazil	Sauropod		5	Adult and juvenile	Leonardi (1989)
Early Cretaceous	Brazil	Sauropod		6 (Serrote do Pimenta)	Adult	Leonardi (1989)
Early Cretaceous	Brazil	Sauropod		5 (São Dominguez)	Adult	Leonardi (1989)
Early Cretaceous	Brazil	Sauropod		7 (Piau)	Adult	Leonardi (1989)
Early Cretaceous	Brazil	Sauropod		4 trackways, 12 tracks (Serrote do Letreiro)	Adult	Carvalho (2000)
Early Cretaceous	Chile	Theropod		2	Adult	Rubilar et al. (2008)
Early Cretaceous	China	Sauropod		4	Juvenile	Lockley et al. (2002b, 2018)
Early Cretaceous	China	Deinonychosaur	*Dromaeopodus shandongensis*	6	Adult	Li et al. (2008)
Early Cretaceous	China	Ornithopod	cf. *Ornithopodichnus*	3	Adult	Xing et al. (2015a)
Early Cretaceous	Texas, USA	Sauropod	*Brontopodus*	12 (Paluxy River)	Adult	Bird (1939, 1941, 1954), Farlow (1987), Farlow et al. (1989)
Early Cretaceous	Texas, USA	Sauropod	*Brontopodus*	23 (Davenport Ranch)	Adult and juvenile	Bird (1944, 1954), Ostrom (1985), Farlow (1987), Farlow et al. (1989)
Early Cretaceous	Texas, USA	Theropod		11	Adult	Albritton (1942, p. 160)
Early Cretaceous	New Mexico, USA	Ornithopod	*Caririchnium*	55 (Mosquero)	Adult	Lockley and Hunt (1995a, Fig. 5.20), Cotton et al. (1988, Fig. 3)
Early Cretaceous	New Mexico, USA	Ornithopod	*Caririchnium*	Several (Clayton Lake)	Adult	Hunt and Lucas (1998b), Lucas and Dalman (2016), Lucas et al. (2024b)
Early Cretaceous	New Mexico, USA	Crocodylian		5	Adult	Bennett (1992)

(Continued)

TABLE 3.1 Putative gregarious transgressive tracksites.—cont'd

Age	Location	Biotaxa	Ichnotaxa	No. of trackways	Age class	References
Early Cretaceous	Colorado, USA	Ornithopod	*Caririchnium*	Multiple sites (e.g., 6 at Dinosaur Ridge)	Adult	Lockley et al. (1992a,Lockley et al. (1992b)), Lockley and Hunt (1995a), Lockley (1995, 1997)
Early Cretaceous	Colorado, USA	Bird	*Ignotornis mcconnelli*	6	Adult	Lockley et al. (2009, Fig. 15)
Early Cretaceous	Arkansas, USA	Sauropod	*Brontopodus*	10	Adult	Pittman and Gillette (1989)
Early Cretaceous	Algeria	Sauropod	*Brontopodus*	3	Adult	Zaagane et al. (2023)
Early Cretaceous	Spain	Sauropod		2	Adult	Castanera et al. (2011)
Early Cretaceous	Spain	Ornithopod		3	Adult	Castanera et al. (2011)
Early Cretaceous	Spain	Sauropod		8	Adult	García-Ortiz and Pérez-Lorente (2014)
Early Cretaceous	Spain	Small ornithopod	*Dinehichnus*	8	Adult	Aguirrezabala et al. (1985)
Early Cretaceous	Spain	Small theropod		3	Adult	Pérez-Lorente et al. (1997)
Early Cretaceous	Spain	Ornithopod (Castanera et al. (2014) or theropod (Barco et al., 2006)	*Iguanodontipus oncalensis*	12	Adult	Barco et al. (2006), Castanera et al. (2014), Díaz-Martínez et al. (2023, Fig. 5B)
Early Cretaceous	England	Ornithopod		4	Adult	Delair and Lander (1973), Delair (1981)
Early Cretaceous	Germany	Sauropod	*Brontopodus (Rotundichnus münchehagensis)*	7	Adult	Hendricks (1981)
Early Cretaceous	Australia	Birds		Several individual tracks	Adult	Martin et al. (2023)
Early Cretaceous	Peru	Theropod		4 and 3	Adult	Moreno et al. (2012, Fig. 4)
Early Cretaceous	Chile	Theropod		2	Adult	Moreno et al. (2012, Fig. 6)

TABLE 3.1 Putative gregarious transgressive tracksites.—cont'd

Age	Location	Biotaxa	Ichnotaxa	No. of trackways	Age class	References
Early Cretaceous/ Late Cretaceous	Australia	Small ornithopod or theropod	*Wintonopus, Skartopus*	156	Adult	Thulborn and Wade (1979, 1984), Romilio et al. (2013), Thulborn (2017)
Late Cretaceous	Bolivia	Sauropod		8 (Toro Toro)	Adult and juvenile	Leonardi (1984, 1989)
Late Cretaceous	Bolivia	Sauropod		11 (Humaca)	Subadult	Lockley et al. (2002a, Fig. 4)
Late Cretaceous	Bolivia	Sauropod		2 (Cal Orck'o)	Adult	Lockley et al. (2002a), Calvo et al. (2022)
Late Cretaceous	Argentina	Abelisaurid theropod	aff. *Asianopodus pulvinicalx*	5	Adult	Heredia et al. (2020, Fig. 5)
Late Cretaceous	Argentina	Sauropod	*Titanopodus mendozensis*	4 2	Adults and juvenile/ subadults Adult and juvenile/ subadult	González Riga and Ortiz (2011), Tomaselli (2014)
Late Cretaceous	Colorado, USA	Ceratopsian	*Ceratopsipes goldenensis*	4	Adult	Lockley and Hunt (1995b)
Late Cretaceous	Canada	Tyrannosaur	*Bellatoripes fredlundi*	3	Adult	McCrea et al. (2014b)
Late Cretaceous	Canada	Hadrosaur	*Hadrosauropodus*	15	Adult	Enriquez et al. (2022)
Late Cretaceous	Morocco	Sauropod		2	Adult	Boudchiche et al. (2017)
Eocene	Utah, USA	Perissodactyl		2	Adult	Lockley and Hunt (1995a, Fig. 6.12)
Eocene	New Mexico, USA	Artiodactyl		3	Adult and juvenile	Lockley and Hunt (1995a, Fig. 6.16)
Oligocene	Spain	Birds		6	Adult	de Raaf et al. (1965)
Miocene	Spain	Artiodactyl	*Fustinianapodus arriazui*	Majority of 309 individual tracks	Adult and juvenile	Díaz-Martínez et al. (2020)

(*Continued*)

TABLE 3.1 Putative gregarious transgressive tracksites.—cont'd

Age	Location	Biotaxa	Ichnotaxa	No. of trackways	Age class	References
Miocene	Italy	Artiodactyl		5	Adult	Dalla Vecchia and Rustioni (1996)
Miocene	California, USA	Camelid		2	Adult	Lockley and Hunt (1995a, Fig. 6.20)
Pleistocene	England	*Homo antecessor*		8	Adult and juvenile	Ashton et al. (2014, Figs. 9−10)
Pleistocene	Nevada, USA	Sloth		7	Adult	Coffin (1889), Lockley and Hunt (1995a, Fig. 6.29)
Pleistocene	Nevada, USA	Mammoth		3	Adult	Coffin (1889), Lockley and Hunt (1995a, Fig. 6.29)
Holocene	Nicaragua	*Homo sapiens*		11	Adult	Lucas et al. (2024a)

2.2.1 *Ichnotaxonomy*

Ichnotaxa do not equate exactly to biotaxa and usually represent a higher taxonomic category (e.g., Farlow, 1992; Day et al., 2004; Lucas, 2007). However, gregarious concentrations should include the same ichnotaxa. Some modern migrations (e.g., annual wildebeest and zebra migrations in East Africa) include multiple taxa progressing in the same direction. Thus, there is potential for an ichnological signature of transgressive sites involving several ichnotaxa, which represents gregarious behavior in several biotaxa. Such sites have yet to be recognized.

2.2.2 *Bedding plane*

Often, vertebrate tracks occur on multiple bedding planes representing different time episodes (e.g., Lockley et al., 1995). Gregarious behavior can only be demonstrated by the presence of tracks on the same bedding plane representing the same time interval.

2.2.3 *Time averaging*

Tracks on the same bedding plane may represent time averaging. Indicators of time averaging may include frequent trackway intersections, overprinting of tracks, and dissimilar preservation of adjacent tracks (size, depth, margin slumping, erosion) (Lockley, 1991b; Day et al., 2004). Tracksites are sensitive to even small amounts of time-averaging that may skew interpretations of gregarious behavior such as that several solitary individuals traveling in a common direction and passing the same point within a short period could produce a set of subparallel trackways (Myers and Fiorillo, 2009, Fig. 1). Time averaging of a lacustrine margin tracksite may be measured in hours or months depending on the environmental circumstances (Cohen et al., 1991). "It may be tempting to view a tracksite as being produced by contemporaneous animals, particularly if tracks appear parallel or associated in some other way, but a sediment may be exposed and susceptible to track formation continuously or sporadically over minutes, days, months or even years" (Falkingham, 2014, p. 225).

2.2.4 Subparallel orientation

The principal indicator of gregarious behavior since Hitchcock (1836) has been subparallel orientation of trackways of the same ichnotaxon, indicating a grouping of individuals moving in the same direction.

2.2.5 Directional barriers

Physical barriers can result in animals walking in the same direction (Ostrom, 1972). The most common example of this is animals walking parallel to lacustrine or marine shorelines (e.g., Getty et al., 2012, 2015, 2017; Myers and Fiorillo, 2009). Ostrom (1972) argued that the presence of trackways with perpendicular orientations may indicate the absence of barriers, but this may not always be the case.

2.2.6 Ecological attractors

Some environmental components attract organisms, notably water (e.g., García-Ortiz and Pérez-Lorente, 2014, Fig. 2), but also possibly sources of foodstuffs such as dense vegetation. In this instance, tracks are parallel and at right angles to the perceived attractor (e.g., lake shore).

2.2.7 Intertrack spacing

Gregarious animals usually move in a structured grouping with a fairly regular spacing. Lockley (1989) emphasized the concept of intertrackway spacing as a parameter for measuring regularity of trackways in parallel sets. A notable exception with multiple overlapping tracks is the Early Cretaceous Davenport Ranch site in Texas in the United States (Bird, 1944, 1954; Farlow, 1987; Farlow et al., 1989: Figs. 1 and 2). However, arguably, the intertrack spacing is mainly relevant to a group that is in a single row with the axis perpendicular to the direction of travel (J. Harris, pers. commun., 2024), In this case, intertrackway spacing would be expected to be somewhat regular because of body width. However, large groups of animals usually don't move in a single row, and there are individuals in front and behind that do not necessarily walk in neat columns. Thus, while animals in front may create trackways that do not overlap, the following animals in the group may step directly onto tracks made by the first animals or in between their trackways (J. Harris, pers. commun., 2024) as seems to be the situation at the Davenport Ranch site (Figs. 3.1 and 3.2). Other large transgressive sites seem to demonstrate a fairly regular intertrack distance (Fig. 3.4).

2.2.8 Age structure

Several groupings of parallel trackways include both large and small tracks of the same ichnotaxon and usually are interpreted to be adults and juveniles of the same biotaxon (Table 3.1). Such a mixing of age classes would be expected in a large social group of a species and may indicate a spatial organization of individuals within a group (e.g., Bakker, 1968; Lockley, 1995). However, track ichnogenera clearly represent biotaxa above the genus level. For example, there is only one common ichnogenus of sauropod track, *Parabrontopodus*, in the Upper Jurassic Morrison Formation of the western United States (Foster and Lockley, 2006), and yet there are several families (and many genera) of sauropod biotaxa of different sizes present in the fauna (e.g., Foster, 2003, 2020). *Brontopus* also occurs in the Morrison at three localites (Foster and Lockley, 2006, 2024). Thus, small "juvenile" tracks could potentially represent different biotaxa instead of juveniles of large "adults."

2.2.9 *Speed and directional change*

Simultaneous changes in direction and speed in a social group should be reflected by the tracks of most individuals in the social group (Lockley, 1991b; González Riga and Ortiz, 2011; Tomaselli, 2014).

2.2.10 *Topography*

Local topography can result in subparallel orientation of trackways. This is particularly evident on foreset beds of eolian dune faces in the lower Permian and Lower Jurassic of western North America on which most trackways are oriented up foresets (e.g., Lockley and Hunt, 1995a; Rowland and Mercadante, 2014; Engelmann and Chure, 2017; Marchetti et al., 2020).

Transgressive sites are influenced by both biological and environmental factors. At one end of the spectrum are tracks on eolian foresets whose parallel orientations are entirely the result of topography and environment (e.g., Coconino, Lyons, and parts of the Navajo/Nugget formations of western North America and other ichnofaunas of the *Chelichnus* Ichnofacies of Hunt and Lucas, 2007: Table 1). At the other end are parallel trackways that include the passing of a probable social group with either standard intertrack distances or a mix of individuals of different sizes/age classes and no indication of directional constraints (e.g., Davenport Ranch site: Bird, 1944, 1954: Figs. 3.1 and 3.2). In the middle of the spectrum are parallel trackways that may indicate gregarious behavior but may also have been influenced by environmental factors, notably lacustrine shorelines and possibly fluvial sand bars, which are important sources of tracks (e.g., Late Jurassic/Early Newark Supergroup, Eocene Green River Formation, and other ichnofaunas of the *Grallator* Ichnofacies of Hunt and Lucas, 2007).

We recognize two scales of congregation sites. Microcongregation sites are small areas (up to tens or hundreds of square meters) with concentrations of the same ichnotaxon focused on a discrete and often recognizable locus. Examples include watering holes (e.g., Bennett et al., 2009), caves (e.g., Lockley and Meyer, 2000; Willey et al., 2005; Pastoors et al., 2015), forest areas (e.g., Parker and Balsley, 1989; Van Allen et al., 2005), and areas on tidal and lacustrine-margin flats (e.g., surfaces with concentrations of *Dromopus* or *Batrachichnus* in the early Permian: MacDonald, 1994; *Rhychosauroides* in the Middle Triassic: Diedrich, 2008; *Aquatilavipes* in the Early Cretaceous: Currie, 1981; Table 2). These sites are characterized by a combination of short trails and individual prints with common overprinting (Morse et al., 2010). Megacongregation sites are large areas (up to hundreds or thousands of square meters) with multiple trackways in which few, if any, demonstrate an obvious preferred orientation with no clear cause for the concentration. Prominent dinosaur examples of megacongregation sites include the Late Jurassic "Dinosaur Disco" of the Swiss Alps (Lockley and Meyer, 2000) and the Middle Jurassic "Stomping Ground" and other portions of the Moab Megatracksite in Utah in the United States (Lockley, 1991a; Lockley, 2021). We list a selection of congregation sites in Table 3.2, but the actual number of localities is significantly larger.

TABLE 3.2 Selected gregarious congregation tracksites.

Age	Location	Biotaxa	Ichnotaxa	Micro	Macro	References
Late Pennsylvanian/ Early Permian	Canada	Seymouriamorph	*Amphisauropus latus*	X		Van Allen et al. (2005, Figs. 12A and 13)
Early Permian	New Mexico, USA	Diapsid	*Dromopus agilis*	X		MacDonald (1994)
Early Permian	New Mexico, USA	Temnospondyl	*Batrachichnus salamandroides*	X		MacDonald (1994)
Middle Triassic	Germany	Archosauromorph lepidosauromorph	*Rhychosauroides*		X	Diedrich (2008)
Early Jurassic	Colorado, USA	Theropods			X	Lockley (1991b, Fig. 4.4)
Early Jurassic	Lesotho	Theropods		X		Ellenberger et al. (1963), Ellenberger (1974)
Late Jurassic	Switzerland	Sauropods			X	Lockley and Meyer (2000, Fig. 7.16)
Middle Jurassic	Utah, USA	Theropods			X	Lockley (1991b, Fig. 12.2A), Lockley and Hunt, 1995a, Fig. 4.36)
Late Jurassic	Utah, USA	Sauropods and theropods			X	Lockley et al. (2023)
Early Cretaceous	Canada	Ornithopods			X	Currie and Sarjeant (1979)
Early Cretaceous	Spain	Quadruped			X	Diaz-;Martinez et al. (2023, Fig. 6)
Early Cretaceous	New Mexico, USA	Ornithopods	*Caririchnium*		X	Hunt and Lucas (1998b), Lucas and Dalman (2016), Lucas et al. (2024b)
Early Cretaceous	Canada	Birds	*Aquatilavipes*	X		Currie (1981)
Early Cretaceous	Korea	Birds	Several			Kim et al. (2012)
Late Cretaceous	Utah, USA	Pterosaurs		X		Parker and Balsley, (1989), Lockley and Hunt (1995a, Fig. 5.36)
Late Cretaceous	Utah, USA	Hadrosaurs		X	X	Balsley and Parker (1983), Lockley and Hunt (1995a, Fig. 5.35)
Late Cretaceous	Korea	Pterosaurs	*Pteraichnus*	X		Jung et al. (2022)
Late Cretaceous/ Paleocene	Colorado, USA	Birds		X		Lockley and Hunt (1995a, Fig. 6.2)

(*Continued*)

TABLE 3.2 Selected gregarious congregation tracksites.—cont'd

Age	Location	Biotaxa	Ichnotaxa	Micro	Macro	References
Miocene	Iran	Artiodactyls, birds and other taxa		X		Abbassi (2010)
Miocene	Iran	Felids and canids		X		Abbassi et al. (2021)
Pleistocene	Kenya	*Homo ergaster/ erectus*		X		Bennett et al. (2009)
Pleistocene	France	*Homo sapiens*		cave		Lockley and Meyer (2000, Fig. 10.12)

Congregation sites provide poor evidence for gregarious behavior and are more likely the result of other behavior, such as feeding. Megacongregation sites may represent "taphotracksites" in that they may be artifacts of time-averaged trackways on a preservationally optimal substrate in certain environments.

Seeking evidence of gregarious behavior from tracks is certainly worthwhile. However, caution is advised in the inference of gregarious behavior from tracksites. As noted earlier, there are many factors that might influence the appearance of group behavior. The most significant influences are directional barriers, ecological attractors, and time averaging.

2.3 Fossil record of ichnology of gregarious or putative gregarious behavior in tetrapods

Trackways of tetrapods only become common in the Early Carboniferous (Lucas, 2019, 2024; Lucas and Stimson, 2024). It is notable that most literature on predinosaur (pre-Late Triassic) vertebrate tracks is devoted to ichnotaxonomy, and much more literature on younger tracksites, notably in the Mesozoic, focuses on paleoecology. Therefore, we would expect that there may be examples of gregarious behavior in older strata that may not have been recognized. The earliest congregation site is from close to the Pennsylvanian/Permian boundary in eastern Canada (Van Allen et al., 2005). The first transgressive and the next congregation sites are in lower Permian redbeds of New Mexico in the United States. The one transgressive site has only two trackways of *Dimetropus,* and it might be the result of an accidental alignment (Hunt et al., 1993b: Fig. 3.5). The congregation sites are among the Robledo Mountains tracksites and feature local concentrations of *Dromopus agilis* and *Batrachichnus salamandroides* in a broadly tidal flat environment (MacDonald, 1994).

Lower Permian eolian strata in western North America, Scotland, Germany, and probably Argentina have parallel trackways of "*Chelichnus*" on foresets that were undoubtedly topography-controlled.

There is only one transgressive tracksite in the Early Triassic of Germany that features dicynodont tracks (Haubold, 1984). However, in the Middle Triassic, there are multiple transgressive sites featuring archosaurs and dicynodonts from North America and Europe and a large congregation site from Germany (Rühle v. Lilienstern, 1939; Peabody, 1948; Courel and Demathieu, 1976; Hunt et al., 1993a; Diedrich, 2008; Klein et al., 2024, Tables 1–2). It is possible that the Early-Middle Triassic witnessed the evolution of complex gregarious behavior in multiple groups of tetrapods, notably dicynodonts and some archosaurs, but as aforementioned, this behavior may remain to be recognized in Permian tracksites. Indeed, we would expect gregarious behavior in earlier therapsid herbivores such as *Lystrosaurus* because these lineages produce bonebeds suggestive of such behavior.

Late Triassic transgressive tracksites occur in North and South America and Europe (Conrad et al., 1987; Lockley et al., 1992a; Furrer, 1993; Lockley and Hunt, 1995a; Cotton et al., 1987, 1988; Lockley et al., 2001; Apesteguía et al., 2021). Trackmakers include early sauropodomorphs and theropods.

Parallel trackways of *Brasilichnium* (tritlyodontid or basal mammaliaform) from the Early Jurassic of western North America occur on forest beds in eolian strata and are topographically controlled (Lockley and Hunt, 1995a, Fig. 4.21; Rowland and Mercadante, 2014; Engelmann and Chure, 2017). Other Early Jurassic transgressive tracksites are widespread in North America, Europe, and eastern Asia and appear to represent gregarious behavior in sauropodomorphs and theropods (Shikama, 1942; de Lapparent and Montenat, 1967; Lockley et al., 1998b, 2006a,b; Lockley, 2011; Milner et al., 2006, 2011). Lacustrine margin tracksites of the Newark Supergroup (e.g., Hitchcock, 1836, 1848; Ostrom, 1972) may have been influenced by shoreline trends (Getty et al., 2012, 2015, 2017). Lockley (1991a, Fig. 4.4) illustrate a congregation site of theropod tracks from Colorado in the United States. Ellenberger (1974; Ellenberger et al., 1963) described several congregation sites, dominated by theropod tracks, from Lesotho.

Middle Jurassic transgressive sites are small in number but widely dispersed in Europe, Africa, and south and eastern Asia and principally pertain to sauropods (Day et al., 2002; Malkani, 2007; Xing et al., 2019; Lallensack et al., 2019). A large congregation site of theropod tracks occurs in Utah in the United States (Lockley, 1991b, Fig. 12.2A; Lockley and Hunt, 1995a, Fig. 4.36).

Transgressive sites of Late Jurassic age are common in western North America and the Iberian peninsula (e.g., Lockley, 1991b, 1995; Santos et al., 1995; Lockley et al., 1986, 1998; Barnes and Lockley, 1994; Piñuela et al., 2016). These represent sauropods, theropods, and the first examples of gregarious behavior in ornithopods.

The largest number of described transgressive tracksites interpreted to indicate gregarious behavior are from the Early Cretaceous (Figs. 3.1–3.4). They occur in North and South America, Europe, north Africa, and eastern Asia and pertain principally to sauropods and ornithopods with a smaller number to theropods (Table 3.1). Congregational sites are known from Canada, the United States, Korea, and Spain that principally preserve ornithopods as well as undetermined quadrupeds, pterosaurs, and birds (e.g., Currie and Sarjeant, 1979; Currie, 1981; Jung et al., 2022; Díaz-Martínez et al., 2023) (Fig. 3.3).

Transgressive tracksites of Late Cretaceous age occur in North and South America and Africa and represent sauropods, hadrosaurs, ceratopsians, and theropods (Leonardi, 1989; Lockley and Hunt, 1995a; Lockley et al., 2002a; McCrea et al., 2014a; Heredia et al., 2020; Enriquez et al., 2022; Zaagane et al., 2023). Congregational sites from North America and Asia preserve tracks of hadrosaurs, pterosaurs, and birds (Balsley and Parker, 1983; Parker and Balsley, 1989; Lockley and Hunt, 1995a; Kim et al., 2012).

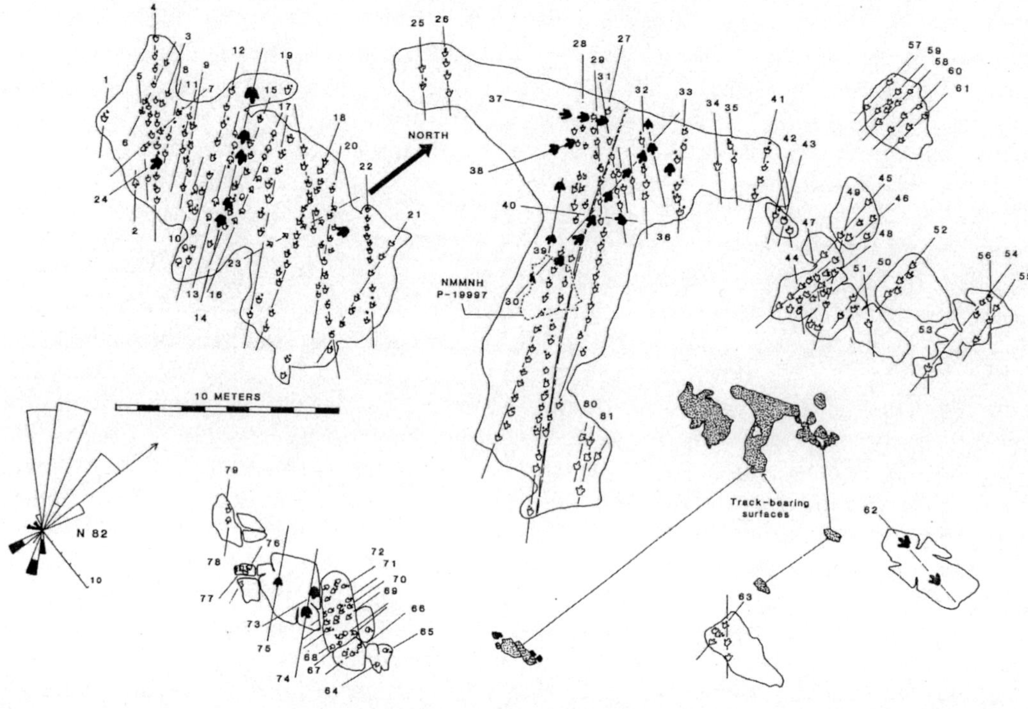

FIGURE 3.4 Multiple parallel trackways of the ornithopod track *Caririchnium leonardii* from the uppermost Mesa Rica Sandstone (Lower Cretaceous) at Mosquero Creek, New Mexico, USA. *After Lockley and Hunt (1995a, Fig. 5.20).*

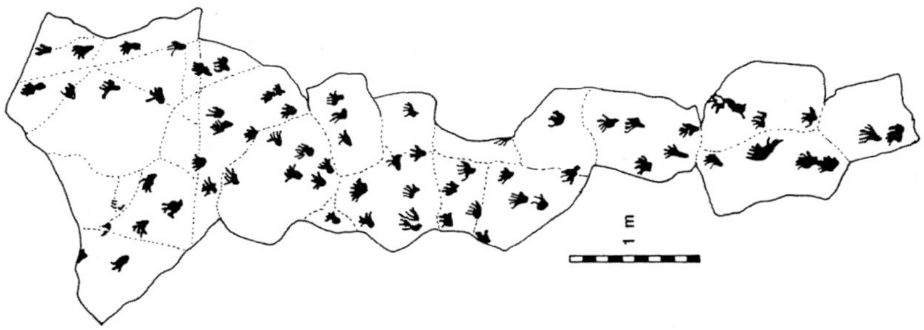

FIGURE 3.5 Two subparallel trackways of *Dimetropus* isp. from the Robledo Mountains Formation (Lower Permian: middle-upper Wolfcampian) from Prehistoric Trackways National Monument southern New Mexico, USA. *From Hunt et al. (1993b, Fig. 7).*

Cenozoic tracks have received much less study with regard to gregarious behavior because fossil mammals have modern analogs that are demonstrably social, so the recognition of this behavior in the fossil record is to be expected and thus is not notable. In Tables 3.1 and 3.2, we list a sample of Cenozoic tracksites, both congregation and transgressive (Fig. 3.6).

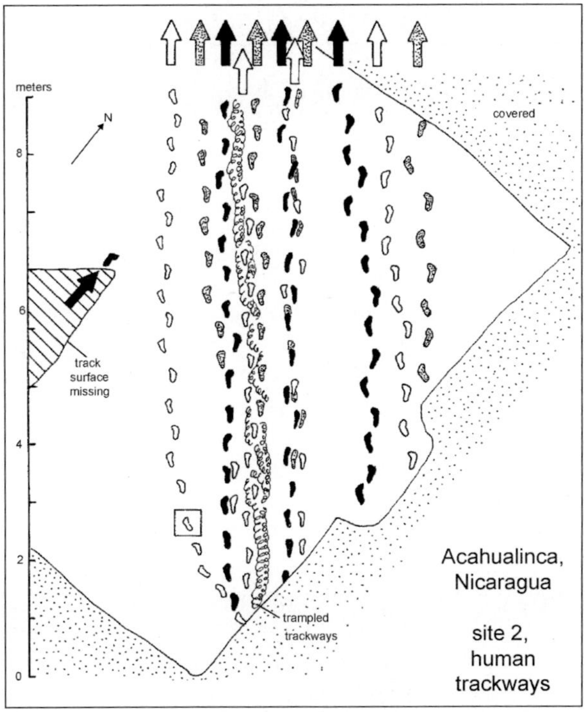

FIGURE 3.6 Parallel trackways of *Homo sapiens* from the Holocene site of Acahualinca, Nicaragua. *From Lucas et al. (2024a, Fig. 3).*

FIGURE 3.7 Reconstructions of the early Permian sphenacodontid *Dimetrodon* with traditional sprawling posture (*left*) and a more upright stance (*right*) that appears to be more consistent with trackways of *Dimetropus* as shown in Fig. 3.5. *From Hunt and Lucas (2024) after MacDonald (1994).*

3. Ichnopathology

Paleopathologies have been recognized from fossil trackways on the basis of: (1) irregular gait; or (2) pathologic foot impressions (Lockley et al., 1994c; Tanke and Rothschild, 2002; McCrea et al., 2015). Most reports are from dinosaurians or associated tracks (Table 3.3). Lockley et al. (1994c) was an influential study that recognized nine examples of alternating long and short steps postulated to be indicative of limping or some other unusual gait in a variety of dinosaurs from the Early Jurassic to the Early Cretaceous. Dantas et al. (1994) defined limping as an irregular gait in which one leg or foot is favored over the other, resulting in consistent differences in the length of alternate steps. However, Farlow (pers comm. in Dantas et al., 1994) pointed out that inconsistency of gait may simply be an indication of a healthy animal favoring one side over the other, as is often seen in the gaits of modern healthy animals. McCrea et al. (2015) distinguished consistencies of irregularities in pace and stride values that they defined as irregular gaits, whereas they termed limping gaits the more extreme examples that are more likely to be cases of ichnopathology. In the latter, there is a visible pathology in the morphology of the track associated with an obvious limp. Thus, McCrea et al. (2015) concluded that the majority of the trackways described by Lockley et al. (1994c) that show no foot pathologies are likely examples of irregular gaits of healthy animals, rather than pathologically influenced limping behavior, although they could be result of other limb or girdle injuries. They classified trackways with one or more visible, repetitive pathologies, but with a nearly normal gait (no or minimal trace of a limp) as having compensated gaits (McCrea et al., 2015). McCrea et al. (2015) further recognized five factors that may cause pathologies affecting the both morphology and the locomotory function of trackmakers' limbs. However, of these, only biomechanical injury due to physical damage to the body affecting the movement of the limb and its elements is identifiable from tracks. Genetic or disease-induced malformations would be indistinguishable from those caused by injuries.

The majority of reported pathologies and putative pathologies relate to bipedal, tridactyl trackmakers, principally theropods (Table 3.3). The narrow digits and long stride of theropods facilitate the recognition of abnormal foot structure and of irregular pacing.

Schulp and Brokx (1999) described Late Cretaceous sauropod trackways found on a steeply inclined (60°) rock face in Spain. The inclined rock is undergoing strain that deforms the tracks and creates asymmetrical dinosaur tracks and apparent limping dinosaur trackways. This is probably a rare example of such geophysical deformation.

The oldest pathologic trackways are of the theropod ichnotaxon *Anchisauripus* from the Late Triassic of Wales (Tucker and Burchette, 1977) and the phytosaur track *Apatopus* from the Late Triassic of Utah (Foster et al., 2000, 2003). In the former, there is a malformation of digit III, and strong angulation of the distal end of the toe impression was repeated in several tracks (Tucker and Burchette, 1977). The trackway configuration is normal, so it represents a compensated gait (*sensu* McCrea et al., 2015). The latter trackway consists of two left manus-pes sets, a tail drag, and a right manus-pes set, with a second right pes skin-impression heel mark of *Apatopus* at the top of the Chinle Group. In the left pes, both tracks in sequence have shortened digit IVs (usually the longest digit) relative to typical *Apatopus*, indicating possible malformation or complete loss of the distal phalanges of the digit in this individual (Foster et al., 2000, 2003).

TABLE 3.3 Ichnopathologies of dinosaur tracks.

Age	Location	Biotaxa	Ichnotaxa	Ichnological expression of pathology	Inferred pathology	References
Late Triassic	Wales	Theropod	*Anchisauripus* isp.	Strong angulation of the distal end of toe impression repeated in several tracks	Distinctive malformation of digit III, compensated gait	Tucker and Burchette (1977), McCrea et al. (2015)
Late Triassic	Utah, USA	Phytosaur	*Apatopus*	In the left pes, both tracks in sequence have shortened digit IVs (usually the longest digit) relative to typical *Apatopus*	Possible malformation or complete loss of the distal phalanges of digit IV	Foster et al. (2000, 2003)
Early Jurassic	Connecticut, USA	Theropod	*Sauroidichnites abnormis*	The midlines of the two left prints were oriented toward the direction of travel, the two right prints were strongly rotated toward the trackway midline by nearly 45°	Injury to the right foot	Hitchcock (1844) (McCrea et al. (2015)
Early Jurassic	Connecticut, USA		*Antipus bifidus*	Quadrupedal animal that exhibited only two digit impressions on the fore-foot	Entire lateral digit is missing – we consider that this ichnotaxon does not represent a vertebrate track	Hitchcock (1858), McCrea et al. (2015)
Early Jurassic	New Jersey, USA	Small theropod	cf. *Grallator sensu lato*	Right to left (R-L) step of 60 cm, and a longer left to right (L-R) step of 76 cm, short/long step ratio of 79%	Injury to the left foot, or left side of the trackmaker or irregular gaits of healthy animals	Lockley et al. (1994c)
Early Jurassic	Northeastern USA	Theropod	*Eubrontes*	Trackway segment in which the right footprints completely lacked any impression of digit II	Digit II missing in right foot, compensated gait	Abel (1935), Thulborn (1990, Fig. 5.8), Lockley (1991b)
Early Jurassic	China	Theropod	*Eubrontes pareschequier*	Possible swelling feature that at the posterior-medial end of one track	Distorted metatarsophalangeal but could be taphonomic artifact	Xing et al. (2009, 2012), Lockley et al. (2013), Xing et al. (2014), McCrea et al. (2015, Fig. 5)
Early Jurassic	China	Theropod	*Eubrontes pareschequier*	Absence of the distal end of digit III on one track	Amputation or severe dislocation of the terminal end of this digit - could be taphonomic artifact	Xing et al. (2009), (2012), Lockley et al. (2013); Xing et al. (2014); McCrea et al. (2015), Fig. 5)

(*Continued*)

TABLE 3.3 Ichnopathologies of dinosaur tracks.—cont'd

Age	Location	Biotaxa	Ichnotaxa	Ichnological expression of pathology	Inferred pathology	References
Middle Jurassic	Morocco	Small theropod		Average length of the short step is 63 cm compared to 79 cm for the long step (= ratio of 80%), right foot abnormal in having the two outer digits (III and IV) held very closely together	A foot injury	Jenny and Josen (1982), Ishigaki (1986), Thulborn (1990), Lockley et al. (1994), Ishigaki and Lockley (2010), McCrea et al. (2015)
Middle Jurassic	Morocco	Sauropod (manus only)		Short/long step ratio of 73% from 0.53 (short step/mean stride) to 0.73 (long step/mean stride)	An injury or affliction to the left foot or left side but probably irregular gait of healthy animals	Ishigaki (1989), Lockley and Rice (1990), Lockley et al. (1994c), McCrea et al. (2015)
Late Jurassic	Utah, USA	Theropod	*Hispanosauropus*	Average short to long step ratio of 79%	Probably irregular gait of healthy animal	Lockley et al. (1994c, Fig. 3A–B), McCrea et al. (2015), Foster (2015)
Late Jurassic	Portugal	Large theropod		Mean short step/long step ratio is 87%	Possibly an injury to the left foot or side but probably irregular gait of healthy animal	Lockley et al. (1994c), Dantas et al. (1994), McCrea et al. (2015)
Late Jurassic	Portugal	sauropod		Mean short/long step ratio is 88%	An injury or affliction to the right side but probably irregular gait of healthy animal	Lockley et al. (1994, Fig. 3C), Dantas et al. (1994), McCrea et al. (2015)
Late Jurassic	Spain	Theropod	*Megalosauripus* or *Hispanosauropus*	Digit IV on both right footprints has digit impression oriented almost perpendicular to the midline footprints resulting in a III–IV divarication value of 80°, over 40° greater than observed from the III–IV divarication values of the left prints	Pathologic digit IV on right foot and compensated gait	Avanzini et al. (2008, Fig. 2)
Early Cretaceous	New Mexico, USA	Ornithopod	*Caririchnium*	Mean short/long step ratio of 80%	Probably irregular gait of healthy animal	Lockley et al. (1994), Lockley and Hunt (1995), McCrea et al. (2015)

Early Cretaceous	Australia	Small theropod	*Skartopus*	Short/long step ratio of 83%	An injury to the left foot or side but probably irregular gait of healthy animal	Wade (1989), McCrea et al. (2015)
Early Cretaceous	Spain	Theropod or ornithopod		Short/long step lengths of 84/96 cm (= ratio 88%).	Probably irregular gait of healthy animal	Moratalla et al. (1988), McCrea et al. (2015)
Early Cretaceous	Spain	Theropod		Front projection of digit II on left footprint is completely missing, and the mark of this digit is likely backwardly displaced	Injury to, or loss of, digit II	Herrera-Castillo et al. (2022)
Early Cretaceous	Spain	Ornithopod		Irregular pace pattern and differences between the right and the left tracks, notably in the relative position of the lateral digits II and IV with respect to the central digit III	An injury or pathology on left digit II	Casanovas et al. (1995), Pérez-Lorente (2003), Razzolini et al. (2016)
Early Cretaceous	China	Theropod		Single track reported with unusually shortened digit II	An injury or congenital deformation, but could simply be an extramorphological variant	Xing et al. (2012), McCrea et al. (2015)
Early Cretaceous	Canada	Theropod	cf. *Irenesauripus mclearni*	Trackway exhibiting an extremely short pace and stride, the left prints consistently strongly rotated toward the midline of the trackway, some of the right prints are rotated inward as well, but not to the same degree	Injury to the soft tissue (e.g., muscle, tendon, ligament) of the foot	McCrea et al. (2014a, 2015)
Early Cretaceous/ Late Cretaceous	Colorado, USA	Theropod		Right theropod footprint, one of the outer digit impressions exhibits a strong curve toward the digit III impression, the curved digit impression also appears to be unusually thickened compared to the other two digit impressions	A significant dislocation of digit II with resultant swelling of the soft tissues	McCrea et al. (2015)

(*Continued*)

TABLE 3.3 Ichnopathologies of dinosaur tracks.—cont'd

Age	Location	Biotaxa	Ichnotaxa	Ichnological expression of pathology	Inferred pathology	References
Late Cretaceous	Canada	Tyrannosaur	*Bellatoripes fredlundi*	The two left footprints each have an obviously truncated second digit, indicating pathology, uninjured foot exhibited a very slight outward rotation	Injury above the level of the feet, somewhere between the ankles and the hips, compensated gait	McCrea et al. (2014b, 2015)
Late Cretaceous	Canada	Theropod		Isolated track with severe curvature of digit III as well as the very high splay of both outer digits	Injury to the trackmaker's third digit	McCrea et al. (2014a, 2015)
Late Cretaceous	Canada	Large theropod		Uncollected isolated track presumed digit IV that appears to show severe rotational twisting, so that nearly half of the digit was folded underneath itself and the anterior portion of the metatarsal pad	A relatively recent dislocation injury to the trackmaker because the affected claw retains its normal shape in spite of being dislocated beneath the foot	McCrea et al. (2015)
Late Cretaceous	Mongolia	Hadrosaur		An uncollected, single footprint exhibited a bulbous expansion under the fourth digit impression	Possible soft tissue swelling	Currie et al. (2003)

Four of the Early Jurassic examples are from the long-studied Newark Supergroup of the northeastern United States, and two are from China. Hitchcock (1844) described the first pathologic trackway from Connecticut of four small theropod tracks of "*Sauroidichnites abnormis.*" The midlines of the two left prints are oriented toward the direction of travel, whereas the two right prints are strongly rotated toward the trackway midline by nearly 45 degrees, indicating an injury to the right foot. Subsequently, Hitchcock (1858, pl. 36, Fig. 8, pl. 48, Fig. 10) reported on another trackway from Connecticut, the holotype of "*Antipus bifidus.*" The validity of this ichnotaxon and indeed of the specimen being a track at all has been considered "extremely doubtful" (Lull, 1904, p. 536), "might be organic" (Coombs, 1996, p. 327), or "enigmatic … V shaped grooves" (Olsen et al., 1998, p. 597). We consider that it is uncertain that "*Antipus bifidus*" even represents a vertebrate track (Hitchcock, 1858) and that interpretations of it as demonstrating a pathologic foot (e.g., McCrea et al., 2015) are unjustified.

The two other putative pathological specimens from the Newark Supergroup are from small (*Grallator*) and large (*Eubrontes*) theropod trackways. Abel (1935) first reported a trackway of *Eubrontes* in which the right footprints have only a stub of an impression of digit II. The inferred pedal digit loss in the right foot of an otherwise normal *Eubrontes* trackway indicates that it is an example of a compensated gait. The *Grallator* trackway from New Jersey has an alternating right to left (R-L) step of 60 cm and a longer left to right (L-R) step of 76 cm and probably represents the irregular gait of a healthy animal (Lockley et al., 1994; McCrea et al., 2015).

Both putative Early Jurassic track pathologies from China pertain to the large theropod *Eubrontes pareschequier* and are preserved on two isolated tracks on the same block (Xing et al., 2009, 2012b, 2014; Lockley et al., 2013; McCrea et al., 2015). One pes track lacks the terminal portion of the digit III impression, and the other exhibits a swelling on the medial edge of the digit II impression. Both isolated tracks possibly exhibit extramorphological or taphonomic artifacts (McCrea et al., 2015). The most convincing pathologies are present on successive tracks in a trackway.

The two described putative examples from the Middle Jurassic are both from Morocco. Ishigaki (1989) first recognized that a manus only trackway of a sauropod has alternating lengths of paces, and this has been interpreted as evidence of limping (Lockley and Rice, 1990; Lockley et al., 1994c), but is probably the irregular gait of a healthy animal (McCrea et al., 2015). Ishigaki (1986) described a theropod trackway, first reported by Jenny and Josen (1982). This trackway shows limping (*sensu* Dantas et al., 1994) and a repetitive malformed foot with digits III and IV closely appressed (McCrea et al., 2015). An artistic rendering of the trackway and the theropod was a frontispiece in Lockley (1991b).

There are four putative occurrences of ichnopathology in the Late Jurassic, one from the United States, two from Portugal, and one from Spain. Three of the instances from Utah in the United States (Lockley et al., 1994c; McCrea et al., 2015) and a theropod and a sauropod from Portugal (Dantas et al., 1994; Lockley et al., 1994c; McCrea et al., 2015) probably represent the irregular gaits of a healthy animal (a theropod). Avanzini et al. (2008, Fig. 2) described a theropod trackway from Spain in which the digit IV on both right footprints had the digit impressions oriented almost perpendicular to the midline footprints, resulting in a digit III–IV divarication value of 80 degrees, over 40 degree greater than observed from the digit III–IV divarication values of the left prints. This resulted from a pathologic digit IV on the right foot, and the trackway demonstrates a compensated gait (*sensu* McCrea et al., 2015).

The majority of putative ichnopathologies are of Early Cretaceous age. There are eight occurrences/areas of occurrence from Spain, the United States, Canada, China, and Australia related to theropods and ornithopods (Table 3.3). Note that García-Ortiz and Pérez-Lorente (2014, Table 1) listed 28 tracksites with evidence of gregarious behavior in the La Rioja area of Spain. Three of the examples pertain to ornithopods or tridactyl tracks that could be ornithopod. One ornithopod trackway of *Caririchnium* from New Mexico in the United States probably represents the irregular gait of a healthy animal (Lockley et al., 1994c; Lockley and Hunt, 1995a; McCrea et al., 2015). An ornithopod trackway from Spain displays an irregular pace pattern and differences between the right and the left tracks, notably in the relative position of the lateral digits II and IV with respect to the central digit III caused by an injury or pathology on left digit II (Casanovas et al., 1995; Pérez-Lorente, 2003; Razzolini et al., 2016). A second Spanish tridactyl trackway of a theropod or ornithopod probably represents the irregular gait of a healthy animal (Moratalla et al., 1988; McCrea et al., 2015).

One Early Cretaceous theropod trackway of *Skartopus* from Australia probably represents the irregular gait of a healthy animal (Wade, 1989; McCrea et al., 2015), but the four other trackways display ichnopathologies. A trackway from Spain exhibits the completely missing projection of digit II on the left footprint, and the mark of this digit is likely displaced backward due to an injury and/or loss of digit II (Herrera-Castillo et al., 2022). Xing et al. (2012b) described a single track from China with an unusually shortened digit II that could be the result of an injury or congenital deformation, but also could simply be an extramorphological variant (McCrea et al., 2015). A Canadian trackway of cf. *Irenesauripus mclearni* exhibits an extremely short pace and stride with the left prints consistently rotated strongly toward the midline of the trackway, and some of the right prints rotated inward as well, but not to the same degree. This probably resulted from an injury to the soft tissue (e.g., muscle, tendon, ligament) of the foot (McCrea et al., 2014a, 2015). A mid-Cretaceous trackway from Colorado in the United States displays a right theropod footprint with one of the outer digit impressions exhibiting a strong curve toward the digit III impression, and the curved digit impression also appears to be unusually thickened compared with the other two digit impressions. This could have resulted from a significant dislocation of digit II, with resultant swelling of the soft tissues (McCrea et al., 2015).

Reported Late Cretaceous occurrences of ichnopathology include three theropods from Canada and a hadrosaur from Mongolia. The Mongolian hadrosaur track is an uncollected, single footprint with a bulbous expansion under the fourth digit impression due to possible soft tissue swelling (Currie et al., 2003). Two of the Canadian theropod tracks are also isolated specimens. One isolated track displays severe curvature of digit III as well as the very high splay of both outer digits, which is interpreted as injury to the third digit of the trackmaker (McCrea et al., 2014b, 2015). The other is an uncollected isolated track, with a presumed digit IV that appears to show severe rotational twisting, so that nearly half of the digit was folded underneath itself and the anterior portion of the metatarsal pad. This is interpreted to be the result of a relatively recent dislocation injury to the trackmaker because the affected claw retains its normal shape in spite of being dislocated beneath the foot (McCrea et al., 2015). While these isolated tracks may indeed demonstrate pathologies, individual footprints typically display a wide range of taphonomic or/and extramorphological variation, so we are reluctant to diagnose ichnopathologies from individual footprints that are not parts of longer trackways. The other Canadian example is much more convincing and consists of a trackway of *Bellatoripes fredlundi* in which the two left footprints each have a truncated second digit, indicating pathology, and the uninjured right foot exhibits a very slight outward rotation. This could have resulted from injury above the level of the feet, somewhere between the ankles and the hips. The trackway demonstrates a compensated gait (McCrea et al., 2014b, 2015).

Gillette and Thomas (1985) noted tail drags in Early Cretaceous ornithopod trackways from Clayton Lake, New Mexico in the United States and considered that they might indicate a pathological gait (also see Hunt and Lucas, 2004; Lucas and Dalman, 2016). However, recent restudy of these features indicates that they are toe drags, not tail drags.

4. Gait

Characteristics of trackways reflect the gait of the track maker and thus have great potential in the study of the evolution of locomotion. However, gait analysis has received relatively little attention. A few examples of the importance or potential importance of gait information derived from trackways include:

1. The oldest tetrapod tracks are from the Devonian (Lucas, 2015, 2019, 2024). These tetrapod footprints indicate an alternating pattern of limb-supported locomotion with a larger hind foot than forefoot. This is the characteristic pattern of quadrupedal locomotion of post-Devonian tetrapods (Lucas, 2015, 2019, 2024).
2. Most Paleozoic tetrapods employed the sprawling gait in which the proximal limb bones (humerus and femur) are held nearly horizontal to the substrate so that the feet hit the substrate well lateral to the body midline. However, there is little study of the variation among Paleozoic tetrapods in their use of the sprawling gait (e.g., Sukhanov, 1968; Bernardi and Avanzini, 2011). More analysis needs to be done to better understand variation in late Paleozoic sprawling gaits based on differences in body size, coupling value, foot and tail structure, and speed of progression (Lucas, 2019).
3. An unresolved issue with Paleozoic tetrapod locomotion is based on the presence of relatively narrow-gauge trackways as early as the Late Carboniferous that suggest that the trackmakers employed a semiupright gait (e.g., Hunt and Lucas, 1998; Voigt et al., 2007). Particularly striking in this regard are long, narrow-gauge trackways of eupelycosaurs (ichnogenus *Dimetropus*) from the lower Permian of New Mexico, USA (Hunt and Lucas, 1998; Lucas, 2019: Fig. 3.5). Three ideas have been proposed to explain such narrow-gauge trackways made by late Paleozoic tetrapods: (1) that they indicate semiupright walking, at least over short distances (Hunt and Lucas, 1998; Abbott et al., 2017: Figs. 3.5 and 3.7); (2) that the trackmaker used considerable lateral bending of the trunk (axial bending) to produce such trackways (Hopson, 2012, 2015); or (3) that the trackmaker swung the limbs far forward over a horizontal arc to plant them facing forward near the trackway midline where limb retraction took place (Berman and Henrici, 2003). More analysis is needed to resolve these disparate explanations (Lucas, 2019).

4. The shift from a sprawling gait to an upright gait has long been considered to have taken place during the Triassic (e.g., Kubo and Benton, 2009). However, late Permian archosauriform footprints could indicate that the shift from sprawling to upright gaits had taken place well in advance of the end of the Permian (Bernardi et al., 2015). The same is true for recently described cynodont and gorgonopsid therapsid footprints (Marchetti et al., 2019b,c). Thus, these older, narrow-gauge trackways may demonstrate that the capability of at least short-term use of semiupright gaits by tetrapods goes back to the Carboniferous (Lucas, 2019).
5. Ceratopsid dinosaurs have been traditionally restored with sprawling forelimbs. The Laramie Formation (Upper Cretaceous: Maastrichtian) of Colorado in the United States yielded the first ceratopsid tracks, which were named *Ceratopsipes goldenensis* (Lockley and Hunt, 1995b). These tracks indicate that models of forelimb posture of ceratopsids that postulated a sprawling stance are incorrect and that ceratopsids had a rhinoceros-like posture with parasagittal forelimb kinematics and the ability to run faster than extant elephants (Lockley and Hunt, 1995b; Paul and Christiansen, 2000). The trackways show that the hands were placed directly beneath the glenoids, and that manual impressions were directed laterally, not medially, as in sprawling reptiles. Pedal impressions in trackways are medial to the manual impressions, due to the slightly averted elbow and to the asymmetrical distal femoral condyles, which directed the crus slightly medially. The limbs of ceratopsians of all sizes display substantial joint flexure, strongly indicating that the elephantine forelimb posture that has sometimes been suggested as the alternative to a sprawling posture is erroneous (Paul and Christiansen, 2000).
6. Horses possess one of the most diverse gait sets known among mammals (Vincelette, 2021). Various breeds have been observed employing 14 out of the 17 major quadrupedal gaits (Renders and Vincelette, 2023b). Determining the gait of quadrupedal fossil species from trackways is challenging (Kienapfel et al., 2014; Renders and Vincelette, 2023b). Quadrupeds are capable of a large variety of gaits, including both laterally and diagonally coordinated walking and running gaits, as well as asymmetrical bounding gaits. These gaits can leave trackways that look superficially similar, as, for example, walks, trots, and slow racks that all form trackways with ipsilateral foot pairs landing near each other (Renders and Vincelette, 2023b). New methodologies provide a framework for helping to identify horse gait and velocity from trackways (Renders and Vincelette, 2023a,b).
7. Bipedalism is a key adaptation in human evolution. The famous Pliocene hominin footprints from Laetoli, Tanzania, preserve bipedal tracks probably produced by *Australopithecus afarensis* (Leakey and Hay, 1979; Tuttle et al., 1991). They have been the focus of much research about locomotion (e.g., Tuttle, 1987; Day and Wickens, 1980; White and Suga, 1987; Meldrum et al., 2011). Hatala et al. (2016) determined that they are morphologically distinct from those of both chimpanzees and habitually barefoot modern humans. The Laetoli hominin probably used a more flexed limb posture at foot strike than modern humans when walking bipedally. The Laetoli footprints provide a clear evidence of an early hominin bipedal gait that probably involved a limb posture that was slightly but significantly different from *Homo sapiens*, and these data support

the hypothesis that important evolutionary changes to hominin bipedalism occurred within the past 3.66 Myr (Meldrum et al., 2011; Hatala et al., 2016).

5. Speed

Alexander (1976) published an elegant and highly influential study of estimating speed from trackways. Although most trackways are produced (unsurprisingly) by plodding animals, some were made by individuals moving rapidly (e.g., Thulborn and Wade, 1979). This became another major ichnological support for the "Dinosaur Renaissance" (Bakker, 1968): dinosaurs were not giant sluggish lizards, they were agile organisms that could run. As with gregarious behavior and ichnopathologies, the vast amount of literature on speed from trackways is on dinosaur tracks and speed.

The original formula for calculating speed from tracks proposed by Alexander (1976) has proved remarkably robust with only a minor amendment by Ruiz and Torices (2013):

$$v = 0.226\, g^{0.5}\, \lambda^{1.67}\, h^{-1.17}$$

(where v is velocity, g is the force of gravity at the Earth's surface, λ is the absolute stride length, and h is the total height of its hind limbs, measured from the ground to the acetabulum/pelvic articulation of the femur and generally $h = 4\mathrm{FL}$, where FL = footprint length according to Alexander (1976).

Observed speeds reported for some recent mammals differ from theoretical predictions by factors as large as 1.5 or 2.5 (Bakker, 1987; Alexander, 2006), but conclusions based on observations of free-ranging animals require assumptions about their size, weight, and proportions that cannot be tested (Alexander et al., 1977; Ruiz and Torices, 2013).

Alexander (1976) hypothesized that animals are running when the relative stride length (RSL, the ratio between stride length SL and the acetabular height (h) is greater than 2, and lower ratios indicate walking. Subsequently, Thulborn (1982) proposed that <1.9 was walking, $> 2 < 2.9$ was trotting (jogging of Weems, 2006), and >3 was running. Weems (2006) assumed that a value of 2.02 represents the transition between walking and trotting, and the trotting to running transition is at 2.96 RSL.

Evidence of running dinosaurs has been reported at several tracksites (e.g., Farlow, 2018; Navarro-Lorbes et al., 2021; Perez-Lorente, 1996). The highest calculated speeds of running theropod dinosaurs are from the Lower Jurassic of Utah (Lockley et al., 2021) and the Lower Cretaceous of Texas (Farlow, 1981) in the United States, with calculated speeds of 10.8–13.8 and 9.4–11.9 ms^{-1}, and the Lower Cretaceous of Spain at 8.8 and 12.4 ms^{-1} (Navarro-Lorbes et al., 2021).

Speed calculations have also been made for mammalian trackways using Alexander's formula and other methods. Renders and Vincelette (2023a) derived a formula for equid speed based on studies of modern horses:

$$v = (0.2 \times \mathrm{SL}/\mathrm{HL}) - 0.09$$

(where v = velocity (m/second), and SL/HL = stride length/front hoof length ratio).

The proposed formulae have been very useful, but they have challenges notably related to the nature of footprint preservation and of calculating dimensions (e.g., Lucas et al., 2024b). Footprints are not simply molds of feet. Track morphology results from an interaction between dynamics, anatomy, and substrate (e.g., Falkingham, 2014). Therefore, the dimensions and shapes of tracks vary, and thus the length of a print and the stride will not be constant for use in calculations. Additionally, hip height is important in the Alexander formula. The ideal situation would be to be able to measure this dimension from a skeleton of the animal known to have produced the tracks. However, complete and undistorted skeletons are relatively uncommon, and it is rare to be able to associate a biotaxon at a low taxonomic level with a trackway. Furthermore, Alexander (1976) postulated that the hip height was four times the length of the foot length, but this is ultimately an approximation. Others have used other numbers with no explanation (e.g., ×5.7 in Kim et al., 2009). Hip heights vary among taxa, and this can introduce inconsistencies. Overall, the Alexander method seems to be a well-tested method, but should probably be considered to give an estimate of speed rather than a truly accurate velocity.

6. Other behaviors

Tetrapod tracks and trackways have been utilized to interpret other behaviors including:

1. There have been multiple studies of traces produced by swimming or underwater walking, principally in the Mesozoic and mainly relating to dinosaurs, crocodylomorphs, turtles, and pterosaurs (e.g., Foster and Lockley, 1997; Lockley and Wright, 2003; Avanzini et al., 2005; Milner et al., 2006, 2016; Ezquerra et al., 2007; Xing et al., 2013, 2016; Lee et al., 2019).
2. There are several reports of dinosaur trackways that highlight a turning of direction of travel (e.g., Lockley, 1990, 1991b; Ishigaki and Matsumoto, 2009; Xing et al., 2015b; Goodell et al., 2021; Lockley et al., 2021). However, while interesting, this behavior is of little paleoecological significance.
3. Lucas et al. (2010) described associated body impressions of a Pennsylvanian temnospondyl oriented in opposite directions that could be evidence of a courtship ritual.
4. There are several examples of dinosaurs crouching, especially in bipedal taxa, often with metapodial impressions and sometimes with ischial or pubic impressions (e.g., Lockley et al., 2003; Milàn et al., 2008; Gierlinski et al., 2009; Milner et al., 2009; Xing et al., 2012a; Romano and Citton, 2017).
5. There are multiple dinosaur trackways that preserve tail drag traces (e.g., Platt and Hasiotis, 2008; Kim and Lockley, 2013), which can provide information about locomotion

7. Conclusions

1. The majority of the work on gregariousness and other behavior, determining speed, and recognizing pathologies from tracks pertains to dinosaurs.

2. Tracksites related to gregarious behavior are either transgressive sites where several individuals moved across an area in subparallel alignment, or congregation sites, where tracks of multiple individuals are concentrated at a location.
3. There are two scales of congregation sites; (1) microcongregation sites (up to tens or hundreds of square meters) that are concentrations of the same ichnotaxon focused on a discrete and often recognizable locus; and (2) megacongregation sites that are large areas (up to hundreds or thousands of square meters) with multiple trackways in which few, if any, demonstrate an obvious preferred orientation with no clear cause for the concentration.
4. Factors to be considered in the evaluation of evidence for gregarious behavior include: (1) ichnotaxonomy; (2) bedding plane; (3) time averaging; (4) subparallel orientation; (5) directional barriers; (6) ecological attractors; (7) intertrack spacing; (8) age structure; (9) speed and directional change; and (10) topography.
5. The first good track evidence for complex gregarious behavior in tetrapods is from the Early-Middle Triassic, but we predict that it may remain to be recognized in Permian tracksites.
6. Paleopathologies have been recognized from fossil trackways on the basis of: (1) irregular gait or (2) pathologic foot impressions. Most reports are from dinosaurs or associated tracks.
7. Consistent irregularities in pace and stride values are defined as irregular gaits, and the more extreme examples are more likely to be cases of ichnopathology.
8. The majority of reported ichnopathologies and putative ichnopathologies relate to bipedal, tridactyl trackmakers, principally theropods.
9. Trackways have great potential in the study of gaits and their evolution but have received relatively little attention.
10. Formulae for calculating speed from tetrapod trackways have been very useful, but they have challenges notably related to the nature of footprint preservation and of calculating dimensions.
11. Ichnological evidence for gregarious behavior and elevated speeds were important in the reevaluation of dinosaur biology in the 1970s and 1980s.

Acknowledgments

The authors thank many of their colleagues for helpful discussions on the topics covered in this chapter over the years including Bill Cotton, Jim Farlow, Martin Lockley, Masaki Matsukawa, Rich McCrea, John Ostrom, and John Rogers. The American Museum of Natural History kindly supplied an image. The authors particularly thank John Foster and Jerry Harris, who provided thoughtful and helpful reviews as well as indicating references that the authors had overlooked.

Declaration of competing interest

The authors declare that they have no known competing financial interests or personal relationships that could have appeared to influence the work reported in this chapter.

Data availability

Only published data were used for the research described in this chapter.

References

Abbassi, N., 2010. Vertebrate footprints from the Miocene Upper Red Formation, Shokorchi Area, Zanjan Province, NW Iran. Ichnos 17 (2), 115–126. https://doi.org/10.1080/10420941003659493.

Abbassi, N., Khoshyar, M., Lucas, S.G., Esmaeili, F., 2021. Extensive vertebrate tracksite from the Upper Red Formation (middle–late Miocene), west Zanjan, northwestern Iran. Fossil Record 24, 101–116. https://doi.org/10.5194/fr-24-101-2021.

Abbott, C.P., Sues, H.-D., Lockwood, R., 2017. The *Dimetrodon* dilemma: reassessing posture in sphenacodonts. Geological Society of America Abstracts with Programs 49. https://doi.org/10.1130/abs/2017AM-307190.

Abel, O., 1935. Vorzeitliche lebensspuren. Gustav Fisher, Jena.

Aguirrezabala, L.M., Torres, J.A., y Viera, L.I., 1985. El weald de Igea (Cameros-La Rioja). Sedimentología, Bioestratigrafía y Paleoicnología de gran des reptiles (Dinosaurios). Munibe 37, 111–138.

Albritton Jr., C.C., 1942. Dinosaur tracks near Comanche, Texas. Field Lab. 10 (2), 160–181.

Alexander, R.M.N., 1976. Estimates of speeds of dinosaurs. Nature 261, 129–130. https://doi.org/10.1038/261129a0.

Alexander, R.M.N., 2006. Dinosaurs biomechanics. Proc. Royal Soc. B 273, 1849–1855.

Alexander, R.M.N., Langman, V.A., Jayes, A.S., 1977. Fast locomotion of some large African ungulates. J. Zool. 183, 291–300.

Apesteguía, S., Riguetti, F., Citton, P., Veiga, G.D., Poiré, D.G., de Valais, S., Gabriel Zacarías, G., 2021. The Ruditayoj-Tunasniyoj fossil area (Chuquisaca, Bolivia): a Triassic chirotheriid megatracksite and reinterpretation of purported thyreophoran tracks. Hist. Biol. 33 (11), 2883–2896.

Ashton, N., Lewis, S.G., De Groote, I., Duffy, S.M., Bates, M., Bates, R., Hoares, P., Lewis, M., Parfitt, S.A., Peglar, S., Williams, C., Stringer, C., 2014. Hominin footprints from early pleistocene deposits at Happisburgh, UK. PLoS One 9 (2), e88329. https://doi.org/10.1371/journal.pone.0088329.

Avanzini, M., Leonardi, G., 1993. Dinosauri nel Trentino: il giacimento dei Lavini di Marco. UCT, Trento 18 (214), 35–41.

Avanzini, M., García–Ramos, J.C., Lires, J., Menegon, M., Piñuela, L., Fernández, L.A., 2005. Turtle tracks from the late Jurassic of Asturias, Spain. Acta Palaeontol. Pol. 50 (4), 743–755.

Avanzini, M., Pinuela, L., Garcia-Ramos, J.C., 2008. Theropod palaeopathology inferred from a Late Jurassic trackway, Asturias (N. Spain). Ocyctos 8, 71–75.

Bakker, R.T., 1968. The superiority of dinosaurs. Discovery, New Haven 3 (2), 11–22.

Bakker, R.T., 1975. Dinosaur renaissance. Scientif. Amer. 232 (4), 58–79.

Bakker, R.T., 1986. The Dinosaur Heresies: New Theories Unlocking the Mystery of the Dinosaurs and Their Extinction. William Morrow and Company, New York, p. 481.

Bakker, R.T., 1987. The return of the dancing dinosaurs. In: Czerkas, S.J., Olson, E.C. (Eds.), Dinosaurs Past and Present, vol. I. Natural History Museum of Los Angeles County, Los Angeles, pp. 38–69.

Balsley, J.K., Parker, L.R., 1983. Cretaceous Wave Dominated Delta, Barrier Island, and Submarine Fan Depositional Systems: Book Cliffs East Central Utah. American Association of Petroleum Geology Field Guide, 279 pp.

Barco, J.L., Canudo, J.I., Ruiz-Omeñaca, J.I., 2006. New Data on *Therangospodus oncalensis* from the Berriasian Fuentesalvo Tracksite (Villar del Río, Soria, Spain): An example of gregarious behaviour in theropod dinosaurs. Ichnos 13 (4), 237–248. https://doi.org/10.1080/10420940600843682.

Barnes, F.A., Lockley, M.G., 1994. Trackway evidence for social sauropods from the Morrison Formation, Eastern Utah (USA). Gaia 10, 37–42.

Bennett, S.C., 1992. Reinterpretation of problematic tracks at Clayton Lake State Park, New Mexico: not one pterosaur, but several crocodiles. Ichnos 2, 37–42.

Bennett, M.R., Harris, J.W., Richmond, B.G., Braun, D.R., Mbua, E., Kiura, P., Olago, D., Kibunjia, M., Omuombo, C., Behrensmeyer, A.K., Huddart, D., 2009. Early hominin foot morphology based on 1.5-million-year-old footprints from Ileret, Kenya. Science 323 (5918), 1197–1201.

Berman, D.S., Henrici, A.C., 2003. Homology of the astragalus and structure and function of the tarsus of Diadectidae. J. Paleontol. 77, 172–188.

Bernardi, M., Avanzini, M., 2011. Locomotor behavior in early reptiles: insights from an unusual *Erpetopus* trackway. J. Paleontol. 85, 925–929.

Bernardi, M., Klein, H., Petti, F.M., Ezcurra, M.D., 2015. The origin and early radiation of archosauriformes: integrating the skeletal and footprint record. PLoS One. https://doi.org/10.1377/journal.pone.0128449.

Bird, R.T., 1939. Thunder in his footsteps. Nat. Hist. 43, 254–261.

Bird, R.T., 1941. A dinosaur walks into the museum. Nat. Hist. 47, 74–81.

Bird, R.T., 1944. Did *Brontosaurus* ever walk on land? Nat. Hist. 53, 61–67.

Bird, R.T., 1954. We captured a "live" brontosaur. Natl. Geogr. 105, 707–722.

Boudchiche, L., Masrour, M., Boutakiout, M., Hamzaoui, L., Gascón, J.,H., Pérez-Lorente, F., 2017. Jarrate Lbel: new Upper Cretaceous continental site in Morocco. A probable ornitischian non-onithopod trackway and three amble gait titanosauriform trackways. Geogaceta 61, 19–22.

Calvo, J.O., González Riga, B.J., Apesteguía, S., Tomaselli, M.B., 2022. Sauropod ichnology: overview and new research lines from a South American perspective. In: Alejandro Otero, A., Carballido, J.L., Pol, D. (Eds.), South American Sauropodomorph Dinosaurs: Record, Diversity and Evolution. Springer Earth System Sciences, pp. 503–540.

Carvalho, I.S., 2000. Huellas de saurópodos de la Formación Antenor Navarro (Cretácico Temprano de la Cuenca de Sousa), Serrote do Letreiro, Paraíba, Brasil. Ameghiniana 37, 353–362.

Casanovas, M.L., Ezquerra, R., Fernández, A., Montero, D., Pérez-Lorente, F., Santafé, J.V., Fernández-Baldor, F.T., Viera, L.V., 1995. El yacimiento de La Canal (Munilla, La Rioja, España). La variación de velocidad en función del tamaño del pie de los ornitópodos. Zubia 13, 55–81.

Castanera, D., Barco, J.L., Díaz-Martínez, I., Gascón, J.H., Perez-Lorente, F., Canudo, J.I., 2011. New evidence of a herd of titanosauriform sauropods from the lower Berriasian of the Iberian range (Spain). Palaeogeogr. Palaeoclimatol. Palaeoecol. 310 (3–4), 227–237.

Castanera, D., Vila, B., Razzolini, N.L., Santos, V.D., Pascual, C., Canudo, J.I., 2014. Sauropod trackways of the Iberian peninsula: palaeogeological and palaeoenvironmental implications. J. Iber. Geol. 40 (1), 49–59. https://doi.org/10.5209/rev_JIGE.2014.v40.n1.44087.

Coffin, A., 1889. Geologic Footprints in Nevada Sandstone. Hadley, Indiana.

Cohen, A., Lockley, M., Halfpenny, J., Michel, A.E., 1991. Modern vertebrate track taphonomy at Lake Manyara, Tanzania. Palaios 6, 371–389.

Conrad, K., Lockley, M.G., Prince, N.K., 1987. Triassic and Jurassic vertebrate-dominated trace fossil assemblages of the Cimarron Valley region: implications for paleoecology and biostratigraphy. New Mexico Geological Society, Guidebook 38, 127–138.

Coombs, W.P., 1996. Redescription of the ichnospecies *Antipus flexiloquus* Hitchcock, from the early Jurassic of the Connecticut valley. J. Paleontol. 70 (2), 327–331.

Cotton, W.D., Cotton, J.E., Hunt, A.P., 1987. Social behavior of Cretaceous ornithopod dinosaurs of New Mexico. In: Lucas, S.G., Estep, J.W., Williamson, T.E., Morgan, G.S. (Eds.), New Mexico's Fossil Record, vol 11. New Mexico Museum of Natural History and Science, Bulletin, pp. 45–50.

Cotton, W.D., Cotton, J.E., Hunt, A.P., 1988. Evidence for social behavior in ornithopod dinosaurs from the Dakota Group of northeastern New Mexico, U.S.A. Ichnos 6 (3), 141–149.

Courel, L., Demathieu, G., 1976. Une ichnofaune reptilienne remarquable dans les gre's Triassiques de Largentie're (Arde'che, France). Palaeontographica Abteilung A Palaontologie. Stratigraphie 151, 194–216.

Currie, P.J., 1981. Bird footprints from the gething formation (Aptian, Lower Cretaceous) of northeastern British Columbia, Canada. J. Vertebr. Paleontol. 1, 257–264.

Currie, P.J., 1983. Hadrosaur trackways from the lower Cretaceous of Canada. Acta Palaeontol. Pol. 28, 63–73.

Currie, P.J., Sarjeant, W.A.S., 1979. Lower Cretaceous dinosaur footprints from the Peace River canyon, British Columbia, Canada. Palaeogeogr. Palaeoclimatol. Palaeoecol. 28, 103–115.

Currie, P.J., Badamgarav, D., Koppelhus, E.B., 2003. The first late Cretaceous footprints from the Nemegt locality in the Gobi of Mongolia. Ichnos 10, 1–13.

Dalla Vecchia, F.M., Rustioni, M., 1996. Mammalian trackways in the Conglomerato di Osoppo (Udine, NE Italy) and their contribution to its age determination. Mem. Sci. Geol. 48, 221–232.

Dantas, P., dos Santos, V.F., Lockley, M.G., Meyer, C.A., 1994. Footprint evidence for limping dinosaurs from the Upper Jurassic of Portugal. Gaia 10, 43–48.

Day, J.J., Norman, D.B., Gale, A.S., Upchurch, P., Powell, H.P., 2004. A Middle Jurassic dinosaur trackway site from Oxfordshire, UK. Palaeontology 47 (2), 319–348.

Day, J.J., Upchurch, P., Norman, D.B., Gale, A.S., Powell, H.P., 2002. Sauropod trackways, evolution, and behavior. Science 296 (5573), 1659.

Day, M.H., Wickens, E.H., 1980. Laetoli Pliocene hominid footprints and bipedalism. Nature 286, 385–387. https://doi.org/10.1038/286385a0.

de Lapparent, A.F., Montenat, C., 1967. Les empreintes de pas de l'Infralias de Veillon (Vendée). Mémoires de la Société géologique de France, nouvelle série 107, 1–44.

Delair, J.B., 1981. Multiple trackways from the Isle of Purbeck. Proc. Dorset Nat. History Archaeol. Soc. 102, 65–68.

Delair, J.B., Lander, A.B., 1973. A short history of the discovery of reptilian footprints in the Purbeck Beds of Dorset, with notes on their distribution. Proc. Dorset Nat. History Archaeol. Soc. 94, 17–20.

Díaz-Martínez, I., Suarez-Hernando, O., Larrasoaña, J.C., Martínez-García, B.M., Baceta, J.I., Murelaga, X., 2020. Multi-aged social behaviour based on artiodactyl tracks in an early Miocene palustrine wetland (Ebro Basin, Spain). Sci. Rep. 10 (1), 1099. https://doi.org/10.1038/s41598-020-57438-4.

Díaz-Martínez, I., Citton, P., Castanera, D., 2023. What do their footprints tell us? Many questions and some answers about the life of non-avian dinosaurs. J. Iber. Geol. 1–22. https://doi.org/10.1007/s41513-023-00226-6.

Diedrich, C., 2008. Millions of reptile tracks—early to Middle Triassic carbonate tidal flat migration bridges of Central Europe—reptile immigration into the Germanic Basin. Palaeogeogr. Palaeoclimatol. Palaeoecol. 259 (4), 410–423.

Ellenberger, P., 1974. Contribution a la classification des pistes de vertebres du Trias: Les types du Stormberg d'Afrique du Sud (II): Le Stormberg Superior. Palaeovertebrata Memoir Extraordinaire, Laboratoire de Paleontologie des Vertebres, Montpellier.

Ellenberger, F., Ellenberger, P., Fabre, J., Mendrez, C., 1963. Deux nouvelles dalles à pistes de Vertébrés fossiles découvertes au Basutoland (Afrique du Sud). Compte Rendus de la Société Géologique de France 1963, 315–317.

Engelmann, G.F., Chure, D.J., 2017. Morphology and sediment deformation of downslope *Brasilichnium* trackways on a dune slipface in the Nugget Sandstone of northeastern Utah, USA. Palaeontol. Electron. 20 (2), 1–21. https://doi.org/10.26879/617/palaeo-electronica.org/content/2017/1845-downslope-trackways.

Enriquez, N.J., Campione, N.E., White, M.A., Fanti, F., Sissons, R.L., Sullivan, C., Vavrek, M.J., Bell, P.R., 2022. The dinosaur tracks of Tyrants Aisle: an upper Cretaceous ichnofauna from Unit 4 of the Wapiti formation (upper Campanian), Alberta, Canada. PLoS One 17 (2), e0262824. https://doi.org/10.1371/journal.pone.0262824.

Ezquerra, R., Doublet, S., Costeur, L., Galton, P.M., Pérez-Lorente, F., 2007. Were non-avian theropod dinosaurs able to swim? Supportive evidence from an Early Cretaceous trackway, Cameros Basin (La Rioja, Spain). Geology 35 (6), 507–510.

Falkingham, P.L., 2014. Interpreting ecology and behaviour from the vertebrate fossil track record. J. Zool. 292, 222–228.

Farlow, J.O., 1981. Estimates of dinosaur speeds from a new trackway site in Texas. Nature 294, 747–748.

Farlow, J.O., 1987. Lower Cretaceous Dinosaur Tracks, Paluxy River Valley, Texas. South- Central Section. Geological Society of America, Baylor University, pp. 1–50.

Farlow, J.O., 1992. Sauropod tracks and trackmakers: Integrating the ichnological and skeletal records. Zubia 10, 89–138.

Farlow, J.O., 2018. Noah's Ravens: Interpreting the Makers of Tridactyl Dinosaur Footprints. Indiana University Press, p. 800.

Farlow, J.O., Pittman, J.G., Hawthorne, J.M., 1989. *Brontopodus birdi*, Lower Cretaceous sauropod footprints from the U.S. Gulf coastal plain. In: Gillette, D.D., Lockley, M.G. (Eds.), Dinosaur Tracks and Traces. Cambridge University Press, Cambridge, pp. 371–394.

Foster, J.R., 2003. Paleoecological analysis of the vertebrate fauna of the Morrison Formation (Upper Jurassic), Rocky Mountain region, U.S.A. New Mexico Museum of Natural History and Science, Bulletin 23, 73.

Foster, J.R., 2015. Theropod dinosaur ichnogenus *Hispanosauropus* identified from the Morrison Formation (Upper Jurassic), western North America. Ichnos 22 (3–4), 183–191.

Foster, J., 2020. Jurassic West: The Dinosaurs of the Morrison Formation and Their World. Indiana University Press, 560 pp.

Foster, J.R., Lockley, M.G., 1997. Probable crocodilian tracks and traces from the Morrison Formation (Upper Jurassic) of eastern Utah. Ichnos 5 (2), 121–129.

Foster, J.R., Lockley, M.G., 2006. The vertebrate ichnological record of the Morrison Formation (Upper Jurassic, North America). New Mexico Museum of Natural History and Science, Bulletin 36, 203–216.

Foster, J.R., Lockley, M.G., 2024. The vertebrate ichnological record of the Morrison Formation (Upper Jurassic, North America) – Part 2. New Mexico Museum of Natural History and Science, Bulletin 96, 339–354.

Foster, J.R., Hamblin, A.H., Lockley, M.G., 2000. The oldest evidence of a sauropod dinosaur in the western United States and other important vertebrate trackways from Grand Staircase-Escalante National Monument, Utah. Ichnos 7, 169–181.

Foster, J.R., Hamblin, A.H., Lockley, M.G., 2003. *Apatopus* trackway and other footprints from the Chinle Group of southern Utah: an update. Ichnos 10, 165–167.

Fuentes Vidarte, C., 1996. Primeras huellas de aves en el Weald de Soria (España). Nuevo icnogénero, *Archeornithipus* y nueva icnoespecie A. meijidei. Estud. Geol. 52, 63–75.

Furrer, H., 1993. Entdeckung und Untersuchung der Dinosaurierfährten im Nationalpark. Cratschla: Ediziuns Specialas 1, 24 pp.

García-Ortiz, E., Pérez-Lorente, F., 2014. Palaeoecological inferences about dinosaur gregarious behaviour based on the study of tracksites from La Rioja area in the Cameros Basin (Lower Cretaceous, Spain). J. Iber. Geol. 40 (1), 113–127. https://doi.org/10.5209/rev_JIGE.2014.v40.n1.44091.

García-Ramos, J.C., Lires, J., Pinuela, L., 2002. Dinosaurios: rutas por el Jurásico de Asturias. La Voz de Asturias, Lugones, Asturias, Spain, 204 pp.

Garcia-Ramos, J.C., Pinuela, L., Lires, J., 2006a. Atlas del Jurásico de Asturias. Ediciones Nobel, Oviedo, 225 pp.

Garcia-Ramos, J.C., Pinuela, L., Ruiz-Omeñaca, J.I., 2006b. El Jurásico de Asturias y sus huellas de dinosaurios. Yacimientos de la playa de La Griega (Colunga) y de los acantilados de Tereñes (Ribadeeslla). In: Garcia-Ramos, J.C., Jimenez-Sanchez, M., Pinuela, L., Dominguez Cuesta, M.J., Lopez Fernandez, C. (Eds.), Patrimonio Geológico en Asturias: la cuenca alta del río Nalón y la Costa de los Dinosaurios. Itinerarios geológicos de la VII reunión de la Comisión de Patrimonio Geológico. Servitec, Oviedo, pp. 31–48.

Getty, P.R., Judge, A., Csonka, J., Bush, A.M., 2012. Were Early Jurassic dinosaurs gregarious? Reexamining the evidence from Dinosaur Footprint Reservation in Holyoke, Massachusetts. In: Thomas, M.A. (Ed.), Guidebook for Field Trips in Connecticut and Massachusetts, Proceeding of the 47th Annual Meeting of the Northeastern Section of the Geological Society of America, Hartford, CT, USA, 18–20 March 2012, State Geological and Natural History Survey of Connecticut: Hartford, CT, USA, 2012, pp. A1–A18.

Getty, P.R., Hardy, L., Bush, A.M., 2015. Was the *Eubrontes* track maker gregarious? Testing the herding hypothesis at Powder Hill Dinosaur Park, Middlefield, Connecticut. Bull. Peabody Mus. Nat. Hist. 56, 95–106.

García-Ramos, J.C., Pinuela, L., Lires, J., 2004. Guía del Jurásico de Asturias. Rutas por los yacimientos de huellas de dinosaurios. Zinco Comunicación, Gijón, 118 pp.

Getty, P.R., Aucoin, C., Fox, N., Judge, A., Hardy, L., Bush, A.M., 2017. Perennial lakes as an environmental control on theropod movement in the Jurassic of the Hartford Basin. Geosciences 7 (1), 13. https://doi.org/10.3390/geosciences7010013.

Gierliński, G., Pieńkowski, G., 1999. Dinosaur track assemblages from the Hettangian of Poland. Geol. Q. 43 (3), 329–346.

Gierliński, G.G., Lockley, M.G., Niedźwiedzki, G., 2009. A distinctive crouching theropod trace from the Lower Jurassic of Poland. Geol. Q. 53 (4), 471–476.

Gillette, D.D., Thomas, D.A., 1985. Dinosaur tracks in the Dakota Formation (Aptian-Albian) at Clayton Lake State Park, Union County, New Mexico. N. M. Geol. Soc. Guidebook 36, 283–288.

Gomes, J.P., 1915-1916. Descoberta de rastos de sáurios gigantescos no Jurássico do Cabo Mondego. Comunicações Comissão Serviços Geológicos de Portugal 11, 132–134.

González Riga, B.J., Ortiz, D.L., 2011. Yacimiento de las icnitas *Titanopodus mendozensis*, Cretácico Superior de Mendoza (Argentina): una síntesis. In: Resúmenes del 7° Encuentro del international center for earth sciences (ICES) E-ICES 7, Malargüe, Mendoza, Argentina, pp. 1–14.

Goodell, Z., Lockley, M.G., Lucas, S.G., Schumacher, B.A., Smith, J.A., Trujillo, R., Xing, L., 2021. A high-altitude sauropod trackway site in the Jurassic of Colorado: the longest known consecutive footprint sequence reveals evidence of sharp turning behavior. New Mexico Museum of Natural History. Bulletin 82, 101–112.

Hatala, K.G., Demes, B., Richmond, B.G., 2016. Laetoli footprints reveal bipedal gait biomechanics different from those of modern humans and chimpanzees. Proc. Royal Soc. B 283, 20160235. https://doi.org/10.1098/rspb.2016.0235.

Haubold, H., 1984. Saurierfährten, second ed. Die Neue Brehm Bücherei, Wittenberg (Ziemsen). 231 pp.

Hendricks, A., 1981. Die Saurierfährten von Münchehagen bei Rehburg Loccum (NW-Deutschland). Abhandlungen aus dem Landesmuseum für Naturkunde Munster Westfalen 43, 1–22.

Heredia, A.M., Díaz-Martínez, I., Pazos, P.J., Comerio, M., Fernández, D.E., 2020. Gregarious behaviour among non-avian theropods inferred from trackways: A case study from the Cretaceous (Cenomanian) Candeleros Formation of Patagonia, Argentina. Palaeogeogr. Palaeoclimatol. Palaeoecol. 538 (2020), 109480. https://doi.org/10.1016/j.palaeo.2019.109480.

Herrera-Castillo, C.M., Moratalla, J.J., Belaústegui, Z., Marugán-Lobón, J., Martín-Abad, H., Nebreda, S.M., López-Archilla, A.I., Angela, D., BuscalioniI, A.D., 2022. A theropod trackway providing evidence of a pathological foot from the exceptional locality of Las Hoyas (upper Barremian, Serranı'a de Cuenca, Spain). PLoS One 17 (4), e0264406. https://doi.org/10.1371/journal.pone.0264406.

Hitchcock, E., 1836. Ornithichnology—Description of the foot marks of birds, (Ornithichnites) on New Red Sandstone in Massachusetts. Am. J. Sci. 29, 307–340.

Hitchcock, E., 1844. Report on ichnolithology, or fossil footmarks, with a description of several new species, and the coprolites of birds, from the valley of Connecticut River, and of a supposed footmark from the valley of Hudson River. Am. J. Sci. 47, 292–322.

Hitchcock, E., 1848. An attempt to discriminate and describe the animals that made the fossil footmarks of the United States, and especially of New England. Am. Acad. Arts Sci. Memoir 3, 129–256.

Hitchcock, E., 1858. Ichnology of New England. A Report on the Sandstone of the Connecticut Valley, Especially its Fossil footmarks. Boston. Commonwealth of Massachusetts. William White, 220 pp.

Hopson, J.A., 2012. The role of foraging mode in the origin of therapsids: Implications for the origin of mammalian endothermy. Fieldiana Life Earth Sci. 5, 126–148.

Hopson, J.A., 2015. Fossils, trackways, and transitions in locomotion: A case study of *Dimetrodon*. In: Dial, K.P., Shubin, N., Brainerd, E.L. (Eds.), Great Transformations in Vertebrate Evolution. The University of Chicago Press, Chicago, pp. 125–141.

Hunt, A.P., Lucas, S.G., 1998a. Vertebrate tracks and the myth of the belly-dragging, tail-dragging tetrapods of the Late Paleozoic: In. In: Lucas, S.G., Estep, J.W., Hoffer, J.M. (Eds.), Permian Stratigraphy and Paleontology of the Robledo Mountains, New Mexico, vol 12. New Mexico Museum of Natural History and Science, Bulletin, pp. 67–69.

Hunt, A.P., Lucas, S.G., 1998b. Tetrapod ichnofaunas from the Lower Cretaceous of northeastern New Mexico, USA. New Mexico Museum of Natural History and Science, Bulletin 14, 163–167.

Hunt, A.P., Lucas, S.G., 2004. Multiple parallel dinosaur tail drags from the Early Cretaceous of New Mexico. J. Vertebr. Paleontol. 24 (Suppl. no. 3), 73A.

Hunt, A.P., Lucas, S.G., 2007. Tetrapod ichnofacies: a new paradigm. Ichnos 14, 59–68.

Hunt, A.P., Lucas, S.G., 2021. The ichnology of vertebrate consumption: Dentalites, gastroliths and bromalites. New Mexico Museum of Natural History. Bulletin 87, 216.

Hunt, A.P., Lucas, S.G., 2024. Bromalites, Pabulites, Digestilites, Micturalites and Other Trace Fossil Evidence of Predation and Consumption, This Volume.

Hunt, A.P., Santucci, V.L., Lockley, M.G., Olson, T.J., 1993a. Dicynodont trackways from the Holbrook Member of the Moenkopi Formation (Middle Triassic: Anisian), Arizona, USA. In: Lucas, S.G., Morales, M. (Eds.), The Nonmarine Triassic, vol. 3. New Mexico Museum of Natural History and Science, Bulletin, pp. 213–218.

Hunt, A.P., Lockley, M.G., Lucas, S.G., MacDonald, J.P., Hotton III, N., Kramer, J., 1993b. Early Permian tracksites in the Robledo Mountains, south-central New Mexico. In: Lucas, S.G., Zidek, J. (Eds.), Vertebrate Paleontology in New Mexico, vol. 2. New Mexico Museum of Natural History and Science, Bulletin, pp. 23–31.

Hunt, A.P., Lucas, S.G., Klein, H., 2018. Late Triassic nonmarine vertebrate and invertebrate trace fossils and the pattern of the Phanerozoic record of vertebrate trace fossils. In: Tanner, L.H. (Ed.), The Late Triassic World. Topics in Geobiology 46. Springer Verlag, New York, pp. 447–543.

Ishigaki, S., 1986. Dinosaur footprints of the Atlas Mountains. Nat. Stud. 32 (1), 6–9.

Ishigaki, S., 1989. Footprints of swimming Sauropods from Morocco. In: Gillette, D.D., Lockley, M.G. (Eds.), Dinosaur Tracks and Traces. Cambridge University Press, Cambridge, pp. 83–86.

Ishigaki, S., Lockley, M.G., 2010. Didactyl, tridactyl and tetradactyl theropod trackways from the Lower Jurassic of Morocco: Evidence of limping, labouring and other irregular gaits. Hist. Biol. 22, 100–108.

Ishigaki, S., Matsumoto, Y., 2009. "Off-tracking"-like phenomenon observed in the turning sauropod trackway from the Upper Jurassic of Morocco. Memoirs of Fukui Prefecture Dinosaur Museum 8, 1–10.

Jenny, J., Josen, J.-A., 1982. Decouverte d'empreintes de pas de dinosaur-iens dans le Jurassique Inferier (Pliensbachien) du Haut–Atlas central (Maroc). Compte Rendus de l'Academie des Sciences, Paris 294 (II), 223–226.

Jung, J., Huh, M., Unwin, D.M., Smyth, R.S., Hwang, K.G., Kim, H.J., Choi, B.D., Xing, L., 2022. Evidence for a mixed-age group in a pterosaur footprint assemblage from the early Upper Cretaceous of Korea. Sci. Rep. 12, 10707. https://doi.org/10.1038/s41598-022-14966-5.

Kienapfel, K., Läbe, S., Preuschoft, H., 2014. Do tracks yield reliable information on gaits? – Part 1: The case of horses. Fossil Record 17, 59–67. https://doi.org/10.5194/fr-17-59-2014.

Kim, J.-Y., Lockley, M.G., 2013. Review of dinosaur tail traces. Ichnos 20, 129–141.

Kim, J.Y., Lockley, M.G., Kim, H.M., Lim, J.D., Kim, K.S., 2009. New dinosaur tracks from Korea, *Ornithopodichnus masanensis* ichnogen. et ichnosp. nov. (Jindong Formation, Lower Cretaceous): implications for polarities in ornithopod foot morphology. Cretac. Res. 30 (6), 1387–1397.

Kim, J.Y., Lockley, M.G., Seo, S.J., Kim, K.S., Kim, S.H., Baek, K.S., 2012. A paradise of Mesozoic birds: The world's richest and most diverse Cretaceous bird track assemblage from the Early Cretaceous Haman Formation of the Gajin Tracksite, Jinju, Korea. Ichnos 19 (1–2), 28–42. https://doi.org/10.1080/10420940.2012.660414.

Klein, H., Niedźwiedzki, G., 2012. Revision of the Lower Triassic tetrapod ichnofauna from Wióry, Holy Cross Mountains, Poland. New Mexico Museum of Natural History and Science, Bulletin 56, 1–62.

Klein, H., Lucas, S.G., Lallensack, J.N., Marchetti, L., 2024. Peabody's legacy: The Moenkopi Formation (Middle Triassic, Anisian) tetrapod ichnofauna— updates from an extensive new tracksite in NE Arizona, USA. Paläontol. Z. 98, 357–389. https://doi.org/10.1007/s12542-023-00680-8.

Kubo, T., Benton, M.J., 2009. Tetrapod postural shift estimated from Permian and Triassic trackways. Palaeontology 52, 1029–1037.

Lallensack, J.N., Ishigaki, S., Lagnaoui, A., Buchwitz, M., Oliver Wings, O., 2019. Forelimb orientation and locomotion of sauropod dinosaurs: Insights from the ?Middle Jurassic Tafaytour Tracksites (Argana Basin, Morocco). J. Vertebr. Paleontol. 38 (5), e1512501. https://doi.org/10.1080/02724634.2018.1512501.

Leakey, M.D., Hay, R.L., 1979. Pliocene footprints in the Laetolil beds at Laetoli, Northern Tanzania. Nature 278, 317–323. https://doi.org/10.1038/278317a0.

Lee, Y.-N., Yang, S.-Y., Seo, S.-J., Baek, K.-S., Lee, D.-J., Park, E.-J., Han, S.-W., 2000. Distribution and paleobiological significance of dinosaur tracks in the Jindong Formation (Albian) in Kosong County, Korea. Paleontol. Soc. Korea Special Publ. 4, 1–12.

Lee, Y.N., Lee, H.J., Kobayashi, Y., Paulina-Carabajal, A., Barsbold, R., Fiorillo, A.R., Tsogtbaatar, K., 2019. Unusual locomotion behaviour preserved within a crocodyliform trackway from the Upper Cretaceous Bayanshiree Formation of Mongolia and its palaeobiological implications. Palaeogeogr. Palaeoclimatol. Palaeoecol. 533, 109239.

Leonardi, G., 1984. Le impronte fossili di dinosauri. In: Bonaparte, J.F., Colbert, E.H., Currie, P.J., De Ricqles, A.J., Kielan-Jaworowska, Z., Leonardi, G., Morello, F., Taquet, P. (Eds.), Sulle orme dei dinosauri. Erizzo (Esplorazioni e ricerche IX), Italia, pp. 161–186.

Leonardi, G., 1989. Inventory and statistics of the South American dinosaurian ichnofauna and its paleobiological interpretation. In: Gillette, D.D., Lockley, M.G. (Eds.), Dinosaur Tracks and Traces. Cambridge University Press, Cambridge, pp. 165–178.

Li, R., Lockley, M.G., Makovicky, P.J., Matsukawa, M., Norell, M.A., Harris, J.D., Liu, M., 2008. Behavioral and faunal implications of Early Cretaceous deinonychosaur trackways from China. Naturwissenschaften 95, 185–191. https://doi.org/10.1007/s00114-007-0310-7.

Lockley, M.G., 1989. Tracks and traces: new perspectives on dinosaurian behavior, ecology and biogeography. In: Padian, K., Chure, D.J. (Eds.), The Age of Dinosaurs. Short Courses in Paleontology, vol. 2. Paleontological Society, Knoxville, Tennessee, pp. 134–145.

Lockley, M.G., 1990. Did "*Brontosaurus*" ever swim out to sea? Ichnos 1, 81–90.

Lockley, M.G., 1991a. The Moab megatracksite: A preliminary description and discussion of millions of Middle Jurassic tracks in eastern Utah. In: Averett, W.R. (Ed.), Guidebook for dinosaur quarries and tracksites tour, western Colorado and eastern Utah. Grand Junction Geological Society, Grand Junction, pp. 59–65.

Lockley, M.G., 1991b. Tracking Dinosaurs: Aa New Look at an Ancient World. Cambridge University Press, Cambridge, 238 pp.

Lockley, M., 1995. Track records. Nat. Hist. 104 (6), 46–54.

Lockley, M.G., 1997. The paleoecological and paleoenvironmental utility of dinosaur tracks. In: Farlow, J.O., Brett-Surman, M.K. (Eds.), The Complete Dinosaur. Indiana University Press, Bloomington=Indianapolis, pp. 554–578.

Lockley, M.G., 2011. Theropod and prosauropod dominated ichnofaunas from the Navajo–Nugget Sandstone (lower Jurassic) at Dinosaur National Monument; Implications for prosauropod behavior and ecology. New Mexico Museum of Natural History. Bulletin 53, 316–320.

Lockley, M.G., 2021. The distribution of theropod-dominated ichnofaunas in the Moab Megatracksite area, Utah: Implications for Late Jurassic palaeobiology along an arid coast. Hist. Biol. https://doi.org/10.1080/08912963.2021.1975279.

Lockley, M.G., Gierlinski, G.D., 2006. Diverse vertebrate ichnofaunas containing *Anomoepus* and other unusual trace fossils from the Lower Jurassic of the western United States: implications for paleoecology and palichnostratigraphy. New Mexico Museum of Natural History and Science, Bulletin 37, 176–191.

Lockley, M.G., Hunt, A.P., 1995a. Dinosaur Tracks and Other Fossil Footprints of the Western United States. Columbia University Press, New York, 338 pp.

Lockley, M.G., Hunt, A.P., 1995b. The first ceratopsid tracks and associated ichnofauna from the Laramie Formation (Upper Cretaceous: Maastrichtian) of Colorado. J. Vertebr. Paleontol. 15, 592–614.

Lockley, M.G., Meyer, C.A., 2000. Dinosaur Tracks and Other Fossil Footprints of Europe. Columbia University Press, New York, 360 pp.

Lockley, M.G., Schumacher, B.A., 2021. The process of ranking the geo-heritage "values" of dinosaur tracksites in the USA and globally. New Mexico Museum of Natural History and Science, Bulletin 82, 177–184.

Lockley, M.G., Wright, J.L., 2003. Pterosaur swim tracks and other ichnological evidence of behaviour and ecology. Geol. Soc. London Special Publ. 217 (1), 297–313.

Lockley, M.G., Houck, K.J., Prince, N.K., 1986. North America's largest dinosaur trackway site: implications for Morrison Formation paleoecology. Geol. Soc. Am. Bull. 97 (10), 1163–1176.

Lockley, M.G., Conrad, K., Paquette, M., Hamblin, A., 1992a. Late Triassic vertebrate tracks in the Dinosaur National Monument area. Utah Geol. Survey Miscell. Publ. 92–3, 383–391.

Lockley, M.G., Holbrook, J., Hunt, A.P., Matsukawa, M., Meyer, C., 1992b. The dinosaur freeway: a preliminary report on the Cretaceous megatracksite, Dakota Group, Rocky Mountain Front Range and Highplains; Colorado, Oklahoma and New Mexico. In: Flores, R. (Ed.), Mesozoic of the Western Interior. Society of Economic Paleontologists and Mineralogists, Tulsa, Klahoma, Midyear Meeting Fieldtrip Guidebook, pp. 39–54.

Lockley, M.G., Meyer, C., dos Santos, V.F., 1994a. Trackway evidence for a herd of juvenile sauropods from the Late Jurassic of Portugal. Gaia 10, 27–35.

Lockley, M.G., Meyer, C.A., Hunt, A.P., Lucas, S.G., 1994b. The distribution of sauropod tracks and trackmakers. Gaia 10, 233–248.

Lockley, M.G., Hunt, A.P., Moratalla, J.J., Matsukawa, M., 1994c. Limping dinosaurs? Trackway evidence for abnormal gaits. Ichnos 3, 193–202.

Lockley, M.G., Hunt, A.P., Haubold, H., Lucas, S.G., 1995. Fossil footprints in the DeChelly Sandstone of Arizona: with paleoecological observations on the ichnology of dune facies. In: Lucas, S.G., Heckert, A.B. (Eds.), Early Permian Footprints and Facies, vol. 6. New Mexico Museum of Natural History and Science, Bulletin, pp. 225–233.

Lockley, M.G., dos Santos, V.F., Meyer, C., Hunt, A.P., 1998a. A new dinosaur tracksite in the Morrison Formation, Boundary Butte, Southeastern Utah. In: Carpenter, K., Chure, D., Kirkland, J. (Eds.), The Upper Jurassic Morrison Formation: An Interdisciplinary Study, vol 23. Modern Geology, pp. 317–330.

Lockley, M.G., Hunt, A.P., Meyer, C.M., Rainforth, E., Schultz, R., 1998b. A survey of fossil footprints at Glen Canyon National Recreation Area (western USA): A case study in documentation of trace fossil resources. Ichnos 5, 177–211.

Lockley, M.G., Wright, J.L., Lucas, S.G., Hunt, A.P., 2001. The late Triassic sauropod track record comes into focus. Old legacies and new paradigms. N. M. Geol. Soc. Guidebook 52, 181–190.

Lockley, M.G., Rice, A., 1990. Did "Brontosaurus" ever swim out to sea? Evidence from brontosaur and other dinosaur footprints. Ichnos 1, 81–90.

Lockley, M., Schulp, A.S., Meyer, C.A., Leonardi, G., Mamani, D.K., 2002a. Titanosaurid trackways from the Upper Cretaceous of Bolivia: evidence for large manus, wide gauge locomotion and gregarious behaviour. Cretac. Res. 23, 383–400.

Lockley, M., Wright, J., White, D., Matsukawa, M., Jianjun, L., Lu, F., Hong, L., 2002b. The first sauropod trackways from China. Cretac. Res. 23, 363–381.

Lockley, M., Matsukawa, M., Jianjun, L., 2003. Crouching theropods in taxonomic jungles: ichnological and ichnotaxonomic investigations of footprints with metatarsal and ischial impressions. Ichnos 10 (2–4), 169–177.

Lockley, M.G., Lucas, S.G., Gaston, R., Hunt, A.P., 2004. Ichnofaunas from the Triassic-Jurassic boundary sequences of the Gateway area, western Colorado: implications for faunal composition and correlations with other areas. Ichnos 11, 89–102.

Lockley, M.G., Lucas, S.G., Hunt, A.P., 2006a. *Eosauropus*, a new name for a Late Triassic track: further observations on the Late Triassic ichnogenus *Tetrasauropus* and related forms, with notes on the limits of interpretation. New Mexico Museum of Natural History and Science, Bulletin 37, 192–198.

Lockley, M., Milner, A.C., Hamblin, D., 2006b. Dinosaur tracksites from the Kayenta Formation (Lower Jurassic), 'Desert Tortoise site,' Washington County, Utah. New Mexico Museum of Natural History and Science, Bulletin 37, 269–275.

Lockley, M., Chin, K., Houck, K., Matsukawa, M., Kukihara, R., 2009. New interpretations of *Ignotornis*, the first-reported Mesozoic avian footprints: Implications for the paleoecology and behavior of an enigmatic Cretaceous bird. Cretac. Res. 30 (4), 1041–1061.

Lockley, M.G., Houck, K., Yang, S.-Y., Matsukawa, M., Lim, S.-K., 2006c. Dinosaur dominated footprint assemblages from the Cretaceous Jindong Formation, Hallayo Haesang National Park, Goseong County, South Korea: Evidence and implications. Cretac. Res. 27, 70–101.

Lockley, M.G., Huh, M., Gwak, S.-G., Koo Geun Hwang, K.G., Paik, I.S., 2012. Multiple tracksites with parallel trackways from the Cretaceous of the Yeosu City area Korea: Implications for gregarious behavior in ornithopod and sauropod dinosaurs. Ichnos 19 (1–2), 105–114. https://doi.org/10.1080/10420940.2011.625793.

Lockley, M.G., Li, J.J., Li, R.H., Matsukawa, M., Harris, J.D., Xing, L.D., 2013. A review of the tetrapod track record in China, with special reference to type ichnospecies: Implications for ichnotaxonomy and paleobiology. Acta Geologica Sinica (English edition) 87 (1), 1–20.

Lockley, M.G., Li, J., Xing, L., Guo, B., Matsukawa, M., 2018. Large theropod and small sauropod trackmakers from the Lower Cretaceous Jingchuan Formation, Inner Mongolia, China. Cretac. Res. 92, 150–167.

Lockley, M.G., Xing, L., Kim, K.S., Meyer, C.A., 2021. Tortuous trackways: evidence and implications of deviations, turns and changes in direction by dinosaurian trackmakers. Hist. Biol. 33, 3326–3339. https://doi.org/10.1080/08912/963.2020.18659/45.

Lockley, M.G., Murphey, P., Romilio, A., 2023. A large, saurischian-dominated, tracksite in the Upper Jurassic Morrison Formation, Grand County, Utah: a spectacular example of extensive dinoturbation. Anat. Rec. 306 (S1), 171–173.

Lucas, S.G., 2007. Tetrapod footprint biostratigraphy and biochronology. Ichnos 14 (1–2), 5–38.

Lucas, S.G., 2015. *Thinopus* and a critical review of Devonian tetrapod footprints. Ichnos 22 (3–4), 136–154.

Lucas, S.G., 2019. An ichnological perspective on some major events of Paleozoic tetrapod evolution. Bolletino della Soc. Paleontol. Ital. 58, 223–266.

Lucas, S.G., 2024. Devonian tetrapod footprints and tetrapod origins. In: Lucas, S.G., Hunt, A.P., Klein, H. (Eds.), Vertebrate Ichnology: Tetrapod Tracks and Trackways. Elsevier, pp. 29–86.

Lucas, S.G., Spielmann, J.A., Klein, H., Lerner, A.J., 2010. Ichnology of the Redonda Formation. New Mexico Museum of Natural History and Science, Bulletin 47, 75.

Lucas, S.G., Dalman, S.G., 2016. The Early Cretaceous Clayton Lake dinosaur tracksite, northeastern New Mexico. New Mexico Museum of Natural History, Bulletin 74, pp. 127–140.

Lucas, S.G., Espinoza, E., Alvarado, G.E., 2024a. Acahualinca, Nicaragua, a scientifically significant site of ancient human footprints in the New World. Geol. Soc. London Special Publications 543 (1), 543–2022.

Lucas, S.G., Rogers, J.B., Kvasnak, M., 2024b. Photogrammetric study of the Clayton Lake Dinosaur tracksite, Lower Cretaceous of Northeastern New Mexico, USA. New Mexico Museum of Natural History and Science, Bulletin.

Lucas, S.G., Stimson, M.R., 2024. Carboniferous terapod tracks. In: Lucas, S.G., Hunt, A.P., Klein, H. (Eds.), Vertebrate Ichnology: Tetrapod Tracks and Trackways. Elsevier, Amsterdam, pp. 39–86.

Lull, R.S., 1904. Fossil footprints of the Jura-Trias of North America. Boston Soc. Nat. Hist. Memoirs 5, 461–557.

MacDonald, J.P., 1994. Earth's First Steps: Tracking Life before the Dinosaurs. Johnson Books, Boulder, 290 pp.

Malkani, M.S., 2007. Trackways evidence of sauropod dinosaurs confronted by a theropod found from Middle Jurassic Samana Suk Limestone of Pakistan. Sind. Univ. Res. J. (Sci. Ser.) 39 (1), 1–14.

Marchetti, L., Voigt, S., Lucas, S.G., 2019b. An anatomy-consistent study of the Lopingian eolian tracks of Germany and Scotland reveals the first evidence of the end-Guadalupian mass extinction at low paleolatitudes of Pangea. Gondwana Res. 73, 32–53.

Marchetti, L., Klein, H., Buchwitz, M., Ronchi, A., Smith, R.M.H., De Klerk, W.J., Sciscio, L., Groenwald, G.H., 2019c. Permian- Triassic vertebrate footprints from South Africa: Ichnotaxonomy producers and biostratigraphy through two major faunal crises. Gondwana Res. 72, 139–168.

Marchetti, L., Francischini, H., Lucas, S.G., Voigt, S., Hunt, A.P., Santucci, V.L., 2020. Paleozoic Vertebrate Ichnology of Grand Canyon National Park. In: Santucci, V.L., Tweet, J.S. (Eds.), Grand Canyon National Park Centennial Paleontological Resource Inventory. Natural Resource Report NPS/GRCA/NRR—2020/2103, pp. 333–379.

Martin, A.J., Lowery, M., Hall, M., Vickers-Rich, P., Rich, T.H., Serrano-Brañas, C.I., Swinkels, P., 2023. Earliest known Gondwanan bird tracks: Wonthaggi Formation (Early Cretaceous), Victoria, Australia. PLoS One 18 (11), e0293308. https://doi.org/10.1371/journal.pone.0293308.

Matsukawa, M., Hamuro, T., Mizukami, T., Fujii, S., 1997. First trackway evidence of gregarious dinosaurs from the Lower Cretaceous Tetori Group of eastern Toyama Prefecture, central Japan. Cretac. Res. 18 (4), 603–619.

McCrea, R.T., 2000. Dinosaur footprints in the Lower Cretaceous (Albian) Gates Formation of Alberta, Canada; their use in palaeobiology and palaeoenvironmental interpretation. J. Paleontol. Soc. Korea 4, 169–178.

McCrea, R.T., 2001. The distribution of vertebrate ichnotaxa from Lower Cretaceous (Albian) Gates Formation track-sites near Grande Cache, Alberta: Implications for habitat preference and functional pedal morphology. In: Alberta Palaeontological Society, Fifth Annual Symposium, pp. 42–46.

McCrea, R.T., 2003. Fossil vertebrate tracksites of Grande Cache, Alberta. Can. Palaeobiol. 9, 11–30.

McCrea, R.T., Buckley, L.G., Plint, A.G., Currie, P.J., Haggart, J.W., Helm, C.W., Pemberton, S.G., 2014a. A review of vertebrate track-bearing formations from the Mesozoic and earliest Cenozoic of western Canada with a description of a new theropod ichnospecies and reassignment of an avian ichnogenus. In: Lockley, M.G., Lucas, S.G. (Eds.), Fossil Footprints of Western North America, vol 62. New Mexico Museum of Natural History and Sciences Bulletin, pp. 5–93.

McCrea, R.T., Buckley, L.G., Farlow, J.O., Lockley, M.G., Currie, P.J., Matthews, N.A., Pemberton, S.G., 2014b. A 'terror of tyrannosaurs': The first trackways of tyrannosaurids and evidenced of gregariousness and pathology in Tyrannosauridae. PLoS One 9 (7), 1–13. https://doi.org/10.1371/journal.pone.0117606 e0117606.

McCrea, R.T., Tanke, D.H., Buckley, L.G., Lockley, M.G., Farlow, J.O., Xing, L., Matthews, N.A., Helm, C.W., Pemberton, S.G., Breithaupt, B.H., 2015. Vertebrate ichnopathology: Pathologies inferred from dinosaur tracks and trackways from the Mesozoic. Ichnos 22 (3–4), 235–260.

Meldrum, D.J., Lockley, M.G., Lucas, S.G., Musiba, C., 2011. Ichnotaxonomy of the Laetoli trackways: The earliest hominin footprints. J. Afr. Earth Sci. 60 (1–2), 1–12.

Milàn, J., Loope, D.B., Bromley, R.G., 2008. Crouching theropod and *Navahopus* sauropodomorph tracks from the Early Jurassic Navajo Sandstone of USA. Acta Palaeontol. Pol. 53 (2), 197–205.

Milner, A.R., Lockley, M.G., Kirkland, J.I., 2006. A large collection of well-preserved theropod dinosaur swim tracks from the Lower Jurassic Moenave Formation, St. George, Utah. New Mexico Museum of Natural History and Science, Bulletin 37, 315–328.

Milner, A.R., Harris, J.D., Lockley, M.G., Kirkland, J.I., Matthews, N.A., 2009. Bird-like anatomy, posture, and behavior revealed by an Early Jurassic theropod dinosaur resting trace. PLoS One 4 (3), e4591.

Milner, A.R.C., Birthisel, T.A., Kirkland, J.I., Breithaupt, B.H., Matthews, N.A., Lockley, M.G., Santucci, V.L., Gibson, S.Z., DeBlieux, D.D., Hurlbut, M., Harris, J.D., Olsen, P.E., 2011. Tracking Early Jurassic dinosaurs across southwestern Utah and the Triassic-Jurassic transition. In: Bonde, J.W., Milner, A.R.C. (Eds.), Field Trip Guide Book, 71st Annual Meeting of the Society of Vertebrate Paleontology, Paris Las Vegas, Las Vegas, Nevada, November 2–5, 2011. Nevada State Museum Paleontological Papers 1, 107 pp.

Milner, A.R., Lockley, M.G., Falkingham, P.L., Marty, D., Richter, A., 2016. Dinosaur swim track assemblages: characteristics, contexts, and ichnofacies implications. In: Falkingham, P.L., Marty, D., Richter, A. (Eds.), Dinosaur Tracks: The Next Steps. Indiana University Press, Bloomington, pp. 152–180.

Moratalla, X., Sanz, X.L., Domínguez, I.M., Jiménez, S., 1988. Yacimentos paleoicnologicoas de La Rioja; Huellas de Dinosaurios. Gobierno de la Rioja, España, 95 pp.

Moreno, K., De Valais, S., Blanco, N.M., Tomlinson, A.J., Jacay, J., Calvo, J., 2012. Large theropod dinosaur footprint associations in western Gondwana: Behavioural and palaeogeographic implications. Acta Palaentologica Polonica 57 (1), 73–83. https://doi.org/10.4202/app.2010.0119.

Morse, S.A., Bennett, M.R., Gonzalez, S., Huddart, D., 2010. Techniques for verifying human footprints: reappraisal of pre-Clovis footprints in Central Mexico. Quat. Sci. Rev. 29 (19–20), 2571–2578.

Myers, T.S., Fiorillo, A.R., 2009. Evidence for gregarious behavior and age segregation in sauropod dinosaurs. Palaeogeogr. Palaeoclimatol. Palaeoecol. 274 (1–2), 96–104.

Navarro-Lorbés, P., Ruiz, J., Díaz-Martínez, I., Isasmendi, E., Sáez-Benito, P., Viera, L., Pereda-Suberbiola, X., Torices, A., 2021. Fast-running theropods tracks from the Early Cretaceous of La Rioja, Spain. Sci. Rep. 11, 1–13. https://doi.org/10.1038/s41598-021-02557-9.

Olsen, P.E., Smith, J.B., McDonald, N.G., 1998. Type material of the type species of the classic theropod footprint genera *Eubrontes, Anchisauripus*, and *Grallator* (Early Jurassic, Hartford and Deerfield basins, Connecticut and Massachusetts, USA). J. Vertebr. Paleontol. 18, 586–601.

Ösi, A., Pálfy, J., Makádi, L., Szentesi, Z., Gulyás, P., Rabi, M., Botfalvai, G., Hips, K., 2011. Hettangian (Early Jurassic) dinosaur tracksites from the Mecsek Mountains, Hungary. Ichnos 18, 79–94. https://doi.org/10.1080/10420940.2011.573603.

Ostrom, J.H., 1972. Were some dinosaurs gregarious? Palaeogeogr. Palaeoclimatol. Palaeoecol. 11, 287–301.

Ostrom, J.H., 1985. Social and unsocial behavior in dinosaurs. Field Mus. Nat. Hist. Bull. 55 (9), 10–21.

Parker, L.R., Balsley, J.K., 1989. Coal mines as localities for studying dinosaur trace fossils. . In Dinosaur tracks and traces. In: Gillette, D.D., Lockley, M.G. (Eds.), Dinosaur Tracks and Traces. Cambridge University Press, Cambridge, pp. 353–359.

Pastoors, A., Lenssen-Erz, T., Ciqae, T., Kxunta, U., Thao, T., Bégouën, R., Biesele, M., Clottes, J., 2015. Tracking in caves: Experience based reading of Pleistocene human footprints in French caves. Camb. Archaeol. J. https://doi.org/10.1017/S0959774315000050.

Paul, G.S., Christiansen, P., 2000. Forelimb posture in neoceratopsian dinosaurs: implications for gait and locomotion. Paleobiology 26 (3), 450–465.

Peabody, F.E., 1948. Reptile and amphibian trackways from the Moenkopi Formation of Arizona and Utah. University of California Publications, Bulletin of the Department of Geological Sciences 27, 295–468.

Perez-Lorente, F.P., 1996. Pistas teropodas en cifras. Zubia 14, 37–55.

Pérez-Lorente, F., 2003. Aportaciones de los yacimientos de La Barguilla, Santisol y Santa Juliana (Hornillos de Cameros, La Rioja. España). In: Pérez-Lorente, F. (Ed.), Dinosaurios y otros reptiles mesozoicos en España, pp. 161–194.

Pérez-Lorente, F., Cuenca-Besos, G., Aurell, M., Canudo, J.I., Soria, A.R., Ruiz-Omeñaca, J.I., 1997. Las Cerradicas tracksite (Berriasian, Galve, Spain): growing evidence for quadrupedal ornithopods. Ichnos 5, 109–120. https://doi.org/10.1080/10420949709386410.

Piñuela, L., Garcia-Ramos, J.C., Lires, J., 2002. Comportamiento gregario de ornitópodos en el Jurásico de Asturias. In: Pérez-Lorente, F. (Ed.), *Dinosaurios y otros reptiles Mesozicos de Espana*.Resúmenes de las comunicaciones ponencias y paneles, Universidad de LaRioja –X aniversario. Universidad de La Rioja, pp. 44–45.

Piñuela, L., García-Ramos, J.C., Romano, M., Ruiz-Omeñaca, J.I., 2016. First record of gregarious behavior in robust medium-sized Jurassic ornithopods: evidence from the Kimmeridgian trackways of Asturias (N Spain) and some general considerations on other medium-large ornithopod tracks in the Mesozoic record. Ichnos 23, 298–311.

Pittman, J.G., Gillette, D.D., 1989. The Briar site: a new sauropod dinosaur tracksite in Lower Cretaceous beds of Arkansas, USA. In: Gillette, D.D., Lockley, M.G. (Eds.), Dinosaur Tracks and Traces. Cam. Univ. Pres, Cambridge, pp. 313–332.

Platt, B.F., Hasiotis, S.T., 2008. A new system for describing and classifying tetrapod tail traces with implications for interpreting the dinosaur tail trace record. Palaios 23 (1), 3–13.

de Raaf, J.F.M., Beets, C., van der Sluijs, G.K., 1965. Lower Oligocene bird-tracks from Northern Spain. Nature, London 207 (July 10), 146–148.

Razzolini, N.L., Vila, B., Díaz-Martínez, I., Manning, P.L., Galobart, À., 2016. Pes shape variation in an ornithopod dinosaur trackway (Lower Cretaceous, NW Spain): new evidence of an antalgic gait in the fossil track record. Cretac. Res. 58, 125–134.

Renders, E., Vincelette, A., 2023a. Methodology for the determination of modern and fossil horse gaits from trackways. J. Paleontol. Techniq. 27, 1–25.

Renders, E., Vincelette, A., 2023b. New Studies on the Gaits Displayed by Miocene, Pliocene, and Pleistocene Fossil Horse Trackways. In: Narayan, E. (Ed.), Animal Science Annual Volume. IntechOpen, London. https://doi.org/10.5772/intechopen.112157.

Romano, M., Citton, P., 2017. Crouching theropod at the seaside. Matching footprints with metatarsal impressions and theropod authopods: a morphometric approach. Geol. Mag. 154 (5), 946–962.

Romilio, A., Tucker, R.T., Salisbury, S.W., 2013. Reevaluation of the Lark Quarry dinosaur Tracksite (late Albian–Cenomanian Winton Formation, central-western Queensland, Australia): no longer a stampede? J. Vertebr. Paleontol. 33 (1), 102–120.

Rowland, S.M., Mercadante, J.M., 2014. Trackways of a gregarious, dunefield-dwelling, Early Jurassic therapsid in the Aztec Sandstone of southern Nevada. Palaios 29 (10), 539–552.

Rubilar, D., Moreno, K., Blanco, N., Calvo, J.O., 2008. Theropod dinosaur trackways from the Lower Cretaceous of the Chacarilla Formation, Chile. Rev. Geol. Chile 35 (1), 175–184.

Rühle, V., Lilienstern, H., 1939. Fährten und Spuren im *Chirotherium*-Sandstein von Südthüringen. Fortschritte der Geologie und Paläontologie 12, 293–387.

Ruiz, J., Torices, A., 2013. Humans running at stadiums and beaches and the accuracy of speed estimations from fossil trackways. Ichnos 20 (1), 31–35.

Santos, V.F., Galopim de Carvalho, A.M., da Silva, C.M., 1995. A jazida da Pedreira da Ribeira do Cavalo (Sesimbra) ou a história das pegadas de dinossáurio que nunca mais poderemos visitar. Almadam, II Série 4, 175–177.

Schulp, A.S., Brokx, W.A., 1999. Maastrichtian sauropod footprints from the Fumanya site, Bergueda, Spain. Ichnos 6 (4), 239–250.

Schumacher, B.A., Lockley, M.G., 2022. North America's largest assemblage of Late Jurassic dinosaur trackways, a new look at the Purgatoire Valley dinosaur tracksite (Dinosaur Lake), Morrison Formation, southeastern Colorado, United States. Hist. Biol. 35, 1146–1161.

Shikama, T., 1942. Footprints from Chinchou, Manchoukuo, of *Jeholosauripus*, the Eo–Mesozoic dinosaur. Bull. Cent. Nat. Mus. Manchoukuo 3, 21–31.

Sukhanov, V.B., 1968. Symmetrical Locomotion of Terrestrial Vertebrates and Some Features of Movement of Lower Tetrapods. Nauka, Leningrad, 186 pp.

Tanke, D.H., Rothschild, B.M., 2002. Dinosores: An annotated bibliography of dinosaur paleopathology and related topics—1838–2001. New Mexico Museum of Natural History and Science, Bulletin 20.

Thulborn, R.A., 1982. Speeds and gaits of dinosaurs. Palaeogeogr. Palaeoclimatol. Palaeoecol. 38, 227–256.

Thulborn, T., 1990. Dinosaur Tracks. Chapman and Hall, London, 410 pp.

Thulborn, T., 2017. Behaviour of dinosaurian track-makers in the Winton Formation (Cretaceous, Albian–Cenomanian) at Lark Quarry, western Queensland, Australia: running or swimming? Ichnos 24 (1), 1–18.

Thulborn, R.A., Wade, M., 1979. Dinosaur stampede in the Cretaceous of Queensland. Lethaia 12, 275–279.

Thulborn, R.A., Wade, M., 1984. Dinosaur trackways in the Winton Formation (mid-Cretaceous) of Queensland. Memoir. Queensl. Mus. 21, 413–517.

Tomaselli, M.B., 2014. Desplazamiento enmanada de saurópodos titanosaurios: Estudio de las icnitas *Titanopodus*, Cretácico Superior de Mendoza, Argentina. Tesis de Licenciatura. Universidad del Aconcagua (Mendoza, Argentina) 70.

Tucker, M.E., Burchette, T.P., 1977. Triassic dinosaur footprints from South Wales: Their context and preservation. Palaeogeogr. Palaeoclimatol. Palaeoecol. 22, 195–208.

Tuttle, R.H., 1987. Kinesiological inferences and evolutionary implications from Laetoli bipedal trails G-1, G-2/3, and A. In: Leakey, M.D., Harris, J.M. (Eds.), Laetoli: A Pliocene Site in Northern Tanzania. Clarendon Press, Oxford, UK, pp. 503–520.

Tuttle, R.H., Webb, D.M., Baksh, M., 1991. Laetoli toes and *Australopithecus afarensis*. Hum. Evol. 6, 193–200. https://doi.org/10.1007/BF02438142.

Van Allen, H.E.K., Calder, J.H., Hunt, A.P., 2005. The trackway record of a tetrapod community in a walchian conifer forest from the Permo-Carboniferous of Nova Scotia. In: Lucas, S.G., Zeigler, K.E. (Eds.), The Nonmarine Permian: New Mexico Museum of Natural History and Science, Bulletin, vol 30, pp. 322–332.

Vincelette, A., 2021. Determining the gait of Miocene, Pliocene, and Pleistocene horses. Foss. Rec. 24, 151–169. https://doi.org/10.5194/fr-24-151-2021.

Voigt, S., Berman, D.S., Henrici, A.C., 2007. First well-established track-trackmaker association of Paleozoic tetrapods based on *Ichniotherium* trackways and diadectid skeletons from the lower Permian of Germany. J. Vertebr. Paleontol. 27, 553–570.

Wade, M., 1989. The stance of dinosaurs and the Cossack dancer syndrome. In: Gillette, D.D., Lockiey, M.G. (Eds.), Dinosaur Tracks and Traces. Cambridge University Press, Cambridge, pp. 73–82.

Weems, R., 2006. Locomotor speeds and patterns of running behavior in non-maniraptoriform theropod dinosaurs. In: Harris, J.D., Lucas, S.G., Spielmann, J.A., Lockley, M.G., Milner, A.R.C., Kirkland, J.I. (Eds.), The Triassic-Jurassic Terrestrial Transition, vol 37. New Mexico Museum of Natural History & Science, Bulletin, pp. 379–389.

White, T.D., Suwa, G., 1987. Hominid footprints at Laetoli: facts and interpretations. Am. J. Phys. Anthropol. 7 (72), 485–514. https://doi.org/10.1002/ajpa.1330720409.

Willey, P., Stolen, J., Crothers, G., Watson, P.J., 2005. Preservation of prehistoric footprints in Jaguar cave, Tennessee. J. Cave Karst Stud. 67, 61–68.

Xing, L.D., Harris, J.D., Toru, S., Masato, F., Dong, Z.M., 2009. Discovery of dinosaur footprints from the Lower Jurassic Lufeng Formation of Yunnan Province, China and new observations on *Changpeipus*. Geol. Bull. China 28 (1), 16–29.

Xing, L., Chen, Q., Wang, D., Lallensack, J.N., 2025. The ichnology of reproduction in fossil tetrapods. In: Lucas, S.G., Hunt, A.P., Klein, H. (Eds.), Vertebrate Ichnology, this volume. Elsevier, Amsterdam.

Xing, L.D., Gierliński, G.G., Harris, J.D., Divay, J.D., 2012a. A probable crouching theropod dinosaur trace from the Tuchengzi Formation in Chichengarea, Hebei Province, China. Geol. Bull. China 31 (1), 20–25.

Xing, L.D., Harris, J.D., Gierlinski, G.D., Gingras, M.K., Divay, J.D., Tang, Y.G., Currie, P.J., 2012b. Early Cretaceous pterosaur tracks from a "buried" dinosaur tracksite in Shandong Province, China. Palaeoworld 21, 50–58.

Xing, L., Lockley, M.G., Zhang, J., Milner, A.R., Klein, H., Li, D., Persons, W.S., Ebi, J., 2013. A new Early Cretaceous dinosaur track assemblage and the first definite non-avian theropod swim trackway from China. Chin. Sci. Bull. 58, 2370–2378.

Xing, L.D., Klein, H., Lockley, M.G., Wetzel, A., Li, Z.D., Li, J.J., Gierliski, G.D., Zhang, J.P., Matsukawa, M., Divay, J.D., Zhou, L., 2014. *Changpeipus* (theropod) tracks from the Middle Jurassic of the Turpan Basin, Xinjiang, Northwest China: Review, new discoveries, ichnotaxonomy, preservation and paleoecology. Vertebrata PalasiAtica 52 (2), 233–259.

Xing, L., Lockley, M.G., Klein, H., Gierliński, G.D., Ye, Y., Zhang, J., Persons IV, W.S., Wang, T., 2019. First thyreophoran type tracks from the Middle Jurassic Chuanjie Formation of Yunnan Province, China. Ichnos 26 (1), 8–15.

Xing, L., Lockley, M.G., Zhang, J., Klein, H., Marty, M., Peng, G., Ye, Y., McCrea, R.T., Persons IV, W.S., Xu, T., 2015a. The longest theropod trackway from East Asia, and a diverse sauropod-, theropod-, and ornithopod-track assemblage from the Lower Cretaceous Jiaguan Formation, southwest China. Cretac. Res. 56, 345–362.

Xing, L., Marty, D., Wang, K., Lockley, M.G., Chen, S., Xu, X., Liu, Y., Kuang, H., Zhang, J., Ran, H., Persons IV, W.S., 2015b. An unusual sauropod turning trackway from the Early Cretaceous of Shandong Province, China. Palaeogeogr. Palaeoclimatol. Palaeoecol. 437, 74–84.

Xing, L., Li, D., Falkingham, P.L., Lockley, M.G., Benton, M.J., Klein, H., Zhang, J., Ran, H., Persons IV, W.S., Dai, H., 2016. Digit-only sauropod pes trackways from China—evidence of swimming or a preservational phenomenon? Sci. Rep. 6 (1), 21138.

Zaagane, M., Hunt, A.P., Bendella, M., Lucas, S.G., Draoui, A., Mouzti, O., 2023. Multiple trackways from the Cenomanian Djoua series, in In Amenas region (southeastern Algeria), represent the first ichnological evidence of gregarious behavior in Cretaceous sauropods from Africa. Hist. Biol. https://www.tandfonline.com/action/showCitFormats?doi=10.1080/08912963.2023.2269173.

CHAPTER

4

Fish trails and imprints

Nicholas J. Minter

School of the Environment and Life Sciences, University of Portsmouth, Portsmouth, United Kingdom

1. Introduction

A wide range of trails and imprints have been attributed to fish. *Undichna* is by far the most common and is interpreted as the locomotory trail of fish (Anderson, 1976; de Gibert et al., 1999; Trewin, 2000; Minter and Braddy, 2006), but an increasing number of trace fossils have been further attributed to fish and have expanded their behavioral repertoire through the geological record. *Parundichna* is also interpreted as a fish locomotory trail (Simon et al., 2003), whereas *Broomichnium*, *Pisquiesichnus*, and *Raederichnus* are interpreted as fish resting traces (Benner et al., 2008; Ayranci et al., 2022; Halpern et al., 2022), and *Piscichnus*, *Osculichnus*, and *Daandavichnus* are interpreted as the nesting or feeding traces of fish (Feibel, 1987; Gregory, 1991; Demircan and Uchman, 2010; Rodriguez-de la Rosa et al., 2021). This review focuses on surface trails and imprints attributed to fish (Appendix 1); however, a range of open nomenclature burrows and the named burrows *Cornulatichnus* and *Redondarefugium* have been attributed to sheltering or aestivating fish (McAllister, 1988; Carroll and Trewin, 1995; Gobetz et al., 2006; Lucas et al., 2010; Marshall and Rogers, 2012). In addition, the fossil record and named ichnotaxa of dentalites, coprolites, and regurgitalites attributed to fish were reviewed recently by Hunt and Lucas (2021).

Fish swimming trails, resting imprints, and feeding and nesting imprints range from the Silurian to Recent (Fig. 4.1), and they have a near global distribution, being reported from all present continents apart from Antarctica (Fig. 4.2). This enables the trace fossil record of fish to (i) complement that of their body fossil record (Friedman and Sallan, 2012) to deduce their evolutionary history; (ii) provide in situ evidence for their palaeoenvironmental and palaeogeographic distribution; and (iii) help deduce the swimming styles of extinct fish and other aspects of their behavioral repertoires (e.g., Demírcan and Uchman, 2010; Martin et al., 2010; Belvedere et al., 2011; Löwemark, 2015; Fan et al., 2019; Baucon et al., 2023).

Vertebrate Ichnology
https://doi.org/10.1016/B978-0-443-38351-9.00002-5

FIGURE 4.1 Stratigraphic distribution of the different ichnogenera and ichnospecies of surface trails and imprints attributed to fish. The thickness of the lines corresponds with the number of occurrences from each period. Solid lines represent known presence, and dashed lines represent inferred presence between two periods of known presence.

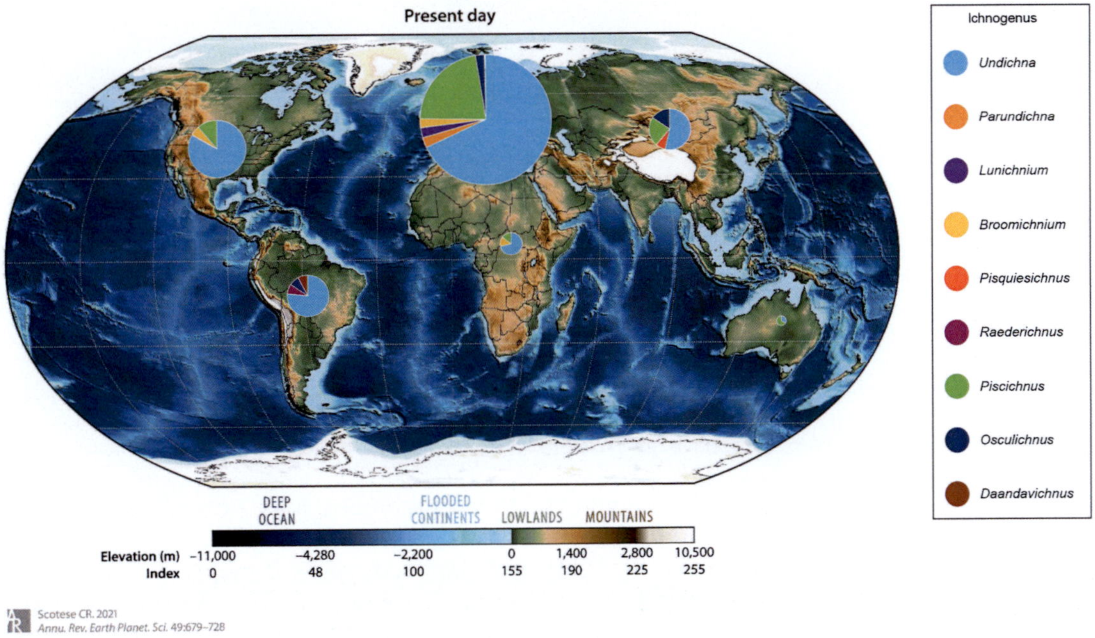

FIGURE 4.2 Geographic distribution of occurrences of ichnogenera of surface trails and imprints attributed to fish. The diameters of the circles correspond with the number of total occurrences from each continent, and the divisions correspond with the proportions of each reported ichnogenus. **Map lines delineate study areas and do not necessarily depict accepted national boundaries**; *Modified after Scotese (2021).*

2. Methods for studying fish trails and imprints

Undichna is the most common trace fossil attributed to fish, and it can comprise unpaired (u) and paired (p) sinusoidal waves, scalloped waves (s), furrows (f), and accessory impressions and imprints (i) (Fig. 4.3; Anderson, 1976; Higgs, 1988; Trewin, 2000; Minter and Braddy, 2006). The range of measurements that could be taken to quantify and characterize such traces were illustrated by Trewin (2000) and include the wavelength (λ = distance between two consecutive crests or troughs of a wave) and amplitude (a = distance, perpendicular to the mid-line, between the crest and trough of a wave) of a sinusoidal wave; the phase difference (pd = distance between the crests or troughs of two intertwined waves); and the width (w = distance, perpendicular to the mid-line, between paired waves) (Fig. 4.3). In addition to such measurements, the use of digital photography and digitized images has facilitated more advanced spatial and mathematical analyses of fish trails. For example, Martin et al. (2010) demonstrated that it is possible to determine the spatial co-ordinates of points along the lengths of the waves of a specimen of *Undichna* and then applied Fourier analysis to interpret the body size and swimming mode of the producer.

Typically, *Undichna* is a very low relief epichnial or hypichnial trail, and so traditionally, it has been documented with photography using low angle lighting to pick out these subtle surficial features. As with other vertebrate trace fossils, photogrammetry (Falkingham et al., 2018) is increasingly being used as a technique to document individual fish trace fossils

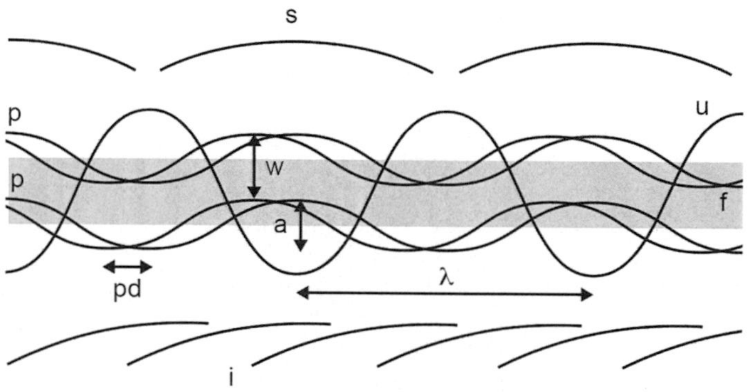

FIGURE 4.3 Fish trail terminology and measurements. Terminology for descriptive elements: u = unpaired wave, p = paired waves, s = scalloped waves, i = imprints, f = furrow. Measurements: λ = wavelength, a = amplitude, w = width between paired waves, pd = phase difference between waves. *Modified after Trewin (2000).*

(Baucon et al., 2023). In addition, aerial surveys and laser scanning have proved beneficial for mapping large surfaces and determining the spatial arrangement of traces, especially of the dish-shaped *Piscichnus* (Belvedere et al., 2011), and the production of mapped surfaces has provided evidence to help test hypotheses as to whether some surfaces record trackways of vertebrates or multiple individual feeding traces of fish (Martinell et al., 2001; Lucas, 2015; Niedzwiedzki and Szrek, 2020).

3. Ichnotaxa

3.1 Fish swimming trails

The most common ichnotaxon interpreted as the swimming trails of fish is the ichnogenus *Undichna* Anderson, 1976. In addition, the monospecific ichnogenus *Parundichna* (Simon et al., 2003) and a particular ichnospecies of *Lunichnium*, *L. westerbergensis* (Schweigert, 2001), have also been interpreted as fish swimming trails.

3.1.1 *Undichna*

The ichnogenus *Undichna* was erected by Anderson (1976), with the type ichnospecies being *U. simplicitas* Anderson, 1976, based on material that had been previously described but not formally classified (Anderson, 1970; Warren, 1972) from the Permian of South Africa. The ichnogenus has subsequently been reviewed on several occasions (de Gibert et al., 1999; Trewin, 2000; Minter and Braddy, 2006) and has been reported from the Silurian to the Recent (Fig. 4.1). *Undichna* comprises sets of horizontal sinusoidal waves, including single waves, unpaired waves, paired waves, and various combinations thereof, which may be intertwined and crossing over one another, or be separated (Anderson, 1976; de Gibert et al., 1999; Trewin, 2000; Minter and Braddy, 2006, Fig. 4.3). Some ichnospecies of *Undichna* have additional elements such as continuous furrows/ridges, and oblique imprints may be present (Minter and Braddy, 2006). The waves of *Undichna* are preserved as continuous epichnial grooves or hypichnial ridges but sometimes they are intermittent and only the troughs/crests of the waves are preserved (Fig. 4.4). Currently, 10 ichnospecies are considered valid (Table 4.1).

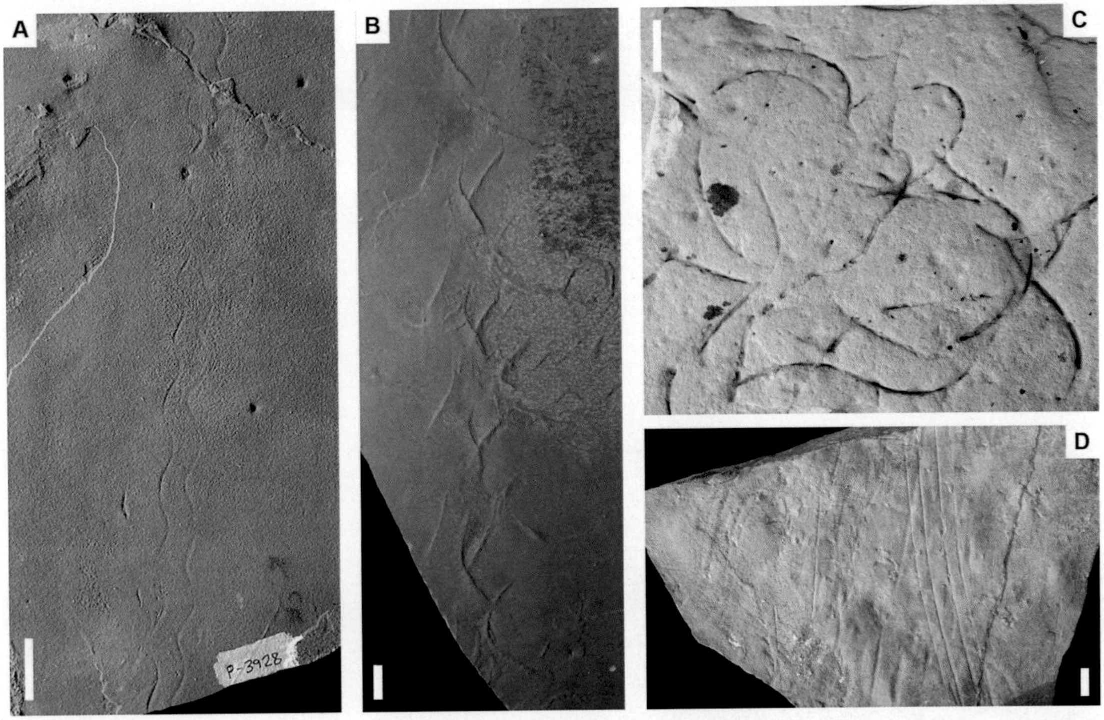

FIGURE 4.4 Examples of fish swimming trails. (A) *Undichna bina* from the Permian of New Mexico, New Mexico Museum of Natural History and Science NMMNH P3928. (B) *U. britannica* from the Carboniferous of Alabama, University of Alabama Museum UAM PV2005.7.67. (C) *U. unisulca* from the Silurian of Sweden, Senckenberg Institute Frankfurt SMF XXX 902. (D) *U. quadrisulcata* from the Eocene of Spitsbergen, Natural History Museum Berlin MB.W 5641. Scale bars 1 cm in A-B, and 2 cm in C-D. *(C) Image courtesy of Dirk Knaust. (D) Image courtesy of Dirk Knaust.*

As with all ichnotaxa, the classification of *Undichna* is governed by the principles of ichnotaxonomy, as laid out by Bertling et al. (2006, 2022), and should therefore be based primarily on morphological ichnotaxobases. A difference in approach between invertebrate and vertebrate ichnologists has been highlighted by Minter et al. (2007), with the former approach being more morphological and the latter more producer-related. However, despite being attributed to vertebrates, *Undichna* has always been treated more like an invertebrate trace in terms of ichnotaxonomic approaches (Minter and Braddy, 2006). An amelioration between the traditional approaches of invertebrate and vertebrate ichnologists to ichnotaxonomy may be made by the fact that the morphology of a trace fossil is a function of the anatomy of a producer, its behavior, and the substrate conditions (Minter et al., 2007). As such, the morphology of a trace fossil encodes aspects of the producer, and these happen to be more prevalent than aspects of behavior and the substrate for vertebrate traces such as trackways. The key distinction is that the morphology of a trace is an objectively observable phenomenon, whereas attributions of a producer, as well as the behavior and substrate conditions, are subjective interpretations.

TABLE 4.1 The valid ichnospecies of *Undichna* and their key distinguishing morphological characteristics.

Ichnotaxon	Key distinguishing morphological characteristics
U. simplicitas Anderson, 1976 	A single unpaired wave that has a high amplitude, an outer pair of parallel waves that have a low amplitude and are often scalloped convex outwards, and an inner pair of parallel waves that have intermediate amplitude. The single unpaired wave may be alongside or intertwined with the inner pair of parallel waves.
U. bina Anderson, 1976 	One pair of parallel waves.
U. insolentia Anderson, 1976 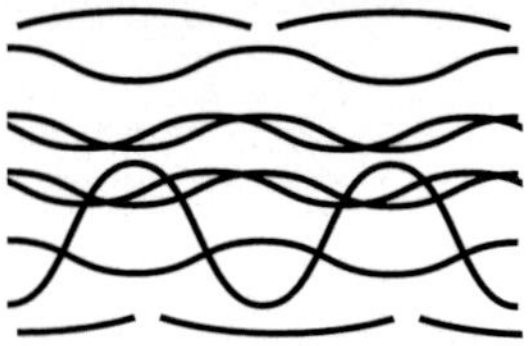	A single unpaired wave that has a high amplitude, an outer pair of parallel waves that have a low amplitude and are often scalloped convex outwards, a middle pair of parallel waves, and two inner pairs of parallel waves that have intermediate amplitudes. The single unpaired wave may be intertwined with all the paired parallel waves, and the two inner pairs of parallel waves are intertwined.
U. britannica Higgs, 1988 	Two unpaired waves that are intertwined and have different amplitudes. An outer pair of parallel waves that have a low amplitude may be present or intermittent; or the outer parts may consist of oblique imprints and be accompanied by a narrow, sinuous, median furrow.
U. consulca Higgs, 1988 	Two unpaired waves that are intertwined, have different amplitudes, and are accompanied by a shallow, broad, continuous impression.
U. unisulca de Gibert et al., 1999 	A single unpaired wave.
U. quina Trewin, 2000 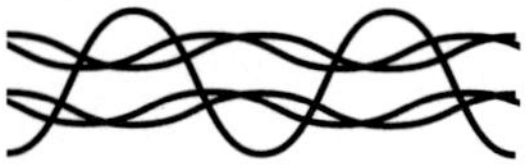	A single unpaired wave that has a high amplitude, and two pairs of parallel waves that have similar but lower amplitudes. The single unpaired is intertwined with the two pairs of parallel waves and the two pairs of parallel waves are also intertwined.
U. trisulcata Morrissey et al., 2004 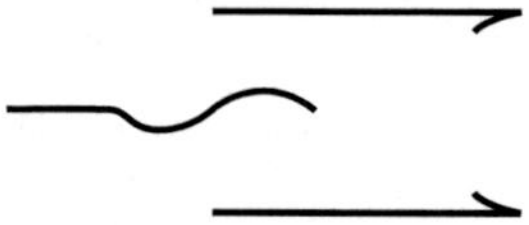	A single straight or slightly sinusoidal imprint, and an outer pair of straight parallel imprints that may have comma or circular-shaped terminations at one end.

TABLE 4.1 The valid ichnospecies of *Undichna* and their key distinguishing morphological characteristics.—cont'd

Ichnotaxon	Key distinguishing morphological characteristics
U. septemsulcata Wisshak et al., 2004 	A single unpaired wave that has a high amplitude, two pairs of outer parallel waves that have low amplitudes, and an inner pair of parallel waves that have intermediate amplitude. All of the waves have the same wavelength and are in parallel and not intertwined.
U. quadrisulcata Knaust, 2019 	Two or three pairs of parallel waves that have the same amplitude and wavelength and are not intertwined.

The key distinguishing morphological characteristics are summarized and simplified from the original and emended diagnoses (Anderson, 1976; de Gibert et al., 1999; Trewin, 2000; Minter and Braddy, 2006; Wisshak et al., 2004; Knaust, 2019). Diagnoses are not presented verbatim here because they differ in format and lack consistency in their use of terminology.

The ichnogenus *Undichna* was reviewed by Trewin (2000) and then again by Minter and Braddy (2006) due to several new ichnospecies having been named in the intervening period. The review by Minter and Braddy (2006) reduced the number of valid ichnospecies from 14 to nine. Only one new ichnospecies, *U. quadrisulcata* (Knaust, 2019), has been erected since. Minter and Braddy (2006) identified that there are subset relationships among the ichnospecies of *Undichna*, with the simpler ichnospecies potentially representing elements of the more complex ichnospecies. These subset relationships could be considered as minor morphological variants among certain ichnospecies that could arise due to minor behavioral or preservational variation. In the case of minor behavioral variation, a producer could swim at different heights above the sediment–water interface and that would affect which fins penetrated into the sediment to leave behind trails. In the case of minor preservational variation, trails from different fins could have been incised to different depths below the sediment–water interface, and then, the observed trail morphology could be that of an undertrail due to erosion prior to or after lithification, or due to the parting plane along which a trail is discovered (Minter and Braddy, 2006). As such, Minter and Braddy (2006) recommended caution before erecting any new ichnospecies of *Undichna*, and that if any potential new ichnospecies represents a subset of the morphology of existing ichnospecies then it needed to be demonstrated that they were a recurrent morphology, based on multiple specimens from a single locality or a morphology that occurs across different localities.

The ichnospecies *Undichna radnicensis* Turek 1989, *U. tricosta* Lu and Chen, 1998, *U. prava* Lu et al., 2004, and *U. gosiutensis* de Gibert, 2001 are all considered as junior synonyms of existing ichnospecies of *Undichna* (Minter and Braddy, 2006). *Undichna radnicensis* was erected for a highly variable range of morphologies that had transitional forms between them (Turek, 1989, 1996), and Minter and Braddy (2006) were able to justify its synonymy with *U. britannica* on the basis of intergrading forms from the Triassic of China reported by Lu et al. (2004). Trewin (2000) proposed that *U. tricosta* should be synonymized with *U. simplicitas*, because it falls within the minimum diagnosis of Anderson (1976) for that ichnospecies, and Minter and Braddy (2006) agreed with this proposal. In addition, Minter and Braddy (2006) regarded *U. prava* as an incomplete *U. tricosta* and therefore further synonymized it with *U. simplicitas*. *Undichna gosiutensis* was synonymized with *U. quina* by Minter and Braddy (2006) as it represents a subset of the morphology of the latter and was based on only one specimen.

One new ichnospecies of *Undichna*, *U. quadrisulcata* (Fig. 4.4D), has been erected since the review of Minter and Braddy (2006), and it represents a morphology that was originally considered to be within *U. insolentia* by Anderson (1976). However, while *U. quadrisulcata* represents a subset of the morphology of *U. insolentia* (*sensu* Anderson, 1976), Knaust (2019) has demonstrated that it is a recurrent morphology that should be considered in its own right because it occurs at numerous different localities of different ages, including the Eocene of Spitsbergen (Knaust, 2019); Triassic of the Netherlands (Knaust, 2019); Permian of South Africa (Anderson, 1976), the Falkland Islands (Trewin, 2000), Uruguay (Piñeiro, 2006), and France (Moreau et al., 2022); and Carboniferous of Argentina (Buatois and Mángano, 1993, 1994; Cardonatto and Melchor, 2014).

The ichnogenus *Cochlichnus* Hitchcock, 1858 and ichnospecies *Undichna unisulca* de Gibert et al., 1999 are both similar in that they each comprise a single sinusoidal wave; however, they may be differentiated because *U. unisulca* may be an order of magnitude larger than *Cochlichnus* (de Gibert et al., 1999), and *Cochlichnus* also tends to be more angular and not as sharply incised (Morrissey et al., 2004). Another ichnotaxon, *Cochlea archimedia* Hitchcock 1858 was reassigned as *Undichna* isp. by Benner and de Gibert (2009). This presents a potential ichnotaxonic issue because *Cochlea* would have priority over *Undichna* but *Cochlea* is an unavailable name because it is pre-occupied (Benner and de Gibert, 2009).

3.1.2 Other fish swimming trails

Two other ichnotaxa, *Parundichna schoelli* Simon et al., 2003 and *Lunichnium westerbergensis* (Schweigert, 2001), have been interpreted as the swimming traces of fish (Table 4.2). The ichnogenus *Parundichna* is monospecific and was erected by Simon et al. (2003) based on material from the Triassic of Germany that comprises alternately symmetrical and overlapping

TABLE 4.2 The valid ichnospecies of other fish swimming trails and their key distinguishing morphological characteristics.

Ichnotaxon	Key distinguishing morphological characteristics
Parundichna schoelli Simon et al., 2003	Alternately symmetrical and overlapping clusters of sigmoidal imprints. Clusters comprise six to eight parallel sigmoidal imprints.
Lunichnium westerbergensis (Schweigert, 2001)	Alternately symmetrical pairs of imprints that are curved concave outwards and are of different lengths.

The key distinguishing morphological characteristics are summarized and simplified from the original and emended diagnoses (Walter, 1983; Turek, 1989, 1996; Schweigert, 2001; Simon et al., 2003; Minter and Braddy, 2006). Diagnoses are not presented verbatim here because they include producer-related features, rather than strict morphological ichnotaxobases, and lack consistency in their use of terminology.

clusters of sigmoidal imprints. *Lunichnium westerbergensis* is known only from the Jurassic of Germany and consists of paired curvilinear imprints with alternate symmetry (Schweigert, 2001; Minter and Braddy, 2006).

The ichnogenus *Lunichnium* Walter, 1983 was originally interpreted as the swimming trace of an arthropod, and then later as a swimming trackway of an amphibian (Turek, 1989). *Lunichnium westerbergensis* was originally placed within *Undichna* by Schweigert (2001), most likely because it was interpreted as being produced by a fish; however, it was re-assigned by Minter and Braddy (2006) based on its morphological pattern of paired imprints with alternate symmetry being more similar to that of *Lunichnium* than *Undichna*. This demonstrates that morphological features take precedent over interpretations of producers when assigning trace fossils to particular ichnogenera (Bertling et al., 2006, 2022; Minter et al., 2007).

3.2 Fish resting imprints

Three ichnogenera have been interpreted as the resting traces of fish, *Broomichnium* Kuhn, 1958, *Pisquiesichnus* Ayranci et al., 2022, and *Raederichnus* Halpern et al., 2022 (Table 4.3). *Broomichnium* consists of nested pairs of imprints that have different angles between them (Fig. 4.5A) and is known from the Permian of South Africa (Kuhn, 1958; Braddy and Briggs, 2002; van Dijk et al., 2002), Quaternary of the USA (Benner et al., 2008, 2009), and Holocene of Sweden (Uchman and Kumpulainen, 2011). *Broomichnium* was originally interpreted as the

TABLE 4.3 The valid ichnospecies of fish resting imprints and their key distinguishing morphological characteristics.

Ichnotaxon	Key distinguishing morphological characteristics
Broomichnium permianum Kuhn, 1958	Small, bilaterally symmetrical structures composed of a nested inner and outer pair of thin linear or curvilinear imprints. Each imprint within a pair is of equal length. The imprints of the inner pair are parallel or slightly diverging from the midline of the trace, and those of the outer pair diverge from the midline at a greater angle.
Broomichnium flirii Benner et al., 2008	Small, bilaterally symmetrical structures composed of a nested inner and outer pair of thin linear or curvilinear imprints, plus a long straight imprint positioned a short distance along the midline of the trace in the direction that the paired imprints diverge. The paired outer and inner imprints may show striations. Commonly occurs in a linear succession of several individual imprints, but may also occur in isolation.
Pisquiesichnus dashtgardi Ayranci et al., 2022	Wing-shaped to ellipsoidal, shallow structures that contain up to five closely spaced and gently curved internal striae. Surrounded by a mud-covered rim.
Raederichnus Halpern et al., 2022	Shallow, teardrop-shaped imprint. An anterior discoid portion makes up one third of the trace length and has paired kidney-shaped lobes on either side toward the anterior end, and a faintly scalloped anterior margin. A posterior spindle-shaped portion makes up two thirds of the trace length and tapers posteriorly to a blunt rounded termination. Multiple traces may be aligned and partially overlap each other.

The key distinguishing morphological characteristics are summarized and simplified from the original and emended diagnoses (Kuhn, 1958; Braddy and Briggs, 2002; van Dijk et al., 2002; Benner et al., 2008; Ayranci et al., 2022; Halpern et al., 2022). Diagnoses are not presented verbatim here because they include producer-related features, rather than strict morphological ichnotaxobases, and lack consistency in their use of terminology.

FIGURE 4.5 Examples of fish resting, feeding, and nesting imprints. (A) *Broomichnium flirii* from the Pleistocene of the USA, Tufts University Geological Collection TUGC 11,251. (B) *Pisquiesichnus dashtgardi* from the Silurian of Saudi Arabia, Saudi Geological Survey SGS-QAL-0020. (C) *Piscichnus waitemata* from the Pliocene of the Azores. (D) *Osculichnus labialis* from the Eocene of Turkey, Museum of Natural History of the General Directorate of Mineral Research and Exploration in Ankara Tr-19. (E) *Daandavichnus batoideum* from the Cretaceous of Mexico, Paleontological Collection of the Laboratorio de Ciencias de la Tierra y Paleontología, Facultad de Estudios Superiores Iztacala, Universidad Nacional Autónoma de Mexico, STA-BAP-2.4 Scale bars in cm in A-B, 5 cm in C, and 1 cm in D-E. *(A) Image courtesy of Jake Benner. (B) Image courtesy of Korhan Ayranci. (C) Image courtesy of Alfred Uchman. (D) Image courtesy of Alfred Uchman. (E) Image courtesy of María Fernández-Barajas and Rubén Rodríguez-de la Rosa.*

tracks of a tetrapod (Kuhn, 1958; van Dijk et al., 2002), and its junior synonym *Quadrispinichna* Braddy and Briggs, 2002 was interpreted as the resting traces of crustaceans, but subsequent neoichnological investigations have found that similar traces can be made by sculpin fish (Benner et al., 2008). The ichnogenus *Pisquiesichnus* is monospecific, with the type ichnospecies *Pisquiesichnus dashtgardi* Ayranci et al., 2022, and was erected relatively recently based on striated wing-shaped structures from the Silurian of Saudi Arabia (Ayranci et al., 2022, Fig. 4.5B). The ichnogenus *Raederichnus* is monospecific, with the type ichnospecies *Raederichnus dondasi* Halpern et al., 2022, and is from probable Silurian strata of Argentina, and consists of shallow, teardrop-shaped imprints that can be aligned and may overlap (Halpern et al., 2022).

The ichnogenus *Broomichnium* was originally erected based on material from the Permian of South Africa (Kuhn, 1958). Anderson (1974), in their unpublished PhD thesis, used one of the same specimens as their holotype for *Quadrispinichna* and, unbeknownst to them, Braddy and Briggs (2002) formalized the erection of *Quadrispinichna* and assigned additional material from the Permian of New Mexico into the ichnogenus. Van Dijk et al. (2002) identified this issue and highlighted that *Broomichnium* takes priority and that *Quadrispinichna* is its junior synonym. In addition, through analysis of large samples of material and transitional specimens, Minter and Braddy (2009) were able to demonstrate that the material referred to *Quadrispinichna* by Braddy and Briggs (2002) represents incomplete examples, most likely due to preservational variation, of the arthropod trace fossil *Tonganoxichnus*. The type ichnospecies of *Broomichnium, B. permianum* Kuhn, 1958, is therefore known only from the Permian of South Africa (Kuhn, 1958; Braddy and Briggs, 2002; van Dijk et al., 2002; Minter and Braddy, 2009). An additional ichnospecies, *B. flirii* Benner et al., 2008, was subsequently erected based on material from the Pleistocene of the USA (Fig. 4.5A), and its recurrence is demonstrated through a report from the Holocene of Sweden (Uchman and Kumpulainen, 2011).

3.3 Fish feeding and nesting imprints

Three ichnogenera have been interpreted as the feeding and nesting traces of fish: *Piscichnus* Feibel, 1987, *Osculichnus* Demircan and Uchman, 2010, and *Daandavichnus* Rodriguez-de la Rosa et al., 2021 (Table 4.4). *Piscichnus* and *Osculichnus* each comprise three ichnospecies, demonstrating that they represent a recurrent form from different localities. *Piscichnus* is a dm-scale dish-shaped to plug-shaped structure (Feibel, 1987; Gregory, 1991; Niedzwiedzki and Szrek, 2020, Fig. 4.5C) that has mostly been reported from the Cenozoic (e.g., Feibel, 1987; Gregory, 1991; Kotake, 2007; Löwemark, 2015; Nielsen et al., 2020, Fig. 4.1) but occurrences also extend back to the Devonian (Lucas, 2015; Niedzwiedzki and Szrek, 2020) and Silurian (Abdlmutalib et al., 2022). *Osculichnus* is an elliptical bilobate imprint with lobes of differing shapes and sizes that are separated by lines with different morphologies (Demircan and Uchman, 2010; Szrek et al., 2016; Rodríguez-de la Rosa et al., 2021, Fig. 4.5D). *Osculichnus* is a relatively recently erected ichnogenus, with the type ichnospecies *Osculichnus labialis* Demircan and Uchman, 2010, but it is evidently recurrent, and reports range from the Eocene-Oligocene (Demircan and Uchman, 2010) to Cretaceous (Rodriguez-de la Rosa et al., 2021) and Devonian (Szrek et al., 2016; Fan et al., 2019). *Daandavichnus* has been erected only recently and is monospecific, comprising the type ichnospecies *Daandavichnus batoideum*

TABLE 4.4 The valid ichnospecies of fish feeding and nesting imprints and their key distinguishing morphological characteristics.

Ichnotaxon	Key distinguishing morphological characteristics
Piscichnus brownii Feibel, 1987	A steep-sided, cylindrical- or plug-shaped, to shallow circular dish-shaped structure of dm-scale. Oriented concave upward, more or less vertical to bedding.
Piscichnus waitemata Gregory, 1991	A bowl-shaped or subcylindrical structure of dm-scale with a steep, smooth margin that can be associated with a concave-up, dish-shaped structure at the top. Oriented vertical to bedding.
Piscichnus sanctacrucensis Niedzwiedzki and Szrek, 2020	A circular- to oval-shaped, very shallow and dish-shaped structure of dm-scale, sometimes slightly irregular in outline and surrounded by a flat rim. Curved concave up in vertical profile.
Osculichnus labialis Demírcan and Uchman, 2010	Bilobate structure, generally elliptical or crescentic in outline, with smaller and larger lobes separated by an undulate furrow.
Osculichnus tarnowskae Szrek et al., 2016	Bilobate structure, mostly elliptical or crescentic in outline, with smaller and larger lobes separated by an undulate furrow. In some specimens a deeply arcuate line or more often a trapezoidal shape is present instead of the line. The surface inside the structure is undulating with additional wrinkles.
Osculichnus tepitsin Rodríguez-de la Rosa et al., 2021	Bilobate structure. lenticular in outline, with two elongate lobes, separated by an M-shaped elongate furrow. The lower lobe has an elongate, nearly subrectangular shape, with fusiform lateral ends; and the upper lobe has a ventral convex projection in its medial portion. Two short, subtriangular impressions are present.
Daandavichnus batoideum Rodríguez-de la Rosa et al., 2021	Subcircular to trapezoid-shaped structure, with one section preserving a curved impression with an M-shaped external boundary, and another section preserving two circular depressions with smooth borders, occasionally connected by striations forming an arch.

The key distinguishing morphological characteristics are summarized and simplified from the original and emended diagnoses (Feibel, 1987; Gregory 1991; Demírcan and Uchman, 2010; Szrek et al., 2016; Niedzwiedzki and Szrek, 2020; Rodríguez-de la Rosa et al., 2021). Diagnoses are not presented verbatim here because they include producer-related features, rather than strict morphological ichnotaxobases, and lack consistency in their use of terminology.

Rodriguez-de la Rosa et al., 2021. *Daandavichnus* is a subcircular to trapezoid-shaped structure with an undulatory boundary on one side and two circular traces preserved within the other side (Rodriguez-de la Rosa et al., 2021, Fig. 4.5E). *Daandavichnus* is known only from the Cretaceous of Mexico but occurs in association with *Osculichnus* (Rodriguez-de la Rosa et al., 2021). Un-named large and elaborate radial structures from the Devonian of China have been compared with modern display structures produced by male pufferfish when attracting a mate and have been interpreted as courtship traces (Zong and Gong, 2018).

The ichnogenus *Piscichnus* was erected by Feibel (1987) based on material from the Plio-Pleistocene of Kenya, with the type ichnospecies *P. brownii* Feibel, 1987. *Piscichnus* is similar to a number of other trace fossils, particularly the plug-shaped burrows *Bergaueria, Conichnus, Conostichnus*, and *Kulindrichnus* (Pemberton et al., 1988), and its distinction was discussed by Gregory (1991). In addition, Gregory (1991) emended the diagnosis of *Piscichnus* to include plug-shaped structures as well as dish-shaped structures, and erected the new ichnospecies, *P. waitemata* Gregory 1991, for material from the Miocene of New Zealand. Gregory (1991) did not provide a formal diagnosis for *P. waitemata*, but this was later rectified by Nielsen et al. (2020).

Traces from the Devonian of Poland that were originally identified as the oldest known tetrapod trackways (Niedzwiedzki et al., 2010) have been re-assigned to *Piscichnus brownii* and re-interpreted as fish feeding or nesting traces (Lucas, 2015). This re-assignment and re-interpretation was based on most of the proposed trackways lacking alternating symmetry with distinct manus and pes pairs, and the imprints themselves have variable morphologies and depths (Lucas, 2015). In a similar situation, Martinell et al. (2001) re-assigned and re-interpreted what was believed to be a surface trampled by dinosaurs as an array of *Piscichnus* traces. As highlighted by Bertling et al. (2022), the arrangement of subunits should be considered as a valid ichnotaxobase, and, in this case, the lack of a regular alternating trackway pattern is important. Subsequently, Niedzwiedzki and Szrek (2020) have agreed in part with Lucas (2015) that some of the Devonian structures from Poland are fish nesting or feeding traces and identified them as *P. brownii* and erected the new ichnospecies *P. sanctacrucensis*; however, they maintain that some of the structures are of tetrapod origin.

3.4 Other fish trace fossils

Burrows (Chapter 12) and coprolites (Chapter 9) have also been attributed to fish and are briefly mentioned here because they are covered in more detail in other chapters. A number of near vertical, cylindrical to flask-shaped burrows, ranging from the Devonian to Cretaceous, have been interpreted as the burrows of lungfish (Marshall and Rogers, 2012). Of the burrows attributed to lungfish, one ichnogenus has been formally named, *Redondarefugium* Lucas et al., 2010. This material was described originally by Gobetz et al. (2006) and later formally named by Lucas et al. (2010) and consists of vertical to oblique burrows that are slightly helical, with a flask-shaped end, and are unlined with an ornamented wall (Gobetz et al., 2006; Lucas et al., 2010). The majority of burrows attributed to lungfish, however, are often described in open nomenclature, and there has been considerable debate over whether they are the burrows of lungfish or crayfish (Dubiel et al., 1987, 1988; McAllister, 1988; Hasiotis et al., 1993). Important criteria established for identifying lungfish burrows, in the first instance consist of associated remains of lungfish body fossils (McAllister, 1988), or that the burrows should be of similar size and shape to known lungfish (Marshall and Rogers, 2012). Furthermore, lungfish burrows may be identified on the basis of clustering of burrows that originate from a single bedding surface, as suggestive of estivation behavior; and burrow walls are generally smooth, although symmetrical projections may also be present (McAllister, 1988; Hasiotis et al., 1993; Marshall and Rogers, 2012). In contrast, it has been suggested that burrows that may be attributed to crayfish have more complex architectures, including branching, and have highly textured walls (Hasiotis et al., 1993). The ichnogenus *Cornulatichnus* was erected from the Devonian of the UK by Carroll and Trewin (1995) for large, subvertical, conical-shaped burrows with subcircular openings that taper toward their base, and have a thin lining and smooth outer wall and massive infill. Carroll and Trewin (1995) noted the similarities with structures that had been interpreted as the estivation burrows of lungfish but, as there were no known lungfish of sufficient size from the same deposits, they instead interpreted the burrows as temporary shelters of eel-like fish.

The fossil record of the evidence for predation by fish, including direct evidence from trace fossils, has been reviewed (McAllister, 2003); and an increase in the prevalence of bite marks on specimens of jawless fish through the Silurian to Devonian has been found to be correlated with the increase in diversity of jawed fish (Randle and Sansom, 2019). Bite marks attributed to fish have also been reported on echinoid spines (Wilson et al., 2015) and, while not formally assigned, similarities were noted with the ichnogenera *Linichnus* Jacobsen and Bromley, 2009, *Machichnus* Mikulas et al., 2006, and *Nihilichnus* Mikulas et al., 2006. Unnamed resting and predation trace fossils from the Cretaceous of Morocco have been attributed to fish (Lkebir et al., 2021). Other evidence of predation by fish comes from the very large number of named ichnogenera of coprolites that have been attributed to fish, especially chondrichthyans: *Bibliocoprus*, *Costacoprus*, *Crassocoprus*, *Crustacoprus*, *Dakyronocopros*, *Daniacopros*, *Elacocoprus*, *Elongatocoprus*, *Eucoprus*, *Heliocoprus*, *Heteropolacopros*, *Huberobromus*, *Hyronocopros*, *Iuloeidocoprus*, *Kalocoprus*, *Kinneybromus*, *Liassocopros*, *Maculacoprus*, *Malericopros*, *Megaheteropolacopros*, *Megakalocoprus*, *Saurocopros*, *Scrollocoprus*, *Speiracoprus*, *Strabelocoprus*, *Strophocopros*, *Tikicopros*, *Vinculostercus*, *Virgacoprus*; and these have been reviewed by Hunt and Lucas (2021).

4. Discussion

4.1 Stratigraphic and geographic distribution of fish trails and imprints

The earliest fish body fossils date back to the Cambrian; however, articulated remains are rare through the Ordovician to Early Devonian (Friedman and Sallan, 2012), and this is mirrored in the fish trace fossil record. *Undichna* is the most common trace fossil attributed to fish and, of them all, it has the most complete stratigraphic range. *Undichna* has been reported from the Silurian (Knaust and Minter, 2018) to the Holocene (Uchman and Kumpulainen, 2011), with most reports coming from the Carboniferous and Permian (Fig. 4.1). To date, *Undichna* has only not been reported from the Paleocene, Oligocene, and Pliocene. Unsurprisingly, the simplest ichnospecies, *U. unisulca*, has the longest range as it can most likely be made by a wide variety of different fish through time. *U. simplicitas*, *U. bina*, and *U. britannica* are all relatively prevalent through time, and *U. britannica* is particularly common from the Carboniferous. Some of the ichnospecies that most likely had very specific producers, such as *U. trisulcata* and *U. septemsulcata*, are known only from the Devonian. The two other swimming traces attributed to fish have very limited stratigraphic ranges, with *Parundichna schoelli* known only from the Triassic (Simon et al., 2003) and *Lunichnium westerbergensis* from the Jurassic (Schweigert, 2001; Minter and Braddy, 2006). The fish resting trace *Broomichnium* is known only from the Permian (Kuhn, 1958; van Dijk et al., 2002) and Quaternary (Benner et al., 2008, 2009; Uchman and Kumpulainen, 2011); and *Pisquiesichnus* and *Raederichnus* are known only from the Silurian (Ayranci et al., 2022; Halpern et al., 2022). Fish nesting or feeding traces have extensive stratigraphic records, ranging from the Silurian to Cenozoic. *Piscichnus* has been reported several times in the literature, mostly from the Cenozoic but it also extends back to the Devonian (Lucas, 2015; Niedzwiedzki and Szrek, 2020) and Silurian (Abdlmutalib et al., 2022). *Osculichnus* and *Daandavichnus* are less common than *Piscichnus*, although the former has a long stratigraphic range from the Devonian (Szrek et al., 2016; Fan et al., 2019) to the Cretaceous (Rodríguez-de la Rosa et al., 2021) and Eocene-Oligocene (Demírcan and Uchman, 2010); while the latter is known only from the Cretaceous (Rodríguez-de la Rosa et al., 2021). Un-named traces interpreted as courtship structures produced by fish have been described from the Devonian of China (Zong and Gong, 2018). Coprolites attributed to early fish date back to the Late Ordovician (Aldridge et al., 2006; Hunt and Lucas, 2021).

Fish trace fossils are known from all continents except Antarctica (Fig. 4.2), and the vast majority of occurrences have been reported from Europe; whereas very few are from Oceania; and, despite *Undichna* being first named based on material from Africa, the continent itself has relatively few reports of fish trace fossils. The combined total of reports from North and South America and Asia are equivalent to those from Europe. This suggests that outside of Europe, the other continents are undersampled and may well yield new and interesting data on fish traces. This has already been the case with recent reports from Asia, extending records of some ichnotaxa back to the Silurian (Abdlmutalib et al., 2022; Ayranci et al., 2022). *Undichna* is the most common ichnogenus, occurring across six continents, followed next by *Piscichnus* across five continents, and *Osculichnus* and *Broomichnium* across three continents, and then *Parundichna*, *Lunichnium westerbergensis*, *Pisquiesichnus*, *Raederichnus*, and *Daandavichnus* are known only from one area each. *Pisquiesichnus*, *Raederichnus*, and *Daandavichnus* have all been erected within the last couple of years (Rodríguez-de la Rosa et al., 2021; Ayranci et al., 2022; Halpern et al., 2022) and so now that such traces have been formally described and named, they may prove recurrent with additional sampling across the globe.

4.2 Palaeoenvironmental distribution of fish trails and imprints

Given the ubiquity of fish within aquatic environments, it might be predicted that their trace fossils would be equally as common across all environments; however, this is not borne out when analyzing their palaeoenvironmental distribution. Fish trace fossils, and particularly *Undichna*, are most common from lacustrine environments, followed by marginal marine, alluvial, and tidal settings; with few occurrences from marine environments through geological time (Fig. 4.6; see caption for definitions of palaeoenvironmental categories). de Gibert et al. (1999) identified taphonomic controls on the preservation of fish trace fossils such as *Undichna*, with them most likely being preserved in environments that were low energy but with relatively rapid burial, and had fine-grained sediment and limited infaunal bioturbation. The predictions of taphonomic controls by de Gibert et al. (1999) are borne out by the predominance of *Undichna* in lacustrine and low energy nonmarine and marginal-marine environments; however, *Undichna* is now also being discovered from low-energy marine environments such as the offshore zone and deep marine (Fig. 4.6). Of the other trace fossils attributed to fish, *Piscichnus* occurs from lacustrine environments like *Undichna*, but also from higher energy shallow-marine settings, and this is most likely related to their inferred producers being rays that would have lived in such environments (Gregory, 1991; Schindler et al., 2005; Kotake, 2007; Löwemark, 2015; Uchman et al., 2018). *Broomichnium* is known only from glaciolacustrine settings (Kuhn, 1958; van Dijk et al., 2002; Benner et al., 2008, 2009; Uchman and Kumpulainen, 2011).

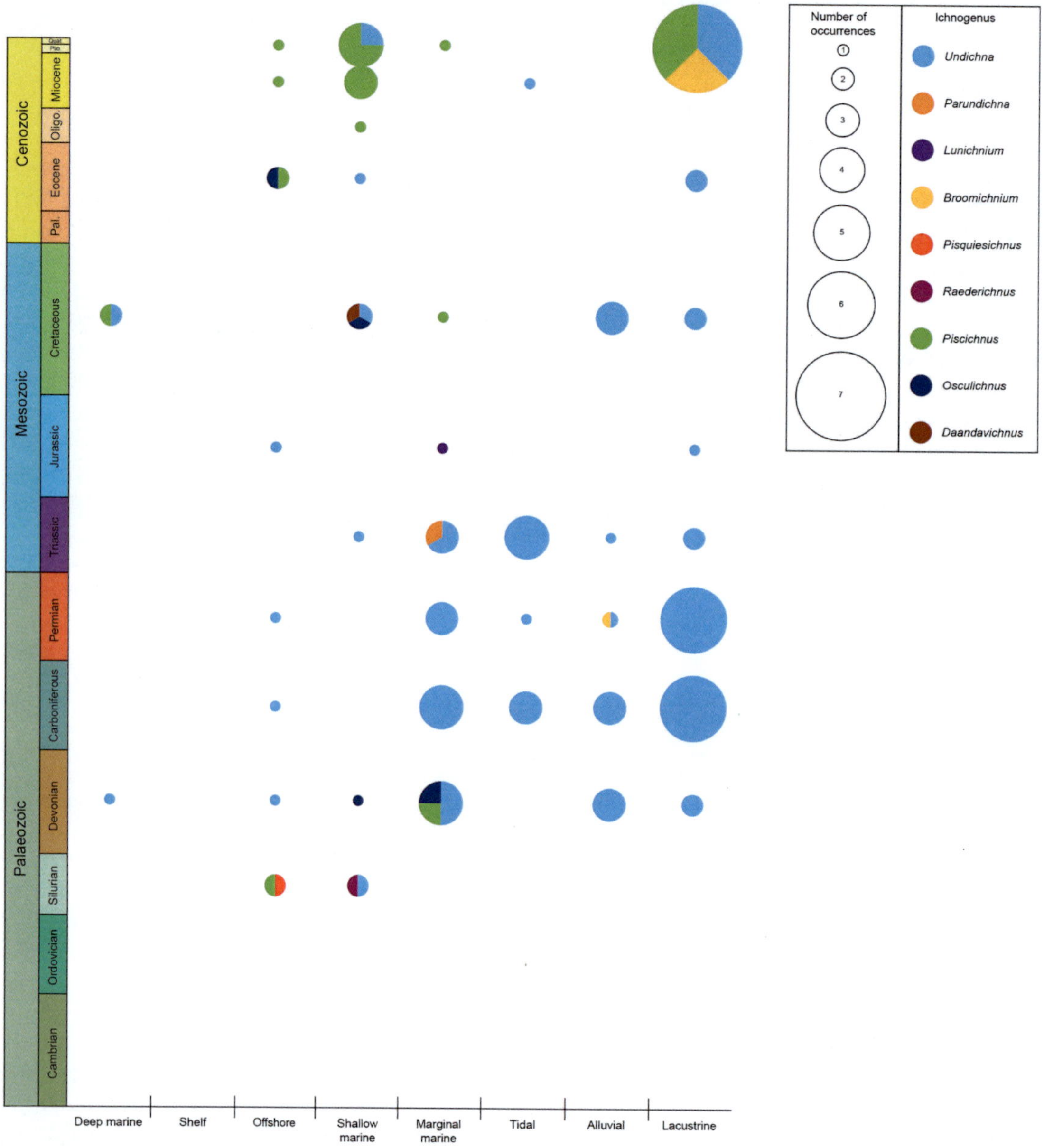

FIGURE 4.6 Temporal and palaeoenvironmental distribution of the different ichnogenera of surface trails and imprints attributed to fish. The diameters of the circles correspond with the number of total occurrences from each period and palaeoenvironment, and the divisions correspond with the proportions of each reported ichnogenus. Data from the Quaternary and Pliocene have been pooled. The palaeoenvironmental categories are modified versions of those employed by Minter et al. (2017) and Buatois et al. (2020) for analyzing macroevolutionary trends of trace fossils: lacustrine (long-lived water bodies and their deep and shallow/margins as well as glaciolacustrine); alluvial (floodplains, ephemeral water bodies, and abandoned and active fluvial channels); tidal flats (supratidal and intertidal above the mean low water level, as well as delta plains that are affected by subaerial exposure); marginal marine (estuaries, bays, and lagoons that are affected by changing salinity conditions); shallow marine (foreshore, shoreface, and subtidal above the fair weather wave base but below the mean low water level, as well as delta fronts); offshore (between fair weather wave base and storm wave base, and includes pro-delta); and deep marine (slope to basin plain).

4.3 Producers of fish trails and imprints

The morphological complexity of fish trace fossils varies and, the more complex a trace, the greater the number of features present that enable producers to be attributed with a higher degree of precision and certainty. In addition, the spatial and temporal co-occurrence with fish body fossils (de Gibert et al., 1999; Costeur and Ezquerra, 2009; Martin et al., 2010), as well as neoichnological observations and experiments (Pearson et al., 2007; Benner et al., 2008; Demírcan and Uchman, 2010; Uchman et al., 2018; Myers et al., 2019) have enabled more precise interpretations of producers.

When considering the producers of fish swimming traces, not only does the anatomy of a fish and their fins come into play, but also their swimming style. In fish with carangiform and subcarangiform locomotion, undulation is confined to the posterior half or third of the body, and the amplitude of undulation increases posteriorly (Breder, 1926; Bainbridge, 1958, 1963), meaning that the amplitude of the impressions will increase from those produced by the pectoral to pelvic and anal and caudal fins. In fish with anguilliform locomotion, the whole body undulates (Breder, 1926; Gray, 1933), so all fins will incise impressions of similar amplitude. Fish swimming traces have been attributed to jawless fish such as cephalaspids, as well as jawed fish such as acanthodians, cartilaginous fish, and bony fish, including members of the Actinopterygii (ray-finned fish) and Sarcopterygii (lobe-finned fish).

One of the most convincing arguments as to the producer of an ichnospecies of *Undichna* is that of *U. trisulcata* being made by a cephalaspid. The attribution of *U. trisulcata* to a cephalaspid is based on the straight parallel imprints being made by the margins of the head shield or pectoral fins, and the single straight or slightly sinusoidal imprint being made by the caudal fin (Morrissey et al., 2004). Ichnotaxa of *Undichna* with multiple paired impressions have generally been attributed to acanthodians. Anderson (1976) was not able to identify the producer of *U. insolentia*, attributing it to an unknown fish; however, Buatois and Mángano (1994) suggested that it was made by an acanthodian, with the multiple paired parallel and scalloped impressions, and unpaired impression being made by the various fin spines of such a fish. *U. septemsulcata* was attributed to an acanthodian on the basis that the single high amplitude unpaired wave was made by the caudal fin, the two pairs of low amplitude outer parallel waves were made by double or bifurcated pectoral fins, and the inner pair of parallel waves that have intermediate amplitude were made by the pelvic fins (Wisshak et al., 2004). In addition, Tindal et al. (2021) interpreted the producer of an example of *U. britannica* from the Devonian Achanarras Limestone Member of Scotland as an acanthodian, such as *Diplacanthus*, *Mesacanthus*, or *Cheiracanthus*, on the basis that their body fossils are well known from this unit; and attributed the intertwined impressions to the pelvic and anal fins, the accessory intertwined imprints to an anal fin membrane, and the outer discontinuous paired imprints to the pectoral fins.

The majority of ichnospecies of *Undichna* have been interpreted to have been produced by actinopterygians. Anderson (1976) attributed *U. simplicitas* to an actinopterygian, with the single high amplitude unpaired wave being produced by the caudal fin, and the outer and inner pairs of parallel waves to the pectoral and pelvic fins, respectively. *Undichna bina* was also attributed to an actinopterygian with subcarangiform locomotion by Anderson (1976), although Trewin (2000) suggested that it can also be made by an eel-like actinopterygian with anguilliform locomotion. Trewin (2000) also suggested that an anguilliform palaeoniscid actinopterygian could have been responsible for *U. quina*, with the single high amplitude unpaired wave being made by the caudal fin, and the two pairs of parallel waves that have similar amplitudes to each other and are intertwined being made by similarly developed pectoral and pelvic fins. *U. quadriculcata*, with its two or three pairs of parallel impressions, is interpreted as being produced by the bony spines of the pectoral and pelvic fins of an amiiform actinopterygian (Knaust, 2019).

Higgs (1988) suggested that *U. britannica* and *U. consulca* were made by palaeoniscid actinopterygians, with the two intertwined waves being made by the caudal and anal fins, and the broad central impression in *U. consulca* being made by the underside of the body; singling out *Acrolepis* and *Cornuboniscus* as the mostly likely genera of fish, respectively, for *U. britannica* and *U. consulca*. *Acrolepis* is contemporaneous with the *Undichna* that Higgs (1988) identified, whereas *Cornuboniscus* occurs in the same formation. de Gibert et al. (1999) also identified a potential producer from the same locality as their example of *U. britannica* and suggested that an elopiform fish was responsible due to it having similarly developed anal and caudal fins that could have incised the two intertwined waves. *Undichna unisulca* is the simplest ichnospecies of *Undichna* and so provides little evidence in itself to correlate it with specific producers other than they must have had one fin down their body axis that extended below all others. de Gibert et al. (1999) suggested that *U. unisulca* was made by the pycnodontiform *Macromesodon*, in which the large anal fin would have been able to leave the single sinusoidal wave. Pycnodontiforms have been identified as the most likely producers of other occurrences of *U. unisulca* (Costeur and Ezquerra, 2009; Herrera-Castillo et al., 2022). Based on contemporaneous body fossils, Lu et al. (2012) suggested that, for their Triassic examples of *U. unisulca*, a fish like the palaeoniscid *Perleidus* was likely a candidate for the tracemaker because it has a large lower lobe to the caudal fin that could have incised the sediment while the other fins did not.

Some fish swimming traces have been attributed to sarcopterygians. *Parundichna* was convincingly attributed to a coelacanth by Simon et al. (2003) on the basis of the swimming patterns of modern individuals and computer simulations, whereby the alternately symmetrical and overlapping clusters of sigmoidal imprints were made by paired pectoral and pelvic fins moving in an alternating, tetrapod-like gait pattern. *Lunichnium westerbergensis* is also interpreted as probably having been produced by a sarcopterygian fish (Schweigert, 2001) that would have been employing an alternating gait pattern (Minter and Braddy, 2006). In addition, Knaust (2019) suggested a coelacanth as a potential alternative producer for *U. quadrisulcata*.

The interpretation of the producer of *Broomichnium* has varied considerably. Originally, *Broomichnium* was interpreted as a tetrapod track by various authors (see van Dijk et al., 2002), although this is unlikely given the sharp nature and symmetry of the imprints; Braddy and Briggs (2002) later interpreted it as a crustacean resting trace. For *B. flirii* at least, neoichnological observations have revealed that similar traces are left behind as the resting impressions of Cottidae fish (Benner et al., 2009). Based on the morphology and spatial arrangement of the wing-shaped impressions of the resting trace *Pisquiesichnus*, Ayranci et al. (2022) suggested that it was most likely made by an ancestor of the Gobiidae. The age and dimensions of *Raederichnus* have been used to suggest that the most likely producer was an early jawless fish as opposed to a xiphosuran or trilobite (Halpern et al., 2022).

Piscichnus is generally interpreted as the feeding trace of elasmobranch rays (Gregory, 1991; Kotake, 2007; Löwemark, 2015), and this has been supported by neoichnological observations (Martinell et al., 2001; Uchman et al., 2018). *Daandavichnus* has also been interpreted as the feeding trace of rays (Rodríguez-de la Rosa et al., 2021). A host of other neoichnological observations, however, have thrown up alternatives for *Piscichnus*. Neoichnological observations by Pearson et al. (2007) identified that sturgeon are capable of making *Piscichnus*-like traces (Pearson et al., 2007), and this interpretation has been followed by some authors (Belvedere et al., 2011), with certain features such as radial imprints supporting bony sturgeons as producers over cartilaginous rays. Further neoichnological observations have demonstrated that other bony fish such as gobies can also create *Piscichnus*-like traces when performing sand-shifting behavior (Myers et al., 2019). As an alternative to production by fish, walruses employing hydraulic jet feeding behavior have been suggested as potential producers of *Piscichnus* (Gingras et al., 2007), and Pleistocene examples from Willapa Bay have been found associated with potential prey as evidenced by bivalve trace fossils. It has also been noted that traces superficially similar to *Piscichnus* have been made by the feeding activities of gulls, but these traces occur as nested mounds and therefore differentiate them from *Piscichnus* and eliminate gulls as potential producers (Melnyk et al., 2022). To explain the behavior demonstrated by *Osculichnus*, Demírcan and Uchman (2010) performed experiments whereby they used the heads of carp and trout, penetrating the sediment with their mouths, to demonstrate that such a trace morphology was a fish-feeding trace. Other authors have suggested that the producers of certain *Osculichnus* were lungfish (Szrek et al., 2016; Fan et al., 2019). Such feeding traces have been found in association with the burrow *Sinusichnus*, which has been attributed to decapod crustaceans, and so may have been the focus of such feeding behavior (Fan et al., 2019). Benthic fish are known to feed on burrowing arthropods such as amphipod crustaceans (McCurdy et al., 2005), although observations of novel mobbing behaviors by fish, firing hydraulic jets, against burrow-dwelling sessile predatory worms (Lachat and Haag-Wackernagel, 2016) suggest an alternative possibility.

5. Summary

Fish surface trails and imprints record the locomotory, resting, feeding, and nesting behavior of fish, and are relatively common in the geological record, ranging in total from the Silurian to Recent, and across all continents except Antarctica. As such, the fish trace fossil record may be used to complement that of their body fossils to help deduce their evolutionary history, provide in situ evidence for their palaeoenvironmental and palaeogeographic distribution, and provide data on the swimming styles of extinct fish and other aspects of their behavioral repertoires.

Acknowledgments

This work benefited from the comments of Spencer Lucas and Dirk Knaust, with valuable suggestions of additional ichnotaxa and references to include within the scope of this review. Spencer Lucas is further thanked for providing access to material at the New Mexico Museum of Natural History and Science. I am indebted to the members of the Alabama Paleontological Society, especially Prescott Atkinson, Ron Buta, and David Kopaska-Merkel for facilitating and hosting me on a research trip to study trace fossils from the Carboniferous of Alabama, and Mark Uhen is thanked for providing access to material at the University of Alabama Museum. Dirk Knaust, Jake Benner, Korhan Ayranci, Alfred Uchman, María Fernández-Barajas, and Rubén Rodríguez-de la Rosa are all thanked for sharing images of fish trace fossils that I could include here.

APPENDIX 1 Occurrences of fish trace fossils.

References	Ichnogenus	Ichnospecies	Period	Environment (lacustrine, alluvial, marginal marine, shallow marine, deep marine)	Location
Uchman and Kumpulainen (2011)	*Undichna*	isp.	Holocene	Lacustrine	Sweden
Fliri et al. (1970), Fliri et al. (1971)	*Undichna*	*britannica* *unisulca*	Pleistocene	Lacustrine	Austria
Benner et al. (2009)	*Undichna*	*simplicitas* *bina* *britannica* *unisulca*	Pleistocene	Lacustrine	USA
Kim and Kang (2018)	*Undichna*	*britannica*	Pleistocene	Shallow marine	South Korea
Hovikoski et al. (2004)	*Undichna*	isp.	Miocene	Tidal	Peru
Loewen and de Gibert (1999)	*Undichna*	*unisulca*	Eocene	Lacustrine	USA
Martin et al. (2010)	*Undichna*	*simplicitas*	Eocene	Lacustrine	USA
Knaust (2019)	*Undichna*	*quadrisulcata*	Eocene	Shallow marine	Norway
de Gibert et al. (1999), (2000)	*Undichna*	*unisulca* *britannica*	Cretaceous	Lacustrine	Spain
Melchor and Cardonatto (1998)	*Undichna*	isp.	Cretaceous	Shallow marine	Argentina
Kim et al. (2022)	*Undichna*	*simplicitas* *bina* *britannica*	Cretaceous	Lacustrine	South Korea
Costeur and Ezquerra (2009)	*Undichna*	*unisulca*	Cretaceous	Alluvial	Spain
Segura et al. (2016)	*Undichna*	*unisulca*	Cretaceous	Alluvial	Spain
Herrera-Castillo et al. (2022)	*Undichna*	*unisulca*	Cretaceous	Alluvial	Spain
Baucon et al. (2023)	*Undichna*	*unisulca*	Cretaceous	Deep marine	Italy
de Gibert (2001)	*Undichna*	*quina*	Jurassic	Offshore	USA
Benner and de Gibert (2009)	*Undichna*	isp.	Jurassic	Lacustrine	USA
Lu and Chen (1998), Lu et al. (2004)	*Undichna*	*bina* *britannica* *simplicitas* *quina*	Triassic	Lacustrine	China

APPENDIX 1 Occurrences of fish trace fossils.—cont'd

References	Ichnogenus	Ichnospecies	Period	Environment (lacustrine, alluvial, marginal marine, shallow marine, deep marine)	Location
Diedrich (2000), (2002)	*Undichna*	isp.	Triassic	Tidal	Germany
Mickelson (2004)	*Undichna*	isp.	Triassic	Tidal	USA
Xing et al. (2023)	*Undichna*	*bina*	Triassic	Lacustrine	China
		britannica			
		unisulca			
		quina			
		isp.			
Ronchi et al. (2018)	*Undichna*	*britannica*	Triassic	Shallow marine	Italy
Lu et al. (2012)	*Undichna*	*unisulca*	Triassic	Marginal marine	China
Sciscio et al. (2020)	*Undichna*	*unisulca*	Triassic	Alluvial	South Africa
		isp.			
Todesco and Avanzini (2008)	*Undichna*	*quina*	Triassic	Tidal	Italy
Knaust (2019)	*Undichna*	*quadrisulcata*	Triassic	Tidal	Netherlands
Feng et al. (2019)	*Undichna*	isp.	Triassic	Marginal marine	China
Haughton (1925), Abel (1935), Plumstead (1970), Anderson (1970), (1976)	*Undichna*	*simplicitas*	Permian	Lacustrine	South Africa
		bina			
		britannica			
		insolentia			
		quadrisulcata			
Smith (1987), (1993)	*Undichna*	*britannica*	Permian	Alluvial	South Africa
Trewin (2000), Trewin et al. (2002)	*Undichna*	*bina*	Permian	Lacustrine	Falkland Islands
		insolentia			
		quadrisulcata			
Trewin (2000), Trewin et al. (2002)	*Undichna*	*quina*	Permian	Offshore	Falkland Islands
Warren (1972)	*Undichna*	*simplicitas*	Permian	Marginal marine	Australia
Walter (1983)	*Undichna*	isp.	Permian	Lacustrine	Germany
Minter and Braddy (2006), (2009)	*Undichna*	*bina*	Permian	Tidal	USA

(Continued)

APPENDIX 1 Occurrences of fish trace fossils.—cont'd

References	Ichnogenus	Ichnospecies	Period	Environment (lacustrine, alluvial, marginal marine, shallow marine, deep marine)	Location
Ronchi et al. (2018)	*Undichna*	*bina*	Permian	Marginal marine	Italy
		quina			
Bordy et al. (2011)	*Undichna*	*britannica*	Permian	Lacustrine	South Africa
Moreau and Gand (2022)	*Undichna*	*britannica*	Permian	Lacustrine	France
Pineiro (2006)	*Undichna*	*quadrisulcata*	Permian	Marginal marine	Uruguay
Moreau et al. (2022)	*Undichna*	*quadrisulcata*	Permian	Lacustrine	France
		isp.			
Buatois et al. (1997), (1998)	*Undichna*	*simplicitas*	Carboniferous	Tidal	USA
		britannica			
		unisulca			
Soler-Gijón and Moratalla (2001)	*Undichna*	*britannica*	Carboniferous	Marginal marine	Spain
		unisulca			
Melchor et al. (1993)	*Undichna*	*insolentia*	Carboniferous	Lacustrine	Argentina
Netto et al. (2003)	*Undichna*	*consulca*	Carboniferous	Lacustrine	Brazil
Higgs (1988)	*Undichna*	*britannica*	Carboniferous	Lacustrine	UK
		consulca			
		bina			
		unisulca			
Walter (1982), Turek (1989), (1996)	*Undichna*	*britannica*	Carboniferous	Lacustrine	Czech Republic
Rindsberg (1990)	*Undichna*	isp.	Carboniferous	Alluvial	USA
Lucas et al. (2004)	*Undichna*	*britannica*	Carboniferous	Tidal	USA
Buatois and Mángano (1993), (1994)	*Undichna*	*britannica*	Carboniferous	Lacustrine	Argentina
		quadrisulcata			
Buatois and Mángano (2003)	*Undichna*	*britannica*	Carboniferous	Marginal marine	Argentina
		isp.			
Keighley and Pickerill (1997)	*Undichna*	*bina*	Carboniferous	Alluvial	Canada
		consulca			
		isp.			

APPENDIX 1 Occurrences of fish trace fossils.—cont'd

References	Ichnogenus	Ichnospecies	Period	Environment (lacustrine, alluvial, marginal marine, shallow marine, deep marine)	Location
Pazos (2002)	*Undichna*	isp.	Carboniferous	Lacustrine	Argentina
Hunt and Lucas (2003), Martin (2003), Hunt et al. (2004), Martin and Rindsberg (2004)	*Undichna*	isp.	Carboniferous	Tidal	USA
Archer and Maples (1984), Archer (1993)	*Undichna*	*britannica*	Carboniferous	Marginal marine	USA
Cameron et al. (1998)	*Undichna*	*bina*	Carboniferous	Marginal marine	Canada
		britannica			
		unisulca			
		insolentia			
		isp.			
Muszer (2020)	*Undichna*	*britannica*	Carboniferous	Offshore	Poland
Fillmore et al. (2011)	*Undichna*	*britannica*	Carboniferous	Alluvial	USA
		quina			
		unisulca			
Sutcliffe et al. (1999)	*Undichna*	isp.	Devonian	Slope to basin plain	Germany
Trewin and Davidson (1996)	*Undichna*	*unisulca*	Devonian	Lacustrine	UK
Morrissey et al. (2004)	*Undichna*	*trisulcata*	Devonian	Alluvial	UK
		unisulca			
		simplicitas			
Wisshak et al. (2004)	*Undichna*	*septemsulcata*	Devonian	Alluvial	Norway
Niedzwiedzki and Szrek (2020)	*Undichna*	isp.	Devonian	Marginal marine	Poland
Klug et al. (2021)	*Undichna*	*britannica*	Devonian	Offshore	Morocco
Trewin et al. (2012)	*Undichna*	*unisulca*	Devonian	Alluvial	Scotland
Tindal et al. (2021)	*Undichna*	*britannica*	Devonian	Lacustrine	Scotland
		isp.			
Mikulas et al. (2013)	*Undichna*	isp.	Devonian	Marginal marine	Russia
Knaust and Minter (2018)	*Undichna*	*unisulca*	Silurian	Shallow marine	Sweden

(*Continued*)

APPENDIX 1 Occurrences of fish trace fossils.—cont'd

References	Ichnogenus	Ichnospecies	Period	Environment (lacustrine, alluvial, marginal marine, shallow marine, deep marine)	Location
Simon et al. (2003)	*Parundichna*	*schoelli*	Triassic	Marginal marine	Germany
Schweigert (2001)	*Lunichnium*	*westerbergensis*	Jurassic	Marginal marine	Germany
Uchman and Kumpulainen (2011)	*Broomichnium*	*flirii*	Quaternary	Lacustrine	Sweden
Benner et al. (2008), (2009)	*Broomichnium*	*flirii*	Pleistocene	Lacustrine	USA
Kuhn (1958), van Dijk et al. (2002)	*Broomichnium*	*permianum*	Permian	Alluvial	South Africa
Ayranci et al. (2022)	*Pisquiesichnus*	*dashtgardi*	Silurian	Offshore	Saudi Arabia
Halpern et al. (2022)	*Raederichnus*	*dondasi*	Silurian	Shallow marine	Argentina
Ekdale and Lewis (1991)	*Piscichnus*	isp.	Quaternary	Offshore	New Zealand
Gingras et al. (2007)	*Piscichnus*	isp.	Pleistocene	Marginal marine	USA
Benner et al. (2009)	*Piscichnus*	isp.	Pleistocene	Lacustrine	USA
Nielsen et al. (2020)	*Piscichnus*	*waitemata*	Pleistocene	Shallow marine	Greece
Feibel (1987)	*Piscichnus*	*brownii*	Plio-Pleistocene	Lacustrine	Kenya
Massari and d'Alessandro (2010)	*Piscichnus*	isp.	Pliocene	Shallow marine	Italy
Uchman et al. (2018)	*Piscichnus*	*waitemata*	Pliocene	Shallow marine	Portugal
Gregory (1991)	*Piscichnus*	*waitemata*	Miocene	Shallow marine	New Zealand
Lowemark (2015)	*Piscichnus*	*waitemata*	Miocene	Offshore	Taiwan
Kotake (2007)	*Piscichnus*	*waitemata*	Miocene	Shallow marine	Japan
Pervesler et al. (2011)	*Piscichnus*	isp.	Miocene	Shallow marine	Austria
Schindler et al. (2005)	*Piscichnus*	isp.	Oligocene	Shallow marine	Germany
Belvedere et al. (2011)	*Piscichnus*	isp.	Eocene	Offshore	Italy
Martinell et al. (2001)	*Piscichnus*	isp.	Cretaceous	Marginal marine	Spain
Baucon et al. (2023)	*Piscichnus*	isp.	Cretaceous	Deep marine	Italy
Lucas (2015), Niedzwiedzki and Szrek (2020)	*Piscichnus*	*sanctacrucensis* *brownii*	Devonian	Marginal marine	Poland

APPENDIX 1 Occurrences of fish trace fossils.—cont'd

References	Ichnogenus	Ichnospecies	Period	Environment (lacustrine, alluvial, marginal marine, shallow marine, deep marine)	Location
Abdlmutalib et al. (2022)	*Piscichnus*	isp.	Silurian	Offshore	Saudi Arabia
Demircan and Uchman (2010)	*Osculichnus*	*labialis*	Eocene	Offshore	Turkey
Rodriguez-de la Rosa et al. (2021)	*Osculichnus*	*tepitsin*	Cretaceous	Shallow marine	Mexico
Szrek et al. (2016)	*Osculichnus*	*tarnowskae*	Devonian	Shallow marine	Poland
Fan et al. (2019)	*Osculichnus*	isp.	Devonian	Marginal marine	China
Rodriguez-de la Rosa et al. (2021)	*Daandavichnus*	*batoideum*	Cretaceous	Shallow marine	Mexico

References

Abdlmutalib, A.J., Ayranci, K., Yassin, M.A., Hussaini, S.R., Abdullatif, O.A., Humphrey, J.D., 2022. Impact of sedimentary fabrics on small-scale permeability variations within fine-grained sediments: Early Silurian Qusaiba Member, northern Saudi Arabia. Mar. Petrol. Geol. 139, 105607.

Abel, O., 1935. Vorzeitliche Lebensspuren. Gustav Fischer, Jena.

Aldridge, R.J., Gabbott, S.E., Siveter, L.J., Theron, J.N., 2006. Bromalites from the Soom Shale Lagerstätte (Upper Ordovician) of South Africa: palaeoecological and palaeobiological implications. Palaeontology 49, 857–871.

Anderson, A.M., 1970. An analysis of supposed fish trails from interglacial sediments in the Dwyka Series, near Vryheid, Natal. In: Proceedings and Papers, Second IUGS Gondwana Symposium, pp. 637–647. South Africa.

Anderson, A.M., 1974. Arthropod Trackways and Other Trace Fossils from the Early Permian Lower Karoo Beds of South Africa. Unpublished PhD dissertation. University of Witwatersrand.

Anderson, A.M., 1976. Fish trails from the early permian of South Africa. Palaeontology 19, 397–409.

Archer, A.W., 1993. Reappraisal of Pennsylvanian trace fossil assemblages in the Eastern Interior Coal Basin, USA. 5.1–5.14. In: Archer, A.W., Feldman, H.R., Lanier, W.P. (Eds.), Incised Paleovalley of the Douglas Group in Northeastern Kansas. Field Guide and Related Contributions. Kansas Geological Survey, Open-File Report, 93–24.

Archer, A.W., Maples, C.G., 1984. Trace-fossil distribution across a marine to nonmarine gradient in the Pennsylvanian of southwestern Indiana. J. Paleontol. 58, 448–466.

Ayranci, K., Wetzel, A., Kaminski, M.A., Kurtulus, B., Rabaoui, L., 2022. Possibly the oldest fish-made resting traces. Ichnos 29, 61–70.

Bainbridge, R., 1958. The speed of swimming fish as related to size and to frequency and amplitude of the tail beat. J. Exp. Biol. 35, 109–144.

Bainbridge, R., 1963. Caudal fin and body movement in the propulsion of some fish. J. Exp. Biol. 40, 23–56.

Baucon, A., Ferretti, A., Fioroni, C., Pandolfi, L., Serpagli, E., Piccinini, A., Neto de Carvalho, C., Cachão, M., Linley, T., Muñiz, F., Belaústegui, Z., Jamieson, A., Lo Russo, G., Guerrini, F., Ferrando, S., Priede, I., 2023. The earliest evidence of deep sea vertebrates. Proc. Nat. Acad. Sci. USA 120, e2306164120.

Belvedere, M., Francheschi, M., Morsilli, M., Zoccarato, P.L., Mietto, P., 2011. Fish feeding traces from Middle Eocene limestones (Gargano Promontory, Apulia, southern Italy). Palaios 26, 693–699.

Benner, J.S., de Gibert, J.M., 2009. *Cochlea archimedia*: Hitchcock's fish trail. Ichnos 16, 274–280.

Benner, J.S., Ridge, J.C., Taft, N.K., 2008. Late Pleistocene freshwater fish (Cottidae) trackways from New England (USA) glacial lakes and a reinterpretation of the ichnogenus *Broomichnium* Kuhn. Palaeogeogr. Palaeoclimatol. Palaeoecol. 260, 375–388.

Benner, J.S., Ridge, J.C., Knecht, R.J., 2009. Timing of post-glacial reinhabitation and ecological development of two New England, USA, drainages based on trace fossil evidence. Palaeogeogr. Palaeoclimatol. Palaeoecol. 272, 212–231.

Bertling, M., Braddy, S.J., Bromley, R.G., Demathieu, G.R., Genise, J., Mikuláš, R., Nielsen, J.K., Nielsen, K.S.S., Rindsberg, A.K., Schlirf, M., Uchman, A., 2006. Names for trace fossils: a uniform approach. Lethaia 39, 265–286.

Bertling, M., Buatois, L.A., Knaust, D., Laing, B., Mángano, M.G., Meyer, N., Mikuláš, R., Minter, N.J., Neumann, C., Rindsberg, A.K., Uchman, A., Wisshak, M., 2022. Names for trace fossils 2.0: theory and practice in ichnotaxonomy. Lethaia 55, 1–19.

Bordy, E.M., Linkermann, S., Prevec, R., 2011. Palaeoecological aspects of some invertebrate trace fossils from the mid- to Upper Permian Middleton formation (Adelaide Subgroup, Beaufort Group, Karoo Supergroup), Eastern Cape, South Africa. J. Afr. Earth Sci. 61, 238–244.

Braddy, S.J., Briggs, D.E.G., 2002. New Lower Permian nonmarine arthropod trace fossils from New Mexico and South Africa. J. Paleontol. 76, 546–557.

Breder, C.M., 1926. The locomotion of fishes. Zoologica 4, 159–297.

Buatois, L.A., Mángano, M.G., 1993. Trace fossils from a Carboniferous turbiditic lake: implications for the recognition of additional nonmarine ichnofacies. Ichnos 2, 237–258.

Buatois, L.A., Mángano, M.G., 1994. Pistas de peces en el Carbonífero de la cuenca Paganzo (Argentina): su significado estratigráfico y paleoambiental. Ameghiniana 31, 33–40.

Buatois, L.A., Mángano, M.G., 2003. Caracterización icnológica y paleoambiental de la localidad tipo de *Orchesteropus atavus* Frenguelli, Huerta de Huachi, provincia de San Juan, Argentina. Ameghiniana 40, 53–70.

Buatois, L.A., Mángano, M.G., Maples, C.G., Lanier, W.P., 1997. The paradox of nonmarine ichnofaunas in tidal rhythmites: integrating sedimentologic and ichnologic data from the Late Carboniferous of eastern Kansas, USA. Palaios 12, 467–481.

Buatois, L.A., Mángano, M.G., Maples, C.G., Lanier, W.P., 1998. Ichnology of an Upper Carboniferous fluvio-estuarine paleovalley: the Tonganoxie Sandstone, Buildex Quarry, eastern Kansas, USA. J. Paleontol. 72, 152–180.

Buatois, L.A., Mángano, M.G., Minter, N.J., Zhou, K., Wisshak, M., Wilson, M.A., Olea, R., 2020. Quantifying ecospace utilization and ecosystem engineering during the early Phanerozoic – the role of bioturbation and bioerosion. Sci. Adv. 6, eabb0618.

Cameron, B., Wood, D., Van Dommelen, R., 1998. Vertebrate and invertebrate trace fossils from the Horton Bluff formation (lower Carboniferous) of Nova Scotia. Min. Matt. Nova Scotia 98, 6.

Cardonatto, M.C., Melchor, R.N., 2014. Biomechanical analysis of fish swimming trace fossils (*Undichna*): preservation and mode of locomotion. In: Hembree, D.I., Platt, B.F., Smith, J.J. (Eds.), Experimental Approches to Understanding Fossils, Topics in Geobiology, vol. 41. Springer, Dordrecht, pp. 265–303.

Carroll, S., Trewin, H.H., 1995. *Cornulatichnus*: a new trace fossil from the old red Sandstone of Orkney. Scot. J. Geol. 31, 37–41.

Costeur, L., Ezquerra, R., 2009. Early Cretaceous fish trails from La Rioja, Spain. Cretac. Res. 30, 1027–1030.

Demírcan, H., Uchman, A., 2010. Kiss of death of a hunting fish: trace fossil *Osculichnus labialis* igen. et isp. nov. from late Eocene – early Oligocene prodelta sediments of the Mezardere Formation, Thrace Basin, NW Turkey. Acta Geol. Pol. 60, 29–38.

Diedrich, C., 2000. New vertebrate tracks from the Lower Muschelkalk (Middle Triassic) of the Osnabrueck Hills and Teutoburg forest (NW-Germany) and their stratigraphic and palaeogeographic significance in the Germanic basin. Neues Jahrbuch Geol. Palaontol. Abhand. 217, 369–395.

Diedrich, C., 2002. Vertebrate track bed stratigraphy at new megatrack sites in the Upper Wellenkalk Member and orbicularis Member (Muschelkalk, Middle Triassic) in carbonate tidal flat environments of the western Germanic Basin. Palaeogeogr. Palaeoclimatol. Palaeoecol. 183, 185–208.

Van Dijk, D.E., Channing, A., Van der Heever, J.A., 2002. Permian trace fossils attributed to tetrapods (Tierberg formation, Karoo basin, South Africa). Palaeontologica Africana 38, 49–56.

Dubiel, R.F., Blodgett, R.H., Bown, T.M., 1987. Lungfish burrows in the upper Triassic Chinle and Dolores formations – Reply. J. Sed. Petrol. 57, 512–521.

Dubiel, R.F., Blodgett, R.H., Bown, T.M., 1988. Lungfish burrows in the Upper Triassic Chinle and Dolores formations – reply. J. Sed. Petrol. 58, 367–369.

Ekdale, A.A., Lewis, D.W., 1991. Trace fossils and paleoenvironmental control of ichnofacies in a Late Quaternary gravel and loess fan delta complex, New Zealand. Palaeogeogr. Palaeoclimatol. Palaeoecol. 81, 253–279.

Falkingham, P.L., Bates, K.T., Belvedere, M., Castanera, D., Citton, P., Díaz-Martinez, I., Farlow, J., Getty, P., Hatala, K.G., Klein, H., Lallensack, J.N., Marty, D., Meyer, C.A., Milàn, J., Minter, N.J., Razzolini, N.L., Romilio, A., Sciscio, L., Tanaka, I., Wiseman, A.L.A., Xing, L., 2018. A standard protocol for documenting modern and fossil ichnological data. Palaeontology 61, 469–480.

Fan, R.-Y., Zong, R.-W., Gong, Y.-M., 2019. Fish hunting trace *Osculichnus* and the oldest *Sinusichnus sinuosus* from the Upper Devonian of South China. Palaeogeogr. Palaeoclimatol. Palaeoecol. 530, 103–112.

Feibel, C.S., 1987. Fossil fish nests from the Koobi Fara Formation (Plio-Pleistocene) of northern Kenya. J. Paleontol. 61, 130–134.

Feng, X., Chen, Z.-Q., Benton, M.J., Wu, S., Bottjer, D.J., Thompson, J.R., 2019. A diverse trackway-dominated marine ichnoassemblage from the Lower Triassic in the northern Paleotethys: ichnology and implications for biotic recovery. Palaeogeogr. Palaeoclimatol. Palaeoecol. 519, 124–140.

Fillmore, D.L., Lucas, S.G., Simpson, E.L., 2011. The fish swimming trace *Undichna* from the Mississippian Mauch Chunk formation, eastern Pennsylvania. Ichnos 18, 27–34.

Fliri, F., Bortenschlager, S., Felber, H., Heissel, W., Hilscher, H., Resch, W., 1970. Der Bänderton von Baumkirchen (Inntal, Tirol): eine neue Schlüsselstelle zür Kenntnis der Würm-Vereisung der Alpen. Z. Gletscherkd. Glazialgeol. 6, 5–35.

Fliri, F., Hilscher, H., Markgraf, V., 1971. Weitere Untersuchungen zur Chronologie der Alpinen Vereisung (Bänderton von Baumkirchen, Inntal, Nordtirol). Z. Gletscherkd. Glazialgeol. 7, 5–24.

Friedman, M., Sallan, L.C., 2012. Five hundred million years of extinction and recovery: a Phanerozoic survey of large-scale diversity patterns in fishes. Palaeontology 55, 707–742.

De Gibert, J.M., 2001. *Undichna gosiutensis*, isp. nov.: a new fish trace fossil from the Jurassic of Utah. Ichnos 8, 15–22.

De Gibert, J.M., Buatois, L.A., Fregenal-Martínez, M.A., Mángano, M.G., Ortega, F., Poyato-Ariza, F.J., Wenz, S., 1999. The fish trace fossil *Undichna* from the Cretaceous of Spain. Palaeontology 42, 409–427.

de Gibert, J.M., Fregenal-Martinez, M.A., Buatois, L.A., Mángano, M.G., 2000. Trace fossils and their palaeoecological significance in Lower Cretaceous lacustrine conservation deposits, El Montsec, Spain. Palaeogeogr. Palaeoclimatol. Palaeoecol. 156, 89–101.

Gingras, M.K., Armitage, I.A., Pemberton, S.G., Clifton, H.E., 2007. Pleistocene walrus herds in the Olympic Peninsula area: trace fossil evidence of predation by hydraulic jetting. Palaios 22, 539–545.

Gobetz, K.E., Lucas, S.G., Lerner, A.J., 2006. Lungfish burrows of varying morphology from the Upper Triassic Redonda formation, Chinle Group, eastern New Mexico. New Mex. Mus. Nat. Hist. Sci. Bull. 37, 140–146.

Gray, J., 1933. Studies in animal locomotion I. The movement of fish with special reference to the eel. J. Exp. Biol. 10, 88–104.

Gregory, M.R., 1991. New trace fossils from the Miocene of Northland, New Zealand: *Rorschachichnus amoeba* and *Piscichnus waitemata*. Ichnos 1, 195–205.

Halpern, K., Gouiric-Cavalli, S., Taglioretti, M.L., Farenga, M., Scaglia, F., Calvo Marcilese, L., Del Río, J.L., 2022. *Raederichnus dondasi* a new trace fossil from the Early Paleozoic of Argentina reveals shoaling behavior in early fish. Palaios 37, 418–432.

Hasiotis, S.T., Mitchell, C.E., Dubiel, R.F., 1993. Application of morphologic burrow interpretations to discern continental burrow architects; lungfish of crayfish? Ichnos 2, 315–333.

Haughton, S.H., 1925. Exhibit: tracks of animals preserved in the Ecca Shales of the Cape Province. Trans. Roy. Soc. S. Afr. 13, 28–29.

Herrera-Castillo, C.M., Moratalla, J.J., Belaústegui, Z., Marugán-Lobón, J., Martín-Abad, H., Nebreda, S.M., López-Archilla, A.I., Buscalioni, A.D., 2022. A theropod trackway providing evidence of a pathological foot from the exceptional locality of Las Hoyas (upper Barremian, Serranía de Cuenca, Spain). PLoS One 17, e0264406.

Higgs, R., 1988. Fish trails in the upper Carboniferous of South-west England. Palaeontology 31, 255–272.

Hitchcock, E., 1858. Ichnology of new England. A report on the Sandstone of the Connecticut valley, especially its Footmarks. W. White, Boston, p. 220.

Hovikoski, J., Roddaz, M., Brusset, S., Antoine, P.-O., Hermoza, W., Baby, P., Pittman, I.R., Räsänen, M., 2004. Miocene brackish and freshwater tidal deposition in Western Amazonia, pp. 40–41. Abstract book – First International Congress on Ichnology, Trelew, Argentina.

Hunt, A.P., Lucas, S.G., 2003. Large tetrapods in the Early Pennsylvanian – ichnological evidence from Alabama. Geol. Soc. Am. Abstr. Progr. 35, 499.

Hunt, A.P., Lucas, S.G., 2021. The ichnology of vertebrate consumption: dentalites, gastroliths, and bromalites. New Mex. Mus. Nat. Hist. Sci. Bull. 87, 1–216.

Hunt, A.P., Lucas, S.G., Lockley, M.G., 2004. Large pelycosaur footprints from the Lower Pennsylvanian of Alabama, USA. Ichnos 11, 39–44.

Jacobsen, A.R., Bromley, R.G., 2009. New ichnotaxa based on tooth impressions on dinosaur and whale bones. Geol. Q. 53, 373–382.

Keighley, D.G., Pickerill, R.N., 1997. Systematic ichnology of the Mabou and Cumberland groups (Carboniferous) of western Cape Breton Island, eastern Canada, 1: burrows, pits, trails, and coprolites. Atl. Geol. 33, 181–215.

Kim, J.Y., Kang, J.H., 2018. Trace fossils from the Late Pleistocene marginal marine deposits of Jeju Island, Korea: implications for the *Psilonichnus* and *Skolithos* ichnofacies. J. Korean Earth Sci. Soc. 39, 23–45.

Kim, K.S., Lockley, M.G., Romilio, A., Bae, S.M., Lim, J.D., 2022. Fish swim traces from the Jindong Formation (Cretaceous) Korea: implications for lake basin ichnofacies and paleoecology. Cretac. Res. 131, 105070.

Klug, C., Lagnaoui, A., Jobbins, M., Bel Haouz, W., Najih, A., 2021. The swimming trace *Undichna* from the latest Devonian Hangenberg Sandstone equivalent of Morocco. Swiss J. Palaeontol. 140, 19.

Knaust, D., 2019. The fish trail *Undichna quadrisulcata* isp. nov. from the Eocene of Spitsbergen. Ichnos 26, 178–186.

Knaust, D., Minter, N.J., 2018. The fish swimming trace *Undichna unisulca* from the Silurian of Sweden: probably the oldest vertebrate locomotion trace. Lethaia 51, 469–472.

Kotake, N., 2007. *Macaronichnus* isp. associated with *Piscichnus waitemata* in the Miocene of Yonaguni-jima Island, southwest Japan. In: Miller, I.I.I.,W. (Ed.), Trace Fossils: Concepts, Problems, Perspectives. Elsevier, Amsterdam, pp. 492–501.

Kuhn, O., 1958. Die Fährten der Vorzeitlichen Amphibien und Reptilien (Bamberg, Meisenbach).

Lachat, J., Haag-Wackernagel, D., 2016. Novel mobbing strategies of a fish population against a sessile annelid predator. Sci. Rep. 6, 33187.

Lkebir, N., Masrour, M., Torices, A., Pérez-Lorente, F., 2021. *Cubichnia* and *Praedichnia* in fossil ichnites of a flat fish from the Lower Cretaceous Talmest palaeoichnological site, western high atlas (Morocco). Ichnos 28, 60–71.

Loewen, M.A., de Gibert, J.M., 1999. The first occurrence of Cenozoic fish trails (*Undichna*) from Eocene fossil lake, Wyoming. J. Vertebr. Paleontol. 19, 59A.

Löwemark, L., 2015. Evidence for targeted elasmobranch predation on thalassinidean shrimp in the Miocene Taliao Formation, NE Taiwan. Lethaia 48, 227–234.

Lu, Z., Chen, B., 1998. Discovery of Late Triassic fish trails (*Undichna*) in Hengshan County, Shaanxi, China. Acta Palaeontol. Sin. 37, 76–84.

Lu, Z., Hou, J., Chen, B., Hao, C., Li, X., 2004. Genetic interpretation of fish swimming trails and calculation of fish-body length in Late Triassic Hengshan, Shaanxi Province, China. Sci. China Earth Sci. 47, 272–279.

Lu, T.Q., Wang, Z.L., Yang, X.Y., Zhang, X.L., 2012. First record of Lower Triassic *Undichna* spp. fish swimming traces from Emei, Sichuan Province, China. Chin. Sci. Bull. 57, 1320–1324.

Lucas, S.G., 2015. *Thinopus* and a critical review of Devonian tetrapod footprints. Ichnos 22, 136–154.

Lucas, S.G., Lerner, A.J., Bruner, M., Shipman, P., 2004. Middle Pennsylvanian ichnofauna from eastern Oklahoma, USA. Ichnos 11, 45–55.

Lucas, S.G., Spielmann, J.A., Klein, H., Lerner, A.J., 2010. Ichnology of the Redonda formation. New Mex. Mus. Nat. Hist. Sci. Bull. 47, 1–75.

Marshall, M.S., Rogers, R.R., 2012. Lungfish burrows from the Upper Cretaceous Maevarano formation, Mahajanga Basin, northwestern Madagasca. Palaios 27, 857–866.

Martin, A.J., 2003. *Undichna*: A Trace Fossil Bridging the Gap between Fish and Tetrapods. University of Alabama, pp. 19–21. Program and abstracts: Workshop on Permo-Carboniferous Ichnology.

Martin, A.J., Rindsberg, A.K., 2004. Ichnological Insights from the Union Chapel Mine, Alabama (USA) on Carboniferous Vertebrate Behaviour, pp. 52–53. Abstract book – First International Congress on Ichnology, Trelew, Argentina.

Martin, A.J., Vazquez-Prokopec, G.M., Page, M., 2010. First-known feeding trace of the Eocene bottom-dwelling fish *Notogoneus osculus* and its paleontological significance. PLoS One 5, e10420.

Martinell, J., de Gibert, J.M., Domènech, R., Ekdale, A.A., Steen, P.P., 2001. Cretaceous ray traces? An alternative interpretation for the alleged dinosaur tracks of La Posa, Isona, NE Spain. Palaios 16, 409–416.

Massari, F., d'Alessandro, A., 2010. Icehouse, cool-water carbonate ramps: the case of the Upper Pliocene Capodarso Fm (Sicily): role of trace fossils in the reconstruction of growth stages of prograding wedges. Facies 56, 47–58.

McAllister, J.A., 1988. Lungfish burrows in the Upper Triassic Chinle and Dolores formations, Colorado Plateau – comments on the recognition criteria of fossil lungfish burrows. J. Sediment. Petrol. 58, 365–367.

McAllister, J., 2003. Predation of fishes in the fossil record. In: Kelley, P.H., Kowalewski, M., Hansen, T.A. (Eds.), Predator-Prey Interactions in the Fossil Record. Plenum, New York, pp. 303–324.

McCurdy, D.G., Forbes, M.R., Logan, S.P., Lancaster, D., Mautner, A.I., 2005. Foraging and impacts by benthic fish on the intertidal amphipod *Corophium volutator*. J. Crustac Biol. 25, 558–564.

Melchor, R.N., Cardonatto, M.C., 1998. Pistas de peces (*Undichna*): rango estratigráfico, paleoambientes y mecanismos de propulsión del productor. Tercera Reunión Argentina de Icnología. In: Primera Reunión de Icnología del Mercosur, pp. 20–21.

Melchor, R.N., Bazán, J., Fernández, M.A., 1993. Asociación de trazas fósiles de la facies pelíticas de la Formación Agua Escondida (Carbonífero Superior?), sureste de Mendoza, Argentina. In: Primera Reunión Argentina de Icnología, Resúmenes y Conferencias Invitadas, p. 15.

Melnyk, S., Lazowski, C.N., Gingras, M.K., 2022. Sedimentological and ecological significance of a biodeformational structure associated with an unusual feeding behavior in gulls (*Larus* sp.). Ichnos 29, 84–92.

Mickelson, D.L., 2004. The diversity and stratigraphic distribution of pre-dinosaurian communities from the Triassic Moenkopi formation, Capitol Reef National Park and Glen Canyon National Recreation area, Utah. Park Paleontol. 8, 1–4.

Mikulas, R., Kedlecova, E., Fejfar, O., Dvorak, Z., 2006. Three new ichnogenera of biting and gnawing traces on reptilian and mammalian bones: a case study from the Miocene of the Czech Republic. Ichnos 13, 1–15.

Mikulas, R., Meskis, S., Ivanov, A., Luksevics, E., Zupins, I., Stinkulis, G., 2013. A rich ichnofossil assemblage from the Frasnian (Upper Devonian) deposits at Andoma Hill, Onega Lake, Russia. Bull. Geosci. 88, 389–400.

Minter, N.J., Braddy, S.J., 2006. The fish and amphibian swimming traces *Undichna* and *Lunichnium*, with examples from the Lower Permian of New Mexico, USA. Palaeontology 49, 1123–1142.

Minter, N.J., Braddy, S.J., 2009. Ichnology of an early permian intertidal flat: the Robledo Mountains formation of southern New Mexico, USA. Spec. Pap. Palaeontol. 82, 5–107.

Minter, N.J., Braddy, S.J., Davies, R.B., 2007. Between a rock and a hard place: arthropod trackways and ichnotaxonomy. Lethaia 40, 365–375.

Minter, N.J., Buatois, L.A., Mángano, M.G., Davies, N.S., Gibling, M.R., MacNaughton, R.B., Labandeira, C.C., 2017. Early bursts of diversification defined the faunal colonization of land. Nat. Ecol. Evol. 1, 0175.

Moreau, J.-D., Gand, G., 2022. New data on the Permian ecosystem of the Rodez Basin: ichnofauna (traces of protostomians, tetrapods and fishes), jellyfishes and plants from Banassac-Canilhac (Lozère, southern France). Geodiversitas 44, 975–987.

Moreau, J.-D., Michelin, A., Fara, E., Gand, G., Galtier, J., Puech, G., Fouché, S., 2022. Ichnofossils and body fossils from the permian of the Sorgue Valley (Saint-Affrique Basin, southern France): palaeoenvironmental implications. Hist. Biol. https://doi.org/10.1080/08912963.2022.2148205.

Morrissey, L.B., Braddy, S.J., Bennett, J.P., Marriott, S.B., Tarrant, P.R., 2004. Fish trails from the Lower Old Red Sandstone of Tredomen Quarry, Powys, southeast Wales. Geol. J. 39, 337–358.

Muszer, J., 2020. First discovery of Mississippian trace fossils in the Swiebodzice Unit from the Witoszow region (SW Poland). Geol. Q. 64, 838–860.

Myers, R.A., Gingras, M.K., Keyes, G., Konhauser, K.O., Zonneveld, J.-P., 2019. Bioresuspension behaviors of the gobiid, *Valenciennea puellaris*, and the biogenic sedimentary structures it produces. Palaios 34, 179–189.

Netto, R.G., Balistieri, P., Paz, C.P., Lavina, E., 2003. Upper Carboniferous–Lower Permian rhythmites from Itararé Group (Paraná Basin, Brazil): are they all varvites?. In: III Simposio Argentino del Paleozoico Superior y Reunión Anual del Grupo de Trabajo Argentino del Proyecto IGCP 471, La Plata, p. 4.

Niedzwiedzki, G., Szrek, P., 2020. Non-tetrapod trace fossils from the Middle Devonian tetrapod tracksite at Zachelmie Quarry, Holy Cross Mountains, Poland. Palaeogeogr. Palaeoclimatol. Palaeoecol. 553, 109763.

Niedzwiedzki, G., Szrek, P., Narkiewicz, K., Narkiewicz, M., Ahlberg, P.E., 2010. Tetrapod trackways from the early Middle Devonian period of Poland. Nature 463, 43–48.

Nielsen, J.K., Nara, M., Jacobsen, A.R., 2020. Bowl-shaped structures in a Pleistocene clastic carbonate wedge on the Island of Rhodes, Greece. Ichnos 27, 326–333.

Pazos, P.J., 2002. Palaeoenvironmental framework of the glacial–postglacial transition (Late Paleozoic) in the Paganzo-Calingasta Basin (southern South America) and the Great Karoo-Kalahari Basin (southern Africa): ichnological implications. Gondwana Res. 5, 619–640.

Pearson, N.J., Gingras, M.K., Armitage, I.A., Pemberton, S.G., 2007. Significance of Atlantic sturgeon feeding excavations, Mary's point, Bay of Fundy, New Brunswick, Canada. Palaios 22, 457–464.

Pemberton, S.G., Frey, R.W., Bromley, R.G., 1988. The ichnotaxonomy of *Conostichus* and other plug-shaped ichnofossils. Can. J. Earth Sci. 25, 866–892.

Pervesler, P., Roetzel, R., Uchman, A., 2011. Ichnology of shallow sublittoral siliciclastics of the Burgschleinitz formation (Lower Miocene, Eggenburgian) in the Alpine-Carpathian foredeep (NE Austria). Austr. J. Earth Sci. 104, 81–96.

Piñeiro, G., 2006. Nuevos aportes a la paleontología del Pérmico de Uruguay. In: Veroslavsky, G., Martınez, S., Ubilla, M. (Eds.), Cuencas Sedimentarias de Uruguay. Geología, Paleontología y Resursos Naturales. DIRAC – Facultad de Ciencias (Universidad de la Republica), Montevideo, pp. 129–140.

Plumstead, E.P., 1970. Section A, post symposium excursion. In: Guide Book, Second IUGS Gondwana Symposium, South Africa, vol. 4, pp. 1–37.

Randle, E., Sansom, R.S., 2019. Bite marks and predation of fossil jawless fish during the rise of the jawed vertebrates. Proc. Royal Soc. B 286, 20191596.

Rindsberg, A.K., 1990. Freshwater to marine trace fossils of the Mary Lee Coal zone and overlying strata (Westphalian A), Pottsville Formation of northern Alabama. 82–95. In: Gastaldo, R.A., Demko, T.M., Liu, Y. (Eds.), Carboniferous Coastal Environments and Palaeocommunities of the Mary Lee Coal Zone, Marion and Walker Counties, Alabama. A Guidebook for Field Trip VI, Southeastern Section, Geological Society of America. Geological Survey of Alabama, Tuscaloosa, 138 pp.

Rodríguez-de la Rosa, R.A., del Rosario Fernández-Barajas, M., Valdes-Vergara, N.A., Prado-Escamilla, E., 2021. Fish trace fossils from the Lower Cretaceous of Puebla, Mexico. J. South Am. Sci. 112, 103547.

Ronchi, A., Santi, G., Marchetti, L., Bernardi, M., Gianolla, P., 2018. First report on swimming trace fossils of fish from the Upper Permian and Lower Triassic of the Dolomites (Italy). Ann. Soc. Geol. Pol. 88, 111–125.

Schindler, T., Poschmann, M., Wuttke, M., 2005. Chondrichthyan feeding depressions in a subtidal coastal environment from the Mainz Basin (Oligocene, SW Germany). Neues Jahrbuch Geol. Palaontol. Abhand. 237, 29–39.

Schweigert, G., 2001. Eine Schwimmspur aus dem Nusplinger Plattenkalk (Oberjura, SW-Deutschland). Neues Jahrbuch Geol. Palaontol. Monatsh. 2001, 77–84.

Sciscio, L., Bordy, E.M., Head, H.V., 2020. A Late Triassic aquatic community: *Undichna*-like and related swimming traces from a freshwater pond in the lower Elliot Formation of South Africa. J. Afr. Earth Sci. 172, 104026.

Scotese, C.R., 2021. An atlas of Phanerozoic paleogeographic maps: the seas come in and the seas go out. Annu. Rev. Earth Planet Sci. 49, 679–728.

Segura, M., Barroso-Barcenilla, F., Berrocal-Casero, M., Castanera, D., García-Hidalgo, J.F., Santos, V.F., 2016. A new Cenomanian vertebrate tracksite at Tamajón (Guadalajara, Spain): palaeoichnology and palaeoenvironmental implications. Cretac. Res. 57, 508–518.

Simon, T., Hagdorn, H., Hagdorn, M.K., Seilacher, A., 2003. Swimming trace of a coelacanth fish from the Lower Keuper of south-west Germany. Palaeontology 46, 911–926.

Smith, R.M.H., 1987. Morphology and depositional history of exhumed Permian point bars in the southwestern Karoo, South Africa. J. Sediment. Petrol. 57, 19–29.

Smith, R.M.H., 1993. Sedimentology and ichnology of floodplain paleosurfaces in the Beaufort Group (Late Permian), Karoo Sequence, South Africa. Palaios 8, 339–357.

Soler-Gijón, R., Moratalla, J.J., 2001. Fish and tetrapod trace fossils from the Upper Carboniferous of Puertollano, Spain. Palaeogeogr. Palaeoclimatol. Palaeoecol. 171, 1–28.

Sutcliffe, O.E., Briggs, D.E.G., Bartels, C., 1999. Ichnological evidence for the environmental setting of the Fossil-Lagerstätten in the Devonian Hunsrück Slate, Germany. Geology 27, 275–278.

Szrek, P., Salwa, S., Niedzwiedzki, G., Dec, M., Ahlberg, P.E., Uchman, A., 2016. A glimpse of a fish trace – an exceptional fish feeding trace fossil from the Lower Devonian of the Holy Cross Mountains, Poland. Palaeogeogr. Palaeoclimatol. Palaeoecol. 454, 113–124.

Tindal, B.H., Shillito, A.P., Davies, N.S., 2021. First report of fish trace fossils (*Undichna*) from the Middle Devonian Achanarras Limestone, Caithness Flagstone Group. Scot. J. Geol. 57, 023.

Todesco, R., Avanzini, M., 2008. First record of the fish trace fossil *Undichna* from the Middle Triassic of Italy. Studi Trentini di Scienze Naturali. Acta Geol. 83, 253–257.

Trewin, N.H., 2000. The ichnogenus *Undichna*, with examples from the Permian of the Falkland Islands. Palaeontology 43, 979–997.

Trewin, N.H., Davidson, R.J., 1996. An Early Devonian lake and its associated biota in the Midland valley of Scotland. Trans. R. Soc. Edinb. Earth Sci. 86, 233–246.

Trewin, N.H., MacDonald, D.I.M., Thomas, C.G.C., 2002. Stratigraphy and sedimentology of the Permian of the Falkland Islands: lithostratigraphic and palaeoenvironmental links with South Africa. J. Geol. Soc. 159, 5–19.

Trewin, N.H., Gurr, P.R., Jones, R.B., Gavin, P., 2012. The biota, depositional environment and age of the Old Red Sandstone of the island of Kerrera, Scotland. Scot. J. Geol. 48, 77–90.

Turek, V., 1989. Fish and amphibian trace fossils from Westphalian sediments of Bohemia. Palaeontology 32, 623–643.

Turek, V., 1996. Fish trace fossil interpreted as a food gathering swimming trail from the Upper Carboniferous (Westphalian) of Bohemia. Cas. Nar. Muzea Rada Prirodoved. 165, 5–8.

Uchman, A., Kumpulainen, R.A., 2011. Trace fossils in Quaternary glacial varved clays near Uppsala, Sweden. GFF 133, 135–140.

Uchman, A., Torres, P., Johnson, M.E., Berning, B., Ramalho, R.S., Rebelo, A.C., Melo, C.S., Baptista, L., Madeira, P., Cordeiro, R., Ávila, S.P., 2018. Feeding traces of recent ray fish and occurrences of the trace fossil *Piscichnus waitemata* from the Pliocene of Santa Maria Island, Azores (northeast Atlantic). Palaios 33, 361–375.

Walter, H., 1982. Neue Arthropodenfährten aus den Oberhöfer Schichten (Rotliegendes, Thüringer Wald) mit Bemerkungen über Ichnia limnisch-terrestricher Tuffite innerhalb der varistischen Molasse. Freiberger Forschungschefte C 375, 87–100.

Walter, H., 1983. Zur Taxonomie,Ökologie und Biostratigraphie der Ichnia limnisch-terrestrischer Arthropoden des mitteleuropäischen Jungpaläozoikums. Freiberger Forschungschefte C382, 146–193.

Warren, A., 1972. Queensland trace fossils pose 230 million years old problem. Aust. Nat. Hist. 17, 160–162.

Wilson, M.A., Borszcz, T., Zaton, M., 2015. Bitten spines reveal unique evidence for fish predation on Middle Jurassic echinoids. Lethaia 48, 4–9.

Wisshak, M., Volohonsky, E., Blomeier, D., 2004. Acanthodian fish trace fossils from the Early Devonian of Spitsbergen. Acta Palaeontol. Pol. 49, 629–634.

Xing, L., Wang, M., Klein, H., Gao, J., 2023. Life on a lake bottom: diverse horseshoe crab and fish ichnofauna from Yanchang Group (Upper Triassic) of northern Shaanxi Province, China – implications for palaeoenvironment. Hist. Biol. https://doi.org/10.1080/08912963.2022.2162398.

Zong, R., Gong, Y., 2018. Possible courtship behaviour of Devonian fish: evidence from large radial trace fossils in northwestern China. Palaeogeogr. Palaeoclimatol. Palaeoecol. 505, 180–186.

PART II

The ichnology of consumption

CHAPTER

5

Analyzing coprolites: Different methods for the study of coprolites

Paula Dentzien-Dias[1] *and Weronika Łaska*[2]

[1]Departamento de Paleontologia e Estratigrafia, Geosciences Institute, Federal University of Rio Grande Do Sul, Rio Grande Do Sul, Brazil; [2]Institute of Evolutionary Biology, Faculty of Biology, Biological and Chemical Research Centre, University of Warsaw, Warsaw, Poland

1. Introduction

The study of coprolites (fossil feces) has improved enormously in the last decades. Several types of dietary remains can be recovered from coprolites, from bones to plant fragments. To decipher part of the food web, classical coprolite analysis (morphology and macroscopic inclusions) is fundamental to identify the producer and reconstruct part of the trophic chain (Barrios-de Pedro and Buscalioni, 2018; Hunt and Lucas, 2019). However, the use of different methodologies can help to identify a variety of biological interactions and microscopic inclusions, leading to a better understanding of coprolite fossilization.

In this way, different techniques can provide evidence of rare fossils (Qvarnström et al., 2016). For example, coprolite thin sections reveal parasite eggs, fungi, and cyanobacteria (Fig. 5.1; Northwood, 2005; Dentzien-Dias et al., 2013, 2018; Nonsrirach et al., 2023). Scanning electron microscopy (SEM) has helped to decipher how coprolites, and their different inclusions, are preserved (Owocki et al., 2012; Barrios-de Pedro et al., 2018; Bajdek and Bienkwska-Wasiluk, 2020). In another way the use of two or more methodologies can fill gaps in the identification of faunal diversity and soft-tissue remains (Chin, 2002; Dentzien-Dias et al., 2018). Different kinds of computer tomography scanning have shown amazing images of the food inclusions in coprolites, such as articulated vertebrate and invertebrate remains (Bravo-Cuevas et al., 2017; Qvarnström et al., 2021). These different analyses have shown how special is the preservation of coprolites and the different inclusions in them. Here we summarize some of the most common and effective ways to analyze coprolites and their principal results (Table 5.1).

Vertebrate Ichnology
https://doi.org/10.1016/B978-0-443-38351-9.00001-3

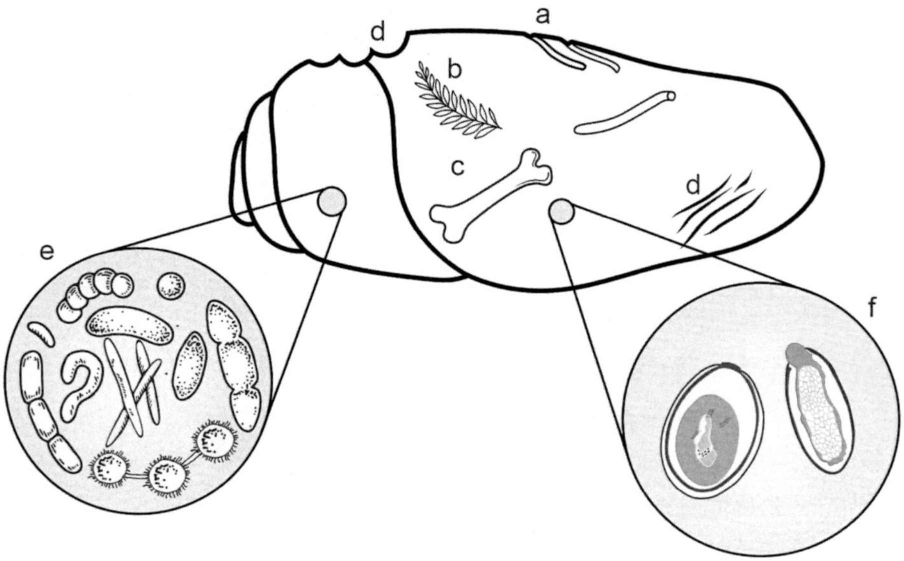

FIGURE 5.1 Line drawing of a coprolite with different biological components and interactions that can be found with different methods. (A) Invertebrate burrows, (B–C) food residues, (D) Bite and scratch traces, (E) Microbial remains and, (F) Parasite eggs.

2. Coprolite analyses

2.1 Nondestructive analyses

The documentation of coprolites is fundamental before any analyses. Besides 2D photos, it is possible to undertake digital 3D modeling by photogrammetry. This methodology has been extensively used in ichnology (Petti et al., 2008; Mujal et al., 2020) and will not be discussed in this chapter (see the chapter by Lallensack in this volume for a discussion of 3D modeling).

2.1.1 External/morphological analyses

The first thing to do when you find a coprolite in the field is to take a photo and make a stratigraphic section. If it is not in situ, try to find the producing level. After the field, the most important thing is to describe the coprolite; however, describing a coprolite is not a simple thing. Usually, the first thing to do is to measure the coprolite, and as some coprolites can be flattened, so it is necessary to measure the length, width, and height (Fig. 5.2).

According to Chin (2002), the volumes of small coprolites can be determined by the simple method of measuring water displacement by submerging samples (Archimedes' principle). However, porous specimens must be put in water to absorb water before measuring the volume (see Chin 2002 for more details).

The morphology is essential to infer the producer because it reflects the intestinal morphology. Generally, coprolites are divided into spiral or nonspiral (Hunt and Lucas, 2012). According to McAllister (1985), some or all agnathans, placoderms, dipnoans,

TABLE 5.1 A selection of coprolite analyses and the information obtained through the methods.

Analyses method	Type of analysis	Type of information	References
External/ morphological analyses	Nondestructive	Intestinal morphology, Probable producer, Dietary residues, Taphonomical modifications, Coprophagy/Dentalites	Hunt and Lucas (2012); Hunt et al. (1994); Dentzien-Dias et al. (2012), 2021.
(Micro-) CT, synchrotron-based phase-contrast tomography	Nondestructive	Dietary residues, Coprophagy/ burrows	Milàn, 2012b; Vršanský et al., (2013), Qvarnström et al. (2017), 2019a; 2019 b, 2021, Byrne et al. (2022)
Petrography	Partially destructive	Dietary residues, Parasite eggs, Cyanobacteria, Taphonomical modifications	Northwood (2005); Dentzien-Dias et al. (2013).
Confocal microscope	Partially destructive	Organic structures (cell structures and collagen fibers)	Hollocher et al. (2001); Jurigan et al. (2023).
SEM-EDS	Partially destructive	Dietary residues (hair, bones, plants); bacteria; chemical composition	Owocki et al. (2012); Dentzien-Dias et al. (2017); Mychajliw et al. (2020); Gross et al. (2023).
Acid dissolution	Destructive	Palynological and parasitological remains	Poinar and Boucot (2006); Hugot et al. (2014); Cardia et al. (2018); Barrios-de Pedro et al. (2020a).
DR-X	Destructive	Mineralogical composition	Hollocher et al. (2005); Rodrigues et al. (2018).
Lipid and isotopic analysis	Destructive	Cholesteroids, sterols, other biomolecules	Gill and Bull (2012); Gill et al. (2009); Barrios-de Pedro et al. (2020c).

actinistians, and chondrichthyans have spiral valves in their intestines. For more morphological and terminological examples of coprolites, see Hunt and Lucas (2012).

Another important external analysis is to search using different light incidences for inclusions (bone and plant fragments), surface marks (desiccation cracks), borings and bites, and adhesions (sand, plant, and shell impressions; Thulborn, 1991).

2.1.2 Computed tomography scanning

To analyze the internal structure of the coprolite—slice it up. Since each coprolite is unique, rather than employing classical true physical serial sectioning (e.g., Sollas, 1904; Bednarz and McIlroy, 2009), consider using noninvasive X-ray tomographic techniques to digitally cut the coprolite into slices.

One such technique is attenuation-based X-ray (micro-) computed tomography (CT). In short, during the CT scanning process, X-rays generated by an X-ray tube (an X-ray source) pass through a specimen, where some of the X-ray photons are absorbed before reaching an

FIGURE 5.2 Schematic drawing of a coprolite and standard measurements. (A) Spiral coprolite with its length (L) and width (W), (B) Spiral coprolite in cross section with an ellipsoidal shape, height (H).

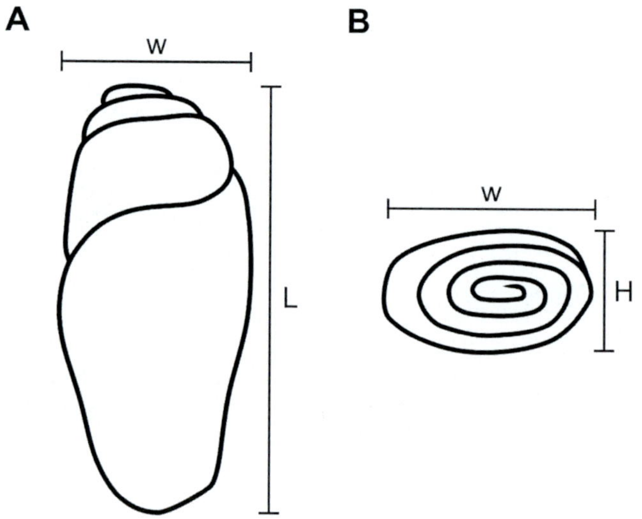

X-ray detector positioned opposite the source. The detector collects the attenuation data in the form of a two-dimensional (2D) radiograph, known as a projection image. A series of these projections from various angles is acquired by rotating the specimen either 180 degrees or 360 degrees, on a manipulation stage. By applying reconstruction mathematical algorithms, these projections are converted into a stack of multiple cross-sectional image slices (Fig. 5.3A–B) that in postprocessing steps are used to construct 3D visualizations of the analyzed coprolite, as in Fig. 5.3C–D (e.g., Kak and Slaney, 2001). The use of CT imaging for coprolite analysis was pioneered by Milàn et al. (2012a), who examined the internal

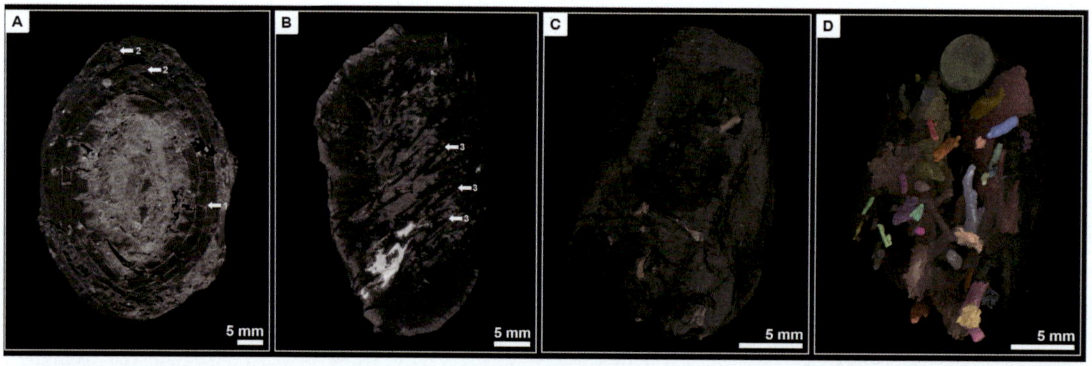

FIGURE 5.3 Examples of micro-CT-based images of different coprolites. (A–B) A single, virtual, and horizontal cross-sectional slice through a coprolite with arrows indicating prominent spiraling (1), cracks of drying (2), and fish scales (3). (A) KB_H21 (B) PV632P_01 (C) 3D visualization of an external surface of a coprolite (PV_0636_01). (D) 3D visualization showing internal morphology of the coprolite (PV_0636_01). Note differently colored inclusions dispersed throughout the coprolite matrix. The external surface of the coprolite is transparent gray.

architecture and bone content of coprolites from the lower Paleocene (Danian) of Denmark, and burrows from coprophageous organisms dug into Lower Cretaceous coprolites (Milàn et al., 2012b). Since that time, the application of CT for examining both the internal and external characteristics of coprolites has gained increasing attention in numerous studies (Milàn et al., 2015; Wang et al., 2018; Rummy et al., 2021; Halaclar et al., 2022; Jacquet et al., 2023; Yao et al., 2023).

While micro-CT provides spatial resolution ranges from below 0.25 mm down to less than 1 μm, it falls short in differentiating finer structures (e.g., Cunningham et al., 2014). For more detailed three-dimensional (3D) analysis of the interior of the coprolite, synchrotron X-ray microtomography (SR-μCT), with its excellent spatial resolution, can be employed instead (e.g., Tafforeau et al., 2006). Although SR-μCT, thanks to the high flux monochromatic X-ray beam, generates high-fidelity sharp images free of beam-hardening artifacts; both methods do not effectively address the problem of very weak-to-no absorption contrast between structures of interest within an analyzed sample, such as between inclusions and the coprolite matrix (Fig. 5.3B).

A real game changer in this context is propagation phase-contrast synchrotron microtomography (PPC-SRμCT) (e.g., Tafforeau et al., 2006; Lak et al., 2008; Sanchez et al., 2012, 2013). This method, among others, reveals structures of similar mineral density that are unresolved with conventional attenuation-based X-ray tomographic methods (e.g., Tafforeau et al., 2006; Qvarnström et al., 2017). Instead of X-ray attenuation, phase-contrast imaging directly detects the phase shift of the wave passing through the scanned sample (e.g., Tafforeau and Smith, 2008). After scanning, the reconstructions of the recorded data using a phase retrieval approach produce image stacks that later are imported into specialized software for 3D analysis and visualizations.

Vršanský et al. (2013) initiated the application of PPC-SRμCT to visualize coprolites extruded from an immature cockroach, both preserved in Cretaceous amber from Lebanon. Qvarnström et al. (2017) applied PPC-SRμCT to obtain high-quality 3D visualizations of inclusions of two coprolites from the Upper Triassic of Krasiejów, Poland, which they recognized as beetle remains (Qvarnström et al., 2019a) and a partly articulated fish and fragments of bivalves. The method has also been applied to study inclusions of Late Devonian coelacanth coprolites of the Holy Cross Mountains with conodont elements preserved inside (Zatoń et al., 2017). Late Jurassic coprolites of pterosaurs include foraminiferans, skeletal elements of ostracods and bivalves, and other crustaceans/arthropods (Qvarnström et al., 2019b), termite coprolites inside Eocene-Oligocene borings in wood from the Malzieu Basin in France (Moreau et al., 2020), and the beetle remains within Triassic coprolite of possible dinosauriform origin (Qvarnström et al., 2021).

2.1.2.1 Image segmentation during post-processing

Accurate segmentation of structures of interest within coprolites (including structural patterns such as spirals or drying cracks) from the image stack acquired via X-ray imaging techniques, is the first and essential step for their correct identification and quantification, enabling further interpretations such as diet of the producer or paleoecological interactions in ancient ecosystems (Qvarnström et al., 2017). Thanks to recent technological advances and increased accessibility, CT has rapidly become a standard method in paleontology, including coprolite analysis, producing high-fidelity data. Moreover, the explosive growth of open-source CT processing software, e.g., Fiji (Schindelin et al., 2012), 3D Slicer (Fedorov et al., 2012), Drishti (Limaye, 2012), and MeVisLab (Egger et al., 2012), along with free access to tens of thousands of video tutorials (mostly hosted on YouTube, e.g., https://youtube.com/playlist?list=PLeaIM0zUlEqswa6Pskg9uMq15LiWWYP39&feature=shared, https://youtube.com/@DrishtiViz?feature=shared, https://youtu.be/FcRQY1PzMJ8?feature=shared, https://youtu.be/oOQATwrTyq0?feature=shared) and text-based guides with step-by-step instructions on how to perform image processing and data analysis of CT data, allow researchers to quickly explore 3D scientific images and generate their own 3D models.

The segmentation and labeling of individual structures within coprolites often present challenges due to their diverse complexity, shapes, sizes, and mineralization. Moreover, the presence of various inclusions, including bones, which share identical image intensities, adds to the complexity of interpretation. Hence, traditional manual segmentation methods—whether fully manual or involving basic threshold-based approaches—tend to be time-consuming and inefficient, especially when dealing with associated inclusions such as hair, fish scales, articulated bones, or invertebrate skeletal elements, which often show small contrasts. To address these challenges, automated segmentation using the state-of-the-art convolutional neural network (CNN) deep learning methods can be employed. Such advanced tools are available, for example, within the software Dragonfly, free for academic users.

2.2 Partially destructive analyses

When a destructive analysis is performed, a small piece of one of the poles of the coprolite should be taken, avoiding destroying the morphology of the entire specimen.

2.2.1 *Petrographic analysis*

This analysis is very efficient for the recognition of inclusions and petrographic fabrics (Fig. 5.4A; Jurigan et al., 2023). If the specimen is elongated, longitudinal and transverse sections should be made to assemble a more complete view of the inclusions (Dentzien-Dias et al., 2013). On the other hand, if the specimen is small or rare, the thin section should be taken from the end of the coprolite to preserve the morphology (Chin, 2002). The techniques for preparing coprolite thin sections are very similar to those used to prepare the standard petrographic sections (Wilson, 1994). Usually, the coprolite, before being cut, should be hardened by embedding into epoxy resin, and then cut with a diamond saw using a standard thin section method (Chin, 2002; Dentzien-Dias et al., 2012, 2013; Nonsrirach et al., 2023). The coprolite slices should be glued to glass slides, and optimal thickness for transmission microscopy was obtained using a grinder/sander. The thin section analyses should be made in a petrographic transmitted light polarizing microscope.

This technique is one of the most used because it is one of the most accessible, and provides very good insight into different inclusions and the inner morphology of the coprolites. For example, Williams (1972) described folded structures inside one of the whorls of a spiral coprolite that were later interpreted as a folded fecal ribbon (McAllister, 1985). Another important contribution is the analyses of coprofabrics, which can be very useful for

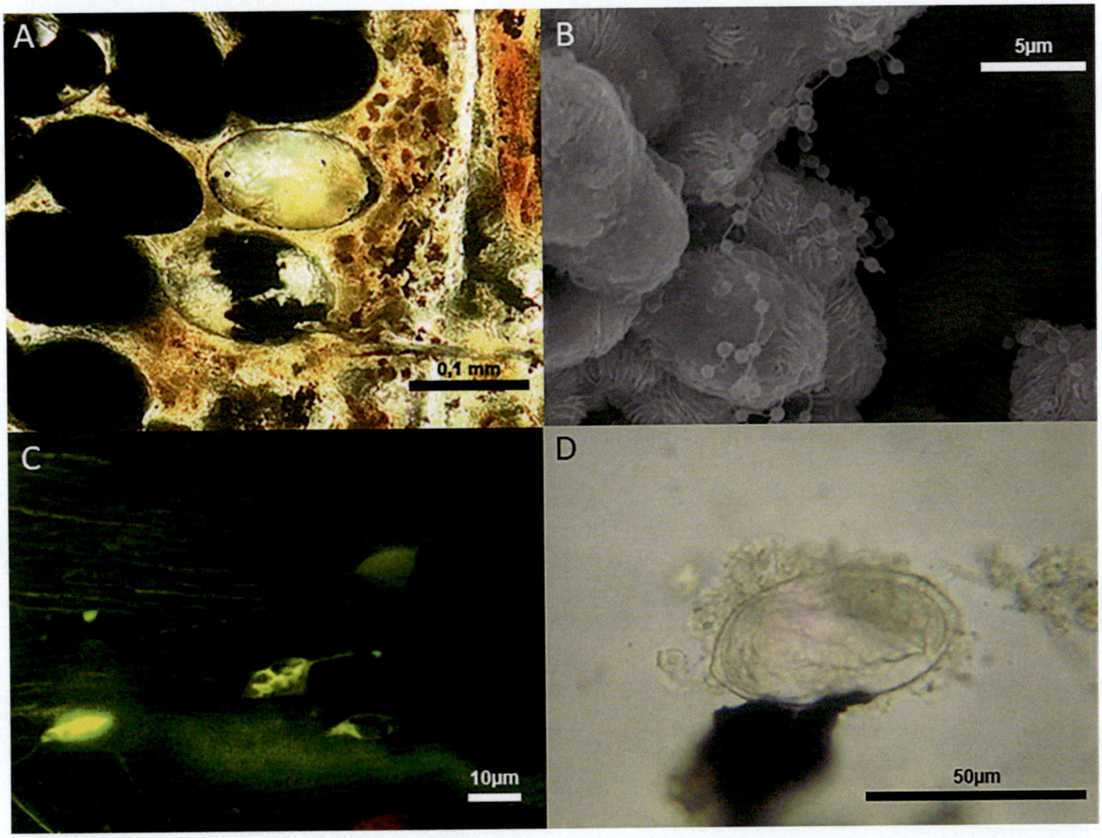

FIGURE 5.4 Different inclusions analyzed in different methodologies. (A) Coprolite thin section analyzed in a petrographic microscope, note the cestode eggs (Dentzien-Dias et al., 2013). (B) SEM image of an actinomycete with aerial spore chains arising from verrucose colonies (Dentzien-Dias et al., 2017). (C) Confocal images of some cell structures with possible nuclei and collagen fibers preserved on a bone surface (Jurigan et al., 2023). (D) Strongylida ovoidal eggs recovered after acid dissolution (Cardia et al., 2021).

classification of coprolite types (Chin, 2007a). Different coprofabrics were described from the Permian (Jurigan et al., 2023) to the Cretaceous (Chin, 2007b; Barrios-de Pedro et al., 2020b).

2.2.2 *Scanning electron microscope and energy-dispersive X-ray spectroscopy*

The SEM has shown exceptionally detailed soft tissues and different types of bacteria in coprolites. Before the analyses, the coprolite must be dried for 24 h at 40°C in a drying oven and broken, to expose a fresh portion and to avoid external contamination by microbes (Dentzien-Dias et al., 2017, Fig. 5.4B). This portion will be mounted on Al stubs, coated with Au, and subjected to analysis under the SEM. The SEM works using a focused beam of high-energy electrons that interacts with the surface of the coprolite to generate a magnified image. Energy-dispersive X-ray spectroscopy (EDS) analyses could be also conducted using the same equipment to determine the chemical composition of the coprolite.

Barrios-de Pedro and Buscalioni (2018) analyzed Cretaceous coprolites through SEM to study the digestion features, such as corrosion lines and shape of the fractures at the ends. The analyses allowed the authors to recognize three digestive strategies. Bajdek and Bienkowska-Wasiluk (2020) described different inclusions on Oligocene coprolites, such as parasite eggs and different bacterial remains, as thin-walled spheres interpreted as pseudomorphs after microbial cell walls. The thin-walled spheres have been reported from coprolites from different producers and periods (Lamboy et al., 1994; Hollocher and Hollocher, 2012; Dentzien-Dias et al., 2018).

Another impressive finding with the use of SEM is the recognition of muscle tissue preserved in coprolites. Chin et al. (2003) reported undigested muscle tissue in a tyrannosaurid coprolite. This finding allowed the authors to conclude that the large predator had a short gut-residence time and that the coprolite had undergone rapid lithification and minimal recrystallization. In the same study, the use of EDS for chemical comparison between the bones and muscle inclusions indicated that even blood cells fossilized.

2.2.3 *Confocal microscopy*

Coprolite thin sections can be analyzed under a confocal microscope to survey preserved organic structures that are less evident in petrographic microscopy (Hollocher et al., 2001; Jurigan et al., 2023). Confocal images acquired on some bone surfaces also revealed some cell structures preserved and possible collagen fibers (Fig. 5.4C). Even though there are different confocal microscopes with different techniques to generate images, all can be used to search organic matter in coprolites. Jurigan et al. (2023) obtained images through an EC Plan-Neofluar 40×/1.30 oil M27 objective, excitation with 405 nm laser at 4% of emission, and resolution of $0.10 \times 0.10 \times 0.60$ μm per pixel (for x, y, and z, respectively), which resulted in an image format of 1024×1024 pixels.

2.3 Destructive analyses

2.3.1 *Acid dissolution*

This method is efficient to find parasite eggs and cysts from the Triassic to Holocene (Fig. 5.4D; Poinar and Boucot, 2006; Silva et al., 2014; Barrios-de Pedro et al., 2020a; Cardia et al., 2021). The coprolites must be scrubbed thoroughly with a toothbrush and abrasive soap powder to remove any superficial material (Poinar and Boucot, 2006). Individual samples of the surface and the interior of the coprolite ($\cong$ 1.0 g each) must be extracted using a punch or an electric drill (Barrios-de Pedro et al., 2020a; Cardia et al., 2021), and then slightly ground using an agate mortar and pestle. The powder from the surface and the interior should be stored individually in properly labeled Falcon 15 mL polypropylene tubes. According to Ferreira et al. (2011), a 10% hydrochloric acid solution should be added to dissociate minerals. To stop the reaction, a double volume of distilled water must be added (Cardia et al., 2021). According to Barrios-de Pedro et al. (2020a), all tools used during the extraction and preparation of the coprolites must be carefully cleaned with 70% ethanol and then rinsed three consecutive times using ultrafiltered (0.45 μm) distilled water between the samples to avoid cross-contamination.

The resulting solution must be washed several times with distilled water, and then sifted through a 500 mesh (25 μm) Tyler sieve, as described by Bouchet et al. (1999). The residual material retained on the sieve must be washed again in distilled water. A drop of this material is then placed on a microscope slide, three drops of glycerin are added, and the slide is covered with a coverslip (Cardia et al., 2021). The slide must be analyzed by optical microscopy at 100× and 400× magnification (Barrios-de Pedro et al., 2020a; Cardia et al., 2021).

2.3.2 *Lipid analysis*

Lipids can be extracted from coprolites, mainly from those of Quaternary herbivores, such as ground sloths and mammoths (Gill and Bull, 2012). However, Chin (1996) has shown that the lipid content of five herbivorous and three carnivorous coprolites from the Mesozoic was preserved.

Umamaheswaran et al. (2019) recovered organic molecules preserved in coprolites from the Triassic that could be related to the dietary habits of the coprolite producer and paleoenvironment. More recently, Tripp et al. (2022) combined analyses of lipids and stable isotopes from Carboniferous coprolites and revealed the dietary spectrum and trophic position of the producers, with an overwhelming abundance of cholesteroids.

According to Gill and Bull (2012), organic solvents are used to extract lipids from dried material, to give a total lipid extract (TLE). Then it is separated into fractions using column chromatography. The column chromatography involves passing the TLE through a glass column packed with a stationary phase (alumina or silica) and eluting using a sequence of solvents of increasing polarity. Specific techniques such as selected ion monitoring gas chromatography-mass spectrometry (SIM-GC/MS) or gas chromatography tandem mass spectrometry (GC/MS) can be employed to lower the limit of quantification (Evershed and Bethell, 1996). Although the details of the extraction methods vary according to the material analyzed, the general approach remains the same.

Nevertheless, lipids are transformed into stable derivatives during fossilization (Tripp et al., 2022). For a detailed description of the lipid and stable isotope analysis, see Gill and Bull (2012) and Tripp et al. (2022).

2.3.3 *X-ray diffraction*

X-ray diffraction is widely used to determine the mineralogical composition of coprolites. According to Hollocher et al. (2005), a ~1.5 fragment of a coprolite must be crushed to a fine powder to be analyzed in an X-ray diffractometer. Rodrigues et al. (2018) sampled three different areas from the same coprolite to better understand its fossilization. For example, if there is a halo (Thulborn, 1991) the area with different colors and textures should be sampled. This sampling technique indicated that four phases occurred during the coprolite fossilization process (Rodrigues et al., 2018).

3. Conclusions

There is no doubt that only the morphological/external analyses of coprolites can indicate the producer and its prey/food. However, to obtain a more complete understanding of the depositional environment, fossilization, and the relationship between microorganisms and vertebrates, it is fundamental to use a large variety of analyses. In this matter, a protocol is proposed (Fig. 5.5), to extract all the potential of the coprolites, the soft tissue, parasites, and perfectly preserved food inclusions. Nevertheless, the different methodologies that are used by researchers will depend on the amount of coprolites they have, to consider a destructive analysis, the funding and accessibility they have to the newest technologies (that usually are expensive).

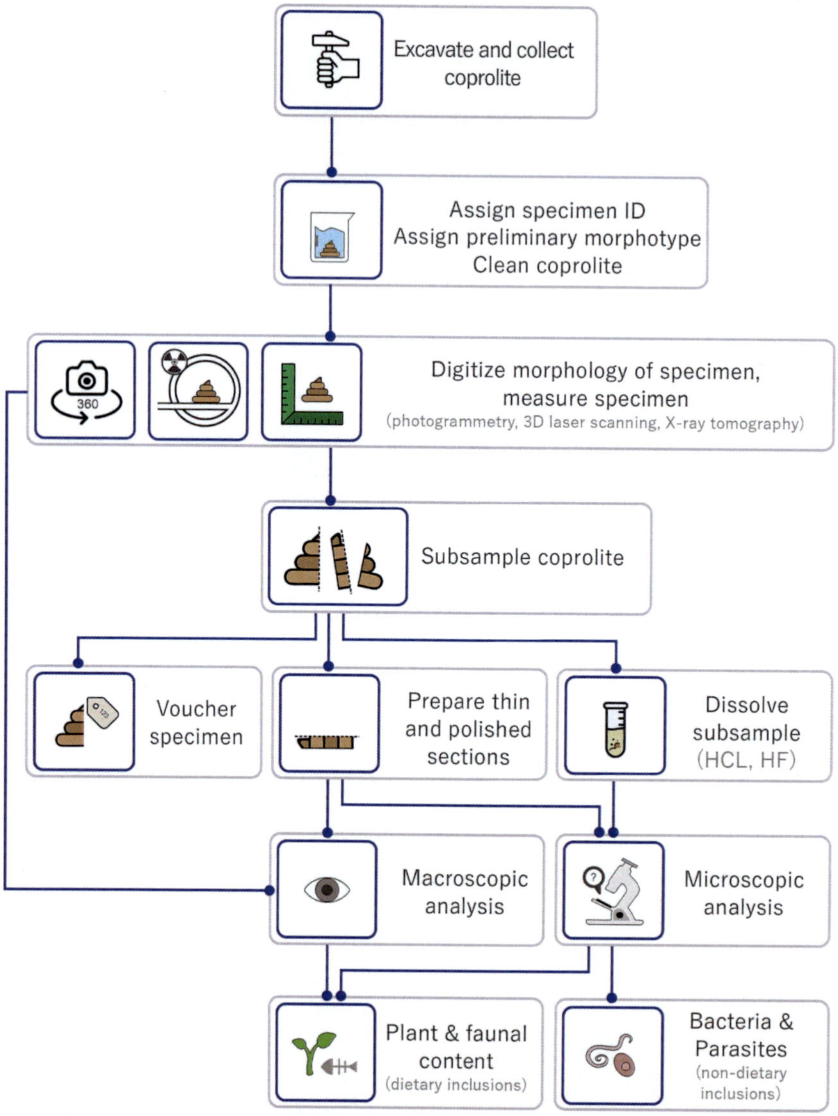

FIGURE 5.5 Protocol flowchart of the standard methods employed in coprolite analysis.

Funding

PD-D was supported by the Brazilian National Council for Scientific and Technological Development (CNPq). WŁ was supported by I.3.4 Action of the Excellence Initiative—Research University Program at the University of Warsaw, funded by the Ministry of Education and Science, Poland.

Acknowledgments

We thank MSc. Sheron Medeiros for Figs. 5.1 and 5.2.

References

Bajdek, P., Bienkowska-Wasiluk, M., 2020. Deep-sea ecosystem revealed by teleost fish coprolites from the Oligocene of Poland. Palaeogeogr. Palaeoclimatol. Palaeoecol. 540, 109546. https://doi.org/10.1016/j.palaeo.2019.109546.

Barrios-de Pedro, S., Buscalioni, Á.D., 2018. Scrutinizing Barremian coprolite inclusions to record digestive strategies. Ann. Soc. Geol. Pol. 88, 20−221. https://doi.org/10.14241/asgp.2018.014.

Barrios-de Pedro, S., Poyato-Ariza, F.J., Moratalla, J.J., Buscalioni, Á.D., 2018. Exceptional coprolite association from the Early Cretaceous continental Lagerstätte of Las Hoyas, Cuenca, Spain. PLoS One 13 (5), e0196982. https://doi.org/10.1371/journal.pone.0196982.

Barrios-de Pedro, S., Osuna, A., Buscalioni, Á.D., 2020a. Helminth eggs from Early Cretaceous faeces. Sci. Rep. 10, 18747. https://doi.org/10.1038/s41598-020-75757-4.

Barrios-de Pedro, S., Chin, K., Buscalioni, Á.D., 2020b. The late Barremian ecosystem of Las Hoyas sustained by fishes and shrimps as inferred from coprofabrics. Cretac. Res. 110, 104409. https://doi.org/10.1016/j.cretres.2020.104409.

Barrios-de Pedro, S., Rogers, K.M., Alcorlo, P., Buscalioni, Á.D., 2020c. Food web reconstruction through isotopic compositions of fossil feces: insights into the ecology of a late Barremian freshwater ecosystem (Las Hoyas, Cuenca, Spain). Cretac. Res. 108, 104343. https://doi.org/10.1016/j.cretres.2019.104343.

Bednarz, M., McIlroy, D., 2009. Three-dimensional reconstruction of "phycosiphoniform" burrows: implications for identification of trace fossils in core. Palaeontol. Electron. 12, 15 pp. http://palaeoelectronica.org/2009_3/195/index.html.

Bouchet, F., Lefèvre, C., West, D., Corbett, D., 1999. First paleoparasitological analysis of a midden in the Aleutian Islands (Alaska): results and limits. J. Parasitol. 85, 369−372.

Bravo-Cuevas, V.K., Morales-Garcia, N.M., Barrõn-Ortiz, C.R., Theodor, J.M., Cabral-Perdomo, M.A., 2017. Canid coprolites from the late pleistocene of hidalgo. Importance for the carnivore record of north America. Ichnos 24 (4), 239−249. https://doi.org/10.1080/10420940.2016.1270209.

Byrne, H.M., Niedźwiedzki, G., Blom, H., Kear, B.P., Ahlberg, P.E., 2022. Coprolite diversity reveals a cryptic ecosystem in an early Tournaisian lake in East Greenland: implications for ecosystem recovery after the end-Devonian extinction. Palaeogeogr. Palaeoclimatol. Palaeoecol. 605, 111215. https://doi.org/10.1016/j.palaeo.2022.111215.

Cardia, D.F.F., Bertini, R.J., Camossi, L.G., Richini-Pereira, V.B., Losnak, D.O., Francischini, H., Dentzien-Dias, P., 2021. Paleoparasitological analysis of a coprolite assigned to a carnivoran mammal from the upper Pleistocene Touro Passo Formation, Rio Grande do Sul, Brazil. An Acad. Bras Ciências 93. https://doi.org/10.1590/0001-3765202120190876.

Cardia, D.F.F., Bertini, R.J., Camossi, L.G., Letizio, L.A., 2018. The first record of Ascaridoidea eggs discovered in Crocodyliformes hosts from the Upper Cretaceous of Brazil. Rev. Bras. Palaontol. 21 (3), 238−244. https://doi.org/10.4072/rbp.2018.3.04.

Chin, K., 1996. The Paleobiological Implications of Herbivorous Dinosaur Coprolites: Ichnologic, Petrographic, and Organic Geochemical Investigations. University of California, Santa Barbara.

Chin, K., 2002. Analyses of coprolites produced by carnivorous vertebrates. In: Kowalewski, M., Kelley, P.H. (Eds.), The Fossil Record of Predation, The Paleontological Society Papers, vol 8, pp. 43−49.

Chin, K., 2007a. Thin section analysis of lithified coprolites (fossil feces). In: Microsc. Microanal, pp. 504–505. https://doi.org/10.1017/S143192760707540X.

Chin, K., 2007b. The paleobiological implications of herbivorous dinosaur coprolites from the Upper Cretaceous Two Medicine Formation of Montana: why eat wood? Palaios 22 (5), 554–566. https://doi.org/10.2110/palo.2006.p06-087r.

Chin, K., Eberth, D.A., Schweitzer, M.H., Rando, T.A., Sloboda, W.J., Horner, J.R., 2003. Remarkable preservation of undigested muscle tissue within a Late Cretaceous tyrannosaurid coprolite from Alberta, Canada. Palaios 18 (3). https://doi.org/10.1669/0883-1351(2003)018<0286:RPOUMT>2.0.CO;2.

Cunningham, J.A., Rahman, I.A., Lautenschlager, S., Rayfield, E.J., Donoghue, P.C., 2014. A virtual world of paleontology. Trends Ecol. Evol. 29 (6), 347–357. https://doi.org/10.1016/j.tree.2014.04.004.

Dentzien-Dias, P.C., de Figueiredo, A.E.Q., Horn, B., Cisneros, J.C., Schultz, C.L., 2012. Paleobiology of a unique vertebrate coprolites concentration from rio do rasto formation (Middle/Upper permian), parana basin, Brazil. J. S. Am. Earth Sci. 40, 53–62. https://doi.org/10.1016/j.jsames.2012.09.008.

Dentzien-Dias, P.C., Poinar, J.,G., de Figueiredo, A.E.Q., Pacheco, A.C.L., Horn, B.L., Schultz, C.L., 2013. Tapeworm eggs in a 270 million-year-old shark coprolite. PLoS One 8 (1), e55007. https://doi.org/10.1371/journal.pone.0055007.

Dentzien-Dias, P.C., Poinar, G., Francischini, H., 2017. A new actinomycete from a Guadalupian vertebrate coprolite from Brazil. Hist. Biol. 29, 770–776. https://doi.org/10.1080/08912963.2016.1241247.

Dentzien-Dias, P., Carrillo-Briceño, J.D., Francischini, H., Sánchez, R., 2018. Paleoecological and taphonomical aspects of the late Miocene vertebrate coprolites (Urumaco Formation) of Venezuela. Palaeogeogr. Palaeoclimatol. Palaeoecol. 490, 590–603. https://doi.org/10.1016/j.palaeo.2017.11.048.

Dentzien-Dias, P., Hunt, A.P., Lucas, S.G., Francischini, H., Gulotta, M., 2021. Coprolites from shallow marine deposits of the Nanjemoy Formation, lower Eocene of Virginia, USA. Lethaia 54 (1), 26–39. https://doi.org/10.1111/let.12380.

Egger, J., Tokuda, J., Chauvin, L., Freisleben, B., Nimsky, C., Kapur, T., Wells, W., 2012. Integration of the OpenIGTLink Network Protocol for image-guided therapy with the medical platform MeVisLab. Int. J. Med. Robot. Comput. Assist. Surg. 8, 282–290. https://doi.org/10.1002/rcs.1415.

Evershed, R.P., Bethell, P.H., 1996. Application of multimolecular biomarker techniques to the identification of fecal material in archaeological soils and sediments. In: Orna, M.V. (Ed.), Archaeological Chemistry. American Chemical Society, pp. 157–172. https://doi.org/10.1021/bk-1996-0625.ch013.

Fedorov, A., Beichel, R., Kalpathy-Cramer, J., Finet, J., Fillion-Robin, J.C., Pujol, S., Bauer, C., Jennings, D., Fennessy, F., Sonka, M., Buatti, J., 2012. 3D slicer as an image computing platform for the quantitative imaging Network. Magn. Reson. Imaging 30, 1323–1341. https://doi.org/10.1016/j.mri.2012.05.001.

Ferreira, L.F., Reinhard, K., Araújo, A., 2011. Fundamentos da Paleoparasitologia, 1a ed. Editora Fiocruz, Rio de Janeiro. 482 pp.

Gill, F.L., Bull, I.D., 2012. Lipid analysis of vertebrate coprolites. New Mexico Museum of Natural History and Science Bulletin 57, 93–98. http://econtent.unm.edu/cdm/ref/collection/bulletins/id/1817.

Gill, F.L., Crump, M.P., Schouten, R., Bull, I.D., 2009. Lipid analysis of a ground sloth coprolite. Quat. Res. 72, 284–288. https://doi.org/10.1016/j.yqres.2009.06.006.

Gross, M., Prieto, J., Grímsson, F., Bojar, H.P., 2023. Hyena and 'false' sabre-toothed cat coprolites from the late Middle Miocene of south-eastern Austria. In: Hist Biol, pp. 1–20.

Halaclar, K., Rummy, P., Deng, T., Van Do, T., 2022. Footprint on a coprolite: a rarity from the Eocene of vietnam. Palaeoworld 31 (4), 723–732. https://doi.org/10.1016/j.palwor.2022.01.010.

Hollocher, T.C., Chin, K., Hollocher, K., Kruge, M.A., 2001. Bacterial residues in coprolite of herbivorous dinosaurs: role of bacteria in mineralization of feces. Palaios 16, 547–565. https://doi.org/10.2307/3515628.

Hollocher, K., Alcober, O.A., Colombi, C.E., Hollocher, T.C., 2005. Carnivore coprolites from the Upper Triassic Ischigualasto Formation, Argentina: chemistry, mineralogy, and evidence for rapid initial mineralization. Palaios 20, 51–63. https://doi.org/10.2110/palo.2003.p03-98.

Hollocher, K., Hollocher, T.C., 2012. Early processes in the fossilization of terrestrial feces to coprolites, and microstructure preservation. New Mexico Museum of Natural History Bull 57, 79–91.

Hugot, J.P., Gardner, S.L., Borba, V., Araujo, P., Leles, D., Da-Rosa, Á.A.S., Ferreira, L.F., Araújo, A., 2014. Discovery of a 240 million year old nematode parasite egg in a cynodont coprolite sheds light on the early origin of pinworms in vertebrates. Parasites Vectors 7, 486. https://doi.org/10.1186/s13071-014-0486-6.

Hunt, A.P., Chin, K., Lockley, M.G., 1994. The paleobiology of vertebrate coprolites. In: Donovan, S. (Ed.), The Palaeobiology of Trace Fossils. John Wiley and Sons, London, pp. 221–240.

Hunt, A.P., Lucas, S.G., 2012. Descriptive terminology of coprolites and recent feces. Nat. Hist. Sci. Bull. 57, 153–160. N. M. Mus.

Hunt, A.P., Lucas, S.G., 2019. Hyena hegemony: biogeography and taphonomy of Pleistocene vertebrate coprolites with description of a new mammoth coprolite ichnotaxon. Ichnos 27 (2), 111–121. https://doi.org/10.1080/10420940.2019.1612393.

Jacquet, S.M., Webb, J.L., Huntley, J.W., Selly, T., Schiffbauer, J.D., 2023. X-Ray tomographic microscopy of Eocene coprolites from pipestone springs main pocket, southwest Montana. Front. Earth Sci. 11, 1130107. https://doi.org/10.3389/feart.2023.1130107.

Jurigan, I., Ricardi-Branco, F., Dentzien-Dias, P., 2023. Permian western Gondwana food chain elucidated by coprolites from the Corumbataí Formation (Paraná Basin, Brazil). J. S. Am. Earth Sci. 127, 104414. https://doi.org/10.1016/j.jsames.2023.104414.

Kak, A.C., Slaney, M., 2001. Principles of Computerized Tomographic Imaging. Society For Industrial And Applied Mathematics, Philadelphia. https://doi.org/10.1137/1.9780898719277.

Lak, M., Néraudeau, D., Nel, A., Cloetens, P., Perrichot, V., Tafforeau, P., 2008. Phase contrast X-ray synchrotron imaging: opening access to fossil inclusions in opaque amber. Microsc. Microanal. 14, 251–259. https://doi.org/10.1017/S1431927608080264.

Lamboy, M., Rao, V.P., Ahmed, E., Azzouzi, N., 1994. Nannostructure and significance of fish coprolites in phosphorites. Mar. Geol. 120, 373–383. https://doi.org/10.1016/0025-3227(94)90068-X.

Limaye, A., 2012. Drishti: a volume exploration and presentation tool. In: Developments in X-Ray Tomography VIII, vol 8506. International Society for Optics and Photonics, p. 85060X.

McAllister, J.A., 1985. Reevaluation of the formation of spiral coprolites. In: Palaeontological Contributions, vol 114. University of Kansas, 1e12.1.

Milàn, J., Rasmussen, B.W., Lynnerup, N., 2012a. A coprolite in the MDCT-scanner–internal architecture and bone contents revealed. In: Hunt, et al. (Eds.), Vertebrate Coprolites. New Mexico Museum of Natural History and Science Bulletin, vol 57, pp. 99–104.

Milàn, J., Rasmussen, B.W., Bonde, N., 2012b. Coprolites with prey remains and traces from coprophagous organisms from the Lower Cretaceous (late Berriasian) Jydegaard Formation of Bornholm, Denmark. New Mexico Museum of Natural History and Science Bulletin 57, 235–240.

Milàn, J., Hunt, A.P., Adolfssen, J.S., Rasmussen, B.W., Morten, B., 2015. First record of a vertebrate coprolite from the upper cretaceous (maastrichtian) chalk of stevns klint, Denmark. Fossil record 4. New Mexico Museum of Natural History and Science Bulletin 67, 227–229.

Moreau, J.D., Trincal, V., Nel, A., Simon-Coinçon, R., Sallé, V., Le Couls, M., Néraudeau, D., Fernandez, V., 2020. Hidden termite coprolites revealed by synchrotron microtomography inside Eocene–Oligocene filled woodborings from the Malzieu Basin, Lozère, southern France. Lethaia 53 (1), 106–117. https://doi.org/10.1111/let.12344.

Mujal, E., Marchetti, L., Schoch, R.R., Fortuny, J., 2020. Upper Paleozoic to lower Mesozoic tetrapod ichnology revisited: photogrammetry and relative depth pattern inferences on functional prevalence of autopodia. Front. Earth Sci. 8, 248. https://doi.org/10.3389/feart.2020.00248.

Mychajliw, A.M., Rice, K.A., Tewksbury, L.R., Southon, J.R., Lindsey, E.L., 2020. Exceptionally preserved asphaltic coprolites expand the spatiotemporal range of a North American paleoecological proxy. Sci. Rep. 10 (1), 5069. https://doi.org/10.1038/s41598-020-61996-y.

Nonsrirach, T., Morand, S., Ribas, A., Manitkoon, S., Lauprasert, K., Claude, J., 2023. First discovery of parasite eggs in a vertebrate coprolite of the Late Triassic in Thailand. PLoS One 18 (8), e0287891. https://doi.org/10.1371/journal.pone.0287891.

Northwood, C., 2005. Early Triassic coprolites from Australia and their palaeobiological significance. Palaeontology 48, 49–68. https://doi.org/10.1111/j.1475-4983.2004.00432.x.

Owocki, K., Niedźwiedzki, G., Sennikov, A.G., Golubev, V.K., Janiszewska, K., Sulej, T., 2012. Upper Permian vertebrate coprolites from Vyazniki and Gorokhovets, Vyatkian regional stage, Russian platform. Palaios 27, 867–877. https://doi.org/10.2110/palo.2012.p12-017r.

Petti, F.M., Avanzini, M., Belvedere, M., De Gasperi, M., Ferretti, P., Girardi, S., Remondino, F., Tomasoni, R., 2008. D Digital 3D modelling of dinosaur footprints by photogrammetry and laser scanning techniques: integrated approach at the Coste dell'Anglone tracksite (lower Jurassic, southern Alps, northern Italy). Acta Geol. 83, 303–315.

Poinar, G., Boucot, A.J., 2006. Evidence of intestinal parasites of dinosaurs. Parasitology 133, 245–249. https://doi.org/10.1017/S0031182006000138.

Qvarnström, M., Niedźwiedzki, G., Živilė, Ž., 2016. Vertebrate coprolites (fossil feces): an underexplored Konservat-Lagerstätte. Earth Sci. Rev. 162, 44–57. https://doi.org/10.1016/j.earscirev.2016.08.014.

Qvarnström, M., Niedźwiedzki, G., Tafforeau, P., Žigaitė, Ž., Ahlberg, P.E., 2017. Synchrotron phase-contrast microtomography of coprolites generates novel palaeobiological data. Sci. Rep. 7, 2723. https://doi.org/10.1038/s41598-017-02893-9.

Qvarnström, M., Wernström, J.V., Piechowski, R., Tałanda, M., Ahlberg, P.E., Niedźwiedzki, G., 2019a. Beetle-bearing coprolites possibly reveal the diet of a Late Triassic dinosauriform. R. Soc. Open Sci. 6, 181042. https://doi.org/10.1098/rsos.181042.

Qvarnström, M., Elgh, E., Owocki, K., Ahlberg, P.E., Niedźwiedzki, G., 2019b. Filter feeding in Late Jurassic pterosaurs supported by coprolite contents. PeerJ 7, e7375. https://doi.org/10.7717/peerj.7375.

Qvarnström, M., Fikáček, M., Wernström, J.V., Huld, S., Beutel, R.G., Arriaga-Varela, E., Ahlberg, P.E., Niedźwiedzki, G., 2021. Exceptionally preserved beetles in a Triassic coprolite of putative dinosauriform origin. Curr. Biol. 31, 3374–3381. https://doi.org/10.1016/j.cub.2021.05.015.

Rodrigues, M.I.C., da Silva, J.H., Santos, F.E.P., Dentzien-Dias, P., Cisneros, J.C., de Menezes, A.S., Freire, P.T.C., Viana, B.C., 2018. Physicochemical analysis of Permian coprolites from Brazil. Spectrochim. Acta Mol. Biomol. Spectrosc. 189, 93–99. https://doi.org/10.1016/j.saa.2017.08.011.

Rummy, P., Halaclar, K., Chen, H., 2021. The first record of exceptionally-preserved spiral coprolites from the Tsagan-Tsab Formation (Lower Cretaceous), Tatal, western Mongolia. Sci. Rep. 11 (1), 7891. https://doi.org/10.1038/s41598-021-87090-5.

Sanchez, S., Ahlberg, P.E., Trinajstic, K.M., Mirone, A., Tafforeau, P., 2012. Three-dimensional synchrotron virtual paleohistology: a new insight into the world of fossil bone microstructures. Microsc. Microanal. 18, 1095–1105. https://doi.org/10.1017/S1431927612001079.

Sanchez, S., Fernandez, V., Pierce, S.E., Tafforeau, P., 2013. Homogenization of sample absorption for the imaging of large and dense fossils with synchrotron microtomography. Nat. Protoc. 8, 1708–1717. https://doi.org/10.1038/nprot.2013.098.

Schindelin, J., Arganda-Carreras, I., Frise, E., Kaynig, V., Longair, M., Pietzsch, T., Cardona, A., 2012. Fiji: an open-source platform for biological-image analysis. Nat. Methods 9, 676–682. https://doi.org/10.1038/nmeth.2019.

Silva, P.A., Borba, V.H., Dutra, J.M., Leles, D., Da-Rosa, A.A., Ferreira, L.F., Araujo, A., 2014. New ascarid species in cynodont coprolite dated of 240 million years. An Acad. Bras Ciências 86, 265–269. https://doi.org/10.1590/0001-3765201320130036.

Sollas, W.J., 1904. A method for the investigation of fossils by serial sections. Philos. Trans. R. Soc. Lond. B Biol. Sci. 196, 257–263. https://doi.org/10.1098/rstb.1904.0008.

Tafforeau, P., Smith, T.M., 2008. Nondestructive imaging of hominoid dental microstructure using phase contrast X-ray synchrotron microtomography. J. Hum. Evol. 54, 272–278. https://doi.org/10.1016/j.jhevol.2007.09.018.

Tafforeau, P., Boistel, R., Boller, E., Bravin, A., Brunet, M., Chaimanee, Y., Cloetens, P., Feist, M., Hoszowska, J., Jaeger, J.J., Kay, R.F., 2006. Applications of X-ray synchrotron microtomography for non-destructive 3D studies of paleontological specimens. Appl. Phys. A 83, 195–202. https://doi.org/10.1007/s00339-006-3507-2.

Thulborn, R.A., 1991. Morphology, preservation and palaeobiological significance of dinosaur coprolites. Palaeogeography, Palaeoclimatology. 83, 341–366.

Tripp, M., Wiemann, J., Brocks, J., Mayer, P., Schwark, L., Grice, K., 2022. Fossil biomarkers and biosignatures preserved in coprolites reveal carnivorous diets in the carboniferous mazon creek ecosystem. Biology 11 (9), 1289. https://doi.org/10.3390/biology11091289.

Umamaheswaran, R., Prasad, G.V.R., Rudra, A., Dutta, S., 2019. Biomarker signatures in Triassic coprolites. Palaios 34, 458–467. https://doi.org/10.2110/palo.2019.023.

Vršanský, P., van de Kamp, T., Azar, D., Prokin, A., Vidlička, L.U., Vagovič, P., 2013. Cockroaches probably cleaned up after dinosaurs. PLoS One 8 (12), e80560. https://doi.org/10.1371/journal.pone.0080560.

Wang, X., White, S.C., Balisi, M., Biewer, J., Sankey, J., Garber, D., Tseng, Z.J., 2018. First bone-cracking dog coprolites provide new insight into bone consumption in *Borophagus* and their unique ecological niche. Elife 7, e34773. https://doi.org/10.7554/eLife.34773.

Williams, M.E., 1972. The origin of 'spiral coprolites'. University of Kansas. Palaeontological Contributions 59, 1–19.
Wilson, J.W., 1994. Histological techniques. In: Leiggi, P., May, P. (Eds.), Vertebrate Paleontological Techniques, Vol 1. Cambridge University Press, Cambridge.
Yao, M.T., Sun, Z.Y., Meng, Q.Q., Chen, S.M., Li, J.C., Jiang, D.Y., 2023. A possible pseudosuchian archosaur coprolite from the Middle Triassic lacustrine Ordos Basin, China: palaeobiological and palaeoecological significance. In: Neues Jahrbuch Geol Palaontol Abhand, pp. 261–275.
Zatoń, M., Broda, K., Qvarnström, M., Niedźwiedzki, G., Ahlberg, P.E., 2017. The first direct evidence of a Late Devonian coelacanth fish feeding on conodont animals. Sci. Nat. 104, 1–5. https://doi.org/10.1007/s00114-017-1455-7.

CHAPTER

6

Vertebrate dentalites

Adrian P. Hunt[1] and Spencer G. Lucas[2]

[1]Flying Heritage and Combat Armor Museum, Everett, WA, United States; [2]New Mexico Museum of Natural History, Albuquerque, NM, United States

1. Introduction

Hunt et al. (2018, p. 500) introduced the term dentalite to encompass all traces produced on a substrate by the teeth or oral cavity of a vertebrate or invertebrate. These trace fossils were generally referred to previously as "bite marks" by paleontologists (e.g., Avilla et al., 2004; Drumheller et al., 2014) and more correctly by ichnologists as "bite traces" (e.g., Mikuláš et al., 2006; Mikuláš and Dvořák, 2010; Zonneveld et al., 2022b). Most examples of dentalites were made by vertebrates, but the term also applies to traces left by the jaw apparatus of invertebrates (e.g., *Bicrescomanducator rolli*—cephalopod dentalites on ammonites: Andrew et al., 2010: *Gnathichnus pentax*—echinoid dentalite on belemnite—Bromley, 1975). The substrate of vertebrate dentalites is most often bone (Binford, 1981; Fernández-Jalvo and Andrews, 2016) but can be vegetation (Lucas, 2016), invertebrate hard parts (Kauffman and Sawdo, 2013), coprolites (Godfrey and Smith, 2010), or sediment (Gingras et al., 2007).

Several ichnologists disagree with usage of the term dentalite (e.g., Zonneveld et al., 2022b; LaBarge and Njau, 2024; Mikuláš, pers. commun., 2024). Their objections are principally:

1. Bite mark, tooth mark, and gnaw mark have historic precedence.
2. Dentalite represents obscure, exclusionary language that excludes nonichnologists.
3. There is no need for a new term.
4. A dentalite arguably refers to the absence of rock where biting occurred and therefore the incorporation of the Greek lithos in the term may not be appropriate.

We believe that historical precedence is not relevant to the continued use of imprecise, vernacular terms. Regurgitalite, by comparison, was only introduced in 1992 but has gained widespread acceptance because it is more useful than older imprecise terms such as regurgitation, gastric residue, and ejecta (Hunt and Lucas, 2012, 2024). New terminology is always initially to a degree exclusionary (cf. regurgitalite), but arguably the meaning of dentalite is easy to discern and will aid in its acceptance. In our opinion, there is a need for precise terminology in regard to traces produced by biting that is filled by dentalite. We deem the

Vertebrate Ichnology
https://doi.org/10.1016/B978-0-443-38351-9.00003-7

derivation of the term dentalite to be appropriate and consonant with others as it refers to a trace on lithified material.

Ultimately, a redundant and irrelevant term will be ignored, and a useful one will be utilized. We note that the term dentalite has been utilized in studies from the Americas to Europe to Asia and whose foci range in age from Devonian to Miocene (Cueille et al., 2020; Jagt et al., 2020; Bosio et al., 2021; da Costa Pereira et al., 2022; Robin et al., 2022; Gross et al., 2023; Mazurek and Antczak, 2023; Sagar et al., 2024).

Hunt and Lucas (2021, Tables A.1–A.5) recently provided an extensive review of the record of vertebrate dentalites that is briefly reprised and extensively updated here. We also include a complete revision of the ichnotaxonomy of vertebrate dentalites and related ichnotaxa. Minter (2024) also reviews fish dentalites on a lithic substrate. Note that we do not consider invertebrate dentalites.

Institutional acronyms: Ah SGDB, Collection "Ahníkov" of the Doly Bílina Company, Doly Bílina, Czech Republic; CMM, Department of Paleontology, Calvert Marine Museum, Solomons, Maryland, USA; coll. Doly Bílina, Collection "Ahníkov" of the Doly Bílina Company, Doly Bílina, Czech Republic; CHMHZ, Chlupáč's Museum of Earth History, Faculty of Science, Charles University, Prague. CUMZ, University Museum of Zoology, University of Cambridge, Cambridge, England; DGM, Coleção de Paleontologia do Museu de Ciências da Terra, Rio de Janeiro, Brazil; DPC, Duke Lemur Center, Durham, USA; IUPC: Indiana University Paleontology Collection, Bloomington, USA; KK, Kronosaurus Korner, Richmond, Australia; MCC, Museu Câmara Cascudo of Universidade Federal do Rio Grande do Norte, Natal city, Rio Grande do Norte State, Brazil; MCT, Museu de Ciências da Terra, Rio de Janeiro city, Rio de Janeiro State, Brazil; MGUH, Statens Naturhistoriske Museum (Museum Geologicum Universitatis Hauniensis), Copenhagen, Denmark; MLP, Museo de La Plata, La Plata, Argentina; MNHN, Muséum national d'Histoire naturelle, Paris, France; MPCA, Museo Provincial Carlos Ameghino, Cipolletti, Argentina; PGI-NRI, Geological Museum of the Polish Geological Institute—National Research Institute, Warszawa, Poland; NMNHU-P, collection AR, Department of Palaeontology of the National Museum of Natural History, National Academy of Sciences of Ukraine, Kyiv, Ukraine; RTMP, The Royal Tyrrell Museum of Palaeontology, Alberta, Canada; SMNS, Staatliche Museum für Naturkunde, Stuttgart, Germany; STA-BAP, Paleontological Collection of the Laboratorio de Ciencias de la Tierra y Paleontología, Facultad de Estudios Superiores Iztacala, Universidad Nacional Autònoma de Mexico; Tr, Museum of Natural History of the General Directorate of Mineral Research and Exploration, Ankara, Turkey; UMMP, University of Michigan Museum of Paleontology, Ann Arbor, Michigan, USA; WHYJ, Laboratory of Geobiology, School of Earth Sciences, China University of Geosciences, Wuhan, China; WNoZ, Museum of Earth Sciences at the University of Silesia, Sosnowiec, Poland; X, Section X—Problematica and trace fossils of type and figured specimen register of Paleontology Collection, School of Environment, University of Auckland, Auckland, New Zealand; ZPAL, Instytut Paleobiologii Polskiej Akademii Nauk, Warsaw, Poland.

2. Actualistic studies

Buckland (1822, 1824) conducted the first actualistic study of dentalites by feeding cow bones to a hyena from a traveling show and comparing the bite marks with those present

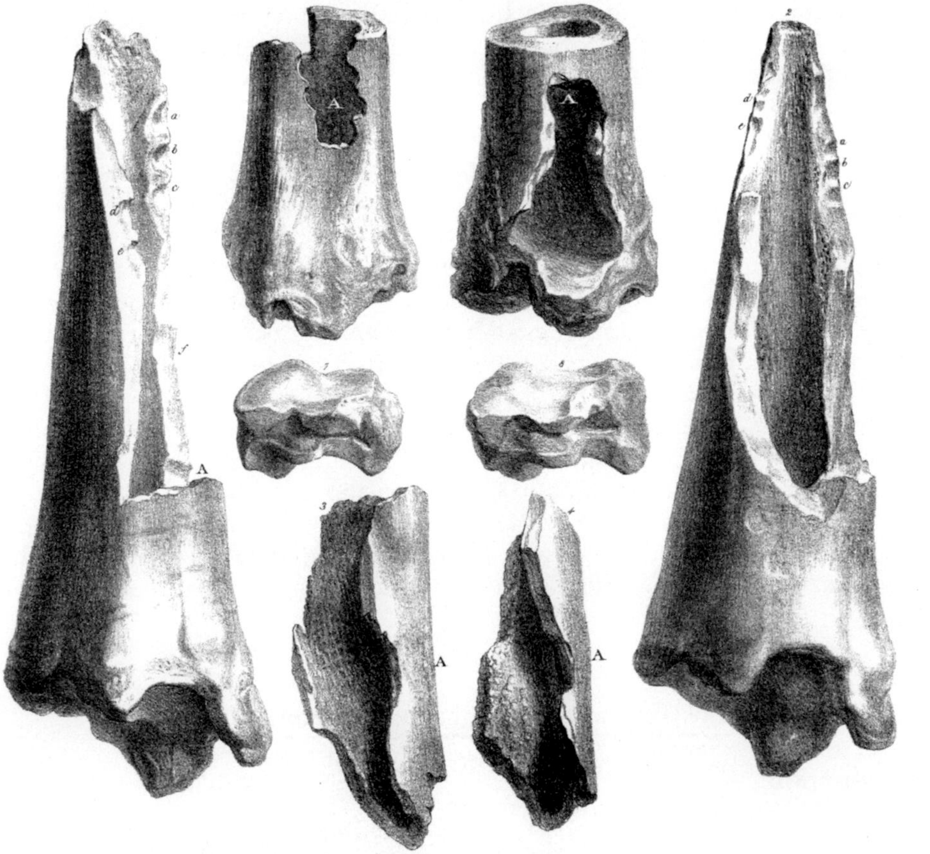

FIGURE 6.1 Comparison of Pleistocene and recent damage to bovid bones by hyena jaws. "1. Residuary part of the lower extremity of the tibia of an ox, which I saw given entire to a Cape hyaena in Mr. Wombell's traveling collection at Oxford, in December 1822: Marks of the teeth are definitely distinctly visible at a, b, c, d, e, f.... 2. Fragment of a similar tibia from Kirkdale cave, broken nearly in the same manner as No. 1, and bearing similar marks of teeth at a, b, c, d, e: In the recent, as in the antediluvian specimen, the lower condyle has, from its hardness, been left unbroken. 3. Splinter from another bone broken by the hyaena at Oxford: the cavity at A. was produced by the hyaena's tooth. 4. Similar splinter, bearing a similar cavity A., from the cave at Kirkdale, and partially incrusted with stalagmite. 5. Inside of the lower extremity of the recent specimen, No. 1, in which the hole A. was produced by the hyaena's bite ... . 6. Lower extremity of another tibia from Kirkdale, in which the form of the cavity A. resembles that in No. 5. 7. Scaphoid bone of the left carpus of an ox, which, with the other component bones of the carpus, lay all night untouched in the hyaena's cage at Oxford. 8. Similar bone from Kirkdale, equally untouched" (Buckland, 1824, p. 276–277). *From Buckland (1824), pl. 23.*

on Pleistocene bison bones from Kirkdale Cave (Fig. 6.1). Lyell (1863) provided porcupines at the London Zoo with bone and antler samples to gnaw upon, and he compared the results with marks on probable Pleistocene bones from Saint Prest in France. Modern actualistic studies were largely inspired by controversies generated by the interpretation of early archeological sites in East Africa and South Africa, including Olduvai Gorge (Tanzania), Koobi Fora (Kenya), and Sterkfontein, Klasies River Mouth and Swartkrans (South Africa) (Arriaza et al.,

2021). This research started in the 1970s but reached a crescendo in the 1980s (e.g., Sutcliffe, 1970; Binford, 1981; Brain, 1970, 1978, 1981; Bunn, 1982, 1983; Haynes, 1980, 1982, 1983; Blumenschine, 1986, 1988). New methodologies are being developed (e.g., Domínguez-Rodrigo and Piqueras, 2003; Domínguez-Rodrigo et al., 2012).

Several studies have focused on large carnivores. For example, Parkinson et al. (2015) studied lion and tiger dentalites. Arriaza et al. (2021) recently studied leopard dentalites, utilizing microphotogrammetry and geometric morphometrics. They demonstrated that tooth marks inflicted by leopards, spotted hyenas and lions can be statistically differentiated based on their morphology, and that the ratio of pits to scores inflicted by the leopard is closer to that of hyenas. Blumenschine et al. (1996) conducted blind tests that indicated that marks inflicted on bone surfaces by carnivore teeth, hammerstone percussion, and metal knife cutting and scraping can be distinguished with near-perfect reliability without scanning electron microscopy or consideration of only conspicuous marks.

Actualistic studies on crocodile dentalites were notably spurred by dissertations and subsequent research by Njau (2006; Njau and Blumenschine, 2006, 2012; Njau and Gilbert, 2016), Drumheller (Drumheller-Horton, 2012; Drumheller and Brochu, 2014; 2016), and LaBarge (LaBarge, 2022; La Barge and Njau, 2024). D'Amore (2009) studied tooth marks of *Varanus komodoensis* as analogs for dentalites of theropods. Domínguez-Rodrigo and Baquedano (2018) used machine learning methods to distinguish butchery cut marks from crocodile bite marks.

3. Fossil record of dentalites on bony substrate

3.1 Introduction

Hunt and Lucas (2021, Tables A.1–A.11) attempted to compile data on all recorded specimens of vertebrate dentalites older than Pleistocene in age. The volume of literature on Pleistocene vertebrate dentalites precludes a complete review, so Hunt and Lucas (2021) only included representative examples. We have characterized traces as marine or nonmarine, again based largely on the original literature. We somewhat arbitrarily included lagoons in the nonmarine category, even though many such environments have marine influences.

We organize our review of the dentalite record by the substrate on which the dentalite is registered, because identifying the producers of dentalites is often problematic. There are several clear patterns to the documented record of dentalites (Hunt and Lucas, 2021: Fig. 6.2).

1. There is a bias toward dentalites of large animals preserved on large pieces of substrate (e.g., cetaceans, dinosaurs, and marine reptiles). This is presumably related to ease of recognition and to the fact that smaller animals may be wholly ingested or totally destroyed by mastication.
2. Hunt et al. (2018) coined the term Taxophile Effect to refer to the disproportionate study of more "popular" taxonomic groups. This is well seen in the marked disparity between the numbers of reports of dentalites on dinosaur bones relative to those on the bones of Cenozoic mammals. The Taxophile Effect may also be at play with invertebrates, as

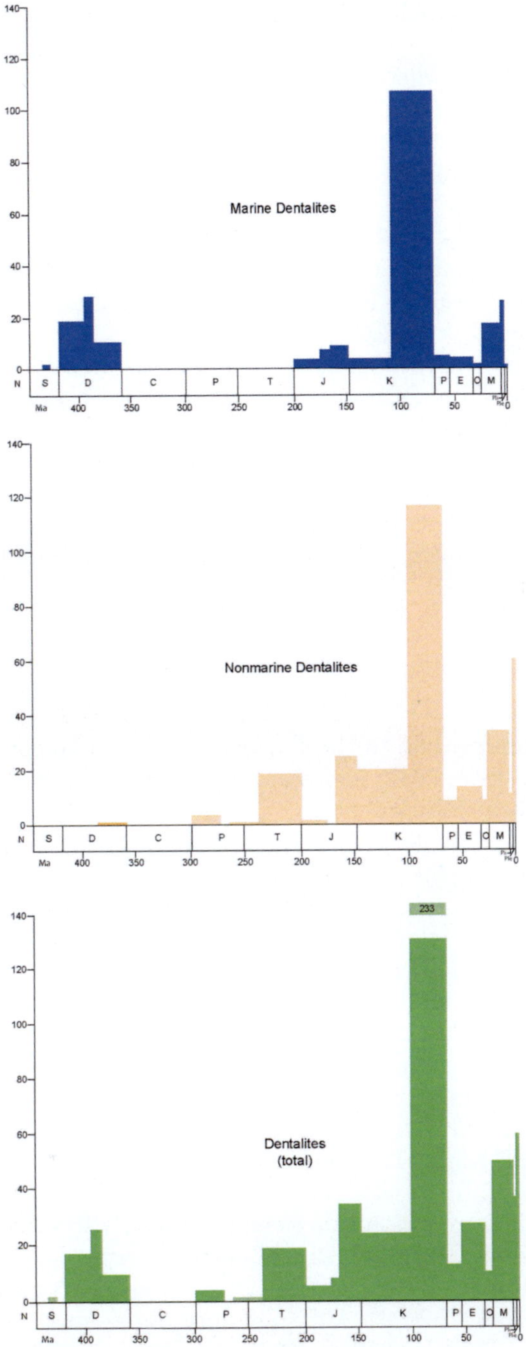

FIGURE 6.2 Temporal distribution of reports of dentalites. *For sources see Hunt and Lucas, (2021, Tables A.1–4).*

there are more reports of dentalites on elegant ammonoids than on any other invertebrate taxonomic group.

3. Dentalites are nonrandomly distributed in time, with notable acmes in the Late Triassic, Late Cretaceous, and late Cenozoic.
4. Prior to the Quaternary, with the exception of the nonmarine Mesozoic, most dentalites are reported from marine facies. In the Quaternary, a large percentage of dentalites are from cave deposits.
5. There is a clear seeker bias (seek and ye shall find), as careful investigation of large, existing collections, at least of vertebrate specimens, has always led to the identification of multiple examples of dentalites (e.g., Jacobsen, 1995, 1998; Konuki, 2008; Drumheller et al., 2020; Dalman and Lucas, 2021).

Dentalites on a bony substrate pertain to the *Cubiculum* ichnofacies of Lucas (2016), those on vegetation to the *Palaeoscolytus* ichnofacies of Lucas (2016), and those on coprolites to the *Gaspeichnus* ichnofacies of Hunt et al. (2018). The *Cubiculum* Ichnofacies (e.g., Ekrt et al., 2016; Collareta et al., 2023b) and *Gaspeichnus* Ichnofacies (Godfrey and Collareta, 2022) have received positive recognition. Dentalites on marine invertebrates could be assigned to the *Gnathichnus* Ichnofacies of Bromley and Asgaard (1993), which may be a tier of the *Trypanites* Ichnofacies (Frey and Seilacher, 1980; MacEachern et al., 2007, 2012).

3.2 Paleozoic

3.2.1 Marine

Putative jawed vertebrates date to the Ordovician (Karatajute-Talimaa and Predtechenskyj, 1995; Sansom et al., 1996, 2012; Smith and Sansom, 1997), but the first definitive body fossil remains are of Early Silurian age (Zhao and Zhu, 2010; Brazeau and Friedman, 2015). The earliest record of a dentalite is on an unidentified jawless cyathaspid from the Middle Silurian (Wenlock) Man on the Hill (MOTH) site in the Mackenzie Mountains, Northwest Territories of Canada (Randle and Sansom, 2019a,b). There are no other recorded Silurian examples, but dentalites become relatively common in the Devonian, and the study of ostracoderms indicates increasing frequency of these traces on aquatic vertebrates through the period (Lebedev et al., 2009; Randle and Sansom, 2019a).

Early Devonian dentalites are present on heterostracans in Canada, the USA, England, Wales, Scotland, Ukraine, Sweden, and Russia (Tarrant, 1991; Lebedev et al., 2009; Elliott and Petriello, 2011; Randle and Sansom, 2019a). Dentalites of Middle Devonian age are known in heterostracans, arthrodires, and antiarchs in Latvia, Estonia, Russia, and the USA (Obruchev and Mark-Kurik, 1965; Lebedev et al., 2009; Johanson et al., 2013; Tuuling, 2015; Randle and Sansom, 2019a).

Late Devonian dentalites are present on heterostracans, arthrodires, antiarchs, and osteolepiforms in Scotland, Estonia, Belgium, Russia, Canada, and Morocco (Capasso et al., 1996; Lebedev et al., 2009; Scott et al., 2012; Cloutier, 2013; Glinskiy and Mark-Kurik, 2016; Randle and Sansom, 2019a; Robin et al., 2022). Although eurypterids might be considered the perpetrators of some putative Devonian dentalites (e.g., Romer, 1933, but see Lamsdell and Braddy, 2009), there are only two probable recorded occurrences (Lebedev et al., 2009; Elliott and Petriello, 2011; Blieck, 2017).

There are very few reports of dentalites on Carboniferous or Permian fish fossils. The marked decline after the Devonian probably relates in large part to the extinction of many of the armored fish, as sheets of dermal armor provide a preferential substrate to preserve dentalites. Zangerl and Richardson (1963) published a classic study of fossiliferous Pennsylvanian black shales that contain fossils that appeared to preserve abundant examples of bite marks. They documented a range of incomplete skeletons of fish, which they interpreted as representing dismembered carcasses and "gastric residues" (regurgitalites). However, subsequent study has made it clear that the partial skeletons represent normal decay processes, most of the putative regurgitalites are not regurgitalites, and there are no preserved dentalites (Elder, 1985; Elder and Smith, 1988).

3.2.2 Nonmarine

There are only a small number of reports of Paleozoic nonmarine dentalites. However, one of the oldest tetrapod fossils, a humerus from the Famennian (Late Devonian), has puncture marks on the dorsal and ventral surfaces that may be dentalites (Shubin et al., 2004). The earliest example in the nonmarine Paleozoic of multiple dentalites is on an articulated skeleton of the early Permian synapsid predator *Varanops*, which preserves traces interpreted to be the result of scavenging by a dissorophoid amphibian (Reisz and Tsuji, 2006, p. 1023). Reisz and Tsuji (2006, p. 1023) stated that "few Paleozoic tetrapods have been prepared with the kind of care required to readily identify bite marks, or studied with these taphonomic features in mind." The early Permian Craddock Bone Bed in Texas yields abundant dentalites produced by *Dimetrodon* that have not been described in detail (Houpt et al., 2020). Dentalites are present on a snout of the amphibian *Diplocaulus* but most commonly on skeletal elements of adult dimetrodonts and subadult diadectids. Damage is concentrated on distal limb elements and has been interpreted as the result of feeding on major muscle masses (Houpt et al., 2020).

Another early Permian example is represented by a puncture dorsal to the caniniform tooth in the skull of *Dimetrodon* (*Bathygnathus*) *borealis* (Langston, 1963; Brink et al., 2013, 2015). The classic late Permian Karoo faunas preserve few reported dentalites, but one example is of dentalites likely made by the gorgonopsian *Aelurognathus* on a therapsid skeleton (Fordyce et al., 2012).

3.3 Mesozoic

3.3.1 Marine

There are a large number of dentalites reported from the marine Mesozoic (Hunt and Lucas, 2021, Table A.1). The majority are related to marine reptiles, and many dentalites in the Cretaceous were produced by sharks. Marine tetrapods provided an important substrate for the preservation of dentalites throughout the Mesozoic and Cenozoic. They generally have larger body sizes and more robust skeletal elements than chondrichthyans and bony fish, both factors that aid in the preservation and recognition of dentalites. The greatest number of occurrences is from the Late Cretaceous, with many from the sedimentary record of the Western Interior Seaway of western North America (Hunt and Lucas, 2022).

There are two dentalite records from the Early Jurassic, both from ichthyosaurs. Five specimens of the ichthyosaur *Temnodontosaurus* from Germany have dentalites attributed to *Temnodontosaurus* and the crocodylomorph *Steneosaurus* (Pardo-Pérez et al., 2018). A skull of the ichthyosaur *Leptonectes* cf. *L. tenuirostris* from England has a healed wound that appears to have been a dentalite (Maisch and Matzke, 2003).

There are a number of dentalites reported from the Middle Jurassic of England, and one from Scotland. One example of the fish *Leedsicthys* preserves a tooth of the marine crocodylomorph *Metriorhynchus* in a dentalite (Martill, 1985). Four specimens of the plesiosaur *Cryptoclidus* or cf. *Cryptoclidus* from England include dentalites, three of which are attributed to a pliosaur maker and one to *Metriorhynchus* (Martill et al., 1994; Forrest, 2000, 2003; Rothschild et al., 2018). Two ichthyosaurs, one from England and one from Scotland, include dentalites, one identified as having been produced by a pliosaur (Martill, 1996; Brusatte et al., 2015). The Middle-Upper Jurassic Rosso Ammonitico Veronese of northeastern Italy yields a sauropsid rib with possible shark dentalites (Serafini et al., 2024).

Late Jurassic dentalites are not numerous but are more broadly distributed than those from earlier parts of the period. Two specimens of turtles from Switzerland assigned to *Plesiochelys* sp. and an indeterminate taxon have dentalites attributed to the crocodyliform *Machimosaurus* (Meyer, 1991). Bones of the marine crocodyliforms *Geosaurus vignaudi* and *G. saltillense* from Mexico include dentalites (Frey et al., 2002; Buchy et al., 2006). Late Jurassic plesiosaurs with dentalites occur in the USA, England, and Mexico, and Late Jurassic ichthyosaurs with dentalites are present in Russia, Poland, and Norway (Martill et al., 1994; Wahl, 2006; Buchy, 2007; Druckenmiller et al., 2012; Novis, 2012; Zverkov et al., 2015; Tyborowski, 2016). One of the plesiosaurs and one of the ichthyosaur occurrences are attributed to bites from pliosaurs (Buchy, 2007; Zverkov et al., 2015).

The majority of Early Cretaceous dentalite records are from Australia (Hunt and Lucas, 2021, Table A.1). These specimens include dentalites on the ichthyodectiform fish *Cooyoo australis* (Wretman and Kear, 2014). The holotype of the "elasmosaur" *Eromangasaurus carinognathus* has dentalites made by the "pliosaur" *Kronosaurus* (Thulborn and Turner, 1993; Kear, 2005; McHenry, 2009). In turn, there are two examples of dentalites on *Kronosaurus*, one produced by intraspecific interaction (McHenry, 2009) and one by a cretoxyrhinid shark (Holland, 2018). Kellner et al. (2010) described shallow punctures on an ornithocheiroid pterosaur wing metacarpal. Early Cretaceous dentalites outside of Australia include specimens of the crocodyliform *Pachycheilosuchus trinquei* from the USA (Rogers, 2003) and a spinosaur tooth within a dentalite on the pterosaur *Anhanguera* sp. from Brazil (Buffetaut et al., 2004; Kellner, 2004).

Late Cretaceous marine dentalites are nearly global in their distribution, being present in New Zealand, Australia, Japan, Italy, France, Germany, Belgium, Netherlands, Sweden, Jordan, Angola, Mali, and Chile. However, the vast majority of occurrences are from the Western Interior of the USA and Canada (Hunt and Lucas, 2021). A wide range of taxonomic groups preserve dentalites, but the most characteristic group from the Late Cretaceous is the marine lizards, the mosasaurs. Dentalites produced by sharks, notably *Squalicorax* and *Cretoxyrhina*, become abundant in the Late Cretaceous (e.g., Schwimmer et al., 1997; Everhart, 2005c), and this trend extends into the Cenozoic.

There are 20 records of dentalites on large chondrichthyans and teleosts from the Late Cretaceous (Hunt and Lucas, 2021, Table A.1), including on specimens of *Cretoxyrhina*

FIGURE 6.3 Late Cretaceous dentalites from marine strata of Kansas, USA. (A) Proximal end of mosasaur rib with characteristic serrated dentalites (*Linichnus serratus*) produced by *Squalicorax* sp. (Everhart, 2017, Fig. 6.13). (B) Left hyoplastron of the marine turtle *Protostega gigas* (FHSM VP-17979), in ventral view, from Santonian chalk, preserving dentalites originally interpreted as produced by a mosasaur (Everhart, 2017, Fig. 6.13). Note that several punctures have an associated scrape mark similar to the "drag-snags" of Njau and Gilbert (2016), which may suggest a crocodilian origin of the dentalites. *Photographs courtesy of Michael Everhart.*

mantelli, *Eumylodus laqueatus*, *Xiphactinus audax*, and *Pachyrhizodus caninus* (Shimada, 1997; Schwimmer et al., 1997; Shimada and Everhart, 2004; Cicimurri et al., 2008; Amalfitano et al., 2019). One of these specimens is from Italy, and the remainder are from the USA. Of the identified tracemakers, 16 belong to *Squalicorax*, and one is *Cretoxyrhina mantelli* (Hunt and Lucas, 2021, Table A.1).

Published Late Cretaceous dentalite records on marine turtles number about 23 (Fig. 6.3B) (Hunt and Lucas (2021), Table A.1). These include single occurrences from Italy, Angola, Jordan, and The Netherlands, with the remainder from the USA (Mulder, 2003; Kaddumi, 2006; Mateus et al., 2012; Amalfitano et al., 2017). All dentalites on Jurassic turtles are attributed to crocodylomorphs. Late Cretaceous dentalites include two or three attributed to crocodylomorphs, four or five to mosasaurs and 17 to sharks, principally *Squalicorax* with a few attributed to *Cretoxyrhina* or other taxa (Hunt and Lucas, 2021, Table A.1). Marine crocodylomorphs are less common in the Late Cretaceous, and there are only three examples with dentalites from Mali, France, and the USA, and the two identified tracemakers are selachians (Hua et al., 2007; Boles and Lacovara, 2013; Hill et al., 2015; O'Leary et al., 2019).

There are more than 50 documented examples of dentalites on Late Cretaceous mosasaurs (Fig. 6.3A: Hunt and Lucas, 2021, Table A.1; Jagt et al., 2020; Polcyn et al., 2023). The majority of the specimens are from the United States, but they also are present in The Netherlands, Belgium, Germany, Spain, and Angola (Bardet et al., 1998; Dortangs et al., 2002; Corral et al., 2004; Lingham-Solar, 2004; Rothschild et al., 2005; Strganac et al., 2015). There are 10 examples of dentalites on mosasaurs produced by mosasaurs, some due to presumed intraspecific head biting (e.g., Rothschild and Martin, 1993; Bell and Martin, 1995; Everhart, 2008; Carr et al., 2020). There are approximately 40 examples attributed to sharks, about a quarter of these produced by *Squalicorax* and several by *Cretoxyrhina*, but many are attributed to undetermined selachians (e.g., Fig. 6.3A: Schwimmer et al., 1997; Everhart, 1999; Konuki, 2008; Polcyn et al., 2023).

There are 14 reports of dentalites on plesiosaur bones, half from the USA and the others from Sweden, Japan, Chile, Argentina, and New Zealand (Sato et al., 2006; Barnes and Hiller, 2010; Einarsson et al., 2010; Shimada et al., 2010; Kubo et al., 2012; Otero et al., 2014; Araújo et al., 2015; Hunt and Lucas, 2021, Table A.1). The majority of specimens preserve dentalites produced by sharks, primarily *Squalicorax*, with two examples of *Cretoxyrhina* (e.g., Schwimmer et al., 1997; Everhart, 2005a; Araújo et al., 2015). There is one report of a mosasaur-produced dentalite on a plesiosaur (Einarsson et al., 2010), and two plesiosaur skeletons that record dentalites derived from both a mosasaur and one or more selachians (Konuki, 2008; Barnes and Hiller, 2010).

Six dinosaur specimens that preserve dentalites have been recovered from marine strata in the USA—four hadrosaurs, a nodosaur, an ornithomimosaur, and another theropod (Schwimmer et al., 1997; Everhart and Hamm, 2005; Everhart and Ewell, 2006; Schein and Poole, 2014; Brownstein, 2018; Brownstein and Bissell, 2021). All but the theropod, whose dentalite is attributed to a crocodylomorph, were bitten by sharks. Two specimens of the pterosaur *Pteranodon* preserve dentalites produced by *Squalicorax* (and a small fish) and *Cretoxyrhina* (Konuki, 2008; Ehret and Harrell, 2018; Hone et al., 2018a,b). There is one record of dentalites on the bird *Hesperornis* produced by a polycotylid plesiosaur (Martin et al., 2016).

3.3.2 *Nonmarine*

Although there is only a small data set, it seems that dentalites are relatively less common in the Paleozoic than the Mesozoic (Hunt and Lucas, 2021, Table A.2). We posit that this may be caused by several factors that distinguish most Mesozoic vertebrate predators from most Paleozoic vertebrate predators, including (1) the evolution of laterally compressed teeth in terrestrial predators, which would penetrate through soft tissue to bone more effectively; (2) the development of a more upright gait among predators that would provide for more three-dimensional predation—more opportunity for biting the dorsal as well as the lateral regions of prey; and (3) increased body size of prey that would invite more extensive scavenging of carcasses than is feasible with small-bodied prey.

There are relatively few unambiguous reports of dentalites from the nonmarine Early and Middle Triassic. Modesto and Botha-Brink (2010) described a possible puncture on the scapula of a juvenile *Lystrosaurus* from the Early Triassic of South Africa. Fernandez et al. (2013) reported possible dentalites on the temnospondyl *Broomistega putterilli* from a burrow, also in the Early Triassic of South Africa. Cruickshank (1986) proposed the first ichnotaxonomic name for a vertebrate dentalite, *Mandaodontites coxi* for the imprint of a partial dental arcade with ovoid tooth impressions (Hunt and Lucas, 2021). This dentalite, from the Middle Triassic of Tanzania, is on a dicynodont bone and is attributed to an archosaur (see Zonneveld, 2022 and discussion below). Schoch et al. (2022) and Mujal et al. (2022) describe dentalites from the Middle Triassic of Germany.

There are a large number of reports of Late Triassic dentalites from a wide range of taxa across a broad geographic area (North and South America and Europe), although they are never abundant in any given fossil assemblage. The extent of the dentalite record thus broadly parallels that of the osseous record, with many more specimens known from the Late Triassic than in the earlier part of the period. The prevalence of dentalites correlates to a marked increase in the diversity of taxa with dorsal armor in Late Triassic nonmarine environments (nonarchosaurian archosauriforms [*Vancleavea*], pseudosuchians [*Revueltosaurus*], aetosaurs, phytosaurs, paracrocodylomorphs, crocodylomorphs, turtles). Hunt et al. (2018) and Hunt and Lucas (2021) recently reviewed the record of Late Triassic dentalites. Sadlok (2022) discussed several specimens in bone fragments from southern Poland.

The only record of dentalites on a Late Triassic amphibian is a clavicle fragment of *Koskinonodon perfectum* from New Mexico that preserves two rows of tooth marks made by blade-like and rounded teeth (Rinehart et al., 2006: Fig. 6.4A and B). Rinehart et al. (2006) named these dentalites *Heterodontichnites hunti* and attributed them to a phytosaur. Dentalites on several phytosaur skulls of *Nicrosaurus* and *Pseudopalatus* in North America and Europe represent intraspecific head biting, as occurs in modern crocodylians (Abel, 1922a,b; Camp, 1930; Ruben, 1990). There are only two reports of dentalites on aetosaurs. Both specimens are on the aetosaurs *Typothorax* and *Desmatosuchus* from the southwestern United States (Zeigler et al., 2002; Drymala and Bader, 2012; Drymala et al., 2021). Battista et al. (2024) report six examples of dentalites representing 0.21% of the >2700 bones sampled from the Late Triassic of Brazil.

Dentalites on terrestrial carnivores include two incomplete femora of paracrocodylomorphs ("Rauisuchia") from northern New Mexico, USA (Drumheller et al., 2014), with the dentalites interpreted to be from phytosaurs. A humerus of the holotype of the paracrocodylomorph *Postosuchus kirkpatricki* preserves dentalites of unknown origin (Weinbaum, 2013). The holotype skeleton of the crocodylomorph *Dromicosuchus grallator* from North Carolina, USA, preserves damage to the cervical area and mandible attributed to the paracrocodylomorph whose skeleton was found immediately above it, and it is inferred that the two animals died and were buried together during the act of predation (Sues et al., 2003). Sereno and Novas (1994) described healed intraspecific bite marks on the saurischian *Herrerasaurus*.

The majority of dentalites described from the Late Triassic occur on dicynodont skeletons, which are always the largest-bodied herbivorous taxa in their faunas (in contrast, they are not often registered on the bones of "prosauropod" dinosaurs, which dominate vertebrate faunas in certain environments/geographic locations). A femur of a dicynodont similar to *Ischigualastia* from New Mexico, USA, has multiple dentalites that are tooth marks of a phytosaur assigned to *Heterodontichnites hunti* (Lucas and Hunt, 1993; Hunt and Lucas, 2014: Fig. 6.4C and D). Budziszewska-Karwowska et al. (2010) described bite marks on a dicynodont tibia from Poland. The bone has longitudinal dentalites on the anterior side of its shaft, as well as a row of small oval pits that are interpreted to represent the scavenging of more than one carnivore.

Elsewhere in southern Poland, there are six small dicynodont bones with dentalites (Hunt and Lucas, 2021, Table A.2). These traces were assigned to the dentalite ichnotaxa *Linichnus serratus*, *Knethichnus parallelum*, and *Nihilichnus nihilicus* and attributed to theropod dinosaurs (Dzik et al., 2008; Niedźwiedzki et al., 2010). Dentalites also occur on ribs of the dicynodont "*Jachaleria*" *candelariensis* and other dicynodonts from Brazil (Braunn et al., 2001; Vega-Dias and Schultz, 2007). Niedźwiedzki et al. (2010) suggested that dentalites on dicynodont bones indicate that an increase in the size of dicynodonts through the Late Triassic may have been driven by selection pressure to reach a size refuge from early dinosaur predators.

Other dentalites on herbivore bones include the topotype of the pseudosuchian *Revueltosaurus callenderi* that preserves dentalites (gnaw marks: Hunt et al., 2005; Hunt and Lucas,

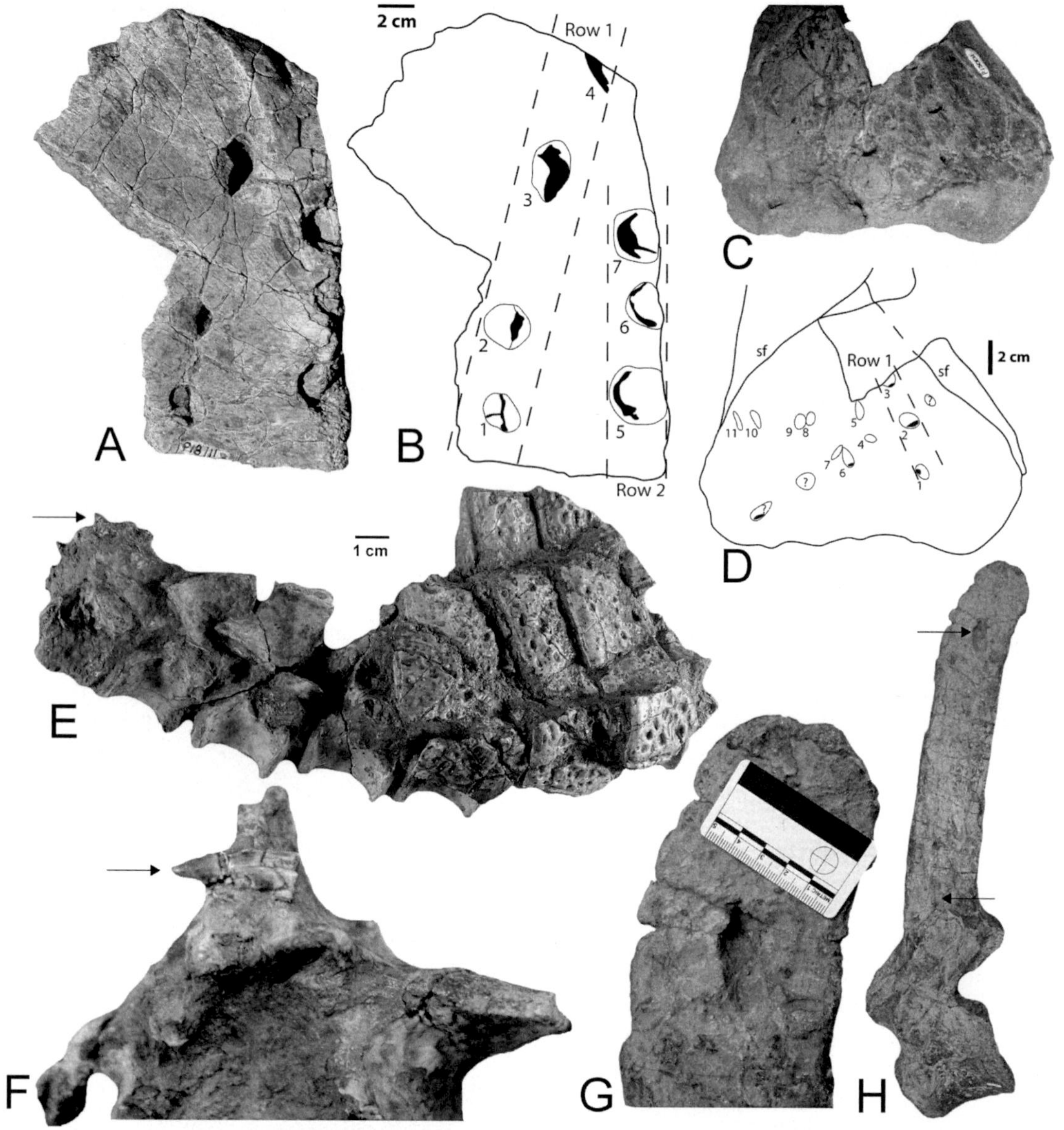

FIGURE 6.4 Dentalites from the Triassic and Cretaceous of New Mexico, USA. (A and B) NMMNH P-18111 from the Lamy amphibian quarry, 12 cm-long *Koskinodon perfectum* clavicle fragment with holotype bite marks of *Heterodontichnites hunti*. (A) Interior surface. (B) Sketch showing the tooth mark numbers and rows (after Rinehart et al., 2006, Fig. 1C). (C and D) NMMNH P-13001 from locality L-1410, a left dicynodont femur of *Ischigualastia* sp. with bite marks of *Heterodontichnites hunti*. (C) Tooth marks on the anterior distal surface. (D) Sketch showing the toothmark numbers and spiral fractures (after Rinehart et al., 2006, Fig. 2C). sf = spiral fracture. (E and F) NMMNH P-16932, part of the topotype skeleton of *Revueltosaurus callenderi* from the Bull Canyon Formation (NMMNH L-467). (E) Dorsal vertebral column in lateral view. Note on left the truncation of the last two neural spines, dentalites on last centrum and isolated carnivore tooth (marked by arrow). (F) Close up of laterally compressed, recurved tooth (arrow). (G and H) NMMNH P-30188, lambeosaurine dorsal vertebra from the De-na-zin Member of the Kirtland Formation (NMMNH locality L-4280). (G) Close up of dentalites on the distal neural spine. (H) Lateral view of dorsal vertebra with dentalites indicated by arrows. From Hunt and Lucas (*2014*, Fig. 17).

2014: Fig. 6.4E and F). Müller et al. (2015) described dentalites on bones of the traversodontid cynodont *Exaeretodon*, from Brazil, which were attributed to an ecteniniid cynodont. Romo-de-Vivar-Martínez et al. (2017) reported a pathology in the rhynchocephalian *Clevosaurus brasiliensis*, also from Brazil, that resulted from injuries (including bites) produced during intraspecific fights. Holgado et al. (2015) recognized dentalites that were produced by fish on a skeleton of a protorosaurian within a regurgitalite from Italy.

The Jurassic and Cretaceous have nonmarine faunas dominated by dinosaurs that yield most of the described dentalites (Hunt and Lucas, 2021, Table A.2). The Jurassic and Cretaceous share with the Triassic a pattern of an increasing volume of dentalites toward the end of each period, reflecting the increase in preservation of nonmarine vertebrate specimens. Throughout the Mesozoic, the majority of nonmarine dentalites are from North America. Hunt (1984, 1987) predicted that dentalites would be rarer on dinosaur bones than on mammal bones because nonavian theropods lacked the dentition or jaw mechanics to manipulate and modify bones in a similar manner. Fiorillo (1991a) validated this hypothesis by demonstrating that dinosaur faunas exhibited 4% or less of bones with dentalites, whereas in the mammal faunas he studied the percentages varied from 13.1% to 37.5% (but see Drumheller et al., 2020, for a notable exception).

Dentalites are rare in the Early Jurassic. Occurrences include a crocodylomorph bite on a "prosauropod" bone from China (Xing et al., 2018) and a theropod bitten by a theropod from Antarctica (Hammer and Hickerson, 1993). There are no reports of dentalites from the Middle Jurassic, which in large part probably reflects the paucity of nonmarine faunas of that age.

The Late Jurassic has a record of dentalites that is fairly diverse taxonomically and geographically, although it is dominated by specimens associated with sauropods from North America (Hunt and Lucas, 2021, Table A.2). Two turtles from Germany preserve dentalites, one produced by a theropod and one by a crocodylomorph or a theropod (Joyce, 2000; Karl, 2012). The remaining Late Jurassic dentalites are registered on dinosaur bones. There are four records of dentalites from large theropods. Specimens of *Sinraptor dongi* from China and *Allosaurus* from the United States preserve cranial or mandibular dentalites that probably resulted from intraspecific combat (Gilmore, 1920; Tanke and Currie, 1998). Another specimen of *Allosaurus* preserves a pubic foot bitten by *Ceratosaurus* or *Torvosaurus* (Chure et al., 1998). Augustin et al. (2020a,b) described dentalites of a metriacanthosaurid and of a mammal, respectively, on different mamenchisaurid sauropod dinosaur specimens from China.

Cetiosauriscus greppini from Switzerland and *Europasaurus holgeri* from Germany are sauropods with dentalites produced by the huge crocodile *Machimosaurus* and a small crocodylomorph or shark, respectively (Meyer and Thüring, 2003; Slodownik and Wings, 2015; Wings, 2015). Osborn (1904) first reported dentalites on a sauropod from the Upper Jurassic Morrison Formation of the western USA. Subsequently, dentalites have been described from many taxa from the Morrison, including the sauropods *Camarasaurus*, *Apatosaurus*, *Diplodocus* and a brachiosaur (e.g., Osborn, 1904; Matthew, 1908; Jensen, 1988; Chure et al., 1998; Jacobsen, 1998; Hunt et al., 1994; Myers, 2004; Kirkland et al., 2005; Jennings and Hasiotis, 2006; Myers and Storrs, 2007; Bader et al., 2009; Storrs et al., 2013; Hone and Chure, 2018). Virtually all of these dentalites are attributed to theropods, in general, and most to *Allosaurus*, in particular (e.g., Osborn, 1904; Matthew, 1908; Chure et al., 1998; Jennings and Hasiotis, 2006). Dentalites on the holotype skeleton of *Camarasaurus lewisi* may pertain to *Torvosaurus* (Jensen, 1988;

Hunt et al., 1994). There are only single occurrences of dentalites on other Morrison herbivorous dinosaurs—the ankylosaur *Mymoorapelta maysi* and the stegosaur *Stegosaurus* (Carpenter et al., 2005; Kirkland et al., 2005).

It is notable that, during the Jurassic, dentalites attributed to crocodylomorphs are restricted to small-bodied tetrapods such as turtles (e.g., Karl and Tichy, 2004), with only two possible associations with sauropods (Meyer and Thüring, 2003; Bader et al., 2009). One of these is attributed to *Machimosaurus*, the largest crocodyliform of the Jurassic, with a body length of up to approximately 6.9 m (Young et al., 2016). While some species of this genus were more marine, the Late Jurassic *M. mosae* was adapted to a semi-aquatic lifestyle (Young et al., 2014).

There is a time interval from the beginning of the Jurassic until the Early Cretaceous, possibly the Aptian, when there are no known large, semi-aquatic predators. In the Late Triassic, phytosaurs such as *Smilosuchus gregorii* (Camp, 1930; Long and Murry, 1995) had a body length longer than that of contemporaneous terrestrial tetrapods. Prosauropods were longer, but apparently were geographically and ecologically separated from phytosaurs. However, after the extinction of the phytosaurs (an earliest Jurassic event: Lucas and Tanner, 2018), no semi-aquatic predators were more than half the length of the largest associated terrestrial predators or herbivores until the Cretaceous. As Young et al. (2016, p. 11) noted "it is unclear whether *Machimosaurus rex* (~7.15 m) was the largest crocodylomorph of the pre-Aptian Early Cretaceous, as the ages and body sizes of the giant pholidosaurids *Sarcosuchus hartii* (Marsh, 1869) and *Chalawan thailandicus* (Buffetaut and Ingavat, 1980) are uncertain." By the Aptian-Albian, large crocodylomorphs such as *Sarcosuchus* (Sereno et al., 2001) and large, semi-aquatic theropods such as *Suchomimus* and *Baryonyx* entered this niche. Hunt and Lucas (2021) termed this interval without large semi-aquatic predators the "SAP (semi-aquatic predator) Gap" The prevalence of theropod footprints on lake margins (Hunt and Lucas, 2007b; Milner and Kirkland, 2007) and biomechanical data (Molnar, 1973; Bakker and Bir, 2004; but see Yun, 2019) suggest that terrestrial nonavian dinosaurs may have partially exploited this niche during the SAP Gap (sensu Hunt and Lucas, 2021). However, the paucity of dinosaur dentalites on either turtles or crocodiles in the Mesozoic suggests otherwise (Hunt and Lucas, 2021, Table A.2).

Early Cretaceous dentalites are less numerous than those from the Late Jurassic and include traces on a turtle from France and on crocodyliforms from Spain and England, all produced by crocodyliforms, with most attributed to goniopholidids (Andrade et al., 2011; Buscalioni et al., 2013; Ristevski et al., 2018; Gônet et al., 2019; Hunt and Lucas, 2021, Table A.2). There are only two reports of Early Cretaceous theropods with dentalites, which are an ornithomimosaur from France and a tetanuran from Australia (Gônet et al., 2019; Poropat et al., 2019). Several ornithopod specimens preserve dentalites, including *Tenontosaurus tilletti* from the USA, *Lurdusaurus arenatus* from Niger and *Camptosaurus* and an iguanodont from Romania (Grigorescu, 1992; Taquet and Russell, 1999; Posmoşanu, 2003; Gignac et al., 2010). There are a smaller number of occurrences on sauropod specimens, including *Dongbeititan dongi* from China, a titanosauriform from Spain and an indeterminate taxon from Korea (Paik et al., 2011; Xing et al., 2012; Alonso et al., 2017). Smith et al. (2023) described five morphotypes of dentalites on pterosaur bones from the Kem Kem Group of Morocco.

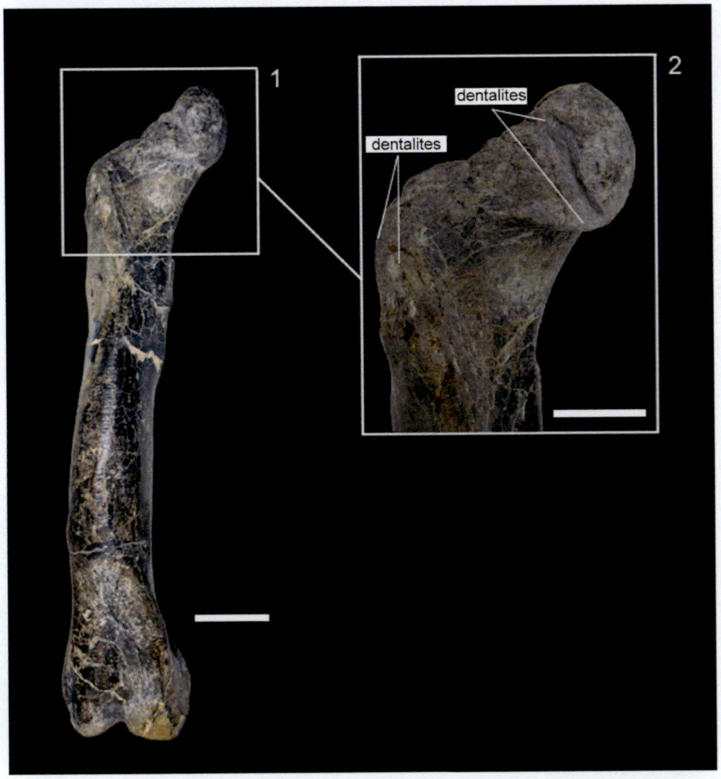

FIGURE 6.5 Dentalites on a right femur of a tyrannosaurid (NMMNH P-25083) from the Upper Cretaceous (Campanian) De-na-zin Member, Kirtland Formation, New Mexico, USA. (1) Femur, in anterior view. (2) Close-up of the anterior surface of the femoral head showing dentalites. Scale bar in (1) equals 10 cm, scale bar in (2) equals 5 cm. *From Dalman and Lucas (2021), Fig. 6.*

The Late Cretaceous has the largest number of recorded dentalites from nonmarine depositional environments before the Pleistocene (Figs. 6.2, 6.3G and H, 6.5, 6.6: Hunt and Lucas, 2021, Table A.2; Reis, 2022; Adrian et al., 2023). Taxa with dentalites include multiple examples of turtles and crocodylomorphs, scores of examples of nonavian dinosaurs (theropods, sauropods, thyreophorans, marginocephalians, ornithopods) as well as a pterosaur and a mammal (Siviero, 2019; Hunt and Lucas, 2021, Table A.2; da Costa Pereira et al., 2022; Brown et al., 2022; Reis et al., 2023; Adrian et al., 2023; Coppock and Currie, 2024). These traces have a broad geographic range in North America (Canada, USA, Mexico), South America (Argentina, Brazil), Europe (Hungary, Romania, Spain), Asia (Russia, China, Mongolia, Pakistan), and Africa (Madagascar). There are several reasons why this dataset is so large: (1) the Late Cretaceous yields the largest sample of excavated vertebrate fossils of large size before the Pleistocene; (2) the Western Interior Basin preserved large volumes of Late Cretaceous vertebrate fossils and is in itself a Megalagerstätte (Hunt and Lucas, 2022), and later tectonics, climate, and human history led to large numbers of collections and studies; (3) the Taxophile Effect is a strong influence, as many bones or skeletons of dinosaurs with dentalites warrant individual papers, which skews the dataset, and the most popular dinosaurs (tyrannosaurs) have a disproportionately large literature; and (4) the large body size

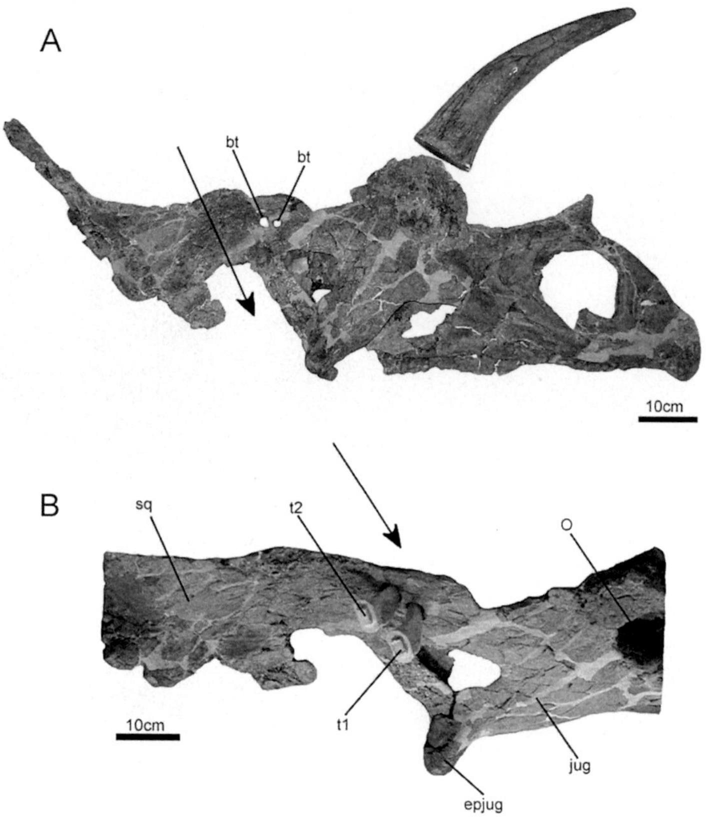

FIGURE 6.6 A nearly complete skull of a chasmosaurine ceratopsian (NMMNH P-50000) from the late Campanian of the Kirtland Formation of northwestern New Mexico, USA showing evidence of an attack by a large tyrannosaurid dinosaur. (A) Skull in right lateral view; (B) Dorsolateral view of the skull and the proximal end of the right squamosal with inserted tyrannosaurid teeth (casts). *Abbreviations*: *bt*, bite trace/dentalite; *epjug*, epijugal; *jug*, jugal; *O*, orbit; *sq*, squamosal; *t1-t2*, teeth (casts). The arrows show the orientation of tyrannosaurid snout with respect to ceratopsian skull. *From Dalman and Lucas (2018, Fig. 4).*

of many Late Cretaceous terrestrial vertebrates results in more recognition and preservation of dentalites. The large dataset demonstrates several patterns:

1. The majority of dentalites on turtles were produced by crocodylomorphs (Carpenter and Lindsay, 1980; Fiorelli, 2010; Noto et al., 2012; Botfalvai et al., 2014).
2. Most dentalites on crocodylomorphs represent intraspecific aggression or predation (Williamson, 1996; Avilla et al., 2004; De Vasconcellos and Carvalho, 2010; Codrea et al., 2010, 2012; de Araújo Júnior and da Silva Marinho, 2013; Botfalvai et al., 2014)
3. Crocodylomorph-produced dentalites are rare on bones of dinosaurs, except for examples that record biting by the giant *Deinosuchus* (Rivera-Sylva et al., 2009, 2012; Schwimmer, 2010).
4. Intraspecific head biting was common among large theropods, including tyrannosaurs (e.g., Bell and Currie, 2010; Currie and Eberth, 2010; Hone and Tanke, 2015; Brown et al., 2022) and abelisaurs (Rogers et al., 2004; Malkani, 2006, 2009, 2010).
5. Cannibalism is demonstrable in tyrannosaurids (e.g., Dalman and Lucas, 2021: Fig. 5.2).
6. The most common dinosaurian substrates of dentalites are bones of hadrosaurs and ceratopsians. Jacobsen (1998) analyzed 1000 dinosaur bones from the Dinosaur Park

Formation in Alberta, Canada, and recorded the highest percentage of dentalites on hadrosaurid bones (14%), whereas only 5% of the ceratopsid bones showed tooth marks, and 2% of the tyrannosaurid bones. One of the most well-known examples of putative evidence of carnivore biting involves the mid-caudal neural spines of the Late Cretaceous hadrosaur *Edmontosaurus*, possibly bitten by *Tyrannosaurus* (Carpenter, 1998), but the damage is more likely due to trampling, with the putative tooth marks actually related to a pyogenic infection in which the openings allowed the discharge of pus (Tanke and Rothschild, 2014).

7. The earliest mammalian-produced dentalites occur in the Late Cretaceous and include traces on *Champsosaurus* from Canada (Longrich and Ryan, 2010), the theropods *Buitreraptor gonzalezorum* from Argentina (Gianechini and de Valais, 2016) and *Velociraptor* from Mongolia (Saneyoshi et al., 2011), the protoceratopsians *Protoceratops* and *Bagaceratops* from Mongolia (Saneyoshi et al., 2011), ornithischians from Canada (Longrich and Ryan, 2010), the mammal *Eodelphis* from Canada (Longrich and Ryan, 2010) and multiple taxa from Argentina (de Valais et al., 2012).
8. Large sample sizes of both dentalites and associated body fossils facilitate study of the producers and their behavior (e.g., Brown et al., 2021).

3.4 Cenozoic

3.4.1 *Marine*

Marine dentalites are less common in the Cenozoic than in the Mesozoic (Hunt and Lucas, 2021, Table A.3). The majority of examples were produced by selachians on a substrate of marine mammal bones in (decreasing order of abundance) cetaceans, pinnipeds, and sirenians.

Perez et al. (2021) described four examples of shark dentalites preserved on cartilaginous carcharhiniform centra. These specimens are all from the Neogene Atlantic Coastal Plain of the United States and are of Miocene and Miocene or Pliocene age. A *Carcharhinus* from the Miocene of the USA contains a conspecific tooth in a dentalite, and a *Carcharodon megalodon* from the Pliocene preserves a shark dentalite (Purdy et al., 2001; Godfrey, 2003). Shark-bitten shark teeth have been documented but they are usually attributed to self-bitten teeth (Purdy et al., 2001; Renz, 2002; Perez, 2020).

Two sawfish from the Pliocene of Italy and the USA preserve dentalites of unknown origin, and one specimen of a tuna (*Thunnus*) from the Pliocene of the USA has dentalites produced by istiophorid billfish (Purdy et al., 2001; Schneider and Fierstine, 2004; Collareta et al., 2017a).

There are only three recorded Paleogene turtles and one crocodylomorph with dentalites. An indeterminate chelonioid from the Paleocene of Denmark preserves three types of dentalites, *Nihilichnus nihilicus*, interpreted as crocodilian, *Machichnus bohemicus*, interpreted as selachian, and small, circular traces from sharks or bony fish (Milàn et al., 2011). Myrvold et al. (2018) described other Paleocene specimens from Denmark consisting of fragments of a carapace of a chelonid turtle and a hypoplastron of *Ctenochelys* cf. *C. stenoporus*, both with marks produced by a crocodilian and the former with shark dentalites. A dryosaurid from the Paleocene of Niger has dentalites of crocodylomorph origin (Martin, 2013). Peters (2018) noted dentalites on the dermochelyid *Psephophorus polygonus* from the Miocene of the Netherlands.

There is a small but widespread record of dentalites on fossil penguin bones. These include the Eocene *Palaeeudyptes klekowskii* and unidentified taxa from Antarctica, cf. *Spheniscus* and an unidentified form from the Miocene of Argentina and Pliocene and *Tereingaomis moisleyi* from New Zealand (McKee, 1987; Walsh and Hune, 2001; Cione et al., 2010; Hospitaleche, 2016; Irazoqui and Acosta Hospitaleche, 2022). The producers of the dentalites were identified as teleost, *Galeocerdo aduncus*, cf. *Carcharhinus* and a dolphin, respectively (McKee, 1987; Walsh and Hune, 2001; Cione et al., 2010; Hospitaleche, 2016). Terrestrial carnivores also produced dentalites on penguin bones (Hunt and Lucas, 2021, Table A.3). A charadriiform bird from the Pliocene of Spain preserves dentalites attributed to a shark (Muñiz et al., 2008).

The largest number (>40) of reports of marine dentalites from the Cenozoic relate to large, nondelphin cetaceans (Hunt and Lucas, 2021, Table A.3; Godfrey, 2021; Benites-Palomino et al., 2022; Collareta et al., 2023a). The earliest example is the basilosaurid *Dorudon atrox* from the Eocene of Egypt that preserves dentalites attributed to *Basilosaurus isis* (Uhen, 2004; Fahlke, 2012: Fig. 6.7). There are currently no Oligocene records. The earliest records of odontocete (e.g., Lambert et al., 2018) and mysticete (e.g., Noriega et al., 2007; Collareta et al., 2017b) dentalites are from the Miocene, and these, and all younger examples, are attributed to sharks. About a third of all records of dentalites are merely attributed to unspecified cetaceans (e.g., Uhen, 2004, Fig. 137; Hunt and Lucas, 2021, Table A.3). The majority of Pliocene dentalites that are attributed to a specific genus are attributed to *Carcharodon* (e.g., Bianucci et al., 2002; Ehret et al., 2009; Govender and Chinsamy, 2013; Govender, 2015, 2019). There are very few reports of Pleistocene marine dentalites (e.g., Cicimurri and Knight, 2009).

Reports of dentalites on dolphins are restricted to the Oligocene, Miocene, and Pliocene (Hunt and Lucas, 2021, Table A.3; Boessenecker and Geisler, 2023; Godfrey, 2024). Pervesler

FIGURE 6.7 Partial cranium of a juvenile of *Dorudon atrox* (UMMP 94814) from the middle-late Eocene of Egypt, in dorsal view, with dentalites highlighted by white circles.

et al. (1995) described a specimen from the Oligocene of Austria, and Boessenecker and Geisler (2023) described others from South Carolina, USA. The four examples from the Miocene are all from the eastern United States. Specimens of *Eurhinodelphis*, a eurhinodelphinid and *Hadrodelphis* and an unidentified dolphin preserve dentalites of sharks (Dawson and Gottfried, 2002; Godfrey, 2003, 2024; Godfrey et al., 2018). There are three Pliocene examples from northern Italy. Specimens of *Astadelphis gastaldii*, *Hemisyntrachelus cortessi* and an indeterminate taxon preserve dentalites attributed to *Cosmopolitodus hastalis* and *Carcharodon carcharias* (*Knethichnus parallelum*), respectively (Portis, 1883; Cigala-Fulgosi, 1990; Bianucci et al., 2010; Jacobsen and Bromley, 2009).

Nine pinnipeds preserve dentalites (Hunt and Lucas, 2021, Table A.3; Godfrey et al., 2024). A pinniped from the Miocene of the USA preserves dentalites attributed to a shark (Bigelow, 1994). Two specimens from the Miocene and Pliocene of the USA have dentalites produced by mammals, possibly a pilot whale or beluga-like cetacean, a terrestrial carnivore, a dusignathine or odobenine walrus, or a conspecific otariid (Boessenecker and Perry, 2011). Collareta et al. (2017b) describe dentalites produced by *Carcharocles megalodon* on a pinniped from the Miocene of Peru. Two specimens of *Homiphoca* from the Pliocene of Spain preserve *Linichnus* produced by teleosts or small sharks (Rahmat et al., 2018; Muñiz et al., 2020). Godfrey et al. (2024) describe a tooth of *Carcharodon carcharias* embedded in a dentalite on an early Pliocene seal calcaneum from Florida, USA. An example of *Eumetopias jubatus* from the Pleistocene of Canada yields dentalites attributed to a bull of that species (Harington et al., 2004).

There are only four records of dentalites on sirenian bones, and all were produced by sharks (Hunt and Lucas, 2021, Table A.3: Merella et al., 2021). The oldest are Oligocene specimens of *Halitherium schinzii* from Germany and Austria (Pervesler et al., 1995; Diedrich, 2008). A Miocene specimen occurs in Venezuela (Aguilera and Rodrigues de Aguilera, 2004). Shark dentalites on ribs of *Metaxytherium subapenninum* from the lower the Pliocene of Italy include two whose morphology matches that of juvenile teeth of the extant bluntnose sixgill shark *Hexanchus griseus,* which is known from the same site (Merella et al., 2021, 2022).

3.4.2 Nonmarine

As mammalian faunas typically yield an order of magnitude more dentalites than reptilian faunas (Fiorillo, 1991a), one would expect a very large literature on Cenozoic dentalites, particularly relative to the Mesozoic, but this is not the case (at least until the Pleistocene). It is clear that the Taxophile Effect is important to the large volume of literature on dinosaur dentalites. Many individual examples of dinosaur dentalites warrant individual papers, which is clearly not the case for Cenozoic mammals (Hunt and Lucas, 2021, Table A.4). The volume of dentalite reports in the Pleistocene also reflects in part the Taxophile Effect relative to hominin evolution, as well as other factors, including (1) the large volume of Pleistocene vertebrate fossils relative to earlier time periods; (2) the widespread preservation of cave faunas—large sample sizes, often accumulated by predators and carefully studied to document this; hominin fossils are sometimes present, and surface preservation of bones is very good; and (3) the ease of using modern analogs, which has promoted study. The dataset for the nonmarine Cenozoic (Hunt and Lucas, 2021, Table A.4) is thus the most incomplete in this study. Many pre-Pleistocene dentalites are not recorded, or the occurrences are buried within taxonomic or faunal discussions and are difficult to find.

Sinclair and Granger (1914, p. 310), who amassed important collections of early Paleocene mammals from New Mexico, USA, noted that "A quarter, at least, of the Puerco specimens, collected by the 1913 expedition, show traces of gnawing, probably by small plagiaulacids [multituberculate mammals]." In marked contrast, we have only found five other published mentions of dentalites on Paleocene mammal bones (Simpson and Elftman, 1928; Gingerich, 1987; Bloch and Boyer, 2001; Secord et al., 2002; Longrich and Ryan, 2010). In addition, the Pleistocene literature on dentalites, particularly with regard to hominins and cave faunas, is very extensive, and we have not captured all of it here. Instead, we have attempted to record the significant literature and a sampling of the remainder.

There is only one dentalite recorded from a lizard, which occurs on *Varanus* sp. from the Miocene of Greece and is interpreted to have been produced by the hyaenid *Protictitherium* (Georgalis et al., 2018).

There are many reports of dentalites on Cenozoic turtles (21) and crocodylomorphs (6), and occurrences are mainly clustered in the Paleocene-Eocene and the Pleistocene (Hunt and Lucas, 2021, Table A.4; Zonneveld et al., 2022a; Yanenko and Kovalchuk, 2023; Zonneveld and Bartels, 2023). The majority of these dentalites are attributed to crocodylomorphs (e.g., Sawyer and Erickson, 1998; Fuentes, 2003; Njau, 2006; Mackness et al., 2010; Böhme et al., 2011; Hastings et al., 2015; Scheyer et al., 2018; Falk et al., 2019).

There are relatively few reports of dentalites on avian specimens. Occurrences include on *Cayaoan brunetti* and a penguin from the Miocene of Argentina (Cione et al., 2010; De Mendoza and Haidr, 2018) and on the ibis *Geronticus* cf. *G. calvus* from the Pliocene of South Africa (Pavia et al., 2017). The sources of the dentalites are identified as mammals or unknown. In the Holocene of New Zealand, moas were subject to predation by the eagle *Harpagornis*, producing dentalites (Farlow and Holtz, 2002).

Eocene mammals with dentalites are known in the USA and Germany, with one occurrence in France and one in England. Specimens from the United States include on the pantodont *Coryphodon*, resulting from intraspecific combat, and on the "condylarth" *Apheliscus chydaeus*, the carnivore *Viverravus* and the primate *Notharctus* (Gingerich, 1987; Lucas and Schoch, 1987; Alexander, 1992; Gingerich, 1994: Hunt and Lucas, Table A.4). Alexander and Burger (2001) described an unusual occurrence of a dentalite on a perissodactyl within a consumulite of the crocodilian *Pristichampsus vorax*. They also listed several primate specimens (*Notharctus*, *Omomys*) that appear to preserve raptor dentalites. The German Eocene includes the two important Lagerstätten of Messel and Geiseltal. Dentalites produced by crocodiles occur on several mammal fossils from these Lagerstätten, including equids, the ceratomorph *Lophiodon*, and the primate *Europolemur* (Franzen and Frey, 1993; Franzen, 2001; Falk et al., 2019). Mammals from Quercy in France preserve dentalites produced by rodents such as *Archaeomys* (Laudet and Fosse, 2001). The rodents *Isoptychus* sp. and *Thalerimys fordi* from southern England preserve dentalites produced by the amphicyonid *Cynodictis* cf. *C. lacustris* (Vasileiadou et al., 2007, 2009, 2020).

There are multiple reports of Oligocene dentalites from the United States and one from Egypt (Hunt and Lucas, 2021, Table A.4). The dentalites from the USA occur on several taxa, including oreodonts, entelodonts, and *Titianotherium*, and include bite marks and evidence of gnawing (Sinclair, 1922; Scott and Jepsen, 1936; Tanke et al., 1992; Effinger, 1998; Tanke and Currie, 1998; Longrich and Ryan, 2010). Four species of primates (*Aegyptopithecus*,

Propliopithecus) from the Fayum Formation of Egypt preserve dentalites produced by carnivore-like mammals (Gebo and Simons, 1984).

Miocene dentalites are much more widely distributed than those from the Paleogene, and they occur in France, Spain, Hungary, Greece, Austria, China, Uganda, South Africa, Kenya and the USA (Domingo et al., 2011; Aiglstorfer et al., 2019; Di Gregorio and de Araújo-Júnior, 2020; Hunt and Lucas, 2021, Table A.4). The substrates for these dentalites include rhinocerotids (Antunes et al., 2006; Deng and Tseng, 2010), a cervid (Havlik et al., 2014), a proboscidean (Havlik et al., 2014), an equid (Diffendal, 2003), a climacoceratid (Pickford, 1996), a tragulid (Sánchez et al., 2015), moschids (Domingo et al., 2011; Aiglstorfer et al., 2019), camelids (Winkler, 1987; Morgan and Williamson, 2000), a primate (Zapfe, 1981) and a hominin (Gommery et al., 2007; Mikulas et al., 2006; also see). All of the above have dentalites attributed to mammalian predators, except for the climatococeratid from South Africa that was bitten by a crocodilian. The most important study of Miocene dentalites was by Mikuláš et al. (Ekrt et al., 2016), who studied multiple dentalites on ungulates, small mammals and carnivores and erected six ichnotaxa for these traces, which have received widespread acceptance—*Nihilichnus nihilicus*, *N. mortalis*, *Machichnus regularis*, *M. multilineatus*, *M. bohemicus* and *Brutalichnus brutalis*. They attributed these ichnotaxa to squirrels, the carnivore *Amphicyon* sp. and crocodilians.

There is a smaller number of records of Pliocene dentalites (Hunt and Lucas, 2021, Table A.4). Ungulates from Italy have traces produced by the hyena *Pachycrocuta brevirostris*, and a glyptodont (cf. *Eosclerocalyptus lineatus*) from Argentina was bitten by the procyonid, *Chapalmalania* (Mazza et al., 2004; Mazza, 2006; de los Reyes et al., 2013). A Canadian mammoth was gnawed by a beaver, and a rodent from Chad was bitten by a mammalian carnivore (Harington, 1996; Denys et al., 2003). Diverse mammals from Tanzania have traces of indeterminate mammalian predators (Su and Harrison, 2008). Sahle et al. (2017) reported crocodile dentalites on diverse large mammals from Tanzania (bovids, equid, ungulate) and on *Australopithecus anamensis* and *A. afarensis* from Ethiopia.

There are large numbers of reports of dentalites from the Pleistocene from every continent, except Antarctica (Hunt and Lucas, 2021, Table A.4). Much of the impetus for the rise of interest in dentalites relates to studies focused on the taphonomy of Pleistocene hominins, in particular, which has expanded more broadly to other bone accumulating agents, notably in caves (e.g., Binford, 1981; Brain, 1981; Andrews, 1990; Blumenschine, 1995; Hart and Sussman, 2005; Domínguez-Rodrigo and Barba, 2006). Indeed, the most well-known examples of dentalites relate to Pleistocene hominins—leopard predation on *Paranthropus robustus* (Brain, 1981), eagle predation on the Taung child, *Australopithecus africanus* (Berger and Clarke, 1995), bites of *Crocodylus anthropophagus* on *Homo habilis* (Davidson and Soloman, 1990) and *Pachycrocuta brevirostris*-produced dentalites on *H. erectus* (Boaz et al., 2000, 2004).

Dentalites occur on a wide range of Pleistocene taxa. It is notable that many dentalites in the Old World are attributable to hyenas, principally *Crocuta crocuta spelaea*, but also *Pachycrocuta brevirostris*. Cave hyena dentalites occur in England, Germany, the Czech Republic and Morocco (Buckland, 1824; Diedrich, 2011, 2012a,c, 2013), and those of the giant, short-faced hyena are present in Spain, China, and Pakistan (Palmqvist et al., 1996; Boaz et al., 2000, 2004; Dennell et al., 2008). There are several occurrences in Australia of dentalites of the marsupial "lion" *Thylacoleo carnifex* (Horton and Wright, 1981; Runnegar, 1983; Camens and Carey, 2013; Dortch et al., 2016). It is also notable that intraspecific dentalites occur in several Pleistocene taxa, including *Ursus spelaeus* in Spain (Capasso, 1998; Pinto Llona and Andrews, 2004; Rabal-Garcés et al., 2012), *Canis lupus* in the United States (Courville, 1953), *Smilodon fatalis* in the United States (Miller, 1980; Akersten, 1985; Rothschild and Martin, 1993), *Panthera leo spelaea* in Germany (Diedrich, 2011) and *Crocuta crocuta spelaea* in the Czech Republic (Diedrich, 2006a).

Sutcliffe (1973) first identified osteophagia in herbivorous mammals as a significant phenomenon, notably ungulates such as giraffes and some deer that chew antlers and bone, presumably for nutritional supplements. This behavior has been recognized in several recent taxa but not yet in fossils.

4. Dentalites on invertebrate hard tissue substrate

The ichnological record of predation on invertebrates by vertebrate predators and scavengers is surprisingly small. There are, in fact, only two main taxa that have been documented in any detail: (1) predation on sessile echinoderms; and (2) predation on nektonic nautiloids and ammonoids (Hunt and Lucas, 2021, Table A.5).

There are three reports of dentalites, possibly produced by vertebrates, on Paleozoic crinoids from the Middle Devonian of Germany (Bohatý, 2008, 2009) and Poland (Gorzelak et al., 2011) and the early Carboniferous of the United States (Gahn, 2004). There are single examples, all from Poland, from the Middle Triassic (Salamon and Gorzelak, 2008), Late Jurassic (Lach et al., 2015), and Late Cretaceous (Salamon and Gorzelak, 2010). Crinoids have relatively massive skeletons and lack a large volume of easily digestible tissue, so they are not attractive to Recent fish (Gahn, 2004). However, since the Ordovician, crinoids have served as substrate hosts to a wide range of organisms, including brachiopods, bryozoans, corals, gastropods, ophiuroids, myzostomid annelids, and clionid sponges. Thus, at least some of the dentalites on crinoids may be "collateral damage" resulting from predation on their epibionts (Gahn, 2004). Cione et al. (2010) suggested a similar origin for some dentalites on Miocene penguin bones.

Echinoids, in contrast to crinoids, appear more attractive as a food source because of their thin test and relatively more abundant soft tissue. The earliest example of a vertebrate dentalite on an echinoid is from the Middle Jurassic of Israel (Wilson et al., 2014) although the enigmatic echinoderm *Bolboporites* from the Ordovician may preserve dentalites (Kushlina, 2006; Gillet et al., 2019). There are 11 reports of Late Cretaceous dentalites on echinoids from Germany and the Netherlands (Hunt and Lucas, 2021, Table A.5). The mosasaur *Prognathodon* is proposed as the producer of a dentalite from Germany (Neumann and Hampe, 2018), and the shark *Squalicorax* for another specimen from that country (Neumann, 2000). Other examples are generally attributed to indeterminate fish (e.g., Thies and Reif, 1985; Jagt et al., 2018). Current records suggest that there is great potential to find additional dentalites on echinoids.

There are three examples of dentalites on nonmarine bivalves from the Late Triassic of Poland (Gorzelak et al., 2010), and the Early Cretaceous of Spain (Bermúdez-Rochas et al., 2013) and Australia (Kear and Godthelp, 2008). Nonmarine invertebrates are much less studied than vertebrates from the same environments, so we expect more dentalites to be present but not yet documented.

A few other sessile invertebrates yield dentalites of vertebrate origin, including a Carboniferous brachiopod (Alexander, 1981), Late Cretaceous rudist (Hattin, 1988), an Early Jurassic brachiopod (Tasnadi-Kubacska, 1962) and marine bivalves from the Middle Devonian (Nagel-Myers et al., 2009), Middle Jurassic (Phipps, 2008) and Late Cretaceous (Kauffman, 1972). There are only two dentalites reported from fossil coral. Galle and Mikuláš (2003) described predation of fish on rugose coral from the Middle Devonian of the Czech Republic, and Kaufman (1981) described gall-like growths on Late Pleistocene coral from Jamaica that resemble Recent examples of predation by threespot damselfish. Given the importance of predation on coral reefs by Recent vertebrates such as parrotfish (Peyrot-Clausade et al., 2000), we expect there to be many more examples of dentalites on scleractinian corals and, by analogy, in extinct coral groups (and other reef-forming invertebrates).

There are only three examples of dentalites on vagrant fossil invertebrates. Zatoń et al. (2007) and Neumann (2000) describe traces on sea stars from the Middle Jurassic of Poland and the Late Cretaceous of Germany, respectively. Bishop (1972) reported a dentalite on the crab *Raninella* from the Late Cretaceous of the United States. A perfectly preserved, fully grown soft-tissue specimen of the octobrachian coleoid *Plesioteuthis subovata* from the Late Jurassic Solnhofen of Germany has an embedded tooth of the pterosaur *Rhamphorhynchus muensteri* (Hoffmann et al., 2020).

Kauffman and Kesling (1960) wrote an influential paper describing putative predation on the Late Cretaceous ammonite *Placenticeras cf. P. whitfieldi* by a platecarpine mosasaur (Fig. 6.8). Subsequently, there has been a large number of reports of dentalites on nautiloids and ammonoids, with the majority being of Late Cretaceous age. Older dentalites on cephalopod fossils are reported from the Late Devonian (Slotta et al., 2011), Late Mississippian (Bond and Saunders, 1989), Early Pennsylvanian (e.g., Mapes and Hansen, 1983), Early Triassic (Hoffmann and Keupp, 2015), Late Triassic (Tichy and Urbanek, 2004), Early Jurassic (e.g., Andrew et al., 2010; Takeda and Tanabe, 2014; Andrew et al., 2015), Middle Jurassic (e.g., Richter, 2009), Early Cretaceous (Hoffmann and Keupp, 2015) and Late Cretaceous (Mazuch et al., 2024). The inferred sources of these dentalites include teleosts (e.g., Martill, 1990), sharks (Vullo, 2011), and marine reptiles, including a nothosaur (Tichy and Urbanek, 2004). Late Cretaceous examples occur in Morocco (Gale et al., 2017), Madagascar (Hoffmann and Keupp, 2015), Mexico (e.g., Ifrim, 2013), and Canada (e.g., Hewitt and Westermann, 1990). There are multiple reports from the USA, and Kauffman (1990b) mentions more than 100 specimens. One report from Mexico identifies the tracemaker as the shark *Ptychodus* (Ifrim, 2013), but the remainder are identified as being produced by mosasaurs (e.g., Saul, 1979; Tsujita and Westermann, 1998; Mazuch et al., 2024). Many of the Late Cretaceous dentalites can be assigned to *Mikulasichnus quadripertitus* igen nov. (see ichnotaxonomy: Fig. 6.9).

Some subcircular traces on Late Cretaceous ammonoids represent limpet (patellogastropod) home scars (e.g., Knsc et al., 1994), and it has been proposed that these caused all of the putative dentalites (Kase et al., 1994, 1998; Johnston et al., 1997; Seilacher, 1998). However, this seems an overstatement, and many putative dentalites are correctly identified as such (Tsujita and Westermann, 2001; King, 2009; Mapes and Chaffin, 2003). Other dentalites occurring on nektonic invertebrates include examples of a foraminiferan with parottfish traces from the Eocene of India (Syed and Sengupta, 2019) and a Late Cretaceous squid with marks of the mosasaur *Tylosaurus proriger* (Stewart and Carpenter, 1990).

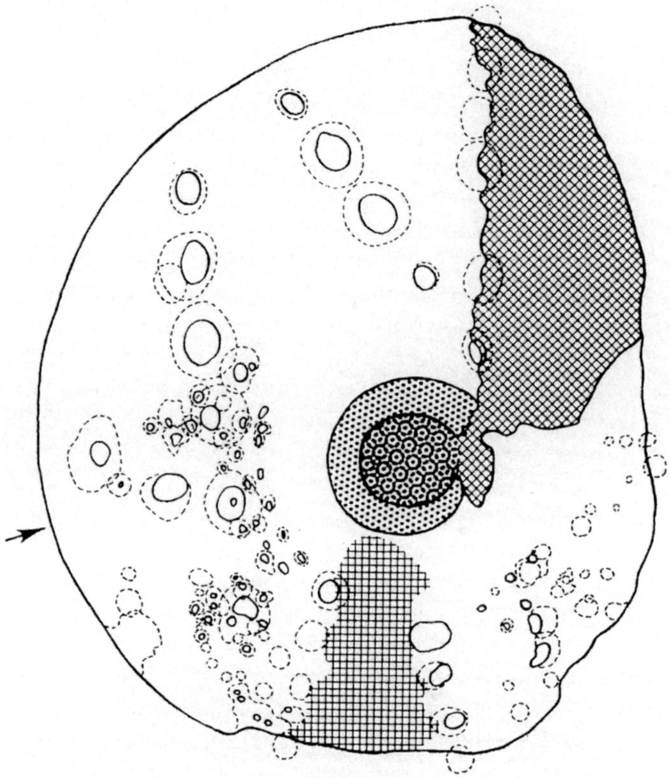

FIGURE 6.8 Left side of UMMP 35484 *Placenticeras* sp. cf. P. *whitfieldi* with *Mikulasichnus quadripertitus* igen. nov. Vertical and horizontal crosshatching—area covered by matrix; 45-degree crosshatching—crushed zone of shell; dotted patterns—umbilical region; solid lines—outlines of perforations; dashed lines—outlines of crushed zones and dents; arrow-adapical limit of the living chamber (approximate). X 3/8 (Kauffman and Kesling, 1960, Fig. 2).

Tischlinger (2001) described the only vertebrate dentalites on flying insects. Two specimens of insects, the odonatan *Cymatophlebia longialata* and the neuropteran *Archegetes neuropterum*, have damage to their wings, including absence of a wingtip that he attributed to pterosaur predation.

There is a distinct paucity of dentalites attributed to teleosts, even though many recent teleost taxa feed on a variety of invertebrates (e.g., Backus, 1964; Brown-Sarracino et al., 2007). However, it is difficult to identify teleost dentalites because there is so little known about fish traces, notably in temperate areas (Cione et al., 2010).

5. Dentalites on coprolite substrate

Coprolites with dentalites have only been recognized in the past decade, principally by Godfrey and co-workers. They are all from aquatic environments, mainly marine, and all but one is definitely Cenozoic in age. Hunt et al. (2018) erected the *Gaspeichnus* ichnofacies for traces on a coprolitic substrate, which also include footprints and borings.

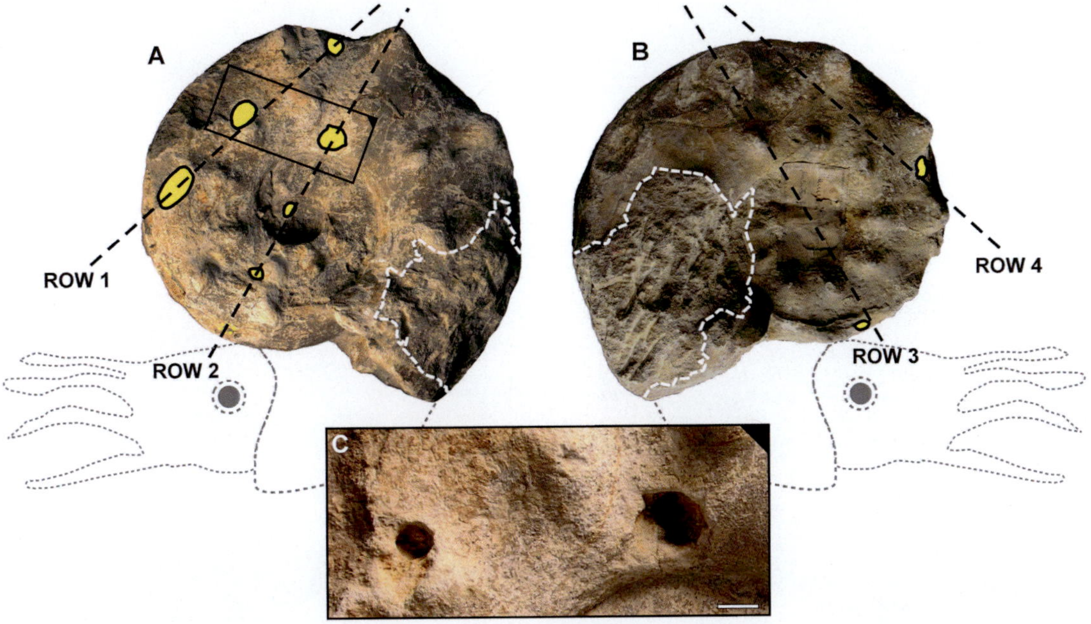

FIGURE 6.9 Holotype of *Mikulasichnus quadripertitus* igen nov. on *Mammites nodosoides* (Schlüter, 1871) (CHMHZ-Mch0001). (A and B) Tooth traces (yellow) arranged in tooth rows (black dashed lines). The area of a putative additional attack on the body chamber is marked by a white dashed line. (C) Enlarged part of the left side of the area indicated by the black rectangle in A, showing oval and square marks. Scale bar equals 10 mm. *From Mazuch et al. (2024, Fig. 3) courtesy of Radek Mikuláš.*

Godfrey and Smith (2010) first described dentalites on coprolites of unknown origin, which derive from the Miocene Calvert Cliffs of Maryland in the United States. They describe two coprolites with dentalites produced by sharks that could have resulted from: (1) aborted coprophagy, (2) benthic or nektonic exploration, or (3) predation. Below we name this trace *Selachiodentichnus godfreyi* igen et isp. nov (Fig. 6.10). Moore (2021) described another coprolite from this location. Godfrey et al. (2020) record a Paleocene crocodile coprolite from the United States with dentalites made by a chondrichthyan, actinopterygian, or small crocodilian.

Godfrey and Palmer (2015) described a coprolite bitten by the gar *Lepidosteus* from South Carolina, USA, from a mixed lag of Late Cretaceous, early Paleocene and Plio-Pleistocene taxa. Lindholm et al. (2022) named a new ichnotaxon, *Machichnus dimorphodon*, for traces of gar dentalites on a coprolite from Virginia. Frandsen (2020, p. 25-26, 28, 31, 33, 36, 39) illustrated seven additional coprolites with putative dentalites. Godfrey and Frandsen (2016) described another dentalite of *Lepidosteus* from another locality in South Carolina in a coprolite that might pertain to a crocodile. Subsequently, Frandsen and Godfrey (2019) illustrated two coprolites within the jaws of an articulated specimen of the gar *Atractosteus simplex* from the lacustrine Eocene Green River Formation of Wyoming, USA. A crocodilian coprolite from the Paleocene of Maryland, USA, exhibits dentalites of unknown origin (Godfrey et al., 2020).

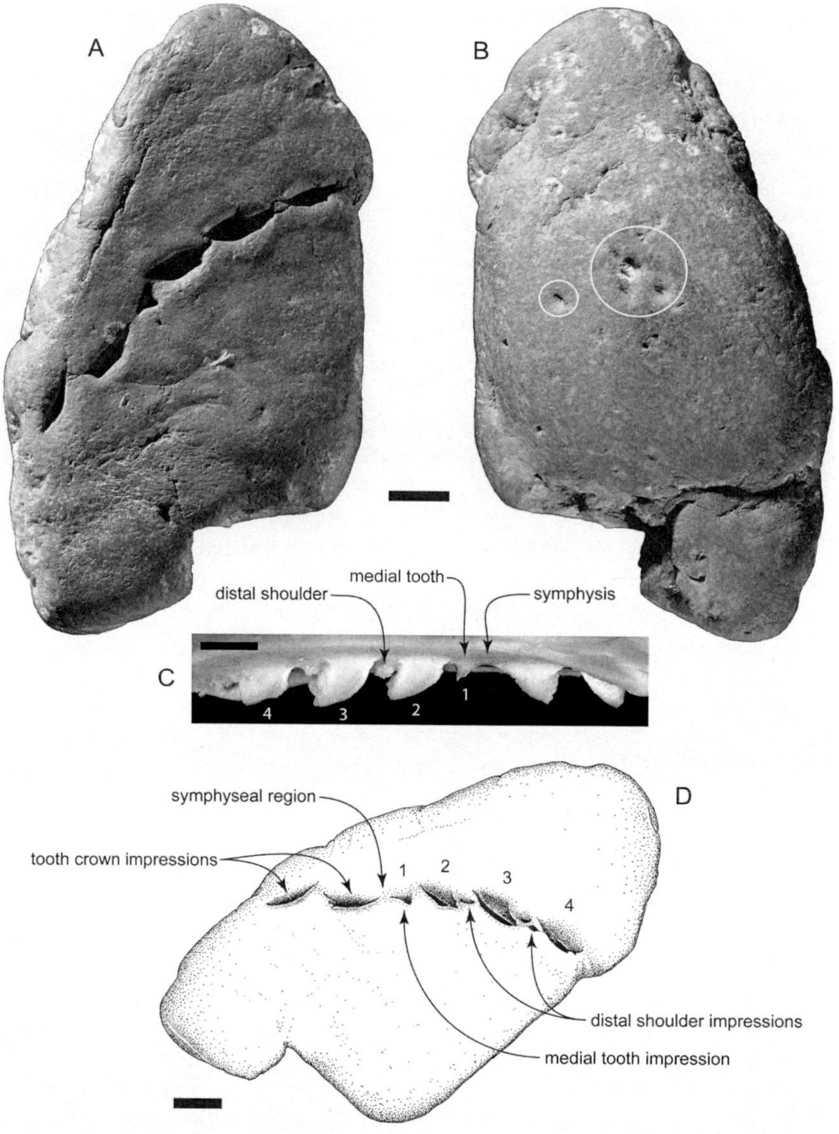

FIGURE 6.10 Holotype of *Selachiodentichnus godfreyi igen* et isp. nov. coprolite (CMM-V-2244) from the Miocene of Maryland, USA with dentalite of six tooth marks. (A) Flattened lower surface with six deeper tooth impressions. (B) Convex upper surface showing at least five shallower tooth impressions. (C) Silicone rubber cast of the tooth impressions in labial view. Cast teeth 1–4 in c correspond to numbered impressions 1–4 in D. (D) Drawing of the flattened lower surface of CMM-V-2244. Scale bars equal 10 mm (Godfrey and Smith, 2010, Fig. 2). *Figure courtesy of Stephen Godfrey.*

Frandsen (2020) illustrated two Pre-Cenozoic coprolites with putative dentalites. However, a specimen from the Late Triassic of New Mexico, USA, has longitudinal striations that we identify not as dentalites but rather the characteristic surface texture of *Alococoprus* (Frandsen, 2020, p. 22). Thus, the oldest of his specimens that we concur is a dentalite is from the Early Jurassic of Lyme Regis, England, with conical depressions (Frandsen, 2020, p. 76).

Dentzien-Dias et al. (2018) described teleost (Serrasalmidae) dentalites on coprolites from the Miocene of Venezuela. A coprolite of a carnivorous fish, likely a shark, from the Miocene of Italy contains dentalites possibly attributed to both selachians and teleosts (Collareta et al., 2019b). Dentzien-Dias et al. (2021) described multiple dentalites on fish coprolites from the Eocene of Virginia, USA. Cueille et al. (2020) described 17 coprolites from the Rhaetian of the United Kingdom with probable dentalites. Rozada et al. (2021) noted a coprolite from the Early Cretaceous of France with a possible shark dentalite.

6. Dentalites on lithic substrate

A variety of vertebrate behaviors involve direct interaction between mouth parts and the lithic substrate, notably feeding by aquatic fish and tetrapods, feeding by aerial birds and pterosaurs and digging by rodents. As Vallim et al. (2024) noted, vertebrates may be the "ignored bioturbators" in marine environments. Vertebrate megafaunal bioturbation can be categorized as pit creators, bulldozers, and trench diggers (Vallim et al., 2024). Recent vertebrates that substantially disturb the substrate during aquatic feeding, including chondricththyans (e.g., Common Skate *Raja erinacea*, Bat Ray *Myliobatis californicus*: Cook, 1971), actinopterygians (e.g., Atlantic Sturgeon *Acipenser oxyrinchus oxyrinchus*: Armitage and Gingras, 2003), osteichthyans (e.g., Goat fish—Mullidae: Geister, 1998; flatfish such as California Halibut *Paralichthys californicus*: Cook, 1971), reptiles (e.g., Loggerhead Turtle: *Caretta caretta*: Preen, 1996), birds (e.g., Yellow-eyed Penguin *Megadyptes antipodes*: Mattern et al., 2007, 2018), pinnipeds (e.g., Walrus *Odobenus rosmarus*: Kastelein et al., 1989), cetaceans (e.g., Gray Whale *Eschrichtius robustus*: Nelson et al., 1987, 1992: Bottlenose Dolphin *Tursiops truncatus*: Rossbach and Herzing, 1997), sirenians (e.g., Dugong *Dugong dugon*: Cossa et al., 2023) and carnivores (e.g., Sea Otter *Enhydra lutris*: Calkins, 1978).

The oldest marine traces attributed to vertebrate feeding were described by Fischer (1978) from the Ordovician Harding Sandstone of Colorado, USA. He erected the ichnogenus *Agnathichnus* for what he interpreted to be the feeding trace of a jawless fish. However, one of us (SGL) is revising this ichnofauna and considers *Agnathichnus* to be a synonym of *Treptichnus*, a zig-zag style of arthropod feeding trace.

Osculichnus tarnowskae is an ichnotaxon from the Early Devonian of Poland. It consists of a bilobate, generally elliptical, epichnial pit in the substrate that was produced by a fish probably feeding on bivalves, polychaetes, and arthropods (Szrek et al., 2016). The producer was probably a lungfish similar to *Dipnorhynchus* (Szrek et al., 2016). *Osculichnus fani* isp. nov. occurs in the Late Devonian of China (Fan et al., 2019: Fig. 6.11), and the type ichnospecies, *O. labialis*, is from the late Eocene-early Oligocene of Turkey (Demircan and Uchman, 2010).

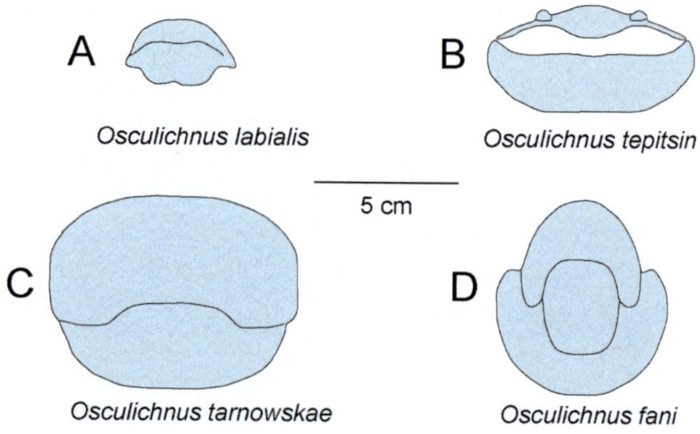

FIGURE 6.11 Morphology of ichnospecies of *Osculichnus. Adapted from Rodríguez-de la Rosa (2021, Fig. 3).*

Geister (1998) described sigmoidal epichnial grooves from Early Devonian limestones of the Czech Republic that are 20–35 cm long, 3–5 cm wide, and less than 3 cm deep. He interpreted these as feeding traces made by jawless or jawed fishes feeding on the muddy sea floor.

Large furrows, up to 60 cm wide, up to 30 cm deep, and up to 9 m long, are present on a Middle Jurassic bedding plane of limestone in northern Switzerland (Geister, 1998). There are three morphologies of trace. Geister (1998) suggested that the traces more than 40 cm wide were produced by the snout of the pliosaur *Liopleurodon*, and the smaller ones, less than 15 cm wide, could have been made by plesiosauroids or by the narrow, pointed snouts of ichthyosaurs.

Calvo et al. (1987) named *Megaplanolites ibericus* for a large tubular trace fossil from the Late Jurassic of Spain. They interpreted it as a feeding or locomotion trace of a large worm, but Geister (1998) suggested that it represents a vertebrate feeding trace.

Thousands of ovate depressions on bedding planes of Late Cretaceous strata from Spain, previously identified as dinosaur tracks, were re-interpreted by Martinelli et al. (2001) as traces of the feeding activity of rays or other fish with similar behavior. Howard et al. (1977) described how rays excavate large pits in shallow estuarine conditions to search for infaunal food sources by flapping the "wings" to hydraulically erode the sediment. Howard et al. (1977) noted Pleistocene and Cretaceous examples of this trace. Subsequently, Gregory et al. (1979) described how the New Zealand eagle ray, *Myliobatis tenuicaudatus*, hydraulically excavates much of its vertical-sided feeding depression by the action of water jetted downwards through the mouth and/or gill clefts and noted that other rays probably use a similar methodology. In this case, the depression would represent a dentalite. Gregory (1991) named this trace *Piscichnus waitemata*, and this ichnotaxonomy has been used by many subsequent workers (see below: Fig. 6.12). *Piscichnus browni* is a bowl-shaped trace described for fish nests (Feibel, 1987), and Gregory (1991) substantially broadened the definition of the ichnogenus to include cylindrical and plug-shaped ichnofossils. Unsurprisingly, the structure of a fish nest differs substantially from that of a pit produced by hydraulic excavation (compare Figs. 6.12 and 6.13). Thus, *Piscichnus browni* differs substantially from *P. waitemata* in many regards: (1) shallow, wide depression versus deep and narrow (2) symmetrical in cross section versus asymmetrical; (3) gentle sides versus steep to undercut sides; (4) infilling is fine and sometimes weakly bedded versus coarse infilling; (5) no organic fill versus common

FIGURE 6.12 Multiple specimens of *Piscichnus sanctacrucensis* from Middle Devonian (Eifelian) dolomites exposed in the Zachełmie Quarry, Holy Cross Mountains, southern Poland. *Image courtesy of Joerg Schneider.*

FIGURE 6.13 Holotype of *Physichnium* (*Piscichnus*) *waitemata* igen. nov. (X39, FR Number R09/f116) from the Cape Rodney formation (early Miocene), Mathesons Bay, North Island, New Zealand. Note the U-shaped excavation infilled by conglomerate. *Image courtesy of Neville Hudson.*

shelly debris; and (6) no overlapping of structures versus some structures cut across previous ones. Therefore, herein we describe the new ichnogenus *Physichnium* for "*P.*" *waitemata* and revise *Piscichnus* (see Ichnotaxonomy). *Physichnium* (*Piscichnus*) *waitemata* is widely distributed in the Late Cretaceous-Holocene (see Ichnotaxonomy). For example, Uchman et al. (2018) described *Physichnium* (=*Piscichnus*) *waitemata* from the Pliocene of Santa Maria Island (Azores Archipelago), which they interpreted to have been produced by rays hunting for polychaetes, crustaceans, and bivalves. Belvedere et al. (2011) described more than 200 crescentic traces from the Middle Eocene of Italy that they assigned to *Physichnium* (=*Piscichnus*) isp. and interpreted as the feeding traces of sturgeon-like fishes rather than rays or flat fishes.

Poropat et al. (2021) describe two morphologies of trace fossil from the Cenomanian–lowermost Turonian Winton Formation of Australia, which they attribute to feeding traces of fish. One morphotype consists of two indentations and superficially resembles feeding traces made by the Recent sturgeon *Acipenser oxyrinchus* (Pearson et al., 2007). There are multiple examples of a U-shaped trace that are interpreted as snout marks of a lungfish (Poropat et al., 2021).

Undichna is a common sinuous trace produced by a swimming fish (Minter and Braddy, 2006; Minter, 2024). Martin et al. (2010) described a specimen of *Undichna* cf. *U. simplicitas* from the lacustrine Green River Formation (early Eocene) of the western United States that preserves overlapping ellipsoids along the midline. These were interpreted as being produced by the ventral mouth of *Notogoneues osculus* during feeding (Martin et al., 2010).

Excavation by hydraulic jetting through the oral cavity is used by walruses. Gingras et al. (2007) described Pleistocene pits from the northwestern United States that they attributed to walruses feeding on deep-burrowing bivalves. Walruses root for prey with their snouts and emit a jet of water that liquefies the bottom sediments where a bivalve has burrowed (Gingras et al., 2007).

Both birds and pterosaurs produced feeding traces on substrate. Erickson (1967) first described dabble (surface grazing) marks associated with a trackway of *Presbyornis* (*Presbyorniformipes feduccii*) from the Eocene of the United States (also see Lockley and Hunt, 1995, p. 253; Yang et al., 1995; Hunt and Lucas, 2007a, Fig. 3; Lockley and Harris, 2010, Fig. 6). These feeding trails show the slow back and forth motion of the beak relative to the trackway midline and the bird's direction of progression (Lockley and Harris, 2010).

Lockley et al. (2009) noted several trackways of the Early Cretaceous *Ignotornis* that exhibit a "shuffling" behavior in which the right and left feet each advanced for very short distances of less than a single footprint length. They interpreted this as a type of "foot stirring" behavior characteristic of modern herons that "stir" up the substrate in search of food. Traces from the Early Cretaceous of Korea are very similar to those produced by the black-faced spoonbill *Platalea*. These traces consist of sets of fine, zigzag, arcuate to semicircular grooves that "sweep" back and forth across the trackway (Swennen and Yu, 2005; Lockley et al., 2008; Lockley and Harris, 2010, Fig. 11). Subsequently, Falk (2004; Falk et al., 2010, 2014) described a wide range of avian feeding interactions with the substrate from the Early Cretaceous of Korea, including pecking, probing, and scything traces. Fiorillo et al. (2010) interpreted Late Cretaceous subcircular to suboval depressions associated with *Ignotornis* and *Aquatilavipes* tracks from the Cantwell Formation of Alaska, USA, as probe marks.

Abbassi et al. (2015) described four types of bird feeding traces from the Neogene of Iran, including (1) bi-lobed *Osculichnus*-like traces which represent tips of the two bills; (2) curviform traces possibly related to the first; (3) ring and crescent tracs representing single probing acts; and (4) chain-shaped traces produced by a dragging beak. Brustur and Jipa (2020) noted examples of the first two types from the Miocene of Romania, including fine examples of the bilobate form.

Melchor et al. (2006) and Genise et al. (2009) described putative peck marks associated with "bird-like" tracks from the Upper Triassic-Lower Jurassic Santo Domingo Formation of Argentina. These traces substantially predate the emergence of avian theropods but could nonetheless be feeding traces.

Feeding traces have also been attributed to pterosaurs. There are several occurrences of shallow, paired, rounded impressions interpreted as beak traces associated with pterosaur tracks from the Middle Jurassic of Utah (Lockley and Wright, 2003), the Early Cretaceous of England (Wright et al., 1997), and the Late Cretaceous of Utah (Parker and Balsley, 1989; Wright et al., 1997; Mazin et al., 2003). Sinuous traces from the Late Cretaceous of Utah and Mexico, some originally interpreted as tail drags, may represent sweeping movements of pterodactyloid jaws (Lockley and Wright, 2003; Rodriguez-de la Rosa, 2003).

Several rodents utilize their dentition, as well as their claws, for the excavation of burrows. Good examples have been described from the Late Oligocene and Miocene of Nebraska and Colorado in the United States. Tooth marks of *Palaeocaster fossor* occur in the spiral *Daimonelix* burrows from the late Oligocene-early Miocene of Nebraska (Martin and Bennett, 1977; Yelinek, 2005). Gobetz and Martin (2006) described dentalites of a gopher-like rodent, possibly *Gregorymys*, from the Early Miocene of that state. Late Miocene burrows from Nebraska produced by marmotine ground squirrels contain paired striations, which are common dentalites in rodent burrows (Joeckel and Tucker, 2013). Gobetz (2006) and Hembree and Hasiotis (2008) described other burrows of mylagaulids with dentalites from the Late Miocene of the adjacent state of Colorado. One of us (SGL) is describing Pliocene rodent burrows from New Mexico with dentalites.

Geophagy is the consumption of soil or minerals (including salt licks) and is practiced by many Recent animals, principally herbivores, including ungulates, seed-eating birds and humans and nonprimates (Jones and Hanson, 1985; Izawa, 1993; Pryce, 1994; Abrahams and Parsons, 1996; Mahaney et al., 1996; Diamond et al., 1999; Gilardi et al., 1999; Abrahams, 2003; Tobler et al., 2009). There are multiple potential advantages to geophagy principally related to mineral supplementation or alleviating gastrointestinal disorders (adsorb toxins, adjust gut pH, antidiarrhoea, counteract the effects of endoparasites, protect the gastrointestinal lining from biological and chemical damage) (Gilardi et al., 1999; Krishnamani and Mahaney, 2000; Abrahams, 2003). Geophagy should result in dentalites in nonprimates, and these could be recognized in the fossil record.

Some Recent terrestrial vertebrates, including bears and raccoons, use mechanical excavation (i.e., digging into the sediment) to feed (Gingras et al., 2007). Currently there is no described fossil record for this behavior.

7. Dentalites on plant substrate

Specialized herbivorous vertebrates date back to the Permian (Lucas et al., 2018), and Paleozoic–Cenozoic invertebrate predation on plants has been widely documented (e.g., Scott, 1991; Labandeira, 1998), but there are very few examples of vertebrate dentalites on vegetation, and all are from the Cenozoic. All dentalites are preserved on the more dense portions of plants such as wood and nuts.

There are few well-documented Mesozoic examples of dentalites on plant material. However, Lewis (2011) described fossil vines associated with hadrosaur bones from the Late

Cretaceous of Texas, USA, with damage indicating that the branches were dominantly removed by shearing consistent with herbivory. In addition, the internal anatomy of the vines shows tracheid-filled false rings that are generally caused by mechanical removal of a portion of the plant during life. Manchester et al. (2010) described palm seeds in the same area associated with bones of juvenile ceratopsian and hadrosaurian dinosaurs and interpreted them as food debris.

Dentalites produced by rodents occur on seeds and nuts from the Eocene to the Pleistocene (Collinson, 1990). Collinson and Hooker (2000) recorded gnaw marks on late Eocene seeds of *Stratiotes* from southern England. Kodaira (1921) and Yoshikawa (2000) described dentalites produced by mice on nuts of *Juglans* spp. from the Pleistocene of Japan. The latter example was attributed to *Apodemus speciosus* (Yoshikawa, 2000). Gregor (1982) described another example of rodent activity in the stripping of Miocene *Araucaria* cones.

Beavers actively modify woody tissue and produce dentalites. Wood-cutting behavior may date back to the Oligocene, but the earliest dentalites on wood are Pliocene in age (Rybczynski, 2008). Dentalites occur with skeletal remains of *Dipoides* sp. from the early Pliocene on Ellesmere Island in Canada (Harington, 1996; Hutchison and Harington, 2002; Tedford and Harington, 2003). Both *Diploides* and the Recent *Castor* employ the same function of their incisors, but the living beaver has a more efficient cutting mechanism (Rybczynski, 2008).

Picidae (woodpeckers) are a well-defined family whose members mostly peck on wood to extract insects and their larvae (Shakya et al., 2017). Woodpeckers occupy a variety of habitats, but are highly specialized ecologically and behaviorally. Currently, 33 genera and 217 species of Recent taxa are recognized on every continent except Antarctica (Dickinson and Remsen, 2013). The Picidae probably evolved in the Paleogene of the Old World (Moore and Miglia, 2009; Shakya et al., 2017; Mayr, 2006, 2022). We would expect that woodpecker peck marks would occur in Paleogene and Neogene wood.

One of the most striking examples of convergent evolution within recent mammals is the suite of anatomical specializations shared by the primate *Daubentonia* of Madagascar and the marsupial *Dactylopsila* of Australia and New Guinea (Beck, 2009). These two genera have independently evolved extremely similar adaptations for feeding on xylophagous (wood—boring) insect larvae including enlarged incisors to gouge holes in wood, cranial modifications to strengthen the skull against the stresses generated by wood gouging and elongate manual digits that are used as probes to extract the larvae (Koenigswald and Schierning, 1987). Thus, these mammals fill the same ecological niche as woodpeckers. Paleogene apatemyids exhibit very similar craniodental and postcranial adaptations to *Daubentonia* and *Dactylopsila* and presumably occupied the woodpecker niche (Koenigswald and Schierning, 1987). The enigmatic Oligo-Miocene Australian metatherian *Yalkaparidon* also has similar adaptations and is probably a fourth "mammalian woodpecker" (Beck, 2009). All these taxa should have produced dentalites on wood.

8. Dentalites on other substrates

Dentalites occur, or potentially could occur, on a number of other substrates.

1. Hirsch et al. (1997) described predatory peck marks on fossil eggshells from the Eocene of Colorado, USA.
2. Possible dentalites on dinosaur skin include concave depressions with radiating stress fractures in *Psittacosaurus* (Lingham-Soliar, 2008) and a healed wound in *Edmontosaurus* (Rothschild and Depalma, 2013).
3. Chumakov et al. (2013) erected three new ichospecies, *Machichnus normani*, *M. harlandi* and *M. jeansi*, for scratch marks on phosphorite nodules and pebbles from the Late Cretaceous of England. They are interpreted as scratches that originated from the teeth of fish scraping bacterial or algal film off the surface of nodules that were covered with phosphate gel.
4. Some rocks are gnawed by rodents and other mammals for minerals or to wear down incisors (e.g., Gobetz and Hattin, 2002), although no paleontological specimens have been identified.
5. Jensen et al. (2019) report human chewing of birch pitch for nonnutritional reasons from the Holocene, and similar behavior might be seen in older hominins.

9. Ichnotaxobases

9.1 Introduction

Bromley and Jacobsen (2008) outlined research designed to produce an ichnotaxonomy of dentalites. They also noted how few ichnotaxa have been named for trace fossils in bone and recommended the development of rigorous ichnotaxobases and ichnotaxonomy. Currently, there are 28 named ichnotaxa of dentalites (Table 6.1).

9.2 Ichnotaxobases

Ichnotaxobases are distinctive morphologic features of a trace fossil that display significant and readily detectable variability and are thus important to ichnotaxonomic classifications (e.g., Bromley, 1996; Buatois and Mángano, 2011; Pirrone et al., 2014).

Pirrone et al. (2014) proposed ichnotaxobases for bioerosional structures of all kinds in bone; the following are particular to dentalites.

9.2.1 Substrate

We follow Bertling et al. (2006) in considering substrate as an important ichnotaxobase if it implies a different behavior of the producer. This is clearly the case if the dentalite is in a bone, wood, a coprolite, or an invertebrate shell (Lucas, 2016).

9.2.2 Orientation

Virtually all vertebrate teeth/dentitions function (move) vertically (dorso-ventrally), though lateral motion of the teeth/dentitions is also significant in some vertebrate taxa. Thus, dentalites can be divided into those produced by a vertical or lateral impact or a combination of the two. Binford's (1981) influential classification scheme of tooth marks recognized two vertical (punctures, pits) and two lateral tooth strike marks (scores, furrows).

TABLE 6.1 Ichnotaxa of dentalites.

Ichnotaxon	Author	Substrate	Producers
Mandaodontites coxi	Cruickshank (1986)	Bone	Pseudosuchian
Physichnium waitemata	Gregory (1991); this paper	Sand	Batoid
Nihilichnus nihilicus	Mikuláš et al. (2006)	Bone	Carnivores, herbivores
Nihilichnus covichi	Rasser et al. (2016)	Shell	*Tinca* spp.
Nihilichnus sulcatus	Trifilio et al. (2023)	Bone	Carnivores
Nihilichnus clavus	LaBarge and Njau (2024)	Bone	Mammal, crocodile
Nihilichnus sicarius	LaBarge and Njau (2024)	Bone	Crocodile
Nihilichnus hastarius	LaBarge and Njau (2024)	Bone	Crocodile
Machichnus regularis	Mikuláš et al. (2006)	Bone	Not specified
Machichnus multilineatus	Mikuláš et al. (2006)	Bone	Not specified
Machichnus bohemicus	Mikuláš et al. (2006)	Bone	Not specified
Machichnus normani	Chumakov et al. (2013)	Pebble	Probably fish
Machichnus harlandi	Chumakov et al. (2013)	Pebble	Probably fish
Machichnus jeansi	Chumakov et al. (2013)	Pebble	Probably fish
Machichnus fatimae	de Araújo-Júnior et al. (2017)	Bone	Large-sized canid *Protocyon troglodytes*
Machichnus dimorphodon	Lindholm et al. (2022)	Coprolite	Gar
Machichnus inrosus	Yanenko and Kovalchuk (2023)	Bone	Rodent *Anomalomys grytsivensis*
Brutalichnus brutalis	Mikuláš et al. (2006)	Bone	Carnivores
Heterodontichnites hunti	Rinehart et al. (2006)	Bone	Phytosaur
Linichnus serratuse	Jacobsen and Bromley (2009)	Bone	Theropods
Linichnus bromleyi	Muñiz et al. (2020)	Bone	Shark
Knethichnus parallelus	Jacobsen and Bromley (2009)	Bone	Theropods, Sharks
Osculichnus labialis	Demircan and Uchman (2010)	Sand	Fish
Osculichnus tarnowskae	Szrek et al. (2016)	Sand	Lungfish
Osculichnus tepitsin	Rodríguez-de la Rosa et al. (2021)	Sand	Benthic-feeding fish
Osculichnus fani	This paper	Sand	Lungfish
Daandavichnus batoideum	Rodríguez-de la Rosa et al. (2021)	Sand	Batoid
Mikulasichnus quadripertitus	This paper	Shell	Mosasaur
Selachiodentichnus godfreyi	This paper	Coprolite	Shark

Vertical features include deep (puncture of Binford, 1981; perforation of Fernández-Jalvo and Andrews, 2016) and shallow (pits) penetrations. Binford (1981) suggested the definitions are based on whether there is collapse of surficial layers (in bone, but this could also be applicable to shells), but a simpler differentiation might be whether the depression is deeper than wide. Binford (1981) considered pits to be more typical of gnawing.

Traces produced by lateral impacts of teeth are generally U-shaped (Fiorillo, 1991a,b; Fernández-Jalvo and Andrews, 2016). Binford's (1981) terms scores and furrows distinguish single and repeated jaw action. Some traces represent both vertical and lateral contact. Many of these include a tooth strike that slipped, but others may represent "torsional forces applied against incompletely gripped bones that slip on clasped jaws during side-to-side head thrusting and clockwise to counterclockwise death roll pivots … …and are common in crocodile-modified assemblages" (Njau and Gilbert, 2016, p. 5).

9.2.3 *Size*

Absolute size and size relative to the dimensions of the substrate are important. Small tooth marks on a large bone (or shell) probably denote postmortem damage, although the opposite is not always the case.

9.2.4 *Evidence of tooth structure*

Tooth structure can be seen in vertical marks by the cross section of the impression (e.g., Cruickshank, 1986; Rinehart et al., 2006). In lateral marks, serrations can produce lineations (Jacobsen and Bromley, 2009). Grooved incisors of rodents can also produce lineations (Fernández-Jalvo and Andrews, 2016). Bicarinate crocodile teeth produce distinctive marks (Njau and Blumenschine, 2006).

9.2.5 *Evidence of dentition structure*

The most useful traces provide not only evidence of behavior but also of the tracemaker's morphology. Bromley and Jacobsen (2008) favored naming bite marks based on the damage of a single tooth and regard multiple tooth marks as compound trace fossils. However, this suggestion fails to recognize the significance of heterodonty and the variation in single dental configurations (arcades). Thus, the ideal ichnotaxobase is the tooth marks of an entire dental arcade (Lucas, 2016). However, imprints of a large section of a dental arcade are uncommon, so most ichnotaxonomy has developed around single tooth marks. In some cases, the pattern of a dentition is preserved (Cruickshank, 1986; Rinehart et al., 2006).

9.2.6 *Pattern of occurrence*

Many dentalites are not isolated (Pirrone et al., 2014). Pits often occur in groupings. Multiple lateral marks can be sinuous, arcuate, or subparallel (Mikuláš et al., 2006; Jacobsen and Bromley, 2009). Gnawing traces are usually subparallel (Fernandez-Jalvo and Andrews, 2016, Figs. A.194–A.201).

9.2.7 *Location*

Buckland (1822, 1824) recognized that the location of dentalites on the substrate was important to identify the behavior and identity of the tracemaker. Subsequently, the pattern of dentalites on skeletons has been recognized as important in assessing predation and scavenging in animals with and without living analogs (e.g., Binford, 1981; Brain, 1981; Hunt et al., 1994).

9.2.8 Summary

There are several ichnotaxobases available for dentalites, but the ideal one is based on the tooth marks of an entire dental arcade. There are 26 named dentalites, which are reviewed below (Table 6.1). In addition, we review the ichnotaxonomy of *Piscichnus* since it was previously utilized for some dentalites.

10. Ichnotaxonomy

10.1 Introduction

The following is a review of dentalite ichnotaxa and a review/revision of *Piscichnus*. The lists of referred specimens include ichnotaxa identified by the authors of the cited papers. However, we have not attempted to review all the literature on dentalites and to identify other specimens.

10.1.1 Ichnofamily Machichnidae Wisshak et al. (2019)

Type ichnogenus: *Machichnus* Mikuláš et al. (2006).

Diagnosis: Punctures to grooves, both of a somewhat irregular outline, often in sets, in bone.

Included ichnogenera: *Knethichnus* Jacobsen and Bromley, 2009; *Linichnus* Jacobsen and Bromley, 2009; *Machichnus* Mikuláš et al., 2006; *Mandaodonites* Cruickshank, 1986; *Nihilichnus* Mikuláš et al., 2006; *Heterodontichnites* Rinehart et al., 2006; *Mikulasichnus* igen. nov.; *Selachiodentichnus* igen. nov.

Distribution: See below. Note also Machichnidae indet on a Late Cretaceous sauropod from Brazil (da Costa Pereira et al., 2022).

Discussion: Several ichnospecies of the ichnogenera listed are not preserved on bone. "The principal type of substrate may serve as an ichnotaxobase, especially in bioerosion traces, but its sole use should be avoided, unless informed by knowledge about the behavior of specialist producers" (Bertling et al., 2022, p. 1). We maintain a conservative position and retain ichnospecies within ichnogenera even if the original diagnosis or intent was to recognize traces on a bony substrate.

10.1.2 **Brutalichnus** *Mikuláš et al. (2006)*

Type ichnospecies: *Brutalichnus brutalis* Mikuláš et al., (2006).

Included species: Known only from the type ichnospecies.

Distribution: Middle Triassic of Germany (Mujal et al., 2022; Schoch et al., 2022), Late Cretaceous of Brazil (Paes-Neto et al., 2018), early Miocene of Czech Republic (Mikuláš et al., 2006), Pliocene of Spain (Muñiz et al., 2009), Pleistocene of USA (Wiest et al., 2016).

Revised diagnosis: Straight to arcuate trough linear crack or trough breakages that penetrate through the cortical layer into cancellous core with some radial breakages within the feature.

Discussion: Wisshak et al. (2019, p, 24) considered that the biogenic nature of the structure remains to be demonstrated as this ichnotaxon does "not exhibit tooth traces. The breakage is at least as likely due to diagenetic effects such as compaction." Thus, they consider this ichnotaxon to be a nomen dubium. However, Mikuláš et al. (2006, p. 126) note that "points where the bone is broken more strongly, showing radial pattern of fractures, are indication that this breakage is due to animal bites, and not by a postburial compaction." This trace has been recognized at several sites by different researchers (e.g., Paes-Neto et al., 2018; Schoch et al., 2022; LaBarge, 2022). Cracks associated with punctures have been described by LaBarge (2022, p. 7A). Pending further documentation of the type material and the radial fractures we consider this ichnotaxon valid.

10.1.3 Brutalichnus brutalis *Mikuláš et al. (2006)*

Holotype: coll. Doly Bílina, Inv. No Ah-1444 SGDB (Mikuláš et al., 2006, Fig. 10A).

Referred specimens: Two other specimens from type locality out of 400 bone fragments selected for study.

Diagnosis: As for the ichnogenus.

Type locality: Ahníkov, Czech Republic.

Type horizon: Early Miocene.

Distribution: As for the ichnogenus.

Discussion: Examples of this ichnotaxon from the Pleistocene of Texas, USA (Wiest et al., 2016, Fig. 5; Wiest, 2018, Fig. 5) are wrongly assigned, as those traces are not associated with cracks.

10.1.4 Heterodontichnites *Rinehart et al. (2006)*

Type ichnospecies: *Heterodontichnites hunti* Rinehart et al., (2006).

Included ichnospecies: Known only from the type ichnospecies.

Distribution: Late Triassic (Carnian: Adamanian) of New Mexico.

Revised diagnosis: Dentalite that differs from others in that the tooth impressions occur in an approximately straight or gently arcuate line, comprise both antero-posteriorly elongate and approximately round impressions in close proximity.

Discussion: Wisshak et al. (2019) doubted the distinction of this ichnogenus from *Mandaodonites* because both have rounded tooth impressions, and it is not apparent that the former has a more curved tooth row. However, the principal difference, as the ichnogeneric name suggests, is that *Heterodontichnites* preserves the impression of a heterodont dentition. Although it is a matter of degree, it is also evident that the tooth impressions of *Mandaodonites* are more ovoid in cross section and arranged in a more arcuate pattern than those of *Heterodontichnites* (compare Cruickshank, 1986, Figs. 1–3 and Rinehart et al., 2006, Figs. 1 and 2). Thus, we consider the ichnogenus to be valid.

10.1.5 Heterodontichnites hunti *Rinehart et al., (2006)*

Holotype: NMMNH P-18111 holotype, a tooth-marked *Koskinonodon perfectum* clavicle fragment (Rinehart et al., 2006, Fig. 1: Fig. 6.4A and B) from the Lamy Amphibian Quarry, near Lamy, NM.

Referred specimens: NMMNH P-13001, a tooth-marked femur of the dicynodont cf. *Ischigualastia* (Rinehart et al., 2006, Fig. 2) from NMMNH locality 1410.

Diagnosis: As for the ichnogenus.

Type locality: Lamy Amphibian Quarry, near Lamy, NM.

Type horizon: Garita Creek Formation (Late Triassic: Carnian: Adamanian).

Distribution: As for the ichnogenus.

Discussion: *Heterodontichnites* best matches the tooth shape and configuration of a phytosaur (Rinehart et al., 2006).

10.1.6 Knethichnus *Jacobsen and Bromley 2009*

Type ichnospecies: *Knethichnus parallelus* Jacobsen and Bromley (2009), emended by Wisshak et al. (2019.).

Included ichnospecies: Known only from the type ichnospecies.

Distribution: Triassic to Pliocene of western North America, Europe, and South America.

Diagnosis: Scraping structures in which serration traces extend as parallel grooves leading in some cases away from an initial groove.

Discussion: *Knethichnus parallelus* consists of serration traces produced when a denticulate tooth, typically theropod or selachian, was dragged across a bone at an angle at which only the denticles on the tooth meet the bone (Jacobsen and Bromley, 2009). "Depending on bone structure, preservation, erosion and preparation, the area showing serration traces may be the only preserved feature of the tooth trace" (Jacobsen and Bromley, 2009, p. 378).

10.1.7 Knethichnus parallelus *Jacobsen and Bromley (2009), emended by Wisshak et al. (2019)*

Holotype: RTMP 88.36.39 (Jacobsen and Bromley 2009, Fig. 6.3C–D).

Referred specimens: Cited by Jacobsen and Bromley (2009)—Pliocene dolphin material (Cigala-Fulgosi, 1990), Late Cretaceous mosasaur bones (Schwimmer et al., 1997; Corral et al., 2004). On Late Triassic dicynodont (Niedźwiedzki et al., 2010). On Late Cretaceous tyrannosaur (McLain et al., 2018). On Middle Triassic bones from Germany (Mujal et al., 2022; Schoch et al., 2022). On Late Cretaceous sauropod from Brazil (Reis et al., 2023). On Late Cretaceous tyrannosaur from USA (McLain, 2016). On Late Cretaceous dryosaurid from Mali (Hill et al., 2015; O'Leary et al., 2019). On Late Cretaceous hadrosaur from USA (Siviero, 2019). On Oligocene dolphin from USA (Boessenecker and Geisler, 2023).

Diagnosis: As for ichnogenus.

Type locality: Alberta, Canada.

Type horizon: Upper Cretaceous Dinosaur Park Formation,

Distribution: As for the ichnogenus.

Discussion: *Knethichnus parallelum* Jacobsen and Bromley (2009) was emended to *K. parallelus* by Wisshak et al. (2019) because the gender of the ichnogenus is masculine.

10.1.8 Linichnus *Jacobsen and Bromley (2009)*

Type ichnospecies: *Linichnus serratus* Jacobsen and Bromley (2009).

Included ichnospecies: *Linichnus serratus* Jacobsen and Bromley (2009), *Linichnus bromleyi* Muñiz et al. (2020).

Distribution: On Middle Triassic bones from Germany (Mujal et al., 2022; Schoch et al., 2022).

Linichnus cf. *L. serratus* on Late Cretaceous mosasaur bones from the Netherlands (Jagt et al., 2020). cf. *Linichnus* isp. on Eocene turtle from USA (Zonneveld and Bartels, 2023). *Linichnus* isp. on Miocene sperm whales from Peru (Benites-Palomino et al., 2022). *Linichnus* isp. on Pliocene whale from Spain (Muñiz et al., 2009). *Linichnus* isp. on Pliocene seal from Spain (Rahmat et al., 2018).

Revised diagnosis: Muñiz et al. (2020) amended the diagnosis to include traces on invertebrate substrates, but we maintain the original intent of Jacobsen and Bromley (2009), which was to restrict the ichnogenus to vertebrate material.

10.1.9 **Linichnus serratus** ***Jacobsen and Bromley (2009)***

Holotype: RTMP 84.82.01 (Jacobsen and Bromley 2009, Fig. 6.2B–C).

Referred specimens: On Late Triassic dicynodont (Niedźwiedzki et al., 2010). On Late Triassic bones from Poland (Sadlok, 2022). On Late Cretaceous sauropod from Brazil (Reis et al., 2023). On Late Cretaceous tyrannosaur bone from USA (McLain, 2016). On Late Cretaceous dryosaurid from Mali (Hill et al., 2015; O'Leary et al., 2019). On Late Cretaceous tyrannosaur in USA (Brown et al., 2022). On Late Cretaceous hadrosaur from USA (Siviero, 2019). On Oligocene dolphin from USA (Boessenecker and Geisler, 2023). On Miocene shark from Peru (Collareta et al., 2023a). On Mio-Pliocene marine vertebrate bones from USA (Boessenecker et al., 2014).

Revised diagnosis: *Linichnus* with a serrated edge.

Type locality: Alberta, Canada.

Type horizon: Upper Cretaceous Horseshoe Canyon Formation.

Distribution: As for the ichnogenus.

Discussion: In some instances, serration patterns and intertooth distances can allow recognition of the dentalite-maker. Some Canadian specimens can be matched with *Saurornitholestes* (Jacobsen, 2001; Jacobsen and Bromley, 2009).

10.1.10 **Linichnus bromleyi** ***Muñiz et al. (2020)***

Holotype: On left rib of mysticte whale (PS/C/001) deposited in the Center for Paleontological Interpretation, Viator (Almería, Spain) (Muñiz et al., 2020, Fig. 6.2A–B).

Paratypes: On right dentary of whale (CpzT/001), Laboratory of Research Group RNM293 Geomofología Ambiental y Recursos Hídricos of the University of Huelva) (Muñiz et al., 2020, Fig. 6.3A–B); on innominate bone of *Homiphoca* sp. (RNM238/PN1/H001), Collection of the Research Group RNM238 Paleontology and Applied Ecology, University of Huelva) (Muñiz et al., 2020, Fig. 6.3C–D).

Referred specimens: On Late Triassic bones from Poland (Sadlok, 2022). On Late Cretaceous sauropod from Brazil (Reis et al., 2023). On Oligocene dolphin from USA (Boessenecker and Geisler, 2023). On Oligocene whale from USA (Boessenecker et al., 2017—identified originally as *Linichnus* isp.). On Miocene whale from USA (Godfrey and Lowry, 2021). On Neogene sharks from eastern USA (Perez et al., 2021).

Diagnosis: *Linichnus* with a nonserrated edge.

Type locality: Monte Palmo de Salas, Viator, Almería, Andalusia, southern Spain.

Type horizon: Pliocene Unit 1.

Distribution: As for the ichnogenus.

Discussion: Boessenecker et al. (2017, p. 23) first suggested "the description of a new ichnotaxon for smooth-edged *Linichnus* traces, typically inferred to reflect the tips of shark teeth scraping into bone." Muñiz et al. (2020) considered that this trace was produced by an undetermined chondrichthyan.

10.1.11 Machichnus *Mikuláš et al. (2006)*

Type ichnospecies: *Machichnus regularis* Mikuláš et al., (2006).

Included ichnospecies: *Machichnus regularis* Mikuláš et al., (2006), *Machichnus multilineatus* Mikuláš et al., (2006), *Machichnus bohemicus* Mikuláš et al., (2006), *Machichnus normani* Chumakov et al., (2013), *Machichnus harlandi* Chumakov et al., (2013), *Machichnus jeansi* Chumakov et al., (2013), *Machichnus fatimae* Araújo-Júnior et al., (2017), *Machichnus dimorphodon* Lindholm et al. (2022), *Machichnus inrosus* Yanenko and Kovalchuk (2023).

Distribution: Triassic-Recent of Europe, South America and Antarctica. Occurrences not identified to the ichnospecies level include: *Machichnus*-like on Middle Triassic bones from Germany (Mujal et al., 2022; Schoch et al., 2022). On Late Cretaceous turtle bones from the Netherlands (Jagt et al., 2020). On Eocene penguin bones from Antarctica (Irazoqui and Acosta Hospitaleche, 2022). On Eocene mammal bones from England (Vasileiadou et al., 2022). On Miocene bird from Argentina (De Mendoza and Haidr, 2018).

Revised diagnosis: Shallow, serial, parallel, or subparallel grooves in hard substrates with groove surface being smooth or longitudinally striated and oriented perpendicular to substrate edge with each usually consisting of dozens of grooves that are typically uniform in shape and dimensions.

Discussion: The ichnogenus has been applied to a "surprisingly heterogeneous" variety of traces (Ekrt et al., 2016, p. 208) including dentalites on freshwater to marine environments with potential producers raning from fish to rodents. The ichnogenus is probably in need of taxonomic revision.

Machichnus represents gnawing or rasping traces (Mikuláš et al., 2006). Mikuláš (2013) referred to "incipient forms" of this ichnogenus on pottery from an archeological site.

10.1.12 Machichnus regularis *Mikuláš et al. (2006)*

Holotype: Coll. Doly Bílina, Inv. No. Ah-1401 SGDB (Mikuláš et al., 2006, Fig. 7A–K).

Referred specimens: 85 specimens from the type locality from approximately 400 studied bone fragments. On Pleistocene mammoth from USA (Wiest et al., 2016).

Type locality: Ahníkov, Czech Republic.

Type horizon: Early Miocene.

Distribution: As for the ichnogenus.

Revised diagnosis: *Machichnus* consisting of grooves that typically less than 1 mm in width and are arcuate in cross-section, and the surface across the grooves is relatively smooth as the margins of the grooves are rounded.

10.1.13 Machichnus multilineatus *Mikuláš et al. (2006)*

Holotype: Coll. Doly Bílina, Inv. No. Ah-1407 SGDB (Mikuláš et al., 2006, Fig. 8C).

Referred specimens: 15 specimens from the type locality from approximately 400 studied bone fragments.

Revised diagnosis: *Machichnus* consisting of grooves that typically 1–2 mm in width and are arcuate in cross-section, and the surface across the grooves is relatively rough as the margins of the grooves are angular.

Type locality: Ahníkov, Czech Republic.

Type horizon: Early Miocene.

Distribution: Early Miocene of Czech Republic.

10.1.14 Machichnus bohemicus *Mikuláš et al. (2006)*

Holotype: Coll. Doly Bílina, Inv. No. Ah-1435 SGDB (Mikuláš et al., 2006, Fig. 9D).

Referred specimens: More than 40 specimens from the type locality from approximately 400 studied bone fragments. On Jurassic bones from USA (McHugh et al., 2020). On Eocene penguin bones from Antarctica (Irazoqui and Acosta Hospitaleche, 2022). On Eocene turtle from Denmark (Milàn et al., 2011). On Late Pleistocene mammals from Brazil (Araújo-Júnior et al., 2017). On Pleistocene mammoth from USA (Wiest et al., 2016).

Diagnosis: Shallow, thin, discrete, parallel to subparallel, smooth-bottomed scratches. The scratches occur in small groups or series of corresponding width and length.

Type locality: Ahníkov, Czech Republic.

Type horizon: Early Miocene.

Distribution: Early Miocene of Czech Republic.

Discussion: *M. bohemicus* was subsequently interpreted as a trace of a crocodile (Mikuláš and Dvořák, 2010).

10.1.15 Machichnus normani *Chumakov et al. (2013)*

Holotype: On pebble in collection of Sedgwick Museum of Earth Sciences, Cambridge, UK (Chumakov et al., 2013, Fig. 6.3A–B).

Referred specimens: The holotype and one additional specimen on the same pebble.

Diagnosis: *Machichnus* composed of two parallel series of short, scratches, one of which series consists of strictly parallel striae at a regular interval, whereas the second series is composed of three times the amount of subparallel scratches that may crosscut.

Type locality: Southern England, north of London.

Type horizon: Upper Cretaceous Cambridge Greensand (Cenomanian).

Distribution: As for the ichnogenus.

Discussion: Chumakov et al. (2013) described three new ichnospecies based on traces on isolated phosphate pebbles from the Cambridge Greensand—*M. normani*, *M. harlandi*, and *M. jeansi*. These pebbles were collected individually over a wide area over a period of 100 years in southern England. Striations were previously ascribed to glacial processes (Chumakov et al., 2013). In addition, the Cambridge Greensand was extensively mined for "coprolites" (= mainly phosphate nodules) in the Nineteenth Century, and some of these mined specimens came to the University of Cambridge (Ford and O'Connor, 2009). Chumakov et al. (2013) hypothesized that the traces were produced by fish scraping a bacterial or algal film on phosphatic nodules on the Cretaceous seafloor. We are skeptical about the validity of the proposed ichnospecies given the various possible sources of the score marks and also because of their poor provenance.

10.1.16 Machichnus harlandi *Chumakov et al. (2013)*

Holotype: On pebble in collection of Sedgwick Museum of Earth Sciences, Cambridge, UK (Chumakov et al., 2013, Fig. 6.3D–E).

Referred specimens: None.

Diagnosis: *Machichnus* consisting of isolated, systematically recurring, both longitudinally striated and smooth scratches.

Type locality: Southern England, north of London.

Type horizon: Upper Cretaceous Cambridge Greensand (Cenomanian).

Distribution: As for the ichnogenus.

10.1.17 Machichnus jeansi *Chumakov et al. (2013)*

Holotype: On pebble (sample no. 145) in collection of Sedgwick Museum of Earth Sciences, Cambridge, UK (Chumakov et al., 2013, Fig. 6.3C).

Referred specimens: None. The holotype and one additional specimen on the same pebble.

Diagnosis: *Machichnus* consisting of deep thin scratches, narrowing to one end. They occur parallel in a simple series.

Type locality: Southern England, north of London.

Type horizon: Upper Cretaceous Cambridge Greensand (Cenomanian).

Distribution: As for the ichnogenus.

10.1.18 Machichnus fatimae *Araújo-Júnior et al. (2017)*

Holotype: MCC 491-V, groove in the right ulna of *Glyptotherium* sp. (Mammalia, Cingulata) (Araújo-Júnior et al., 2017, Fig. 4, see arrow).

Referred specimens: MCC 242-V, caudal vertebra of *Glyptotherium* sp. (Mammalia, Cingulata); MCC 411-V, right tibia of *Smilodon populator* (Mammalia, Carnivora); MCC 476-V, right ectocuneiform of *Eremotherium laurillardi* (Mammalia, Pilosa); MCC 482-V, left radius of *Glyptotherium* sp. (Mammalia, Cingulata) (Araújo-Júnior et al., 2017). On Late Pleistocene-early Holocene *Eremotherium* from Brazil (da Costa et al., 2023).

Diagnosis: *Machichnus* consisting of regularly arcuate, large-sized grooves with smooth-bottomed surface that are enlarged and U-shaped in cross section with some branching.

Type locality: Lajedo da Escada, municipality of Baraúna, Rio Grande do Norte.
State, Brazil.

Type horizon: Late Pleistocene-early Holocene age.

Distribution: Pleistocene to Holocene of Brazil.

Discussion: *Machichnus fatimae* is interpreted as gnawing traces resulting from the feeding behavior of a carnivore, possibly *Protocyon troglodytes* (Araújo-Júnior et al., 2017).

10.1.19 Machichnus dimorphodon *Lindholm et al. (2022)*

Holotype: A set of seven parallel, gently curving gouges, including five larger/deeper (primary) gouges and two finer (secondary) gouges on CMM-V-8998, a vertebrate (probably crocodilian) coprolite (Lindholm et al., 2022, Fig. 1A–B).

Paratype: A set of several primary and secondary gouges on the external surface of CMM-V-4480 (Lindholm et al., 2022, Fig. 2A).

Referred specimen: A set of primary and secondary gouges on the external surface of CMM-V-6615 from Summerville, South Carolina, USA, of unknown specific age but probably Oligocene-Pleistocene (Lindholm et al., 2022, Fig. 6.2B).

Diagnosis: *Machichnus* ichnospecies with parallel linear gouges of any length in nearly straight or variably curved lines, preserved together in two forms as primary bold gouges and fine secondary gouges, the latter being adjacent, variably interspersed, and parallel to the primary gouges.

Type locality: Horseshoe, right bank of the Pamunkey River, New Kent County, Virginia, USA.

Type horizon: Miocene Calvert Formation.

Distribution: Miocene and possibly Late Cretaceous to Plio-Pleistocene (Lindholm et al., 2022).

Discussion: There are dentalites on several coprolites of different ages, but this is the first named ichnotaxon from this kind of substrate (Hunt and Lucas, 2021). The preservation of dentalites on coprolites indicates rapid solidification of feces.

10.1.20 Machichnus inrosus *Yanenko and Kovalchuk (2023)*

Holotype: NMNHU-P AR 407, a series of scratches on the upper right margin of the nuchal plate of *Testudo chernovi* (Yanenko and Kovalchuk, 2023, Fig. 3).

Diagnosis: *Machichnus* ichnospecies consisting of narrow, subparallel grooves with a double longitudinal striation exposed on a small area of the substrate.

Type locality: Grytsiv locality, Khmelnytskyi Oblast, Ukraine.

Type horizon: Upper Miocene karst filling.

Distribution: Miocene of Ukraine.

Discussion: *Machichnus inrosus* is similar to other ichnospecies (*M. regularis*, *M. multilineatus*, *M. bohemicus*) and may not be a valid ichnotaxon. These traces are identified as gnawing marks of the rodent *Anomalomys grytsivensis*, as evidenced by the width of in lower incisors and the presence of two characteristic longitudinal ridges on their labial surface (Yanenko and Kovalchuk, 2023).

10.1.21 Mandaodonites *Cruickshank (1986)*

Type ichnospecies: *Mandaodonites coxi* Cruickshank (1986).

Included ichnospecies: Known only from the type ichnospecies.

Revised diagnosis: Conical impressions of varying size arranged in a sigmoidal curve with each row being at least 16 cm long, the impressions may occur in triplets and average 6.12 mm in diameter on one row and 7.47 mm in the other with maximum and minimums of 10.5 and 3.5 mm.

Distribution: Triassic of Tanzania.

Discussion: Zonneveld (2022) considered that *Mandaodonites coxii* is a combination of nonbiogenic, fracture-related pits that were inadvertently enhanced during fossil preparation. We find Zonneveld's (2022) analysis unconvincing, particularly because it is based on his clear misinterpretation of how an archosaur would make such tooth marks in bone.

10.1.22 Mandaodontites coxi *Cruickshank (1986)*

Holotype: In CUMZ T1125, dicynodont right femur (Cruickshank, 1986, Fig. 1) and silicone rubber cast (Cruickshank, 1986, Fig. 2).

Referred specimens: None.

Diagnosis: As for ichnogenus.

Type locality: Ruhuhu Valley, Songea District, Southern Province, Tanzania.

Type horizon: Middle Triassic Manda Formation.

Distribution: As for the ichnogenus.

Discussion. This ichnospecies was probably produced by a medium-sized pseudosuchian.

10.1.23 Nihilichnus *Mikuláš et al. (2006)*

Type ichnospecies: *Nihilichnus nihilicus* Mikuláš et al., (2006).

Included ichnospecies: *Nihilichnus nihilicus* Mikuláš et al. (2006); *Nihilichnus covichi* Rasser et al. (2016); *Nihilichnus sulcatus* Trifilio et al. (2023).

Distribution: See below under ichnospecies. Occurrences that did not discriminate an ichnospecies include: On Late Triassic bones from Poland (Sadlok, 2022). On Eocene turtle from USA (Zonneveld and Bartels, 2023). On Pliocene whale from Spain (Muñiz et al., 2009).

Revised diagnosis: Roughly triangular, circular, elliptical, or lenticular holes, or external pits, found on skeletal material, occurring solitarily or in groups, which may show recurring patterns. Outer margin of the concavity is either smooth and continuous with surrounding skeletal material or shows minute irregular jags. Lacks organized boring patterns (LaBarge and Njau, 2024).

Discussion: *Nihilichnus covichi* is the only ichnospecies of the ichnogenus not found on a bony substrate.

10.1.24 Nihilichnus nihilicus *Mikuláš et al. (2006)*

Holotype: Coll. Doly Bílina, Inv. No. Ah-1439 SGDB (Mikuláš et al., 2006, Fig. 6J).

Amended diagnosis: Roughly triangular, circular, or elliptical holes in vertebrate bone, passing through the cortical bone (LaBarge and Njau, 2024).

Referred specimens: On Middle Triassic bones from Germany (Mujal et al., 2022; Schoch et al., 2022). On Late Triassic dicynodont from Poland (Niedźwiedzki et al., 2010). On Eocene penguin bones from Antarctica (Irazoqui and Acosta Hospitaleche, 2022). On Early Cretaceous fish from Australia (Wretman and Kear, 2014). On Late Cretaceous tyrannosaur in USA (Brown et al., 2022). On Late Cretaceous hadrosaur from USA (Siviero, 2019). On Late Cretaceous turtle from USA (Adrian et al., 2023). On Miocene turtle from the Netherlands (Peters, 2018). On Late Miocene crocodilian from Brazil (Di Gregorio and de Araújo-Júnior, 2020). On Pleistocene mammoth from USA (Wiest et al., 2016). On Pleistocene mammal from Brazil (Andrade, 2019).

Type locality: Lower Miocene, Ahníkov, Czech Republic.

Type horizon: Lower Miocene,

Distribution: Triassic-Pleistocene of Europe and North and South America.

10.1.25 Nihilichnus covichi *Rasser et al. (2016)*

Holotype: Hole in the shell of *Gyraulus trochiformis* (SMNS 101527; Rasser et al., 2016, Fig. 2A–B).

Paratypes: SMNS 101524, 101525, 101526, 101527, 101528, 101529, 101530, 101531, 01532, 101533, holes in specimens of *Gyraulus trochiformis* and *Bania pseudoglobulus* (Rasser and Covich, 2014).

Referred specimens: Holes in shells of *Nematurella bavarica*: LfU SPR2014-029 and SPR2014-051 (Kirchberg Formation: Lauingen, SE Germany); SPR2014-041; SPR2014-043; SPR2014-042 and SPR2014-050 (Kirchberg Formation, Germany).

Diagnosis: Solitary to rarely paired *Nihilichnus* with mainly ellipsoidal to sometimes circular outline; the hole margin is straight and perpendicular to the inner and outer surfaces, and very few small scratches may be present on the shell surface in close vicinity to the holes.

Type locality: Steinheim am Albuch, Baden-Württemberg, Germany.

Type horizon: Steinheimer Seeschichten (middle Miocene).

Distribution: Miocene of Germany.

Discussion: *Nihilichnus covichi* was probably produced by the pharyngeal teeth of the durophagous tench fish (*Tinca* spp) (Rasser et al., 2016).

10.1.26 Nihilichnus sulcatus *Trifilio et al. (2023)*

Holotype: Holes and furrows on a right lunate (MCC 197-V) of *Eremotherium laurillardi* (Trifilio et al., 2023, Fig. 9).

Referred specimens: A rib fragment (MCC 3264.107-V) of *Holmesina paulacoutoi* from the Quaternary sediments of Complexo Suíço cave, Felipe Guerra municipality, Rio Grande do Norte State, Brazil, left ulna (MCT 3981-M) of *Valgipes bucklandi* from Gruna das Três Cobras cave, Serra do Ramalho municipality, Bahia State, Brazil.

Diagnosis: *Nihilichnus* with ellipsoidal holes with branching furrows. Both holes and furrows penetrate through the cortical bone into the cancellous bone, but neither contains filling or bioglyphs. The holes are perpendicular to the bone surface and have sharp edges, whereas the furrows have a smooth bottom, U-shaped cross-sectional morphology, and smooth edges.with mainly ellipsoidal to sometimes circular outline; the hole margin is straight and perpendicular to the inner and outer surfaces and very few small scratches may be present on the shell surface in close vicinity to the holes.

Type locality: F3 cave, Baraúna municipality, Rio Grande do Norte State, Brazil.

Type horizon: Quaternary sediments.

Distribution: Quaternary of Brazil.

Discussion: The presence of furrows connected to the perforations is the primary distinction between this and other ichnospecies of *Nihilichnus* (Trifilio et al., 2023, Fig. 9).

10.1.27 Nihilichnus hastarius *LaBarge and Njau (2024)*

Holotype: On IUPC 101303, which is a femur of *Equus oldowayensis*, likely modified by *Crocodylus anthropophagus* (LaBarge and Njau, 2024, Fig. 3B).

Referred specimens: Ah-1428-SGDB, Ah-1429-SGDB. Ah-1431-SGDB, Ah-1434-SGDB (Mikuláš et al., 2006). KK R0457 (Wretman and Kear, 2014). RTMP-uncatalogued (Jacobsen et al., 2015). MPCA 470-75 (Gianechini and de Valais, 2016). MNHN Sa 3661 (Aiglstorfer et al., 2019). WNoZ/S/7/170, WNoZ/S/7/267, WNoZ/S/7/268 (Sadlok, 2022). SMNS 81169, SMNS 81361, SMNS 84244, SMNS 84256, SMNS 97010/1, SMNS 97012 (Mujal et al., 2022). MCC 90-V, MCC 977-V, MCC 1250-V M (Trifilio et al., 2023).

Diagnosis: Lenticular holes in vertebrate bone, passing through the cortical bone. Outline presents two vertices on opposing sides of trace.

Type locality: Olduvai Gorge, Tanzania.

Type horizon: Olduvai Lowermost Bed II (Quaternary: ~1.8 Ma).

Distribution: Early Cretaceous to Quaternary of Australia, Argentina, Brazil, France, Hungary, and Tanzania.

Discussion: LaBarge and Njau (2024) noted that *N. hastarius* is similar in formation mechanics to *N. nihilicus* but possesses a distinct cross section. Bisected punctures are are diagnostic of crocodile feeding (Njau and Blumenschine, 2006; LaBarge and Njau, 2024).

10.1.28 Nihilichnus clavus *LaBarge and Njau (2024)*

Holotype: On IUPC 101303 that is a femur of *Equus oldowayensis*, likely modified by *Crocodylus anthropophagus* (LaBarge and Njau, 2024, Fg. 4A).

Referred specimens: IUPC 101304 (LaBarge and Njau, 2024, Fig. 3.4). Ah-1428-SGDB, Ah-1429-SGDB, Ah-1434-SGDB, Ah-1455-SGDB (Mikuláš et al., 2006). MGUH 29293 (Milàn et al., 2011). ZPAL V.33⁄464 (Niedźwiedzki et al., 2010). KK R0457 (Wretman and Kear, 2014). RTMP-uncatalogued (Jacobsen et al., 2015). MPCA 470-73 (Gianechini and de Valais, 2016). MNHN Sa 3660, MNHN Sa 3661, MNHN Sa 3675, MNHN Sa 3728, MNHN Sa 3966, MNHN Sa 3967, MNHN Sa 3969, MNHN Sa 3977 (Aiglstorfer et al., 2019). DGM 1173-R (Di Gregorio and de Araújo-Júnior, 2020). MLP 12-I-20-307t3, MLP 12-I-20-308t1 (Irazoqui and Acosta Hospitaleche, 2022). DPC 12545 (Zonneveld et al., 2022a). WnoZ/S/7/268 (Sadlok, 2022). SMNS 81342, SMNS 84253 (Mujal et al., 2022).

Diagnosis: Circular to elliptical pits on the surface of vertebrate bone, not passing through cortical bone. Interior of pit evenly slopes to singular point with no obvious facets. Concavity may either be conical or hemispherical.

Type locality: Olduvai Gorge, Tanzania.

Type horizon: Olduvai Lowermost Bed II (Quaternary: ~1.8 Ma).

Distribution: Middle Triassic to Quaternary of Australia, France, Germany, Poland, Hungary, Argentina, Brazil, Canada, Egypt, Tanzania, and Antarctica.

Discussion: Rounded pits can be produced by a wide range of predators from mammals to crocodiles (LaBarge and Njau, 2024).

10.1.29 Nihilichnus sicarius *LaBarge and Njau (2024)*

Holotype: On IUPC 101303 that is a femur of *Equus oldowayensis*, likely modified by *Crocodylus anthropophagus* (LaBarge and Njau, 2024, Fig. 5A).

Referred specimens: Ah-1431-SGDB (Mikuláš et al., 2006). RTMP67.15.21; RTMP85.36.314 (Jacobsen and Bromley, 2009). KK R0457 (Wretman and Kear, 2014). WnoZ/S/7/267, WnoZ/S/7/268 (Sadlok, 2022). SMNS 81169, SMNS 81342, SMNS 97010/5 (Mujal et al., 2022).

Diagnosis: Lenticular pits on the surface of vertebrate bone, not passing through the cortical bone. Width to length ratio is greater than 0.25, and sides slope to an internal vertex dividing pit, resulting in a "V-shaped" cross section.

Type locality: Olduvai Gorge, Tanzania.

Type horizon: Olduvai Lowermost Bed II (Quaternary: ~1.8 Ma).

Distribution: Middle Triassic to Quaternary of Australia, Germany, Hungary, Poland, and Canada.

Discussion: Bisected pits are commonly associated with crocodiles but records date back to the Missile Triassic (LaBarge and Njau, 2024).

10.1.30 Mikulasichnus *igen. nov.*

Type ichnospecies: *Mikulasichnus* (*Nihilichnus*) *quadripertitus* (Mikuláš, 2024 in Mazuch et al., 2024).

Included species: Known only from the type ichnospecies.

Etymology: To honor Radek Mikuláš for his pioneering, and ongoing, contributions to the study of bioerosion.

Diagnosis: Dentalite that differs from others in that the tooth impressions occur in an approximately straight line, comprise round impressions that are spaced regularly in close proximity with the distance between the pits being approximately an order of magnitude greater than their diameter, and if two lines occur they converge at an angle of about 20 degrees.

Distribution: Late Cretaceous of North America, Europe and Africa.

Discussion: *Mikulasichnus* igen. nov. is similar to *Mandaoichnites* and *Heterodontichnites* and dissimilar to *Nihilichnus* in that it preserves the impression of a partial dentition. Kauffman and Kesling (1960: Fig. 6.7) first interpreted this Late Cretaceous trace as having been produced by a mosasaur. Subsequently, Kauffman (1990) mentioned 100 specimens from the United States, and other such mosasaur dentalites are described from Canada, Mexico, Morocco and Europe (Hunt and Lucas, 2021, Table A.5).

10.1.31 Mikulasichnus (Nihilichnus) quadripertitus *(Mikuláš, 2024 in Mazuch et al., 2024)*

Holotype: CHMHZ-Mch0001, four rows of holes on both sides of *Mammites nodosoides* (Mazuch et al., 2024; Figs. 2–3; Fig. 6.9).

Referred specimens: Specimens from the Late Cretaceous of Czech Republic, of USA (e.g., Kauffman and Kesling, 1960; Mapes and Hansen, 1983; Kauffman, 1990, 2004), Morocco (Gale et al., 2017), Madagascar (Hoffmann and Keupp, 2015), Mexico (e.g., Saul, 1979; Ifrim, 2013) and Canada (e.g., Hewitt and Westermann, 1990; Tsujita and Westermann, 1998).

Diagnosis: As for ichnogenus.

Type locality: Měcholupy, Czech Republic.

Type horizon: *Mammites nodosoides* ammonite Zone, spongilitic calcareous, silty marlstones to limestones, Bílá Hora Formation (lower to middle Turonian), lower Turonian "Malnice beds."

Distribution: As for ichnogenus.

Discussion: This ichnospecies is widely distributed in the Late Cretaceous. As discussed above, there has been some debate about the origin of this trace.

10.1.32 Selachiodentichnus *igen. nov.*

Type ichnospecies: *Selachiodentichnus godfreyi* isp. nov.

Included species: Known only from the type ichnospecies.

Etymology: From the Greek *selachos* shark, Latin *dentis* tooth and Greek *ichnos* trace.

Diagnosis: Dentalite that differs from others in that the multiple tooth impressions occur along an arc, comprising lanceolate impressions 1.5–3 mm in width and approximately 10 mm in length that are strongly asymmetrical, and that are closely spaced.

Distribution: Miocene of North America.

Discussion: Godfrey and Smith (2010) hypothesized that the trace was made by *Galeocerdo* cf. *aduncus* from among the sharks from Calvert Cliffs on the basis of asymmetrical anteroposterior or parasymphysial teeth, the presence of accessory conules on the teeth, the rounded mesial tooth margins, strongly oblique anteroposterior tooth impressions and the labiolingual narrowness and the lack of twisting along the mesial margin of the crown impressions.

10.1.32.1 Selachiodentichnus godfreyi isp. nov.

Holotype: CMM-V-2244, coprolite preserving six shark tooth impressions and silicone rubber cast (Fig. 6.10).

Referred specimens: Shark-bitten coprolite, CMM-V-3245, also from Calvert Cliffs, Maryland, USA.

Etymology: To honor Stephen J. Godfrey who described the first, and many subsequent, dentalites on coprolites.

Diagnosis: As for ichnogenus.

Type locality: Calvert Cliffs, Maryland, USA.

Type horizon: Miocene Chesapeake Group.

Distribution: As for ichnogenus.

Discussion: The holotype preserves a curved pattern of six impressions from a single functional tooth row on the flattened lower surface of the coprolite (Fig. 6.6). The convex upper surface is marked by about five shallow tooth impressions, which are interpreted as representing impressions from the opposing dental arcade (Godfrey and Smith, 2010). A silicone rubber cast of the impressions on the lower surface demonstrates the general morphology of the teeth. The impressions were made by one medial tooth and five anteroposterior teeth, three on one side of the medial tooth and two from the contralateral side of the mandible (Godfrey and Smith, 2010). The medial tooth impression is much smaller than those of the anteroposterior teeth. Except for the medial (symphysial) tooth, the crowns are all strongly asymmetrical (the apex of each crown points distally) (Godfrey and Smith, 2010). The mesial margins of the crowns are convex (evenly rounded), and they exhibit a slight labial recurvature. The mesial margin of each tooth overlaps its mesial neighbor labially (Godfrey and Smith, 2010). The tooth impressions are nearly uniform, both in size and morphology, except for the small medial tooth, suggesting a gradationally homodont dentition (Godfrey and Smith, 2010).

10.1.33 *Ichnofamily* incertae sedis

10.1.33.1 *Daandavichnus* Rodríguez-de la Rosa in Rodríguez-de la Rosa et al. (2021)

Type ichnospecies: *Daandavichnus batoideum* Rodríguez-de la Rosa in Rodríguez-de la Rosa et al. (2021).

Included ichnospecies: Known only from the type ichnospecies.

Distribution: Early Cretaceous of Mexico.

Diagnosis: Trace fossil circumscribed within a subcircular to trapezoid outline; the anterior area preserves a curved impression with a M-shaped external boundary; posterior area preserves two circular depressions with smooth borders, occasionally connected by striations forming an arch.

Discussion: *Daandavichnus* is interpreted as the impression of the distal part of the jaw apparatus of a benthic-feeding organism, a small-sized member of Batoidea. It originated during the symmetrical protrusion of the jaw apparatus, while feeding (Rodríguez-de la Rosa et al., 2021).

10.1.33.2 *Daandavichnus batoideum* Rodríguez-de la Rosa in Rodríguez-de la Rosa et al. (2021)

Holotype: STA-BAP-2.4 is an epichnial impression on slab STA-BAP-2 and a silicone cast is cataloged under the same number (Rodríguez-de la Rosa et al., 2021, Fig. 4A—B).

Referred specimens: STA-BAP-1.1 to 1.3, STA-BAP-2.4 to 2.8.

Type locality: Puebla, Central Mexico.

Type horizon: Lower Cretaceous San Juan Raya Formation (upper Valanginian to lower Hauterivian).

Diagnosis: As for ichnogenus.

Distribution: As for the ichnogenus.

Discussion: Some benthic feeders evolved buccal specializations, such as jaw protrusion, an adaptation for feeding that evolved within several actinopterygian lineages (Hernandez and Staab, 2015).

10.1.33.3 *Osculichnus* Demírcan and Uchman (2010)

Type ichnospecies: *Osculichnus labialis* Demírcan and Uchman (2010).

Included ichnospecies: *Osculichnus labialis* Demírcan and Uchman, 2010, *Osculichnus tarnowskae* Szrek et al., 2016, *Osculichnus tepitsin* Rodríguez-de la Rosa et al., 2021, *Osculichnus fani* isp. nov. (Fig. 6.11).

Distribution: Upper Devonian of China (Fan et al., 2019), Early Jurassic of Poland (Pieńkowski, 1985), Late Jurassic of Spain (Rodriguez-de la Rosa et al., 2021; Santos et al., 2022). Eocene-Oligocene of Turkey (Demírcan and Uchman, 2010), *Osculichnus* cf. *labialis*, late Eocene-early Oligocene of Turkey (Demircan and Uchman, 2016), Lower Devonian of Poland (Szrek et al., 2016),

Diagnosis: Hypichnial, bilobate mounds, generally elliptical or crescentic in outline, having a smaller and a larger, lip-like lobe separated by undulate furrow.

Discussion: *Osculichnus* is interpreted as a hunting trace (praedichnion) of a fish penetrating the seafloor through a sand layer into underlying mud (in the best-preserved specimens) or directly in surficial mud (poorly preserved specimens) (Demírcan and Uchman, 2010).

The ichnogenus *Osculichnus* shows notable morphological variation, indicating different fish trace-makers (Rodríguez-de la Rosa et al., 2021). Rodríguez-de la Rosa et al. (2021) note that morphology ranges from the typical lip-like traces of *Osculichnus labialis* (Demírcan and Uchman, 2010) to the rather "square" mouth of *Osculichnus tarnowskae*, attributed to a lungfish (Szrek et al., 2016). Other *Osculichnus* traces show lip-like lobes of rather equal size and shape, such as *O. fani* isp. nov. and those specimens reported by Demírcan and Uchman (2016).

10.1.33.4 *Osculichnus labialis* Demírcan and Uchman (2010)

Holotype: Tr19 (Demírcan and Uchman, 2010, Fig. 3A).

Paratype: Tr22 (Demírcan and Uchman, 2010, Fig. 3B).

Referred specimens: 56 specimens on 49 slabs (Tr1-Tr32, Tr93-112).

Revised diagnosis: *Oscularis* ichnospecies that differs from others in being comprised of rounded bilobate mounds, and maxillary barbel impressions, if present, are large and rounded in shape (Fig. 6.11A).

Type locality: Thrace Basin near Malkara, Turkey.

Type horizon: Mezardere Formation (late Eocene—early Oligocene).

Distribution: Late Jurassic of Spain (Rodríguez-de la Rosa et al., 2021), late Eocene-early Oligocene of Turkey (Demírcan and Uchman, 2010), *Osculichnus* cf. *O. labialis* in late Eocene-early Oligocene of Turkey (Demircan and Uchman, 2016).

Discussion: There are relatively few examples of described modern analogs (Demircan and Uchman, 2010). Examples include elongate depressions ("snout-markings") off the eastern coast of the USA produced by fishes of the families Moridae and Halosauridae (Stanley, 1971), and Jones et al. (1992) described "feeding scars" produced by three genera of teleosts in areas in One Tree Lagoon on the Great Barrier Reef.

10.1.33.5 *Osculichnus tarnowskae* Szrek et al. (2016)

Holotype: Muz. PGI-NRI 1733.II.406 (original and silicone cast: Szrek et al., 2016, Fig. 4),

Paratype: Muz. PGI-NRI 1733.II.418 (silicone cast, original left in the field; Szrek et al., 2016, Fig. 5).

Referred specimens: 41 silicone casts (Muz. PGI-NRI 1733.II.391–432).

Revised diagnosis: *Oscularis* ichnospecies that differs from others in having a recurrent shape of a closed mouth that is U-shaped or often trapezoidal, in being slightly smaller and in preserving more subtle details, such as imprints of wrinkles, folded skin, and tuberculated tooth plates (Fig. 6.11C).

Type locality: Ujazd village Holy Cross Mountains, southern Poland.

Type horizon: Lower Devonian of Poland.

Distribution: Lower Devonian of Poland.

Discussion: As Szrek et al. (2016, p. 116) acknowledged, *Osculichnus tarnowskae* "is very similar to *O. labialis*." However, the different shape of the mouth imprint suggests that it does indeed represent a different ichnospecies (Fig. 6.7). Szrek et al. (2016) postulated that this ichnospecies was produced by a short-snouted dipnoan similar to *Dipnorhynchus* based on overall morphology, trapezoidal shape of the mouth, the probable presence of tuberculate tooth-plates, and the imprints of folding skin.

10.1.33.6 *Osculichnus tepitsin* Rodríguez-de la Rosa et al. (2021)

Holotype: STA-BAP-3, slab and associate silicone cast, cataloged under the same number (Fig. 6.2E).

Referred specimens: Five, medium-to fine-grained greyish sandstone slabs were collected (STA-BAP-1 to STA-BAP-5; CP.UAZ-015, CP. UAZ-016).

Diagnosis: *Osculichnus* ichnospecies that differs from others in having an elongate, nearly subrectangular, lower jaw lobe, with fusiform lateral ends, an upper jaw lobe with a ventral convex projection in its medial portion and the presence of two short, subtriangular, maxillary barbel impressions (Fig. 6.11B).

Type locality: Puebla, Central Mexico.

Type horizon: Lower Cretaceous San Juan Raya Formation (upper Valanginian to lower Hauterivian).

Distribution: Lower Cretaceous of Mexico.

Discussion: *Osculichnus tepitsin* is interpreted as a benthic-feeding trace of an unknown fish (Rodríguez-de la Rosa et al., 2021).

10.1.33.7 *Osculichnus fani* isp. nov.

Holotype: WHYJ-1, with enlarged lower lobe and an irregular central cavity (Fan et al., 2019, Fig. 5B).

Paratype: Uncollected specimen with prominent central lobe (Fan et al., 2019, Fig. 5A, central specimen indicated by white arrow).

Etymology: Named for Ruo-ying Fan, who first described this trace.

Referred specimens: Huangjintang locality, lowermost sandstone, 15 specimens (Fan et al., 2019, Fig. 6.4A–E), middle sandstone, five specimens (Fan et al., 2019, Fig. 4F), upper sandstone, six specimens. Guodingshan hill locality, nine specimens (Fan et al., Fig. 5A).

Diagnosis: *Osculichnus* ichnospecies that differs from others in having either a central groove or cavity or a prominent central lobe of rounded rectangular shape that may extend over the lower lobe (Fig. 6.11D).

Type locality: Guodingshan hill locality, Wuhan city, Hubei Province, China.

Type horizon: Upper Devonian Wutong.

Distribution: Devonian of China.

Discussion: *Osculichnus fani* isp. nov. demonstrates a range of morphological variation (Fan et al., 2019, Fig. 6). All specimens have two raised outer lobes, some specimens have a lateral central groove or cavity and others have a central lobe that is oriented dorsoventrallly and has the shape of a rounded rectangle. The arched upper lobe is indicative of a trapezoidal outline of the snout and a pair of arches in the ventral margin of the upper lip which is similar to extant lungfishes (Fan et al., 2019).

10.1.33.8 *Physichnium* igen. nov.

Type ichnospecies: *Physichnium* (*Pischichnus*) *waitemata* (Gregory, 1991).

Included species: Known only from the type ichnospecies.

Etymology: From the Greek *physa* (bellows) in allusion to the inferred production by projecting fluid and *ichnos* (trace).

Revised diagnosis: A cylindrical or plug-like structure with a steep, smooth margin which may have an undercut, asymmetrical in cross section with a steeper slope at one end of the depression, oriented vertical to bedding, averaging 15 to 20 cm in diameter and 15 cm in depth maximum, and infilling consists of coarse material (rarely up to conglomerate), which is typically structureless and often contains shelly debris.

Distribution: Cretaceous-Pleistocene of New Zealand, United States, and Antarctica. Occurrences include the Upper Cretaceous, Utah, USA (Howard et al., 1977), Upper Cretaceous Blackhawk Formation, Utah, USA (Kamola, 1984), Upper Cretaceous Santa Marta Formation, James Ross Island, Antarctica (Scasso et al., 1991), Upper Cretaceous Tremp La Posa

Formation, Group, Spain (Martinelli et al., 2001), middle Eocene Peschici Formation, Italy (Belvedere et al., 2011), basal Miocene of New Zealand (Gregory, 1991), lower Miocene Cape Rodney and Puketi beds of New Zealand (Brook, 1983; Gregory, 1991), Miocene sandstone of Floras Lake, Oregon, USA (Leithold and Bourgeois, 1984), Miocene Taliao Formation of Taiwan (Löwemark, 2015), middle Miocene Tougane Formation of Japan (Kotake et al., 2004), upper Miocene-lower Pliocene Tokomaru Formation of New Zealand (Blom, 1982, 1984), lower Pliocene, Santa Maria, Azores Archipelago (Uchman et al., 2018), lower Pliocene Paeoneone Formation of New Zealand (Ballance et al., 1977), upper Pliocene Capodarso Formation of Italy (Massari, and D'Alessandro, 2010), Pleistocene Merced Formation of California, USA (Hunter et al., 1984), Pleistocene of Georgia–Florida and North Carolina, USA (Howard et al., 1977), Pleistocene Clifton Sand, New Zealand (Gregory, 1991), Pleistocene Cape Arkhangelos Formation of Rhodes (Nielsen et al., 2020) and Holocene of Georgia, USA (Howard et al., 1977).

Discussion: This trace has been interpreted as a feeding trace produced by rays excavating feeding holes by hydraulic jetting (Gregory et al., 1979; Gregory, 1991; Uchman et al., 2018). This ichnogenus is quite common from the Cretaceous until the Recent, which is consistent with the postulated producer being a member of the Batoidea (e.g., Gregory, 1991; Scasso et al., 1991; Martinelli et al., 2001; Uchman et al., 2018; but see Belvedere et al., 2011). Most previously reported occurrences of *Piscichnus* pertain to *Physichnium* ig. nov.

10.1.34 Physichnium waitemata *(Gregory, 1991)* igen. nov.

Holotype: Specimen X39 (FR Number R09/f116) Gregory, 1991, Fig. 6; Fig. 6.13).

Referred specimens: None.

Diagnosis: As for ichnogenus.

Type locality: Mathesons Bay, North Island, New Zealand.

Type horizon: Cape Rodney Formation (early Miocene).

Distribution: As for the ichnogenus.

Discussion: Gregory (1991) considered that Devonian traces from the Catskill deltaic complex of New York, USA described by Miller (1979) pertain to this ichnospecies. However, his "large vertical structures" are not similar to this ichnotaxon (Miller, 1979, pl. 1E), so all occurrences are late Mesozoic or younger. Gregory (1991) did not list a diagnosis for this ichnospecies, and Nielsen et al. (2020) were the first to do so.

10.1.35 Piscichnus *Feibel (1987)*

Type ichnospecies: *Piscichnus brow*ni Feibel (1987).

Included ichnospecies: *Piscichnus brow*ni Feibel (1987), *Piscichnus sanctacrucensis* Niedźwiedzski and Szrek (2020).

Revised diagnosis: A shallow, circular, dish-shaped structure, symmetrical in cross section, oriented concave upward on a horizontal plane with gentle sides and a flat or rounded base, and the edge may be surrounded by a low rim and infilling consists of a single structureless fill or several weakly bedded depositional units.

Distribution: Devonian to Recent of Africa and Europe.

Discussion: The structures typically occur in scattered pattern with variable diameter, but they seldom intersect. The surface of the structure may be marked by a lamina of grains slightly coarser than the underlying material. *Piscichnus* is a relatively uncommon ichnogenus.

10.1.36 Piscichnus browni *Feibel (1987)*

Lectotype: Unnumbered, uncollected specimen (Feibel, 1987, Fig. 2.1).

Referred specimens: Approximately 70 specimens at type locality.

Diagnosis: Differs from other ichnospecies of *Piscichnus* in large size (diameter 0.45–1.35 m, depth 5–15 cm), in lacking a prominent rim, and the diameter of the structure is far greater than the maximum depth, producing a diameter to depth ratio of up to 10:1.

Type locality: Kolom Guo sand river, south-central Koobi Fora region, Kenya.

Type horizon: Upper portion of the KBS Member of the Koobi Fora Formation (Plio-Pleistocene).

Distribution: Plio-Pleistocene of Africa.

Discussion: Feibel (1987, p. 131) denoted the type material as "approximately 70 structures occur, exposed over an area of about 2250 m" but did not designate a holotype. We consider the type material of Feibel (1987) to be syntypes and we designate a lectotype (Feibel, 1987, Fig. 3). Feibel (1987) interpreted this trace as being a nest produced by cichlid fish excavating using their mouth, and he described Recent traces of this type from the same area.

10.1.37 Piscichnus sanctacrucensis *Niedźwiedzski and Szrek (2020)*

Holotype: Muz PGI 1728.II.22 (Niedźwiedzski and Szrek. 2020, Figs. 9A and 12E) represented by a resin cast (Niedźwiedzski and Szrek 2020, Fig. 8B).

Referred specimens: Numerous uncollected specimens from the type locality Fig. 6.12).

Revised diagnosis: Differs from other species of *Piscichnus* in being small (15–20 cm in diameter) to medium (20–30 cm in diameter) size and relatively deep (1–5 cm) possessing a prominent rim that is flat and may fully encircle the structure.

Type locality: Zachełmie Quarry (Holy Cross Mountains, southern Poland).

Type horizon: Wojciechowice Formation (lower Middle Devonian).

Distribution: Devonian of Poland.

Discussion: Lucas (2015) first identified *Piscichnus* from the type locality. Niedźwiedzski and Szrek (2020) also identified *Piscichnus browni* from the locality, but these specimens are smaller (10–50 cm in diameter) and relatively deeper (2–15 cm deep) than most specimens of this ichnospecies, and we suspect that future work will result in these specimens being assigned to *P. sanctacrucensis*.

11. Utility of dentalites

Dentalites have the potential to elucidate a wide range of behaviors, and ecological and evolutionary scenarios, including (1) predation, including hunting strategies; (2) bite method and force; (3) dietary selection; (4) feeding; (5) scavenging strategies; (6) methodologies of bone accumulation; (7) trophic patterns; (8) intraspecific (agonistic) interactions; (9) tooth sharpening; and (10) bone and rock utilization for other purposes, including mineral extraction (Binford, 1981; Hunt et al., 1994; Drumheller-Horton, 2012; Lucas, 2016). They thus are of diverse paleoethological significance.

12. Conclusions

Dentalites encompass all traces produced on a substrate by the teeth or oral cavity of a vertebrate or invertebrate.

The majority of vertebrate dentalites are on a bony substrate. Their fossil record is biased by the Taxophile Effect and toward large animals and certain time periods with acmes in the Late Triassic, Late Cretaceous, and Late Cenozoic. Prior to the Quaternary, with the exception of the nonmarine Mesozoic, most dentalites are reported from marine facies.

The ichnological record of predation on invertebrates by vertebrate predators and scavengers is surprisingly small. There are only two main taxa documented in detail: (1) predation on sessile echinoderms; and (2) predation on nektonic nautiloids and ammonoids.

Coprolites with dentalites have only been recognized in the past decade. They are all from aquatic environments, mainly marine, and all but one is definitely Cenozoic in age.

A variety of vertebrate behaviors involve direct interaction between mouth parts and the lithic substrate, including feeding by aquatic fish and tetrapods, feeding by aerial birds and pterosaurs and digging by terrestrial rodents. Vertebrate ichnotaxa include *Osculichnus*, *Daandavichnus*, and *Physichnium* gen. nov. *Piscichnus* is the trace of a fish nest.

There are very few examples of vertebrate dentalites on vegetation, and all are from the Cenozoic. All dentalites are preserved on the more dense portions of plants such as wood and nuts.

Ichnotaxobases for dentalites include substrate, orientation, size, evidence of tooth structure, evidence of dentition structure, pattern of occurrence, and location.

The ichnofamily Machichnidae Wisshak et al., 2019 includes *Machichnus* Mikuláš et al., 2006, *Knethichnus* Jacobsen and Bromley, 2009; *Linichnus* Jacobsen and Bromley, 2009; *Machichnus* Mikuláš et al., 2006, *Mandaodonites* Cruickshank, 1986, *Nihilichnus* Mikuláš et al., 2006, *Heterodontichnites* Rinehart et al., 2006, *Mikulasichnus* igen. nov. and *Selachiodentichnus* igen. nov.

Buckland conducted the first actualistic study of dentalites by feeding cow bones to a hyena. Modern actualistic studies were largely triggered by controversies generated by the interpretation of early archeological sites in East Africa and South Africa. Recent studies have focused on large mammals and crocodiles.

Dentalites have diverse potential in paleoethology, including (1) predation, including hunting strategies; (2) bite method and force; (3) dietary selection; (4) feeding; (5) scavenging strategies; (6) methodologies of bone accumulation; (7) trophic patterns; (8) intraspecific (agonistic) interactions; (9) tooth sharpening; and (10) bone and rock utilization for other purposes, including mineral extraction.

In general, ichnologists utilize an ichnotaxonomy for dentalites, but vertebrate paleontologists do not. There is a need for a refined dentalite ichnology beginning with compilation and synthesis of the entire fossil record, rigorous ichnotaxonomy and determination of analytical criteria for establishing inferences about the behaviors archived by dentalites.

Declaration of competing interest

The authors declare that they have no known competing financial interests or personal relationships that could have appeared to influence the work reported in this paper.

Data availability

No data were used for the research described in this article.

Acknowledgments

We thank many colleagues who have graciously allowed us access to collections, including Michael Brett-Surman, Thomas Lehmann, Krister Smith, Anika Vogel, and others for helpful discussions over the years, including Lewis Binford, Brooks Britt, Ken Carpenter, Sebastian Dalman, Christian Meyer, Jesper Milàn, and Larry Rinehart. Neville Hudson provided important collection information. Mike Everhart, Stephen Godfrey, Neville Hudson, Radek Mikuláš, and Joerg Schneider kindly supplied images. Radek Mikuláš provided a thoughtful and knowledgeable review although he disagrees with usage of the term dentalite. Hendrik Klein also provided an insightful review.

References

Abbassi, N., Kundrát, M., Mirzaie Ataabadi, M., Ahlberg, P.E., 2015. Avian ichnia and other vertebrate trace fossils from the Neogene Red Beds of Tarom valley in north-western Iran. Hist. Biol. 28 (8), 1075–1089.

Abel, O., 1922a. Die Schnauzenverletzungen der Parasuchier und ihre biologische Bedeutung. Päleontol. Z. 5, 25–57.

Abel, O., 1922b. Discussion of Krausel, 1922. Paläontol. Z. 4, 87.

Abrahams, P.W., 2003. Human geophagy: a review of its distribution, causes, and implications. In: Skinner, H.C., Berger, A.R. (Eds.), Geology and Health: Closing the Gap. Oxford University Press, Oxford, pp. 31–36.

Abrahams, P.W., Parsons, J.A., 1996. Geophagy in the tropics—a literature review. Geogr. J. 162, 63–72.

Adrian, B., Smith, H.F., Kelley, K., Wolfe, D.G., 2023. A new baenid, *Edowa zuniensis* gen. et sp. nov., and other fossil turtles from the Upper Cretaceous Moreno Hill Formation (Turonian), New Mexico, USA. Cretac. Res. 144, 105422. https://doi.org/10.1016/j.cretres.2022.105422.

Aguilera, O., Rodrigues de Aguilera, D.R., 2004. Giant-toothed white sharks and wide-toothed mako (Lamnidae) from the Venezuela Neogene: their role in the Caribbean, shallow-water fish assemblage. Caribb. J. Sci. 40, 368–382.

Aiglstorfer, M., Heizmann, E.P.J., Peigné, S., 2019. Who killed *Micromeryx flourensianus*? A case study of taphonomy and predation on ruminants in the middle Miocene of France. Lethaia 52 (3), 429–444. https://doi.org/10.1111/let.12322.

Akersten, W.A., 1985. Canine function in *Smilodon* (Mammalia; Felidae; Machairodontinae). Contributions Science, Natural History Museum Los Angeles County, 356, pp. 1–22.

Alexander, J., 1992. Alas, poor *Notharctus*: natural history, 101 (8), 54–59.

Alexander, J.P., Burger, B.J., 2001. Stratigraphy and taphonomy of Grizzly buttes, bridger formation, and the middle Eocene of Wyoming. In: Gunnell, G.F. (Ed.), Eocene Biodiversity: Unusual Occurrences and Rarely Sampled Habitats. Kluwer Academic/Plenum Publishers, New York, pp. 165–196.

Alexander, R.R., 1981. Predation scars preserved in Chesterian brachiopods: probable culprits and evolutionary consequences for the articulates. J. Paleontol. 55, 192–203.

Alonso, A., Canudo, J.I., Fernández-Baldor, F.T., Huerta, P., 2017. Isolated theropod teeth associated with sauropod remains from El Oterillo II (Early Cretaceous) site of Salas de los Infantes (Burgos, Spain). J. Iber. Geol. 43, 193–215.

Amalfitano, J., Dalla Vecchia, F.M., Giusberti, L., Fornaciari, E., Luciani, V., Roghi, G., 2017. Direct evidence of trophic interaction between a large lamniform shark, *Cretodus* sp., and a marine turtle from the Cretaceous of northeastern Italy. Palaeogeogr. Palaeoclimatol. Palaeoecol. 469, 104–121.

Amalfitano, J., Giusberti, L., Fornaciari, E., Dalla Vecchia, F.M., Luciani, V., Kriwet, J., Carnevale, G., 2019. Large deadfalls of the 'ginsu' shark *Cretoxyrhina mantelli* (Agassiz, 1835) (Neoselachii, Lamniformes) from the upper Cretaceous of northeastern Italy. Cretac. Res. 98, 250–275.

Andrade, L.C.D., 2019. Cenário paleoambiental, tafonômico e paleopatológico dos mamíferos fósseis do depósito de tanque Zabelê, Capoeiras, Pernambuco, Brasil. Ph. D Dissertation. Universidade Federal de Pernambuco, Pernannuco, 164 p.

Andrade, M.B., Edmonds, R., Benton, M.J., Schouten, R., 2011. A new Berriasian species of *Goniopholis* (Mesoeucrocodylia, Neosuchia) from England, and a review of the genus. Zool. J. Linn. Soc. 163, S66–S108. https://doi.org/10.1111/j.1096-3642.2011.00709.x, 66e108.

Andrew, C., Howe, P., Paul, C., 2015. Fatally bitten ammonites from septarian concretions of the 'Marston Marble' (lower Jurassic, Sinemurian), Somerset, UK, and their taphonomy. Proc. Geologists' Assoc. 126, 119–129.

Andrew, C., Howe, P., Paul, C.R., Donovan, S.K., 2010. Fatally bitten ammonites from the lower Lias group (lower Jurassic) of Lyme Regis, Dorset. Proc. Yorks. Geol. Soc. 58, 81–94.

Andrews, P., 1990. Owls, Caves and Fossils: Predation, Preservation and Accumulation of Small Mammal Bones in Caves, with an Analysis of the Pleistocene Cave Faunas from Westbury-sub-Mendip. University of Chicago Press, Somerset, UK. Chicago, 239 p.

Antunes, M.T., Balbino, A.C., Ginsburg, L., 2006. Ichnological evidence of a Miocene rhinoceros bitten by a bear-dog (*Amphicyon giganteus*). Ann. Paleontol. 92, 31–39.

Araújo, R., Polcyn, M.J., Lindgren, J., Jacobs, L.L., Schulp, A.S., Mateus, O., Gonçalves, A.O., Morais, M.L., 2015. New aristonectine elasmosaurid plesiosaur specimens from the Early Maastrichtian of Angola and comments on paedomorphism in plesiosaurs. Neth. J. Geosci. 94, 93–108.

Araújo Júnior, H.I., da Silva Marinho, T., 2013. Taphonomy of a *Baurusuchus* (Crocodyliformes, Baurusuchidae) from the adamantina formation (upper Cretaceous, Bauru basin), Brazil: implications for preservational modes, time resolution and paleoecology. J. S. Am. Earth Sci. 47, 90–99.

Araújo-Júnior, H.I., de Souza Barbosa, F.H., Medeiros da Silva, L.H., 2017. Overlapping paleoichnology, paleoecology and taphonomy: analysis of tooth traces in a Late Pleistocene-early Holocene megafaunal assemblage of Brazil and description of a new ichnotaxon in hard substrate. Palaeogeogr. Palaeoclimatol. Palaeoecol. 468, 122–128.

Armitage, I.A., Gingras, M.K., 2003. Sedimentologic and environmental implications of Atlantic sturgeon (*Acipenser oxyrhynchus*) feeding traces, Bay of Fundy, new Brunswick, Canada. Geol. Soc. Am. 35 (6), 578. Abstracts with Programs.

Arriaza, M.C., Aramendi, J., Maté-González, M.A., Yravedra, J., Stratford, D., 2021. Characterising leopard as taphonomic agent through the use of micro-photogrammetric reconstruction of tooth marks and pit to score ratio. Hist. Biol. 33 (2), 176–185.

Augustin, F.J., Matzke, A.T., Maisch, M.W., Pfretzschner, H.-U., 2020a. A theropod dinosaur feeding site from the Upper Jurassic of the Junggar Basin, NW China. Palaeogeogr. Palaeoclimatol. Palaeoecol. 560. https://doi.org/10.1016/j.palaeo.2020.109999.

Augustin, F.J., Matzke, A.T., Maisch, M.W., Hinz, J.K., Pfretzschner, H.U., 2020b. The smallest eating the largest: the oldest mammalian feeding traces on dinosaur bone from the Late Jurassic of the Junggar Basin (northwestern China). Sci. Nat. 107, 1–5.

Avilla, L.S., Fernandes, R., Ramos, D.F., 2004. Bite marks on a crocodylomorph from the Upper Cretaceous of Brazil: evidence of social behavior? J. Vertebr. Paleontol. 24, 971–973.

Backus, G.J., 1964. The effects of fish-grazing on invertebrate evolution in shallow tropical waters. Occasional Papers of the Allan Hancock Foundation, 27, pp. 1–34.

Bader, K.S., Hasiotis, S.T., Martin, L.D., 2009. Application of forensic science techniques to trace fossils on dinosaur bones from a quarry in the Upper Jurassic Morrison Formation, northeastern Wyoming. Palaios 24, 140–158.

Bakker, R.T., Bir, G., 2004. Dinosaur crime scene investigations: theropod behaviour at Como Bluff, Wyoming, and the evolution of birdness. In: Currie, P.J., Koppelhus, E.B., Shugar, M.A., Wright, J.L. (Eds.), Feathered Dragons: Studies on the Transition from Dinosaurs to Birds. Indiana University Press, Indianapolis, pp. 301–342.

Ballance, P.F., Hayward, B.W., Wakefield, L.L., 1977. Group nomenclature of the late Oligocene and early Miocene rocks in Auckland and Northland, New Zealand; and an Akarana supergroup. N. Z. J. Geol. Geophys. 20, 673–686.

Bardet, N., Jagt, J.W.M., Kuypers, M.M.M., Dorttmgs, R.W., 1998. Shark tooth marks on a vertebra of the mosasaur *Plioplatecarpus marshi* from the Late Maastrichtian of Belgium. Publ. Natuurhistorisch Genootschap Limburg 41, 52–55.

Barnes, K.M., Hiller, N., 2010. The taphonomic attributes of a Late Cretaceous plesiosaur skeleton from New Zealand. Alcheringa 34, 333–344.

Battista, F., Ribeiro, A.M., Melo, T.P., Schultz, C.L., 2024. A reinterpretation of taphonomy and palaeoecology of the early Carnian *Santacruzodon* assemblage zone type-locality (Santa Maria Supersequence), Brazil. Palaeogeogr. Palaeoclimatol. Palaeoecol., 112080 https://doi.org/10.1016/j.palaeo.2024.112080.

Beck, R.M., 2009. Was the Oligo-Miocene Australian metatherian Yalkaparidon a 'mammalian woodpecker'? Biol. J. Linn. Soc. 97 (1), 1–7.

Bell Jr., G.L., Martin, J.E., 1995. Direct evidence of aggressive intraspecific competition in *Mosasaurus conodon* (Mosasauridae: Squamata). J. Vertebr. Paleontol. 15 (Suppl. 3), 18A.

Bell, P.R., Currie, P.J., 2010. A tyrannosaur jaw bitten by a confamilial: scavenging or fatal agonism? Lethaia 43, 278–281.

Belvedere, M., Franceschi, M., Morsilli, M., Zoccarato, P.L., Mietto, P., 2011. Fish feeding traces from middle Eocene limestones (Gargano Promontory, Apulia, southern Italy). Palaios 26, 693–699.

Benites-Palomino, A., Velez-Juarbe, J., Altamirano-Sierra, A., Collareta, A., Carrillo-Briceño, J.D., Urbina, M., 2022. Sperm whales (Physeteroidea) from the Pisco Formation, Peru, and their trophic role as fat sources for late Miocene sharks. Proc. R. Soc. B 289 (1977), 20220774.

Berger, L.R., Clarke, R.J., 1995. Eagle involvement in accumulation of the Taung child fauna. J. Hum. Evol. 29, 275–299.

Bermúdez-Rochas, D.D., Delvene, G., Ruiz-Omeñaca, J.I., 2013. Evidence of predation in Early Cretaceous unionoid bivalves from freshwater sediments in the Cameros Basin, Spain. Lethaia 46, 57–70.

Bertling, M., Braddy, S.J., Bromley, R.G., Demathieu, G.R., Genise, J., Mikuláš, R., Nielsen, J.K., Nielsen, K.S.S., Rindsberg, A.K., Schlirf, M., Uchman, A., 2006. Names for trace fossils: a uniform approach. Lethaia 39, 265–286.

Bertling, M., Buatois, L.A., Knaust, D., Laing, B., Mángano, M.G., Meyer, N., Mikuláš, R., Minter, N.J., Neumann, C., Rindsberg, A.K., Uchman, A., 2022. Names for trace fossils 2.0: theory and practice in ichnotaxonomy. Lethaia 55 (3), 1–19.

Bianucci, G., Sorce, B., Storai, T., Landini, W., 2010. Killing in the Pliocene: shark attack on a dolphin from Italy. Palaeontology 53, 457–470.

Bianucci, G., Bisconti, M., Landini, W., Storai, T., Zuffa, M., Giuliani, S., Mojetta, A., 2002. Trophic interaction between white shark (*Carcharodon carcharias*) and cetaceans: a comparison between Pliocene and recent data from central Mediterranean Sea. In: Vacchi, M., La, M.G., Serena, F., Séret, B. (Eds.), Proceedings of the IV European Elasmobranch Association Meeting, Livono Italy. Abbeville, Imprimerie F. Paillart, pp. 33–48.

Bigelow, P.K., 1994. Occurrence of a squaloid shark (Chondrichthyes: Squaliformes) with the pinniped *Allodesmus* from the upper Miocene of Washington. J. Paleontol. 68, 680–684.

Binford, L.R., 1981. Bones: Ancient Men and Modern Myths. Academic Press, New York, 320 p.

Bishop, G.A., 1972. Crab bitten by a fish from the upper Cretaceous Pierre shale of South Dakota. Bull. Geol. Soc. Am. 83, 3823–3826.

Blieck, A., 2017. Heterostracan vertebrates and the great Eodevonian biodiversification event—an essay. Palaeobiodivers. Palaeoenviron. 97, 375–390.

Bloch, J.I., Boyer, D.M., 2001. Taphonomy of Small Mammals in Freshwater Limestones from the Paleocene of the Clarks Fork Basin, vol 33. University of Michigan Museum of Paleontology, pp. 185–198. Papers.

Blom, W.M., 1982. Sedimentology of the Tokumaru Formation, Waiupu Subdivision, Raukumara Peninsula. M. Sc. Thesis, Auckland University, Auckland, p. 75.

Blom, W.M., 1984. Stratigraphy and sedimentology of Tokomaru Formation (late Miocene—early Pliocene), eastern Raukumara Peninsula, New Zealand. N. Z. J. Geol. Geophys. 27, 125–137.

Blumenschine, R.J., 1986. Carcass consumption sequences and the archaeological distinction of scavenging and hunting. J. Hum. Evol. 15 (8), 639–659.

Blumenschine, R.J., 1988. An experimental model of the timing of hominid and carnivore influence on archaeological bone assemblages. J. Archaeol. Sci. 15, 483–502.

Blumenschine, R.J., 1995. Percussion marks, tooth marks and the experimental determinations of the timing of hominid and carnivore access to long bones at FLK *Zinjanthropus*, Olduvai Gorge, Tanzania. J. Hum. Evol. 29, 21–51.

Blumenschine, R.J., Marean, C.W., Capaldo, S.D., 1996. Blind tests of inter-analyst correspondence and accuracy in the identification of cut marks, percussion marks, and carnivore tooth marks on bone surfaces. J. Archaeol. Sci. 23 (4), 493–507.

Boaz, N.T., Ciochon, R.L., Xu, Q., Liu, J., 2000. Large mammalian carnivores as a taphonomic factor in the bone accumulation at Zhoukoudian. Acta Anthropol. Sin. 19 (Suppl.), 224–234.

Boaz, N.T., Ciochon, R.L., Xu, Q., Liu, J., 2004. Mapping and taphonomic analysis of the *Homo erectus* loci at Locality 1 Zhoukoudian, China. J. Hum. Evol. 46, 519–549.

Boessenecker, R.W., Geisler, J.H., 2023. New skeletons of the ancient dolphin *Xenorophus sloanii* and *Xenorophus simplifies* sp. nov. (Mammalia, Cetacea) from the Oligocene of South Carolina and the ontogeny, functional anatomy, asymmetry, pathology, and evolution of the earliest Odontoceti. Diversity 15, 1154. https://doi.org/10.3390/d15111154.

Boessenecker, R.W., Perry, F.A., 2011. Mammalian bite marks on juvenile fur seal bones from the Late Neogene Purisima formation of central California. Palaios 26, 115–120.

Boessenecker, R.W., Perry, F.A., Schmitt, J.G., 2014. Comparative taphonomy, taphofacies, and bonebeds of the Mio-Pliocene Purisima Formation, Central California: strong physical control on marine vertebrate preservation in shallow marine settings. PLoS One 9 (3), e91419.

Boessenecker, R.W., Ahmed, E., Geisler, J.H., 2017. New records of the dolphin *Albertocetus meffordorum* (Odontoceti: Xenorophidae) from the lower Oligocene of South Carolina: encephalization, sensory anatomy, postcranial morphology, and ontogeny of early odontocetes. PLoS One 12 (11), e0186476. https://doi.org/10.1371/journal.pone.0186476.

Bohatý, J., 2008. Classification of pre- and postmortem ossicular modifications of the cupressocrinitid skeletons (Crinoidea), Devonian 4. Arbeitstreffen deutschsprachiger Echinodermenforscher 4th. Workshop of German and Austrian Echinoderm Research Abstracts, Andreas Kroh and Brigitta Schmid Naturhistorisches Museum, Wien, pp. 3–13.

Bohatý, J., 2009. Pre-and postmortem skeletal modifications of the Cupressocrinitidae (Crinoidea, Cladida). J. Paleontol. 83, 45–62.

Böhme, M., Prieto, J., Schneider, S., Hung, N.V., Tran, D.N., 2011. The Cenozoic on-shore basins of Northern Vietnam: biostratigraphy, vertebrate and invertebrate faunas. J. Asian Earth Sci. 40, 672–687.

Boles, Z., Lacovara, K., 2013. Evidence of extensive scavenging/predation by Late Cretaceous marine organisms from the basal Hornerstown Formation, New Jersey, USA. J. Vertebr. Paleontol. (Suppl. 3), 89A.

Bosio, G., Collareta, A., Di Celma, C., Lambert, O., Marx, F.G., de Muizon, C., Gioncada, A., Gariboldi, K., Malinverno, E., Malca, R.V., Urbina, M., 2021. Taphonomy of marine vertebrates of the Pisco formation (Miocene, Peru): insights into the origin of an outstanding fossil-Lagerstätte. PLoS One 16 (7), e0254395.

Botfalvai, G., Prondvai, E., Ősi, A., 2014. Inferred bite marks on a Late Cretaceous (Santonian) bothremydid turtle and a hylaeochampsid crocodilian from Hungary. Cretac. Res. 50, 304–317.

Bond, P.N., Saunders, W.B., 1989. Sublethal injury and shell repair in Upper Mississippian ammonoids. Paleobiology 15, 414–428.

Brain, C.K., 1970. New finds at the Swartkrans australopithecine site. Nature 225, 1112–1119.

Brain, C.K., 1978. Interpreting the bone accumulations from the Sterkfontein Valley caves: a case study in Quaternary research. South. Afr. Humanit. 23, 465–468.

Brain, C.K., 1981. The Hunters or the Hunted? University of Chicago Press, Chicago, 384 p.

Braunn, P.R., Fonseca, R.C., Ferigolo, J., 2001. Evidências de possível necrofagia em costelas de *Jachaleria candelariensis* (Therapsida, Dicynodontia) do Triássico Superior do Estado do Rio Grande do Sul, Brasil. In: XVII Congresso Brasieiro Paleontologia, Resumos, Rio Branca, vol 49. Univeridade Federale do Acre.

Brazeau, M.D., Friedman, M., 2015. The origin and early phylogenetic history of jawed vertebrates. Nature 520, 490–497.

Brink, K.S., Campione, N.E., Hawthorn, J.R., 2013. Amniote faunal revision of the Pictou group (Permo-Carboniferous), Prince Edward Island, Canada. C. R. Palevol. 12, 473–485.

Brink, K.S., Maddin, H.C., Evans, D.C., Reisz, R.R., 2015. Re-Evaluation of the historic Canadian fossil *Bathygnathus borealis* from the early Permian of Prince Edward Island. Can. J. Earth Sci. 52, 1109–1120.

Bromley, R.G., 1975. Comparative analysis of fossil and recent echinoid bioerosion. Palaeontology 18 (4), 725–739.

Bromley, R.G., 1996. Trace Fossils: Biology and Taphonomy, second ed. Unwin Hyman, London. 361 p.

Bromley, R.G., Asgaard, U., 1993. Two bioerosion ichnofacies produced by early and later burial associated with sea-level change. Geol. Rundsch. 82, 872–874.

Bromley, R.G., Jacobsen, A.R., 2008. Ichnotaxa for bite traces of tetrapods: a new area of research or a total waste of time?. In: Uchman, A. (Ed.), Abstract Book and the Intra-Congress Field Trip Guidebook, Second International Congress on Ichnology, vol 20.

Brook, F.J., 1983. Lower Miocene Geology of North Kaipara, Northern New Zealand, with Emphasis on the Paleontology. Unpublished Doctoral Dissertation. University of Auckland, 390 p.

Brown, C.M., Tanke, D.H., Hone, D.W.E., 2021. Rare evidence for 'gnawing-like' behavior in a small-bodied theropod dinosaur. PeerJ 9, e11557. https://doi.org/10.7717/peerj.11557.

Brown, C.M., Currie, P.J., Therrien, F., 2022. Intraspecific facial bite marks in tyrannosaurids provide insight into sexual maturity and evolution of bird-like intersexual display. Paleobiology 48 (1), 12–43.

Brown-Sarracino, J., Peckol, P., Curran, A., Robbart, M., 2007. Spatial variation in sea urchins, fish predators, and bioerosion rates on coral reefs of Belize. Coral Reefs 26, 71–78.

Brownstein, C.D., 2018. Trace fossils on dinosaur bones reveal ecosystem dynamics along the coast of eastern North America during the latest Cretaceous. PeerJ 6, e4973. https://doi.org/10.7717/peerj.4973.

Brownstein, C.D., Bissell, I., 2021. An elongate hadrosaurid forelimb with biological traces informs the biogeography of the Lambeosaurinae. J. Paleontol. 95, 367–375.

Brusatte, S.L., Young, M.T., Challands, T.J., Clark, N.D., Fischer, V., Fraser, N.C., Liston, J.J., MacFadyen, C.C., Ross, D.A., Walsh, S., Wilkinson, M., 2015. Ichthyosaurs from the Jurassic of Skye, Scotland. Scot. J. Geol. 51, 43–55.

Brustur, T., Jipa, D.C., 2020. Possible Middle Pontian avian and mammalian trace fossils from the Dacian Basin (Slănicul de Buzău Valley, Romania). Geo-Eco-Marina 25, 193–202.

Buchy, M.C., 2007. Mesozoic Marine Reptiles from North-East Mexico: Description, Systematics, Assemblages and Palaeobiogeography. Ph.D. dissertation. University of Karlsruhe, Karlsruhe, 87 p.

Buchy, M.-C., Vignaud, P., Frey, E., Stinnesbeck, W., González González, A.H., 2006. A new thalattosuchian crocodyliform from the Tithonian (Upper Jurassic) of north-eastern Mexico. C. R. Palevol. 5, 785–794.

Buckland, W., 1822. Account of an assemblage of fossil teeth and bones of elephant, rhinoceros, hippopotamus, bear, tiger and hyaena, and sixteen other animals; discovered in a cave at Kirkdale, Yorkshire, in the year 1821; with a comparative view of five similar caverns in various parts of England, and others on the continent. Phil. Trans. Roy. Soc. Lond. 112, 171–236.

Buckland, W., 1824. Reliquiae Diluvianae; or, Observations on the Organic Remains Contained in Caves, FIssures, and Diluvial Gravel, and on Other Geological Phenomena, Attesting the Action of an Universal Deluge. J. Murray, London, 303 p.

Budziszewska-Karwowska, E., Bujok, A., Sadlok, G., 2010. Bite marks on an upper Triassic dicynodontid tibia from Zawiercie, Kraków-Czestochowa Upland, southern Poland. Palaios 25, 415–421.

Buatois, L.A., Mángano, M.G., 2011. Ichnology: Organism-Substrate Interactions in Space and Time. Cambridge University Press, Cambridge, 358 p.

Buffetaut, E., Ingavat, R., 1980. A new crocodilian from the Jurassic of Thailand, *Sunosuchus thailandicus* n. sp. (Mesosuchia, Goniopholididae), and the palaeogeographical history of South-east Asia in the Mesozoic. Geobios 13, 879–889.

Buffetaut, E., Martill, D., Escuillié, F., 2004. Pterosaurs as part of a spinosaur diet. Nature 430, 33-33.

Bunn III., H.T., 1982. Meat-eating and Human Evolution: Studies on the Diet and Subsistence Patterns of Plio-Pleistocene Hominids in East Africa. Ph.D. dissertation. University of California, Berkeley, 514 p.

Bunn, H.T., 1983. Comparative analysis of modern bone assemblages from a San huntergatherer camp in the Kalahari Desert, Botswana, and from a spotted hyena den near Nairobi, Kenya. In: Clutton-Brock, J., Grigson, C. (Eds.), Animals and Archaeology. Volume I: Hunters and Their Prey, BAR International Series 283. British Archaeological Reports, Oxford, pp. 143–148.

Buscalioni, A.D., Alcalá, L., Espílez, E., Mampel, L., 2013. European Goniopholididae from the early Albian Escucha Formation in Ariño (Teruel, Aragon, España). Spanish J. Paleontol. 28, 103–122.

Calkins, D.G., 1978. Feeding behavior and major prey species of the sea otter, *Euhydra lutris*, in Montague Strait, Prince William Sound, Alaska. Fish. Bull. 76, 125–131.

Calvo, J.M., Gil, E., Meléndez, G., 1987. *Megaplanolites ibericus* (ichnogen. et ichnosp. nov.), a new trace fossil from the Upper Jurassic (uppermost Oxfordian) of Bueña (Teruel Province, Iberian Chain, Spain). Palaeogeogr. Palaeoclimatol. Palaeoecol. 61, 199–204.

Camens, A.B., Carey, S.P., 2013. Contemporaneous trace and body fossils from a Late Pleistocene Lakebed in Victoria, Australia, allow assessment of bias in the fossil record. PLoS One 8 (1), e52957. https://doi.org/10.1371/journal.pone.0052957.

Camp, C.L., 1930. A Study of Phytosaurs, vol 10. Memoirs of the University of California, 174 p.

Capasso, L., 1998. Cranial pathology of *Ursus spelaeus* Rosenmüller & Heinroth from Chateau Pignon, Basque Territories (Spain). Int. J. Osteoarchaeol. 8, 107–115.

Capasso, L., Bacchia, F., Rabottini, N., Rothshild, B.M., Mariani-Costantini, R., 1996. Possible intraspecific aggressive behavior of Devonian giant fishes (Arthrodira, Dinichthydae). J. Paleopathol. 8, 153–160.

Carpenter, K., 1998. Evidence of predatory behavior by carnivorous dinosaurs. Gaia 15, 135–144.

Carpenter, K., Lindsey, D., 1980. The dentary of *Brachychampsa montana* Gilmore (Alligatorinae; Crocodylidae), a late Cretaceous turtle-eating alligator. J. Paleontol. 54, 1213–1217.

Carpenter, K., Sanders, F., McWhinney, L.A., Wood, L., 2005. Evidence for predator-prey relationships: examples for *Allosaurus* and *Stegosaurus*. In: Carpenter, K. (Ed.), The Carnivorous Dinosaurs. Indiana University Press, Bloomington, pp. 325–350.

Carr, R., Lindgren, J., Schweitzer, M., Woodward, H., Scannella, J., 2020. Soft tissue preservation and paleoecology of a new giant plioplatecarpine mosasaur from the Bearpaw Shale of Montana (U.S.A.). Society of Vertebrate Paleontology 80th Annual Meeting Virtual 2020 October 12-16 2020 (unnumbered page). www.vertpaleo.orgConferenceProgram.

Chumakov, N.M., Dronov, A.V., Mikuláš, R., 2013. New ichnospecies of scratching traces from phosphatic nodules (Cenomanian, England). Stratigr. Geol. Correl. 21, 291–299.

Chure, D.J., Fiorillo, A.R., Jacobsen, A., 1998. Prey bone utilization by predatory dinosaurs in the Late Jurassic of North America, with comments on prey bone use by dinosaurs throughout the Mesozoic. Gaia 15, 227–232.

Cicimurri, D.J., Knight, J.L., 2009. Two shark-bitten whale skeletons from Coastal Plain deposits of South Carolina. SE. Nat. 8, 71–83.

Cicimurri, D.J., Parris, D.C., Everhart, M.J., 2008. Partial dentition of a chimaeroid fish (Chondrichthyes, Holocephali) from the Upper Cretaceous Niobrara Chalk of Kansas, USA. J. Vertebr. Paleontol. 28, 34–40.

Cigala-Fulgosi, F., 1990. Predation (or possible scavenging) by a great white shark on an extinct species of bottlenosed dolphin in the Italian Pliocene. Tert. Res. 12, 17–36.

Cione, A.L., Hospitaleche, C.A., Pérez, L.M., Laza, J.H., César, I., 2010. Trace fossils on penguin bones from the Miocene of Chubut, southern Argentina. Alcheringa 34, 433–454.

Cloutier, R., 2013. Great Canadian Lagerstätten 4. The Devonian Miguasha biota (Québec): UNESCO World Heritage Site and a time capsule in the early history of vertebrates. Geosci. Can. 40, 149–163.

Codrea, V., Barbu, O., Jipa-Murzea, C., 2010. Upper Cretaceous (Maastrichtian) land vertebrate diversity in Alba District (Romania). Bull. Geol. Soc. Greece 43, 594–601.

Collareta, A., Caasati, S., Di Cencio, A., 2017a. A pristid sawfish from the lower Pliocene of Lucciolabella (Radicofani basin, Tuscany, central Italy). Atti della Società Toscana di Scienze Naturali. Mem. Ser. A 124, 49–55.

Collareta, A., Casati, S., Di Cencio, A., Bianucci, G., 2023a. The deep past of the White Shark, *Carcharodon carcharias*, in the Mediterranean sea: a synthesis of its palaeobiology and palaeoecology. Life 13, 2085. https://doi.org/10.3390/life13102085.

Collareta, A., Gemelli, M., Varola, A., Bianucci, G., 2019b. Trace fossils on a trace fossil: a vertebrate-bitten vertebrate coprolite from the Miocene of Italy. Neues Jahrbuch Geol. Palaontol. Abhand. 293, 117–126.

Collareta, A., Merella, M., Casati, S., Di Cencio, A., Tinelli, C., Bianucci, G., 2023b. Polyplacophoran feeding traces on Mediterranean Pliocene sirenian bones: insights on the role of grazing bioeroders in shallow-marine vertebrate falls. Life 13 (2), 327.

Collareta, A., Lambert, O., Landini, W., Di Celma, C., Malinverno, E., Varas-Malca, R., Urbina, M., Bianucci, G., 2017b. Did the giant extinct shark *Carcharocles megalodon* target small prey? Bite marks on marine mammal remains from the late Miocene of Peru. Palaeogeogr. Palaeoclimatol. Palaeoecol. 469, 84–91.

Collinson, M.E., 1990. Plant evolution and ecology during the Early Cainozoic diversification. Adv. Bot. Res. 17, 1–98.

Collinson, M.E., Hooker, J.J., 2000. Gnaw marks on Eocene seeds: evidence for early rodent behavior. Palaeogeogr. Palaeoclimatol. Palaeoecol. 157, 127–149.

Cook, D.O., 1971. Depressions in shallow marine sediment made by benthic fish. J. Sediment. Res. 41, 577–578.

Coppock, C., Currie, P.J., 2024. Additional *Albertosaurus sarcophagus* (Tyrannosauridae, Albertosaurinae) material from the Danek bonebed of Edmonton, Alberta, Canada with evidence of cannibalism. Can. J. Earth Sci. 1–7. https://doi.org/10.1139/cjes-2023-0055.

Corral, J.C., Pereda Suberbiola, X., Bardet, N., 2004. Marcas de ataque atribuidas a un selacio en una vertebra de mosasaurio del Cretácico Superior de Álava (Region Vasco-Cantábrica). Rev. Esp. Palaontol. 19, 23–32.

Cossa, D., Cossa, M., Timba, I., Nhaca, J., Macia, A., Infantes, E., 2023. Drones and machine-learning for monitoring dugong feeding grounds and gillnet fishing. Mar. Ecol. Prog. Ser. 716, 123–136. https://doi.org/10.3354/meps14361.

Courville, C.B., 1953. Cranial injuries in prehistoric animals; with special notes on a healed wound of the skull in the dire wolf (*Canis* [*Aenocyon*] *dirus* [Leidy] and a mortal wound in the California black bear (*Ursus americanus*). Bull. Los Ang. Neurol. Soc. 18, 117–126.

Cruickshank, A.R.I., 1986. Archosaur predation on an East African Middle Triassic dicynodont. Palaeontology 29, 415–422.

Cueille, M., Green, E., Duffin, C.J., Hildebrand, C., Benton, M.J., 2020. Fish and crab coprolites from the latest Triassic of the UK: From Buckland to the Mesozoic Marine Revolution. Proc. Geologists' Assoc. 131, 699–721.

Currie, P.J., Eberth, D.A., 2010. On gregarious behavior in *Albertosaurus*. Can. J. Earth Sci. 47, 1277–1289.

da Costa, J.P., Trifilio, L.H.M.D.S., de Araújo-Júnior, H.I., Ximenes, C.L., 2023. Trace fossils on megafaunal bone remains from Quaternary natural tank deposits of Brazil: A case study in João Cativo Paleontological site, Megafauna Valley, Brazil. Ichnos 1–10. https://doi.org/10.1080/10420940.2023.2204232.

da Costa Pereira, G., Luiz, P.V., Bogado, J.P., Ribeiro, T.B., Belfort, L.P., de Valais, S., dos Anjos Candeiro, C.R., 2022. Dino on the menu: Tooth traces in a sauropod epiphysis from the Presidente Prudente Formation (Campanian-Maastrichtian), Bauru Group, Brazil-palaeobiological and palaeoecological implications. Hist. Biol. 34 (11), 2079–2088.

Dalman, S.G., Lucas, S.G., 2018. New evidence for predatory behavior in tyrannosaurid dinosaurs from the Kirtland Formation (Late Cretaceous, Campanian), Northwestern New Mexico. N. M. Museum Nat. Hist. Sci. Bull. 79, 113–124.

Dalman, S.G., Lucas, S.G., 2021. New evidence for cannibalism in tyrannosaurid dinosaurs from the Upper Cretaceous (Campanian/Maastrichtian) San Juan Basin of New Mexico. N. M. Museum Nat. Hist. Sci. Bull. 82, 39–56.

D'Amore, D.C., 2009. A Functional, Behavioral, and Taphonomic Analysis of Ziphodont Dentition: Novel Methodology for the Evaluation of Carnivorous Dinosaur Feeding Paleoecology. Ph.D. dissertation. Rutgers University-New Brunswick, New Brunswick, 221 p.

Davidson, I., Soloman, S., 1990. Was OH7 the victim of a crocodile attack? In: Solomon, S., Davidson, I., Watson, D. (Eds.), Problem Solving in Taphonomy: Archaeological and Palaeontological Studies from Europe, Africa and Oceania. Tempus, St. Lucia, pp. 197–206.

Dawson, S.D., Gottfried, M.D., 2002. Paleopathology in a Miocene kentriodontid dolphin (Cetacea: Odontoceti). Smithsonian Contribut. Paleontol. 93, 263–270.

de los Reyes, M., Poiré, D., Soibelzon, L., Zurita, A.E., Arrouy, M.J., 2013. First evidence of scavenging of a Glyptodont (Mammalia, Glyptodontidae) from the Pliocene of the Pampean region (Argentina): taphonomic and paleoecological remarks. Palaeontol. Electron. 16 (2), 15A; 13p; palaeo-electronica.org/content/2013/434-scavenging-of-a-glyptodont.

De Mendoza, R.S., Haidr, N.S., 2018. Predation trace fossils in a new specimen of *Cayaoa bruneti* Tonni (Aves, Anseriformes) from the Gaiman Formation (early Miocene, Chubut, Argentina). Ameghiniana 55, 483–488.

Demircan, H., Uchman, A., 2010. Kiss of death of a hunting fish: Trace fossil *Osculichnus labialis* igen. et isp. nov. from late Eocene–early Oligocene prodelta sediments of the Mezardere Formation, Thrace Basin, NW Turkey. Acta Geol. Pol. 60, 29–38.

Demircan, H., Uchman, A., 2016. Ichnology of prodelta deposits of the Mezardere Formation (late Eocene—early Oligocene) in the Gökçeada Island, western Turkey. Geodin. Acta 28 (1–2), 86–100. https://doi.org/10.1080/09853111.2015.1113720.

Deng, T., Tseng, Z.J., 2010. Osteological evidence for predatory behavior of the giant percrocutid (*Dinocrocuta gigantea*) as an active hunter. Chin. Sci. Bull. 55, 1790–1794.

Dennell, R.W., Coard, R., Turner, A., 2008. Predators and scavengers in early Pleistocene southern Asia. Quat. Int. 192, 78–88.

Dentzien-Dias, P., Carrillo-Briceño, J.D., Francischini, H., Sánchez, R., 2018. Paleoecological and taphonomical aspects of the Late Miocene vertebrate coprolites (Urumaco Formation) of Venezuela. Palaeogeogr. Palaeoclimatol. Palaeoecol. 490, 590–603.

Dentzien-Dias, P., Hunt, A.P., Lucas, S.G., Francischini, H., Gulotta, M., 2021. Coprolites from shallow marine deposits of the Nanjemoy Formation, Lower Eocene of Virginia, USA. Lethaia 54, 26–39.

Denys, C., Viriot, L., Daams, R., Pelaez-Campomanes, P., Vignaud, P., Andossa, L., Brunet, M., 2003. A new Pliocene xerine sciurid (Rodentia) from Kossom Bougoudi, Chad. J. Vertebr. Paleontol. 23, 676–687.

de Valais, S., Apesteguía, S., Garrido, A.C., 2012. Cretaceous small scavengers: feeding traces in tetrapod bones from Patagonia, Argentina. PLoS One 7 (1), e29841. https://doi.org/10.1371/journal.pone.0029841.

De Vasconcellos, F.M., Carvalho, I.S., 2010. Paleoichnological assemblage associated with *Baurusuchus salgadoensis* remains, a Baurusuchidae Mesoeucrocodylia from the Bauru Basin, Brazil (Late Cretaceous). N. M. Museum Nat. Hist. Sci. Bull. 51, 227–237.

Diamond, J., Bishop, K.D., Gilardi, J.D., 1999. Geophagy in New Guinea birds. Ibis 141 (2), 181–193.

Dickinson, E.C., Remsen, J.V., 2013. Non-passerines. The Howard and Moore Complete Checklist of the Birds of the World, 4th. Edition, vol 1. 461 p.

Diedrich, C., 2006. By Ice Age spotted hyenas protracted, cracked, nibbled and chewed skeleton remains of *Coelodonta antiquitatis* (Blumenbach 1799) from the Lower Weichselian (Upper Pleistocene) freeland prey deposit site Bad Wildungen-Biedensteg (Hesse, NW Germany). J. Taphon. 4, 95–127.

Diedrich, C., 2008. The food of the miosiren *Anomotherium langenwieschei* (Siegfried)—Indirect proof of seaweed or seagrass by xenomorphic oyster fixation structures in the Upper Oligocene (Neogene) of the Doberg, Bünde (NW Germany) and comparisons to modern *Dugong dugon* (Müller) feeding strategies. Senckenberg. Maritima 38, 59–73.

Diedrich, C.G., 2011. Late Pleistocene steppe lion *Panthera leo spelaea* (Goldfuss, 1810) footprints and bone records from open air sites in northern Germany–evidence of hyena-lion antagonism and scavenging in Europe. Quat. Sci. Rev. 30, 1883–1906.

Diedrich, C.G., 2013. Extinctions of late Ice Age cave bears as a result of climate/habitat change and large carnivore lion/hyena/wolf predation stress in Europe. ISRN Zool. 2013. ArticleID138319,25 pages. https://doi: org/10.1155/2013/138319.

Diffendal Jr., R., 2003. Field trip Ash Fall Historic Park: Field Trip 5. In: Association of American State Geologists, 95th Annual Meeting. Lincoln Nebraska, 37 p.

Di Gregorio, E.D.B., de Araújo Júnior, H.I., 2020. Ocorrência da Icnoespécie *Nihilichnus nihilicus* Mikulás, Kadlecová, Fejfar & Dvorák (2006) em Vértebra de Crocodylia da Formação Solimões (Mioceno Superior da Bacia do Acre). Anu. do Inst. Geociencias 43 (1), 408–413.

Domingo, M.S., Sanchez, I.M., Alberdi, M.T., Azanza, B., Morales, J., 2011. Evidence of predation/scavenging on Moschidae (Mammalia, Ruminantia) from the Late Miocene of Spain. Lethaia 45, 386–400.

Domínguez-Rodrigo, M., Barba, R., 2006. New estimates of tooth mark and percussion mark frequencies at the FLK Zinj site: The carnivore-hominid-carnivore hypothesis falsified. J. Hum. Evol. 50 (2), 170–194.

Domínguez-Rodrigo, M., Baquedano, E., 2018. Distinguishing butchery cut marks from crocodile bite marks through machine learning methods. Sci. Rep. 8 (1), 5786. https://doi.org/10.1038/s41598-018-24071-1.

Domínguez-Rodrigo, M., Piqueras, A., 2003. The use of tooth pits to identify carnivore taxa in tooth-marked archaeofaunas and their relevance to reconstruct hominid carcass processing behaviours. J. Archaeol. Sci. 30, 1385–1391.

Domínguez-Rodrigo, M., Gidna, A., Yravedra, J., Musiba, C., 2012. A comparative neotaphonomic study of felids, hyenids and canids: an analogical framework based on long bone modification patterns. J. Taphon. 10, 147–164.

Dortangs, R.W., Schulp, A.S., Mulder, E.W., Jagt, J.W., Peeters, H.H., De Graaf, D.T., 2002. A large new mosasaur from the Upper Cretaceous of The Netherlands. Neth. J. Geosci. 81, 1–8.

Dortch, J., Cupper, M., Grün, R., Harpley, B., Lee, K., Field, J., 2016. The timing and cause of megafauna mass deaths at Lancefield Swamp, south-eastern Australia. Quat. Sci. Rev. 145, 161–182.

Druckenmiller, P.S., Hurum, J.H., Knutsen, E.M., Nakrem, H.A., 2012. Two new ichthyosaurs (Ichthyosauria: Ophthalmosauridae) from the Agardhfjellet Formation (Upper Jurassic: Volgian), Svalbard, Norway. Norw. J. Geol. 92, 311–339.

Drumheller, S.K., Brochu, C.A., 2014. A diagnosis of *Alligator mississippiensis* bite marks with comparisons to existing crocodilian datasets. Ichnos 21, 131–146.

Drumheller, S.K., Brochu, C.A., 2016. Phylogenetic taphonomy: A statistical and phylogenetic approach for exploring taphonomic patterns in the fossil record using crocodylians. Palaios 31, 463–478.

Drumheller, S.K., Stocker, M.R., Nesbitt, S.J., 2014. Direct evidence of trophic interactions among apex predators in the Late Triassic of western North America. Naturwissenschaften 101, 975–987.

Drumheller, S.K., McHugh, J.B., Kane, M., Riedel, A., D'Amore, D.C., 2020. High frequencies of theropod bite marks provide evidence for feeding, scavenging, and possible cannibalism in a stressed Late Jurassic ecosystem. PLoS One 15 (5), e0233115. https://doi.org/10.1371/journal.pone.0233115.

Drumheller-Horton, S.K., 2012. An Actualistic and Phylogenetic Approach to Identifying and Interpreting Crocodylian Bite Marks. Ph.D. dissertation. University of Iowa, Iowa City, 170 p.

Drymala, S.M., Bader, K., 2012. Assessing predator–prey interactions through the identification of bite marks on an aetosaur (Archosauria, Suchia) osteoderm from the Upper Triassic (Norian) Chinle Formation in Petrified Forest. J. Vertebr. Paleontol. 32 (Suppl. to no. 3), 89.

Drymala, S.M., Bader, K., Parker, W.G., 2021. Bite marks on an aetosaur (Archosauria, Suchia) osteoderm: Assessing Late Triassic predator-prey ecology through ichnology and tooth morphology. Palaios 36, 28–37.

Dzik, J., Sulej, T., Niedźwiedzki, G., 2008. A dicynodont-theropod association in the latest Triassic of Poland. Acta Palaeontol. Pol. 53, 733–738.

Effinger, J.A., 1998. Entelodontidae. In: Janis, C.M., Scott, K.M., Jacobs, L.L. (Eds.), Evolution of Tertiary Mammals of North America, Volume 1, Terrestrial Carnivores, Ungulate and Ungulate-like Mammals. Cambridge University Press, Cambridge, pp. 375–380.

Ehret, D.J., Harrell Jr., T.L., 2018. Feeding traces on a *Pteranodon* (Reptilia: Pterosauria) bone from the Late Cretaceous (Campanian) Mooreville Chalk in Alabama, USA. Palaios 33, 414–418.

Einarsson, E., Lindgren, J., Kear, B.P., Siverson, M., 2010. Mosasaur bite marks on a plesiosaur propodial from the Campanian (Late Cretaceous) of southern Sweden. GFF 132, 123–138.

Ehret, D.J., MacFadden, B.J., Salas-Gismondi, R., 2009. Caught in the act: Trophic interactions between a 4-million-year-old white shark (*Carcharodon*) and mysticete whale from Peru. Palaios 24, 329–333.

Ekrt, B., Mikuláš, R., Wagner, J., Čermák, S., Procházková, K., Kadlecová, E., Fejfar, O., 2016. New contribution to the palaeoichnology and taphonomy of the Ahníkov fossil site, Early Miocene, Most Basin (the Czech Republic). Fossil Imprint 72, 202–214.

Elder, R.L., 1985. Principles of Aquatic Taphonomy with Examples from the Fossil Record. Ph.D. dissertation. University of Michigan, Ann Arbor, 336 p.

Elder, R.L., Smith, G.R., 1988. Fish taphonomy and environmental inference in paleolimnology. Palaeogeogr. Palaeoclimatol. Palaeoecol. 62, 577–592.

Elliott, D.K., Petriello, M.A., 2011. New poraspids (Agnatha, Heterostraci) from the Early Devonian of the western United States. J. Vertebr. Paleontol. 31, 518–530.

Erickson, B.R., 1967. Fossil bird tracks from Utah. Museum Observor (Science Museum of Minnesota), 5, pp. 1–7.

Everhart, M.J., 1999. Evidence of feeding on mosasaurs by the Late Cretaceous lamniform shark, Cretoxyrhina mantelli. J. Vertebr. Paleontol. 17 (Suppl. 3), 43–44.

Everhart, M.J., 2005a. Bite marks on an elasmosaur (Sauropterygia; Plesiosauria) paddle from the Niobrara Chalk (Upper Cretaceous) as probable evidence of feeding by the lamniform shark, Cretoxyrhina mantelli. PalArch Vertebr. Paleontol. 2, 14–24.

Everhart, M.J., 2005c. Elasmosaurid remains from the Pierre Shale (Upper Cretaceous) of western Kansas. Possible missing elements of the type specimen of *Elasmosaurus platyurus* Cope, 1868? PalArch Vertebr. Palaeontol. 4, 19–32.

Everhart, M.J., 2008. A bitten skull of *Tylosaurus kansasensis* (Squamata: Mosasauridae) and a review of mosasaur-on-mosasaur pathology in the fossil record. Trans. Kans. Acad. Sci. 111, 251–262.

Everhart, M.J., 2017. Oceans of Kansas—A Natural History of the Western Interior Sea, second ed. Indiana University Press, Bloomington. 427 p.

Everhart, M.J., Ewell, K., 2006. Shark-bitten dinosaur (Hadrosauridae) caudal vertebrae from the Niobrara Chalk (Upper Coniacian) of western Kansas. Trans. Kans. Acad. Sci. 109, 27–36.

Everhart, M.J., Hamm, S.A., 2005. A new nodosaur (Dinosauria: Nodosauridae) from the Smoky Hill Chalk (Upper Cretaceous) of western Kansas. Kansas Acad. Sci. Trans. 108, 15–21.

Fahlke, J.M., 2012. Bite marks revisited—evidence for middle-to-late Eocene *Basilosaurus isis* predation on *Dorudon atrox* (both Cetacea, Basilosauridae). Palaeontol. Electron. 15 (3), 32A, 16p; palaeo-electronica.org/content/2012-issue-3-articles/339-archaeocete-predation.

Falk, A.R., 2004. Foot and Hindlimb Morphology, Soft Tissues, and Tracemaking Behaviors of Early Cretaceous Birds from China and the Republic of Korea with a Comparison to Modern Avian Morphology and Behavior. Ph.D. dissertation. Lawrence, University of Kansas, 750 p.

Falk, A.R., Hasiotis, S.T., Martin, L.D., 2010. Feeding traces associated with bird tracks from the Lower Cretaceous Haman Formation, Republic of Korea. Palaios 25, 730–741.

Falk, A.R., Lim, J.D., Hasiotis, S.T., 2014. A behavioral analysis of fossil bird tracks from the Haman Formation (Republic of Korea) shows a nearly modern avian ecosystem. Vertebr. Palasiat. 52, 129–152.

Falk, D., Mccoy, V., Wings, O., 2019. Ichnofossils of the Eocene Fossillagerstätte Geiseltal (Saxony-Anhalt, Germany). Hallesches Jahrbuch für Geowissenschaften, Beiheft, 46, pp. 15–16.

Fan, R.Y., Zong, R.W., Gong, Y.M., 2019. Fish hunting trace *Osculichnus* and the oldest *Sinusichnus sinuosus* from the Upper Devonian of South China. Palaeogeogr. Palaeoclimatol. Palaeoecol. 530, 103–112.

Farlow, J.O., Holtz, T.R., 2002. The fossil record of predation in dinosaurs. Paleontol. Soc. Pap. 8, 251–266.

Feibel, C.S., 1987. Fossil fish nests from the Koobi Fora Formation (Plio-Pleistocene) of northern Kenya. J. Paleontol. 61, 130–134.

Fernandez, V., Abdala, F., Carlson, K.J., Cook, D.C., Rubidge, B.S., Yates, A., Tafforeau, P., 2013. Synchrotron reveals Early Triassic odd couple: Injured amphibian and aestivating therapsid share burrow. PLoS One 8 (6), e64978. https://doi.org/10.1371/journal.pone.0064978.

Fernandez-Jalvo, Y., Andrews, P., 2016. Atlas of Taphonomic Identifications: 1001+ Images of Fossil and Recent Mammal Bone Modification. Springer, New York, 359 p.

Fiorelli, L.E., 2010. Predation bite-marks on a peirosaurid crocodyliform from the Upper Cretaceous of Neuquén Province, Argentina. Ameghiniana 47, 387–400.

Fiorillo, A.R., 1991a. Prey bone utilization by predatory dinosaurs. Palaeogeogr. Palaeoclimatol. Palaeoecol. 88, 157–166.

Fiorillo, A.R., 1991b. Taphonomy and depositional setting of Careless Creek Quarry (Judith River Formation), Wheatland County, Montana, USA. Palaeogeogr. Palaeoclimatol. Palaeoecol. 81, 281–311.

Fiorillo, A.R., McCarthy, P.J., Flaig, P.P., 2010. Taphonomic and sedimentologic interpretations of the dinosaur-bearing Upper Cretaceous strata of the Prince Creek Formation, northern Alaska: insights from an ancient high-latitude terrestrial ecosystem. Palaeogeogr. Palaeoclimatol. Palaeoecol. 295, 376–388.

Fischer, W.A., 1978. The habitat of the early vertebrates: trace and body fossil evidence from the Harding Formation (Middle Ordovician), Colorado. Mt. Geol. 15, 1–26.

Ford, T.D., O'Connor, B., 2009. A vanished industry: Coprolite mining. Mercian Geol. 17, 93–100.

Fordyce, N., Smith, R., Chinsamy, A., 2012. Evidence of a therapsid scavenger in the Late Permian Karoo Basin, South Africa. South Afr. J. Sci. 108, 114–118.

Forrest, R., 2000. A large rhomaleosaurid pliosaur from the Upper Lias of Rutland. Mercian Geol. 15, 37–40.

Forrest, R., 2003. Evidence for scavenging by the marine crocodile Metriorhynchus on the carcass of a plesiosaur. Proc. Geologists' Assoc. 114, 363–366.

Frandsen, G., 2020. Coprolites of the World. Monee, Poozeum, 92 p.

Frandsen, G., Godfrey, S.J., 2019. A gar-bitten coprolite from the Eocene Green River Formation near Kemmerer, Wyoming, U.S.A. Ecphora 34 (1), 2–4.

Franzen, J.L., 2001. Taphonomic analysis of the Messel Formation (Germany). In: Gunnell, G.F. (Ed.), Eocene Biodiversity: Unusual Occurrences and Rarely Sampled Habitats. Kluwer Academic/Plenum Press, Boston, pp. 197–214.

Franzen, J.L., Frey, E., 1993. *Europolemur* completed. Kaupia 3, 130.

Frey, E., Buchy, M.C., Stinnesbeck, W., López-Oliva, J.G., 2002. *Geosaurus vignaudi* n. sp. (Crocodyliformes: Thalattosuchia), first evidence of metriorhynchid crocodilians in the Late Jurassic (Tithonian) of central-east Mexico (State of Puebla). Can. J. Earth Sci. 39, 1467–1483.

Frey, R.W., Seilacher, A., 1980. Uniformity in marine invertebrate ichnology. Lethaia 13, 183–207.

Fuentes, E.J., 2003. Predación crocodiliana a quelonios. Un Neochelys (Pelomedusidae), del Eoceno de Zemora, lisiando por un Asiatosuchus. Stud. Geol. Salmant. 39, 11–23.

Gahn, F.J., 2004. Parasitism and Predation on Paleozoic Crinoids. Ph.D. dissertation. The University of Michigan, Ann Arbor, 120 p.

Gale, A.S., Kennedy, W.J., Martill, D., 2017. Mosasauroid predation on an ammonite—*Pseudaspidoceras*—from the Early Turonian of south-eastern Morocco. Acta Geol. Pol. 67, 31–46.

Galle, A., Mikuláš, R., 2003. Evidence of predation on the rugose coral *Calceola sandalina* (Devonian, Czech Republic). Ichnos 10, 41–45.

Gebo, D.L., Simons, E.S., 1984. Puncture marks on early African anthropoids. Am. J. Phys. Anthropol. 63, 31–35.

Geister, J., 1998. Lebensspuren von meersauriern und ihren beutetieren im mittleren Jura (Callovien) von Liesberg, Schweiz. Facies 39, 105–124.

Genise, J.F., Melchor, R.N., Archangelsky, M., BalaL, O., Straneck, R., de Valais, S., 2009. Application of neoichnological studies to behavioural and taphonomic interpretation of fossil bird-like tracks from lacustrine settings: the

Late Triassic-Early Jurassic? Santo Domingo Formation, Argentina. Palaeogeogr. Palaeoclimatol. Palaeoecol. 272, 143–161.

Georgalis, G.L., Rage, J.C., de Bonis, L., Koufos, G.D., 2018. Lizards and snakes from the late Miocene hominoid locality of Ravin de la Pluie (Axios Valley, Greece). Swiss J. Geosci. 111, 169–181.

Gianechini, F.A., de Valais, S., 2016. Bioerosion trace fossils on bones of the Cretaceous South American theropod *Buitreraptor gonzalezorum* Makovicky, Apesteguía and Agnolín, 2005 (Deinonychosauria). Hist. Biol. 28, 533–549.

Gignac, P.M., Makovicky, P.J., Erickson, G.M., Walsh, R.P., 2010. A description of *Deinonychus antirrhopus* bite marks and estimates of bite force using tooth indentation simulations. J. Vertebr. Paleontol. 30, 1169–1177.

Gilardi, J.D., Duffey, S.S., Munn, C.A., Tell, L.A., 1999. Biochemical functions of geophagy in parrots: detoxification of dietary toxins and cytoprotective effects. J. Chem. Ecol. 25, 897–922.

Gillet, E., Lefebvre, B., Gardien, V., Steimetz, E., Durlet, C., Marin, F., 2019. Reinterpretation of the enigmatic Ordovician genus *Bolboporites* (Echinodermata). Zoosymposia 15 (1), 44–70 ff10.11646/zoosymposia.15.1.7ff. ffhal-02333918f.

Gilmore, C.W., 1920. Osteology of the carnivorous Dinosauria in the United States National museum: with special reference to the genera *Antrodemus* (*Allosaurus*) and *Ceratosaurus*. Proc. U. S. Natl. Mus. 60, 1–154.

Gingerich, P.D., 1987. Early Eocene bats (Mammalia, Chiroptera) and other vertebrates in freshwater limestones of the Willwood Formation, Clark's Fork Basin, Wyoming. In: Contributions from the Museum of Paleontology, vol 27. The University of Michigan, pp. 275–320.

Gingerich, P.D., 1994. New species of *Apheliscus, Haplomylus*, and *Hyopsodus* (Mammalia, Condylarthra) from the late Paleocene of southern Montana and early Eocene of northwestern Wyoming. In: Contributions from the Museum of Paleontology, vol 29. The University of Michigan, pp. 119–134.

Gingras, M.K., Armitage, I.A., Pemberton, S.G., Clifton, H.E., 2007. Pleistocene walrus herds in the Olympic Peninsula area: Trace-fossil evidence of predation by hydraulic jetting. Palaios 22, 539–545.

Glinskiy, V.N., Mark-Kurik, E., 2016. Revision of *Psammosteus livonicus* Obruchev (Agnatha, Heterostraci) from the Devonian Amata Regional Stage of the NW of the East European Platform. Est. J. Earth Sci. 65, 1–18.

Gobetz, K.E., 2006. Possible burrows of mylagaulids (Rodentia: Aplodontoidea: Mylagaulidae) from late Miocene (Barstovian) Pawnee Creek Formation of northeastern Colorado. Palaeogeogr. Palaeoclimatol. Palaeoecol. 237, 119–136.

Gobetz, K.E., Hattin, D.E., 2002. Rodent-gnawed carbonate rocks from Indiana. Proc. Indiana Acad. Sci. 111, 1–8.

Gobetz, K.E., Martin, L.D., 2006. Burrows of a gopher-like rodent, possibly Gregorymys (Geomyoidea: Geomyidae: Entoptychtinae), from the early Miocene Harrison Formation, Nebraska. Palaeogeogr. Palaeoclimatol. Palaeoecol. 237, 305–314.

Godfrey, S.J., 2003. Miocene sharks and smoking guns… Bugeye Times. Quart. Newslett. Calvert Marine Museum 28 (1), 5–7.

Godfrey, S.J., 2021. Shark-bitten Miocene baleen whale ulna. Ecphora 36 (3), 11.

Godfrey, S.J., 2024. Miocene crocodile-bitten dolphin vertebra. Ecphora 39 (1), 1.

Godfrey, S.J., Collareta, A., 2022. A new ichnotaxonomic name for burrows in vertebrate coprolites from the Miocene Chesapeake Group of Maryland, USA. Swiss J. Palaeontol. 141, 9. https://doi.org/10.1186/s13358-022-00250-6.

Godfrey, S.J., Frandsen, G., 2016. Vertebrate-bitten coprolite from South Carolina. Ecphora 31 (1), 12–14.

Godfrey, S.J., Lowry, A.J., 2021. The ichnospecies *Linichnus bromleyi* on a Miocene baleen whale radius preserving multiple shark bite-shake traces suggests scavenging. Carnets Géol. 21 (17), 91–398.

Godfrey, S.J., Palmer, B.T., 2015. Gar-bitten coprolite from South Carolina, USA. Ichnos 22, 103–108.

Godfrey, S.J., Smith, J., 2010. Shark-bitten vertebrate coprolites from the Miocene of Maryland. Naturwissenschaften 97, 461–467.

Godfrey, S.J., Murdoch, P., Dewaele, L., Perez, V.J., Schumaker, C., 2024. First report in the fossil record of a shark tooth embedded in a pinniped bone. In: Laurin, M., Modesto, S.P., Reisz, R.R. (Eds.), The Importance of Scientific Illustrations in Paleontology: A Tribute to Diane Scott. Comptes Rendus Palevol, vol 23, pp. 107–118. https://doi.org/10.5852/cr-palevol2024v23a8 (8).

Godfrey, S.J., Alford, A., Collareta, A., Weems, R.E., 2020. A Paleocene vertebrate-bitten crocodilian coprolite from Liverpool Point, Maryland, USA. Neues Jahrbuch Geol. Palaontol. Abhand. 2020, 237–244.

Godfrey, S.J., Ellwood, M., Groff, S., Verdin, M.S., 2018. *Carcharocles*-bitten odontocete caudal vertebrae from the Coastal Eastern United States. Acta Palaeontol. Pol. 63, 463–468.

Gommery, D., Pickford, M., Senut, B., 2007. A case of carnivore-inflicted damage to a fossil femur from Swartkrans, comparable to that on a hominid femur representing *Orrorin tugenensis,* BAR 1003'00 (Kenya). Ann. Transvaal Mus. 44, 215–218.

Gônet, J., Rozada, L., Bourgeais, R., Allain, R., 2019. Taphonomic study of a pleurosternid turtle shell from the Early Cretaceous of Angeac-Charente, southwest France. Lethaia 52, 232–243.

Gorzelak, P., Niedźwiedzki, G., Skawina, A., 2010. Pathologies of non-marine bivalve shells from the Late Triassic of Poland. Lethaia 43, 285–289.

Gorzelak, P.L., Rakowicz, L., Salamon, M.A., Szrek, P., 2011. Inferred placoderm bite marks on Devonian crinoids from Poland. Neues Jahrbuch Geol. Palaontol. Abhand. 259, 105–112.

Govender, R., 2015. Shark-cetacean trophic interaction, Duinefontein, Koeberg, (5 Ma), South Africa. South Afr. J. Sci. 111, 11–12.

Govender, R., 2019. Early Pliocene fossil cetaceans from Hondeklip Bay, Namaqualand, South Africa. Hist. Biol. 1–20. https://doi.org/10.1080/08912963.2019.1650273.

Govender, R., Chinsamy, A., 2013. Early Pliocene (5 Ma) shark-cetacean interaction from Langebaanweg, western coast of South Africa. Palaios 28, 27–277.

Gregor, H.J., 1982. *Pinus aurimontana* N. Sp.-Eine Neue Kiefernart aus dem Jungtertiaer des Goldbergs (Ries.). Stutt. Beitr. Naturk. Ser. B Geol. Palontol. 83, 1–19.

Gregory, M.R., 1991. New trace fossils from the Miocene of Northland, New Zealand: *Rorschachichnus amoeba* and *Piscichnus waitemata*. Ichnos 1, 195–205.

Gregory, M.R., Ballance, P.F., Gibson, G.W., Ayling, A.M., 1979. On how some rays (Elasmobranchia) excavate feeding depressions by jetting water. J. Sediment. Petrol. 49, 1125–1130.

Grigorescu, D., 1992. Nonmarine Cretaceous formations of Romania. In: Mateer, N.J., Pei-Ji, C. (Eds.), Aspects of Nonmarine Cretaceous Geology, Special Volume of ICGP Project 245. China Ocean Press, Beijing, pp. 142–164.

Gross, M., Prieto, J., Grímsson, F., Bojar, H.P., 2023. Hyena and 'false'sabre-toothed cat coprolites from the late Middle Miocene of south-eastern Austria. Hist. Biol. 1–20. https://doi.org/10.1080/08912963.2023.2237979.

Hammer, W.R., Hickerson, W.J., 1993. A brief review of Mesozoic terrestrial vertebrates from the Transantarctic Mountains including evidence of scavenging theropod. Antarct. J. U. S. 28, 33–35.

Harington, C.R., 1996. Paleoecology of a Pliocene beaver-pond site in the Canadian Arctic Islands. In: 30th International Geological Congress, Beijing, China, Abstracts, vol 2, p. 114.

Harington, C.R., Ross, R.L., Mathewes, R.W., Stewart, K.M., Beattie, O., 2004. A late Pleistocene Steller sea lion (*Eumetopias jubatus*) from Courtenay, British Columbia: Its death, associated biota, and paleoenvironment. Can. J. Earth Sci. 41, 1285–1297.

Hart, D., Sussman, R.W., 2005. Man the Hunted: Primates, Predators, and Human Evolution. Westview Press, 336 p.

Hastings, A.K., Bloch, J.I., Jaramillo, C.A., 2015. A new blunt-snouted dyrosaurid, *Anthracosuchus balrogus* gen. et sp. nov. (Crocodylomorpha, Mesoeucrocodylia), from the Palaeocene of Colombia. Hist. Biol. 27, 998–1020.

Hattin, D.E., 1988. Rudists as historians. Fort Hays Studies, Third Series, 10, pp. 4–22.

Havlik, P., Aiglstorfer, M., Beckmann, A.K., Gross, M., Böhme, M., 2014. Taphonomical and ichnological considerations on the late Middle Miocene Gratkorn locality (Styria, Austria) with focus on large mammal taphonomy. Palaeobiodivers. Palaeoenviron. 94, 171–188.

Haynes, G., 1980. Evidence of carnivore gnawing on Pleistocene and recent mammalian bones. Paleobiology 6, 341–351.

Haynes, G., 1982. Utilization and skeletal disturbances of North American prey carcasses. Arctic 35, 266–281.

Haynes, G., 1983. A guide for differentiating mammalian carnivore taxa responsible for gnaw damage to herbivore long bones. Paleobiology 9, 164–172.

Hembree, D.I., Hasiotis, S.T., 2008. Miocene vertebrate and invertebrate burrows defining compound paleosols in the Pawnee Creek Formation, Colorado, U.S.A. Palaeogeogr. Palaeoclimatol. Palaeoecol. 270, 349–365.

Hernandez, L.P., Staab, K.L., 2015. Bottom feeding and beyond: how the premaxillary protrusion of Cypriniforms allowed for a novel kind of suction feeding. Integr. Compar. Biol. 1–11. https://doi.org/10.1093/icb/icv038.

Hewitt, R.A., Westermann, G.E.G., 1990. Mosasaur tooth marks on the ammonite *Placenticeras* from the Upper Cretaceous of Alberta, Canada. Can. J. Earth Sci. 27, 469–472.

Hill, R.V., Roberts, E.M., Tapanila, L., Bouare, M.L., Sissoko, F., O'Leary, M.A., 2015. Multispecies shark feeding in the Trans-Saharan seaway: Evidence from Late Cretaceous dyrosaurid (Crocodyliformes) fossils from northeastern Mali. Palaios 30, 589–596.

Hirsch, K.F., Kihm, A.J., Zelenitsky, D.K., 1997. New eggshell of ratite morphotype with predation marks from the Eocene of Colorado. J. Vertebr. Paleontol. 17, 360–369.

Hoffmann, R., Keupp, H., 2015. Definition of (paleo-) pathological phenomena. In: King, C., Korn, D., De Baets, K., Kruta, I., Mapes, R.H. (Eds.), Ammonoid Paleobiology: From Anatomy to Ecology. Topics in Geobiology 43. Springer, New York, pp. 877–926.

Hoffmann, R., Bestwick, J., Berndt, G., Berndt, R., Fuchs, D., Klug, C., 2020. Pterosaurs ate soft-bodied cephalopods (Coleoidea). Sci. Rep. 10 (1), 1230.

Holgado, B., Dalla Vecchia, F.M., Fortuny, J., Bernardini, F., Tuniz, C., 2015. A reappraisal of the purported gastric pellet with pterosaurian bones from the Upper Triassic of Italy. PLoS One 10 (11), e0141275. https://doi.org/10.1371/journal.pone.0141275.

Holland, T., 2018. The mandible of *Kronosaurus queenslandicus* Longman, 1924 (Pliosauridae, Brachaucheniinae), from the Lower Cretaceous of Northwest Queensland, Australia. J. Vertebr. Paleontol. 38, e1511569.

Hone, D.W., Chure, D.J., 2018. Difficulties in assigning trace makers from theropodan bite marks: An example from a young diplodocoid sauropod. Lethaia 51, 456–466.

Hone, D.W.E., Tanke, D.H., 2015. Pre- and postmortem tyrannosaurid bite marks on the remains of *Daspletosaurus* (Tyrannosaurinae: Theropoda) from Dinosaur Provincial Park, Alberta, Canada. PeerJ 3, e885. https://doi.org/10.7717/peerj.885.

Hone, D.W., Tanke, D.H., Brown, C.M., 2018a. Bite marks on the frill of a juvenile *Centrosaurus* from the Late Cretaceous Dinosaur Provincial Park Formation, Alberta, Canada. PeerJ 6, e5748. https://doi.org/10.7717/peerj.5748.

Hone, D.W.E., Witton, M.P., Habib, M.B., 2018b. Evidence for the Cretaceous shark *Cretoxyrhina mantelli* feeding on the pterosaur *Pteranodon* from the Niobrara Formation. PeerJ 6, e6031. https://doi.org/10.7717/peerj.6031.

Horton, D.R., Wright, R.V., 1981. Cuts on Lancefield bones: Carnivorous *Thylacoleo*, not humans, the cause. Archaeol. Ocean. 16, 73–80.

Hospitaleche, C.A., 2016. Paleobiological remarks on a new partial skeleton of the Eocene Antarctic penguin *Palaeeudyptes klekowskii*. Ameghiniana 53, 269–281.

Houpt, T.J., Flis, C.J., Bakker, R.T., Temple, D.T., 2020. *Dimetrodon*s disassembling prey: Multiple taphonomic signatures at Craddock Bone Bed, Seymour, TX. In: Society of Vertebrate Paleontology 80th Annual Meeting Virtual 2020 October 12-16 2020 unnumbered page). www.vertpaleo.orgConferenceProgram.

Howard, J.D., Mayou, T.V., Heard, R.W., 1977. Biogenic sedimentary structures formed by rays. J. Sediment. Petrol. 47, 339–346.

Hua, S., Buffetaut, E., Legall, C., Rogron, P., 2007. *Oceanosuchus boecensis* n. gen, n. sp., a marine pholidosaurid (Crocodylia, Mesosuchia) from the Lower Cenomanian of Normandy (western France). Bull. Soc. Geol. Fr. 178, 503–513.

Hunt, A.P., 1984. Fluvial vertebrate taphonomy: historical perspectives. N. M. J. Sci. 24 (2), 26–27.

Hunt, A.P., 1987. Phanerozoic trends in nonmarine taphonomy: implications for Mesozoic vertebrate taphonomy and paleoecology. Geol. Soc. Am. Abstr. Progr. 19, 171.

Hunt, A.P., Lucas, S.G., 2007a. Cenozoic vertebrate trace fossils of North America: ichnofaunas, ichnofacies and biochronology. N. M. Museum Nat. Hist. Sci. Bull. 42, 17–41.

Hunt, A.P., Lucas, S.G., 2007b. Tetrapod ichnofacies: A new paradigm. Ichnos 14, 59–68.

Hunt, A.P., Lucas, S.G., 2012. Classification of vertebrate coprolites and related trace fossils. N. M. Museum Nat. Hist. Sci. Bull. 57, 137–146.

Hunt, A.P., Lucas, S.G., 2014. Vertebrate trace fossils from New Mexico and their significance. N. M. Museum Nat. Hist. Sci. Bull. 64, 9–40.

Hunt, A.P., Lucas, S.G., 2021. The ichnology of vertebrate consumption: Dentalites, gastroliths and bromalites. N. M. Museum Nat. Hist. Sci. Bull. 87, 215 p.

Hunt, A.P., Lucas, S.G., 2022. The Upper Cretaceous fossil record of the Western Interior Basin of North America is a Megalagerstätte. Geol. Soc. Am. Abstr. Progr. 54 (5). https://2022doi:0.1130/abs/2022AM-383304.

Hunt, A.P., Lucas, S.G., 2024. Regurgitalites. In: Lucas, S.G., Hunt, A.P., Klein, H. (Eds.), Vertebrate Ichnology. Elsevier, Amsterdam (this volume).

Hunt, A.P., Lucas, S.G., Klein, H., 2018. Late Triassic nonmarine vertebrate and invertebrate trace fossils and the pattern of the Phanerozoic record of vertebrate trace fossils. In: Tanner, L.H. (Ed.), The Late Triassic World. Topics in Geobiology 46. Springer Verlag, New York, pp. 447–543.

Hunt, A.P., Lucas, S.G., Spielmann, J.A., 2005. The postcranial skeleton of *Revueltosaurus callenderi* (Archosauria: Crurotarsi) from the Upper Triassic of Arizona and New Mexico, USA. N. M. Museum Nat. Hist. Sci. Bull. 29, 67–76.

Hunt, A.P., Meyer, C.A., Lockley, M.G., Lucas, S.G., 1994. Archaeology, toothmarks and sauropod taphonomy. In: Lockley, M.G., dos Santos, V.F., Meyer, C.A., Hunt, A.P. (Eds.), Aspects of Sauropod Palaeobiology, vol 10. Gaia, pp. 225–232.

Hunter, R.E., Clifton, H.E., Hall, N.T., Chin, J.L., 1984. Pleistocene shoreline and shelf deposits at Fort Funston and their relation to sea-level changes. In: Society of Economic Paleontologists and Mineralogists, 1984 Midyear Meeting, San Jose, California, vol 3, pp. 1–30. Field Trip Guidebook.

Hutchison, J.H., Harington, C.R., 2002. A peculiar new fossil shrew (Lipotyphla, Soricidae) from the High Arctic of Canada. Can. J. Earth Sci. 39, 439–443.

Ifrim, C., 2013. Paleobiology and paleoecology of the early Turonian (Late Cretaceous) ammonite *Pseudaspidoceras flexuosum*. Palaios 28, 9–22.

Irazoqui, F., Acosta Hospitaleche, C., 2022. Bioerosive traces in fossil penguin bones (Aves, Sphenisciformes) from the Eocene of Marambio/Seymour Island (West Antarctica). Hist. Biol. 34 (12), 2341–2349.

Izawa, K., 1993. Soil-eating in *Aouatta* and *Ateles*. Int. J. Primatol. 14, 229–242.

Jacobsen, A.R., 1995. Predatory Behaviour of Carnivorous Dinosaurs: Ecological Interpretations Based on Tooth Marked Dinosaur Bones and Wear Patterns of Theropod Teeth. M.Sc. thesis. University of Copenhagen, Copenhagen, 79 p.

Jacobsen, A.R., 1998. Feeding behavior of carnivorous dinosaurs as determined by tooth marks on dinosaur bones. Hist. Biol. 13, 17–26.

Jacobsen, A.R., 2001. Tooth-marked small theropod bone: an extremely rare trace. In: Tanke, D.H., Carpenter, K. (Eds.), Mesozoic Vertebrate Life. Indiana University Press, Bloomington, pp. 58–63.

Jacobsen, A.R., Bromley, R.G., 2009. New ichnotaxa based on tooth impressions on dinosaur and whale bones. Geol. Q. 53, 373–382.

Jacobsen, A.R., Lauridsen, H., Fiirgaard, B., Boel, L.W.T., Hansen, K., 2015. Healed or non-healed? Computed tomography (CT) visualisation of morphology of bite trace ichnotaxa on a dinosaur bone. Ann. Soc. Geol. Pol. 85 (3), 457–464. https://doi.org/10.14241/asgp.2015.022.

Jagt, J.W., Deckers, M.J., De Leebeeck, M., Donovan, S.K., Nieuwenhuis, E., 2020. Episkeletozoans and bioerosional ichnotaxa on isolated bones of Late Cretaceous mosasaurs and cheloniid turtles from the Maastricht area, the Netherlands. Geologos 26, 39–49.

Jagt, J.W., van Bakel, B.W., Deckers, S.K., Donovan, R.H.B., Fraaije, E.A., Jagt-Yazykova, J.L., Nieuwenhuis, E., Thijs, B., 2018. Late Cretaceous echinoderm 'odds and ends' from the Low Countries. Contemp. Trends Geosci. 7, 255–282.

Jennings, D.S., Hasiotis, S.T., 2006. Taphonomic analysis of a dinosaur feeding site using geographic information systems (GIS), Morrison Formation, southern Bighorn Basin, Wyoming, USA. Palaios 21, 480–492.

Jensen, J.A., 1988. A fourth new sauropod dinosaur from the Upper Jurassic of the Colorado Plateau and sauropod bipedalism. Great Basin Nat. 48, 121–145.

Jensen, T.Z., Niemann, J., Iversen, K.H., Fotakis, A.K., Gopalakrishnan, S., Vågene, Å.J., Pedersen, M.W., Sinding, M.H.S., Ellegaard, M.R., Allentoft, M.E., Lanigan, L.T., 2019. A 5700 year-old human genome and oral microbiome from chewed birch pitch. Nat. Commun. 10, 5520. https://doi.org/10.1038/s41467-019-13549-9.

Joeckel, R.M., Tucker, S.T., 2013. Exceptionally well preserved latest Miocene (Hemphillian) rodent burrows from the eastern Great Plains, United States, and a review of the burrows of North American rodents. Palaios 28, 793–824.

Johanson, Z., Smith, M., Kearsley, A., Pilecki, P., Mark-Kurik, E., Howard, C., 2013. Origins of bone repair in the armour of fossil fish: response to a deep wound by cells depositing dentine instead of dermal bone. Biol. Lett. 9. https://doi.org/10.1098/rsbl.2013.0144.

Johnston, P.A., Kase, T., Seilacher, A., 1997. Mosasaurs, Limpets and Diagenesis: Alternative Pathways to Punctured Ammonite Shells. CSPG-SEPM Joint Convention, Program with. https://doi.org/10.1098/rsbl.2013.0144.

Jones, R.L., Hanson, H.C., 1985. Mineral Licks, Geophagy, and Biogeochemistry of North American Ungulates. The Iowa State University Press, Ames, 301 p.

Jones, G.P., Ferrell, D.J., Sale, P.F., 1992. Fish feeding and dynamics of soft-sediment mollusc populations in a coral reef lagoon: Source. Mar. Ecol. Prog. Ser. 80 (2/3), 175–190.

Joyce, W.G., 2000. The first complete skeleton of *Solnhofia parsonsi* (Cryptodira, Eurysternidae) from the Upper Jurassic of Germany and its taxonomic implications. J. Paleontol. 74, 684–700.

Kaddumi, H.F., 2006. A new genus and species of gigantic marine turtles (Chelonioidea: Cheloniidae) from the Maastrichtian of the Harrana Fauna-Jordan. PalArch 3, 1–13.

Kamola, D.L., 1984. Trace fossils from marginal-marine facies of the Spring Canyon Member, Blackhawk Formation (Upper Cretaceous), East-Central Utah. J. Paleontol. 58, 529–541.

Karatajute-Talimaa, V., Predtechenskyj, N., 1995. The distribution of the vertebrates in the Late Ordovician and Early Silurian palaeobasins of the Siberian platform. Bulletin du Muséum national d'Histoire naturelle, 4ème série–section C–Sciences de la Terre, Paléontologie, Géologie, Minéralogie, 17, pp. 39–55.

Karl, H.V., 2012. Bite traces in a turtle shell fragment from the Kimmeridgian (Upper Jurassic) of northern Germany. Stud. Geol. Salmant. 9, 25–30.

Karl, H.V., Tichy, G., 2004. The structure of fossil teeth of chelonophagous crocodiles (Diapsida: Crocodylia). Stud. Geol. Salmant. 40, 115–124.

Kase, T., Shigeta, Y., Futakami, M., 1994. Limpet home depressions in Cretaceous ammonites. Lethaia 27, 49–58.

Kase, T., Johnston, P.A., Seilacher, A., Boyce, J.B., 1998. Alleged mosasaur bite marks on Late Cretaceous ammonites are limpet (patellogastropod) home scars. Geology 26, 947–950.

Kastelein, R.A., Stephens, S., Mousterd, P., 1989. The excavation technique for mollusks of Pacific walruses (*Odobenus rosmarus divergens*) under controlled conditions. Aquat. Mamm. 15, 3–5.

Kaufman, L., 1981. There are biological disturbances on Pleistocene coral reefs. Paleobiology 7, 527–532.

Kauffman, E.G., 1972. *Ptychodus* predation upon a Cretaceous *Inoceramus*. Palaeontology 15, 439–444.

Kauffman, E.G., 1990a. Cretaceous fish predation on a large squid. In: Boucot, A.J. (Ed.), Evolutionary Paleobiology of Behavior and Coevolution. Elsevier, Amsterdam, pp. 195–196.

Kauffman, E.G., 1990b. Mosasaur predation on ammonites during the Cretaceous—an evolutionary history. In: Boucot, A.J. (Ed.), Evolutionary Paleobiology of Behavior and Coevolution. Elsevier, Amsterdam, pp. 184–189.

Kauffman, E.G., 2004. Mosasaur predation on Upper Cretaceous nautiloids and ammonites from the United States Pacific coast. Palaios 19, 96–100.

Kauffman, E.G., Kesling, R.V., 1960. An Upper Cretaceous ammonite bitten by a mosasaur. Contrib. Mus. Paleontol. Univ. Mich. 15, 193–248.

Kauffman, E.G., Sawdo, J.K., 2013. Mosasaur predation on a nautiloid from the Maastrichtian Pierre Shale, Central Colorado, Western Interior Basin, United States. Lethaia 46, 180–187.

Kear, B.P., 2005. A new elasmosaurid plesiosaur from the Lower Cretaceous of Queensland, Australia. J. Vertebr. Paleontol. 25, 792–805.

Kear, B.P., Godthelp, H., 2008. Inferred vertebrate bite marks on an Early Cretaceous unionoid bivalve from Lightning Ridge, New South Wales, Australia. Alcheringa 32, 65–71.

Kellner, A.W.A., 2004. On a pterosaur neck with a dinosaur tooth: scavenging or predation. Nat. Nascosta 29, 41–43.

Kellner, A.W.A., Rich, T.H., Costa, F.R., Vickers-Rich, P., Kear, B.P., Walters, M., Kool, L., 2010. New isolated pterodactyloid bones from the Albian Toolebuc Formation (western Queensland, Australia) with comments on the Australian pterosaur fauna. Alcheringa 34, 219–230.

King, S.D., 2009. The Ability of Mosasaurs to Produce Unique Puncture Marks on Ammonite Shells. M.S. thesis. Bowling Green State University, 144 p.

Kirkland, J.I., Scheetz, R.D., Foster, J.R., 2005. Jurassic and Lower Cretaceous Dinosaur Quarries of western Colorado and eastern Utah. compiler. In: Gigi, R. (Ed.), 2005 Rocky Mountain Section of the Geological Society of America Field Trip Guidebook, vol 402. Grand Junction Geological Society, Field Trip, pp. 1–26.

Knsc, T., Shigeta, Y., Futakami, M., 1994. Limpet home depressions in Cretaceous ammonites. Lethaia 27, 49–58.

Kodaira, R., 1921. Fossil nut-shells of *Juglans sieboldiana*, Maximovicz in the Lignite of Asahiyama, near Nagano City, Province of Shinano. J. Geol. Soc. Jpn. 28, 1–5.

Koenigswald, W.V., Schierning, H.P., 1987. The ecological niche of an extinct group of mammals, the early Tertiary apatemyids. Nature 326, 595–597.

Konuki, R., 2008. Biostratigraphy of Sea Turtles and Possible Bite Marks on a *Toxochelys* (Testudine, Chelonioidea) from the Niobrara Formation (Late Santonian), Logan County, Kansas, and Paleoecological Implications for Predator–Prey Relationships Among Large Marine Vertebrates. M.S. thesis. Fort Hays State University, 169 p.

Kotake, N., Kondo, Y., Fujii, T., Shibasaki, H., 2004. Paleontologic and ichnologic interpretation for a peculiar mode of occurrence of shell concentrations in the lower part of the Tatamigaura Sandstone Member of the Middle Miocene Tougane Formation, Hamada, Western part of Shimane, Japan. J. Geol. Soc. Jpn. 110 (12), 733–745.

Krishnamani, R., Mahaney, W.C., 2000. Geophagy among primates: adaptive significance and ecological consequences. Anim. Behav. 59, 899–915.

Kubo, T., Mitchell, M.T., Henderson, D.M., 2012. *Albertonectes vanderveldei*, a new elasmosaur (Reptilia, Sauropterygia) from the Upper Cretaceous of Alberta. J. Vertebr. Paleontol. 32, 557–572.

Kushlina, V.B., 2006. Biting traces on echinoderms from the Ordovician of the St. Petersburg region 1053 (Russia). In: Mikuláš, R., Rindsberg, A.K. (Eds.), Abstract Book: Workshop on Ichnotaxonomy—III, 1054 Prague and Moravia, Czech Republic. Institute of Geology and Academy of Sciences of the Czech 1055 Republic, Prague, pp. 12–14.

Labandeira, C.C., 1998. Early history of arthropod and vascular plant associations. Annu. Rev. Earth Planet Sci. 26, 329–377.

LaBarge, T.W., 2022. Taphonomy and Ichnology of Nile Crocodile Feeding Behavior. Doctoral dissertation, Indiana University, 59 p.

LaBarge, T.W., Njau, J.K., 2024. Taxonomic reappraisal of *Nihilichnus* from taphonomic perspectives of crocodile predatory ecology. Ichnos. https://doi.org/10.1080/10420940.2024.2353199.

Lach, R., Brom, K., Leśko, K., 2015. Bite marks and overgrowths on crinoids from the Štramberkite marks and overgrow-type limestones in Poland. Neues Jahrbuch Geol. Palaontol. Abhand. 276, 151–154.

Lambert, O., de Muizon, C., Malinverno, E., Celma, C.D., Urbina, M., Bianucci, G., 2018. A new odontocete (toothed cetacean) from the Early Miocene of Peru expands the morphological disparity of extinct heterodont dolphins. J. Syst. Palaeontol. 16, 981–1016.

Lamsdell, J.C., Braddy, S.J., 2009. Cope's Rule and Romer's theory: Patterns of diversity and gigantism in eurypterids and Palaeozoic vertebrates. Biol. Lett. 6, 265–269.

Langston, W.J., 1963. Fossil Vertebrates and the Late Palaeozoic Redbeds of Prince Edward Island, vol 187. National Museum of Canada Bulletin, pp. 1–36.

Laudet, F., Fosse, P., 2001. Un assemblage d'os grignote par les rongeurs au Paleogene (Oligocene superieur, phosphorites du Quercy). Comptes Rendus Acad. Sci. Ser. IIA Earth Planet. Sci. 333, 195–200.

Lebedev, O.A., Mark-Kurik, E., Karatajūtė-Talimaa, V.N., Lukševičs, E., Ivanov, A., 2009. Bite marks as evidence of predation in early vertebrates. Acta Zool. 90 (Suppl. 1), 344–356.

Leithold, E.L., Bourgeois, J., 1984. Characteristics of coarse-grained sequences deposited in nearshore, wave-dominated environments—examples from the Miocene of south-west Oregon. Sedimentology 31 (6), 749–775.

Lewis, C., 2011. Evidence of Dinosaur Herbivory in the Upper Cretaceous Aguja Formation, Big Bend National Park. M.S. thesis. Texas Tech University, Lubbock, 144 p.

Lindholm, A., Godfrey, S.J., Ward, L.W., Collareta, A., 2022. A gar-bitten vertebrate coprolite from the Neogene of the Atlantic Coastal Plain, USA, and a new ichnospecies of Machichnus Mikuláš et al., 2006. Ichnos 29 (3–4). https://doi.org/10.1080/10420940.2023.2182300.

Lingham-Soliar, T., 2004. Palaeopathology and injury in extinct mosasaurs (Lepidosauromorpha, Squamata) and implications for modern reptiles. Lethaia 37, 255–262.

Lingham-Soliar, T., 2008. A unique cross section through the skin of the dinosaur *Psittacosaursus* from China showing a complex fibre architecture. Proc. R. Soc. Ser. B 275, 775–780.

Lockley, M.G., Harris, J.D., 2010. On the trail of early birds: a review of the fossil footprint record of avian morphological and behavioral evolution. In: Ulrich, P.K., Willet, J.H. (Eds.), Trends in Ornithology Research. Nova Science Publications, Inc, pp. 1–63.

Lockley, M.G., Hunt, A.P., 1995. Dinosaur Tracks and Other Fossil Footprints of the Western United States. Columbia University Press, New York, 338 p.

Lockley, M.G., Wright, J.L., 2003. Pterosaur swimtracks and other ichnological evidence of behaviour and ecology. In: Buffetaut, E., Mazin, J.M. (Eds.), Evolution and Palaeobiology of Pterosaurs, vol 217. Geological Society of London, Special Publications, pp. 297–313.

Lockley, M., Harris, J., Chin, K., Matsukawa, M., 2008. Ichnological evidence for morphological and behavioral convergence between Mesozoic and extant birds: paleobiological implications. In: Uchman, A. (Ed.), Second International Congress on Ichnology Abstract Book, vol 71. Polish Geological Institute, Warszawa.

Lockley, M.G., Chin, K., Houck, M., Matsukawa, M., Kukihara, R., 2009. New interpretations of *Ignotornis* the first-reported Mesozoic avian footprints: Implications for the ecology and behavior of an enigmatic Cretaceous bird. Cretac. Res. 30, 1041–1061.

Long, R.A., Murry, P.A., 1995. Late Triassic (Carnian and Norian) Tetrapods from the Southwestern United States. N. M. Museum Nat. Hist. Sci. Bull. 4, 1–254.

Longrich, N.R., Ryan, M.J., 2010. Mammalian tooth marks on the bones of dinosaurs and other Late Cretaceous vertebrates. Palaeontology 53, 703–709.

Löwemark, L., 2015. Evidence for targeted elasmobranch predation on thalassinidean shrimp in the Miocene Taliao Formation, NE Taiwan. Lethaia 48 (2), 227–234.

Lucas, S.G., 2015. *Thinopus* and a critical review of Devonian tetrapod footprints. Ichnos 22 (3–4), 136–154. https://doi.org/10.1080/10420940.2015.1063491.

Lucas, S.G., 2016. Two new, substrate-controlled nonmarine ichnofacies. Ichnos 23, 248–261.

Lucas, S.G., Hunt, A.P., 1993. A dicynodont from the Upper Triassic of New Mexico and its biochronological significance. N. M. Museum Nat. Hist. Sci. Bull. 3, 321–325.

Lucas, S.G., Schoch, R.M., 1987. Paleopathology of Early Cenozoic *Coryphodon* (Mammalia; Pantodonta). J. Vertebr. Paleontol. 7, 145–154.

Lucas, S.G., Tanner, L.H., 2018. The missing mass extinction at the Triassic-Jurassic boundary. In: Tanner, L. (Ed.), The Late Triassic World. Springer, New York, pp. 721–785.

Lucas, S.G., Rinehart, L.F., Celeskey, M.D., 2018. The oldest specialized tetrapod herbivore: A new eupelycosaur from the Permian of New Mexico, USA. Palaeontol. Electron. 21 (3). https://doi.org/10.26879/899palaeo-electronica.org/content/2018/2343-new-eupelycosaur, 39A 1-42.

Lyell, C., 1863. Principles of Geology. John Murray, London, 398 p.

MacEachern, J.A., Bann, K.L., Gingras, M.K., Zonneveld, J.P., Dashtgard, S.E., Pemberton, S.G., 2012. The ichnofacies paradigm. In: Knaust, D., Bromley, R.G. (Eds.), Trace Fossils as Indicators of Sedimentary Environments, Developments in Sedimentology, vol 64. Elsevier, pp. 103–138.

MacEachern, J.A., Bann, K.L., Pemberton, S.G., Gingras, M.K., 2007. The ichnofacies paradigm: high resolution paleoenvironmental interpretation of the rock record. In: MacEachern, J.A., Bann, K.L., Gingras, M.K., Pemberton, S.G. (Eds.), Applied Ichnology. SEPM Short Course Notes, vol 52, pp. 27–64.

Mackness, B.S., Cooper, J.E., Wilkinson, C.E.C., Wilkinson, D., 2010. Palaeopathology of a crocodile femur from the Pliocene of eastern Australia. Alcheringa 34, 515–521.

Mahaney, W.C., Hancock, R.G.V., Aufreiter, S., Huffman, M.A., 1996. Geochemistry and clay mineralogy of termite mound soil and the role of geophagy in chimpanzees of the Manhale Mountains, Tanzania. Primates 37 (2), 121–134.

Maisch, M.W., Matzke, A.T., 2003. The cranial osteology of the ichthyosaur *Leptonectes* cf. *tenuirostris* from the Lower Jurassic of England. J. Vertebr. Paleontol. 23, 116–127.

Malkani, M.S., 2006. First rostrum of carnivorous *Vitakridrinda* (abelisaurid theropod dinosaur) found from the Latest Cretaceous Dinosaur beds (Vitakri) Member of Pab Formation, Alam Kali Kakor Locality of Vitakri area, Barkhan District, Balochistan, Pakistan. Sind. Univ. Res. J. (Sci. Ser.) 38, 5–24.

Malkani, M.S., 2009. New *Balochisaurus* (Balochisauridae, Titanosauria, Sauropoda) and Vitakridrinda (Theropoda) remains from Pakistan. Sind. Univ. Res. J. (Sci. Ser.) 41, 65–92.

Malkani, M.S., 2010. *Vitakridrinda* (Vitakrisauridae, Theropoda) from the Latest Cretaceous of Pakistan. J. Earth Sci. 21, 204–212.

Manchester, S.R., Lehman, T.M., Wheeler, E.A., 2010. Fossil palms (Arecaceae, Coryphoideae) associated with juvenile herbivorous dinosaurs in the upper Cretaceous Aguja formation, Big Bend National Park, Texas. Int. J. Plant Sci. 171, 679–689.

Mapes, R., Chaffin, D.T., 2003. Predation on cephalopods: A general overview with a case study from the upper Carboniferous of Texas. In: Kelley, P.H., Kowalewski, M., Hansen, T.A. (Eds.), Predator-prey Interactions in the Fossil Record. Kluwer Academic/Plenum Publishers, New York, pp. 177–213.

Mapes, R.H., Hansen, M.C., 1983. Pennsylvanian shark-cephalopod predation: A case study. Lethaia 17, 175–183.

Marsh, O.C., 1869. Notice of some new reptilian remains from the Cretaceous of Brazil. Amer. J. Sci. 2 (141), 390–392.

Martill, D.M., 1985. The preservation of marine vertebrates in the Lower Oxford Clay (Jurassic) of central England. Philos. Trans. R. Soc. London, Ser. A B 311, 155–165.

Martill, D.M., 1990. Predation on *Kosmoceras* by semionotid fish in the Middle Jurassic Oxford Clay of England. Palaeontology 33, 739–742.

Martill, D.M., 1996. Fossils explained 17: Ichthyosaurs. Geol. Today 12, 194–196.

Martill, D.M., Taylor, M.A., Duff, K.L., Riding, J.B., Bown, P.R., 1994. The trophic structure of the biota of the Peterborough Member, Oxford Clay Formation (Jurassic), UK. J. Geol. Soc. Lond. 151, 173–194.

Martin, L.D., Bennett, D.K., 1977. The burrows of the Miocene beaver *Palaeocastor*, western Nebraska, USA. Palaeogeogr. Palaeoclimatol. Palaeoecol. 22, 173–193.

Martin, A.J., Vazquez-Prokopec, G.M., Page, M., 2010. First known feeding trace of the Eocene bottom-dwelling fish *Notogoneus osculus* and its paleontological significance. PLoS One 5 (5), e10420. https://doi.org/10.1371/journal.pone.0010420.

Martin, J.E., 2013. Surviving a potentially lethal injury? Bite mark and associated trauma in the vertebra of a dyrosaurid crocodilian. Palaios 28, 6–8.

Martin, L.D., Rothschild, B.M., Burnham, D.A., 2016. *Hesperornis* escapes plesiosaur attack. Cretac. Res. 63, 23–27.

Martinelli, J., De Gibert, J.M., Domènech, R., Ekdale, A.A., Steen, P.P., 2001. Cretaceous ray traces?: An alternative interpretation for the alleged dinosaur tracks of La Posa, Isona, NE Spain. Palaios 16, 409–416.

Massari, F., D'Alessandro, A., 2010. Icehouse, Cool-Water Carbonate Ramps: The Case of the Upper Pliocene Capodarso Fm (Sicily): Role of trace fossils in the reconstruction of growth stages of prograding wedges. Facies 56 (1), 47–58.

Mateus, O., Polcyn, M.J., Jacobs, L.L., Araújo, R., Schulp, A.S., Marinheiro, J., Pereira, B., Vineyard, D., 2012. Cretaceous amniotes from Angola: dinosaurs, pterosaurs, mosasaurs, plesiosaurs, and turtles. In: Hurtado, P.H., Fernández-Baldor, F.T., Sanagustín, J.I.C. (Eds.), Actas de V Jornadas Internacionales sobre Paleontología de Dinosaurios y su Entorno, pp. 71–105.

Mattern, T., Ellenberg, U., Houston, D.M., Davis, L.S., 2007. Consistent foraging routes and benthic foraging behaviour in yellow-eyed penguins. Mar. Ecol. Prog. Ser. 343, 295–306. https://doi.org/10.3354/meps06954.

Mattern, T., McPherson, M.D., Ellenberg, U., van Heezik, Y., Seddon, P.J., 2018. High definition video loggers provide new insights into behaviour, physiology, and the oceanic habitat of a marine predator, the yellow-eyed penguin. PeerJ 6, e5459. https://doi.org/10.7717/peerj.5459.

Matthew, W., 1908. *Allosaurus,* a carnivorous dinosaur, and its prey. Am. Mus. J. 8, 3–5.

Mayr, G., 2006. First fossil skull of a Palaeogene representative of the Pici (woodpeckers and allies) and its evolutionary implications. Ibis 148 (4), 824–827.

Mayr, G., 2022. Paleogene Fossil Birds, second ed. Springer, Cham. 239 p.

Mazin, J., Billon-Bruyat, J., Hantzepergue, P., Larauire, G., 2003. Ichnological evidence for quadrapedal locomotion in pterodactyloid pterosaurs: trackways from the late Jurassic of Crayssac. In: Buffetaut, E., Mazin, J.M. (Eds.), Evolution and Palaeobiology of Pterosaurs, vol 217. Geological Society of London, Special Publications, pp. 283–296.

Mazuch, M., Košťák, M., Mikuláš,, R., Culka, A., Kohout, O., Jagt, J.M.W., 2024. Bite traces of a large, mosasaur-type(?) vertebrate predator in the lower Turonian ammonite *Mammites nodosoides* (Schlüter, 1871) from the Czech Republic. Cretac. Res. 153. https://doi.org/10.1016/j.cretres.2023.105714.

Mazurek, D., Antczak, M., 2023. Late Cretaceous coprolite from the Opole area (southern Poland) as evidence for a variable diet in shell-crushing shark *Ptychodus* (Elasmobranchii: Ptychodontidae). PeerJ 11, e16598.

Mazza, P.P., 2006. Poggio Rosso (Upper Valdarno, central Italy), a window on latest Pliocene wildlife. Palaios 21, 493–498.

Mazza, P.P., Bertini, A., Magi, M., 2004. The late Pliocene site of Poggio Rosso (central Italy): Taphonomy and paleoenvironment. Palaios 19, 227–248.

McHenry, C.R., 2009. 'Devourer of Gods'—The Palaeoecology of the Cretaceous Pliosaur *Kronosaurus queenslandicus*. Ph.D. dissertation. University of Newcastle, Newcastle, 616 p.

McHugh, J.B., Drumheller, S.K., Riedel, A., Kane, M., 2020. Decomposition of dinosaurian remains inferred by invertebrate traces on vertebrate bone reveal new insights into Late Jurassic ecology, decay, and climate in western Colorado. PeerJ 8, e9510. https://doi.org/10.7717/peerj.9510.

McKee, J.W.A., 1987. The occurrence of the Pliocene penguin *Tereingaomis moisleyi* (Sphenisciformes: Spheniscidae) at Hawera, Taranaki, New Zealand. N. Z. J. Zool 14, 557–561.

McLain, M.A., 2016. Taphonomy of a Lance Formation (Maastrichtian, WY) Dinosaur Bonebed with a Focus on Tooth Traces. Ph.D. dissertation. Loma Linda University, Loma Linda, 149 p.

McLain, M.A., Nelsen, D., Snyder, K., Griffin, C.T., Siviero, B., Brand, L.R., Chadwick, A.V., 2018. Tyrannosaur cannibalism: a case of a tooth-traced tyrannosaurid bone in the Lance Formation (Maastrichtian), Wyoming. Palaios 33, 164–173.

Melchor, R.N., Bedatou, E., de Valais, S., Genise, J.F., 2006. Lithofacies distribution of invertebrate and vertebrate trace-fossil assemblages in an Early Mesozoic ephemeral fluvio-lacustrine system from Argentina: implications for the *Scoyenia* ichnofacies. Palaeogeogr. Palaeoclimatol. Palaeoecol. 239 (3–4), 253–285.

Merella, M., Collareta, A.,, Casati, S.,, Di Cencio, A., Bianucci, G., 2021. An unexpected deadly meeting: deep-water (hexanchid) shark bite marks on a sirenian skeleton from Pliocene shoreface deposits of Tuscany (Italy). Neues Jahrbuch Geol. Palaontol. Abhand. 301 (3), 295–305. https://doi.org/10.1127/njgpa/2021/1012.

Merella, M., Collareta, A., Casati, S., Di Cencio, A., Bianucci, G., 2022. An unexpected deadly meeting: deep-water (hexanchid) shark bite marks on a sirenian skeleton from Pliocene shoreface deposits of Tuscany (Italy). Neues Jahrbuch Geol. Palaontol. Abhand. 301 (3), 295–305. https://doi.org/10.1127/njgpa/2022/1040.

Meyer, C.A., 1991. Burial experiments with marine turtle carcasses and their paleoecological significance. Palaios 6, 89–96.

Meyer, C.A., Thüring, B., 2003. Dinosaurs of Switzerland. C. R. Palevol. 2, 103–117.

Mikuláš, R., 2013. Ichnotaxonomy and interpretation of "incipient" insect trace fossils in the archaeological context of Abusir (Holocene, Egypt). Stratigr. Geol. Correl. 21, 323–336.

Mikuláš, R., Dvorák, Z., 2010. Possible crocodylian bite traces, Miocene of the Most Basin (Czech Republic). N. M. Museum Nat. Hist. Sci. Bull. 51, 191–194.

Mikuláš, R., Kadlecová, E., Fejfar, O., Dvořák, Z., 2006. Three new ichnogenera of biting and gnawing traces on reptilian and mammalian bones: A case study from the Miocene of the Czech Republic. Ichnos 13, 113–127.

Milàn, J., Lindow, B.E., Lauridsen, B.W., 2011. Bite traces in a turtle carapace fragment from the middle Danian (Lower Paleocene) bryozoan limestone, Faxe, Denmark. Bull. Geol. Soc. Den. 59, 61–67.

Miller, M.F., 1979. Paleoenvironmental distribution of trace fossils in the Catskill deltaic complex, New York State. Palaeogeogr. Palaeoclimatol. Palaeoecol. 28, 117–141.

Miller, G.J., 1980. Some new evidence in support of the stabbing hypothesis for *Smilodon californicus* Bovard. Carnivore 3, 8–26.

Milner, A.R.C., Kirkland, J.I., 2007. The case for fishing dinosaurs at the St. George Dinosaur Discovery Site at Johnson Farm. Utah Geol. Surv. Notes 39, 1–3.

Minter, N.J., 2024. Fish trails and imprints. In: Lucas, S.G., Hunt, A.P., Klein, H. (Eds.), Vertebrate Ichnology. Elsevier, Amsterdam (this volume).

Minter, N.J., Braddy, S.J., 2006. The fish and amphibian swimming traces *Undichna* and *Lunichnium*, with examples from the Lower Permian of New Mexico, USA. Palaeontology 49, 1123–1142.

Modesto, S.P., Botha-Brink, J., 2010. A burrow cast with *Lystrosaurus* skeletal remains from the Lower Triassic of South Africa. Palaios 25, 274–281.

Molnar, R.E., 1973. The Cranial Morphology and Mechanics of *Tyrannosaurus rex* (Reptilia: Saurischia). Ph.D. dissertation. University of California Los Angeles, Los Angeles, 450 p.

Moore, M., 2021. Bitten coprolite: A dentalite. Ecphora 36 (3), 10.

Moore, W.S., Miglia, K.J., 2009. Woodpeckers, toucans, barbets, and allies (Piciformes). In: Hedges, S.B., Kumar, S. (Eds.), The Timetree of Life. Oxford Biology, pp. 445–450.

Morgan, G.S., Williamson, T.E., 2000. Middle Miocene (late Barstovian) vertebrates from the Benavidez Ranch local fauna, Albuquerque Basin, New Mexico. N. M. Museum Nat. Hist. Sci. Bull. 16, 195–208.

Mujal, E., Foth, C., Maxwell, E.E., Seegis, D., Schoch, R.R., 2022. Feeding habits of the Middle Triassic pseudosuchian *Batrachotomus kupferzellensis* from Germany and palaeoecological implications for archosaurs. Palaeontology 65 (3), e12597.

Mulder, E.W.A., 2003. Comparative osteology, palaeoecology and systematics of the Late Cretaceous turtle *Allopleuron hofmanni* (Gray 1831) from the Maastrichtian type area. Publicaties van het Natuurhistorisch Genootschap in Limburg 44, pp. 23–92.

Müller, R.T., de Araujo-Junior, H.I., Aires, A.S.S., Roberto-da-Silva, L., Dias-da-Silva, S., 2015. Biogenic control on the origin of a vertebrate monotypic accumulation from the Late Triassic of southern Brazil. Geobios 48 (4), 331–340.

Muñiz, F., Toscano, A., Bromley, R.G., Esperante, R., 2009. Excepcional caso de interacción trófica entre tiburones hexanchiformes y una ballena Balaenoptera en el Plioceno inferior de Huelva (SO de España). In: Libro de Actas de las XXV Jornadas de la Sociedad española de Paleontología, Ronda (Málaga). Sociedad Española de Paleontología, pp. 242–244.

Muñiz, F., Belaústegui, Z., Toscano, A., Ramirez-Cruzado, S., Gámez Vintaned, J.A., 2020. New ichnospecies of *Linichnus* Jacobsen & Bromley, 2009. Ichnos 27, 344–351.

Muñiz, F., Toscano, A., Sánchez, A., Esperante, R., Korestky, I., 2008. Bioerosión producida por necrófagos marinos en restos óseos del Plioceno Inferior de Huelva (SO España). In: Libro de Actas XXIV Jornadas de la Sociedad Española de Paleontología. Sociedad Española de Paleontología, Sevilla, pp. 47–48.

Myers, T.S., 2004. Taphonomy of the Mother's Day Quarry: Implications for Gregarious Behavior in Sauropod Dinosaurs. M.S. thesis. University of Cincinnati, Cincinnati, 119 p.

Myers, T.S., Storrs, G.W., 2007. Taphonomy of the Mother's Day Quarry, Upper Jurassic Morrison Formation, south-central Montana, USA. Palaios 22, 652–666.

Myrvold, K.S., Milàn, J., Rasmussen, J.A., 2018. Two new finds of turtle remains from the Danian and Selandian (Paleocene) deposits of Denmark with evidence of predation by crocodilians and sharks. Bull. Geol. Soc. Den. 66, 211–218.

Nagel-Myers, J.J., Dietl, G.P., Brett, C.E., 2009. First report of sublethal breakage-induced predation on Devonian bivalves. Palaios 24, 460–465.

Nielsen, J.K., Nara, M., Jacobsen, A.R., 2020. Bowl-shaped structures in a Pleistocene clastic carbonate wedge on the Island of Rhodes, Greece. Ichnos 27, 326–333.

Nelson, C.H., Johnson, K.R., Barber Jr., J.H., 1987. Gray whale and walrus feeding excavations on the Bering Shelf, Alaska. J. Sediment. Petrol. 57, 419–430.

Nelson, C.H., Johnson, K.R., Barber Jr., J.H., Phillips, R.L., 1992. Gray Whale and walrus feeding excavation on the Bering Shelf, Alaska. In: Alaska OCS Region, Fourth Information Transfer Meeting Conference Proceedings. Anchorage. U.S. Department of the Interior Minerals Management Service, pp. 87–104.

Neumann, C., 2000. Evidence of predation on Cretaceous sea stars from north-west Germany. Lethaia 33, 65–70.

Neumann, C., Hampe, O., 2018. Eggs for breakfast? Analysis of a probable mosasaur biting trace on the Cretaceous echinoid *Echinocorys ovata* Leske, 177. Mitteilungen aus dem Museum für Naturkunde in Berlin, Fossil Record 21, 55–66.

Niedźwiedzki, G., Gorzelak, P., Sulej, T., 2010. Bite traces on dicynodont bones and the early evolution of large terrestrial predators. Lethaia 44, 87–92.

Niedźwiedzki, G., Szrek, P., 2020. Non-tetrapod trace fossils from the Middle Devonian tetrapod tracksite at Zachełmie Quarry, Holy Cross Mountains, Poland. Palaeogeogr. Palaeoclimatol. Palaeoecol. 553, 109763.

Njau, J.K., 2006. The Relevance of Crocodiles to Oldowan Hominin Paleoecology at Olduvai Gorge, Tanzania. Ph.D. dissertation. Rutgers University, New Brunswick, 325 p.

Njau, J.K., Blumenschine, R.J., 2006. A diagnosis of crocodile feeding traces on larger mammal bone, with fossil examples from the Plio-Pleistocene Olduvai Basin, Tanzania. J. Hum. Evol. 50, 142–162.

Njau, J.K., Blumenschine, R.J., 2012. Crocodylian and mammalian carnivore feeding traces on hominid fossils from FLK 22 and FLK NN 3, Plio-Pleistocene, Olduvai Gorge, Tanzania. J. Hum. Evol. 63, 408–417.

Njau, J.K., Gilbert, H., 2016. Standardizing terms for crocodile-induced bite marks on bone surfaces in light of the frequent bone modification equifinality found to result from crocodile feeding behavior, stone tool modification, and trampling. FOROST For. Osteol. Occasion. Publicat. 3, 1–13.

Noriega, J.I., Cione, A.L., Aceñolaza, F.G., 2007. Shark tooth marks on Miocene balaenopterid cetacean bones from Argentina. Neues Jahrbuch für Geologie und Paläontologie, 245, pp. 185–192.

Noto, C.R., Main, D.J., Drumheller, S.K., 2012. Feeding traces and paleobiology of a Cretaceous (Cenomanian) crocodyliform: Example from the Woodbine Formation of Texas. Palaios 27, 105–115.

Novis, L.K., 2012. A Taphonomic Study of Marine Reptiles from the Upper Jurassic of Svalbard. M.S. thesis. University of Oslo, Oslo, 51 p.

Obruchev, D.V., Mark-Kurik, E.Y., 1965. Psammosteidy (Agnatha, Psammosteidae) Devona SSSR. Institute of Geology, Academy of Sciences of the Estonian SSR, Tallinn, 304 p.

O'Leary, M.A., Bouaré, M.L., Claeson, K.M., Heilbronn, K., Hill, R.V., Mccartney, J., Sessa, J.A., Sissoko, F., Tapanila, L., Wheeler, E., Roberts, E.M., 2019. Stratigraphy and paleobiology of the Upper Cretaceous-Lower Paleogene sediments from the trans-Saharan seaway in Mali. Bull. Am. Mus. Nat. Hist. 436, 1–183.

Osborn, H.F., 1904. Fossil Wonders of the West: The Dinosaurs of the Bone-Cabin Quarry: being the first description of the greatest "find" of extinct animals ever made. Century Illustrated Magazine (New Series), 168, pp. 680–694.

Otero, R.A., Soto-Acuna, S., Vargas, A.O., Rubilar-Rogers, D., 2014. A new postcranial skeleton of an elasmosaurid plesiosaur from the Upper Cretaceous of central Chile and reassessment of *Cimoliasaurus andium* Deecke. Cretac. Res. 50, 318–331.

Paes Neto, V.D., Francischini, H., Martinelli, A.G., Marinho, T.D.S., Ribeiro, L.C.B., Soares, M.B., Schultz, C.L., 2018. Bioerosion traces on titanosaurian sauropod bones from the Upper Cretaceous Marília Formation of Brazil. Alcheringa 42, 415–426.

Paik, I.S., Kim, H.J., Lim, J.D., Huh, M., Lee, H.I., 2011. Diverse tooth marks on an adult sauropod bone from the Early Cretaceous, Korea: Implications in feeding behaviour of theropod dinosaurs. Palaeogeogr. Palaeoclimatol. Palaeoecol. 309, 342–346.

Palmqvist, P., Martínez-Navarro, B., Arribas, A., 1996. Prey selection by terrestrial carnivores in a Lower Pleistocene paleocommunity. Paleobiology 22, 514–534.

Pardo-Pérez, J.M., Kear, B.P., Mallison, H., Gómez, M., Moroni, M., Maxwell, E.E., 2018. Pathological survey on *Temnodontosaurus* from the Early Jurassic of southern Germany. PLoS One 13 (10), e0204951. https://doi.org/10.1371/journal.pone.0204951.

Parker, L.R., Balsley, J.K., 1989. Coal mines as localities for studying dinosaur trace fossils. In: Gillette, D.D., Lockley, M.G. (Eds.), Dinosaur Tracks and Traces. Cambridge University Press, Cambridge, pp. 353–359.

Parkinson, J.A., Plummer, T., Hartstone-Rose, A., 2015. Characterizing felid tooth marking and gross bone damage patterns using GIS image analysis: an experimental feeding study with large felids. J. Hum. Evol. 80, 114–134.

Pavia, M., Davies, G.B., Gommery, D., Kgasi, L., 2017. Mid-Pliocene bald ibis (*Geronticus* cf. *calvus*; Aves: Threskiornithidae) from the Cradle of Humankind, Gauteng, South Africa and its environmental and evolutionary implications. Paläontol. Z. 91, 237–243.

Pearson, N.J., Gingras, M.K., Armitage, I.A., Pemberton, S.G., 2007. Significance of Atlantic sturgeon feeding excavations, Mary's Point, Bay of Fundy, New Brunswick, Canada. Palaios 22 (5), 457–464.

Perez, V.J., 2020. Self-bitten snaggletooth. Ecphora 35, 12.

Perez, V.J., Godfrey, S.J., Chapman, P.F., 2021. Rare evidence of shark-on-shark trophic interactions in the fossil record. Acta Palaeontol. Pol. 66 (4), 847–856.

Pervesler, P., Roetzel, R., Steininger, F., 1995. Taphonomie der Sirenen in den marinen Flachwasserablagerungen (Burgschleinitz-Formation, Eggenburgium, Unter- miozän) der Gemeindesandgrube Kühnring (Niederösterreich). Jahrbuch der Geologischen Bundesandesanstallten 138, 89–121.

Peters, M.E., 2018. Identification and Taphonomy of a Miocene Leatherback Turtle (Testudines: Dermochelyidae) from the Westerschelde. Master's thesis, Utrecht University, The Netherlands, 32 p.

Peyrot-Clausade, M., Chabanet, P., Conand, C., Fontaine, M., Letourneur, Y., Harmelinvivien, M., 2000. Sea urchin and fish bioerosion on La Réunion and Moorea reefs. Bull. Mar. Sci. 66, 477–485.

Phipps, K.J., 2008. Evidence of predation on *Gryphaea* (*Bilobissa*) *lituola* Lamarck, 1819, from the Oxford Clay Formation of South Cave Station Quarry, Yorkshire. Proc. Geol. Assoc. 119, 277–285.

Pickford, M., 1996. Fossil crocodiles (*Crocodylus lloydi*) from the Lower and Middle Miocene of southern Africa. Ann. Paleontol. 82, 235–250.

Pieńkowski, G., 1985. Early Liassic trace fossil assemblages from the Holy Cross Mountains, Poland: their distribution in continental and marginal marine environments. In: Curran, H.A. (Ed.), Biogenic Structures: Their Use in Interpreting Depositional Environments, The Society of Economic Paleontologists and Mineralogists, vol 35. Special Publication, pp. 37–51.

Pinto Llona, A.C., Andrews, P., 2004. Taphonomy and paleoecology of Ursus spelaeus from northern Spain. Publications du musée des Confluences, 2, pp. 163–170.

Pirrone, C.A., Buatois, L.A., Bromley, R.G., 2014. Ichnotaxobases for bioerosion trace fossils in bones. J. Paleontol. 88, 195–203.

Polcyn, M.J., Schulp, A.S., Gonçalves, A.O., 2023. Remarkably well-preserved in-situ gut-content in a specimen of *Prognathodon kianda* (Squamata: Mosasauridae) reveals multispecies intrafamilial predation, cannibalism, and a new Mosasaurine taxon. In: Lee, Y.-N. (Ed.), Windows into Sauropsid and Svolution: Essays in Honor of Prof. Louis L. Jacobs. Seoul Dinosaur Science Press, pp. 66–98.

Poropat, S.F., White, M.A., Vickers-Rich, P., Rich, T.H., 2019. New megaraptorid (Dinosauria: Theropoda) remains from the Lower Cretaceous Eumeralla Formation of Cape Otway, Victoria, Australia. J. Vertebr. Paleontol. 39, 4. https://doi.org/10.1080/02724634.2019.1666273.

Poropat, S.F., White, M.A., Ziegler, T., Pentland, A.H., Rigby, S.L., Duncan, R.J., Sloan, T., Elliott, D.A., 2021. A diverse Late Cretaceous vertebrate tracksite from the Winton Formation of Queensland, Australia. PeerJ 9, e11544. https://doi.org/10.7717/peerj.11544.

Portis, A., 1883. Nuovi studi sulle tracce attribuite all'uomo pliocenico. Memorie della Reale Accademia delle Scienze di Torino, 35, pp. 2–35.

Posmoşanu, E., 2003. Iguanodontian dinosaurs from the lower Cretaceous Bauxite site from Romania. Acta Paleontol. Roman. 4, 431–439.

Preen, A.R., 1996. Infaunal mining: A novel foraging method of loggerhead turtles. J. Herpetol. 30 (1), 94–96.

Pryce, E., 1994. Grey louries *Corythaixoides concolor* feeding on clay. Babbler 26–27, 23–24.

Purdy, R., Schneider, V.P., Applegate, S.P., McClellan, J.H., Meyer, R.L., Slaughter, B.H., 2001. The Neogene sharks, rays, and bony fishes from Lee Creek Mine, Aurora, North Carolina. In: Ray, C.E., Bohaska, D.J. (Eds.), Geology and Paleontology of the Lee Creek Mine, North Carolina, Volume 3, vol 90. Smithsonian Contributions to Paleobiology, pp. 71–202.

Rabal-Garcés, R., Cuenca-Bescós, G., Ignacio Canudo, J., De Torres, T., 2012. Was the European cave bear an occasional scavenger? Lethaia 45, 96–108.

Rahmat, S., Muñiz, F., Toscano, A., Esperante, R., Koretsky, I., 2018. First European record of *Homiphoca* (Phocidae: Monachinae: Lobodontini) and its bearing on the paleobiogeography of the genus. Hist. Biol. 1–9. https://doi.org/10.1080/08912963.2018.1507030.

Randle, E., Sansom, R.S., 2019a. Bite marks and predation of fossil jawless fish during the rise of jawed vertebrates. Proc. R. Soc. Ser. B 286, 1–7. https://doi.org/10.1098/rspb.2019.1596.

Randle, E., Sansom, R.S., 2019b. Data from: Bite marks and predation of fossil jawless fish during the rise of jawed vertebrates. Proc. R. Soc. Ser. B. https://doi.org/10.5061/dryad.f32p5g7 v5, Dataset.

Rasser, M.W., Covich, A.P., 2014. Predation patterns on Gyraulus shells from Miocene Lake Steinheim and their potential for intralacustrine coevolution. Lethaia 47 (4), 524–532.

Rasser, M.W., Vallon, L.H., Salvador, R.B., 2016. Perforations of freshwater snail shells from the Miocene of Germany: *Nihilichnus covichi* n. isp. Ichnos 23 (3–4), 222–227.

Reis, L.F.S., 2022. Traços de mordidas em osso de dinossauro da Formação São José do Rio Preto, Bacia Bauru, Cretáceo Superior, Brasil. Universidade Federal de São Carlos MS Thesis 1–65.

Reis, L.F.S., Ghilardi, A.M., Fernandes, M.A., 2023. Bite traces in a sauropod rib from the Upper Cretaceous São José do Rio Preto Formation (Bauru Basin), Brazil. Hist. Biol. 35 (8), 1329–1340.

Reisz, R.R., Tsuji, L.A., 2006. An articulated skeleton of *Varanops* with bite marks: The oldest known evidence of scavenging among terrestrial vertebrates. J. Vertebr. Paleontol. 26, 1021–1023.

Renz, M., 2002. Megalodon: Hunting the Hunter. PaleoPress, Lehigh Acres, 159 p.

Richter, A.E., 2009. Ammoniten-Gehäuse mit Bissspuren. Berliner Paläobiologische Abhandlungen 10, 297–305.

Rinehart, L.F., Lucas, S.G., Spielmann, J.A., 2006. Bite marks on tetrapod bones from the Upper Triassic Chinle Group representing a new genus. N. M. Museum Nat. Hist. Sci. Bull. 37, 160–163.

Ristevski, J., Young, M.T., De Andrade, M.B., Hastings, A.K., 2018. A new species of *Anteophthalmosuchus* (Crocodylomorpha, Goniopholididae) from the Lower Cretaceous of the Isle of Wight, United Kingdom, and a review of the genus. Cretac. Res. 84, 340–383.

Rivera-Sylva, H.E., Frey, E., Guzmán-Gutiérrez, J.R., 2009. Evidence of predation on the vertebra of a hadrosaurid dinosaur from the Upper Cretaceous (Campanian) of Coahuila, Mexico. Carnets Géol. Letter 2, 1–5.

Rivera-Sylva, H.E., Hone, D.W.E., Dodson, P., 2012. Bite marks of a large theropod on a hadrosaur limb bone from Coahuila, Mexico. Bol. Soc. Geol. Mex. 64, 155–159.

Robin, N., Noirit, F., Chevrinais, M., Clément, G., Olive, S., 2022. Vertebrate predation in the Late Devonian evidenced by bite traces and regurgitations: implications within an early tetrapod freshwater ecosystem. Pap. Palaeontol. 8 (4), e1460. https://doi.org/10.1002/spp2.14601.

Rodriguez-de la Rosa, R.A., 2003. Pterosaur tracks from the latest Campanian Cerro del Pueblo Formation of southeastern Coahuila, Mexico. In: Buffetaut, E., Mazin, J.M. (Eds.), Evolution and Palaeobiology of Pterosaurs, vol 217. Geological Society of London, Special Publications, pp. 275–282.

Rodriguez-de la Rosa, R.A., del Rosario Fernandez-Barajas, M., Valdes-Vergara, N.A., Prado-Escamilla, E., 2021. Fish trace fossils from the Lower Cretaceous of Puebla, Mexico. J. S. Am. Earth Sci. 112, 103547. https://doi.org/10.1016/j.jsames.2021.103547.

Rogers, J.V., 2003. *Pachycheilosuchus trinquei*, a new procoelous crocodyliform from the Lower Cretaceous (Albian) Glen Rose Formation of Texas. J. Vertebr. Paleontol. 23, 128–145.

Rogers, R.R., Krause, D.W., Curry Rogers, K., Rasoamiaramanana, A.H., Rahantrisoa, L., 2004. Paleoenvironment and paleoecology of *Majungasaurus crenatissimus* (Theropoda: Abelisauridae) from the Late Cretaceous of Madagascar. In: Sampson, S.D., Krause, D.W. (Eds.), *Majungasaurus crenatissimus* (Theropoda:

Abelisauridae) from the Late Cretaceous of Madagascar. Journal of Vertebrate Paleontology, Memoir, vol 8, pp. 21–31.

Romer, A.S., 1933. Eurypterid influence on vertebrate history. Science 78, 114–117.

Romo-de-Vivar-Martínez, P.R., Martinelli, A.G., Paes Neto, V.D., Soares, M.B., 2017. Evidence of osteomyelitis in the dentary of the late Triassic rhynchocephalian *Clevosaurus brasiliensis* (Lepidosauria: Rhynchocephalia) from southern Brazil and behavioural implications. Hist. Biol. 29, 320–327.

Rossbach, K.A., Herzing, D.L., 1997. Underwater observations of benthic-feeding bottlenose dolphins (*Tursiops truncatus*) near Grand Bahama Island, Bahamas. Mar. Mamm. Sci. 13, 498–504.

Rothschild, B.M., DePalma, R., 2013. Skin pathology in the Cretaceous: Evidence for probable failed predation in a dinosaur. Cretac. Res. 42, 44–47.

Rothschild, B.M., Martin, L.D., 1993. Paleopathology: Disease in the Fossil Record. CRC Press, Baton Rouge, 386 p.

Rothschild, B.M., Clark, N.D., Clark, C.M., 2018. Evidence for survival in a Middle Jurassic plesiosaur with a humeral pathology: What can we infer of plesiosaur behaviour? Palaeontol. Electron. 21, 1–11.

Rothschild, B.M., Martin, L.D., Schulp, A.S., 2005. Sharks eating mosasaurs, dead or alive? Netherlands J. Geosci. Geol. Mijnbouw 84, 335–340.

Rozada, L., Allain, R., Vullo, R., Goedert, J., Augier, D., Jean, A., Marchal, J., Peyre de Fabrègues, C., Qvarnström, M., Royo-Torres, R., 2021. A Lower Cretaceous Lagerstätte from France: a taphonomic overview of the Angeac-Charente vertebrate assemblage. Lethaia 54, 141–165.

Ruben, J.A., 1990. Evidence of convergent behavioral patterns in male crocodilians and phytosaurs. In: Boucot, A.J. (Ed.), Evolutionary Paleobiology of Behavior and Coevolution. Elsevier, Amsterdam, pp. 427–428.

Runnegar, B., 1983. A *Diprotodon* ulna chewed by the marsupial lion, *Thylacoleo carnifex*. Alcheringa 7, 23–25.

Rybczynski, N., 2008. Woodcutting behavior in beavers (Castoridae, Rodentia): Estimating ecological performance in a modern and a fossil taxon. Paleobiology 34, 389–402.

Sadlok, G., 2022. Carnivorous reptile feeding strategies and postmortem food-processing behaviors: tooth traces on bones from the Upper Triassic Grabowa Formation (southern Poland). J. Paleontol. 96 (5), 1189–1208.

Sagar, R., Kapur, V.V., Kumar, K., Morthekai, P., Sharma, A., Shukla, S.K., Ghosh, A.K., Chauhan, G., Thakkar, M.G., 2024. First record of Chelonian coprolites from the Early-Middle Miocene Kutch Basin, western India, and their palaeodietary and palaeobiological implications. Geobios 84, 83–101. https://doi.org/10.1016/j.geobios.2023.12.004.

Sahle, Y., El Zaatari, S., White, T.D., 2017. Hominid butchers and biting crocodiles in the African Plio–Pleistocene. Proc. Natl. Acad. Sci. USA 114, 13164–13169.

Salamon, M.A., Gorzelak, P., 2008. Signs of predation on Middle Triassic crinoids form Poland. In: Kroh, A., Schmid, B. (Eds.), 4. Arbeitstreffen Deutschsprachiger Echinodermenforscher, 4th Workshop of German and Austrian Echinoderm Research Abstracts. Naturhistorisches Museum Wien, Wien, pp. 43–44.

Salamon, M., Gorzelak, P., 2010. Late Cretaceous crinoids (Crinoidea) from Eastern Poland. Palaeontograph. Abteilung 291, 1–43.

Sánchez, I.M., Quiralte, V., Ríos, M., Morales, J., Pickford, P., 2015. First African record of the Miocene Asian mouse-deer Siamotragulus (Mammalia, Ruminantia, Tragulidae): implications for the phylogeny and evolutionary history of the advanced selenodont tragulids. J. Syst. Palaeontol. 13, 543–556.

Saneyoshi, M., Watabe, M., Suzuki, S., Tsogtbaatar, K., 2011. Trace fossils on dinosaur bones from Upper Cretaceous eolian deposits in Mongolia: taphonomic interpretation of paleoecosystems in ancient desert environments. Palaeogeogr. Palaeoclimatol. Palaeoecol. 311, 38–47.

Sansom, I.J., Davies, N.S., Coates, M.I., Nicoll, R.S., Ritchie, A., 2012. Chondrichthyan-like scales from the Middle Ordovician of Australia. Palaeontology 55, 243–247.

Sansom, I.J., Smith, M.M., Smith, M.P., 1996. Scales of thelodont and shark-like fishes from the Ordovician of Colorado. Nature 379, 628–630.

Santos, A., Piñuela, L., Mayoral, E., García-Ramos, J.C., 2022. Primer registro de *Osculichnus* del Jurásico. In: Ejemplos del Kimmeridigiense de Asturias (N España) XXXVII Jornadas de Paleontología V Congreso Ibérico de Paleontología, Libro de Resúmenes, p. 177.

Sato, T., Hasegawa, Y., Manabe, M., 2006. A new elasmosaurid plesiosaur from the Upper Cretaceous of Fukushima, Japan. Palaeontology 49, 467–484.

Saul, L.R., 1979. A hollow spined *Anapachydiscus peninsularis* with possible mosasaur bite impressions. Contr. Sci. Nat. Hist. Museum L. A. County 304, 1–8.

Sawyer, G.T., Erickson, B.R., 1998. Paleopathology of the Paleocene crocodile Leidyosuchus. Sci. Mus. Minn. Monogr. 4, 38 p.

Scasso, R.A., Olivero, E.B., Buatois, L.A., 1991. Lithofacies, biofacies and ichnoassemblage evolution of a shallow submarine volcaniclastic fan-shelf depositional system (Upper Cretaceous, James Ross Island, Antarctica). J. S. Am. Earth Sci. 4, 239–260.

Schein, J.P., Poole, J.C., 2014. A shark-bitten dinosaur (Hadrosauridae) femur from the latest Maastrichtian basal Hornerstown Formation, New Jersey, U.S.A. Mosasaur 8, 15–22.

Scheyer, T.M., Delfino, M., Klein, N., Bunbury, N., Fleischer-Dogley, F., Hansen, D.M., 2018. Trophic interactions between larger crocodylians and giant tortoises on Aldabra Atoll, Western Indian Ocean, during the Late Pleistocene. R. Soc. Open Sci. 5, 171800. https://doi.org/10.1098/rsos.171800.

Schlüter, C., 1871. Cephalopoden der oberen deutschen Kreide. Palaeontographica 21, 1–24.

Schneider, V.P., Fierstine, H.L., 2004. Fossil tuna vertebrae punctured by istiophorid billfishes. J. Vertebr. Paleontol. 24, 253–255.

Schoch, R.R., Seegis, D., Mujal, E., 2022. The Middle Triassic vertebrate deposits of Kupferzell (Germany): Palaeoenvironmental evolution of complex ecosystems. Palaeogeogr. Palaeoclimatol. Palaeoecol. 603, 111181.

Schwimmer, D.R., 2010. Bite marks of the giant crocodilian *Deinosuchus* on Late Cretaceous (Campanian) bones. N. M. Museum Nat. Hist. Sci. Bull. 51, 183–190.

Schwimmer, D.R., Stewart, J.D., Williams, G.D., 1997. Scavenging by sharks of the genus *Squalicorax* in the Late Cretaceous of North America. Palaios 12, 71–83.

Scott, A.C., 1991. Evidence for plant-arthropod interactions in the fossil record. Geol. Today 7, 58–61.

Scott, E.E., Clemens, M.E., Ryan, M.J., Jackson, G., Boyle, J.T., 2012. A *Dunkleosteus* suborbital from the Cleveland Shale, northeastern Ohio, showing possible arthrodire-inflicted bite marks: evidence for agonistic behavior, or postmortem scavenging? Geol. Soc. Am. Abstr. Progr. 61.

Scott, W.B., Jepsen, G.L., 1936. The mammalian fauna of the White River Oligocene. Part I. Insectivora and Carnivora. Trans. Am. Phil. Soc. 28, 1–153.

Secord, R., Gingerich, P.D., Bloch, J.I., 2002. *Mylanodon rosei,* a new metacheiromyid (Mammalia, Palaeanodonta) from the late Tiffinian (late Paleocene) of northwestern Wyoming. In: Contributions of the Museum of Paleontology, vol 30. University of Michigan, pp. 385–399.

Seilacher, A., 1998. Mosasaurs, limpets or diagenesis: how *Placenticeras* shells got punctured. Mitteilungen aus dem Museum für Naturkunde in Berlin. Fossil Rec. 1, 93–102.

Serafini, G., Danise, S., Maxwell, E.E., Martire, L., Amalfitano, J., Miriam, C., Hohenstein, U.T., Giusberti, L., 2024. Of his bones are crinoid made: taphonomy and deadfall ecology of marine reptiles from a pelagic setting (Middle-Upper Jurassic of Northeastern Italy). Riv. Ital. Paleontol. Stratigr. 130 (1), 97–128. https://doi.org/10.54103/2039-4942/22314.

Sereno, P.C., Novas, F.E., 1994. The skull and neck of the basal theropod *Herrerasaurus ischigualastensis*. J. Vertebr. Paleontol. 13, 451–476.

Sereno, P.C., Larsson, H.C., Sidor, C.A., Gado, B., 2001. The giant crocodyliform *Sarcosuchus* from the Cretaceous of Africa. Science 294, 1516–1519.

Shakya, S.B., Fuchs, J., Pons, J.M., Sheldon, F.H., 2017. Tapping the woodpecker tree for evolutionary insight. Mol. Phylogenet. Evol. 116, 182–191.

Shimada, K., 1997. Paleoecological relationships of the Late Cretaceous lamniform shark *Cretoxyrhina mantelli* (Agassiz). J. Paleontol. 71, 926–933.

Shimada, K., Everhart, M.J., 2004. Shark–bitten *Xiphactinus audax* (Teleostei: Ichthyodectiformes) from the Niobrara Chalk (Upper Cretaceous) of Kansas. Mosasaur 7, 35–39.

Shimada, K., Tsuihiji, T., Sato, T., Hasegawa, Y., 2010. A remarkable case of a shark-bitten elasmosaurid plesiosaur. J. Vertebr. Paleontol. 30, 592–597.

Shubin, N.H., Daeschler, E.B., Coates, M.I., 2004. The early evolution of the tetrapod humerus. Science 304, 90–93.

Simpson, G.G., Elftman, H.O., 1928. Hind limb musculature and habits of a Palaeocene multituberculate. Am. Mus. Novit. 333, 19 p.

Sinclair, W.J., 1922. Entelodonts from the Big Badlands of South Dakota in the Geological Museum of Princeton University. Proc. Am. Phil. Soc. 60, 467–495.

Sinclair, W.J., Granger, W., 1914. Palaeocene deposits of the San Juan Basin, New Mexico. Bull. Am. Mus. Nat. Hist. 33, 297–316.

Siviero, B.C., 2019. Behavioral Causes of Paleopathologies in Cretaceous Edmontosaurus Annectens, Lance Formation, Wyoming. Doctoral dissertation, Loma Linda University, 211 p.

Slodownik, M., Wings, O., 2015. Bite marks on *Europasaurus* bones from the Langenberg Quarry near Goslar (Lower Saxony, Germany). In: 13th Annual Meeting of the European Association of Vertebrate Palaeontologists Opole, Poland, 8-12 July 2015, Abstracts, p. 141.

Slotta, F., Korn, D., Klug, C., Kröger, B., Keupp, H., 2011. Sublethal shell injuries in Late Devonian ammonoids (Cephalopoda) from Kattensiepen (Rhenish Mountains). Neues Jahrbuch Geol. Palaontol. Abhand. 261, 321–336.

Smith, M.M., Sansom, I.J., 1997. Exoskeletal micro-remains of an Ordovician fish from the Harding Sandstone of Colorado. Palaeontology 40, 645–658.

Smith, R.E., Martill, D.M., Longrich, N., Unwin, D.M., Ibrahim, N., Zouhri, S., 2023. Comparative taphonomy of Kem Kem Group (Cretaceous) pterosaurs of southeast Morocco. Evol. Earth 1, 100006. https://doi.org/10.1016/j.eve.2023.100006.

Stanley, D.J., 1971. Fish-produced markings on the outer continental margin. East of the Middle Atlantic States. J. Sediment. Petrol. 41, 159–170.

Stewart, J.D., Carpenter, K., 1990. Examples of vertebrate predation on cephalopods in the Late Cretaceous of the Western Interior. In: Boucot, A.J. (Ed.), Evolutionary Paleobiology of Behavior and Coevolution. Elsevier, Amsterdam, pp. 203–207.

Storrs, G.W., Oser, S.E., Aull, M., 2013. Further analysis of a Late Jurassic dinosaur bone-bed from the Morrison Formation of Montana, USA, with a computed three-dimensional reconstruction. Earth Environ. Sci. Trans. R. Soc. Edinburgh 103, 443–458. https://doi.org/10.1017/S1755691013000248.

Strganac, C., Jacobs, L.L., Polcyn, M.J., Mateus, O., Myers, T.S., Salminen, J., May, S.R., Araújo, R., Ferguson, K.M., Gonçalves, A.O., Morais, M.L., 2015. Geological setting and paleoecology of the Upper Cretaceous Bench 19 marine vertebrate bonebed at Bentiaba, Angola. Neth. J. Geosci. 94, 121–136.

Su, D.F., Harrison, T., 2008. Ecological implications of the relative rarity of fossil hominins at Laetoli. J. Hum. Evol. 55, 672–681.

Sues, H.D., Olsen, P.E., Carter, J.G., Scott, D.M., 2003. A new crocodylomorph archosaur from the Upper Triassic of North Carolina. J. Vertebr. Paleontol. 23, 329–343.

Sutcliffe, A.J., 1970. Spotted hyaena: crusher, gnawer, digester and collector of bones. Nature 227, 1110–1113.

Sutcliffe, A.J., 1973. Similarity of bones and antlers gnawed by deer to human artifacts. Nature 246, 428–430.

Swennen, C., Yu, Y.T., 2005. Food and feeding behavior of the blackfaced spoonbill. Waterbirds 28, 19–27.

Syed, R., Sengupta, S., 2019. First record of parrotfish bite mark on larger foraminifera from the Middle Eocene of Kutch, Gujarat, India. Curr. Sci. 116, 363–365.

Szrek, P., Salwa, S., Niedźwiedzki, G., Dec, M., Ahlberg, P.E., Uchman, A., 2016. A glimpse of a fish face—an exceptional fish feeding trace fossil from the Lower Devonian of the Holy Cross Mountains, Poland. Palaeogeogr. Palaeoclimatol. Palaeoecol. 454, 113–124.

Takeda, Y., Tanabe, K., 2014. Low durophagous predation on Toarcian (Early Jurassic) ammonoids in the northwestern Panthalassa shelf basin. Acta Palaeontol. Pol. 60, 781–794.

Tanke, D., Currie, P.J., 1998. Head-biting behavior in theropod dinosaurs: paleopathological evidence. Gaia 15, 167–184.

Tanke, D.H., Rothschild, B.M., 2014. Paleopathology in Late Cretaceous Hadrosauridae from Alberta, Canada with comments on a putative *Tyrannosaurus* bite injury on an *Edmontosaurus* tail. In: Eberth, D., Evans, D. (Eds.), Hadrosaurs. Bloomongton. Indiana University Press, pp. 540–571.

Tanke, D.H., Currie, P.J., Larson, P.L., 1992. Once bitten twice shy: Predator tooth marks on oreodont (Mammalia: Merycoidodontoidae) skulls, Middle ND Upper Oligocene, Brule Formation of South Dakota and Nebraska, USA. J. Vertebr. Paleontol. 12 (Suppl. to no. 3), 54–55A.

Taquet, P., Russell, D.A., 1999. A massively-constructed Iguanodont from Gadoufaoua, Lower Cretaceous of Niger. Ann. Palaeontol. 85, 5–96.

Tarrant, P.R., 1991. The ostracoderm *Phialaspis* from the Lower Devonian of the Welsh Borderland and South Wales. Palaeontology 34 (2), 399–438.

Tasnadi-Kubacska, A., 1962. Palaopathologie. Gustav Fischer Verlag, Jena, 269 p.

Tedford, R.H., Harington, C.R., 2003. An Arctic mammal fauna from the early Pliocene of North America. Nature 425, 388–390.

Thies, D., Rief, W.-E., 1985. Phylogeny and evolutionary ecology of Mesozoic Neoselachii. Neues Jahrbuch Geol. Palaontol. Abhand. 169, 333–361.

Thulborn, T., Turner, S., 1993. An elasmosaur bitten by a pliosaur. Mod. Geol. 18, 489–501.

Tichy, G., Urbanek, E., 2004. Biss-Spuren eines Sauriers an *Pinacoceras parma* Mojsisovics, ein Ammonit der Halleiner Obetrias. GeoAlp 1, 87–90.

Tischlinger, H., 2001. Bemerkungen zur Insekten Taphonomie der Solnhofer Plattenkalke. Archaeopteryx 19, 29–44.

Tobler, M.W., Carrillo-Percastegui, S.E., Powell, G., 2009. Habitat use, activity patterns and use of mineral licks by five species of ungulate in south-eastern Peru. J. Trop. Ecol. 25, 261–270.

Trifilio, L.H.M.D.S., de Araújo Júnior, H.I., Porpino, K.D.O., 2023. The paleoichnofauna in bones of Brazilian Quaternary cave deposits and the proposition of two new ichnotaxa. Ichnos 30 (3), 207–234.

Tsujita, C.J., Westermann, G.E., 1998. Ammonoid habitats and habits in the Western Interior Seaway: a case study from the Upper Cretaceous Bearpaw Formation of southern Alberta, Canada. Palaeogeogr. Palaeoclimatol. Palaeoecol. 144 (1–2), 135–160.

Tsujita, C.J., Westermann, G.E., 2001. Were limpets or mosasaurs responsible for the perforations in the ammonite *Placenticeras*? Palaeogeogr. Palaeoclimatol. Palaeoecol. 169, 245–270.

Tuuling, T., 2015. A Specimen of *Psammolepis* in the Gauja Formation (Estonia). B.Sc. thesis. University of Tartu, Tartu, 31 p.

Tyborowski, D., 2016. A new ophthalmosaurid ichthyosaur from the Late Jurassic of Owadów-Brzezinki Quarry, Poland. Acta Palaeontol. Pol. 61, 791–804.

Uchman, A., Torres, P., Johnson, M.E., Berning, B., Ramalho, R.S., Rebelo, A.C., Melo, C.S., Baptista, L., Madeira, P., Cordeiro, R., Avila, S.P., 2018. Feeding traces of recent ray fish and occurrences of the trace fosil *Piscichnus waitemata* from the Pliocene of Santa Maria Island, Azores (Northeast Atlantic). Palaios 33, 361–375.

Uhen, M.D., 2004. Form, function, and anatomy of Dorudon atrox (Mammalia, Cetacea): An archaeocete from the middle to late Eocene of Egypt. Univ. Michigan Pap. Paleontol. 34, 1–222.

Vallim, A.L., Schenone, S., Thrush, S.F., 2024. Megafauna: the ignored bioturbators. Mar. Ecol. Prog. Ser. 733, 137–144. https://doi.org/10.3354/meps14556.

Vasileiadou, K., Hooker, J.J., Collinson, M.E., 2007. Taphonomic evidence of a Paleogene mammalian predator–prey interaction. Palaeogeogr. Palaeoclimatol. Palaeoecol. 243, 1–22.

Vasileiadou, K., Hooker, J.J., Collinson, M.E., 2009. Paleocommunity reconstruction and accumulation of micromammalian remains (late Eocene, southern England). Palaios 24, 553–567.

Vasileiadou, K., Hooker, J.J., Collinson, M.E., 2022. Taphonomy of an Eocene micromammal assemblage in a lake-margin depositional setting elucidates an ancient food web. Palaeontol. Electron. 25 (2), a16. https://doi.org/10.26879/1214.

Vega-Dias, C., Schultz, C.L., 2007. Evidence of archosauriform feeding on dicynodonts in the Late Triassic of southern Brazil. PaleoBios 27, 62–67.

Vullo, R., 2011. Direct evidence of hybodont shark predation on Late Jurassic ammonites. Naturwissenschaften 98, 545–549.

Wahl, W.R., 2006. A juvenile plesiosaur (Reptilia: Sauropterygia) assemblage from the Sundance Formation (Jurassic), Natrona County, Wyoming. PALUDICOLA 5, 255–261.

Walsh, S.A., Hume, J.P., 2001. A new Neogene marine avian assemblage from north-central Chile. J. Vertebr. Paleontol. 21, 484–491.

Weinbaum, J.C., 2013. Postcranial skeleton of *Postosuchus kirkpatricki* (Archosauria: Paracrocodylomorpha), from the upper Triassic of the United States. Geol. Soc. Lond. Spec. Publicat. 379, 525–553.

Wiest, L.A., 2018. Taphonomic and Mass-Extinction Research from an Ichnological Perspective [Doctoral Dissertation]. Baylor University, 87 p.

Wiest, L.A., Esker, D., Driese, S.G., 2016. The Waco Mammoth National Monument may represent a diminished watering-hole scenario based on preliminary evidence of post-mortem scavenging. Palaios 31 (12), 592–606.

Williamson, W.E., 1996. ?*Brachychampsa sealeyi*, sp. nov., (Crocodylia, Alligatoroidea) from the Upper Cretaceous (Lower Campanian) Menefee Formation, northwestern New Mexico. J. Vertebr. Paleontol. 16, 421–431.

Wilson, M.A., Borszcz, T., Zatoń, M., 2014. Bitten spines reveal unique evidence for fish predation on Middle Jurassic echinoids. Lethaia 48, 4–9.

Wings, O., 2015. The Langenberg Quarry near Goslar: Unique window into a terrestrial Late Jurassic ecosystem in Northern Germany. Abstract Volume of the 12th Symposium on Mesozoic Terrestrial Ecosystems, pp. 99–100.

Winkler, D.A., 1987. Vertebrate-bearing eolian unit from the Ogallala Group (Miocene) in northwestern Texas. Geology 15, 705–708.

Wisshak, M., Knaust, D., Bertling, M., 2019. Bioerosion ichnotaxa: review and annotated list. Facies 65 (2), 24. https://doi.org/10.1007/s10347-019-0561-8.

Wretman, L., Kear, B.P., 2014. Bite marks on an ichthyodectiform fish from Australia: possible evidence of trophic interaction in an Early Cretaceous marine ecosystem. Alcheringa 38, 170–176.

Wright, J.L., Unwin, D.M., Lockley, M.G., Rainforth, E.C., 1997. Pterosaur tracks from the Purbeck Limestone Formation of Dorset, England. PGA (Proc. Geol. Assoc.) 108, 39–48.

Xing, L., Bell, P.R., Currie, P.J., Shibata, M., Tseng, K., Dong, Z., 2012. A sauropod rib with an embedded theropod tooth: direct evidence for feeding behaviour in the Jehol group, China. Lethaia 45, 500–506.

Xing, L., Rothschild, B.M., Randolph-Quinney, P.S., Wang, Y., Parkinson, A.H., Ran, H., 2018. Possible bite-induced abscess and osteomyelitis in *Lufengosaurus* (Dinosauria: sauropodomorph) from the Lower Jurassic of the Yimen Basin, China. Sci. Rep. 8, 5045. https://doi.org/10.1038/s41598-018-23451-x.

Yanenko, V., Kovalchuk, O., 2023. Late Miocene turtles of Grytsiv (western Ukraine) with rodent gnaw marks on the carapace surface. Zoodiversity 57 (4), 311–322. https://doi.org/10.15407/zoo2023.04.311.

Yang, S.-Y., Lockley, M.G., Greben, R., Erickson, B.R., Lim, S.-K., 1995. Flamingo and duck-like bird tracks from the Late Cretaceous and early Tertiary: evidence and implications. Ichnos 4, 21–34.

Yelinek, K.D., 2005. Taphonomy of *Daemonelix*: Implications for the Preservation of Terrestrial Vertebrate Burrows and Description of Associated Invertebrate Burrows. M.S. thesis. University of Colorado, Boulder, 50 p.

Yoshikawa, H., 2000. *Juglans ailanthifolia* Carr gnawed by large Japanese field mice, *Apodemus speciosus* (Temminck), from the middle Pleistocene Sahama Mud in Hanamatsu City, Shizuoka Prefecture, central Japan. Sci. Rep. Toyohashi Mus. Nat. Hist. 10, 23–30.

Young, M.T., Rabi, M., Bell, M.A., Foffa, D., Steel, L., Sachs, S., Peyer, K., 2016. Big-headed marine crocodyliforms and why we must be cautious when using extant species as body length proxies for long-extinct relatives. Palaeontol. Electron. 19 (3). https://doi.org/10.26879/648.palaeo-electronica.org/content/2016/1554-teleosaurid-size-estimation, 30A: 1-14.

Young, M.T., Hua, S., Steel, L., Foffa, D., Brusatte, S.L., Thüring, S., Mateus, O., Ruiz-Omeñaca, J.I., Havlik, P., Lepage, Y., De Andrade, M.B., 2014. Revision of the Late Jurassic teleosaurid genus *Machimosaurus* (Crocodylomorpha, Thalattosuchia). R. Soc. Open Sci. 1, 140222. https://doi.org/10.1098/rsos.140222.

Yun, C., 2019. Comments on the ecology of Jurassic theropod dinosaur *Ceratosaurus* (Dinosauria: Theropoda) with critical reevaluation for supposed semiaquatic lifestyle. Volumina Jurassica 17, 111–116.

Zangerl, R., Richardson, E.S., 1963. The paleoecological history of two Pennsylvanian black shales. Fieldiana, Geological Memoirs 4, 352 p.

Zapfe, H., 1981. Ein Schädel von Mesopithecus mit Biss-Spuren. Folia Primatol. 35, 248–258.

Zatoń, M., Villier, L., Salamon, M.A., 2007. Signs of predation in the Middle Jurassic of south-central Poland: Evidence from echinoderm taphonomy. Lethaia 40, 139–151.

Zeigler, K.E., Heckert, A.B., Lucas, S.G., 2002. A new species of Desmatosuchus (Archosauria: Aetosauria) from the Upper Triassic of the Chama Basin, north-central New Mexico. N. M. Museum Nat. Hist. Sci. Bull. 21, 215–219.

Zhao, W.-J., Zhu, M., 2010. Siluro-Devonian vertebrate biostratigraphy and biogeography of China. Palaeoworld 19, 4–26.

Zonneveld, J.P., 2022. On the status of the ichnotaxon *Mandaodonites coxii* Cruickshank, 1986. Ichnos 1–9. https://doi.org/10.1080/10420940.2022.2138378.

Zonneveld, J.P., Bartels, W.S., 2023. The occurrence and significance of ulcerative shell disease (USD) on geoemydid turtles, Middle Eocene, Southwestern Wyoming, United States. Palaios 38 (11), 453–473. https://doi.org/10.2110/palo.2023.009.

Zonneveld, J.-P., AbdelGawad, M.K., Miller, E.R., 2022a. Ectoparasite borings, mesoparasite borings, and scavenging traces in early Miocene turtle and tortoise shell: Moghra formation, Wadi Moghra, Egypt. J. Paleontol. 96 (2), 304–322. https://doi.org/10.1017/jpa.2021.92.

Zonneveld, J.P., Hasiotis, S., Gingras, M.K., 2022b. Tooth marks, gnaw marks, claw-marks, bite marks, scratch marks, etc: terminology in ichnology. Ichnos. https://doi.org/10.1080/10420940.2022.2058937.

Zverkov, N.G., Arkhangelsky, M.S., Stenshin, I.M., 2015. A review of Russian Upper Jurassic ichthyosaurs with an intermedium/humeral contact: Reassessing Grendelius McGowan, 1976. Proc. Zool. Inst. RAS 319, 558–588.

Further reading

Araújo-Júnior, H.I., Porpino, K.O., Bergqvist, L.P., 2011. Marcas de dentes de carnívoros/carniceiros em mamíferos pleistocênicos do Nordeste do Brasil. Rev. Bras. Palaontol. 14, 291–296.

Averianov, A., Dyke, G., Danilov, I., Skutschas, P., 2015. The paleoenvironments of azhdarchid pterosaurs localities in the Late Cretaceous of Kazakhstan. ZooKeys 483, 59–80.

BJorn, W.M., 1982. Sedimentology of the Tokomaru Formation, Waiapu Subdivision, Raukumara Peninsula. M.Sc. thesis. Auckland University, Auckland, 120 p.

Buffetaut, E., 1983. Wounds on the jaw of an Eocene mesosuchian crocodilian as possible evidence for the antiquity of crocodilian intraspecific fighting behavior. Paläontol. Z. 57 (1–2), 143–145.

Collareta, A., Cigala Fulgosi, F., Bianucci, G., 2019a. A new kogiid sperm whale from northern Italy supports psychrospheric conditions in the early Pliocene Mediterranean Sea. Acta Palaeontol. Pol. 64, 609–626.

Everhart, M.J., Everhart, P.A., Shimada, K., 1995. New specimen of shark bitten mosasaur vertebrae from the Smoky Hill Chalk (Upper Cretaceous) in western Kansas. Kansas Acad. Sci. Trans. 14, 19.

Gilbert, C.C., McGraw, W.S., Delson, E., 2009. Plio-Pleistocene eagle predation on fossil cercopithecids from the Humpata Plateau, southern Angola. Am. J. Phys. Anthropol. 139, 421–429.

Gorzelak, P., Salamon, M.A., Baumiller, T.K., 2012. Predator-induced macroevolutionary trends in Mesozoic crinoids. Proc. Natl. Acad. Sci. USA 18, 7004–7007.

Rivera-Sylva, H.E., Frey, E., Guzmán-Gutierrez, J.R., Palomino-Sánchez, F., Stinnesbeck, W., 2011. Un *Deinosuchus riograndensis* (Eusuchia: Alligatoroidea) de Coahuila, norte de México. Rev. Mex. Ciencias Geol. 28, 267–274.

CHAPTER

7

Regurgitalites

Adrian P. Hunt[1] *and Spencer G. Lucas*[2]

[1]Flying Heritage and Combat Armor Museum, Everett, WA, United States; [2]New Mexico Museum of Natural History, Albuquerque, NM, United States

1. Introduction

Regurgitalites are bromalites egested through the mouth (Hunt, 1992). They are the least studied bromalites, although there has been a notable increase in publications on them during the past few years. Regurgitalites have the least extensive fossil record among bromalites. Hunt and Lucas (2021b) recently provided an extensive review of the record of regurgitalites that is briefly reprised and updated here.

Hunt and Lucas (2012a) (Fig. 7.1) introduced the term purgolite for an accumulation of regurgitalites (accretionary, if concentrated by physical processes, and ethological, if by behavior) and others for specific types, including ejectalite (deriving from oral cavity or gastrointestinal tract anterior to the stomach), ekrhexalite (derives from the stomach), ornithoregurgitalite (produced by a bird), and strigilite (fossil owl pellets). Some of these cannot be readily discriminated in the fossil record (Gordon et al., 2020). Myhrvold (2011) coined the term emetolite for fossilized gastric pellets produced by routine vomiting (emesis). Bertling et al. (2022) recently revised their previous classification of trace fossils (Bertling et al., 2006) to explicitly include regurgitalites.

Abbreviations are:

FHPR, Utah Field House of Natural History State Park Museum, Vernal, Utah, USA; NMMNH, New Mexico Museum of Natural History, Albuquerque, New Mexico, USA; OUM, Oxford University Museum of Natural History, Oxford, England; SMF, Senckenberg Research Institute Frankfurt, Frankfurt, Germany; SMNS, Staatliches Museum für Naturkunde, Stuttgart, Germany.

2. Regurgitation

The vomiting reflex is common in vertebrates and is largely a protective mechanism for the bulk ejection of noxious material (Sims et al., 2000). Habitual regurgitation occurs in a variety

Vertebrate Ichnology
https://doi.org/10.1016/B978-0-443-38351-9.00009-8

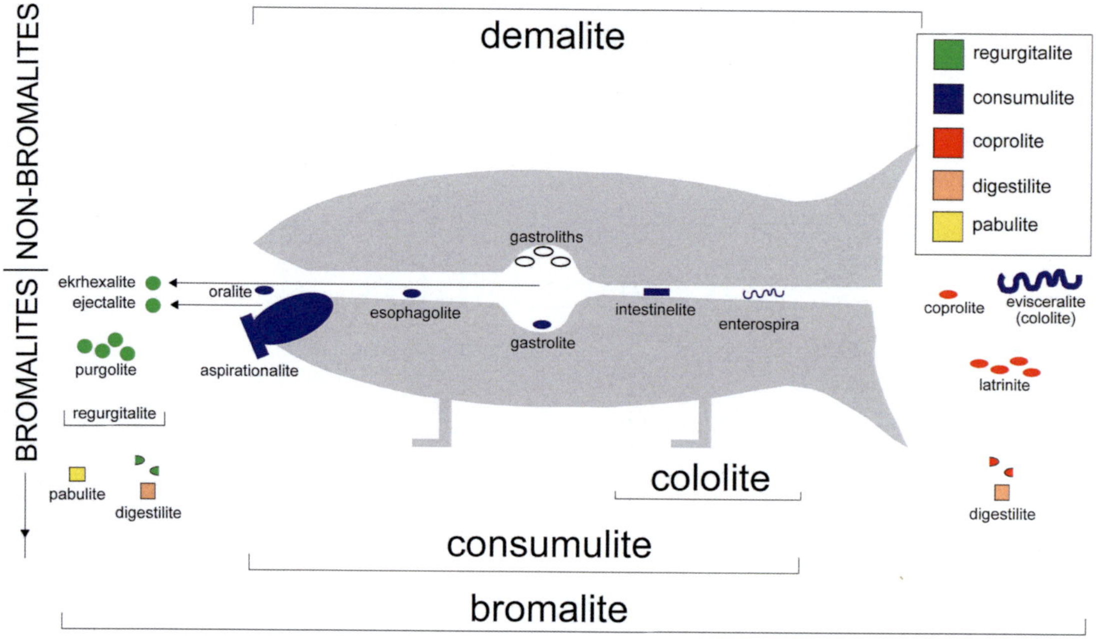

FIGURE 7.1 Selected classification terms for regurgitalites and associated trace fossils. *From Hunt and Lucas (2021) (Fig. 3).*

of recent carnivorous, piscivorous and insectivorous species that routinely ingest food with a high proportion of indigestible material that is egested through the mouth (Schäfer, 1972; Myhrvold, 2011). The other principal expulsion mechanism is stomach eversion, which occurs in sharks and rays and, possibly, some amphibians (e.g., Sims et al., 2000; Brunnschweiler et al., 2005). The range of reasons for regurgitation include: (1) eradicating poison; (2) removing mechanically dangerous matter; (3) feeding offspring; (4) removing parasites; (5) eliminating indigestible material; and (6) getting rid of sloughed gastric mucosa and mucus (Gudger, 1949; Beintema, 1991; Sims et al., 2000; Klug and Vallon, 2018; Hoffmann et al., 2019).

Many bony fish and chrondrichthyans regurgitate indigestible material (e.g., Hattin, 1988; Bowman, 1986; Andrews et al., 1998; Zatoń and Salamon, 2008; Aas et al., 2017). Most carnivorous snakes and many lizards egest gastric pellets and other materials such as eggshell fragments and bird remains (Gans, 1952; Reinhard and Vogel, 1980; Fathinia et al., 2009; Myhrvold, 2011; Bury and Drohobycka-Wawryka, 2020). Long et al. (2010) reported regurgitation in turtles. Varanids produce gastric pellets (Petzold, 1967; Auffenberg, 1981), as do many crocodilians (Dolowy et al., 1960; Scherpner, 1980; Fisher, 1981a, 1981b; Chabreck, 1996; Andrews et al., 2000; Myhrvold, 2011). Virtually all birds that eat invertebrate or vertebrate animals egest pellets, and the majority of birds regurgitate to feed their young (e.g., Rea, 1973; Duke et al., 1976; Andrews, 1990; Zijlstra and Van Eerden, 1995; Myhrvold, 2011). Some marine mammals, whales and most pinnipeds produce gastric pellets (e.g., Fea and Harcourt, 1997; Goodman-Lowe, 1998; Clarke et al., 1998).

3. Actualistic studies

In the early 1970s, it was recognized that avian regurgitalites are important in the development of microvertebrate accumulations (Dodson, 1973; Dodson and Wexlar, 1979; Mellett, 1975; Mayhew, 1977), and since that time, there have been extensive actualistic studies of recent regurgitations. Peter Andrews has been prominent in this work with authorship of two extensive volumes (Andrews, 1990; Fernández-Jalvo and Andrews, 2016). Selected recent works that provide access to this large literature include Fernández-Jalvo and Andrews (2016), Fernández-Jalvo et al. (2016), Denys et al. (2018) and Montalvo and Fernández (2019).

3.1 Identification of regurgitalites

Hunt (1992) originally proposed the encompassing term bromalite in part because it can be difficult to distinguish a regurgitalite from a coprolite. The majority of studies have identified bromalites as one or the other, and some regurgitalites may have been mistaken for coprolites (Myhrvold, 2011; Vallon, 2012). Recent raptor pellets can be identified as to originator (e.g., Moon, 1940; Terry, 2007, 2010). Regurgitalites have no unique characteristics, but they can be recognized on the basis of a suite of features (Wilson, 1987; Myhrvold, 2011; Thies and Hauff, 2012; Hunt et al., 2015; Klug and Vallon, 2018; Hoffmann et al., 2019; Gordon et al., 2020; Freimuth, 2020; Freimuth et al., 2021; Hunt and Lucas, 2021; Foster et al., 2022):

1. Geometry. Putative regurgitalites range from three-dimensional cylindrical or amorphous bodies to splatters of associated organic matter in a discrete area on a bedding plane (e.g., Hunt et al., 2012a,b; Hoffmann et al., 2019; Gordon et al., 2020; Foster et al., 2022).
2. Nonbiotic content. Regurgitalites contain a relative (to coprolites) lack of phosphatic matrix, and some specimens have none (Hunt et al., 2015; Klug and Vallon, 2018; Gordon et al., 2020; Freimuth et al., 2021; Foster et al., 2022).
3. Biotic content.
   a. Physicochemical characteristics. The principal inclusions are elements that are not easy to digest, so they may be preferentially evacuated (e.g., vertebrate skeletal elements, invertebrate valves and cuticle, dermal elements—fur and feathers) (Myhrvold, 2011; Hunt et al., 2015; Hoffmann et al., 2019). It is important to note that some hard parts, such as invertebrate shells, can pass through the digestive tracts of durophagous fish and some birds (Cate and Evans, 1994; Zuschin et al., 2003). Nevertheless, the reverse could also be true, and regurgitalites could be characterized by materials that are easy to digest and that would not be present in coprolites (Gordon et al., 2020).
   b. Size. Hard parts are relatively larger and articulated more often in regurgitalites than in coprolites (Hoffmann et al., 2019).
   c. Evidence of digestion. Regurgitalites should demonstrate evidence of digestion, but less than in coprolites (Bochenski et al., 1993, 1998; Hockett, 1996; Hunt et al., 2015; Hoffmann et al., 2019; Gordon et al., 2020).
      i. Physical—broken or rounded elements.
      ii. Chemical—etched or pitting.

d. Dentalites. Evidence of predation is likely to be better preserved in elements in regurgitalites that have shorter residence time in the digestive tract than in coprolites (Hunt et al., 2015).
e. Element composition. Recent and fossil avian pellets can be characterized by a prevalence of skulls, articulated elements, and the association of one or more discrete skeletons (Myhrvold, 2011; Hoffmann et al., 2019; Freimuth et al., 2021).
f. Element orientation. Elements are often aligned about their long axes and closely packed (Myhrvold, 2011; Holgado et al., 2015).

4. Fossil record

Most identified regurgitalites occur in marine or lacustrine environments characterized by low energy and fine-grained sediments (Hoffmann et al., 2019; Hunt and Lucas, 2021b). Overviews of the fossil record of regurgitalites were provided by Thies and Hauff (2012), Klug and Vallon (2018), Hoffmann et al. (2019) and Hunt and Lucas (2021b).

4.1 Paleozoic

The earliest putative regurgitalites are from the Late Ordovician Soom Shale Lagerstätte of South Africa (Aldridge et al., 2006). Aldridge et al. (2006) described compact, pellet-like clusters of fragmented conodont elements that they tentatively interpreted as regurgitalites produced by conodont animals.

A regurgitalite, likely produced by a nautiloid, occurs in the Silurian of Poland (Brachaniec et al., 2015). This specimen consists of fragmented and intermingled angular and nonabraded mollusc, brachiopod, and crinoid remains (Brachaniec et al., 2015).

A definitive vertebrate regurgitalite is from the Early Devonian locality of Duntrune in Scotland (Burrow and Turner 2010). From this locality, Burrow and Turner (2010) described and illustrated a bromalite that includes four tooth whorls, two fin spines, a scapulocoracoid, flank scales, tesserae, and umbellate scales of one individual of the acanthodian fish *Nostolepis scotica*. They also mentioned several hundred similar specimens from the important fish localities at both Duntrune and Tillywhandland, each of which includes elements from a single specimen of *Ischnacanthus gracilis* or *Mesacanthus mitchelli*. Klug and Vallon (2018) reported three regurgitalites containing cephalopod material from the latest Devonian of Morocco. They contain dissolved ammonite shell fragments and jaws of *Mimimitoceras* and are tentatively ascribed to a gnathostome.

Robin et al. (2022) analyzed vertebrate predation in the Late Devonian (late Famennian) freshwater ecosystem of Strud in Belgium by documenting both dentalites and regurgitalites. They identified 12 regurgitalites, which were chemically analyzed and imaged using scanning electron microscopy. Acanthodian spines and scales occur in almost all of the regurgitalites, but actinopterygian and sarcopterygian elements never cooccur in the bromalites.

Salamon et al. (2014; Figs. 6 and 7) reported one regurgitalite from the Devonian and nine from the Mississippian composed of shell fragments. Angular shell fragment debris may indicate the presence of durophagous fish, so the distribution of such material should parallel that of shell-rich regurgitalites (Oji et al., 2003; Salamon et al., 2014). This seems to be the case in the Devonian to the Mississippian (Salamon et al., 2014; Fig. 7). Thus, based on the work of Oji et al. (2003), we predict that there was a rise in such regurgitalites during the Paleogene and a major increase in the Neogene.

Regurgitalites are notable in several Pennsylvanian units of the United States. The Mississippian Bear Gulch Lagerstätte of Montana includes the ostracode-bearing regurgitalite *Ostracobromus snowyensis* (Hunt et al., 2012a, Fig. 7.2A), which is also present in the Pennsylvanian Hamilton Lagerstätte (Cunningham et al., 1993, p. 230). Zangerl and Richardson (1963) described faunas from the Early Pennsylvanian Carbondale Formation of Indiana. They distinguished two types of regurgitalites based on the degree of processing in the digestive tract—ejected prey (mastication, minor digestion) and gastric residues (more extensive digestion) (Zangerl and Richardson, 1963). However, the putative regurgitalites containing vertebrate material probably mainly represent decayed fish instead of regurgitalites (Elder, 1985; Elder and Smith, 1988). Nevertheless, some of the accumulations of invertebrate debris (e.g., of the bivalve *Myalina*) do probably represent vertebrate regurgitalites (e.g., Zangerl and Richardson, 1963, pl. 44D).

Soler-Gijón and Ruiz (2023) described a rare specimen of a regurgitalite (or coprolite) containing a fragmentary insect wing. This specimen derives from Carboniferous (Stephanian) strata of the Puertollano basin in Spain.

The Missourian Atrasado Formation at the Kinney Brick Quarry and Tinajas Lagerstätte in New Mexico, USA, yield large samples of bromalites, including the regurgitalite *Conchobromus kinneyensis*, which contains abundant conchostracans (Hunt et al., 2012a,b,c; Hunt and Lucas, 2021a: Fig. 7.2B).

FIGURE 7.2 Holotypes of regurgitalites from the Pennsylvanian of the western United States. (A) NMMNH P-54134, holotype of *Ostracobromus snowyensis* (Hunt et al., 2012a; Fig. 2D and E) Bear Gulch Limestone Member of the Tyler Formation (Mississippian: Chesterian), Bear Gulch Lagerstätte, Montana, USA. (B) NMMNH P-19157, holotype of *Conchobromus kinneyensis* (Hunt et al., 2012b), from Upper Pennsylvanian Tinajas Member of the Atrasado Formation (Missourian: Kasimovian), Kinney Brick Quarry, New Mexico, USA (Hunt, 1992, Fig. 1H; Hunt et al., 2012b, Fig. 3J).

4.2 Mesozoic

Salamon et al. (2012) described a number of putative regurgitalites from the Middle Triassic Gogolin Formation in southern Poland. These are discrete accumulations composed primarily of angular bivalve shell fragments with sharp, nonabraded margins and crinoid ossicles with many breaks. They are interpreted to represent regurgitalites of durophagous vertebrates, including sharks, colobodontid fish, and placodont and pachypleurosaur reptiles (Salamon et al., 2012). Another possible regurgitalite from the same formation is fusiform in shape, composed of fragmented bivalve shells and, based on its size, is attributed to the sharks *Acrodus* or *Palaeobates* (Niedźwiedzki et al., 2021).

The Dingxiao area of Xingyi City in Guizhou Province, South China, yields abundant fossils of Ladinian (Middle Triassic) age including specimens of the pachypleurosaur sauropterygian *Keichousaurus hui*. Ye et al. (2024) describe disarticulated bones of at least three individuals of this taxon as a regurgitalite attributed to the sauropterygians *Nothosaurus* or *Lariosaurus*.

Another Ladinian occurrence of regurgitalites is in the lacustrine Erfurt Formation, exposed in a commercial limestone quarry near Vellberg in Germany. Small tetrapod bones are often found disarticulated but in discrete associations. Schoch and Seegis (2016) interpreted such accumulations as regurgitalites, which were possibly produced by temnospondyls living in the lake.

The Reingraben Shales in Austria yield a Late Triassic (early Carnian) Konservat-Lagerstätte that contains regurgitalites (Lukeneder et al., 2020; Lukeneder and Lukeneder, 2021) of invertebrate debris, no matrix, and evidence of acid etching (Lukeneder et al., 2020). The constituents are entire shells and fragments of the ammonoid *Austrotrachyceras* and rare teuthid arm hooks and buccal cartilage of *Phragmoteuthis* (Lukeneder et al., 2020). The regurgitalites are attributed to durophagous sharks such as *Acrodus* (Lukeneder et al., 2020; Lukeneder and Lukeneder, 2021).

Dalla Vecchia et al. (1989) described a regurgitalite from the Late Triassic of northern Italy as containing pterosaur skeletal elements, but they were subsequently reinterpreted as pertaining to a protorosaurian similar to *Langobardisaurus* (Holgado et al., 2015). The producer is hypothesized to be a large fish such as *Saurichthys*, *Birgeria* or a coelacanthiform (Holgado et al., 2015). Gordon et al. (2020) described another Late Triassic regurgitalite from Arizona that is composed of a compact mass of skeletal material of the pseudosuchian archosaur *Revueltosaurus*. They interpret the producer as a phytosaur, rauisuchid, or temnospondyl.

The Lower Jurassic Posidonienschiefer Formation of southern Germany yields a number of regurgitalites. Keller (1977; also see Vallon, 2012) considered a coiled-up skeleton of a 1.6-m-long specimen of the ichthyosaur *Stenopterygius quadriscissus* as a regurgitalite produced by the large predatory ichthyosaur *Temnodontosaurus* (=*Leptopterygius*). Jäger (2001, 2005) interpreted disarticulated bones of a small ichthyosaur to be another regurgitalite produced by *Temnodontosaurus*. The type specimen of the small ornithischian *Emausaurus ernsti* may also represent a regurgitalite produced by a large ichthyosaur or marine crocodile (Haubold, 1990; Thies and Hauff, 2012). Thies and Hauff (2012) described a fourth regurgitalite that contains the remains of four specimens of the actinopterygian *Dapedium* sp. and a lower jaw of *Lepidotes* sp. It may have been produced by a chondrichthyan (*Hybodus*), actinopterygians (pachycormiforms), or marine reptiles (crocodilians, ichthyosaurs, plesiosaurs).

Cooper et al. (2022) figured three additional fish-bearing regurgitalites with single inclusions of *Dapedium* sp., *Pholidophorus* sp., and a juvenile *Pachycormus macropterus*. A large accumulation of disarticulated and brecciated bones containing the anterior portion of the *Saurostomus esocinus* and remains referable to *Saurorhynchus* sp. and *Leptolepis* sp. is interpreted as a regurgitated slurry (Cooper and Maxwell., 2022). Cooper et al. (2024) note a specimen of an undescribed coccolepidid, which consists of an articulated skull, with some disarticulated elements of the postcranial skeleton in a concentration behind the operculum. They interpret it as a regurgitalite produced by a small ichthyosaur, a hybodontiform shark, or even a large coleoid cephalopod.

Jäger (1991) described a "Seeball" from Holzmaden, composed of spines of the echinoid *Diademopsis*, that Hoffmann et al. (2019) considered to be a regurgitalite. Böttcher (1989, 1990) also noted regurgitalites from Holzmaden, and Frentzen (1936) described the ammonite *Amaltheus* from what he considered to be a fish consumulite, but it is more likely a regurgitalite.

Pinna et al. (1985) and Garassino and Donovan (2000) described regurgitalites from the Early Jurassic of Italy. They contain fish centra and scales and coleoid hooks, and they were ascribed them to thylacocephalan crustaceans, but they could pertain to vertebrates.

Zatoń and coworkers (Zatoń et al., 2007; Zatoń and Salamon, 2008) described regurgitalites from the Middle Jurassic of Poland. Nine specimens are composed of mollusks (scaphopods, gastropods, bivalves, ammonites, and belemnites), articulate brachiopods and echinoderms (asteroids, crinoids, and echinoids), and the probable producers are palaeospinacid sharks, although other vertebrates such as durophagous pycnodontiform fish, cannot be excluded (Zatoń et al., 2007). Zatoń and Salamon (2008) described another 11 specimens that contain remains of diverse taxa, including gastropods, scaphopods, bivalves, belemnites, ammonites, brachiopods, abryozoan, crinoids, echinoids, ophiuroids, and even wood, which they interpreted to have been produced by an opportunistic generalist, most likely a fish.

The Middle Jurassic of the Bielefeld area of northern Germany has yielded possible regurgitalites of marine reptiles (Wetzel, 1953, 1964; Horstmann and Maier, 1957). These contain juvenile ammonites, belemnites, cephalopods, echinoderms, scaphopods and foraminiferans with smaller quantities of fish scales, crustacean limbs, and reptile teeth (Wetzel, 1953, 1964; Horstmann and Maier, 1957).

The majority of regurgitalites from the Late Jurassic derive from the Nusplingen and Solnhofen Lagerstätten of Germany (e.g., Schweigert, 2023). Stevens et al. (2014) described four closely associated belemnites from Nusplingen that they interpreted as a regurgitalite produced by an elasmobranch, holocephalan, or marine reptile. Other Nusplingen regurgitalites contain echinoid (*Pseudodiadema*, *Nenoticidaris*, *Plegiocidaris*) and sea star (*Sphaeraster*) remains that do not occur elsewhere in these limestones, and they are attributed to the pycnodont fish *Gyrodus* (Dietl and Schweigert, 2001). Dietl and Schweigert (2001) and Schweigert et al. (2001) reported a regurgitalite that contained bones of the pterosaur *Rhamphorhynchus* (Fig. 7.3), possibly produced by a crocodile, *Cricosaurus* (*Geosaurus*) or *Dakosaurus*, or large predatory fish such as *Caturus* (Hoffmann et al., 2019) or possibly *Aspidorhynchus* (Frey and Tischlinger, 2012). Other Nusplingen regurgitalites contain the echinoid *Plegiocidaris* and the fishes *Tharsisdubius* and *Caturus* (Grawe-Baumeister et al., 2000; Vallon, 2012; Albersdörfer and Häckel, 2015; Viohl, 2015; Hoffmann et al., 2019).

FIGURE 7.3 Regurgitalite from the Upper Jurassic Nusplingen Lithographic Limestone of Germany. It (SMNS Inv. Nr. 63990) contains bones of a pterosaur, probably *Rhamphorhynchus*, and is interpreted as having been produced by a large fish or a marine crocodile (Schweigert et al., 2001). The regurgitalite is approximately 4 cm tall. *Image courtesy of Günter Schweigert.*

Six coiled vertebral columns of the actinopterygian *Leptolepis sprattiformis* from the Solnhofen Lagerstätte represent a regurgitalite (Janicke and Schairer, 1970). Janicke (1970) and Barthel (1978) described regurgitalites from Solnhofen containing the actinopterygian *Caturus* and interpreted them as produced by a large fish or crocodile. Hoffmann et al. (2019) listed other fish regurgitalites from the Solnhofen strata, including: (1) multiple mollusks, among them ammonites, from *Lepidotes* (Wellnhofer, 1964; Barthel and Janicke, 1970); (2) lamellaptychi from sharks (Barthel and Janicke, 1970); and (3) lamellaptychi from Holostei, Amiiformes (e.g., *Caturus*) or Aspidorhynchiformes (Mehl, 1978).

There are two Solnhofen pterosaurs associated with possible regurgitalites. The first is represented by a loose spiral of an articulated vertebral column of *Leptolepides* preserved adjacent to the mandibular tip of a specimen of *Scaphognathus crassirostris* (Bennett, 2014). This fish skeleton may have been a regurgitalite, or it may have been ejected from the animal during the decay process (Bennett, 2014). The second is a specimen of "*Pterodactylus propinquus*" from the Late Jurassic of Germany (now lost) that preserves disarticulated fish debris around the gular pouch, which could have been regurgitated from the gut (Broili, 1938; Wellnhofer, 1970; Witton, 2018).

Two specimens of the pterosaur *Kunpengopterus sinensis* from the Late Jurassic Tiaojishan Formation of China are associated with regurgitalites (Jiang et al., 2022). Both regurgitalites contain scales of an unnamed palaeonisciform fish.

The Upper Jurassic lithographic limestone Lagerstätte of the Causse Méjean in southern France has only been extensively sampled very recently (Moreau et al., 2022). Regurgitalites include concentrations of echinoid elements produced by fish (Moreau et al., 2022; Fig. 15).

Foster et al. (2022) (Fig. 7.4) described a regurgitalite from the Upper Jurassic Morrison Formation from Utah in the USA. This small bromalite was probably produced by a fish or semiaquatic mammal and consists of an accumulation of a small concentration of lissamphibian bones that are otherwise rarely preserved in the Morrison.

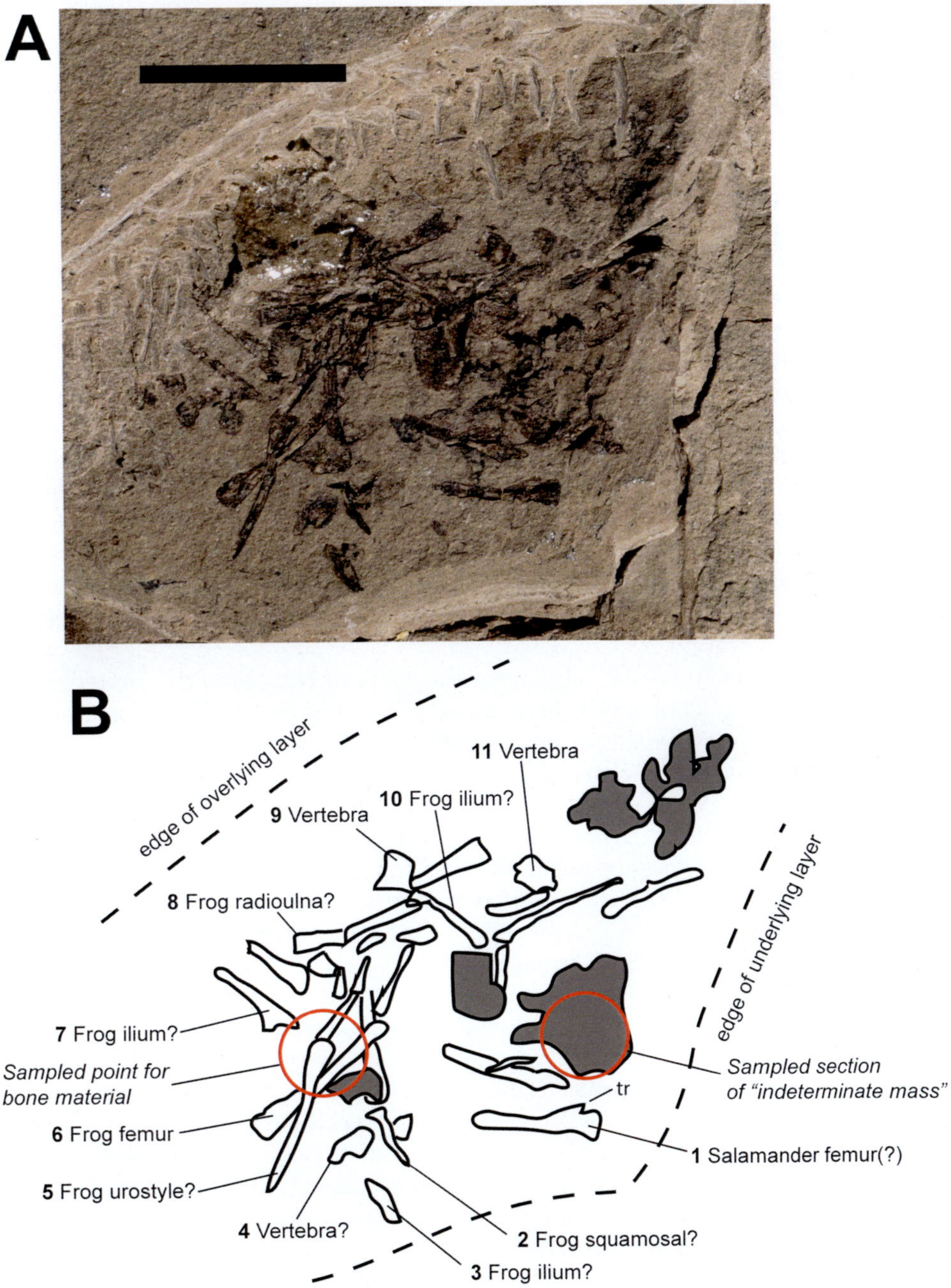

FIGURE 7.4 Regurgitalite (FHPR 16914) from the Upper Jurassic Salad Bar locality, Brushy Basin Member of the Morrison Formation, San Juan County, Utah, USA. (A) Regurgitalite bone cluster, scale bar is 5 mm with lighting from the lower right. (B) Diagram of elements in A, with possible identifications labeled. Gray areas are indeterminate, nonbone material. Red circles indicate areas scanned with XRF for geochemical analysis. Scale is the same as for A. Abbreviation in B, tr is trochanter. *From Foster et al. (2022) (Fig. 2).*

A regurgitalite from the latest Kimmeridgian or earliest Tithonian of the Rosso Ammonitico Veronese of northeastern Italy contains the postcranial remains of a small teleosauroid (Serafini et al., 2022). Its preservational condition (grouping of skeletal elements, acid etching) is distinct from other thalattosuchians from this unit (Serafini et al., 2023).

Sanz et al. (2001) described a regurgitated pellet from the Early Cretaceous Los Hoyas Lagerstätte in Spain. This bromalite consists of partial skeletons of four juvenile birds and may have derived from a nonavian theropod or a pterosaur (Sanz et al., 2001). Pellets occur in several birds from the Jehol Lagerstätte in China (Wang et al., 2016; O'Connor, 2019). All appear to be incorporeal pelletites and not egested. Probable regurgitalites composed of fish scales and bones also occur with three specimens of the Jehol troodontid *Anchiornis huxleyi* (Zheng et al., 2018b).

The majority of Late Cretaceous regurgitalites contain vertebrate elements, but Neumann (2000) described subrounded or elliptical lenses consisting of densely packed asteroid ossicles and/or other echinoderm remains from chalk in Germany. These pellets are interpreted as possible regurgitalites produced by bony fishes or sharks (Neumann, 2000). Similar bromalites have been found in the Late Cretaceous and Danian of northern Europe, although some have been attributed to large asteroideans (Wright and Wright, 1940; Rasmussen, 1950; Müller, 1953; Gale, 1987; Neumann, 2000).

Wetzel (1960, 1964) described possible regurgitalites from the Late Cretaceous of Chile. They contain ammonite larvae and adult *Baculites* as well as bivalves and plankton, and he ascribed them to plesiosaurs.

Hattin (1996) described a regurgitalite from the chalk of Kansas in the United States. This bromalite includes not only bone, but also cirriped plates and coccoliths and was attributed to the chondrichthyan *Ptychodus*. Capasso (2019) described a regurgitalite just exterior to the oral cavity of the pycnodont *Acrorhinichthys poyato*i from the Middle Cenomanian of Lebanon.

Everhart (2017) noted the occurrence of similar bromalites in the Kansas Cretaceous. Marine deposits of the Late Cretaceous of the Western Interior Seaway have yielded a variety of other regurgitalites. Bishop (1975) illustrated two regurgitalites, one containing disarticulated fragments of the crab *Dakoticancer* and the other of the mud shrimp *Callianassa*. Regurgitalites from Kansas include heavily-pitted bones of fish, marine reptiles (several mosasaurs, one plesiosaur) and the dinosaur *Niobrarasaurus* (Everhart, 1999, 2003, 2004, 2017). A regurgitalite consisting of the front part of a juvenile mosasaur skull has teeth that are eroded down to the roots (Everhart, 2017; Fig. 3). An isolated mandible of the ornithocheiroid *Pteranodon* preserves a pellet between the mandibular rami, probably held in place by throat tissues before burial (Brown, 1943; Bennett, 2001; Witton, 2018). This specimen preserves several fish vertebrae (Bennett, 2001). Stewart and Carpenter (1990) noted two other possible regurgitalites from Kansas: (1) a juvenile specimen of the ammonite ?*Clioscaphites choteauensis* with apatitic matrix in the outer whorl; and (2) a mass of teleost bones, batoid denticles and fragments of a teuthidid gladius. Martin and Tate (1976) described material associated with *Baptornis advenus* that could represent a regurgitalite (or a coprolite or evisceralite).

Longrich et al. (2022) examined a large sample from the Maastrichtian phosphate deposits of Morocco of acid-etched specimens of vertebrates including the mosasaurids *Halisaurus arambourgi*, *Gavialimimus almaghrebensis*, *Eremiasaurus heterodontus*, an elasmosaurid plesiosaur and fish including *Enchodus libycus* and *E. elegans*. They interpreted these specimens, that also include decalcified teeth, as regurgitalites produced by *Thalassotitan atrox*. Cooper et al. (2022) described three jaw fragments of the pycnodont *Phacodus punctatus* from the same strata that display severe pitting, corrosive marking and enamel discoloration and interpreted these specimens as regurgitalites probably produced by a large mosasaur such as *Prognathodon*.

Dalla Vecchia and Chiappe (2002) described a bird from the Late Cretaceous Nammoûra Lagerstätte of Lebanon. This skeleton is preserved in a ball-like mass of powdered bone mixed with carbonized feathers and represents a regurgitalite (Dalla Vecchia and Chiappe, 2002; Myhrvold, 2011).

Three multiindividual aggregates of mammalian skeletons from the Late Cretaceous nesting locality of Egg Mountain in the United States represent regurgitalites (Freimuth, 2020; Freimuth et al., 2021). One yields two individuals of the multituberculate *Filikomys primaevus*, and the other two are composed of the remains of 3 and 11 individuals, primarily of the marsupiali *Alphadon halleyi*. These regurgitalites are interpreted as being produced by *Troodon formosus* on the basis of abundant shed teeth and nesting evidence, and this would be consistent with previous inferences of this predator as having a diet of small-bodied prey, manipulating prey during feeding, heightened metabolic processes, and potential nocturnality (Freimuth et al., 2021).

4.3 Cenozoic

The only described specimen from the Paleocene is an ornithoregurgitalte (*sensu* Hunt and Lucas, 2012a) from the Salamanca Formation of Argentina at Punta Peligro (Muzzopappa et al., 2021). This orithoregurgitalite contains a single skeleton of the anuran *Calyptocephalella sabrosa*.

Eocene regurgitalites occur in shallow lacustrine facies of the western United States and Canada. Wilson (1977a, b, 1980, 1987) studied Middle Eocene freshwater vertebrate fossils from the interior of British Columbia and northern Washington. In 25 vertebrate fossil assemblages he noted that up to 69% of fish remains occurred in pellets ("fish-bone coprolites"), which he interpreted as principally avian regurgitalites (Wilson, 1987). Given the large number of these pellets, we deem it much more likely that these regurgitalites derive from fish. Buskirk et al. (2015, 2016) described several morphotypes of bromalites from the Middle Eocene Florissant Formation of Colorado, including regurgitalites that are ovoid (rarely sinusoidal) in shape with little to no groundmass and with a high content of crushed invertebrate shells—these are likely regurgitalites.

The Eocene Messel Lagerstätte in Germany yielded a regurgitalite that resembles Recent owl pellets, and, thus, it is considered to represent the oldest strigilite (Mayr and Schaal, 2016; Fig. 7.5C). The strigilite has a maximum length of 22.5 mm and is composed largely of bones of a wren-sized bird, some of which are pitted, and it also includes remains of feathers (Mayr and Schaal, 2016). Fossils of unambiguous raptorial birds are rare in the Messel avifauna, but one species of small owl, *Palaeoglaux artophoron*, is known from two individuals (Mayr and Schaal, 2016). The regurgitalite is in the lower size spectrum to be expected for a gastric pellet of this owl (Mayr and Schaal, 2016). Early Eocene owls were probably diurnal, and the fact that the Messel strigilite contains a bird rather than a small mammal may be support for this scenario (Mayr and Schaal, 2016; Mayr and Kitchener, 2023).

FIGURE 7.5 Regurgitalites from the early Eocene Messel Lagerstätte of Germany. (A) Anterior of the skeleton of the hyaenodontid mammal *Lesmesodon edingeri* (SMF-ME 3843a) from a regurgitalite probably produced by a large booid snake (*Eoconstrictor fischeri*). (B) Regurgitalite (SMF-ME, 813a) of a snake containing the skeleton of a small adult bird. (C) Strigilite (SMF-ME 11309a) that may have been produced by the owl *Palaeoglaux artophoron* containing a small bird? *Images B and C courtesy of Krister Smith and Anika Vogel.*

Two other gastric pellets with bird remains from Messel have more elongated shapes and probably derive from snakes or other squamates (Mayr and Schaal, 2016; Gunnell et al., 2018: Fig. 7.5A–B). The only other regurgitalite reported from Messel includes a skeleton of the hyaenodontid mammal *Lesmesodon edingeri* and was probably produced by a large boid snake (Morlo et al., 2012; Schaal, 2016: Fig. 7.5A).

The Eocene *Omomys* Quarry in Wyoming in the United States consists almost exclusively of bones of the eponymous primate as well as bones of birds, including owls. It may be a strigilite purgolite (Murphey et al., 2001; Alexander and Burger, 2001).

A specimen of the owl *Stihanus* from the latest Eocene Peanut Peak Member of the Chadron Formation in South Dakota is a possible strigilite (Hunt and Lucas, 2007a). Lucas et al. (2012) (Fig. 7.6) described a purgolite composed of strigilites from the early Oligocene of eastern Wyoming, first mentioned by Walton (1990) and Korth and Emry (1991). The strigilites are preserved as calcareous nodules full of fossil mammal bones and yielded the type specimens of the sciurid rodent *Cedromus wilsoni* and the todid bird *Palaeotodus emryi*. The strigilite assemblage also includes the skeletons, bearing skulls, of at least three individual owls, which strongly reinforces the interpretation of these small masses of bone as owl pellets (Lucas et al., 2012). Other possible Oligocene occurrences from Wyoming include a skeleton of the insectivore *Centenodon chadronensis* that exhibits features suggestive of it originating in

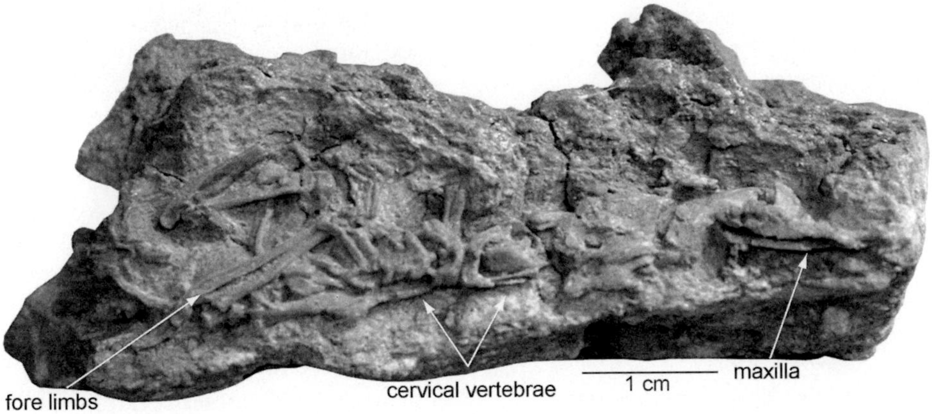

FIGURE 7.6 Strigilite with anterior portion of skeleton of a small rodent from the Orella Member of the White River Formation (early Oligocene), Wyoming, USA (Lucas et al., 2012, Fig. 4A).

a strigilite (Lillegraven et al., 1981), and pellets and possible owl bones from another locality (Walton, 1990). The late Oligocene of Germany also preserves presumed strigilites that contain rodent specimens (Engesser and Storch, 1999; Smith and Wuttke, 2015).

Gawne (1975) distinguished a strigilite from the early Miocene of central New Mexico. This specimen is a small pellet containing tightly packed cranial and postcranial specimens of two rodents (*Proheteromys cejanus* and *P.* aff. *P. floridanus*) (Gawne, 1975).

Two fossil regurgitalites from the Andalhuala Formation (Upper Miocene), Santa María of Argentina are composed of articulated and disassociated bones and teeth of octodontid and cricetid rodents that show evidence of corrosion (Nasif et al., 2009). Nasif et al. (2009) interpret these as ornithoregurgitalites (*sensu* Hunt and Lucas, 2012a) of "terror birds" (Phorusrhacidae, Psilopterinae). This is the oldest record of a cricetid rodent in South America.

An early Miocene lake in eastern Spain's Iberian Peninsula yields "abundant clumps of scales and bones" of unidentified teleosts (Álvarez-Parra et al., 2021, p. 10). These specimens are flat accumulations that lack phosphatic matrix. Álvarez-Parra et al. (2021) recognized them as bromalites, and they clearly represent regurgitalites.

The Miocene Pisco Formation of Peru is a marine Lagerstätte that includes multiple vertebrate specimens. One skeleton of the beaked whale *Messapicetus gregarius* has an accumulation of skeletons of *Sardinops* cf. *S. sagax*, adjacent to its skull, which are interpreted as comprising a regurgitalite (Bosio et al., 2021; Fig. 22F–H).

The best example of Pliocene strigilites is a putative purgolite of slightly disaggregated strigilites from Arizona in the United States (Walton, 1990; Czaplewski, 2011). A less well documented occurrence is in the adjacent state of New Mexico (Walton, 1990).

Many Pleistocene cave deposits and some archeological sites yield bone accumulations derived from raptor, notably owl, regurgitations (e.g., Andrews, 1990; Hunt and Lucas, 2007a). Rodent fossils from Pleistocene cave deposits in Poland may derive from strigilites (Kowalski, 1960; Sulimski, 1964).

Withnell et al. (2022) conducted a notable study of a purgolite of rodent-bearing strigilites from Utah in the United States. They utilized high-resolution X-ray computed tomography

(CT) to segment, digitally prepare, and describe teeth belonging to *Allophaiomys pliocaenicus*, *Mictomys kansasensis*, or *Mictomys meltoni*, and a previously recognized but unnamed morphotype of *Phenacomys*.

There are very few convincing examples, prior to the Pleistocene, of paleontological microvertebrate accumulations that comprise bones derived from raptor regurgitations (Lucas et al., 2012). However, many Pleistocene cave deposits and some archeological sites yield bone accumulations derived from raptor, notably owl, regurgitations (e.g., Andrews, 1990; Hunt and Lucas, 2007a; Montalvo et al., 2012). Indirect evidence of strigilites (or other ornithoregurgitalites) is based on analysis of microvertebrate accumulations (e.g., Mellett, 1975; Mayhew, 1977; Kusmer, 1990). There is strong evidence that some cave microvertebrate accumulations, particularly of bats and lizards, resulted from accumulation by owls (e.g., Williams, 1952; Morgan, 1994).

Cenozoic marine regurgitalites are rare. Schrøder et al. (2022) described a bromalite from the early Eocene Fur Formation of Denmark. This specimen is located anterior and superior to a skeleton of the stromateiform fish *Butyrumichthys henricii*; it contains fish debris and may represent a regurgitalite. Milner (in Hunt and Lucas, 2007a) noted fish regurgitalites from the Late Pleistocene calcareous concretions from the Champlain Sea Clays of eastern Canada. Baldanza et al. (2013) described fossil ambergris from the Pleistocene of Italy, but this is probably fecal in origin (Clarke, 2006).

5. Ichnotaxobases

Currently, there are only two named regurgitalite ichnotaxa, but there is great potential to construct an enlarged ichnotaxonomy that will further the study of these traces. Potential ichnotaxobases include:

1. Composition
   a. Biologic
      i. Taxonomy of inclusions
      ii. Physical or chemical modification of inclusions
   b. Nonbiologic
      i. Volume of matrix
      ii. Composition of matrix
2. Morphology
   a. Geometry
   b. Size

5.1 Regurgitalite ichnotaxa

Currently, there are only two named regurgitalite ichnotaxa, and both are from the Carboniferous of the United States (Hunt et al., 2012a, 2012b).

5.1.1 Ostracobromus *(Hunt et al., 2012a)*

Type ichnospecies: *Ostracobromus snowyensis* (Hunt et al., 2012a).

Included species: Known only from the type ichnospecies.

Distribution: Mississippian of Montana and Pennsylvanian of Kansas, USA.

Revised diagnosis: Bromalite that differs from other ichnogenera in being ovoid in lateral view and thin in cross-section with maximum width and length of 26 mm and is characterized by inclusions of multiple valves of ostracods set in a fine-grained phosphatic groundmass.

Discussion: The ichnogenus was originally described as a possible regurgitalite, but we are confident of its assignment given the paucity of groundmass relative to comparable coprolites and the concentration of indigestible ostracod valves.

5.1.2 Ostracobromus snowyensis *(Hunt et al., 2012a)*

Holotype: NMMNH P-54134 (Hunt et al., 2012a; Fig. 7.2A, D and E).

Referred specimens: None.

Type locality: Bear Gulch Lagerstätte, Big Snowy Mountains, Montana.

Type horizon: Bear Gulch Limestone Member of the Tyler Formation (Mississippian: Chesterian).

Distribution: As for the ichnogenus.

Diagnosis: As for ichnogenus.

Discussion: This ichnotaxon not only occurs at its type locality, Bear Gulch Lagerstätte, but also in the Pennsylvanian Hamilton Lagerstätte of Kansas, USA (Cunningham et al., 1993, p. 230).

5.1.3 Conchobromus *(Hunt et al., 2012b)*

Type ichnospecies: *Conchobromus kinneyensis* (Hunt et al., 2012b).

Included species: Known only from the type ichnospecies.

Distribution: Pennsylvanian of New Mexico.

Revised diagnosis: Bromalite that differs from other ichnogenera in being flattened and ovoid in shape and composed predominantly of conchostracan valves (*Pseudestheria* sp.) with little groundmass, varies in width and length from 14 by 10 mm to 26 by 21 mm.

Discussion: *Conchobromus* consists of densely packed and disarticulated valves of the conchostracan *Pseudestheria* sp. with little groundmass and is common both at Kinney and at the Tinajas Lagerstätte in central New Mexico, USA, both of which are in the same stratigraphic unit (Hunt et al., 2012a, 2012c).

5.1.4 Conchobromus kinneyensis *(Hunt et al., 2012b)*

Holotype: NMMNH P-19157 (Hunt, 1992; Fig. 1H; Hunt et al., 2012b; Fig. 3J; Fig. 7.2B).

Referred specimens: Kinney Brick Quarry, Bernalillo County, New Mexico (NNMNH L-345); NMMNH P-12994, NMMNH P-19156 (Hunt et al., 2012b, Fig. 3K), NMMNH P-19166; Tinajas Lagerstätte, Socorro County, New Mexico (NNMNH L-4667): NMMNH P-37668, NMMNH P-37673, NMMNH P-37667, NMMNH P-37675, NMMNH uncatalogued (Hunt and Lucas, 2021a, Fig. 13A–C).

Type locality: Kinney Brick Quarry, Bernalillo County, New Mexico.

Type horizon: Upper Pennsylvanian Tinajas Member of the Atrasado Formation (Missourian: Kasimovian).

Distribution: As for ichnogenus.

Diagnosis: As for ichnogenus.

Discussion: Huber (1992), p. 9 (Fig. 2) and Scholze et al. (2021) considered that the holotype of *Conchobromus kinneyensis* represented a "current-deposited grouping of disarticulated carapaces." However, this seems unlikely since there is no other evidence for swirling currents at the Kinney Brick Quarry, and no other fossil specimens occur in such groupings in this Lagerstätte (Hunt and Lucas, 2021b).

6. Utility

Regurgitalites have diverse utility, including (Hunt and Lucas, 2021b):

1. Providing evidence of the evolution of predation (Mayr and Schaal, 2016).
2. Analysis of taphonomy and sedimentary environments (Wilson, 1987).
3. Proxy for the presence of biotaxa.
4. Loci for exceptional preservation (Gordon et al., 2020).
5. Biogeographic studies (Darroch et al., 2021).
6. Evaluating digestive processes of producer (Gordon et al., 2020).
7. Evidence of the evolution of durophagy (Lucas and Hunt, 2023).

7. Conclusions

Regurgitalites are not only the least studied bromalites, but they can also be the most difficult to identify. Bromalite pellets can represent coprolites or regurgitalites, and two-dimensional examples could also be taphonomic artifacts (decayed specimens or physical concentrations). Many identified vertebrate regurgitalites were produced by fish or birds and are preserved in a limited range of environmental settings (e.g., principally aquatic low energy). Thus, the fossil record of regurgitalites is strongly controlled by both taxonomic and taphonomic factors. Regurgitalites need a robust ichnotaxonomy based on sound ichnotaxobases rooted in their composition and morphology.

Declaration of competing interest

The authors declare that they have no known competing financial interests or personal relationships that could have appeared to influence the work reported in this chapter.

Data availability

No data were used for the research described in this chapter.

Acknowledgments

We thank many colleagues who have graciously allowed us access to collections, including Michael Brett-Surman, Thomas Lehmann, Krister Smith, Anika Vogel, and others for helpful discussions including Bret Buskirk, Bob Emry, John Foster and Chris Duffin. Günter Schweigert, Krister Smith, Anika Vogel, and John Foster kindly supplied images, and John Foster and Hendrik Klein provided helpful reviews.

References

Aas, T.S., Sixten, H.J., Hillestad, M., Sveier, H., Ytrestøyl, T., Hatlen, B., Åsgård, T., 2017. Measurement of gastrointestinal passage rate in Atlantic salmon (*Salmo salar*) fed dry or soaked feed. Aquacult. Rep. 8, 49–57.

Albersdörfer, R., Häckel, W., 2015. Die Kieselplattenkalke von Painten. In: Arratia, G., Schultze, H.P., Tischlinger, H., Viohl, G. (Eds.), Solnhofen—ein Fenster in die Jurazeit. Verlag Dr. F. Pfeil, München, pp. 126–133.

Aldridge, R.J., Gabbott, S.E., Siveter, L.J., Theron, J.N., 2006. Bromalites from the Soom Shale Lagerstätte (Upper Ordovician) of South Africa: palaeoecological and palaeobiological implications. Palaeontology 49, 857–871.

Alexander, J.P., Burger, B.J., 2001. Stratigraphy and taphonomy of Grizzly Buttes, Bridger Formation, and the middle Eocene of Wyoming. In: Gunnell, G.F. (Ed.), Eocene Biodiversity: Unusual Occurrences and Rarely Sampled Habitats. Kluwer Academic/Plenum Publishers, New York, pp. 165–196.

Álvarez-Parra, S., Albesa, J., Gouiric-Cavalli, S., Montoya, P., Peñalver, E., Sanjuan, J., Crespo, V.D., 2021. The early Miocene lake of Foieta la Sarra-A in eastern Iberian Peninsula and its relevance for the reconstruction of the Ribesalbes–Alcora Basin palaeoecology. Acta Palaeontol. Pol. 66 (Suppl). https://doi.org/10.4202/app.00842.2020.

Andrews, P., 1990. Owls, Caves and Fossils: Predation, Preservation and Accumulation of Small Mammal Bones in Caves, with an Analysis of the Pleistocene Cave Faunas from Westbury-sub-Mendip. University of Chicago Press, Somerset, UK. Chicago, 239p.

Andrews, P.L.R., Sims, D.W., Young, J.Z., 1998. Induction of emesis by the sodium channel activator veratrine in the lesser spotted dogfish, *Scyliorhinus canicula* (Chondrichthyes: Elasmobranchii). J. Mar. Biol. Assoc. U. K. 78, 1269–1279.

Andrews, P.L.R., Axelsson, M., Franklin, C., Holmgren, S., 2000. The emetic reflex in a reptile (*Crocodylus porosus*). J. Exp. Biol. 203, 1625–1632.

Auffenberg, W., 1981. The Behavioral Ecology of the Komodo Monitor. University Presses of Florida, Gainesville, 406 p.

Baldanza, A., Bizzarri, R., Famiani, F., Monaco, P., Pellegrino, R., Sassi, P., 2013. Enigmatic, biogenically induced structures in Pleistocene marine deposits: a first record of fossil ambergris. Geology 41, 1075–1078.

Barthel, K.W., 1978. Solnhofen: Ein Blick in die Erdgeschichte. Ott Verlag, Thun, 393 p.

Barthel, K.W., Janicke, V., 1970. Aptychen als Verdauungsrückstand—Ein Fund aus den Solnhofener Plattenkalken, unteres Untertithon, Bayern. Neues Jahrbuch Geol. Palaontol. Monatsh. 1970, 65–68.

Beintema, A.J., 1991. Penguins shed stomach lining. Nature 352, 480–481.

Bennett, S.C., 2001. The osteology and functional morphology of the Late Cretaceous pterosaur *Pteranodon*. Palaeontograph. Abteilung 260, 1–153.

Bennett, S.C., 2014. A new specimen of the pterosaur *Scaphognathus crassirostris*, with comments on constraint of cervical vertebrae number in pterosaurs. Neues Jahrbuch Geol. Palaontol. Abhand. 271, 327–348.

Bertling, M., Braddy, S.J., Bromley, R.G., Demathieu, G.R., Genise, J., Mikuláš, R., Nielsen, J.K., Nielsen, K.S.S., Rindsberg, A.K., Schlirf, M., Uchman, A., 2006. Names for trace fossils: A uniform approach. Lethaia 39, 265–286.

Bertling, M., Buatois, L.A., Knaust, D., Laing, B., Mángano, M.G., Meyer, N., Mikuláš, R., Minter, N.J., Neumann, C., Rindsberg, A.K., Uchman, A., 2022. Names for trace fossils 2.0: theory and practice in ichnotaxonomy. Lethaia 55 (3), 1–19.

Bishop, G.A., 1975. Traces of predation. In: Frey, R.W. (Ed.), The Study of Trace Fossils. Springer, New York, pp. 261–287.

Bochenski, Z.M., Tomek, T., Boev, Z., Mitev, I., 1993. Patterns of birdbone fragmentation in pellets of the tawny owl (*Strix aluco*) and the eagle bowl (*Bubo bubo*) and their taphonomic implications. Acta Zool. Cracov. 36, 313–328.

Bochenski, Z.M., Huhtala, K., Jussila, P., Pulliainen, E., Tornberg, R., Tunkkari, P.S., 1998. Damage to bird bones in pellets of Gyrfalcon *Falco rusticolus*. J. Archaeol. Sci. 25, 425–433.

Bosio, G., Collareta, A., Di Celma, C., Lambert, O., Marx, F.G., de Muizon, C., Gioncada, A., Gariboldi, K., Malinverno, E., Malca, R.V., Urbina, M., Bianucci, G., 2021. Taphonomy of marine vertebrates of the Pisco Formation (Miocene, Peru): insights into the origin of an outstanding Fossil-Lagerstätte. PLoS One 16 (7), e0254395. https://doi.org/10.1371/journal.

Böttcher, R., 1989. Über die Nahrung eines *Leptopterygius* (Ichthyosauria, Reptilia) aus dem süddeutschen Posidonienschiefer (Unterer Jura) mit Bemerkungen über den Magen der Ichthyosaurier. Stuttgarter Beiträge zur Naturkunde, Serie B 155, 1–19.

Böttcher, R., 1990. Neue Erkenntnisse über die Fortpflanzungsbiologie der Ichthyosaurier (Reptilia). Stuttgarter Beiträge zur Naturkunde, Serie B 164, 1–51.

Bowman, R.E., 1986. Effect of regurgitation on stomach content data of marine fisheries. Environ. Biol. Fish. 16, 171–181.

Brachaniec, T., Niedźwiedzki, R., Surmik, D., Krzykawski, T., Szopa, K., Gorzelak, P., Salamon, M.A., 2015. Coprolites of marine vertebrate predators from the Lower Triassic of southern Poland. Palaeogeogr. Palaeoclimatol. Palaeoecol. 435, 118–126.

Broili, E., 1938. Beobachtungen an *Pterodactylus*. Sitzungberichte der Bayerischen Akademie der Wissenschaften. Math. Nat. Abteilung 1938, 139–154.

Brown, B., 1943. Flying reptiles. Nat. Hist. 52 (3), 104–111.

Brunnschweiler, J.M., Andrews, P.L.R., Southall, E.J., Pickering, M., Sims, D.W., 2005. Rapid voluntary stomach eversion in a free-living shark. J. Mar. Biol. 85, 1141–1144.

Burrow, C.J., Turner, S., 2010. Reassessment of "*Protodus*" *scoticus* from the early Devonian of Scotland. In: Elliott, K., Maisey, J.G., Yu, X., Miao, D. (Eds.), Morphology, Phylogeny and Paleobiogeography of Fossil Fishes. Verlag Dr. Friedrich Pfeil, Munich, pp. 123–144.

Bury, S., Drohobycka-Wawryka, A., 2020. Pellet egestion in modern carnivorous snakes. Curr. Zool. 66, 593–595. https://doi.org/10.1093/cz/zoaa009.

Buskirk, B.L., Hunt, A.P., Lucas, S.G., 2015. Who's eating who? Preliminary analysis of enigmatic bromalites from the Eocene Florissant Formation, Colorado. Geol. Soc. Am. Abstr. Progr. 47 (7), 346.

Buskirk, B.L., Bourgeois, J., Meyer, H.W., Nesbitt, E.A., 2016. Freshwater molluscan fauna from the Florissant Formation, Colorado: paleohydrologic reconstruction of a latest Eocene lake. Can. J. Earth Sci. 53 (6), 630–643.

Capasso, L., 2019. Palaentological evidence of piscivorous habits of some pycnodonts from the Middle Cenomanian of Lebanon. Thalass. Salentina 41, 91–110.

Cate, A.S., Evans, I., 1994. Taphonomic significance of the biomechanical fragmentation of live molluscan shell material by a bottom-feeding fish (*Pogonias cromis*) in Texas coastal bays. Palaios 9, 254–274.

Chabreck, R.H., 1996. Regurgitation by the American alligator. Herpetol. Rev. 27 (4), 185–186.

Clarke, M.R., Pailiza, O., Aguayo, A., 1998. Sperm whales of the southeast Pacific. Part IV: fatness, food and feeding. In: Pilleri, G. (Ed.), Investigations on Cetacea. University of Berne, Brain Anatomy Institute, Berne, pp. 53–195.

Clarke, R., 2006. The origin of ambergris. Lat. Am. J. Aquatic Mamm. 5, 7–21.

Cooper, S.L.A., Maxwell, E.E., 2022. Revision of the pachycormid fish *Saurostomus esocinus* Agassiz from the Early Jurassic (Toarcian) of Europe with new insight into the origins of suspension-feeding in Pachycormidae. Pap. Palaeontol. 8, e1467. https://doi.org/10.1002/spp2.1467.

Cooper, S.L., Marson, K.J., Smith, R.E., Martill, D., 2022. Contrasting preservation in pycnodont fishes reveals first record of regurgitalites from the Upper Cretaceous (Maastrichtian) Moroccan phosphate deposits. Cretac. Res. 131, 105111. https://doi.org/10.1016/j.cretres.2021.105111.

Cooper, S.L.A., López-Arbarello, A., Maxwell, E.E., 2024. First occurrence of a †coccolepidid fish (?Chondrostei: †Coccolepididae) from the upper Lias (Toarcian, early Jurassic) of southern Germany. Palaeontol. Electron. 27 (1), a23. https://doi.org/10.26879/1326.palaeo-electronica.org/content/2024/5197-new-toarcian-coccolepidid-fish.

Cunningham, C.R., Feldman, H.R., Franseen, E.K., Gastaldo, R.A., Mapes, G., Maples, C.G., Schultze, H.P., 1993. The Upper Carboniferous Hamilton Fossil-Lagerstätte in Kansas: a valley-fill, tidally influenced deposit. Lethaia 26, 225–236.

Czaplewski, N.J., 2011. An owl-pellet accumulation of small Pliocene vertebrates from the Verde Formation, Arizona, USA. Palaeontol. Electron. 14 (3), 1–33.

Dalla Vecchia, F.M., Chiappe, L.M., 2002. First avian skeleton from the Mesozoic of northern Gondwana. J. Vertebr. Paleontol. 22, 856–860.

Dalla Vecchia, F.M., Muscio, G., Wild, R., 1989. Pterosaur remains in a gastric pellet from the upper Triassic (Norian) of Rio Seazza Valley (Udine, Italy). Gortania Atti Museo Friuli Storia Naturale, vol 10, pp. 121–132.

Darroch, S.A.F., Fraser, D., Casey, M.M., 2021. The preservation potential of terrestrial biogeographic patterns. Proc. R. Soc. Ser. B 288, 20202927. https://doi.org/10.1098/rspb.2020.2927.

Denys, C., Stoetzel, E., Andrews, P., Bailon, S., Rihane, A., Huchet, J.B., Fernández-Jalvo, Y., Laroulandie, V., 2018. Taphonomy of small predators multi-taxa accumulations: palaeoecological implications. Hist. Biol. 30, 868–881.

Dietl, G., Schweigert, G., 2001. Im Reich der Meerengel. Pfeil, München, 144 p.

Dodson, P., 1973. The significance of small bones in paleoecological interpretation. Rocky Mt. Geol. 12, 15–19.

Dodson, P., Wexlar, D., 1979. Taphonomic investigations of owl pellets. Paleobiology 5, 275–284.

Dolowy, W.C., Fisher, L.E., Williamson, W., 1960. Some observations on vomiting in animals. J. Am. Vet. Med. Assoc. 137, 426–427.

Duke, G.E., Evanson, O.A., Redig, P.T., Rhoades, D.D., 1976. Mechanism of pellet egestion in great horned owls (*Bubo virginianus*). Am. J. Physiol. 231, 1824–1829.

Elder, R.L., 1985. Principles of Aquatic Taphonomy with Examples from the Fossil Record [Ph.D. Dissertation]. University of Michigan, Ann Arbor, 336 p.

Elder, R.L., Smith, G.R., 1988. Fish taphonomy and environmental inference in paleolimnology. Palaeogeogr. Palaeoclimatol. Palaeoecol. 62, 577–592.

Engesser, B., Storch, G., 1999. Eomyiden (Mammalia, Rodentia) aus dem Oberoligozän von Enspel im Westerwald (Westdeutschland). Eclogae Geol. Helv. 92, 483–493.

Everhart, M.J., 1999. Evidence of feeding on mosasaurs by the Late Cretaceous lamniform shark, *Cretoxyrhina mantelli*. J. Vertebr. Paleontol. 17 (Suppl. 3), 43–44.

Everhart, M.J., 2003. First records of plesiosaur remains in the lower Smoky Hill Chalk Member (Upper Coniacian) of the Niobrara Formation in western Kansas. Kansas Acad. Sci. Trans. 106, 139–148.

Everhart, M.J., 2004. Late Cretaceous interaction between predators and prey. Evidence of feeding by two species of shark on a mosasaur. PalArch Vertebr. Palaeontol. Ser. 1, 1–7.

Everhart, M.J., 2017. Oceans of Kansas—A Natural History of the Western Interior Sea, Second ed. Indiana University Press, Bloomington. 427 p.

Fathinia, B., Anderson, S.C., Rastegar-Pouyani, N., Jahani, H., Mohamadi, H., 2009. Notes on the natural history of *Pseudocerastes urarachnoides* (Squamata: Viperidae). Russ. J. Herpetol. 16 (2), 134–138.

Fea, N., Harcourt, R., 1997. Assessing the use of fecal and regurgitate analysis as a means of determining the diet of New Zealand Fur Seals. In: Hindell, M., Kemper, C. (Eds.), Marine Mammal Research in the Southern Hemisphere. Volume 1: Status, Ecology and Medicine. Surrey Beatty, Chipping Norton, pp. 143–150.

Fernandez-Jalvo, Y., Andrews, P., 2016. Atlas of Taphonomic Identifications: 1001+ Images of Fossil and Recent Mammal Bone Modification. Springer, New York, 359 p.

Fernandez-Jalvo, Y., Andrews, P., Denys, C., Sese, C., Stoetzel, E., Marin-Monfort, D., Pesquero, D., 2016. Taphonomy for taxonomists: implications of predation in small mammal studies. Quat. Sci. Rev. 139, 138–157.

Fisher, D.C., 1981a. Crocodilian scatology, microvertebrate concentrations, and enamel-less teeth. Paleobiology 7, 262–275.

Fisher, D.C., 1981b. Taphonomic interpretation of enamel-less teeth in the Shotgun local fauna (Paleocene, Wyoming). In: Contributions of the Museum of Paleontology, vol 25. University of Michigan, pp. 259–275.

Foster, J.R., Hunt, A.P., Kirkland, J.I., 2022. Significance of a small regurgitalite containing lissamphibian bones, from the Morrison Formation (Upper Jurassic), within a diverse plant locality deposit in southeastern Utah, USA. Palaios 37 (8), 433–442.

Freimuth, W.J., 2020. Paleoecological Utility of Feeding Traces at Egg Mountain, a Rich Terrestrial Vertebrate Locality of the Upper Cretaceous Two Medicine Formation, Montana, U.S.A. MS Thesis. Montana State University, Bozeman, 209 p.

Freimuth, W.J., Varricchio1, D.J., Brannick, A.L., Weaver, L.N., Wilson Mantilla, G., 2021. Mammal-bearing gastric pellets potentially attributable to *Troodon formosus* at the Cretaceous Egg Mountain locality, two Medicine formation, Montana, USA. Palaeontology. https://doi.org/10.1111/pala.12546.

Frentzen, K., 1936. Ein fossiler Mageninhalt aus dem Lias Delta (Amaltheen-Schichten) von Reichenbach, Aalen. Beitrag Naturkund. Forsch. S. W. Deutsch. 1, 153–156.

Frey, E., Tischlinger, H., 2012. The Late Jurassic pterosaur *Rhamphorhynchus*, a frequent victim of the ganoid fish *Aspidorhynchus*? PLoS One 7, e31945. https://doi.org/10.1371/journal.pone.0031945.

Gale, A.S., 1987. Goniasteridae (Asteroidea, Echinodermata) from the Late Cretaceous of north-west Europe. 1. Introduction. The genera *Metopaster* and *Recurvaster*. Mesozoic Research 1, 1–69.
Gans, C., 1952. The functional morphology of the egg-eating adaptations in the snake genus: *Dasypeltis*. Zoologica 37, 209–244.
Garassino, A., Donovan, D.T., 2000. A new family of coleoids from the Lower Jurassic of Osteno, northern Italy. Palaeontology 43, 1019–1038.
Gawne, C.E., 1975. Rodents from the Zia Sand Miocene of New Mexico. Am. Mus. Novit. 2586, 25 p.
Goodman-Lowe, G.D., 1998. Diet of the Hawaiian monk seal (*Monachus schauinslandi*) from the Northwestern Hawaiian Islands during 1991 to 1994. Mar. Biol. 132, 535–546.
Gordon, C.M., Roach, B.T., Parker, W.G., Briggs, D.E., 2020. Distinguishing regurgitalites and coprolites: a case study using a Triassic bromalite with soft tissue of the pseudosuchian archosaur *Revueltosaurus*. Palaios 35, 111–121.
Grawe-Baumeister, J., Schweigert, G., Dietl, G., 2000. Echiniden aus dem Nusplinger Plattenkalk (Ober-Kimmeridgium, Südwestdeutschland). Stuttgarter Beiträge zur Naturkunde B 286, 1–39.
Gudger, E.W., 1949. Natural history notes on tiger sharks, *Galeocerdo tigrinus*, caught at Key West, Florida, with emphasis on food and feeding habits. Copeia 1949, 39–47.
Gunnell, G.F., Lehmann, T., Ruf, I., Habersetzer, J., Morlo, M., Rose, K.D., 2018. Ferae—animals that eat animals. In: Smith, K.T., Schaal, S.F.K., Habersetzer, J. (Eds.), Messel—An Ancient Greenhouse Ecosystem. Schweizerbart, Stuttgart, pp. 270–283.
Hattin, D.E., 1988. Rudists as historians. Fort Hays Studies, Third Series, vol 10, pp. 4–22.
Hattin, D.E., 1996. Fossilized regurgitate from Smoky Hill Member of Niobrara chalk (upper Cretaceous) of Kansas, USA. Cretac. Res. 17, 443–450.
Haubold, H., 1990. Ein neuer Dinosaurier (Ornithischia, Thyreophora) aus dem unteren Jura des nördlichen Mitteleuropa. Rev. Paléobiol. 9, 149–177.
Hockett, B.S., 1996. Corroded, thinned and polished bones created by golden eagles (*Aquila chrysaetos*): taphonomic implications for archaeological interpretation. J. Archaeol. Sci. 23, 587–591.
Hoffmann, R., Stevens, K., Keupp, H., Simonsen, S., Schweigert, G., 2019. Regurgitalites–a window into the trophic ecology of fossil cephalopods. J. Geol. Soc. 177, 82–102.
Holgado, B., Dalla Vecchia, F.M., Fortuny, J., Bernardini, F., Tuniz, C., 2015. A reappraisal of the purported gastric pellet with pterosaurian bones from the Upper Triassic of Italy. PLoS One 10 (11), e0141275. https://doi.org/10.1371/journal.pone.0141275.
Horstmann, G., Maier, D., 1957. Neue Sauriergewölle aus dem Bielefelder Dogger. Aufschluss 8, 17–21.
Huber, P., 1992. Faunal distribution, seasonal deposition, and fish taphonomy at the upper Pennsylvanian (Missourian) Kinney Brick Company quarry, central New Mexico. In: Zidek, J. (Ed.), Geology and Paleontology of the Kinney Brick Quarry, Late Pennsylvanian, Central New Mexico. New Mexico Bureau of Mines and Mineral Resources, Bulletin, vol 138, pp. 37–48.
Hunt, A.P., 1992. Late Pennsylvanian coprolites from the Kinney Brick Quarry, central New Mexico, with notes on the classification and utility of coprolites. In: Zidek, J. (Ed.), Geology and Paleontology of the Kinney Brick Quarry, Late Pennsylvanian, Central New Mexico. New Mexico Bureau of Mines and Mineral Resources, Bulletin, vol 138, pp. 221–229.
Hunt, A.P., Lucas, S.G., 2007. Cenozoic vertebrate trace fossils of North America: ichnofaunas, ichnofacies and biochronology. New Mexico Museum of Natural History and Science, Bulletin 42, pp. 17–41.
Hunt, A.P., Lucas, S.G., 2012a. Classification of vertebrate coprolites and related trace fossils. New Mexico Museum of Natural History and Science, Bulletin 57, pp. 137–146.
Hunt, A.P., Lucas, S.G., 2021b. The ichnology of vertebrate consumption: dentalites, gastroliths and bromalites. New Mexico Museum of Natural History and Science Bulletin 87, 215 p.
Hunt, A.P., Lucas, S.G., 2021a. A review of the bromalite ichnofauna from the Kinney Brick Quarry Lagerstätte (Late Pennsylvanian) of New Mexico, USA with descriptions of new consumulites and coprolites. New Mexico Museum of Natural History and Science, Bulletin 84, pp. 217–236.
Hunt, A.P., Lucas, S.G., Spielmann, J.A., Lockley, M.G., 2012a. Bromalites from the Mississippian bear Gulch Lagerstätte of central Montana, USA. New Mexico Museum of Natural History and Science, Bulletin 57, pp. 171–174.
Hunt, A.P., Lucas, S.G., Spielmann, J.A., Suazo, T., Cantrell, A., 2012b. A re-evaluation of late Pennsylvanian bromalites from the Kinney Brick quarry Lagerstätte, New Mexico, USA. New Mexico Museum of Natural History and Science, Bulletin 57, pp. 185–192.

Hunt, A.P., Lucas, S.G., Spielmann, J.A., Cantrell, A., Suazo, T., Lerner, A.J., 2012c. Bromalites from the Tinajas Lagerstätte (late Pennsylvanian: late Missourian), central New Mexico, USA. New Mexico Museum of Natural History and Science, Bulletin 57, pp. 175–183.

Hunt, A.P., Lucas, S.G., Buskirk, B.L., 2015. The value of vomit: the fossil record of vertebrate regurgitalites and its significance. Geol. Soc. Am. Abstr. Progr. 47 (7), 346.

Jäger, M., 1991. Lias epsilon von Dotternhausen 2. Teil. Fossilien 1/91, 33–36.

Jäger, M., 2001. Das Fossilienmuseum im Werkforum. Self-Published, Dotternhausen, 149 p.

Jäger, M., 2005. The Museum of Fossils in the Werkforum: Guidebook of the Exhibition of Jurassic Fossils. Holcim, Dotternhausen.

Janicke, V., 1970. Ein Strobilodus als Speiballen im Solnhofener Plattenkalk (Tiefes Untertithon, Bayern). Neues Jahrbuch Geol. Palaontol. Monatsh. 1970, 61–64.

Janicke, V., Schairer, G., 1970. Fossilerhaltung und Problematica aus den Solnhofener Plattenkalken. Neues Jahrbuch Geol. Palaontol. Monatsh. 1970, 452–464.

Jiang, S., Wang, X., Zheng, X., Cheng, X., Wang, X., Wei, G and, Kellner, A.W.A., 2022. Two emetolite-pterosaur associations from the Late Jurassic of China: showing the first evidence for antiperistalsis in pterosaurs. Philos. Trans. R. Soc. B 377, 20210043. https://doi.org/10.1098/rstb.2021.0043.

Keller, T., 1977. Frassreste im süddeutschen Posidonienschiefer. Jahrshefte der Gesellschaft für Naturkunde in Württemberg 132, 117–134.

Klug, C., Vallon, L.H., 2018. Regurgitated ammonoid remains from the latest Devonian of Morocco. Swiss J. Palaeontol. 138, 87–97.

Korth, W.W., Emry, R.J., 1991. The skull of *Cedromus* and a review of the Cedromurinae (Rodentia, Sciuridae). J. Paleontol. 65, 984–994.

Kowalski, K., 1960. Cricetidae and Microtidae (Rodentia) from the Pliocene of Weze (Poland). Acta Zool. Cracow 5, 447–505.

Kusmer, K.D., 1990. Taphonomy of owl pellet deposition. J. Paleontol. 64, 629–637.

Lillegraven, J.A., McKenna, M.C., Krishtalka, L., 1981. Evolutionary Relationships of Middle Eocene and Younger Species of Centetodon (Mammalia, Insectivora, Geolabidae) with a Description of the Dentition of *Ankylodon* (Adapisoricidae), vol 45. University of Wyoming Publication, 115 p.

Long, C.T., Page, R.B., Howard, A.M., McKeon, G.P., Felt, S.A., 2010. Comparison of gastrografin to barium sulphate as a gastrointestinal contrast agent in red-eared slider turtles (*Trachemys scripta elegans*). Vet. Radiol. Ultrasound 51, 42–47.

Longrich, N.R., Jalil, N.E., Khaldoune, F., Yazami, O.K., Pereda-Suberbiola, X., Bardet, N., 2022. *Thalassotitan atrox*, a giant predatory mosasaurid (Squamata) from the upper Maastrichtian phosphates of Morocco. Cretac. Res. 140, 105315. https://doi.org/10.1016/j.cretres.2022.105315.

Lucas, S.G., Hunt, A.P., 2023. There was no Mesozoic marine revolution. In: Proceedings the 4th International Electronic Conference on Geosciences Proceedings, Proceedings 2023, vol 87, p. 37. https://doi.org/10.3390/IECG2022-14819.

Lucas, S.G., Emry, R.J., Krainer, K., Hunt, A.P., Spielmann, J.A., 2012. Strigilites (fossilized owl pellets) from the Oligocene of Wyoming. New Mexico Museum of Natural History and Science, Bulletin 57, pp. 325–335.

Lukeneder, A., Lukeneder, P., 2021. The upper Triassic Polzberg palaeobiota from a marine Konservat-Lagerstätte deposited during the Carnian Pluvial Episode in Austria. Sci. Rep. 11, 16644. https://doi.org/10.1038/s41598-021-96052-w.

Lukeneder, A., Surmik, D., Gorzelak, P., Niedźwiedzki, R., Brachaniec, T., Salamon, M.A., 2020. Bromalites from the Upper Triassic Polzberg section (Austria); insights into trophic interactions and food chains of the Polzberg palaeobiota. Sci. Rep. 10, 20545. https://doi.org/10.1038/s41598-020-77017-x.

Martin, L.D., Tate, J., 1976. The skeleton of *Baptornis advenus* (Aves: Hesperornithformes). Smithsonian Contrib. Paleobiol. 27, 35–66.

Mayhew, D.F., 1977. Avian predators as accumulators of fossil mammal material. Boreas 6, 25–31.

Mayr, G., Kitchener, A.C., 2023. Early Eocene fossil illuminates the ancestral (diurnal) ecomorphology of owls and documents a mosaic evolution of the strigiform body plan. Ibis 165 (1), 231–247.

Mayr, G., Schaal, S.F.K., 2016. Gastric pellets with bird remains from the early Eocene of Messel. Palaios 31, 447–451.

Mehl, J., 1978. Ein Koprolith mit Ammoniten-Aptychen aus den Solnhofener Plattenkalken. Jahresberichte der Wetterauischen Gesellschaft für Naturkunde 1978, 85–89.

Mellett, J.S., 1975. Scatological origins of microvertebrate fossils. Science 185, 349–350.

Montalvo, C.I., Fernández, F.J., 2019. Review of the actualistic taphonomy of small mammals ingested by South American predators. Its importance in the interpretation of the fossil record. Publicación Electrónica de la Asociación Paleontológica Argentina 19 (1), 18–46.

Montalvo, C.I., Tomassini, R.L., Visconti, G., Tiranti, S., 2012. Análisis tafonómico de micromamíferos del Pleistoceno superior del Quequén Salado, provincia de Buenos Aires, Argentina Revista Brasileira de Paleontologia. Revist Brasil de Paleontolog 15, 345–358.

Moon, E.L., 1940. Notes on hawk and owl pellet formation and identification. Trans. Kans. Acad. Sci. 43, 457–466.

Moreau, J.-D., Vullo, R., Charbonnier, S., Jattiot, R., Trincal, V., Néraudeau, D., Fara, E., Baret, L., Garassino, A., Gand, G., Lafaurie, G., 2022. Konservat-Lagerstätten from the Upper Jurassic Lithographic Limestone of the Causse Méjean (Lozère, Southern France): Palaeontological and Palaeoenvironmental Synthesis. Geological Magazine. https://doi.org/10.1017/S0016756821001382.

Morgan, G.S., 1994. Late Quaternary fossil vertebrates from the Cayman Islands. In: Brunt, M.A., Davies, J.E. (Eds.), The Cayman Islands: Natural History and Biogeography. Kluwer Academic Publishers, Dordrecht, pp. 465–508.

Morlo, M., Gunnell, G.F., Smith, K.T., 2012. Mammalian carnivores from Messel and a comparison of non-volant predator guilds from the middle Eocene of Europe and North America. In: Lehmann, T., Schaal, S.K.F. (Eds.), The World at the Time of Messel: Puzzles in Palaeobiology, Palaeoenvironment, and the History of Early Primates. Senckenberg Gesellschaft für Naturforschung, Frankfurt am Main, pp. 120–121.

Müller, A.H., 1953. Die isolierten Skelettelemente der Asteroidea (Asterozoa) aus der obersenonen Schreibkreide von Rügen. Beihefte zur Zeitschrift Geologie 8, 1–66.

Murphey, P.C., Torick, L.L., Bray, E.S., Chandler, R., Evanoff, E., 2001. Taphonomy, fauna, and depositional environment of the *Omomys* Quarry, an unusual accumulation from the Bridger Formation (middle Eocene) of southwestern Wyoming (USA). In: Gunnell, G.F. (Ed.), Eocene Biodiversity: Unusual Occurrences and Rarely Sampled Habitats. Kluwer Academic/Plenum Publishers, New York, pp. 361–401.

Muzzopappa, P., Martinelli, A.G., Garderes, J.P., Rougier, G.W., 2021. Exceptional avian pellet from the Paleocene of Patagonia and description of its content: a new species of calyptocephalellid (Neobatrachia) anuran. Pap. Palaeontol. 7 (2), 1133–1146.

Myhrvold, N.P., 2011. A call to search for fossilised gastric pellets. Hist. Biol. 24, 505–517.

Nasif, N.L., Esteban, G.I., Ortiz, P.E., 2009. Novedoso hallazgo de egagrópilas en el Mioceno tardío, Formación Andalhuala, provincia de Catamarca, Argentina. Serie Correlación Geológica. Temas de Paleontología I 25, 105–114.

Neumann, C., 2000. Evidence of predation on Cretaceous sea stars from north-west Germany. Lethaia 33, 65–70.

Niedźwiedzki, R., Surmik, D., Chećko, A., Salamon, M.A., 2021. A regurgitalite of the Middle Triassic (Muschelkalk) from Upper Silesia (Poland). Geology Geophys. Environ. 47 (1), 33–40.

O'Connor, J.K., 2019. The trophic habits of early birds. Palaeogeogr. Palaeoclimatol. Palaeoecol. 513, 178–195.

Oji, T., Ogaya, C., Sato, T., 2003. Increase of shell-crushing predation recorded in fossil shell fragmentation. Paleobiology 29, 520–526.

Petzold, H.G., 1967. Notizen zur Gewöllbildung be einen Bindenwarem (*Varanus salvator*) und einige allgemeine Bemerkungen über Repetilliengewölle. Zool. Gart. 34, 134–138.

Pinna, G., Arduini, P., Pesarini, C., Teruzzi, G., 1985. Some controversial aspects of the morphology and anatomy of *Ostenocaris cypriformis* (Crustacea, Thylacocephala). In: Transactions of the Royal Society of Edinburgh, Earth Sciences, vol 76, pp. 373–379.

Rasmussen, H.W., 1950. Cretaceous Asteroidea and Ophiuroidea with special reference to the species found in Denmark. Danmarks Geologiske Undersùgelse 77, 1–134.

Rea, A.M., 1973. Turkey vultures casting pellets. Auk 90, 209–210.

Reinhard, W., Vogel, Z., 1980. Die Nattern. In: Grzimek, B., Hediger, H., Klemmer, K., Kuhn, O., Wermuth, H. (Eds.), Grzimeks Tierleben, 6, Kriechtiere. Deutscher Taschenbuch-Verlag, München, pp. 390–423.

Robin, N., Noirit, F., Chevrinais, M., Clément, G., Olive, S., 2022. Vertebrate predation in the Late Devonian evidenced by bite traces and regurgitations: implications within an early tetrapod freshwater ecosystem. Pap. Palaeontol. 8 (4), e1460. https://doi.org/10.1002/spp2.1460.

Salamon, M.A., Gorzelak, P., Niedźwiedzki, R., Trzesiok, D., Baumiller, T.K., 2014. Trends in shell fragmentation as evidence of mid-Paleozoic changes in marine predation. Paleobiology 40, 14–23.

Salamon, M.A., Niedźwiedzki, R., Gorzelak, P., Lach, R., Surmik, D., 2012. Bromalites from the Middle Triassic of Poland and the rise of the Mesozoic Marine revolution. Palaeogeogr. Palaeoclimatol. Palaeoecol. 321, 142–150.

Sanz, J.L., Chiappe, L.M., Fernádez-Jalvo, Y., Ortega, F., Sánchez-Chillón, B., Poyatoariza, F.J., Pérez-Moreno, B.P., 2001. An Early Cretaceous pellet. Nature 409, 998–1000.

Schaal, S.K.F., 2016. The Messel pit fossil site. In: Fraser, N., Sues, H.-D. (Eds.), Terrestrial Conservation Lagerstätten: Windows into the Evolution of Life on Land. Dunedin Academic Press, Dunedin, 450 p.

Scherpner, C., 1980. Die Krokodile. In: Grzimek, B., Hediger, H., Klemmer, K., Kuhn, O., Wermuth, H. (Eds.), Grzimeks Tierleben, 6, Kriechtiere. Deutscher Taschenbuch-Verlag, München, pp. 128–147.

Schoch, R.R., Seegis, D., 2016. A Middle Triassic palaeontological gold mine: the vertebrate deposits of Vellberg (Germany). Palaeogeogr. Palaeoclimatol. Palaeoecol. 459, 249–267. https://doi.org/10.1016/j.palaeo.2016.07.002.

Schrøder, A.E., Rasmussen, J.A., Møller, P.R., Carnevale, G., 2022. *Butyrumichthys henricii* gen. et sp. nov.: a new stromateiform fish from the lower Eocene Fur Formation, Denmark. J. Vertebr. Paleontol. 42 (3), e2171798. https://doi.org/10.1080/02724634.2023.2171798.

Schweigert, G., Dietl, G., Wild, R., 2001. Miscellanea aus dem Nusplinger Plattenkalk (Ober-Kimmeridgium, Schwäbische Alb). 3. Ein Speiballen mit Flugsaurierresten: Jahresberichte und Mitteilungen des Oberrheinischen Geologischen Vereins Band, 83, pp. 357–364.

Schweigert, G., 2023. The Nusplingen Plattenkalk—observations, assumptions and enigmas of a Late Jurassic fossil Konservat-Lagerstätte. In: Reitner, J., Reich, M., Jan-Peter Duda, J.-P. (Eds.), Fossillagerstätten and Taphonomy: New Ideas and Approaches. Abstracts of the Symposium November 23–25, vol 2023, pp. 72–80.

Serafini, G., Gordon, C.M., Foffa, D., Cobianchi, M., Giusberti, L., 2022. Tough to digest: first record of Teleosauroidea (Thalattosuchia) in a regurgitalite from the Upper Jurassic of northeastern Italy. Pap. Palaeontol. 8 (6), e1474. https://doi.org/10.1002/spp2.1474.

Serafini, G., Foffa, D., Young, M.T., Friso, G., Cobianchi, M., Giusberti, L., 2023. Reappraisal of the thalattosuchian crocodylomorph record from the Middle-Upper Jurassic Rosso Ammonitico Veronese of northeastern Italy: age calibration, new specimens and taphonomic biases. PLoS One 18 (10), e0293614.

Schäfer, W., 1972. Ecology and Paleoecology of Marine Environments. University of Chicago Press, Chicago, 568 p.

Scholze, F., Schneider, J.W., Lucas, S.G., 2021. Conchostracans from the Late Pennsylvanian Kinney Brick Quarry Konservat Lagerstätte, New Mexico (USA)—taxonomy, biostratigraphy, and paleoecology: New Mexico Museum of Natural History. Bulletin 84, 287–299.

Sims, D.W., Andrews, P.L.R., Young, J.Z., 2000. Fish behaviour: stomach rinsing in rays. Nature 404, 566.

Smith, K.T., Wuttke, M., 2015. Avian pellets from the late Oligocene of Enspel, Germany-ecological interactions in deep time. Palaeobiodivers. Palaeoenviron. 95, 103–113.

Soler-Gijón, R., Ruiz, A.D., 2023. "Carbonífero de Puertollano" Natural Monument (Puertollano basin, Spain): a window for the knowledge of early vertebrates. Spanish J. Palaeontol. 38 (1), 81–94. https://doi.org/10.7203/sjp.26788.

Stevens, K., Mutterlose, J., Schweigert, G., 2014. Belemnite ecology and the environment of the Nusplingen Plattenkalk (Late Jurassic, southern Germany): evidence from stable isotope data. Lethaia 47, 512–523.

Stewart, J.D., Carpenter, K., 1990. Examples of vertebrate predation on cephalopods in the Late Cretaceous of the Western Interior. In: Boucot, A.J. (Ed.), Evolutionary Paleobiology of Behavior and Coevolution. Elsevier, Amsterdam, pp. 203–207.

Sulimski, A., 1964. Pliocene Lagomorpha and Rodentia of Weze 1 (bone breccia). Acta Palaeontol. Pol. 9, 149–261.

Terry, R.C., 2007. Inferring predator identity from skeletal damage of small-mammal prey remains. Evol. Ecol. Res. 9, 199–219.

Terry, R.C., 2010. On raptors and rodents: testing the ecological fidelity and spatiotemporal resolution of cave death-assemblages. Paleobiology 36, 137–160.

Thies, D., Hauff, R.B., 2012. A Speiballen from the Lower Jurassic Posidonia Shale of South Germany. Neues Jahrbuch Geol. Palaontol. Abhand. 267, 117–124.

Vallon, L.R., 2012. Digestichnia (Vialov, 1972)—an almost forgotten ethological class for trace fossils. New Mexico Museum of Natural History and Science, Bulletin 57, pp. 131–135.

Viohl, G., 2015. Die lithographischen Plattenkalke im engeren Sinne. In: Arratia, G., Schultze, H.P., Tischlinger, H., Viohl, G. (Eds.), Solnhofen—ein Fenster in die Jurazeit. Verlag Dr. F. Pfeil, München, pp. 78–100.

Walton, A.H., 1990. Owl pellets and the fossil record. In: Boucot, A.J. (Ed.), Evolutionary Paleobiology of Behavior and Coevolution. Elsevier, Amsterdam, pp. 233–241.

Wang, M., Zhou, Z., Sullivan, C., 2016. A fish-eating enantiornithine bird from the Early Cretaceous of China provides evidence of modern avian digestive features. Curr. Biol. 26, 1170–1176.

Wellnhofer, P., 1964. Zur Pelecypodenfauna der Neuburger Bankkalke (Mitteltithon). Abhandlungen der Bayerischen Akademie der Wissenschaften, Mathematisch-Naturwissenschaftliche Klasse. Neue Folge 119, 1–143.
Wellnhofer, P., 1970. Die Pterodacyloidea (Pterosauria) der Oberjura-Plattenkalke Süddeutschlands. Bayerische Akademie der Wissenschaften, Mathematisch Wissenschaftlichen Klasse. Abhandlugen 141, 1–133.
Wetzel, W., 1953. Die Bielefelder Garantianen-Knollen als palaobiologisches Objekt. Z. Dtsch. Geol. Ges. 104, 499–505.
Wetzel, W., 1960. Nachtrag zum Fossilarchiv der *Fuiriquina*-Schichten. Neues Jahrbuch fur Geologie und Palaontologie Monatshefte 1960, 439–446.
Wetzel, W., 1964. Mikroorganismen aus jurassischen und kretazischen Saurier-Gewöllen. Z. Dtsch. Geol. Ges. 1964, 867–874.
Withnell, C.B., Bell, C.J., Jass, C.N., 2022. Extraordinary accumulation of arvicoline rodents from little Dell Dam (Pleistocene), Utah. SW. Nat. 66 (2), 91–101.
Williams, E.E., 1952. Additional notes on fossil and subfossil bats from Jamaica. J. Mammal. 33, 171–179.
Wilson, M.V.H., 1977a. Paleoecology of Eocene lacustrine varves at Horsefly, British Columbia. Can. J. Earth Sci. 14, 953–962.
Wilson, M.V.H., 1977b. Middle Eocene freshwater fishes from British Columbia. Royal Ontario Museum. Life Sci. Contrib. 113, 1–61.
Wilson, M.V.H., 1980. Eocene lake environments: depth and distance- from-shore variation in fish, insect, and plant assemblages. Palaeogeogr. Palaeoclimatol. Palaeoecol. 32, 21–44.
Wilson, M.V.H., 1987. Predation as a source of fish fossils in Eocene lake sediments. Palaios 2, 497–504.
Witton, M.P., 2018. Pterosaurs in Mesozoic food webs: a review of fossil evidence. In: Hone, D.W.E., Witton, M.P., Martill, D.M. (Eds.), New Perspectives on Pterosaur Palaeobiology, vol 455. Geological Society of London, Special Publications, pp. 7–23.
Wright, C.W., Wright, E.V., 1940. Notes on Cretaceous Asteroidea. Geol. Soc. London Quart. J. 94, 231–248.
Ye, X.J., Sun, Z.Y., Yao, M.T., 2024. A *Keichousaurus*-bearing regurgitalite from the Middle Triassic Xingyi Fauna, Dingxiao of Xingyi city, Guizhou, South China. Palaeoworld 33 (2), 363–373. https://doi.org/10.1016/j.palwor.2023.03.009.
Zangerl, R., Richardson, E.S., 1963. The paleoecological history of two Pennsylvanian black shales. Fieldiana, Geological Memoirs 4, 352 p.
Zatoń, M., Salamon, M.A., 2008. Durophagous predation on Middle Jurassic molluscs, as evidenced from shell fragmentation. Palaeontology 51, 63–70.
Zatoń, M., Villier, L., Salamon, M.A., 2007. Signs of predation in the Middle Jurassic of south-central Poland: evidence from echinoderm taphonomy. Lethaia 40, 139–151.
Zheng, X., Wang, X., Sullivan, C., Zhang, X., Zhang, F., Wang, Y., Li, F., Xu, X., 2018. Exceptional dinosaur fossils reveal early origin of avian-style digestion. Sci. Rep. 8, 14217. https://doi.org/10.1038/s41598-018-32202-x.
Zijlstra, M., Van Eerden, M.R., 1995. Pellet production and the use of otoliths in determining the diet of cormorants *Phalacrocorax carbo sinensis*: Trials with captive birds. Ardea 83, 123–131.
Zuschin, M., Stachowitsch, M., Stanton Jr., R.J., 2003. Patterns and processes of shell fragmentation in modern and ancient marine environments. Earth Sci. Rev. 63, 33–82.

Further reading

Hunt, A.P., Lucas, S.G., 2012b. Descriptive terminology of coprolites and recent feces. New Mexico Museum of Natural History and Science, Bulletin 57, pp. 153–160.
Kner, R., 1868. Über *Conchopoma gadiforme* nov. gen. et spec. Und *Acanthodes* aus dem Rothliegenden (der untern Dyas) von Lebach bei Saarbriicken in Rheinprcussen. Wiener Akademie Wissenschaften Sitzungberichte, 57, pp. 278–305.
Korth, W., 1979. Taphonomy of microvertebrate fossil assemblages. Ann. Carnegie Mus. 48, 235–285.
Lukeneder, A., Lukeneder, P., 2023. New data on the marine Upper Triassic palaeobiota from the Polzberg Konservat-Lagerstätte in Austria. Swiss J. Palaeontol. 142 (1), 9. https://doi.org/10.1186/s13358-023-00269-3.
McAllister, J.A., 1988. Preliminary description of the coprolitic remains from Hamilton quarry, Kansas. In: Mapes, G., Mapes, R.H. (Eds.), Regional Geology and Paleontology of Upper Paleozoic Hamilton Quarry Area, vol. 6. Kansas Geological Survey Guidebook, pp. 195–202.

CHAPTER

8

Consumulites

Adrian P. Hunt[1] and Spencer G. Lucas[2]

[1]Flying Heritage and Combat Armor Museum, Everett, WA, United States; [2]New Mexico Museum of Natural History, Albuquerque, NM, United States

1. Introduction

Hunt and Lucas (2012) introduced the term consumulite to refer to material ingested and preserved within the gastrointestinal tract of a fossil vertebrate. Hunt and Lucas (2021b) recently provided an extensive review of the record of consumulites that is briefly reprised and updated here. Abbreviations are as follows: AMNH, American Museum of Natural History, New York City, New York, USA; FMNH, Field Museum of Natural History, Chicago, USA; FHSM, Sternberg Museum of Natural History, Fort Hays State University, Hays, Kansas, USA; GPIT, Geologisch-Paläontologisches Institut Tübingen, Tübingen, Germany; NHMS, Naturhistorisches Museum Schloss Bertholdsburg, Schleusingen, Germany; HMV, Hunterian Museum, Glasgow, Scotland; NMMNH, New Mexico Museum of Natural History, Albuquerque, New Mexico, USA; OUM, Oxford University Museum of Natural History, Oxford, UK; SMF, Senckenberg Research Institute Frankfurt, Frankfurt, Germany; SMNS, Staatliches Museum für Naturkunde, Stuttgart, Germany; USNM, United States National Museum, Smithsonian, Washington, DC, USA; YPM, Yale Peabody Museum of Natural History, New Haven, Connecticut, USA.

2. Terminology

Consumulites are the bromalites with the most complicated terminological history (Fig. 8.1; Table 8.1: Hunt and Lucas, 2012; 2021b). Hunt and Lucas (2012) reviewed the long history of terms applied to bromalites preserved within the body cavity and introduced the term consumulite to encompass all such trace fossils. Hunt and Lucas (2012; 2021b) redefined existing terms and introduced new ones to provide a refined terminology, including aspirationalite (protruding from mouth), oralite (within the oral cavity), esophogalite (in the gastrointestinal tract anterior to the stomach), gastrolite (in the stomach), cololite (in

Vertebrate Ichnology
https://doi.org/10.1016/B978-0-443-38351-9.00005-0

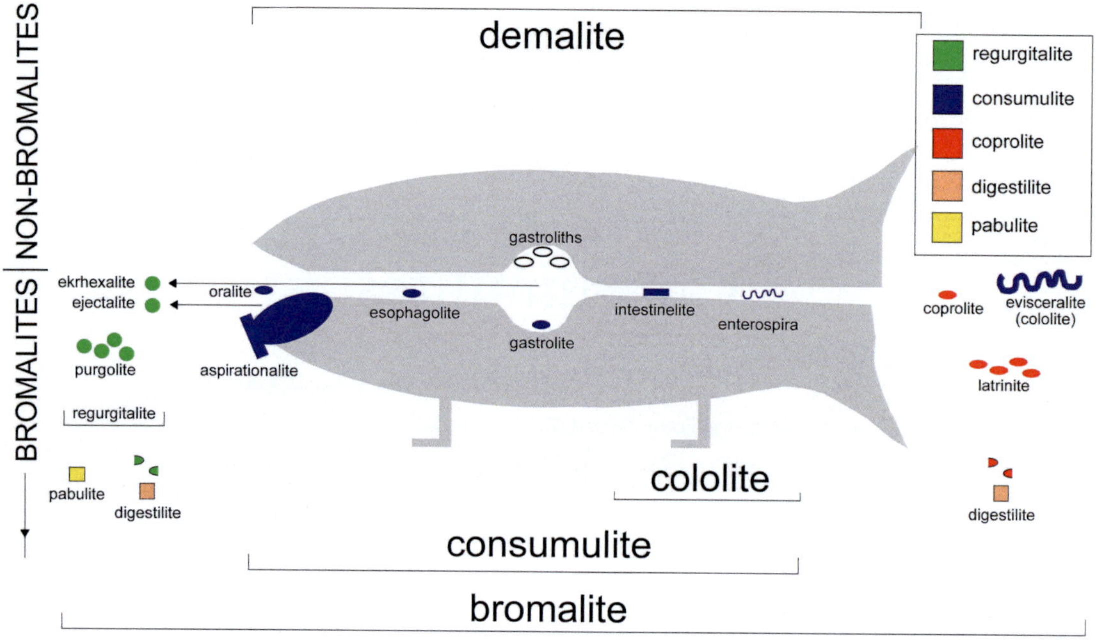

FIGURE 8.1 Classification terms for consumulites, other bromalites, and associated trace fossils. *From Hunt and Lucas (2021b).*

TABLE 8.1 Consumulite terminology.

Placement of consumulite		Agassiz (1833)	Hoernes (1904)	Fritsch (1907)	Northwood (2005)	Hunt and Lucas (2012, 2021b)	
Mouth	Protruding from oral cavity						Aspirationalite
	Within oral cavity						Oralite
Esophagus	Esophagus					Consumulite	Esophagalite
	Avian crop						Cropalite
Stomach/ ventriculus	Avian proventriculus				Gastrolite		Proventrilite
	Stomach						Gastrolite
Intestines	Nonspiral intestines	Cololite				Cololite	Intestinilite
	Spiral valve		Enterolite	Enterospira			Enterospira
Preserved in absence of body		Cololite					Evisceralite

the gastrointestinal tract posterior to the stomach), intestinelite (cololite in intestines), enterospira (cololite in a spiral valve), and incorporeal pelletite (pelletite preserved within the body cavity). The majority of consumulites represent gastrolites. An evisceralite is a cololite that is a preserved segment of infilled fossilized intestines preserved independent of, or exterior to, a carcass (Hunt and Lucas 2012: Table 8.1). In rare examples, a consumulite occurs within a consumulite, and we term this a matryoshkalite in reference to the stacked Russian dolls that Lomax (2021) alluded to in reference to such specimens.

3. Actualistic studies

The preservation of consumulites is favored by a longer gut retention time and a lower digestive acidity (Miller and Pittman, 2021). In birds, gut retention times are influenced by overall ecology and short-term behaviors (Miller and Pittman, 2021). In Recent birds, fully terrestrial species tend to have much longer gut retention times than flying taxa, with the most active having the lowest time (Jackson, 1992; Hilton et al., 1999; Caviedes-Vidal et al., 2007; Frei et al., 2014). Gut retention is also related to aspects of diet, including fiber content, lipid content, particle size, and energy density (Balch and Campling, 1965; Warner, 1981; Hilton et al., 1998, 2000a; McWhorter and Martínez del Rio, 2000; Levey and Del Rio, 2001). Short-term behavioral changes affecting retention include dietary switching (Hilton et al., 2000b), migration (McWilliams et al., 1999), and raising young (Thouzeau et al., 2004). Stomach acidities vary among Recent raptors, with less bone remaining in the pellets of raptors with lower stomach pH (Duke et al., 1975; Duke, 1997; Miller and Pittman, 2021). However, enzymatic activity and time, and not only low pH, control bone corrosion (Denys et al., 1995).

There is a wide range of experimental work related to mammalian digestion ranging from a human test subject ingesting insectivores (Crandall and Stahl, 1995) to laboratory studies of the chemistry of digestion (Denys et al., 1995). Many studies focus on digestion as it relates to the preservation of microvertebrate accumulations (e.g., Fisher, 1981b; Andrews and Evans, 1983, 1990).

Milàn (2012) conducted experiments on feeding in a variety of extant crocodylians. Crocodylians have a very effective digestive system with a hydrochloric acid concentration that exceeds that of mammalian carnivores by a factor of 50 (Coulson et al., 1989; Milàn, 2012). This environment decalcifies and dissolves all bone tissue completely before excretion and removes the enamel from teeth (Fisher, 1981a,b; Coulson et al., 1989; Trutnau and Sommerlad, 2006; Milàn, 2012).

4. Recognition and taphonomy

Not all foreign matter lying within, or apparently within, a fossil body represents a consumulite. Demalites are biogenic material putatively preserved within the body cavity of an animal that does not pertain to that individual (Hunt and Lucas, 2012). They can have five principal sources: (1) apparent preservation as the result of stacked carcasses with parts of one skeleton seeming to be within the body outline of a superposed specimen, as has been proposed for specimens of the Late Triassic theropod dinosaur *Coelophysis* (Colbert, 1989; Gay, 2005; Nesbitt et al., 2006; Rinehart et al., 2009); (2) gignolites, such as embryos or eggs; (3) postmortem utilization of the carcass (e.g., shed carnivore teeth); (4) material introduced by sedimentological processes; and (5) true consumulites. Thus, consumulites are a subset of demalites referring specifically to fossilized, ingested food material preserved within the body cavity (Hunt and Lucas, 2012). In much of the vertebrate paleontological literature, consumulites are referred to informally as "stomach contents" or "gut contents." Most consumulites represent gastrolites. They are easiest to recognize due to the size of the stomach (volume of the bromalite) and the fact that the contents still often preserve morphological details, relative to the contents of the posterior digestive tract. Thus, the contents of a decomposed stomach are likely to be still identifiable.

Consumulites, with the notable exception of evisceralites, cannot be unambiguously recognized unless they occur within an articulated or at least substantially complete skeleton. Voss et al. (2019) adapted identification criteria proposed for the recognition of gastroliths by Wings (2004) to apply to stomach contents, and these criteria are relevant to all consumulites (Hunt and Lucas, 2021b):

1. In association with the skeleton of "an appropriate consumer"—this can be useful, but some consumulites preserve prey that would be unexpected in the consumer (e.g., a fish within the apparently terrestrial theropod dinosaur *Baryonyx*: Charig and Milner, 1986, 1997) or an absence of expected prey (e.g., no insects within the pangolin *Eomanis*, a putative anteater: von Koenigswald et al., 1981; Richter, 1988).
2. In an anatomically correct position, though there are definitely exceptions related to rupture or displacement of the gastrointestinal tract (e.g., Eocene *Amia*: Boreske, 1974; Grande, 1980).
3. Discrete clusters of bromalitic material—often the case, but it can be disseminated, for example, clustered and separated by distance from other such clusters, and, in some cases, the putative consumulite could be an associated concretion (e.g., fish associated with *Elasmosaurus*: Everhart, 2017).
4. Found in a low-energy depositional setting in sediment finer in clast size than in the consumulite.
5. Dentalites present—dentalites are actually rarely reported in consumulites (an exception is the consumulite of *Postosuchus alisonae*: Sues et al., 2003; Peyer et al., 2008), and their absence can provide evidence of prey swallowed whole (e.g., conchostracan valves in the salamander *Jeholotriton*: Dong et al., 2012).
6. A further criterion is evidence of mechanical or chemical processing, which is relatively uncommon (e.g., etched headshield of *Cephalapis* in *Ptomacanthus*: Denison, 1956; Allen and Tarlo, 1963; Miles, 1973).

Consumulites are principally associated with, or at least recognized in the context of, articulated skeletons (Hunt and Lucas, 2021b). The majority of articulated skeletons are from aquatic environments, as are many consumulites. Articulated skeletons are common in Lagerstätten that preserve vertebrate fossils (Hunt and Lucas, 2021b). Thus, the record of consumulites is biased toward Lagerstätten such as the Cleveland Shale (Devonian), Holzmaden (Jurassic), Solnhofen (Jurassic), Jehol (Cretaceous), Messel (Eocene), and Green River (Eocene). Beyond Lagerstätten, consumulites are most common in fine-grained deposits of low-energy environments (Hunt and Lucas, 2021b). The vast majority of consumulites represent carnivorous animals because plant material is usually finely macerated during digestion, and it is much easier to recognize a bone as a foreign object than carbonaceous debris (Hunt and Lucas, 2021b).

Large body size favors the recognition of consumulites (Hunt and Lucas, 2021b). The combination of large body size and an aquatic lifestyle results in an extensive record of consumulites in large Mesozoic marine vertebrates, notably ichthyosaurs, plesiosaurs, and mosasaurs (Hunt and Lucas, 2021b).

5. Fossil record

Hunt and Lucas (2021b, Tables A.6—A.11) attempted to compile data on all recorded specimens of consumulites older than Pleistocene in age. However, most references to consumulites, particularly in the older literature, are within publications that have a different focus and are difficult to search for, so the compilation is undoubtedly incomplete. The volume of literature on Pleistocene consumulites precludes a complete review, so herein we only include representative examples of this age. We have characterized traces as marine or nonmarine, based largely on the original literature. We somewhat arbitrarily included lagoons in the nonmarine category, even though many such environments have marine influences. Significant publications since, or uncited in Hunt and Lucas (2021b) include Cooper et al. (2022), Cooper and Maxwell (2022, 2023), Druckenmiller et al. (2020), Dudgeon et al. (2023), Ebert, and Kölbl-Ebert (2008), Eriksson et al. (2022), Harper (2022), Hoffmann and Stevens (2020), Hu et al. (2022), Hua et al. (2024), Jamison-Todd et al. (2022), Maxwell et al. (2021), Polcyn et al. (2023), Sheffield et al. (2001), Spiekman et al. (2024), Therrien et al. (2023), Wang et al. (2022), Weis et al. (2024), White et al. (2022, 2023) and Wu et al. (2023).

The earliest invertebrate consumulites occur in the mollusc-like *Kimberella* from the Ediacaran (Bobrovskiy et al., 2022). Vertebrate consumulites have an extensive fossil record starting in the Devonian (Hunt and Lucas, 2021b). We review the fossil record of consumulites by era and environment. Evisceralites are considered separately because of their limited fossil record. Significant sources of data on the fossil record of consumulites include Massare (1987), Boucot (1990), Viohl (1990), Cicimurri and Everhart (2001), Konuki (2008), Boucot and Poinar (2010), Naish (2014), Everhart (2017), O'Connor (2019), and Hunt and Lucas (2021b).

5.1 Eviscerallites

Agassiz (1833, p. 676) introduced the term cololite to refer to the fossilized "more or less stuffed intestines of fish." He used the term in relation to both infilled intestines preserved within a body cavity and also those preserved in isolation and not directly associated with a body cavity; the latter are now termed eviscerallites (Hunt and Lucas, 2012). Eviscerallites are not found in association with skeletons but "M. Agassiz has ingeniously explained this fact by observing the process of decomposition of dead fishes in the lakes of Switzerland. The dead fish floats on the surface with its belly upwards, until the abdomen is so distended with putrid gas, that it bursts: through the aperture thus formed the bowels come forth into the water, still adhering together in their natural state of convolution" (Buckland, 1835, p. 156). The best evidence for this type of preservation may derive from the Middle Jurassic (late Callovian) fish fauna of the Luciano Mesa Member of the Todilto Formation in eastern New Mexico, USA (Koerner, 1930; Schaeffer and Patterson, 1984; Lucas et al., 1985: Fig. 8.2).

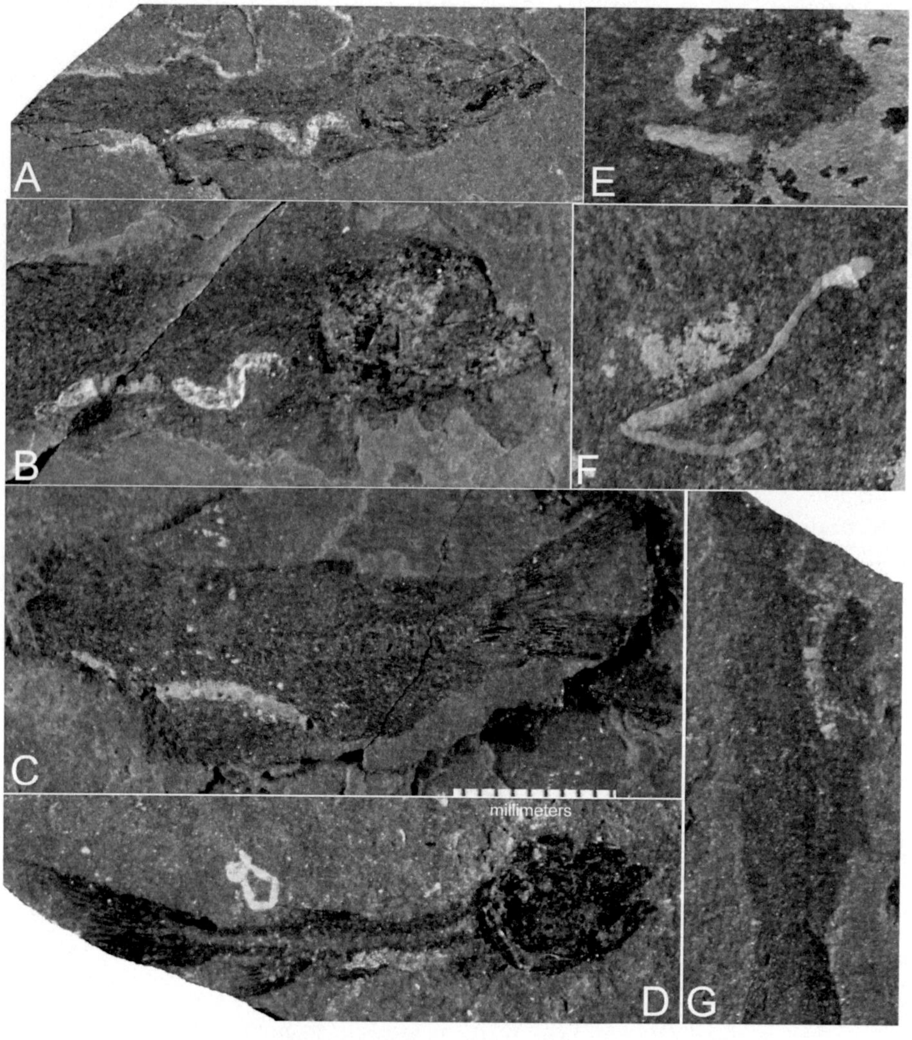

FIGURE 8.2 Middle Jurassic consumulites and evisceralites in "holostean" fishes from the Luciano Mesa Member of the Todilto Formation at NMMNH locality L-1476, Bull Canyon (A-D, G), and NMMNH locality L-3520, Warm Springs (E–F) in New Mexico, USA. **A,** NMMNH P-32715, *Todiltia schoewei* in lateral view with consumulite. **B,** NMMNH P-32713, *Todiltia schoewei* in lateral view with consumulite. **C,** NMMNH P-32718, *Todiltia schoewei* in lateral view with consumulite. **D,** NMMNH P-32714, *Todiltia schoewei* in lateral view with adjacent evisceralite. **E,** NMMNH P-28700, evisceralite in lateral view. **F,** NMMNH P-28700, evisceralite in lateral view. **G,** NMMNH P-32712, *Todiltia schoewei* in lateral view with consumulite. All to same scale (Hunt and Lucas, 2014, Fig. 12).

There, multiple specimens of the holostean fish *Todiltia schoewei* preserve consumulites, and the surrounding matrix also preserves isolated sections of intestinal tract (Schaeffer and Patterson, 1984; Lucas et al., 1985; Hunt and Lucas, 2014, Fig. 8.1). This type of evisceralite, in which a portion of the gastrointestinal tract is physically removed from a carcass, is dislocational (Hunt and Lucas, 2021b).

Buckland (1835, 1836) recognized that spiral coprolites were derived from fish with valvular intestines. Fritsch (1895) and Neumayer (1904) considered that some of the spiral bromalites actually represented fossilized valvular intestines rather than coprolites. Subsequently, Hoernes (1904) and Fritsch (1907) introduced terms for bromalites that represent infilled spiral-valved intestines, enterolite and enterospira, respectively. Hunt and Lucas (2012) accepted the term enterospira over enterolite because of widespread usage. The idea that spiral bromalites represent infilled intestines has been reexamined several times (e.g., Williams, 1972; McAllister, 1985; Ward et al., 2020). A 300–500 μm separation between whorls could represent a thin mucusoal membrane and be used as a criterion to distinguish intestinal infillings (Ward et al., 2020).

There are few other recorded examples of dislocational evisceralites. Clark (1989) described the earliest such evisceralite from the Late Mississippian (Serpukhovian) of Bearsden, Scotland (Fig. 8.3).

A second type of evisceralite results from preferential fossilization of the gastrointestinal tract, and is termed preservational (Hunt and Lucas, 2012, 2021b). Geologists of the U.S. Geological Survey and others first described unusual sideritic specimens from the Paleocene of North Dakota and the Miocene of Washington (Amstutz, 1958; Roberts, 1958; Brown, 1962). Subsequently, similar specimens were described from the Permian of China and the Late Cretaceous of Canada and Madagascar (Broughton et al., 1977; Broughton, 1981; Seilacher et al., 2001; Hunt and Lucas, 2016). Hunt et al. (2012a) designated one of the preservational evisceralites from North Dakota as the holotype of *Hiabromus seilacheri* (Fig. 8.3).

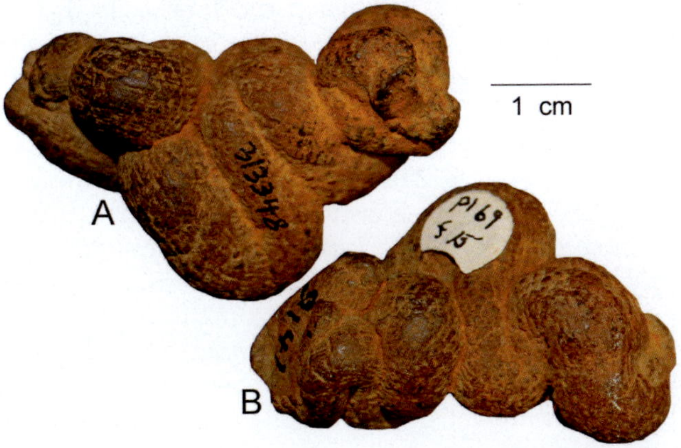

FIGURE 8.3 Holotype of evisceralite *Hirabromus seilacheri*. A-B, USNM 313348, Holotype bromalite in lateral views from the Miocene of Salmon Creek, Washington.

5.2 Other consumulites

5.2.1 Paleozoic

5.2.1.1 Marine

The earliest vertebrates were jawless, so they are considered to have been primarily filter feeders. Thus, it is not surprising that the earliest consumulite known consists of a gastrolite infilled by sediment within the furcacaudiform thelodont *Pezopallichthys ritchie* from the Middle Silurian of Canada (Wilson and Caldwell, 1993, 1998: Fig. 8.4). This strongly suggests that stomachs evolved before jaws in early vertebrates (Wilson and Caldwell, 1993). Other furcacaudiforms from the Early Devonian of Canada have similar arenaceous consumulites (Wilson and Caldwell, 1993, 1998), as do the endeiolepidiform *Endeiolepis aneri* and the euphanerid *Euphanerops longaevus*, both from the Late Devonian of Canada (Stensiö, 1939; Arsenault and Janvier, 1991; Janvier, 1996a,b; Janvier and Arsenault, 2007).

The oldest known consumulites that contain recognizable prey are from the Early Devonian of England and include a *Cephalaspis* head shield within the shark *Ptomacanthus* and a small ostracoderm within an acanthodian (Denison, 1956; Allen and Tarlo, 1963; Miles, 1973). Middle Devonian examples from Scotland occur in several specimens of *Coccosteus* and *Glyptolepis* as well as in acanthodians, actinopterygians, and placoderms (Heintz, 1938; Miles and Westoll, 1968; Ahlberg, 1992; Davidson and Trewin, 2005). One *Glyptolepis* contains a small *Glyptolepis* (Ahlberg, 1992). This is both the earliest record of cannibalism and the first to demonstrate a fish swallowing another head-first. This orientation of prey acquisition is prevalent in gape-limited predatory fish throughout the Phanerozoic. Swallowing prey head-first significantly reduces: (1) esophageal abrasion; (2) escape rate of the prey; and (3) the prey manipulation period (Reimchen, 1991). These advantages diminish with small prey. Head-first prey acquisition also occurs in most, but not all (e.g., Braz and Marques, 2016) snakes and squamates. A significant literature demonstrates analogous causation: (1) reduce the resistance offered by the appendages and body covering of the prey; (2) reduce the prey-handling time, which decreases the period of vulnerability to other predators; and (3) advantageous to ophiophagous snakes because it decreases the resistance from posteriorly projecting, overlapping ventral scales (Diefenbach and Emslie, 1971; Loop and Bailey, 1972; Loop, 1974; Greene, 1976; de Queiroz and de Queiroz, 1987; Wiseman et al., 2019). As with fish, the probability of head-first ingestion increases with increasing prey size (de Queiroz and de Queiroz, 1987).

The geographic distribution and volume of consumulite specimens increase substantially in the Late Devonian (Hunt and Lucas, 2021b, Table A.6: Fig. 8.4). More than 50 fish from the Cleveland Shale in Ohio, USA, preserve consumulites (Williams, 1990). These include more than 40 sharks representing four species of *Cladoselache*, *Ctenacanthus compressus*, and unidentified taxa. The majority of consumulites consist of paleoniscoid debris, but they also include a couple of complete fish (*Kentuckia blavini*), several contain arthropods and four yield conodonts. Three specimens contain the crustacean *Concavicaris* cf. *C. bradleyi*, which were swallowed tail first (Williams, 1990). Whereas most fish are swallowed head-first, most invertebrates with appendages in the Devonian and later are swallowed tail first, presumably to avoid damage to the predator during ingestion. Other Late Devonian consumulites are from Russia (Zakharenko, 2008), Latvia (Upeniece, 2001, 2011), Canada (Arsenault, 1982;

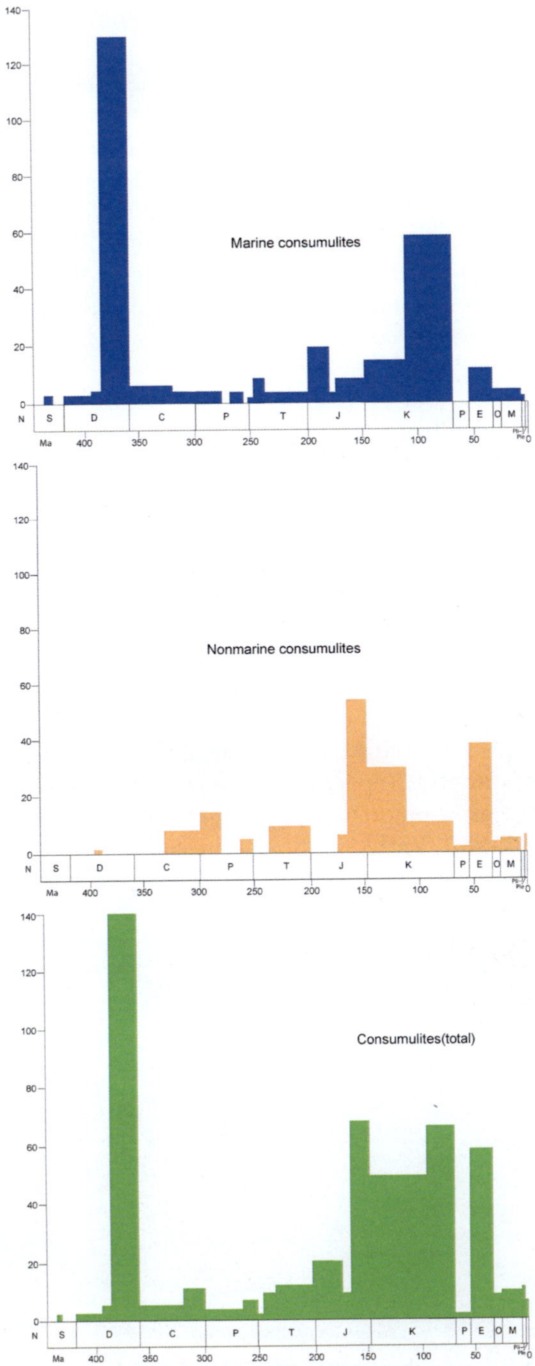

FIGURE 8.4 Temporal distribution of number of reported specimens of consumulites (Hunt and Lucas (2021b, Fig. 12)). The Devonian peak is the result of the Cleveland Shale Lagerstätte.

Janvier, 1996b; Chevrinais et al., 2017) and Australia (Dennis and Miles, 1981; Long, 1991, 1995; Choo et al., 2009).

Notably fewer records of consumulites come from the later Paleozoic (Hunt and Lucas, 2021b: Fig. 8.4). There are reports of specimens from the early Carboniferous of England and Scotland (Traquair, 1879; Watson, 1937; Viohl, 1990; Coates and Sequeira, 2001). The selachian *Akmonistan zangerli* from Scotland preserves the oldest incorporeal pelletite (*Chondripilula zideki*), and a second occurs in another shark (*Cobelodus aculeatus)* from the Late Pennsylvanian of the United States (Coates and Sequeira, 2001; Hunt, 1992; Hunt et al., 2012b; Hunt and Lucas, 2021a). *Typhloesus wellsi* from the Bear Gulch Lagerstätte in the Lower Mississippian of Montana, USA, is a rare example of an invertebrate that preserves consumulites of vertebrates (fish) as well as conodont apparatuses and worm teeth (Conway Morris, 1976, 1990).

Most Permian consumulites derive from Germany (Hunt and Lucas, 2021b, Table A.6). Three specimens are known from the late Permian Kupferschiefer (Weigelt, 1928a,b, 1930a,b; Malzahn, 1968; Schaumberg, 1979), and the remainder are from earlier in the period (Heidtke, 2007; Kriwet et al., 2008). Lohmann and Sachs (2001) reported on consumulites through four ontogenetic stages of the temnospondyl amphibian *Sclerocephalus haeuseri*. The only reptilian consumulites from the marine Paleozoic pertain to the mesosaur *Brazilosaurus sanpauloensis* from the lower Permian of Brazil and include crustacean fragments and bones of very young mesosaurids (Silva et al., 2017) (Fig. 8.5).

5.2.1.2 Nonmarine

The pattern of nonmarine Paleozoic consumulites is a reverse of that of the marine, with increasing numbers through time (Hunt and Lucas, 2021b, Fig. 12, Table A.7). There are no Devonian examples. Carboniferous specimens are currently only known from two Lagerstätten in the United States, Mazon Creek in Illinois (Richardson and Johnson, 1971; Richardson, 1980; Milner, 1982; Viohl, 1990; Godfrey, 1997) and the Kinney Brick Quarry

FIGURE 8.5 Holotype of intestinelite *Chondripilula zideki* within NMMNH P-19182, skeleton of the symmoriid chondrichthyan *Cobelodus aculeatus*, from the Late Pennsylvanian Kinney Brick Quarry Lagerstätte of New Mexico in the United States.

in New Mexico (Hunt, 1992; Hunt et al., 2012b; Werneburg et al., 2013, 2021; Hunt and Lucas, 2021a). One host is a shark, but all the remaining pertain to small amphibians (body lengths less than 50 mm). The "branchiosaur" *Milnerpeton huberi* from New Mexico and an unidentified amphibian from Illinois each have ostracods as consumulites (Fig. 8.6). The dvinosaurian *Bermanerpeton kinneyi* from Kinney preserves syncarid arthropods (*Uronectes* cf. *kinniensis*, *Aenigmacaris* cf. *minima)*, spines of *Acanthodes* cf. *kinneyi*, actinopterygian scales and ribs and neural arches of amphibians in consumulites (Figs. 8.7 and 8.8). Hunt and Lucas (2021a) named the ostracod-bearing consumulite *Werneburgichnus kinneyensis* (Fig. 8.6), and that containing more diverse content is named *W. varius* (Figs. 8.7 and 8.8). *Chondripilula zideki* is a pelletal body preserved within the posterior intestines

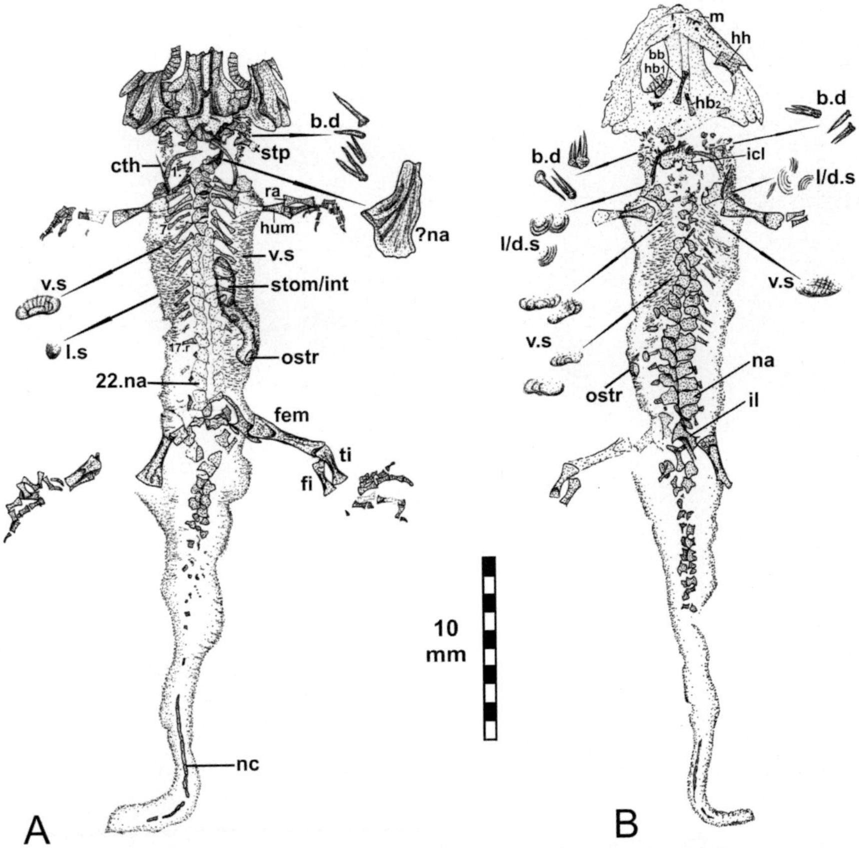

FIGURE 8.6 Holotype of *Werneburgichnus kinneyensis* igen et isp. nov. within skeleton of *Milnererpeton huberi*. **A,** Holotype NMMNH P-3696a. **B,** counterpart of the holotype NMMNH P-3696b. Abbreviations: bb, basibranchial; b.d, branchial denticles; cth, cleithrum; d.s, dorsal scales; fem, femur; fi, fibula; hb1, 2, hypobranchials 1 and 2; hh, hypohyal; hum, humerus; icl, interclavicle; il, ilium; l.s, lateral scales; m, maxilla; na, neural arch; nc, notochord; ostr, ostracodes; ra, radius; stp, stapes; stom/int, stomach or intestine; ti, tibia; v.s, ventral scales. *(from Werneburg et al., 2013, Fig. 5).*

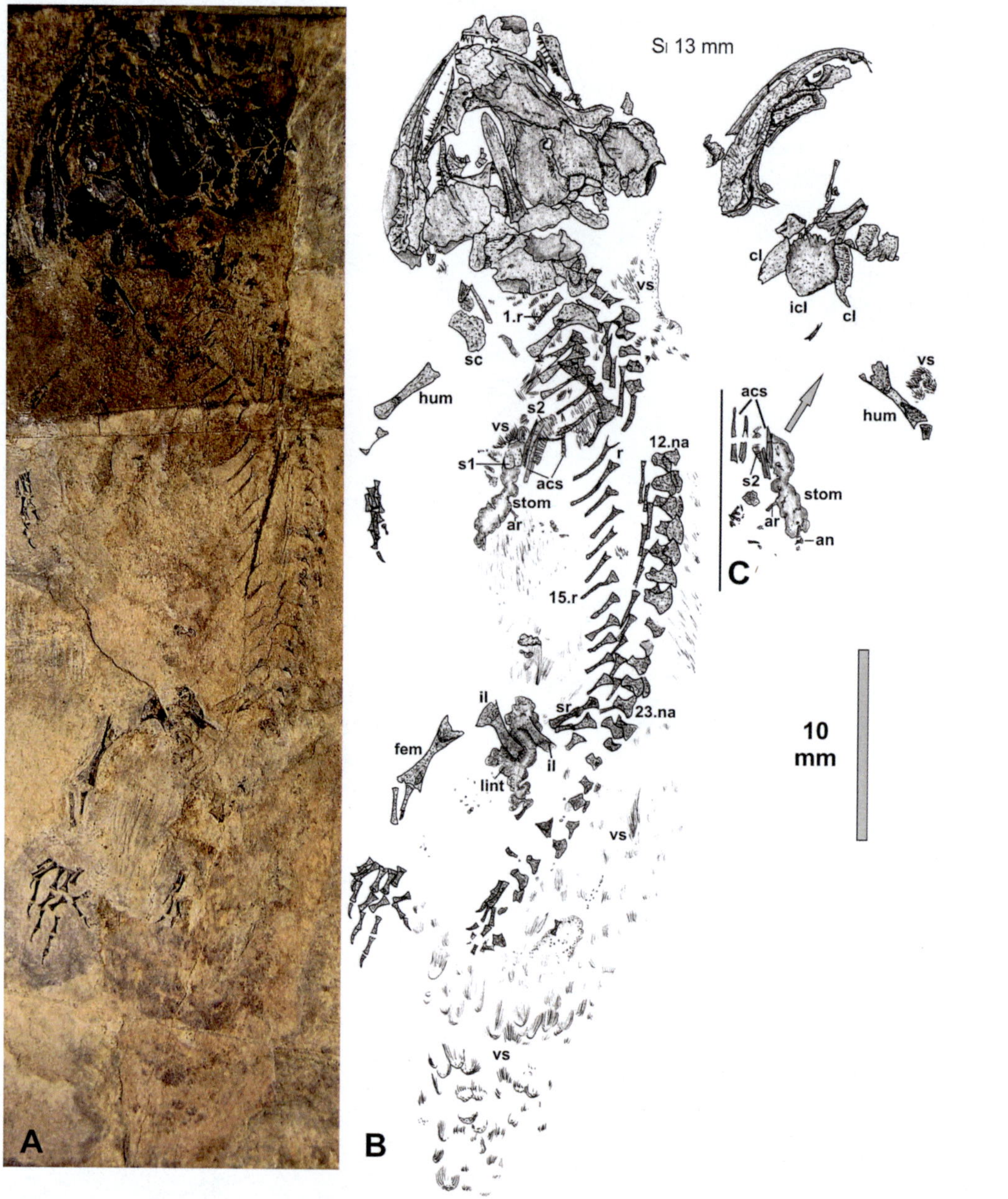

FIGURE 8.7 Holotype of *Werneburgichnus varius* within holotype skeleton of *Bermanerpeton kinneyi*—CM 41708. Part in ventral view (**A–B**), and counterpart in dorsal view (**C**). Abbreviations: acs = spine pairs of *Acanthodes*, an = amphibian neural arches, ar = small amphibian rib, cl = clavicle, fem = femur, hum = humerus, icl = interclavicle, il = ilium, lint = large intestine (intestinilite), na = neural arch, r = rib, s1 = syncarid arthropod 1, s2 = syncarid arthropod 2, sc = scapulocoracoid, stom = stomach (gastrolite), versus = ventral scales. *(from Werneburg et al., 2021, Fig. 3).*

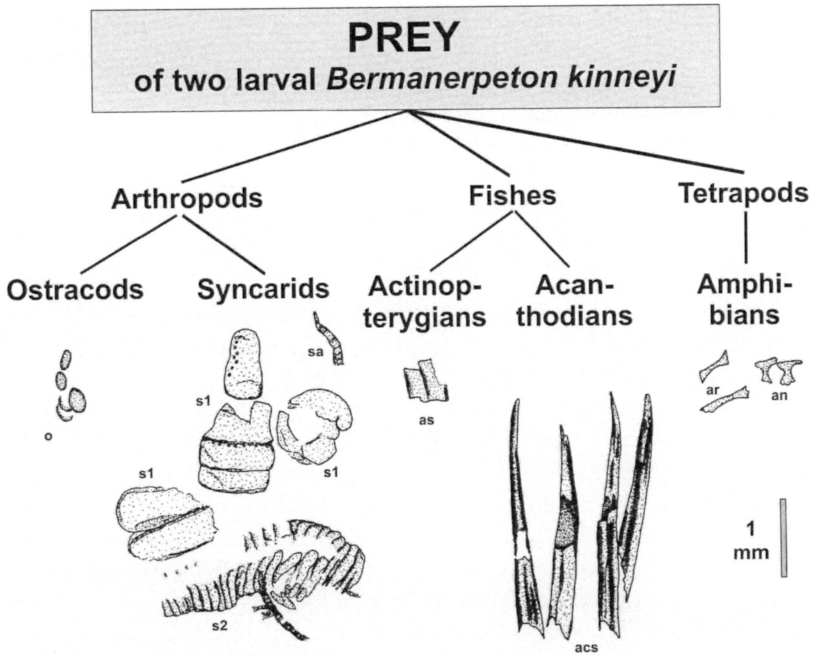

FIGURE 8.8 Gastrolite and cololite contents of two larval specimens of *Bermanerpeton kinneyi* (NMMNH P.67679a, b, and NHMS—WC 231) from the Kinney Brick Quarry Lagerstätte of New Mexico, USA. Abbreviations: **acs** = spine pairs of *Acanthodes*, **an** = amphibian neural arches, **ar** = amphibian rib, **as** = actinopterygian scales, **o** = ostracods, **r** = rib, **s1** = syncarid arthropod 1, **s2** = syncarid arthropod 2, **sa** = syncarid antennule. *(from Werneburg et al., 2021, Fig. 13).*

of sharks (Hunt and Lucas, 2021a: Fig. 8.5). No identifiable food items occur in the specimens from Mazon Creek (Godfrey, 1997).

Small "branchiosaur" amphibians from the lower Permian of Germany preserve a variety of consumulites that range from conchostracans to other amphibians (e.g., Werneburg, 1986, 1988, 1989, 2020; Witzmann, 2009). These strata also yielded the first matryoshkalite (consumulite within a consumulite), which is the shark *Triodus sessilis*, which ingested two amphibians, *Archegaurus decheni* and *Cheliderpeton latirostre*; the latter had swallowed a specimen of *Acanthodes bronni* (Kriwet et al., 2008). Two early Permian reptile skeletons from the United States contain oralites. A specimen of the captorhinimorph *Romeria* sp. from Texas contains a smaller skull and partial skeleton of probably the same species (Case, 1911; Eaton, 1964). Two skulls of the parareptile *Delorhynchus priscus* from Oklahoma preserve fragments of arthropod cuticle between the palatal teeth (Modesto et al., 2009; Reisz et al., 2014).

Late Permian reptile skeletons with consumulites occur in Tanzania and Germany (Hunt and Lucas, 2021b, Table A.7). A specimen of the gorgonopsid "*Aelurognathus*" *parringtoni* from Tanzania includes a mandibular symphysis of the dicynodont *Katumbia parringtoni* as a gastrolite (Maisch, 2009). Both Weigelt (1928a) and Munk and Sues (1993) described ovules of the conifer *Ullmannia frumentaria* associated with different specimens of the early archosauromorph reptile *Protorosaurus speneri* from Germany. The first report had been regarded with skepticism (e.g., Haubold and Schaumberg, 1985) because the plant material was scattered within a disarticulated skeleton, but the latter demonstrated the presence of plant material within an articulated ribcage. Munk and Sues (1993) also described coarse clastics with interspersed carbonized material (that they interpreted as macerated plant matter) in a consumulite of the pareisaurian parareptile *Parasaurus geinitzi.*

5.2.2 *Mesozoic*

5.2.2.1 Marine

The Mesozoic marine record of consumulites is extensive and dominated by sharks, bony fish, and the three main clades of marine reptiles (ichthyosaurs, plesiosaurs, and mosasaurs). The frequency of occurrences increases through the Mesozoic and reaches its maximum during the Late Cretaceous (Hunt and Lucas, 2021b, Table A.8: Fig. 8.4).

An Early Triassic specimen of the ray-finned fish *Birgeria* from Madagascar yields conspecifics as a demalite. This was originally interpreted as an example of ovoviparity by Beltan (1977), but these specimens more likely represent a consumulite (Viohl, 1990). Middle Triassic fish consumulites occur in specimens of *Saurichthys* in China and in the Monte San Giorgio Lagerstätte in Switzerland (Wu et al., 2015; Argyriou et al., 2016). Monte San Giorgio also yields consumulites in a nothosaur and a pachypleurosaur (Tschanz, 1989; Diedrich, 2015). Ichthyosaur consumulites occur not only at this Swiss locality (Rieber, 1970; Brinkmann, 2004) but also in the Early Triassic of Norway (Buchy et al., 2004) and the Late Triassic of China (Cheng et al., 2006) and the United States (Camp, 1930; Druckenmiller et al., 2014). The gastrointestinal tracts of Early and Middle Triassic ichthyosaurs yield only cephalopod hooklets, but the Late Triassic examples include both vertebrate remains and mollusk shell fragments (Rieber, 1970; Camp, 1980; Brinkmann, 2004; Buchy et al., 2004; Cheng et al., 2006; Druckenmiller et al., 2014).

Spiekman et al. (2024) describe a consumulite in the archosauromorph *Dinocephalosaurus orientalis* from the Middle Triassic of China that contains fish including a complete specimen of *Lashanichthys* sp. and a partial specimen referred to *Colobodus* sp. A Late Triassic thalattosauroid from Alaska preserves consumulite material in the vicinity of the pelvis (Druckenmiller et al., 2020).

The Early Jurassic (Toarcian) Holzmaden Lagerstätte (Posidonienschiefer Formation) of Germany has produced many specimens of consumulites (Hunt and Lucas, 2021b, Table A.8). These records include about 250 belemnite guards in a specimen of the shark *Hybodus* (Brown, 1900; Pollard, 1990; Doyle and McDonald, 1993; Hoffmann and Stevens, 2020) and coleoid remains in the bony fishes *Saurostomus* and *Pachycormus* (Urlichs et al., 1994; Přikryl et al., 2012). Cooper and Maxwell (2022, Table 1) revised *Saurostomus* and listed several specimens with consumulites that indicate a pelagic teuthophagous ecology. Cooper et al. (2022) described two specimens of the pachycormiform *Germanostomus pectopteri* with consumulites. The gastrointestinal tract of the holotype is very well preserved and preserves details of the soft tissue anatomy including the alimentary canal as well as infilling consumulites. A very unusual specimen of the actinopterygian fish *Pachycormus macropterus* preserves a large ammonite inside its gut (Cooper and Maxwell, 2023).

Contemporary strata in Luxembourg yield six specimens of pachycormid fish with esophagalites and gastrolites with octobrachian coleoid gladii (Weis et al., 2024). The fish represent

Pachycormus macropterus and the gladii are ascribed to *Teudopsis bollensis*, *Teudopsis* sp. indet. and Loligosepiidae indet. These specimens, together with others from Germany indicate that teuthophagy was widespread at the base of the clade that contains the suspension-feeding pachycormid giants of the Jurassic–Cretaceous (Weis et al., 2024).

Mateer (1974) documented an indeterminate organic area above the sacral area in a specimen of the Jurassic marine crocodylomorph *Steneosaurus bollensis*. However, the majority of consumulites from the Jurassic Posidonienschiefer pertain to ichthyosaurs. These include more than 35 specimens found in skeletons of *Stenopterygius* spp. and *Leptopterygius* spp. (von Wurstemberger, 1876; Keller, 1976; Massare, 1987; Böttcher, 1989; Bürgin, 2000). These consumulites principally contain cephalopod hooklets, but also include an actinopterygian fish, juvenile *Stenopterygius*, and rare wood fragments (e.g., Keller, 1976; Massare, 1987; Böttcher, 1989; Bürgin, 2000). Dick et al. (2016) studied a large sample of consumulites in *Stenopterygius quadriscissus* and demonstrated that ontogenetic shifts in the size and shape of the dentition coincide with dietary changes. The smallest specimens fed on small, burst-swimming fishes, with a progressive shift with increasing body size toward consuming faster-moving fish and, ultimately, cephalopods.

Ichthyosaur consumulites also occur in the Lower Jurassic of England in several species of *Ichthyosaurus* (Buckland, 1858; Pollard, 1968; Lomax et al., 2019, Fig. 8.9). The majority of

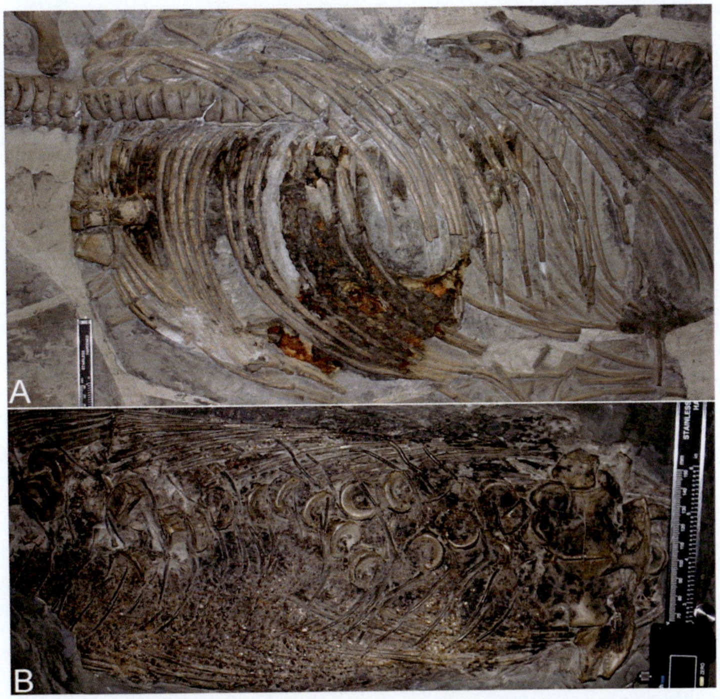

FIGURE 8.9 Consumulites in ichthyosaur skeletons from the lower Lias (Hettangian-lower Pliensbachian) of Lyme Regis, UK. **A,** Consumulite in lateral view (OUM J10320). **B,** Consumulite in lateral view (OUM J13593), also see Buckland, 1836, pl. 14) *(from Hunt and Lucas, 2021b, Fig. 14).*

specimens yield only cephalopod hooklets, but Buckland (1858) reported some *Pholidophorus* fish scales and spines. Middle Jurassic consumulites also derive from England from the Oxford Clay. They include cephalopods in the plesiosaurs *Simolestes vorax* and *Peloneustes* sp. (Andrews, 1910; Martill, 1992; Cicimurri and Everhart, 2001) and the crocodyliform *Metriorhynchus* sp. (Martill, 1985).

Hua et al. (2024) described a metriorhynchid specimen with stomach contents from the Middle Jurassic of France. This skeleton of *Metriorhynchus* cf. *superciliosus* preserves a gastrolite containing the gill apparatus remains of the large *Leedsichthys* and also invertebrate remains. Hua et al. (2024) postulate that the *Leedsichthys* was not the direct prey but rather that it was scavenged. Thus, longirostrine metriorhynchids were piscivorous but also probably opportunistic scavengers.

Late Jurassic consumulites are more geographically widespread but are not numerous (Hunt and Lucas, 2021b: Fig. 8.4). They include amorphous gut contents in the ichthyodectiform fish *Allothrissops* in Europe (Patterson and Rosen, 1977), small teleosts in the pachycormid *Hypsocormus posterodorsalis* from Nusplingen, Germany (Maxwell et al., 2021) and cephalopod remains in the ichthyosaurs *Pantosaurus striatus* and *Ophthalmosaurus natans* from the United States (Massare and Young, 2005; Wahl, 2012). One American plesiosaur (cf. *Pantosaurus striatus)* has ichthyosaur embryos in its digestive tract (O'Keefe et al., 2009), and another (cf. *Tatenectes laramiensis)* contains hybodont shark teeth, spines and denticles in addition to coleoid cephalopod hooklets (Wahl, 1998; Cicimurri and Everhart, 2001). Late Jurassic plesiosaur skeletons in Russia and England also contain consumulites with hooklets as well as fish (Zhuravlev, 1943a,b; Gekker and Gekker, 1955; Tarlo, 1959). Two skeletons of Metriorhynchidae from Germany preserve consumulites (Spindler et al., 2021). Taylor et al. (1993) described a specimen of *Pliosaurus brachyspondylus* that contains ornithischian dinosaur osteoderms, the oldest example of clearly terrestrial vertebrate remains in a marine consumulite.

Consumulites are geographically widespread in the Lower Cretaceous, but the majority of the specimens come from two countries, Brazil and Australia (Hunt and Lucas, 2021b, Table A.8; White et al., 2023). The Aptian Santana Formation of Brazil yields at least 10 examples of fish consumulites that contain identifiable fish that were ingested head-first (Patterson and Rosen, 1977; Case, 1982; Viohl, 1990; Wilby and Martill, 1992; Maisey, 1994, 1996; Mulder, 2013). The Early Cretaceous of Australia has produced consumulites in the turtle cf. *Notochelone* sp. (inoceramids), the ichthyosaur *Platypterygius longmani* (turtle, bird, fish), two specimens of the pliosaur *Kronosaurus* (fish, turtle, elasmosaur) and three plesiosaurs (principally invertebrates, teleost) (Kear et al., 2003; McHenry et al., 2005; Kear, 2006; McHenry, 2009; White et al., 2023). The plesiosaur *Nichollsia borealis* from Canada contains fish in a consumulite (Druckenmiller and Russell, 2009).

The very extensive record of Late Cretaceous consumulites is heavily biased toward the well excavated and researched vertebrate fauna of the Western Interior seaway of North America, notably in the United States, and particularly the state of Kansas (Hunt and Lucas, 2021b, Table A.8). There are six occurrences of shark consumulites from the United States, and one each from Italy and Lebanon, which preserve a wide range of prey including fish, turtle, plesiosaur, and mosasaur (Sternberg, 1917; Martin and Rothschild, 1989; Viohl, 1990; Druckenmiller et al., 1993; Shimada, 1997; Everhart, 2003; Amalfitano et al., 2017).

Arguably the most famous consumulite is the *Gillicus arcuatus* swallowed by *Xiphactinus audax*, the "fish within a fish" from Kansas that is on display at the Sternberg Museum of Natural History (e.g., Walker, 2006; Everhart, 2017, Fig. 1: Fig. 8.10). The Kansas chalk yields at least half a dozen more specimens of *Xiphactinus* with consumulites as well as several other bony fishes that exhibit similar examples (Hunt and Lucas, 2021b, Table A.8; see discussions or/and listings in Bardack, 1965; Konuki, 2008; Everhart, 2017). There is also a significant record of osteicthyans with consumulites from Lebanon (Davis, 1887; Woodward, 1901; Viohl, 1990; Konuki, 2008; Amalfitano et al., 2017). Other examples are known from Brazil (Viohl, 1990), Morocco (Arambourg, 1954; Cavin, 1999), France (Poplin, 1986) and Germany (Diedrich, 2012).

A large number of mosasaur skeletons preserve consumulites (Hunt and Lucas, 2021b, Table A.8). Most examples are from the United States (e.g., Konuki, 2008; Everhart, 2017), but specimens also occur in Canada (Konishi et al., 2011, 2014), Belgium (Dollo, 1887a,b, 1913) and Angola (Strganac et al., 2015; Polcyn et al., 2023). Mosasaur consumulites contain a large range of vertebrate prey items, including turtles (Dollo, 1887a,b; Konishi et al., 2011), sharks (Martin and Bjork, 1987), bony fishes (Williston, 1899, 1914; Camp, 1942; Martin and Bjork, 1987; Konishi et al., 2014; Everhart, 2017), plesiosaurs (Everhart, 2004), mosasaurs (Martin and Bjork, 1987; Bell and Barnes, 2007; Strganac et al., 2015) and birds (Martin and Bjork, 1987). The invertebrate content of mosasaur consumulites includes belemnites (Dollo, 1913), echinoids (Dollo, 1913), and bivalves (Martin and Fox, 2004). Polcyn et al. (2023) described a notable skeleton of *Prognathodon kianda* from the lower Maastrichtian of Angola that contains three partial mosasaurs of different taxa, including a cannibalized specimen of the host species.

Late Cretaceous plesiosaur consumulites occur in the United States (Cope, 1868, 1872; Brown, 1904; Martin and Kennedy, 1988; Storrs, 1999; Cicimurri and Everhart, 2001), Canada (Nicholls, 1988), and Japan (Matsumoto et al., 1982; Sato and Tanabe, 1998; Sato and Storrs, 2000). Most of these consumulites contain cephalopods (Brown, 1904; Sato and Tanabe, 1998; Sato and Storrs, 2000; Cicimurri and Everhart, 2001), fishes (Cope, 1868; Brown, 1904; Martin and Kennedy, 1988; Nicholls, 1988; Cicimurri and Everhart, 2001), mosasaurs (Cope, 1872; Storrs, 1999) and pterosaurs (Brown, 1904).

There is a single example of a Cretaceous pterosaur consumulite. This bromalite occurs in the throat of *Pteranodon* sp. and includes partially digested fish material, so it may be a regurgitalitic mass in the process of being egested (Brown, 1943; Bennett, 2001).

FIGURE 8.10 *Xiphactinus audax* (FHSM VP-333) from the Upper Cretaceous Smoky Hill Chalk of Kansas, USA, with a consumulite of *Gillicus arcuatus* (Everhart, 2017, fig. 5.1). Length of *Xiphactinus* is approximately 4 m long and *Gillicus* is 1.8 m long. *Photograph courtesy of Michael Everhart (from Hunt and Lucas, 2021b, Fig. 15).*

5.2.2.2 Nonmarine

There are many instances of Mesozoic nonmarine consumulites (Fig. 8.4), but a large percentage are associated with two Lagerstätten, the Late Jurassic Solnhofen Limestone of Germany, which is dominated by fish, and the Early Cretaceous Jehol biota of China in which most bromalites occur in tetrapods (Hunt and Lucas, 2021b, Fig. 12, Table A.9).

There are no reported Early or Middle Triassic consumulites. There are eight Late Triassic examples in terrestrial reptiles and one in a pterosaur. Three phytosaurs from the United States and India preserve consumulites that indicate consumption of a range of tetrapods, including the metoposaurid *Apachesarus* and reptiles, including *Vancleavea*, *Malerisaurus*, a phytosaur and a rhynchosaur (Chatterjee, 1978, 1980; Hunt, 1991, 1994, 2001; Hunt and Lucas, 2014). The paracrocodylomorph *Postosuchus alisonae* from the eastern United States contains a gastrolite with a diverse assortment of prey, including a partial skeleton of a small aetosaur (cf. *Aetosaurus*), a snout, left coracoid, and left humerus of the traversodont cynodont *Plinthogomophodon herpetairus*, two articulated phalanges of a large dicynodont, and a fragment of an unidentified ?temnospondyl bone. Some of the bones have dentalites, and others display periosteal erosion likely caused by digestion (Sues et al., 2003; Peyer et al., 2008).

Three skeletons of the Late Triassic theropod *Coelophysis bauri* from the Ghost Ranch quarry in New Mexico in the western United States yield several specimens of consumulites (Nesbitt et al., 2006; Rinehart et al., 2009; Hunt and Lucas, 2014): (1) AMNH 7224 contains a gastrolite with elements of the crocodylomorph *Hesperosuchus* (Nesbitt et al., 2006); (2) NMMNH P-44552 has an ovoid mass located between the proximal caudals and the ischia that is probably a cololite but could be an expelled coprolite (Rinehart et al., 2009: Fig. 8.11); and (3) NMMNH P -44551 includes an oralite that contains *Coelophyis* elements indicating cannabilism. The Late Triassic Swiss theropod *Notatesseraeraptor frickensis* has a consumulite that includes a maxilla of the rhynchocephalian *Clevosaurus* (Unterrassner, 2009).

There are no reported Early Jurassic nonmarine consumulites, but Middle Jurassic specimens include fish, amphibians, and a dinosaur (Hunt and Lucas, 2021b, Fig. 12, Table A.9). Many specimens of the bony fish *Todiltia schoewei* and a few of *Hulettia americana* from the Middle Jurassic Todilto Formation in New Mexico contain gastrolites and intestinilites (Lucas et al., 1985b; Hunt and Lucas, 2014: Fig. 8.2). Post-Paleozoic amphibian consumulites are rare, but there are three examples from the Middle Jurassic of China, including *Jeholotriton paradoxus* and *Chunperpeton tianyiensis*, which contain conchostracans, and a specimen of *C. tianyiensis* that yields insects (Gao and Shubin, 2003; Dong et al., 2012). The theropod *Poekilopleuron bucklandii* from France contains a tooth of the hybodont shark *Polyacrodus* sp. and bone fragments that resemble skeletal elements of cartilaginous fishes (Eudes-Deslongchamps, 1838).

There are more than 120 consumulites from the Upper Jurassic of the Solnhofen area in Germany (including the Ettling Lagerstätte of Ebert et al., 2015), which principally relate to fish and pterosaur skeletons. More than 110 fish skeletons representing more than 20 taxa yield consumulites that contain fishes (e.g., Viohl, 1990; Ebert, and Kölbl-Ebert, 2008; Ebert et al., 2015; Hunt and Lucas, 2021b, Table A.9). The coelacanth *Undina penicillata* is the only taxon from Solnhofen to have undoubtedly ingested a conspecific, and this was initially interpreted as evidence of ovoviparity but more recently of predation (Watson, 1927;

FIGURE 8.11 Skeleton of *Coelophysis bauri* (NMMNH P-42352) associated with a probable cololite. **A**, Overview of cololite lying between the proximal caudals and ischia (cm scale). **B**, Close-up of the cololite showing numerous small, bony, rod-like inclusions (cm scale). **C**, Photomicrograph of the rod-like inclusions (fish scale material?) (mm scale). **D**, Photomicrograph of the surface of a fish scale from the same quarry showing wavy, microscopic, rodlike structure that may be the source of the rod-like structures in the cololite *(from Rinehart et al., 2009; Fig. 118).*

Schultze, 1972). Five specimens of the pterosaurs *Rhamphorhynchus* and two of *Pterodactylus* contain consumulites composed of fish material (Broili, 1938; Wellnhofer, 1970, 1975a,b; Kellner, 2006; Frey and Tischlinger, 2012; Witton, 2013, 2018; Hone et al., 2013, 2015). The holotype of the theropod *Compsognathus longipes* contains a skeleton of the lepidosaur *Bavarisaurus* cf. *B. macrodactylus* (Ostrom, 1978). Outside of Solnhofen, there are four Late Jurassic fishes and a turtle that also contain consumulites (Eastman, 1911; De Saint Seine, 1949; Joyce et al., 2021).

The majority of nonmarine consumulites from the Lower Cretaceous derive from the Jehol Lagerstätte in northeastern China and include specimens from a frog, a choristodere, 10 theropods, 13 birds, a pterosaur, and a mammal (Hunt and Lucas, 2021b, Table A.9; Wang et al.,

2022; Wu et al., 2023). The frog *Genibatrachus baoshanensis* contains as a consumulite a skeleton of a salamander, cf. *Nuominerpeton* (Xing et al., 2019). Seven juvenile skulls of the choristodere *Monjurosuchus splendens* were found within the skeleton of an adult, indicating cannibalism in this taxon (Wang et al., 2005). The diversity of small, derived theropods and birds and their consumulites from Jehol provides important information about the development of the crop (cropalite), proventriculus (proventrilite), and ventriculus (O'Connor et al., 2019; Hu et al., 2022), discussed below. Mayr et al. (2020) demonstrated that several putative ovarian follicles in some Jehol birds actually represent propagules assignable to *Carpolithes multiseminalis*: (1) carbonaceous or originally carbonaceous; (2) putative follicles have similar dimensions in differently sized birds; (3) simultaneous maturing of multiple follicles unlikely; (4) no evidence of medullary bones; and (5) unlikely in specimens with little or no soft tissue preservation. O'Connor (2019) and O'Connor and Zhou (2020) recognized other paravian consumulites from Jehol (Hunt and Lucas, 2021b, Table A.9). The gastrolite of the mammal *Repenomamus robustus* is significant in that it provides direct evidence of predation on dinosaurs, as it contains remains of a juvenile of the ceratopsian *Psittacosaurus* (Hu et al., 2005). Surprisingly, Ji et al. (2016) reported fish remains in the presumably herbivorous ankylosaur *Liaoningosaurus paradoxus*.

A salamander and a bird from the Early Cretaceous of Spain yield consumulites (Sanz et al., 1996; Evans, 2016; O'Connor, 2019). The Australian crocodyliform *Confractosuchus sauroktonos* has abdominal contents of an ornithopod dinosaur (White et al., 2022). Two theropods contain consumulites, the large English theropod *Baryonyx walkeri* contains fish and *Iguanodon* bones, and the Italian *Scipionyx samniticus* yields fish and lepidosaurian bones (Charig and Milner, 1986, 1997; Dal Sasso and Signore, 1998; Dal Sasso and Maganuco, 2011). The ornithischian ankylosaur *Mimni* sp. from Australia contains a consumulite with plant debris and seeds (Molnar and Clifford, 2000, 2001). *Ludodactylus sibbicki* is a pterosaur from Brazil with the unusual oralite of a leaf between mandibular rami (Frey et al., 2003; Witton, 2018).

There are only about a dozen reports of consumulites from the Late Cretaceous, and they all represent tetrapods (Hunt and Lucas, 2021b, Table A.9: Fig. 8.4). Two examples that do not pertain to dinosaurs or birds are both from Brazil. These pertain to an unidentified frog and the baurusuchid crocodyliform *Aplestosuchus sordidus* (Leal and Martill, 2007; Godoy et al., 2014). A third example is in a champsosaur from Montana, USA (Dudgeon et al., 2023). Theropods are represented by three specimens from Mongolia and two from the United States. One American example is acid-etched vertebrae and a fragmentary dentary from a juvenile hadrosaur found in association with *Daspletosaurus* sp. (Varricchio, 2001). The second is a juvenile of *Gorgosaurus libratus* that contains two yearling caenagnathids (Therrien et al., 2023). The Mongolian (Hone et al., 2012; Lee et al., 2014), or putative Mongolian (Sereno et al., 2009; Fowler et al., 2011), specimens are *Deinocheirus mirificus* (fish), the enigmatic ?*Raptorex kriegsteini* (*Lycoptera* or Ellimmichthyiformes) and *Velociraptor mongoliensis* (azhdarchid pterosaur). At least four hadrosaurs from Canada and the United States contain plant debris that has been interpreted as consumulites (Kräusel, 1922; Ostrom, 1964; Taggart and Cross, 1997; Currie et al., 1995; Tweet et al., 2008, 2016). The sole Late Cretaceous bird consumulite of *Enantiophoenix electrophyla* is from Lebanon and is interesting because blebs of amber were interpreted to be evidence of feeding on sap (Dalla Vecchia and Chiappe, 2002; Cau and Arduini, 2008).

5.2.3 *Cenozoic*

5.2.3.1 Marine

There are relatively few consumulite reports from the Paleogene-Quaternary. The majority pertain to fish, with four examples from whales and one from a bird (Hunt and Lucas, 2021b, Table A.10: Fig. 8.4).

The early Eocene Pesciara di Bolca Konservat–Lagerstätte in Italy has yielded nine examples of consumulites from fish—two from sharks and the remainder from bony fish (Agassiz, 1833–1845a,b; Sorbini, 1972; Fanti et al., 2016; Friedman and Carnevale, 2018; Hunt and Lucas, 2021b, Table A.10). The presence of remains of the barracuda *Sphyraena bolcensis* within the shark *Galeorhinus cuvieri* is interesting because there is a predator: prey relationship between these genera in modern reef systems (Fanti et al., 2016). Other consumulite examples occur in a Miocene shark and a whale from Peru (Lambert et al., 2015; Collareta et al., 2017) and bony fish in the Oligocene of Germany (Weiler, 1934) and Romania (Paucă, M., 1933), the Miocene of Algeria (Carnevale and Pietsch, 2006), Peru (Collareta et al., 2017) and the United States (Crane, 1996) and the Pliocene of Italy (Sorbibi in Boucot, 1990; Viohl, 1990).

Three late Eocene whales from Egypt contain consumulites. *Basilosaurus isis* yields bones of a juvenile whale (*Durodon atrox*) and large fishes (*Pycnodus mokattamensis*), *B. cetoides* contains fishes and sharks, and *Durodon atrox* contains fishes (Swift and Barnes, 1996; Uhen, 2004; Voss et al., 2019). Martínez-Cáceres et al. (2017) described another example from Peru. An undescribed late Miocene whale contains a mass of bones of the sardine *Sardinops*, which provides the first evidence of piscivory in an edentulous mysticete (Collareta et al., 2015). The volume of the fish argues against them being accidently ingested. Lambert et al. (2015) described specimens of the clupeiform *Sardinops* sp. cf. *S. sagax* in the odontocete *Messapicetus gregarius* from the same strata in Peru. An early Oligocene loon (*Colymboides metzleri*) from Germany has a gastrolite consisting of fragments of small fish (Mayr, 2004).

5.2.3.2 Nonmarine

The consumulite record from nonmarine Cenozoic rocks is skewed by large samples from several Lagerstätten, notably the Eocene Green River Formation of the United States and the Middle Eocene Messel Lagerstätte of Germany (Hunt and Lucas, 2021b, Fig. 12, Table A.11). The Paleocene Menat Lagerstätte in France yields the oldest Cenozoic consumulites, with a fish within a crocodylomorph and a gastrolite within an insectivore (Guth, 1962; Wedmann et al., 2018).

The lacustrine Green River Formation includes a number of fish consumulites (Jepsen, 1967; Boreske, 1974; Ulrich, 1978—cited by Boucot, 1990; Grande, 1980). A notable specimen is an example of a matryoshkalite (consumulite within a consumulite) as *Amia uintaensis* consumed *Diplomystus*, which, in turn, had consumed *Knightia* (Boreske, 1974; Grande, 1980). The Green River yields several aspirationalites (Fig. 8.12).

The Middle Eocene of Messel has yielded many consumulites, notably from multiple mammals (Franzen, 1976, 1977, 1984, 1985; Richter and Storch, 1980; von Koenigswald et al., 1981, 1992; Maier et al., 1986; Richter, 1988; Collinson, 1990; Storch, 2001) and birds (Mayr and Peters, 1998; Mayr and Mourer-Cauvré, 2000; Mayr, 2002, 2009, 2013; Mayr et al., 2004, 2006; Morlo, 2004; Mayr and Richter, 2011; Hunt and Lucas, 2021b, Table A.11:

FIGURE 8.12 Aspirationalite from the early Eocene Green River Formation of Wyoming, USA. *Mioplosus labracoides* swallowing *Knightia eocaena.* The specimen is approximately 50 cm long. *Photograph courtesy of Vincent Santucci, National Park Service (from Hunt and Lucas, 2021b, Fig. 11).*

Fig. 8.13A). These consumulites pertain to both carnivores and herbivores. Multiple specimens of *Rhenanoperca minuta* and *Thaumaturus intermedius* yield insects, crustaceans, and fish (Richter and Baszio, 2001, 2006). Two lizards and two snakes contain gastrolites (Greene, 1983; Schaal, 2004; Weber, 2004; Smith and Scanferla, 2016; Scanferla and Smith, 2020). One specimen of the booid *Eoconstrictor fisheri* has the lizard *Geiseltaliellus maarius* in its stomach that, in turn, contains an insect in its digestive tract, so it represents a matryoshkalite (Smith and Scanferla, 2016, 2020; Scanferla and Smith, 2020, Fig. 8.14). This is only the second example of a consumulite within a consumulite. Five specimens of the putative pangolin *Eomanis waldi* have gastrolites of coarse sand and plant debris rather than the insects that might be expected (von Koenigswald et al., 1981; Richter, 1988). Other terrestrial mammals with consumulites represent four families (Pantolestidae, Dichobunidae, Equidae, and Amphilemuridae: Fig. 8.13B). The bat *Palaeochiropteryx tupaiodon* principally yields lepidopteran wings but also some hair that presumably was derived from grooming (Richter and Storch, 1980; Richter, 1988). *Archaeonycteris* yields beetles, and other bats also contain moths and caddis flies and beetles (Habersetzer et al., 1994: Fig. 8.13B). Several birds yield consumulites. Eleven taxa of birds yield gastrolites and cropalites (Mayr and Peters, 1998; Mayr and Mourier-Chauviré, 2000; Mayr, 2002, 2009, 2013, 2015; Morlo, 2004; Mayr et al., 2006; Mayr and Richter, 2011; Mayr and Wilde, 2014). For example, the mousebird *Masillacolius brevidactylus* yields a large seed (Mayr, 2015).

Alexander and Burger (2001) described a crocodilian, *Pristiochampsus vorax*, from the middle Eocene Grizzly Buttes locality in Wyoming in the United States, which had ingested a hindlimb of the perissodactyl *Helaletes* sp. Oligocene specimens include the frog *Palaeobatrachus* sp. from Germany with ingested fish (Wuttke and Poschmann, 2010), tadpoles of *Pelobates decheni* and *Eopelobates anthracinus* from Germany with granular gut contents, the tortoises *Stylemys* sp. and *Stylemys nebrascensis* from the United States with hackberry seeds (Marron and Moore, 2013) and the bird *Oligocolius psittacocephalon* from Germany with fruit stones in its crop (Mayr, 2013).

Miocene consumulites include the fish *Lepidocottus* from Germany, which fed on planorbid snails (Wuttke and Poschmann, 2010), and *Rama pueyoi* frogs and their tadpoles from Spain, respectively, which preserve gastrolites composed of gastropod shells (with rarer plants, arthropods and larval anurans) and fine debris that includes diatoms (McNamara et al., 2009, 2010; Wuttke and Poschmann, 2010). The only recorded Miocene mammalian consumulite is

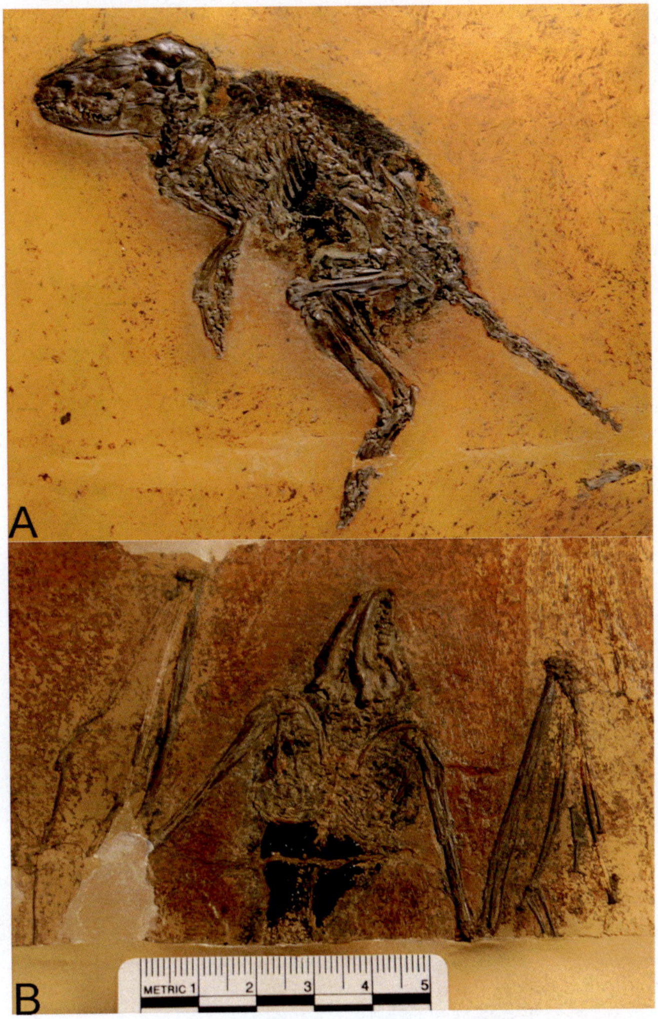

FIGURE 8.13 Gastrolites in tetrapods from the middle Eocene Messel Lagerstätte of Germany. A, Amphilemurid mammal *Macrocranion tupaiodon* with approximate length of 15 cm. B, SMF-ME 11369, archaeonycterid bat *Archaeonycteris* sp.

from the canid *Carpocyon* from the United States, which ingested the rabbit *Hypolagus* and other small mammals (Green, 1948; Stirton, 1959; Tedford in Boucot, 1990). We have found no records of Pliocene consumulites.

There are many examples of Pleistocene consumulites. The frozen mummies of large herbivores that contain consumulites, notably *Mammuthus*, *Coelodonta*, *Equus*, and *Bison*, have been extensively studied (e.g., Ukraintseva, 1981, 1993; Guthrie, 1990; Harington, 2007; van Geel et al., 2008; Boeskorov et al., 2011, 2014; Kosintsev et al., 2012; Potapova and Potapova, 2024).

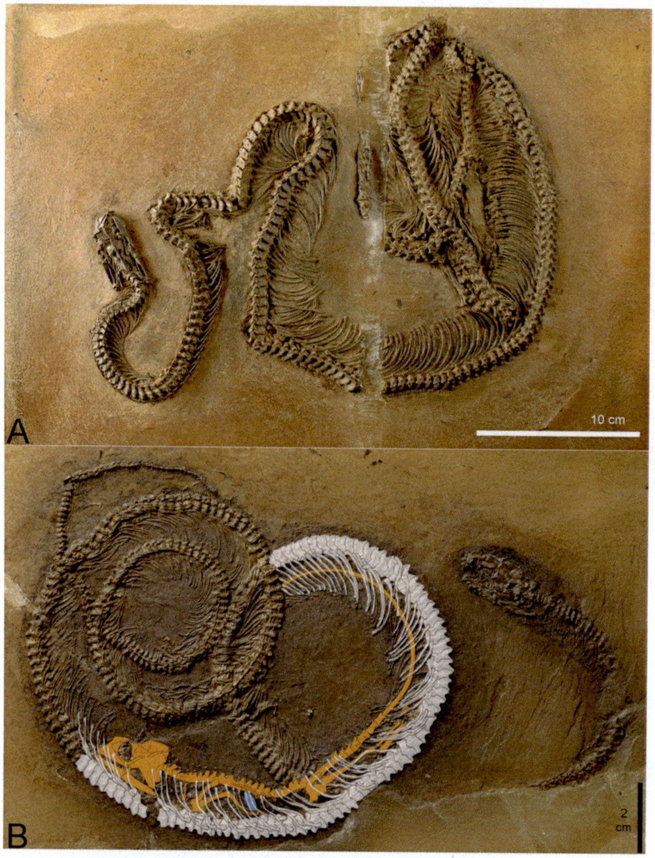

FIGURE 8.14 A matryoshkalite from the Middle Eocene Messel Lagerstätte of Germany. **A**. Skeleton of booid *Eoconstrictor fischeri* (SMF ME 11398) with no consumulite. **B.** Skeleton of *Eoconstrictor fischeri* with matryoshkalite. Interpretive drawing of SMF ME 11332a overlaid on a photograph. The lizard, *Geiseltaliellus maarius* (orange), is preserved in the stomach of the snake *Eoconstrictor fisheri* (white). The lizard was swallowed headfirst, and the tail does not appear to have been lost during the encounter with the snake. The position of an unidentified insect in the abdominal cavity of the lizard is indicated in outline (blue). *Published with kind permission of ©Krister T. Smith 2016.*

Several studies use the terms "feces" and "dung" to refer to fragments of consumulites of *M. primigenius* (e.g., Van Geel et al., 2004, 2011; Mol et al., 2006).

6. Ichnotaxonomy

Knaust (2020) and Bertling et al. (2022) have both expressed the opinion that consumulites "should not be named because no part of an animal should be named separately" to prevent an "ichnotaxon from entering into synonymy with the producer biotaxon" (Bertling et al., 2022, p. 4). They do, however, consider evisceralites (their cololites) as nameable because of the disassociation of the trace fossil and the body fossil. Furthermore, they note that the definition of a trace fossil (as given by Bertling et al., 2006) is that it results from the life activity of an individual organism that has modified its substrate. In the case of an evisceralite this modification is taken to be the "the separation and uptake (outside of the body) followed by the transport and compositional changes (inside the body) of the nutrient particles" (Bertling et al., 2022, p. 4). Even though gastroliths typically lie within a body cavity "the trace is not the piece of rock; rather it is the microscopic surface alteration" and thus by inference they could be given ichnotaxonomic nomenclature (Bertling et al., 2022, p. 4).

Nevertheless, these lines of reasoning strike us as illogical. First, if a named gastrolith within a skeleton does not risk a confusing synonymy with the host animal why does this apply to consumulites? Second, "compositional changes (inside the body) of the nutrient particles" occur in consumulites. Third, the surfaces of elements within consumulites often display evidence of physical and chemical erosion that is directly analogous to "the microscopic surface alteration" of gastroliths. Fourth, it seems unreasonable that a pellet within the posterior gastrointestinal tract cannot have a name but if it is located 1 centimeter outside the anus it can, and what of a fecal mass that is preserved in mid-excretion? We believe that consumulites should be named for the same reason as other traces—to facilitate recognition, description, and comparison.

7. Ichnotaxobases

There are currently only four named consumulites, three preserved within the body cavity, *Werneburgichnus kinneyensis*, *W. varius*, and *Chondripilula zideki* (Hunt and Lucas, 2021a), and the evisceralite *Hiabromus seilacheri* (Hunt et al., 2012a). Evisceralites are discrete bodies, so ichnotaxobases include shape, size, content, composition, and surface texture. However, there are distinct morphologies and content that should allow the erection of other ichnotaxa of consumulites preserved within body cavities. Bromalite content representing different diets would represent different ichnotaxa because they represent different behaviors. Potential ichnotaxobases include: (1) location of bromalite in digestive tract; (2) morphology, for example, pellet, infilling that follows form of digestive tract; (3) nature of content and inclusions (including taxonomic identification of inclusions); and (4) degree of mechanical and biochemical processing.

8. Systematic ichnotaxonomy

8.1 *Werneburgichnus* Hunt and Lucas (2021a)

Type ichnospecies: *Werneburgichnus kinneyensis* Hunt and Lucas (2021).

Included ichnospecies: The type ichnospecies and *Werneburgichnus varius* (Hunt and Lucas (2021)).

Distribution: Pennsylvanian and early Permian of the United States and Germany.

Diagnosis: Invertebrate or vertebrate specimens within gastrolites and/or intestinilites of small, branchiosaur-like amphibians.

Discussion: Consumulites occur in several Pennsylvanian and early Permian branchiosaurs and other small, pedomorphic amphibians (Hunt and Lucas, 2021a,b). There is clearly an ontogenetic change in diet in some taxa, which is reflected in the different composition of the consumulites (Werneburg, 1986, 1988, 1989).

8.2 *Werneburgichnus kinneyensis* Hunt and Lucas (2021a)

Holotype: NMMNH P-3696a, b, within holotype of *Milnerpeton huberi* (Hunt et al., 1996; Werneburg et al., 2013, Figs. 4 and 5, 8–10; Hunt and Lucas, 2021a, Figs. 2–4: Fig. 8.6).

Type locality: NMMNH locality 345, Kinney Brick Quarry, Bernalillo County, New Mexico, USA.

Type horizon: Upper Pennsylvanian Tinajas Member of the Atrasado Formation (Missourian: Kasimovian).

Distribution: Middle and Late Pennsylvanian of New Mexico and Illinois, USA.

Referred specimens: Within NMMNH P-15019a-b, referred specimen of *Milnerpeton huberi* (Werneburg et al., 2013) from NMMNH locality 345, Kinney Brick Quarry, New Mexico, USA5); within unidentified amphibian from Mazon Creek fauna at Peabody Coal Pit 11, Illinois, USA (Godfrey, 1997).

Diagnosis: Ichnospecies of *Werneburgichnus* consisting of scattered ostracods, usually isolated, within a gastrolite and possibly intestinilite.

Description: The part of the holotype of *Milnerpeton huberi*, NMMNH P-369a, preserves a sinuous body with rounded ends that is on the anatomical left side (right of holotype part, which preserves the skeleton in ventral view) and extends from about halfway down the dorsal vertebral series close to the midline to about three quarters of the way along the lateral margin (Werneburg et al., 2013, Fig. 4-4; Hunt and Lucas, 2021a, Figs. 2–3). This consumulite is 7.25 mm in length, with widths ranging from 0.6 to 1 mm. It consists primarily of fine-grained matrix but contains ostracods, notably at the posterior lateral margin (Werneburg et al., 2013, Fig. 5; Hunt and Lucas, 2021a, Fig. 3). The counterpart does not show the elongate consumulite, but also preserves an ostracod on the posterolateral margin (Werneburg et al., 2013, Figs. 5 and 8; Hunt and Lucas, 2021a, Figs. 2–3), and ostracods more anteromedially and anterolaterally (Werneburg et al., 2013, Fig. 8A–C; Hunt and Lucas, 2021a, Fig. 4A–C).

NMMNH P-15019a, b is a second specimen of *Milnerpeton huberi* from Kinney that preserves a consumulite of *Werneburgichnus kinneyensis* (Werneburg et al., 2013, Fig. 10; Hunt and Lucas, 2021a, Fig. 5). This specimen includes several specimens of ostracods on the counterpart (NMMNH P-15019a: Werneburg et al., 2013, Fig. 10A; Hunt and Lucas, 2021a, Fig. 5A).

Discussion: Hunt et al. (1992, 1996) first described specimens of *Milnererpeton*, but they did not recognize the consumulites. Werneburg et al. (2013) recognized the cylindrical consumulite containing ostracods. They interpreted it as stomach or intestines and the ostracods as specimens of *Carbonita* (Kietzke and Kaesler, 1992). The anterior third of the consumulite is wider than the posterior portion, and it is possible that the posterior segment represents the intestines. *Milnererpeton huberi* had a system of branchial ossicles, bearing one or two needle-like points that would have allowed filter feeding on planktonic prey (Werneburg et al., 2013). Thus, the fine-grained material between the ostracods may result from consumption of such prey.

8.2.1 Werneburgichnus varius *Hunt and Lucas (2021a)*

Holotype: CM 41708a, b, within holotype of *Bermanerpeton kinneyi* (Werneburg et al., 2021, Figs. 3, 7, 13; Hunt and Lucas, 2021a, Figs. 6–7: Fig. 8.7).

Type locality: NMMNH locality 345, Kinney Brick Quarry, Bernalillo County, New Mexico, USA.

Type horizon: Upper Pennsylvanian Tinajas Member of the Atrasado Formation (Missourian: Kasimovian).

Distribution: Late Pennsylvanian of New Mexico, USA.

Referred specimens: Within two paratype skeletons of *Bermanerpeton kinneyi*, NMMNH P–67679a, b and NHMS-WC 231 (Werneburg et al., 2021).

Diagnosis: Ichnospecies of *Werneburgichnus* with specimens of diverse invertebrates and vertebrates within the stomach and intestines.

Description: The holotype is preserved on both the part and counterpart of CM 41708a, b. It consists of two segments of the gastrointestinal tract, one identified as a gastrolite posterior to the pectoral girdle and the other as an intestinilite associated with the pelvis (Hunt and Lucas, 2021a, Figs. 6–7). The gastrolite has irregular borders and is about 6 mm long. There are several identifiable specimens within or in the immediate vicinity of the gastrolite, including several carapace fragments and an antennule of a syncarid crustacean, two pairs of acanthodian spines, and a small amphibian rib and neural arch (Werneburg et al., 2021).

A second specimen of *Werneburgichnus varius* occurs within another skeleton of *Bermanerpeton kinneyi* (NMMNH P-67679; Werneburg et al., 2021, Fig. 7; Fig. 8.7). The gastrolite is subround in shape, with a diameter of about 2 mm. Immediately posterior to the gastrolite are two syncarid carapace segments and an ostracod. The linear intestinilite is associated with a syncarid antennule, a small amphibian rib and two actinoperygian scales (Fig. 8.8).

Discussion: The ostracods have smooth shells and are tentatively assigned to *Carbonita* (Kietzke and Kaesler, 1992; Werneburg et al., 2021). Two syncarid crustacean carapace fragments are preserved, and they may pertain to *Uronectes* cf. *kinniensis* (smaller carapace elements) and to *Aenigmacaris* cf. *minima* (larger carapace elements) (Werneburg et al., 2021). The vertebrate remains in the consumulite are pairs of spines from *Acanthodes* cf. *kinneyi*, actinopterygian scales and small ribs and neural arches of an indeterminate amphibian (Werneburg et al., 2021).

8.3 *Chondripilula* Hunt and Lucas (2021a)

Type ichnospecies: *Chondripilula zideki* Hunt and Lucas (2021a).

Included ichnospecies: Known only from the type ichnospecies.

Distribution: Mississippian and Pennsylvanian of the USA and Scotland.

Diagnosis: Large (15–30 mm in length) spiral pellet with broader posterior termination located in the hind skeleton of a chondrichthyan adjacent or posterior to the pelvic plate.

Discussion: This ichnotaxon is known from the Kinney specimen from the Pennsylvanian and two others, from the Mississippian of Scotland (Coates and Sequeira, 2001) and the Middle Pennsylvanian of Indiana, USA (Zangerl and Richardson, 1963).

8.4 *Chondripilula zideki* Hunt and Lucas (2021a)

Holotype: NMMNH P-19182, within skeleton of the symmoriid chondrichthyan *Cobelodus aculeatus* (Zidek, 1992, Fig. 3A; Hunt, 1992, Fig. 2; Hunt et al., 2012b, Fig. 4; Hunt and Lucas, 2021a, Fig. 9: Fig. 8.5).

Type locality: NMMNH locality 345, Kinney Brick Quarry, Bernalillo County, New Mexico, USA.

Type horizon: Upper Pennsylvanian Tinajas Member of the Atrasado Formation (Missourian: Kasimovian).

Distribution: As for ichnogenus.

Referred specimens: Within HMV8246, the holotype of *Akmonistion zangerli* (Coates and Sequeira, 2001, Fig. 14) from the Mississippian Manse Burn Formation (Mississippian: Serpukhovian) of Scotland. Within FMNH PF 2207, skeleton of *Stethacanthus* from the Middle Pennsylvanian (Desmoinesian) of Indiana, USA (Zangerl and Richardson, 1963, p. 197).

Diagnosis: As for ichnogenus.

Description: The holotype intestinelite is an incorporeal pelletite *sensu* Hunt and Lucas (2012). It is located posterior to the pelvic plate and immediately dorsal to the first series of hemal arches (Hunt and Lucas, 2021a, Fig. 9). It is broadly oval in lateral view, with a length of 28 mm and a maximum diameter of 14 mm. It tapers anteriorly and has a broad and rounded posterior termination. The overall shape is consistent with a heteropolar morphology, but the surface was slightly damaged during preparation, so this cannot be confirmed. It contains macerated fish debris that appears to be acanthodian.

The holotype of *Akmonistion zangerli* cololites preserves an incorporeal pelletite directly above the pelvic plates and ventral to the dorsal fin and thus anterior to the position in the holotype of *Chondripilula zideki* (Coates and Sequeira, 2001). This specimen is subrectangular in lateral view with a length of 15 mm and a maximum diameter of 10 mm. The posterior end is blunt. Three spirals are preserved in this specimen in addition to finer striae. It appears to be amphipolar in morphology. Zangerl and Richardson (1963) noted a spiral coprolite in the pelvic area of a specimen of *Stethacanthus* from Indiana.

Discussion: *Chondripilula zideki* represents a formed pellet, so it must be located within the rectum or cloaca, posterior to the spiral valve, which gives some indication of the internal anatomy of the host sharks. Thus, the cloaca of *Cobelodus aculeatus* is more posterior than that of *Akmonistion* or *Stethacanthus.*

8.5 *Hirabromus* Hunt et al. (2012b)

Type ichnospecies: *Hirabromus seilacheri* Hunt et al. (2012b).

Included ichnospecies: Known only from the type ichnospecies.

Distribution: Permian of China, Cretaceous of Canada and Madagascar, Paleocene and Miocene of United States.

Diagnosis: Bromalite that differs from other ichnogenera in consisting of an elongate, convoluted, longitudinally striated cylinder with tapering terminations composed of siderite.

Discussion: These ichnofossils have been identified from the Permian (China), the Cretaceous (Canada, Madagascar) and the Paleocene and Miocene (United States) and have variously been interpreted as coprolites, pseudofossils or casts of internal organs (Amstutz, 1958;

Broughton et al., 1977; Broughton, 1981; Schmitz and Benda, 1991; Spencer, 1993; Mustoe, 2000; Seilacher et al., 2001). Seilacher et al. (2001) argued that they are both ichnofossils and cololites and represent what Hunt and Lucas (2012) termed evisceralites. Distinctive features of these bromalites that indicate that they are cololites include: (1) longitudinal striations that represent taenial muscle bands; (2) sinuous shape; (3) prominent tapering at both ends; and (4) more complex specimens (e.g., Hunt et al., 2012b, Fig. 3K–N) consist of segments whose opposing ends taper in opposite directions (Seilacher et al., 2001).

8.6 *Hirabromus seilacheri* Hunt et al. (2012b)

Holotype: USNM 313350 (Hunt et al., 2012b, Fig. 3O–R; Brown, 1962, pl. 69, Fig. 17: Fig. 8.3).

Type locality: 8 miles south of Rhame, North Dakota.

Type horizon: Fort Union Formation, Paleocene (Torrejonian).

Distribution: As for ichnogenus.

Referred specimens: Upper Permian, China (Seilacher et al., 2001, Fig. 1A): Upper Miocene Wilkes Formation, WA: USNM 313348 (Fig. 8.2; Brown, 1962, pl. 69, Fig. 15); GPIT 1860/1 (Seilacher et al., 2001, Fig. 1B); GPIT 1860/2 (Seilacher et al., 2001, Fig. 1C); GPIT 1860/3 (Seilacher et al., 2001, Fig. 3A): GPIT 1860/4 (Seilacher et al., 2001, Fig. 3B): GPIT 1860/5 (Seilacher et al., 2001, Fig. 3C); YPM56257 (Seilacher et al., 2001, Fig. 4): Paleocene Fort Union Formation, ND: USNM 313349 (Brown, 1962, pl. 69, Fig. 16); USNM 313723 (Brown, 1962, pl. 69, Fig. 18); USNM 313724 (Brown, 1962, pl. 69, Fig. 19): Paleocene Golden Valley Formation, ND, YPM 56410 (Seilacher et al., 2001, Fig. 5 center); YPM 56411 (Seilacher et al., 2001, Fig. 5 right).

Diagnosis: As for ichnogenus.

Description: USNM 313350 is an elongate (54 mm-long), longitudinally striated cylinder with acute tips at either end (Brown, 1962, pl. 69, fig.17l; Hunt et al., 2012a, Fig. 3O–R). The specimen is only flexed in one dimension, with two tight curves in lateral view.

Discussion: Given its size and preservation in a Paleocene fluvial environment, the holotype probably represents a crocodylian. In contrast, large specimens from the upper Miocene of Washington presumably pertain to a mammal (Hunt et al., 2012a). Broughton (2017) proposed an ad hoc hypothesis that multidecimeter-long specimens may be evisceralites of an unknown taxon of giant terrestrial earthworm (Oligochaeta) that existed from the Late Cretaceous to the Neogene. We consider this hypothesis unlikely.

9. Utility

Consumulites provide compelling evidence of diet, as they are direct evidence of what a vertebrate animal consumed. However, consumulites do not necessarily reflect the full diet of a predator, and the recognizable remains of different prey taxa/elements may have different residence times within the gut (Harper, 2022). This is well illustrated by experiments to interpret the stomach contents of modern walrus whereby evidence of much of the prey ingested is not recognizable because of rapid digestion (Sheffield et al., 2001; Harper, 2022). And, of course there can be temporal differences in diet, notably observed in *Ursus arctos horribilis* seasonally gorging on *Oncorhynchus nerka.*

Consumulites provide evidence for assessing the chemistry of digestive systems by examining the etching and erosion of included materials. Strganac et al. (2015) and Polcyn et al. (2023) described the skeleton of the large mosasaur *Prognathodon kianda* that contained partial skulls and skeletons of the smaller of three other mosasaurs. Strganac et al. (2015) noted that the skull roof, snout and other areas covered by thin flesh in life are etched, and tooth crowns are dissolved above the gum line.

Previously, the Extant Phylogenetic Bracket approach was often utilized to reconstruct the digestive tract (e.g., Bryant and Russell, 1992; Witmer, 1995), but consumulites can provide direct evidence of the structure. Studying the evolution of the components of the digestive system has been done very successfully with early birds (e.g., O'Connor et al., 2019) and also with fish (Janvier and Arsenault, 2007).

Dietary changes through ontogeny can be demonstrated by consumulites. Werneburg (1986, 1988, 1989; Werneburg et al., 2007) has demonstrated ontogenetic change in diet in several Pennsylvanian and early Permian branchiosaurs and other small, pedomorphic amphibians. Similarly, Dick et al. (2016) analyzed trophic niche ontogeny in the Jurassic ichthyosaur *Stenopterygius* utilizing consumulites and Richter and Baszio (2006) in the Eocene fish *Rhenanoperca*.

Consumulites provide direct information for evaluating the evolution of diets within clades. Thus, Druckenmiller et al. (2014) utilized consumulites of Triassic ichthyosaurs to assess the diversity of their predatory guilds relative to Jurassic taxa.

Consumulites can provide strong evidence for environmental tolerances. Werneburg et al. (2021) noted that freshwater/brackish water ostracods (*Carbonita*) in consumulites in the small Pennsylvanian amphibians *Milnerpeton* and *Bermanerpeton* demonstrated that they were nonmarine and not marine. (e.g., nonmarine ostracods in larval amphibians indicate that the amphibians were not marine in lifestyle).

Consumulites preserve a wide range of organic elements with a poor fossil record, ranging from lepidopteran wings and hair (Richter and Storch, 1980; Richter, 1988) to embryos (O'Keefe et al., 2009) and thus can be Micro-Lagerstätten as first proposed by Wilby and Martill (1992). In addition, consumulites can also preserve tissues of the gastrointestinal tract (Wilby and Martill, 1992). The systematic study of consumulites will undoubtedly yield significant records of contained fossils as has the recent focus on the content of coprolites (e.g., Qvarnström et al., 2016).

10. Conclusions

There is an extensive and under-appreciated fossil record of vertebrate consumulites starting in the Middle Silurian. Marine and nonmarine strata from the Cretaceous Period have a notable abundance of consumulites.

The study of consumulites will be greatly enhanced by further description of specimens and by the application of more robust ichnotaxonomy. There are currently only four named consumulites, three preserved within the body cavity, *Werneburgichnus kinneyensis*, *W. varius* and *Chondripilula zideki* and the evisceralite *Hiabromus seilacheri*.

Ichnotaxobases of consumulites include (1) location of bromalite in digestive tract; (2) morphology, for example, pellet, infilling that follows form of digestive tract; (3) nature of content and inclusions (including taxonomic identification of inclusions); and (4) degree of mechanical and biochemical processing.

Recognition criteria for consumulites include a combination of: (1) in an anatomically correct position; (2) discrete clusters of bromalitic material; (3) found in a low-energy depositional setting in sediment finer in clast size than in the consumulite; (4). dentalites present; and (5) evidence of mechanical or chemical processing.

Consumulites are principally associated with, or at least recognized in the context, of articulated skeletons. The majority of articulated skeletons are from aquatic environments, as are many consumulites. Articulated skeletons are common in Lagerstätten, and the record of consumulites is biased toward them.

The vast majority of consumulites represent carnivorous animals because plant material is usually finely macerated during digestion, and it is much easier to recognize a bone as a foreign object than carbonaceous debris.

Large body size favors the recognition of consumulites. The combination of large body size and an aquatic lifestyle results in an extensive record of consumulites in large Mesozoic marine vertebrates, notably ichthyosaurs, plesiosaurs, and mosasaurs.

Consumulites provide evidence for: (1) assessing the chemistry of digestive systems by examining the etching and erosion of included materials; (2) the structure and evolution of the digestive tract; (3) dietary changes through ontogeny; (4) evaluating the evolution of diets within clades; and (5) environmental tolerances.

Consumulites preserve a wide range of organic elements with a poor fossil record, ranging from lepidopteran wings to hair, and thus represent Micro-Lagerstätten. The majority of consumulites are gastrolites.

Food stuffs that pass through the digestive tract are subject to chemical and physical processes that result in characteristic damage. Crocodilian-digested bones most closely resemble bones exposed to hydrochloric acid as opposed to the damage due to a mix of acid and digestive enzymes exhibited by bones consumed by mammalian and avian predators.

Matryoshkalite is a new term for a consumulite within a consumulite. There are only three currently known examples, from the Permian of Germany and the Eocene of the United States and Germany.

Declaration of competing interest

The authors declare that they have no known competing financial interests or personal relationships that could have appeared to influence the work reported in this paper.

Data availability

No data were used for the research described in this article.

Acknowledgments

We thank many colleagues who have graciously allowed us access to collections, including Michael Brett-Surman, Paul Jeffery, Thomas Lehmann, Krister Smith, and Anika Vogel. Mike Everhart, Vince Santucci, Krister Smith, and Anika Vogel kindly supplied images, and Jesper Milàn and Hendrik Klein provided insightful reviews.

References

Agassiz, L., 1833–1845a. Recherches sur les Poissons Fossiles. Tome 2. Neuchatel. Imprimerie de Petitpierre et Prince, p. 288.

Agassiz, L., 1833–1845b. Recherches sur les Poissons Fossiles. Tome 4. Neuchatel. Imprimerie de Petitpierre et Prince, p. 188.

Agassiz, L., 1833. Letter to Prof. Bronn of Nov. 8, 1833. Neue Entedeckungen über fossile Fische. Neues Jahr. Mineral. Geol. Petrefaktend. 1833, 675–677.

Ahlberg, P.E., 1992. The palaeoecology and evolutionary history of the porolepiform sarcopterygians. In: Mark-Kurik, E. (Ed.), Fossil Fishes as Living Animals, vol 1. Estonian Akademia, Tallinn, pp. 71–90.

Alexander, J.P., Burger, B.J., 2001. Stratigraphy and taphonomy of Grizzly Buttes, Bridger Formation, and the middle Eocene of Wyoming. In: Gunnell, G.F. (Ed.), Eocene Biodiversity: Unusual Occurrences and Rarely Sampled Habitats. Kluwer Academic/Plenum Publishers, New York, pp. 165–196.

Allen, J.R.L., Tarlo, L.B., 1963. The Downtonian and Diffonian facies of the Welsh borderland. J. Geol. Soc. Lond. 138, 375–405.

Amalfitano, J., Dalla Vecchia, F.M., Giusberti, L., Fornaciari, E., Luciani, V., Roghi, G., 2017. Direct evidence of trophic interaction between a large lamniform shark, *Cretodus* sp., and a marine turtle from the Cretaceous of northeastern Italy. Palaeogeogr. Palaeoclimatol. Palaeoecol. 469, 104–121.

Amstutz, G., 1958. Coprolites: a review of the literature and a study of specimens from southern Washington. J. Sediment. Petrol. 28, 498–508.

Andrews, C.W., 1910. A Descriptive Catalogue of the Marine Reptiles of the Oxford Clay. Part I. British Museum (Natural History), London, 205 p.

Andrews, P., Evans, E.N., 1983. Small mammal bone accumulations produced by mammalian carnivores. Paleobiology 9, 289–307.

Arambourg, C., 1954. Les poisons Crétaces de Jebel Tselfat (Maroc). In: Éditions du Service géologique du Maroc Mémoire, vol 118, 188 p.

Argyriou, T., Clauss, M., Maxwell, E.E., Furrer, H., Sánchez-Villagra, M.R., 2016. Exceptional preservation reveals gastrointestinal anatomy and evolution in early actinopterygian fishes. Sci. Rep. 6, 18758. https://doi.org/10.1038/srep18758.

Arsenault, M., 1982. *Eusthenopteron foordi*, a predator on *Homalacanthus concinnus* from the Escuminac Formation, Miguasha, Quebec. Can. J. Earth Sci. 19, 2214–2217.

Arsenault, M., Janvier, P., 1991. The anaspidlike craniates of the Escuminac Formation (Upper Devonian) from Miguasha (Quebec, Canada) with remarks on anaspid-petromyzontid relationships. In: Chang, M.M., Liu, Y.H., Zhang, G.R. (Eds.), Early Vertebrates and Related Problems of Evolutionary Biology. Science Press, Beijing, pp. 19–44.

Balch, C.C., Campling, R.C., 1965. Rate of passage of digesta through the ruminant digestive tract. In: Dougherty, R.W., Allen, R.S., Burroughs, W., Jacobson, N.L., McGillard, A.D. (Eds.), Physiology of Digestion in the Ruminant. Butterworth, London, pp. 108–123.

Bardack, D., 1965. Anatomy and evolution of chirocentrid fishes. Paleontological Contributions of the University of Kansas, Vertebrata, 10, pp. 1–88.

Bell Jr., G.L., Barnes, K.R., 2007. First record of stomach contents in *Tylosaurus nepaeolicus* and comments on predation among mosasauridae. In: Second Mosasaur Meeting Abstract Booklet and Field Guide. Hays Kansas, Sternberg Museum of Natural History, pp. 9–10.

Beltan, L., 1977. La parturition d'un Actinopterygian de l'Eotrias du nord-ouest de Madagascar: Comptes Rendus Hebdomadaires des Séances de l'Académie des Sciences, Paris, Série D, 284, pp. 2223–2225.

Bennett, S.C., 2001. The osteology and functional morphology of the Late Cretaceous pterosaur *Pteranodon*. Palaeontograph. Abteilung 260, 1–153.

Bertling, M., Braddy, S.J., Bromley, R.G., Demathieu, G.R., Genise, J., Mikuláš, R., Nielsen, J.K., Nielsen, K.S.S., Rindsberg, A.K., Schlirf, M., Uchman, A., 2006. Names for trace fossils: a uniform approach. Lethaia 39, 265–286.

Bertling, M., Buatois, L.A., Knaust, D., Laing, B., Mángano, M.G., Meyer, N., Mikuláš, R., Minter, N.J., Neumann, C., Rindsberg, A.K., Uchman, A., 2022. Names for trace fossils 2.0: theory and practice in ichnotaxonomy. Lethaia 55 (3), 1–19.

Bobrovskiy, I., Nagovitsyn, A., Hope, J.M., Luzhnaya, E., Brocks, J.J., 2022. Guts, gut contents, and feeding strategies of Ediacaran animals. Curr. Biol. 32, 5382–5389. https://doi.org/10.1016/j.cub.2022.10.051.

Boeskorov, G.G., Potapova, O.R., Mashchenko, E.N., Protopopov, A.V., Kuznetsova, T.V., Agenbroad, L., Tikhonov, A.N., 2014. Preliminary analyses of the frozen mummies of mammoth (*Mammuthus primigenius*), bison (*Bison priscus*) and horse (*Equus* sp.) from the Yana-Indigirka Lowland, Yakutia, Russia. Integr. Zool. 9, 471–480.

Boeskorov, G.G., Lazarev, P.A., Sher, A.V., Davydov, S.P., Bakulina, N.T., Shchelchkova, M.V., Binladen, J., Willerslev, E., Buigues, B., Tikhonov, A.N., 2011. Woolly rhino discovery in the lower Kolyma River. Quat. Sci. Rev. 30, 2262–2272.

Boreske Jr., J.R., 1974. A review of North American amiid fishes. Museum Comp. Zool. Bull. 146, 1–87.

Böttcher, R., 1989. Über die Nahrung eines Leptopterygius (Ichthyosauria, Reptilia) aus dem süddeutschen Posidonienschiefer (Unterer Jura) mit Bemerkungen über den Magen der Ichthyosaurier. Stutt. Beitr. Naturkd. B 155, 1–19.

Boucot, A.J., 1990. Evolutionary Paleobiology of Behavior and Coevolution. Elsevier, Amsterdam, 735 p.

Boucot, A.J., Poinar Jr., G.O., 2010. Fossil Behavior Compendium. CRC Press, Boca Raton, 391 p.

Braz, H.B., Marques, O.A., 2016. Tail-first ingestion of prey by the false coral snake, *Erythrolamprus aesculapii*: Does it know where the tail is? Salamandra 52, 211–214.

Brinkmann, W., 2004. Mixosaurier (Reptilia, Ichthyosauria) mit Quetschzähnen aus der Grenzbitumenzone (Mitteltrias) des Monte San Giorgio (Schweiz, Kanton Tessin). Schweiz. Paläontol. Abh. 124, 1–84.

Broili, E., 1938. Beobachtungen an *Pterodactylus*. Sitzungber. Bayer. Akad. Wissensch. Math.-Naturwiss. Abt. 1938, 139–154.

Broughton, P.L., 1981. Casts of vertebrate internal organs from the Upper Cretaceous of western Canada. J. Geol. 89, 741–749.

Broughton, P.L., 2017. Enigmatic origin of massive Late Cretaceous-to-Neogene coprolite-like deposits in North America: a novel palaeobiological alternative to inorganic morphogenesis. Lethaia 50, 194–216.

Broughton, P.L., Simpson, F., Whitaker, S.H., 1977. Late Cretaceous coprolites from southern Saskatchewan: comments on excretion, plasticity and nomenclature. Bull. Can. Petrol. Geol. 25, 1097–1099.

Brown, B., 1904. Stomach stones and the food of plesiosaurs. Science 20, 184–185.

Brown, C., 1900. Ober das Genus *Hybodus* und seine systematische Stellung. Palaeontographica 46, 149–174.

Brown, B., 1943. Flying reptiles. Nat. Hist. 52, 104–111.

Brown, R.W., 1962. Paleocene Flora of the Rocky Mountains and Great Plains. United States Geological Survey. Professional Paper 375, 119 p.

Bryant, H.N., Russell, A.P., 1992. The role of phylogenetic analysis in the inference of unpreserved attributes of extinct taxa. Philos. Trans. R. Soc. Lond. Ser. B Biol. Sci. 337, 405–418.

Buchy, M.C., Taugourdeau, P., Janvier, P., 2004. Stomach contents of a Lower Triassic ichthyosaur from Spitzbergen. Oryctos 5, 47–55.

Buckland, W., 1835. On the discovery of coprolites, or fossil faeces, in the Lias at Lyme Regis, and in other formations. Trans. Geol. Soc. Lond. 3 (series 2), 223–238.

Buckland, W., 1836. Geology and mineralogy considered with reference to natural theology. In: The Bridgewater Treatises on the Power, Wisdom and Goodness of God as Manifested in the Creation, 2 volumes. Pickering, London, 599 p. and 128 p.

Buckland, W., 1858. Geology and mineralogy considered with reference to natural theology. Treatise vol I. In: The Bridgewater Treatises on the Power, Wisdom and Goodness of God as Manifested in the Creation, third ed., 2 volumes. Pickering, London. 552 p. and 143 p.

Burgin, T., 2000. *Euthynotus* cf. *incognitus* (Actinopterygii, Pachycormidae) als Mageninhalt eines Fischsauriers aus dem Posidonienschiefer Süddeutschlands (Unterer Jura, Lias epsilon). Eclogae Geol. Helv. 93, 491–496.

Camp, C.L., 1930. A study of phytosaurs. In: Memoirs of the University of California, vol 10, 174 p.

Camp, C.L., 1942. California mosasaurs. In: Memoirs of the University of California, vol 13, 68 p.

Camp, C.L., 1980. Large ichthyosaurs from the Upper Triassic of Nevada. Palaeontograph. Abteilung 170, 139–200.

Carnevale, G., Pietsch, T.W., 2006. Filling the gap: a fossil frogfish, genus *Antennarius* (Teleostei, Lophiiformes, Antennariidae), from the Miocene of Algeria. J. Zool. 270, 448–457.

Case, E.C., 1911. A Revision of the Cotylosauria of North America, vol 145. Publications of the Carnegie Institution of Washington, pp. 1–122.

Case, G.R., 1982. A Pictorial Guide to Fossils. Van Nostrand Reinhold, New York, 514 p.

Cau, A., Arduini, P., 2008. *Enantiophoenix electrophyla* gen. et sp. nov. (Aves, Enantiornithes) from the Upper Cretaceous (Cenomanian) of Lebanon and its phylogenetic relationships. Atti della Soc. Ital. Sci. Nat. Mus. Civico Storia Nat. Milano 149 (2), 293–324.

Caviedes-Vidal, E., McWhorter, T.J., Lavin, S.R., Chediack, J.G., Tracy, C.R., Karasov, W.H., 2007. The digestive adaptation of flying vertebrates: high intestinal paracellular absorption compensates for smaller guts. Proc. Natl. Acad. Sci. U.S.A. 104, 19132–19137.

Cavin, L., 1999. Occurrence of a juvenile teleost, *Enchodus* sp., in a fish gut content from the Upper Cretaceous of Goulmima, Morocco. Palaeontology 60, 57–72.

Charig, A.J., Milner, A.C., 1986. *Baryonyx*, a remarkable new theropod dinosaur. Nature 324, 359–361.

Charig, A.J., Milner, A.C., 1997. *Baryonyx walkeri*, a fish-eating dinosaur from the Wealden of Surrey. Bull. Br. Mus. (Nat. Hist.) Geol. 53, 11–70.

Chatterjee, S., 1978. A primitive parasuchid (phytosaur) reptile from the Upper Triassic Maleri Formation of India. Palaeontology 21, 83–127.

Chatterjee, S., 1980. *Malerisaurus*, a new eosuchian reptile from the Late Triassic of India. Philos. Trans. R. Soc. London, Ser. A B 291, 163–200.

Cheng, L., Wings, O., Chen, X., Sander, P.M., 2006. Gastroliths in the Triassic ichthyosaur *Panjiangsaurus* from China. J. Paleontol. 80, 583–588.

Chevrinais, M., Jacquet, C., Cloutier, R., 2017. Early establishment of vertebrate trophic interactions: foodweb structure in Middle to Late Devonian fish assemblages with exceptional fossilization. Bull. Geosci. 92, 491–510.

Choo, B., Long, J.A., Trinajstic, K., 2009. A new genus and species of basal actinopterygian fish from the Upper Devonian Gogo Formation of Western Australia. Acta Zool. 90, 194–210.

Cicimurri, D.J., Everhart, M.J., 2001. An elasmosaur with stomach contents and gastroliths from the Pierre Shale (Late Cretaceous) of Kansas. Trans. Kans. Acad. Sci. 104, 129–143.

Clark, N.D.L., 1989. A Study of a Namurian Crustacean-Bearing Shale from the Western Midland Valley of Scotland. Ph.D. dissertation. University of Glasgow, Glasgow, 372 p.

Coates, M.I., Sequeira, S.E.K., 2001. A new stethacanthid chondrichthyan from the lower Carboniferous of Bearsden, Scotland. J. Vertebr. Paleontol. 21, 438–459.

Colbert, E.H., 1989. The Triassic dinosaur *Coelophysis.* Museum of Northern Arizona. Bulletin 57, 160.

Collareta, A., Landini, W., Augusto, C., Budiel, C., Vera, W.V., Sierra, A., Ali, J., Schmitt, M.U., Bianucci, G., 2017. A well preserved skeleton of the fossil shark *Cosmopolitodus hastalis* from the late Miocene of Peru, featuring fish remains as fossilized stomach contents. Riv. Ital. Paleontol. Stratigr. 123, 11–22.

Collareta, A., Landini, W., Lambert, O., Post, K., Tinelli, C., Di Celma, C., Panetta, D., Tripodi, M., Salvadori, P.A., Caramella, D., Marchi, D., 2015. Piscivory in a Miocene Cetotheriidae of Peru: first record of fossilized stomach content for an extinct baleen-bearing whale. Sci. Nat. 102, 1–12. https://doi.org/10.1007/s00114-015-1319-y.

Collinson, M.E., 1990. Plant evolution and ecology during the Early Cainozoic diversification. Adv. Bot. Res. 17, 1–98.

Conway Morris, S., 1976. A new Cambrian lophophorate from the Burgess Shale of British Columbia. Palaeontology 19, 199–222.

Conway Morris, S., 1990. *Typhloesus wellsi* (Melton and Scott, 1973), a bizarre metazoan from the Carboniferous of Montana, U.S.A. Philos. Trans. R. Soc. London, Ser. A B 327, 595–624.

Cooper, S.L.A., Maxwell, E.E., 2022. Revision of the pachycormid fish *Saurostomus esocinus* Agassiz from the Early Jurassic (Toarcian) of Europe, with new insight into the origins of suspension-feeding in Pachycormidae. Pap. Palaeontol. 8 (6), e1467. https://doi.org/10.1002/spp2.1467.

Cooper, S.L.A., Maxwell, E.E., 2023. Death by ammonite: Fatal ingestion of an ammonoid shell by an Early Jurassic bony fish. Geol. Mag. 160, 1254–1261. https://doi.org/10.1017/S0016756823000456.

Cooper, S.L., Giles, S., Young, H., Maxwell, E.E., 2022. A new large† Pachycormiform (Teleosteomorpha:† Pachycormiformes) from the Lower Jurassic of Germany, with affinities to the suspension-feeding clade, and comments on the gastrointestinal anatomy of pachycormid fishes. Diversity 14, 1026. https://doi.org/10.3390/d14121026.

Cope, E.D., 1868. [Remarks on a new enaliosaurian, *Elasmosaurus platyurus*.]. Proc. Acad. Nat. Sci. Phila. 20, 92–93.

Cope, E.D., 1872. [On a species of *Clidastes* and on *Plesiosaurus gulo* Cope]. Proc. Acad. Nat. Sci. Phila. 24, 127–129.

Coulson, R.A., Herbert, J.D., Coulson, T.D., 1989. Biochemistry and physiology of alligator metabolism in vivo. Am. Zool. 29, 921–934.

Crandall, B.D., Stahl, P.W., 1995. Human digestive effects on a micromammalian skeleton. J. Archaeol. Sci. 22, 789–797.

Crane Jr., J.M., 1996. Late Tertiary radiation of viperfishes (Chauliodontidae) based on a comparison of Recent and Miocene species. Contrib. Sci. LA Countr. Museum 115, 29.

Currie, P.J., Koppelhus, E.B., Muhammad, A.F., 1995. Stomach contents of a hadrosaur from the Dinosaur Park Formation (Campanian, Upper Cretaceous) of Alberta, Canada. In: Sun, A., Wang, Y. (Eds.), Sixth Symposium on Mesozoic Terrestrial Ecosystems and Biota, Short Papers. China Ocean Press, Beijing, pp. 111–114.

Dalla Vecchia, F.M., Chiappe, L.M., 2002. First avian skeleton from the Mesozoic of northern Gondwana. J. Vertebr. Paleontol. 22, 856–860.

Dal Sasso, C., Signore, M., 1998. Exceptional soft-tissue preservation in a theropod dinosaur from Italy. Nature 392, 383–387.

Dal Sasso, C., Maganuco, S., 2011. *Scipionyx samniticus* (Theropoda: Compsognathidae) from the Lower Cretaceous of Italy. Osteology, ontogenetic assessment, phylogeny, soft tissue anatomy, taphonomy and palaeobiology. Mem. della Soc. Ital. Sci. Nat. del Mus. Civico Storia Nat. Milano 37, 1–281.

Davidson, R.G., Trewin, N.H., 2005. Unusual preservation of the internal organs of acanthodian and actinopterygian fish in the Middle Devonian of Scotland. Scot. J. Geol. 41, 129–134.

Davis, J.W., 1887. The fossil fish of the Chalk of Mount Lebanon in Syria. Sci. Trans. R. Soc. Dublin Ser. 2 3, 457–636.

Denison, R.H., 1956. A Review of the Habitat of the Earliest Vertebrates: Fieldiana Geology, vol 11, pp. 361–457.

Dennis, K., Miles, R.S., 1981. A pachyosteomorph arthrodire from Gogo, Western Australia. Zool. J. Linn. Soc. 73, 213–258.

Denys, C., Fernández-Jalvo, Y., Dauphin, Y., 1995. Experimental taphonomy: Preliminary results of the digestion of micromammal bones in laboratory. Compt. Rend. Acad. Sci. Sér. II A (Paris) 321, 803–809.

de Queiroz, A., de Queiroz, K., 1987. Prey handling behavior of *Eumeces gilberti* with comments on headfirst ingestion in squamates. J. Herpetol. 21, 57–63.

de Saint Seine, P., 1949. Les poisons des calcaires lithographiques de Cerin (Ain). Nouvelles Archives de Muséum d'histoire naturelle de Lyon, 2, pp. 1–351.

Dick, D.G., Schweigert, G., Maxwell, E.E., 2016. Trophic niche ontogeny and palaeoecology of early Toarcian *Stenopterygius* (Reptilia: Ichthyosauria). Palaeontology 59, 423–431.

Diedrich, C.G., 2012. Stomach and gastrointestinal tract contents in Late Cenomanian (Upper Cretaceous) teleosts from black shales of Germany and analysis of fish mortality and food chains in the upwelling-influenced Pre-North Sea Basin of Europe. New Mexico Museum Nat. Hist. Sci. Bull. 57, 241–256.

Diedrich, C.G., 2015. The vertebrates from the Lower Ladinian (Middle Triassic) bonebed of Lamerden (Germany) as palaeoenvironment indicators in the Germanic Basin. Open Geosci. 7 (1). https://doi.org/10.1515/geo-2015-0062 id.62, 28pp.

Diefenbach, C.O., Emslie, S.G., 1971. Cues influencing the direction of prey ingestion of the Japanese snake, *Elaphe climacophora*. Herpetologica 27, 461–466.

Dollo, L., 1887a. Le hainosaure et les nouveaux vertébrés fossiles du Musée de Bruxelles. Rev. Des. Quest. Sci. 21, 504–539.

Dollo, L., 1887b. On some Belgian fossil reptiles. Geol. Mag. 4, 392–396.

Dollo, L., 1913. *Globidens fraasi*, mosasaurien mylodonte nouveau du Maestrichtien (Crétacé supérieur) du Limbourg, et l'éthologie de la nutrition chez les mosasauriens: Archives de Biologie, vol 28, pp. 609–626.

Dong, L.P., Huang, D.Y., Wang, Y., 2012. Two Jurassic salamanders with stomach contents from Inner Mongolia, China. Chin. Sci. Bull. 57, 72–76.

Doyle, P., MacDonald, D.I., 1993. Belemnite battlefields. Lethaia 26, 65–80.

Druckenmiller, P.S., Russell, A.P., 2009. Earliest North American occurrence of Polycotylidae (Sauropterygia: Plesiosauria) from the Lower Cretaceous (Albian) Clearwater Formation, Alberta, Canada. J. Paleontol. 83, 981–989.

Druckenmiller, P.S., Daun, A.J., Skulan, J.L., Pladziewicz, J.C., 1993. Stomach contents in the Upper Cretaceous shark *Squalicorax falcatus*. J. Vertebr. Paleontol. 13, 33A–34A.

Druckenmiller, P.S., Kelley, N., Whalen, M.T., Mcroberts, C., Carter, J.G., 2014. An Upper Triassic (Norian) ichthyosaur (Reptilia, Ichthyopterygia) from northern Alaska and dietary insight based on gut contents. J. Vertebr. Paleontol. 34, 1460–1465.

Druckenmiller, P.S., Kelley, N.P., Metz, E.T., Baichtal, J., 2020. An articulated Late Triassic (Norian) thalattosauroid from Alaska and ecomorphology and extinction of Thalattosauria. Sci. Rep. 10 (1), 1746. https://doi.org/10.1038/s41598-020-57939-2.

Dudgeon, T.W., Mallon, J.C., Evans, D.C., 2023. The first report of *Champsosaurus lindoei* (Choristodera: Champsosauridae) from the Campanian of the United States: anatomical, phylogenetic, and palaeoecological significance. Zool. J. Linn. Soc. 1–20. https://doi.org/10.1093/zoolinnean/zlad087.

Duke, G.E., 1997. Gastrointestinal physiology and nutrition in wild birds. Proc. Nutr. Soc. 56, 1049–1056.

Duke, G.E., Jegers, A.A., Loff, G., Evanson, 0.A., 1975. Gastric digestion in some raptors. Comp. Biochem. Physiol. A Physiol. 50, 649–656.

Eaton, T.H., 1964. A captorhinomorph predator and its prey (Cotylosauria). Am. Mus. Novit. 2169, 1–3.

Eastman, C.R., 1911. Jurassic saurian remains ingested within fish. Ann. Carnegie Mus. 8, 182–187.

Ebert, M., Kölbl-Ebert, M., 2008. Räuber-Beute-Beziehungen bei *Orthogonikleithrus hoelli* Arratia 1997. Archaeopteryx 26, 11–18.

Ebert, M., Kölbl-Ebert, M., Lane, J.A., 2015. Fauna and predator-prey relationships of Ettling, an actinopterygian fish-dominated Konservat Lagerstätte from the Late Jurassic of southern Germany. PLoS One 10 (1), e0116140. https://doi.org/10.1371/journal.pone.0116140.

Eriksson, M.E., De La Garza, R., Horn, E., Lindgren, J., 2022. A review of ichthyosaur (Reptilia, Ichthyopterygia) soft tissues with implications for life reconstructions. Earth Sci. Rev. 226, 103965. https://doi.org/10.1016/j.earscirev.2022.103965.

Eudes-Deslongchamps, J.A., 1838. Mémoire sur le *Poikilopleuron bucklandii*, grand saurien fossile, intermédiaire entre les crocodiles et les lézards. Mém. Soc. Lin. Normandie 6, 37–146.

Evans, S.E., 2016. Urodela. In: Poyato-Ariza, F.J., Buscalioni, Á.D. (Eds.), Las Hoyas: A Cretaceous wetland: A multidisciplinary synthesis after 25 years of research on an exceptional fossil Lagerstätte from Spain. Volume dedicated to the late Jordi María de Gibert. Verlag Dr. Friedrich Pfeil, Munich, pp. 138–142.

Everhart, M.J., 2003. First records of plesiosaur remains in the lower Smoky Hill Chalk Member (Upper Coniacian) of the Niobrara Formation in western Kansas. Kansas Acad. Sci. Trans. 106, 139–148.

Everhart, M.J., 2004. Late Cretaceous interaction between predators and prey. Evidence of feeding by two species of shark on a mosasaur. PalArch Vertebr. Palaeontol. Ser. 1, 1–7.

Everhart, M.J., 2017. Oceans of Kansas—A Natural History of the Western Interior Sea, second ed. Indiana University Press, Bloomington. 427 p.

Fanti, F., Minelli, D., Conte, G.L., Miyashita, T., 2016. An exceptionally preserved Eocene shark and the rise of modern predator–prey interactions in the coral reef food web. Zool. Lett. 2 (1), 9. https://doi.org/10.1186/s40851-016-0045-4.

Fisher, D.F., 1981a. Crocodilian scatology, microvertebrate concentrations, and enamel-less teeth. Paleobiology 7, 262–275.

Fisher, D.C., 1981b. Taphonomic interpretation of enamel-less teeth in the Shotgun local fauna (Paleocene, Wyoming). In: Contributions of the Museum of Paleontology, vol 25. University of Michigan, pp. 259–275.

Fowler, D.W., Woodward, H.N., Freedman, E.A., Larson, P.L., Horner, J.R., 2011. Reanalysis of "*Raptorex kriegsteini*": A juvenile tyrannosaurid dinosaur from Mongolia. PLoS One 6 (6), e21376. https://doi.org/10.1371/journal.pone.0021376.

Franzen, J.L., 1976. Senckenburgs Grabungskampagne 1975 in Messel: Erste Ergebnisse und Ausblick. Nat. Museum Frankfurt 106, 217–223.

Franzen, J.L., 1977. Die Entstehung der Fossilfundstelle Messel. Ber Naturforsch. Ges. Freib. 67, 53–58.

Franzen, J.L., 1984. Die Stammesgeschichte der Pferde in ihrer wissenschafthistorischen Entwickung. Nat. Museum Frankfurt 114, 149–162.

Franzen, J.L., 1985. Exceptional preservation of Eocene vertebrates in the lake deposit of Grube Messel (West Germany). Philos. Trans. R. Soc. Lond. Ser. B Biol. Sci. 311, 181–186.

Frei, S., Ortmann, S., Reutlinger, C., Kreuzer, M., Hatt, J.M., Clauss, M., 2014. Comparative digesta retention patterns in ratites. Auk 132, 119–131.

Frey, E., Martill, D.M., Buchy, M.C., 2003. A new crested ornithocheirid from the Lower Cretaceous of northeastern Brazil and the unusual death of an unusual pterosaur. In: Buffetaut, E., Mazln, J.-M. (Eds.), Evolution and Palaeobiology of Pterosaurs, vol 217. Geological Society of London, Special Publications, pp. 55–63.

Frey, E., Tischlinger, H., 2012. The Late Jurassic pterosaur *Rhamphorhynchus*, a frequent victim of the ganoid fish *Aspidorhynchus*? PLoS ONE 7. https://doi.org/10.1371/journal.pone.0031945.

Friedman, M., Carnevale, G., 2018. The Bolca Lagerstätten: shallow marine life in the Eocene. J. Geol. Soc. 175, 569–579.

Fritsch, A., 1895. Fauna der Gaskohle und der Kalksteine der Permformation Böhmens, Band 3. F. Řivnáč, Prague, 132 p.

Fritsch, A., 1907. Miscellanea Palaeontologica. I Palaeozoica. Rivnac, Prague, 23 p.

Gao, K.Q., Shubin, N.H., 2003. Earliest known crown-group salamanders. Nature 422, 424–428.

Gay, R., 2005. Sexual dimorphism in the Early Jurassic theropod dinosaur *Dilophosaurus* and a comparison with other related forms. In: Carpenter, K. (Ed.), The Carnivorous Dinosaurs. Indiana University Press, Bloomington, pp. 277–283.

Gekker, E.L., Gekker, R.F., 1955. Remains of Teuthoidea from the White Lias and Lower Cretaceous of the Volga region. Voprosi Palaeontologii 1955 (2), 36–44 [In Russian].

Godfrey, S.J., 1997. Tetrapods (amphibians and reptiles). In: Shabica, C.W., Hay, A.A. (Eds.), Richardson's Guide to the Fossil Fauna of Mazon Creek. Northeastern Illinois University, Chicago, pp. 256–269.

Godoy, P.L., Montefeltro, F.C., Norell, M.A., Langer, M.C., 2014. An additional baurusuchid from the Cretaceous of Brazil with evidence of interspecific predation among Crocodyliformes. PLoS One 9 (5), e97138. https://doi.org/10.1371/journal.pone.0097138.

Grande, L., 1980. Paleontology of the Green River Formation, with a review of the fish fauna. Geol. Surv. Wyoming Bull. 63, 333.

Green, M., 1948. A new species of dog (*Tomarctus*) from the lower Pliocene of California. Univ. Calif. Dept. Geol. Sci. Bull. 28, 81–90.

Greene, H.W., 1976. Scale overlap, a directional sign stimulus for prey ingestion by ophiophagous snakes. Z. Tierpsychol. 41, 113–120.

Greene, H.W., 1983. Dietary correlates of the origin and radiation of snakes. Am. Zool. 23, 431–441.

Guth, C., 1962. Un insectivore de Meant. Ann. Paleontol. 48, 1–10.

Guthrie, R.D., 1990. Frozen Fauna of the Mammoth Steppe: The Story of Blue Babe. University of Chicago Press, Chicago, 323 p.

Habersetzer, J., Richter, G., Storch, G., 1994. Paleoecology of early middle Eocene bats from Messel, FRG: Aspects of flight, feeding and echolocation. Hist. Biol. 8, 235–260.

Harington, C.R., 2007. Late Pleistocene mummified mammals. In: Elias, S.A. (Ed.), Encyclopedia of Quaternary Science. Elsevier, Amsterdam, pp. 3197–3202.

Harper, E.M., 2022. Hunting evidence for the Mesozoic marine revolution: Progress and challenges. Boll. Soc. Paleontol. Ital. 61 (1), 1–18.

Haubold, H., Schaumberg, G., 1985. Die Fossilien des Kupferschiefers: Pflanzen-und Tierwelt zu Beginn des Zechsteins-eine Erzlagerstätte und ihre Paläonthologie. A. Ziemsen Verlag, Wittenberg, 223 p.

Heidtke, U.H.J., 2007. Ein ausgestorbenes Fischkonzept: Acanthodier, schwache Krieger im Kettenhemd. In: Schindler, T., Heidtke, U.H.J. (Eds.), Kohlesümpfe, Seen und Halbwüsten-Dokumente einer rund 300 Millionen Jahre alten Lebewelt zwischen Saarbrücken und Mainz). Pollichia Sonderveröffentlichung, vol 10. Pollichia Sonderveröffentlichung, Bad Dürkheim, pp. 230–238.

Heintz, A., 1938. Notes on Arthrodira. Nor. Geografisk Tidsskr. 18, 1–27.

Hilton, G.M., Furness, R.W., Houston, D.C., 2000a. A comparative study of digestion in North Atlantic seabirds. J. Avian Biol. 31, 36–46.

Hilton, G.M., Furness, R.W., Houston, D.C., 2000b. The effects of diet switching and mixing on digestion in seabirds. Funct. Ecol. 14, 145–154.

Hilton, G.M., Houston, D.C., Furness, R.W., 1998. Which components of diet quality affect retention time of digesta in seabirds? Funct. Ecol. 12, 929–939.

Hilton, G.M., Houston, D.C., Barton, N.W.H., Furness, R.W., Ruxton, G.D., 1999. Ecological constraints on digestive physiology in carnivorous and piscivorous birds. J. Exp. Zool. A: Ecol. Genet. Physiol. 283, 365–376.

Hoernes, R., 1904. Über Koprolithen und Enterolithen. Biologie Centralblatte, Leipzig, 24, pp. 566–576.

Hoffmann, R., Stevens, K., 2020. The palaeobiology of belemnites—foundation for the interpretation of rostrum geochemistry. Biol. Rev. 95 (1), 94–123.

Hone, D.W.E., Habib, M.B., Lamanna, M.C., 2013. An annotated and illustrated catalogue of Solnhofen (Upper Jurassic, Germany) pterosaur specimens at Carnegie Museum of Natural History. Ann. Carnegie Mus. 82, 65–191.

Hone, D., Henderson, D.M., Therrien, F., Habib, M.B., 2015. A specimen of *Rhamphorhynchus* with soft tissue preservation, stomach contents and a putative coprolite. PeerJ 3, e1191.

Hone, D., Tsuihiji, T., Watabe, M., Tsogtbaatr, K., 2012. Pterosaurs as a food source for small dromaeosaurs. Palaeogeogr. Palaeoclimatol. Palaeoecol. 331, 27–30.

Hu, Y., Meng, J., Wang, Y., Li, C., 2005. Large Mesozoic mammals fed on young dinosaurs. Nature 433, 149–152.

Hu, H., Wang, Y., McDonald, P.G., Wroe, S., O'Connor, J.K., Bjarnason, A., Bevitt, J.J., Yin, X., Zheng, X., Zhou, Z., Benson, R.B., 2022. Earliest evidence for fruit consumption and potential seed dispersal by birds. Elife 11, e74751. https://doi.org/10.7554/eLife.74751.

Hua, S., Liston, J., Tabouelle, J., 2024. The diet of *Metriorhynchus* (Thalattosuchia, Metriorhynchidae): Additional discoveries and paleoecological implications. Fossil Stud. 2, 66–76. https://doi.org/10.3390/fossils2010002.

Hunt, A.P., 1991. Two phytosaur (Reptilia: Archosauria) skeletons from the Bull Canyon Formation (Late Triassic) of east-central New Mexico with preserved stomach contents. N. M. Geol. 13, 93.

Hunt, A.P., 1992. Late Pennsylvanian coprolites from the Kinney Brick Quarry, central New Mexico, with notes on the classification and utility of coprolites. In: Zidek, J. (Ed.), Geology and Paleontology of the Kinney Brick Quarry, Late Pennsylvanian, Central New Mexico: New Mexico Bureau of Mines and Mineral Resources, Bulletin, vol 138, pp. 221–229.

Hunt, A.P., 1994. Vertebrate Paleontology and Biostratigraphy of the Bull Canyon Formation (Chinle Group: Norian), East-Central New Mexico with Revisions of the Families Metoposauridae (Amphibia: Temnospondyli) and Parasuchidae (Reptilia: Archosauria). Ph.D. dissertation. University of New Mexico, Albuquerque, p. 403.

Hunt, A.P., 2001. The vertebrate fauna, biostratigraphy and biochronology of the type Revueltian land-vertebrate faunachron, Bull Canyon Formation (Upper Triassic), east-central New Mexico. In: Lucas, S.G., Ulmer-Scholle, D.S. (Eds.), Geology of the Llano Estacado, vol 52. New Mexico Geological Society Guidebook, Socorro, pp. 123–151.

Hunt, A.P., Lucas, S.G., 2012. Classification of vertebrate coprolites and related trace fossils. N. M. Museum Nat. Hist. Sci. Bull. 57, 137–146.

Hunt, A.P., Lucas, S.G., 2014. Vertebrate trace fossils from New Mexico and their significance. N. M. Museum Nat. Hist. Sci. Bull. 64, 9–40.

Hunt, A.P., Lucas, S.G., 2016. The most significant record of "middle"-late Paleozoic vertebrate coprolites: Silurian-Carboniferous ichnoassemblages from Scotland. Geol. Soci. America Abstr. Progr. 48 (7). https://doi.org/10.1130/abs/2016AM-284166.

Hunt, A.P., Lucas, S.G., 2021a. A review of the bromalite ichnofauna from the Kinney Brick Quarry Lagerstätte (Late Pennsylvanian) of New Mexico, USA with descriptions of new consumulites and coprolites. In: Lucas, S.G., DiMichele, W.A., Allen, B.D. (Eds.), Kinney Brick Quarry Lagerstätte. New Mexico Museum of Natural History and Science Bulletin, vol 84, pp. 209–228.

Hunt, A.P., Lucas, S.G., 2021b. The ichnology of vertebrate consumption: Dentalites, gastroliths and bromalites. N. M. Museum Nat. Hist. Sci. Bull. 87, 215.

Hunt, A.P., Lucas, S.G., Berman, D.S., 1992. The Late Pennsylvanian amphibian fauna of the Kinney Quarry, central New Mexico. In: Zidek, J. (Ed.), Geology and Paleontology of the Kinney Brick Quarry, Late Pennsylvanian, Central New Mexico. New Mexico Bureau of Mines and Mineral Resources, Bulletin, vol 138, pp. 211–220.

Hunt, A.P., Lucas, S.G., Berman, D.S., 1996. A new amphibamid (Amphibia: Temnospondyli) from the Late Pennsylvanian (Middle Stephanian) of central New Mexico, USA. Paläontol. Zeitsch. 70, 555–565.

Hunt, A.P., Lucas, S.G., Spielmann, J.A., 2012a. The bromalite collection at the National Museum of Natural History (Smithsonian Institution), with descriptions of new ichnotaxa and notes on other significant coprolite collections. N. M. Museum Nat. Hist. Sci. Bull. 57, 105–114.

Hunt, A.P., Lucas, S.G., Spielmann, J.A., Suazo, T., Cantrell, A., 2012b. A re-evaluation of Late Pennsylvanian bromalites from the Kinney Brick Quarry Lagerstätte, New Mexico, USA. N. M. Museum Nat. Hist. Sci. Bull. 57, 185–192.

Jackson, S., 1992. Do seabird gut sizes and mean retention times reflect adaptation to diet and foraging method? Physiol. Zool. 65, 674–697.

Jamison-Todd, S., Moon, B.C., Rowe, A.J., Williams, M., Benton, M.J., 2022. Dietary niche partitioning in Early Jurassic ichthyosaurs from Strawberry Bank. J. Anat. 241 (6), 1409–1423.

Janvier, P., 1996a. Early Vertebrates. Oxford University Press, Oxford, 408 p.

Janvier, P., 1996b. The Miguasha "Anapsida". In: Schultze, H.-P., Cloutier, R. (Eds.), Devonian Fishes and Plants of Miguasha. Verlag Dr. Friedrich Pfeil, Quebec, Canada. Munich, pp. 134–140.

Janvier, P., Arsenault, M., 2007. The anatomy of *Euphanerops longaevus* Woodward, 1900, an anaspid-like jawless vertebrate from the Upper Devonian of Miguasha, Quebec, Canada. Geodiversitas 29, 143–216.

Jepsen, G.L., 1967. Notable geobiologic moments. Geotimes 12, 16–18.

Ji, Q., Wu, X., Cheng, Y., Ten, F., Wang, X., Ji, Y., 2016. Fish hunting ankylosaurs (Dinosauria, Ornithischia) from the Cretaceous of China. J. Geol. 40, 183–190.

Joyce, W.G., Mäuser, M., Evers, S.W., 2021. Two turtles with soft tissue preservation from the platy limestones of Germany provide evidence for marine flipper adaptations in Late Jurassic thalassochelydians. PLoS One 16 (6), e0252355. https://doi.org/10.1371/journal.pone.0252355.

Kear, B.P., 2006. First gut contents in a Cretaceous sea turtle. Biol. Lett. 2, 113–115.

Kear, B.P., Boles, W.E., Smith, E.T., 2003. Unusual gut contents in a Cretaceous ichthyosaur. Proc. R. Soc. Lond. Ser. B Biol. Sci. 270 (Suppl. 2), S206–S208.

Keller, T., 1976. Magen- und Darminhalte von Ichthyosaurien des süddeutschen Posidonienschiefers. Neues Jahrbuch Geol. Palaontol. Monatsh. 5, 266–283.

Kellner, A.W.A., 2006. Pterosauros—os Senhores Do Céu Do Brasil. Vieira and Lent, Rio de Janeiro, 176 p.

Kietzke, K.K., Kaesler, R.L., 1992. Late Pennsylvanian Ostracoda from the Kinney Brick Quarry, Bernalillo County, New Mexico, with notes on other microfossils. N. M. Bureau Mines Miner. Resour. Bull. 138, 127–133.

Knaust, D., 2020. Invertebrate coprolites and cololites revised. Pap. Palaeontol. 6 (3), 385–423.

Koerner, H.E., 1930. Jurassic fishes from New Mexico. Am. J. Sci. Ser. 5 (19), 463.

Konishi, T., Newbrey, M.G., Caldwell, M.W., 2014. A small, exquisitely preserved specimen of *Mosasaurus missouriensis* (Squamata, Mosasauridae) from the upper Campanian of the Bearpaw Formation, western Canada, and the first stomach contents for the genus. J. Vertebr. Paleontol. 34, 802–819.

Konishi, T., Brinkman, D., Massare, J.A., Caldwell, M.W., 2011. New exceptional specimens of *Prognathodon overtoni* (Squamata, Mosasauridae) from the upper Campanian of Alberta, Canada, and the systematics and ecology of the genus. J. Vertebr. Paleontol. 31, 1026–1046.

Konuki, R., 2008. Biostratigraphy of Sea Turtles and Possible Bite Marks on a *Toxochelys* (Testudine, Chelonioidea) from the Niobrara Formation (Late Santonian), Logan County, Kansas, and Paleoecological Implications for Predator–Prey Relationships Among Large Marine Vertebrates. M.S. thesis. Fort Hays State University, Fort Hays, 169 p.

Kosintsev, P.A., Lapteva, E.G., Trofimova, S.S., Zanina, O.G., Tikhonov, A.N., Van der Plicht, J., 2012. Environmental reconstruction inferred from the intestinal contents of the Yamal baby mammoth Lyuba (*Mammuthus primigenius* Blumenbach, 1799). Quat. Int. 255, 231–238.

Kräusel, R., 1922. Die Nahrung von *Trachodon*. Paläontol. Z. 4, 80.

Kriwet, J., Witzmann, F., Klug, S., Heidtke, U.H., 2008. First direct evidence of a vertebrate three-level trophic chain in the fossil record. Proc. R. Soc. Ser. B Biol. Sci. 275, 181–186.

Lambert, O., Collareta, A., Landini, W., Post, K., Ramassamy, B., Di Celma, C., Urbina, M., Bianucci, G., 2015. No deep diving: evidence of predation on epipelagic fish for a stem beaked whale from the Late Miocene of Peru. Proc. R. Soc. Ser. B 282, 20151530. https://doi.org/10.1098/rspb.2015.1530.

Leal, M.E.C., Martill, D., 2007. Anurans of the Crato Formation. In: Martill, D., Bechly, G., Loveridge, R.F. (Eds.), The Crato Fossil Beds of Brazil: Window into an Ancient World. Cambridge University Press, Cambridge, pp. 444–451.

Lee, Y.N., Barsbold, R., Currie, P.J., Kobayashi, Y., Lee, H.J., Godefroit, P., Escuillié, F., Chinzorig, T., 2014. Resolving the long-standing enigmas of a giant ornithomimosaur *Deinocheirus mirificu*. Nature 515, 257–260.

Levey, D.J., del Rio, C.M., 2001. It takes guts (and more) to eat fruit: lessons from avian nutritional ecology. Auk 118, 819–831.

Lohmann, U., Sachs, S., 2001. Observations on the postcranial morphology, ontogeny and palaeobiology of *Sclerocephalus haeuseri* (Amphibia: Actinodontidae) from the lower Permian of southwest Germany. Memoir. Queensl. Mus. 46, 771–782.

Lomax, D.R., 2021. Locked in Time: Animal Behavior Unearthed in 50 Extraordinary Fossils. Columbia University Press, New York, pp. 1–296.

Lomax, D.R., Larkin, N.R., Boomer, I., Dey, S., Copestake, P., 2019. The first known neonate skeleton: a rediscovered specimen from the Lower Jurassic, UK. Hist. Biol. 31, 600–609.

Long, J.A., 1991. Arthrodire predation by *Onychodus* (Pisces, Crossopterygii) from the Late Devonian Gogo Formation, Western Australia. West. Austr. Mus. Rec. 15, 503–516.

Long, J.A., 1995. The Rise of Fishes. The Johns Hopkins University Press, Baltimore, 223 p.

Loop, M.S., 1974. The effect of relative prey size on the ingestion behavior of the Bengal monitor, *Varanus bengalensis* (Sauria: Varanidae). Herpetologica 30, 123–127.

Loop, M.S., Bailey, L.G., 1972. The effect of relative prey size on the ingestion behavior of rodent-eating snakes. Psychonomic Sci. 28 (3), 167–169.

Lucas, S.G., Kietzke, K., Hunt, A.P., 1985. The Jurassic System in east-central New Mexico. In: Lucas, S.G., Zidek, J. (Eds.), Santa Rosa-Tucumcari Region, vol 36. New Mexico Geological Society Guidebook, Socorro, pp. 213–242.

Maier, W., Richter, R., Storch, G., 1986. *Leptictidium nasutum*—ein archisches Saugetier aus Messel mit aussergewohnlichen Anpassungen. Nat. Mus. 116, 1–19.

Maisch, M.W., 2009. The small dicynodont *Katumbia parringtoni* (von Huene 1942)(Therapsida: Dicynodontia) from the upper Permian Kawinga Formation of Tanzania as gorgonopsian prey. Palaeodiversity 2, 279–282.

Maisey, J.G., 1994. Predator relationships and trophic level reconstruction in a fossil fish community. Environ. Biol. Fish. 40, 1–22.

Maisey, J.G., 1996. Discovering Fossil Fishes. Henry Holt and Company, New York, pp. 1–223.

Malzahn, E., 1968. Uber neue Funde von *Janassa bitminosa* (Schloth.) im niederrheinischen Zechstein. Geologidches Jahrbuch 85, 67–96.

Marron, A.O., Moore, J.R., 2013. Evidence of frugivory and seed dispersal in Oligocene tortoises from South Dakota. Geol. Mag. 150, 1143–1149.

Martill, D.M., 1985. The preservation of marine vertebrates in the Lower Oxford Clay (Jurassic) of central England. Phil. Trans. R. Soc. Lond. B 311, 155–165.

Martill, D.M., 1992. Pliosaur Stomach Contents from the Oxford Clay: Mercian Geologist, vol 13, pp. 37–42.

Martin, J.E., Bjork, P.R., 1987. Gastric residues associated with a mosasaur from the Late Cretaceous (Campanian) Pierre Shale in South Dakota. Dakoterra 3, 68–72.

Martin, J.E., Fox, J.E., 2004. Molluscs in the stomach contents of *Globidens*, a shell crushing mosasaur, from the Late Cretaceous Pierre Shale, Big Bend area of the Missouri River, central South Dakota. Geol. Soc. Am. Abstr. Progr. 36, 80.

Martin, J.E., Kennedy, L.E., 1988. A plesiosaur with stomach contents from the Late Cretaceous (Campanian) Pierre Shale of South Dakota: a preliminary report. Proc. S. D. Acad. Sci. 67, 76–79.

Martin, L.D., Rothschild, B.M., 1989. Paleopathology and diving mosasaurs. Am. Sci. 77, 460–467.

Martínez-Cáceres, M., Lambert, O., de Muizon, C., 2017. The anatomy and phylogenetic affinities of *Cynthiacetus peruvianus*, a large *Dorudon*-like basilosaurid (Cetacea, Mammalia) from the late Eocene of Peru. Geodiversitas 39 (1), 7–163.

Massare, J.A., 1987. Tooth morphology and prey preference of Mesozoic marine reptiles. J. Vertebr. Paleontol. 7, 121–137.

Massare, J.A., Young, H.A., 2005. Gastric contents of an ichthyosaur from the Sundance Formation (Jurassic) of central Wyoming. PALUDICOLA 5, 20–27.

Mateer, N.J., 1974. Three Mesozoic crocodiles in the collections of the Palaeontological Museum, Uppsala. Bulletin of the Geological Institutions of the University of Uppsala. New Ser. 4, 53–72.

Matsumoto, T., Obata, I., Okazaki, Y., Kani, Y., 1982. An interesting occurrence of a fossil reptile in the Cretaceous of the Obira area, Hokkaido. Proc. Jpn. Acad. Ser. Bibliogr. 58, 109–113.

Maxwell, E.E., Lambers, P., López-Arbarello, A., Schweigert, G., 2021. Re-evaluation of pachycormid fishes from the Late of Southwestern Germany. Acta Palaeontol. Pol. 65 (3), 429–453.

Mayr, G., 2002. A new specimen of *Salmila robusta* (Aves: Gruiformes: Salmilidae n. fam.) from the Middle Eocene of Messel. Paläontol. Z. 76, 305–316.

Mayr, G., 2004. A partial skeleton of a new fossil loon (Aves, Gaviiformes) from the early Oligocene of Germany with preserved stomach content. J. Ornithol. 145, 281–286.

Mayr, G., 2009. Paleogene Fossil Birds. Springer, Berlin, 262 p.

Mayr, G., 2013. Late Oligocene mousebird converges on parrots in skull morphology. Ibis 155, 384–396.

Mayr, G., 2015. A new specimen of the Early Eocene *Masillacolius brevidactylus* and its implications for the evolution of feeding specializations in mousebirds (Coliiformes). Comptes Rendus Palevol 14, 363–370.

Mayr, G., Kaye, T.G., Pittman, M., Saitta, E.T., Pott, C., 2020. Reanalysis of putative ovarian follicles suggests that Early Cretaceous birds were feeding not breeding. Scientif. Rep. 10, 1–10. https://doi.org/10.1038/s41598-020-76078-2.

Mayr, G., Mourer-Chauviré, C., 2000. Rollers (Aves. Coraciiformes s.s.) from the Middle Eocene of Messel (Germany) and the Upper Eocene of the Quercy (France). J. Vertebr. Paleontol. 20, 533–546.

Mayr, G., Peters, D.S., 1998. The mousebirds (Aves: Coliiformes) from the Middle Eocene of Grube Messel (Hessen, Germany). Senckenberg. Lethaea 78, 179–197.

Mayr, G., Richter, G., 2011. Exceptionally preserved plant parenchyma in the digestive tract indicates a herbivorous diet in the middle Eocene bird *Strigogyps sapea* (Ameghinornithidae). Paläontol. Z. 85, 303–307.

Mayr, G., Wilde, V., 2014. Eocene fossil is earliest evidence of flower-visiting by birds. Biol. Lett. 10 (5), 20140223. https://doi.org/10.1098/rsbl.2014.0223.

Mayr, G., Mourer-Chauviré, C., Weidig, I., 2004. Osteology and systematic position of the Eocene Primobucconidae (Aves, Coraciiformes sensu stricto), with first records from Europe. J. Syst. Paleontol. 2, 1–12.

Mayr, G., Poschmann, M., Wuttke, M., 2006. A nearly complete skeleton of the fossil galliform bird *Palaeortyx* from the late Oligocene of Germany. Acta Ornithol. (Warszaw) 41, 129–135.

McAllister, J.A., 1985. Reevaluation of the origin of spiral coprolites. Univ. Kansas Palaeontol. Contrib. 114, 1–12.

McHenry, C.R., 2009. 'Devourer of Gods'–The Palaeoecology of the Cretaceous Pliosaur *Kronosaurus queenslandicus*. Ph.D. dissertation. University of Newcastle, Newcastle, 616 p.

McHenry, C.R., Cook, A.G., Wroe, S., 2005. Bottom-feeding plesiosaurs. Science 310, 75.

McNamara, M.E., Orr, P.J., Kearns, S.L., Alcalá, L., Anadon, P., Peñalver-Mollá, E., 2009. Soft-tissue preservation in Miocene frogs from Libros, Spain: insights into the genesis of decay microenvironments. Palaios 24, 104–117.

McNamara, M.E., Orr, P.J., Kearns, S.L., Alcalá, L., Anadón, P., Peñalver-Mollá, E., 2010. Exceptionally preserved tadpoles from the Miocene of Libros, Spain: Ecomorphological reconstruction and the impact of ontogeny upon taphonomy. Lethaia 43, 290–306.

McWhorter, T.J., Martínez del Rio, C., 2000. Does gut function limit hummingbird food intake? Physiol. Biochem. Zool. 73, 313–324.

McWilliams, S.R., Caviedes-Vidal, E., Karasov, W.H., 1999. Digestive adjustments in cedar waxwings to high feeding rate. J. Exp. Zool. 283, 394–407.

Milan, J., 2012. Crocodylian scatology—a look into morphology, internal architecture, inter- and intraspecific variation and prey remains in extant crocodylian feces. N. M. Museum Nat. Hist. Sci. Bull. 57, 65–71.

Miles, R.S., 1973. Articulated acanthodian fishes from the Old Red Sandstone of England, with a review of the structure and evolution of the acanthodian shoulder-girdle. Bull. Br. Mus. (Nat. Hist.) Geol. 24, 113–213.

Miles, R.S., Westoll, T.S., 1968. The placoderm fish *Coccosteus cuspidatus* Miller ex Agassiz from the Middle Old Red Sandstone of Scotland. Part I. Descriptive morphology. Trans. R. Soc. Edinb. 67, 373–476.

Miller, C.V., Pittman, M., 2021. The diet of early birds based on modern and fossil evidence and a new framework for its recognition. Biol. Rev. https://doi.org/10.1111/brv.12743.

Milner, A.R., 1982. Small temnospondyl amphibians from the Middle Pennsylvanian of Illinois. Palaeontology 25, 635–664.

Modesto, S.P., Scott, D.M., Reisz, R.R., 2009. Arthropod remains in the oral cavities of fossil reptiles support inference of early insectivory. Biol. Lett. 5, 838–840.

Mol, D., Shoshani, J., Tikhonov, A., Van Geel, B., Sano, S.I., Lazarev, P., Boeskorov, G., Agenbroad, L.D., 2006. The Yukagir Mammoth: Brief history, 14C dates, individual age, gender, size, physical and environmental conditions and storage. Sci. Ann. Sch. Geol. Arist. Univ. Thessaloniki 98, 299–314. Special Volume.

Molnar, R.E., Clifford, H.T., 2000. Gut contents of a small ankylosaur. J. Vertebr. Paleontol. 20, 194–196.

Molnar, R.E., Clifford, H.T., 2001. An ankylosaurian cololite from the Lower Cretaceous of Queensland, Australia. In: Carpenter, K. (Ed.), The Armored Dinosaurs. Indiana University Press, Bloomington, pp. 399–412.

Morlo, M., 2004. Diet of *Messelornis* (Aves: Gruiformes), an Eocene bird from Germany. Cour. Forschungsinst. Senckenberg 252, 29–33.

Mulder, E.W.A., 2013. On the piscivorous behaviour of the Early Cretaceous amiiform neopterygian fish *Calamopleurus cylindricus* from the Santana Formation, northeast Brazil. Netherlands J. Geosci. Geol. Mijnbouw 92, 119–122.

Munk, W., Sues, H.D., 1993. Gut contents of *Parasaurus* (Pareiasauria) and *Protorosaurus* (Archosauromorpha) from the Kupferschiefer (Upper Permian) of Hessen, Germany. Paläontol. Z. 67, 169–176.

Mustoe, G.E., 2000. Enigmatic origin of ferruginous "coprolites": evidence from the Miocene Wilkes Formation, southwestern Washington. Geol. Soc. Am. Bull. 113, 673–681.

Naish, D., 2014. The fossil record of bird behavior. J. Zool. 292, 268–280.

Nesbitt, S.J., Turner, A.H., Erickson, G.M., Norell, M.A., 2006. Prey choice and cannibalistic behavior in the theropod *Coelophysis*. Biol. Lett. 2, 611–614.

Neumayer, L., 1904. Die Koprolithen des Perm von Texas. Palaeontographica 51, 121–128.

Nicholls, E.L., 1988. Marine Vertebrates of the Pembina Member of the Pierre Shale (Campanian, Upper Cretaceous) of Manitoba and Their Significance to the Biogeography of the Western Interior Seaway. Ph.D. dissertation. University of Calgary, Calgary, 317 p.

Northwood, C., 2005. Early Triassic coprolites from Australia and their palaeobiological significance. Palaeontology 48, 49–68.

O'Connor, J.K., 2019. The trophic habits of early birds. Palaeogeogr. Palaeoclimatol. Palaeoecol. 513, 178–195.

O'Connor, J., Zheng, X., Dong, L., Wang, X., Wang, Y., Zhang, X., Zhou, Z., 2019. Microraptor with ingested lizard suggests non-specialized digestive function. Curr. Biol. 29, 2423–2429.

O'Connor, J.K., Zhou, Z., 2020. The evolution of the modern avian digestive system: Insights from paravian fossils from the Yanliao and Jehol biotas. Palaeontology 63, 13–27.

O'Keefe, F.R., Street, H.P., Cavigelli, J.P., Socha, J.J., O'Keefe, R.D., 2009. A plesiosaur containing an ichthyosaur embryo as stomach contents from the Sundance Formation of the Bighorn Basin, Wyoming. J. Vertebr. Paleontol. 29, 306–1310.

Ostrom, J.H., 1964. A reconsideration of the paleoecology of hadrosaurian dinosaurs. Am. J. Sci. 262, 975–997.

Ostrom, J.H., 1978. The osteology of *Compsognathus longipes* Wagner. Zitteliana 4, 73–118.

Paucă, M., 1933. Die fossile Fauna und Flora aus dem Oligozän von Suslaneşti-Muscel in Rumänien. In: Eine systematische und palaeobiologische Studie, vol 16. Anuarul Institutului Geologic al României, Bucureşti [for 1931), 99 p.

Patterson, C., Rosen, D.E., 1977. Review of the icthyodectiform and other Mesozoic teleost fishes and the theory and practice of classifying fossils. Bull. Am. Mus. Nat. Hist. 158, 83–172.

Peyer, K., Carter, J.G., Sues, H.D., Novak, S.E., Olsen, P.E., 2008. A new suchian archosaur from the Upper Triassic of North Carolina. J. Vertebr. Paleontol. 28, 363–381.

Polcyn, M.J., Schulp, A.S., Gonçalves, A.O., 2023. Remarkably well-preserved in-situ gut-content in a specimen of *Prognathodon kianda* (Squamata: Mosasauridae) reveals multispecies intrafamilial predation, cannibalism, and a new Mosasaurine taxon. In: Lee, Y.-N. (Ed.), Windows into Sauropsid and Synapsid Evolution: Essays in Honor of Prof. Louis L. Jacobs: Seoul Dinosaur Science Press, pp. 66–98.

Pollard, J.E., 1968. The gastric contents of an ichthyosaur from the lower Lias of Lyme Regis, Dorset. Palaeontology 11, 376–388.

Pollard, J.E., 1990. Evidence for diet. In: Briggs, D.E.G., Crowther, P.R. (Eds.), Palaeobiology: A Synthesis. Blackwell, Oxford, pp. 362–367.

Poplin, C., 1986. Taphocoenoses et restes alimentaires de Vertébrés carnivores. Bulletin du Muséum National d'Histoire Naturelle, Paris, Série C., 8 Muséum National d'Histoire Naturelle, Paris, pp. 257–267.

Potapova, A., Potapova, E., 2024. Late Pleistocene mummified mammals. In: Elias, S.A. (Ed.), Encyclopedia of Quaternary Science, third ed. Elsevier, Amsterdam, pp. 541–568 https://doi.org/10.1016/B978-0-323-99931-1.00275-0.

Přikryl, T., Koštk, M., Mazuch, M., Mikuláš, R., 2012. Evidence for fish predation on a coleoid cephalopod from the Lower Jurassic Posidonia Shale of Germany. Neues Jahrbuch Geol. Palaontol. Abhand. 263, 25–33.

Qvarnström, M., Niedźwiedzki, G., Žigaitė, Ž., 2016. Vertebrate coprolites (fossil faeces): an underexplored Konservat-Lagerstätte. Earth Sci. Rev. 162, 44–57.

Reimchen, T., 1991. Evolutionary attributes of head first manipulation and swallowing in piscivores. Can. J. Zool. 69, 2912–2916.

Reisz, R.R., MacDougall, M.J., Modesto, S.P., 2014. A new species of the parareptile genus *Delorhynchus*, based on articulated skeletal remains from Richards Spur, lower Permian of Oklahoma. J. Vertebr. Paleontol. 34, 1033–1043.

Richardson, E.S., 1980. Life at Mazon Creek. In: Langenheim, R.L., Mann, C.J. (Eds.), Middle and Late Pennsylvanian Strata on Margin of Illinois Basin. 10th Annual Field Conference of the Society of Economic Paleontologists and Mineralogists, Great Lakes Section. University of Illinois Urbana, Urbana, pp. 217–224.

Richardson Jr., E.S., Johnson, R.G., 1971. The Mazon Creek fauna. In: Proceedings of the North American Paleontology Convention II. Lawrence, pp. 1222–1235.

Richter, G., 1988. Problems in the analysis of stomach contents of Eocene mammals from the Messel Oil Shale-layers. Cour. Forschung. Sencken. 107, 105–117.

Richter, G., Baszio, S., 2001. First proof of planctivory/insectivory in a fossil fish: *Thaumaturus intermedius* from the Eocene Lake Messel (FRG). Palaeogeogr. Palaeoclimatol. Palaeoecol. 173, 75–85.

Richter, G., Baszio, S., 2006. First evidence of size-related change of diet ("switching") in a fossil fish. Palaeogeogr. Palaeoclimatol. Palaeoecol. 237, 270–279.

Richter, R., Storch, G., 1980. Beiträge zur Ernährungs-biologie eozäner Fledermause aus der "Grube Messel.". Nat. Mus. Frankfurt 110, 353–367.

Rieber, H., 1970. *Phragmoteuthis? ticinensis* n. sp., ein Coleoidea-Rest aus der Grenzbitumenzone (Mittlere Trias) des Monte San Giorgio (Kt. Tessin, Schweiz). Paläontol. Z. 44, 32–40.

Rinehart, L.F., Lucas, S.G., Heckert, A.B., Spielmann, J.A., Celeskey, M.D., 2009. The paleobiology of *Coelophysis bauri* (Cope) from the Upper Triassic (Apachean) Whitaker quarry, New Mexico, with detailed analysis of a single quarry block. N. M. Museum Nat. Hist. Sci. Bull. 45, 260 p.

Roberts, A.E., 1958. Geology and Coal Resources of the Toledo-Castle Rock District, Cowlitz and Lewis Counties, vol 1062. U.S. Geological Survey, Bulletin, Washington, 71 p.

Sanz, P.M., Chiappe, L.M., Pérez-Moreno, B., Buscalioni, A.D., Moratalla, J.J., Ortega, F., Poyata-Ariza, F.J., 1996. An Early Cretaceous bird from Spain and its implications for the evolution of avian flight. Nature 382, 442–445.

Sato, T., Storrs, G.W., 2000. An early polycotylid plesiosaur (Reptilia: Sauropterygia) from the Cretaceous of Hokkaido, Japan. J. Paleontol. 74, 907–914.

Sato, T., Tanabe, K., 1998. Cretaceous plesiosaurs ate ammonites. Nature 394, 629–630.

Scanferla, A., Smith, K.T., 2020. Exquisitely preserved fossil snakes of Messel: insight into the evolution, biogeography, habitat preferences and sensory ecology of early boas. Diversity 12, 100. https://doi.org/10.3390/d12030100.

Schaal, S., 2004. *Palaeopython fischeri* n. sp. (Serpentes: Boidae), eine Riesenschlange aus dem Eozän (MP 11) von Messel. Cour. Forschungin. Senckenber. 252, 35–45.

Schaeffer, B., Patterson, C., 1984. Jurassic fishes from the western United States, with comments on Jurassic fish distribution. Am. Mus. Novit. 2796, 86.

Schaumberg, G., 1979. Neue Nachweise von Bryozoen und Brachiopoden als Nahrung des permischen Holocephalen *Janassa bituminosa* (Schlotheim): Ein Beitrag zur Ökologie des Kupferschiefers. Philippia 4, 3–11.

Schmitz, M., Benda, P.L., 1991. Coprolites from the Maastrichtian Whitemud Formation of southern Saskatchewan: Morphological classification and interpretation of diagenesis. Paläontol. Z. 65, 199–211.

Schultze, H.-P., 1972. Early growth stages in coelacanth fishes. Nat. New Biol. 236, 90–91.

Seilacher, A., Marshall, C., Skinner, H.C.W., Tsuihiji, T., 2001. A fresh look at sideritic "coprolites.". Paleobiology 27, 7–13.

Sereno, P.C., Tan, L., Brusatte, S.L., Kriegstein, H.J., Zhao, X., Cloward, K., 2009. Tyrannosaurid skeletal design first evolved at small body size. Science 26, 418–422.

Sheffield, G., Fay, F.H., Feder, H., Kelly, B.P., 2001. Laboratory digestion of prey and interpretation of walrus stomach contents. Mar. Mamm. Sci. 17, 310–330.

Shimada, K., 1997. Paleoecological relationships of the Late Cretaceous lamniform shark *Cretoxyrhina mantelli* (Agassiz). J. Paleontol. 71, 926–933.

Silva, R.R., Ferigolo, J., Bajdek, P., Piñeiro, G., 2017. The feeding habits of Mesosauridae. Front. Earth Sci. 5, 23. https://doi.org/10.3389/feart.2017.00023.

Smith, K.T., Scanferla, A., 2016. Fossil snake preserving three trophic levels and evidence for an ontogenetic dietary shift. Palaeobiodiv. Palaeoenv. 96, 589–599.

Smith, K.T., Scanferla, A., 2020. More than one large constrictor snake lurked around paleolake Messel. Palaeontograph. Abteilung: Palaeozool. Stratigr. 323 (1), 75–103. https://doi.org/10.1127/pala/2021/0119.

Sorbini, L., 1972. I fossili di Bolca. Edizioni Corev, Verona, 132 p.

Spencer, P.K., 1993. The "coprolites" that aren't: the straight poop on specimens from the Miocene of southwestern Washington. Ichnos 2, 231–236.

Spiekman, S.N., Wang, W., Zhao, L., Rieppel, O., Fraser, N.C., Chun, L.I., 2024. *Dinocephalosaurus orientalis* Li, 2003: A remarkable marine archosauromorph from the Middle Triassic of southwestern China. In: Earth and

Environmental Science Transactions of the Royal Society of Edinburgh, pp. 1–33. https://doi.org/10.1017/S175569102400001X.

Spindler, F., Lauer, R., Tischlinger, H., Mäuser, M., 2021. The integument of pelagic crocodylomorphs (Thalattosuchia: Metriorhynchidae). Palaeontol. Electron. 24 (2), a25. https://doi.org/10.26879/1099/palaeo-electronica.org/content/2021/3399-metriorhynchid-skin.

Stensiö, E.A., 1939. A new anaspid from the Upper Devonian of Scaumenac Bay in Canada, with remarks on the other anaspids. K. Sven. Vetenskapsakademiens Handl. 18 (3), 3–25.

Sternberg, C.H., 1917. Hunting Dinosaurs in the Badlands of the Red Deer River, Alberta, Canada. Lawrence. Charles H. Sternberg, San Diego, 232 p.

Stirton, R.A., 1959. Time, Life and Man. Wiley, New York, 558 p.

Storch, G., 2001. Paleobiological implications of the Messel mammalian assemblage. In: Gunnell, G.F. (Ed.), Eocene Biodiversity: Unusual Occurrences and Rarely Sampled Habitats. Kluwer Academic/Plenum, New York, pp. 215–235.

Storrs, G.W., 1999. An examination of Plesiosauria (Diapsida: Sauropterygia) from the Niobrara Chalk (upper Cretaceous) of central North America. Univ. Kansas Paleontol. Contrib. New Ser. 11, 15 p.

Strganac, C., Jacobs, L.L., Polcyn, M.J., Mateus, O., Myers, T.S., Salminen, J., May, S.R., Araújo, R., Ferguson, K.M., Gonçalves, A.O., Morais, M.L., 2015. Geological setting and paleoecology of the Upper Cretaceous Bench 19 marine vertebrate bonebed at Bentiaba, Angola. Neth. J. Geosci. 94, 121–136.

Sues, H.D., Olsen, P.E., Carter, J.G., Scott, D.M., 2003. A new crocodylomorph archosaur from the Upper Triassic of North Carolina. J. Vertebr. Paleontol. 23, 329–343.

Swift, C.C., Barnes, L.G., 1996. Stomach contents of *Basilosaurus cetoides*: Implications for the evolution of cetacean feeding behavior, and the evidence for vertebrate fauna of epicontinental Eocene seas. Paleontol. Soc. Spec. Publ. 8, 380.

Taggart, R.E., Cross, A.T., 1997. The relationship between land plant diversity and productivity and patterns of dinosaur herbivory Dinofest International. In: Wolberg, D.L., Stump, E., Rosenberg, G. (Eds.), Dinofest International: Proceedings of a symposium held at Arizona State University. National Academy of Sciences, Philadelphia, pp. 403–416.

Tarlo, L.B., 1959. *Pliosaurus brachyspondylus* Owen from the Kimmeridge Clay. Palæontology 1, 83–291.

Taylor, M.A., Norman, D.B., Cruickshank, A.R.I., 1993. Remains of an ornithischian dinosaur in a pliosaur from the Kimmeridgian of England. Palæontology 36, 357–360.

Therrien, F., Zelenitsky, D.K., Tanaka, K., Voris, J.T., Erickson, G.M., Currie, P.J., DeBuhr, C.L., Kobayashi, Y., 2023. Exceptionally preserved stomach contents of a young tyrannosaurid reveal an ontogenetic dietary shift in an iconic extinct predator. Sci. Adv. 9 (49), eadi0505. https://doi.org/10.1126/sciadv.adi0505.

Thouzeau, C., Peters, G., Le Bohec, C., Le Maho, Y., 2004. Adjustments of gastric pH, motility and temperature during long-term preservation of stomach contents in free-ranging incubating king penguins. J. Exp. Biol. 207, 2715–2724.

Traquair, R.H., 1879. Evidence as to the predaceous habits of the larger Palaeoniscidae. Proc. R. Phys. Soc. Edinburgh 5, 128–130.

Trutnau, L., Sommerlad, R., 2006. Crocodilians. Their Natural History and Captive Husbandry. Edition Chimaira, Frankfurt am Main, 646 p.

Tschanz, K., 1989. *Lariosaurus buzzii* n. sp. from the Middle Triassic of Monte San Giorgio (Switzerland) with comments on the classification of nothosaurs. Palaeontogr. Abteil. A Paläozool. Stratigr. 208, 153–179.

Tweet, J., Chin, K., Ekdale, A.A., 2016. Trace fossils of possible parasites inside the gut contents of a hadrosaurid dinosaur, Upper Cretaceous Judith River Formation, Montana. J. Paleontol. 90, 279–287.

Tweet, J.S., Chin, K., Murphy, N.L., Braman, D.R., 2008. Probable gut contents within a specimen of *Brachylophosaurus canadensis* (Dinosauria: Hadrosauridae) from the Upper Cretaceous Judith River Formation of Montana. Palaios 28, 624–635.

Uhen, M.D., 2004. Form, function, and anatomy of *Dorudon atrox* (Mammalia, Cetacea): an archaeocete from the middle to late Eocene of Egypt. Univ. Michigan Pap. Paleontol. 34, 1–222.

Ukraintseva, V.V., 1981. Vegetation of warm late Pleistocene intervals and the extinction of some large herbivorous mammals. Polar Geogr. Geol. 5, 189–203.

Ukraintseva, V.V., 1993. Vegetation cover and environment of the Mammoth Epoch in Siberia. In: Agenbroad, L.D., Mead, J.I., Hevly, R.H. (Eds.), Rapid City, the Mammoth Site of Hot Springs, p. 309.

Unterrassner, L., 2009. The Anterior Appendicular Morphology and the Stomach Content of the First Theropod Skeleton (Saurischia, Dinosauria) of Switzerland (Late Triassic; Frick, Canton Aargau). M.S. thesis. University of Zurich, Zurich, 123 p.

Upeniece, I., 2001. The unique fossil assemblage from the Lode quarry (Upper Devonian, Latvia). Mitteil. Mus. Nat. Berl. Geowissensch. Reihe 4, 101–119.

Upeniece, I., 2011. Palaeoecology and juvenile individuals of the Devonian placoderm and acanthodian fishes from Lode site, Latvia [Ph.D. dissertation]. Disert. Geol. Univ. Latviens. 21, 1–221.

Urlichs, M., Wild, R., Ziegler, N., 1994. Der Posidonien-Schiefer des unteren Juras und seine Fossilien. Stutt. Beitr. Nat. Ser. C 36, 1–95.

Van Geel, B., Pals, J.P., Van Reenen, G., Mol, D., 2004. Plant remains in the Yukagir mammoth dung and an environmental reconstruction; in the Yukagir Mammoth: Outcome of the first stage of research work. In: Proceedings of International Scientific Practical Conference. SB RAS Publishing House, Yakutsk, pp. 58–59.

Van Geel, B., Guthrie, R.D., Altmann, J.G., Broekens, P., Bull, I.D., Gill, F.L., Jansen, B., Nieman, A.M., Gravendeel, B., 2011. Mycological evidence of coprophagy from the feces of an Alaskan late glacial mammoth. Quat. Sci. Rev. 30, 2289–2303.

Van Geel, B., Aptroot, A., Baittinger, C., Birks, H.H., Bull, I.D., Cross, H.B., Evershed, R.P., Gravendeel, B., Kompanje, E.J., Kuperus, P., Mol, D., 2008. The ecological implications of a Yakutian mammoth's last meal. Quat. Res. 69, 361–376.

Varricchio, D.J., 2001. Gut contents from a Cretaceous tyrannosaurid: implications for theropod dinosaur digestive tracts. J. Paleontol. 75, 401–406.

Viohl, G., 1990. Piscivorous fishes of the Solnhofen lithographic limestone. In: Boucot, A.J. (Ed.), Evolutionary Paleobiology of Behavior and Coevolution. Elsevier, Amsterdam, pp. 285–303.

Von Koenigswald, W., Richter, G., Storch, G., 1981. Nacvhweis von Hornshuppen bei *Eomanis waldi* aus der "Grube Messel" bei Darmstadt (Mammalia, Pholidota). Senckenberg. Lethaea 61, 291–298.

von Wurstemberger, A.R., 1876. Über Lias epsilon: Schweizerbart. Jahresh. Ver. Vaterländis. Nat. Württemberg 32, 192–233.

Voss, M., Antar, M.S.M., Zalmout, I.S., Gingerich, P.D., 2019. Stomach contents of the archaeocete *Basilosaurus isis*: Apex predator in oceans of the late Eocene. PLoS One 14 (1), e0209021. https://doi.org/10.1371/journal.pone.0209021.

Wahl, W., 1998. Plesiosaur gastric contents from the upper Redwater Shale (lower Oxfordian) of the Sundance Formation (Jurassic) of Wyoming. J. Vertebr. Paleontol. 18 (Suppl. to no. 3), 84A.

Wahl, W.R., 2012. Gastric contents of a plesiosaur from the Sundance Formation of Hot Springs County, Wyoming, and implications for the paleobiology of cryptocleidid plesiosaurs. PALUDICOLA 9, 32–39.

Walker, M.V., 2006. The impossible fossil—revisited. Kansas Acad. Sci. Trans. 109, 87–96.

Wang, X., Miao, D., Zhang, Y., 2005. Cannibalism in a semi-aquatic reptile from the Early Cretaceous of China. China Science Bulletin 50, 281–283.

Wang, X., Cau, A., Guo, B., Ma, F., Qing, G., Liu, Y., 2022. Intestinal preservation in a birdlike dinosaur supports conservatism in digestive canal evolution among theropods. Sci. Rep. 12 (1), 19965. https://doi.org/10.1038/s41598-022-24602-x.

Ward, D.J., Duffin, C.J., Ward, A., 2020. Are spiral bromalites enterospirae or coprolites? A brief review with some additional evidence: Society of Vertebrate Paleontology 80th Annual Meeting Virtual 2020 October 12–16 2020. http://www.vertpaleo.org. Conference Program, unnumbered page.

Warner, A.C.I., 1981. Rate of passage of digesta through the gut of mammals and birds. Nutr. Abstr. Rev. Ser. B 51, 789–820.

Watson, D.M.S., 1927. The reproduction of the coelacanth. Proc. Zool. Soc. Lond. 28, 453–457.

Watson, D.M.S., 1937. The acanthodian fishes. Philos. Trans. R. Soc. London, Ser. A B 228, 49–146.

Weber, S., 2004. *Ornatocephalus metzleri* gen. et spec. nov. (Lacertilia, Scincoida): Taxonomy and paleobiology of a-basal scincoid lizard from the Messel Formation (Middle Eocene: basal Lutetian, Geiseltalium). Abh. Senckenb. Natforsch. Ges. 561, 1–159.

Wedmann, S., Uhl, D., Lehmann, T., Garrouste, R., Nel, A., Gomez, B., Smith, K., Schaal, S.F.K., 2018. The Konservat-Lagerstätte Menat (Paleocene; France)-an overview and new insights. Geol. Acta 16, 189–213.

Weigelt, J., 1928a. Die Pflanzenreste des mitteldeutschen Kupferschiefers und ihre Einschaltung ins Sediment: eine palökologische Studie. Fortsch. Geol. Paläontol. 6, 395–592.

Weigelt, J., 1928b. Ganoidfischleichen in Kupferschiefer und in der Gegenwart. Palaeobiologica 1, 323–356.

Weigelt, J., 1930a. Über die vermutliche Nahrung von *Protorosaurus* und über einen körperlich erhaltenen Fruchtstand von *Archaeopodocarpus germanicus* aut. Leopoldina 6, 269–280.

Weigelt, J., 1930b. Wichtige Fischreste aus dem Mansfelder Kupferschiefer. Leopoldina 6, 601–624.

Weiler, W., 1934. Ein fossiler Cybiide mit Magenhalt. Palaeobiologica 6, 17–24.

Weis, R., Delsate, D., Klug, C., Argyriou, T., Fuchs, D., 2024. Pachycormid fish fed on octobrachian cephalopods: new evidence from the '*Schistes bitumineux*'(early Toarcian) of southern Luxembourg. Swiss J. Palaeontol. 143 (1), 1–16.

Wellnhofer, P., 1970. Die Pterodacyloidea (Pterosauria) der Oberjura-Plattenkalke Süddeutschlands: Bayerische Akademie der Wissenschaften, Mathematisch Wissenschaftlichen Klasse. Abhandlugen 141, 1–133.

Wellnhofer, P., 1975a. Die Rhamphorhynchoidea (Pterosauria) der Oberjura Plattenkalke Süddeutschlands, vol 148. Palaeontographica, pp. 1–186.

Wellnhofer, P., 1975b. Die Rhamphorhynchoidea (Pterosauria) der Oberjura Plattenkalke Süddeutschlands, vol 149. Palaeontographica, pp. 1–30.

Werneburg, R., 1986. Die Stegocephalen (Amphibia) der Goldlauterer Schichten (Unterrotliegendes, Perm) des Thüringer Waldes, Teil I: *Apateon flagrifer* (Whitt.). Freiberger Forschungshefte—A C410, 87–100.

Werneburg, R., 1988. Die Stegocephalen der Goldlauterer Schichten (Unterrotliegendes, Unterperm) des Thüringer Waldes, Teil II: *Apateon kontheri* n. sp., *Melanerpeton eisfeldi* n. sp. und andere. Freiberger Forschungshefte—A C427, 7–29.

Werneburg, R., 1989. Die Amphibienfauna der Manebacher Schichten (Unterrotliegendes, Unterperm) des Thüringer Waldes. Veröffentlich. Naturhist. Mus. Schleus. 4, 55–68.

Werneburg, R., 2020. On the morphology of large branchiosaurids (Amphibamiformes) from the Rotliegend (Lower Permian) of the Saar-Nahe basin, Germany. Semana 35, 39–54.

Werneburg, R., Ronchi, A., Schneider, J.W., 2007. The early Permian branchiosaurids (Amphibia) of Sardinia (Italy): systematic palaeontology, palaeoecology, biostratigraphy and palaeobiogeographic problems. Palaeogeog. Palaeoclimat. Palaeoecol. 252, 383–404.

Werneburg, R., Schneider, J.W., Lucas, S.G., 2013. The dissorophoid *Milnererpeton huberi* (Temnospondyli) from the Late Pennsylvanian Kinney Brick Quarry in New Mexico restudied—paleontology, paleoenvironment, and age. N. M. Museum Nat. Hist. Sci. Bull. 59, 349–369.

Werneburg, R., Schneider, J.W., Lucas, S.G., 2021. The new dvinosaurian *Bermanerpeton kinneyi* (Temnospondyli), with a few "branchiosaurid" characters, from the Late Pennsylvanian Kinney Brick Quarry. N. M. Museum Nat. Hist. Sci. Bull. 84, 433–451.

White, J.M., Barron, A., McCurry, M.R., Tim Denham, T., 2023. Investigating gut contents of the leptocleidian plesiosaur *Umoonasaurus demoscyllus* using micro-CT imaging. Alcheringa 47 (2), 206–210. https://doi.org/10.1080/03115518.2023.2194944.

White, M.A., Bell, P.R., Campione, N.E., Sansalone, G., Birch, S.A., Bevitt, J.J., Molnar, R.E., Cook, A.G., Wroe, S., Elliott, D.A., 2022. Abdominal contents reveal Cretaceous crocodyliforms ate dinosaurs. Gondwana Res. 106, 281–302. https://doi.org/10.1016/j.gr.2022.01.016.

Wilby, P.R., Martill, D.M., 1992. Fossil fish stomachs: a micoenvironment for exceptional preservation. Hist. Biol. 6, 25–36.

Williams, M.E., 1972. The origin of "spiral coprolites.", vol 59. University of Kansas Palaeontological Contributions, pp. 1–19.

Williams, M.E., 1990. Feeding behavior in Cleveland Shale fishes. In: Boucot, A.J. (Ed.), Evolutionary Paleobiology of Behavior and Coevolution. Elsevier, Amsterdam, pp. 273–287.

Williston, S.W., 1899. Some additional characters of the mosasaurs, vol 8. Kansas University Quarterly, pp. 39–41.

Williston, S.W., 1914. Water Reptiles of the Past and Present. University of Chicago Press, Chicago, pp. 1–251.

Wilson, M.V.H., Caldwell, M.W., 1993. New Silurian and Devonian fork-tailed 'thelodonts' are jawless vertebrates with stomachs and deep bodies. Nature 361, 442–444.

Wilson, M.V.H., Caldwell, M.W., 1998. The Furcacaudiformes, a new order of jawless vertebrates with thelodont scales, based on articulated Silurian and Devonian fossils from northern Canada. J. Vertebr. Paleontol. 18, 10–29.

Wings, O., 2004. Identification, Distribution, and Function of Gastroliths in Dinosaurs and Extant Birds with Emphasis on Ostriches (*Struthio camelus*). Ph.D. dissertation. Rheinische Friedrich-Wilhelms-Universität, Bonn, 187 p.

Wiseman, K.D., Greene, H.W., Koo, M.S., Long, D.J., 2019. Feeding ecology of a generalist predator, the California kingsnake (*Lampropeltis californiae*): why rare prey matter. Herpetol. Conserv. Biol. 14, 1–30.

Witmer, L.M., 1995. The extant phylogenetic bracket and the importance of reconstructing soft tissues in fossils. In: Thomason, J. (Ed.), Functional Morphology in Vertebrate Paleontology. Cambridge University Press, Cambridge, pp. 19–33.

Witton, M.P., 2013. Pterosaurs: Natural History, Evolution, Anatomy. Princeton University Press, Princeton, pp. 1–291.

Witton, M.P., 2018. Pterosaurs in Mesozoic food webs: a review of fossil evidence. In: Hone, D.W.E., Witton, M.P., Martill, D.M. (Eds.), New Perspectives on Pterosaur Palaeobiology, vol 455. Geological Society of London, Special Publications, pp. 7–23.

Witzmann, F., 2009. Cannibalism in a small growth stage of the early Permian branchiosaurid *Apateon gracilis* (Credner, 1881) from Saxony. Fossil Rec. 12, 7–11.

Woodward, A.S., 1901. Catalog of fossil fishes in the British Museum (Natural History). Part VI. In: Containing the Actinopterygian Teleostomi of the Suborders Isospondyli (In Part), Ostariophysi, Apodes, Percesoces Hemibranchii, Acanthopterygii, and Anacanthini. British Museum (Natural History), London, 636 p.

Wu, F.X., Sun, Y.L., Hao, W.C., Jiang, D.Y., Sun, Z.Y., 2015. A new species of *Saurichthys* (Actinopterygii; Saurichthyiformes) from the Middle Triassic of southwestern China, with remarks on pattern of the axial skeleton of saurichthyid fishes. Neues Jahrbuch Geol. Palaontol. Abhand. 275, 249–267.

Wu, Y., Ge, Y., Hu, H., Stidham, T.A., Li, Z., Bailleul, A.M., Zhou, Z., 2023. Intra-gastric phytoliths provide evidence for folivory in basal avialans of the Early Cretaceous Jehol Biota. Nat. Commun. 14 (1), 4558. https://doi.org/10.1038/s41467-023-40311-z.

Wuttke, M., Poschmann, M., 2010. First finding of fish in the diet of a water-dwelling extinct frog *Palaeobatrachus* from the Upper Oligocene Fossil-Lagerstätte Enspel (Westerwald Mountains, Western Germany). Palaeobiodivers. Palaeoenviron. 90, 59–64.

Xing, L., Niu, K., Evans, S.E., 2019. Inter-amphibian predation in the Early Cretaceous of China. Sci. Rep. 9, 7751. https://doi.org/10.1038/s41598-019-44247-7.

Zakharenko, G.V., 2008. Possible evidence of predation in placoderms (Pisces) of the Evlanovo basin of central Russia. Paleontol. J. 42, 522–525.

Zangerl, R., Richardson, E.S., 1963. The paleoecological history of two Pennsylvanian black shales. Field. Geol. Mem. 4, 352 p.

Zhuravlev, K.N., 1943a. [The remains of Upper Jurassic sea reptiles at the Saveljevka Shale Mine site]. Bull. Acad. Sci. URSS Class. Sci. Biol. 5, 293–306 (In Russian, with English summary, p. 306.

Zhuravlev, K.I., 1943b. Nakhodki ostatkov verkhneoorskikh reptilii v savelevskom slantysevom rudniki [Finds of the remains of the upper orian reptile in the Savelievsky schantose mines]. Izvestiya Akad Nauk Sooza SSR Otdelenie Biologich Nauk 5, 293–306 (In Russian).

Zidek, J., 1992. Late Pennsylvanian Chondrichtyes, Acanthodii, and deep bodied Actinopterygii from the Kinney Quarry, Manzanita Mountains, New Mexico. New Mex. Bur. Mines Mineral Res. Bull. 138, 145–182.

Further reading

Andrews, P., 1990. Owls, Caves and Fossils: Predation, Preservation and Accumulation of Small Mammal Bones in Caves, with an Analysis of the Pleistocene Cave Faunas from Westbury-sub-Mendip, Somerset, UK. University of Chicago Press, Chicago, pp. 1–239.

Broughton, P.L., Simpson, F., Whitaker, S.H., 1978. Late Cretaceous coprolites from western Canada. Palaeontology 21, 443–453.

Buchy, M.-C., Vignaud, P., Frey, E., Stinnesbeck, W., González González, A.H., 2006. A new thalattosuchian crocodyliform from the Tithonian (Upper Jurassic) of north-eastern Mexico. Comptes Rendus Palevol 5, 785–794.

Druckenmiller, P.S., Russell, A.P., 2008. Skeletal anatomy of an exceptionally complete specimen of a new genus of plesiosaur from the Early Cretaceous (early Albian) of northeastern Alberta, Canada. Palaeontogr. Am. 283, 1–3.

Fernandez-Jalvo, Y., Andrews, P., 2016. Atlas of Taphonomic Identifications: 1001+ Images of Fossil and Recent Mammal Bone Modification. Springer, New York, 359 p.

Gripshover, N.D., Jayne, B.C., 2023. Using natricine snakes to test how prey type and size affect predatory behaviors and performance. Front. Behav. Neurosci. 17, 1134131. https://doi.org/10.3389/fnbeh.2023.1134131.

Hunt, A.P., Lucas, S.G., Klein, H., 2018. Late Triassic nonmarine vertebrate and invertebrate trace fossils and the pattern of the Phanerozoic record of vertebrate trace fossils. In: Tanner, L.H. (Ed.), The Late Triassic World. Topics in Geobiology 46. Springer Verlag, New York, pp. 447–543.

Martin, J.E., Fox, J.E., 2007. Stomach contents of *Globidens*, a shell-crushing mosasaur (Squamata), from the Late Cretaceous Pierre Shale Group, Big Bend area of the Missouri River, central South Dakota. In: Martin, J.E., Parris, D.C. (Eds.), The Geology and Paleontology of the Late Cretaceous Marine Deposits of the Dakotas, vol 427. Geological Society of America, Special Paper, pp. 167–176.

O'Connor, J.K., Zhou, Z., 2020. The evolution of the modern avian digestive system: insights from paravian fossils from the Yanliao and Jehol biotas. Palaeontology 63, 13–27.

Richter, G., 1992. Fossilized gut contents: analysis and interpretation. In: Schaal, S., Ziegler, W. (Eds.), Messel: An Insight into the History of Life and of the Earth. Oxford University Press, Oxford, pp. 285–289.

Sander, P.M., Pérez de Villar, P.R., Furrer, H., Wintrich, T., 2022. Giant Late Triassic ichthyosaurs from the Kössen Formation of the Swiss Alps and their paleobiological implications. J. Vertebr. Paleontol. https://doi.org/10.1080/02724634.2021.2046017.

Van Geel, B., Protopopov, A., Bull, I., Duijm, E., Gill, F., Lammers, Y., Nieman, A., Rudaya, N., Trofimova, S., Tikhonov, A.N., Vos, R., 2014. Multiproxy diet analysis of the last meal of an early Holocene Yakutian bison. J. Quat. Sci. 29, 261–268.

Van Nest, J., 2002. The good earthworm: How natural processes preserve upland Archaic archaeological sites of western Illinois, U. S. A. Geoarchaeology. Int. J. 17, 53–90.

CHAPTER

9

Vertebrate coprolites

Adrian P. Hunt[1] *and Spencer G. Lucas*[2]

[1]Flying Heritage and Combat Armor Museum, Everett, WA, United States; [2]New Mexico Museum of Natural History, Albuquerque, NM, United States

1. Introduction

William Buckland first recognized fossil vertebrate feces in the 1820s (Buckland, 1822, 1824). Subsequently, he coined the term coprolite for these trace fossils (Buckland, 1829). There has since been diverse and extensive study of vertebrate coprolites (see reviews in: Reinhard and Bryant, 1992; Duffin, 2009, 2012a,b; Bryant and Reinhard, 2012; Hunt et al., 2012d; Pemberton, 2012; Shillito et al., 2020; Hunt and Lucas, 2021e). After footprints, they are the most studied vertebrate trace fossils, and coprolites are the subject of an extensive and rapidly growing literature. Much recent work is focused on new methodologies, notably related to inclusions in coprolites, which is reviewed elsewhere in this volume by Dentzien-Dias and Łaska (2025). Hunt and Lucas (2021e) recently provided an extensive review of the fossil record of coprolites that is reprised and updated here. The principal goals of this paper are to provide an overview of the fossil record of vertebrate coprolites and its utility and also to review their ichnotaxonomy. Elsewhere we will more fully revise the ichnotaxonomy.

Abbreviations are as follows: GLCA, Glen Canyon National Recreation Area fossil collection reposited at Museum of Northern Arizona, Flagstaff, Arizona, USA; IITKGP, Indian Institute of Technology, Kharagpur, India; IVPP, Institute of Vertebrate Paleontology and Paleoanthropology, Chinese Academy of Sciences, Beijing, Peoples Republic of China; LACM, Natural History Museum of Los Angeles County, Los Angeles, California; LF, Lauer Foundation for Paleontology, Science and Education, Wheaton, Illinois, USA; MCYNAP, Museum of Chuxiong Yi Nationality Autonomous Prefecture, Chuxiong, Peoples Republic of China; NMMNH, New Mexico Museum of Natural History, Albuquerque, New Mexico, USA; OUM, Oxford University Museum of Natural History, Oxford, UK; USNM, United States National Museum, Smithsonian, Washington, DC, USA; YPM, Yale Peabody Museum of Natural History, New Haven, Connecticut, USA.

Vertebrate Ichnology
https://doi.org/10.1016/B978-0-443-38351-9.00008-6

2. Terminology

Buckland (1829; see review in Hunt and Lucas, 2012a) introduced the term coprolite for fossilized feces. The term coprolite has been consistently utilized for expelled fossil feces, although there are a few examples of the term being utilized for consumulites (Hunt and Lucas, 2012a, 2020, 2021e). Hunt and Lucas (2012a) reviewed all usage of terminology related to coprolites and introduced other related terms including latrinite (accretionary and ethological) for accumulations of coprolites and guanolite (ornithoguanolite, pinnipedaguanolite, chiropteraguanolite) for fossil guano (Fig. 9.1: Glossary of Terms).

Neumayer (1904) first introduced a terminology for spiral coprolites on the basis of specimens from the early Permian of Texas (Fig. 9.2). He recognized that some coprolites have spirals that are distributed roughly evenly along the long axis in lateral view (amphipolar), whereas others have the spirals concentrated at one end (heteropolar). Heteropolar coprolites are much more common throughout the fossil record than amphipolar forms. Hunt et al. (2007) recognized two forms of heteropolar morphology—microspiral, in which the posterior spire constitutes less than 50% of the length of the coprolite in lateral view, and macrospiral, in which it represents 50%—75% of the length of the coprolite (Fig. 9.3). Several authors proposed different descriptive terminologies for spiral coprolites (e.g., Jain, 1983; McAllister, 1985; Laojumpon et al., 2012), and Hunt and Lucas (2012b) synthesized aspects of these schemes (coils, lip) and defined new terms (anterior coil, posterior spire). Scroll coprolites represent a third major morphology of spiral coprolites that have the structure of a rolled

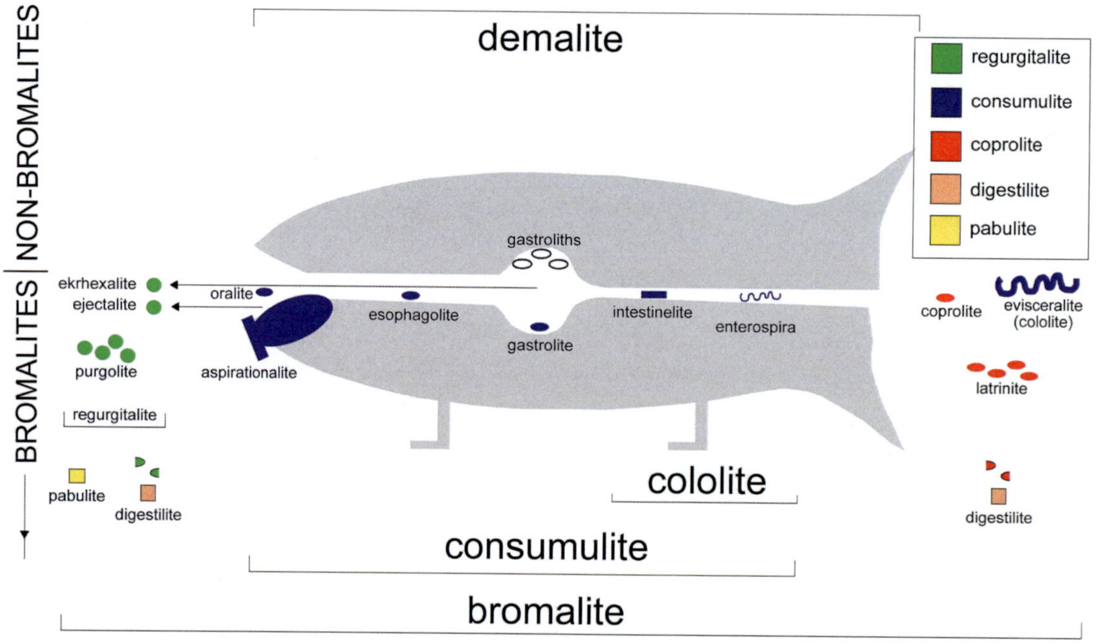

FIGURE 9.1 Selected classification terms for coprolites, other bromalites and associated trace fossils. *From Hunt and Lucas (2021e).*

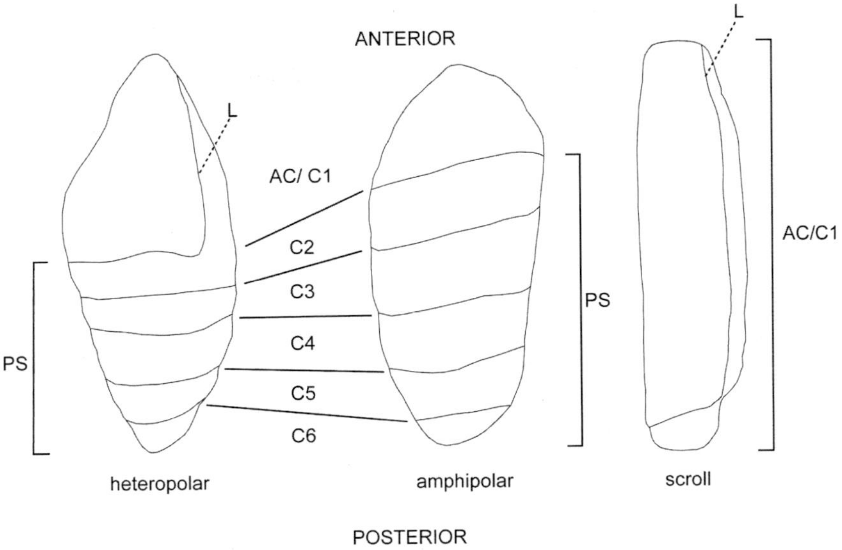

FIGURE 9.2 Descriptive terminology of spiral coprolites. **AC** is anterior coil, **C** is coil, **L** is lip, **PS** is posterior spire. *From Hunt and Lucas (2012b).*

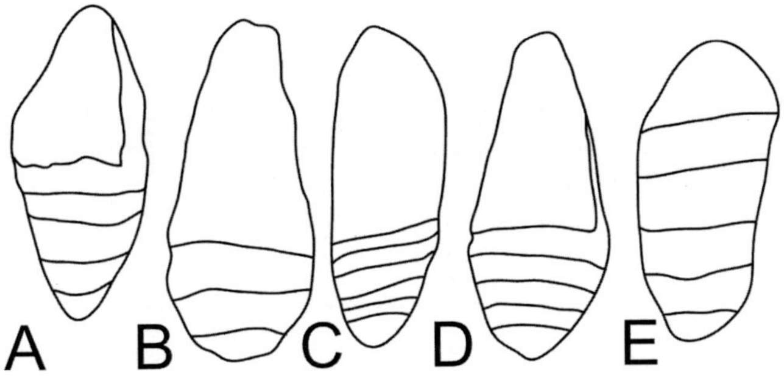

FIGURE 9.3 Principal morphotypes of spiral coprolites: Macrospiral heteropolar (A), microspiral heteropolar (B–D) and amphipolar coprolites (E). (A) *Liassocopros*. (B) *Saurocopros*. (C) *Malericopros*. (D) *Heteropolacopros*. (E) *Hyronocopros*. Not to scale. *From Hunt and Lucas (2021e).*

sheet of paper that consists essentially of a single coil (Hunt and Lucas, 2012b). Coprolite terminology is listed in the Glossary of Terms.

Nonspiral coprolites display a wide range of morphologies (e.g., Häntzschel et al., 1968, Fig. 9.1). Thulborn (1991) noted that many coprolites have terminations of different shapes. He applied the term anisopolar to this type of coprolite and the term isopolar to forms in which the two ends are of the same shape. Typically, the posterior (distal) end of an anisopolar coprolite that emerges first from the anus/cloaca is broadly rounded, and the anterior (proximal) end is tapered to a point (mucro of Thulborn, 1991). The trailing end is pinched by the constriction of the cloaca/anal margins as it closes (Thulborn, 1991: Fig. 9.4). Hunt and

FIGURE 9.4 Descriptive terminology of nonspiral coprolites. *From Hunt and Lucas (2012b).*

Lucas (2012b) proposed the term segments for the discrete longitudinal elements of a coprolite (pellets of Diedrich, 2012a)

3. Actualistic studies

Buckland (1822, 1824) conducted the first actualistic studies of coprolites when he fed cow bones to a hyena and studied the resulting feces to assist in the identification of specimens from a Pleistocene cave in northern England. Subsequently, Buckland recognized spiral coprolites, which he hypothesized to be produced by chondrichthyans. He dissected extant rays and scyliorhinid sharks (dogfish) to study the spiral valves of their intestinal tracts, and, subsequently, he injected the intestines with Roman cement (Buckland, 1835; Duffin, 2009). The injection process produced "artificial coprolites that in form are exactly similar to many of our fossil specimens" (Buckland, 1835, p. 234). Zangerl and Richardson (1963) conducted analogous experiments.

Modern feces have been the subject of extensive study, but actualistic studies in relation to coprolites have been mainly limited to anecdotal comments (e.g., Jain, 1983; McAllister, 1996), although there are a few notable exceptions (e.g., Esteban-Nadal et al., 2010; Milàn and Hedergaard, 2010; Milàn, 2012; Wings, 2012). However, many neontological studies provide important data for the study of coprolites such as the particle size of ingesta (Fritz, 2007; Fritz et al., 2009), fecal dimensions relative to body size (Platt et al., 2020) and fecal composition, diet and digestive physiology (Canfield and Fairburn, 1983; Doherty, 2009). There is clear potential for more actualistic studies related to coprolites.

4. Fossil record

4.1 Introduction

We have previously reviewed the fossil record of vertebrate animal coprolites (Hunt et al., 2012d; Hunt and Lucas, 2021e) and, specifically, coprofaunas from the Carboniferous-Permian (Hunt and Lucas, 2013), Triassic (Hunt et al., 2007, 2013b, 2018), Jurassic (Hunt and Lucas, 2014a), Cretaceous-Paleogene (Hunt and Lucas, 2007a, 2016b; Hunt et al., 2015b), Cenozoic (Hunt and Lucas, 2007a), and Pleistocene (Hunt and Lucas, 2018c, 2020). Here, we expand and update the overview provided by Hunt and Lucas (2021e) and add a section on human coprolites.

The fossil record of vertebrate coprolites is influenced by megabiases that deserve further study, including:

1. Most coprolites represent carnivores, except in rare instances (e.g., desiccated Pleistocene ichnofaunas: Hunt and Lucas, 2021c). Not only are herbivore feces preferentially degraded as they represent still "useable" plant debris, but also carnivore coprolites feces can fossilize quickly because of their high phosphate content (e.g., Hunt et al., 1994; Hollocher and Hollocher, 2012).
2. Most coprolites in nonmarine environments, at least through the Paleogene, represent aquatic or semi-aquatic animals. Thus, for example, many larger coprolites in the later Mesozoic may represent crocodiles rather than dinosaurs (Hunt and Lucas, 2017).
3. There are apparently significant size biases in the preservation of coprolites, and coprolites longer than 10 cm are rare before the Neogene, with rare examples (e.g., Chin et al., 1998). Coprolites are three-dimensional, unlike the majority of trace fossils, and are presumably subject to a variety of taphonomic processes usually associated with body fossils.
4. Coprolites can fossilize quickly, and, thus, they are common in hydrodynamically accumulated bone beds in marine and nonmarine environments (e.g., Hunt and Lucas, 2010). Coprolites are the only trace fossils that can be reworked. Thus, coprolites should a priori be considered as potentially physically robust fossils that are commonly transported.
5. Spiral coprolites, many produced by chondrichthyans, seem to be disproportionally represented in marine coprofaunas (Hunt et al., 2015b).
6. Coprolites occurrences are nonrandomly distributed through the Phanerozoic with notable acmes in Permian and Triassic red beds, Late Cretaceous marine and nonmarine strata, Cretaceous to Paleogene phosphorites and diverse Pleistocene environments.

5. Paleozoic

The oldest vertebrate body fossils are from the lower Cambrian of China (e.g., Shu et al., 1999, 2003; Shu, 2008). Cambrian coprolites occur at a number of localities, but all have been attributed to invertebrates (e.g., Chen and Chen, 1980; Peel, 2015; Shen et al., 2014; Kimmig and Strotz, 2017; Kimmig and Pratt, 2018). Some larger forms could have arguably been produced by vertebrates, such as round masses of fragmented skeletal material from the middle Cambrian Spence Shale of Utah. These were identified as coprolites by Conway Morris and Robison (1986) and attributed either to very large individuals of the arthropod predator *Anomalocaris* or to another, unidentified large predator.

There are several occurrences of clusters of conodont elements from the Early Ordovician of Kazakhstan, Sweden, and Australia that probably represent coprolites (Tolmacheva, 1996; Tolmacheva and Purnell, 2002; Stewart and Nicoll, 2003). These coprolites were most likely produced by conodont animals (Tolmacheva and Purnell 2002), and thus, these are the oldest "vertebrate" coprolites.

The Middle Ordovician Winneshiek Lagerstätte of Iowa in the United States and the Late Ordovician Soom Shale Lagerstätte of South Africa both preserve diverse bromalites that probably include some of vertebrate origin (Aldridge et al., 2006; Hawkins et al., 2018). The Winneshiek Shale yields a variety of vermiform bromalites that are principally preserved three-dimensionally and composed of calcium phosphate with a minority preserved as carbonaceous compressions (Briggs et al., 2015; Liu et al., 2017; Hawkins et al., 2018). Hawkins et al. (2018) identified five morphotypes that may have been produced by eurypterids, agnathans, and/or conodonts. Inclusions identified in thin sections are principally conodonts, but also include phyllocarids, other small arthropods such as ostracods, and linguloids. Some Winneshiek bromalites preserve a concentrically layered internal structure, which is characteristic of the coprolites of animals with a valvular intestinal tract such as chordates (Hawkins et al., 2018). The abundance and diversity of the Winneshiek bromalites may provide independent evidence of predation in the fauna during the Great Ordovician Biodiversification Event (Hawkins et al., 2018).

Aldridge et al. (2006) described five principal morphotypes of Ordovician bromalites, including coprolites. Brachiopod-bearing coprolites could pertain to conodont animals or several groups of invertebrates. Some coprolites, or possibly regurgitalites, are composed of conodont elements more fragmented than those in the Early Ordovician specimens, and they were probably produced by conodont animals (Aldridge et al., 2006). Corrugated/spiral bromalites, coiled coprolites, and wrinkled coprolites could all have been produced by vertebrates. The most likely vertebrate coprolites are those that are amphipolar spiral in morphology (Aldridge et al., 2006, pl. 1, Figs. 9.9 and 9.10). This corresponds to the intestinal morphology of many early fish (McAllister, 1987). While eurypterids also possessed a partially spiraled gut, there is no evidence that they produced spiral coprolites.

Caster and Kjellesyig-Waering (1964) described a putative eurypterid coprolite that is unstructured and contained fragments of the exoskeleton of the eurypterid *Megalograptus ohioensis* and a trilobite cephalon, in association with body fossils of *M. ohioensis* from the Ordovician of Ohio, as evidence of cannibalism. Unstructured masses containing disarticulated agnathan fragments are common in the Monks Water fish bed in the Silurian Hagshaw Hills inlier in Scotland, together with the eurypterid *Lanarkopterus dolichoschelus,* and these were interpreted by Selden (1979) as eurypterid coprolites. These examples suggest eurypterid coprolites consist of unstructured material. Thus, we consider the spiral morphology to be diagnostic of vertebrates.

There are few putative Early or Middle Silurian vertebrate coprolites. There are Early Silurian coprolites from Scotland and Nova Scotia of such a size that they could have been produced by vertebrates, but they have been attributed to large predatory eurypterids such as *Lanarkopterus* (Gilpin, 1886; Rolfe, 1973; Selden, 1984; Turner, 1999). Late Silurian coprolites include two unusual morphologies. Gilmore (1992) described more than 150 scroll and nonspiral coprolites from western Ireland and southern Scotland. The coprolites are composed of argillaceous clastic material. These coprolites occur with body fossils of agnathans, to which they are attributed. Scroll coprolites are uncommon in the fossil record (e.g., Hunt et al., 2012h; Stringer and King, 2012), and the only other examples of nonphosphatic spiral coprolites are from the Permian of Antarctica (Retallack and Krull, 1999). Bicknell et al. (2023) described Silurian coprolites from Australia with abundant trilobite fragments that are unlikely to have been made by fish and were probably produced by eurypterids.

The Telychian/Llandovery agnathan locality of Birk Knowes, Lanarkshire, Scotland, yields a large number of coprolites. They are composed principally of agnathan fragments with little matrix. Specimens at the Hunterian Museum in Glasgow preserve two principal morphologies. The majority of bromalites are flattened ovoids averaging 30 mm in length and 40 mm in height. A small number of specimens are parallel-sided linear ribbons with a maximum length of 105 mm and a width of 24 mm. The lack of matrix could suggest that these bromalites represent regurgitalites. However, the abundance of specimens and the regular size of the pellets suggest they are coprolites, albeit unusual ones.

The Ludlow bone bed (Ludlow Bone Bed Member of Downton Castle Sandstone Formation) is a Late Silurian source of vertebrates in the area along the border between England and Wales. Coprolites have long been known to be locally common in the Ludlow bone bed (Strickland and Hooker, 1853; Murchison, 1867). They include spiral morphologies and are phosphatic, and thus are similar to many geologically younger vertebrate coprolites (Murchison, 1839, p. 607, pl. 2, Figs. 46, 47, 54, 55).

Vertebrate coprolites become more common in the Devonian, with the most important assemblages derived from fossil-fish localities in Scotland, and Lagerstätten elsewhere. The most significant Early Devonian coprolites from Scotland derive from the Lochkovian Tillywhandland Quarry, Forfarshire (Trewin and Davidson, 1996). Many coprolites contain spines and scales of the acanthodian fish *Mesacanthus* and were probably produced by *Ischnacanthus* (Trewin and Davidson, 1996; Trewin, 2008; Newman and Davidson, 2010). The *Mesacanthus* spines are usually oriented anteriorly, which indicates head-first ingestion of spiny acanthodians (Trewin, 2008). Most coprolites are flattened. A second important assemblage of coprolites occurs at Duntrune Quarry in Angus. Spiral coprolites are uncommon in the nonmarine assemblages of the Early Devonian of Scotland.

The Early Devonian Hunsrück Slate Lagerstätte of Germany yields pyritized coprolites of various morphologies and includes local concentrations (Bartels et al., 1998; Wagner and Boyce, 2006; Kühl et al., 2012). Small strings of feces that represent deposit feeders are widespread, but larger coprolites attributed to fish are only common in the northern Hunsrück where body fossils are rare (Bartels et al., 1998). These coprolites range from ovoids to spirals in morphology and locally are concentrated in thin layers (Bartels et al., 1998, Fig. 220; Kühl et al., 2012, Fig. 119).

The Achanarras Quarry in the Upper Flagstone Group of the Middle Old Red Sandstone of Caithness (Eifelian–Givetian) yields a large sample of Middle Devonian coprolites (Rayner,

1963; Hamilton and Trewin, 1988, 1994; Trewin, 1986). Other Scottish Middle Devonian coprolites derive from Orkney and Moray. Most coprolites are preserved three dimensionally and include several examples of spiral coprolites, and one with a scroll morphology.

McAllister (1996) published a detailed study of more than 120 bromalites, principally coprolites, from the Upper Devonian Escuminac Formation of the Miguasha Lagerstätte, Quebec, Canada. These coprolites are typical of Devonian assemblages in that spiral forms are relatively uncommon (only three to four in the Escuminac) as are specimens over 4–5 cm in length, and larger forms are very rare. Small vertebrate coprolites occur in the Late Devonian of Poland, and one larger spiral specimen contains conodont elements (Zatoń and Rakociński, 2014; Zatoń et al., 2017). Devonian coprolites have been reported from other areas, including Brazil (Maisey and Melo, 2005), Latvia/Estonia (Hunt et al., 2012b), the United States (e.g., Branson, 1914), and South Africa (Harris and Gess, 2017).

Carboniferous strata record the first large samples of vertebrate coprolites (Turner, 1999; Hunt and Lucas, 2021e: Fig. 9.5). This is a strictly Laurussian record that increases in diversity and abundance from the Mississippian into the Pennsylvanian. Mansky et al. (2012) described coprolites from the Early Mississippian (Tournaisian) Horton Bluff Formation of Nova Scotia. A large and diverse coprolite ichnofauna occurs in the Tournaisian of Greenland and represents tetrapods and a variety of fish (Byrne et al., 2022). In Scotland, there are Early Mississippian coprolites (e.g., Pollard, 1985) as well as large collections of coprolites from three Middle Mississippian (Viséan) localities: (1) East Kirkton, West Lothian; (2) Wardie, Midlothian; and (3) Anstruther, Fife (Buckland, 1836; Sumner, 1991, 1994). The Wardie and Anstruther localities yield the oldest coprofaunas dominated by spiral coprolites. These coprolites occur in concretions and average about 5 cm in length. Very large coprolites from East Kirkton measuring more than 17 cm in length have been attributed to eurypterids (Sumner, 1994). The Late Mississippian (Serpukhovian) of Bearsden, East Dunbartonshire, Scotland, yields a large sample of coprolites, as does the similar-age Bear Gulch Lagerstätte of Montana, USA (Zidek, 1980; Clark, 1989; Hunt et al., 2012e). Other Mississippian coprolites occur in the USA (e.g., Fayetteville Shale of Arkansas and Oklahoma, Michigan Formation of Michigan, Buffalo Wallow Formation of Kentucky) and England (Hunt and Lucas, 2013; Greb et al., 2015; Duffin and Ward, 2020).

The first abundant samples of vertebrate coprolites occur in the Pennsylvanian (Hunt and Lucas, 2021e: Fig. 9.5). Coprolites are found in marine and lagoonal shales at many localities as well as in some nonmarine coal successions (Hunt and Lucas, 2013). Four bromalite faunas of Late Pennsylvanian (Missourian/Kasimovian) age in New Mexico, USA, represent an ecological transect from lacustrine to basinal marine: (1) Tinajas Lagerstätte—lacustrine; (2) Kinney Brick Quarry Lagerstätte—lagoonal; (3) Erickson site—nearshore marine; and (4) Sacramento Mountains—offshore marine (Harris et al., 2017; Hunt and Lucas, 2017a; Hunt and Lucas, 2021e: Fig. 9.6). There are clear trends through these ichnofaunas from lacustrine to offshore marine in terms of a trend of flattened preservation to three-dimensional preservation, decreasing diversity of kinds of bromalites (regurgitalites, consumulites) and increasing proportions of spiral coprolites (Hunt and Lucas, 2017a).

There are several Pennsylvanian examples of stratigraphically narrow but geographically widespread zones rich in spiral coprolites in marine shales in Colorado (Weber Formation, lower Belden Formation) and West Virginia (Conemaugh Group) in the USA, and in Germany (Göttelborner Beds) (Price, 1927; Johnson, 1934; Guthörl, 1959; Kneuper and

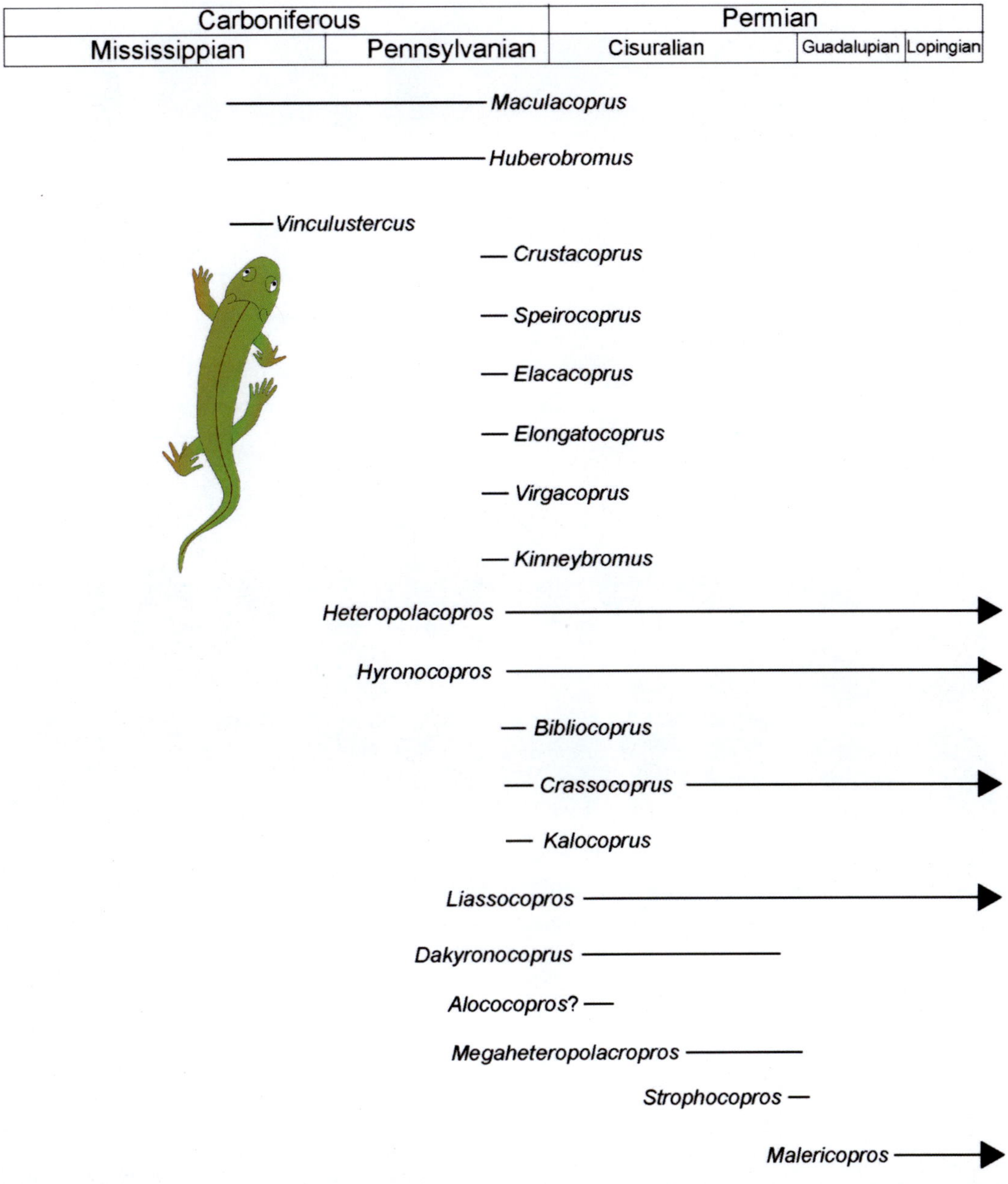

FIGURE 9.5 Stratigraphic distribution of Paleozoic coprolite ichnogenera. *Modified from Hunt and Lucas (2021e).*

Schönenberg, 1962; Hunt et al., 2012b; Hunt and Lucas, 2013). A variety of Pennsylvanian Lagerstätten in the United States preserve coprolites, including Mazon Creek in Illinois, the Hamilton Quarry of Kansas and the Kinney Brick Quarry and Tinajas localities of New

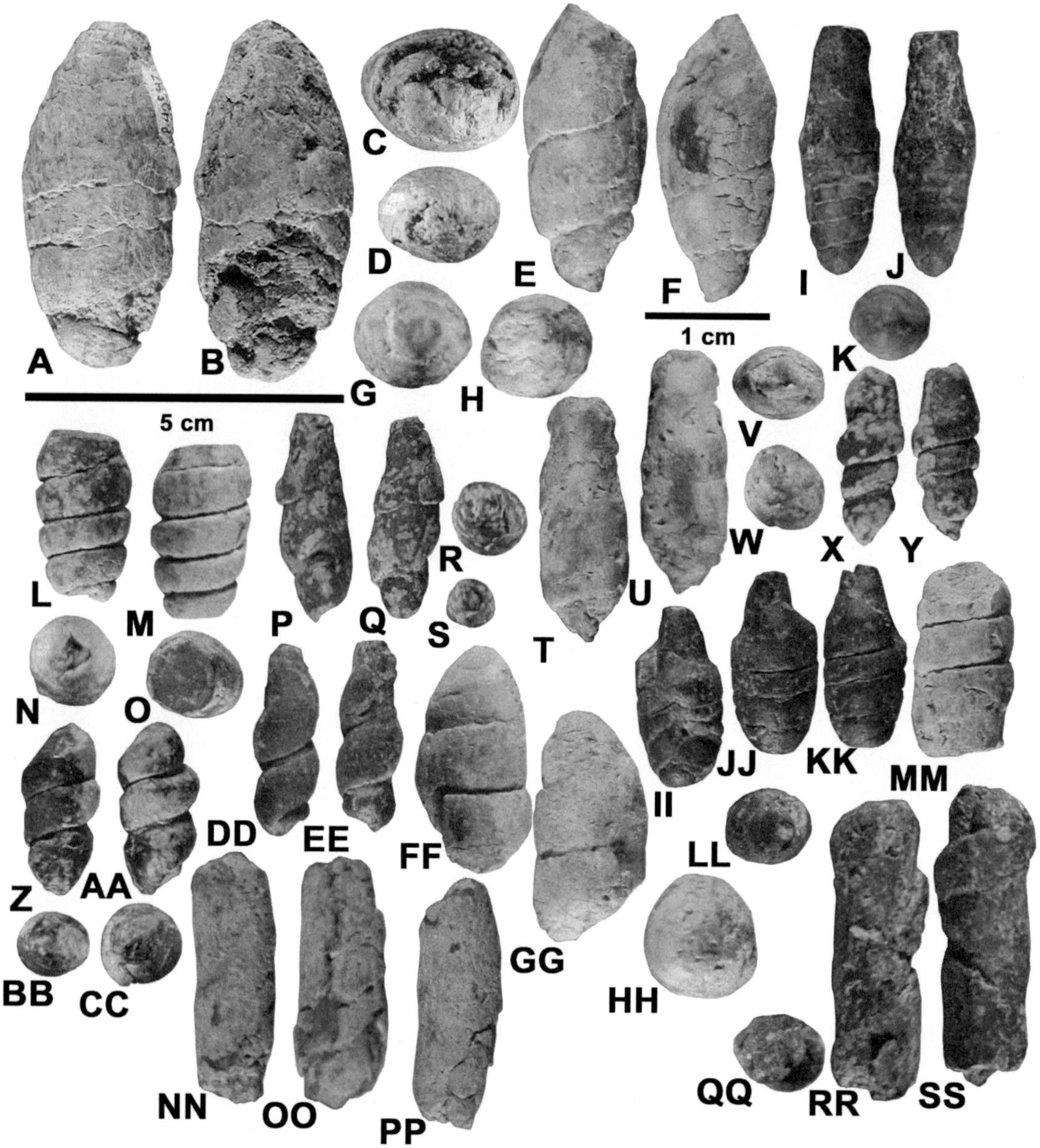

FIGURE 9.6 An example of a Paleozoic ichnofauna. Selected spiral coprolites from the Late Pennsylvanian (Missourian) Beeman Formation, Otero County, New Mexico, USA. (A–D) *Liassocoprus hawkinsi*, NMMNH P-42547, coprolite in A-B, axial and C-D, polar views. (E–H) *Hyronocoprus* ichno sp., NMMNH P-63521, coprolite in E-F, axial and G-H, polar views. (I–K) *Heteropolacoprus texaniensis*, NMMNH P-63514, coprolite in I-J, axial and K, polar views. (L–O) *Hyronocoprus amphipola*, NMMNH P63520, coprolite in L-M, axial and N–O, polar views. (P–S) *Kalocoprus oteroensis*, NMMNH P-63511, coprolite in P-Q, axial and R–S, polar views. (T–W) *Heteropolacoprus ichno sp.*, NMMNH P-63517, coprolite in T-U, axial and V–W, polar views. (X–CC) *Kalocoprus oteroensis*. X–Y, NMMNH P-63527, coprolite in axial views. Z-CC, NMMNH P-63523, coprolite in Z-AA, axial and BB-CC, polar views. (DD-EE) *Kalocoprus oteroensis*, NMMNH P-63513, coprolite (holotype) in axial views. (FF–HH) *Hyronocoprus amphipola*, NMMNH P-63519, coprolite in FF-GG, axial and HH, polar views. (II–LL) *Heteropolacoprus texaniensis*, NMMNH P-63524, coprolite in II-KK, axial and LL, polar views. (MM) *Hyronocoprus amphipola*, NMMNH P-63518, coprolite in axial view. (NN–SS) *Bibliocoprus beemanensis*. NN-PP, NMMNH P-63515, coprolite in axial views. QQ-SS, NMMNH P-63503, coprolite (holotype) in QQ, polar and RR-SS, axial views. *From Hunt et al. (2012h).*

Mexico (Richardson and Johnson, 1971; Richardson, 1980; McAllister, 1988; Shabica and Godfrey, 1997; Hunt et al., 2012i,j; Harris et al., 2017; Hunt and Lucas, 2017a, 2021e; Tripp et al., 2021, 2022). Marine invertebrates in coprolites from Hamilton aid in paleoenvironmental interpretation (Cunningham et al., 1993). Zangerl and Richardson (1963) conducted a classic and extensive study of the paleoecology of the Middle Pennsylvanian (Desmoinesian/Moscovian) Mecca Quarry Shale and Logan Quarry members of the Carbondale Formation in western Indiana, USA, and described multiple bromalites (coprolites and regurgitalites: Elder, 1985; Elder and Smith, 1988).

Dawson (1854, 1862) first noted common coprolites in the fossil-bearing tree stumps of the Early Pennsylvanian Joggins Formation of Nova Scotia, and they have been described from various localities in eastern Canada (Brown and Lyell, 1845; Keighley and Pickerill, 1997; Falcon-Lang et al., 2006; Ó Gogáin et al., 2016; Chipman et al., 2020; Bingham-Koslowski et al., 2021; Hunt, 2022). Other coal-bearing Pennsylvanian strata in England, Scotland, Belgium, Germany, and Poland yield coprolites (e.g., Buckland, 1836, 1858; Bayer, 1934; Anderson et al., 1997; Hunt et al., 2012b,c; Hunt and Lucas, 2013; Krzykawski et al., 2014, Lomax et al., 2016). Hodnett and Lucas (2018) described an unusual, large coprolite that contained body fossils of fishes and a tetrapod from the Upper Pennsylvanian of New Mexico that may represent a fecal cloud derived from a large shark.

Many Pennsylvanian coprolites have a spiral morphology, and most of these are heteropolar in form, including *Crassocoprus* and *Heteropolacopros* (Hunt et al., 2012h,i,j, 2013c; Hunt and Lucas, 2017a; Chipman et al., 2020; Hunt and Lucas, 2021e, Figs. 25 and 26). Amphipolar morphologies, such as *Kalocoprus* and *Hyronocoprus,* are less common, and scroll coprolites (e.g., *Bibliocoprus*) are only locally present (Hunt et al., 2012h; Hunt and Lucas, 2017a; Chipman et al., 2020; Hunt and Lucas, 2021e, Fig. 26). The Benxi Formation of Mississippian/Pennsylvanian (Serpukhovian to Bashkirian) age in northern China yields small coprolites that could have been produced by fish (Gong et al., 2010).

Whereas the Carboniferous coprolite record is strictly Laurussian, the Permian coprolite record globalizes to include records from Gondwana (Hunt and Lucas, 2013). A similar pattern is seen in the bone record, and this is due, at least in part, to the glaciations in Gondwana during parts of the Carboniferous-earliest Permian, which restricted the distribution of vertebrates in the southern continents (Lucas, 2006). Permian coprolites are even more abundant than those of the Pennsylvanian and are particularly common in red beds.

Early Permian coprolites occur in red beds of the American Southwest in New Mexico (Hunt et al., 2005b, 2013a; Cantrell et al., 2012), Texas (Neumayer, 1904; Olson, 1966; Olson and Mead, 1982; Sander, 1989; Hunt and Lucas, 2005a,b, 2023b, Hunt et al., 2005a, 2012c), and Oklahoma (Olson, 1971, 1977; Hunt et al., 2012b), and in marine strata in Kansas (Williams, 1972; McAllister, 1985). These ichnofaunas principally yield spiral coprolites that are commonly heteropolar (*Heteropolacopros, Megaheteropolacopros, Liassocoprus, Malericoprus, Saurocoprus*), although amphipolar (*Hyronocoprus*) and nonspiral forms (*Alococopros, Dakyronocopros, Strophocoprus*) are also present (Neumayer, 1904; Hunt and Lucas, 2005a,b, 2023b; Hunt et al., 2005a, 2012a,b, 2013a; Cantrell et al., 2012). Similar age coprolites are present in the lower Permian Dunkard Group of West Virginia in the eastern United States (Bowen, 2013; Lucas, 2013; Hembree and Blair, 2016; Hembree and Bowen, 2017).

Coprolites are common in the Rotliegend strata of Western and Central Europe, including France (Gaudry, 1887), Germany (Lohmann and Sachs, 2001; Eichler and Werneburg, 2010), and the Czech Republic (Reuss, 1856; Lemke and Weiler, 1942; Zajíc, 2014). Other early Permian coprolites are found in Brazil and Namibia and include specimens associated with skeletons of *Mesosaurus* (Ruedemann, 1929; Horsthemke et al., 1990; Warren et al., 2001; Silva et al., 2017). Middle Permian coprolites are less common, but they are present in Brazil (Dentzien-Dias et al., 2012, 2013, 2017), Russia (Ochev, 1974), and Antarctica (Retallack and Krull, 1999). The Antarctic specimens are of note because they are composed of arenaceous matrix (Retallack and Krull, 1999). Late Permian coprolites have been reported from England (Bell et al., 1979) and South Africa (Smith, 1993; Smith and Botha-Brink, 2011), but they are most common in Russia where one specimen contains possible premammalian hair (Owocki et al., 2012; Sennikov and Golubev, 2012; Bajdek et al., 2016, 2017; Niedźwiedzki et al., 2016b).

5.1 Mesozoic

There is an abundant record of vertebrate coprolites in the Triassic that increases in abundance and diversity through that period (Hunt et al., 2018: Fig. 9.7). As in the Permian, Triassic coprolites are particularly abundant in nonmarine red beds. Early Triassic coprolites are geographically widely dispersed but relatively small in number. The largest sample is from the Arcadia Formation (Induan) in Queensland, northeastern Australia (Northwood, 1997, 2005). This ichnofauna includes *Hyronocopros amphipola*, *Alococopros triassicus*, *Eucoprus cylindratus* and indeterminate coprolites (Hunt et al., 2007). A similar ichnofauna occurs in the Burgersdorp Formation (Olenekian) of South Africa with over 30,000 specimens and including *Alococopros triassicus*, *Eucoprus cylindratus*, *Heteropolacopros* sp. and *Hyronocopros amhipola* (Bender and Hancox, 2003, 2004; Yates et al., 2012; Hancox et al., 2020; Hancox and Rubidge, 2023; Montgomery, 2022; Hoffman et al., 2023; Wolvaardt et al., 2023; Zipfel et al., 2023). Other significant ichnofaunas occur in the Vokhama Formation in Russia (Niedźwiedzki et al., 2016b), and the Bulgo Sandstone in Australia (Niedźwiedzki et al., 2016a). Other coprolites also occur in units of the Buntsandstein in France (Gall, 1971; Gall and Grauvogel-Stamm, 1993) and Germany (e.g., Dachroth, 1985) and the Beaufort Group in South Africa (Smith and Botha-Brink, 2011).

Early Triassic coprolites from marine strata of Japan and Poland have been utilized to address changing ecosystems after the PTB extinctions and in relation to the "Mesozoic Marine Revolution" (Nakajima and Izumi, 2014; Brachaniec et al., 2015). Similarly, a diverse coprolite fauna from the Middle Triassic of Luoping, China has been used as evidence of the emergence of complex trophic ecosystems in the Anisian as part of the biotic recovery after the end-Permian extinctions (Hu et al., 2010, 2011; Luo et al., 2017). Yao et al. (2022) and You et al. (2020) described coprolites from the early Ladinian lacustrine ecosystem of the Yandang Formation in the Ordos Basin as derived from the earliest known Mesozoic-type, trophically multileveled lacustrine ecosystem after the end-Permian mass extinction.

Middle Triassic nonmarine coprolite localities are widely dispersed in North (USA) and South America (Argentina, Brazil), Europe (England, France, Germany, Switzerland, Italy, Poland, and Slovenia), Asia (Russia, Kazakhstan, India, China) and Africa (South Africa) (Hunt et al., 2013b). The best described Middle Triassic ichnofaunas are from Brazil (Santa Maria Formation: Souto, 2001; Hunt et al., 2013b; Francischini et al., 2018), Argentina

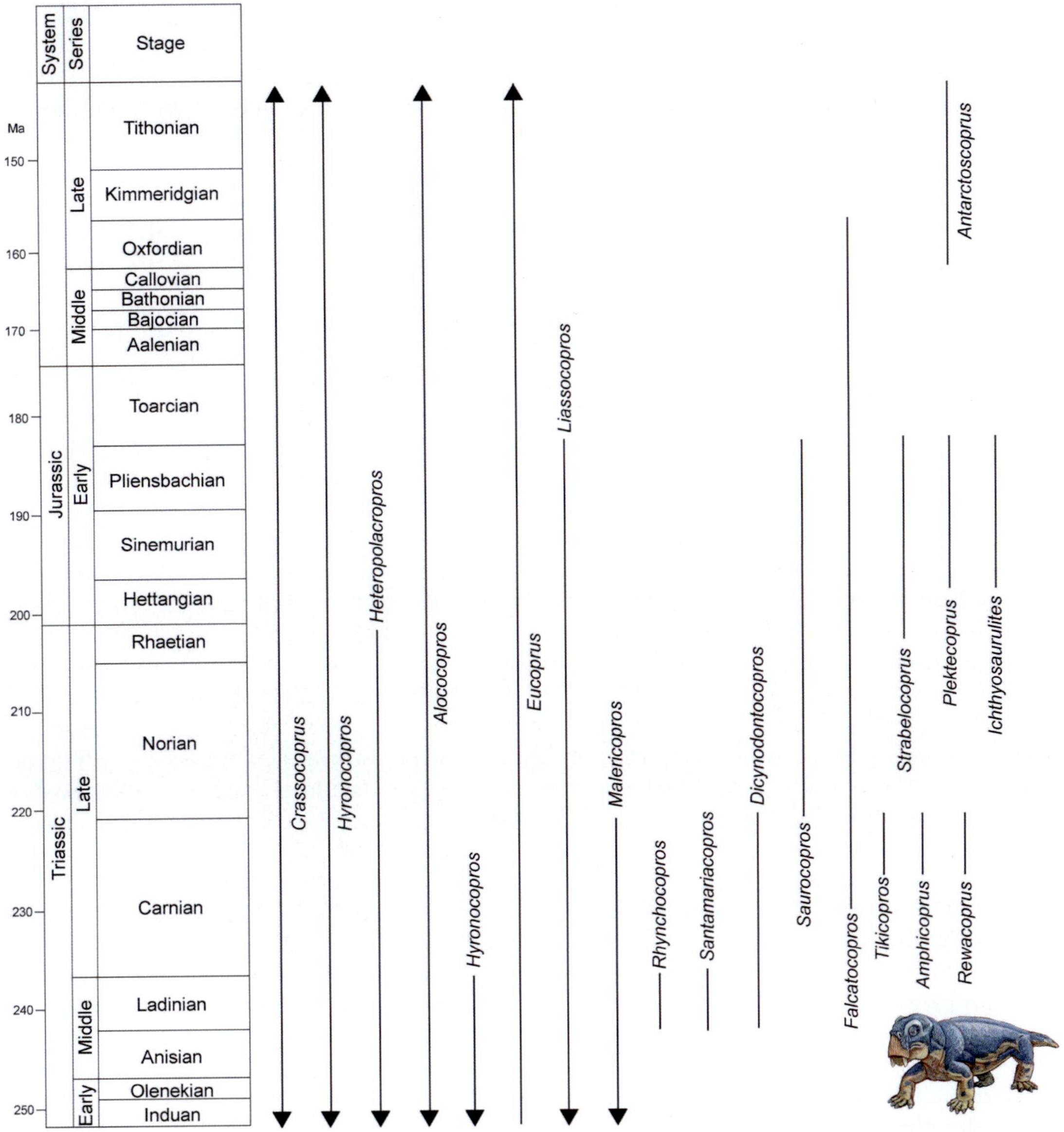

FIGURE 9.7 Stratigraphic distribution of Triassic and Jurassic coprolite ichnogenera. *Modified after Hunt and Lucas (2021e).*

(Potrerillos, Cacheuta, Río Blanco and Chañares formations: Rusconi, 1947, 1949; Mancuso et al., 2004; Fiorelli et al., 2013; Loinaze et al., 2018; Mancuso et al., 2018), the United States (Moenkopi Formation: Benz, 1980; Morales, 1987; Boy et al., 2001; Schoch et al., 2010), Russia (Karagachka, Donguz I and Bukobay localities: Ochev, 1974), Kazakhstan (Mollo-Khara-Bala-

Kantemir locality: Ochev, 1974) and Poland (Muschelkalk: Chrzastek, 2008; Kowal-Linka and Bodzioch, 2012). The Santa Maria Formation of Brazil yields the earliest unambiguous vertebrate herbivore coprolites, *Rhynchocopros* and *Santamariacopros* (Hunt et al., 2013b; but see Francischini et al., 2018). The oldest latrinite is from the Chañares Formation of Argentina (Fiorelli et al., 2013).

The Late Triassic sample of coprolites is by far the largest from this period, and the majority of specimens are from nonmarine strata (Hunt and Lucas, 2021e, Figs. 27 and 28). There are records in the United States (Lucas et al., 1985; Wahl et al., 1998; Hunt et al., 1998, 2007, 2013b, 2018), Greenland (Milàn et al., 2012b, 2021; Hansen, 2014; Hansen et al., 2016), England (Duffin, 1979; Swift and Duffin, 1999; Hunt et al., 2013b; Cross et al., 2018; Cueille et al., 2020; Williams et al., 2022), The Netherlands (Klompmaker et al., 2010), Switzerland (Fluckiger, 1861), Germany (Fraas, 1891; Schoch, 2012), Poland (Badjek et al., 2014, 2019; Zatoń et al., 2015; Qvarnström et al., 2019a,b, 2021), India (Matley, 1939a,b; Sohn and Chatterjee, 1979; Jain, 1983; Vijaya et al., 2009; Rakshit et al., 2019; Umamaheswaran et al., 2019; Rakshit and Ray, 2021), Thailand (Laojumpon et al., 2012), Argentina (Contreras, 1995; Holocher et al., 2005), Brazil (Langer, 2005; Battista et al., 2024), Morocco (Hunt et al., 2013b; Zouheir et al., 2023), Madagascar (Burmeister et al., 2006) and South Africa (Anderson et al., 1998). The majority of specimens are from the nonmarine Chinle Group and Newark Supergroup of western and eastern North America, respectively, and the marine Rhaetic bone beds in Europe (Duffin, 1979; Swift and Duffin, 1999; Hunt et al., 2013b).

The largest and most studied Late Triassic coprolite assemblages are from about 20 stratigraphic units of the Chinle Group of the western and southwestern United States (Arizona, New Mexico, Colorado, Texas, Utah, Wyoming: Hunt et al., 1998, 2007, 2013b, 2018: Fig. 9.8). Late Triassic nonmarine coprofaunas include a wide diversity of morphologies representing large herbivores (*Dicynodontocopros*) and carnivores with heterospiral (*Heteropolacopros*), scroll and nonspiral (*Alococoprus, Eucoprus*) coprolite morphologies (Hunt et al., 2013b, 2018; Bajdek et al., 2014; Zatoń et al., 2015; Rakshit et al., 2019; Bajdek et al., 2019; Qvarnström et al., 2019a, 2021). The most abundant marine coprolites are from Rhaetian bone beds of western Europe, notably the United Kingdom, and morphotypes include spiral and nonspiral forms (Duffin, 1979; Swift and Duffin, 1999; Hunt et al., 2013b; Allard et al., 2015; Norden et al., 2015; Mears et al., 2016; Slater et al., 2016; Cavicchini et al., 2018; Cross et al., 2018; Cueille et al., 2020; Ronan et al., 2020; Evans et al., 2024; Williams et al., 2022). In England, two types of bone beds can be distinguished in the Westbury Formation on the basis of taphonomy and rare earth element geochemistry, distal beds with a high degree of transport and proximal beds with significantly less evidence of high energy transport (Trueman and Benton, 1997; Evans et al., 2024). Coprolites also occur in the marine Carnian of Austria (Lukeneder et al., 2020; Lukeneder and Lukeneder, 2021).

Jurassic coprolites are much less common than those in the Triassic and understudied, with the notable exception of those from the Lias of England, which yields several ichnotaxa, including multiple spiral forms such as *Liassocoprus, Strabelocoprus* and *Saurocoprus* (Hunt et al., 2007, 2012a; Duffin, 2009, 2010, 2012a; Hunt and Lucas, 2021e, Fig. 27). Other Early Jurassic marine units in Europe yield relatively few coprolites, for example, the Posidonienschiefer of Germany (e.g., Hauff, 1921). Garassino and Donovan (2000) describe possible coprolites of marine reptiles, some containing coleoid hooks, which are common in ichthyosaur consumulites, from the Early Jurassic of Italy.

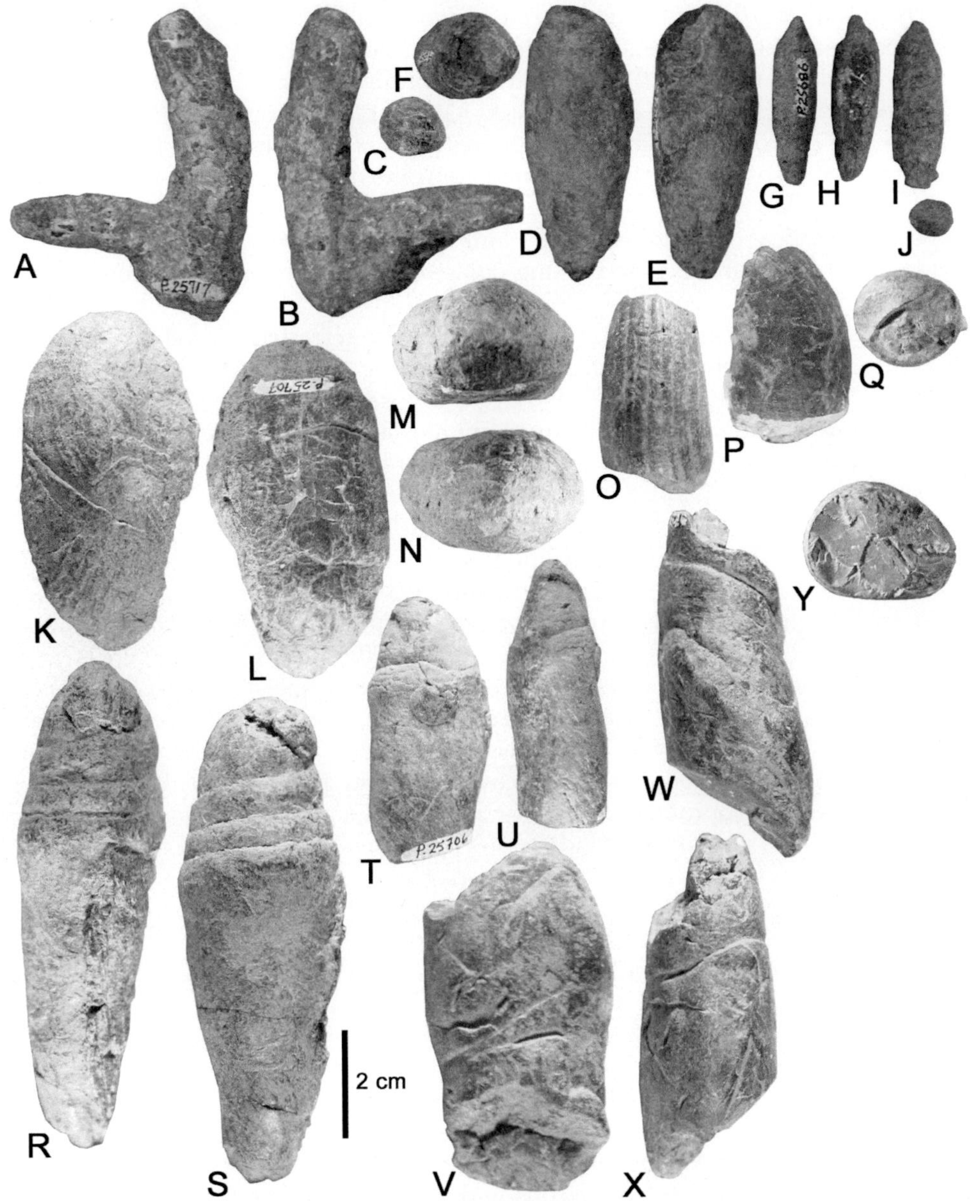

FIGURE 9.8 An example of an early Mesozoic ichnofauna. Late Triassic (Adamanian) coprolites from the Ciniza pond locality, Bluewater Creek Formation, New Mexico, USA (NMMNH locality 1864). (A–C) NMMNH P-25717, two conjoined coprolites in A-B, axial and C, polar views. (D–F) NMMNH P-25692, heterospiral coprolite in D-E, axial and F, polar views. (G–J) NMMNH P-25686 from NMMNH locality 1864, heterospiral coprolite in G-I, axial and J, polar views. (K–N) NMMNH P-25707, large ovoid coprolite in K-L, axial and M-N, polar views. (O–Q) NMMNH P-25706, *Alococopros triassicus* in O–P, axial and Q, polar views. (R and S), NMMNH P-1864, large *Heteropolacopros texaniensis* in axial views. (T and U) NMMNH P-25706, partial *Heteropolacopros texaniensis* in axial views. (V–Y) NMMNH P25709, segment of *Heteropolacopros texaniensis* in V-X, axial and Y, polar views. *From Hunt et al. (2013b).*

The Early Jurassic portion of the nonmarine Newark Supergroup in eastern North America contains coprolites that are little studied (e.g., Hitchcock, 1844; Dana, 1845; Gilfillian and Olsen, 2000; Lanzirotti et al., 2000), and a smaller number occur in the Glen Canyon Group of the southwestern USA (Clark and Fastovsky, 1986). Górecki et al. (2019) described plant material from the coprolite of a large predatory dinosaur from the Early Jurassic of Poland. Subsequently, Barbacka et al. (2022) examined coprolites of herbivorous sauropodomorphs, ornithischians, and large carnivorous theropods from the Hettangian of the Holy Cross Mountains in Poland and described their plant content. In general, there is no significant turnover of coprolite ichnotaxa across the Triassic/Jurassic boundary (Hunt and Lucas, 2018a, 2021e).

Nonmarine Middle Jurassic coprolites are rare (as are terrestrial vertebrate body fossils). Hill (1976) described a mass of small pellets containing the plant *Ptilophyllum* in England and tentatively assigned them to an ornithopod dinosaur or mammal, although the latter is highly improbable given their size (about 1 cm in diameter) and age. The Natural History Museum in London has coprolite collections from the marine Purbeck Limestone and Oxford Clay formations of England, which are largely unstudied (Hunt et al., 2007, 2012c). The Oxford Clay is notable for yielding many vertebrate coprolites (Martill, 1985; Martill et al., 1994).

Several lithographic limestone Lagerstätten of Late Jurassic age in Germany contain coprolites. Schweigert and Dietl (2012) described specimens from the Nusplingen Lithographic Limestone, and specimens are also common in the Solnhofen and Ettling Lagerstätten (Barthel, 1978; Barthel et al., 1990; Kemp and Trueman, 2003; Röper, 2005; Ebert et al., 2015). *Medusites capillaris is a* problematic fossil from the Solnhofen Limestones that consists of white or pinkish, coiled parallel strings of homogenous phosphatic matter that could be the coprolite of a coleoid or possibly a small fish, although we consider a vertebrate origin unlikely given its morphology (Schweigert, 2001; Schweigert and Dietl, 2012). Hone et al. (2015) described rare pterosaur coprolites from Solnhofen, and Qvarnström et al. (2019b) described others from Poland. A Late Jurassic lithographic limestone Lagerstätte in the Causse Méjean (southern France) also contains coprolites (Moreau et al., 2022). The Talbragar Fish Bed Lagerstätte in New South Wales, Australia, is of Kimmeridgian-Tithonian age and yields abundant coprolites (Beattie and Avery, 2012; Schwarzhans et al., 2018).

Despite the local abundance of dinosaur body fossils in the Late Jurassic, their coprolites are rare. Hunt and Lucas (2014a) reviewed the record of Late Jurassic coprolites from the United States. It includes probable theropod specimens and putative specimens from the Morrison Formation supposedly produced by herbivorous dinosaurs (Chin and Kirkland, 1998; Chin and Bishop, 2004, 2007). Bigurrarena Ojeda et al. (2017, 2023) named *Antarctoscoprus longinensis* for a coprolite produced by a nonteleostean actinopterygian from the Late Jurassic of Antarctica (see also Doyle and Whitham, 1991; Gouiric-Cavalli et al., 2017, 2019).

Cretaceous coprolites are more numerous and more widespread than in any earlier time period. Early Cretaceous coprolites are mainly known from nonmarine settings and are relatively poorly documented, although their occurrences are widespread in North and South America, Africa, Australasia, and Europe (Hunt and Lucas, 2016b: Fig. 9.9). Bertrand (1903) authored the first monograph on vertebrate coprolites, based on specimens that had been found in strata that yielded multiple skeletons of *Iguanodon* from Belgium. He ascribed them to theropod dinosaurs (see also Casier, 1960, 1978), but they more likely pertain to crocodiles (Abel, 1935). Multiple coprolites also occur in an *Iguanodon* bone bed at Nehden in

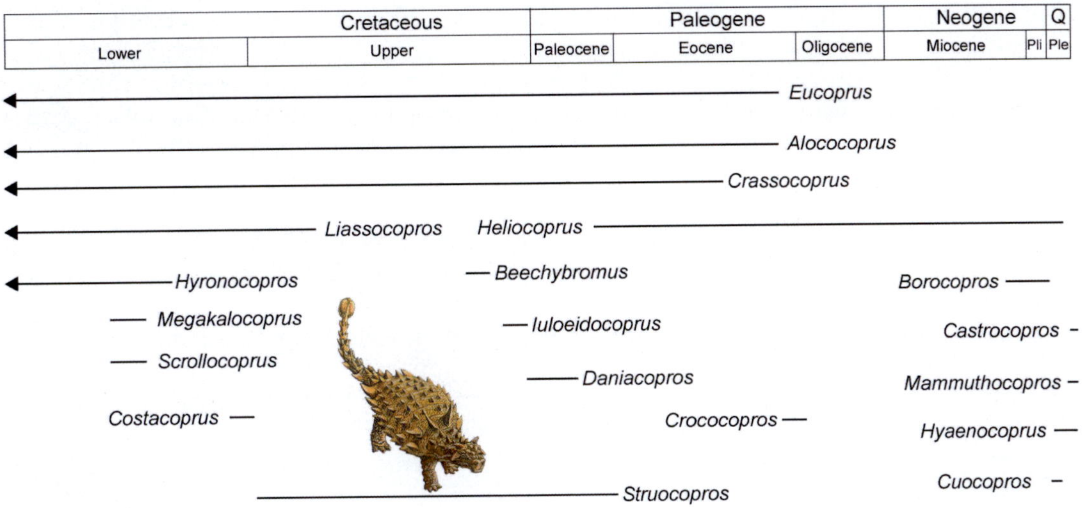

FIGURE 9.9 Stratigraphic distribution of Cretaceous-Quaternary bromalite ichnogenera. *Modified from Hunt and Lucas (2021e).*

Germany (Norman, 1987). Other notable coprofaunas from Early Cretaceous nonmarine strata include the Maceió and São Sebastião formations of Brazil (Souto, 2008; Souto and Schwanke, 2010), the Wealden Group of England (Buckland, 1835; Martill and Naish, 2001; Goldring et al., 2005), the Escucha Formation (Alcalá et al., 2012; Vajda et al., 2016) and La Huérguina Formation (Las Hoyas Konservat-Lagerstätte: Barrios-de Pedro et al., 2018, 2020a,c) of Spain, the Angeac-Charente bone bed of France (Rozada, 2019; Rozada et al., 2021), the Aïn El Guettar Formation of Tunisia (Dridi, 2022) and the Jehol Lagerstätte of China (Burnham, 2008; Pan et al., 2011, 2013). Rummy et al. (2021) described diverse coprolites from Mongolia. Other nonmarine Early Cretaceous coprolites are present in the United States, Mexico, Denmark, Tunisia, and Mongolia (Hunt and Lucas, 2016b).

Marine reptile coprolites are uncommon. Early Cretaceous ichthyosaur coprolites occur with a large concentration of skeletons in the Zapata Formation at Torres del Paine National Park in Chile (Stinnesbeck et al., 2014). The lagoonal Civita di Pietraroja Lagerstätte in Italy has yielded coprolites of a herbivorous fish, possibly *Notagogus* (Russo et al., 2015).

The record of Late Cretaceous coprolites is much more extensive and more widespread than for the earlier part of the period (Hunt and Lucas, 2016b: Fig. 9.9). In addition, there are much larger samples from marine units, reflecting the great extent of continental flooding. And, Late Cretaceous-Paleogene phosphorites yield coprolites over a wide area from northwest Africa to the Middle East (Hunt and Lucas, 2016b).

Woodward (1729) was the first to describe Late Cretaceous specimens that were later recognized as coprolites, from the marine Gray Chalk Group (Cenomanian) in England (Duffin, 2009; Hunt et al., 2015b). Subsequently, Mantell (1822) described and illustrated similar specimens and others from the Turonian-Maastrichtian White Chalk Group (Turonian-Maastrichtian), which Buckland (1835) ultimately identified as coprolites. Later, several coprolites, principally heteropolar in form, were described from the Chalk of southern England

(Longbottom and Patterson, 2002; Hunt et al., 2015b). Elsewhere, Late Cretaceous marine coprolites are widely distributed in three principal areas: (1) Europe; (2) Middle East/North Africa; and (3) North America. Many of the coprolites from Europe are spiral forms from chalk facies in the Netherlands, Belgium, Sweden, Germany, the Czech Republic, Austria, and Poland (e.g., Hunt et al., 2015b; Milàn et al., 2015; Hunt and Lucas, 2016b). Nonspiral forms are locally common (Hunt and Lucas, 2018b). Large samples demonstrate a diversity of morphotypes (e.g., Eriksson et al., 2011; Anagnostakis, 2013), and some coprolites from Austria and Poland are composed of *Inoceramus* shell fragments (Götzinger and Becker, 1932; Bajdek, 2013), and others from Poland contain brachiopod shell fragments and foraminiferan shells (Mazurek and Antczak, 2023).

Phosphorites are nonrandomly distributed in space and time (e.g., Cook and McElhinny, 1979; Dornbos, 2011). The North Africa/Middle East phosphogenic province ranges in age from Late Cretaceous to Eocene and is part of the South Tethyan Phosphogenic Province (STPP) (e.g., Sheldon, 1964; Klemme, 1985). It yields reports of marine coprolites from Togo to Iraq, including Late Cretaceous specimens from Jordan, Israel, Syria, Egypt, Mauritania, Tunisia, Senegal, Morocco, Mali, and Niger (Tapanila et al., 2008; Hunt and Lucas, 2016b; O'Leary et al., 2019). Duffin and Ward (2023a,b) described two new ichnotaxa, *Alococopros milnei* and *Struocopros catapitostromata*, from the latest Cretaceous phosphorite of Morocco. Many coprolites from the phosphorites are correctly identified, but some of the putative coprolites are probably concretions (cf. Horton, 2012; Hunt and Lucas, 2016b). Capasso (2019) described a coprolite of a pycnodont from the Cenomanian of Lebanon that is not associated with phosphorites and that appears to have been expelled just before fossilization.

The Western Interior Seaway in North America yields abundant marine coprolites. The first published record of a North American coprolite was a heteropolar specimen from the Upper Cretaceous of New Jersey (DeKay, 1830a,b; Buckland, 1835). Subsequently, coprolites have been described from Upper Cretaceous deposits elsewhere in the eastern United States, in New Jersey (Hunt and Lucas, 2016b), North Carolina (Hunt et al., 2012b), South Carolina (Hunt et al., 2012b; Schwimmer et al., 2015b), Georgia (Harrell and Schwimmer, 2010; Hunt et al., 2012b) and Alabama (Hunt et al., 2012b). The majority of the coprolites from the western United States derive from Kansas (e.g., Stewart, 1978; Hattin, 1996; Everhart, 2005, 2017), but they also occur in Texas (Friedman, 2012), Colorado (Shimada et al., 2006; Hunt et al., 2012b) and Wyoming (Hunt and Lucas, 2016b). Mosasaur coprolites occur in southern Saskatchewan (Mahaney et al., 2013; Hunt and Lucas, 2018b) and in Cenomanian bone beds in Saskatchewan and Manitoba (Schröder-Adams et al., 2001; Cumbaa et al., 2006; Phillips, 2008; Underwood and Cumbaa, 2010).

The majority of vertebrate coprolites from the nonmarine Late Cretaceous derive from North America, and there are also specimens from Europe, India, and Africa (Hunt and Lucas, 2022b). Many of the nonmarine units in the Western Interior of North America (Canada, United States, Mexico) that produce vertebrate body fossils also yield coprolites, but there are relatively few published descriptions. The most extensive described samples in North America are from the San Juan Basin in New Mexico (Suazo et al., 2012; Sullivan and Jasinski, 2012; Hunt et al., 2012b), and the largest individual specimens are putative tyrannosaur coprolites from Saskatchewan and Alberta (Chin et al., 1998, 2003). Apart from putative coprolites of herbivorous dinosaurs from Montana (Chin, 1996, 1997, 2007; Chin and Gill, 1996; Hollocher et al., 2001; Chin et al., 2009) and possibly Coahuila, Mexico

(Rodriguez de la Rosa et al., 1998) and Utah (Ridgwell et al., 2015), all other occurrences represent carnivores and include specimens from Texas, Utah, Wyoming, Montana and Alaska in the United States (Rowe et al., 1992; Baghai-Riding and DiBenedetto, 2001; Hollocher et al., 2010; Hunt et al., 2012b; Hunt and Lucas, 2016b), and Alberta in Canada (Waldman and Hopkins, 1970; Coy, 1995; Hunt and Lucas, 2016b). Many of these coprolites may not represent dinosaurs (Hunt and Lucas, 2017b). Other nonmarine Late Cretaceous coprolites occur in India (e.g., Matley, 1939a,b; Ghosh et al., 2003; Hunt et al., 2007; Khosla et al., 2015, 2016), France (Robertson, 1834), Hungary (Segesdi et al., 2017), Austria (Ősi et al., 2021), Brazil (Souto, 2010; Souto and Fernandes, 2015; Ferreira et al., 2019; de Oliveira et al., 2021), Madagascar (Rogers et al., 2013) and Kazakhstan (Nesov, 1995, 1997; Averianov et al., 2015).

5.2 Paleogene

The majority of Paleogene coprolite localities are in nonmarine strata, although the largest individual sample sizes are from marine deposits (Hunt and Lucas, 2016b). North America has the best-described record of Cenozoic vertebrate coprolites (Hunt and Lucas, 2007, 2016b), with nonmarine specimens from every epoch and notably large samples from the Eocene Green River Formation (e.g., Edwards, 1976; Grande, 1980, 1984, 2013; Wilson, 1987; Wells et al., 1993; Hunt et al., 2012b; Hunt and Lucas, 2024), Eocene Renova Formation (Lofgren et al., 2017; Jacquet et al., 2023) and Eocene/Oligocene White River Group, including a latrinite (e.g., Sinclair, 1922; Wanless, 1923; Abel, 1926, 1935; Stovall and Strain, 1936; Vogeltanz, 1965, 1967; Clark et al., 1967; Lemley, 1971; Edwards, 1973a,b; Edwards and Yatkola, 1974; Retallack, 1983; LaGarry, 1997; DiBenedetto, 2002; Hembree and Hasiotis, 2004, 2007; Meehan, 2007; Hunt and Lucas, 2007a; Hunt et al., 2012b,g: Fig. 9.9).

Nonmarine Paleogene coprolites are present at several localities in South America, including the Paleocene of Brazil (Souto, 2007), Paleocene/Eocene of Argentina (Krause et al., 2007; Krause and Piña, 2012), and Oligocene of Brazil (Castro et al., 1988). Lucas et al. (2012) described a large coprofauna from the late Eocene of northeastern Kazakhstan, and other Asian localities are in the Paleocene of China (Young, 1964; Meng and Wyss, 1997) and the middle Eocene of Pakistan (Gingerich, 1977). Paleocene coprolites occur in the Menat Lagerstätte in France (Wedmann et al., 2018).

Coprolites, principally from crocodiles and fish, are known from both of the famous German Eocene Lagerstätten of Geiseltal (Voigt, 1934) and Messel (Fikentscher, 1933; Nürnberger, 1934; Schmitz, 1991; Richter and Baszio, 2001a,b; Richter and Wedmann, 2005; Hunt and Lucas, 2024). Baranov et al. (2021) described an unusual occurrence of fly larvae in a specimen of a mammalian coprolite within Eocene Baltic amber. Other nonmarine coprolites from Europe occur in the Eocene London Clay (Buckland, 1835; Henslow, 1845; Clouter et al., 2000; Rayner et al., 2009) and the Oligocene of France (Buckland, 1835; Duffin, 2009). Halaçlar et al. (2023) described crocodile coprolites from the late Eocene of Vietnam and named them *Crococopros naduongensis*. One specimen has yielded a partial footprint of a crocodile (Halaclar et al., 2022).

Paleogene phosphorites yield marine coprolites over a wide area of the South Tethyan Phosphogenic Province (STPP) from northwest Africa to the Middle East, including Iraq, Jordan, Egypt, Algeria, Senegal, Mali, Morocco, Algeria, Togo and Nigeria (Hunt and Lucas,

2016b). The marine Eocene is characterized by very large samples (>1K specimens) of marine coprolites, notably the early Eocene Potapaco Member of the Nanjemoy Formation, Virginia (Dentzien-Dias et al., 2021) and the upper Eocene Yazoo Clay, Louisiana (Stringer and King, 2012), both in the United States, and from the middle Eocene Fürstenau Formation of Germany (Diedrich and Felker, 2012). The latter two coprofaunas consist principally of heteropolar spiral coprolites. Other marine coprolites occur in the Paleocene of Denmark (Milàn, 2010; Milàn et al., 2012b; Milàn and Hunt, 2016), the Oligocene of Germany (Weiler, 1922; Zotz, 1928; Martini, 1965), Oligocene of Denmark (Milàn et al., 2018), Oligocene of Poland (Bajdek and Bienkowska-Wasiluk, 2020) and the Paleocene and Eocene of France (Meillet, 1842) and Belgium (Buckland, 1835).

5.3 Neogene

The Neogene coprolite record is dominated by nonmarine assemblages with relatively few marine examples. Miocene coprolites have an almost worldwide distribution and occur in South, North and Central America, Europe, Asia, Africa and Australasia. Virtually all examples are from nonmarine strata, and they include an increased number of coprolites attributed to herbivores relative to older time intervals.

Some of the most significant Miocene coprofaunas from Europe are from Spain and yield the earliest hyena coprolites (Pesquero et al., 2011, 2013a,b), which were to become prevalent during the latest Neogene and Quaternary (Hunt and Lucas, 2020). Abella et al. (2021) described two morphotypes of coprolites of bone-consuming carnivores from Spain and Gross et al. (2023) two from Austria. Antunes et al. (2006a,b) reported the unusual occurrence of mammalian footprints in coprolites from Portugal. Miocene tetrapod coprolites occur in the Czech Republic (Fejfar et al., 2003; Mikuláš and Dvorák, 2010), and aquatic coprolites occur in Hungary, Spain, Malta, and Russia (Pedley, 1978; Peñalver and Gaudant, 2010; Román et al., 2024; Tarasenko and Sivkova, 2023). Brachaniec et al. (2022) analyzed ferruginous masses from the Miocene of Poland that may represent coprolites of turtles and snakes.

One of the largest coprolite ichnofaunas from the Miocene of South America is from the Uranco Formation of Venezuela, and it includes specimens produced by herbivorous and carnivorous mammals and reptiles (Royo and Gomez, 1960; Dentzien-Dias et al., 2018). There are several localities yielding coprolites in Argentina (Tauber et al., 2007; Verzi et al., 2008; Tomassini and Montalvo, 2010; Aceñolaza, 2012; Montalvo et al., 2016, 2019; Tomassini et al., 2019), and specimens have been reported from Colombia (Carlini et al., 1997) and in northwestern Amazonia (Hoorn, 1994). In Central America, coprolites occur in Miocene strata in Panama (Whitmore and Stewart, 1965; Hastings et al., 2013).

Wang et al. (2018) described coprolites of the bone-cracking dog *Borophagus* from the Miocene of California in the western United States that Hunt and Lucas (2021d) subsequently named *Borocopros wangi*. Other borophagine coprolites occur in New Mexico and Texas (Hunt and Lucas, 2021d). Martin (1981) documented other carnivore coprolites to the north in Oregon. There are other undescribed Miocene coprolites from North America, including rhinoceros coprolites from Mexico (Hunt and Lucas, 2007a).

There are relatively few reports of Neogene coprolites from Africa and Asia, but they include specimens from terrestrial and lacustrine environments in Uganda and Kenya, respectively (Greenwood, 1951; Bishop, 1964), and fish, turtle, and crocodilian coprolites from India (Sharma and Patnaik, 2010; Kapur et al., 2019; Sagar et al., 2024). Australasian Miocene coprolites occur in a fluvial deposit in central Australia (Megirian et al., 1996) and a lacustrine maar Lagerstätte in New Zealand (Lindqvist and Lee, 2009; Lee et al., 2016).

Miocene marine coprolites are uncommon, and most derive from the Calvert Cliffs of Maryland in the United States (Godfrey and Smith, 2010; Godfrey and Collareta, 2022; Godfrey et al., 2022). Specimens of note include one yielding feather impressions and another that includes dentalites produced by a shark and others with burrows (Wetmore, 1943; Godfrey and Smith, 2010; Godfrey and Collareta, 2022; Godfrey et al., 2022). Collareta et al. (2019b, 2022) described another dentalite-bearing coprolite and one with the echinoid trace *Gnathichnus pentax* from northern and southern Italy, respectively.

The record of Pliocene coprolites is principally from nonmarine strata in Africa, Europe, and South America. Most Pliocene African coprolites were discovered during studies of early hominins in the eastern and central parts of the continent, including Ethiopia (Jacobs, 1985), Tanzania (Harrison, 2011; Njau and Blumenschine, 2012), and Zaire (Harris et al., 1987). Other Pliocene coprolites occur in Namibia (Morales et al., 2011) and Libya (Muftah, 2020; Muftah et al., 2020). Pliocene or Villafranchian (late Pliocene-early Pleistocene) coprolites, principally derived from hyenas, are known from multiple localities in Europe, including Spain (Arribas et al., 2009; Madurell-Malapeira et al., 2010, 2011), France (Delson et al., 2006; Argant and Bonifay, 2011) and Italy (Girotti et al., 2004). Farlow et al. (2010) described coprolites from the late Neogene of Indiana in the United States.

South American Pliocene coprolite records are principally from Argentina (Aceñolaza, 2012; Tomassini and Montalvo, 2013; Cenizo et al., 2016). The North American Pliocene has produced few coprolites. Hunt and Lucas (2019, 2021d) described a specimen of the borophagine coprolite *Borocopros* from southwestern New Mexico in the United States. There are possible Pliocene coprolites from China (Kao, 1962).

There are few records of Pliocene marine coprolites. Hunt et al. (2015a) named *Helicoprus clarki* for a spiral chondrichthyan coprolite from the Red Crag Formation (upper Pliocene-lower Pleistocene) of England.

5.4 Quaternary

There are numerous Quaternary coprolites in paleontological and archeological sites worldwide, which include hominin specimens. The largest Quaternary coprolite fossil record is in North America (Figs. 9.10 and 9.11), where the caves of the arid Southwest have yielded large accumulations of coprolites, including latrinites (e.g., Rampart Cave, Bechan Cave: Fig. 9.12) of megaherbivore coprolites such as those of ground sloth and mammoth (*Castrocopros, Mammuthocopros*) as well as of smaller herbivores (Martin et al., 1961; Hansen, 1980; Mead et al., 1984, 1986a,b, 1993, 2020; Glowiak, 2007; Mead and Swift, 2012; Hunt and Lucas, 2018c, 2020; Hunt and Lucas, 2021e, Figs. 29 and 30). The most extensive record is from small rodents such as *Neotoma* (Tweet et al., 2012). *Neotoma* coprolites also occur in the Rancho La Brea (tar pits) Lagerstätte (Rice et al., 2019; Mychajliw et al., 2020).

There is a smaller Quaternary coprolite record in South America (e.g., Spillmann, 1929; Ferreira et al., 1991; Verde and Ubilla, 2002; Kerber and Oliveira, 2008; Hunt and Lucas,

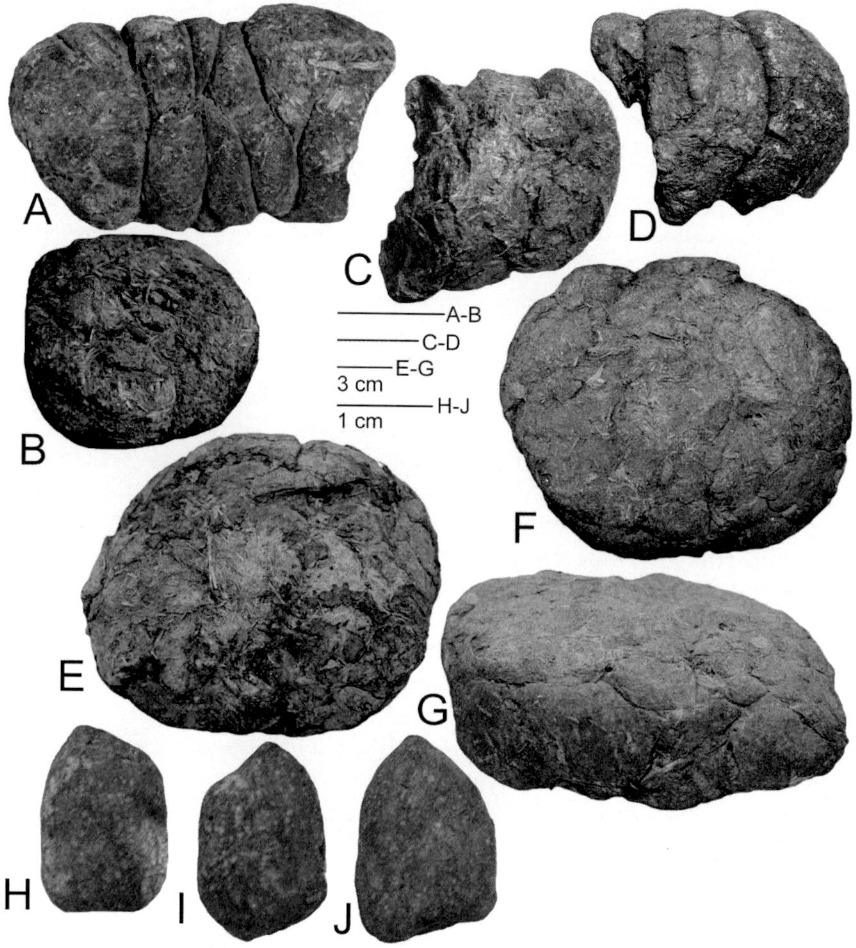

FIGURE 9.10 Examples of Cenozoic coprolites. Late Pleistocene herbivore coprolites. (A and B) USNM PAL 720155, *Castrocopros martini*, holotype coprolite from Rampart Cave, Arizona, USA, in lateral (A) and terminal (B) views. (C and D) YPM 10574, *Castrocopros hauthali*, holotype coprolite from Cueva de Milodón, Ultima Esperanza Province, Chile in lateral views. (E–G) GLCA 381, *Mammuthocopros allenorum*, holotype coprolite from Bechan Cave, Utah, USA, in terminal (E–F) and lateral (G) views. (H–J) GLCA 2379, three coprolites of *Suaviocopros harrisi* from Hooper's Hollow, Utah, USA, in lateral view. *From Hunt and Lucas (2020).*

2018c; Petrigh et al., 2019; Tietze et al., 2019; Agustín et al., 2021; Cardia et al., 2021; Oyarzún-Ruiz et al., 2021). Jacob et al. (2024) described a particularly interesting carnivore coprolite that contains ground sloth osteoderms from the late Pleistocene of Brazil. The Old World is characterized by caves with hyena coprolites and latrinites (Buckland, 1822, 1824, 1827; Chow, 1955; Kao, 1962; Musil, 1962; Mitzopoulos and Zapfe, 1963; Mohr, 1964; Scott, 1987; Ferrereira et al., 1993; Fernandéz-Rodrìguez et al., 1995; Tournepiche and Couture, 1999; Boaz et al., 2000; Berger et al., 2009; Diedrich, 2012a; Perri et al., 2017; Hunt and Lucas, 2020; Espigares et al., 2023; Zipfel et al., 2023). Other types of coprolites also occur such as

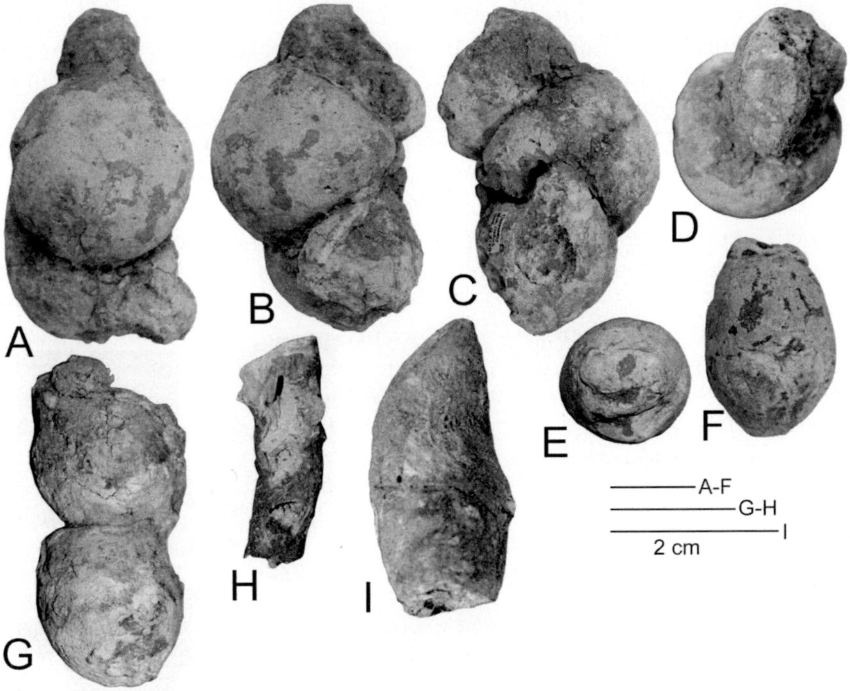

FIGURE 9.11 Examples of Cenozoic coprolites. Late Pleistocene carnivore coprolites. (A–F), *Hyaenacoprus bucklandi* (Hunt et al., 2012a) from the Kirkdale Cave, Yorkshire, England (Hunt et al., 2012a, Fig. 9.5). (A–D) OUM Q6168, holotype partial coprolite in A–C, lateral and D, polar views. (E and F), OUM Q6167 in lateral (E) and terminal (F) views. (G) USNM, 509327, large felid coprolite from Rampart Cave, Arizona, USA, in lateral view. (H) USNM uncatalogued, coprolite of *Bassariscus* sp. from Rampart Cave, Arizona, USA, in lateral view. (I) GLCA 2382, large felid coprolite from Bechan Cave, Utah, USA, in lateral view. *From Hunt and Lucas (2020).*

FIGURE 9.12 Late Pleistocene latrinite of *Castrocopros martini* from Rampart Cave, Grand Canyon National Park, Arizona, USA. *Image courtesy of Vincent Santucci.*

the canid coprolite *Cuocopros yuanmouensis* from the Early Pleistocene of China (Farjand et al., 2024). New Zealand and other Pacific islands yield locally abundant and important records of bird coprolites that are Quaternary and Holocene in age (James and Burney, 1997; Horrocks et al., 2004; Wood et al., 2008, 2012a,b).

Davis (Davis et al., 1977; Davis, 1987; Davis and Shafer, 2006) first recognized that spores of the dung fungus *Sporormiella* became abundant following the historic introduction of grazing herbivores in the western United States. During the Holocene this fungus is rare. However, *Sporormiella* spores are abundant before the extinction of Pleistocene megaherbivores, and they have been recovered from coprolites of *Mammuthocopros* from Bechan Cave in Arizona (Davis, 1987; Gill et al., 2013). Thus, *Sporormiella* spores are a proxy for an abundance of large terrestrial herbivores and can provide evidence for their extinction (Davis, 1987). *Sporormiella* spores have also been widely utilized to study patterns of extinction or/and the immigration or introduction of moas into New Zealand (Wood et al., 2011), giant tortoises into the Galapagos Islands (Froyd et al., 2013), giant lemurs, elephant birds and the pygmy hippopotamus into Madagascar (Burney et al., 2003) and cattle into Brazil (Raczka et al., 2016).

5.5 Human coprolites

Hominid coprolites have been described exclusively in the archeological literature (Reinhard and Bryant, 1992; Bryant and Reinhard, 2012; Shillito et al., 2020). The known fossil record of human coprolites and its study is strongly biased by a number of factors:

1. A scarcity of human skeletal remains in some areas, notably in early prehistoric North America, has led to a greater focus on collecting and studying coprolites as evidence for human presence (Jenkins et al., 2012; Shillito et al., 2020). In later periods, and most other regions, an abundance of other types of evidence has contributed to a relative lack of interest in human coprolites (Shillito et al., 2020).
2. Much coprolite research has been conducted in the United States where work has focused in the arid Southwest or Great Basin regions, where there are preferential conditions for desiccation and numerous caves and rock shelters (Reinhard, 1988; Reinhard and Bryant, 2008; Shillito et al., 2020).
3. Reinhard and Bryant (Reinhard, 2006; Reinhard and Bryant, 2008; Bryant and Reinhard, 2012) noted an acme of coprolite studies ("Golden Age of Coprolite Analysis") in the United States in the 1960s through the early 1990s and a subsequent decline in the 1990s and 2000s.
4. Much early work in Europe focused on latrines and cesspits rather than coprolites, and study in Asia only started in the past couple of decades, so the majority of known human coprolite sites are from North America (Shillito et al., 2020).

The record of hominin coprolites has most recently been reviewed by Shillito et al. (2020). Most known localities yield coprolites of *Homo sapiens*. However, De Lumley (1969) and Trevor-Deutsch and Bryant (1978) studied coprolites of *Homo erectus,* and Callen (1969) those of Neanderthals, all from France. Rampelli et al. (2021) reconstructed part of the Neanderthal gut microbiome based on sediment samples.

The majority of human coprolites have been documented from North America and notably the United States. The Great Basin of the western United States yields coprolites from multiple localities including Danger Cave, Hogup Cave, and Cowboy Cave in Utah and Hidden Cave and Lovelock Cave in Nevada (Heizer, 1967; Fry, 1970, 1977, 1978; Napton and Heizer, 1970; Kelso, 1971). Localities in Oregon include Paisley Caves (Gilbert et al., 2008; Goldberg et al., 2009; Shillito et al., 2018, 2020; Blong et al., 2020), Dirty Shame Rockshelter (Hall, 1969, 1977) and Connley Caves (McDonough, 2019).

Further south, on the Colorado Plateau, human coprolites were recovered from many sites in National Park Service areas including Mug House, Step House and Hoy House, Mesa Verde National Park, Antelope House, Canyon de Chelly National Monument, Inscription House, Turkey Cave, and Keet Seel Ruin, Navajo National Monument, Pueblo Alto and Pueblo Bonito, Chaco Culture National Historical Park and Oakleaf Alcove, Glen Canyon National Recreation Area, and Aztec Ruins National Monument (Williams-Dean, 1986; Williams-Dean and Bryant, 1975; Stiger, 1977; Clary, 1984; Fry and Hall, 1986; Mead and Agenbroad,1992; Brand, 1994; Hunt et al., 2012g; Camacho and Reinhard, 2020; Shillito et al., 2020). Thousands of coprolites also occur at Salmon Ruin in New Mexico and a few at Navajo Mountain (Irwin-Williams and Shelley, 1980; Wormington and Agogino, 1994; Reed, 2006; Reinhard, 2007; Reinhard and LeRoy-Toren, 2006; Shillito et al., 2020). Holocene human coprolites occur at Antelope Cave in northwest Arizona (Reinhard et al., 2012; Fig. 9.13B) and Turkey Pen Ruin rock-shelter in Cedar Mesa, Utah (Battillo, 2019).

Another area with abundant human coprolites is the Pecos region of southwest Texas (Bryant, 1969). The most prominent site is Hinds Cave, which has yielded more than 2000 coprolites and has been studied extensively (e.g., Williams-Dean, 1978; Sobolik, 1991, 1994; Danielson, 1993; Danielson and Reinhard, 1998; Riley, 2008, 2012: Fig. 9.13C and D). Other localities include Baker Cave and Conejo Shelter (Bryant, 1974b; Sobolik, 1988; Reinhard, 1990; Beck et al., 2019: Fig. 9.13A). Other human coprolites occur in the southeastern United States (e.g., Young, 1910; Watson and Yarnell, 1966; Bryant, 1974c; Schoenwetter, 1974; Watson and Yarnell, 1966) and Puerto Rico (Cano et al., 2014). Mexican sites include the Ocampo Caves in Tamaulipas, Cueva de los Muertos Chiquitos in Durango, and sites in Tehuacan and Coahuila (MacNeish, 1958; Callen, 1963, 1965, 1967a, 1967b, 1968; Bryant, 1975; Jiménez et al., 2012; Borry et al., 2020; Pucu et al., 2020).

Human coprolites occur in South America, notably Peru. Peruvian sites include Pikimachay, Ayacucho, Ayacucho, Atacama, and the Desert Norte Chico region (Callen and Cameron, 1955, 1960; Callen, 1965; Vinton et al., 2009; Haas et al., 2013).

A notable individual coprolite from Europe is the "Lloyds Bank coprolite" from the Viking occupation of York, England, which is 20 cm in length and 5 cm in width (Jones, 1983). Many European sites are latrines and cesspits in urban settings that have different formation processes and taphonomy from discrete coprolite deposits and which do not yield individual specimens (Shillito et al., 2020). Sites that have yielded individual coprolites include the Louvre in Paris and other sites in France (Bouchet, 1995; Bouchet et al., 2003). Many coprolites in Europe and the Middle East have been recognized from thin sections or biochemical analysis, for example in Switzerland (Pichler et al., 2014), Crete (Bull et al., 1999), Turkey (Shillito et al., 2011), Iraq (Matthews, 1995) and Iran (Matthews et al., 2014). There are relatively few coprolite localities in Asia, but they include the Chinese sites of Tianluoshan, Zhejiang Province (Zhang et al., 2020) and Yuhuicun and Houtieying, Anhui Province (Zhang et al., 2019;

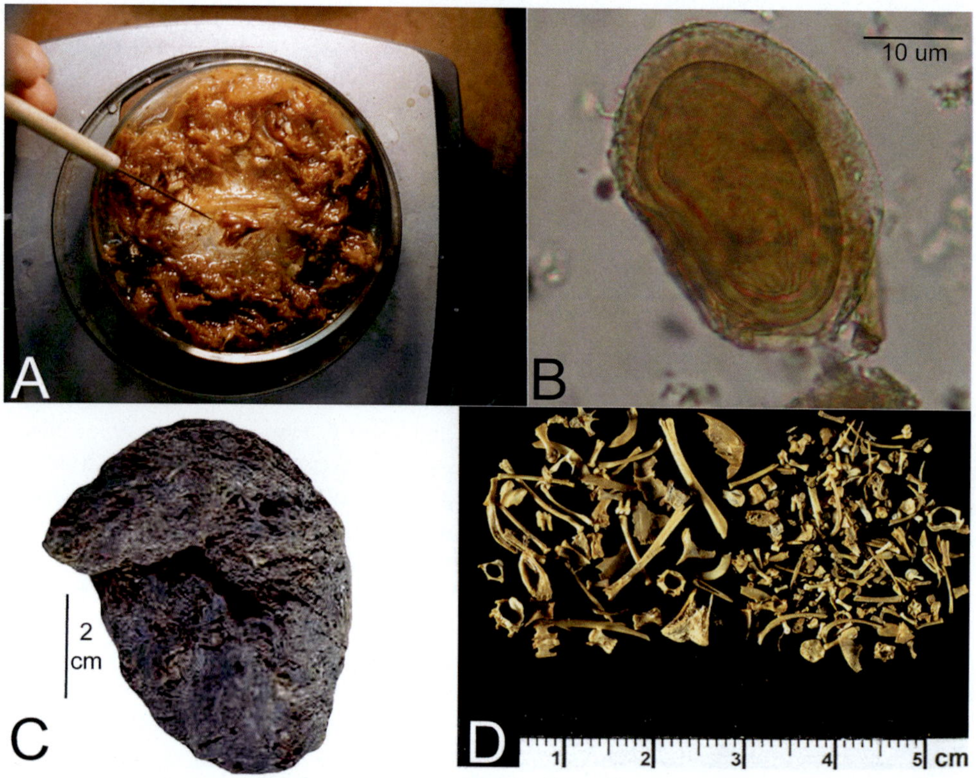

FIGURE 9.13 Coprolites of *Homo sapiens* and their contents. (A) Dried coprolite (Upper Archaic Period) from Conejo Shelter, Lower Pecos region of Texas, USA, hydrated by soaking in a weak solution of trisodium phosphate with probe pointing to a wild onion (*Allium*) bulb, one of many found in these coprolites that contained high amounts of plant fiber. (B) A parasitic acanthocephalan (thorny-headed worm) from a human coprolite, Antelope Cave, Arizona, USA, dating to about 1100 years old, demonstrating parasitism due to the dietary habit of eating insects. Scale bar is 10 μm. (C) One of many coprolites found at Hinds Cave (Middle Archaic Period), Texas, USA. Most coprolites were amorphous and flattened, caused by loose stools from eating excessive amounts of fiber, but this one represents a more typical morphology of a human coprolite. (D) Small mammal, reptile, bird, and fish bones recovered from coprolites from Hinds Cave. Hunt and Lucas (2021b, Fig. 9.2). *Photographs courtesy of Vaughn M. Bryant, Jr. and Karl J. Reinhard.*

Borry et al., 2020) and in the Torihama shell midden in Fukui Prefecture, Japan (Matsui et al., 2003).

6. Ichnotaxonomy

6.1 Introduction

Bertling et al. (2006, p. 265) noted that "The need to name trace fossils … has unambiguously been accepted for decades." Initially, vertebrate ichnologists were slow to name coprolites. Bian et al. (1991) and Hunt et al. (1998) named the first ichnotaxa of vertebrate coprolites. The description and naming of vertebrate coprolites has allowed detailed analyses of distribution patterns, including a robust biostratigraphy and biochronology for the Triassic, utilizing 14 ichnogenera (Hunt et al., 2018, Fig. 12.11: Hunt and Lucas, 2021e, Fig. 27) and construction of paleoenvironmental transects in the Pennsylvanian using seven ichnogenera (Hunt and Lucas, 2017a, Fig. 9.1). Initially relatively few other workers used a formal ichnotaxonomy for vertebrate coprolites, but that has changed significantly in the last few years (e.g., Milàn, 2018; Rakshit et al., 2019; Rummy et al., 2021; Rakshit and Ray, 2021; Zouheir et al., 2023; Bendella et al., 2023; Bigurrarena Ojeda et al., 2023; Duffin and Ward, 2023a,b; Halaçlar et al., 2023; Hancox and Rubidge, 2023; Farjand et al., 2024, Table 9.1).

Several authors have explicitly advocated the use of morphotypes rather than ichnotaxa for coprolites because they consider that: (1) different animals may produce similar fecal morphologies; (2) differing morphologies and contents of feces are driven by seasonal or ontogenetic factors; and (3) there is variation in preservation (Chin in Hunt et al., 1994; Bajdek et al., 2014; Francischini et al., 2018). The first two arguments are largely based on the misconception that modern feces are not distinguishable because of variability, even though wildlife biologists routinely track and study the distribution of extant taxa based on the distinct morphologies of their feces (e.g., Murie, 1974; Jenkins and Burrows, 1980; Stuart and Stuart, 2000; Chame, 2003). In an arguably extreme case, Taglioretti et al. (2014) used pellet size and shape and were able to discriminate droppings of guanaco from those of huemul with 100% accuracy and with identical results in identifying camelid coprolites. Whereas primates (such as ourselves) produce unusually variable morphologies of feces (e.g., Lewis and Heaton, 1977; Hunt et al., 2012d), the majority of vertebrates (carnivores and herbivores) produce a consistent morphology of feces, which, as just noted, is apparent from studies of Recent animal traces.

Pleistocene coprolites can usually be assigned to genera and species (e.g., Hunt and Lucas, 2020, 2022a). However, many older coprolite ichnotaxa do not represent fossil genera and species but rather higher taxonomic groups. However, this is not an argument against naming them. Authors who decline to erect ichnotaxa discriminate morphotypes (e.g., Eriksson et al., 2011; Hansen et al., 2016; Francischini et al., 2018), which are ichnotaxa without names. A nomenclature that recognizes repetitive morphologies is essential for synthetic ichnology be it biochronology, biogeography, or paleoecology. Footprint ichnotaxa also do not relate to low taxonomic levels of biotaxa, and yet they have a proven utility for 200 years, and their use has, for example, resulted in the construction of robust biochronologies (e.g., Lucas, 2007; Klein and Lucas, 2024; Lucas and Stimson, 2024; Marchetti et al., 2024). Thus, we strongly advocate the application of a rigorous ichnotaxonomy to vertebrate coprolites for both philosophical and practical reasons.

7. Ichnotaxobases

Vallon (2012) first explicitly discussed ichnotaxobases for coprolites. The most utilized ichnotaxobases for coprolites are:

TABLE 9.1 Named coprolite ichnotaxa.

Ichnotaxon	Author	Producer	Age	Notes
Alococopros indicus	Hunt et al. (2007)	Crocodylomorph	Late Cretaceous	Valid
Alococopros triassicus	Hunt et al. (2007)	Archosauromorph	Early Permian-Late Eocene	Valid
Alococopros milnei	Duffin and Ward (2023a)	Crocodilian or chelonian	Late Cretaceous	Valid
Amphicopros tikiensis	Rakshit and Ray (2021)	Saurichthyids or lungfish	Late Triassic	Valid
Antarctoscoprus longinensis	Bigurrarena Ojeda et al. (2023)	Nonteleostean actinopterygian	Late Jurassic	Valid
Beechybromus wellschi	Hunt and Lucas (2018b)	Mosasaur	Late Cretaceous	Valid
Bibliocopros beemanensis	Hunt et al. (2012h)	Chondrichthyan	Late Pennsylvanian	Valid
Borocopros wangi	Hunt and Lucas (2021c)	*Borophagus* spp.	Latest Miocene - latest Pliocene	Valid
Castrocopros martini	Hunt and Lucas (2018c)	*Nothrotheriops shastensis*	Late Pleistocene	Valid
Castrocopros hauthali	Hunt and Lucas (2018c)	*Mylodon darwinii*	Late Pleistocene	Valid
Costacoprus chinae	Hunt et al. (2012c)	Fish	Early Cretaceous	Valid
Crassocoprus mcallisteri	Hunt et al. (2012i)	Chondrichthyan	Late Pennsylvanian-Eocene	Valid
Crococopros naduongensis	Halaçlar et al. (2023)	Crocodilians	Late Eocene	Valid
Crustacoprus tinajaensis	Hunt et al. (2012j)	Acanthodian, platysomid	Late Pennsylvanian	Valid
Cuocopros yuanmouensis	Rummy and Halaçlar (2024) (in Farjand et al., 2024)	*Sinocuon yuanmouensis*	Early Pleistocene	Valid
Dakyronocopros arroyoensis	Hunt and Lucas (2005a)	Fish or tetrapod	Early Permian	Valid
Daniacopros hofstedtii	Milàn and Hunt (2016)	Fish less derived than teleosts or advanced actinopterygian	Early Paleocene	Valid
Dicynodontocopros maximus	Hunt et al. (1998)	Dicynodont	Late Triassic	Valid
Elacocoprus williamsi	Hunt et al. (2012j)	Chondrichthyan or sarcopterygian	Late Pennsylvanian	Valid

TABLE 9.1 Named coprolite ichnotaxa.—cont'd

Ichnotaxon	Author	Producer	Age	Notes
Elongatocoprus amadoensis	Hunt et al. (2012j)	Chondrichthyan or sarcopterygian	Late Pennsylvanian	Valid
Eucoprus cylindratus	Hunt and Lucas (2012b)	Fish, amphibians, tortoises, iguanas, crocodylians, theropods, birds, mammalian carnivores	Early Triassic –late Eocene	Valid
Falcatocopros oxfordiensis	Hunt et al. (2007)	Vertebrate	Late Triassic-Late Jurassic	Valid
Heliocoprus clarki	Hunt et al. (2015b)	Rajidae (skates)	Late Pliocene-Pleistocene	Valid
Heteropolacopros texaniensis	Hunt et al. (1998)	Chondrichthyan	Early Permian-Late Triassic	Valid
Huberobromus ovatus	Hunt and Lucas (2021a)	Chondrichthyan	Mississippian-Pennsylvanian	Valid
Hyaenacoprus bucklandi	Hunt et al. (2012a)	Hyena	Pliocene-Recent	Valid
Hyronocopros amphipola	Hunt et al. (2005b)	Sarcopterygian or chondrichthyan	Late Pennsylvanian-early Permian	Valid
Hyronocopros hunti	Rummy et al. (2021)	Asipenceriformes	Early Cretaceous	Valid
Hyronocopros tsagantsabensis	Rummy et al. (2021)	Asipenceriformes	Early Cretaceous	Valid
Ichthyosaurolites duffini	Hunt et al. (2012a)	Plesiosaur or ichthyosaur	Early Jurassic	Valid
Iuloeidocoprus mantelli	Hunt et al. (2012b)	Chondrichthyan or Osteichthyan	Late Cretaceous	Valid
Kalocoprus oteroensis	Hunt et al. (2012h)	Chondrichthyan	Late Pennsylvanian	Valid
Kinneybromus jurgenai	Hunt and Lucas (2021a)	Chondrichthyan	Late Pennsylvanian	Valid
Liassocopros hawkinsi	Hunt et al. (2007)	Chondrichthyan	Late Pennsylvanian-Late Cretaceous	Valid
Maculacoprus ateri	Hunt and Lucas (2021a)	Acanthodians, palaeonisciforms or platysomids	Mississippian-Pennsylvanian	Valid
Malericopros matleyi	Hunt et al. (2007)	Chondrichthyan	Permian-Late Triassic	Valid
Megaheteropolacopros sidmacadami	Hunt et al. (2005a)	Large chondrichthyan	Early Permian (Wolfcampian-middle Leonardian)	Valid

(Continued)

TABLE 9.1 Named coprolite ichnotaxa.—cont'd

Ichnotaxon	Author	Producer	Age	Notes
Mammuthocopros allenorum	Hunt and Lucas (2020)	*Mammuthus columbi*	Late Pleistocene	Valid
Megakalocoprus barremianensis	Rummy et al. (2021)	Acipenceriformes	Early Cretaceous	Valid
Plektecoprus whitbyensis	Hunt et al. (2012a)	Ichthyosaur or plesiosaur	Early Jurassic	Valid
Revueltobromus complexus	Hunt and Lucas (2014b)	Crocodylomorphs	Late Triassic	Valid, possibly evisceralite
Rewacopros jaini	Rakshit and Ray (2021)	Xenacanthids	Late Triassic	Valid
Rhynchocopros soutoi	Hunt et al. (2013b)	Rhynchosaur	Middle Triassic	Valid
Rummycopros yini	This paper; Bian et al. (1991)	Fish	Unknown	Valid
Santamariacopros elongatus	Hunt et al. (2013b)	Dicynodont	Middle Triassic	Valid
Saurocopros bucklandi	Hunt et al. (2007)	Chondrichthyan	Late Triassic-Early Jurassic	Valid
Scrollocoprus tatalensis	Rummy et al. (2021)	Sarcopterygian	Early Cretaceous	Valid
Speiracoprus socorroensis	Hunt et al. (2012j)	Chondrichthyan or sarcopterygian	Late Pennsylvanian	Valid
Strabelocoprus pollardi	Hunt et al. (2012a)	Chondrichthyan or sarcopterygian	Late Triassic-Early Jurassic	Valid
Strophocopros valensis	Hunt and Lucas (2005b)	Chondrichthyan	Early Permian	Valid
Struocopros friedmanae	Hunt and Lucas (2023a)	Chondrichthyan	Late Cretaceous-early Eocene	Valid
Struocopros catapitostromata	Duffin and Ward (2023b)	Medium to large lamniform sharks, rays myliobatiformes	Late Cretaceous	Valid
Suaviocopros harrisi	Hunt and Lucas (2022a)	*Euceratherium collinum*	Late Pleistocene	Valid
Tikicopros triassicus	Rakshit et al. (2019)	Hybodontid sharks	Late Triassic	Valid
Vinculostercus vermiformis	Duffin and Ward (2020)	Chondrichthyans or actinopterygians	Early Carboniferous	Valid
Virgacoprus brevis	Hunt and Lucas (2021a)	Haplolepid a small palaeonisciform or an acanthodian	Late Pennsylvanian	Valid

TABLE 9.1 Named coprolite ichnotaxa.—cont'd

Ichnotaxon	Author	Producer	Age	Notes
Yufenites lufengensis	Bian et al. (1991)	Fish	Early Jurassic	Nomen dubium
Yufenites helicoides	Bian et al. (1991)	Fish	Early Jurassic	Nomen dubium
Yufenites tongziensis	Bian et al. (1991)	Fish	Middle Jurassic	Nomen dubium
Yufenites fuxianensis	Bian et al. (1991)	Fish	Late Jurassic	Nomen dubium
Yufenites tongchuanensis	Bian et al. (1991)	Fish	Middle Jurassic	Nomen dubium

1. External morphology—Coprolites and Recent feces display a range of morphotypes that are useful for ichnotaxonomic definitions (Häntzschel et al., 1968; Chame, 2003; Hunt and Lucas, 2012b).
2. Internal structure—Many coprolites have a spiral or scroll internal structure (Hunt and Lucas, 2012b). Spiral coprolites are either amphipolar or heteropolar (microspiral or macrospiral) in form. Many ichnotaxa are based on these structures (e.g., *Heteropolacopros:* Hunt et al., 1998; *Scrollocoprus*: Rummy et al., 2021).
3. Surface texture—Some coprolites such as *Strophocopros* have distinct surface textures (Hunt and Lucas, 2005b).
4. Size—Size by itself is not usually an ideal ichnotaxobase except in cases where there are very distinct size classes (e.g., *Megaheteropolacopros*: Hunt et al., 2005a; Hunt and Lucas, 2023b).
5. Inclusions—Certain coprolites contain distinct or/and abundant inclusions, such as conchostracan valves in *Crustacoprus* (Hunt et al., 2012j). Quaternary herbivore coprolites often contain well-preserved plant material whose size and taxonomic identity can be ichnotaxobases as in *Suaviocopros* and *Mammuthocopros* (Hunt and Lucas, 2020, 2021c).
6. Mineralogy—Most coprolites are phosphatic in composition, but others have different mineralogies such as *Rhynchocopros soutoi* and *Santamariacopros elongatus*, which are calcareous (Hunt et al., 2013).

7.1 Systematic ichnotaxonomy

7.1.1 Introduction

There are currently 61 named ichnotaxa of coprolites, of which we consider 56 valid (Table 9.1). We intend later to provide a full review of this ichnotaxonomy, but herein we list basic information about the named ichnotaxa. They are grouped by major morphological types (heteropolar, amphipolar, etc.).

Bian et al. (1991) proposed an ichnotaxonomic scheme in which coprolite ichnogenera were based on the source organism as Hitchcock had originally done for vertebrate tracks. Thus, *Yufenites* (spiral morphology) was produced by fish, *Wafenites* by amphibians, *Lonfenites* by reptiles, *Niaofenites* by birds, and *Shoufenites* by mammals. However, naming ichnotaxa based on originator rather than morphology is contradictory to all recent ichnotaxonomic methodology. Bian et al. (1991) named five ichnospecies of *Yufenites* and You et al. (2020) a sixth (Fig. 9.14). *Wafenites, Lonfenites, Niaofenites,* and *Shoufenites* have no type ichnospecies, diagnoses (apart from the putative source organism), or type specimens and are therefore *nomina nuda*. The ichnospecies of *Yufenites* named by Bian et al. (1991) are three amphipolar forms (*Y. lufengensis, Y. helicoides, Y. yini*) and two heteropolar (*Y. tongziensis, Y. fuxianensis*). Bian et al. (1991) did not explicitly assign a type species but we, as first revisers, consider that reasonably the first named ichnospecies, *Yufenites lufengensis* is the

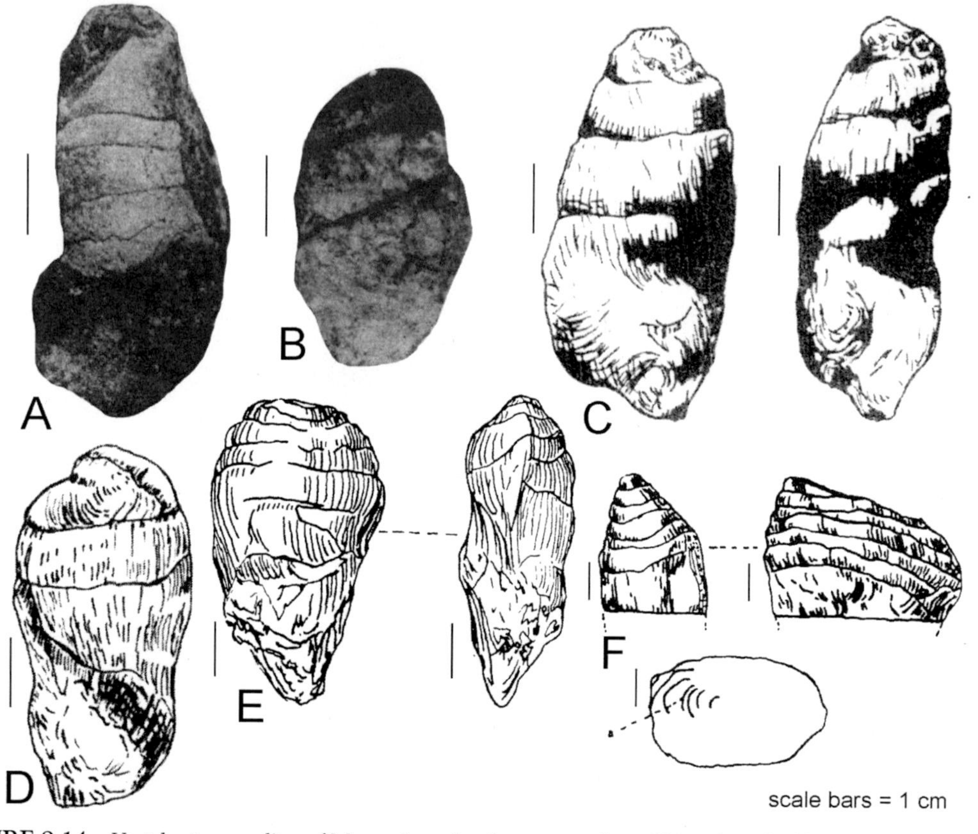

FIGURE 9.14 Vertebrate coprolites of Mesozoic and unknown age from China described by Bian et al. (1991). (A) Holotype coprolite of *Yufenites lufengensis,* lectotype ichnospecies of *nomen dubium* ichnogenus *Yufenites* in lateral view. (B) Holotype coprolite of *nomen dubium Yufenites helicoides* in lateral view. (C) Holotype coprolite of *Rummycopros yini* igen. nov. in lateral views. (D) Holotype coprolite of nomen dubium *Yufenites tongziensis* in lateral view. (E) Holotype coprolite of *nomen dubium Yufenites fuxianensis* in lateral views. (F) partial coprolite of *Yufenites* isp. in lateral and posterior views. *Modified from Bian et al. (1991), Figs. 9.1–9.6.*

type lectotype ichnospecies (Bian et al., 1991, Fig. 9.1). We consider the holotype of *Y. lufengensis* to be an indeterminate, probably amphipolar coprolite. Thus, we consider the ichnogenus *Yufenites,* of which *Y. lufengensis* is the lectotype ichnospecies, to be a nomen dubium *or* nomen nudum. We also consider the holotypes of the ichnospecies *Y. helicoid*es, *Y. tongziensis,* and *Y. fuxianensis* to be indeterminate, and thus, these ichnotaxa are nomina dubia (Bian et al., 1991, Figs. 9.1, 9.3, 9.5). We consider the holotype of *Y. yini* to be diagnosable and thus we assign it to a new ichnogenus below (*Rummycopros* igen. nov.). You et al. (2020) named *Yufenites tongchuanensis* but did not designate or illustrate a holotype or provide a clear differential diagnosis. You et al. (2020, p. 6) note a "Positive model: a complete fish manure fossil specimen (Fig. 9.3C)" but this figure illustrates a partial fish and not a coprolite. You et al. (2020, Fig. 3B, E, G–I) do illustrate coprolites (and inclusions: You et al., 2020, Fig. 4B–G), but we do not consider any distinctive. Thus, we consider *Yufenites tongchuanensis* to be a nomen dubium or *nomen nudum.*

7.1.2 Heterospiral coprolites

7.1.2.1 *Heteropolacopros* (Hunt et al., 1998)

Type ichnospecies: *Heteropolacopros texaniensis* (Hunt et al., 1998).

Included ichnospecies: Known only from the type ichnospecies.

Distribution: Late Pennsylvanian (Missourian) to Late Triassic (Rhaetian) of North America, Europe, Africa, and Asia.

Diagnosis: A microspiral heteropolar coprolite that is small (usually less than 4 cm long) with three to four coils forming less than 50% of the length. The posterior spire is wider than the first coil (always less than one third of the total length) and gradually decreases in diameter both up and down the long axis (the tip comes to a rounded point), and the posterior spire gradually tapers to a rounded tip (Hunt et al., 2018: Figs. 9.6I–K, II–LL and 9.8R–Y).

Tracemaker: Heterospiral coprolites are produced by valvular intestines. The phylogenetic distribution of valvular intestines is not totally understood. McAllister (1987) presented evidence that some or all agnathans, placoderms, dipnoans, actinistians, and chondrichthyans have valvular intestines. Only two fossil actinopterygians have evidence of this structure, and McAllister (1987) hypothesized that this group progressively reduced the valvular intestine; teleosts do not exhibit this kind of structure. Dipnoans probably produced amphipolar coprolites (Hunt and Lucas, 2012b), and Pennsylvanian and Triassic heteropolar coprolites probably represent chondrichthyans.

Discussion: This ichnospecies is very long-ranging, somewhat variable in morphology and in need of an ichnotaxonomic revision.

7.1.2.2 *Megaheteropolacopros* Hunt et al., 2005a

Type ichnospecies: *Megaheteropolacopros sidmcadamsi* Hunt et al., 2005a.

Included ichnospecies: Known only from the type ichnospecies.

Distribution: Early Permian of Texas and New Mexico (Coyotean-Redtankian).

Diagnosis: Phosphatic microspiral heteropolar coprolite that differs from *Heteropolacopros* and other heteropolar coprolites in being large (length typically >120 mm) with a short posterior spire that is about 20%–24% of the length and that consists of five or less coils and has an acute apical posterior tip.

Tracemaker: Chondrichthyan.

Discussion: Many specimens of this ichnotaxon exhibit evidence of flattening and are ovoid in cross section.

7.1.2.3 *Liassocopros* Hunt et al., 2007

Type ichnospecies: *Liassocopros hawkinsi* Hunt et al., 2007.

Included ichnospecies: Known only from the type ichnospecies.

Distribution: Late Pennsylvanian (Missourian) to Early Jurassic (Pliensbachian) of North and South America, Europe, and Asia.

Diagnosis: A microspiral heteropolar coprolite that tends to be large (>5 cm long) with a small number (<3) of wide coils (Fig. 9.6A–D).

Tracemaker: This heteropolar coprolite was produced by a chondrichthyan.

Discussion: This is a very long-ranging ichnospecies that may be in need of taxonomic revision.

7.1.2.4 *Saurocopros* Hunt et al., 2007

Type ichnospecies: *Saurocopros bucklandi* Hunt et al., 2007.

Included ichnospecies: Known only from the type ichnospecies.

Distribution: Late Triassic (Carnian) to Early Jurassic (Pliensbachian) of Europe and Asia.

Diagnosis: Microspiral heteropolar coprolite that differs from *Malericopros* in being tapered below the spiral demarcation and from *Heteropolacopros* in having a small number of wide spirals (typically 3) at the anterior end.

Tracemaker: Chondrichthyan.

Discussion: Hunt et al. (2007) named this coprolite in recognition of the fact that Buckland (1835: 227, pl. 28, Figs. 9.6, 9.7, 9.9) termed this morphology "Sauro-coprolites," and they noted that it does not pertain to a reptile.

7.1.2.5 *Malericopros* Hunt et al., 2007

Type ichnospecies: *Malericopros matleyi* Hunt et al., 2007.

Included ichnospecies: Known only from the type ichnospecies.

Distribution: Late Permian (Lopingian) to Late Triassic (Carnian) of Europe and Asia.

Diagnosis: Microspiral heteropolar coprolite with a maximum diameter that is posterior to the anterior coil.

Tracemaker: Chondrichthyan.

Discussion: This ichnospecies was originally described from the Upper Triassic of India.

7.1.2.6 *Strabelocoprus* Hunt et al., 2012a

Type ichnospecies: *Strabelocoprus pollardi* Hunt et al., 2012a.

Included ichnospecies: Known only from the type ichnospecies.

Distribution: Late Triassic (Rhaetian) to Early Jurassic (Pliensbachian) of Europe.

Description: Heteropolar, microspiral coprolite with a small number of coils (<3) in lateral view, exhibiting very wide spirals in posterior view and has a width that exceeds half of its length.

Tracemaker: Chondrichthyan.

Discussion: This ichnospecies seems to be a good index ichnotaxon for the Rhaetian-Pliensbachian as it has a restricted range and a distinctive morphology.

7.1.2.7 *Crassocoprus* Hunt et al., 2012j

Type ichnospecies: *Crassocoprus mcallisteri* Hunt et al., 2012j.

Included ichnospecies: Known only from the type ichnospecies.

Distribution: Late Pennsylvanian (Missourian) to Eocene of North America and Europe.

Diagnosis: Heteropolar macrospiral coprolite that differs from others in consisting of approximately 10 tight, irregularly spaced coils that constitute 75% of the length of the posterior spire of the coprolite and includes irregular striations parallel to the long axis.

Tracemaker: Chondrichthyan.

Discussion: This ichnotaxon is common in the shallow marine Pennsylvanian strata of the western United States but also occurs in a similar environment in the Eocene of Germany (Diedrich and Felker, 2012; Hunt et al., 2012j; Hunt and Lucas, 2017a).

7.1.2.8 *Speiracoprus* Hunt et al., 2012j

Type ichnospecies: *Speiracoprus socorroensis* Hunt et al., 2012j.

Included ichnospecies: Known only from the type ichnospecies.

Distribution: Late Pennsylvanian (Missourian) of New Mexico, USA.

Diagnosis: Heteropolar, macrospiral coprolite that differs from others in being proportionally short and wide with three coils with deep sulci between them, a narrow, acutely tipped posterior end, and a rounded, conical anterior termination.

Tracemaker: Chondrichthyan or sarcopterygian.

Discussion: This distinctive coprolite is only known from one specimen.

7.1.2.9 *Daniacopros* Milàn and Hunt, 2016

Type ichnospecies: *Daniacopros hofstedtii* Milàn and Hunt, 2016.

Included ichnospecies: Known only from the type ichnospecies.

Distribution: Early Paleocene, Denmark.

Diagnosis: Heteropolar, microspiral coprolite in which the posterior spine constitutes considerably less than 50% of the length in lateral view. The anterior end is narrow with a rounded tip. Irregular longitudinal striations extend from the anterior end for about one third the length of the coprolite. It differs from other described coprolites in the combination of a short posterior spire and a blunt anterior termination.

Tracemaker: Fish less derived than teleosts or an advanced actinopterygian.

7.1.2.10 *Rewacopros* Rakshit and Ray, 2021

Type ichnospecies: *Rewacopros jaini* Rakshit and Ray, 2021.

Included ichnospecies: Known only from the type ichnospecies.

Distribution: Late Triassic (Carnian-Norian) of India.

Diagnosis: Small (<35 mm in length), cylindrical coprolites; ovoid in shape; elliptical cross-section; heteropolar microspiral; rough surface texture; internally the layers converge anteriorly; clusters/pockets of dense inclusions of bones and fish scales; overall mineralogy calcium phosphate (Fig. 9.15C and D).

Tracemaker: Xenacanthids.

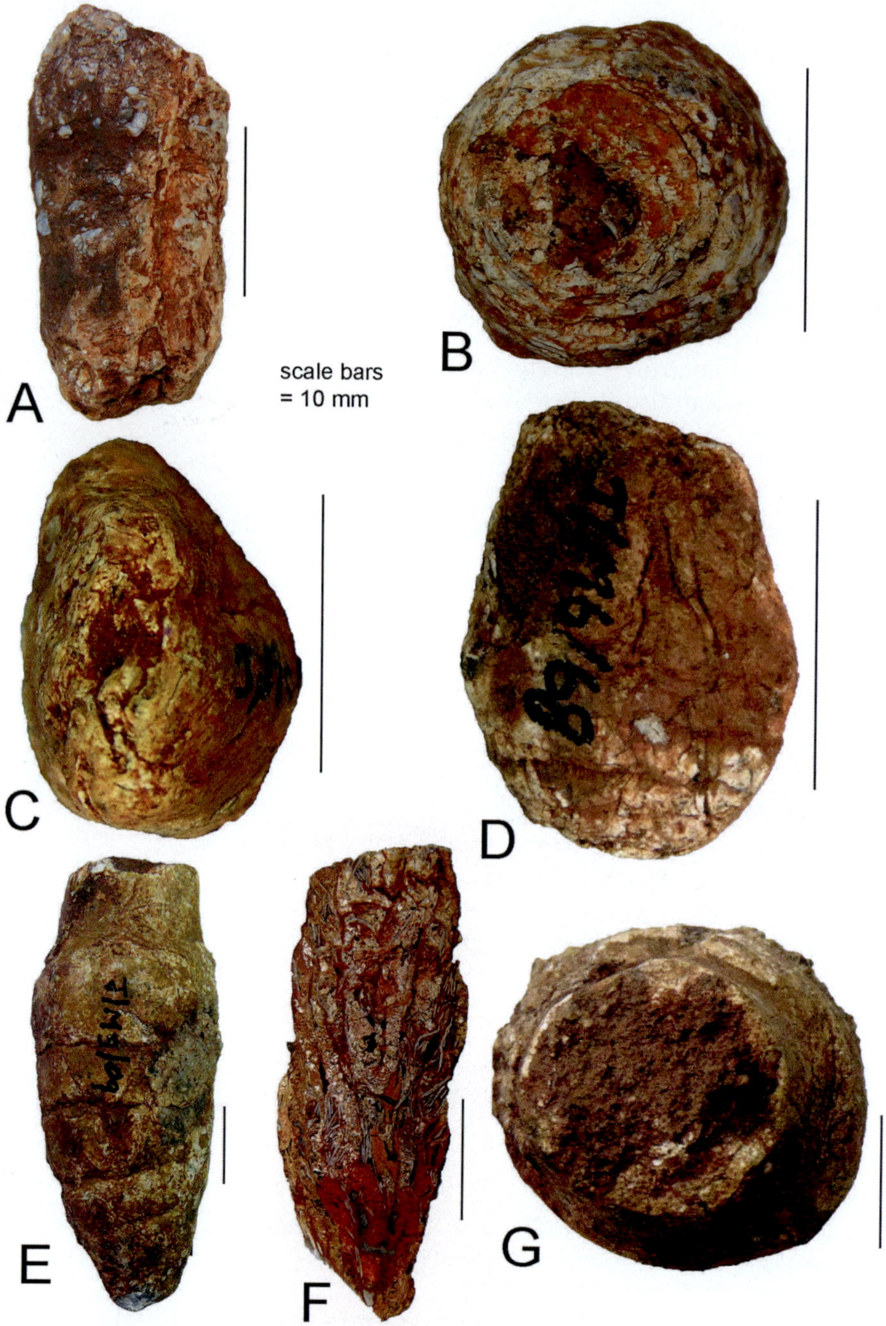

FIGURE 9.15 Triassic coprolite ichnotaxa named from the Late Triassic of India. (A and B), *Tikicopros triassicus*. A, IITKGPC64 in lateral view (Rakshit et al., 2019, Fig. 4D). B, IITKGPC109 in terminal view (Rakshit et al., 2019, Fig. 5A). (C and D) *Rewacopros jaini* in terminal (C) and lateral (D) views (Rakshit and Ray, 2021, Fig. 9.6A and C). (E–G) *Amphicopros tikiensis* in lateral external (E), lateral cross section (F) and terminal (G) views (Rakshit and Ray, 2021, Fig. 9.3K, M, N). *Images courtesy of Nibedita Rakshit.*

Discussion: Numerous fish scales of different sizes and shapes are present on the external surface of *Rewacopros*, and internally they are arranged along the layers. These are mostly ganoid fish scales, which are elongated and rhomboidal (Rakshit and Ray, 2021).

7.1.3 Amphipolar coprolites

7.1.3.1 *Rummycopros* igen. nov.

Type ichnospecies: *Rummycopros yini* (Bian et al., 1991).

Included ichnospecies: Known only from the type ichnospecies.

Etymology: Named for Paul Rummy of the Key Laboratory of Vertebrate Evolution and Human Origins, Institute of Vertebrate Paleontology and Paleoanthropology, Chinese Academy of Sciences, Beijing, for his contributions to the study of coprolites.

Revised diagnosis: Heterospiral microspiral coprolite that differs from all others except *Malericopros* in that the maximum diameter is in the anterior coil and from *Malericopros* in that the posterior spire is comprised of three wide coils and comprises approximately 50% of the length of the coprolite.

Distribution: Unknown age in China.

Discussion: This specimen was originally described by Yin (1945).

7.1.3.2 *Rummycopros yini* (Bian et al., 1991)

Holotype: Coprolite in the collection of the Palaeontological Research Office of the former central Geological Survey (Bian et al., 1991, Fig. 9.4: Fig. 9.14C).

Referred specimens: None.

Diagnosis: As for ichnogenus.

Description: The holotype is 56 mm long with a maximum width of 26.4 mm. The coprolite is laterally flattened. The coils decrease in width from anterior to posterior, the widest being 1.25 mm. In lateral view, there is a sulcus at the base of the posterior spire.

Type locality: Unknown.

Type horizon: Unknown.

Distribution: As for ichnogenus.

7.1.3.3 *Hyronocopros* Hunt et al., 2005b

Type ichnospecies: *Hyronocopros amphipola* Hunt et al., 2005b.

Included ichnospecies: *Hyronocopros amphipola* Hunt et al., 2005b, *H. tsagantsabensis* Rummy et al., 2021, *H. hunti* Rummy et al., 2021.

Distribution: Early Permian (Wolfcampian) to Early Cretaceous (Barremian) of North America and Asia.

Diagnosis: Coprolite that differs from others in having an amphipolar morphology with up to five coils of even width in lateral view and a paucity of inclusions.

Discussion: Neumayer (1904) first described these coprolites, including specimens of *Hyronocopros amphipola*, from the early Permian of Texas.

7.1.3.4 *Hyronocopros amphipola* Hunt et al., 2005b

Diagnosis: Small ichnospecies of *Hyronocopros* that is less than 2 cm in length (Fig. 9.6L–O, FF–HH, MM).

Distribution: Late Pennsylvanian to Middle Triassic of North and South America and Australia.

Tracemaker: Fish.

Discussion: *Hyronocoprus amphipolar* is the most widespread ichnospecies of this ichnotaxon.

7.1.3.5 *Hyronocopros tsagantsabensis* Rummy et al., 2021

Diagnosis: *Hyronocoprus* is similar in morphology but twice the size of *H. amphipola* and half the size of *H. hunti*; all specimens are smooth with inclusions and have up to three coils (Fig. 9.16A–C, H–J, K–N).

Distribution: Early Cretaceous of Mongolia.

Tracemaker: Acipenseriformes.

7.1.3.6 *Hyronocopros hunti* Rummy et al., 2021

Diagnosis: *Hyronocopros hunti* is distinctively large in size and is generally four times longer than the holotype of *H. amphipola* with four coils and a small nubbin on the anterior end (Fig. 9.16V–Z).

Distribution: Early Cretaceous of Mongolia.

Tracemaker: Acipenseriformes.

Discussion: *Hyronocopros hunti* is currently the largest known amphipolar coprolite ichnotaxon.

7.1.3.7 *Iuloeidocoprus* Hunt et al., 2012b

Type ichnospecies: *Iuloeidocoprus mantelli* Hunt et al., 2012b.

Included ichnospecies: Known only from the type ichnospecies.

Distribution: Upper Cretaceous of North America and Europe.

Diagnosis: Amphipolar coprolite that differs from other ichnogenera in being characterized by a cylindrical shape with a subrounded cross section with multiple, closely spaced spirals and a posterior termination that has a radial pattern and an acute, raised center.

Tracemaker: Chondrichthyan.

Discussion: Mantell (1822, p. 103, pl. 9, Figs. 9.4–9.11) described "aments or cones of unknown vegetables" from the Gray Chalk Marl of Sussex, England. These specimens encompassed both heteropolar (*Liassocoprus*: Mantell, 1822, pl. 9, Figs. 9.4 and 9.5) and amphipolar (Mantell, 1822, pl. 9, Figs. 9.6–9.11) coprolites. Buckland (1829a, p. 143) noted that "the supposed fir-cones or Iuli in the chalk and chalk marl .... are also of fecal origin ... .for these the provisional name of Coprusiuloides is proposed." Subsequently, Buckland (1835, p. 234) utilized the term Iulo-eido-coprolites. Later, Mantell acknowledged that these fossils are coprolites, possibly of sharks (Buckland, 1836, p. 155).

7.1.3.8 *Elacacoprus* Hunt et al., 2012j

Type ichnospecies: *Elacacoprus williamsi* Hunt et al., 2012j.

Included ichnospecies: Known only from the type ichnospecies.

Distribution: Late Pennsylvanian of New Mexico.

Diagnosis: Elongated amphipolar coprolite that differs from other amphipolar ichnogenera (e.g., *Hyronocoprus*, *Kalocoprus*) in being narrow, elongate, and tapering posteriorly.

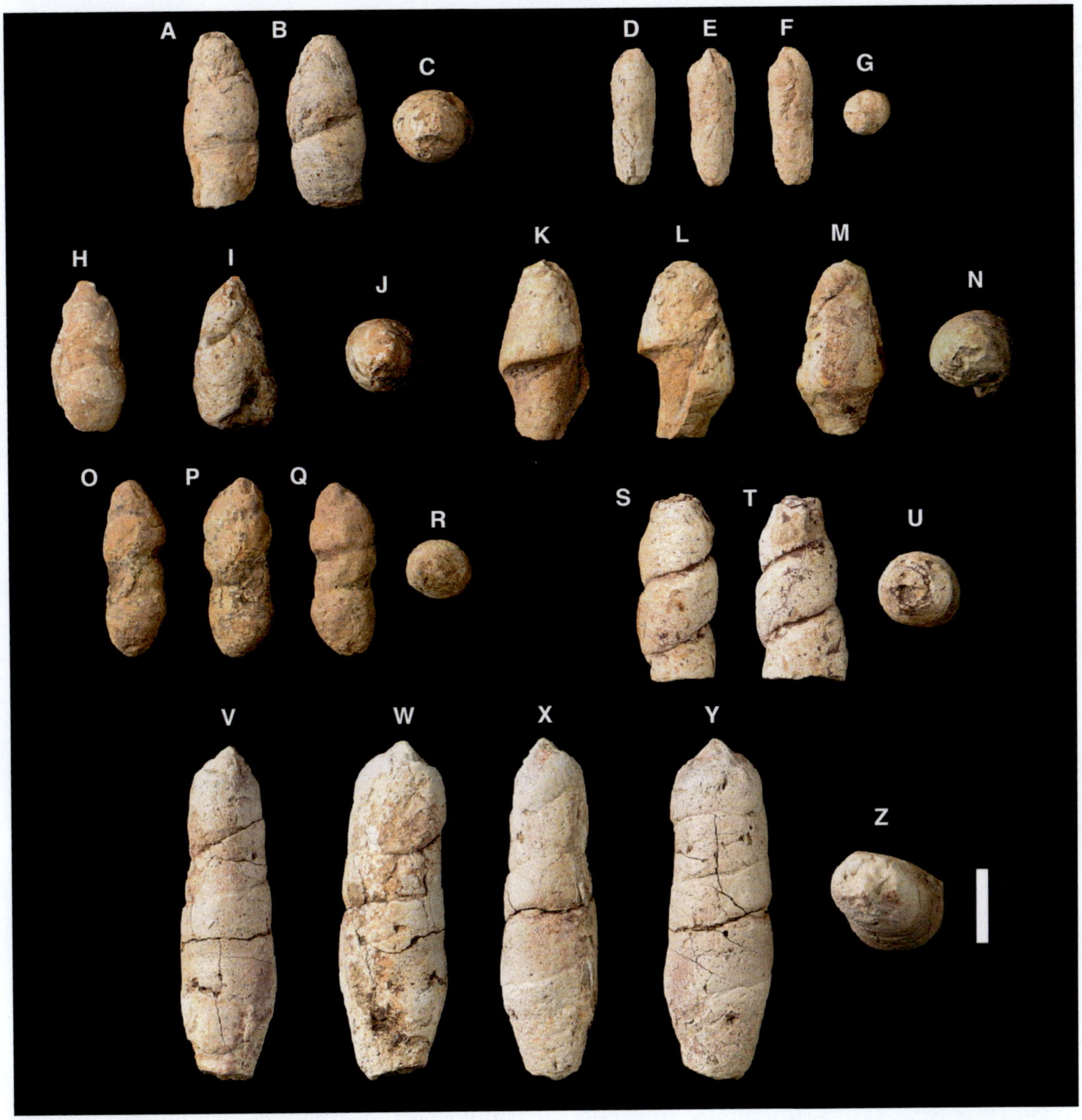

FIGURE 9.16 Type and referred specimens of vertebrate coprolites from the Lower Cretaceous Tsagan-Tsab formation, Tatal, western Mongolia (Rummy et al., 2021, Fig. 9.2). (A–C) *Hyronocopros tsagantsabensis,* holotype coprolite IVPP V 27,544 in lateral (A–B) and terminal C) views. (D–G) *Scrollocoprus tatalensis,* holotype IVPP V 27,545, coprolite in lateral (D–F) and terminal (G) views. (H–J) *Hyronocopros tsagantsabensis,* referred specimen IVPP V 27,546 in lateral (H–I) and terminal (J) views. (K–N) *Hyronocopros tsagantsabensis,* referred specimen IVPP V 27,547 in lateral (K–M) and terminal (N) views. (O–R) *Megakalocoprus barremianensis,* referred specimen IVPP V-27548 in lateral O-Q) and terminal (R) views. (S–U) *Megakalocoprus barremianensis,* holotype IVPP V 27,549 in lateral (S–T) and terminal (U) views. (V–Z) *Hyronocopros hunti,* holotype IVPP V 27,550 in lateral (V–Y) and terminal (Z) views. Scale bar equals 2 cm. *From Rummy et al. (2021, Fig. 9.2), courtesy of Paul Rummy.*

Tracemaker: Chondrichthyan or a sarcopterygian.

Discussion: It is composed of fine-grained phosphatic groundmass with no inclusions.

7.1.3.9 *Kalocoprus* Hunt et al., 2012h

Type ichnospecies: *Kalocoprus oteroensis* Hunt et al., 2012h.

Included ichnospecies: Known only from the type ichnospecies.

Distribution: Late Pennsylvanian (Missourian) of New Mexico.

Diagnosis: Dominantly amphipolar coprolite (the spirals do not extend all the way to the posterior end) that differs from *Hyronocoprus* in that there are a small number of coils (about three) along the length of the coprolite and that the spirals are separated from each other by deep sulci so that a large percentage of the circumference is visible in lateral view, and, as such, the entire spiral coprolite has the appearance of being slightly "unwound" (it is a loose spiral) (Fig. 9.66X–CC, DD and EE).

Discussion: It could be hypothesized that *Kalocoprus* is actually a degraded and uncoiled variant of *Hyronocoprus*, but we dismiss this idea because *Hyronocoprus*: (1) has more (>4) spirals in lateral view that are proportionally shorter than in *Kalocoprus*; (2) lacks a tapering termination; and (3) the spirals are more rounded in cross section (Hunt et al., 2012h).

7.1.3.10 *Helicocoprus* Hunt et al., 2015a

Type ichnospecies: *Helicocoprus clarki* Hunt, Lucas and Lichtig, 2015a.

Included ichnospecies: Known only from the type ichnospecies.

Distribution: Late Pliocene-Pleistocene of England.

Diagnosis: Amphipolar coprolite that differs from others in the wide separation of the coils, from *Hyronocoprus* in the small number of coils (3) and from *Kalococoprus* in that coils are separated by deep sulci with a V-cross section and spirals with an acute cross section.

Tracemaker: Rajidae (skates).

Discussion: All specimens are from the Red Crag Formation and are polished and pitted, which could indicate reworking.

7.1.3.11 *Vinculostercus* Duffin and Ward, 2020

Type ichnospecies: *Vinculostercus vermiformis* Duffin and Ward, 2020.

Included ichnospecies: Known only from the type ichnospecies.

Diagnosis: Amphipolar spiral coprolites measuring up to 3 mm long. Anterior end tapered; posterior end squared off. Up to four slightly overlapping coils, the anterior coil (C1) being the largest. Inclusions absent. Fecal material consists of a series of narrow, flattened ribands that are simple and unevenly distributed and sometimes overlapping anteriorly but becoming progressively more twisted and parallel to the walls of the coils anteriorly.

Distribution: Early Carboniferous (Dinantian) of England.

Tracemaker: Chondrichthyan or actinopterygian.

Discussion: This is the only named vertebrate coprolite less than 5 mm in length.

7.1.3.12 *Amphicopros* Rakshit and Ray, 2021

Type ichnospecies: *Amphicopros tikiensis* Rakshit and Ray, 2021.

Included ichnospecies: Known only from the type ichnospecies.

Distribution: Late Triassic (Carnian-Norian) of India.

Diagnosis: Amphipolar spiral coprolites, circular cross-section, differentiated from others by a high abundance of inclusions of bones and fish scales; overall mineralogy calcium phosphate (Fig. 9.15E–G).

Tracemaker: Actinopterygians.

Discussion: *Amphicopros* contains a high abundance of inclusions.

7.1.3.13 *Megakalocoprus* Rummy et al., 2021

Type ichnospecies: *Megakalocoprus barremianensis* Rummy et al., 2021.

Included ichnospecies: Known only from the type ichnospecies.

Distribution: Lower Cretaceous of western Mongolia.

Diagnosis: An amphipolar coprolite with the spirals is spaced along the longitudinal section of the specimen and is not concentrated at one end. *Megakalocoprus* is similar to *Kalocoprus* in the manner of having the same number of coils, with three situated along the lateral length but differing from *Hyronocoprus* in that there are more than four spirals in lateral view that lack tapering ends. Also, the spirals of *Megakalocoprus* are seen to be separated by a deep sulcus, thus, giving an "unwound" appearance. *Megakalocoprus* is three times larger than *Kalacoprus* (Fig. 9.16O–U).

Tracemaker: Acipenseriformes.

Discussion: The largest specimen of *Kalocoprus* is 18.8 mm long, while the holotype specimen of *Megakalacoprus* is at least 49.86 mm long (Rummy et al., 2021).

7.1.4 Scroll coprolites

7.1.4.1 *Bibliocoprus* Hunt et al., 2012h

Type ichnospecies: *Bibliocoprus beemanensis* Hunt et al., 2012h.

Included ichnospecies: Known only from the type ichnospecies.

Distribution: Late Pennsylvanian (middle to upper Missourian) of New Mexico.

Diagnosis: Scroll coprolite that differs from microspiral heteropolar coprolites (e.g., *Heteropolacopros*, *Malericoprus*, *Saurocoprus*, *Crassocoprus*, *Speiracoprus*, and *Strabelocoprus*) in being parallel-sided and cylindrical in lateral view, and in having coils in lateral view that constitute less than 20% of the length and that are relatively thin, so that the coprolite has the general form of a rolled-up scroll of parchment (Fig. 9.6NN–SS).

Tracemaker: Chondrichthyan

Discussion: *Bibliocoprus* is one of the most common ichnotaxa of coprolites at its type locality in the Beeman Formation.

7.1.4.2 *Tikicopros* Rakshit et al., 2019

Type species: *Tikicopros triassicus* Rakshit et al., 2019.

Included ichnospecies: Known only from the type ichnospecies.

Diagnosis: Small, cylindrical coprolites; rounded anterior and tapering posterior ends; circular or elliptical in cross-section; scrolled multiple layers like rolled sheets of paper; linear/straight free edge, dense inclusions of bones and scales; overall mineralogy calcium phosphate (Fig. 9.15A and B).

Distribution: Late Triassic (Carnian) of India.

Tracemaker: Hybodontid sharks such as *Lonchidion* and *Pristrisodus.*

Discussion: Inclusions include ganoid fish scales, teeth, lower jaw and skeletal remains of various osteichthyans, chondrichthyans, archosauriformes and indeterminate reptiles (Rakshit et al., 2019).

7.1.4.3 *Scrollocoprus* Rummy et al., 2021

Type ichnospecies: *Scrollocoprus tatalensis*, Rummy et al., 2021.

Included ichnospecies: Known only from the type ichnospecies.

Distribution: Lower Cretaceous of Western Mongolia.

Diagnosis: Small, anisopolar, cylindrical rod-like scroll coprolite that differs from *Tikicopros* and *Bibliocoprus* in having a posterior end with a distinct tip and a rounded anterior end. Also, a thin coil that is barely visible can be seen on the pointed tip. *Scrollocoprus* differs from *Eucoprus* in that *Eucoprus* is perfectly cylindrical in shape, while *Scrollocoprus* has one end that is wider than the other (Fig. 9.16D–G).

Tracemaker: Unknown fish.

Discussion: *Scrollocoprus* is the only the second named Mesozoic ichnogenus of scroll coprolite after *Tikicopros*.

7.1.5 Other spiral coprolites

7.1.5.1 *Elongatocoprus* Hunt et al., 2012j

Type ichnospecies: *Elongatocoprus amadoensis* Hunt et al., 2012j.

Included ichnospecies: Known only from the type ichnospecies.

Distribution: Late Pennsylvanian (middle Missourian) of New Mexico.

Diagnosis: Coprolite that differs from other ichnogenera in consisting of an elongate, gently coiled cylinder with a round cross section that is tightly spiraled in cross section.

Tracemaker: Chondrichthyan or sarcopterygian.

Discussion: The coprolite consists of fine-grained phosphatic groundmass with no inclusions. This ichnospecies is only known from a single specimen. Because of its narrow, elongate morphology, it is likely that specimens of *Elongatocoprus* that are free of matrix might have poor preservation potential.

7.1.5.2 *Struocopros* Hunt and Lucas, 2023a

Type ichnospecies: *Struocopros friedmanae*, Hunt and Lucas, 2023a.

Included ichnospecies: *Struocopros friedmanae* Hunt and Lucas, 2023a; *S. catapitostromata* Duffin and Ward, 2023b.

Diagnosis: Coprolite that differs from all others in appearing to be composed of stacked thick, rounded layers that are folded dorsally (see orientation discussion below) like a series of hard taco shells (folded corn tortillas) but that has a coiled internal structure and with few inclusions.

Distribution: Late Cretaceous (Cenomanian) to Paleogene (early Eocene) of North America and Africa.

Tracemaker: Chondrichthyan.

Discussion: The orientation of this coprolite is difficult to interpret because of the unusual morphology. Thus, we somewhat arbitrarily designate the ventral side to be that at the apices of the folds of the layers.

Discussion: These Eocene coprolites are of similar size to those from the Cretaceous but lack inclusions.

7.1.5.3 *Struocopros friedmanae* Hunt and Lucas, 2023a

Diagnosis: Ichnospecies of *Struocopros* that differs from *S. catapitostromata* in containing inclusions, having thicker individual coils that are fewer in number, and lacking lateral offset and variation in inter-coil junctions.

Distribution: Late Cretaceous (Cenomanian) to Paleogene (early Eocene) of North America.

Tracemaker: Chondrichthyan.

Discussion: *Struocopros friedmanae* is rare in the ichnofaunas in which it occurs.

7.1.5.4 *Struocopros catapitostromata* Duffin and Ward, 2023b

Diagnosis: Coprolites with three to five coils, an anterior coil with well-preserved lip and nubbin, foreshortened antero-posterior axis and expanded lateral axis, rounded to flattened suboval cross-section and occasionally laterally offset adjacent coils declined at 45° to the coprolite long axis. Specimens of *S. catapitostromata* differ from those of *S. friedmanae* in that the latter may contain inclusions, have thicker individual coils that are fewer in number, and lack the lateral offset and variation in inter-coil junctions shown by the Moroccan ichnospecies (Fig. 9.17).

Distribution: Latest Cretaceous of Morocco.

Tracemaker: Medium to large lamniformes sharks such as *Squalicorax* and *Serratolamna maroccana*, rays such as *Schizorhiza stromeri* or myliobatiformes such as *Rhombodus binkhorsti* (Duffin and Ward, 2023b).

Discussion: This ichnospecies is found about 2 m below the K-Pg boundary (Duffin and Ward 2023b).

7.1.6 *Cylindrical coprolites*

7.1.6.1 *Eucoprus* Hunt and Lucas, 2012b

Type ichnospecies: *Eucoprus cylindratus* Hunt and Lucas, 2012b.

Distribution: Early Triassic (Lootsbergian) to late Eocene of North America, Australasia, and Europe.

Description: Coprolite that is cylindrical in shape, with rounded ends, and contains no osseous inclusions.

Discussion: This ichnospecies is probably much more widespread than has been currently documented.

Tracemaker: *Eucoprus* was produced by a wide variety of carnivorous tetrapods based on its long stratigraphic range.

7.1.6.2 *Crustacoprus* Hunt et al., 2012j

Type ichnospecies: *Crustacoprus tinajaensis* Hunt et al., 2012j.

FIGURE 9.17 Holotype and paratypes of *Struocopros catapitostromata* from the Couche III Upper Bone Bed (Late Maastrichtian) of Sidi Chennane Quarry, Ouled Abdoun Phosphate Basin, Morocco. (A–F) LF 5649: Holotype. Coprolite in terminal (A), posterior (B), anterior (C) lateral (D–E) and terminal view (F). K-L. LF 5650: Paratype. Coprolite in terminal (A), posterior (B), anterior (C) lateral (D–E) and terminal view (F). All coprolite images with anterior at the top. The position of the nubbin is indicated by (**n**). *From Duffin and Ward (2023b, Fig. 9.2), courtesy of Chris Duffin.*

Included ichnospecies: Known only from the type ichnospecies.

Diagnosis: Bromalite that differs from other ichnogenera except *Conchobromus kinneyensis* in being composed predominantly of conchostracan valves and is distinct from *Conchobromus* in that it contains abundant, fine-grained phosphatic groundmass and is cylindrical in shape.

Distribution: Late Pennsylvanian (Missourian) of New Mexico, USA.

Tracemaker: Acanthodian or platysomid.

Discussion: The abundant groundmass indicates that *Crustacoprus* is a coprolite, whereas the lack of groundmass indicates that *Conchobromus* is a regurgitalite. The holotype contains abundant specimens of the conchostracan *Lioestheria carinacurvata* (Martens and Lucas, 2005; Scholze et al., 2021).

7.1.6.3 *Costacoprus* Hunt et al., 2012c

Type ichnospecies: *Costacoprus chinae* Hunt et al., 2012c.

Included ichnospecies: Known only from the type ichnospecies.

Diagnosis: Coprolite that differs from other ichnogenera in being cylindrical in shape with closely spaced lateral ridges along its length that average 2–3 mm in width and that are at right angles to the long axis.

Distribution: Early Cretaceous of England.

Discussion: This ichnogenus is currently only known from the Early Cretaceous of Cambridgeshire, England.

7.1.6.4 *Crococopros* Halaçlar et al., 2023

Type species: *Crococopros naduongensis* Halaçlar et al., 2023.

Included ichnospecies: Known only from the type ichnospecies.

Diagnosis: *Crococopros* coprolites are fusiform with cross sections oval in shape, featuring smoothly concave ends and circumferential zones of constriction. They lack longitudinal striations, distinguishing them from *Alococopros*. In addition, *Crococopros* is larger than *Alococopros* and lacks the narrowly spaced "ribs" characteristic of *Costacoprus*. Compared with *Eucoprus*, *Crococopros* exhibits larger size, more rounded apices, and possesses circumferential zones of constriction, thus distinguishing it from all three mentioned ichnogenera (Fig. 9.18).

Distribution: Paleogene of China and Vietnam.

Tracemaker: Crocodilian.

FIGURE 9.18 Holotype of *Crococopros naduongensis* (IVPP V 27941/46) in lateral view.

Discussion: The type material is from the middle-late Eocene of Vietnam, but very similar specimens occur in the Paleocene of China (Halaçlar et al., 2023).

7.1.7 Striated coprolites

7.1.7.1 *Alococopros* Hunt et al., 2007

Type ichnospecies: *Alococopros triassicus* Hunt et al., 2007.

Included ichnospecies: *Alococopros triassicus* Hunt et al., 2007; *A. indicus* Hunt et al., 2007; *A. milnei* Duffin and Ward, 2023a.

Diagnosis: Differs from other coprolite ichnogenera in often being arcuate in lateral view and subrounded in cross-section with regularly spaced, thin, longitudinal grooves.

Distribution: Early Triassic (Lootsbergian) to late Eocene of North America, Asia, Africa, and Australasia.

Tracemaker: Archosauromorphs, crocodylomorphs, crocodylians.

Discussion: It is common for specimens to be conjoined (Hunt et al., 2018, Fig. 12.9). This ichnospecies is currently only known from the Triassic to the Eocene. The specimens referred to this ichnospecies from the early Permian of New Mexico have widely spaced, coarse grooves (Cantrell et al., 2012: Fig. 2C–J) and probably represent a different ichnotaxon.

7.1.7.2 *Alococopros triassicus* Hunt et al., 2007

Diagnosis: Differs from *Alococopros indicus* in being less than one-fourth as long (typically 2 cm in length) and from *A. milnei* in lacking a well-defined nubbin anteriorly and longitudinal ridges that show sinistral or dextral drift along the length of the coprolite (Fig. 9.8O–Q).

Distribution: Early Triassic (Lootsbergian) to late Eocene of North America, Asia, Africa, and Australasia.

Tracemaker: Archosauromorphs, crocodylomorphs, crocodylians.

Discussion: Northwood (2005) noted that longitudinal intestinal rugae, which would produce striated feces occur in both amphibians and reptiles, but argued that *Alococopros triassicus* (her "longitudinally striated coprolites") represent archosauromorphs, because: (1) this ichnotaxon first occurs in the Early Triassic; (2) some extant reptiles have longitudinal rugae; and (3) they resemble extant crocodile feces (Young 1964). This is a reasonable hypothesis with Late Triassic forms produced by crocodylomorphs and later examples by crocodilians.

7.1.7.3 *Alococopros indicus* Hunt et al., 2007

Diagnosis: Ichnospecies that differs from *Alococopros triassicus* in being more than four times as long (typically 10 cm in length) and from *A. milnei* in arcuate shape and in lacking a well-defined nubbin anteriorly and longitudinal ridges that show sinistral or dextral drift along the length of the coprolite.

Distribution: Late Cretaceous (Maastrichtian) of India.

Tracemaker: Crocodilian.

Discussion: The presence of this ichnotaxon suggests the presence of a large crocodilian in the Lameta Formation of India.

7.1.7.4 *Alococopros milnei* Duffin and Ward, 2023a

Diagnosis: Coprolites belonging to *Alococopros milnei* can be distinguished from those of *A. triassicus* and *A. indicus* on the basis of their more rounded cross section, generally straight shape (as opposed to commonly arcuate), the presence of a well-defined nubbin anteriorly and longitudinal ridges that may show sinistral or dextral drift along the length of the structure. Occasional ridge bifurcation, zig-zag ridge arrangement, and intercalary ridges are also confined to *A. milnei* (Fig. 9.19).

Distribution: Latest Cretaceous (Maastrichtian) of Morocco.

Tracemaker: Crocodilian or chelonian (Duffin and Ward, 2023a).

Discussion: Inclusions are usually absent on the coprolite surface, although some specimens do show polished phosphatic debris measuring approximately 250 μm in diameter embedded over the whole coprolite surface. Sparse angular bone and scale debris may be present internally (Duffin and Ward, 2023a).

7.1.7.5 *Revueltobromus* Hunt and Lucas, 2014b

Type ichnospecies: *Revueltobromus complexus* Hunt and Lucas, 2014b.

Included ichnospecies: Known only from the type ichnospecies.

Distribution: Late Triassic (Norian) of New Mexico, USA.

Diagnosis: Bromalite that differs from others in being composed of small, striated, and lunate pellets within a thin cylindrical sheath.

Tracemaker: Given the similarity to *Alococoprus*, it is likely that *Revueltobromus* was produced by a crocodylomorph.

Description: *Revueltobromus complexus* is a complex specimen that consists of small *Alococopros*-like specimens (lunate, striated) encased in an apatitic sheath that covers one end (Hunt et al., 2013b: Fig. 9.8N–Q). This ichnotaxon apparently comprises multiple coprolite-like pellets (*Alococoprus triassicus*) within a structure, so it may not represent an evacuated object but rather an evisceralite or a section of the gastrointestinal tract preserved outside the body cavity. Conjoined specimens of *Alococopros* (usually twice the size of the components of *Revueltobromus*) occur at other localities (e.g., Suazo et al., 2012: Fig.7A–G).

7.1.8 *Segmented coprolites*

7.1.8.1 *Strophocopros* Hunt and Lucas, 2005b

Type ichnospecies: *Strophocopros valensis* Hunt and Lucas, 2005b.

Included ichnospecies: Known only from the type ichnospecies.

Diagnosis: Coprolite that differs from other ichnogenera in having a surface texture with a fibrous, ropy texture and in being over 10 cm in length.

Distribution: Early Permian (middle Leonardian) of Texas, USA.

Tracemaker: Unknown.

Discussion: Hunt and Lucas (2005b) termed the distinctive surface texture "strophic" from the Greek for twisted cord or rope.

7.1.8.2 *Hyaenacoprus* Hunt et al., 2012a

Type ichnospecies: *Hyaenacoprus bucklandi* Hunt et al., 2012a.

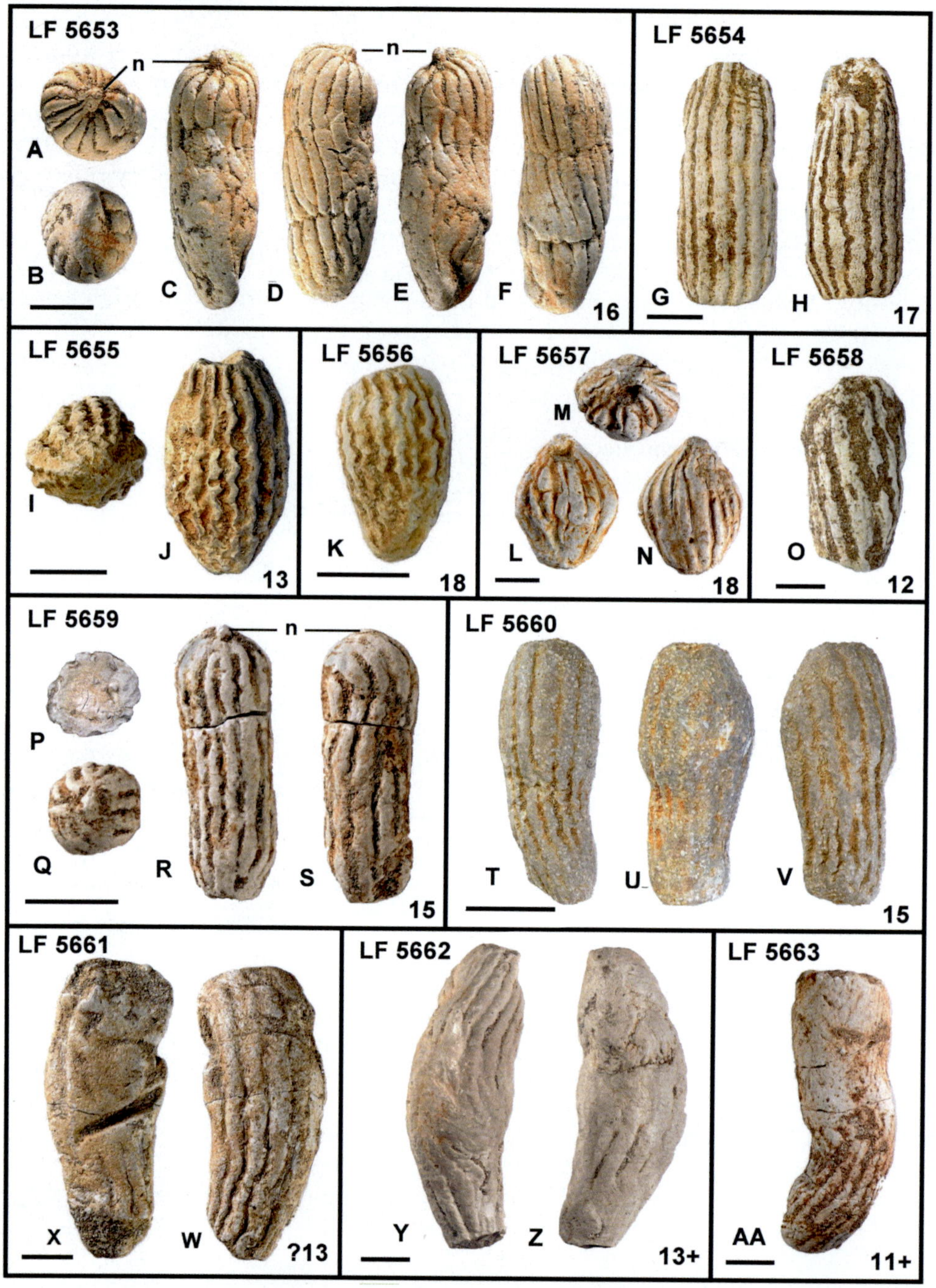

FIGURE 9.19 Specimens of *Alococopros milnei* from Couche III Upper Bone Bed (Late Maastrichtian) of Sidi Chennane Quarry, Ouled Abdoun Phosphate Basin, Morocco. (A–F) LF 5653 (holotype). (G and H) LF5654. (I and J) LF 5655. (K) LF 5656. (L–N) LF 5657. (O) LF 5658. (P–S) LF 5659. (T–V) LF 5660. (W and X) LF 5661. (Y and Z) LF 5662. (AA) LF, 5663. A, M, Q in anterior view; B, I in terminal view; C–H, J–K, L, N, O, R–AA in lateral view; P in section. The nubbin is indicated by n. The number of longitudinal striations is shown in the bottom right corner of each box. Scale bars are 10 mm. *From Duffin and Ward (2023b, Fig. 9.3), courtesy of Chris Duffin.*

Diagnosis: Phosphatic coprolite that differs from other ichnogenera in being composed of a series of rounded segments (pellets of Diedrich, 2012b), some of which are subspherical and are white in color with many small, angular bone fragments (Fig. 9.11A–F).

Included ichnospecies: Known only from the type ichnospecies.

Distribution: Upper Pliocene-Pleistocene of Europe, Asia, and Africa.

Tracemaker: *Crocuta crocuta spelaea* and possibly *Pachycrocuta.*

Discussion: Buckland (1822, 1824) described the geology and paleontology of Kirkdale Cave in Yorkshire, which he interpreted to be a Late Pleistocene hyena den. He discovered "many small balls of the solid calcareous excrement of an animal that had fed on bones, resembling the substance known in the old Materia Medica by the name of album graecum .... its external form is that of a sphere, irregularly compressed, as in the feces of sheep, and varying from half an inch to an inch in diameter; its color is yellowish white, its fracture is usually earthy and compact, resembling steatite, and sometimes granular; when compact, it is interspersed with minute cellular cavities: it was at first sight recognized by the keeper of the Menagerie at Exeter Change, as resembling, both in form and appearance, the feces of the spotted or Cape Hyaena, which he stated to be greedy of bones, beyond all other beasts under his care" (Buckland, 1822, p. 186–187). Subsequently, Buckland was able to conduct actualistic studies to confirm his hypothesis: "I have had an opportunity of seeing a Cape Hyaena at Oxford, in the traveling collection of Mr. Wombwell, the keeper of which confirmed in every particular the evidence given to Dr. Wollaston by the keeper at Exeter Change ... ...The keeper pursuing this experiment to its final result [the feeding of bones to the hyena], presented me the next morning with a large quantity of album graecum, disposed in balls, that agree entirely in size, shape, and substance with those found in the den at Kirkdale" (Buckland, 1824, p. 38). This was the first recognition of vertebrate coprolites.

7.1.8.3 *Castrocopros* Hunt and Lucas, 2018c

Type ichnospecies: *Castrocopros martini* Hunt and Lucas, 2018c.

Included ichnospecies: *Castrocopros martini* Hunt and Lucas, 2018c; *C. hauthali* Hunt and Lucas, 2018c.

Diagnosis: Subcylindrical coprolite with rounded cross section and apices that is typically 5 cm in diameter, composed of anterior-posteriorly thick plates that may be parallel-sided to wedge-shaped in cross section with a composition predominantly of short twigs with maximum widths of 1–4 mm and maximum lengths of 10–28 mm.

Distribution: Late Pleistocene of western and southwestern United States and Chile.

Discussion: The preservation of plant material in *Castrocopros* is similar to that of coprolites and feces of *Equus* spp. (Mead and Agenbroad, 1989). However, in *Castrocopros*, the maximum width of twigs is much larger (4 mm in *Castrocopros*, 2 mm in coprolites of *Equus*) and the maximum lengths are shorter (about 38 mm in coprolites of *Equus*, 28 mm in *Castrocopros*), even when the horse had eaten woody shrubs (Mead and Agenbroad, 1989, Fig. 9.6).

7.1.8.4 *Castrocopros martini* Hunt and Lucas, 2018c

Diagnosis: Ichnospecies of *Castrocopros* that differs from *C. hauthali* in that segments are broad (antero-postero length is typical close to one half of the width of the coprolite) and the constituents do not include grasses, sedges, and fruits (Fig. 9.10A,B and 6.12).

Distribution: Late Pleistocene (late Rancholabrean) of western and southwestern United States (Nevada, Arizona, Utah, New Mexico, and Texas).

Tracemaker: *Nothrotheriops shastensis*, Pleistocene ground sloth from North America.

Discussion: The first specimens of *Castrocopros martini* were excavated in the early 1930s at Gypsum Cave in Nevada, USA, and they were immediately identified as pertaining to *Nothrotheriops shastensis* on the basis of associated osteological specimens and similarity to sloth coprolites from Chile (Harrington, 1933, 1940; Laudermilk and Munz, 1938).

7.1.8.5 *Castrocopros hauthali* Hunt and Lucas, 2018c

Diagnosis: Ichnospecies of *Castrocopros* that differs from *C. martini* in that segments are narrow (antero-postero length is typically close to one quarter, and always less than one half of the width of the coprolite), and the constituents include grasses, sedges and fruits (Fig. 9.10C and D).

Distribution: Late Cretaceous of Chile.

Tracemaker: *Mylodon darwinii*, Pleistocene ground sloth from South America.

Discussion: Hauthal et al. (1899, pl. 4, Figs. 9.5–9.7; Hunt and Lucas, 2018, Fig. 9.1) provided the earliest and best illustrations of this ichnospecies, which clearly demonstrate the narrow segments that typify this ichnospecies.

7.1.8.6 *Beechybromus* Hunt and Lucas, 2018b

Type ichnospecies: *Beechybromus wellschi* Hunt and Lucas, 2018b.

Included ichnospecies: Known only from the type ichnospecies.

Diagnosis: Large, elongate subcylindrical coprolite with rounded cross section that is composed of up to 12 discoidal segments; the posterior termination is rounded, and the anterior consists of a triangular apex.

Distribution: Late Cretaceous of Saskatchewan, Canada.

Tracemaker: Mosasaur.

Discussion: There is a range of morphological variation in this ichnospecies from subcylindrical to bi-lobed.

7.1.8.7 *Borocopros* Hunt and Lucas, 2021d

Type ichnospecies: *Borocopros wangi* Hunt and Lucas, 2021d.

Included ichnospecies: Known only from the type ichnospecies.

Diagnosis: Light-colored, segmented phosphatic coprolite containing abundant large bone fragments that differs from *Hyaenacoprus* in being much shorter in overall length and ovoid in lateral section with less rounded segments (Fig. 9.19).

Distribution: Latest Miocene to latest Pliocene of the western United States.

Tracemaker: Borophagine canid *Borophagus*.

Discussion: Coprolites of bone-crushing carnivores are predicted to be more common than those of other predators because they are composed almost entirely of hydroxyapatite powder, which facilitates their preservation (Hunt and Lucas, 2020).

7.1.8.8 *Cuocopros* Rummy and Halaçlar, 2024 (in Farjand et al., 2024)

Type ichnospecies: *Cuocopros yuanmouensis* Rummy and Halaçlar, 2024 (in Farjand et al., 2024).

Included ichnospecies: Known only from the type ichnospecies.

Diagnosis: Displays a length range of 15–55 mm and a maximum width range of 17–35 mm. The overall shape is rounded, with one end appearing more tapered than the other with no disc shape. The coprolites exhibit a color variation from white to yellowish-brown (HUE 10 YR 8/1 to 5/4) and contain fewer bone fragments than *Borocopros* and possible pollen (Figs. 9.20 and 9.21).

Distribution: Early Pleistocene (2.6–1.7 Ma).

Tracemaker: Dhole canid *Sinocuon yuanmouensis.*

Discussion: The specimens of this ichnotaxon occur in latrinite inclusions of hair and all contain bone fragments and inclusions of hair (Farjand et al., 2024).

7.1.9 Ovoid coprolites

7.1.9.1 *Rhynchocopros* Hunt et al., 2013b

Type ichnospecies: *Rhynchocopros soutoi* Hunt et al., 2013b.

Included ichnospecies: Known only from the type ichnospecies.

Diagnosis: Calcareous coprolite that differs from other ichnogenera in having the shape of a well-rounded ovoid with a length:width ratio of approximately 1:0.5, in being composed primarily of calcite and often occurring in masses of up to six.

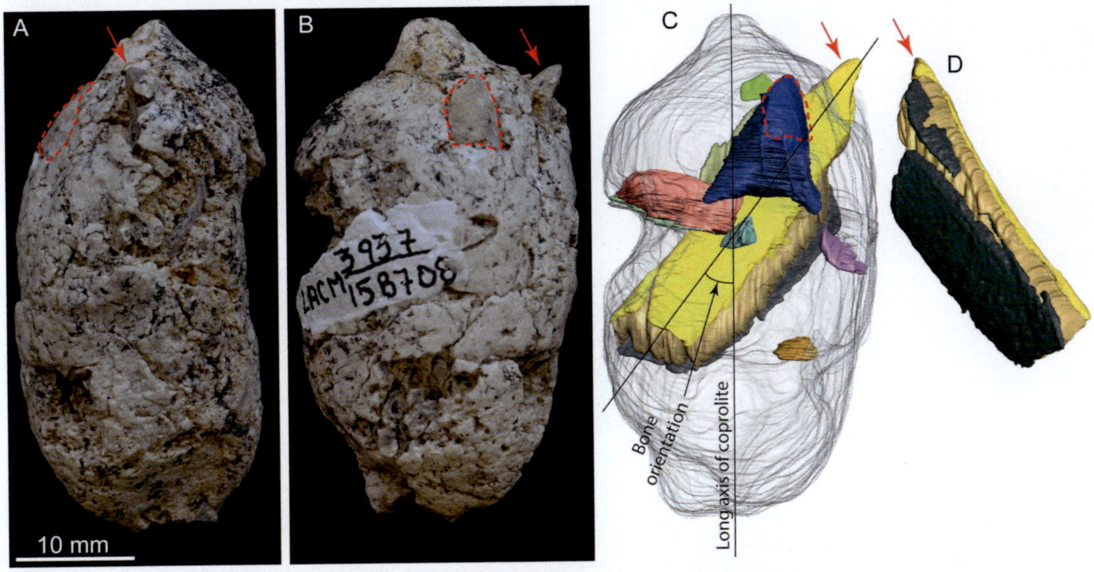

FIGURE 9.20 Holotype of *Borocopros wangi* LACM 158708, a complete coprolite from the upper Miocene Mehrten Formation, Stanislaus County, California. (A and B) Lateral views about 90° rotations, anterior is up. (C) digitally separated individual bones (in different colors) within coprolite matrix (light gray), identical view as B. (D) rotated view of a rib fragment seen in C, showing the convex (external) side, yellow and dark green shapes representing internal (toward chest cavity) and external cortical bone respectively, and yellowish brown sandwiched between the cortical bones being cancellous bone. Red arrows indicate the same protruded tip of rib fragment, and red dashed lines define the exposed outlines of a flat bone (mostly buried within coprolite matrix; dark blue piece in C shows the full extent of this bone within the coprolite). The rib is the only identifiable bone fragment. Wang et al. (2018, Fig. 9.3) and Hunt and Lucas (2021d, Fig. 9.1). *Courtesy of Xiaoming Wang.*

FIGURE 9.21 Holotype and referred specimens of *Cuocopros yuanmouensis* Rummy and Halaçlar, 2024 (in Farjand et al., 2024) from the early Pleistocene member three of the Yuanmou Formation, Juna village, Yuanmou County, Yunnan, Southwest China. Top left with red asterisk, holotype. MCYNAP 2015-LCZN-T1-135.1, a complete coprolite pellet with constriction mark. Remaining images are referred specimens in lateral views, MCYNAP 2015-LCZN-T1-135.2 to 2015-LCZN-T1-135.38. *From Farjand et al. (2024, Fig. 9.3), courtesy of Paul Rummy.*

Distribution: Middle Triassic (Berdyankian) of Brazil.

Tracemaker: Rhynchosaur.

Discussion: The prevalence of this ichnogenus in co-occurrence with rhynchosaur bones is a strong indication of the probable source animal. Francischini et al. (2018) contested the validity of this ichnotaxon.

7.1.9.2 *Santamariacopros* Hunt et al., 2013b

Type ichnospecies: *Santamariacopros elongatus* Hunt et al., 2013b.

Included ichnospecies: Known only from the type ichnospecies.

Diagnosis: Coprolite that differs from other ichnogenera in being an elongate, rounded cylinder with a length:width ratio of 1: >0.5 which is composed principally of calcite and lacks inclusions and often occurs in masses of up to four.

Distribution: Middle Triassic (Berdyankian) of Brazil.

Tracemaker: Dicynodont.

Discussion: It is likely that this herbivore coprolite pertains to the dicynodonts in the Santa Maria fauna. Francischini et al. (2018) disputed the validity of this ichnotaxon.

7.1.9.3 *Huberobromus* Hunt and Lucas, 2021a

Type ichnospecies: *Huberobromus ovatus* Hunt and Lucas, 2021a.

Included ichnospecies: Known only from the type ichnospecies.

Distribution: Mississippian and Pennsylvanian of the United States.

Diagnosis: Large (up to 60 mm by 45 mm), flattened, and subovoid bromalite with groundmass and abundant fish debris.

Tracemaker: Chondrichthyan.

Discussion: *Huberobromus ovatus* is widespread throughout the Carboniferous in North America in ichnofaunas in fissile shale.

7.1.9.4 *Maculacoprus* Hunt and Lucas, 2021a

Type ichnospecies: *Maculacoprus ateri* Hunt and Lucas, 2021a.

Included ichnospecies: Known only from the type ichnospecies.

Distribution: Mississippian and Pennsylvanian of the United States.

Diagnosis: Small, dense, dark flattened ovoid with macerated fish material and matrix.

Tracemaker: Acanthodians and palaeonisciformes and larger ones by platysomids.

Discussion: *Maculacoprus* differs from *Huberobromus* in: (1) smaller size; (2) more subcircular shape; (3) maceration of fish material; (4) larger proportion of fine-grained matrix; and (5) dark color. *Maculacoprus ovatus* specimens typically contain only palaeonisciform debris, although some also yield some acanthodian material.

7.1.9.5 *Virgacoprus* Hunt and Lucas, 2021a

Type ichnospecies: *Virgacoprus brevisa* Hunt and Lucas, 2021a.

Included ichnospecies: Known only from the type ichnospecies.

Distribution: Pennsylvanian (Missourian: Kasimovian) of New Mexico, USA.

Diagnosis: Flattened, small (5–11 mm in length), narrow (1–2 mm in width) coprolite with rounded terminations, composed of macerated fish debris in fine-grained matrix.

Tracemaker: A small fish, possibly a haplolepid such as *Pyritocephalus* (Huber, 1992), a small palaeonisciform (Bardack, 1992), or an acanthodian (Zidek, 1992).

Discussion: *Virgacoprus brevis* is considered to represent a coprolite rather than a regurgitalite because of its fusiform shape, finely macerated fish material, and fine-grained matrix.

7.1.10 Pelletoidal coprolites

7.1.10.1 *Suaviocopros* Hunt and Lucas, 2021c

Type ichnospecies: *Suaviocopros harrisi* Hunt and Lucas, 2021c.

Included ichnospecies: Known only from the type ichnospecies.

Distribution: Late Pleistocene (late Rancholabrean) of Utah, USA.

Diagnosis: Short coprolite, usually subround in cross section but sometimes subtriangular with the anterior end forming a rounded point and the posterior often forming a concave pit, typically 15–25 mm long with a minimum diameter of 11–17 mm, and composed of more than 95% tree or shrub foliage (principally *Artemisia tridentata*, *Acacia* sp., *Oenothera* sp., *Quercus* sp., *Rhus* sp., and *Chrysothamnus* sp.) and less than 5% grass (Fig. 9.10H–J).

Tracemaker: Shrub-ox *Euceratherium collinum.*

Discussion: All specimens are from Glen Canyon National Recreation Area, Utah, USA at Bechan Cave, Cottonwood Alcove, Cowboy Cave, Grobot Grotto, Hooper's Hollow, Oak Haven, and Shrub-ox Alcove. These coprolites are assigned to *Euceratherium* on the basis of: (1) association with a tooth of *Euceratherium collinum* in Bechan Cave; (2) composition indicates production by a browser; and (3) larger than feces of other artiodactyls known from fossils on the Colorado Plateau (Kropf, 2005; Kropf et al., 2007; Hunt and Lucas, 2021c). There are four radiocarbon dates on *Suaviocopros harrisi* that range from 20,930 ± 400 to 11,630 ± 150 yr BP (Hunt and Lucas, 2021, Table 2). These ages are all Rancholabrean (late Pleistocene).

7.1.11 Elongate coprolites

7.1.11.1 *Dicynodontocopros* Hunt et al., 1998

Type ichnospecies: *Dicynodontocopros maximus* Hunt et al., 1998.

Distribution: Middle Triassic to Late Triassic of North America, Africa, and Asia.

Description: Large vertebrate coprolites (up to 10 cm long), often dark gray in color, preserving up to four loose coils, often containing large (up to 4 mm in diameter) blebs of carbonaceous plant debris, having a slightly arcuate long axis with one end broadly rounded and the other coming to an acute tip, with a maximum width-to-length ratio of about 45%–50%.

Tracemaker: Dicynodont (principally *Placerias* during the Late Triassic).

Discussion: This ichnotaxon is most abundant at the *Placerias* quarry in northeastern.

Arizona, USA (Hunt et al., 1998, 2013b). It is a proxy for several localities where dicynodont bones are lacking (e.g., in western New Mexico and Texas, USA, and Morocco).

7.1.11.2 *Dakyronocopros* Hunt and Lucas, 2005a

Type species: *Dakyronocopros arroyoensis* Hunt and Lucas, 2005a.

Included species: Known only from the type species.

Distribution: Early Permian of Texas (early Wolfcampian-middle Leonardian).

Diagnosis: Coprolite that differs from other coprolite ichnogenera in being teardrop-shaped, the widest point is approximately a third of the way from the broadest end, it tapers at both ends, and the broader end is almost flat, whereas the apical end is a rounded point, laterally flattened, with no spiraling or inclusions.

Tracemaker: Aquatic or terrestrial vertebrate.

Discussion: This ichnogenus from the early Wolfcampian Nocona Formation to the middle Leonardian Arroyo Formation is remarkably uniform in morphology (Hunt and Lucas, 2005a).

7.1.11.3 *Falcatocopros* Hunt et al., 2007

Type ichnospecies: *Falcatocopros* oxfordensis Hunt et al., 2007.

Included ichnospecies: Known only from the type ichnospecies.

Distribution: Late Triassic (Carnian) to Late Jurassic (Oxfordian) of North America and Europe.

Diagnosis: Differs from other coprolite ichnogenera in being long, narrow, and arcuate in lateral view, and rounded to subrounded in cross section with a width that gradually decreases from one end to the other.

Tracemaker: Unidentified tetrapod.

Discussion: This ichnospecies is uncommon, and probably, at least in part, the result of a taphonomic artifact related to its slender morphology that would render it susceptible to mechanical breakage.

7.1.11.4 *Kinneybromus* Hunt and Lucas, 2021a

Type ichnospecies: *Kinneybromus jurgenai* Hunt and Lucas, 2021a.

Included ichnospecies: Known only from the type ichnospecies.

Diagnosis: Large (>80 cm long), flat (thickness <2 mm) bromalite with elongate ovoid shape, scalloped margin, and fine-grained matrix.

Distribution: Late Pennsylvanian (Missourian) of New Mexico, USA.

Tracemaker: Chondrichthyan.

Discussion: *Kinneybromus* is identified as a coprolite rather than a regurgitalite because of its abundant matrix and lack of large skeletal elements. This distinctive coprolite is only known from the holotype.

7.1.11.5 *Antarctoscoprus* Bigurrarena Ojeda et al., 2023

Type ichnospecies: *Antarctoscoprus longinensis* Bigurrarena Ojeda et al., 2023.

Included ichnospecies: Known only from the type ichnospecies.

Diagnosis: Roughly rounded to elongate small coprolite (7–62 mm long) with a well-defined outline, typically preserved in 2D, with an internal content composed of actinopterygian remains and a groundmass or mucus.

Distribution: Late Jurassic of Antarctica.

Tracemaker: Nonteleostean actinopterygians (i.e., pachycormiformes and amiids).

Discussion: The contents of the coprolites consist of isolated and fragmentary actinopterygian cranial bones, teeth, vertebrae, fin rays, and scales. These remains show signs of consumption and digestion (e.g., cracking, abrasion, corrosion) (Bigurrarena Ojeda et al., 2023, Fig. 9.3).

7.1.12 *Coiled coprolites*

7.1.12.1 *Plektecoprus* Hunt et al., 2012a

Type ichnospecies: *Plektecoprus whitbyensis* Hunt et al., 2012a.

Included ichnospecies: Known only from the type ichnospecies.

Distribution: Early Jurassic of England.

Diagnosis: Coprolite that differs from other ichnogenera in being elongated and rounded in cross section with a loose spiral coil and having a conical posterior end and a broad, rounded anterior end.

Tracemaker: Marine reptile.

Discussion: This ichnogenus is currently only known from the Lower Lias of Yorkshire, England. The elongate shape of this ichnospecies and its loose coil may have made it susceptible to mechanic destruction, so *Plektecoprus* may be rarely preserved.

7.1.13 *Massive coprolites*

7.1.13.1 *Ichthyosaurolites* Hunt et al., 2012a

Type ichnospecies: *Ichthyosaurolites duffini* Hunt et al., 2012a.

Included ichnospecies: Known only from the type ichnospecies.

Distribution: Early Jurassic of England.

Diagnosis: Bromalite that differs from other ichnogenera in consisting of a wide, flattened rectangle with a rounded tip with abundant phosphatic groundmass and inclusions of multiple skeletal elements of juvenile ichthyosaurs (vertebral diameters typically about 12–15 mm).

Tracemaker: Large ichthyosaur such as *Temnodontosaurus*.

Discussion: This distinct morphology of this bromalite was first recognized by Buckland (1836, pl. 15, Fig. 9.18) and later by Duffin (2010, pl. 77, Fig. 9.1).

7.1.13.2 *Mammuthocopros* Hunt and Lucas, 2020

Type ichnospecies: *Mammuthocopros allenorum* Hunt and Lucas, 2020.

Included ichnospecies: Known only from the type ichnospecies.

Distribution: Late Pleistocene of North America.

Diagnosis: Coprolite that differs from other ichnogenera in being characterized by a flattened sphere shape, large size (maximum diameter >220 mm), >60% size fraction of contents greater than 1 mm, and individual twigs (principally grasses and sedges) mainly longer than 5 mm (Fig. 9.10E–G).

Discussion: Mead et al. (1986a) first assigned this morphotype tentatively to *Mammuthus columbi*, and this was confirmed through DNA study by Karpinski et al. (2017). This ichnogenus is only known from seven caves in southeastern Utah, USA.

8. Producers

Before Buckland even coined the term coprolite in 1829, he was studying these trace fossils with an interest in their origin (Buckland, 1822). His early actualistic work on hyena dung and bone-processing was a century ahead of its time relative to any subdiscipline of paleontology (Pemberton and Frey, 1991; Duffin, 2006). From Buckland on there has been a strong interest in associating vertebrate coprolites not just with a general category of animals but with a particular genus. In many ways, this interest in the animal that produced the trace mirrors the history of vertebrate track studies and contrasts with the invertebrate ichnological study of both kinds of traces (Hunt and Lucas, 2010). As in the study of vertebrate tracks, vertebrate coprolite workers have treated the trace as a substitute for a biotaxon, whereas invertebrate ichnologists have been more interested in the behavioral aspects of the ichnofossils, not in the identity of their producer (Hunt and Lucas, 2007b; Lucas, 2007). Another similarity exists between vertebrate tracks and vertebrate coprolites. Tracks rarely allow identification of a genus or species known from the body fossil record, and the same is true of coprolites. Indeed, almost all tetrapod footprint ichnogenera are equivalent to a body fossil-based family or a higher taxon (Lucas, 2007), and it is likely that most pre-Pleistocene coprolite ichnotaxa, with some notable exceptions, also correspond to supraspecific taxa. This means that both footprint and coprolite ichnogenera necessarily have much longer temporal ranges and therefore slower apparent evolutionary turnover rates than do body fossil genera (Lucas, 2007).

The majority of vertebrate coprolites represent those of carnivorous animals, because of two main factors: (1) herbivores defecate a large proportion of undigested vegetable material that is the basis of a diverse microecosystem of bacterial and other scavengers that diminish and/or destroy the bolus, whereas carnivore feces are largely composed of unattractive waste products; and (2) the calcium phosphate in consumed bone facilitates the lithification of most carnivore feces (Hunt et al., 1994; Hollocher and Hollocher, 2012).

The main lines of evidence to identify the specific producer of a vertebrate coprolite type are (Hunt and Lucas, 2010, 2021e; Chame and Sianto, 2014):

1. Morphology—Some coprolites have a very distinctive morphology and structure. Notably, spiral coprolites that are produced by sharks, rays, or less advanced fish with spiral valves have been studied extensively to investigate their origins (Hunt and Lucas, 2012b).
2. Analogy—In the case of late Cenozoic animals, it is often possible to compare the morphology and content of coprolites with those of living animals. Thus, comparison of putative mammoth coprolites with elephant dung has been useful in identification of the producers (Mead et al., 1986a; Hunt and Lucas, 2020).
3. Association in digestive tract—In a few rare instances, coprolites have been found within the body cavity of Paleozoic fish, Mesozoic dinosaurs, and Cenozoic crocodilians, thus establishing their producers.
4. Association with skeletal remains (articulated carcass or bone bed)—Some coprolites are found in close association with skeletons but not within them, such as *Dicynodontocopros* at the Late Triassic *Placerias* quarry bone bed, which is dominated by the bones of dicynodonts (Hunt et al., 1998). The association thus suggests the producer.
5. Biochemistry—Pleistocene and Holocene coprolites are often identifiable to species level because of the preservation of complex organic molecules, including DNA (Höss et al., 1996; Poinar et al., 1998, 2003; Karpinksi et al., 2017). Borry et al. (2020) applied Shotgun metagenomics to archeological coprolites. They were able to distinguish morphologically similar human and canine coprolites.

6. Inclusions—Inclusions in coprolites can provide information about diet that may help identify the producer (e.g., Qvarnström et al. (2019a)).
7. Size—Size, notably diameter, can be of some use in the identification of the producer of coprolites, as small animals cannot produce large coprolites, although the reverse is not true. In cases of very large dimensions, size may allow attribution. For example, very large Late Cretaceous coprolites have reasonably been attributed to tyrannosaurid dinosaurs, the only vertebrate taxon at that time large enough to produce such large coprolites (Chin et al., 1998, 2003).

9. Utility of coprolites

9.1 Introduction

Coprolites can serve as proxies for biotaxa (the animals that produced them). The producer of individual vertebrate coprolites can usually only be identified at a high taxonomic level (order or higher) with very few exceptions before the Pleistocene, such as dicynodont therapsids from the Late Triassic of the United States and hyperodapedontidae rhynchosaurs from the Middle Triassic of Brazil (Hunt et al., 2013b). Many Pleistocene vertebrate coprolites from nonmarine environments can be identified at the specific level on the basis of morphology and DNA (e.g., *Hyaenacoprus* derives from *Crocuta*: Bon et al., 2012). Regardless of the known taxonomic level of the producer, the distribution of a coprolite ichnotaxon is a proxy for a biological taxonomic unit. Vertebrate coprolites provide some biochronology in the Pennsylvanian to the Quaternary (Hunt and Lucas, 2021e, Figs. 25, 27, 29; Figs. 9.5, 9.7, 9.9), and invertebrate coprolites, although represented by far fewer ichnotaxa, provide a biochronology through the entire Phanerozoic (Knaust, 2020).

Vertebrate coprolites can delimit biogeographic regions. For example, there are two distinct biogeographic and taphonomic provinces defined by vertebrate coprolites in the Pleistocene (Hunt and Lucas, 2020, Fig. 9.1). The *Castrocopros* province of the New World is characterized by a dominance of herbivore coprolites, which are preserved almost exclusively in caves, and the *Hyaenacoprus* province in the Old World is dominated by hyena coprolites that occur mainly in caves (Hunt and Lucas, 2020; Hunt and Lucas, 2021e, Fig. 31).

All trace fossils, including coprolites, are facies fossils. Spiral vertebrate coprolites in shallow marine environments define the *Crassocoprus* ichnofacies that ranges from the Mississippian to the Eocene. On a finer scale, vertebrate coprolites allow the discrimination of three ichnocoenoses in the nonmarine Late Triassic and four in an ecological transect from lacustrine to shallow marine conditions in the Late Pennsylvanian (Hunt and Lucas, 2017a).

The first important inclusions to be recognized in coprolites were spores in arthropod coprolites and pollen in vertebrate herbivore coprolites. Subsequently, a wide range of organic materials, including soft tissues, have been identified in vertebrate coprolites, particularly muscle tissues, eggs, hair, parasites, bacteria, fungi, and feathers. Thus, some coprolites represent Lagerstätten (Qvarnström et al., 2016).

Coprolites obviously represent the end point of digestion, so they can provide evidence about the evolution of feeding, diet, and digestion. For example, vertebrate coprolites of the Silurian to Carboniferous demonstrate changing patterns of predation and food processing among carnivorous fishes as demonstrated by an increase in groundmass, increase in size, and the diversification of spiral morphologies (Hunt and Lucas, 2016a).

Coprolites provide ecological information about their producers, and Niedźwiedzki and co-workers have utilized them to reconstruct ecosystems and analyze diversity (Niedźwiedzki et al., 2016a,b). Coprolites are also of utility in a number of other areas from the study of the evolution of human diseases to sedimentology to providing mineable resources (guanolite) (Hunt, 1992; Bryant and Reinhard, 2012; Hunt et al., 2012d).

9.2 Ichnofacies

Coprolites are trace fossils and thus facies fossils. Hunt and co-workers (Hunt et al., 1994, 1998, 2007, 2013a; Hunt and Lucas 2007b) first recognized discrete associations of vertebrate coprolites and sedimentary facies in the Late Triassic of western North America. Hunt and Lucas (2017a) recognized that four bromalite ichnofaunas of Missourian (Late Pennsylvanian) age in New Mexico that represent an ecological transect from lacustrine to basinal marine: (1) Tinajas Lagerstätte—lacustrine; (2) Kinney Brick Quarry Lagerstätte—lagoonal; (3) Erickson site—nearshore marine; and (4) Sacramento Mountains—offshore marine. There are clear trends through these ichnofaunas (Tinajas-Kinney-Erickson-Sacramento): (1) flattened preservation in matrix to isolated three dimensional; (2) diverse bromalites to only coprolites; and (3) increasing proportion of spiral coprolites so that the most basinward ichnofauna is dominated by spiral coprolites. In addition, the Kinney ichnofauna is typical of lagoonal/estuarine ichnofaunas of the Paleozoic and Mesozoic. Hunt et al. (2015b) had identified the presence of abundant spiral coprolites as being characteristic of certain marine environments.

Buckland (1829, 1835) largely based his definition of coprolites on the recognition that heteropolar "bezoar stones" from the Early Jurassic of Lyme Regis in southwestern England represent fossil feces. Heteropolar coprolites occur in large numbers in shallow marine strata from the Mississippian to the Eocene (e.g., Hunt et al., 2015b). Hunt et al. (2018) defined the *Crassocoprus* ichnofacies to include marine trace fossil ichnocoenoses dominated by heteropolar coprolites and that include coprolites of low to moderate ichnodiversity. Shale substrates typify the *Crassocoprus* ichnofacies. The name is for *Crassocoprus*, a macrospiral heteropolar coprolite from the Pennsylvanian that is attributed to a chondrichthyan (Hunt et al., 2012j).

Heteropolar coprolites date back to the Devonian and first became abundant during the Pennsylvanian (Hunt and Lucas, 2013). The principal large samples of heteropolar coprolites (shallow marine setting unless indicated otherwise) occur in the: 1. Middle-Late Mississippian Wardie, Midlothian, Scotland (Middle Mississippian: Viséan) (Buckland, 1835; Sumner, 1991) Anstruther, Fife, Scotland (Middle Mississippian: Viséan) (Sumner, 1991). Bearsden, East Dunbartonshire, Scotland (Late Mississippian: Serpukhovian) (Clark, 1989). 2. Late Pennsylvanian of Park and Chaffee counties, Colorado, USA (Johnson, 1934); Bassam Park, Colorado, USA (Houck et al., 2004); Morgantown, West Virginia, USA (Price, 1927); and Sacramento Mountains, New Mexico, USA (Hunt et al., 2012h). 3. Early Permian Manhattan, Kansas, USA (Williams, 1972; McAllister, 1985). 4. Middle/late Permian southern Brazil—lacustrine (Dentzien-Dias et al.,

2012). 5. Late Permian/Early Triassic European Russia—nonmarine (Niedźwiedzki et al., 2016b). 6. Early Jurassic Lyme Regis, England (Buckland, 1835; Hunt et al., 2012a). 7. Late Cretaceous southern England, Western Europe (Hunt et al., 2015b). 8. Middle-Late Eocene northern Germany (Diedrich and Felker, 2012) and Alabama, USA (Stringer and King, 2012).

Hunt et al. (2018) also named the *Gaspeichnus* Ichnofacies for traces developed on the substrate of a coprolite. Feces provide a source of food for some insects and vertebrates. Coprolites also undergo bioerosion and are the subject of accidental contact by vertebrates. Therefore, coprolites preserve a variety of traces including vertebrate dentalites, vertebrate footprints, and invertebrate borings. The *Gaspeichnus* Ichnofacies was thus defined to include marine and nonmarine trace fossil ichnocoenoses of low diversity dominated by macroscopic borings and vertebrate dentalites and footprints that utilize coprolites as a substrate. The name is for *Gaspeichnus*, a sinuous coprolite boring from the Devonian (Hunt et al., 2018). Other examples of the *Gaspeichnus* Ichnofacies include (1) flask-shaped borings in both bone and coprolites from the Cretaceous-Eocene of Mali that represent the putative bivalve boring named *Gastrochaenolites ornatus* (Tapanila et al., 2004); (2) an amphipolar coprolite with an invertebrate boring from the Rio do Rasto Formation (middle/upper Permian) in southern Brazil (Dentzien-Dias et al., 2012); (3) small holes in Late Triassic coprolites from Arizona, USA, produced by insects, such as dipteran larvae (Wahl et al., 1998); (4) putative examples of coprophagy traces of dung beetles in the Cretaceous and Paleogene of the USA (Bradley, 1946; Chin and Gill, 1996); (5) Miocene mammalian footprints preserved in coprolites from Portugal, including a tridactyl footprint that can be ascribed to a right foot of the rhinoceros *Hispanotherium matritense*, a tridactyl, left foot impression of a perissodactyl, possibly an *Anchitherium*-like equid, and a didactyl track from a small-sized ruminant, most probably a cervid, genus *Procervulus* (Antunes et al., 2006b); (6) Miocene aquatic coprolites from the eastern United States and Italy containing dentalites produced by gars and sharks (Godfrey and Smith, 2010; Godfrey and Palmer, 2015; Collareta et al., 2019;Godfrey et al., 2020); (7) Miocene crocodile coprolites with burrows (Godfrey and Collareta, 2022; Godfrey et al., 2022); and (8) a Miocene coprolite from Italy with the echinoid trace *Gnathichnus pentax* from southern Italy (Collareta et al., 2022).

9.3 Biostratigraphy and biochronology

Vertebrate coprolites can serve as proxies for biotaxa and thus are of biostratigraphic and biochronological utility (e.g., Hunt, 1992; Hunt et al., 1998, 2005a, 2007, 2013b, 2018; Hunt and Lucas, 2021e, Figs. 25, 27, 29). However, vertebrate ichnotaxa almost always correspond to higher-level taxonomic groups than do body fossils. Thus, footprint ichnogenera are often equivalent to the family (or higher) taxonomic level of body fossils (Lucas, 2007). Pre-Pleistocene coprolites probably represent, in most cases, even higher-level taxonomic groups ("order" or above) (Hunt et al., 2007, 2013b). Examples of lower-level identifications would be coprolites attributed to dicynodont therapsids from the Late Triassic of the United States (Hunt et al., 1998) and to tyrannosaurid theropods from the Late Cretaceous of Canada (Chin et al., 1998, 2003).

Most Pleistocene vertebrate coprolites can be identified at the specific level on the basis of morphology (e.g., Mead and Swift, 2012) and more recently by DNA. For example, *Hyaena-coprus* derives from *Crocuta* (Bon et al., 2012), and *Mammuthocoprus* derives from *Mammuthus columbi* (Karpinski et al., 2017). However, all coprolite ichnotaxa do have defined stratigraphic ranges that parallel the stratigraphic ranges of the producing animals, so the coprolites have potential utility in biostratigraphy and biochronology (Hunt et al., 2005a; Hunt et al., 2007, 2013a,b; Hunt and Lucas, 2021e, Figs. 25, 27, 28; Hunt and Lucas, 2023b). Indeed, coprolites have demonstrated biochronologic utility, particularly in the Late Triassic-Early Jurassic (Hunt et al., 2018, Fig. 12.11: Hunt and Lucas, 2021e, Figs. 25, 27, 29; Figs. 9.5, 9.7, 9.9).

9.4 Biogeography

Vertebrate coprolites can delimit biogeographic regions as proxies for biotaxa. For example, there are two distinct biogeographic and taphonomic provinces for vertebrate coprolites in the Pleistocene (Hunt and Lucas, 2020). The *Castrocopros* province of the New World is characterized by a dominance of herbivore coprolites, which are preserved almost exclusively in caves, and the *Hyaenacoprus* province in the Old World is dominated by hyena coprolites that occur mainly in caves (Hunt and Lucas, 2018c, 2019, 2020; Hunt and Lucas, 2021: Fig. 9.22). In the Late Triassic, northern Gondwana (North Africa, India) has a vertebrate

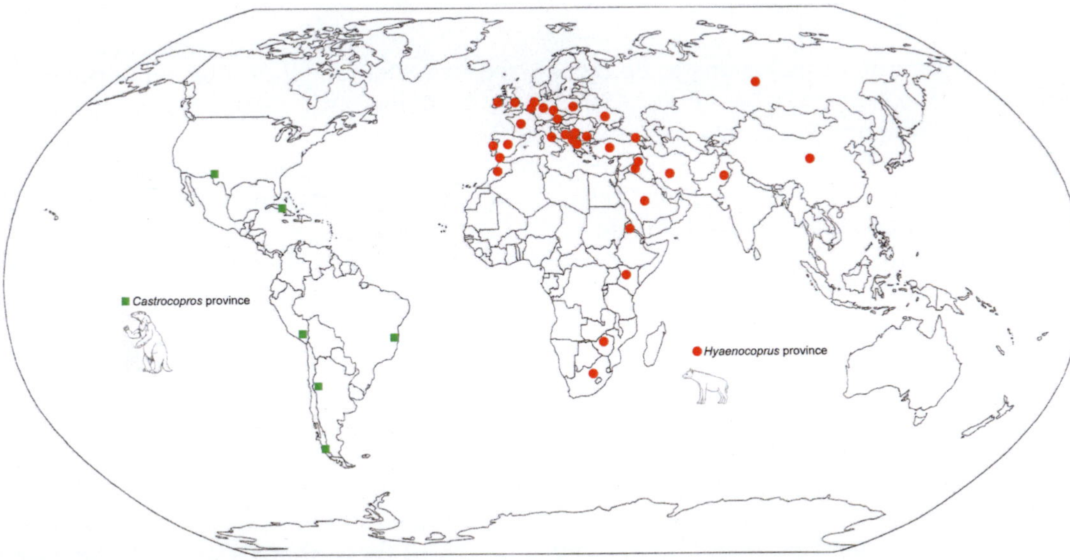

FIGURE 9.22 Late Pleistocene biogeographic provinces based on vertebrate coprolites. The *Castrocopros* province (green squares) characterized by *Castrocopros* spp. produced by ground sloths and the *Hyaenacoprus* province (red circles) by *Hyaenacoprus bucklandi* produced by hyenas. For details of localities see Hunt and Lucas (2020, Table 1 and Supplemental data file). ***Map lines delineate study areas and do not necessarily depict accepted national boundaries;*** *From Hunt and Lucas (2021e).*

fauna characterized by phytosaurs and metoposaurs that is similar to that of Laurussia. Coprolites as proxy for biotaxa can record the presence of a taxon in the absence of bones (e.g., dicynodonts in the Late Triassic: Hunt et al., 2013b).

Late Triassic (late Carnian) coprofaunas from Laurussia occur in the USA, Greenland, Germany, and Thailand (Hunt et al., 2018). *Alococopros triassicus* is abundant in North America. In Gondwana, coprofaunas occur in Argentina, South Africa, India, Madagascar, Brazil, and Morocco. *Alococopros triassicus* is absent from all these coprofaunas except for an ichnofauna from the Timezgadiouine Formation, northern Argana Basin, Morocco (Zouheir et al., 2023). The northern osteological faunas of the late Carnian of Gondwana (e.g., India) contain similar phytosaur-metoposaur associations as in Laurussia and differ from those in South America, which are dominated by dicynodonts and cynodonts (Zouheir et al., 2023). There are late Carnian coprolites in the upper Santa Maria Supersequence of Brazil (Francischini et al., 2018) and the Ischigualasto Formation of Argentina (Holocher et al., 2005). These coprolites are relatively uncommon but are apparently distinct from those in the Moroccan coprofaunas in lacking *Alococopros*, *Dicynodontocopros*, and spiral forms. Because coprolites are proxies for the taxa that produced them, this suggests a closer similarity between the vertebrate faunas of Morocco and Laurussia, particularly North America, than is the case for India, Madagascar, Thailand, and South America (Zouheir et al., 2023). This may relate to the timing of rifting between North Africa and North America and a relatively late paleogeographic connection between the two areas (Zouheir et al., 2023).

9.5 Lagerstätten

The first important inclusions to be recognized in vertebrate coprolites were pollen in herbivore coprolites (Martin et al., 1961; Hunt and Fiacconi, 2018). Subsequently, a wide range of organic materials, including soft tissues, have been identified in vertebrate coprolites, particularly muscle tissues, eggs, hair, parasites, bacteria, fungi, and feathers (e.g., Chin et al., 1998, 2003; Cardia et al., 2019). Recent studies have also highlighted that coprolites contain diverse body fossils of delicate and rare organisms (Dentzien-Dias et al., 2013, 2017; Bajdek et al., 2016; Chin et al., 2017; Qvarnström et al., 2016, 2017, 2019, 2021). Thus, Qvarnström et al. (2016) concluded that the term Lagerstätten is applicable to some coprolites.

9.6 Diet

Coprolites can provide diverse information about the diet of their producers (e.g., Hansen, 1978; McAllister, 1988; Qvarnström et al., 2019a,b). Qvarnström et al. (2019a) identified abundant beetle and arthropod inclusions in coprolites attributed to the dinosauriform *Silesaurus opolensis* and hypothesized that it specialized in a diet of these animals using its beak-like jaws to peck small insects off the ground.

9.7 Digestion

Coprolites can provide direct evidence of digestive processes. For example, Chin et al. (1998) argued that the distinct fragmentation and angularity of bones within a putative coprolite of *Tyrannosaurus rex* reflected extensive biting rather than gizzard-induced breakage. Subsequently, Chin et al. (2003) identified the fossilized remains of undigested muscle tissue in another tyrannosaurid coprolite that indicated a relatively short gut-residence time.

9.8 Parasites

Vertebrate coprolites provide important information on the presence and taxonomy of parasites whose body fossils are rare. Remains of endoparasites extend as far back as the Carboniferous. Paleoparasitological research on coprolites has largely focused on archeological specimens (De Baets et al., 2021). However, older coprolites provide evidence of parasites including from the Pleistocene (e.g., Jouy-Avantin et al., 1999; Schmidt et al., 1992), Eocene (Robin et al., 2016), Cretaceous (Poinar and Boucot, 2006; Poinar, 2021), Triassic (Da Silva et al., 2014; Francischini et al.,2018), and Permian (Dentzien-Dias et al., 2013).

9.9 Trophic relationships

Coprolites have long been recognized as having utility in the reconstruction of ancient trophic relationships (e.g., Hunt et al., 1994; Thulborn, 1991; Smith and Both-Brink, 2011; Zaton et al., 2015; Bajdek et al., 2016; Qvarnström et al., 2019b; Qvarnström, 2020a,b; Dentzien-Dias et al., 2021).

9.10 Health

The analysis of human coprolites can address health issues. For example, Bryant and Reinhard (2012) noted that the analysis of groups of coprolites from specific regions of the American Southwest reveals ancient diets that were high in fiber, rich in food diversity, and relied on calories from plants containing inulin, not starch. Those data are now being used to gain a better understanding of why certain Native American ethnic groups suffer high rates of obesity and type II diabetes when eating traditional Western diets rather than the types of ancient diets eaten by their ancestors (Bryant and Reinhard, 2012).

9.11 Faunal evolution and extinction

Coprolites can also be used as biotaxa proxies to analyze ecosystem evolution, extinction, and diversity patterns (e.g., Martin et al., 1985; Nakajima and Izumi, 2014; Niedźwiedzki et al., 2016a,b; Hunt and Lucas, 2021e; Fig. 9.23).

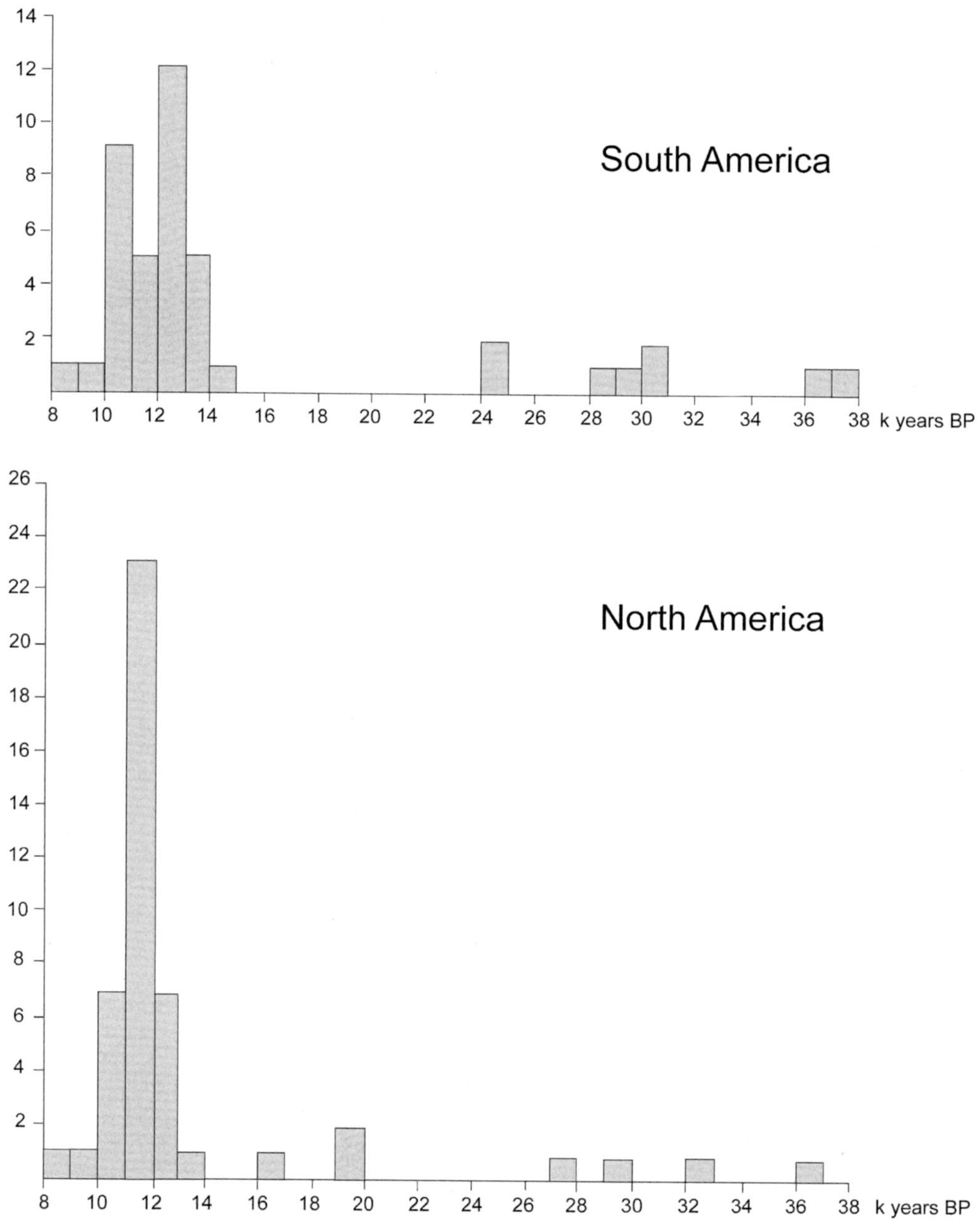

FIGURE 9.23 Age distribution of radiometric ages on ground sloth coprolites from caves in North and South America. This data demonstrates increased cave utilization during the latest Pleistocene, which was likely climate driven. There is virtually no evidence of humans hunting sloths, which would be another explanation for this behavioral shift. Thus, the giant sloth demise is more likely related to climatic factors rather than human overkill, which are the two most prevalent hypotheses for their extinction. *From Hunt and Lucas (2018c).*

10. Conclusions

1. Vertebrate coprolites are second only to tracks in the volume of research on vertebrate trace fossils. Coprolites have been studied for almost 200 years, and there is a robust terminology for their description.
2. The fossil record of vertebrate coprolites is influenced by megabiases that deserve further study, including a preponderance produced in different environments by carnivores, aquatic or semiaquatic organisms and sharks, nonrandom distribution through time and distinct size selection.
3. The first known vertebrate coprolites are from the Early or Middle Ordovician and they become common in the Devonian with the first large ichnofaunas in the Carboniferous. There are notable acmes of abundance and diversity in the Triassic, Cretaceous, and Pleistocene.
4. Hominid coprolites have been described exclusively in the archeological literature, and the known fossil record of human coprolites and its study is strongly biased by a number of factors including more study where human skeletal remains are uncommon, in arid environments, and elevated levels of study from 1960 to 1990; much early work outside North America has focused on cesspits.
5. Ichnotaxobases for coprolites include external morphology, internal structure, surface texture, size, inclusions, and mineralogy.
6. Bian et al. (1991) and Hunt et al. (1998) named the first ichnotaxa of vertebrate coprolites. There are currently 61 named ichnotaxa of coprolites, of which we consider 56 valid.
7. Coprolite ichnotaxa do not represent biotaxa genera and species but rather higher taxonomic groups as is the case with tracks.
8. There is need for a rigorous ichnotaxonomy of vertebrate coprolites. The use of ichnotaxonomy in regard to coprolites has increased significantly in the past few years.
9. Coprolites have diverse utility in the study of ichnofacies, biostratigraphy and biochronology, biogeography, Lagerstätten, diet, digestion, parasites, trophic relations, human health, faunal evolution, and extinction.
10. Internally, coprolites can be Lagerstätten that preserve a wide range of organisms with an otherwise poor fossil record.

Glossary of terms

Accretionary latrinite Latrinite that results from accumulation due to physical, rather than biological, processes (Hunt and Lucas, 2012a)

Amphipolar Spiral coprolite that preserves coils that extend for the majority of the length of the coprolite (>75%) in lateral view (Neumayer, 1904; Hunt and Lucas, 2012b: Fig. A.1)

Anisopolar Non-spiral coprolite with terminations of different shapes (Thulborn, 1991)

Anterior coil The anterior end of a spiral coprolite consisting of a single, antero-posteriorly elongate coil (Hunt and Lucas, 2012b: Fig. A.1)

Chiropteraguanolite Guanolite produced by bats (Hunt and Lucas, 2012a)

Coil The external expression of the spiral, conular elements of a spiral coprolite (Hunt and Lucas, 2012b: Fig. A.1)

Coprolite Fossil fecal material that has been ejected from the posterior end of the gastrointestinal tract (Buckland, 1829; Hunt and Lucas, 2012a)

Ethological latrinite Latrinite that results from behavior of an organism (Hunt and Lucas, 2012a)
Excorporeal pelletite Pelletite preserved outside body cavity, coprolite (Hunt and Lucas, 2012a)
Guanolite Fossil guano deposit (Hunt and Lucas, 2012a)
Heteropolar Spiral coprolite with the coils concentrated at one end in lateral view (Neumayer, 1904: Fig. A.1)
Incorporeal pelletite Pelletite preserved within the body cavity (Hunt and Lucas, 2012a)
Isopolar Non-spiral coprolite with terminations of the same shape (Thulborn, 1991)
Latrinite An accumulation of coprolites (Hunt and Lucas, 2012a)
Lip An exposed edge of a flap on the anterior coil of a heteropolar coprolite (Jain, 1983: Fig. A.1)
Macrospiral Heteropolar coprolite in which the posterior spire constitutes 50%–75% of the length of the coprolite in lateral view (Hunt and Lucas, 2012b)
Microspiral Heteropolar coprolite in which the posterior spire constitutes less than 50% of length
Mucro Anterior end of an anisopolar coprolite that is tapered to a point (Thulborn, 1991: Fig. A.2)
Neotomalite Fossil packrat midden (Hunt and Lucas, 2012a)
Ornithoguanolite Guanolite produced by birds (Hunt and Lucas, 2012a)
Pelletite Fossil fecal pellets preserved inside (incorporeal pelletite) or outside (excorporeal pelletite) the body cavity (Hunt and Lucas, 2012a)
Posterior spire The posterior segment of heteropolar and amphipolar coprolites that consists of a number of closely spaced coils (Hunt and Lucas, 2012b: Fig. A.1)
Scroll Spiral coprolite with the structure of a rolled sheet of paper that consists essentially of a single coil (Hunt and Lucas, 2012b: Fig. A.1)
Segments Discrete longitudinal elements of a non-spiral coprolite (Hunt and Lucas, 2012b: Fig. A.2)

Acknowledgments

We thank many colleagues who have graciously allowed us access to collections, including Kim Besom, Michael Brett-Surman, Sandra Chapman, David and Janet Gillette, Colleen Hyde, Paul Jeffery, Thomas Lehmann, Jordan Mallon, Timothy Rowe, Andrey Sennikov, Krister Smith, Alan Tennyson and Anika Vogel, and others for helpful discussions, including Paula Dentzien-Dias, Chris Duffin, and Jesper Milàn. Vaughn M. Bryant, Jr., Chris Duffin, Nibedita Rakshit, Karl J. Reinhard, Paul Rummy, and Vince Santucci kindly supplied images, and Paul Rummy and Hendrik Klein provided insightful reviews.

Declaration of competing interest

The authors declare that they have no known competing financial interests or personal relationships that could have appeared to influence the work reported in this paper.

Data availability

No data were used for the research described in this article.

References

Abel, O., 1926. Amerikafahrt: Beobachtungen und Studeien eines Naturforschers auf einer Reise nach Nordamerika und Westindien. G. Fischer, Jena, p. 462.

Abel, O., 1935. Vorzeitliche Lebensspuren. Verlag von Gustav Fischer, Jena, p. 644.

Abella, J., Martín-Perea, D.M., Valenciano, A., Hontecillas, D., Montoya, P., Morales, J., 2021. Coprolites in natural traps: direct evidence of bone-eating carnivorans from the Late Miocene Batallones-3 site, Madrid, Spain. Lethaia. https://doi.org/10.1111/let.12438.

Aceñolaza, F.G., 2012. Neogene coprolites (Upper Miocene-Lower Pliocene) of Entre Ríos province, northeastern Argentina. Universidad Autónoma de Entre Ríos. Secretaría de Ciencia y Técnica. Scientia Interfluvius 3, 41–53.

Agustín, B., Julián, F.F., Ornela, B.M., 2021. Carnivore coprolites from "Gruta del Indio" site as source of paleoparasitological and paleoecological evidences (late Pleistocene-Holocene, Mendoza, Argentina). Archaeol. Anthropol. Sci. 13, 1–14. https://doi.org/10.1007/s12520-021-01272-w.

Alcalá, L., Espílez, E., Mampel, L., Kirkland, J.I., Ortiga, M., Rubio, D., González, A., Ayala, D., Cobos, A., Royo-Torres, R., Gascó, F., Pesquero, M.D., 2012. A new Lower Cretaceous vertebrate bonebed near Ariño (Teruel, Aragón, Spain); Found and managed in a joint collaboration between a mining company and a palaeontological park. Geoheritage. https://doi.org/10.1007/s12371-012-0068-y.

Aldridge, R.J., Gabbott, S.E., Siveter, L.J., Theron, J.N., 2006. Bromalites from the Soom Shale Lagerstätte (Upper Ordovician) of South Africa: palaeoecological and palaeobiological implications. Palaeontology 49, 857–871.

Allard, H., Carpenter, S.C., Duffin, C.J., Benton, M.J., 2015. Microvertebrates from the classic Rhaetian bone beds of Manor Farm quarry, near Aust (Bristol, UK). Proc. Geologists' Assoc. 126 (6), 762–776.

Anagnostakis, S., 2013. Upper Cretaceous Coprolites From the Münster Basin (Northwestern Germany) – a Glimpse into the Diet of Extinct Animals. M.S. thesis. Lund University, Lund, p. 28.

Anderson, J.M., Anderson, H.M., Cruickshank, A.R.I., 1998. Late Triassic ecosystems of the Molteno lower Elliot biome of southern Africa. Palaeontology 41, 387–421.

Anderson, L.I., Dunlop, J.A., Horrocks, C.A., Winkelmann, H.M., Eagar, R.M.C., 1997. Exceptionally preserved fossils from Bickershaw, Lancashire UK (upper Carboniferous, Westphalian A (Langsettian). Geol. J. 3, 197–210.

Antunes, M.T., Balbino, A.C., Ginsburg, L., 2006a. Ichnological evidence of a Miocene rhinoceros bitten by a bear-dog (*Amphicyon giganteus*). Ann. Paleontol. 92, 31–39.

Antunes, M.T., Balbino, A.C., Ginsburg, L., 2006b. Miocene Mammalian footprints in coprolites from Lisbon, Portugal. Ann. Paleontol. 92, 13–30.

Argant, J., Bonifay, M.F., 2011. Hyaena coprolites (*Pachycrocuta brevirostris*) from the Ceyssaguet Villafranchian site (Lavoûte-sur-Loire, Haute-Loire, France): pollen analyses and palaeoenvironmental indications. Quaternaire 22, 3–11.

Arribas, A., Garrido, G., Viseras, C., Soria, J.M., Pla, S., Solano, J.G., Garcés, M., Beamud, E., Carrión, J.S., 2009. A mammalian lost world in southwest Europe during the Late Pliocene. PLoS One 4 (9), e7127. https://doi.org/10.1371/journal.pone.0007127.

Averianov, A., Dyke, G., Danilov, I., Skutschas, P., 2015. The paleoenvironments of azhdarchid pterosaurs localities in the Late Cretaceous of Kazakhstan. ZooKeys 483, 59–80.

Baghai-Riding, N.L., DiBenedetto, J.N., 2001. An unusual dinosaur coprolite from the Campanian Aguja Formation, Texas. Gulf Coast Association of Geological Societies Transactions 51, 9–20.

Bajdek, P., 2013. Coprolite of a durophagous carnivore from the Upper Cretaceous Godula Beds, Outer Western Carpathians, Poland. Geol. Q. 57, 361–364.

Bajdek, P., Bienkowska-Wasiluk, M., 2020. Deep-sea ecosystem revealed by teleost fish coprolites from the Oligocene of Poland. Palaeogeogr. Palaeoclimatol. Palaeoecol. 540. https://doi.org/10.1016/j.palaeo.2019.109546.

Bajdek, P., Owocki, K., Niedźwiedzki, G., 2014. Putative dicynodont coprolites from the Upper Triassic of Poland. Palaeogeogr. Palaeoclimatol. Palaeoecol. 411, 1–17.

Bajdek, P., Szczygielski, T., Kapuścińska, A., Sulej, T., 2019. Bromalites from a turtle-dominated fossil assemblage from the Triassic of Poland. Palaeogeogr. Palaeoclimatol. Palaeoecol. 520, 214–228.

Bajdek, P., Owocki, K., Sennikov, A.G., Golubev, V.K., Niedźwiedzki, G., 2017. Residues from the Upper Permian carnivore coprolites from Vyazniki in Russia-key questions in reconstruction of feeding habits. Palaeogeogr. Palaeoclimatol. Palaeoecol. 482, 70–82.

Bajdek, P., Qvarnström, M., Owocki, K., Sulej, T., Sennikov, A.G., Golubev, V.K., Niedźwiedzki, G., 2016. Microbiota and food residues including possible evidence of pre-mammalian hair in Upper Permian coprolites from Russia. Lethaia 49, 455–477.

Baranov, V.A., Engel, M.S., Hammel, J., Hörnig, M.K., van de Kamp, T., Zuber, M., Haug, J.T., 2021. Synchrotron-radiation computed tomography uncovers ecosystem functions of fly larvae in an Eocene forest. Palaeontol. Electron. 24 (1), a07. https://doi.org/10.26879/1129.

Barbacka, M., Górecki, A., Pacyna, G., Pieńkowski, G., Philippe, M., Bóka, K., Ziaja, J., Jarzynka, A., Qvarnström, M., Niedźwiedzki, G., 2022. Early Jurassic coprolites: insights into palaeobotany and the feeding behaviour of dinosaurs. Papers Palaeontol. 8 (2), e1425. https://doi.org/10.1002/spp2.1425.

Bardack, D., 1992. Late Pennsylvanian paleonisciform fish from the Kinney Quarry, New Mexico: New Mexico Bureau Mines and Mineral Resources. Bulletin 138, 197–203.

Barrios-de Pedro, S., Chin, K., Buscalioni, Á.D., 2020a. The late Barremian ecosystem of Las Hoyas sustained by fishes and shrimps as inferred from coprofabrics. Cretac. Res. 110, 104409. https://doi.org/10.1016/j.cretres.2020.104409.

Barrios-de Pedro, S., Poyato-Ariza, F.J., Moratalla, J.J., Buscalioni, Á.D., 2018. Exceptional coprolite association from the Early Cretaceous continental Lagerstätte of Las Hoyas, Cuenca, Spain. PLoS One 13 (5), e0196982. https://doi.org/10.1371/journal.pone.0196982.

Barrios-de Pedro, S., Rogers, K.M., Alcorlo, P., Buscalioni, Á.D., 2020c. Food web reconstruction through isotopic compositions of fossil faeces: insights into the ecology of a late Barremian freshwater ecosystem (Las Hoyas, Cuenca, Spain). Cretac. Res. 108, 104343. https://doi.org/10.1016/j.cretres.2019.104343.

Bartels, C., Briggs, D.E.G., Brassel, G., 1998. The fossils of the Hunsrück slate. In: Marine Life in the Devonian. Cambridge University Press, Cambridge, p. 309.

Barthel, K.W., 1978. Solnhofen: Ein Blick in der Erdgeschichte. Thun, Ott Verlag 393.

Barthel, K.W., Swinburne, N.H.M., Conway Morris, S., 1990. Solnhofen: A Study in Mesozoic Palaeontology. Cambridge University Press, Cambridge, p. 236.

Battillo, J., 2019. Farmers who forage: interpreting paleofecal evidence of wild resource use by early corn farmers in the north American Southwest. Archaeol. Anthropol. nce 11, 5999–6016. https://doi.org/10.1007/s12520-019-00944-y.

Battista, F., Ribeiro, A.M., Melo, T.P., Schultz, C.L., 2024. A reinterpretation of taphonomy and palaeoecology of the early Carnian *Santacruzodon* assemblage zone type-locality (Santa Maria Supersequence), Brazil. Palaeogeogr. Palaeoclimatol. Palaeoecol., 112080 https://doi.org/10.1016/j.palaeo.2024.112080.

Bayer, B., 1934. Koprolity z kounovských vrestev na Plzensku. Príroda Brno 27, 21–22.

Beattie, R.G., Avery, S., 2012. Palaeoecology and palaeoenvironment of the Jurassic Talbragar fossil fish bed, Gulgong, New South Wales, Australia. Alcheringa 36, 453–468.

Beck, C.W., Bryant, V.M., McDonough, K.N., 2019. Evidence for non-random distribution of pollen in human coprolites. Archaeol. Anthropol. Sci. 11, 5983–5998. https://doi.org/10.1007/s12520-019-00839-y.

Bell, J., Holden, J., Pettigrew, T.H., Sedman, K.W., 1979. The Marl Slate and basal Permian breccia at Middridge, Co. Durham. Proc. Yorks. Geol. Soc. 42, 439–460.

Bendella, M., Hunt, A.P., Benyoucef, M., Bouchemla, I., 2023. The first Mesozoic vertebrate coprolites from Algeria. Ichnos 30 (1), 27–38. https://doi.org/10.1080/10420940.2023.2187387.

Bender, P.A., Hancox, P.J., 2003. Fossil fishes of the *Lystrosaurus* and *Cynognathus* assemblage zones, Beaufort Group, South Africa: correlative implications. Council Geosci. Bull. 136, 1–27.

Bender, P.A., Hancox, P.J., 2004. Newly discovered fish faunas from the Early Triassic, Karoo Basin, South Africa, and their correlative implications. Gondwana Res. 7, 185–192.

Benz, S., 1980. The Stratigraphy and Paleoenvironment of the Triassic Moenkopi Formation at Radar Mesa, Arizona. M.S. thesis. Northern Arizona University, Flagstaff, p. 43.

Berger, L.R., Pickering, R., Kuhn, B., Backwell, L., Hancox, P.J., Kramers, J.D., Boshoff, P., 2009. A Mid-Pleistocene in situ fossil brown hyaena (*Parahyaena brunnea*) latrine from Gladysvale Cave, South Africa. Palaeogeogr. Palaeoclimatol. Palaeoecol. 279 (3–4), 131–136.

Bertling, M., Braddy, S.J., Bromley, R.G., Demathieu, G.R., Genise, J., Mikuláš, R., Nielsen, J.K., Nielsen, K.S.S., Rindsberg, A.K., Schlirf, M., Uchman, A., 2006. Names for trace fossils: a uniform approach. Lethaia 39, 265–286.

Bertrand, C.E., 1903. Les coprolithes de Bernissart. I. Partie: Les coprolithes qui ont ete attribues aux Iguanodons, vol. 1. Mémoire du Musée Royal d'Histoire Naturelle de Belgique, Brussels, pp. 1–154.

Bian, Z.X., Wang, J.Y., Wang, Z.X., 1991. Yufenites in Lufeng, Yunnan, Tongzi, Guizhou, and fuxian, Shaanxi. J. Chengdu Coll. Geol. 18 (2), 20–24.

Bicknell, R.D., Smith, P.M., Kimmig, J., 2023. Novel coprolitic records from the Silurian (Přídolí) Wallace Shale of New South Wales. Alcheringa Austr. J. Palaeontol. 47 (1), 24–30.

Bigurrarena Ojeda, M.A., Gouiric-Cavalli, S., Pérez, L.M., Reguero, M., 2017. Preliminary Study of the Coprolites From the Ameghino Formation (Late Jurassic) of the Antarctic Peninsula. IX Congreso Latinoamericano de Ciencia Antártica. Visiones sobre Ciencia Antártica, Punta Arenas, pp. 90–93.

Bigurrarena Ojeda, M.A., Gouiric-Cavalli, S., Pérez, L.M., Marcelo, A., Reguero, M.A., 2023. Bromalites from the Ameghino (= Nordenskjöld) formation Upper Jurassic of Antarctic Peninsula. Publicación Electrónica de la Asociación Paleontológica Argentina 23 (2), 45–64. https://doi.org/10.5710/PEAPA.11.04.2023.446.

Bingham-Koslowski, N., Grey, M., Ehrman, J.M., Pufahl, P., 2021. Investigating the relationship between the elemental composition of coprolites and trophic level: a brief digest of Late Carboniferous coprolites from the Joggins Formation, Nova Scotia, Canada. Atl. Geol. 57, 107–108.

Bishop, W.W., 1964. More fossil primates and other Miocene mammals from north-east Uganda. Nature 203, 1327–1331.

Blong, J.C., Adams, M.E., Sanchez, G., Jenkins, D.L., Bull, I.D., Shillito, L.M., 2020. Younger Dryas and early Holocene subsistence in the northern Great Basin: multiproxy analysis of coprolites from the Paisley caves, Oregon, USA. Archaeol.Anthropol. Sci. 12, 224. https://doi.org/10.1007/s12520-020-01160-9.

Boaz, N.T., Ciochon, R.L., Xu, Q., Liu, J., 2000. Large mammalian carnivores as a taphonomic factor in the bone accumulation at Zhoukoudian. Acta Anthropol. Sin. 19 (Suppl. ment), 224–234.

Bon, C., Berthonaud, V., Maksud, F., Labadie, K., Poulain, J., Artiguenave, F., Wincker, P., Aury, J.M., Elalouf, J.M., 2012. Coprolites as a source of information on the genome and diet of the Cave Hyena. Proc. Biol. Sci. 279, 2825–2830.

Borry, M., Cordova, B., Perri, A., Wibowo, M., Honap, T.P., Ko, J., Yu, J., Britton, K., Girdland-Flink, L., Power, R.C., Stuijts, I., Salazar-García, D.C., Hofman, C., Hagan, R., Samdapawindé Kagoné, T., Meda, N., Carabin, H., Jacobson, D., Reinhard, K., Lewis, C., Kostic, A., Jeong, C., Herbig, A., Hübner, A., Warinner, C., 2020. CoproID predicts the source of coprolites and paleofeces using microbiome composition and host DNA content. PeerJ 8, e9001. https://doi.org/10.7717/peerj.9001.

Bouchet, F., 1995. Recovery of helminth eggs from archeological excavations of the Grand Louvre (Paris, France). J. Parasitol. 81, 785–787.

Bouchet, F., Harter, S., Le Bailly, M., 2003. The State of the Art of Paleoparasitological Research in the Old World, vol. 98. Memórias do Instituto Oswaldo Cruz, pp. 95–101.

Bowen, J.J., 2013. The Neoichnology of Juliform Millipedes and Upper Monongahela to Lower Dunkard Group Paleosols: A Multi-Proxy Approach to Paleolandscape Variability. M.S. thesis. Ohio University, Athens, p. 305.

Boy, J.A., Schoch, R.R., Lucas, S.G., 2001. The Moenkopi Formation in east-central New Mexico: stratigraphy and vertebrate fauna. New Mex. Geol. Soc. Guidebook 52, 103–108.

Brachaniec, T., Niedźwiedzki, R., Surmik, D., Krzykawski, T., Szopa, K., Gorzelak, P., Salamon, M.A., 2015. Coprolites of marine vertebrate predators from the Lower Triassic of southern Poland. Palaeogeogr. Palaeoclimatol. Palaeoecol. 435, 118–126.

Brachaniec, T., Środek, D., Surmik, D., Niedźwiedzki, R., Georgalis, G.L., Płachno, B.J., Duda, P., Lukeneder, A., Gorzelak, P., Salamon, M.A., 2022. Comparative actualistic study hints at origins of alleged Miocene coprolites of Poland. PeerJ 10, e13652. https://doi.org/10.7717/peerj.13652.

Bradley, W.H., 1946. Coprolites from the Bridger Formation of Wyoming: their composition and microorganisms. Am. J. Sci. 244, 215–239.

Brand, M.J., 1994. Prehistoric Anasazi Diet: A Synthesis of Archaeological Evidence. M.A. thesis. The University of British Columbia, Vancouver, p. 113.

Branson, E.B., 1914. The Devonian fishes of Missouri. The University of Missouri Bulletin. Science 2, 59–74.

Briggs, D.E.G., Liu, H.P., Mckay, R.M., Witzke, B.J., 2015. Bivalved arthropods from the Middle Ordovician Winneshiek Lagerstätte, Iowa, USA. J. Paleontol. 89, 991–1006.

Brown, R., Lyell, C., 1845. On the geology of Cape Breton. Quart. J. Geol. Soc. 1, 207–213.

Bryant, V.M., 1969. Late Full-Glacial, and Post-glacial Pollen Analysis of Texas Sediments. Ph.D. dissertation. Austin, University of Texas, Austin, p. 148.

Bryant, V.M., 1974b. The role of coprolite analysis in archeology. Bull. Tex. Archeol. Soc. 74, 1–28.

Bryant, V.M., 1974c. Pollen analysis of prehistoric human feces from Mammoth Cave. In: Watson, P.J. (Ed.), Archaeology of the Mammoth Cave Area. New York, Academic Press, New York, pp. 203–249.

Bryant, V.M., 1975. Pollen as an indicator of prehistoric diets in Coahuila, Mexico. Bull. Tex. Archeol. Soc. 46, 87–106.

Bryant Jr., V.M., Reinhard, K.J., 2012. Coprolites and archaeology: the missing links in understanding human health. New Mex. Mus. Nat. Hist. Sci. Bull. 57, 379–387.

Buckland, W., 1822. Account of an Assemblage of Fossil Teeth and Bones of Elephant, Rhinoceros, hippopotamus, Bear, Tiger and Hyaena, and Sixteen Other Animals; Discovered in a Cave at Kirkdale, Yorkshire, in the Year 1821; With a Comparative View of FIve Similar Caverns in Various Parts of England, and Others on the Continent, vol. 112. Philosophical Transactions of the Royal Society of London, pp. 171–236.

Buckland, W., 1824. Reliquiae Diluvianae; or, Observations on the Organic Remains Contained in Caves, FIssures, and Diluvial Gravel, and on Other Geological Phenomena, Attesting the Action of an Universal Deluge. J. Murray, London, p. 303.

Buckland, W., 1827. Observations on the bones of hyaenas and other animals on the cavern of Lunel near Montpelier, and in the adjacent strata of marine formation. Proc. Geol. Soc. Lond. 1, 3–6.

Buckland, W., 1829. A Paper by Dr. Buckland, vol. 1. Proceedings of the Geological Society of London, pp. 142–143.

Buckland, W., 1835. On the discovery of coprolites, or fossil faeces, in the Lias at Lyme Regis, and in other formations. Trans. Geol. Soc. London 3 (series 2), 223–238.

Buckland, W., 1836. Geology and mineralogy considered with reference to natural theology. The Bridgewater Treatises on the Power. Wisdom Goodness God Manif. Creat. London 599, 128.

Buckland, W., 1858. Geology and Mineralogy Considered With Reference to Natural Theology. The Bridgewater Treatises on the Power, Wisdom and Goodness of God as Manifested in the Creation, third ed. Treatise VI. London, Pickering, 2 volumes. 552 p. and 143 p.

Bull, I.D., Betancourt, P.P., Evershed, R.P., 1999. Chemical evidence supporting the existence of a structured agricultural manuring regime on Pseira Island, Crete during the Minoan Age. Aegaeum 20, 69–75.

Burmeister, K.C., Flynn, J.J., Parrish, J.M., Wyss, A.R., 2006. Paleogeographic and biostratigraphic implications of new early Mesozoic vertebrates from Poamay, central Morondava basin, Madagascar. New Mex. Mus. Nat. Hist. Sci. Bull. 37, 457–475.

Burney, D.A., Robinson, G.S., Burney, L.P., 2003. *Sporormiella* and the late Holocene extinctions in Madagascar. Proc. Natl. Acad. Sci. U.S.A. 100 (19), 10800–10805.

Burnham, D.A., 2008. A review of the Early Cretaceous Jehol Group on northeastern China and a revision of concerning the origin of flight paradigm. Oryctos 7, 27–43.

Byrne, H.M., Niedźwiedzki, G., Blom, H., Kear, B.P., Ahlberg, P.E., 2022. Coprolite diversity reveals a cryptic ecosystem in an early Tournaisian lake in East Greenland: implications for ecosystem recovery after the end-Devonian extinction. Palaeogeogr. Palaeoclimatol. Palaeoecol. 605, 111215. https://doi.org/10.1016/j.palaeo.2022.111215.

Callen, E.O., 1963. Diet as revealed by coprolites. In: Brothwell, D., Higgs, E. (Eds.), Science in Archeology. Basic Books, London, pp. 186–194.

Callen, E.O., 1965. Food habits of some pre-Columbian Indians. Econ. Bot. 19, 335–343.

Callen, E.O., 1967a. Analysis of the Tehuacan coprolites. In: Byers, D. (Ed.), The Prehistory of the Tehuacan Valley: Environment, and Subsistence, vol.1. University of Texas Press, Austin, pp. 261–289.

Callen, E.O., 1967b. The first new world cereal. Am. Antiq. 32, 535–538.

Callen, E.O. Plants, diet, and early agriculture of some cave dwelling pre-Columbian Mexican Indians.1968 Actas y Memorias del 37 Congreso Internacional de Americanistas, 2, 641-656.

Callen, E.O., 1969. Les coprolithes de la cabane acheuleenne du Lazaret: Analyse et diagnostic. Memoires de la Societe Prehistorique Francaise 7, 123–124.

Callen, E.O., Cameron, T.W.M., 1955. The diet and parasites of prehistoric Huaca Prieta Indians as determined by dried coprolites. Proc. Roy. Soc. Can. 5, 51–52.

Callen, E.O., Cameron, T.W.M., 1960. A prehistoric diet revealed in coprolites. New Sci. 8 (190), 35–40.

Camacho, M., Reinhard, K.J., 2020. Pinworm research in the Southwest USA: five decades of methodological and theoretical development and the epidemiological approach. Archaeol. Anthropol. Sci. 12 (2). https://doi.org/10.1007/s12520-019-00994-2.

Canfield, P.J., Fairburn, A.J., 1983. Effect of various diets on faecal analysis in normal dogs. Res. Vet. Sci. 34, 24–27.

Cano, R.J., Rivera-Perez, J., Toranzos, G.A., Santiago-Rodriguez, T.M., Narganes-Storde, Y., Chanlatte-Baik, L., Garcia-Roldan, E., Bunkley-Williams, L., Massey, S.E., 2014. Paleomicrobiology: revealing fecal microbiomes of ancient indigenous cultures. PLoS One 9 (9), e106833. https://doi.org/10.1371/journal.pone.0106833.

Cantrell, A.K., Suazo, T.L., Spielmann, J.A., Lucas, S.G., 2012. Vertebrate coprolites from the lower Permian (lower Wolfcampian) Gallina well locality, Joyita Hills, Socorro County, New Mexico. New Mex. Mus. Nat. Hist. Sci. Bull. 57, 197–201.

Capasso, L., 2019. Palaentological evidence of piscivorous habits of some pycnodonts from the Middle Cenomanian of Lebanon. Thalass. Salentina 41, 91–110.

Cardia, D.F., Bertini, R.J., Camossi, L.G., Letizio, L.A., 2019. First record of *Acanthocephala* parasites eggs in coprolites preliminary assigned to Crocodyliformes from the Adamantina Formation (Bauru Group, upper Cretaceous), Sao

Paulo, Brazil. Anais da Academia Brasileira de Ciências. Earth Sci. 91 (Suppl. 2). https://doi.org/10.1590/0001-3765201920170848.

Cardia, D.F.F., Bertini, R.J., Camossi, L.G., Richini-Pereira, V.B., Losnak, D.O., Francischini, H., Dentzien-Dias, P., 2021. Paleoparasitological analysis of a coprolite assigned to a carnivoran mammal from the Upper Pleistocene Touro Passo formation, Rio Grande do Sul, Brazil. An Acad. Bras Ciências 93 (Suppl. 2), 1–13. https://doi.org/10.1590/0001-3765202120190876 e20190876.

Carlini, S.C., Vizcaíno, S.F., Scillato-Yané, G.J., 1997. Armored xenathrans: a unique taxonomic and taphonomic assemblage. In: Kay, R.F., Madden, R.D., Cifelli, R.L., Flynn, J.J. (Eds.), Vertebrate Paleontology in the Neotropics: The Miocene Fauna of La Venta, Colombia. Smithsonian Institution Press, Washington, pp. 213–226.

Casier, E., 1960. Les Iguanodons de Bernissart. Institute Royal d'Histoire Naturelle de Belgique, Brussels, p. 134.

Casier, E., 1978. Les Iguanodons de Bernissart. Institute Royal d'Histoire Naturelle de Belgique, Brussels, p. 166.

Caster, K.E., Kjellesyig-Waering, E.N., 1964. Upper Ordovician eurypterids from Ohio. Palaeontogr. Am. 4, 297–358.

Castro, A.C.J., Fernandes, A.C.S., Carvalho, I.S., 1988. Coprólitos de aves da Bacia de Taubaté, SP. Anais do Congresso Brasileiro do Geologia. Belém 35, 2358–2370.

Cavicchini, I., Heyworth, H.C., Duffin, C.J., Hildebrandt, C., Benton, M.J., 2018. A Rhaetian microvertebrate fauna from Stowey quarry, Somerset, UK. Proc. Geologists' Assoc. 129 (2), 144–158.

Cenizo, M., Soibelzon, E., Magnussen Saffer, M., 2016. Mammalian predator–prey relationships and reoccupation of burrows in the Pliocene of the Pampean Region (Argentina): new ichnological and taphonomic evidence. Hist. Biol. 28, 1026–1040.

Chame, M., 2003. Terrestrial mammal feces: a morphometric summary and description. Memórias do Instituto Oswaldo Cruz 98 (Suppl. 1), 71–94.

Chame, M., Sianto, L., 2014. Coprolite diagnosis, or who made the coprolite? In: Ferreira, L.F., Reinhard, K.J., Araújo, A. (Eds.), Foundations of Paleoparasitology. Rio de Janeiro, Editora Fiocruz, pp. 255–272.

Chen, M., Chen, Y., 1980. Note on a possible coprolites from the lowest Cambrian strata of Yangtze Gorge. Sci. Geol. Sin. 10, 406–407.

Chin, K., 1996. The Paleobiological Implications of Herbivorous Dinosaur Coprolites: Ichnologic, Petrographic, and Organic Geochemical Investigations. Ph.d. dissertation. University of California, Santa Barbara, p. 162.

Chin, K., 1997. What did dinosaurs eat? Coprolites and other direct evidence of dinosaur diets. In: Farlow, J.O., Brett-Surman, M.K. (Eds.), The Complete Dinosaur. Indiana University Press, Bloomington, pp. 371–382.

Chin, K., 2007. The paleobiological implications of herbivorous dinosaur coprolites from the Upper Cretaceous Two Medicine Formation of Montana: why eat wood? Palaios 22, 554–566.

Chin, K., Bishop, J., 2004. Traces within Traces: Evidence for Coprophagy in a Probable Theropod Coprolite From the Jurassic Morrison Formation of Utah, USA, vol. 26. First International Congress on Ichnology.

Chin, K., Bishop, J.R., 2007. Exploited Twice: Bored Bone in a Theropod Coprolite From the Jurassic Morrison Formation of Utah, vol. 88. U.S.A. SEPM Special Publication, pp. 379–387.

Chin, K., Gill, B.D., 1996. Dinosaurs, dung beetles, and conifers: participants in a Cretaceous food web. Palaios 11, 280–285.

Chin, K., Kirkland, J.I., 1998. Probable herbivore coprolites from the Upper Jurassic Mygatt-Moore Quarry, western Colorado. Mod. Geol. 23, 249–275.

Chin, K., Feldmann, R.M., Tashman, J.N., 2017. Consumption of crustaceans by megaherbivorous dinosaurs: dietary flexibility and dinosaur life history strategies. Sci. Rep. 7, 11163. https://doi.org/10.1038/s41598-017-11538-w.

Chin, K., Hartman, J.H., Roth, B., 2009. Opportunistic exploitation of dinosaur dung: fossil snails in coprolites from the Upper Cretaceous Two Medicine Formation of Montana. Lethaia 42, 185–198.

Chin, K., Tokaryk, T.T., Erickson, G.M., Calk, L.C., 1998. A king sized theropod coprolite. Nature 393, 680–682.

Chin, K., Eberth, D.A., Schweitzer, M.H., Rando, T.A., Sloboda, W.J., Horne, J.R., 2003. Remarkable preservation of undigested muscle tissue within a Late Cretaceous tyrannosaurid coprolite from Alberta, Canada. Palaios 18, 286–294.

Chipman, M., Grey, M., Pufahl, P.K., 2020. Coprolites from a brackish ecosystem in the Pennsylvanian Joggins Formation, Nova Scotia, Canada and their palaeoecological implications. Palaeogeogr. Palaeoclimatol. Palaeoecol. 538, 109407. https://doi.org/10.1016/j.palaeo.2019.109407.

Chow, M.M., 1955. Two specimens of mammalian coprolites from the Sanmen Rapids, Shanshi and Singtsai, Honan. Acta Paleontologica Sinica 3, 283–286.

Chrzastek, A., 2008. Vertebrate remains from the Lower Muschelkalk of Raciborowice Górne (North-SudeticBasin, SW Poland). Geol. Q. 52, 225–238.

Clark, J., Beerbower, J.R., Kietzke, K.K., 1967. Oligocene sedimentation, stratigraphy, paleoecology and paleoclimatology in the Big Badlands of South Dakota. Fieldiana Geol. Mem. 5, 157.

Clark, J.M., Fastovsky, D.E., 1986. Vertebrate biostratigraphy of the Glen Canyon Group in northern Arizona. In: Padian, K. (Ed.), The Beginning of the Age of Dinosaurs: Faunal Change across the Triassic-Jurassic Boundary. Cambridge University Press, Cambridge, pp. 285–301.

Clark, N.D.L., 1989. A Study of a Namurian Crustacean-Bearing Shale From the Western Midland Valley of Scotland. Ph.D. dissertation. University of Glasgow, Glasgow, p. 372.

Clary, K.H., 1984. Anasazi diet and subsistence as revealed by coprolites from Chaco Canyon. In: Judge, W.J., Schelberg, J.D. (Eds.), Recent Research on Chaco Prehistory: Albuquerque, Division of Cultural Resources, Reports of the Chaco Center, vol. 8. National Park Service, pp. 265–279.

Clouter, F., Mitchell, T., Rayner, D., Rayner, M., 2000. London Clay fossils of the Isle of Sheppey: a collectors guide to the fossil animals of the London Clay between Minster and Warden Point. Sheppey Medway Medway Fossil Mineral Soc. 100.

Collareta, A., Gemelli, M., Varola, A., Bianucci, G., 2019. Trace fossils on a trace fossil: a vertebrate-bitten vertebrate coprolite from the Miocene of Italy. Neues Jahrbuch Geol. Palaontol. Abhand. 293, 117–126.

Collareta, A., Peri, E., Godfrey, S.J., Bianucci, G., 2022. Just a different place to graze? An unusual occurrence of the echinoid feeding trace *Gnathichnus pentax* on a marine vertebrate coprolite (Miocene, Italy) and its palaeoethological implications. Carnets Geologie 22 (20), 847–855. https://doi.org/10.2110/carnets.2022.2220.

Contreras, V.H., 1995. Datos preliminares de estrucuturas biodeposicionales (coprólitos) de La Formacion Ischigualasto (Triásico superior, Temprano), en su localidad tipo (Ischigualasto, San Juan, Argentina). 2 Reunión Argentina de Icnologia. San Juan, pp. 7–8.

Conway Morris, S., Robison, R.A., 1986. Middle Cambrian priapulids and other softbodied fossils from Utah and Spain. Univ. Kans. Paleontol. Contrib. Pap. 117, 22.

Cook, P.J., McElhinny, M.W., 1979. A re-evaluation of the spatial and temporal distribution of phosphorites in the light of plate tectonics. Econ. Geol. 74, 315–330.

Coy, C.E., 1995. The first record of spiral coprolites from the Dinosaur Park Formation (Judith River Group, Upper Cretaceous) southern Alberta, Canada. J. Paleontol. 69, 1191–1194.

Cross, S.R., Ivanovski, N., Duffin, C.J., Hildebrandt, C., Parker, A., Benton, M.J., 2018. Microvertebrates from the basal Rhaetian bone bed (latest Triassic) at Aust Cliff, SW England. Proc. Geol. Assoc. 129 (5), 635–653. https://doi.org/10.1016/j.pgeola.2018.06.002.

Cueille, M., Green, E., Duffin, C.J., Hildebrand, C., Benton, M.J., 2020. Fish and crab coprolites from the latest Triassic of the UK: from Buckland to the Mesozoic Marine revolution. Proc. Geol. Assoc. 131, 699–721.

Cumbaa, S.L., Schröder-Adams, C., Day, R.G., Phillips, A.J., 2006. Cenomanian bonebed faunas from the northeastern margin, Western Interior Seaway, Canada. New Mex. Mus. Nat. Hist. Sci. Bull. 35, 139–155.

Cunningham, C.R., Feldman, H.R., Franseen, E.K., Gastaldo, R.A., Mapes, G., Maples, C.G., Schultze, H.P., 1993. The Upper Carboniferous Hamilton Fossil-Lagerstätte in Kansas: a valley-fill, tidally influenced deposit. Lethaia 26, 225–236.

Dachroth, W., 1985. Fluvial sedimentary styles and associated depositional environments in the Buntsandstein west of River Rhine in Saar area and Pfalz (F.R. Germany) and Vosges (France). In: Mader, D. (Ed.), Aspects of Fluvial Sedimentation in the Lower Triassic Buntsandstein of Europe, Lecture Notes in Earth Sciences, vol. 4. Springer, Berlin, pp. 197–248.

Dana, S.L., 1845. Analysis of coprolites from the New Red Sandstone Formation of New England. Am. J. Sci. 48, 46–60.

Danielson, D.R., 1993. Phytolith Analysis of Coprolites From the Prehistoric Southwest. M.S. thesis. University of Nebraska, Lincoln, p. 111.

Danielson, D.R., Reinhard, K.J., 1998. Human dental microwear caused by calcium oxalate phytoliths in prehistoric diet of the lower Pecos region, Texas. Am. J. Phys. Anthropol. 107 (3), 297–304. https://doi.org/10.1002/(SICI)1096-8644(199811) 107:3<297::AID-AJPA6>3.0.CO;2-M.

Da Silva, P.A., Borba, V.H., Dutra, J.M.F., Leles, D., Da-Rosa, A.A.S., Ferreira, L.F., Araujo, A., 2014. A new ascarid species in cynodont coprolite dated of 240 million years. An Acad. Bras Ciências 86, 265–270.

Davis, O.K., 1987. Spores of the dung fungus *Sporormiella*: increased abundance in historic sediments and before Pleistocene megafaunal extinction. Quat. Res. 28, 290–294.

Davis, O.K., Kolva, D.A., Mehringer, P.J., 1977. Pollen analysis of Wildcat Lake, Whitman County, Washington: the last 1000 years. Northwest Sci. 51, 13–30.

Davis, O.K., Shafer, D.S., 2006. *Sporormiella* fungal spores, a palynological means of detecting herbivore density. Palaeogeogr. Palaeoclimatol. Palaeoecol. 237, 40–50.

Dawson, J.W., 1854. On the coal measures of the South Joggins. Nova Scotia Quart. J. Geol. Soc. London 10, 1–42.

Dawson, J.W., 1862. Notice of the discovery of additional remains of land animals in the coal-measures of the South Joggins. Nova Scotia Quart. J. Geol. Soc. London 18, 5–7.

De Baets, K., Dentzien-Dias, P., Harrison, G.W.M., Littlewood, D.T.J., Parry, L.A., 2021. Fossil constraints on the timescale of parasitic helminth evolution. In: De Baets, K., Huntley, J.W. (Eds.), The Evolution and Fossil Record of Parasitism. Topics in Geobiology, vol. 49. Springer, Cham. https://doi.org/10.1007/978-3-030-42484-8_7.

DeKay, J.E., 1830a. On the discovery of coprolites in North America. Phil. Magaz. 10, 321–322.

DeKay, J.E., 1830b. On the extinct reptiles of the genera *Mosasaurus* and *Geosaurus* found in the secondary formation of New Jersey; and on the occurrence of a substance recently named coprolite by Dr. Buckland in the same locality. Ann. Lyceum Nat. Hist. N. Y. 3, 134–141.

Delson, E., Faure, M., Guérin, C., Aprile, L., Argant, J., Blackwell, B.A.B., Debard, E., Harcourt-Smith, W., Martin-Suarez, E., Monguillon, A., Parenti, F., Pastre, J.F., Sen, S., Skinner, A.R., Swisher, C.C., Valli, A.M.F., 2006. Franco-American renewed research at the late Villafranchian locality of Senèze (Haute-Loire, France); in late Neogene and Quaternary biodiversity and evolution: regional developments and interregional correlations. In: Proceedings of the 18th International Senckenberg Conference (VI IPCW), vol. I, pp. 275–290.

de Lumley, H., 1969. A Paleolithic camp at Nice. Sci. Am. 220, 42–50.

Dentzien-Dias, P.C., Poinar, G., Francischini, H., 2017. A new actinomycete from a Guadalupian vertebrate coprolite from Brazil. Hist. Biol. 29, 770–776.

Dentzien-Dias, P., Carrillo-Briceño, J.D., Francischini, H., Sánchez, R., 2018. Paleoecological and taphonomical aspects of the Late Miocene vertebrate coprolites (Urumaco Formation) of Venezuela. Palaeogeogr. Palaeoclimatol. Palaeoecol. 490, 590–603.

Dentzien-Dias, P., Hunt, A.P., Lucas, S.G., Francischini, H., Gulotta, M., 2021. Coprolites from shallow marine deposits of the Nanjemoy Formation, Lower Eocene of Virginia, USA. Lethaia 54, 26–39.

Dentzien-Dias, P.C., Figueiredo, A.E.Q., Horn, B., Cisneros, J.C., Schultz, C.L., 2012. Paleobiology of a unique vertebrate coprolites concentration from Rio do Rasto Formation (middle/upper Permian), Paraná Basin, Brazil. J. S. Am. Earth Sci. 40, 53–62.

Dentzien-Dias, P., Łaska, W., 2025. Analyzing Coprolites: Different Methods for the Study of Coprolites. Vertebrate Ichnology. Elsevier, Amsterdam. In this issue.

Dentzien-Dias, P.C., Poinar, G., De Figueiredo, A.E.Q., Pacheco, A.C.L., Horn, B.L.D., Schultz, C.L., 2013. Tapeworm eggs in a 270 million-year-old shark coprolite. PLoS One 8. https://doi.org/10.1371/journal.pone.0055007.

de Oliveira, F.A., Santucci, R.M., de Oliveira, C.E.M., de Andrade, M.B., 2021. Morphological and compositional analyses of coprolites from the Upper Cretaceous Bauru Group reveal dietary habits of notosuchian fauna. Lethaia. https://doi.org/10.1111/let.12431.

DiBenedetto, J.N., 2002. Phytoliths extracted from Oligocene carnivore coprolites. Geol. Soc. Am. Abstr. Progr. 34, 47.

Diedrich, C.G., 2012a. An Ice Age spotted hyena *Crocuta crocuta spelaea* (Goldfuss 1823) population, their excrements and prey from the late Pleistocene hyena den of the Sloup Cave in the Moravian Karst, Czech Republic. Hist. Biol. 24, 61–185.

Diedrich, C.G., 2012b. Typology of Ice Age spotted hyena *Crocuta crocuta spelaea* (Goldfuss, 1823) coprolite aggregate pellets from the European Late Pleistocene and their significance at dens and scavenging sites. New Mex. Mus. Nat. Hist. Sci. Bull. 57, 369–377.

Diedrich, C.G., Felker, C.D.H., 2012. Middle Eocene shark coprolites from shallow marine and deltaic coasts of the pre-North Sea Basin in central Europe. New Mex. Mus. Nat. Hist. Sci. Bull. 57, 311–318.

Doherty, J.M., 2009. Comparison of Digestion Through Examination of Fecal Contents over Differing Feeding Regimens in *Elaphe Guttata, Elaphe Obsoleta*, and *Chelydra serpentina*. M.S. thesis. University of Colorado, Boulder, p. 42.

Dornbos, S.Q., 2011. Phosphatization through the Phanerozoic. In: Allison, P.A., Bottjer, D.J. (Eds.), Taphonomy: Process and Bias through Time. Topics in Geobiology 32. Springer, New York, pp. 435–456.

Doyle, P., Whitham, A.G., 1991. Paleoenvironments of the Nordenskjold Formation: an Antarctic Late Jurassic–Early Cretaceous black shale–tuff sequence. In: Tyson, R.V., Pearson, T.H. (Eds.), Modern and Ancient Continental Shelf Anoxia, vol. 58. Geological Society of London, Special Publication, pp. 397–414.

Dridi, J., 2022. First report of diverse vertebrate coprofaunas from the Lower Cretaceous of Tunisia. Cretac. Res. https://doi.org/10.1016/j.cretres.2022.105192.

Duffin, C.J., 1979. Coprolites: a brief review with reference to specimens from the Rhaetic bone beds of England and South Wales. Mercian Geol. 7, 191–204.

Duffin, C.J., 2006. William Buckland (1984-1856). Geol. Today 22, 104–108.

Duffin, C.J., 2009. Records of Warfare…embalmed in the Everlasting Hills:" A History of Early Coprolite Research. Mercian Geologist, pp. 101–111.

Duffin, C.J., 2010. Coprolites. In: Lord, A.R., Davis, P.G. (Eds.), Fossils from the Lower Lias of the Dorset Coast, vol. 13. Palaeontological Association, Field Guides to Fossils, pp. 395–400.

Duffin, C.J., 2012a. Coprolites and characters in victorian britain. New Mex. Mus. Nat. Hist. Sci. Bull. 57, 45–59.

Duffin, C.J., 2012b. The earliest published records of coprolites. New Mex. Mus. Nat. Hist. Sci. Bull. 57, 25–29.

Duffin, C.J., Ward, D.J., 2020. Coprolites in an early Carboniferous microvertebrate fauna from Derbyshire, UK. Neues Jahrbuch Geol. Palaontol. Abhand. 297, 1–10.

Duffin, C.J., Ward, D.J., 2023a. Coprolites of the ichnogenus *Alococopros* from the late Cretaceous of Morocco. Proc. Geologists' Assoc. https://doi.org/10.1016/j.pgeola.2023.11.005.

Duffin, C.J., Ward, D.J., 2023b. First record of the coprolite ichnogenus *Struocopros* from the late Cretaceous of Morocco. New Mex. Mus. Nat. Hist. Bull. 94, 11–16.

Ebert, M., Kölbl-Ebert, M., Lane, J.A., 2015. Fauna and predator-prey relationships of Ettling, an actinopterygian fish-dominated Konservat Lagerstätte from the Late Jurassic of Southern Germany. PLoS One 10 (1), e0116140. https://doi.org/10.1371/journal.pone.0116140.

Edwards, P., 1973a. A qualitative comparison of the chemical composition of recent feces and coprolites. In: Proceedings of the 83rd Nebraska Academy of Sciences, vol. 12.

Edwards, P., 1973b. Qualitative x-ray diffraction fluorescence analysis of some Oligocene coprolites. Contrib. Geol. 12, 25.

Edwards, P., 1976. Fish coprolites from Fossil Butte, Wyoming. Rocky Mt. Geol. 14, 115–117.

Edwards, P., Yatkola, D., 1974. Coprolites of White River (Oligocene) carnivorous mammals: origin and paleoecologic and sedimentologic significance. Contrib. Geol. 13, 67–73.

Eichler, B., Werneburg, R., 2010. Neufunde von Branchiosauriern aus dem Rotliegend von Frietal bei Dresden im UV-Licht. Geol. Saxonica 56, 137–157.

Elder, R.L., 1985. Principles of Aquatic Taphonomy With Examples From the Fossil Record. Ph.D. dissertation. University of Michigan, Ann Arbor, p. 336.

Elder, R.L., Smith, G.R., 1988. Fish taphonomy and environmental inference in paleolimnology. Palaeogeogr. Palaeoclimatol. Palaeoecol. 62, 577–592.

Eriksson, M.E., Lindgren, J., Chin, K., Månsby, U., 2011. Coprolite morphotypes from the Upper Cretaceous of Sweden: novel views on an ancient ecosystem and implications for coprolite taphonomy. Lethaia 44, 455–468.

Esteban-Nadal, M., Cáceres, I., Fosse, P., 2010. Characterization of a current coprogenic sample originated by *Canis lupus* as a tool for identifying a taphonomic agent. J. Archaeol. Sci. 37, 2959–2970.

Espigares, M.P., Palmqvist, P., Rodríguez-Ruiz, M.D., Ros-Montoya, S., Pérez-Ramos, A., Rodríguez-Gómez, G., Guerra-Merchán, A., García-Aguilar, J.M., Granados, A., Campaña, I., Martínez-Navarro, B., 2023. Sharing food with hyenas: a latrine of *Pachycrocuta brevirostris* in the early Pleistocene assemblage of Fuente Nueva-3 (Orce, Baza basin, SE Spain). Archaeol. Anthropol. Sci 15 (81). https://doi.org/10.1007/s12520-023-01784-7.

Evans, O., Duffin, C.J., Hildebrandt, C., Benton, M.J., 2024. Microvertebrates from the basal Rhaetian bone bed (late triassic) at Lavernock, South Wales. Proc. Geol. Assoc. https://doi.org/10.1016/j.pgeola.2024.05.001.

Everhart, M.J., 2005. Oceans of Kansas—A Natural History of the Western Interior Sea: Bloomington. Indiana University Press, p. 322.

Everhart, M.J., 2017. Oceans of Kansas—A Natural History of the Western Interior Sea, second ed. Indiana University Press, Bloomington, p. 427.

Falcon-Lang, H.J., Benton, M.J., Braddy, S.J., Davies, S.J., 2006. The Pennsylvanian tropical biome reconstructed from the Joggins Formation of Nova Scotia, Canada. J. Geol. Soc. 163, 561–576.

Farjand, A., Fu, L., Rummy, P., Halaçlar, K., Wang, J., You, Q., Su, H., Bi, S., 2024. First record of well-preserved canid coprolites from Eurasia: new insights into the unique ecological niche of Yuanmou Basin. Heliyon 10 (9), e30072. https://doi.org/10.1016/j.heliyon.2024.e30072.

Farlow, J.O., Chin, K., Argast, A., Poppy, S., 2010. Coprolites from the Pipe Creek Sinkhole (late Neogene, Grant County, Indiana, USA). J. Vertebr. Paleontol. 30 (3), 959–969.

Fejfar, O., Dvorák, Z., Kadlecová, E., 2003. New record of early Miocene (MN3a) mammals in the open brown coal pit Merkur, North Bohemia, Czech Republic. Deinsea 10, 163–182.

Fernandéz-Rodrìguez, C., Ramil, P., Martìnez Cortizas, A., 1995. Characterization and depositional evolution of hyaena (*Crocuta crocuta*) coprolites from La Valina Cave (northwest Spain). J. Archaeol. Sci. 22, 597–607.

Ferreira, D.F., Bertini, R.J., Camossi, L.G., Letizio, L.A., 2019. The first record of ascaridoidea eggs discovered in crocodyliformes hosts from the Upper Cretaceous of Brazil. Rev. Bras. Palaontol. 21, 238–244.

Ferreira, L.F., Araújo, A., Confalonieri, U., Chame, M., Gomes, D.C., 1991. *Trichuris* eggs in animal coprolites dated from 30,000 years ago. J. Parasitol. 77, 491–493.

Fikentscher, R., 1933. Koproporphyrin in Tertiären Krokodilkot: ZoologischeAnzeiger, vol. 103, pp. 289–295.

Fiorelli, L.E., Ezcurra, M.D., Hechenleitner, E.M., Arganaraz, E., Taborda, J.R., Trotteyn, M.J., Von Baczko, M.B., Desojo, J.B., 2013. The oldest known communal latrines provide evidence of gregarism in Triassic megaherbivores. Sci. Rep. 3, 3348. https://doi.org/10.1038/srep03348.

Fluckiger, F.A., 1861. Über Koprolithen aus Baselland. Schweizer Zeitschrift Pharmacie 3, 189–195.

Fraas, E., 1891. Die Ichthyosaurier der Süddeutschen Trias- und Jura-Ablagerungen, p. 81. Tübingen.

Francischini, H., Dentzien-Dias, P., Schultz, C.L., 2018. A fresh look at ancient dungs: the Brazilian Triassic coprolites revisited. Lethaia 51, 389–405.

Friedman, V., 2012. Vertebrate coprolites from the lower Eagle Ford Group of north central Texas and their paleoecological significance. New Mex. Mus. Nat. Hist. Sci. Bull. 57, 221–228.

Fritz, F., Hummel, J., Kienzle, E., Arnold, C., Nunn, C., Clauss, M., 2009. Comparative chewing efficiency in mammalian herbivores. Oikos 11, 1623–1632.

Fritz, J., 2007. Allometrie der Kotpartikelgröße von pflanzenfressenden Säugern, Reptilien und Vögeln [Inaugural-Dissertation]. München, Ludwig-Maximilians-Universität, p. 272.

Froyd, C.A., Coffey, E.E.D., Van Der Knaap, W.O., Van Leeuwen, J.F.N., Tye, A., Willis, K.J., 2013. The ecological consequences of megafaunal loss: giant tortoises and wetland biodiversity. Ecol. Lett. https://doi.org/10.1111/ele.12203.

Fry, G., 1970. Preliminary analysis of the Hogupcave coprolites. In: Aikens, C.M. (Ed.), Hogup Cave. University of Utah Anthropological Papers 93. University of Utah Press, Salt Lake City, pp. 247–250.

Fry, G.F., 1977. Analysis of prehistoric coprolites from Utah. In: University of Utah Anthropological Papers, 97, pp. 1–45.

Fry, G.F., 1978. Prehistoric Diet at Danger Cave, Utah, as Determined by the Analysis of Coprolites. Miscellaneous Paper 23, vol. 99. University of Utah Anthropological Papers, Salt Lake City.

Fry, G., Hall, H.J., 1986. Human coprolites. In: Morris, D.P. (Ed.), Archeological Investigations at Antelope House, vol. 19. National Park Service Publications in Archaeology, pp. 165–188.

Gall, J.-C., 1971. Faunes et paysages du Grès à Voltzia du nord des Vosges. Essai paleoecologique sur le Buntsandstein superieur. Mémoires du Service de la Carte Géologique d'Alsace et de Lorraine 34, 1–138.

Gall, J.-C., Grauvogel-Stamm, L., 1993. Paleoecology of terrestrial ecosystems from the Buntsandstein (Lower Triassic) of eastern France. New Mexico Museum of Natural History. Bulletin 3, 141–145.

Garassino, A., Donovan, D.T., 2000. A new family of coleoids from the Lower Jurassic of Osteno, northern Italy. Palaeontology 43, 1019–1038.

Gaudry, A., 1887. L'Actinodon. Mémoire extrait des Nouvelles Archives du Muséum National d'Histoire naturelle, Serie 2 (10), 1–30.

Ghosh, P., Bhattacharya, S.K., Sahni, A., Kar, R.K., Mohabey, D.M., Ambwani, K., 2003. Dinosaur coprolites from the Late Cretaceous (Maastrichtian) Lameta Formation of India. Isotopic and other markers suggesting a C3 plant diet. Cretac. Res. 24, 743–750.

Gilbert, M.T., Jenkins, D.L., Götherstrom, A., Naveran, N., Sanchez, J.J., Hofreiter, M., Thomsen, P.F., Binladen, J., Higham, T.F., Yohe, R.M., Parr, R., Cummings, L.S., Willerslev, E., 2008. DNA from pre-Clovis human coprolites in Oregon, North America. Science 320, 786–789.

Gilfillian, A.M., Olsen, P.E., 2000. The coelacanth *Diplurus longicaudatus* as the origin of the large coprolites occurring in the Triassic-Jurassic lacustrine strata of eastern North America. Geol. Soc. Am. Abstr. Progr. 32 (1), 20.

Gill, J.L., Mclauchlan, K.K., Skibbe, A.M., Goring, S., Zirbel, C.R., Williams, J.W., 2013. Linking abundances of the dung fungus *Sporormiella* to the density of bison: implications for assessing grazing by megaherbivores in palaeorecords. J. Ecol. 101, 1125–1136.

Gilmore, B., 1992. Scroll coprolites from the Silurian of Ireland and the feeding of early vertebrates. Palaeontology 35, 319–333.

Gilpin, E., 1886. The geology of Cape Breton Island, Nova Scotia. Quart. J. Geol. Soc. 42, 515–526.

Gingerich, P.D., 1977. A small collection of fossil vertebrates from the Middle Eocene Kuldana and Kohat formations of Punjab (Pakistan). Contrib. Mus. Paleontol. Univ. Mich. 24, 190–203.

Girotti, O., Capasso Barbato, L., Esu, D., Kotsakis, A., Martinetto, E., Petronio, C., Sardella, R., Squazzini, E., 2004. The section of Torre Picchio (Terni, Umbria, Central Italy): aVillafranchian site rich in mammals, molluscs, ostracods and plants. Revista Italiana di Paleontologia e Stratigrafia 109, 77–98.

Glowiak, E.M., 2007. Gypsum Cave Revisited: Faunal and Taphonomic Analysis of a Rancholabrean to Holocene Fauna in Southern Nevada. M.S. thesis. University of Nevada at Las Vegas, Las Vegas, p. 132.

Godfrey, S.J., Collareta, A., 2022. A new ichnotaxonomic name for burrows in vertebrate coprolites from the Miocene Chesapeake Group of Maryland, USA. Swiss J. Palaeontol. 141 (9). https://doi.org/10.1186/s13358-022-00250-6.

Godfrey, S.J., Palmer, B.T., 2015. Gar-bitten coprolite from South Carolina, USA. Ichnos 22, 103–108.

Godfrey, S.J., Smith, J., 2010. Shark-bitten vertebrate coprolites from the Miocene of Maryland. Naturwissenschaften 97, 461–467.

Godfrey, S.J., Collareta, A., Nance, J.R., 2022. Coprolites from Calvert Cliffs: Miocene fecal pellets and burrowed crocodilian droppings from the Chesapeake Group of Maryland, U.S.A. Riv. Ital. Paleontol. Stratigr. 128 (1), 69–79.

Godfrey, S.J., Alford, A., Collareta, A., Weems, R.E., 2020. A Paleocene vertebrate-bitten crocodilian coprolite from Liverpool Point, Maryland, USA. Neues Jahrbuch Geol. Palaontol. Abhand. 296, 237–244.

Goldberg, P., Berna, F., Macphail, R.I., 2009. Comment on "DNA from Pre-Clovis human coprolites in Oregon, North America". Science 325. https://doi.org/10.1126/science.1167531.148-c.

Goldring, R., Pollard, J.E., Radley, J.D., 2005. Trace fossils and pseudofossils from the Wealden strata (non-marine Lower Cretaceous) of southern England. Cretac. Res. 26, 665–685.

Gong, Y., Zhang, L., Wu, Y., 2010. Carboniferous coprolites from Qinhuangdao, North China. Sci. China Earth Sci. 53, 213–219.

Górecki, A., Barbacka, M., Niedźwiedzki, G., Jarzynka, A., Pacyna, G., Ziaja, J., 2019. Plant remains from fossil faeces (coprolites) of a large predator from the Early Jurassic of Poland (Sołtyków). In: Fekete, K., Michalík, J., Reháková, D. (Eds.), XIV Jurassica Conference and Workshop of the ICS Berriasian Group, Field Trip Guide and Abstracts Book, Earth Science Institute. Slovak Academy of Sciences and Faculty of Natural Sciences, Comenius University, Bratislava, pp. 105–106.

Götzinger, G., Becker, H., 1932. Zur geologischen Gliederung des Wienerwaldflysches (Neue Fossilfunde). Jahrbuch der Geologischen Bundesanstalt, Vienna 82, 343–395.

Grande, L., 1980. Paleontology of the Green River Formation, with a review of the fish fauna. Geol. Surv. Wyom. Bull 63, 333.

Grande, L., 1984. Paleontology of the Green River Formation, with a Review of the Fish Fauna, second ed. vol. 63. Wyoming State Geological Survey Bulletin, p. 333.

Grande, L., 2013. The Lost World of Fossil Lake: Snapshots From Deep Time. University of Chicago Press, Chicago, p. 425.

Greb, S.F., Storrs, G.W., Garcia, W.J., Eble, C.F., 2015. Late Mississippian vertebrate palaeoecology and taphonomy, Buffalo Wallow Formation, western Kentucky, USA. Lethaia 49, 99–218.

Greenwood, P.H., 1951. Fish remains from Miocene deposits of Rusinga Island and Kavirondo Province, Kenya. Ann. Mag. Nat. Hist. 4, 1192–1201.

Gross, M., Prieto, J., Grímsson, F., Hans-Peter Bojar, H.-P., 2023. Hyena and 'false' sabre-toothed cat coprolites from the late Middle Miocene of south-eastern Austria. Hist. Biol. https://doi.org/10.1080/08912963.2023.2237979.

Gouiric-Cavalli, S., Acosta Burllaile, L., Iglesias, A., Moly, J.J., O'Gorman, J.P., Reguero, M., Santillana, S., Talevi, M., Vieytes, C., Bigurrarena Ojeda, M.A., Lusky, J., 2017. Late Mesozoic marine Antarctic fishes: future perspectives based on the newly collections recovered in the Ameghino and López de Bertodano Formations. Res. Knowl. 3 (1), 16–61.

Gouiric-Cavalli, S., Rasia, L.L., Marquez, G.J., Rosato, V.G., Scasso, R.A., Reguero, M.A., 2019. First pachycormiform (Actinopterygii, Pachycormiformes) remains from the Late Jurassic of the Antarctic Peninsula and remarks on bone alteration by recent bioeroders. J. Vertebr. Paleontol. 38 (5), e1524384. https://doi.org/10.1080/02724634.2018.1524384.

Guthörl, P., 1959. Hundert Jahre Pälaontologie und Karbonstratigraphie im Saarbrückener Steinkohlengbiet. Bergfreiheit 24, 111–129.

Haas, J., Creamer, W., Mesiam, L.H., Goldstein, D., Reinhard, K., Rodriguez, C.V., 2013. Evidence for maize (*Zea mays*) in the late Archaic (3000-1800 B.C.) in the Norte Chico region of Peru. Proc. Natl. Acad. Sci. USA 110 (13), 4945–4949.

Halaclar, K., Rummy, P., Deng, T., Van Do, T., 2022. Footprint on a coprolite: A rarity from the Eocene of Vietnam. Palaeoworld 31 (4), 723–732. https://doi.org/10.1016/j.palwor.2022.01.010.

Halaçlar, K., Rummy, P., Liu, J., Hunt, A.P., Do, Minh, N.T., Deng, T., 2023. Exceptionally well-preserved crocodilian coprolites from the Late Eocene of Northern Vietnam: Ichnology and paleoecological significance. iScience 26, 107607. https://doi.org/10.1016/j.isci.2023.107607.

Hall, H.J., 1969. Rehydration and Concentration of Parasite Ova in Human Coprolites from the Great Basin. Honor's Thesis, Department of Anthropology. University of Utah, Salt Lake City, p. 68.

Hall, H.J., 1977. A paleoscatalogical study of diet and disease at Dirty Shame Rockshelter, Southeast Oregon. Tebiwa Misc. Pap. Idaho State Univ. Mus. Nat. Hist. 8, 1–14.

Hamilton, R.F.M., Trewin, N.H., 1988. Environmental controls on fish faunas of the Middle Devonian Orcadian Basin. In: McMillan, N.J., Embry, A.F., Glass, D.J. (Eds.), Devonian of the World, vol. 14. Canadian Society of Petroleum Geologists, Memoir, pp. 589–600.

Hamilton, R.F.M., Trewin, N.H., 1994. Taphonomy of fish beds from the Upper Flagstone Group of the Middle Old Red Sandstone, Caithness. Scot. J. Geol. 30, 175–181.

Hancox, P.J., Rubidge, B.S., 2023. The Beaufort-Stormberg group contact—Implications for Karoo basin development in the Triassic. J. Afr. Earth Sci. 198, 104767. https://doi.org/10.1016/j.jafrearsci.2022.104767.

Hancox, P.J., Neveling, J., Rubidge, B.S., 2020. Biostratigraphy of the *Cynognathus* Assemblage Zone (Beaufort Group, Karoo Supergroup), South Africa. S. Afr. J. Geol. 123 (2), 217–238.

Hansen, B.B., 2014. Description and Analysis of a Rhaetian Vertebrate Coprolite Assemblage of the Kap Stevard Formation, Jameson Land, East Greenland. M.Sc. thesis. Copenhagen, University of Copenhagen, p. 97.

Hansen, B.B., Milàn, J., Clemmensen, L.B., Adolfssen, J.S., Estrup, E.J., Klein, N., Mateus, O., Wings, O., 2016. Coprolites from the Late Triassic Kap Stewart Formation, Jameson Land, East Greenland: morphology, classification and prey inclusions. In: Kear, B.P., Lindgren, J., Hurum, J.H., Milàn, J., Vajda, V. (Eds.), Mesozoic Biotas of Scandinavia and its Arctic Territories, vol. 434. Geological Society of London, Special Publications, pp. 49–69.

Hansen, R.M., 1978. Shasta ground sloth food habits, Rampart Cave, Arizona. Paleobiology 4, 302–319.

Hansen, R.M., 1980. Late Pleistocene plant fragments in the dungs of herbivores at Cowboy Cave. In: Jennings, J.D. (Ed.), Cowboy Cave. University of Utah, Anthropological Papers, vol. 104, pp. 179–189.

Häntzschel, W., El-Baz, F., Amstutz, G.C., 1968. Coprolites: An Annotated Bibliography, vol. 108. Geological Society of America, Memoir, p. 132.

Harrell, S.D., Schwimmer, D.R., 2010. Coprolites of *Deinosuchus* and other crocodylians from the Upper Cretaceous of western Georgia, USA. New Mex. Mus. Nat. Hist. Sci. Bull. 51, 209–213.

Harrington, M.R., 1933. Gypsum Cave, Nevada. Southwest Mus Papers 8, 197.

Harrington, M.R., 1940. Man and Beast in Gypsum Cave: The Desert Magazine, vol. 3, p. 34.

Harris, C., Gess, R.W., 2017. A trace fossil assemblage from 'lingulid mud beds' in the Upper Devonian Witpoort Formation, Witteberg Group, South Africa. Palaeontol. Afr. 52, 158–159.

Harris, J.W., Williamson, P.G., Verniers, J., Tappen, M.J., Stewart, K., Helgren, D., de Heinzelin, J., Boaz, N.T., Bellomo, R.V., 1987. Late Pliocene hominid occupation in Central Africa: the setting, context, and character of the Senga 5A site, Zaire. J. Hum. Evol. 16, 701–728.

Harris, S.K., Lucas, S.G., Lueth, V.W., 2017. Lacustrine coprofauna from the Upper Pennsylvanian (Missourian) Tinajas Member of the Altrasado Formation, Socorro County, New Mexico. New Mexico Museum of Natural History. Bulletin 77, 113–132.

Harrison, T., 2011. Coprolites: taphonomic and paleoecological implications. In: Harrison, H.T. (Ed.), Paleontology and Geology of Laetoli: Human Evolution in Context Geology, Geochronology, Paleoecology and Paleoenvironment, vol. 1. Springer, Dordrecht, pp. 279–292.

Hastings, A.K., Bloch, J.I., Jaramillo, C.A., Rincon, A.F., MacFadden, B.J., 2013. Systematics and biogeography of crocodylians from the Miocene of Panama. J. Vertebr. Paleontol. 33, 239–263.

Hattin, D.E., 1996. Fossilized regurgitate from Smoky Hill Member of Niobrara Chalk (Upper Cretaceous) of Kansas, USA. Cretac. Res. 17, 443–450.

Hauff, B., 1921. Untersuchung der Fossilfundstatten in Posidonienschiefer des oberen Lias Württemberg. Palaeontographica 64, 1–42.

Hauthal, R., Roth, S., Lehmann-Nitsche, R. (Eds.), 1899. El mamífero misterioso de la Patagonia: Grypotherium domesticum, vol. 9. Revista del Museo de La Plata, pp. 411–472.

Hawkins, A.D., Liu, H.P., Briggs, D.E., Muscente, A.D., McKay, R.M., Witzke, B.J., Xiao, S., 2018. Taphonomy and biological affinity of three-dimensionally phosphatized bromalites from the Middle Ordovician Winneshiek Lagerstätte, northeastern Iowa, USA. Palaios 33, 1–15.

Heizer, R.F., 1967. Analysis of human coprolites from a dry Nevada cave. In: Papers on Great Basin Archaeology, Reports of the University of California Survey, vol. 70, pp. 1–120.

Hembree, D.I., Blair, M.G., 2016. A paleopedological and ichnological approach to interpreting spatial and temporal variability in early Permian fluvial deposits of the lower Dunkard Group, West Virginia, USA. Palaeogeogr. Palaeoclimatol. Palaeoecol. 454, 246–266.

Hembree, D.I., Bowen, J.J., 2017. Paleosols and ichnofossils of the Upper Pennsylvanian–lower Permian Monongahela and Dunkard groups (Ohio, USA): A multi-proxy approach to unraveling complex variability in ancient terrestrial landscapes. Palaios 32, 295–320.

Hembree, D.I., Hasiotis, S.T., 2004. Using paleosols and ichnofossils to interpret the changing paleoecology, paleoenvironments and paleoclimate of the Eocene-Oligocene White River Formation, northeastern Colorado. Geol. Soc. Am. Abstr. Progr. 36 (5), 63.

Hembree, D.I., Hasiotis, S.T., 2007. Paleosols and ichnofossils of the White River Formation of Colorado: Insight into soil ecosystems of the North American midcontinent during the Eocene-Oligocene transition. Palaios 22, 123–142.

Henslow, J.S., 1845. On Nodules, Apparently Coprolitic, From the Red Crag, London Clay, and Greensand, vol. 15. Notices and abstracts of the British Association for the Advancement of Science, pp. 51–52.

Hill, C.R., 1976. Coprolites of *Ptilophyllum* cuticles from the Middle Jurassic of North Yorkshire. Bull. Brit. Mus. Nat. Hist. Geol. 27, 289–294.

Hitchcock, E., 1844. Report on ichnolithology or fossil footmarks, with a description of several new species, and the coprolites of birds from the valley of Connecticut River, and of a supposed footmark from the valley of the Hudson River. Am. J. Sci. 47, 292–322.

Hodnett, J.-P., Lucas, S.G., 2018. A nonmarine Late Pennsylvanian vertebrate assemblage in a marine bromalite from the Manzanita Mountains, Bernalillo County, New Mexico. New Mex. Mus. Nat. Hist. Sci. Bull. 79, 251–260.

Hoffman, D.K., Hancox, J.P., Nesbitt, S.J., 2023. A diverse diapsid tooth assemblage from the Early Triassic (Driefontein locality, South Africa) records the recovery of diapsids following the end-Permian mass extinction. PLoS One 18 (5), e0285111. https://doi.org/10.1371/journal.pone.0285111.

Hollocher, K., Hollocher, T.C., 2012. Early processes in the fossilization of terrestrial feces to coprolites, and microstructure preservation. New Mex. Mus. Nat. Hist. Sci. Bull. 57, 79–92.

Hollocher, K.T., Alcober, O.A., Colombi, C.E., Hollocher, T.C., 2005. Carnivore coprolites from the Upper Triassic Ischigualasto Formation, Argentina: Chemistry, mineralogy, and evidence for rapid initial mineralization. Palaios 20, 51–63.

Hollocher, T.C., Chin, K., Hollocher, K.T., Kruge, M.A., 2001. Bacterial residues in coprolite of herbivorous dinosaurs: Role of bacteria in mineralization of feces. Palaios 16, 547–565.

Hone, D., Henderson, D.M., Therrien, F., Habib, M.B., 2015. A specimen of *Rhamphorhynchus* with soft tissue preservation, stomach contents and a putative coprolite. PeerJ 3, e1191. https://doi.org/10.7717/peerj.1191.

Hoorn, C., 1994. An environmental reconstruction of the palaeo-Amazon river system (Middle–Late Miocene, NW Amazonia). Palaeogeogr. Palaeoclimatol. Palaeoecol. 112, 187–238.

Horrocks, M., D'Costa, D., Wallace, R., Gardner, R., Renzo Kondo, R., 2004. Plant remains in coprolites: Diet of a subalpine moa (Dinornithiformes) from southern New Zealand. Emu 104, 149–156.

Horsthemke, E., Ledendecker, S., Porada, H., 1990. Depositional environments and stratigraphic correlation of the Karoo sequence in northwestern Damaraland. Commun. Geol. Surv. South West Afr. Namibia 6, 63–73.

Horton, A., 2012. The occurrence of calcium phosphate in the Mesozoic and Tertiary of eastern England. Mercian Geol. 18, 60–68.

Höss, M., Dilling, A., Currantt, A., Pääbo, S., 1996. Molecular phylogeny of the extinct ground sloth *Mylodon darwinii*. Proc. Natl. Acad. Sci. USA 93, 181–185.

Houck, K.J., Fleming, J., Guerrero, R., Heimink, N., Heberton, A., Itano, W., Titus, A., Barrick, J.E., 2004. Paleontology of the Bassam Park fossil beds (Pennsylvanian), San Isabel National Forest, Colorado. Geol. Soc. Am. Abstr. Progr. 36, 413.

Hu, S.X., Zhang, Q.Y., Zhou, C.Y., 2010. Fossil coprolites from the Middle Triassic Luoping biota and ecological implication. J. Earth Sci. 21 (Special Issue), 191–193.

Hu, S.X., Zhang, Q.Y., Chen, Z.Q., Zhou, C.Y., Lv, T., Xie, T., Wen, W., Huang, J.Y., Benton, M.J., 2011. The Luoping biota: exceptional preservation, and new evidence on the Triassic recovery from end-Permian mass extinction. Proc. Royal Soc. B 278, 2274–2283.

Huber, P., 1992. Faunal distribution, seasonal deposition, and fish taphonomy at the Upper Pennsylvanian (Missourian) Kinney Brick Company Quarry, central New Mexico. In: Zidek, J. (Ed.), Geology and Paleontology of the Kinney Brick Quarry, Late Pennsylvanian, Central New Mexico. New Mexico Bureau of Mines and Mineral Resources, Bulletin, vol. 138, pp. 37–48.

Hunt, A.P., 1992. Late Pennsylvanian coprolites from the Kinney Brick Quarry, central New Mexico, with notes on the classification and utility of coprolites. In: Zidek, J. (Ed.), Geology and Paleontology of the Kinney Brick Quarry, Late Pennsylvanian, Central New Mexico. New Mexico Bureau of Mines and Mineral Resources, Bulletin, vol. 138, pp. 221–229.

Hunt, A.P., 2022. Vertebrate coprolites from the Late Devonian to Pennsylvanian of eastern Canada and their significance. GAC-MAC-IAH-CNC-CSPG, Abstr. 45, 115.

Hunt, C.O., Fiacconi, M., 2018. Pollen taphonomy of cave sediments: What does the pollen record in caves tell us about external environments and how do we assess its reliability? Quat. Int. 485, 68–75.

Hunt, A.P., Lucas, S.G., 2005a. A new coprolite ichnotaxon from the early Permian of Texas. New Mex. Mus. Nat. Hist. Sci. Bull. 30, 121–122.

Hunt, A.P., Lucas, S.G., 2005b. The origin of large vertebrate coprolites from the early Permian of Texas. New Mex. Mus. Nat. Hist. Sci. Bull. 30, 125–126.

Hunt, A.P., Lucas, S.G., 2007a. Cenozoic vertebrate trace fossils of North America: ichnofaunas, ichnofacies and biochronology. New Mex. Mus. Nat. Hist. Sci. Bull. 42, 17–41.

Hunt, A.P., Lucas, S.G., 2007b. Tetrapod ichnofacies: A new paradigm. Ichnos 14, 59–68.

Hunt, A.P., Lucas, S.G., 2010. Crocodylian coprolites and the identification of the producers of coprolites. New Mex. Mus. Nat. Hist. Sci. Bull. 51, 219–226.

Hunt, A.P., Lucas, S.G., 2012a. Classification of vertebrate coprolites and related trace fossils. New Mex. Mus. Nat. Hist. Sci. Bull. 57, 137–146.

Hunt, A.P., Lucas, S.G., 2012b. Descriptive terminology of coprolites and recent feces. New Mex. Mus. Nat. Hist. Sci. Bull. 57, 153–160.

Hunt, A.P., Lucas, S.G., 2013. The fossil record of Carboniferous and Permian vertebrate coprolites. New Mex. Mus. Nat. Hist. Sci. Bull. 60, 121–127.

Hunt, A.P., Lucas, S.G., 2014a. Jurassic vertebrate bromalites of the western United States in the context of the global record. Volumina Jurassica 12 (2), 151–158.

Hunt, A.P., Lucas, S.G., 2014b. Vertebrate trace fossils from New Mexico and their significance. New Mex. Mus. Nat. Hist. Sci. Bull. 64, 9–40.

Hunt, A.P., Lucas, S.G., 2016a. The most significant record of "Middle"- Late Paleozoic vertebrate coprolites: Silurian-Carboniferous Ichnoassemblages from Scotland. Geol. Soc. Am. Abstr. Progr. 48 (7). https://doi.org/10.1130/abs/2016AM-284166.

Hunt, A.P., Lucas, S.G., 2016b. The record of vertebrate coprolites across the Cretaceous/Paleogene boundary. New Mex. Mus. Nat. Hist. Sci. Bull. 74, 79–93.

Hunt, A.P., Lucas, S.G., 2017a. A new vertebrate coprolite locality from the Late Pennsylvanian of Central New Mexico, USA, and the environmental context of Carboniferous bromalites. New Mex. Mus. Nat. Hist. Sci. Bull. 77, 139–150.

Hunt, A.P., Lucas, S.G., 2017b. On the probable rarity of small dinosaur coprolites. Geol. Soc. Am. Abstr. Progr. 49 (6). https://doi.org/10.1130/abs/2017AM-303525.

Hunt, A.P., Lucas, S.G., 2018a. Biochronology of Triassic vertebrate coprolites and the lack of a significant turnover in coprolite ichnofaunas at the Triassic/Jurassic boundary. Geol. Soc. Am. Abstr. Progr. 50 (5). https://doi.org/10.1130/abs/2018RM-314409.

Hunt, A.P., Lucas, S.G., 2018b. Mosasaur coprolites from the Bearpaw Formation (Upper Cretaceous) of Saskatchewan, Canada. New Mex. Mus. Nat. Hist. Sci. Bull. 79, 271–275.

Hunt, A.P., Lucas, S.G., 2018c. The record of sloth coprolites in North and South America: Implications for terminal Pleistocene extinctions. New Mex. Mus. Nat. Hist. Sci. Bull. 79, 277–298.

Hunt, A.P., Lucas, S.G., 2019. Coprolites of bone-crushing carnivores: Abundant in the Old World, rare in the New World. Geol. Soc. Am. Abstr. Progr. 51 (5). https://doi.org/10.1130/abs/2019AM-336095.

Hunt, A.P., Lucas, S.G., 2020. Hyena hegemony: Biogeography and taphonomy of Pleistocene vertebrate coprolites with description of a new mammoth coprolite ichnotaxon. Ichnos 27, 111–121.

Hunt, A.P., Lucas, S.G., 2021a. A review of the bromaliteichnofauna from the Kinney Brick Quarry Lagerstätte (Late Pennsylvanian) of New Mexico, USA with descriptions of new consumulites and coprolites. New Mex. Mus. Nat. Hist. Sci. Bull. 84, 217–236.

Hunt, A.P., Lucas, S.G., 2021b. Coprolites. In: Alderton, D., Elias, S. (Eds.), Encyclopedia of Geology, second ed. 3. Elsevier, London, pp. 532–544.

Hunt, A.P., Lucas, S.G., 2021c. Coprolites in caves: Late Pleistocene coprofaunas of The American Southwest and their significance. New Mex. Mus. Nat. Hist. Sci. Bull. 88, 343–359.

Hunt, A.P., Lucas, S.G., 2021d. Coprolites of *Borophagus* (Mammalia: Canidae) From Western North America and the distribution of the coprolites of bone-cracking carnivores in the Late Cenozoic. New Mex. Mus. Nat. Hist. Sci. Bull. 82, 151–159.

Hunt, A.P., Lucas, S.G., 2021e. The ichnology of vertebrate consumption: Dentalites, gastroliths and bromalites. New Mex. Mus. Nat. Hist. Sci. Bull. 87, 215.

Hunt, A.P., Lucas, S.G., 2022a. Coprolites in caves: Late Pleistocene coprofaunas of the American Southwest and their significance. New Mex. Mus. Nat. Hist. Sci. Bull. 88, 343–359.

Hunt, A.P., Lucas, S.G., 2022b. The Upper Cretaceous fossil record of the Western Interior Basin of North America is a Megalagerstätte. Geol. Soc. Am. Abstr. Progr. 54 (5). https://2022doi: 0.1130/abs/2022AM-383304.

Hunt, A.P., Lucas, S.G., 2023a. An unusual new coprolite ichnotaxon from the Cretaceous-Paleogene of North America. New Mex. Mus. Nat. Hist. Sci. Bull. 94, 247–250.

Hunt, A.P., Lucas, S.G., 2023b. Ichnotaxonomy, distribution and biochronological significance of the early Permian Coprolite *Megaheteropolacoprossidmcadamsi* from the Southwestern United. New Mex. Mus. Nat. Hist. Sci. Bull. 94, 239–245.

Hunt, A.,P., Lucas, S.G., 2024. Vertebrate bromalites from Eocene lacustrine strata – Comparison of ichnofaunas from the Green River Formation of the Western United States and the Messel Lagerstätten in Germany. Geol. Soc. Am. Abstr. Progr. 56 (4). https://doi.org/10.1130/abs/2024CD-399234.

Hunt, A.P., Chin, K., Lockley, M.G., 1994. The paleobiology of coprolites. In: Donovan, S.K. (Ed.), The Paleobiology of Trace Fossils. John Wiley, London, pp. 221–240.

Hunt, A.P., Lucas, S.G., Klein, H., 2018. Late Triassic nonmarine vertebrate and invertebrate trace fossils and the pattern of the Phanerozoic record of vertebrate trace fossils. In: Tanner, L.H. (Ed.), The Late Triassic World, Topics in Geobiology 46. Springer Verlag, New York, pp. 447–543.

Hunt, A.P., Lucas, S.G., Lichtig, A.J., 2015a. A helical coprolite from the Red Crag Formation (Plio-Pleistocene) of England. New Mex. Mus. Nat. Hist. Sci. Bull. 67, 59–61.

Hunt, A.P., Lucas, S.G., Lockley, M.G., 1998. Taxonomy and stratigraphic and facies significance of vertebrate coprolites of the Upper Triassic Chinle Group, western United States. Ichnos 5, 225–234.

Hunt, A.P., Lucas, S.G., Spielmann, J.A., 2005a. Biochronology of early Permian vertebrate coprolites of the American Southwest. New Mex. Mus. Nat. Hist. Sci. Bull. 31, 43–45.

Hunt, A.P., Lucas, S.G., Spielmann, J.A., 2005b. Early Permian vertebrate coprolites from north-central New Mexico with description of a new ichnogenus. New Mex. Mus. Nat. Hist. Sci. Bull. 31, 39–42.

Hunt, A.P., Lucas, S.G., Spielmann, J.A., 2012a. New coprolite ichnotaxa from the Buckland Collection at the Oxford University Museum of Natural History. New Mex. Mus. Nat. Hist. Sci. Bull. 57, 115–124.

Hunt, A.P., Lucas, S.G., Spielmann, J.A., 2012b. The bromalite collection at the National Museum of Natural History (Smithsonian Institution), with descriptions of new ichnotaxa and notes on other significant coprolite collections. New Mex. Mus. Nat. Hist. Sci. Bull. 57, 105–114.

Hunt, A.P., Lucas, S.G., Spielmann, J.A., 2012c. The vertebrate coprolite collection at the Natural History Museum (London). New Mex. Mus. Nat. Hist. Sci. Bull. 57, 125–129.
Hunt, A.P., Lucas, S.G., Spielmann, J.A., 2013a. A large heteropolar coprolite from the Early Permian of New Mexico. New Mex. Mus. Nat. Hist. Sci. Bull. 59, 327–329.
Hunt, A.P., Lucas, S.G., Spielmann, J.A., 2013b. Triassic vertebrate coprolite ichnofaunas. New Mex. Mus. Nat. Hist. Sci. Bull. 61, 237–258.
Hunt, A.P., Lucas, S.G., Milàn, J., Spielmann, J.A., 2012d. Vertebrate coprolite studies: Status and prospectus. New Mex. Mus. Nat. Hist. Sci. Bull. 57, 5–24.
Hunt, A.P., Lucas, S.G., Spielmann, J.A., Lerner, A.J., 2007. A review of vertebrate coprolites of the Triassic with descriptions of new Mesozoic ichnotaxa. New Mex. Mus. Nat. Hist. Sci. Bull. 41, 88–107.
Hunt, A.P., Lucas, S.G., Spielmann, J.A., Lockley, M.G., 2012e. Bromalites from the Mississippian Bear Gulch Lagerstätte of central Montana, USA. New Mex. Mus. Nat. Hist. Sci. Bull. 57, 171–174.
Hunt, A.P., Santucci, V.L., Tweet, J.S., Lucas, S.G., 2012g. Vertebrate coprolites and other bromalites in National Park Service areas. New Mex. Mus. Nat. Hist. Sci. Bull. 57, 343–353.
Hunt, A.P., Lucas, S.G., Milàn, J., Lichtig, A.J., Jagt, J.W.M., 2015b. Vertebrate coprolites from Cretaceous Chalk in Europe and North America and the Shark Surplus Paradox. New Mex. Mus. Nat. Hist. Sci. Bull. 67, 63–68.
Hunt, A.P., Spielmann, J.A., Lucas, S.G., Durney, K., 2013c. Late Pennsylvanian coprofauna from the Finis Shale of north-central Texas. New Mex. Mus. Nat. Hist. Sci. Bull. 60, 129–131.
Hunt, A.P., Lucas, S.G., Spielmann, J.A., Cantrell, A., Suazo, T., 2012h. A new marine coprofauna from the Beeman Formation (Late Pennsylvanian: Late Missourian), Sacramento Mountains, New Mexico, USA. New Mex. Mus. Nat. Hist. Sci. Bull. 57, 193–196.
Hunt, A.P., Lucas, S.G., Spielmann, J.A., Suazo, T., Cantrell, A., 2012i. A re-evaluation of Late Pennsylvanian bromalites from the Kinney Brick Quarry Lagerstätte, New Mexico, USA. New Mex. Mus. Nat. Hist. Sci. Bull. 57, 185–192.
Hunt, A.P., Lucas, S.G., Spielmann, J.A., Cantrell, A., Suazo, T., Lerner, A.J., 2012j. Bromalites from the Tinajas Lagerstätte (Late Pennsylvanian: Late Missourian), central New Mexico, USA. New Mex. Mus. Nat. Hist. Sci. Bull. 57, 175–183.
Irwin-Williams, C., Shelley, P.H., 1980. Investigations at the Salmon Site: The Structure of Chacoan Society in the Northern Southwest. Portales Eastern New Mex. Univ. Public. Anthropol. 4 (3), 144.
Jacquet, S.M., Webb, J.-L., Huntley, J.W., Selly, T., Schiffbauer, J.D., 2023. X-ray tomographic microscopy of Eocene coprolites from Pipestone Springs Main Pocket, southwest Montana. Front. Earth Sci. 11, 1130107. https://doi.org/10.3389/feart.2023.1130107.
Jacobs, L.L., 1985. Review: The Omo micromammals: Systematics and paleoecology of early man sites from Ethiopia by Henry B. Wesselman. J. Vertebr. Paleontol. 5, 281–283.
Jacob, T., Diniz, D., Kerber, L., Dentzien-Dias, P., Francischini, H., 2024. A large-sized mammalian coprolite containing ground sloth osteoderms from the Upper Pleistocene Touro Passo Formation of Brazil. J. S. Am. Earth Sci. 133, 104715. https://doi.org/10.1016/j.jsames.2023.104715.
Jain, S., 1983. Spirally coiled 'coprolites' from the Upper Triassic Maleri Formation, India. Palaeontology 26, 813–829.
James, H.F., Burney, D.A., 1997. The diet and ecology of Hawaii's extinct flightless waterfowl: Evidence from coprolites. Biol. J. Linn. Soc. 62, 279–297.
Jenkins, D., Burrows, G.O., 1980. Ecology of otters in northern Scotland. III. The use of faeces as indicators of otter (*Lutra lutra*) density and distribution. J. Anim. Ecol. 49, 755–774.
Jenkins, D.L., Davis, L.G., Stafford, T.W., Campos, P.F., Hockett, B., Jones, G.T., Cummings, L.S., Yost, C., Connolly, T.J., Yohe, R.M., Gibbons, S.C., Raghavan, M., Rasmussen, M., Paijmans, J.L.A., Hofreiter, M., Kemp, B.M., Barta, J.L., Monroe, C., Gilbert, M.T.P., Willerslev, E., 2012. Clovis age western stemmed projectile points and human coprolites at the Paisley Caves. Science 337 (6091), 223–228.
Jiménez, F.A., Gardner, S.L., Araújo, A., Fugassa, M., Brooks, R.H., Racz, E., Reinhard, K.J., 2012. Zoonotic and human parasites of inhabitants of Cueva de Los Muertos Chiquitos, Rio Zape Valley, Durango, Mexico. J. Parasitol. 98, 304–309.
Johnson, J.H., 1934. A coprolite horizon in the Pennsylvanian of Chafee and Park counties, Colorado. J. Paleontol. 8, 477–479.

Jones, A., 1983. A coprolite from 6–8 pavement. In: Hall, A.R., Kenward, H.K., Williams, D., Greig, J.R.A. (Eds.), Environment and Living Conditions at Two Anglo-Scandinavian Sites. Council of British Archaeology, York, pp. 225–229.

Jouy-Avantin, F., Combes, C., Lumley, H., Miskovsky, J.C., Moné, H., 1999. Helminth eggs in animal coprolites from a middle Pleistocene site in Europe. J. Parasitol. 85, 376–379.

Kao, F.-T.K., 1962. Notes on coprolites from the Nihowan series. Vertebr. Palasiat. 4, 390–397.

Kapur, V.V., Kumar, K., Morthekai, P., Chaddha, A.S., 2019. Palaeodiet of Miocene producer(s) and depositional environment(s): Inferences from the first evidence on microcoprolites from India. Acta Geol. Sin. 94 (5), 1574–1590.

Karpinski, E., Mead, J.I., Poinar, H.N., 2017. Molecular identification of paleofeces from Bechan Cave, Southeastern Utah, USA. Quat. Int. 443, 140–146.

Keighley, D.G., Pickerill, R.K., 1997. Systematic ichnology of the Mabou and Cumberland groups (Carboniferous) of western Cape Breton Island, eastern Canada, 1: Burrows, pits, trails, and coprolites. Atl. Geol. 33, 181–215.

Kelso, G.K., 1971. Hogup Cave, Utah: Comparative Pollen Analysis of Human Coprolites and Cave Fill. Unpublished MA thesis. University of Arizona.

Kemp, R.A., Trueman, C.N., 2003. Rare earth elements in Solnhofen biogenic apatite: geochemical clues to the palaeoenvironment. Sediment. Geol. 155, 109–127.

Kerber, L., Oliveira, E.V., 2008. Fósseis de vertebrados da Formação Passo del Touro (Pleistoceno Superior), Rio Grande do Sul, Brasil. Atualização dos dados e novas contribuições. Gea. J. Geosci. 4, 49–64.

Khosla, A., Chin, K., Alimohammadin, H., Dutta, D., 2015. Ostracods, plant tissues, and other inclusions in coprolites from the Late Cretaceous Lameta Formation at Pisdura, India: Taphonomical and palaeoecological implications. Palaeogeogr. Palaeoclimatol. Palaeoecol. 418, 90–100.

Khosla, A., Chin, K., Verma, O., Kaur, J., Alimohammadin, H., Pal, S., Dutta, D., Rashid, M., Sharma, S., 2016. Systematics and paleobiogeographical implications of the fresh water Late Cretaceous ostracods, charophyte and distinctive residues from coprolites of the Lameta Formation at Pisdura, Chandrapur District (Maharashtra), Central India. New Mex. Mus. Nat. Hist. Sci. Bull. 71, 173–183.

Kimmig, J., Pratt, B.R., 2018. Coprolites in the Ravens Throat River Lagerstätte of northwestern Canada: Implications for the middle Cambrian food web. Palaios 33, 125–140.

Kimmig, J., Strotz, L.C., 2017. Coprolites in mid-Cambrian (Series 2-3) Burgess Shale-type deposits of Nevada and Utah and their ecological implications. Bull. Geosci. 92, 297–309.

Klein, H., Lucas, S.G., 2024. Triassic tetrapod footprints. In: Lucas, S.G., Hunt, A.P., Klein, H. (Eds.), Vertebrate Ichnology. Elsevier, Amsterdam, pp. 179–264.

Klemme, H.D., 1985. Regional geology of Circum-Mediterranean region. AAPG (Am. Assoc. Pet. Geol.) Bull. 42, 477–512.

Klompmaker, A.A., Herngreen, G.F.W., Oosterink, H.W., 2010. Biostratigraphic correlation, paleoenvironment stress, and subrosion pipe collapse: Dutch Rhaetian shales uncover their secrets. Facies 56, 597–613.

Knaust, D., 2020. Invertebrate coprolites and cololites revised. Papers Palaeontol. 6 (3), 385–423.

Kneuper, G., Schönenberg, R., 1962. Das Stefan A (Untere Ottweiler Gruppe) auf der Nordwestflanke des Saarbrücker Haupsattels. Bergfreiheit 12, 1–12.

Kowal-Linka, M., Bodzioch, A., 2012. Warstwykooecionooene z *Dactylosaurus* (Reptilia, Sauropterygia) z retu (Trias dolny, Olenek) Opolszczyzny. Przeglad Geol. 60, 646–649.

Krause, J.M., Piña, C.I., 2012. Reptilian coprolites in the Eocene of central Patagonia: Argentina. J. Paleontol. 86, 527–538.

Krause, J.M., Verde, M., Diz, R., 2007. Coprolitos del Grupo Rio Chico (Paleoceno Médio) en el sudeste de Chubut. Congresso Argentino de Paleontologia y Bioestratigrafia 9, 287. Resúmenes.

Kropf, M., 2005. The Extinct Shrub-Ox (*Euceratherium collinum*) and its Late Quaternary Environment on the Colorado Plateau. M.S. thesis. Northern Arizona University, Flagstaff, p. 179.

Kropf, M., Mead, J.I., Anderson, R.S., 2007. Dung, diet, and the paleoenvironment of the extinct shrub-ox (*Euceratherium collinum*) on the Colorado Plateau, USA. Quat. Res. 67, 143–151.

Krzykawski, T., Szopa, K., Brachaniec, T., Salamon, M.A., 2014. Carboniferous coprolite in the siderite concretion from Sosnowiec-Zagórze Lagerstätte, Poland—preliminary data. Freiberger Forschungshefte - A C 548, 101–107.

Kühl, G., Bartels, C., Briggs, D.E.G., Rust, J., 2012. Visions of a Vanished World: The Extraordinary Fossils of the Hunsrück Slate. Yale University Press, New Haven, p. 128.

LaGarry, H.E., 1997. Geology of the Toadstool Park Region of Northwestern Nebraska, With the Lithostratigraphic Revision and Redescription of the Brule Formation and Remarks on Oligocene Bone Processing. Ph.D. dissertation. University of Nebraska, Lincoln, p. 350.

Langer, M.C., 2005. Studies on continental Late Triassic tetrapod biochronology. I. The type locality of *Saturnalia tupiniquim* and the faunal succession in south Brazil. J. S. Am. Earth Sci. 19, 205–218.

Laojumpon, C., Matkhammee, T., Wathanapitaksakul, A., Suteethorn, V., Suteethorn, S., Lauprasert, K., Srisuk, P., Le Loeuff, J., 2012. Preliminary report on coprolites from the Late Triassic of Thailand. New Mexico Museum of Natural History. Bulletin 57, 207–213.

Lanzirotti, A., Becker, M., Hanson, G.N., Sutton, S.R., 2000. Uranium behavior in Jurassic fish coprolites: Combined Microbeam Synchrotron X-Ray Fluorescence, Diffraction, and XANES Analysis. NSLS Activ. Rep. 2, 48–52.

Laudermilk, J.D., Munz, P.A., 1938. Plants in the Dung of *Nothrotherium* from Rampart and MuavCaves, vol. 487. Carnegie Institute of Washington Publication, Arizona, pp. 271–281.

Lee, D.E., Kaulfuss, U., Conran, J.G., Bannister, J.M., Lindqvist, J.K., 2016. Biodiversity and palaeoecology of Foulden Maar: An early Miocene Konservat-Lagerstätte deposit in southern New Zealand. Alcheringa 40, 525–541.

Lemke, E., Weiler, W., 1942. Die Fischreste ause den permischen Brandschiefern am Fusse des Reisengebirges (Sudetengau). Mitteilungen des Reichsamtes für Bodenforschung Zweigstelle 22, 24.

Lemley, R.E., 1971. Notice of new finds in the Badlands. Proc. S. D. Acad. Sci. 50, 70–74.

Lewis, S.J., Heaton, K.W., 1997. Stool form scale as a useful guide to intestinal transit time. Scandanavian J. Gastroenterol. 32 (9), 920–924.

Lindqvist, J.K., Lee, D.E., 2009. High-frequency paleoclimate signals from Foulden Maar, Waipiata Volcanic Field, southern New Zealand: An Early Miocene varved lacustrine diatomite deposit. Sediment. Geol. 222, 98–110.

Liu, H.P., Bergström, S.M., Witzke, B.J., Briggs, D.E.G., Mckay, R.M., Ferretti, A., 2017. Exceptionally preserved conodont apparatuses with giant elements from the Middle Ordovician Winneshiek Konservat-Lagerstätte, Iowa, USA. J. Paleontol. 91, 493–511.

Lofgren, D.L., Shen, C.Y., Buday, N.N., Ylagan, C.A., Lofgren, K.K., Lai, R., Santana-Grace, D.D., Tabrum, A.R., 2017. Coprolites and mammalian carnivores from Pipestone Springs, Montana, and their paleoecological significance. Ann. Carnegie Mus. 84 (4), 265–285. https://doi.org/10.2992/0.

Lohmann, U., Sachs, S., 2001. Observations on the postcranial morphology, ontogeny and palaeobiology of *Sclerocephalus haeuseri* (Amphibia: Actinodontidae) from the Lower Permian of Southwest Germany. Memoir. Queensl. Mus. 46, 771–782.

Loinaze, V.S.P., Vera, E.I., Fiorelli, L.E., Desojo, J.B., 2018. Palaeobotany and palynology of coprolites from the Late Triassic Chañares Formation of Argentina: implications for vegetation provinces and the diet of dicynodonts. Palaeogeogr. Palaeoclimatol. Palaeoecol. 502, 31–51.

Lomax, D.R., Robinson, P., Cleal, C.J., Bowden, A., Larkin, N.R., 2016. Exceptional preservation of Upper Carboniferous (lower Westphalian) fossils from Edlington, Doncaster, South Yorkshire, UK. Geol. J. 51, 42–50.

Longbottom, A.E., Patterson, C., 2002. Fishes. In: Smith, A.B., Batten, D.J. (Eds.), Fossils of the Chalk, second ed., vol. 2. The Palaeontological Association Field Guides to Fossil, London, pp. 296–324. The Palaeontological Association.

Lucas, S.G., 2006. Global Permian tetrapod biostratigraphy and biochronology. In: Lucas, S.G., Cassinis, G., Schneider, J.W. (Eds.), Non-marine Permian Biostratigraphy and Biochronology, vol. 265. Geological Society of London, Special Publications, pp. 65–93.

Lucas, S.G., 2007. Tetrapod footprint biostratigraphy and biochronology. Ichnos 14, 5–38.

Lucas, S.G., 2013. Vertebrate biostratigraphy and biochronology of the upper Paleozoic Dunkard Group, Pennsylvania–West Virginia–Ohio, USA. Int. J. Coal Geol. 119, 79–87.

Lucas, S.G., Stimson, M.R., 2024. Carboniferous tetrapod footprints. In: Lucas, S.G., Hunt, A.P., Klein, H. (Eds.), Vertebrate Ichnology. Elsevier, Amsterdam, pp. 39–86.

Lucas, S.G., Hunt, A.P., Bennett, S.C., 1985. Triassic vertebrates from east-central New Mexico in the Yale Peabody Museum. In: Lucas, S.G., Zidek, J. (Eds.), Santa Rosa-Tucumcari Region, vol. 36. New Mexico Geological Society Guidebook, Socorro, pp. 199–204.

Lucas, S.G., Emry, R.J., Krainer, K., Hunt, A.P., Spielmann, J.A., 2012. Strigilites (fossilized owl pellets) from the Oligocene of Wyoming. New Mex. Mus. Nat. Hist. Sci. Bull. 57, 325–335.

Lukeneder, A., Lukeneder, P., 2021. The Upper Triassic Polzberg palaeobiota from a marine Konservat-Lagerstätte deposited during the Carnian Pluvial Episode in Austria. Sci. Rep. 11, 16644. https://doi.org/10.1038/s41598-021-96052-w.

Lukeneder, A., Surmik, D., Gorzelak, P., Niedźwiedzki, R., Brachaniec, T., Salamon, M.A., 2020. Bromalites from the Upper Triassic Polzberg section (Austria); Insights into trophic interactions and food chains of the Polzberg palaeobiota. Sci. Rep. 10, 20545. https://doi.org/10.1038/s41598-020-77017-x.

Luo, M., Hu, S., Benton, M.J., Shi, G.R., Zhao, L., Huang, J., Song, H., Wen, W., Zhang, Q., Fang, Y., Huang, Y., 2017. Taphonomy and palaeobiology of early Middle Triassic coprolites from the Luoping biota, southwest China: Implications for reconstruction of fossil food webs. Palaeogeogr. Palaeoclimatol. Palaeoecol. 474, 232–246.

MacNeish, R.S., 1958. Preliminary archeological investigations in the Sierra de Tamaulipas, Mexico, vol. 48. American Philosophical Society Transactions, pp. 235–359.

Madurell-Malapeira, J., Alba, D.M., Aurell-Garrido, P., Moyà-Solà, S., 2011. Taphonomic approach to the last Villafranchian faunas of Europe: The layer 7 of the Vallparadís Estació local section (Vallès-Penedès Basin, NE Iberian Peninsula). IX Encuentro de Jóvenes Investigadores en Paleontología, Morella (Castellón). Proc. Vol. 203–210.

Madurell-Malapeira, J., Minwer-Barakat, R., Alba, D.M., Garcés, M., Gómez, M., Aurell-Garrido, J., Ros-Montoya, S., Moyà-Solà, S., Berástegui, X., 2010. The Vallparadís section (Terrassa, Iberian Peninsula) and the latest Villafranchian faunas of Europe. Quat. Sci. Rev. 29, 3972–3982.

Mahaney, W.C., Barendregt, R.W., Allen, C.C.R., Milner, M.W., Bay, D., 2013. Coprolites from the Cretaceous Bearpaw Formation of Saskatchewan. Cretac. Res. 14, 31–38.

Maisey, J.G., Melo, J.H.G., 2005. Some Middle Devonian (Eifelian–Givetian) fossil fish remains from the Pimenteira Formation of the Parnaíba Basin, northeast Brazil. Arq. Mus. Nac. 63, 495–505.

Mancuso, A.C., Marsicano, C., Palma, R., 2004. Vertebrate coprolites from the Triassic of Argentina (Cuyana Basin). Ameghiniana 41, 347–354.

Mancuso, A.C., Benavente, C.A., Previtera, E., Arcucci, A.B., Irmis, R.B., 2018. Carnivore coprolites from the lower Carnian (Upper Triassic) Chañares Formation, northwestern Argentina. Palaeogeogr. Palaeoclimatol. Palaeoecol. 489, 15–28.

Mansky, C., Lucas, S.G., Hunt, A.P., 2012. Mississippian bromalites from Blue Beach, Nova Scotia, Canada: Vertebrate Coprolites. New Mexico Museum of Natural History. Bulletin 57, 161–170.

Mantell, G.A., 1822. The Fossils of the South Downs: Or Illustrations of the Geology of Sussex, p. 305. London, Lupton Relfe.

Marchetti, L., Mujal, E., Logghe, A., Buchwitz, M., Klein, H., Lucas, S.G., 2024. Permian vertebrate tracks. In: Lucas, S.G., Hunt, A.P., Klein, H. (Eds.), Vertebrate Ichnology. Elsevier, Amsterdam, pp. 87–178.

Martens, T., Lucas, S.G., 2005. Taxonomy and biostratigraphy of Conchostraca (Branchiopoda, Crustacea) from two nonmarine Pennsylvanian and Lower Permian localities in New Mexico. New Mex. Mus. Nat. Hist. Bull. 30, 208–213.

Martill, D.M., 1985. The preservation of marine vertebrates in the Lower Oxford Clay (Jurassic) of central England. Phil. Trans. 311, 155–165.

Martill, D.M., Naish, D., 2001. Dinosaur trace fossils: Footprints, coprolites and gastroliths. In: Martill, D.M., Naish, D. (Eds.), Dinosaurs of the Isle of Wight, vol. 10. Palaeontological Association, Field Guides to Fossils, pp. 310–323.

Martill, D.M., Taylor, M.A., Duff, K.L., Riding, J.B., Bown, P.R., 1994. The trophic structure of the biota of the Peterborough Member, Oxford Clay Formation (Jurassic), UK. J. Geol. Soc. London 151, 173–194.

Martin, J.E., 1981. Contents of coprolites from Hemphillian sediments in northern Oregon and their importance in paleoecological interpretations. Proc. S. D. Acad. Sci. 60, 105–115.

Martin, P.S., Sabela, B.E., Shutler, D., 1961. Rampart Cave coprolites and ecology of the Shasta ground sloth. Am. J. Sci. 259, 102–127.

Martin, P.S., Thompson, R.S., Long, A., 1985. Shasta ground sloth extinction: A test of the blitzkrieg model. In: Mead, J.I., Meltzer, D.J. (Eds.), Environments and Extinction: Man in Late Glacial North America. Orono. University of Maine, pp. 5–14.

Martini, E., 1965. Die Fischfauna von Sieblos/Rhön (Oligozän): 2. Fischreste aus Koprolithen. Senckenbergiana 46, 307–314.

Matthews, W., 1995. Micromorphological characterization and interpretation of occupation deposits and microstratigraphic sequences at Abu Salabikh, Iraq. In: Barham, A.J., Macphail, R.I. (Eds.), Archaeological Sediments and Soils: Analysis. Interpretation and Management. Institute of Archaeology, London, pp. 41–76.

Matthews, W., Shillito, L.-M., Elliot, S., Bull, I.D., Williams, J., 2014. Neolithic lifeways: microstratigraphic traces within houses, animal pens and settlements. In: Whittle, A., Bickle, P. (Eds.), Early Farmers: The View from Archaeology and Science. British Academy and OUP, pp. 251–279.

Matley, C., 1939a. On some coprolites from the Maleri Beds of India. Record Geol. Surv. India 74, 535–547.

Matley, C., 1939b. The coprolites of Pijdura, Central Province. Record Geol. Surv. India 74, 530–534.

Matsui, A., Kanehara, M., Kanehara, M., 2003. Palaeoparasitology in Japan: Discovery of Toilet Features, vol. 98. Memórias do Instituto Oswaldo Cruz, pp. 127–136.

Mazurek, D., Antczak, M., 2023. Late Cretaceous coprolite from the Opole area (southern Poland) as evidence for a variable diet in shell-crushing shark *Ptychodus* (Elasmobranchii: Ptychodontidae). PeerJ 11, e16598. https://doi.org/10.7717/peerj.16598.

McAllister, J.A., 1985. Reevaluation of the origin of spiral coprolites. Univ. Kansas Palaeontol. Contrib. 114, 1–12.

McAllister, J.A., 1987. Phylogenetic distribution and morphological reassessment of the intestines of fossil and modern fishes. Zool. Jahrb. - Abt. Anat. Ontog. Tiere 115, 281–294.

McAllister, J.A., 1988. Preliminary description of the coprolitic remains from Hamilton quarry, Kansas. In: Mapes, G., Mapes, R.H. (Eds.), Regional Geology and Paleontology of Upper Paleozoic Hamilton Quarry Area, vol. 6. Kansas Geological Survey Guidebook, pp. 195–202.

McAllister, J.A., 1996. Coprolites. In: Schultze, H.-P., Cloutier, R. (Eds.), Devonian Fishes and Plants of Miguasha. Verlag Dr Friedrich Pfeil, Quebec, Canada. München, pp. 328–347.

McDonough, K.N., 2019. Middle Holocene menus: dietary reconstruction from coprolites at the Connley Caves, Oregon, USA. Archaeol. Anthropol. Sci.

Mead, J.I., Agenbroad, L.D., 1989. Pleistocene dung and the extinct herbivores of the Colorado Plateau, southwestern USA. Cranium 6 (1), 29–44.

Mead, J.I., Swift, S.L., 2012. Late Pleistocene (Rancholabrean) dung deposits of the Colorado Plateau, Western North America. New Mex. Mus. Nat. Hist. Sci. Bull. 57, 337–342.

Mead, J.I., Agenbroad, L.D., Stuart, A., 1993. Late Pleistocene vertebrates from Bechan Cave, Colorado Plateau, Utah. In: Santucci, V.L., McClelland, L. (Eds.), National Park Service Paleontological Research Abstract Volume. Technical Report NPS/NRPEFO/NRTR-93/11, 69.

Mead, J.I., Agenbroad, L.D., Davis, O.K., Martin, P.S., 1986a. Dung of *Mammuthus* in the arid Southwest, North America. Quat. Res. 25, 121–127.

Mead, J.I., Agenbroad, L.D., Martin, P.S., Davis, O.K., 1984. The mammoth and sloth dung from Bechan Cave in Southern Utah. Curr. Res. Pleistoc. 1, 79–80.

Mead, J.I., Tweet, J.S., Santucci, V.L., Tobin, B., Chambers, C.L., Thomas, S.C., Carpenter, M.C., 2020. Pleistocene/Holocene cave fossils from Grand Canyon National Park: Ice Age (Pleistocene) flora, fauna, environments, and climate of the Grand Canyon, Arizona. In: Santucci, V.L., Tweet, J.S. (Eds.), Grand Canyon National Park: Centennial Paleontological Resource Inventory. Natural Resource Report NPS/GRCA/NRR—2020/2103. National Park Service, Fort Collins, pp. 403–463.

Mead, J.I., Martin, P.S., Euler, R.C., Long, A., Jull, A.T., Toolin, L.T., Donahue, D.J., Linick, T.W., 1986b. Extinction of Harrington's Mountain Goat. Proc. Natl. Acad. Sci. USA 83, 836–839.

Mears, E.M., Rossi, V., MacDonald, E., Coleman, G., Davies, T.G., Arias-Riesgo, C., Hildebrandt, C., Thiel, H., Duffin, C.J., Whiteside, D.I., Benton, M.J., 2016. The Rhaetian (Late Triassic) vertebrates of Hampstead Farm Quarry, Gloucestershire, UK. Proc. Geol. Assoc. 127 (4), 478–505.

Meehan, T.J., 2007. Hair traces in carnivorous coprolites. Geol. Soc. Am. Abstr. Progr. 39 (3), 18.

Megirian, D., Murray, P., Wells, R., 1996. The late Miocene Ongeva local fauna of central Australia. Beagle Rec. Mus. Art Galleries Northern Territory 13, 9–37.

Meillet, A., 1842. Minéralogie de basin de Paris: Analyse de l'Apatélite et des coprolithes du Calcaire grossier et de la craie. Revue de science et industrie, Paris 11, 254–257.

Meng, J., Wyss, A.R., 1997. Multituberculate and other mammal hair recovered from Palaeogene excreta. Nature 385, 712–714.

Mikuláš, R., Dvorák, Z., 2010. Possible crocodylian bite traces, Miocene of the Most Basin (Czech Republic). New Mex. Mus. Nat. Hist. Sci. Bull. 51, 191–194.

Milàn, J., 2010. Coprolites from the Danian Limestone (lower Paleocene) of Faxe Quarry, Denmark. New Mex. Mus. Nat. Hist. Sci. Bull. 51, 215–218.

Milàn, J., 2012. Crocodylian scatology – a look into morphology, internal architecture, inter- and intraspecific variation and prey remains in extant crocodylian feces. New Mex. Mus. Nat. Hist. Bull. 37, 65–71.

Milàn, J., 2018. First find of the coprolite ichnotaxon *Daniacopros hofstedtae*, from the Middle Danian Faxe Formation of Faxe Quarry, Denmark: New Mexico Museum of Natural History. Bulletin 79, 499–501.

Milàn, J., Hedegaard, R., 2010. Interspecific variation in tracks and trackways from extant crocodylians. New Mex. Mus. Nat. Hist. Sci. Bull. 51, 15–29.

Milàn, J., Hunt, A.P., 2016. *Daniacopros hofstedtae*, ichnogen. et ichnosp. nov., a new vertebrate coprolite ichnotaxon from the Lower Danian Stevns Klint Formation of the Hammelev limestone quarry, Denmark. New Mex. Mus. Nat. Hist. Sci. Bull. 74, 159–161.

Milàn, J., Rasmussen, E.S., Dybkjær, K., 2018. A crocodilian coprolite from the lower Oligocene Viborg Formation of Sofienlund Lergrav, Denmark. Bull. Geol. Soc. Den. 66, 181–187.

Milàn, J., Hunt, A.P., Adolfssen, J.S., Rasmussen, B.W., Bjerager, M., 2015. First record of a vertebrate coprolite from the Upper Cretaceous (Maastrichtian) Chalk of Stevns Klint, Denmark. New Mex. Mus. Nat. Hist. Sci. Bull. 67, 227–229.

Milàn, J., Mateus, O., Mau, M., Rudra, A., Sanei, H., Clemmensen, L.B., 2021. A possible phytosaurian (Archosauria, Pseudosuchia) coprolite from the Late Triassic Fleming Fjord Group of Jameson Land, central East Greenland. Bull. Geol. Soc. Den. 69, 71–80.

Milàn, J., Clemmensen, L.B., Adolfssen, J.S., Estrup, E.J., Frobøse, N., Klein, N., Mateus, O., Wings, O., 2012b. A preliminary report on coprolites from the Late Triassic part of the Kap Stewart Formation, Jameson Land, East Greenland. New Mex. Mus. Nat. Hist. Sci. Bull. 57, 203–206.

Mitzopoulos, M.K., Zapfe, H., 1963. Fossile Hyäniden-Koprolithen au Pikermi. Annales Géologie Pays Hellèniques. Athénes 14, 405–407.

Mohr, E., 1964. Bermerkungen uber Hyänenkot und Koprolithen. In: Festschrift, K. (Ed.), Hamburger Zoologische Institut und Museum Mittelung, pp. 107–111. Hamburg.

Montgomery, C., 2022. Application of Micro Computed-Tomography in the Characterisation of Early Triassic Coprolite Microstructures and Microfossils From the Burgersdorp Formation (Tarkastad Subgroup, Beaufort Group) of South Africa. Unpublished MSc dissertation, University of the Witwatersrand, Johannesburg, p. 99.

Montalvo, C.I., Tomassini, R.L., Sostillo, R., 2016. Leftover prey remains: A new taphonomic mode from the Late Miocene Cerro Azul Formation of Central Argentina. Lethaia 49, 219–230.

Montalvo, C.I., Raigemborn, M.S., Tomassini, R.L., Zapata, L., Bargo, M.S., Uncal, M.C.M., Vizcaíno, S.F., 2019. Floodplain taphonomic mode of early Miocene vertebrates of Southern Patagonia, Argentina. Palaios 34, 105–120.

Morales, J., Senut, B., Pickford, M., 2011. *Crocuta dietrichi* from Meob, Namibia: Implications for the age of the Tsondab Sandstone in the coastal part of the Namib Desert. Estud. Geol. 67, 207–215.

Morales, M., 1987. Terrestrial fauna and flora from the Triassic Moenkopi Formation of the southwestern United States. J. Arizona Nevada Acad. Sci. 22, 1–20.

Moreau, J.-D., Vullo, R., Charbonnier, S., Jattiot, R., Trincal, V., Néraudeau, D., Fara, E., Baret, L., Garassino, A., Gand, G., Lafaurie, G., 2022. Konservat-Lagerstätten from the Upper Jurassic lithographic limestone of the Causse Méjean (Lozère, southern France): Palaeontological and palaeoenvironmental synthesis. Geol. Mag. https://doi.org/10.1017/S0016756821001382.

Muftah, A.M., 2020. Panoramic of paleohabitats of Pliocene Sahabi Formation and Recent As-Sahabi Area, Northern Libya. Iraqi Geol. J. 53, 1–11.

Muftah, A.M., El-Shawaihdi, M.H., Al Riaydh, M.H., Boaz, N.T., 2020. Coprolites from the Neogene Sahabi Formation, northeastern Sirt Basin of Libya. Arabian J. Geosci. 13, 1–10.

Murchison, R.I., 1867. Siluria, fourth ed. John Murray, London, p. 566.

Murie, O.J., 1974. Animals Tracks, second ed. Peterson Field Guild Series. Boston, Houghton Mifflin, p. 375.

Musil, R., 1962. Die Höhle "Sveduv stul," ein typischer Höhlenhyänenhors. Anthropos NS 5, 97–260.

Mychajliw, A.M., Rice, K.A., Tewksbury, L.R., Southon, J.R., Lindsey, E.L., 2020. Exceptionally preserved asphaltic coprolites expand the spatiotemporal range of a North American paleoecological proxy. Scientific Rep. 10, 5069. https://doi.org/10.1038/s41598-020-61996-y.

Nakajima, Y., Izumi, K., 2014. Coprolites from the upper Osawa Formation (upper Spathian), northeastern Japan: Evidence for predation in a marine ecosystem 5 Myr after the end-Permian mass extinction. Palaeogeogr. Palaeoclimatol. Palaeoecol. 414, 225–232.

Napton, L.K., Heizer, R.F., 1970. Analysis of human coprolites from archaeological contexts, with primary reference to Lovelock Cave, Nevada. Contrib. Univ. Calif. Archaeol. Res. Facility 10, 87–129.

Nesov, L.A., 1995. Dinosaurs of northern Eurasia: New data about assemblages, ecology and paleobiogeography. Saint Petersburg, Izdatel'stvo Sankt-2 Peterburgskogo Universiteta, p. 156.

Nesov, L.A., 1997. [Cretaceous nonmarine vertebrates of northern Eurasia] (Posthumous Edition). In: Golovneva, L.B., Averianov, A.O. (Eds.), St. Petersburg, Izdatel'stvo Sankt-PeterburgskogoUniversiteta, p. 218.

Neumayer, L., 1904. Die Koprolithen des Perm von Texas. Palaeontographica 51, 121–128.

Newman, M., Davidson, R., 2010. Early Devonian fish from the Midland Valley of Scotland. In: IPC3 Pre-conference Field Trip. Aberdeen. University of Aberdeen, p. 38.

Niedźwiedzki, G., Bajdek, P., Owocki, K., Kear, B.P., 2016a. An Early Triassic polar predator ecosystem revealed by vertebrate coprolites from the Bulgo Sandstone (Sydney Basin) of southeastern Australia. Palaeogeogr. Palaeoclimatol. Palaeoecol. 464, 5–15.

Niedźwiedzki, G., Bajdek, P., Qvarnström, M., Sulej, T., Sennikov, A.G., Golubev, V.K., 2016b. Reduction of vertebrate coprolite diversity associated with the end-Permian extinction event in Vyazniki region, European Russia. Palaeogeogr. Palaeoclimatol. Palaeoecol. 450, 77–90.

Njau, J.K., Blumenschine, R.J., 2012. Crocodylian and mammalian carnivore feeding traces on hominid fossils from FLK 22 and FLK NN 3, Plio-Pleistocene, Olduvai Gorge, Tanzania. J. Hum. Evol. 63, 408–417.

Norden, K.K., Duffin, C.J., Benton, M.J., 2015. A marine vertebrate fauna from the Late Triassic of Somerset, and a review of British placodonts. Proc. Geologists' Assoc. 126 (4–5), 564–581.

Norman, D.B., 1987. A mass-accumulation of vertebrates from the Lower Cretaceous of Nehden (Sauerland), West Germany. Proc. R. Soc. Lond. B Biol. Sci. 230, 215–255.

Northwood, C., 1997. Palaeontological Interpretations of the Early Triassic Arcadia Formation, Queensland. Ph.D. dissertation. La Trobe University, Melbourne, p. 479.

Northwood, C., 2005. Early Triassic coprolites from Australia and their palaeobiological significance. Palaeontology 48, 49–68.

Nürnberger, L., 1934. Koproporhyrin im tertiären Krocodilkot. Nova Acta Leopoldina, Halle 1, 324–325.

Ochev, V.G., 1974. Some remarks on coprolites of Triassic vertebrates: Paleontological. Journal 1974, 253–255.

Ó Gogáin, A., Falcon-Lang, H.J., Carpenter, D.K., Miller, R.F., Benton, M.J., Pufahl, P.K., Ruta, M., Davies, T.G., Hinds, S.J., Stimson, M.R., 2016. Fish and tetrapod communities across a marine to brackish salinity gradient in the Pennsylvanian (early Moscovian) Minto Formation of New Brunswick, Canada, and their palaeoecological and palaeogeographical implications. Palaeontology 59, 689–724.

Olson, E.C., 1966. Community evolution of the origin of mammals. Ecology 47, 291–302.

Olson, E.C., 1971. A skeleton of *Lysorophus tricarinatus* (Amphibia: Lepospondyli) from the Hennessey Formation (Permian) of Oklahoma. J. Paleontol. 45, 443–449.

Olson, E.C., 1977. Permian lake faunas: A study in community evolution. J. Palaeontol. Soc. India 20, 146–163.

Olson, E.C., Mead, J.G., 1982. The Vale Formation (Lower Permian): Its vertebrates and paleoecology. Texas Memorial Mus. Bull. 29, 46.

Ősi, A., Szabó, M., Tóth, E., Bodor, E., Lobitzer, H., Kvaček, J., Svobodová, M., Szente, I., Wagreich, M., Trabelsi, K., Sames, B., 2021. A brackish to non-marine aquatic and terrestrial fossil assemblage with vertebrates from the lower Coniacian (Upper Cretaceous) Gosau Group of the Tiefengraben locality near St. Wolfgang im Salzkammergut, Austria. Cretac. Res. 127. https://doi.org/10.1016/j.cretres.2021.104938.

Owocki, K., Niedzwiedzki, G., Sennikov, A.G., Golubev, V.K., Janiszewska, K., Sulej, T., 2012. Upper Permian vertebrate coprolites from Vyazniki and Gorokhovets, Vyatkian regional stage, Russian Platform. Palaios 27, 867–877.

Oyarzún-Ruiz, P., Pérez-Espinoza, S.A., González-Saldías, F., Martin, F., Lucila Moreno, L., 2021. Paleoparasitological survey of coprolites of Darwin's ground sloth *Mylodon darwini* (Xenarthra, Folivora: Mylodontidae) from Cueva del Milodón Natural Monument, Chilean Patagonia. Archaeol. Anthropol. Sci. 13 (8), 138. https://doi.org/10.1007/s12520-021-01383-4.

Pan, Y., Sha, J., Zhou, Z., Fürsich, F.T., 2013. The Jehol Biota: Definition and distribution of exceptionally preserved relicts of a continental Early Cretaceous ecosystem. Cretac. Res. 44, 30e38. https://doi.org/10.1016/j.cretres.2013.03.007.

Pan, Y., Sha, J., Fürsich, F.T., Wang, Y., Zhang, X., Yao, X., 2011. Dynamics of the lacustrine fauna from the Early Cretaceous Yixian Formation, China: Implications of volcanic and climatic factors. Lethaia. https://doi.org/10.1111/j.1502-3931.2011.00284.x.

Pedley, H.M., 1978. A new fish horizon from the Maltese Miocene and its palaeoecological significance. Palaeogeogr. Palaeoclimatol. Palaeoecol. 24, 73–83.

Peel, J.S., 2015. Bromalites from the Cambrian (Series 2 and 3) of North Greenland. GFF 137, 181–194. https://doi.org/10.1080/11035897.2014.995217.

Pemberton, S.G., 2012. William Buckland (1784-1856) and Henry De La Beche (1796-1855): The early history of coprolites. New Mex. Mus. Nat. Hist. Sci. Bull. 57, 29–43.

Pemberton, S.G., Frey, R.W., 1991. William Buckland and his 'coprolitic vision'. Ichnos 1, 317–325.

Peñalver, E., Gaudant, J., 2010. Limnic food web and salinity of the Upper Miocene Bicorb palaeolake (eastern Spain). Palaeogeogr. Palaeoclimatol. Palaeoecol. 297, 683–696.

Perri, A.R., Heinrich, H., Gur-Arieh, S., Saunders, J.J., 2017. Earliest evidence of *Toxocara* sp. in a 1.2-Million-Yr-old extinct hyena (*Pachycrocuta brevirostris*) coprolite from Northwest Pakistan. J. Parasitol. 103 (1), 138–141. https://doi.org/10.1645/16-71.

Pesquero, M.D., Alcalá, L., Fernández-Jalvo, Y., 2013a. Taphonomy of the reference Miocene vertebrate mammal site of Cerro de la Garita, Spain. Lethaia 46, 378–398.

Pesquero, M.D., Souza-Egipsy, V., Alcalá, L., Ascaso, C., Fernández-Jalvo, Y., 2013b. Calcium phosphate preservation of faecal bacterial negative moulds in hyaena coprolites. Acta Palaeontol. Pol. 59, 997–1005.

Pesquero, M.D., Salesa, M.J., Espílez, E., Mampel, L., Siliceo, G., Alcalá, L., 2011. An exceptionally rich hyaena coprolites concentration in the Late Miocene mammal fossil site of La Roma 2 (Teruel, Spain): Taphonomical and palaeoenvironmental inferences. Palaeogeogr. Palaeoclimatol. Palaeoecol. 311, 30–37.

Petrigh, R.S., Martínez, J.G., Mondini, M., Fugassa, M.H., 2019. Ancient parasitic DNA reveals *Toxascaris leonina* presence in final Pleistocene of South America. Parasitology 146 (10), 284–1288.

Phillips, A., 2008. A Late Cretaceous (Cenomanian) Marine Vertebrate-Rich Bioclastic Horizon From the Northeastern Margin of the Western Interior Seaway, Canada. Ph.D. dissertation. Carleton University, Ottawa, p. 237.

Pichler, S.L., Pumpin, C., Bronnimann, D., Rentzel, P., 2014. Life in the proto-urban style: the identification of parasite eggs in micromorphological thin sections from the Bance,el-Gasfabrik Late Iron Age settlement, Switzerland. J. Archaeol. Sci. 43, 55–65.

Platt, S.G., Elsey, R.M., Bishop, N.D., Rainwater, T.R., Thongsavath, O., Labarre, D., McWilliam, A.G., 2020. Using scat to estimate body size in crocodilians: Case studies of the Siamese crocodile and American alligator with practical applications. Herpetol. Conserv. Biol. 15, 325–334.

Poinar, G., 2021. Fossil record of viruses, parasitic bacteria and parasitic protozoa. In: De Baets, K., Huntley, J.W. (Eds.), The Evolution and Fossil Record of Parasitism: Identification and Macroevolution of Parasites. Topics in Geobiology, vol. 49. Springer, pp. 29–68.

Poinar, G., Boucot, A.J., 2006. Evidence of intestinal parasites of dinosaurs. Parasitology 133 (02), 245–249. https://doi.org/10.1017/S0031182006000138.

Poinar, H.N., Kuch, M., McDonald, G., Martin, P., Pääbo, S., 2003. Nuclear gene sequences from a late Pleistocene sloth coprolite. Curr. Biol. 13, 1150–1152.

Poinar, H.N., Hofreiter, M., Spaulding, W.G., Martin, P.S., Stankiewicz, B.A., Bland, H., Evershed, R.P., Possnert, G., Pääbo, S., 1998. Molecular coproscopy: Dung and diet of the extinct ground sloth *Nothrotheriops shastensis*. Science 281, 402–406.

Pollard, J.E., 1985. Coprolites and ostracods from the Dinantian of Foulden, Berwickshire, Scotland. Trans. R. Soc. Edinb. Earth Sci. 76, 49–51.

Price, P.H., 1927. The coprolite horizon of the Conemaugh Series in and around Morgantown, West Virginia. Ann. Carnegie Mus. 17, 211–254.

Pucu, E., Russ, J., Reinhard, K., 2020. Diet analysis reveals pre-historic meals among the Loma San Gabriel at La Cueva de Los Muertos Chiquitos, Rio Zape, Mexico (600–800 CE). Archaeol. Anthropol. Sci. 12, 25. https://doi.org/10.1007/s12520-019-00950.

Qvarnström, M., 2020a. Food web analysis of Early Mesozoic ecosystems shed new light on early dinosaur evolution. Soc. Vertebr. Paleontol. 80th Ann. Meet. Virt. 2020 October 12-16, unnumbered page.

Qvarnström, M., 2020b. Who Ate Whom? Paleoecology Revealed Through Synchrotron Microtomography of Coprolites (Fossil Feces). Ph.D. dissertation. Acta Universitatis Upsaliensis, Upsalla, p. 49.

Qvarnström, M., Ahlberg, P.E., Niedźwiedzki, G., 2019a. Tyrannosaurid-like osteophagy by a Triassic archosaur. Sci. Rep. 9, 925. https://doi.org/10.1038/s41598-018-37540-4.

Qvarnström, M., Niedźwiedzki, G., Žigaitė, Ž., 2016. Vertebrate coprolites (fossil faeces): an underexplored Konservat-Lagerstätte. Earth Sci. Rev. 162, 44–57.

Qvarnström, M., Elgh, E., Owocki, K., Ahlberg, P.E., Niedźwiedzki, G., 2019b. Filter feeding in Late Jurassic pterosaurs supported by coprolite contents. PeerJ 7, e7375. https://doi.org/10.7717/peerj.7375.

Qvarnström, M., Niedźwiedzki, G., Tafforeau, P., Žigaitė, Ž., Ahlberg, P.E., 2017. Synchrotron phase-contrast microtomography of coprolites generates novel palaeobiological data. Sci. Rep. 7, 2723. https://doi.org/10.1038/s41598-017-02893-9.

Qvarnström, M., Wernström, J.V., Piechowski, R., Tałanda, M., Ahlberg, P.E., Niedźwiedzki, G., 2019c. Beetle-bearing coprolites possibly reveal the diet of a Late Triassic dinosauriform. R. Soc. Open Sci. 6, 181042. https://doi.org/10.1098/rsos.181042.

Qvarnström, M., Anagnostakis, S., Lindskog, A., Scheer, U., Vajda, V., Rasmussen, B.W., Lindgren, J., Eriksson, M.E., 2019d. Multi-proxy analyses of Late Cretaceous coprolites from Germany. Lethaia 52, 550–569.

Qvarnström, M., Fikáček, M., Wernström, J.V., Huld, S., Beutel, R.G., Arriaga-Varela, E., Ahlberg, P.E., Niedźwiedzki, G., 2021. Exceptionally preserved beetles in a Triassic coprolite of putative dinosauriform origin. Curr. Biol. https://doi.org/10.1016/j.cub.2021.05.0.

Raczka, M.F., Bush, M.B., Folcik, A.M., McMichael, C.H., 2016. *Sporormiella* as a tool for detecting the presence of large herbivores in the Neotropics. Biota Neotr. 16 (1), e20150090. https://doi.org/10.1590/1676-0611-BN-2015-0090.

Rakshit, N., Ray, S., 2021. Bone-bearing coprolites from the Upper Triassic of India: ichnotaxonomy, probable producers and predator–prey relationships. Papers Palaeontol. 8 (1), e1418. https://doi.org/10.1002/spp2.1418.

Rakshit, N., Bhat, M.S., Mukherjee, D., Ray, S., 2019. First record of Mesozoic scroll coprolites: classification, characteristics, elemental composition and probable producers. Palaeontology 62, 451–471.

Rampelli, S., Turroni, S., Mallol, C., Hernandez, C., Galván, B., Sistiaga, A., Biagi, E., Astolfi, A., Brigidi, P., Benazzi, S., Lewis, J.,C.M., 2021. Components of a Neanderthal gut microbiome recovered from fecal sediments from El Salt. Commun. Biol. 4, 169. https://doi.org/10.1038/s42003-021-01689-y.

Rayner, D., Mitchell, T., Rayner, M., Clouter, F., 2009. London Clay Fossils of Kent and Essex. Sevenoaks, Caxton and Holmesdale Press and Medway Fossil and Mineral Society, pp. 1–228.

Rayner, D.H., 1963. The Achanarras Limestone of the Middle Old Red Sandstone, Caithness, Scotland. Proc. Yorks. Geol. Soc. 34, 117–138.

Reed, P.F., 2006. Thirty-Five Years of Archaeological Research at Salmon Ruins Volume 3. Tucson. Center for Desert Archaeology and Salmon Ruins Museum, pp. 875–888.

Reinhard, K.J., 1988. Diet, Parasitism, and Anemia in the Prehistoric Southwest. Ph.D. dissertation. Texas A&M University, College Station, pp. 1–201.

Reinhard, K.J., 1990. Archaeoparasitology in North America. Am. J. Phys. Anthropol. 82, 145–163.

Reinhard, K.J., 2006. A coprological view of ancestral Pueblo cannibalism: debate over a single fecal fossil offers a cautionary tale of the interplay between science and culture. Am. Sci. 94, 254–261.

Reinhard, K.J., 2007. Pathoecology of two Anasazi villages. In: Reitz, E.J. (Ed.), Case Studies in Environmental Archaeology, second ed. Plenum Press, New York, pp. 191–210.

Reinhard, K.J., Bryant Jr., V.M., 1992. Coprolite analysis: A biological perspective on archaeology. University of Nebraska, Lincoln. Papers Nat. Res. 46, 245–288.

Reinhard, K., Bryant, V.M., 2008. Pathoecology and the future of coprolite studies in bioarchaeology. In: Stodder, A.L.W. (Ed.), Reanalysis and Reinterpretation in Southwestern Bioarcheology, vol. 59. Arizona State University Anthropological Research Papers, 191-126.

Reinhard, K.J., LeRoy-Toren, S., 2006. Salmon Ruin coprolites: San Juan Diet. In: Reed, P.F. (Ed.), Thirty-five Years of Archaeological Research at Salmon Ruins Volume 3. Tucson, Center for Desert Archaeology and Salmon Ruins Museum, pp. 875–888.

Reinhard, K.J., Johnson, K.L., LeRoy-Toren, S., Wieseman, K., Teixeira-Santos, I., Vieira, M., 2012. Understanding the pathoecological relationship between ancient diet and modern diabetes through coprolite analysis: a case example from Antelope Cave, Mojave County, Arizona. Curr. Anthropol. 53 (4), 506–512.

Retallack, G.J., 1983. Late Eocene and Oligocene Paleosols from Badlands National Park, South Dakota. Geological Society of America Special Paper 193, 82p.

Retallack, G.J., Krull, E.S., 1999. Permian coprolites from Graphite Peak, Antarctica. U.S. Antarctic J. 30, 7–9.
Reuss, A.E., 1856. Über Koprolithen im Rothliegen Böhmens. Sitzsungberichte Berlin Akademie der Wissenschaft Wien, mathematische-naturhistorische Klasse Vienna 18, 124–132.
Rice, K., Mychajliw, A., Tewksbury, L., Lindsey, E., 2019. Stable isotope analysis of asphaltic coprolites reveals stability of woodrat diet (*Neotoma* sp.) from Rancho La Brea, Los Angeles, California. AGU Fall Meeting Abstr. Abstract. PP42C-17.
Richardson Jr., E.S., 1980. Life at Mazon Creek. In: Langenheim, R.L., Mann, C.J. (Eds.), Middle and Late Pennsylvanian Strata on Margin of Illinois Basin. 10th Annual Field Conference of the Society of Economic Paleontologists and Mineralogists, Great Lakes Section. University of Illinois, Urbana, pp. 217–224.
Richardson Jr., E.S., Johnson, R.G., 1971. The Mazon Creek fauna. In: Proceedings of the North American Paleontology Convention II Lawrence, pp. 1222–1235.
Richter, G., Baszio, S., 2001a. First proof of planctivory/insectivory in a fossil fish: *Thaumaturus intermedius* from the Eocene Lake Messel (FRG). Palaeogeogr. Palaeoclimatol. Palaeoecol. 173, 75–85.
Richter, G., Baszio, S., 2001b. Traces of a limnic food web in the Eocene Lake Messel—a preliminary report based on fish coprolite analyses. Palaeogeogr. Palaeoclimatol. Palaeoecol. 166, 345–368.
Richter, G., Wedmann, S., 2005. Ecology of the Eocene Lake Messel revealed by analysis of small fish coprolites and sediments from a drilling core. Palaeogeogr. Palaeoclimatol. Palaeoecol. 223, 147–161.
Ridgwell, N.M., Chin, K., Upchurch, G., Sertich, J., 2015. Rare direct evidence of angiosperm consumption by dinosaurs based on coprolites from the Kaiparowits Formation of Utah. J. Vertebr. Paleontol. 24 (Suppl. to no. 3), 203.
Riley, T., 2008. Diet and seasonality in the Lower Pecos: evaluating coprolite data sets with cluster analysis. J. Archaeol. Sci. 35 (10), 2726–2741.
Riley, T., 2012. Assessing diet and seasonality in the Lower Pecos canyonlands: an evaluation of coprolite specimens as records of individual dietary decisions. J. Archaeol. Sci. 39, 145–162.
Robertson, D., 1834. Bulletin de la Société Géologique de France, series, vol. 1, p. 415.
Robin, N., Foldi, I., Godinot, M., Petit, G., 2016. Scale insect larvae preserved in vertebrate coprolites (Le Quesnoy, France, Lower Eocene): Paleoecological insights. Sci. Nat. 103 (9), 85. https://doi.org/10.1007/s00114-016-1412-x.
Rodriguez De La Rosa, R.A., Cevallo-Ferriz, S.R.S., Silva-Pineda, A., 1998. Paleobiological implications of Campanian coprolites. Palaeogeogr. Palaeoclimatol. Palaeoecol. 142, 231–254.
Rogers, R.R., Krause, D.W., Kast, S.C., Marshall, M.S., Rahantarisoa, L., Robins, C.R., Sertich, J.J., 2013. A new, richly fossiliferous member comprised of tidal deposits in the Upper Cretaceous Maevarano Formation, northwestern Madagascar. Cretac. Res. 44, 12–29.
Rolfe, W.D.I., 1973. Excursion 15: Silurian arthropod and fish assemblages from Lesmahagow, Lanarkshire. In: Bluck, B.J. (Ed.), Excursion Guide to the Geology of the Glasgow District. Geological Society of Glasgow, Glasgow, pp. 119–126.
Román, Z., Segesdi, M., Sebe, K., Földes, T., Bakrač, K., Virág, A., Botfalvai, G., 2024. Palaeontological and taphonomical investigations of the exceptionally rich concentration of Miocene vertebrate coprolites from Pécs-Danitzpuszta (Hungary, Mecsek Mts.). Hist. Biol. https://doi.org/10.1080/08912963.2024.2324435.
Ronan, J., Duffin, C.J., Hildebrandt, C., Parker, A., Hutchinson, D., Copp, C., Benton, M.J., 2020. Beginning of Mesozoic marine overstep of the Mendips: the Rhaetian and its fauna at Hapsford Bridge, Vallis Vale, Somerset, UK. Proc. Geol. Assoc. 131 (5), 578–594.
Röper, M., 2005. Field trip C: lithographic limestones and Plattenkalk deposits of the Solnhofen and Mörnsheibm formations near Eichstätt and Solnhofen. Zitteliana 26, 71–85.
Rowe, T., Cifelli, R.L., Lehman, T.M., Weil, A., 1992. The Campanian Terlingua local fauna, with a summary of other vertebrates from the Aguja Formation, Trans-Pecos Texas. J. Vertebr. Paleontol. 12, 472–493.
Royo, J., Gomez, J., 1960. Los vertebratos de la formation Uranco, Estado Falcon, vol. 11. Tercero Congreso Geológico Venezoelano, Memoir, p. 509.
Rozada, L., 2019. Taphonomie de l'assemblage de macrorestes fossiles de vertébrés du Lagerstätte d'Angeac-Charente (Crétacé inférieur, Charente, France). Ph. D. Dissertation, Paris, Sorbonne Université, p. 285.
Rozada, L., Allain, R., Vullo, R., Goedert, J., Augier, D., Jean, A., Marchal, J., Peyre de Fabrègues, C., Qvarnström, M., Royo-Torres, R., 2021. A Lower Cretaceous Lagerstätte from France: A taphonomic overview of the Angeac-Charente vertebrate assemblage. Lethaia 54, 141–165.
Ruedemann, R., 1929. Fossils from the Permian Tillite of San Paulo, Brazil, and their bearing on the origin of the tillite. Geol. Soc. Am. Bull. 40, 417–426.

Rummy, P., Halaclar, K., Chen, H., 2021. The first record of exceptionally-preserved spiral coprolites from the Tsagan-Tsab Formation (Lower Cretaceous), Tatal, western Mongolia. Sci. Rep. 11, 7891. https://doi.org/10.1038/s41598-021-87090-5.

Rusconi, C., 1947. Primeros hallazgos de coprolitos de reptile en el Triásico de El Challao, Mendoza. Ciencia e Investigación 10, 521–523.

Rusconi, C., 1949. Coprolitos Triassicos de Mendoza. Rev. Mus. Hist. Nat. Mendoza 3, 241–251.

Russo, G., Raia, P., Lauteri, M., 2015. Coprolite specimens from Pietraroja (Lower Cretaceous, Southern Italy): Morphological analysis by scanning electron microscopy. Microscopie 12, 53–59.

Sagar, R., Kapur, V.V., Kumar, K., Morthekai, P., Sharma, A., Shukla, S.K., Ghosh, A.K., Chauhan, G., Thakkar, M.G., 2024. First record of Chelonian coprolites from the Early-Middle Miocene Kutch Basin, western India, and their palaeodietary and palaeobiological implications. Geobios. https://doi.org/10.1016/j.geobios.2023.12.004.

Sander, P.M., 1989. Early Permian depositional environments and pond bonebeds in central Archer County, Texas. Palaeogeogr. Palaeoclimatol. Palaeoecol. 69, 1–21.

Schmidt, G.D., Duszynski, D.W., Martin, P.S., 1992. Parasites of the extinct Shasta Ground Sloth, *Nothrotheriops shastensis*, in Rampart Cave, Arizona. J. Parasitol. 78, 811–816.

Schmitz, M., 1991. Die Koprolithen mitteleozäner Vertebraten der Grube Messel bei Darmstadt. Courier Forschungsinstitut Senckenberg, Band 137, 1–159.

Schoch, R.R., 2012. Dicynodont mandible from the Triassic of Germany forms the first evidence of large herbivores in the Central European Carnian. Neues Jahrbuch Geol. Palaontol. Abhand. 263, 119–123.

Schoch, R.R., Nesbitt, S., Müller, J., Lucas, S.G., Boy, J.A., 2010. The reptile assemblage from the Moenkopi Formation (Middle Triassic) of New Mexico. Neues Jahrbuch Geol. Palaontol. Abhand. 255, 345–369.

Schoenwetter, J., 1974. Pollen analysis of prehistoric human feces from Salts Cave, Kentucky. In: Watson, P. (Ed.), The Archeology of the Mammoth Cave Area. Academic Press, New York, pp. 49–58.

Scholze, F., Schneider, J.W., Lucas, S.G., 2021. Conchostracans from the Late Pennsylvanian Kinney Brick Quarry Konservat Lagerstätte, New Mexico (USA) – Taxonomy, biostratigraphy, and paleoecology. New Mex. Mus. Nat. Hist. Bull. 84, 287–299.

Schröder-Adams, C., Cumba, S.L., Bloch, A.J., Leckie, D.A., Craig, J., Seif El-Dein, S.A., Simons, D.J., Kenig, D.E., 2001. Late Cretaceous (Cenomanian to Campanian) paleoenvironmental history of the eastern Canadian margin of the Western Interior Seaway: Bonebeds and anoxic events. Palaeogeogr. Palaeoclimatol. Palaeoecol. 17, 311–365.

Schwarzhans, W.W., Murphy, T.D., Frese, M., 2018. Otoliths *in situ* in the stem teleost *Cavenderichthys talbragarensis* (Woodward, 1895), otoliths in coprolites, and isolated otoliths from the Upper Jurassic of Talbragar, New South Wales, Australia. J. Vertebr. Paleontol. 38 (6), e1539740. https://doi.org/10.1080/02724634.2018.1539740.

Schweigert, G., 2001. Über *Medusites* Germar (Koprolith) und echte Algen aus oberjurassischen Plattenkalken Süddeutschlands. Neues Jahrbuch Geol. Palaontol. Monatsh. 2001, 237–249.

Schweigert, G., Dietl, G., 2012. Vertebrate coprolites from the Nusplingen lithographic limestone (Upper Jurassic, SW-Germany). New Mex. Mus. Nat. Hist. Sci. Bull. 57, 215–220.

Schwimmer, D.R., Sanders, A.E., Erickson, B.R., Weems, R.E., 2015b. A Late Cretaceous Dinosaur and Reptile Assemblage From South Carolina, vol. 105. U.S.A. Transactions of the American Philosophical Society, p. 157.

Scott, L., 1987. Pollen analysis of hyena coprolites and sediments from Equus Cave Taung, Southern Kalahari (South Africa). Quat. Res. 28, 144–156.

Segesdi, M., Botfalvai, G., Bodor, E.R., Ősi, A., Buczkó, K., Dallos, Z., Tokai, R., Földes, T., 2017. First report on vertebrate coprolites from the Upper Cretaceous (Santonian) Csehbánya Formation of Iharkút, Hungary. Cretac. Res. 74, 87–99.

Selden, P.A., 1979. Functional Morphology of *Baltoeurypterus*. Ph.D. dissertation. University of Cambridge, Cambridge, pp. 1–190.

Selden, P.A., 1984. Autecology of Silurian eurypterids. Spec. Pap. Palaeontol. 32, 39–54.

Sennikov, A.G., Golubev, V.K., 2012. On the faunal verification of the Permo-Triassic boundary in continental deposits of Eastern Europe: 1. Gorokhovets–Zhukov Ravine. Paleontol. J. 46, 313–323.

Shabica, C.W., Godfrey, S.J., 1997. Coprolites and trace fossils (biogenic structures). In: Shabica, C.W., Hay, A.A. (Eds.), Richardson's Guide to the Fossil Fauna of Mazon Creek. Northeastern Illinois University, Chicago, pp. 275–280.

Sharma, M.K., Patnaik, R., 2010. Coprolites from the lower Miocene Baripada beds of Orissa. Curr. Sci. 99, 804–808.

Sheldon, R.P., 1964. Paleolatitudinal and Paleogeographic Distribution of Phosphorite, vol. 501. US Geological Survey Professional Paper, pp. C106–C113.

Shen, C., Pratt, B.R., Zhang, X.-G., 2014. Phosphatized coprolites from the middle Cambrian (Stage 5) Duyun fauna of China. Palaeogeogr. Palaeoclimatol. Palaeoecol. 410, 104–112. https://doi.org/10.1016/j.palaeo.2014.05.035.

Shillito, L.-M., Bull, I.D., Matthews, W., Almond, M.J., Williams, J., Evershed, R.P., 2011. Biomolecular and micromorphological analysis of suspected faecal deposits at Neolithic Catalhoyuk, Turkey. J. Archaeol. Sci. 38, 1869–1877.

Shillito, L.-M., Blong, J., Jenkins, D., Stafford, T.W., Whelton, H., McDonough, K., Bull, I., 2018. New Research at Paisley Caves: Applying new integrated analytical approaches to understanding stratigraphy, taphonomy and site formation processes. PaleoAmerica 4 (1), 82–86.

Shillito, L.-M., Blong, J.C., Green, E.J., Van Asperen, E., 2020. The what, how and why of archaeological coprolite analysis. Earth Sci. Rev. 207, 103196. https://doi.org/10.1016/j.earscirev.2020.103196.

Shimada, K., Schumacher, B.A., Parkin, J.A., Palermo, J.M., 2006. Fossil marine vertebrates from the lowermost Greenhorn Limestone (Upper Cretaceous: Middle Cenomanian) in southeastern Colorado. Memoir of the Paleontological Society. J. Paleontol. 80 (2), 1–45.

Shu, D., 2008. Cambrian explosion: Birth of tree of animals. Gondwana Res. 14, 219–240.

Shu, D.-G., Conway Morris, S., Han, J., Zhang, Z.-F., Yasui, K., Janvier, P., Chen, L., Zhang, X.-L., Liu, J.-N., Li, Y., Liu, H.-Q., 2003. Head and backbone of the Early Cambrian vertebrate *Haikouichthys*. Nature 421, 526–529.

Shu, D.-G., Conway Morris, S., Zhang, Z.-F., Liu, J.N., Han, J., Chen, L., Shu, D.-G., Luo, H.-L., Conway Morris, S., Zhang, X.-L., Hu, S.-X., Chen, L., Han, J., Zhu, M., Li, Y., Chen, L.-Z., 1999. Lower Cambrian vertebrates from South China. Nature 402, 42–46.

Silva, R.R., Ferigolo, J., Bajdek, P., Piñeiro, G., 2017. The feeding habits of Mesosauridae. Front. Earth Sci. 5, 23. https://doi.org/10.3389/feart.2017.00023.

Sinclair, W.J., 1922. Entelodonts from the Big Badlands of South Dakota in the Geological Museum of Princeton University. Proc. Amer. Phil. Soc. 60, 467–495.

Slater, T.S., Duffin, C.J., Hildebrandt, C., Davies, T.G., Benton, M.J., 2016. Microvertebrates from multiple bone beds in the Rhaetian of the M4–M5 motorway junction, South Gloucestershire, UK. Proc. Geologists' Assoc. 127 (4), 464–477.

Smith, R.M.H., 1993. Vertebrate taphonomy of late Permian floodplain deposits in the southwestern Karoo Basin of South Africa. Palaios 8, 45–67.

Smith, R.M.H., Botha-Brink, J., 2011. Morphology and composition of bone-bearing coprolites from the Late Permian Beaufort Group, Karoo Basin, South Africa. Palaeogeogr. Palaeoclimatol. Palaeoecol. 312, 40–53.

Sobolik, K., 1988. The importance of pollen concentration values from coprolites: An analysis of Southwest Texas samples. Palynology 12, 201–214.

Sobolik, K.D., 1991. Paleonutrition of the Lower Pecos Region of the Chihuahuan Desert. Unpublished PhD dissertation, Texas A&M University.

Sobolik, K.D., 1994. Paleonutrition of the Lower Pecos region of the Chihuahuan Desert. In: Sobolik, K.D. (Ed.), Paleonutrition: The Diet and Health of Prehistoric Americans. Carbondale, Southern Illinois University, Center for Archaeological Investigations Occasional Paper No. 22, pp. 247–264.

Sohn, I.G., Chatterjee, S., 1979. Freshwater ostracodes from Late Triassic coprolite in central India. J. Paleontol. 53, 578–586.

Souto, P.R.F., 2001. Tetrapod coprolites from the Middle Triassic of Southern Brazil. GAIA 16, 51–57.

Souto, P.R.F., 2007. Os coprólitos da Bacia de São José de Itaboraí. Estado do Rio de Janeiro, Brasil. In: Ismar de Souza Carvalho, I.D.S., Fernandes, A.C.S., Rodrigues, M.A.C., Carvalho, M.S.S., Arai, M., Oliveira, M.E.Q. (Eds.), Paleontologia: Cenários da vida. 1. Interciência, Rio de Janeiro, pp. 811–818.

Souto, P.R.F., 2008. Coprólitos do Brasil: Principais ocorrências e estudo. Rio de Janeiro, Publit, pp. 1–89.

Souto, P.R.F., 2010. Crocodylomorph coprolites from the Bauru Basin, Upper Cretaceous, Brazil. New Mex. Mus. Nat. Hist. Sci. Bull. 51, 201–208.

Souto, P.R.F., Fernandes, M.A., 2015. Fossilized excreta associated to dinosaurs in Brazil. J. S. Am. Earth Sci. 57, 32–38.

Souto, P.R.F., Schwanke, C., 2010. Ocorrência de coprólitos de vertebrados na bacia de Alagoas, Cretáceo Inferior. Boletim Do Museu Paraense Emílio Goeldi. Cienc. Nat. 5, 189–194.

Spillmann, F., 1929. Das südamerikanische *Mastodon* als Zeitgenose dês Menschen majoiden Kulturkeises. Paläontol. Z. 11, 170–177.

Stewart, I., Nicoll, R.S., 2003. Multi-element apparatus structure of the Early Ordovician conodont *Oepikodus evae* Lindström from Australia and Sweden. In: Mawson, R., Talent, J.A. (Eds.), Contributions to the Second Australian Conodont Symposium (AUSCOS II) Held in Conjunction with Palaeontology Down under 2000 in Orange, Australia, 3–7 July 2000, vol. 2. Courier Forschungsinstitut Senckenberg, pp. 361–387, 45.

Stewart, J.D., 1978. Enterospirae (fossil intestines) from the Upper Cretaceous Niobrara Formation of western Kansas. In: Chorn, J., Reavis, E.A., Stewart, J.D., Whetstone, K.N. (Eds.), Fossil Fish Studies. The University of Kansas Paleontological Contributions, pp. 9–16. Paper 89.

Stiger, M.A., 1977. Anasazi Diet: The Coprolite Evidence. M.A. thesis. University of Colorado, Boulder, pp. 1–125.

Stinnesbeck, W., Frey, E., Rivas, L., Pérez, J.P., Cartes, M.L., Soto, C.S., Lobos, P.Z., 2014. A Lower Cretaceous ichthyosaur graveyard in deep marine slope channel deposits at Torres del Paine National Park, southern Chile. Geol. Soc. Am. Bull. 126, 1317–1339.

Stovall, J.W., Strain, W.S., 1936. A hitherto undescribed coprolite from the White River badlands of South Dakota. J. Mammal. 17, 27–28.

Strickland, H.,E., Hooker, J., 1853. On the distribution and organic contents of the "Ludlow Bone Bed" in the Districts of Woolhope and May Hill with a note on the seed-like bodies found in it. Quart. J. Geol. Soc. 9, 8–12.

Stringer, G.L., King, L., 2012. Late Eocene shark coprolites from the Yazoo Clay in northeastern Louisiana. New Mex. Mus. Nat. Hist. Sci. Bull. 57, 275–309.

Stuart, C., Stuart, T., 2000. A field guide to the tracks and signs of Southern and East African wildlife. Cape Town. In: Struick Nature. Random House Struick), pp. 1–310.

Suazo, T.L., Cantrell, A.K., Lucas, S.G., Spielmann, J.A., Hunt, A.P., 2012. Coprolites across the Cretaceous/Tertiary boundary, San Juan Basin, New Mexico. New Mex. Mus. Nat. Hist. Sci. Bull. 57, 263–274.

Sullivan, R.M., Jasinski, S.E., 2012. Coprolites from the Upper Cretaceous Fruitland, Kirtland and Ojo Alamo formations, San Juan Basin, New Mexico. New Mex. Mus. Nat. Hist. Sci. Bull. 57, 255–262.

Sumner, D., 1991. Palaeobiology, Taphonomy and Diagenesis of a Lower Carboniferous Fish Fauna. Ph.D. dissertation. University of St. Andrews, St Andrews, p. 165.

Sumner, D., 1994. Coprolites from the Visean of East Kirkton, West Lothian, Scotland. Trans. R. Soc. Edinb. Earth Sci. 84, 413–416.

Swift, A., Duffin, C.J., 1999. Trace fossils. In: Swift, A., Martill, D.M. (Eds.), Fossils of the Rhaetian Penarth Group, vol. 9. Palaeontological Association, Field Guides to Fossils, pp. 239–250.

Taglioretti, V., Sardella, N.H., Fugassa, M.H., 2014. Morphometric analysis of modern faeces as a tool to identify artiodactyls' coprolites. Quat. Int. 352, 64–67.

Tapanila, L., Roberts, E.M., Bouaré, M.L., Sissoko, F., O'Leary, M.A., 2004. Bivalve borings in phosphatic coprolites and bone, Cretaceous-Paleogene, northeastern Mali. Palaios 19, 565–573.

Tapanila, L., Roberts, E.M., Bouaré, M.L., Sissoko, F., O'Leary, M.A., 2008. Phosphate taphonomy of bone and coprolite conglomerates: A case study from the Eocene of Mali, NW Africa. Palaios 23, 139–152.

Tarasenko, K.K., Sivkova, T.N., 2023. Metacercariae in coprolites of seals of the genus *Pachyphoca* (Cystophorinae, Phocidae) from the Late Miocene of the North Caucasus. Dokl. Biol. Sci. 511 (1), 277–279.

Tauber, A.A., Palacios, M.E., Cardozo, S.A., 2007. Coprólitos da Formación Santa Cruz (Mioceno Inferior-Medio), Patagônia, Republica Argentina. Congresso Argentino De Paleontologia Y Bioestratigrafia Resúmenes 9, 294.

Thulborn, R.A., 1991. Morphology, preservation and palaeobiological significance of dinosaur coprolites. Palaeogeogr. Palaeoclimatol. Palaeoecol. 83, 341–366.

Tietze, E., Barberena, R., Beltrame, M.O., 2019. Parasite assemblages from feline coprolites through the Pleistocene-Holocene transition in Patagonia: Cueva Huenul 1 Archaeological Site (Argentina). Environ. Archaeol. 28 (5), 367–377. https://doi.org/10.1080/14614103.2019.1689893.

Tolmacheva, T.Y., 1996. New conodont clusters from the Lower Ordovician of central Kazakhstan. In: Dzik, J. (Ed.), Sixth European Conodont Symposium, Abstracts. Instytut Paleobiologii PAN, Warszawa, p. 57.

Tolmacheva, T.Y., Purnell, M.A., 2002. Apparatus composition, growth, and survivorship of the Lower Ordovician conodont *Paracordylodus gracilis* Lindström, 1955. Palaeontology 45, 209–228.

Tomassini, R.L., Montalvo, C.I., 2010. Coprolitos en la Formación Monte Hermoso (Mioceno tardío-Plioceno temprano), Farola Monte Hermoso, Buenos Aires, Argentina. Ameghiniana 47 (1), 111–115.

Tomassini, R.L., Montalvo, C.I., 2013. Taphonomic modes on fluvial deposits of the Monte Hermoso Formation (early Pliocene), Buenos Aires Province, Argentina. Palaeogeogr. Palaeoclimatol. Palaeoecol. 369, 282–294.

Tomassini, R.L., Montalvo, C.I., Bargo, M.S., Vizcaíno, S.F., Cuitiño, J.I., 2019. *Sparassodonta* (Metatheria) coprolites from the early-mid Miocene (Santacrucian age) of Patagonia (Argentina) with evidence of exploitation by coprophagous insects. Palaios 34, 639–651.

Tournepiche, J.F., Couture, C., 1999. The hyena den of Rochelot Cave (Charente, France). Monographien des Römisch-Germanischen Zentralmuseums 42, 89–101.

Trevor-Deutsch, B., Bryant Jr., V.M., 1978. Analysis of suspected human coprolites from Terra Amata, Nice, France. J. Archaeol. Sci. 5, 387–390.

Trewin, N.H., 1986. Palaeoecology and sedimentology of the Achanarras fishbed of the Middle Old Red Sandstone, Scotland. Trans. R. Soc. Edinb. Earth Sci. 77, 21–46.

Trewin, N.H., 2008. Fossils Alive! or New Walks in an Old Field. Dunedin Academic Press, Edinburgh, p. 211.

Trewin, N.H., Davidson, R.G., 1996. An early Devonian lake and its associated biota in the Midland Valley of Scotland: Transactions of the Royal Society of Edinburgh. Earth Sci. 86, 233–246.

Tripp, M., Wiemann, J., Brocks, J., Mayer, P., Schwark, L., Grice, K., 2022. Fossil biomarkers and biosignatures preserved in coprolites reveal carnivorous diets in the carboniferous Mazon Creek ecosystem. Biology 11 (9). https://doi.org/10.3390/biology11091289.

Tripp, M., Wiemann, J., Hope, J.M., Brocks, J.J., Mayer, P., Lidgard, S., Grice, K., 2021. Molecular biomarkers in coprolites illuminate dietary interactions in the Carboniferous Mazon Creek ecosystem. In: IMOG 2021 (European Association of Geoscientists and Engineers) Conference Proceedings, pp. 1–2. https://doi.org/10.3997/2214-4609.202134169.

Trueman, C.N., Benton, M.J., 1997. A geochemical method to trace the taphonomic history of reworked bones in sedimentary settings. Geology 25, 263–266.

Turner, S., 1999. The importance of coprolite events. Ichthyolith Issues 20, 37–38.

Tweet, J.S., Santucci, V.L., Hunt, A.P., 2012. An inventory of packrat (*Neotoma* spp.) middens in National Park Service areas. New Mex. Mus. Nat. Hist. Sci. Bull. 57, 355–368.

Umamaheswaran, R., Prasad, G.V.R., Rudra, A., Dutta, S., 2019. Biomarker signatures in Triassic coprolites. Palaios 34, 458–467.

Underwood, C.J., Cumbaa, S.L., 2010. Chondrichthyans from a Cenomanian (Late Cretaceous) bonebed, Saskatchewan, Canada. Palaeontology 53, 903–944.

Vajda, V., Fernández, M.D.P., Villanueva-Amadoz, U., Lehsten, V., Alcalá, L., 2016. Dietary and environmental implications of Early Cretaceous predatory dinosaur coprolites from Teruel, Spain. Palaeogeogr. Palaeoclimatol. Palaeoecol. 464, 134–142.

Vallon, L.R., 2012. Digestichnia (Vialov, 1972) - An almost forgotten ethological class for trace fossils. New Mex. Mus. Nat. Hist. Sci. Bull. 57, 131–135.

Verde, M., Ubilla, M., 2002. Carnivore mammal coprolites from the Sopas Formation (Upper Pleistocene, Lujanian Stage) of Uruguay. Ichnos 9, 77–80.

Verzi, D.H., Montalvo, C.I., Deschamps, C.M., 2008. Biostratigraphy and biochronology of the Late Miocene of central Argentina: Evidence from rodents and taphonomy. Geobios 41, 145–155.

Vijaya, Prasad, G.V.R., Singh, K., 2009. Late Triassic palynoflora from the Pranhita–Godavari Valley, India: Evidence from vertebrate coprolites. Alcheringa 33, 91–111.

Vinton, S.D., Perry, L., Reinhard, K.J., Santoro, C.M., Teixeira-Santos, I., 2009. Impact of empire expansion on household diet: the Inka in Northern Chile's Atacama Desert. PLoS One 4 (11), e8069. https://doi.org/10.1371/journal.pone.0008069.

Vogeltanz, R., 1965. Austrocknungsstruckturen bei Koprolithen. Neues Jahrbuch Geol. Palaontol. Monatsh. 3, 362–371.

Vogeltanz, R., 1967. Erganzends Mitteilung über Koprolithen Untersuchungen aus dem Unteroligozän von Nebraska. Neues Jahrbuch Geol. Palaontol. Monatsh. 7, 188–191.

Voigt, E., 1934. Die Fische aus der mitteleozänen Braunkohle des geiseltales mit besonderer Berücksichtigung der ehaltenen eichteile. Nova Acta Leopoldina Halle 2, 21–146.

Wagner, T., Boyce, A.J., 2006. Pyrite metamorphism in the Devonian Hunsrück Slate of Germany: Insights from laser microprobe sulfur isotope analysis and thermodynamic modeling. Am. J. Sci. 306, 525–552.

Wahl, A.M., Martin, A.J., Hasiotis, S.T., 1998. Vertebrate coprolites and coprophagy traces, Chinle Formation (Late Triassic), Petrified Forest National Park. In: Santucci, V.L., McClelland, L. (Eds.), National Park Service

Paleontological Research. National Park Service Geological Resources Division Technical Report NPS/NRGRD/GRDTR-98/01, pp. 144–148.
Waldman, M., Hopkins Jr., W.S., 1970. Coprolites from the Upper Cretaceous of Alberta, Canada, with a description of their microflora. Can. J. Earth Sci. 7, 1295–1303.
Wang, X., White, S.C., Balisi, M., Biewer, J., Sankey, J., Garber, D., Tseng, Z.J., 2018. First bone-cracking dog coprolites provide new insight into bone consumption in *Borophagus* and their unique ecological niche. Elife 7, e34773. https://doi.org/10.7554/eLife.347737.
Wanless, H.R., 1923. Stratigraphy of the White River beds of South Dakota. Proc. Am. Phil. Soc. 62, 190–269.
Warren, A.A., Rubidge, B.S., Stanistreet, I.G., Stollhofen, H., Wanke, A., Latimer, E.M., Marsicano, C.A., Damiani, R.J., 2001. Oldest known stereospondylous amphibian from the early Permian of Namibia. J. Vertebr. Paleontol. 21, 34–39.
Watson, P.J., Yarnell, R.A., 1966. Archaeological and paleoethnobotanical investigations in Salts Cave, Mammoth Cave National Park, Kentucky. Am. Antiq. 31, 842–849.
Wedmann, S., Uhl, D., Lehmann, T., Garrouste, R., Nel, A., Gomez, B., Smith, K., Schaal, S.F.K., 2018. The Konservat-Lagerstätte Menat (Paleocene; France)-an overview and new insights. Geol. Acta 16, 189–213.
Weiler, W., 1922. Beiträge zur Kenntnis der Tertiären Fische des Mainzer Beckens I. Abhandlungen der Hessischen Geologischen Landesanstalt zu Darmstadt 6, 71–135.
Wells, N.A., Ferber, C.T., Ohman, S.C., 1993. Discriminant analysis of fish-bearing deposits in the Eocene Green River Formation of Utah and Wyoming. Palaios 8, 81–100.
Wetmore, A., 1943. The occurrence of feather impressions in the Miocene deposits of Maryland. Auk 60, 440–441.
Whitmore, F.C., Stewart, R.H., 1965. Miocene mammals and Central American seaways: Fauna of the Canal Zone indicates separation of Central and South America during most of the Tertiary. Science 148, 180–185.
Williams, H., Duffin, C.J., Hildebrandt, C., Parker, A., Hutchinson, D., Benton, M.J., 2022. Microvertebrates from the Rhaetian bone beds at Westbury Garden Cliff, near Gloucester, UK. Proc. Geologists' Assoc. 133 (2), 119–136.
Williams, M.E., 1972. The Origin of "Spiral Coprolites, University of Kansas Palaeontological Contributions 56, 1–19.
Williams-Dean, G.J., 1978. Ethnobotany and Cultural Ecology of Prehistoric Man in Southwest Texas. Ph.D. dissertation. Texas A&M University, College Station, p. 287.
Williams-Dean, G., 1986. Pollen analysis of human coprolites. In: Morris, D.P. (Ed.), Archaeological Investigations at Antelope House. National Park Service, Washington, pp. 189–205.
Williams-Dean, G., Bryant, V.M., 1975. Pollen analysis of human coprolites from Antelope House, Canyon de Chelly National Monument, Arizona. Kiva 41, 97–111.
Wilson, M.V.H., 1987. Predation as a source of fish fossils in Eocene lake sediments. Palaios 2, 497–504.
Wings, O., 2012. Gastroliths in coprolites—a call to search. New Mex. Mus. Nat. Hist. Bull. 57, 73–77.
Wolvaardt, F.P., Hancox, P.J., Browning, C., Strong, M., 2023. Biostratigraphic and palaeoenvironmental significance of a rich tetrapod fossil accumulation in the Lower to Middle Burgersdorp Formation of the Karoo Basin. J. Afr. Earth Sci. 198, 104776.
Wood, J.R., Wilmshurst, J.M., Worthy, T.H., Cooper, A., 2011. *Sporormiella* as a proxy for non-mammalian herbivores in island ecosystems. Quat. Sci. Rev. 30, 915–920.
Wood, J.R., Wilmshurst, J.M., Worthy, T.H., Cooper, A., 2012a. First coprolite evidence for the diet of *Anomalopteryx didiformis*, an extinct forest ratite from New Zealand: New Zealand. J. Ecol. 36, 164–170.
Wood, J.R., Rawlence, N.J., Rogers, G.M., Austin, J.J., Worthy, T.H., Cooper, A., 2008. Coprolite deposits reveal the diet and ecology of the extinct New Zealand megaherbivore moa (Aves, Dinornithiformes). Quat. Sci. Rev. 27, 2593–2602.
Wood, J.R., Wilmshurst, J.M., Wagstaff, S.J., Worthy, T.H., Rawlence, N.J., Cooper, A., 2012b. High-resolution coproecology: using coprolites to reconstruct the habits and habitats of New Zealand's extinct upland moa (*Megalapteryx didinus*). PLoS One 7 (6), e40025. https://doi.org/10.1371/journal.pone.0040025.
Woodward, J., 1729. An Attempt towards a Natural History of the Fossils of England in a Catalogue of the English Fossils in the Collection of John Woodward. M.D., Volume 2. London, J. Senex, J. Osborn and T. Longman, pp. 1–115.
Wormington, H.M., Agogino, G., 1994. Cynthia Irwin-Williams 1936-1990. Am. Antiq. 59, 667–671.
Yao, M., Sun, Z., Meng, Q., Li, J., Jiang, D., 2022. Vertebrate coprolites from Middle Triassic Chang 7 Member in Ordos basin, China: palaeobiological and palaeoecological implications. Palaeogeogr. Palaeoclimatol. Palaeoecol. 600, 111084. https://doi.org/10.1016/j.palaeo.2022.111084.

Yates, A.M., Neumann, F.H., Hancox, P.J., 2012. The earliest Post-Paleozoic freshwater bivalves preserved in coprolites from the Karoo Basin, South Africa. PLoS One 7, e30228. https://doi.org/10.1371/journal.pone.0030228.
Yin, Z.X., 1945. The fish manure fossil. Geol. Rev. 10 (3–4), 89–96.
You, J., Liu, Y., Zhang, X., Song, S., Zhou, D., Yang, Y., Li, J., 2020. Establishment and significance of ancient lake ecosystems in the Mesozoic—Evidence from coprolite from the Chang 7 section of the Upper Triassic in the Ordos Basin, China. Hist. Biol. 33 (11), 2989–2997. https://doi.org/10.1080/08912963.2020.1840564.
Young, B.H., 1910. The Prehistoric Men of Kentucky. Louisville, Filson Club Publications, p. 25.
Young, C.C., 1964. New fossil crocodiles from China. Vertebrata PalAsiatica 8, 189–210.
Zajíc, J., 2014. Permian fauna of the Krkonoše Piedmont Basin (Bohemian Massif, Central Europe). Acta Musei NationalisPragae. Ser. B Hist. Nat. 70, 131–142.
Zangerl, R., Richardson, E.S., 1963. The paleoecological history of two Pennsylvanian black shales. Fieldiana Geol. Mem. 4, 352.
Zatoń, M., Rakociński, M., 2014. Coprolite evidence for carnivorous predation in a Late Devonian pelagic environment of southern Laurussia. Palaeogeogr. Palaeoclimatol. Palaeoecol. 394, 1–11.
Zatoń, M., Broda, K., Qvarnström, M., Niedźwiedzki, G., Ahlberg, P.E., 2017. The first direct evidence of a Late Devonian coelacanth fish feeding on conodont animals. Sci. Nat. 104, 26. https://doi.org/10.1007/s00114-017-1455-7.
Zatoń, M., Niedźwiedzki, G., Marynowski, L., Benzerara, K., Pott, C., Cosmidis, J., Krzykawski, T., Filipiak, P., 2015. Coprolites of Late Triassic carnivorous vertebrates from Poland: an integrative approach. Palaeogeogr. Palaeoclimatol. Palaeoecol. 430, 21–46.
Zhang, Y., Zhang, D., Yang, Y., Wu, X., 2020. Pollen and lipid analysis of coprolites from Yuhuicun and Houtieying, China: Implications for human habitats and diets. J. Archaeol. Sci. Rep. 29, 102135. https://doi.org/10.1016/j.jasrep.2019.102135.
Zhang, Y., van Geel, B., Gosling, W.D., McMichael, C.N.H., Jansen, B., Absalah, S., Sun, G., Wu, X., 2019. Local vegetation patterns of a Neolithic environment at the site of Tianluoshan, China, based on coprolite analysis. Rev. Palaeobot. Palynol. 271, 104101. https://doi.org/10.1016/j.revpalbo.2019.104101.
Zidek, J., 1980. *Acanthodes lundi*, new species (Acanthodii) and associated coprolites from uppermost Mississippian Heath Formation of central Montana. Ann. Carnegie Mus. 49, 49–78.
Zidek, J., 1992. Late Pennsylvanian Chondrichtyes, Acanthodii, and deep bodied Actinopterygii from the Kinney Quarry, Manzanita Mountains, New Mexico. New Mexico Bureau of Mines and Mineral Resources. Bulletin 138, 145–182.
Zipfel, B., Montgomery, C., Neumann, F.H., Scott, L., Choiniere, J., Hancox, J.P., 2023. Overlooked or unimportant? An overview of the coprolite collections at the University of the Witwatersrand, Johannesburg, South Africa. Curator 66 (1), 149–164.
Zouheir, T., Hunt, A.P., Hminna, A., Saber, H., Schneider, J.W., Lucas, S.G., 2023. Laurussian-aspect of the coprolite association from the Upper Triassic (Carnian) of the Argana Basin, Morocco. Palaios 38 (6), 281–303. https://doi.org/10.2110/palo.2022.022.
Zotz, L., 1928. Die Fische der typischen Fischschiefer aus den Schächten der Kaliwerke Buggingen. Mitteilungen der Badischen Geologische Landesanstalt zu Heidelberg, 10, 145–174.

Further reading

Barrios-de Pedro, S., Osuna, A., Buscalioni, Á.D., 2020b. Helminth eggs from early Cretaceous faeces. Sci. Rep. 10 (1), 18747. https://doi.org/10.1038/s41598-020-75757-4.
Bryant, V.M., 1974a. Prehistoric diet in southwest Texas. The coprolite evidence. Am. Antiq. 39, 407–420.
Ferreira, L.F., Araújo, A., Duarte, A.N., 1993. Nematode larvae in fossilized animal coprolites from lower and middle Pleistocene sites, Central Italy. J. Parasitol. 79, 440–442.
Fry, G.F., 1969. Prehistoric Diet at Danger Cave, Utah as Determined by the Analysis of Coprolites. M.A. Thesis. University of Utah, Salt Lake City, pp. 1–69.
Fry, G.F., 1977. Analysis of Prehistoric Coprolites from Utah. University of Utah Anthropological Papers, vol. 97. University of Utah Press, Salt Lake City, pp. 1–45.
Heizer, R.E., Napton, L.K., 1969. Biological and cultural evidence from Prehistoric human coprolites. Science 165, 563–568.

Hollocher, K.T., Hollocher, T.C., J. Keith Rigby Jr., J.K., 2010. A phosphatic coprolite lacking diagenetic permineralization from the Upper Cretaceous Hell Creek Formation, northeastern Montana: Importance of dietary calcium phosphate in preservation. Palaios 25, 132–140.

Hunt, A.P., Milàn, J., Lucas, S.G., Spielmann, J.A. (Eds.), 2012f. Vertebrate Coprolites. New Mexico Museum of Natural History and Science, Bulletin, vol. 57, p. 387.

Laudermilk, J.D., Munz, P.A., 1934. Plants in the Dung of *Nothrotherium* from Gypsum Cave, vol. 453. Carnegie Institution of Washington, Nevada, pp. 31–37.

Mead, J.I., Agenbroad, L.D., 1992. Isotope dating of Pleistocene dung deposits from the Colorado Plateau, Arizona and Utah. Radiocarbon 34, 1–19.

Milàn, J., Rasmussen, B.W., Lynnerup, N., 2012a. A coprolite in the MDCT-scanner – internal architecture and bone contents revealed. New Mex. Mus. Nat. Hist. Sci. Bull. 57, 99–103.

Murchison, R.I., 1839. The Silurian System, Founded on Geological Researches in the Counties of Salop, Hereford, Radnor, Montgomery, Caermarthen, Brecon, Pembroke, Monmouth, Gloucester, Worcester, and Stafford: With Descriptions of the Coalfields and Overlying Formations. John Murray, London, pp. 1–768.

O'Leary, M.A., Bouaré, M.L., Claeson, K.M., Heilbronn, K., Hill, R.V., Mccartney, J., Sessa, J.A., Sissoko, F., Tapanila, L., Wheeler, E., Roberts, E.M., 2019. Stratigraphy and paleobiology of the Upper Cretaceous-Lower Paleogene sediments from the trans-Saharan seaway in Mali. Bull. Am. Mus. Nat. Hist. 436, 1–183.

Schwimmer, D.R., Weems, R.E., Sanders, A.E., 2015a. A Late Cretaceous shark coprolite with baby freshwater turtle vertebrae inclusions. Palaios 30, 707–713.

Shillito, L.-M., Matthews, W., Bull, I.D., Williams, J., 2013b. Biomolecular investigations of faecal biomarkers at Sheik-e Abad and Jani. In: Matthews, R., Matthews, W., Mohammadifar, Y. (Eds.), The earliest Neolithic of Iran: 2008 Excavations at Sheikh e Abad and Jani. Oxbow and British Institute for Persian Studies, Oxford, pp. 105–115.

Ward, D.J., Duffin, C.J., Ward, A., 2020. Are spiral bromalites enterospirae or coprolites? A brief review with some additional evidence. Soc. Vertebr. Paleontol. 80th Ann. Meet. Virt. 2020 Oct. 12, unnumbered page.

CHAPTER

10

Gastroliths

Spencer G. Lucas[1] *and Adrian P. Hunt*[2]

[1]New Mexico Museum of Natural History, Albuquerque, NM, United States; [2]Flying Heritage and Combat Armor Museum, Everett, WA, United States

1. Introduction

Gastroliths are clasts of sand and/or gravel swallowed by an animal and retained in the digestive tract. The term gastrolith (Greek *gastro*, stomach and *lithos*, stone) has been in frequent use since the early 1900s, and gastroliths are usually referred to in the vernacular as "stomach stones" and "gizzard stones." Hunt and Lucas (2021) recently provided an extensive review of the record of gastroliths that is briefly reprised here. The history of gastrolith research is reviewed in the chapter on the history of vertebrate ichnology in this volume.

2. Terminology

A somewhat confusing and imprecise terminology has long been applied by various workers to gastroliths, and we have clarified and made that terminology more precise (Hunt and Lucas, 2021). Thus, we define the term gastrolith as clasts of sand and/or gravel swallowed by an animal and retained in the digestive tract, and we use the Wentworth Scale of grain sizes to describe the clasts in gastroliths. Wings (2007) presented a somewhat more complex terminology that we do not use.

The terms lithophagy and geophagy have been applied by some to refer to the process of acquiring gastroliths, and some use them interchangeably (e.g., Sokol, 1971; Skoczylas, 1978). Lithophagy has a standard definition as the ingesting of "stones" to aid in digestion. But geophagy refers to eating earth, particularly chalk or clay, and especially by humans, though some nonhuman animals (such as some bats) do engage in geophagy (e.g., Abrahams, 2003). The consumption of earth is either as a dietary supplement or is an eating disorder (Geophagia). Geophagy should not be used in reference to the ingestion of "stones," which is correctly called lithophagy.

Vertebrate Ichnology
https://doi.org/10.1016/B978-0-443-38351-9.00011-6

The term grit is widely used to refer to the rock particles swallowed by birds (e.g., Gionfriddo and Best, 1999). However, grit, generally defined as a hard and sharp (rock) granule, has no size specificity except in industrial applications (sandpaper, etc.). If we follow the Wentworth scale used by most geologists for particle size, gastroliths in living birds, which are about 1–10 mm in size range, are very fine to coarse sand to granules and pebbles. Gastroliths of plesiosaurs and some dinosaurs are larger, as large as cobbles on the Wentworth scale. We do not advocate continued use of the imprecise term "grit" in reference to avian gastroliths.

"Stone" is another imprecise term widely used in the gastrolith literature. It has vernacular definitions such as a hard, solid, nonmetallic mineral matter that rocks are made of, or a hard earthen substance. Indeed, the terms gastrolith and stone are often used interchangeably. However, like grit, stone has no specific size limits, and the sand particles that comprise many gastroliths, particularly in birds, would not be called stones. Thus, Hunt and Lucas (2021) abandoned the imprecise term stone when referring to gastroliths and replaced it with clast, or with more precise terms based on the Wentworth scale, namely sand, granules, pebbles, or cobbles.

3. Actualistic basis

Extensive reviews of gastroliths in extant vertebrates have been published and provide a basis for actualistic interpretation (Wickes, 1908; Baker, 1956; Taylor, 1993; Whittle and Everhart, 2000; Wings, 2004, 2007; Hunt and Lucas, 2021). Gastroliths are rare among living fishes (a handful of osteichthyans swallow sand and/or gravel to grind food: Arnette et al., 2023), and none are known among extant amphibians. Among mammals, some pinnipeds swallow gravel, likely for buoyancy control. Modern pangolins have a muscular stomach and swallow fine gravel that is retained in the stomach. Many living reptile and bird species swallow sand and/or gravel. Thus, some living turtles, snakes, lizards, and crocodilians swallow sand/gravel, mostly to grind food, and many birds (in particular, most herbivorous and omnivorous birds) use gastroliths to aid digestion.

Nevertheless, the distribution of gastroliths is very irregular in extant vertebrates. For example, a few species of crocodilians swallow gravel, whereas others do not, and, in a population of crocodilians that does swallow gravel, some individuals apparently do not. As Whittle and Everhart (2000, p. 77) well observed, "lithophagy in extant turtles, lizards and marine Mesozoic diapsids appears to be species-specific." We would add that the literature suggests that swallowing or not swallowing sand/gravel can be specific to some individuals within a species as well.

The literature on extant crocodilian and bird gastroliths is most extensive and was reviewed by Hunt and Lucas (2021). They concluded that there seems to be little if any direct evidence that extant crocodiles use their gastroliths in digestion (see Wings, 2004), so their use in buoyancy appears to be the best supported function. Most of the extant species of vertebrates that swallow sand/gravel are birds, who use them in a gizzard to aid digestion, particularly herbivorous birds (e.g., Pough et al., 2002; Hu et al., 2022). After birds swallow food, if necessary, they store it in the crop, a chamber connected to the esophagus (a ventral pouch

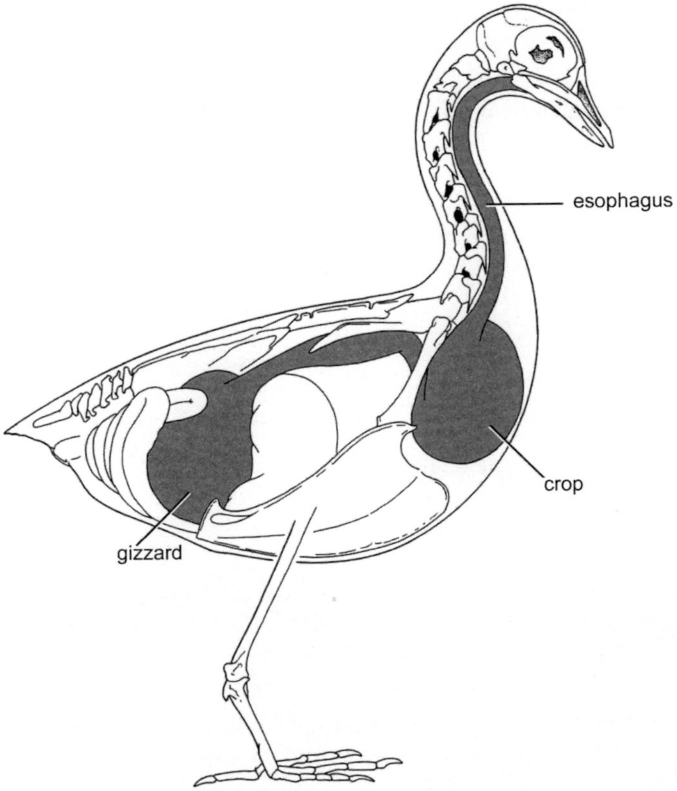

FIGURE 10.1 A modern bird, indicating the position of the crop and the gizzard. *Modified from Zheng et al. (2011).*

anterior to the furcula) (Fig. 10.1). They then pass food into the gizzard (ventriculus or muscular stomach) where it can be ground by sand/gravel that has been swallowed and stored in that chamber. Once processed in the gizzard, the food is passed to the true stomach. This differs from crocodilians, whose swallowed pebbles go directly to the stomach and, if retained, are then stored in the posterior (pyloric) portion of the stomach cavity.

A diverse literature exists on the gastroliths of living birds, much of it summarized by Gionfriddo and Best (1999), who concluded: (1) most birds swallow sand/gravel to mechanically grind and pulverize food in their gizzards, though some sand/gravel is swallowed for mineral extraction (especially calcium) or to stimulate/facilitate digestion; (2) the size of the swallowed clasts correlates to body size and ranges in diameter from <0.1 mm to > 25 mm (in ostriches); (3) the larger the sand/gravel swallowed, the harder and coarser the diet; (4) most birds selectively swallow rough and angular sand/gravel that can become more rounded in the gizzard; and (5) some birds prefer sand/gravel of a certain color (also see Milton et al., 1994; Tryon, 2002).

Fritz et al. (2011) demonstrated that gizzard-based reduction in particle size in herbivorous birds is as efficient as dental mastication by herbivorous mammals (also see Moore, 1999). Thus, the evolution of the avian gizzard presented a solution for birds to the problem of how to grind vegetation to produce more effective digestion (and secondarily compensating for the loss of teeth).

One problem little addressed is the taphonomy of gastroliths. Does a decomposing vertebrate carcass lose its gastroliths rapidly (Everhart, 2005a, 2005b), and could gastroliths be all that remain of a carcass otherwise disintegrated? To our knowledge, the only taphonomic study of gastroliths is Wings (2003), whose actualistic studies suggest that gastroliths are rapidly released from carcasses in water. Wings (2003) also noted that the integrity of a vertebrate animal's integument matters, as there are numerous crocodiles with gastroliths, and almost no birds with gastroliths at the Eocene Messel Lagerstätte in Germany. Further studies of gastrolith taphonomy are needed.

4. Function of gastroliths

A variety of functions have been suggested for gastroliths, only two of which are of significance: use in digestion to grind, pulverize, and/or disintegrate food or use as ballast for buoyancy control (see Wings, 2004, 2007; Hunt and Lucas, 2021, for reviews). Some gastroliths are swallowed while feeding, either by accident (mostly by herbivores browsing close to the ground) or incidentally as the gastroliths inside of a prey item that was consumed. The function of gastroliths in terrestrial tetrapods seems clear—they are used in digestion. But, in aquatic tetrapods, a digestive function and/or use as ballast for buoyancy control have been advocated and have long been a source of disagreement. Based on their review, Hunt and Lucas (2021) concluded that some crocodiles and marine mammals use/used gastroliths for buoyancy control, but all other gastrolith-bearing vertebrates appear to have used them in digestion.

5. High polish does not characterize gastroliths

Identification of fossil gastroliths is only certain when the sand/gravel is found as a concentrated mass in an anatomically plausible position within the abdominal region of a fossil skeleton. This conclusion has long been advocated by various authors (e.g., Bryan, 1931; Dorr, 1966; Lucas, 2000; Wings, 2004, 2007; Schmeisser and Flood, 2008). Nevertheless, many pebbles and granules are identified as gastroliths that lack such a skeletal association.

The idea that high polish characterizes gastroliths, first articulated by Wieland (1906) and Brown (1907), has dominated the quest to recognize gastroliths that lack a skeletal association. Indeed, many authors state a priori that gastroliths are highly polished. This, despite the fact that the gastroliths of living birds are not highly polished nor are most bona fide fossil gastroliths, especially those of plesiosaurs. Indeed, actualistic studies and an evaluation of many bona fide gastroliths indicate that high polish is not characteristic of gastroliths. Despite these conclusions, the literature abounds in diverse identifications of polished "stones" as gastroliths where there is no skeletal association.

Darby and Ojakangas (1980, p 549) expressed skepticism about the diagnostic utility of high polish and used the term "gastromyth" to refer to stones thought to be gastroliths of

FIGURE 10.2 Comparison of polished fluvially transported pebbles and supposed gastroliths. A. Siliceous pebbles in sandstone, Jackpile Member of Morrison Formation, central New Mexico. B. Some of the supposed gastroliths of the Morrison sauropod dinosaur "*Seismosaurus*."

uncertain origin (also see Lucas, 2000). Wings (2009) simulated a bird gastric mill, and it does not polish stones.

The important point, echoing Bryan (1931), is that there are various ways to polish stones, notably by the wind to make them ventifacts (there is even tectonic polish). Polished clasts identified as gastroliths are siliceous, mostly quartz or chert (Fig. 10.2). These clasts could have been polished by wind and/or water, both before ingestion by an animal and after excretion or other removal from an animal's digestive tract. Furthermore, how a stomach or gizzard would polish siliceous clasts is also unclear, and such polishing does not occur in extant birds. Only a small minority of bona fide gastroliths are highly polished (Hunt and Lucas, 2021; Lucas and Hunt, 2021).

There is thus an inability to establish when and where polished clasts were polished and inferring that they acquired their polish while gastroliths is not supported by actualistic studies of gastroliths. The idea that any highly polished clast is a gastrolith needs to be abandoned. Nevertheless, we do acknowledge that some bona fide gastroliths are highly polished (see examples discussed by Wings, 2015; Hunt and Lucas, 2021), but we stress that isolated, polished clasts cannot be assumed to be gastroliths.

6. Gastroliths in fossil vertebrates

Diverse fossil vertebrates from late Permian to Pleistocene age are associated with gastroliths. The most extensive and well-documented record comes from plesiosaurs, the long-necked marine reptiles of the Mesozoic. Though there are relatively few bona fide gastrolith-dinosaur associations, a diverse literature has been published on dinosaur gastroliths. Some fossil birds have gastroliths and, importantly, exceptionally preserved specimens from China document that both the crop and the gizzard had already evolved in birds during the Early Cretaceous. Here, we briefly reprise a more extensive review published by Hunt and Lucas (2021).

6.1 Plesiosaur gastroliths

Plesiosaurs have the most extensive and unambiguous gastrolith record of any fossil vertebrates. Since the 1870s (e.g., Seeley, 1877; Mudge, 1877), numerous articulated or semiarticulated skeletons have been collected, most with closely associated clasts of granule to cobble size, often aggregated as masses in the abdominal cavity (Fig. 10.3). There is now an extensive literature on plesiosaur gastroliths, mostly from the long-necked, but also from the short-necked forms. Gastroliths have thus been documented in the plesiosaur families Pliosauridae, Rhomaleosauridae, Elasmosauridae, and Polycotylidae. Clearly, there are more elasmosaurid records than of other families, but the pattern of gastrolith possession in plesiosaurs is complex, as it is in other tetrapod groups. Hunt and Lucas (2021) concluded that: (1) many plesiosaurs used gastroliths, especially elasmosaurids; (2) plesiosaur gastroliths can be numerous in one individual (usually 50–200 pebbles, but as many as 793 in one individual), show a wide range in size (up to 170 mm maximum diameter), and are mostly siliceous pebbles of varied shapes; (3) most plesiosaur gastroliths are not highly polished; and (4) plesiosaur gastroliths are mostly found in the posterior part of the abdominal region, but it is not certain whether the plesiosaur held them in a separate structure ("gizzard") or in the posterior region of the stomach, like modern crocodiles.

Gastroliths have provided important inferences about plesiosaur behavior. Thus, in the American Late Cretaceous, plesiosaurs generally swam above muddy or chalky sea floors that offered no clasts of the same lithology as their gastroliths. Analysis indicates that such clasts as bedrock were often hundreds of km (maximum estimate = 600 km: Cicimurri and Everhart, 2001) distant from the location of the plesiosaur fossil skeleton. Furthermore, as was first noted by Darby and Ojakangas (1980), the gastroliths of plesiosaurs resemble river cobbles. Thus, O'Gorman et al. (2013, 2014) noted that the sphericity metrics of plesiosaur

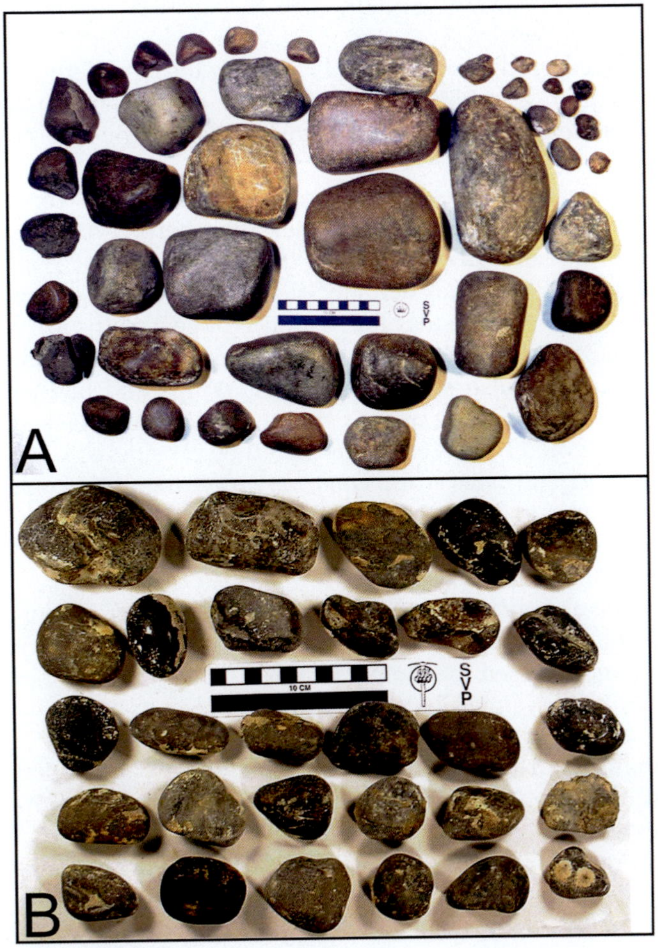

FIGURE 10.3 Two masses of gastroliths from Cretaceous elasmosaurids. *Photographs courtesy of Mike Everhart.*

gastroliths show a close similarity to fluvial pebbles. This, and the fact that many plesiosaur skeletons with gastroliths are found in marine shale and chalk units that are hundreds of kilometers from a possible source of pebbles, indicates that plesiosaurs were prodigious swimmers that likely went very close to shore or ashore to swallow stones in estuaries or along river courses.

The function of plesiosaur gastroliths has been debated for more than a century. The current "consensus" favors their use in digestion, not in buoyancy control. However, most other tetrapods (especially birds) that use gastroliths for dietary purposes are herbivores. Plesiosaurs, like crocodiles, were not herbivorous, so whether or not these predatory marine reptiles would use gastroliths to grind their food is not certain. The possibility that plesiosaurs used gastroliths for both digestion and some amount of buoyancy control remains plausible. More research is needed here.

6.2 Dinosaur gastroliths

An extensive literature has been published on dinosaur gastroliths, even though there are relatively few records of bona fide dinosaurian gastroliths, only a few taxa of dinosaurs had demonstrable gastroliths. These taxa are some prosauropods, sauropods, theropods, an ornithopod, and some ceratopsians.

The case for gastroliths in sauropod dinosaurs is a weak one, despite the fact that a diverse literature identifies gastroliths in these, the largest of all terrestrial tetrapod herbivores (see, for example, Christiansen, 1996). Calvo (1994) first drew attention to this, noting that Wieland's (1906) original report of sauropod gastroliths was actually of clasts near a stegosaur skeleton (Brown, 1907). Calvo (1994, p. 206) concluded that the evidence of gastroliths associated with sauropod bones is "very restricted to fragmentary sauropod remains; thus, these evidences should be interpreted with caution" (also see Lucas, 2000). Wings and Sander (2007) accepted that some sauropods had gastroliths but argued that these dinosaurs lacked a gastric mill as is found in birds. They thus noted that in examples that they considered bona fide gastroliths in sauropods ("*Seismosaurus*," *Cedarosaurus*, *Dinheirosaurus*) the stones are much less than 0.1% of body mass, whereas in herbivorous birds, they are 1% of body mass. They also noted that autochthonous sauropod bone beds lack gastroliths, as did Calvo (1994), Lucas (2000) and Wings (2015).

Wings (2015) reviewed the classic dinosaur bonebeds in the Upper Jurassic Morrison Formation (Bone Cabin quarry, Cleveland–Lloyd quarry, Como Bluff, Dinosaur National Monument, Dry Mesa quarry, Howe quarry) and stressed the almost total lack of any evidence of sauropod gastroliths, even from the articulated/semiarticulated sauropod skeletons found in these bonebeds.

Some paleontologists have argued that sauropod heads and dentitions were so small that they needed a gastric mill to digest the vast amounts of vegetation such behemoths consumed. This found its fullest expression in Bakker (1986, p. 126–138), who presented a lengthy argument for gastroliths in sauropods largely based on his idea that they were endothermic and lacked the masticatory apparatus to process the large quantities of vegetation they ate. He envisioned huge gizzards in sauropods but admitted that direct evidence of sauropod gastroliths was sparse. Indeed, sauropods likely used fermentation to process large amounts of consumed vegetation according to Wings and Sander (2007), as the ability to process vegetation in the mouth was minimal, and there is no compelling evidence of a gastric mill in almost all sauropods.

Among theropod dinosaurs, gastroliths are most common among the Maniraptoriformes, which are the theropods closest to bird ancestry. These theropods have an impressive record of gastroliths that indicates that the crop and gizzard had evolved in some dinosaurs during the Early Cretaceous (Liu et al., 2023; Fig. 10.4).

There are relatively few records of gastroliths in ornithischian dinosaurs. The most convincing are well-described gastroliths from skeletons of the Early Cretaceous ceratopsian *Psittacosaurus* (e.g., Osborn, 1924; Xu, 1997) and an exceptional specimen of the nodosaur *Borealopelta* (Brown et al., 2020).

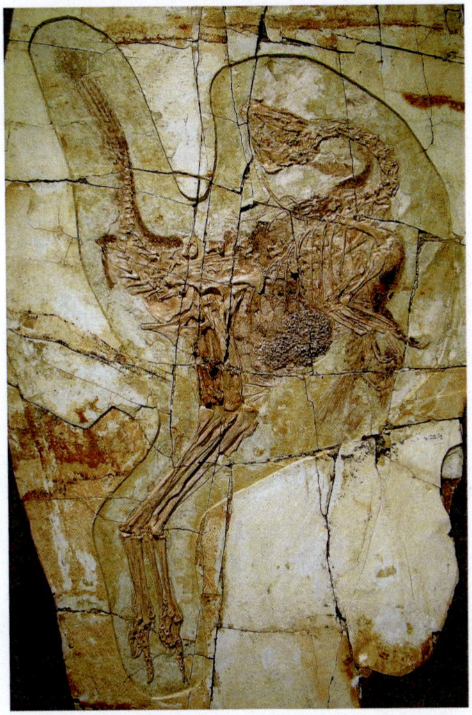

FIGURE 10.4 Skeleton of the maniraptoriform theropod dinosaur *Caudipteryx* (about 1 m long), with a large mass of gastroliths in its abdominal area. *Photograph courtesy of the late Stephen Czerkas.*

6.3 Fossil birds

The fossil record of birds is mostly of isolated and/or incomplete bones, not complete skeletons. This likely explains why there are relatively few documented cases of fossil avian gastroliths. These are mainly of late Cenozoic age. Notable among them is the extensive subfossil record of moas, extinct giant birds from New Zealand, which famously contains many gastroliths that are mostly semiround, white quartz pebbles. These clasts range in size from "gravel" to 100 mm in diameter. Importantly, the clasts are all of lithologies local to the skeletons that contain them. This suggests that moas were localized, only moving tens of kilometers during their lifetimes, not the long distances traveled by some other large, flightless birds, such as ostriches (Anderson, 1989).

Importantly, the spectacular record of fossil birds from the Early Cretaceous strata of northeastern China demonstrates that the avian crop and gizzard had already evolved in the Early Cretaceous. We note that the avian gizzard could have been inherited from nonavian theropods or evolved independently. Thus, the phylogenetic bracket of theropods with gastroliths suggests that birds inherited it from dinosaurian ancestors, but note that the oldest bird, Late Jurassic *Archaeopteryx*, lacks gastroliths.

Louchart and Virot (2011) stated that during the Jurassic-Cretaceous, dental reduction occurred at least six times independently among Aves. They concluded that the horny beak (rhamphotheca) and gizzard replaced the dentition, so the loss of teeth in birds is correlated to the evolution of a muscular gizzard (also see Dilger, 1957).

O'Connor (2019) regarded fossil birds with gizzards as herbivores, and those with crops as granivores. She also noted that the role of gizzards in the tooth loss of birds is complicated by the presence of gastroliths in some toothed birds (*Jeholornis*, *Sapeornis*). So, tooth loss and gastrolith acquisition are not clearly correlated according to O'Connor (2019).

6.4 Other fossil vertebrates

Various records of gastroliths in fossil vertebrates other than plesiosaurs, dinosaurs, and birds are mostly single reports, many of which are problematic. For example, Trewin (1986) documented a placoderm fish (*Coccosteus*) from the Old Red Sandstone with pebbles in the visceral area.

Two Late Cretaceous records are in a large lamniformes shark with apparent black chert gastroliths and a mosasaur with apparent gastroliths. Everhart (2005a) stated that both the shark and the mosasaur likely acquired the gastroliths from preying on plesiosaurs.

The oldest tetrapod gastroliths are in a late Permian pareiasaur from Germany (Munk and Sues, 1993). The abdominal cavity of this pareiasaur had a mass of unsorted granules and pebbles of quartz and chert of varied roundness up to 10 mm in diameter interspersed with carbonized particles that Munk and Sues interpreted as macerated organic matter. They suggested that herbivorous pareiasaurs may have ingested the stones while foraging close to the ground.

Many skeletons of tangasaurid eosuchians (aquatic reptiles from the upper Permian of Madagascar) have masses of gravel in their abdominal regions. The most extensive analysis of these pebbles by Currie (1981) led him to conclude that they were used for buoyancy control. These clasts in *Hovasaurus* are of two sizes, large pebbles and gravel up to 200 mm in diameter and much smaller sand and granules that are 0.5–2.0 mm in diameter. They are mostly quartz and not polished. These clasts are located between the ribs and gastralia of multiple specimens of different body sizes, though two articulated specimens lack any gastroliths.

Fossil crocodiles have a diverse but not prolific record of gastroliths. These include a few records in nonmarine crocodilians and marine crocodiles. However, Hunt and Lucas (2021) questioned the gastrolith identity of some of these records.

There are a few records of gastroliths in coprolites, presumably excreted by the animal that swallowed the gastroliths. These are records primarily in fossil crocodile coprolites (see Hunt and Lucas, 2021).

Codorniú et al. (2013) reported gastroliths from the pterosaur *Pterodaustro* from the Lower Cretaceous of Argentina. They noted the rarity of gastroliths in pterosaurs (many complete skeletons lack gastroliths) and suggested that this pterosaur used the gastroliths to break up hard-shelled crustaceans in its food.

Gastrolith records in fossil mammals are almost unknown. The only record of which we are aware is a skeleton of an Eocene pangolin, *Eomanis*, from Germany that had coarse sand in the visceral area (Von Koenigswald et al., 1981). Extant pangolins similarly swallow sand and gravel as an aid to digestion.

7. Gastroliths as trace fossils

Bertling et al. (2006, p. 266), in their effort to standardize the study and terminology of trace fossils, defined a trace fossil as "a morphologically recurrent structure resulting from the life activity of an individual organism (or homotypic organism) modifying the substrate." They regarded gastroliths as trace fossils, but they classified them as in a "gray zone" with coprolites, "regurgitaliths" (sic), and bite and gnaw structures that have not been considered trace fossils by some workers. Bertling et al. (2022) subsequently clearly stated that a gastrolith is a trace fossil. However, they consider that for a gastrolith, the trace is not the piece of rock; rather, it is the microscopic surface alteration. Bertling et al. (2006) thus list only the etched sculpture of some gastroliths as a possible morphological character that could be of use in an ichnotaxobase.

Nevertheless, gastroliths receive no mention in a very comprehensive book on ichnology by Buatois and Mángano (2011). But, much earlier, Abel (1935), in a classic treatise on trace fossils, devoted 13 pages to reviewing the then published records of gastroliths. And, in their treatise on fossil behavior, Boucot and Poinor (Boucot, 1990; Boucot and Poinar, 2010) regard gastroliths primarily as of use in digestion and assigned them as fossilized evidence of behavior to their highest category 1, or "frozen behavior."

As is currently the case in ichnology, Bertling et al. (2006) argued that all trace fossils need ichnotaxonomic nomenclature governed by the International Commission on Zoological Nomenclature. We advocate development of an ichnotaxonomy for gastroliths. Ichnotaxonomic names, however, should not be assigned to individual grains/clasts of sand/gravel of fossil gastroliths. As the review earlier indicates, the numerous sand grains/gravel found in a fossil vertebrate digestive tract are often diverse in number, size, shape and composition. Naming single sand grains/gravel clasts would thus produce a bewildering plethora of names. Instead, we favor naming the entire gastrolith mass from a single abdominal cavity. Ichnotaxobases could be the number of gastroliths in the mass, their general petrographic composition, and their overall size, shape, surface texture, and other shared morphological features. This likely would produce a workable ichnotaxonomy that recognizes ichnotaxa that are readily distinguished by morphological differences that are a direct reflection of varied behavior.

8. Utility of gastroliths

Gastroliths are a limited and somewhat perplexing part of the ichnological record of vertebrate consumption. Other than birds, relatively few living vertebrate taxa utilize/utilized gastroliths in digestion to pulverize and disintegrate food. Most of these nonavian vertebrates and the gastrolith-using birds are/were herbivores or omnivores. But a compelling case has been made that plesiosaurs used gastroliths for digestion, yet they were predators.

We view gastroliths as a behavioral strategy primarily to grind food, mostly plant matter. Use for ballast in buoyancy control may have been the function of gastroliths in some animals, notably some crocodiles and marine mammals, but this is a little employed strategy for the use of gastroliths. Nevertheless, these strategies for gastrolith use have only been adopted idiosyncratically. Many taxa that could use a "gastric mill" lack it. Some members of taxonomic groups have gastroliths, whereas close relatives do not. And, in populations of some vertebrate species that use gastroliths, not all individuals have gastroliths. These complexities of gastrolith distribution await further explanation, if explanation is possible.

Gastroliths not only provide important information about feeding and buoyancy control, but the need for the animal to collect them provides insight into habitat. Thus, some plesiosaurs must have traveled hundreds of kilometers to acquire their gastroliths, whereas the gastroliths of moas are all of local origin, indicating little travel by these avian giants.

Gastroliths are trace fossils in need of ichnotaxonomy. They provide important insights into various behaviors, notably diet, digestion, buoyancy control, and habitat preferences.

Acknowledgments

Reviews by Hendrik Klein and Robert Weems improved the clarity and content of the manuscript. We also thank Mike Everhart for providing some images used here that were first used by us in Hunt and Lucas (2021).

Declaration of competing interest

The authors declare that they have no known competing financial interests or personal relationships that could have appeared to influence the work reported in this chapter.

Data availability

No data were used for the research described in this chapter.

References

Abel, O., 1935. Vorzeitliche Lebensspuren. Jena, Verlag von Gustav Fischer, 644 p.

Abrahams, P.W., 2003. Human geophagy: a review of its distribution, causes, and implications. In: Skinner, H.C., Berger, A.R. (Eds.), Geology and Health: Closing the Gap. Oxford University Press, Oxford, 179 p.

Anderson, A., 1989. Prodigious Birds: Moas and Moa Hunting in Prehistoric New Zealand. Cambridge University Press, 238 p.

Arnette, S.D., Simonitis, L.E., Egan, J.P., Cohen, K.E., Kolmann, M.A., 2023. True grit?: comparative anatomy and evolution of gizzards in fishes. J. Anat. 244, 260–273.

Baker, A., 1956. The swallowing of stones by animals. Victorian Nat. 73, 82–95.

Bakker, R.T., 1986. The Dinosaur Heresies. William Morrow and Company, New York, 481 p.

Bertling, M., Braddy, S.J., Bromley, R.G., Demathieu, G.R., Genise, J., Mikuláš, R., Nielsen, J.K., Nielsen, K.S.S., Rindsberg, A.K., Schlirf, M., Uchman, A., 2006. Names for trace fossils: a uniform approach. Lethaia 39, 265–286.

Bertling, M., Buatois, L.A., Knaust, D., Laing, B., Mángano, M.G., Meyer, N., Mikuláš, R., Minter, N.J., Neumann, C., Rindsberg, A.K., Uchman, A., Wissak, M., 2022. Names for trace fossils 2.0: theory and practice in ichnotaxonomy. Lethaia 55, 1–19.

Boucot, A.J., 1990. Evolutionary Paleobiology of Behavior and Coevolution. Elsevier, Amsterdam, 735 p.

Boucot, A.J., Poinar Jr., G.O., 2010. Fossil Behavior Compendium. CRC Press, Boca Raton, 391 p.

Brown, B., 1907. Gastroliths. Science 25, 392.

Brown, C.M., Greenwood, D.R., Kalyniuk, J.E., Braman, D.R., Henderson, D.M., Greenwood, C.L., Basinger, J.F., 2020. Dietary palaeoecology of an Early Cretaceous armoured dinosaur (Ornithischia; Nodosauridae) based on floral analysis of stomach contents. R. Soc. Open Sci. 7, 200305. https://doi.org/10.1098/rsos.200305.

Bryan, K., 1931. Wind-worn stones or ventifacts—a discussion and bibliography. In: Reprint and Circular Series National Research Council [Report on Sedimentation 1929-1930], vol 98, pp. 29–50.

Buatois, L.A., Mángano, M.G., 2011. Ichnology: Organism-Substrate Interactions in Space and Time. Cambridge University Press, Cambridge, 358 p.

Calvo, J.O., 1994. Gastroliths in sauropod dinosaurs. Gaia 10, 205–208.

Christiansen, P., 1996. The evidence for and implications of gastroliths in sauropods (Dinosauria, Sauropoda). Gaia 12, 1–7.

Cicimurri, D.J., Everhart, M.J., 2001. An elasmosaur with stomach contents and gastroliths from the Pierre Shale (Late Cretaceous) of Kansas. Trans. Kans. Acad. Sci. 104, 129–143.

Codorniú, L., Chiappe, L.M., Cid, F.D., 2013. First occurrence of stomach stones in pterosaurs. J. Vertebr. Paleontol. 33, 647–654.

Currie, P.J., 1981. *Hovasaurus boulei*: an aquatic eosuchian from the upper Permian of Madagascar. Palaeontol. Afr. 24, 99–168.

Darby, D., Ojakangas, R., 1980. Gastroliths from an Upper Cretaceous plesiosaur. J. Paleontol. 54, 548–556.

Dilger, W.C., 1957. The loss of teeth in birds. Auk 74, 103–104.

Dorr Jr., J., 1966. Wind-polished Stones: Two Similar Sites: Papers of the Michigan Academy of Science, Arts and Letters, vol 51, pp. 265–273.

Everhart, M.J., 2005a. Earliest record of the genus *Tylosaurus* (Squamata: Mosasauridae) from the Fort Hays Limestone (Lower Coniacian) of western Kansas. Kansas Acad. Sci. Trans. 108, 149–155.

Everhart, M.J., 2005b. Elasmosaurid remains from the Pierre shale (Upper Cretaceous) of western Kansas. Possible missing elements of the type specimen of *Elasmosaurus platyurus* Cope, 1868? PalArch Vertebr. Palaeontol. 4, 19–32.

Fritz, J., Hummel, J., Kienzle, E., Wings, O., Streich, W.J., Clauss, M., 2011. Gizzard vs. teeth; it's a tie: food-processing efficiency in herbivorous birds and mammals and implications for dinosaur feeding strategies. Paleobiology 37, 577–586.

Gionfriddo, J.P., Best, L.B., 1999. Grit use by birds: a review. In: Current Ornithology, vol 15, pp. 89–148.

Hu, H., Wang, Y., McDonald, P.G., Wroe, S., O'Connor, J.K., Bjarnason, A., Bevitt, J.J., Yin, X., Zheng, X., Zhou, Z., Benson, R.B.J., 2022. Earliest evidence for fruit consumption and potential seed dispersal by birds. Elife 11, e74751. https://doi.org/10.7554/eLife.74751.

Hunt, A.P., Lucas, S.G., 2021. The ichnology of vertebrate consumption: dentalites, gastroliths and bromalites: New Mexico Museum of Natural History and Science. Bulletin 87, 216.

Liu, S., Li, Z., Liu, D., O'Connor, J.K., 2023. Quantifying the gastral mass in Early Cretaceous ornithuromorphs (Aves, Ornithothoraces) from the Jehol avifauna. Palaeontology 66 (5), e12677.

Louchart, A., Virot, L., 2011. From snout to beak: the loss of teeth in birds. Trends Ecol. Evol. 26, 663–673.

Lucas, S.G., 2000. The gastromyths of "*Seismosaurus*." a Late Jurassic dinosaur from New Mexico: New Mexico Museum of Natural History and Science. Bulletin 17, 61–67.

Lucas, S.G., Hunt, A.P., 2021. Gastroliths in archaeology: a note of caution. Int. J. Osteoarchaeol. https://doi.org/10.1002/oa.3042.

Milton, S.J., Dean, W.H.J., Siegfried, W.R., 1994. Food selection by ostrich in southern Africa. J. Wildl. Manag. 58, 234–248.

Moore, S.J., 1999. Food breakdown in an avian herbivore: who needs teeth? Aust. J. Zool. 47, 625–632.

Mudge, B., 1877. Notes on the Tertiary and Cretaceous periods of Kansas. In: Part I (Geology) U. S. Geological and Geographical Survey of the Territories [Hayden] for 1875. Ninth Annual Report, pp. 277–294.

Munk, W., Sues, H.D., 1993. Gut contents of *Parasaurus* (Pareiasauria) and *Protorosaurus* (Archosauromorpha) from the Kupferschiefer (Upper Permian) of Hessen, Germany. Paläontol. Z. 67, 169–176.

O'Connor, J.K., 2019. The trophic habits of early birds. Palaeogeogr. Palaeoclimatol. Palaeoecol. 513, 178–195.

O'Gorman, J.P., Salgado, L., Cerda, I.A., Gasparini, Z., 2013. First record of gastroliths associated with elasmosaur remains from La Colonia Formation (Campanian-Maastrichtian), Chubut, Patagonia, Argentina, with comments on the probable depositional paleo-environment of the source of the gastroliths. Cretac. Res. 40, 212–217.

O'Gorman, J.P., Oliveiro, E.B., Santillana, S., Everhart, M.J., Reguero, M., 2014. Gastroliths associated with an *Aristonectes* specimen (Plesiosauria, Elasmosauriodae), López de Bertodano Formation (upper Maastrichtian) Seymour Island (*Is. Marambio*). Antarctic Peninsula: Cretac. Res. 50, 228–237.

Osborn, H.F., 1924. *Psittacosaurus* and *Protiguanodon*, two Lower Cretaceous iguanodonts from Mongolia. Am. Mus. Novit. 127, 16.

Pough, F.H., Janis, C.M., Heiser, J.B., 2002. Vertebrate Life, sixth ed. Prentice-Hall Inc., Upper Saddle River. 699 p.

Schmeisser, R.L., Flood, T.P., 2008. Recognition of paleogastroliths from the Lower Cretaceous Cedar mountain formation, Utah using a scanning electron microscope. Ichnos 15, 72–77.

Seeley, H.G., 1877. On *Mauisaurus gardneri* Seeley, an elasmosaur from the base of the Gault of Folkestone. Q. J. Geol. Soc. Lond. 33, 541–546.

Skoczylas, R., 1978. Physiology of the digestive tract. In: Gans, C., Gans, K.A. (Eds.), Biology of the Reptilia, vol. 8. Academic Press, London, pp. 589–717. Physiology B.

Sokol, O., 1971. Lithophagy and geophagy in reptiles. J. Sci. 5, 69–71.

Taylor, M.A., 1993. Stomach stones for feeding or buoyancy? The occurrence and function of gastroliths in marine tetrapods. Philos. Trans. R. Soc. London, Ser. A B 341, 163–175.

Trewin, N.H., 1986. Palaeoecology and sedimentology of the Achanarras fishbed of the Middle Old Red Sandstone, Scotland. Trans. R. Soc. Edinb. Earth Sci. 77, 21–46.

Tryon, C., 2002. Exotic minerals or ostrich gastroliths? An alternative explanation for some evidence of hominin non-utilitarian behavior at Wonderwerk Cave, South Africa. J. Taphon. 8, 235–242.

Von Koenigswald, W., Richter, G., Storch, G., 1981. Nachweis von Hornschuppen bei *Eomanis waldi* aus der "Grube Messel" bei Darmstadt (Mammalia, Pholidota). Senckenberg Lethaea 61, 291–298.

Whittle, C.H., Everhart, M.A., 2000. Apparent and implied evolutionary trends in lithophagic vertebrates from New Mexico and elsewhere: New Mexico Museum of Natural History and Science. Bulletin 17, 75–82.

Wickes, W.H., 1908. Pebble swallowing animals. A sequel to "The Rhaetic Bone Beds". Proc. Bristol Nat. Soc. Ser. 4 11, 25–31.

Wieland, G., 1906. Dinosaurian gastroliths. Science 23, 819–821.

Wings, O., 2003. Observations on the release of gastroliths from ostrich chick carcasses in terrestrial and aquatic environments. J. Taphon. 1, 97–103.

Wings, O., 2004. Identification, Distribution, and Function of Gastroliths in Dinosaurs and Extant Birds with Emphasis on Ostriches (*Struthio camelus*). Ph.D. thesis. Rheinische Friedrich-Wilhelms-Universität, Bonne, 187 p.

Wings, O., 2007. A review of gastrolith function with implications for fossil vertebrates and a revised classification. Acta Palaeontol. Pol. 52, 1–16.

Wings, O., 2009. A simulated bird gastric mill and its implications for fossil gastrolith authenticity. Fossil Rec. 12, 91–97.

Wings, O., 2015. The rarity of gastroliths in sauropod dinosaurs—a case study in the Late Jurassic Morrison Formation, western USA. Fossil Rec. 18, 1–16.

Wings, O., Sander, P.M., 2007. No gastric mill in sauropod dinosaurs: new evidence from analysis of gastrolith mass and function in ostriches. Proc. R. Soc. Lond. Ser. B 274, 635–640.

Xu, X., 1997. A new psittacosaur (*Psittacosaurus mazongshanensis* sp. nov.) from Mazongshan area, Gansu Province, China. In: Dong, Z. (Ed.), Sino-Japanese Silk Road Dinosaur Expedition. China Ocean Press, Beijing, pp. 48–67.

Zheng, X., Martin, L.D., Zhou, Z., Burnham, D.A., Zhang, F., Miao, D., 2011. Fossil evidence of avian crops from the Early Cretaceous of China. Proc. Natl. Acad. Sci. USA 108, 15904–15907.

CHAPTER

11

Pabulites, digestilites, micturalites and other trace fossil evidence of predation and consumption

Adrian P. Hunt[1] *and Spencer G. Lucas*[2]

[1]Flying Heritage and Combat Armor Museum, Everett, WA, United States; [2]New Mexico Museum of Natural History, Albuquerque, NM, United States

1. Introduction

Dentalites, common forms of bromalites (regurgitalites, consumulites, and coprolites) and gastroliths are the principal source of ichnological evidence about vertebrate consumption. However, less common bromalites and some other trace fossils also provide significant information. This ichnological evidence was reviewed by Hunt and Lucas (2021), and their work is reviewed, revised, and updated here. Abbreviations are as follows: FHSM, Sternberg Museum of Natural History, Fort Hays State University, Hays, Kansas, USA; MPA, Museum of Paleontology, Araraquara, Brazil; NMMNH, New Mexico Museum of Natural History, Albuquerque, New Mexico, USA; PW, Generaldirektion Kulturelles Erbe RLP, Mainz, Germany; SMF, Senckenberg Research Institute Frankfurt, Frankfurt, Germany.

2. Pabulites

2.1 Introduction

Many fossil animal remains are potentially the residue remaining after predation or scavenging. Montalvo et al. (2016) introduced the term "leftover prey remains," which represent accumulations of uneaten prey remains discarded by a predator. They based the term on late Miocene microvertebrate accumulations from Argentina.

Subsequently, Klug et al. (2021a, p. 9) coined the formal term pabulite for fossils "which provide evidence for incomplete predation" and which are "leftover fall." They further

Vertebrate Ichnology
https://doi.org/10.1016/B978-0-443-38351-9.00010-4

note that pabulites represent "fossilized food that never entered the digestive tract beyond the buccal cavity" (Klug et al. 2021a, p. 9). Evidence for entry into the buccal cavity is provided by dentalites.

Klug et al. (2021a) noted that pabulites are normally incomplete animals that preserve traces of predation. They describe the example of the belemnite *Passaloteuthis laevigata* from the Toarcian Posidonienschiefer Lagerstätte. Most of the soft parts are missing, but the arm crown is one of the best preserved that is known. Klug et al. (2021a) suggest that the belemnite represents the remnant of the food of a predatory fish, possibly the shark *Hybodus hauffianus*.

Klug et al. (2021a) emphasized that pabulites are numerous. They listed a variety of principally European Jurassic examples including belemnites in which the rostrum cavum is fragmented (Klug et al., 2010; Schweigert, 1999, 2018), isolated soft parts of ammonites and abundant crushed shells (Klug et al., 2021b), incompletely preserved vertebrate remains from the Plattenkalk strata, including isolated heads, fins, or tails of fishes (e.g., Dietl and Schweigert, 2011; Chellouche, 2016; Maxwell et al., 2021), lethally bitten fish (Schweigert et al., 2016), or even large parts of marine crocodiles (Dietl and Schweigert, 2011; Mäuser, 2015), and abundant, crushed ammonite shells (Klug et al., 2021b). Subsequently, Klug et al. (2021b) described another possible pabulite of the cymatoceratid cephalopod *Syrionautilus libanoticus* from the Cenomanian of Lebanon.

Another specimen identified as a probable pabulite from the Posidonienschiefer Formation is a skull of the coccolepidid *Toarcocephalus morlok* (Cooper et al., 2024). This specimen consists of an almost fully articulated skull lacking the body. The incomplete but articulated lateral line scales end abruptly within the vicinity of the pectoral girdle, suggesting decapitation rather than disarticulation due to decay (Cooper et al., 2024).

Pabulites are very common in the fossil record. For example, most specimens preserving dentalites are probably pabulites, although dentalites may also be preserved in some scavenged specimens (e.g., Serafini et al., 2024), regurgitalites, digestilites, consumulites, and coprolites.

2.2 Recognition

No unique characteristic distinguishes pabulites, but they can be recognized and distinguished from other bromalites based on the possession of two or more of the following features (Hunt and Lucas, 2024):

1. Lack of even diffuse matrix around preserved elements (Klug et al., 2021a).
2. Body is often incomplete but with obvious evidence of biological causation (e.g., dentalites present or associated coprolites, regurgitalites, or tracks).
3. Representation and preservation of remaining elements are consistent with partial predation (e.g., Mayr, 2016; Montalvo et al., 2016).
4. Some elements of the body are articulated in their original configuration, still in their primary order (Klug et al., 2021a).
5. Spatial arrangement of debris (Klug et al., 2021a).
6. Dentalites are present (Montalvo and Fernández, 2019).
7. Incomplete preservation of soft tissues (Klug et al., 2021a).
8. High degree of element breakage (Montalvo and Fernández, 2019).
9. Association of carnivore teeth (e.g., Mayr and Poschmann, 2009).

2.3 Utility

Pabulites have utility for addressing a wide range of paleoecological and taphonomic issues at varying levels of synthesis (Hunt and Lucas, 2024). They also provide a clarity of nomenclature in the discussion of fossil faunas. The use of the term pabulite would be useful in the discussion of fish faunas that include both pabulites and skeletons subject to other means of disintegration, for example in the classic Pennsylvanian black shales of Indiana, USA (Zangerl and Richardson, 1963; Elder, 1985; Elder and Smith, 1988) and Plattenkalk faunas from Germany and China (Chellouche, 2016). Examples of studies or potential studies of pabulites are briefly discussed below.

Mayr and Poschmann, 2009 described an isolated foot of a loon from the late Oligocene Enspel fossil site in Germany that represents a lacustrine maar deposit. A crocodile tooth lies on the upper foot, and the taphonomy of the locality indicates minimal water movement. Thus, Mayr and Poschmann (2009) hypothesize that the loon was dismembered by vigorous shaking by a crocodilian with the tooth having been embedded in the skin or musculature (see also Frey and Monninger, 2010).

Subsequently, Mayr (2016) studied isolated feet of multiple birds from the middle Eocene Messel Lagerstätte of Germany that also formed in a lacustrine maar setting. Messel yields a rich and diverse avifauna dominated by well-preserved skeletons of small-sized birds. Larger birds are rare at Messel, and, in eight instances, they are principally only represented by isolated feet (Mayr, 2016: Fig. 11.1). Most of these specimens exhibit broken leg bones with missing ends, suggesting predation (Fig. 11.1). The only nonpiscivorous carnivores preserved at Messel are represented by abundant examples of at least seven taxa of crocodilians (Morlo et al., 2004; Mayr, 2016). A partial skeleton of the primate *Europolemur* from Messel preserves a broken crocodilian tooth stuck in one bone (Franzen and Frey, 1993). There are no direct signs of predation or scavenging by a carnivore on the isolated feet, but the similarity to the Enspel foot and the evidence of possible predators at Messel strongly suggests that they are pabulites produced by crocodilians. We base the first pabulite ichnotaxon, *Frangocuralis mayri* igen. et isp. nov., on this distinct and paleoecologically important trace fossil (see below).

Montalvo et al. (2016) were the first to explicitly recognize pabulites without using that term. They studied three rich vertebrate faunas of small mammals from the upper Miocene Cerro Azul Formation of Argentina. These localities exhibit a high density of specimens in a stratigraphically limited unit, an abundance of small mammals and a high diversity of taxa, so they represent micro-fossil bonebeds (Eberth et al., 2007; Rogers and Kidwell, 2007). Montalvo et al. (2016) posit that the presence of dentalites and association with coprolites in the absence of chemical corrosion suggests that they represent pabulites. There has been a long debate about the origin of small mammal bonebeds (e.g., Dodson, 1973; Korth, 1979), and it is possible that some are largely composed of pabulites.

The concept of pabulites provides a new perspective on examining fossil faunas. For example, in the Upper Jurassic Morrison Formation of western North America, there has been a focus on documenting dentalites and their prevalence (e.g., Drumheller et al., 2020 and references therein), but there is other evidence that could allow the recognition of pabulites (e.g., Jennings and Hasiotis, 2006) that provide a different perspective on predation, scavenging, food webs, and other aspects of paleoecology of the Late Jurassic ecosystem.

FIGURE 11.1 Holotype and referred specimens of the pabulite *Frangocuralis mayri* igen et isp. nov. from the middle Eocene Messel Lagerstätte, Germany. (A-B) Holotype, right foot of unknown genus and species D of bird of Mayr (2016). (A) SMF-ME 707, part A, (B) SMF-ME 707 A, counterpart. (C) SMF-ME 778, referred specimen, right foot of the unknown genus and species A (Gruiformes) of Mayr (2016), SMF-ME 778). (D) SMF-ME 11606, referred specimen, right leg of the unknown genus and species B (?Gruiformes) of Mayr (2016). (E) SMF-ME 11607, referred specimen, right foot of the unknown genus and species C of Mayr (2016). Scale bars are in mm. *Images courtesy of Krister Smith and Anika Vogel.*

2.4 Ichnotaxobases

Ichnotaxobases for pabulites include (1) anatomical representation (Montalvo et al., 2016); (2) degree of bone breakage (Montalvo et al., 2016); (3) presence of dentalites (Montalvo et al., 2016); (4) spatial arrangement of debris (Klug et al., 2021a); (5) incomplete preservation of soft tissue (Klug et al., 2021a); and (6) evidence of particular methodology of predation and/or scavenging (e.g., Mayr, 2016). Herein, we name *Frangocuralis mayri* igen et isp. nov., and we encourage other workers to discriminate and provide an ichnotaxonomic framework for other pabulites.

2.5 Ichnotaxonomy

2.5.1 Frangocuralis, *igen. nov*

Type ichnospecies: *Frangocuralis mayri* Hunt and Lucas, isp. nov.

Included ichnospecies: Known only from the type ichnospecies.

Etymology: From the Latin *frango* (to break) and *curalis* (of the leg), in allusion to the apparent origin of this pabulite.

Distribution: Middle Eocene to late Oligocene of Germany.

Diagnosis: Isolated foot of medium-large terrestrial bird with broken leg bones with missing ends.

Discussion: There is strong evidence that *Frangocuralis mayri* igen. is the result of crocodilian predation (Mayr, 2016; May and Poschman, 2009; and see discussion above). All specimens are preserved in low energy lacustrine environments.

Mayr (2016) described eight feet of different bird taxa, all but one of which were terrestrial rather than aquatic in habit. Most of these taxa are otherwise unknown at Messel. Most birds found in the intestinal tract of modern crocodilians represent aquatic-semiaquatic taxa such as ducks, cormorants, and herons. However, crocodilians have been observed catching low-flying, nonaquatic birds (Mayr, 2016).

2.5.1.1 *Frangocuralis mayri isp. nov*

Holotype: SMF-ME 707 A-B, part and counterpart, right foot of unknown genus and species D of Mayr (2016, Fig. 3D and E; Fig. 11.1A and B).

Etymology: Honoring Gerald Mayr for recognizing and documenting the significance of this pabulite.

Type locality: Messel Lagerstätte.

Type horizon: Messel oil shale (middle Eocene).

Distribution: As for ichnogenus.

Referred specimens: Messel Lagerstatte (middle Eocene): SMF-ME 1577, Left foot of an idiornithid (cf. *Dynamopterus* sp., Cariamiformes) (Mayr, 2016, Fig. 11.1A–D); SMFME, Right foot of the putative cariamiform *Strigogyps* sp. (Mayr, 2016, Fig. 2A and B); SMF-ME 778, Right foot of unknown genus and species A of Mayr (2016) (Gruiformes) (Mayr, 2016, Fig. 2C: Fig. 11.1C); SMF-ME 11606, Right leg of unknown genus and species B of Mayr (2016) (Gruiformes) (Mayr, 2016, Fig. 2D: Fig. 11.1D). SMF-ME 11,607, Right foot of unknown genus and species C of Mayr (2016) (Mayr, 2016, Fig. 3A–C: Fig. 11.1E). SMF-ME 1231, Right foot of the unknown genus and species E of Mayr Mayr (2016) (Mayr, 2016, Fig. 4A) SMF-ME 11286, Left foot of the unknown genus and species F of Mayr (2016) (cf. *Vastanavis*), a semi-zygodactyl bird of uncertain affinities (Mayr, 2016, Fig. 4C).

Enspel site (late Oligocene): PW 2007/5157-LS, Right foot of a loon (Gaviiformes) (Mayr, 2016, Fig. 4B).

Description: SMF-ME 707 A-B is the right hindlimb of the unknown genus and species D of Mayr (2016) that consists of an articulated but dislocated proximal tarsometatarsus and distal tibiotarsus (Fig. 11.1A and B). The tibiotarsus and tarsometatarsus have fractured ends (Mayr, 2016, Fig. 3E).

2.6 Future work

The concept of pabulites was only recently recognized, and it has not yet been widely appreciated. However, it has great potential to aid in many aspects of paleoecology and taphonomy. Nevertheless, pabulites may be difficult to recognize in the absence of other traces. For example, a partial skeleton with some bite marks was likely scavenged and hence is probably a pabulite, whereas an incomplete carcass with no dentalites could have other causation. Many vertebrate specimens exposed on an oxygenated substrate are probably subject to scavenging and may represent pabulites, but in dysoxic or anoxic settings they may be rare (e.g., Holzmaden Lagerstätte). Pabulites have utility in assessing microvertebrate accumulations to distinguish the genesis of the assemblage that result from concentrations produced by digestion (regurgitalites or coprolites) versus physical processes. The development of the study of pabulites requires focus and documentation—for example, actualistic studies, documentation of new examples, re-examination of the taphonomy of faunas, and the establishment of a robust descriptive and ichnotaxonomic protocol.

3. Digestilites

Food stuffs that pass through the digestive tract are subject to chemical and physical processes that result in characteristic damage that can be recognized in regurgitated or defecated material (e.g., Crandall and Stahl, 1995; Fernández-Jalvo et al., 2002, 2014, 2016; Esteban-Nadal et al., 2010; Cohen, 2003; Cohen and Kibii, 2015; Fernandez-Jalvo and Andrews, 2016; Meador, 2017; Montalvo and Fernández, 2019). For example, crocodilian-digested bones most closely resemble bones exposed to hydrochloric acid as opposed to the damage due to a mix of hydrochloric acid and digestive enzymes exhibited by bones consumed by mammalian and avian predators (Fernández-Jalvo et al., 2014; Meador, 2017). This exposure to digestive acid produces a number of unique and identifiable characteristics on bones and teeth, including: (1) surface etching, (2) corrosive pitting, (3) rounding, (4) enamel erosion, (5) foramen excavation, (6) reduction in cortical thickness, (7) corrosive holes, (8) undulations, (9) desquamation, (10) cupules, and (11) polishing (e.g., Lyman, 1994; Crandall and Stahl, 1995; Esteban-Nadal et al., 2010; Cohen, 2003; Cohen and Kibii, 2015; Brown et al., 2016; Meador, 2017; Montalvo and Fernández, 2019). Such specimens represent morphologically recurrent structures resulting from the life activity of an individual organism modifying the substrate, and thus are trace fossils (Bertling et al., 2006). Hunt and Lucas (2021) proposed the term digestilite for fossil specimens that preserve evidence of digestion.

The investigation of damage to ingested vertebrate hard tissue was initially stimulated by the study of the accumulation mechanisms of micromammal taphocoenoses (Dodson, 1973; Dodson and Wexlar, 1979; Korth, 1979; Fisher, 1981a,b; Andrews and Evans, 1983; Andrews, 1990). However, larger bones, such as a hominin specimen from Zhoukoudian in China, also show clear evidence of digestion (Boaz et al., 2000). Thus, Zhoukoudian *Homo erectus* Femur V exhibits a breakage pattern characteristic of hyenas and also has rounded edges and pitting associated with ingestion and partial digestion (Weidenreich, 1941; Boaz et al., 2000).

Digestilites record evidence of mechanical and/or chemical digestion and may derive from regurgitation (e.g., Boaz et al., 2000) or defecation (e.g., Fisher, 1981a,b) or disintegration of a consumulite. Perhaps the best known digestilite is the Denisova 11 hominid, which is the fragmentary limb bone of a Denisovan and Neanderthal hybrid from Denisova Cave, Siberia (Brown et al., 2016; Slon et al., 2018).

FIGURE 11.2 Digestilite composed of etched skeletal elements of a hatchling turtle (FHSM VP-17,572) from lower Santonian chalk, Kansas, USA. *Modified from Hunt and Lucas (2021), Fig. 33. Scale bar in mm. Images courtesy of Michael Everhart.*

Ingested hard tissue constitutes an important contribution to the fossil record. It is widely recognized that many terrestrial microvertebrate faunas are composed of digestilites, the majority from bird regurgitalites, but also from carnivore and crocodile coprolites (Dodson, 1973; Mellett, 1975; Dodson and Wexlar, 1979; Korth, 1979; Fisher, 1981a,b; Andrews and Evans, 1983; Andrews, 1990; Fernández-Jalvo and Andrews, 1992; Williams, 2003; Denys, 2011; Fernández-Jalvo et al., 2016; Denys et al., 2018; Montalvo and Fernández, 2019). Thus, for example, Coco et al. (2020) described late Pleistocene specimens of owls (*Tyto, Athene, Glaucidium*) that exhibit breakage and weathering, consistent with digestion by the giant strigiform *Asio ecuadoriensis.*

The other major importance of digestilites is in the processing of invertebrate hard parts into the sedimentological record. Oji et al. (2003) and Salamon et al. (2014) conducted actualistic studies that demonstrated that predation by fish on bivalves produces shell fragments with sharp, angular margins, in contrast to nonbiogenic physical processes that produce abraded and rounded shell fragments. They thus utilized angular shell fragments as a proxy for the feeding of durophagous fish in the Paleozoic (and later). Digestilites composed of invertebrate material may encompass a significant portion of shell beds from the Mississippian to the Recent (Oji et al., 2003; Salamon et al., 2014). Vertebrate predation is also important in the production of Recent coral sands from reefs, notably by scarid Parrotfish (Peyrot-Clausade et al., 2000).

Digestilites are also important in other taphonomic settings. Wilson (1987) suggested that fish bones in some Eocene lake deposits of British Columbia and Washington are partly dissolved due to possible ingestion, probably by fish (Hunt and Lucas, 2025). Digestilites also occur in the Late Cretaceous marine strata of western North America, representing fish, turtles, mosasaurs, and dinosaurs (Everhart, 2017, Figs. 3.19, 4.3, 6.12, 12.2; Fig. 11.2) and may be very common in the Maastrichtian phosphate deposits of Morocco (Longrich et al., 2022; Cooper et al., 2022).

4. Micturalites

Duvernoy (1844) introduced the term urolite to differentiate the fossil, nonliquid urinary secretions produced by some groups of reptiles, from coprolites. Leydig (1896a,b) and Voigt

(1960) identified urolites from the Triassic and Cretaceous, respectively, which were three-dimensional objects. Fernandes et al. (2004, p. 266) utilized the same term for the trace produced by liquid waste impacting the sediment. Hunt and Lucas (2012) thought that the trace of the excavation pit should have a different name from the three-dimensional ichnofossil and so they introduced the term micturalite for the rare trace fossils produced by the interaction between liquid urine and the substrate (Hunt and Lucas, 2012). McCarville and Bishop (2002) interpreted a subcircular depression (scour?) at a Late Jurassic tracksite in Colorado, USA, as having been eroded by liquid urine of a sauropod dinosaur. However, there is no evidence that this trace fossil resulted from urine impacting the substrate. Fernandes and Souto and co-authors, described a more convincing ichnofossil from the Early Cretaceous of Brazil (Fernandes et al., 2004; Souto and Fernandes, 2015) that we name below.

4.1 Ichnotaxonomy

4.1.1 Urinaichnus, *igen. nov*

Type ichnospecies: *Urinaichnus soutoi* Hunt and Lucas, isp. nov.

Included ichnospecies: Known only from the type ichnospecies.

Etymology: From the Latin *urina* (urine) and the Greek *ichnos* (trace).

Distribution: Early Cretaceous of Brazil.

Diagnosis: Elongate depression with a wide anterior which has a deep pit and has a rounded margin and that tapers posteriorly to a shallower end which has a more acute tip with narrow converging ridges that extend posteriorly from the pit to the rear of the trace (Fig. 11.3).

Discussion: This trace occurs on eolian forests. Fernandes et al. (2004, Fig. 11.5) compared this ichnofossil to erosional depressions produced in the substrate by liquid waste of the Recent ratite *Struthio camelus* (ostrich). They also conducted a simple experiment and poured 2 liters of water from a height of 80 cm onto a slope of loose sand and this also produced a very similar morphology trace (Fernandes et al., 2004, Fig. 2D: Fig. 11.4). Thus, it seems highly probable that this trace was produced by liquid waste of possibly a dinosaur.

4.1.2 Urinaichnus soutoi *isp. nov*

Holotype: MPA-001 is in negative epirelief (Fernandes et al., 2004, Fig. 2C: Fig. 11.3).

Etymology: Honoring Paulo Roberto de Figueiredo Souto for his diverse contributions to vertebrate ichnology.

Type locality: São Bento Quarry, city of Araraquara, São Paulo State, Brazil.

Type horizon: Botucatu Formation, São Bento Group (Lower Cretaceous, Neocomian).

Distribution: As for ichnogenus.

Referred specimens: MPA-002 and MPA-003 (Fernandes et al., 2004, Fig. 2A and B: Fig. 11.5A and B).

Description: MPA-001 is a negative epirelief (Fernandes et al., 2004, Fig. 2C: Fig. 11.3). The elliptical concave portion of the pit (excavation crater of Fernandes et al., 2004) of MPA-001 is 16 cm in length, 11 cm in width and 1.8 cm in depth. The flagstone that includes the specimen is 75 by 45 cm. Fernandes et al. (2004) termed the posteriorly converging ridges as gravitational stream lines.

FIGURE 11.3 Holotype of micturalite *Urinaichnus soutoi* igen. et isp. nov. MPA-001 is in negative epirelief. Scale bar is in cm. *Image courtesy of Paulo Roberto de Figueiredo Souto.*

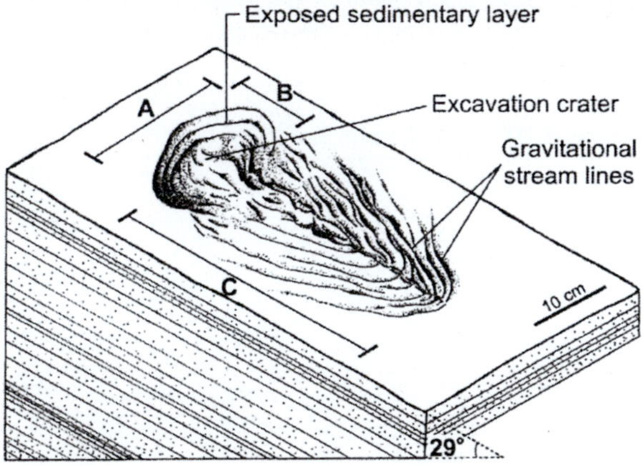

FIGURE 11.4 Diagrammatic representation of morphology of *Urinaichnus soutoi* igen. et isp. nov. *From Fernandes et al. (2004), Fig. 3.*

MPA-002 and MPA-003 respectively are the negative epirelief and positive hyporelief of the same trace fossil (Fernandes et al., 2004, Fig. 2A and B: Fig. 11.5A and B). The flagstones that include the specimens are respectively 50 by 45 cm and 60 by 50 cm. MPA-002 is 19 cm in length, 12 cm wide with a depth of 1.8 cm (Fernandes et al., 2004).

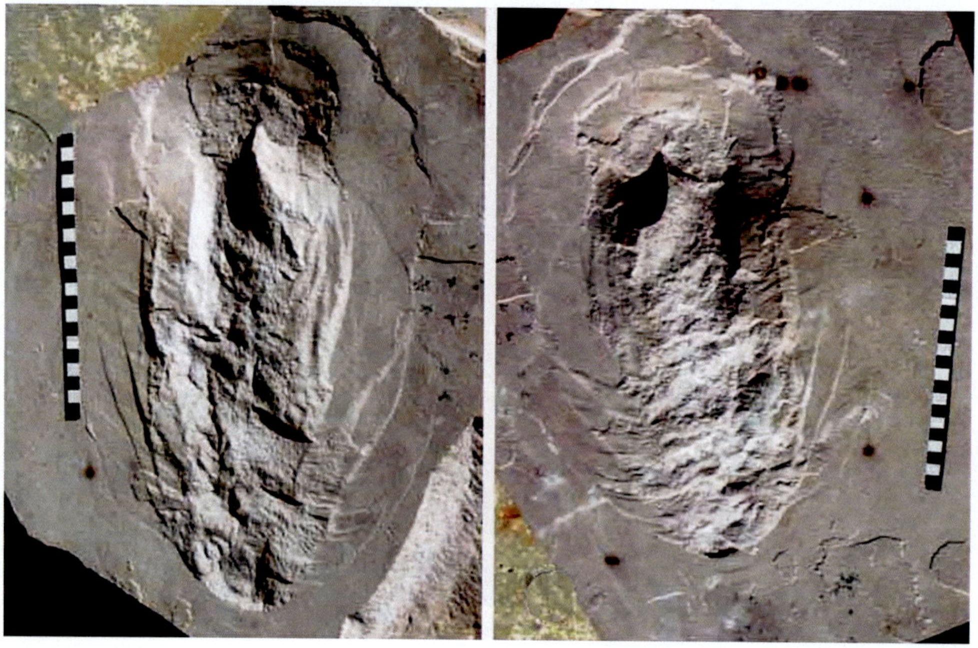

FIGURE 11.5 Referred specimens of micturalite *Urinaichnus soutoi* igen. et isp. nov. (A) MPA-002 in negative epirelief. (B) Sample MPA-003, in positive hyporelief. (C) MPA-001 in negative epirelief. Scale bars are in cm. *Fig. 2A and B from Fernandes et al. (2004) courtesy of Paulo Roberto de Figueiredo Souto.*

Discussion: It is likely that a broadly similar trace consisting of pit with a posteriorly shallowing and tapering termination would also be produced on a flat surface (Fernandes et al., 2004, Figs. 11.4 and 11.5).

5. Other trace fossil evidence of vertebrate feeding

5.1 Introduction

Dentalites, gastroliths and bromalites are not the only trace fossils that provide evidence of vertebrate feeding. Footprints, some nests and mololites also can record evidence of vertebrate predation and consumption.

5.2 Footprints

There are several examples of vertebrate trackways that are purported to capture an act of predation. The first specimen to be so recognized as such was from the Early Cretaceous of Texas and is interpreted to show a theropod attacking a sauropod (Bird, 1954, 1985; Thomas and Farlow, 1997). The theropod repeatedly stepped in the prints made by the sauropod, and the trackways of both made a turn at the same point, which may suggest that the carnivore was following the herbivore (Farlow, 1987a,b; Thomas and Farlow, 1997, Fig. 11.6). Bird

FIGURE 11.6 Putative pursuit trackways from the lower member of the Glen Rose Formation (Lower Cretaceous) at Dinosaur Valley State Park, Texas, USA. The excavation was carried out by 1938 by R. T. Bird (pictured) to collect display trackways for the American Museum of Natural History. The prominent trackway proceeding from bottom left to top right is *Brontopodus* that represents a sauropod with a large U-shaped pes print in the rear and a smaller lunate manus print in front. The theropod trackway of *Irenesauripus* is to the left of the *Brontopodus* and is recognizable by three acute toe prints. Asset # AN1384813, American Museum of Natural History Library.

(1954, 1985) believed that the theropod actually attacked the sauropod, as indicated by a missing left carnivore footprint, which was interpreted to represent the animal being dragged off its feet by the forward motion of the much larger herbivore (Bird, 1985, p. 173; Thomas and Farlow, 1997). However, the subsequent footprint's location and morphology do not support that hypothesis (Lockley and Hunt, 1995; Hunt and Lucas, 1998; Farlow et al., 2012).

The Lark Quarry tracksite from the Early Cretaceous of Australia was originally interpreted to represent a large predatory theropod causing a stampede of small ornithopods and theropods (Thulborn and Wade, 1979, 1984). However, there is a lack of clarity about the taxonomy of the trace makers, the contemporaneity of the trackways, and the speed of the smaller dinosaurs, so the predatory theropod hypothesis is problematic (Romilio and Salisbury, 2011; Romilio et al., 2013; Thulborn, 2013, 2017).

Weems (2018, 2020) described a theropod trackway of *Kayentapus minor* from the Late Triassic of the Culpeper Quarry in Virginia, USA, which is purported to be a possible record of predation. This trackway preserves changing locomotion of an individual—acceleration, running, slowing, slight skid, walking, stopping, and slight shifting of position. This is interpreted as the behavior associated with the capture of prey, but it is not a convincing interpretation, particularly given the lack of evidence of a prey item or its track. It is possible, but unlikely, that the prey was flying and left no tracks.

Malkani (2007) described converging sauropod and theropod trackways from the Middle Jurassic of Pakistan. He concluded that the sauropods were confronted by a theropod but the trackways do not meet and there is no evidence of evasive action by the sauropods, so the trackways are probably not synchronous. Paik et al. (2010) noted the possibility that a solitary theropod was stalking a herd of ornithopods in the Late Cretaceous of Korea.

A Pleistocene hunt of sloths by humans is hypothesized by Bustos et al. (2018) on the basis of trackways at White Sands National Monument in southern New Mexico in the United States. This is of interest as there is a surprising lack of evidence for human hunting of ground sloths (Hunt and Lucas, 2018). The principal pieces of evidence at White Sands are human footprints superimposed on sloth tracks and enigmatic "flailing circles" interpreted as being made by a sloth reaching forward with its forelimbs and leaving knuckle and claw impressions in a defensive mode (Bustos et al., 2018). While these tracks clearly provide evidence of contemporaneity, the evidence for hunting requires more study, particularly with regard to the "flailing circles." These tracks were formed on the margin of a large Pleistocene lake where trackways would be expected to be parallel (e.g., Getty et al., 2017), which should be considered in any analysis.

There are five examples from the Permian of trackway terminations of vertebrates and invertebrates that have been interpreted as evidence of predation (Lockley and Madsen, 1993; Kramer et al., 1995; Hunt and Lucas, 1998; Santi and Stoppini, 2005; Citton et al., 2012: Fig. 11.7). In each case, one trackway appears to terminate at another with the predator

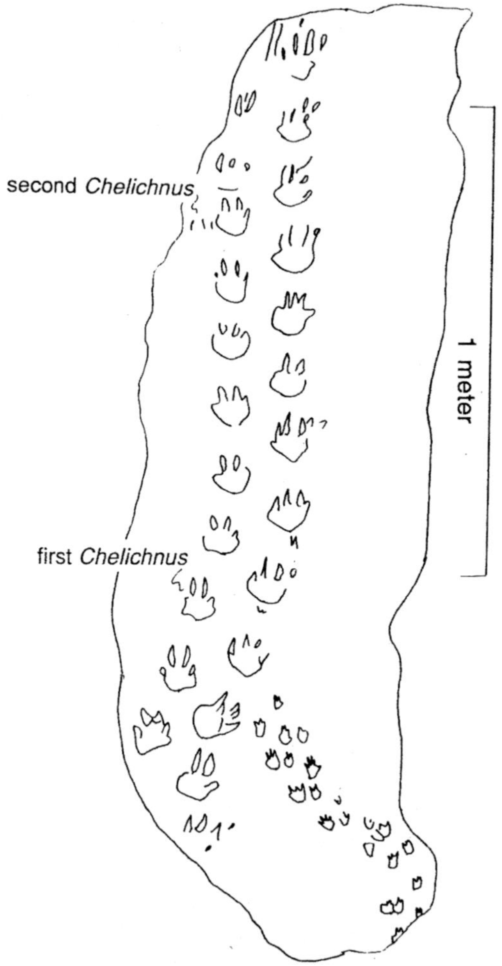

FIGURE 11.7 Putative trackway evidence of predation from Permian Cedar Mountain Formation of Utah. Note the termination of the small "*Chelichnus*" (lower right) trackway at the larger one (probably *Ichniotherium*). *From Hunt and Lucas (1998, Fig. 1); Modified after Lockley and Madsen (1993).*

showing no change in locomotion. None of these examples satisfy more than one of the five (one extra is added herein) criteria established by Hunt and Lucas (1998) to evaluate trackway evidence of predation in such cases: (1) in intersecting trackways, one trackway should terminate at the other; (2) in the case of parallel pursuit trackways, one trackway should end; (3) if the trackway is long enough, there should be evidence of a pursuer adjusting length of stride to sync with prey stride length; (4) one or both trackways should demonstrate change in direction or/and speed—evasion and compensation; (5) the trackway of the prey should show evidence of a struggle—slowing, terminating or lateral movement; and (6) we herein add a sixth criterion of the tracks being on the same bedding plane and hence likely synchronous.

Theropod tracks and inferred trampling have been found associated with large bone accumulations, and this could be evidence of scavenging (e.g., Lockley et al., 1998; Jennings and Hasiotis, 2006; Augustin et al., 2020). An absence of tracks but a presence of broken bones that could have been impacted by trampling together with shed teeth has also been considered as evidence of scavenging (Eberth and Getty, 2005; Snyder et al., 2020). Simpson et al. (2010) described putative digging traces produced by a maniraptoran theropod dinosaur in association with mammalian den complexes from the Late Cretaceous of Utah, which they interpreted as evidence of predation.

5.3 Nests

Lucas and Hunt (2006) defined a nest as the structure made by, or the place chosen by, an animal for spawning, breeding, and/or laying eggs and sheltering young. Carnivore nests, notably dens of Pleistocene hyenas in caves of the Old World, preserve abundant evidence of predation in terms of trace fossils such as dentalites and coprolites, but also body fossils that demonstrate aspects of prey selection, processing, and utilization (e.g., Boaz et al., 2000; Marra et al., 2004; Stiner, 2004; Sala et al., 2012). The most abundant nests in the fossil record are of rodents, notably pack rats (*Neotoma*), which are widely distributed in the Late Pleistocene–Holocene of North America and other areas (e.g., Betancourt et al., 2010; Tweet et al., 2012). Regurgitalites and pabulites can also commonly be found below raptor nests, as may be the case for the South African Taung Child (Berger and Clarke, 1995).

5.4 Mololites

Tooth wear is a type of trace fossil because it falls within "bite and gnaw structures" and reflects the interaction of an organism and a substrate (Bertling et al., 2006, Table 1). Thus, Hunt and Lucas (2021) proposed the term mololite (from the Latin *molo*—to grind) for wear features, both mesowear and microwear (Fortelius and Solounias, 2000; Green and Croft, 2018), on fossil teeth. Fossil teeth preserve wear that has long been recognized as evidence of the nature of the food that is being processed. The study of tooth wear as it relates to diet became a robust tool with the examination of mammalian microwear (particularly with the SEM) in the late 1970s and early 1980s (e.g., Walker et al., 1978; Covert and Kay, 1981; Kay and Covert, 1983; Teaford and Walker, 1984). The majority of wear studies have been on the teeth of terrestrial mammals, but the same methodology has been subsequently applied to other groups such as marine mammals (e.g., Lambert and Bianucci, 2019) and reptiles (e.g., Fiorillo, 1998, 2011). The study of microwear is now widespread, particularly with regard to hominins and other mammals (e.g., Merceron et al., 2005; Scott et al., 2005; Williams and Patterson, 2010; DeSantis, 2016; Ungar et al., 2016; Pappa et al., 2019), but also of fish (e.g., McLennan and Purnell, 2021) and of reptiles (e.g., Bestwick et al., 2019; 2020a,b; Winkler et al., 2019).

6. Conclusions

Diverse kinds of bromalites, as well as other trace fossils, can provide information on vertebrate consumption, and these include pabulites, digestilites, micturalites, mololites, tracks, and nests. Pabulites and digestilites are notably in need of more intense study that requires focus and documentation—actualistic studies, documentation of new examples, reexamination of the taphonomy of faunas and the establishment of a robust descriptive and ichnotaxonomic protocols.

Acknowledgments

We thank many colleagues who have graciously allowed us access to collections, including Michael Brett-Surman, Paul Jeffery, Thomas Lehmann, Gerald Mayr, Krister Smith, Anika Vogel, and Gerald Mayr for helpful discussions. The American Museum of Natural History Library, Mike Everhart, Krister Smith, Paulo Roberto de Figueiredo Souto and Anika Vogel kindly supplied images, and Christian Klug and Henrik Klein provided very helpful reviews.

Declaration of competing interest

The authors declare that they have no known competing financial interests or personal relationships that could have appeared to influence the work reported in this paper.

Data availability

Only published data were used for the research described in this article.

References

Andrews, P., 1990. Owls, Caves and Fossils: Predation, Preservation and Accumulation of Small Mammal Bones in Caves, with an Analysis of the Pleistocene Cave Faunas from Westbury-sub-Mendip. University of Chicago Press, Somerset, UK. Chicago, p. 239.

Andrews, P., Evans, E.N., 1983. Small mammal bone accumulations produced by mammalian carnivores. Paleobiology 9, 289–307.

Augustin, F.J., Matzke, A.T., Maisch, M.W., Pfretzschner, H.-U., 2020. A theropod dinosaur feeding site from the Upper Jurassic of the Junggar Basin, NW China. Palaeogeogr. Palaeoclimatol. Palaeoecol. 560. https://doi.org/10.1016/j.palaeo.2020.109999.

Berger, L.R., Clarke, R.J., 1995. Eagle involvement in accumulation of the Taung child fauna. J. Hum. Evol. 29, 275–299.

Bertling, M., Braddy, S.J., Bromley, R.G., Demathieu, G.R., Genise, J., Mikuláš, R., Nielsen, J.K., Nielsen, K.S.S., Rindsberg, A.K., Schlirf, M., Uchman, A., 2006. Names for trace fossils: a uniform approach. Lethaia 39, 265–286.

Bestwick, J., Unwin, D.M., Purnell, M.A., 2019. Dietary differences in archosaur and lepidosaur reptiles revealed by dental microwear textural analysis. Sci. Rep. 9 (1), 11691.

Bestwick, J., Jones, A.S., Purnell, M.A., Butler, R.J., 2020a. Dietary constraints of phytosaurian reptiles revealed by dental microwear textural analysis. Palaeontology 64, 119–136.

Bestwick, J., Unwin, D.M., Butler, R.J., Purnell, M.A., 2020b. Dietary diversity and evolution of the earliest flying vertebrates revealed by dental microwear texture analysis. Nat. Commun. 11, 5293. https://doi: org/10.1038/s41467-020-19022-2.

Betancourt, J.L., Van Devender, T.R., Martin, P.S. (Eds.), 2010. Packrat Middens: The Last 40,000 Years of Biotic Change. The University of Arizona Press, Tucson, p. 467.

Bird, R.T., 1954. We captured a "live" brontosaur. Natl. Geogr. 105, 707–722.

Bird, R.T., 1985. Bones for Barnum Brown. Texas Christian University Press, Fort Worth, p. 225.

Boaz, N.T., Ciochon, R.L., Xu, Q., Liu, J., 2000. Large mammalian carnivores as a taphonomic factor in the bone accumulation at Zhoukoudian. Acta Anthropol. Sin. 19 (Suppl. ment), 224–234.

Brown, S., Higham, T., Slon, V., Pääbo, S., Meyer, M., Douka, K., Brock, F., Comeskey, D., Procopio, N., Shunkov, M., Derevianko, A., 2016. Identification of a new hominin bone from Denisova Cave, Siberia using collagen fingerprinting and mitochondrial DNA analysis. Sci. Rep. 6 (1–8), 23559. https://doi: org/10.1038/srep23559.

Bustos, D., Jakeway, J., Urban, T.M., Holliday, V.T., Fenerty, B., Raichlen, D.A., Budka, M., C. Reynolds, S.C., Allen, B.D., Love, D.W., Santucci, V.L., Odess, D., Willey, P., McDonald, H.G., Bennett, M.R., 2018. Footprints preserve terminal Pleistocene hunt? Human-sloth interactions in North America. Sci. Adv. 4, eaar7621.

Chellouche, P., 2016. Comparative Actinopterygian Fish Taphonomy of Laminated Fossil Lagerstätten: Case Studies from the Late Mesozoic of Eurasia. Doctoral Thesis at Friedrich. Universität Erlangen-Nürnberg, Alexander, p. 360.

Citton, P., Sacchi, E., Nicosia, U., 2012. Sometimes they come back: recovery and reinterpretation of a trackway slab from the Permian Coconino Sandstone of the Southwestern United States. Ichnos 9 (3), 165–174.

Coco, G.E.L., Agnolín, F.L., Carrión, J.L.R., 2020. Late Pleistocene owls (Aves, Strigiformes) from Ecuador, with the description of a new species. J. Ornithol. 161, 713–721.

Cohen, B.F., Kibii, J.M., 2015. Experimental study of bone modification by captive caracal (*Caracal caracal*); a model for fossil assemblage analysis. J. Archaeol. Sci. 62, 92–104.

Cohen, B., 2003. Actualistic Investigation of Bone Modification on Leporids by Caracal (*Caracal caracal*) and Honey Badger (*Mellivora capensis*); an Insight to the Taphonomy of Cooper's Cave, South Africa. M.S. thesis. University of the Witwatersrand, Johannesburg, p. 160.

Cooper, S.L., Marson, K.J., Smith, R.E., Martill, D., 2022. Contrasting preservation in pycnodont fishes reveals first record of regurgitalites from the Upper Cretaceous (Maastrichtian) Moroccan phosphate deposits. Cretac. Res. 131, 105111. https://doi.org/10.1016/j.cretres.2021.105111.

Cooper, S.L.A., López-Arbarello, A., Maxwell, E.E., 2024. First occurrence of a †coccolepidid fish (?Chondrostei: †Coccolepididae) from the Upper Lias (Toarcian, Early Jurassic) of southern Germany. Palaeontol. Electron. 27 (1), a23. https://doi.org/10.26879/1326.

Covert, H.H., Kay, R.F., 1981. Dental microwear and diet: implications for determining the feeding behaviors of extinct primates, with a comment on the dietary pattern of *Sivapithecus*. Am. J. Phys. Anthropol. 55, 331–336.

Crandall, B.D., Stahl, P.W., 1995. Human digestive effects on a micromammalian skeleton. J. Archaeol. Sci. 22, 789–797.

Denys, C., 2011. Des référentiels en taphonomie des petits vertébrés: bilan et perspectives. In: Laroulandie, V., Mallye, J.B., Denys, C. (Eds.), Taphonomie des Petits vertébrés référentiels et transferts aux fossils, vol. 2269. British Archaeological Reports International Series, pp. 7–22.

Denys, C., Stoetzel, E., Andrews, P., Bailon, S., Rihane, A., Huchet, J.B., Fernández-Jalvo, Y., Laroulandie, V., 2018. Taphonomy of small predators multi-taxa accumulations: palaeoecological implications. Hist. Biol. 30, 868–881.

DeSantis, L.R.G., 2016. Dental microwear textures: reconstructing diets of fossil mammals. Surface Topography. Metrol. Propert. 4. https://iopscience.iop.org/article/10.1088/2051-672X/4/2/023002.

Dietl, G., Schweigert, G., 2011. Im Reich der Meerengel—Fossilien aus dem Nusplinger Plattenkalk, second ed. Pfeil.

Dodson, P., 1973. The significance of small bones in paleoecological interpretation. Rocky Mt. Geol. 12, 15–19.

Dodson, P., Wexlar, D., 1979. Taphonomic investigations of owl pellets. Paleobiology 5, 275–284.

Drumheller, S.K., McHugh, J.B., Kane, M., Riedel, A., D'Amore, D.C., 2020. High frequencies of theropod bite marks provide evidence for feeding, scavenging, and possible cannibalism in a stressed Late Jurassic ecosystem. PLoS One 15 (5), e0233115. https://doi.org/10.1371/journal.pone.0233115.

Duvernoy, G., 1844. Sur l'existence des urolithes fossiles, et sur l'utilité que la science des ossiles organiques pourra tirer de leur distinction d'avec lês coprolithes, pour la détermination de restes fossiles de Sauriens et d'Ophidiens. Comptes Rendus Hebdomadaires des Séances de l'Académie des Sciences 19, 255–260.

Eberth, D.A., Getty, M.A., 2005. Ceratopsian bonebeds: occurrence, origins, and significance. In: Currie, P.J., Koppelhus, E.B. (Eds.), Dinosaur Provincial Park: A Spectacular Ancient Ecosystem Revealed. Indiana University Press, Bloomington, pp. 501–536.

Eberth, D.A., Shannon, M., Noland, B.G., 2007. A bonebeds database: Classification, biases, and patterns of occurrence. In: Rogers, R.R., Eberth, D.A., Fiorillo, A.R. (Eds.), Bonebeds Genesis, Analysis, and Paleobiological Significance. The University of Chicago Press, Chicago and London, pp. 103–220.

Elder, R.L., 1985. Principles of Aquatic Taphonomy With Examples From the Fossil Record. Ph.D. dissertation. University of Michigan, Ann Arbor, p. 336.

Elder, R.L., Smith, G.R., 1988. Fish taphonomy and environmental inference in paleolimnology. Palaeogeogr. Palaeoclimatol. Palaeoecol. 62, 577–592.

Esteban-Nadal, M., Cáceres, I., Fosse, P., 2010. Characterization of a current coprogenic sample originated by *Canis lupus* as a tool for identifying a taphonomic agent. J. Archaeol. Sci. 37, 2959–2970.

Everhart, M.J., 2017. Oceans of Kansas—A Natural History of the Western Interior Sea, second ed. Indiana University Press, Bloomington, p. 427.

Farlow, J.O., 1987a. Lower Cretaceous dinosaur tracks, Paluxy River Valley, Texas. In: Field Trip Guidebook, South Central Section, Geological Society of America, Waco, Texas. Baylor University, p. 50.

Farlow, J., 1987b. Speculations about the diet and digestive physiology of herbivorous dinosaurs. Paleobiology 13, 60–72.

Farlow, J.O., O'Brien, M., Kuban, G.J., Dattilo, B.F., Bates, K.T., Falkingham, P.L., Piñuela, L., Rose, A., Freels, C., Kumagai, C., Libben, C., Smith, J., Whitcraft, J., 2012. Dinosaur tracksites of the Paluxy River Valley (Glen Rose Formation, Lower Cretaceous), Dinosaur Valley State Park, Somervell County, Texas: Actas de V Jornadas Internacionales sobre Paleontologia de Dinosaurios y su Entorno, Salas de los Infantes, Burgos. In: Proceedings of the V International Symposium on Dinosaur Palaeontology and Their Environment, pp. 41–69.

Fernandes, M.A., Fernandes, L.B., dos, R., Souto, P.R. de F., 2004. Occurrence of urolites related to dinosaurs in the Lower Cretaceous of the Bocatu Formation, Paraná basin, São Paulo state, Brazil. Rev. Bras. Palaontol. 7, 263–268.

Fernández-Jalvo, Y., Andrews, P., 1992. Small mammal taphonomy of Gran Dolina, Atapuerca (Burgos), Spain. J. Archaeol. Sci. 19, 407–428.

Fernandez-Jalvo, Y., Andrews, P., 2016. Atlas of Taphonomic Identifications: 1001+ Images of Fossil and Recent Mammal Bone Modification. Springer, New York, p. 359.

Fernandez-Jalvo, Y., Andrews, P., Sevilla, P., Requejo, V., 2014. Digestion versus abrasion features in rodent bones. Lethaia 4, 323–336.

Fernández–Jalvo, Y., Sánchez–Chillón, B., Andrews, P., Fernández–López, S., Alcalá Martínez, L., 2002. Morphological taphonomic transformations of fossil bones in continental environments, and repercussions on their chemical composition. Archaeometry 44, 353–361.

Fernandez-Jalvo, Y., Andrews, P., Denys, C., Sese, C., Stoetzel, E., Marin-Monfort, D., Pesquero, D., 2016. Taphonomy for taxonomists: implications of predation in small mammal studies. Quat. Sci. Rev. 139, 138–157.

Fisher, D.C., 1981a. Crocodilian scatology, microvertebrate concentrations, and enamel-less teeth. Paleobiology 7, 262–275.

Fisher, D.C., 1981b. Taphonomic interpretation of enamel-less teeth in the Shotgun local fauna (Paleocene, Wyoming). In: Contributions of the Museum of Paleontology, vol. 25. University of Michigan, pp. 259–275.

Fiorillo, A.R., 1998. Dental micro wear patterns of the sauropod dinosaurs *Camarasaurus* and *Diplodocus*: evidence for resource partitioning in the Late Jurassic of North America. Hist. Biol. 13, 1–16.

Fiorillo, A.R., 2011. Microwear patterns on the teeth of northern high latitude hadrosaurs with comments on microwear patterns in hadrosaurs as a function of latitude and seasonal ecological constraints. Palaeontol. Electron. 14 (3).

Fortelius, M., Solounias, N., 2000. Functional characterization of ungulate molars using the abrasion-attrition wear gradient: a new method for reconstructing paleodiets. Am. Mus. Novit. 3301, 1–36.

Franzen, J.L., Frey, E., 1993. *Europolemur* completed. Kaupia 3, 113–130.
Frey, E., Monninger, S., 2010. Lost in action—the isolated crocodilian teeth from Enspel and their interpretive value. In: Wuttke, M., Uhl, D., Schindler, T. (Eds.), Fossil-lagerstätte Enspel—Exceptional Preservation in an Upper Oligocene Maar, vol. 90, pp. 65–81. Palaeobiodiversity and Palaeoenvironments.
Getty, P.R., Aucoin, C., Fox, N., Judge, A., Hardy, L., Bush, A.M., 2017. Perennial lakes as an environmental control on theropod movement in the Jurassic of the Hartford Basin. Geosciences 7, 13. https://doi.org/10.3390/geosciences7010013.
Green, J.L., Croft, D.A., 2018. Using dental mesowear and microwear for dietary inference: a review of current techniques and applications. In: Croft, D.A., Su, F., Simpson, S.W. (Eds.), Methods in Paleoecology: Reconstructing Cenozoic Terrestrial Environments and Ecological Communities. Springer International Publishing, Cham, pp. 53–73.
Hunt, A.P., Lucas, S.G., 1998. Ichnological evidence for vertebrate predation in the Paleozoic: is there any? New Mex. Mus. Nat. Hist. Sci. Bull. 12, 59–62.
Hunt, A.P., Lucas, S.G., 2012. Classification of vertebrate coprolites and related trace fossils. New Mex. Mus. Nat. Hist. Sci. Bull. 57, 137–146.
Hunt, A.P., Lucas, S.G., 2018. The record of sloth coprolites in North and South America: implications for terminal Pleistocene extinctions. New Mex. Mus. Nat. Hist. Sci. Bull. 79, 277–298.
Hunt, A.P., Lucas, S.G., 2021. The ichnology of vertebrate consumption: dentalites, gastroliths and bromalites. New Mex. Mus. Nat. Hist. Sci. Bull. 87, 215.
Hunt, A.P., Lucas, S.G., 2024. Pabulites: the legacy of "leftovers". Geol. Soc. Am. Abstr. Progr. 56 (4). https://doi.org/10.1130/abs/2024CD-399235.
Hunt, A.P., Lucas, S.G., 2025. Regurgitalites. In: Lucas, S.G., Hunt, A.P., Klein, H. (Eds.), Vertebrate Ichnology. Elsevier, Amsterdam. In this issue.
Jennings, D.S., Hasiotis, S.T., 2006. Taphonomic analysis of a dinosaur feeding site using geographic information systems (GIS), Morrison Formation, southern Bighorn Basin, Wyoming, USA. Palaios 21, 480–492.
Kay, R.F., Covert, H.H., 1983. True grit: a microwear experiment. Am. J. Phys. Anthropol. 61, 33–38.
Klug, C., Schweigert, G., Dietl, G., Fuchs, D., 2010. First record of a belemnite preserved with beaks, arms and ink sac from the Nusplingen Lithographic Limestone (Kimmeridgian, SW Germany). Lethaia 43, 445–456.
Klug, C., Schweigert, G., Hofmann, R., Weis, R., De Baets, K., 2021a. Fossilized leftover falls as sources of palaeoecological data: a 'pabulite' comprising a crustacean, a belemnite and a vertebrate from the Early Jurassic Posidonia Shale. Swiss J. Paleontol. 140, 10. https://doi.org/10.1186/s13358-021-00225-z.
Klug, C., Schweigert, G., Tischlinger, H., Pochmann, H., 2021a. Failed prey or peculiar necrolysis? Isolated ammonite soft body from the Late Jurassic of Solnhofen (Germany). Swiss J. Palaeontol. 140, 15. https://doi.org/10.1186/s13358-020-00215-7.
Klug, C., Pohle, A., Roth, R., Hofmann, R., Wani, R., Tajika, A., 2021b. Preservation of nautilid soft parts inside and outside the conch interpreted as central nervous system, eyes, and renal concrements from the Lebanese Cenomanian. Swiss J. Palaeontol. 140, 15. https://doi.org/10.1186/s13358-021-00229-9.
Korth, W., 1979. Taphonomy of microvertebrate fossil assemblages. Ann. Carnegie Mus. 48, 235–285.
Kramer, J.M., Erickson, B.R., Lockley, M.G., Hunt, A.P., Braddy, S.J., Lucas, S.G., Heckert, A.B., 1995. Pelycosaur predation in the Permian: evidence from *Laoporus* trackways from the Coconino Sandstone with description of a new species of *Permichnium*. New Mex. Mus. Nat. Hist. Sci. Bull. 6, 245–249.
Lambert, O., Bianucci, G., 2019. How to break a sperm whale's teeth: dental damage in a large Miocene physeteroid from the North Sea basin. J. Vertebr. Paleontol. 39, e1660987. https://doi.org/10.1080/02724634.2019.1660987.
Leydig, F., 1896a. Koprolithen und Urolithen. Neues Jahrbuch Geol. Palaontol. Monatsh. 2, 139–140.
Leydig, F., 1896b. Koprolithen und Urolithen. Biologisches Centralblatt 16, 101–103.
Lockley, M.G., Hunt, A.P., 1995. Dinosaur Tracks and Other Fossil Footprints of the Western United States. Columbia University Press, New York, p. 338.
Lockley, M.G., Madsen Jr., J.H., 1993. Early Permian vertebrate trackways from the Cedar Mesa Sandstone of eastern Utah: evidence of predator-prey interaction. Ichnos 2, 147–153.
Lockley, M.G., Meyer, C., Siber, H.-J., Pabst, B., 1998. Theropod tracks from the Howe Quarry, Morrison Formation, Wyoming. Mod. Geol. 23, 309–316.

Longrich, N.R., Jalil, N.E., Khaldoune, F., Yazami, O.K., Pereda-Suberbiola, X., Bardet, N., 2022. *Thalassotitan atrox*, a giant predatory mosasaurid (Squamata) from the upper Maastrichtian phosphates of Morocco. Cretac. Res. 140, 105315. https://doi.org/10.1016/j.cretres.2022.105315.

Lucas, S.G., Hunt, A.P., 2006. Reappraisal of "reptile nests" from the Upper Triassic Chinle Group, Petrified Forest National Park, Arizona. New Mex. Mus. Nat. Hist. Sci. Bull. 37, 155–159.

Lyman, R.L., 1994. Vertebrate taphonomy. In: Cambridge Manuals in Archaeology. Cambridge University Press, Cambridge, p. 552.

Malkani, M.S., 2007. Trackways evidence of sauropod dinosaurs confronted by a theropod found from Middle Jurassic Samana Suk Limestone of Pakistan. Sind. Univ. Res. J. (Sci. Ser.) 39 (1), 1–14.

Marra, A.C., Villa, P., Beauval, C., Bonfiglio, L., Goldberg, P., 2004. Same predator, variable prey: taphonomy of two Upper Pleistocene hyena dens in Sicily and SW France. Revue de paléobiologie 23, 787–801.

Maxwell, E.E., Lambers, P., López-Arbarello, A., Schweigert, G., 2021. Re-evaluation of pachycormid fishes from the Late of Southwestern Germany. Acta Palaeontol. Pol. 65 (3), 429–453.

Mäuser, M., 2015. Das Wattendorfer Etappenkrokodil. Fossilien 32 (6), 31–37.

Mayr, G., 2016. Avian feet, crocodilian food and the diversity of larger birds in the early Eocene of Messel. Palaeobiodivers. Palaeoenviron. 96, 601–609. https://doi.org/10.1007/s12549-016-0243-2.

Mayr, G., Poschmann, M., 2009. A loon leg (Aves, Gaviidae) with crocodilian tooth from the late Oligocene of Germany. Waterbirds 32 (3), 468–471.

McCarville, K., Bishop, G.A., 2002. To pee or not to pee: evidence for liquid urination in sauropod dinosaurs. Jour. Vert. Paleontol. Supplement 22 (3), 85–85.

McLennan, L.J., Purnell, M.A., 2021. Dental microwear texture analysis as a tool for dietary discrimination in elasmobranchs. Sci. Rep. 11, 2444. https://doi.org/10.1038/s41598-021-81258-9.

Meador, L., 2017. Who Ate the Subfossil Lemurs? A Taphonomic and Community Study of Raptor, Crocodylian and Carnivoran Predation of the Extinct Quaternary Lemurs of Madagascar. Ph.D. dissertation. University of Massachusetts, Amherst, p. 388.

Mellett, J.S., 1975. Scatological origins of microvertebrate fossils. Science 185, 349–350.

Merceron, G., de Bonis, L., Viriot, L., Blondel, C., 2005. Dental microwear of fossil bovids from northern Greece: paleoenvironmental conditions in the eastern Mediterranean during the Messinian. Palaeogeogr. Palaeoclimatol. Palaeoecol. 217, 173–185.

Montalvo, C.I., Fernández, F.J., 2019. Review of the actualistic taphonomy of small mammals ingested by South American predators. Its importance in the interpretation of the fossil record. Publicación Electrónica de la Asociación Paleontológica Argentina 19 (1), 18–46.

Montalvo, C.I., Tomassini, R.L., Sostillo, R., 2016. Leftover prey remains: a new taphonomic mode from the Late Miocene Cerro Azul Formation of Central Argentina. Lethaia 49, 219–230.

Morlo, M., Schaal, S., Mayr, G., Seiffert, C., 2004. An annotated taxonomic list of the Middle Eocene (MP 11) Vertebrata of Messel. Cour. Forschungsinst. Senckenberg 252, 95–108.

Oji, T., Ogaya, C., Sato, T., 2003. Increase of shell-crushing predation recorded in fossil shell fragmentation. Paleobiology 29, 520–526.

Paik, I.S., Kim, H.J., Huh, M., 2010. Impressions of dinosaur skin from the Cretaceous Haman Formation in Korea. J. Asian Earth Sci. 39 (4), 270–274. https://doi.org/10.1016/j.jseaes. 2010.02.015.

Pappa, S., Schreve, D.C., Rivals, F., 2019. The bear necessities: a new dental microwear database for the interpretation of palaeodiet in fossil Ursidae. Palaeogeogr. Palaeoclimatol. Palaeoecol. 514, 168–188.

Peyrot-Clausade, M., Chabanet, P., Conand, C., Fontaine, M., Letourneur, Y., Harmelinvivien, M., 2000. Sea urchin and fish bioerosion on La Réunion and Moorea reefs. Bull. Mar. Sci. 66, 477–485.

Rogers, R.R., Kidwell, S.M., 2007. A conceptual framework for the genesis and analysis of vertebrate skeletal concentrations. In: Rogers, R.R., Eberth, D.A., Fiorillo, A.R. (Eds.), Bonebeds Genesis, Analysis, and Paleobiological Significance. The University of Chicago Press, Chicago and London, pp. 1–64.

Romilio, A., Salisbury, S.W., 2011. A reassessment of large theropod dinosaur tracks from the mid-Cretaceous (late Albian–Cenomanian) Winton Formation of Lark Quarry, central-western Queensland, Australia: a case for mistaken identity. Cretac. Res. 32 (2), 135–142.

Romilio, A., Tucker, R.T., Salisbury, S.W., 2013. Reevaluation of the Lark Quarry dinosaur tracksite (late Albian–Cenomanian Winton formation, central-western Queensland, Australia): no longer a stampede? J. Vertebr. Paleontol. 33, 102–120.

Sala, N., Algaba, M., Arsuaga, J.L., Aranburu, A., Pantoja, A., 2012. A taphonomic study of the Búho and Zarzamora caves. Hyenas and humans in the Iberian Plateau (Segovia, Spain) during the Late Pleistocene. J. Taphon. 10, 477–497.

Salamon, M.A., Gorzelak, P., Niedźwiedzki, R., Trzesiok, D., Baumiller, T.K., 2014. Trends in shell fragmentation as evidence of mid-Paleozoic changes in marine predation. Paleobiology 40, 14–23.

Santi, G., Stoppini, M., 2005. Predator-prey interaction in the Permian of the Orobic basin (North Italy). PalArch Vertebr. Palaeontol. 4, 7–18.

Scott, R., Ungar, P., Bergstrom, T., Brown, C.A., Grine, F.E., Teafordd, M.F., Walker, A., 2005. Dental microwear texture analysis shows within-species diet variability in fossil hominins. Nature 436, 693–695.

Schweigert, G., 1999. Erhaltung und Einbettung von Belemniten im Nusplinger Plattenkalk (Ober-Kimmeridgium, Beckeri-Zone, Schwäbische Alb). Stuttgarter Beiträge zur Naturkunde Serie B 273, 1–35.

Schweigert, G., 2018. Miscellanea aus dem Nusplinger Plattenkalk (Ober-Kimmeridgium, Schwäbische Alb). 19. Zwei Fallbeispiele besonderer Belemnitenerhaltung. Jahresberichte und Mitteilungen des oberrheinischen geologischen Vereins. Neue Folge 100, 509–516. https://doi.org/10.1127/jmogv/100/0015.

Schweigert, G., Maxwell, E., Dietl, G., 2016. First record of a true mortichnium produced by a fish. Ichnos 23, 71–76.

Serafini, G., Danise, S., Maxwell, E.E., Martire, L., Amalfitano, J., Cobianchi, M., Thun Hohenstein, U., Giusberti, L., 2024. Of his bones are crinoid made: taphonomy and deadfall ecology of marine reptiles from a pelagic setting (Middle-Upper Jurassic of northeastern Italy). Riv. Ital. Paleontol. Stratigr. 130 (1), 97–128.

Simpson, E.L., Hilbert-Wolf, H.L., Wizevich, M.C., Tindall, S.E., Fasinski, B.R., Storm, L.P., Needle, M.D., 2010. Predatory digging behavior by dinosaurs. Geology 38, 699–702.

Slon, V., Mafessoni, F., Vernot, B., de Filippo, C., Grote, S., Viola, B., Hajdinjak, M., Peyrégne, S., Nagel, S., Brown, S., Douka, K., 2018. The genome of the offspring of a Neanderthal mother and a Denisovan father. Nature 561, 113–116.

Snyder, K., McLain, M., Wood, J., Chadwick, A., 2020. Over 13,000 elements from a single bonebed help elucidate disarticulation and transport of an *Edmontosaurus* thanatocoenosis. PLoS One 15 (5), e0233182. https://doi.org/10.1371/journal.pone.0233182.

Souto, P.R.F., Fernandes, M.A., 2015. Fossilized excreta associated to dinosaurs in Brazil. J. S. Am. Earth Sci. 57, 32–38.

Stiner, M.C., 2004. Comparative ecology and taphonomy of spotted hyenas, humans, and wolves in Pleistocene Italy. Rev. Paléobiol. 23, 771–785.

Teaford, M.F., Walker, A., 1984. Quantitative differences in dental microwear between primate species with different diets and a comment on the presumed diet of *Sivapithecus*. Am. J. Phys. Anthropol. 64, 191–200.

Thomas, D.A., Farlow, J.O., 1997. Tracking a dinosaur attack. Sci. Am. 277, 74–79.

Thulborn, R.A., 2013. Lark Quarry revisited: a critique of methods used to identify a large dinosaurian track-maker in the Winton Formation (Albian–Cenomanian), western Queensland, Australia. Alcheringa 37, 312–330.

Thulborn, T., 2017. Behaviour of dinosaurian track-makers in the Winton Formation (Cretaceous, Albian–Cenomanian) at Lark Quarry, western Queensland, Australia: running or swimming? Ichnos 24, 1–18.

Thulborn, R.A., Wade, M., 1979. Dinosaur stampede in the Cretaceous of Queensland. Lethaia 12, 275–279.

Thulborn, R.A., Wade, M., 1984. Dinosaur trackways in the Winton Formation (mid-Cretaceous) of Queensland. Memoir. Queensl. Mus. 21, 413–517.

Tweet, J.S., Santucci, V.L., Hunt, A.P., 2012. An inventory of packrat (*Neotoma* spp.) middens in National Park Service areas. New Mex. Mus. Nat. Hist. Sci. Bull. 57, 355–368.

Ungar, P.S., Scott, J.R., Steininger, C.M., 2016. Dental microwear differences between eastern and southern African fossil bovids and hominins. South Afr. J. Sci. 112, 1–5.

Voigt, E., 1960. Über einen mutmasslichen fossilen Harnstein (Urolith) aus der Oberen Kreide. Mitteilungen dem Geol. Staatsinst. Hamburg 29, 85–95.

Walker, A., Hoeck, H.N., Perez, L., 1978. Microwear of mammalian teeth as an indicator of diet. Science 201, 908–910.

Weems, R.E., 2018. A synopsis of the vertebrate fauna from the Culpeper Basin (Upper Triassic-Lower Jurassic, Maryland and Virginia). New Mex. Mus. Nat. Hist. Sci. Bull. 79, 749–768.

Weems, R.E., 2020. Behavioral patterns of the Late Triassic *Kayentapus minor* trackmakers at the Culpeper Quarry near Stevensburg, Virginia USA. New Mex. Mus. Nat. Hist. Sci. Bull. 82, 439–454.

Weidenreich, F., 1941. The extremity bones of *Sinanthropus pekinensis*. Palaeontologica Sinica New Ser D 5, 1–50.

Williams, J.P., 2003. Bones of comprehension: the analysis of small mammal predator-prey interactions. In: Kelley, P.H., Kowalewski, M., Hansen, T.A. (Eds.), Predator-prey Interactions in the Fossil Record. Kluwer Academic/Plenum Publishers, New York, pp. 341–358.
Williams, F.L.E., Patterson, J.W., 2010. Reconstructing the paleoecology of Taung, South Africa from low magnification of dental microwear features in fossil primates. Palaios 25, 439–448.
Wilson, M.V.H., 1987. Predation as a source of fish fossils in Eocene lake sediments. Palaios 2, 497–504.
Winkler, D.E., Schulz-Kornas, E., Kaiser, T.M., Tütken, T., 2019. Dental microwear texture reflects dietary tendencies in extant Lepidosauria despite their limited use of oral food processing. Proc. R. Soc. Ser. B, 20190544. https://doi.org/10.1098/rspb.2019.0544.
Zangerl, R., Richardson, E.S., 1963. The paleoecological history of two Pennsylvanian black shales. Fieldiana Geol Mem. 4, 352.

Further reading

Farlow, J.O., 1985. Notes. In: Bird, R.T. (Ed.), Bones for Barnum Brown. Texas Christian University Press, Fort Worth, pp. 209–211.
Hunt, A.,P., Lucas, S.G., 2024c. Vertebrate bromalites from Eocene lacustrine strata – comparison of ichnofaunas from the Green River Formation of the western United States and the Messel Lagerstätten in Germany. Geol. Soc. Am. Abstr. Progr. 56 (4). https://doi.org/10.1130/abs/2024CD-399234.

PART III

The ichnology of digging and reproduction

CHAPTER

12

Vertebrate burrows

Ricardo N. Melchor[1,2] *and M. Cristina Cardonatto*[1]

[1]Departamento de Geología, Universidad Nacional de La Pampa, Santa Rosa, La Pampa, Argentina; [2]Instituto de Ciencias de la Tierra y Ambientales de La Pampa, Universidad Nacional de La Pampa and Consejo Nacional de Investigaciones Científicas y Técnicas, Santa Rosa, La Pampa, Argentina

1. Introduction

Vertebrates dig to live under a controlled microhabitat, to search for food and for food hoarding, socialization, communication, breeding or for shelter, either under unfavorable surface conditions (e.g., high temperatures, fires, desiccation), and/or to escape from predators (e.g., Kinlaw, 1999; Vizcaíno et al., 2016). The definition of burrow in invertebrate ichnology (in contrast to boring, which is used for excavations in hard substrates) usually refers to an excavation made and maintained by an animal in unconsolidated or soft sediment (e.g., Bromley, 1970, 1996; Frey, 1973; Ekdale et al., 1984). Voorhies (1975) adopted a more comprehensive definition for vertebrate burrows that is followed here: "any hole in the ground made by a vertebrate-whether dug in hard or soft sediment, and whether serving for concealment, shelter, feeding, etc.-is considered a burrow." This definition is more adequate for vertebrate burrowing that can include excavation in a consolidated and/or partially cemented substrate such as hardened soil horizons and even rock with different induration degree (e.g., Lopes et al., 2017; Cardonatto et al., 2018; Guerrero-Arenas et al., 2020).

Vertebrate burrows can be well-defined subterranean systems both permanent and semipermanent, sometimes of considerable complexity; temporary passages through the surficial part of the soil or shallow foraging excavations. It is possible to distinguish the following types of vertebrate burrows: 1) foraging pits, 2) burrow systems *sensu stricto*, 3) nesting burrows or nests, 4) and aestivation burrows. Foraging pits or food probes are shallow burrows for searching for food that are not used for shelter (e.g., Platt, 2014; Cardonatto and Melchor, 2023b). Burrow systems are excavations with different degree of complexity that are mainly used for permanent or transient domicile and for reproduction and care of offspring. Nests are excavations with different complexity that have the primary purpose of laying the eggs and mostly are excavated by reptiles, amphibians, and birds (e.g., Gallardo, 1957; Butler,

Vertebrate Ichnology
https://doi.org/10.1016/B978-0-443-38351-9.00012-8

1995; Doody et al., 2018). Aestivation burrows are mostly of some fish and amphibians and are essentially subvertical structures (Hembree, 2010).

Burrowing organisms are divided into primary excavators, secondary modifiers, and simple occupants (Kinlaw, 1999). Primary excavators are strong and accomplished burrowers who dig a tunnel down into the soil or into a bank, with different degrees of complexity, and can have major environmental impacts. The digging of permanent burrows creates a setting where other species can live, modifies the physical parameters and resource availability, and in this way, burrow systems are related to an increase in biodiversity (Kent and Snell, 1994; Kinlaw and Grasmueck, 2012). Secondary modifiers are animals that inhabit and modify a burrow dug by a primary excavator. Simple occupants of a burrow are those who played no role in digging the structure but simply take advantage of its existence.

Although vertebrates dig burrows under different climatic conditions and biogeographic regions, those inhabiting semiarid or arid zones dig to ameliorate extreme environmental conditions. For example, to avoid high temperatures by building extensive systems with several vertically stacked levels, as is the case for diurnal rodents, and to avoid dehydration by digging deep burrows (Doody et al., 2018; Kinlaw, 2006).

Burrowing mammals exhibit different degree of fossoriality that commonly is divided into semifossorial and fossorial (Shimer, 1903; Ebensperger and Bozinovic, 2000) or between fossorial and subterranean (Ellerman, 1956; Nevo, 1979; Begall et al., 2007; Lessa et al., 2008; Vassallo and Echeverría, 2009; Vizcaíno et al., 2016). Both sets of terms are rough equivalents. We will refer to fossorial habit (when foraging is aboveground and a significant fraction of their lives is spent outside their burrows) and subterranean (when foraging is mostly underground or foraging excursions are made near the entrances, and most of their time is spent in their burrows) (Lessa et al., 2008).

Both the morphology of the structures used in digging by the producer (especially when using teeth, claws or both) and diet have some influence on the excavation mechanism. The bones and muscles of the forearms of burrowing tetrapods are adapted for digging. Muscle-skeletal design is shaped to increase the force applied to the substrate by increasing musculature but also shortening the output arm (distal segments of limbs) or increasing the length of the input arm (muscle insertion far away from joint). The Index of Fossorial Ability (IFA) is a way to quantify this mechanical advantage, which was defined as the ratio between the length of the olecranon process and total ulnar length (Vizcaíno et al., 1999). This index is considered a good indicator of the fossorial adaptation of a particular species (Vizcaíno et al., 1999, 2016; Vizcaíno and Milne, 2002).

The scientific study of vertebrate burrows apparently started with the description of the "gigantic fossils" from the Miocene of Nebraska that Barbour (1892) named *Daimonelix*, after the popular name "Devil's corkscrew" used by ranchmen. This was the first ichnogenus of vertebrate burrows that provoked an extended debate about the origin and nature of the structure. Studies on vertebrate burrows were revived about 50 years later with the discovery and description of subvertical burrows, especially in the Permian of USA, with lungfish remains, even articulated ones, which were interpreted as aestivation burrows (Romer and Olson, 1954; Carlson, 1968; Benton, 1988).

The seminal chapter by Voorhies (1975) summarized the present knowledge on extant and fossil vertebrate burrows and related the apparent dearth of fossil vertebrate burrows to a bias in the description, suggesting that they were much more common than perceived. In the last 2 decades of the 20th century, three new ichnogenera were proposed: *Ichnogyrus*, *Piscichnus*, and *Cornulatichnus* (Bown and Kraus, 1983; Feibel, 1987; Carroll and Trewin, 1995). About 70% of the ichnotaxa of vertebrate burrows were erected in the last 20 years, starting with the description of amphibian aestivation burrows under the ichnogenus *Torridorefugium* Hembree et al., 2005. A recent review of extant and fossil vertebrate burrows can be found in Cardonatto and Melchor (2021b), and some aspects were also covered by Hasiotis et al. (2007) and Hembree (2010). Marchetti et al. (2024) revised the early evolution (Devonian-Triassic) of vertebrate burrowing considering both fossil burrows and burrowing vertebrates and proposed a number of key events and their possible interpretation.

2. Terminology for vertebrate burrow systems

We follow the terminology proposed for vertebrate burrow systems by Cardonatto and Melchor (2020, 2021a), which was mostly adapted from Hickman (1990). This terminology is summarized in Table 12.1, and some key aspects are discussed further. A ramp is a low-angle excavation (inclination >5 degrees) that is distinguished by its inclination from tunnels or cylindrical subhorizontal burrow segments (<5 degrees) (Fig. 12.1A). The term shaft is reserved for a subvertical subcylindrical excavation of a burrow system that can connect two levels or a tunnel with the ground surface (Bromley, 1990, 1996; Frey, 1973; Hickman, 1990). A spiral is a corkscrew-shaped burrow having at least one complete whorl. A chamber is an element having a diameter exceeding that of the associated tunnels, ramps, and/or shafts that is commonly subspherical or ellipsoidal in shape (Fig. 12.1A). The complexity of a burrow system (Hembree and Hasiotis, 2006) is an integer resulting from the summation of the number of entrances + number of segments (ramps, tunnels, shafts) + number of chambers of a burrow system.

Discrete and repetitive traces of biogenic origin observed in the burrow wall/margin or the surface of burrow casts are distinguished as surface ornamentation. Surface features produced during diagenesis or artifacts are not considered surface ornamentation. In particular, primary surface ornamentation (Cardonatto and Melchor, 2020) is referred to sets of curved parallel ridges (in the cast) or grooves (in the burrow margin) representing excavation by the producer, which are interpreted as claw or tooth traces (bioglyphs *sensu* Bromley, 1996) (Fig. 12.1B). The number of traces in a set and the set width are commonly useful for identification of the burrow producer (Cardonatto, 2019; Ponomarenko and Ponomarenko, 2019; Cardonatto and Melchor, 2020, 2021a, 2023a).

Two important measurements of a fossil burrow are the horizontal diameter (Dh: maximum horizontal distance between the sides of a tunnel or chamber taken perpendicular to the burrow axis of the element, Fig. 12.1C) and vertical diameter (Dv; maximum vertical distance between the roof and bottom of the tunnel or chamber, taken perpendicular to the burrow axis, Fig. 12.1C). The cross section of a tunnel or ramp can be characterized by the relative diameter index (RDI), which is the ratio of vertical to horizontal diameter (RDI = Dv/Dh) taken at the same point in the burrow cast (Elissamburu et al., 2011; Cardonatto and Melchor, 2020). The shape of the cross section can also be characterized as vertical elliptical (RDI >1), horizontal elliptical and plano-convex (both with RDI <1), or subcircular (RDI ~ 1).

TABLE 12.1 Compilation of definitions of elements, measurements, and parameters of burrow systems.

Elements of a burrow system (Fig. 12.1A)

Entrance (e). Aperture at the ground surface.

Ramp (Ra). Low-angle excavation (inclination >5 degrees). It is possible to distinguish between entrance ramps (***eRa***) and internal ramps (***iRa***) or those not connected to an entrance.

Tunnel (T). A cylindrical subhorizontal burrow (inclination <5 degrees).

Chamber (C). Element having a diameter exceeding that of the associated tunnels and is commonly subspherical or ellipsoidal in shape. The chamber can be lateral (**Cl**), bilateral (**Cb**), terminal (**Ct**) or located at a ramification (**Cr**).

Ramification (R). Divisions of tunnels into two or three tunnels of similar diameter including bifurcation (**Rb**) and trifurcation (**Rt**).

Blind tunnel (Tb). Short lateral tunnel not related to another element having length greater than horizontal diameter.

Knob (K). Short lateral tunnel not related to another element having length equal or smaller than horizontal diameter.

Shaft (S). Subvertical excavation of a burrow system that can connect two levels or with the ground surface (Frey, 1973; Hickman, 1990; Bromley, 1990, 1996).

Spiral (Sp). Corkscrew-shaped burrow having at least one whorl.

Closed circuit (CC). Interconnected tunnels/ramps and chambers forming a loop.

Segment. Tract of a burrow system limited by two consecutive elements, including entrance, chamber, ramification, and blind tunnel.

Measurements of the burrow systems

Surface ornamentation. Discrete and repetitive traces of biogenic origin observed in the burrow wall/margin or the surface of burrow casts. Surface features produced during diagenesis or artifacts are excluded.

Primary surface ornamentation. Sets of curved parallel ridges (in the cast) or grooves (in the burrow margin) representing excavation by the producer that are interpreted as claw or tooth traces (bioglyphs sensu Bromley 1996).

Secondary surface ornamentation. In vertebrate burrow casts, represented by numerous cylindrical protuberances that are considerably smaller than the hosting burrow, and cannot be linked to digging by the vertebrate producer. They are essentially interpreted as invertebrate burrows. In fossil vertebrate burrows, rhizoliths can also occur (Cardonatto and Melchor, 2020).

Horizontal diameter (Dh). Maximum horizontal distance between the sides of a tunnel or chamber taken perpendicular to the burrow axis of the element (Fig. 12.1C).

Vertical diameter (Dv). Maximum vertical distance between the roof and bottom of the tunnel or chamber, taken perpendicular to the burrow axis (Fig. 12.1C).

Entrance horizontal diameter (e_{Dh}). Maximum horizontal distance between the sides of an entrance.

Entrance vertical diameter (e_{Dv}). Maximum vertical distance between the roof and bottom of an entrance.

Curved length of a burrow system (Lc). Sum of the distance between consecutive elements, measured parallel to the burrow axis.

Straight length of a burrow system (Ls). Straight-line distance between the most distant elements of a system.

Maximum depth of the burrow (Dm). Vertical distance from the highest point of the ground surface to the bottom of an element of the system (i.e., tunnel, chamber, etc.).

TABLE 12.1 Compilation of definitions of elements, measurements, and parameters of burrow systems.—cont'd

Angle of ramification (ar). Angle formed between the two branches in the case of bifurcations. For trifurcations, four angles of ramification measured counterclockwise starting from the main tunnel.
Claw/incisor trace set width (Ws). Distance between the external margins of a set of claw traces taken perpendicular to the long axis of the traces. Also applied for the width of paired incisor traces (Fig. 12.1B).
Claw/incisor trace width (Wt). Distance taken perpendicular to the long axis of each claw trace.
Angle between sets of claw traces (As). Angle formed between two sets of claw or incisor traces taken approximately along the tunnel axis.
Parameters of burrow systems
Relative diameter index (RDI). Ratio of vertical to horizontal diameter (RDI = Dv/Dh) taken at the same point in the tunnel or ramp (Cardonatto and Melchor, 2020).
Complexity of a burrow system (Co). Summation of the number of entrances + number of segments + number of chambers of a burrow system (Hembree and Hasiotis, 2006).
Tortuosity (To). Ratio between the sum of the curved length of each segment and the straight length of the system (To = ΣLc/Ls) (Cardonatto and Melchor, 2020).
Total chamber volume (VC = Σvc). Summation of individual chamber volume (vc), which was estimated using the formula of the rotating ellipsoid: vc = $C_L \times C_{Dh} \times C_{Dv} \times \pi \times 1/6$ (Scheibler et al., 2006; Cardonatto and Melchor, 2020), where C_L: chamber length, C_{Dh}: chamber horizontal diameter, and C_{Dv}: chamber vertical diameter.
Total tunnel volume (VT). Estimated as the cylinder volume: VT = $\pi \times Dh/2 \times Dv/2 \times (Lc\text{-}\Sigma C_L)$.
Ratio of total chamber to total tunnel volume (VC/VT). Index of relative proportion of the chambers in the whole burrow system (Cardonatto and Melchor, 2020).

Modified from Cardonatto and Melchor (2020, 2021a. See also Fig. 12.1.

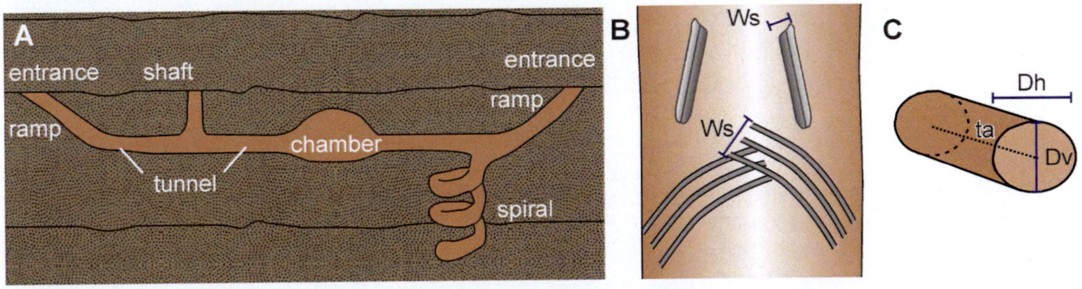

FIGURE 12.1 Parts of a burrow system and some basic measurements. See also Table 12.1. (A) Elements of a burrow system. (B) Sector of a cylindrical burrow cast with sets of three parallel ridges interpreted as claw traces (below) and sets of two closely spaced ridges that are interpreted as incisor traces (above). Ws: width of set of traces. (C) Measurements on a cylindrical burrow segment. *Dv*, vertical diameter; *Dh*, horizontal diameter; *ta*, tunnel axis.

3. Description of selected extant vertebrate burrow systems

This section contains a compilation of the morphology of burrow systems of the main groups of vertebrates, mostly based on descriptions from the biological literature. These valuable descriptions commonly do not consider the surface ornamentation of the burrows and other morphological features that are significant for ichnology.

3.1 Fish burrows

Lungfish (Dipnoi) build two types of burrows: estivation and nesting burrows, although the modern African species *Protopterus annectens* uses the same burrow for estivate and breeding (Greenwood, 1986). Aestivation burrows of the African species *P. annectens* and the South American species *Lepidosiren paradoxa* are similar. Both consist of a subvertical burrow with a terminal widening, where the producer generally remains folded in on itself, sometimes with a mucus cocoon (Kerr, 1898; Johnels and Svensson, 1954; Greenwood, 1986). These aestivation burrows are closed by a clay plug and surrounded by a surficial mound in the ground (Fig. 12.2A). Kerr (1898) noted that some aestivation burrows of *L. paradoxa* can have an upper tract of low inclination and then turn subvertical, as well as several clay plugs in the upper part of the burrow (Fig. 12.2B). *Protopterus annectens* nesting burrows commonly consist of a U-shaped structure with a chamber in the lower part

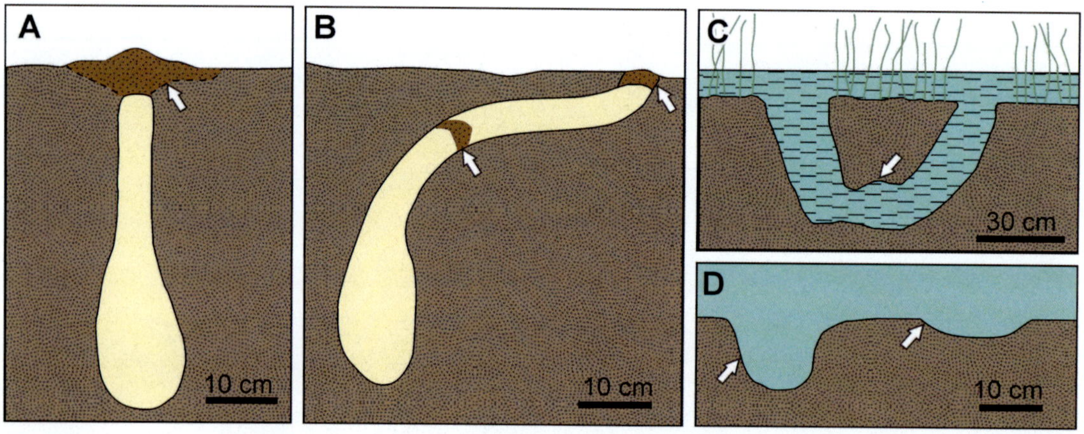

FIGURE 12.2 Cross-sectional views of selected examples of fish burrows. (A). Estivation burrow of *Protopterus annectens* (Dipnoi: Protopteridae). The arrow points to the sedimentary plug and surficial mound. (B) Estivation burrow of *Lepidosiren paradoxa* (Dipnoi: Lepidosirenidae) with two sedimentary plugs (arrows), and a segment of shallow slope. This species also can build estivation burrows similar to those represented in A. (C) One of the more common morphologies of nesting burrows of *P. annectens*. Note the overall U-shape and the central chamber (arrow). (D) Fish burrows with deep subcylindrical shape (left, more typical of ray feeding burrows) and shallow concave shape (right, typical of freshwater fish nests). *(A) Modified from Greenwood (1986). (B) Modified from Kerr (1898). (C) Modified from Greenwood (1986). (D) Drawn from illustrations of Feibel (1987) (rigth) and Kotake and Nara (2002) (left).*

where the eggs are laid down (Fig. 12.2C). Entrances of nesting burrows commonly are left open and underwater (Greenwood, 1986). *Protopterus dolloi* seemingly uses open burrows with the form of a ramp that reach the water table both for aestivation and breeding (Greenwood, 1986).

The remaining fish burrows are essentially related to feeding or breeding (see Muñiz et al., 2015 for a compilation of fish bioturbation structures). Some of these fish burrows are feeding structures of rays and sturgeons. Feeding burrows of several species of rays (Elasmobranchia) have been recorded in estuaries and tidal flats of the Georgia coast (USA) and consist of conical depressions of subcircular to elliptical outline of a few cm to 1 m wide (Howard et al., 1977; Gregory et al., 1979). Sturgeon (*Acipenser oxyrinchus*) feeding structures have been documented in the Bay of Fundy (Canada) and are 5–15 cm wide cylindrical depressions (Pearson et al., 2007). Both rays and sturgeon feeding burrows occur in large numbers and are visible during low tide.

3.2 Amphibian burrows

Some amphibians are obligate burrowers that exhibit digging adaptations, and others are opportunist burrowers that use cracks, natural holes, or burrows of other species. Caecilians (Gymnophiona) are among the more specialized fossorial amphibians, although their burrows have apparently not been described. Most salamanders (Caudata) are not well adapted for digging; *Ambystoma tigrinum* is one of the exceptions, and their excavations have been studied in the terrarium (Dzenowski and Hembree, 2014). The burrows produced under experimental conditions are quite variable, including ramps (Fig. 12.3D) and U-, W-, and J-shaped burrows (Dzenowski and Hembree, 2014).

Anurans display a wide array of burrowing behaviors including avoiding desiccation, reproduction, and/or escape from predation. Anurans commonly burrow in loose sediments, frequently when they are damp, to escape cold winters, hot summers, and long dry seasons (Wells, 2007). Backward hindfeet digging (with or without rotation) is exclusive to frogs and has been developed repeatedly in different families (Emerson, 1976; Sanders and Davies, 1984). Toad burrows were documented by field observations (Gallardo, 1957; Nascimento et al., 2024) and neoichnological experiments (Johnson and Hembree, 2015). *Rhinella granulosa* burrows (Fig. 12.3A) have been described as simple ramps with an entrance of elliptical section, sometimes with slight lateral curvature and occasionally with a terminal chamber (Gallardo, 1957). Sometimes, the initial segment of the burrow forms a high angle with the ground (Fig. 12.3A). Burrows dug by *Scaphiopus holbrookii* in a terrarium were high-angle ramps to subvertical shafts with an ovoid terminal chamber or a simple ovoidal burrow (Johnson and Hembree, 2015). Burrowing frogs also construct simple burrows, sometimes producing a cocoon that protects them from desiccation (McClanahan et al., 1976; Sanders and Davies, 1984). Nesting burrows of *Leptodactylus bufonius* are pyriform (Fig. 12.3C) and comprise a subspherical chamber (containing the eggs) connected by a shaft with the surface (Cei, 1949). This vertical conduit is plugged with sediment and usually includes a chimney or ring-shaped mud mound (Fig. 12.3C).

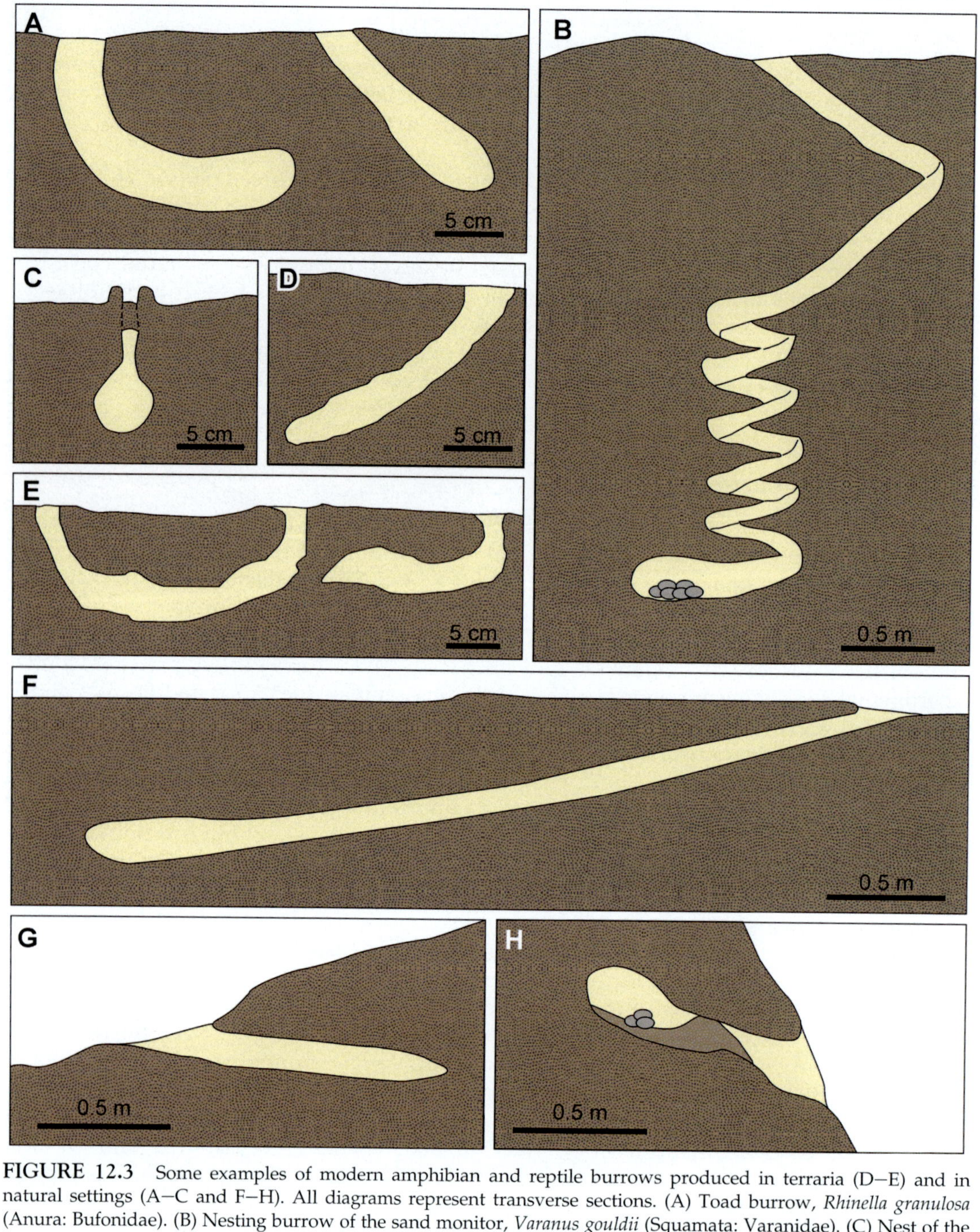

FIGURE 12.3 Some examples of modern amphibian and reptile burrows produced in terraria (D–E) and in natural settings (A–C and F–H). All diagrams represent transverse sections. (A) Toad burrow, *Rhinella granulosa* (Anura: Bufonidae). (B) Nesting burrow of the sand monitor, *Varanus gouldii* (Squamata: Varanidae). (C) Nest of the frog *Leptodactylus bufonius* (Anura: Leptodactylidae). (D) One of the burrow morphologies produced by the tiger salamander, *Ambystoma tigrinum* (Caudata: Ambystomatidae). (E) Two types of burrows produced by skinks, *Mabuya multifaciata* (Squamata: Scincidae). (F) Burrow of the gopher tortoise, *Gopherus polyphemus* (Testudines: Testudinidae). (G–H) Burrow and nest of the Bengal monitor, *Varanus bengalensis* (Squamata: Varanidae). *(A) Modified from Gallardo (1957). (B) Modified from Doody et al. (2018). (C) Based on Cei (1949). (D) Modified from Dzenowski and Hembree (2014). (E) Modified from Catena and Hembree (2014a). (F) Modified from Jones and Franz (1990). (G–H) Modified from Auffenberg (1983).*

3.3 Reptile burrows

Burrowing reptiles include representatives of chelonians, crocodiles, and squamates, especially lizard and snakes, and the subterranean amphisbaenian. In this section, we will mention burrows produced by scincids in the terrarium, by tortoises in natural settings, and monitor burrows and nests.

Skinks (Scincidae) can dig their own burrows or move in the substrate by sand swimming; some species can dig collaboratively a burrow system with multiple entrances that are restricted to a mound (Fenner et al., 2012). Descriptions of skink burrows were conducted on the basis of experiments in terraria (Catena and Hembree, 2014a, 2014b). Sand swimming structures of *Chalcides ocellatus* in a laminated sandy substrate are scarcely distinctive and include subsurface trails, V or U-shaped burrows, and empty burrows (Catena and Hembree, 2014b). Specimens of the skink *Mabuya multifaciata* produced ramps and U- and J-shaped burrows under experimental conditions (Fig. 12.3E) (Catena and Hembree, 2014a).

The gopher tortoise, *Gopherus polyphemus*, builds several meter-long burrows with the shape of a simple ramp and a terminal chamber (Fig. 12.3F). These burrows can be reoccupied or shared with numerous vertebrate and invertebrate species; it has been documented that about 300 species can use gopher tortoise burrows (Jones and Franz, 1990; Kinlaw and Grasmueck, 2012).

Some monitor species (Varanidae) build nesting and/or shelter burrows. The sand monitor, *Varanus gouldii*, digs a complex nesting burrow (Fig. 12.3B) composed by a ramp that then becomes a spiral ending in a chamber where eggs are laid (Doody et al., 2018). This nesting burrow can reach 3 m deep, apparently to avoid dehydration of the eggs (Doody et al., 2018). The Bengal monitor, *Varanus bengalensis*, digs shelter and nesting burrows with the form of a simple ramp (lacking a terminal chamber). The former are excavated in sediment mounds, ravines of below logs or rocks (Fig. 12.3G), whereas nesting burrows are built below abandoned termite nests or in ravines (Fig. 12.3H) (Auffenberg, 1983).

3.4 Bird burrows

Burrowing habits are not common in birds, being more common in some colonial seabirds, and are the normal mode of nesting for petrels, shearwaters, storm petrels, diving petrels, and some auks and penguins (Butler, 1995). Among terrestrial species that excavate or reuse burrows are the burrowing parrots, burrowing owls, bee eaters (*Merops* spp.) (Schmidt and Branch, 2005), kingfishers of the Subfamily Alcedininae, and bank swallows (*Riparia riparia*) (Gallardo, 1972; Butler, 1995). Birds commonly use burrows for breeding, which has several advantages, including avoiding predation, thermal stability, reduced heat loss by chicks, and decreased demand for nest material (Butler, 1995). Bird nesting burrows are commonly constructed in cliffs or ravines (although some species excavate in flat terrain) and are of simple architecture, commonly a tunnel or ramp ending in a chamber (Fig. 12.4).

Woods (1970) described the geometry of nesting burrows of two seabird species from the Malvinas/Falkland Islands, which were located between clumps of tall grasses. These burrows were excavated through a thick and friable peat layer by the great shearwater, *Puffinus gravis*, and the magellanic penguin, *Spheniscus magellanicus* (Fig. 12.4A), those of the last species being deeper.

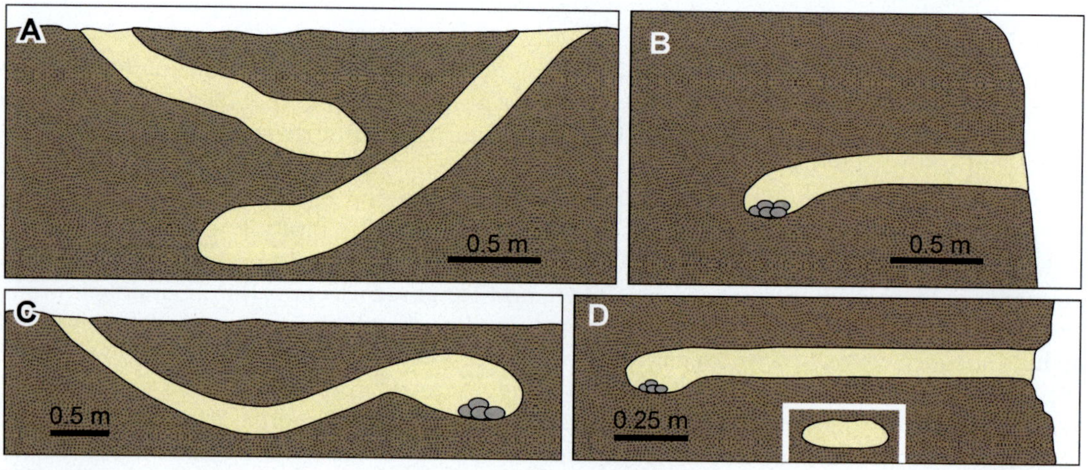

FIGURE 12.4 Examples of avian burrows in transverse section. (A) Burrow of great shearwater (*Puffinus gravis*, Procellariidae, above) and the magellanic penguin (*Spheniscus magellanicus*, Spheniscidae, below). (B) Nesting burrow of the ringed kingfisher (*Megaceryle torquata*, Alcedinidae) in a river ravine. (C) Nesting burrow of the burrowing owl (*Athene cunicularia*, Strigidae). (D) Nesting burrow of the Patagonian burrowing parrot (*Cyanoliseus patagonus*, Psittacidae). The inset displays the transverse section of a tunnel. *(A) Modified from Woods (1970). (B) Modified from de la Peña (2013). (C) Modified from Walker (1952). (D) Modified from Leonardi and Oporto (1983).*

The nesting burrows of the Patagonian burrowing parrot, *Cyanoliseus patagonus*, are tunnels of flattened elliptical cross section ending in a chamber (Fig. 12.4D), arranged essentially perpendicular to the cliff face (Leonardi and Oporto, 1983; Masello et al., 2006). More complex nesting burrows have been documented in this species, including those with two entrance tunnels converging in a chamber and others with a single entrance tunnel connecting two chambers. Burrowing parrots use the beak and claws to excavate (Leonardi and Oporto, 1983). The burrowing owl, *Athene cunicularia*, usually reoccupies abandoned burrows of fossorial mammals, including the plains vizcacha (*Lagostomus maximus*), armadillos (*Chaetophractus villosus*), and prairie dogs (*Cynomys* sp.) (Machicote et al., 2004). Burrowing owl nests have not been apparently studied in detail, although Walker (1952) described them as simple ramps that can be curved in the horizontal plane, eventually ascending before descending and ending in a chamber (Fig. 12.4C). The ringed kingfisher, *Megaceryle torquata*, nests in river ravines digging subhorizontal tunnels with a terminal chamber (Fig. 12.4B) having a diameter about two to three times that of the tunnel (de la Peña, 2013).

3.5 Mammal burrows

The main groups of mammals with fossorial habits are the rodents and armadillos, although there are also examples among the carnivores and the aardvark (*Orycteropus afer*). Burrow diameter is particularly diagnostic for armadillos and rodents, commonly being smaller for the latter. Juvenile armadillos do not excavate their own burrows but share the burrow with adults (McDonough et al., 2000); as a consequence, its size (horizontal and vertical diameter) is useful to distinguish different burrowing species. For armadillos, it is possible to distinguish two types of structures: complex ones used for resting and reproduction, and the simpler and shorter ones that are foraging pits (Platt, 2014; Cardonatto and Melchor, 2023b). The former are more or less permanent, although each individual commonly constructs more than one burrow during its life, changing location following the available food resource. Armadillo burrows commonly have a plano-convex, elliptical, or subcircular entrance that led to a ramp that may widen and form a chamber (Fig. 12.5A); less common branched ramps have also been described (Abba et al., 2005; Cardonatto and Melchor, 2023b), Armadillo burrows with more than one entrance (Fig. 12.5A) are uncommon (e.g., Clark, 1951; Greegor, 1974). Among armadillos, there is a larger group with fossorial habits that build burrows open to the surface (Euphractinae) and those with subterranean habits (Chlamyphorinae) whose tunnels display an active meniscate backfill produced by displacement of the organism in search of food (Melchor et al., 2012b; Cardonatto and Melchor, 2023b).

Rodent burrow systems have several functions in addition to protection against adverse climatic conditions or predators, and to use them for reproduction. These additional functions include food hoarding, socialization, and foraging (for subterranean species) (Kinlaw, 1999). Burrow systems with open entrances (for example, burrows of Caviidae, Octodontidae, Chinchillidae, Sciuridae) are typical of fossorial organisms, whereas burrow systems with closed entrances (namely of Ctenomyidae, Geomyidae, Bathyergidae) are related to subterranean species (Nevo, 1979; Kinlaw, 1999, 2006). Rodent burrow systems in arid or semiarid regions are commonly composed of branched subhorizontal tunnels and have several entrances, ramps, and chambers for different purposes (food storage, latrine, breeding, resting), and one or more levels (e.g., Voorhies, 1975; Reichman and Smith, 1990; Joeckel and Tucker, 2013; Cardonatto and Melchor, 2020; Cardonatto and Melchor, 2021a). Joeckel and Tucker (2013) offered a compilation of the morphology of the burrow systems of living North American rodents.

Rodent burrow systems can be very complex, both in the case of solitary and of colonial species (Fig. 12.5B–E, G). The complexity sometimes arise from reuse by several generations of the same burrow system and from confinement of the burrow system to a sediment mound (commonly a vegetated dune) (Cardonatto and Melchor, 2020, 2021a, 2023a). There are some differences in the complexity of rodent burrow systems related to their solitary or colonial habits, as exemplified by burrows of octodontids belonging to *Spalacopus cyanus* (colonial species) and *Tympanoctomys barrerae* (solitary species) (Fig. 12.5B and D). The systems of *S. cyanus* (Fig. 12.5B) described by Reig (1970) formed a network of subhorizontal interconnected and shallow tunnels with several open entrances. *Tympanoctomys barrerae* burrow systems occur in vegetated sand dunes (nebkhas) with tunnels arranged in up to three levels and having closed circuits (Ojeda et al., 1996; Mares et al., 1997; Cardonatto and Melchor, 2020), where there is a single inhabitant, although the systems can be used by several generations (Fig. 12.5D).

The burrow systems of the colonial rodent *Cynomys* sp. (Sciuridae) described by Verdolin et al. (2008) typically exhibited numerous entrances, nest chambers and widened sectors interpreted as turnarounds (Fig. 12.5C). The ecological equivalent of *Cynomys* in South America (Roig et al., 2009), *Lagostomus maximus* (Chinchillidae), builds similar systems (Fig. 12.5G), with several entrance ramps that converge in a large chamber from which subhorizontal or slightly inclined tunnels arise (Llanos and Crespo, 1952; Jackson et al., 1996; Rafuse et al., 2017). Vizcacheras (burrows of *L. maximus*) commonly are below shrubs or trees and are located in a mound several meters wide that is made of the excavated material.

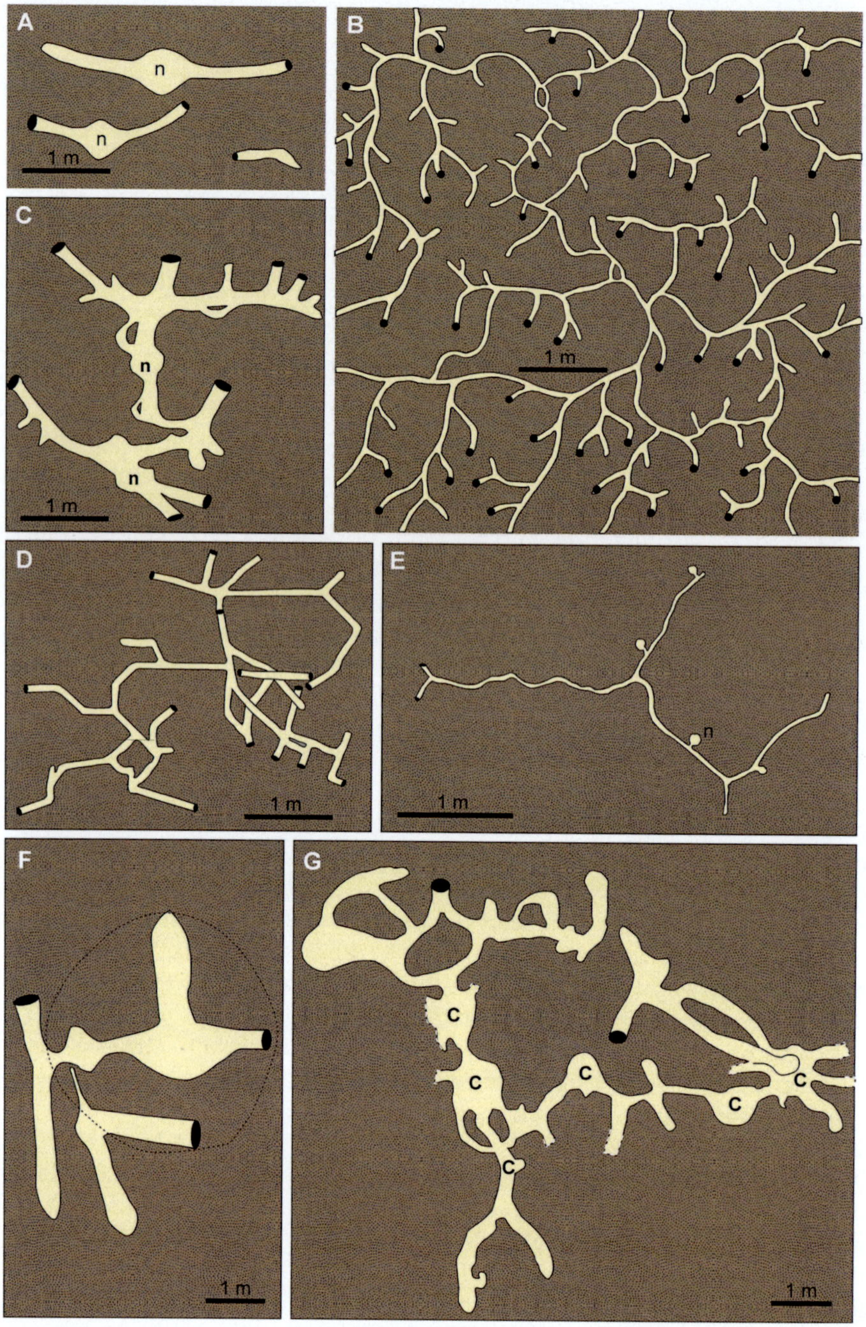

FIGURE 12.5 Selected examples of mammal burrows in plan view. *C*, chamber; n, nest chamber. Black ellipses are entrances. (A) Armadillo burrows. The two upper diagrams belong to the nine-banded armadillo (*Dasypus novemcinctus*, Dasypodidae). The lower diagram belongs to the southern long-nosed armadillo (*Dasypus hybridus*, Dasypodidae). (B) Part of the burrow system of coruros (*Spalacopus cyanus*, Octodontidae). (C) Burrow system of the prairie dog (*Cynomys* sp., Sciuridae). (D) Burrow system of the red vizcacha rat (*Tympanoctomys barrerae*, Octodontidae). (E) Burrow system of the talas tuco-tuco (*Ctenomys talarum*, Ctenomyidae). (F) Burrow system of the aardvark (*Orycteropus afer*, Orycteropodidae) below a termite nest (dashed line), (G) Burrow system of the plain vizcacha ("vizcachera") (*Lagostomus maximus*, Chinchillidae). *(A) Redrawn from González et al. (2001); modified from Clark (1951). (B) Redrawn from Reig (1970). (C) Modified from Verdolin et al. (2008). (D) Modified from Ojeda et al. (1996). (E) Redrawn from Bruch (1937). (F) redrawn from Melton (1976). (G) Modified from Llanos and Crespo (1952).*

Microcavia australis (Caviidae) has social organization, with communal care of youngs, collective detection of predators and cooperation in excavation (Rood, 1972; Ebensperger et al., 2006). Colonies are generally composed of several females and one or two males, in addition to pups and juveniles. Their burrow systems are relatively complex, with several entrances, and tunnels arranged in one or two levels with a significant volume of chambers (Rood, 1970; Cardonatto and Melchor, 2021a).

Extant ctenomyids are mostly solitary and subterranean, and for this reason, the burrow systems are extensive with subhorizontal foraging tunnels (Genise, 1989; Vassallo and Echevería, 2009; Cardonatto and Melchor, 2023a). Ctenomyid burrow systems are easily spotted by the presence of sediment mounds that are located above a plugged entrance. These entrances are connected to ramps and a main or secondary tunnel. Bruch (1937) characterized *Ctenomys talarum* systems as composed of few plugged entrances that led to a main subhorizontal tunnel, with a circular section; with Y forks that led to lateral tunnels, some of which had blind endings; and indicated the presence of three to five lateral chambers (considered nests by the author), which have scarce plant remains or were filled with fresh or dry grass (Fig. 12.5E).

Aardvarks (Orycteropodidae) dig foraging pits and larger burrows for shelter and breeding. Many burrows are located below termite nests (Fig. 12.5F), have one to three entrances, and commonly are a simple ramp with a terminal chamber or more complex tunnel systems (Melton, 1976).

4. Identification of fossil vertebrate burrows

Positive identification of fossil vertebrate burrows usually relies on size and/or the presence of remains of the presumed producer or occupant (e.g., Voorhies, 1975; Zárate et al., 1998; Hasiotis et al., 2007; Genise, 2016). In addition to faunal content, Zárate et al. (1998) emphasized the utility of stratigraphic relationships, cross-sectional shape, and fill to recognize fossil tetrapod burrows. Lucas et al. (2006) discussed 11 criteria to recognize fossil vertebrate burrows, some of these criteria are considered useful herein. The main features that allow a positive identification of fossil tetrapod burrows are as follows (the certainty increasing with the number of features exhibited for a given structure).

- Subcylindrical structure of approximately uniform diameter with a fill that commonly contrasts with the host rock.
- Primary surface ornamentation (bioglyhs), commonly in the form of sets of two to four roughly parallel ridges or furrows (this is a key feature that may be not preserved).
- Tunnel/ramp horizontal diameter commonly larger than a few centimeters (commonly larger than 3–5 cm).
- The structure cuts bedding and/or primary sedimentary structure of the host strata, commonly forming a low angle to the paleohorizontal (except for fish and some amphibian burrows).
- The structure occurs in a continental sedimentary succession, commonly associated with a paleosol (although the latter is not mandatory).

- The presence of articulated or associated bone remains is strongly suggestive of preservation in a burrow fill.

Fossil vertebrate burrows are commonly filled with sediment (burrow casts), except some examples from the Quaternary that may exhibit a partial fill (Quintana, 1992; Buchmann et al., 2009; Dondas et al., 2009; Frank et al., 2012, 2013, 2015; Genise and Farina, 2012). For empty burrows with a diameter compatible with the local fossorial fauna, its identification as a fossil structure is uncertain. Fossil vertebrate burrows should display a difference in color, grain size, and/or sedimentary structure with the host rock, otherwise their identification is difficult. Burrow casts may be more cemented than the host rock, which is related to enhanced porosity of the burrow fill; however, it is uncommon that they are heavily cemented and/or replaced. This is related to the fact that tetrapods commonly do not incorporate organic matter to the fill or wall of a burrow system, except in the case of chambers used as latrines, food caches or nests, or faunal aggregations in chambers.

In cases where the structure potentially interpreted as a tetrapod fossil burrow displays a marked replacement, mineralization, and/or concretionary cementation, it is very likely that it is actually an inorganic structure (for example, some of the examples figured by Tanner and Lucas, 2008) or a concretionary structure probably formed around a former root (for example, most of the structures figured as *Daimonelix* by Raisanen and Hasiotis, 2018). When the boundary between the burrow fill and hosting rock is exposed, the presence of primary surface ornamentation that can be interpreted as digging traces is a key feature for the identification as a tetrapod burrow, although some fossil burrows lack this feature (Cardonatto and Melchor, 2018). This primary surface ornamentation is commonly in the form of two to four parallel ridges on a burrow cast (or furrows in the wall) that can be linked to chisel or scratch digging. It is important to highlight that only discrete and repetitive traces of biogenic origin (bioglyphs) are considered as part of the surface ornamentation of the burrow. Nonrepetitive surface features, including isolated ridges, furrows, or knobs are not indicative (e.g., fig. 12 in Raisanen and Hasiotis, 2018). Imperfect preservation or diagenetic effects are not acceptable as arguments for the lack of distinctiveness or the repetitive character of surface features.

The values of tunnel and ramp diameter (especially the horizontal diameter) should range in a limited interval because most animals construct their burrows that are only slightly larger than their cross sections (Andersen, 1982). Marked changes in the diameter of tunnels and ramps would require alternative interpretations of the structure (other than a fossil burrow). Chambers can be several times the size of an associated tunnel or ramp. Ichnologic studies on modern rodent burrows from central Argentina have documented that a burrow system constructed by a single species may display 10%–20% variation in tunnel/ramp diameter (Cardonatto and Melchor, 2020, 2021a, 2023a). There is no exact diameter that allows distinction between vertebrate and invertebrate burrows; the former are larger, but there is an overlap with invertebrate structures in the range of roughly 3–5 cm (e.g., Krapovickas et al., 2013).

A fossil burrow is an approximately cylindrical structure that cuts the bedding or the internal sedimentary structure of the host rock. In two-dimensional exposures, it may be confused with inorganic structures such as clastic dykes or sills, which are planar bodies. Sedimentary facies-bearing tetrapod fossil burrows are essentially continental and especially represent fluvial, eolian, and shallow lacustrine settings where the burrows are associated with intervals having evidence of soil development (Melchor et al., 2012a).

5. Field and laboratory methods

Field surveys of fossil vertebrate burrows depend upon the type and quality of outcrops and the features of the fossil burrows. Cross-sectional exposures of burrow fills are bidimensional and allow measurement of horizontal and vertical diameter, description of the sedimentary fill and obtaining orthogonal photos to compare the burrow outline (Fig. 12.6A). Cross sections can be compared using geometric morphometric techniques (Cardonatto and Melchor, 2018). Plan view exposures allow description of the tridimensional architecture of the burrow system and, eventually, the slope and length of the system (Fig. 12.6C). Tridimensional exposures can be documented by photogrammetric techniques (e.g., Cardonatto et al., 2023). Burrow fills can be described by measuring detailed sedimentary logs (Fig. 12.6B) that can be interpreted in terms of the taphonomy of the burrows, including the inferred processes of burrow fill (wind or water currents, roof collapse, etc.) (Cardonatto and Melchor, 2018). Oblique exposures are less informative (Fig. 12.6D), although they can be used to complement the field survey. Fossil burrows or ichnofabrics preserved in unconsolidated sedimentary sequences can be documented by sedimentary peel techniques (Voigt and Gittins, 1977; Melchor et al., 2012b) and destructive tomography (Sutton et al., 2014). The construction of a sedimentary peel involves production of a smooth and planar surface (Fig. 12.6E), application of chemical substances to fix the surface (Fig. 12.6F), extraction, and eventual mounting on a rigid fiberboard (Fig. 12.6G) (Voigt and Gittins, 1977).

The presence of bone remains in the burrow fill can be useful for identification of potential producers, especially if they are articulated and appear in a massive interval located in the lower part of the fill or at the end of the burrow fill (Groenewald et al., 2001; Cardonatto and Melchor, 2018). In these cases, they can be preserved as a result of burial when alive or by rapid burial after death (Groenewald et al., 2001; Woodruff and Varricchio, 2011). Disarticulated, or abraded remains or those with signs of subaerial exposure, commonly entered the burrow along with other sedimentary particles, and they have no clear link with the producer (Cardonatto and Melchor, 2018).

When a cross-sectional exposure of the burrow fill is available, it is possible to calculate the area of the cross section and to use this information to estimate the body mass of the producer using allometric equations for fossorial organisms (White, 2005; Wu et al., 2015). The candidates for producing the burrow should be searched in the known fossorial fauna of the lithostratigraphic unit or in contemporaneous units, using the constraint of the body mass (e.g., Cardonatto and Melchor, 2018; Cardonatto et al., 2023; Francischini et al., 2023; Melchor et al., 2023).

Institutional abbreviations. **ABDUG Pal**: Palaeontology Collection, Geology Department, Aberdeen University, UK. **UM-TM**: ISEM (Institut des Sciences de l'Evolution de Montpellier), the University of Montpellier, France. **KUVP**: Division of Vertebrate Paleontology, Biodiversity Institute, Kansas University, Kansas, USA. **UMMP-VP**: Vertebrate Paleontology Collection, University of Michigan Museum of Paleontology, Ann Arbor, Michigan, USA. **GHUNLPam**: Colección Paleontológica, Facultad de Ciencias Exactas y Naturales, Universidad Nacional de La Pampa, Santa Rosa, La Pampa, Argentina. **NMMNH-P**: Paleontology Collection, New Mexico Museum of Natural History & Science, Albuquerque, New Mexico, USA. **BP**: Evolutionary Science Institute, University of Witwatersrand, Johannesburg, South

FIGURE 12.6 Preservation of fossil vertebrate burrows and some study methods. The examples are from Upper Miocene Cerro Azul Formation, La Pampa Province (A–D) and the Holocene deposits of the Gran Salitral saline lake, La Pampa Province, Argentina (E–G). (A) Two-dimensional preservation of cross-section of fossil burrow fill (arrowed). (B) Detailed sedimentary log of the burrow fill illustrated in A, showing a lower laminated interval and a massive upper interval. (C) Three-dimensional preservation of fossil burrow fill showing an enlarged end (arrow). (D) Oblique section of a fossil burrow (arrowed). (E) Clean trench face of Holocene sediments showing a strongly bioturbated interval that was selected for peel construction (arrow). Individual trace fossils are *Nagtuichnus meuleni* (see also Melchor et al., 2012b). (F) Finished peel before extraction. (G) Peel mounted in medium-density fiberboard panel.

Africa. **NRM X**: Department of Paleobiology, Swedish Museum of Natural History, Stockholm, Sweden. **UMPLIC**: Colección de Icnología, Laboratorio de Paleobiología, campus Puerto Escondido, Universidad del Mar, Oaxaca, Mexico.

6. Fossil record of vertebrate burrows

Fossil vertebrate burrows are relatively common biogenic structures, and vertebrate burrowing can be traced back to the Middle Devonian. The oldest tetrapod burrows are probably from the Middle Mississippian of North America (Storm et al., 2010); however, the oldest known convincing evidence of tetrapod burrows are early Permian (Asselian−Artinskian) lysorophid amphibian burrows from Kansas, USA (Hembree et al., 2004). Devonian vertebrate burrows include probable fish feeding/nesting trace fossils from the Middle Devonian (Eifelian) of Poland (Lucas, 2015; Niedźwiedzki and Szrek, 2020); the examples attributed to eel-like fish from the Middle Devonian of Orkney, Scotland (Carroll and Trewin, 1995); and lungfish burrows from the Upper Devonian (Famennian) of the USA (Friedman and Daeschler, 2006). A Middle Mississippian structure from the USA was tentatively interpreted as amphibian burrow (Storm et al., 2010), whereas more recent discoveries from Canada are apparently slightly older (Stimson et al., 2022), but await detailed description. The early (Devonian) fish burrows were subvertical structures with subcylindrical or elliptical cross section, which are similar to the oldest (Lower Permian) tetrapod burrows produced by lysorophid amphibians (Hembree et al., 2004, 2005). These amphibians have reduced or vestigial limbs and an elongated body, resembling the body plan of fishes, which can be an explanation for the simple design and subvertical orientation of the fossil burrows containing articulated specimens of the producer.

Most pre-Cretaceous tetrapod burrows have been attributed to therapsids, in part on the basis of the finding of articulated skeletons in a few late Permian-Early Triassic burrows (Smith, 1987; Groenewald et al., 2001; Damiani et al., 2003; Modesto and Botha-Brink, 2010). A common architecture for late Permian to Jurassic tetrapod burrows is a shallowly inclined ramp with a rounded and not enlarged end, of reduced horizontal diameter, with discrete scratch traces, always showing a flattened elliptical cross section, and commonly with a bilobed bottom (e.g., Damiani et al., 2003; Sidor et al., 2008; Riese et al., 2011; Liu and Li, 2013; Melchor and Loope, 2016; Krummeck and Bordy, 2018). Most Jurassic tetrapod burrows occur in eolian sequences, including the oldest subhorizontal burrow systems that have been assigned to primitive mammals from the Early Jurassic Navajo Sandstone of the USA (Riese et al., 2011). In contrast, there is a dearth of reports of Cretaceous tetrapod burrows, which could possibly be due to the more equable climates that existed for most of this period. An exceptional occurrence for the Late Cretaceous is the unique dinosaur burrow containing an adult and two juvenile remains of ornithopods, further suggesting denning behavior and parental care (Varricchio et al., 2007).

Most Cenozoic tetrapod burrows have been attributed to mammals, mainly to Rodentia and Xenarthra (e.g., Voorhies, 1975; Martin and Bennett, 1977; Zárate et al., 1998; Gobetz and Martin, 2006), whereas a few examples are related to Carnivora (e.g., Hunt et al., 1983; Hembree and Hasiotis, 2008). The record of Paleogene tetrapod burrows is meager and may also be linked to dominantly warm and humid climate conditions (Zachos et al., 2001). The Miocene record of tetrapod burrows is more varied and abundant, with a diversification of the architectural patterns and behavioral strategies that, commonly, appeared under stressed volcaniclastic and eolian environments. The Early Miocene volcaniclastic floodplains of Nebraska, USA, witnessed the appearance of: 1) the celebrated vertical helical burrows ending in a ramp and terminal chamber named *Daimonelix* (e.g., Barbour, 1892; Martin and Bennett, 1977), as well as 2) smaller, complex subhorizontal rodent burrow systems with terminal chambers and subcircular cross section (Gobetz and Martin, 2006), and 3) the first carnivore den (Hunt et al., 1983; Hunt, 1990). Also, in the Early Miocene, the coastal dunes of Germany preserved the oldest fossil food cache (Gee et al., 2003).

The main tetrapod burrowing innovation during the Middle Miocene is represented by cylindrical, subhorizontal, unbranched tunnels with a meniscate backfill interpreted as foraging tunnels of small Chlamyphoridae from southern South America (Melchor et al., 2012b, 2016). Late Miocene tetrapod burrows from central Argentina reflect a remarkable increase in the average size of the fossorial fauna including Dasypodidae, Glyptodontidae, and Mylodontidae (Cardonatto and Melchor, 2018). Pliocene burrows are small (horizontal diameter less than 250 mm), and those from the Atlantic coast of the Buenos Aires Province, Argentina have received a considerable attention, because they are abundant and a significant number of the burrows contains bone remains (Genise, 1989; Quintana, 1992; Fernández et al., 2000; Elissamburu et al., 2011). These are subcircular burrows assigned to rodents and notoungulates. The Pleistocene megafauna of South America is also reflected in the burrow trace fossil record in the form of huge tunnels (up to 2 m or more in width), with flattened elliptical cross sections from Argentina and Brazil (Quintana, 1992; Vizcaíno et al., 2001; Buchmann et al., 2009; Genise and Farina, 2012; Lopes et al., 2017). The smaller burrows are attributed to Dasypodidae and Pampatheriidae and the larger to ground sloths. It has been suggested that the adoption or generalization of burrowing behavior by large Pleistocene herbivorous mammals may reflect the arrival of large carnivorans after the Great American Biotic Interchange, just before the Pliocene-Pleistocene boundary (Soibelzon et al., 2009).

6.1 Ichnotaxobases

It is desirable to adopt a set of morphologic features to be used in the taxonomy of fossil vertebrate burrows, and herein the proposal of Cardonatto and Melchor (2021a) is followed. Generic ichnotaxobases include overall architecture of the burrow system (distinguishing between ramps, subhorizontal ramified systems, and spiral systems), number of levels, presence/absence of closed circuits, morphology of the cross-sectional shape of tunnels (distinguishing between subcircular, flattened elliptical, and vertical elliptical; and including RDI as a useful measure, see Table 12.1), type of burrow filling (active vs. passive), presence/absence of bilobed bottom, and presence/absence of chambers (including VC/VT, if applicable). Burrow system complexity (Co), tortuosity (To), and fractal dimension (FD) are of potential application but require a preservation of the complete burrow system, which is uncommon.

Proposed specific ichnotaxobases are surface ornamentation (distinguishing between primary and secondary surface ornamentation, overall pattern of claw/tooth traces, number of claw/tooth traces on sets, width of sets of claw/tooth traces, type and density of secondary surface ornamentation), details of the burrow system architecture (for example, type of ramification, angle of ramification, ramp inclination), location of chambers (lateral, terminal, bilateral, at a ramification), chamber filling (droppings, plant material, fur, sediment), and tunnel or ramp size, especially as expressed by the horizontal diameter (Dh).

6.2 Ichnotaxa of vertebrate burrows

A probable vertebrate origin has been suggested for ichnofossils comparable with *Megaplanolites* Calvo et al., 1987 by Geister (1998), Knaust (2012), and Abbassi et al. (2021). This ichnogenus was defined as a giant subhorizontal burrow from the Upper Jurassic of the Iberian Chain, Spain. The type material of *Megaplanolites* is burrows preserved in an open marine sequence composed by silty marlstones with marly limestone beds and sandstones (Calvo et al., 1987; Reolid et al., 2023). The producer of *Megaplanolites* is unknown and a vertebrate origin is not favored (Calvo et al., 1987; Reolid et al., 2023). Geister (1998) studied large linear grooves from the Middle Jurassic of Switzerland, which were interpreted as feeding structures of marine vertebrates, and assigned it to *Megaplanolites*. However, the latter ichnogenus is proposed for tubular structures (burrows) (F. Rodriguez-Tovar, written communication, 2024). Abbassi et al. (2021) also described large grazing or locomotion traces from the continental deposits of the Middle—Upper Miocene Upper Red Formation in northwestern Iran, which were compared with *Megaplanolites* and interpreted as produced by a snake-like animal. Both examples compared with *Megaplanolites* and tentatively referred to vertebrates (Geister, 1998; Abbassi et al., 2021) are grooves (preserved as negative epirelief or positive hyporelief) instead of burrows, as the type material of *Megaplanolites*.

Alezichnos Gobetz, 2006

- *Alezichnos chelecharatos* Gobetz, 2006
- *Alezichnos trogodont* Gobetz and Martin, 2006
- *Alezichnos gongylochora* Platt and Smith, 2023

Cornulatichnus Carroll and Trewin, 1995

- *Cornulatichnus edayensis* Carroll and Trewin, 1995

☒ *Cynodontipus* Ellenberger, 1976 (nomen dubium)

- ☒ *Cynodontipus polythrix* Ellenberger, 1976 (nomen dubium)

Daimonelix Barbour, 1892

- *Daimonelix circumaxilis* Barbour, 1892
- ☒ *Daimonelix anaxilis* Barbour, 1892 (junior synonym of *D. circumaxilis*)
- ☒ *Daimonelix bispiralis* Barbour, 1892 (junior synonym of *D. circumaxilis*)
- *Daimonelix carinata* Barbour, 1892
- ☒ *Daimonelix robusta* Barbour, 1892 (junior synonym of *D. circumaxilis*)
- ☒ *Daimonelix petalichnus* Hembree and Hasiotis, 2008 (nomen nudum)
- ☒ *Daimonelix martini* Raisanen and Hasiotis, 2018 (nomen dubium)

☒ *Fractisemita* Raisanen and Hasiotis, 2018 (nomen dubium)

- ☒ *Fractisemita henrii* Raisanen and Hasiotis, 2018 (nomen dubium)

☒ *Ichnogyrus* Bown and Kraus, 1983 (nomen nudum)

- ☒ *Ichnogyrus nididens* Bown and Kraus, 1983 (nomen nudum)

Katarrhedrites Hembree and Hasiotis, 2008

- *Katarrhedrites athesphatichnus* Hembree and Hasiotis, 2008
- ☒ *Katarrhedrites atyphichnus* Hembree and Hasiotis, 2008 (nomen nudum)

☒ *Kladosystemites* Hembree and Hasiotis, 2008 (nomen dubium)

- ☒ *Kladosystemites homocylindrichnus* Hembree and Hasiotis, 2008 (nomen dubium)

Maneraichnus Cardonatto et al., 2023

- *Maneraichnus pampeanum* Cardonatto et al., 2023

Megaichnus Lopes et al., 2017

- *Megaichnus minor* Lopes et al., 2017
- *Megaichnus major* Lopes et al., 2017

Megapermichnus Jenny and Jenny-Deshusses, 1978

- *Megapermichnus aliabadensis* Jenny and Jenny-Deshusses, 1978

☒ *Megatambichnus* Martens, 2001a (nomen nudum)
Nagtuichnus Melchor et al., 2012b

- *Nagtuichnus meuleni* Melchor et al., 2012b

Natatorichnus Lockley et al., 2021

- *Natatorichnus subarenosa* Lockley et al., 2021
- *Natatorichnus sulcatus* Lockley et al., 2021

Piscichnus Feibel, 1987

- *Piscichnus brownii* Feibel, 1987
- *Piscichnus waitemata* Gregory, 1991
- *Piscichnus sanctacruzensis* Niedźwiedzki and Szrek, 2020

☒ *Polychoredrites* Hembree and Hasiotis, 2008 (junior synonym of *Alezichnos*)

- ☒ *Polychoredrites tetrachelichnus* Hembree and Hasiotis, 2008 (junior synonym of *Alezichnos chelecharatos*)
- ☒ *Polychoredrites cetinensis* Mueller and McCann, 2014 (nomen nudum)

Redondarefugium Lucas et al., 2010

- *Redondarefugium abercrombieorum* Lucas et al., 2010

Reniformichnus Krummeck and Bordy, 2018

Reniformichnus katikatii Krummeck and Bordy, 2018
Reniformichnus australis McLoughlin et al., 2020

Tambia Müller, 1969

Tambia spiralis Müller, 1969

Torridorefugium Hembree et al., 2005

Torridorefugium eskridgensis Hembree et al., 2005

Yaviichnus Guerrero-Arenas et al., 2020

Yaviichnus iniyooensis Guerrero-Arenas et al., 2020

7. Description of vertebrate burrow ichnotaxa

Ichnogenus *Alezichnos* Gobetz, 2006.
Fig. 12.7E–F.
Type ichnospecies: *Alezichnos chelecharatos* Gobetz, 2006.

Diagnosis. Natural casts of sinuous, slightly ovate burrows up to at least 7.0 m in length, consisting of nearly horizontal primary tunnels from which secondary tunnels and shafts arise. The asymmetrical, sinuous shape of these tunnels indicates that they form part of a rambling, labyrinthine system, which designates them as the ichnogenus *Alezichnos*. Asymmetry, varying directionality, and irregular branching patterns distinguish ichnofossils of the *Alezichnos* type from more distinctively shaped (strongly helical) terrestrial burrows, such as *Daimonelix*, and from other previously described continental vertebrate burrows (Gobetz, 2006).

Discussion. The ichnogenus was erected for subhorizontal branched tunnel systems with clear surface ornamentation and was related to rodent producers. Two ichnospecies from the Early and Middle Miocene of Nebraska and Colorado (USA) were initially recognized: *A. chelecharatos* and *Alezichnos trogodont* (Gobetz, 2006; Gobetz and Martin, 2006). A further ichnospecies from the Upper Miocene Ogallala Formation (Kansas, USA) was recently proposed: *A. gongylochora* (Platt and Smith, 2023). The type ichnospecies includes burrow casts with a flattened elliptical cross section ($Dv/Dh = 0.88$), horizontal diameter 139–220 mm, and sets of three claw traces as surface ornamentation (Fig. 12.7F). The type material also exhibits a tunnel bifurcation with a strongly ornamented surface (Fig. 12.7E). *Alezichnos trogodont* is smaller than the type ichnospecies ($Dh = 54–74$ mm), with an average cross-section that is subcircular ($Dv/Dh = 1.08$), and displays both claw and incisor traces. *Alezichnos chelecharatos* was related to the extinct mylagaulid rodent *Pterogaulus laevis* that occurs profusely in the host formation. *Alezichnos trogodont* was probably excavated by the extinct geomyid rodent *Gregorymys*. *Alezichnos gongylochora* differs from the remaining ichnospecies in the lack of surface ornamentation and includes terminal or lateral chambers (Platt and Smith, 2023). This lack of surface ornamentation departs from the previous ichnospecies and will probably

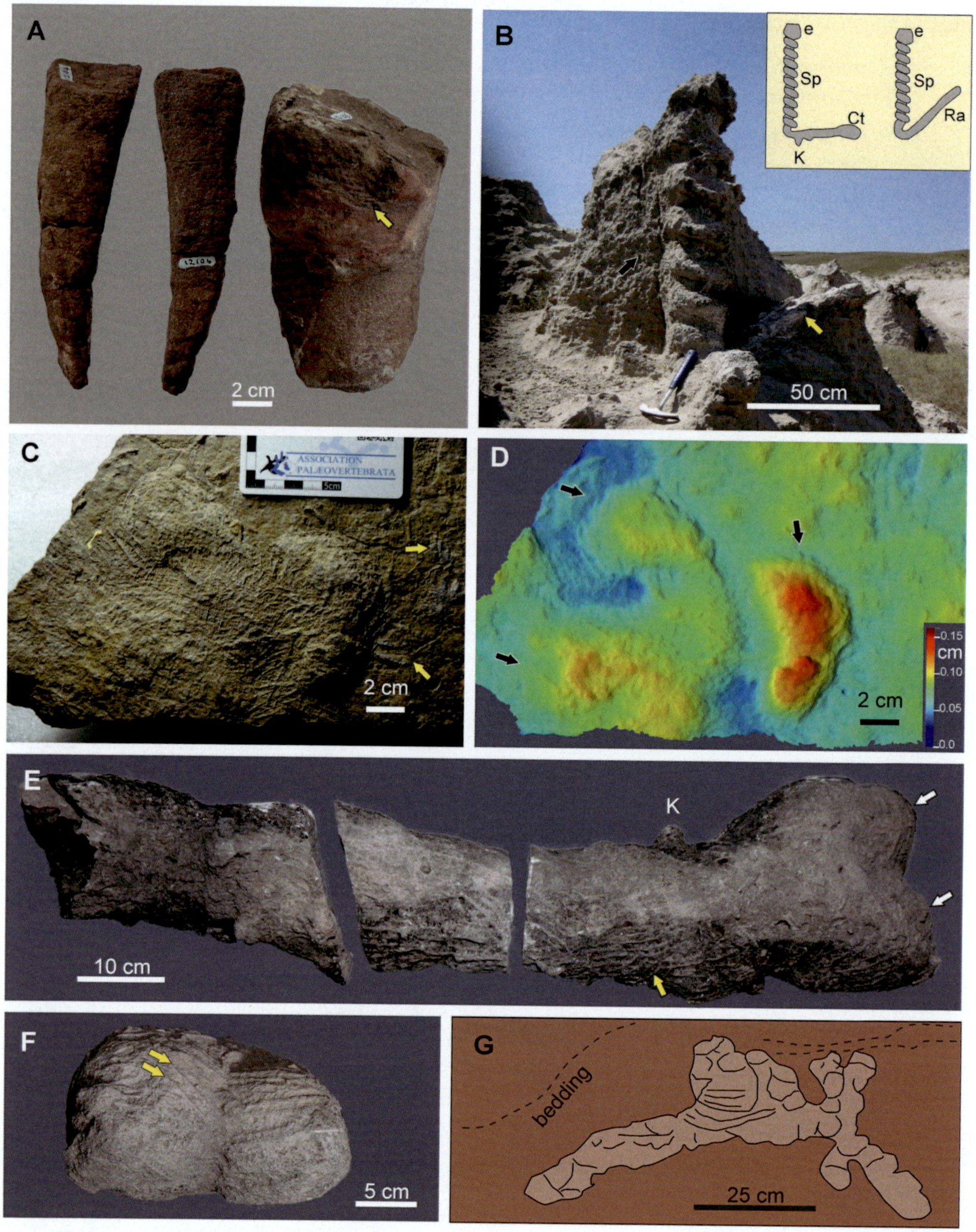

FIGURE 12.7 Holotype or topotype material of vertebrate burrow ichnotaxa. (A) *Cornulatichnus edayensis*. Holotype on left and center (ABDUG Pal. 12104), paratype on right side (ABDUG Pal. 12106). The arrow on the paratype indicates the laminated wall. Devonian Eday Marls and Eday Sandstone (Old Red Sandstone) of the UK. (B) Field specimen of *Diamonelix circumaxilis* from the type locality where Erwin Barbour first discovered them on Pine Ridge, Sioux Co., six miles N of Harrison, Nebraska, USA. Early Miocene Harrison Formation. Note the subvertical spiral (black arrow) and the ascending terminal ramp (yellow arrow). The inset contains diagrams illustrating the salient

boost future revision of the ichnotaxonomic allocation of this ichnospecies. *Alezichnos gongylochora* was tentatively related to ground squirrels.

Ichnogenus *Cornulatichnus* Carroll and Trewin, 1995.

Fig. 12.7A.

Type ichnospecies. *Cornulatichnus edayensis* Carroll and Trewin, 1995.

Diagnosis. Large, subvertical, conical burrow, having a wide, subcircular mouth ranging in diameter from 38 to 90 mm. The burrow tapers downward to a blade-like termination, which may be up to 400 mm from the mouth. A thin lining is developed, and a smooth outer wall is present. Burrow infill is massive (Carroll and Trewin, 1995).

Discussion. This ichnogenus was recognized in the Middle Devonian Eday Marls and Eday Sandstone (Old Red Sandstone) of the UK and was tentatively identified in the Upper Silurian Moor Cliffs Formation, UK (Marriott et al., 2009), although the latter occurrence lacks the diagnostic features of the ichnogenus. The most useful criteria for identifying *Cornulatichnus* are its large size (up to 400 mm long), subvertical orientation, subcircular to elliptical cross section, distinct lining, massive infill, and blade-like termination (Carroll and Trewin, 1995). The authors suggested that was constructed subaqueously as an open shelter, possibly by an eel-like fish. Knaust (2012) suggested that *Cornulatichnus* is a potential subjective synonym of *Lithoplaision* Diblin et al., 1991. However, the latter ichnogenus is distinguished by a thick central tube (about half of the diameter) composed of skeletal grains and an orientation with respect to bedding ranging from near-vertical to subhorizontal (Diblin et al., 1991).

Ichnogenus *Cynodontipus* Ellenberger, 1976.

Fig. 12.7C–D.

Type ichnospecies. *Cynodontipus polythrix* Ellenberger, 1976.

Diagnosis. Vertebrate of large size characterized by the following five salient features: a) Hairy autopod, pentadactyl in appearance, with a sole noticeably shortened longitudinally. Width of foot reaching at least 25 cm. b) The middle digits II-III-IV are subequal, thick, rounded. They diverge by about 50 degrees and curve distally toward the inner edge of

features of *D. circumaxilis* including the entrance (e), also interpreted as turnaround, the spiral (Sp), and the terminal or living chamber (Ct), in addition to occasional knobs (K). In other burrows, the terminal chamber is missing and only a ramp (Ra) is found. (C and D) Holotype of *Cynodontipus polythrix* (UM-TM 34700_1), preserved in positive hyporelief. Lower to Middle Triassic of Hérault, France. (C) Detail of a sector of the slab containing the holotype. Note sets of three parallel ridges (one indicated by a bracket). Arrows indicate two footprints preserved in positive hyporelief on the same surface, the lower footprint is cut by *C. polythrix*. (D) Digital elevation model of the holotype. The arrows point to burrow segments. (E and F) Holotype of *Alezichnos chelecharatos* (KUVP 131601). Late Miocene Pawnee Creek Formation of Colorado, USA. (E) Tunnel segment with distal bifurcation (white arrows). Note marked surface ornamentation in the form of sets of parallel ridges (especially, in the area indicated by a yellow arrow). Note also a knob (K). (F) View of distal bifurcated end showing sets of parallel ridges (yellow arrows). (G) Diagram of the holotype of *Fractisemita henrii* (a field specimen) from the Upper Jurassic Morrison Formation, southeastern Utah, USA. *(A) Photos provided by the University of Aberdeen Museums and Special Collections, licensed under CC by 4.0. (B) Photo courtesy of Robert Hunt. Inset modified from Martin and Bennett (1977). (C and D) Photos provided by ISEM (Institut des Sciences de l'Evolution de Montpellier), the University of Montpellier, France. (E and F) Photos courtesy of Megam Sims (University of Kansas, USA). (G) Drawn after fig. 9E of Raisanen and Hasiotis (2018).*

the foot. They are linked proximally by intermediate hairy zones. c) The extreme fingers I and V are set back behind the other three, they are more divergent and can be guessed about the type by the orientation of a few bristles (incomplete back of the foot). d) No distal claw traces touch the ground. On the contrary, we see the take-off point, upward, with the tips of the fingers and their bristles (effect of expanders). e) The hairs are generally rectilinear, sometimes curved, 10–30 mm long, 0.9 mm thick, slightly tapered at their ends. These bristles radiate around the most proximal part of digit II and especially III, but distally align itself more or less regularly with the edges of each finger with some final convergence (III), and a clear crisscross in the intervals separating digits II-III and III-IV. The deepest parts of the sole of the foot are devoid of it. The morphology of such an autopod is close to that of the most classic Premammalians, especially therapsids (translated of the French from Ellenberger, 1976).

Discussion. This ichnogenus has been proposed to represent a supposed isolated and incomplete fossil footprint of a Triassic hairy therapsid, from the Lower to Middle Triassic of Hérault, France. The erection of the ichnogenus has been criticized due to the incompleteness of the material (Gand et al., 2007), doubtful presence of hair impressions in the footprint (Lockley and Meyer, 2010) and considered a *nomen dubium* probably representing an incomplete chirothere footprint (Klein and Lucas, 2021).

The ichnogenus has also been reinterpreted as representing a procolophonid burrow (Olsen et al., 2012; Sues and Olsen, 2015). This interpretation as a tetrapod burrow is consistent with the existence of sets of three parallel ridges (Fig. 12.7C) that are similar to scratch traces (i.e., the "hairy" appearance in the diagnosis) in other modern and fossil burrows (e.g., Gobetz, 2006; Cardonatto and Melchor, 2021a). The holotype of *Cynodontipus* is also similar to bifurcated burrow terminations described by Voigt et al. (2011) and also shares some morphological and preservational features with *Megatambichnus* (Martens, 2001a). The slab containing the holotype exhibit tridactyl lacertoid footprints, some of them are cut by one of the burrows (Fig. 12.7C). The bottom of three burrow segments can be identified in the holotype (Fig. 12.7D), which are about 35 mm wide. Accepting the vertebrate burrow interpretation, the surface ornamentation is not enough to describe a burrow system, but also features such as overall architecture are necessary. As a consequence, *Cynodontipus* is considered a *nomen dubium*.

Ichnogenus *Daimonelix* Barbour, 1892.

Fig. 12.7B.

Type ichnospecies. *Daimonelix circumaxilis* Barbour, 1892.

Diagnosis. Large, vertical, open cylindrical helical structure, regular in form, with a low-to high-angle, lateral ramp at or near the base, which may contain an enlargement (chamber) and additional tunnels (modified from Raisanen and Hasiotis, 2018).

Discussion. This is the first described tetrapod burrow ichnotaxon. In the original description, from the Lower Miocene Harrison Formation of Nebraska, Barbour (1892) recognized five (ichno)species (*D. circumaxilis*, *D. bispiralis*, *D. anaxilis*, *D. robusta*, and *D. carinata*), that were described and illustrated. Most of these ichnospecies are considered junior synonyms of *C. circumaxilis*, designated as the type ichnospecies by Häntzschel (1975). *Daimonelix bispiralis* results from subsequent burrowing paralleling a previous burrow, *D. anaxilis* lacks the central nonburrowed sediment portion, and *D. robusta* displays a thicker wall but is otherwise morphologically similar. *Daimonelix carinata* displays a "keel" or cylindrical protuberance running parallel and in the lower part of the spiral whorls and is probably distinctive. Since its description the ichnogenus has been loosely applied to more or less helical or circular structures that depart from the morphology of the type ichnospecies. No modern redescription of the type material with an ichnotaxonomical purpose is available. It is herein suggested that the ichnogenus be restricted to large and regular helical structures with a terminal ramp. Raisanen and Hasiotis (2018) proposed an emended diagnosis that is herein modified to fit the original description of the type material. The surface ornamentation in the form of incisor and claw traces was documented in detail by Martin and Bennett (1977).

Barbour (1897) used loosely the name "Daemonelix" (not the correct original spelling) to include a set of structures from the outcrops of the Harrison Formation that mostly represent different trace fossils. These include "Daemonelix fibers" (herein interpreted as rhizoliths), "Daemonelix cakes" (comparable with chambers of the ant nest ichnogenus *Daimoniobarax* Smith et al., 2011), "Daemonelix balls" (probable dung beetle balls), "Daemonelix cigars or fibers" (probable invertebrate burrows, see Yelinek and Chin, 2007), and "Daemonelix irregular and regular" (that correspond to *D. circumaxilis*). Barbour (1897) also described "Daemonelix buds" that are terminal portions of burrow fills with sets of parallel ridges (scratch traces), which are probably comparable with *Alezichnos*. Two additional ichnospecies were proposed recently: *Daimonelix petalichnus* and *Daimonelix martini* (Hembree and Hasiotis, 2008; Raisanen and Hasiotis, 2018). *Daimonelix petalichnus* was erected for a large subcylindrical and vertical weathered structure that was interpreted as representing a tight helical burrow and a terminal chamber (Hembree and Hasiotis, 2008). However, the spiral structure is not clearly defined. This is considered a *nomen dubium*, and the material can be described as probably comparable with *Daimonelix*, but cannot be used as type material of a new ichnospecies. The holotype and all the figured material of *D. martini* lack the regular helical structure that typifies the ichnogenus, display abrupt changes in diameter, lack a cylindrical or elliptical cross section, and lack bioglyphs. *Daimonelix martini* is considered herein a nomen dubium.

The occurrence of *Daimonelix* described by Wood and Wood (1933) and Joeckel and Tucker (2013) includes only the supposed terminal ramp but lacks the helical portion and should not be referred to the ichnogenus. Smith (1987) assigned helical burrow casts with a terminal chamber from the Permian of South Africa to *Daimonelix*. A detailed morphological comparison was offered (Smith, 1987: Table I) that highlights that the main differences with the type material are the lower number of whorls (typically two), the marked downward increase of burrow diameter, the constant dextral coiling, and the horizontal terminal chamber. As the overall architecture is different, the material from the Permian of South Africa should be recognized under a different ichnogenus or, at least, a different ichnospecies within *Daimonelix*.

A fucoid, root, and burrow origin was mentioned, but the favored interpretation by Barbour (1892, 1894, 1895, 1897) was a freshwater sponge. Lugn (1941) also hypothesized that these burrows can be interpreted as fossil lianas. The interpretations by Erwin H. Barbour were strongly biased by the assumption that the host beds were of lacustrine origin and from the observation that thin sections of *Daimonelix* revealed plant cellular structure (Barbour, 1894). *Daimonelix* was covered by siliceous millimeter tubes (rhizoliths or root structures) that preserved plant tissues. Barbour (1892, 1894) documented partially articulated rodent remains inside *Daimonelix* but, assuming that it was impossible that a rodent burrowed underwater, explained its occurrence as the nucleus for growth of the plant tissue of *Daimonelix*. The rodent burrow origin was advanced by Cope (1893), Fuchs (1893) and confirmed by Peterson (1906), Abel (1927), Schultz (1942), and Martin and Bennett (1977).

Peterson (1906) also confirmed that the plant tissues belong to angiosperm roots, compared *Daimonelix* with examples of modern rodent burrows, and suggested that the producer was the beaver *Palaeocastor fossor*. The latter inference was confirmed by Martin and Bennett (1977) by comparing incisor width with putative incisor traces in the burrows and considering the burrow diameter and size of the presumed producer.

Ichnogenus *Fractisemita* Raisanen and Hasiotis, 2018.
Fig. 12.7G.
Type ichnospecies. *Fractisemita henrii* Raisanen and Hasiotis, 2018.
Diagnosis. Unlined, sparse maze of Y-branched tunnels and shafts of varying diameter, angle from the horizontal, and length, with no distinguishable chambers (Raisanen and Hasiotis, 2018).

Discussion. This ichnogenus was erected for structures of the Salt Wash Member of the Upper Jurassic Morrison Formation, Utah, USA. The diagnosis and original descriptions and illustrations show no discrete pattern or morphology. As recognized by Raisanen and Hasiotis (2018), the structures are distorted by pedoturbation and carbonate cementation. The organic origin is doubtful as no repetitive diameter and bioglyphs, and/or, smooth burrow surface is observed. On the contrary, they are similar to coalescent carbonate mazes and/or concretionary growth of pedogenic carbonate around former roots. Although uncertain, the structure probably corresponds to some burrows (or ichnofabric, as suggested by the sparse maze of tunnels in the diagnosis) overprinted by carbonate cementation. *Fractisemita* is herein considered a nomen dubium.

Ichnogenus *Ichnogyrus* Bown and Kraus, 1983.
Fig. 12.8A.
Type ichnospecies. *Ichnogyrus nididens* Bown and Kraus, 1983. See a tridimensional model of the holotype at the University of Michigan Online Repository of Fossils (https://umorf.ummp.lsa.umich.edu/wp/specimen-data/?Model_ID=1833).
Diagnosis. Differs from *Daimonelix* Barbour in its much smaller size and from *Daimonelix* and *Gyrolithes* De Saporta in its tightly closed spiral with contacting whorls (Bown and Kraus, 1983).

Discussion. The ichnogenus was erected on the basis of a single loose specimen with closely spaced whorls composing a straight structure. The slight enlargement toward one of the extremes led Bown and Kraus (1983) to suggest that it was probably connected to a chamber. The ichnogenus was the second to be formally proposed among vertebrate burrows and probably marks the revival of interest in this type of structures. The authors identified a discrete structure with a differential diagnosis, after the suggestions of the ICZN (Ride et al., 2012). The interpretation of the structure as the filling of once open burrow is contradicted by to the closeness of the whorls and absence of bioglyphs. Successive whorls use part of the space of the former, as a consequence, there would be no space for the burrow wall. There is no known similar structure produced by an extant animal, although some analogous remains were described from Cretaceous rocks of England and Australia (Woodward, 1922; Whitehouse, 1934), as *Dinocochlea*. The latter was reinterpreted as a concretion around a former helical burrow (Taylor and Sendino, 2011) and, more recently, as a helical concretion, on the basis of findings of similar structures in the Mississippian Kuna Formation of Alaska (Locock et al., 2020). As a consequence, the most parsimonious interpretation is that

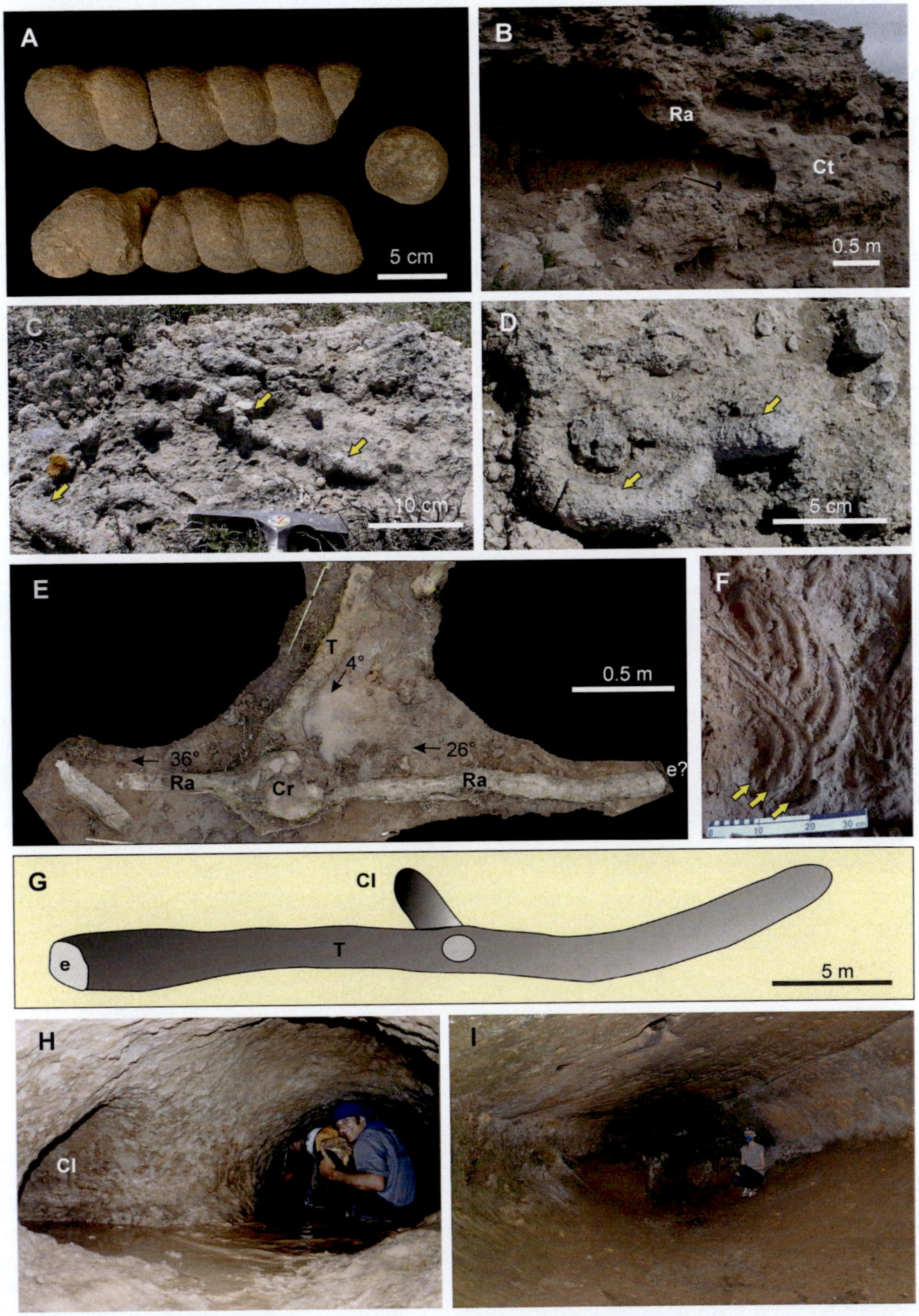

FIGURE 12.8 Holotype and topotype material of vertebrate burrow ichnotaxa (cont.). (A) *Ichnogyrus nididens* (UMMP VP 76862) on lateral (left) and cross-sectional view (right). Lower Eocene Willwood Formation, Wyoming, USA. (B) Field specimen of *Katarrhedrites athesphatichnus* including a ramp (Ra) and a poorly defined terminal chamber (Ct). Middle

Ichnogyrus is an inorganic structure of concretionary origin, and the name is considered a nomen nudum.

Ichnogenus *Katarrhedrites* Hembree and Hasiotis, 2008.

Fig. 12.8B.

Type ichnospecies. *Katarrhedrites athesphatichnus* Hembree and Hasiotis, 2008.

Diagnosis. Subhorizontal tunnel inclined 10–50 degrees. Opening at paleosurface. Terminates in enlarged terminal chamber (Hembree and Hasiotis, 2008).

Discussion. The ichnogenus was created to encompass burrow systems composed by a ramp plus a terminal chamber from the Middle Miocene Pawnee Creek Formation, Colorado, USA. The burrow cast used as the holotype has irregular boundaries and lacks surface ornamentation (Fig. 12.8B), and the diagnosis referred to a general morphology. Two ichnospecies were recognized: *Katarrhedrites athesphatichnus* and *Katarrhedrites atyphichnus*. The second ichnospecies is diagnosed as having a 2–5 mm ramp (not illustrated) ending in a 50–80 mm wide enlargement. The holotype of the latter ichnospecies is a poorly preserved structure interpreted as a terminal chamber. Due to poor preservation, *K. atyphichnus* is considered a nomen dubium. The proposed probable producer for *K. athesphatichnus* was a mammal, likely a carnivore (Hembree and Hasiotis, 2008). The ichnogenus was tentatively recognized (cf. *Katarrhedrites*) in the Mississippian of Canada (Stimson et al., 2022), Early Triassic of South Africa (Modesto and Botha-Brink, 2010), Lower Cretaceous and Late Miocene of USA (Yoshida and Carpenter, 2020; Platt and Smith, 2023), and Middle Miocene of Bolivia (Catena et al., 2017). Knaust (2012) suggested that *Katarrhedrites* is a potential subjective synonym of *Katbergia* Gastaldo and Rolerson, 2008. Potential differences of *Katbergia*, when compared with *Katarrhedrites* are a cylindrical cross section, overall sigmoidal configuration, and presence of paired scratch traces (Gastaldo and Rolerson, 2008).

Ichnogenus *Kladosystemites* Hembree and Hasiotis, 2008.

Fig. 12.8C–D.

Type ichnospecies. *Kladosystemites homocylindrichnus* Hembree and Hasiotis, 2008.

Diagnosis. Boxwork of interconnected, elongate, horizontal tunnels and short subvertical shafts; extend laterally up to 1 m and up to 200 mm deep. Tunnels and shafts straight to

Miocene Pawne Creek Formation, Colorado, USA. (C and D) Field views of *Kladosystemites homocylindrichnus*, including a small-diameter cylindrical burrow network (some arrowed) in C, and detail of two burrows (yellow arrows) surrounding a rhizolith (black arrow) in D. Middle Miocene Pawne Creek Formation, Colorado, USA. (E) Orthomosaic of the holotype (field specimen) of *Maneraichnus pampeanum* showing a large main chamber located at a ramification (Cr), ramps (Ra), possible entrance (e) and a tunnel (T). Sense and inclination of ramps and tunnels are also indicated. Upper Miocene Cerro Azul Formation, Buenos Aires Province, Argentina. (F–H) Holotype of *Megaichnus minor*, from the Quaternary of Rio Grande do Sul, Brazil. (F) Detail of three parallel grooves on the wall of an empty burrow (arrowed). (G) Diagram of the holotype of *M. minor* including a subhorizontal tunnel (T) and a lateral chamber (Cl). (H) View of the main tunnel with a person and lateral chamber (Cl), see also G. (I) Empty structure assigned to *Megaichnus major*, person for scale. This structure was dug in the Jurassic-Cretaceous Botucatu Formation, Brazil. *(A) Photo provided by the University of Michigan Museum of Paleontology. (B) Photo courtesy of Daniel Hembree. (C and D) Photos courtesy of Daniel Hembree. (E) Modified from Cardonatto et al. (2023). (F–H) Photos F, H and I are courtesy of Renato Pereira Lopes. (G) Redrawn from fig. 2A of Lopes et al. (2017). (I) Photo courtesy of Heinrich Theodor Frank.*

sinuous; U- and S-shaped turns common. No chambers present (Hembree and Hasiotis, 2008).

Discussion. The ichnogenus was defined as a boxwork, which is appropriate for an ichnofabric (e.g., Bedatou et al., 2009) and not for a discrete trace fossil (for example, a burrow) that can be defined as an ichnogenus. The composing cylindrical burrow fills lack any bioglyphs but can be covered by thin rhizoliths. They are described as 15–30 mm burrows, but the figured material (fig. 6D–F of Hembree and Hasiotis, 2008) averages 13.5 mm wide (range 11–17 mm, n = 11). This is a very small diameter for a vertebrate burrow (e.g., Krapovickas et al., 2013; Cardonatto and Melchor, 2021b), and no key feature for identification as of vertebrate origin is proposed. Hembree and Hasiotis (2008) suggested a small reptile as potential producer and indicate that they are similar to extant amphisbaenian burrows. *Kladosystemites* is herein interpreted as an ichnofabric of uncertain vertebrate origin and is considered a nomen dubium.

Ichnogenus *Maneraichnus* Cardonatto et al., 2023.

Fig. 12.8E.

Type ichnospecies. *Maneraichnus pampeanum* Cardonatto et al., 2023.

Diagnosis. Interconnected burrow system composed of long entrance ramps, leading to a main chamber that is connected by tunnels or ramps to secondary chamber/s. Chambers commonly located at ramifications, and chamber volume at least one-third of the tunnel and ramp volume. Burrow cross section subcircular and ramifications commonly at straight angle. Occasional presence of surface ornamentation in the form of three claw traces, paired incisor traces and/or bilobed bottom. Distinguished from *Alezichnos* Gobetz, 2006 by the presence of chambers at ramifications and entrance ramps; from *Kladosystemites* Hembree and Hasiotis, 2008 by the presence of chambers and lack of shafts; and from *Yaviichnus* Guerrero-Arenas et al., 2020 because of the presence of a long entrance ramp and lack of shafts (Cardonatto et al., 2023).

Discussion. The ichnogenus was described from the Upper Miocene-Lower Pliocene Cerro Azul Formation of southwestern Buenos Aires Province, Argentina. *Maneraichnus* is distinguishable from other similar ichnogenera by a combination of features. This is particularly true regarding the significant development of entrance ramps, the presence of chambers at bifurcations, the significant volume occupied by chambers in the burrow system and the absence of shafts. A rodent producer is indicated by the presence of interconnected burrow systems with a subcircular cross section and by paired, blunt, incisor traces. The large ratio of chamber volume to tunnel/ramp volume (0.45) is indicative of a communal species, and a body mass of 0.9–4.6 kg is inferred for the producer. Considering the mentioned constraints, the fossorial mammals of the Cerro Azul Formation, and comparison with extant plains vizcacha burrows systems, *Maneraichnus* was likely produced by *Lagostomus* (Late Miocene-Recent) (Cardonatto et al., 2023).

Ichnogenus *Megaichnus* Lopes et al., 2017.

Fig. 12.8F–I.

Type ichnospecies. *Megaichnus minor* Lopes et al., 2017.

Diagnosis. Large (at least 0.6 m in diameter), subhorizontal straight to slightly sinuous tunnels, single or branched, with subcircular to subelliptical outline. Restricted to continental settings, in areas with shallow to relatively high slope. Produced on firmgrounds and hardgrounds that include consolidated sands, weathered volcanic rocks, and sandstones.

Digging marks in the form of long and shallow grooves along the walls and ceiling are common features (Lopes et al., 2017).

Discussion. The Quaternary ichnogenus was erected for flattened elliptical burrows (either with its sediment infill or empty) with type material with an average Dv/Dh ratio of 0.63 (Buchmann et al., 2009). Two ichnospecies were recognized: *M. minor* (Fig. 12.8F–H) and *Megaichnus major* (Fig. 12.8I). The former is 0.6–1.5 m wide, and the latter can reach up to 4 m wide (although no detailed measurements were offered). When the fossil burrows are empty or partially filled with sediment, two or three parallel grooves were identified in the burrow walls, which are interpreted as digging traces (Fig. 12.8F). Also, possible osteoderm and fur impressions were identified. The smaller burrows were probably produced by large dasypodids and pampatheriids (e.g., *Propraopus* and *Pampatherium*), whereas larger burrows were attributed to ground sloths (Lopes et al., 2017). These structures occur in consolidated sands, sandstones, and weathered granitic and basaltic rocks ranging in age from the Precambrian to the late Cenozoic (Lopes et al., 2017). Excavation in sandstones and volcanic rocks is at first sight puzzling, however, only *M. major* was found in hardgrounds. Some of these substrates were weakly cemented sandstones, strongly weathered rocks or rocks having structural weakness planes that likely facilitated excavation by large animals (R. Lopes, written communication, 2024). The large size of unfilled specimens of *M. major* is probably the result of roof collapse, modification by running water, enlargement of existing cavities by ground sloths, or even human modification.

Ichnogenus *Megapermichnus* Jenny and Jenny-Deshusses, 1978.

Type ichnospecies: *Megapermichnus aliabadensis* Jenny and Jenny-Deshusses, 1978.

Description. Slightly flattened and sinuous tube having average diameter of 30 cm, a few meters long, forming a low angle with the bottom of the beds. "Y"-shaped ramifications and a "living chamber" were recorded. Trace ends that are not broken disappear regularly toward the bottom of the bed. The surface of the tubes does not present either striation or segmentation (translated of the French from Jenny and Jenny-Deshusses, 1978).

Discussion. The ichnogenus was defined for Upper Permian Ruteh Formation of northern Iran. The unit is composed of limestones and marls deposited in a shallow marine environment (Jenny and Jenny-Deshusses, 1978). The authors suggested that the probable producers were amphibians or crossopterygian fishes. The attribution to a vertebrate is based on the large size and the inferred environmental setting; however, no comparison with any modern biogenic structure is offered. The bifurcations were not illustrated, and the living chamber is an interpretation of a localized enlargement in the tunnel diameter.

Ichnogenus *Megatambichnus* Martens, 2001a.

Fig. 12.9D.

Type ichnospecies. None proposed.

Description. The name was used by Martens (2001a, b) to refer to burrows or digging traces found in the lower Permian Tambach Formation of Germany. On bedding planes, curved digging traces appear in spiral-shaped "digging passages" with diameters from 20 to 30 cm. Martens (2005) again used the name to refer to burrow cast with a spiral architecture and a terminal chamber, although it was not illustrated. Nyakatura et al. (2015) mentioned the name to refer a burrow structure and considered that the reason for very

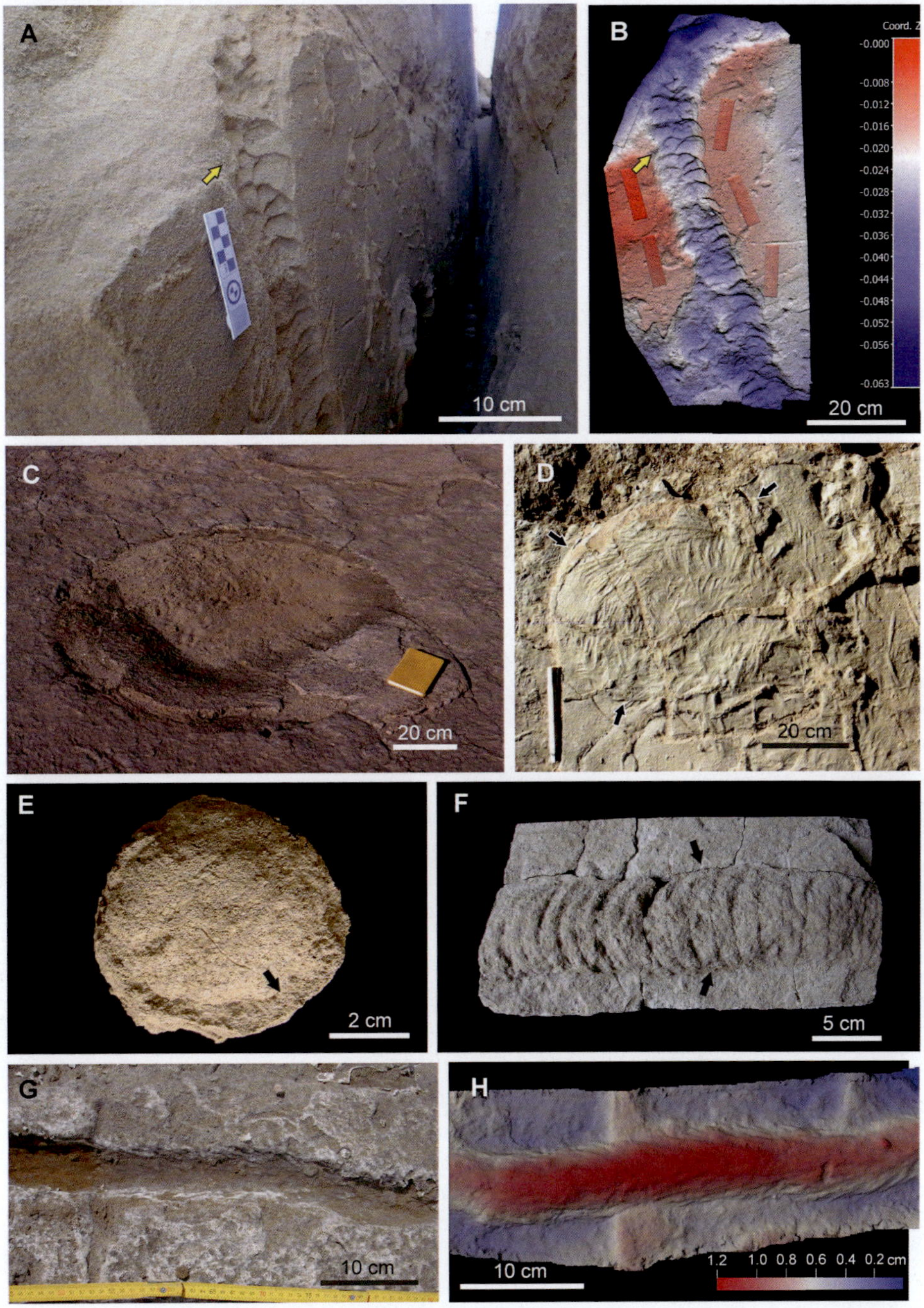

FIGURE 12.9 Holotype and topotype material of vertebrate burrow ichnotaxa (cont.). (A and B). Holotype of *Natatorichnus subarenosa* preserved as negative hyporelief in Pleistocene calcarenites from the of the Bredasdorp Group, Cape coast of South Africa. Oblique field photo (A) and digital elevation model (B). (C) Holotype of *Piscichnus brownii* from the Koobi Fora Formation (Plio-Pleistocene) of northern Kenya. (D) Specimen of *Megatambichnus* from the Lower Permian Tambach Formation of Germany. (E–H) *Nagtuichnus meuleni* from the Holocene sediments of Gran Salitral La Pampa Province (E) and the Upper Miocene-Lower Pliocene Río Negro Formation of the Río Negro Province, both from Argentina (F–H). (E) Holotype specimen (GHUNLPam 27390), cross-sectional view of a burrow fill preserved in full relief. The arrow points to the boundary between the massive external layer and the meniscate internal fill. (F) Paratype (GHUNLPam 3144) exhibiting the meniscate fill. Arrows point the boundary of the burrow fill. (G–H) Surface ornamentation on the burrow wall, including the lower surface in a field specimen (G) and digital elevation model of the polyester resin cast of the same specimen (GHUNLPam 27408). Note the chevron-like ornamentation (digging traces). *(A and B) Images courtesy of Charles Helm. (C) Photo courtesy of Craig Feibel. (D) Source of image https://www.ursaurier.de.*

good preservation of articulated specimens of tetrapods in the Tambach Formation was its preservation in burrows. Voigt (2005) considered *Megatambichnus* a nest structure.

Discussion. This is a digging/burrowing structure of uncertain architecture and morphology assigned to digging diadectids (*Diadectes absitus* or *Orobates pabsti*) due to the close association with the footprint ichnotaxon *Ichniotherium cottae* from the same locality (Martens, 2005; Voigt, 2005). This structure was recognized from the lower Permian Tambach Formation of Germany. The name is considered a nomen nudum as no type ichnospecies and complete description (including illustration) was published to date. Marchetti et al. (2024: fig. 10D–E) illustrated a couple of subhorizontal burrows with scratched surface from the Tambach Formation, but they were not assigned ichnotaxonomically. Considering the pattern of scratch traces in a bedding plane, probably part of the specimens described as *Megatambichnus* reflects a similar preservation to the type material of *Cynodontipus*. However, the architecture of these specimens from Germany is unclear to date.

Ichnogenus *Nagtuichnus* Melchor et al., 2012b.
Fig. 12.9E–H.
Type ichnospecies. *Nagtuichnus meuleni* Melchor et al., 2012b.

Diagnosis. Variably oriented, large, sinuous, unbranched, unlined, cylindrical burrow with a two-zoned backfill composed by a larger central meniscate core and an outer massive, discontinuous and thinner layer. Moderately arcuate menisci of uniform thickness, averaging one-fifth of burrow diameter. External surface of burrow fill with sets of two or three parallel ridges arranged at an oblique angle to the burrow axis. Concave surface of menisci showing groups of paired, rounded pits. In transverse section, outline of menisci composed by a semicircular upper half and a lower half composed of three straight segments; the middle segment is horizontal and the lateral segments are oblique. Burrow fill lithologically similar to the surrounding rock matrix (emended from Melchor et al., 2012b).

Discussion. *Nagtuichnus* can be distinguished from other meniscate trace fossils by the discontinuous, massive layer between a central meniscate core and the excavation boundary (Fig. 12.9E–F), and the presence of sets of parallel ridges composing a chevron pattern in the external wall (Fig. 12.9G–H). The particular cross-sectional outline and the presence of paired pits in the concave surface of *Nagtuichnus* menisci are further key differences (Melchor et al., 2012b). The ichnospecies was assigned to armadillos and particularly to relatives of Chlamyphorinae. This is the first published fossil evidence for this group of animals, and only recently body fossil material of the group has been described (Barasoain et al., 2019), this scarcity of bone remains is probably due to the poor ossification of the carapace. The ichnogenus was originally proposed for material from interdune deposits of the Holocene of the southwest La Pampa Province and the Upper Miocene-Lower Pliocene Río Negro Formation, Río Negro Province (both from Argentina). Further occurrences are from Miocene eolian sequences from Argentina, including the Middle Miocene (Langhian) Santa Cruz Formation, Chubut Province (Melchor et al., 2016), and the Lower Miocene (Burdigalian) Angastaco Formation, Salta Province (Espinoza, 2024).

Ichnogenus *Natatorichnus* Lockley et al., 2021.
Fig. 12.9A–B.
Type ichnospecies. *Natatorichnus subarenosa* Lockley et al., 2021.

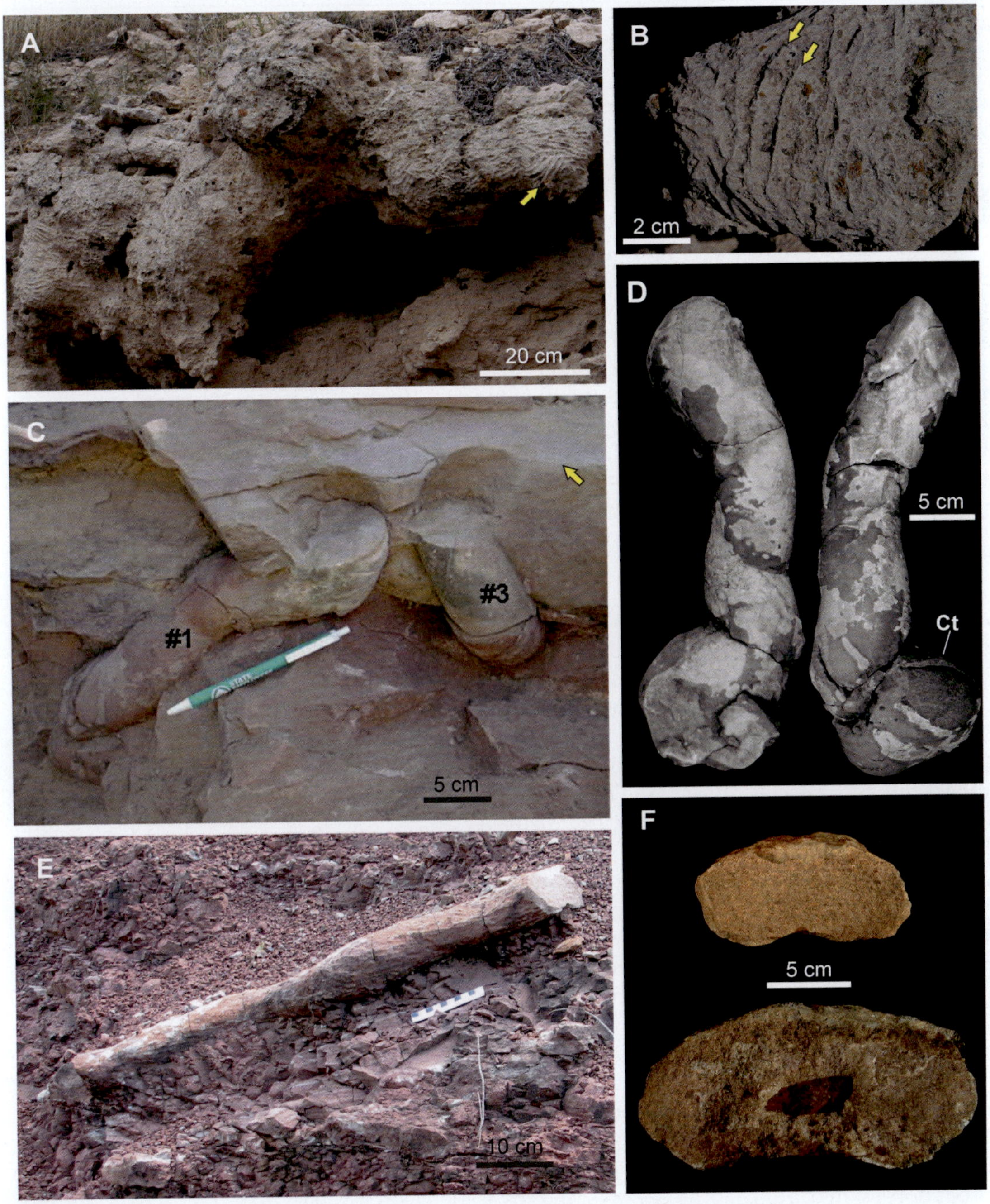

FIGURE 12.10 Holotype and topotype material of vertebrate burrow ichnotaxa (cont.). (A and B) Field views of *Polychoredrites tetrachelichnus* from Middle Miocene Pawnee Creek Formation, Colorado, USA. (A) Cast of part of a burrow system with significant surface ornamentation in the form of sets of up to three claw traces (one arrowed). (B) Detail of the surface ornamentation on the burrow cast showing two parallel ridges (arrows). (C and D) Type material of *Redondarefugium abercrombieorum* from the Montoya Point Member of the Upper Triassic Redonda Formation, New Mexico, USA. Ct: terminal chamber. (C) Field view of holotype (#1) and additional material (#3). Note position of burrows with respect to bedding (arrow). (D) Holotype specimen (NMMNH P-50409) in lateral view showing the terminal chamber (Ct). (E and F) Holotype of *Reniformichnus katikatii* (BP/6/698) from the Lower Triassic Katberg Formation, Eastern Cape, South Africa. (E) Field view of burrow cast (ramp) at 30 degrees to bedding, note surface ornamentation in the form of subparallel ridges. (F) Cross-sectional views of parts of the holotype. Note bilobed bottom. *(A and B) Photos courtesy of Daniel Hembree. (C and D) Photos courtesy of Spencer Lucas. (E and F) Photos courtesy of Emese Bordy.*

Diagnosis. Flattened, sand filled, subhorizontal, linear to meandering, transmitted trail or shallow burrow trace, up to ~10 cm wide, with local width irregularities. Traces are generally upwardly and anteriorly convex into a low roof or ridge, or with ridge or roof mid-line collapsed into a central furrow or sulcus. They are transversely divided into short *spreite* expressed as low amplitude, *en échelon*, anteriorly convex bulges, which tend to be arranged sequentially in alternating left-right patterns (Lockley et al., 2021).

Discussion. Both proposed ichnospecies differ in preservation and are interpreted to represent sand swimming at shallow depths where the roof of the burrow is preserved (*N. subarenosa*; see Fig. 12.9A–B) or collapsed (*N. sulcatus*). The ichnogenus was identified in Late Pleistocene coastal dunes from the Cape coast of South Africa. The authors use the term "burrow trace" or trail to refer to the structure, to highlight the preservation as a transmitted over trace. After common usage in ichnology (Frey, 1973) an intrastratal trail (as can be the case with *Natatorichnus*) is restricted to repichnial structures. However, the authors interpret the structure as fodichnial. By comparison with similar modern structures, the ichnogenus was interpreted to represent foraging behavior by relatives of the golden mole (*Emeritalpa*, Chrysochloridae). A previous mention of a similar structure that can be included in the ichnogenus corresponds to those figured by Ward (1988: fig. 12) from the Cenozoic Tsondab Sandstone of Namibia.

Ichnogenus *Piscichnus* Feibel, 1987.

Fig. 12.9C.

Type ichnospecies. *Piscichnus brownii* Feibel, 1987.

Diagnosis. A shallow-to-deep, steep- or gentle-sided, circular or oval in top view, dish-shaped, plug- or bowl-shaped in vertical cross section, depression or pit-like structure. The edge of the structure/excavation may be surrounded by a low/flat rim. The size ranges reflect different species or size/age of trace maker, preservational and behavioral differences (emended by Niedźwiedzki and Szrek, 2020).

Discussion. The ichnogenus was proposed for shallow, circular, dish-shaped structure, oriented concave upward in a horizontal plane for lacustrine deposits of the Pliocene Koobi Fora Formation of Kenya (Feibel, 1987). There are currently three recognized ichnospecies. *Piscichnus brownii* is a shallow concave upward structure, *Piscichnus waitemata* is a deeper subcylindrical burrow perpendicular to bedding, and *P. sanctacruzensis* is similar to *P. brownii* with a marginal rim. The ichnogenus has been recorded from the Middle Devonian to the Holocene (Table 12.2). *Piscichnus brownii* is particularly recorded in continental settings (a few cases from Middle Devonian to Holocene); *P. waitemata* was commonly found in shallow marine to offshore settings of siliciclastic and carbonate systems (common ichnospecies recorded from the Lower Triassic to Holocene), and *P. sanctacruzensis* was only recorded in ephemeral lakes and/or shallow-water lagoons (single occurrence in the Middle Devonian) (Table 12.2). *Piscichnus brownii* was interpreted as a fish nest; *P. waitemata* was preferentially assigned to predation by hydraulic jetting by rays and, in some cases, by walruses (Gingras et al., 2007 compiled a listing of possible modern producers of similar structures); and *P. sanctacruzensis* was tentatively considered a nesting or feeding structure by fishes. Shorebirds are also potential producers of *Piscichnus*, as some feeding structures of modern gulls produced by foot stamping resemble this ichnogenus (Melnyk et al., 2022). Small specimens (1–3 cm wide) assigned to *Piscichnus* by Abdlmutalib et al. (2022) are probably comparable with *Skolithos* or Siphonichnidae Knaust, 2015.

TABLE 12.2 Compilation of occurrences of the ichnogenus *Piscichnus*.

Ichnospecies	Age	Formation/ locality	Paleoenvironment	Interpretation	Source
Piscichnus brownii					
Pisichnus isp. (= *Piscichnus brownii*)	Pleistocene	Unnamed/New England, USA	Glaciolacustrine	Fish nest	Benner et al. (2009)
P. brownii	Plio-Pleistocene	Koobi Fora Fm/ Kenya	Lacustrine	Fish nest	Feibel (1987)
P. brownii	Middle Devonian (Eifelian)	Wojciechowice Fm/Holy Cross Mountains, Poland	Ephemeral lakes and/or shallow-water lagoons	Fish nest or feeding structure	Lucas (2015), Niedźwiedzkia and Szrekb (2020)
Piscichnus waitemata					
Piscichnus isp. (=*waitemata)*	Late Quaternary	Conway Flat sequence/New Zealand	Distal embayment facies in a fan delta	Ray feeding structure	Ekdale and Lewis (1991)
Piscichnus waitemata(?)	Late Pleistocene	Fuwayrit Fm/Abu Dhabi, U.A.E.	Carbonate foreshore	Giant burrows of unknown producer	Kirkham and Evans (2008)
Piscichnus isp. (=*waitemata*)	Pleistocene	Gravel terrace/ Willapa Bay, Washington, USA	Flood tidal delta	Predation using hydraulic jetting by walruses	Gingras et al. (2007)
P. waitemata	Late Pliocene	Capodarso Fm/ Sicily	Cool water carbonate ramp	Ray feeding trace (¿)	Massari and D'Alessandro (2010)
P. waitemata	Pleistocene	Cape Arkhangelos Fm/Rhodes Is., Greece	Lower shoreface in bioclastic-dominated carbonate wedge	Ray foraging structures	Nielsen et al. (2020)
P. waitemata	Pliocene	Touril Volcano-sedimentary complex/Azores Archipielago	Intertidal zone to fair-weather wave base in a tropical climate	Feeding strutures by ray fishes	Uchman et al. (2018)
P. waitemata	Early-Middle Miocene	Yonaguni Fm/ Yonaguni-jima Is., Japan	Inner shelf and lower shoreface	Ray feeding trace	Kotake and Nara (2002), Kotake (2007)
P. waitemata	Early Miocene	Taliao Fm/ Taiwan	Shoreface to offshore environment	Stingray feeding/ predation on thalassinidean crustaceans	Löwemark (2015)
P. waitemata	Early Miocene	Cape Rodney/ New Zealand	Shallow marine, high energy	Hydraulic jetting by rays	Gregory (1991)

(*Continued*)

TABLE 12.2 Compilation of occurrences of the ichnogenus *Piscichnus*.—cont'd

Ichnospecies	Age	Formation/ locality	Paleoenvironment	Interpretation	Source
Piscichnus isp. (=*waitemata*)	Oligocene	Alzey Fm/ Germany	Coastal subtidal setting	Ray feeding traces	Schindler et al. (2005)
P. waitemata ("rayhole structure")	Late Cretaceous	Blackhawk Fm/ Utah, USA	Shallow marine deltaic	Ray feeding traces	Howard et al. (1977), Kamola (1984), Nielsen et al. (2020)
Piscichnus isp (=*waitemata*)	Lower Triassic	Montney Fm/ Alberta and British Columbia, Canada	Proximal-offshore through upper-shoreface clastic ramp	Fish feeding	Zonneveld et al. (2010)
Piscichnus sanctacrucensis					
P. sanctacrucensis	Middle Devonian (Eifelian)	Wojciechowice Fm/Holy Cross Mountains, Poland	Ephemeral lakes and/or shallow-water lagoons	Fish nest or feeding structure	Niedźwiedzki and Szrek (2020)
Piscichnus isp.					
Piscichnus isp.	Middle Eocene	Peschici Fm/ Apulia Carbonate Platform, Italy	Submarine apron in carbonate platform	Bony fish (¿) traces	Belvedere et al. (2011)
Piscichnus isp.	Late Cretaceous (Maastrichtian)	Arén Fm/Spain	Coastal peritidal setting	Ray traces?	Martinell et al. (2001)
Piscichnus isp.	Late Cretaceous	Wall Creek Fm/ Wyoming, USA	Delta-front turbidite	a fish feeding structure	MacEachern et al. (2005)
Piscichnus isp	Early Cretaceous	Palombini Shale Fm/Italy	Abisal plain deposits	Fish feeding	Baucon et al. (2023)

Ichnogenus *Polychoredrites* Hembree and Hasiotis, 2008.

Fig. 12.10A–B.

Type ichnospecies. *Polychoredrites tetrachelichnus* Hembree and Hasiotis, 2008.

Diagnosis. Interconnected, subhorizontal to subvertical tunnels. Cross-section elliptical. Main tunnel curves downward to large terminal chamber. Side branches end in smaller chambers (Hembree and Hasiotis, 2008).

Discussion. The ichnogenus was proposed on the basis of field material composed of a burrow system preserved in full relief and with scratch traces on the burrow fill surface (Fig. 12.10A–B). Hembree and Hasiotis (2008) argued that the difference with *Alezichnos* was a more regular pattern in *Polychoredrites*, but this was not illustrated or otherwise supported by comparison of any descriptive parameter. Both ichnogenera were erected in the same unit (Miocene Pawnee Creek Formation) and geographic area (northern Colorado,

USA) (Gobetz, 2006; Hembree and Hasiotis, 2008). Regarding the surface ornamentation, Hembree and Hasiotis (2008) recognized sets of four parallel scratch traces, although a maximum of three was identified in the illustrated material. *Polychoredrites* is herein considered a junior synonym of *Alezichnos*. A second ichnospecies, *Polychoredrites cetinensis*, was erected for material from the Miocene Jaraba Formation of NE Spain. This ichnospecies was proposed on the basis of tubular holes in a massive sandstone bed, which display an irregular outline and a wall lacking bioglyphs. There is no filling, and the origin and age of the structure cannot be ascertained. No type material was designated. As a consequence, *P. cetinensis* is considered a nomen nudum.

Ichnogenus *Redondarefugium* Lucas et al., 2010.

Fig. 12.10C–D.

Type ichnospecies. *Redondarefugium abercrombieorum* Lucas et al., 2010.

Diagnosis. Vertical to obliquely oriented, slightly helical, unlined, passively filled hypichnial burrows with a flask-shaped terminus. Cross section round; some burrow walls have transverse, wrinkle-like ornamentation, and burrows weather into checker-like segments (Lucas et al., 2010).

Discussion. The ichnogenus was recognized in the Upper Triassic Redonda Formation of New Mexico, USA. The terminal flask-shaped chamber and cylindrical cross section distinguish *Redondarefugium* from *Torridorefugium* from the Permian of Kansas (Hembree et al., 2004, 2005) (compare Figs. 12.10D and 12.11D). *Capayanichnus* Melchor et al., 2010 from the Miocene of Northwestern Argentina is a subcylindrical burrow of subvertical orientation and similar diameter to *Redondarefugium*, but lacks chambers and displays a marked surface ornamentation. *Capayanichnus* is considered a freshwater crab burrow (Melchor et al., 2010). *Redondarefugium* was tentatively interpreted as a lungfish estivation burrow. The name "Hyperoeuthys teichonomos" proposed by Jones (2008) in an unpublished thesis was based on fossil burrows from the Upper Devonian Catskill Formation (Pennsylvania, USA), which were interpreted as a lungfish burrows. This name was not formally defined, although the material is probably comparable with *Redondarefugium*.

Ichnogenus *Reniformichnus* Krummeck and Bordy, 2018.

Fig. 12.10E–F, 12.11A–B.

Type ichnospecies. *Reniformichnus katikatii* Krummeck and Bordy, 2018.

Diagnosis. Subhorizontal burrows with reniform outlines; inclined at an angle of <30 degrees; straight to very gently sinusoidal form; with a slightly curved/arched dorsum (ceiling) and asymmetrically bilobate ventrum (base) in cross section, as well as an average aspect ratio of ~2. The wall of the burrow cast may be covered by scratches (Krummeck and Bordy, 2018).

Discussion. The ichnogenus was proposed for burrow fills in the form of simple ramps with a reniform cross section, which may present parallel surface ridges representing scratch traces. In addition to the type ichnospecies from South Africa, a second ichnospecies, *Reniformichnus australis* from Australia (Fig. 12.11A–B), was recognized (McLoughlin et al., 2020). The latter ichnospecies is distinguished by a less consistent reniform shape (Fig. 12.11B), the presence of ramps that can reach a high angle in the upper part (Fig. 12.11A), and for poorly developed surface ornamentation (Table 12.3). The original description of

FIGURE 12.11 Holotype and topotype material of vertebrate burrow ichnotaxa (cont.). (A and B) Type material of *Reniformichnus australis* from the Lower Triassic Dooralong Shale, New South Wales, Australia. (A) Field view of burrow casts (arrowed). (B) Cross-sectional view of in situ burrow cast designated as holotype (NRM X 9101a) by McLoughlin et al. (2020). (C) Topotype material of *Tambia spiralis* from the lower Permian Tambach Formation of Germany. Bromacker Quarry near Tambach-Dietharz, central Germany. Bottom of sandstone bed with mudcracks and numerous specimens of *T. spiralis* (some arrowed). The inset displays a close-up of a specimen on a slab exhibited in 1997 outside the Museum der Natur Gotha, Germany. Note cross-cutting by a mudcrack fill. (D) Side view of topotype material of *Torridorefugium eskridgensis* showing the two morphotypes included in the ichnotaxon by Hembree et al. (2005). Type I (upper right) and type II (lower left). Lower Permian Speiser Shale, Kansas, USA. (E, F and G) Topotype material of *Yaviichnus iniyooensis* from the Oligocene Chilapa Formation, Oaxaca, Mexico. (E and F) Oblique view of part of a burrow system (E) and interpretative diagram (F). Note a main chamber (C) that led to a descending and bifurcated ramp (Ra), with a lateral or secondary chamber (Cl). (G) Burrow termination (UMPLIC-381) with well-preserved, paired incisor traces (some arrowed). *(A and B) Photos courtesy of Stephen McLoughlin. (C) Photos by Ricardo Melchor. (D) Photos courtesy of Daniel Hembree. (E and G) Photos courtesy of Rosalía Guerrero-Arenas.*

TABLE 12.3 Comparison of morphological features of ichnotaxonomic value between the recognized ichnospecies of *Reniformichnus*.

Ichnotaxobase	*Reniformichnus katikatii*	*Reniformichnus australis*
Overall architecture	Simple ramp (<30 degrees), slight curvature in plan view	Single ramp (entrance up to 70 degrees passing to < 20 degrees to horizontal)
Cross-sectional shape (RDI)	0.48 ± 0.04 (n = 6)	0.43
Burrow filling	Massive	Massive to laminated
Presence of bilobed bottom	yes	Inconsistently bilobed
Presence of chamber	no	no
Primary surface ornamentation (PSO)	Scratch traces	Absent to subtly preserved scratch traces
Pattern of PSO	Ceiling uncommon sets of 2 or 3 traces, longer in laterals, two sets forming a rhomboid pattern in bottom	Crescentic traces on ceiling, none on bottom
Number of claw traces in set	2 or 3	not recognised
Average width of claw trace set (mm)	16.23	no data
Secondary surface ornamentation	not recognised	not recognised
Ramp Dh (mm)	120.8 ± 8.4 mm (n = 5)	87.0 mm

Data of type material compiled from Krummeck and Bordy (2018) and McLoughlin et al. (2020).

Reniformichnus included several examples of the ichnogenus that were restricted to the latest Permian to earliest Triassic (Krummeck and Bordy, 2018; McLoughlin et al., 2020). However, the ichnogenus is represented in late Permian (Smith et al., 2021), Jurassic (Riese et al., 2011; Melchor and Loope, 2016; Hasiotis et al., 2019), Early Cretaceous and even some extant lizards exhibit a morphology that is compatible with the ichnogenus (Melchor et al., 2023). The ichnogenus was related to cynodonts (Krummeck and Bordy, 2018; McLoughlin et al., 2020), whereas the Early Cretaceous example was tentatively assigned to sphenodontians (Melchor et al., 2023).

Ichnogenus *Tambia* Müller, 1969.

Fig. 12.11C.

Type ichnospecies. *Tambia spiralis* Müller, 1969.

Diagnosis. Shallow vertical, inclined to horizontal burrows with a semi-circular, circular, crescentic, or spiral shape, covered with sets of parallel striae (emended by Knaust et al., 2022).

Discussion. The ichnogenus currently has two ichnospecies, the type ichnospecies and *Tambia gregaria*, both considered of arthropod affinity (Knaust et al., 2022). *Tambia spiralis* has been considered an arthropod trace fossil. Seilacher (2007: plate 3) interpreted *Tambia* as an insect pupation chamber on the basis of similar structures documented for the Oligocene of Wyoming, USA. Martens (2001a,b, 2005) agreed with the arthropod origin and also suggested a probable tetrapod origin on the basis of the occurrence in the same beds (lower Permian Tambach Formation of Germany) of a basal reptile of small size (*Thuringothyris mahlendorffae*) (Müller et al., 2006). It is herein considered that a tetrapod origin is unlikely due to the presence of sets of five–seven parallel scratch traces, which is not expected for a tetrapod autopodium (see also Marchetti et al., 2024). In addition, the forelimb skeleton of the presumed producer displays three longer digits that correspond to II-IV (Müller et al., 2006); as a consequence, the sets of claw traces should be composed by three ridges if produced by the mentioned tetrapod.

Ichnogenus *Torridorefugium* Hembree et al., 2005.

Fig. 12.11D.

Type ichnospecies. *Torridorefugium eskridgensis* Hembree et al., 2005.

Diagnosis. Predominantly vertical, slightly sinuous, unlined, downward-tapering, passively filled, hypichnial burrow with an elliptical cross section (Hembree et al., 2005).

Discussion. This ichnogenus was recognized in the lower Permian Speiser Shale of Kansas, USA. *Torridorefugium* is similar to *Cornulatichnus* in overall shape, elliptical cross section and size. A difference is the lining of *Cornulatichnus*, which is absent in *Torridorefugium*. It is also distinguished from *Redondarefugium* by its cross-sectional shape (elliptical vs. subcircular) and lack of a distal enlargement. Hembree et al. (2005) distinguished two types that share a similar cross-sectional shape and downward tapering: type I and II (Fig. 12.10D). The type I is longer (average length 82.3 mm, range 30–320 mm), maximum/minimum width ratio is 1.46, and displays an average maximum width/length ratio of 0.56. Type II is shorter (average length 29.9 mm, range 20–50 mm), maximum/minimum width ratio is 1.73, and has an average maximum width/length ratio of 1.7. *Torridorefugium* is assigned to estivation by lysorophid amphibians due to the presence of partial and articulated skeletons of *Brachydectes elongatus* in some of the burrow fills (Hembree et al., 2004, 2005). The presence of remains of this amphibian in both types of *Torridorefugium*, in subvertical position in the longest (type I) and horizontally coiled in the shortest (type II) lend support to the weak morphological similarity between both types.

Ichnogenus *Yaviichnus* Guerrero-Arenas et al., 2020.

Fig. 12.11E–F.

Type ichnospecies. *Yaviichnus iniyooensis* Guerrero-Arenas et al., 2020.

Diagnosis. Interconnected burrow system composed of shafts, tunnels and two types of chambers. Large- to medium-sized superior chambers are connected to descending, radiating and inclined shafts, or to horizontal tunnels. Smaller secondary chambers are present at the end of these burrows or lateral to them. Horizontal burrows are straight, sinuous or show "C" or "H" paths. Vertical to subvertical burrows are straight, curved, sinuous or show consecutive arches resembling roughly a helical design. Tunnels and shafts are branched or simple. Horizontal burrows are wider than they are tall, whereas vertical ones are almost circular in cross section. Surface morphology of some burrows includes short, straight, paired traces on the external surface of the burrow fill (Guerrero-Arenas et al., 2020).

Discussion. This ichnogenus presents a complex morphology that includes a variety of elements and distinctive bioglyphs in the form of paired incisor traces (Fig. 12.11G). Guerrero-Arenas et al. (2020) described *Yaviichnus* as burrow systems composed of a large chamber near the top of a paleosol with shafts and radiating tunnels and ramps that may be connected to lateral or terminal (smaller) chambers (Fig. 12.11E–F). Horizontal tunnels are flattened elliptical (average Dh = 98 mm, average Dv = 84 mm, Dv/Dh = 0.86), and shafts are subcircular. No scratch traces were identified, which is a distinction with *Alezichnos*, which also lack chambers. Based on the illustrations, the large chambers appear as roughly stratiform, and it is not clear what is the original chamber outline and to what extent the original shape was distorted by carbonate cementation in the paleosol (Fig. 12.11E–F). The burrow systems were identified in the lower Oligocene Chilapa Formation of Oaxaca, southern Mexico. The unit was interpreted as fluvio-lacustrine with several paleosols. The incisor traces (Fig. 12.11G) are indicative of a rodent origin, and the most likely producer is a geomyid, in particular a species of *Gregorymys*. This is the most common fossorial rodent of the unit, their remains were found in the burrow fills, and the width of incisors matches incisor traces in the burrow walls (Guerrero-Arenas et al., 2020).

8. Concluding remarks

There are 15 ichnogenera and 23 ichnospecies of vertebrate burrows that are considered available, and 6 ichnogenera and 7 ichnospecies that are regarded as unavailable. The latter include three *nomina nuda* (*Ichnogyrus*, *Megatambichnus*, *Katarrhedrites atyphichnus*), three *nomina dubia* (*Fractisemita*, *Cynodontipus*, *Kladosystemites*), and a junior synonym (*Polychoredrites*). An analysis of the stratigraphic distribution of proposed available ichnogenera (Fig. 12.12) suggests that the Oligocene-Miocene (especially the Miocene) is the interval with most of the named ichnogenera, essentially related to mammals and, in many cases, to rodents (including *Alezichnos*, *Daimonelix*, *Katarrhedrites*, *Maneraichnus*, *Nagtuichnus* and *Yaviichnus*). *Piscichnus* is the ichnogenus with the greatest stratigraphic range: from the Middle Devonian to the Holocene, although the record is not continuous (Fig. 12.12).

The renewal of interest in the study of these biogenic sedimentary structures in the last 20 years leads to a remarkable increase in the number of proposed ichnotaxa. In addition to the formal description and naming of ichnotaxa, the number of studies that include fossil vertebrate burrows has increased significantly. It is estimated that, since 1970, approximately 180 papers dealing with vertebrate burrows were published, with about 80% released in the last 20 years. The guidelines for the study of fossil vertebrate burrows and basic techniques are becoming established. Photogrammetric techniques (or other scanning techniques) are useful for documenting type material, especially when the field specimens are large and remain in the field. Tridimensional models can be printed and deposited in a paleontological repository to comply with the recommendations of the ICZN (Ride et al., 2012). A set of ichnotaxobases is proposed herein for application in the erection of new ichnotaxa. Interpretation of new ichnotaxa should consider the criteria suggested herein for the identification of vertebrate

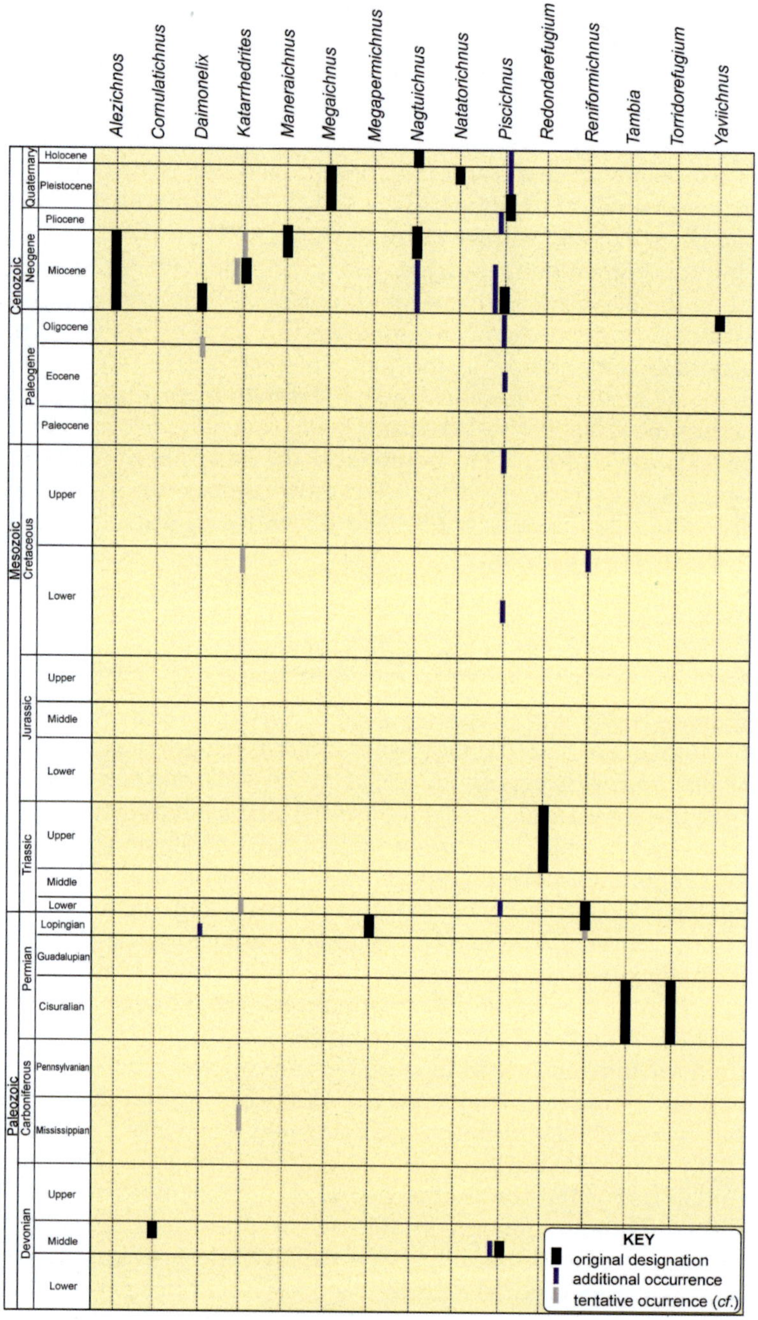

FIGURE 12.12 Compilation of the stratigraphic distribution of ichnotaxa of vertebrate burrows considered available herein. *Source of data:* Alezichnos, *Gobetz (2006), Gobetz and Martin (2006), and Platt and Smith (2023);* Cornulatichnus, *Carroll and Trewin (1995);* Daimonelix, *Martin and Bennett (1977);* Katarrhedrites, *Hembree and Hasiotis (2008), Modesto and Botha-Brink (2010), Catena et al. (2017), Yoshida and Carpenter (2020), Stimson et al. (2022), and Platt and Smith (2023);* Maneraichnus, *Cardonatto et al. (2023);* Megaichnus *Lopes et al. (2017);* Megapermichnus, *Jenny and Jenny Deshusses (1978);* Nagtuichnus, *Melchor et al. (2012b, 2016), and Espinoza (2024);* Natatorichnus, *Lockley et al. (2021);* Piscichnus, *see Table 12.2;* Redondarefugium, *Lucas et al. (2010);* Reniformichnus, *Krummeck and Bordy (2018), McLoughlin et al. (2020), Smith et al. (2021), Melchor et al. (2023);* Tambia, *Martens (2001a);* Torridorefugium, *Hembree et al. (2005);* Yaviichnus, *Guerrero-Arenas et al. (2020).*

burrows. Diagenetic effects, mainly cementation, can obscure the original shape and surface ornamentation of burrow systems, as a consequence it is recommended that this type of material is not used to erect new ichnotaxa. Interpretation of the producer and burrow taphonomy are greatly aided by neoichnologic observations and detailed descriptions of the nature and organization of the burrow fill. For the next decade, we envisage the continued recognition of new vertebrate burrow morphologies and advances in the stratigraphic, sedimentologic, and paleoecological utility of this type of trace fossils.

Acknowledgments

We are indebted to the volume conveners Spencer G. Lucas, Adrian P. Hunt, and Hendrik Klein for the invitation to contribute this chapter. Many persons kindly provided information and/or images of fossil burrows that were used in this chapter. We thank the following people and institutions: Katrina Gobetz (James Madison University, Virginia, USA), Abeer Eladany (Museums and Special Collections, University of Aberdeen, Scotland), Bruce S. Lieberman (Department of Ecology & Evolutionary Biology, Biodiversity Institute, University of Kansas, USA), Anne-Lise Charruault and Mehdi Mouana (Institut des Sciences de l'Evolution de Montpellier, Université de Montpellier, France), David Loope and Robert Hunt Jr. (University of Nebraska–Lincoln, Nebraska, USA), Stephen Hasiotis (University of Kansas, Kansas, USA), Daniel Hembree (Department of Earth and Planetary Sciences, University of Tennessee, Tennessee, USA), Megan Sims (Collection Manager, Vertebrate Paleontology, KU Biodiversity Institute & Natural History Museum, University of Kansas, Kansas, USA), Adam N. Rountrey (University of Michigan Museum of Paleontology, Michigan, USA), Renato Pereira Lopes and Heinrich Theodor Frank (Universidade Federal do Rio Grande do Sul, Rio Grande do Sul, Brazil), Joern Geletneky (Thüringischen Geologischen Vereins, Germany), Charles Helm (African Center for Coastal Palaeoscience, Nelson Mandela University, Gqeberha, South Africa), Craig S. Feibel (Department of Earth and Planetary Sciences, Rutgers University, New Jersey, USA), Spencer G. Lucas (New Mexico Museum of Natural History and Science, New Mexico, USA), Emese Bordy (University of Cape Town, Cape Town, South Africa), Stephen McLoughlin (Swedish Museum of Natural History, Stockholm, Sweden), Francisco Rodríguez-Tovar (Department of Stratigraphy and Paleontology. Universidad de Granada, Spain) and Rosalía Guerrero-Arenas (Laboratorio de Paleobiología, Universidad del Mar, Oaxaca, Mexico). Víctor E. Castillo (INCITAP, La Pampa, Argentina) produced the digital elevation model of *Cynodontipus polythrix* (Fig. 12.7D). This work was funded by project PICT-2019 0114 from the Agencia Nacional de Promoción Científica y Tecnológica, project PIP 2021–23 0146 from the Consejo Nacional de Investigaciones Científicas y Técnicas, and project G22 from Universidad National de La Pampa to RNM.

References

Abba, A.M., Udrizar Sauthier, D.E., Vizcaíno, S.F., 2005. Distribution and use of burrows and tunnels of *Chaetophractus villosus* (Mammalia, Xenarthra) in the eastern Argentinean pampas. Acta Theriol. 50, 115–124.

Abbassi, N., Khoshyar, M., Lucas, S.G., Esmaeili, F., 2021. Extensive vertebrate tracksite from the upper Red Formation (middle-late Miocene), west Zanjan, northwestern Iran. Fossil Rec. 24, 101–116.

Abdlmutalib, A.J., Ayranci, K., Yassin, M.A., Hussaini, S.R., Abdullatif, O.A., Humphrey, J.D., 2022. Impact of sedimentary fabrics on small-scale permeability variations within fine-grained sediments: early Silurian Qusaiba Member, Northern Saudi Arabia. Mar. Petrol. Geol. 139.

Abel, O., 1927. Das daemonelix-problem. Palaeontol. Z. 9, 179–180.

Andersen, D.C., 1982. Below-ground herbivory: the adaptive value of geomyid burrows. Am. Nat. 119, 18–28.

Auffenberg, W., 1983. The burrows of *Varanus bengalensis*: characteristic and use. Record Zool. Surv. India 80, 375–385.

Barasoain, D., Tomassini, R.L., Zurita, A.E., Montalvo, C.I., Superina, M., 2019. A new fairy armadillo (Cingulata, Chlamyphorinae) from the upper Miocene of Argentina: first fossil record of the most enigmatic Xenarthra. J. Vertebr. Paleontol. 39, e1716778.

Barbour, E.H., 1892. Notes on a New Order of Gigantic Fossils, vol. 1. University Studies (University of Nebraska), pp. 301–313.

Barbour, E.H., 1894. Additional notes on the new fossil, *Daimonelix*. Its Mode of Occurrence, its Gross and Minute Structure. In: University Studies, II. University of Nebraska, pp. 1–124.

Barbour, E.H., 1895. Is *Daemonelix* a burrow? A reply to Dr. Theodor Fuchs. Am. Nat. XXIX, 517–527.

Barbour, E.H., 1897. Nature, structure, and phylogeny of *Daemonelix*. GSA Bulletin 8, 305–314.

Baucon, A., Ferretti, A., Fioroni, C., Pandolfi, L., Serpagli, E., Piccinini, A., de Carvalho, C.N., Cachão, M., Linley, T., Muñiz, F., Belaústegui, Z., Jamieson, A., Lo Russo, G., Guerrini, F., Ferrando, S., Priede, I., 2023. The earliest evidence of deep-sea vertebrates. Proc. Natl. Acad. Sci. USA 120, e2306164120.

Bedatou, E., Melchor, R.N., Genise, J.F., 2009. Complex palaeosol ichnofabrics from late Jurassic-early Cretaceous volcaniclastic successions of central Patagonia, Argentina. Sediment. Geol. 218, 74–102.

Begall, S., Burda, H., Schleich, C.E., 2007. Subterranean rodents: news from underground (introduction). In: Begall, S., Burda, H., Schleich, C.E. (Eds.), Subterranean Rodents- News from Underground. Springer-Verlag Berlin Heidelberg, pp. 3–9.

Belvedere, M., Franceschi, M., Morsilli, M., Zoccarato, P.L., Mietto, P., 2011. Fish feeding traces from middle Eocene limestones (Gargano Promontory, Apulia, Southern Italy). Palaios 26, 693–699.

Benner, J.S., Ridge, J.C., Knecht, R.J., 2009. Timing of post-glacial reinhabitation and ecological development of two New England, USA, drainages based on trace fossil evidence. Palaeogeogr. Palaeoclimatol. Palaeoecol. 272, 212–231.

Benton, M.J., 1988. Burrowing by vertebrates. Nature 331, 17–18.

Bown, T.M., Kraus, M.J., 1983. Ichnofossils of the alluvial Willwood Formation (lower Eocene), Bighorn Basin, northwest Wyoming, U.S.A. Palaeogeogr. Palaeoclimatol. Palaeoecol. 43, 95–128.

Bromley, R.G., 1970. Borings as trace fossils and *Entobia cretacea* Portlock, as an example. In: Crimes, P.T., Harper, A.S. (Eds.), Trace Fossils, Geological Journal Special Issue 3, pp. 49–90.

Bromley, R.G., 1990. Trace Fossils. Biology and Taphonomy. Unwin Hyman, Boston, 280 pp.

Bromley, R.G., 1996. Trace Fossils: Biology, Taphonomy and Applications, second ed. Chapman & Hall, London. 361 pp.

Bruch, C., 1937. Notas etológicas acerca del "tuco-tuco". Notas del Museo de La Plata 2, 1–11.

Buchmann, F.S., Lopes, R.P., Caron, F., 2009. Icnofósseis (paleotocas e crotovinas) atribuídos a mamíferos extintos no sudeste e sul do Brasil. Rev. Bras. Palaontol. 12, 247–256.

Butler, D.R., 1995. Zoogeomorphology. Animals as Geomorphic Agents. Cambridge University Press, 240 pp.

Calvo, J.M., Gil, E., Meléndez, G., 1987. *Megaplanolites ibericus* (ichnogen. et ichnosp. nov.), a new trace fossil from the Upper Jurassic (uppermost Oxfordian) of Bueña (Teruel Province, Iberian Chain, Spain). Palaeogeogr. Palaeoclimatol. Palaeoecol. 61, 199–204.

Cardonatto, M.C., 2019. Neoicnología de cuevas de mamíferos y su aplicación a ejemplos fósiles. Unpublished Doctoral Thesis. Centro Regional Universitario Bariloche, Universidad Nacional del Comahue, San Carlos de Bariloche, 360 pp.

Cardonatto, M.C., Feola, S., Melchor, R.N., 2023. Neogene communal rodent burrow systems from central Argentina. Hist. Biol. 1–15.

Cardonatto, M.C., Melchor, R.N., 2018. Large mammal burrows in late Miocene calcic paleosols from central Argentina: paleoenvironment, taphonomy and producers. PeerJ 6, e4787.

Cardonatto, M.C., Melchor, R.N., 2020. Recognition of fossil nebkha deposits: Clues from neoichnology and sedimentology. Palaios 35, 277–291.

Cardonatto, M.C., Melchor, R.N., 2021a. Environmental influence on burrow system features of a colonial and fossorial rodent. Implications for interpreting fossil tetrapod burrows. Palaios 36, 225–245.

Cardonatto, M.C., Melchor, R.N., 2021b. Tocas de vertebrados atuais e fósseis. In: Sedorko, D., Francischini, H. (Eds.), Icnologia: interações entre organismos e substratos. Editora CRV, Curitiba, Brazil, pp. 249–278.

Cardonatto, M.C., Melchor, R.N., 2023a. Burrow systems of modern subterranean rodents (Ctenomyidae): key neoichnologic features and recognition of fossil examples. Ichnos 30, 124–149.

Cardonatto, M.C., Melchor, R.N., 2023b. Neoichnology of armadillos: Keys for the paleoecological and behavioral interpretation of fossil tetrapod burrows. Palaios 38, 57–75.

Cardonatto, M.C., Melchor, R.N., Montalvo, C.I., Bedatou, E., 2018. Pleistocene? megamammal burrows in late Miocene calcareous paleosols from central Argentina, Simposio Latinoamericano de Icnología (SLIC2018). Memorias, Santa Marta, Colombia, p. 74.

Carlson, K.J., 1968. The skull morphology and estivation burrows of the Permian lungfish *Gnathorhiza serrata*. J. Geol. 76, 641–663.

Carroll, S., Trewin, N.H., 1995. *Cornulatichnus*: a new trace fossil from the Old Red Sandstone of Orkney. Scot. J. Geol. 31, 37–41.

Catena, A., Hembree, D.I., 2014a. Biogenic structures of burrowing skinks: Neoichnology of *Mabuya multifasciata* (Squamata:Scincidae). In: Hembree, D., Platt, B.F., Smith, J. (Eds.), Experimental Approaches to Understanding Fossil Organisms: Lessons from the Living- Topics in Geobiology, vol 41, pp. 343–369.

Catena, A.M., Hembree, D.I., 2014b. Swimming through the substrate: the neoichnology of *Chalcides ocellatus* and biogenic structures of sand-swimming vertebrates. Palaeontol. Electron. 17 (37A), 1–19.

Catena, A.M., Hembree, D.I., Saylor, B.Z., Anaya, F., Croft, D.A., 2017. Paleosol and ichnofossil evidence for significant Neotropical habitat variation during the late middle Miocene (Serravallian). Palaeogeogr. Palaeoclimatol. Palaeoecol. 487, 381–398.

Cei, J.M., 1949. Costumbres nupciales y reproducción de un batracio característico chaqueño (*Leptodactylus bufonius* Boul.). Acta Zool. Lilloana 8, 105–110.

Clark, W.K., 1951. Ecological life history of the Armadillo in the Eastern Edwards Plateau region. Am. Midl. Nat. 46, 337–358.

Cope, E., 1893. A supposed new order of gigantic fossils from Nebraska. Am. Nat. 27, 559–560.

Damiani, R., Modesto, S., Yates, A., Neveling, J., 2003. Earliest evidence of cynodont burrowing. Proc. R. Soc. Lond. B Biol. Sci. 270, 1747–1751.

de la Peña, M., 2013. Nidos y Reproducción de las Aves Argentinas. Museo Provincial de Ciencias Naturales "Florentino Ameghino", Santa Fe, 593 pp.

Diblin, M.C., Randazzo, A.F., Jones, D.S., 1991. *Lithoplaision ocalae*: A new trace fossil from the Ocala Limestone (Eocene), Florida. Ichnos 1, 255–260.

Dondas, A., Isla, F.I., Carballido, J.L., 2009. Paleocaves exhumed from the Miramar Formation (Ensenadan Stage-age, Pleistocene), Mar del Plata, Argentina. Quat. Int. 210, 44–50.

Doody, J.S., McHenry, C., Brown, M., Canning, G., Vas, G., Clulow, S., 2018. Deep, helical, communal nesting and emergence in the sand monitor: ecology informing paleoecology? J. Zool. 305, 88–95.

Dzenowski, N., Hembree, D.I., 2014. The neoichnology of two terrestrial Ambystomatid salamanders: Quantifying amphibian burrows using modern analogs. In: Hembree, D., Platt, B.F., Smith, J. (Eds.), Experimental Approaches to Understanding Fossil Organisms, Topics in Geobiology, vol 41. Springer, pp. 305–342.

Ebensperger, L.A., Bozinovic, F., 2000. Energetics and burrowing behaviour in the semifossorial degu *Octodon degus* (Rodentia: Octodontidae). J. Zool. 252, 179–186.

Ebensperger, L.A., Taraborelli, P., Giannoni, S.M., Hurtado, M.J., León, C., Bozinovic, F., 2006. Nest and space use in a highland population of the southern mountain cavy (*Microcavia australis*). J. Mammal. 87, 834–840.

Ekdale, A.A., Bromley, R.G., Pemberton, S., 1984. Ichnology: The Use of Trace Fossils in Sedimentology and Stratigraphy. In: Short Course Notes 15. SEPM (Society for Sedimentary Geology), Tulsa, 317 pp.

Ekdale, A.A., Lewis, D.W., 1991. Trace fossils and paleoenvironmental control of ichnofacies in a late Quaternary gravel and loess fan delta complex, New Zealand. Palaeogeogr. Palaeoclimatol. Palaeoecol. 81, 253–279.

Elissamburu, A., Dondas, A., De Santis, L., 2011. Morfometría de las paleocuevas de la "Fm." Chapadmalal y su asignación a *Actenomys* (Rodentia), *Paedotherium* (Notoungulata) y otros mamíferos fósiles hospedantes. Mastozool. Neotrop. 18, 227–238.

Ellenberger, P., 1976. Une piste avec traces de soies épaisses dans le Trias inférieur a moyen de Lodéve (Hérault, France): *Cynodontipus polythrix* nov. gen., nov. sp. les Cynodontes en France. Geobios 9, 769–787.

Ellerman, J.R., 1956. The subterranean mammals of the World. Trans. Roy. Soc. S. Afr. 35, 11–20.

Emerson, S.B., 1976. Burrowing in frogs. J. Morphol. 149, 437–458.

Espinoza, N., 2024. Icnología aplicada al refinamiento de la interpretación paleoambiental en sucesiones fluvio-eólicas. Unpublished Doctoral Thesis. Departamento de Geología, Universidad Nacional de Córdoba, 256 pp.

Feibel, C.S., 1987. Fossil fish nests from the Koobi Fora Formation (Plio-Pleistocene) of northern Kenya. J. Paleontol. 61, 130–134.

Fenner, A.L., Pavey, C.R., Michael Bull, C., 2012. Characteristics of the burrows of Slater's skink, *Liopholis slateri*. Herpetol. J. 22, 115–121.

Fernández, M.E., Vassallo, A.I., Zárate, M., 2000. Functional morphology and palaeobiology of the Pliocene rodent *Actenomys* (Caviomorpha: Octodontidae): the evolution to a subterranean mode of life. Biol. J. Linn. Soc. 71, 71–90.

Francischini, H., Dentzien-Dias, P., Battista, F., Sipp, G.S., Melo, T.P., Scherer, C.M.S., Schultz, C.L., 2023. Burrows provided shelter for tetrapods in a Permo-Triassic desert. Papers in Palaeontology 9, e1490.

Frank, H.T., Althaus, C.E., Dario, E.M., Tramontina, F.R., Adriano, R.M., Almeida, M.L., Ferreira, G.F., Nogueira, R., Breier, R., 2015. Underground chamber systems excavated by Cenozoic ground sloths in the state of Rio Grande do Sul, Brazil. Rev. Bras. Paleontol. 18, 273–284.

Frank, H.T., Buchmann, F.S., Gonçalves de Lima, L., Fornari, M., Caron, F., 2012. Cenozoic vertebrate tunnels in southern Brazil. In: Netto, R.G., Carmona, N.B., Tognoli, F.M. (Eds.), Ichnology of Latin America: selected papers, Monografías da Sociedade Brasileira de Paleontologia 2, pp. 141–158.

Frank, H.T., Lima, L.G., Gerhard, N.P., Caron, F., Buchmann, F.S.C., Fornari, M., Lopes, R.P., 2013. Description and interpretation of Cenozoic vertebrate ichnofossils in Rio Grande do Sul State, Brazil. Rev. Bras. Paleontol. 16, 83–96.

Frey, R.W., 1973. Concepts in the study of biogenic sedimentary structures. J. Sediment. Petrol. 43, 6–19.

Friedman, M., Daeschler, E.B., 2006. Late Devonian (Famennian) lungfishes from the Catskill Formation of Pennsylvania, USA. Palaeontology 49, 1167–1183.

Fuchs, T., 1893. Ueber die natur von *Daemonelix*, Barbour. Annalen Kaiserlich und Königlich Naturhistorischen Hofmuseums 1893, 91–94.

Gallardo, J.M., 1957. Las subespecies argentinas de *Bufo granulosus*. Revista del Museo Argentino de Ciencias Naturales, Zoología 3, 337–374.

Gallardo, J.M., 1972. Los vertebrados cavícolas de la provincia de Buenos Aires. IDIA suplemento 29, 55–62.

Gand, G., Demathieu, G., Montenat, C., 2007. Les traces de pas d'amphibiens, de dinosaures et autres reptiles du Mesozoïque Français : inventaire et interprétations. Palaeovertebrata 35, 1–149.

Gastaldo, R.A., Rolerson, M.W., 2008. *Katbergia* gen. nov., a new trace fossil from Upper Permian and Lower Triassic rocks of the Karoo Basin: Implications for palaeoenvironmental conditions at the P/Tr extinction event. Palaeontology 51, 215–229.

Gee, C.T., Sander, P.M., Petzelberger, B.E.M., 2003. A Miocene rodent nut cache in coastal dunes of the Lower Rhine Embayment, Germany. Palaeontology 46, 1133–1149.

Geister, J., 1998. Lebensspuren made by marine reptiles and their prey in the Middle Jurassic (Callovian) of Liesberg, Switzerland. Facies 105–124.

Genise, J.F., 1989. Las cuevas con *Actenomys* (Rodentia, Octodontidae) de la Formación Chapadmalal (Plioceno superior) de Mar del Plata y Miramar (Provincia de Buenos Aires). Ameghiniana 26, 33–42.

Genise, J.F., 2016. Ichnoentomology: Insect Traces in Soils and Paleosols. Springer International Publishing, Switzerland.

Genise, J.F., Farina, J.L., 2012. Ants and xenarthrans involved in a Quaternary food web from Argentina as reflected by their fossil nests and palaeocaves. Lethaia 45, 411–422.

Gingras, M.K., Armitage, I.A., Pemberton, S.G., Clifton, H.E., 2007. Pleistocene walrus herds in the Olympic Peninsula area: Trace-fossil evidence of predation by hydraulic jetting. Palaios 22, 539–545.

Gobetz, K.E., 2006. Possible burrows of mylagaulids (Rodentia: Aplodontoidea: Mylagaulidae) from the late Miocene (Barstovian) Pawnee Creek Formation of northeastern Colorado. Palaeogeogr. Palaeoclimatol. Palaeoecol. 237, 119–136.

Gobetz, K.E., Martin, L.D., 2006. Burrows of a gopher-like rodent, possibly *Gregorymys* (Geomyoidea: Geomyidae: Entoptychtinae), from the early Miocene Harrison Formation, Nebraska. Palaeogeogr. Palaeoclimatol. Palaeoecol. 237, 305–314.

González, E.M., Soutullo, A., Altuna, C.A., 2001. The burrow of *Dasypus hybridus* (Cingulata: Dasypodidae). Acta Theriologica 46, 53–59.

Greegor, D.H., 1974. Comparative Ecology and Distribution of Two Species of Armadillos, *Chaetophractus vellerosus* and *Dasypus novemcinctus*. Unpublished PhD Thesis. Department of Biological Sciences, University of Arizona, 114 pp.

Greenwood, P.H., 1986. The natural history of African lungfishes. J. Morphol. 190, 163–179.

Gregory, M.R., 1991. New trace fossils from the Miocene of Northland, New Zealand: *Rorschachichnus amoeba* and *Piscichnus waitemata*. Ichnos 1, 195–205.

Gregory, M.R., Ballance, P.F., Gibson, G.W., Ayling, A.M., 1979. On how some rays (Elasmobranchia) excavate feeding depressions by jetting water. J. Sediment. Res. 49, 1125–1129.

Groenewald, G.H., Welman, J., MacEachern, J.A., 2001. Vertebrate burrow complexes from the Early Triassic *Cynognathus* Zone (Driekoppen Formation, Beaufort Group) of the Karoo Basin, South Africa. Palaios 16, 148–160.

Guerrero-Arenas, R., Jiménez-Hidalgo, E., Genise, J.F., 2020. Burrow systems evince non-solitary geomyid rodents from the Paleogene of southern Mexico. PLoS One 15, e0230040.

Häntzschel, W., 1975. Part W, Miscellanea, Trace Fossils and Problematica. The Geological Society of America, University of Kansas, Boulder, Colorado; Lawrence, Kansas, 269 pp.

Hasiotis, S., Parrish, J.T., Chan, M.A., 2019. Vertebrate burrows in deposits of an eolian system, Lower Jurassic Navajo Sandstone, Moab, Utah, area, U.S.A. Hallesches Jahrbuch für Geowissenschaften 46, 32–34.

Hasiotis, S.T., Platt, B.F., Hembree, D.I., Everhart, M.J., 2007. The trace-fossil record of vertebrates. In: Trace Fossils: Concepts, Problems, Prospects, pp. 196–219.

Hembree, D.I., 2010. Aestivation in the fossil record: evidence from ichnology. Prog. Mol. Subcell. Biol. 49, 245–262.

Hembree, D.I., Hasiotis, S.T., 2006. The identification and interpretation of reptile ichnofossils in paleosols through modern studies. J. Sediment. Res. 76, 575–588.

Hembree, D.I., Hasiotis, S.T., 2008. Miocene vertebrate and invertebrate burrows defining compound paleosols in the Pawnee Creek Formation, Colorado, U.S.A. Palaeogeogr. Palaeoclimatol. Palaeoecol. 270, 349–365.

Hembree, D.I., Hasiotis, S.T., Martin, L.D., 2005. *Torridorefugium eskridgensis* (new ichnogenus and ichnospecies): Amphibian aestivation burrows from the Lower Permian Speiser Shale of Kansas. J. Paleontol. 79, 583–593.

Hembree, D.I., Martin, L.D., Hasiotis, S.T., 2004. Amphibian burrows and ephemeral ponds of the Lower Permian Speiser Shale, Kansas: evidence for seasonality in the midcontinent. Palaeogeogr. Palaeoclimatol. Palaeoecol. 203, 127–152.

Hickman, G.C., 1990. Adaptiveness of tunnel system features in subterranean mammal burrows. In: Nevo, E., Reig, O.A. (Eds.), Evolution of Subterranean Mammals at the Organismal and Molecular Levels. Wiley-Liss, New York, pp. 185–210.

Howard, J.D., Mayou, T.V., Heard, R.W., 1977. Biogenic sedimentary structures formed by rays. J. Sediment. Res. 47, 339–346.

Hunt, R.M., 1990. Taphonomy and sedimentology of Arikaree (lower Miocene) fluvial, eolian, and lacustrine paleoenvironments, Nebraska and Wyoming; A paleobiota entombed in fine-grained volcaniclastic rocks. In: Lockley, M.G., Rice, A. (Eds.), Volcanism and Fossil Biotas. Geological Society of America, Boulder, Colorado, pp. 69–111. Special Paper 244.

Hunt, R.M.J., Xiang-Xu, X.U.E., Kaufman, J., 1983. Miocene burrows of extinct bear dogs: Indication of early denning behavior of large mammalian carnivores. Science 221, 364–366.

Jackson, J.E., Branch, L.C., Villarreal, D., 1996. *Lagostomus maximus*. Mamm. Species 543, 1–6.

Jenny, J., Jenny-Deshusses, C., 1978. Sur la présence de *Megapermichnus* ichnogen nov., nouvel ichnogenre de taille géante dans le Permien de l'Elbourz (Iran). Eclogae Geol. Helv. 71, 313–319.

Joeckel, R.M., Tucker, S.T., 2013. Exceptionally well preserved latest Miocene (Hemphillian) rodent burrows from the Eastern Great Plains, United States, and a review of the burrows of North American rodents. Palaios 28, 793–824.

Johnels, A.G., Svensson, G.S.O., 1954. On the biology of *Protopterus annectens* (Owen). Arkiv för Zoologi 7, 131–158.

Johnson, L.M., Hembree, D.I., 2015. Neoichnology of the eastern spadefoot toad, *Scaphiopus holbrookii* (Anura: Scaphiopodidae): criteria for recognizing anuran burrows in the fossil record. Palaeontol. Electron. 18 (2), 1–29, 43A.

Jones, C.A., Franz, R., 1990. Use of gopher tortoise burrows by Florida mice (*Podomys floridanus*) in Putnam County, Florida. Fla. Field Nat. 18, 45–68.

Jones, W.T., 2008. Ichnotaxonomy and Paleoenvironmental Analysis of Trace Fossils in the Late Devonian Catskill Formation, North-Central Pennsylvania, USA. Unpublished MSc Thesis. Department of Geology, University of Kansas, Kansas, 251 pp.

Kamola, D.L., 1984. Trace fossils from marginal-marine facies of the Spring Canyon Member, Blackhawk Formation (Upper Cretaceous), east-central Utah. J. Paleontol. 58, 529–541.

Kent, D.M., Snell, E., 1994. Vertebrates associated with gopher tortoise burrows in Orange County, Florida. Fla. Field Nat. 22, 8–10.

Kerr, G., 1898. On the habits of *Lepidosiren*. Proc. Zool. Soc. Lond. 1898, 41–44.

Kinlaw, A., 1999. A review of burrowing by semi-fossorial vertebrates in arid environments. J. Arid Environ. 41, 127–145.

Kinlaw, A., 2006. Burrows of Semi-fossorial Vertebrates in Upland Communities of Central Florida: Their Architecture, Dispersion and Ecological Consequences. Unpublished PhD Thesis. University of Florida, 194 pp.

Kinlaw, A., Grasmueck, M., 2012. Evidence for and geomorphologic consequences of a reptilian ecosystem engineer: The burrowing cascade initiated by the Gopher Tortoise. Geomorphology 157–158, 108–121.

Kirkham, A., Evans, G., 2008. Giant burrows in the Quaternary Limestones of Futaysi Island and Al Dabb'iya, Abu Dhabi Emirate. Palaeogeogr. Palaeoclimatol. Palaeoecol. 270, 324–331.

Klein, H., Lucas, S., 2021. The Triassic Tetrapod Footprint Record. New Mexico Museum of Natural History & Science. Bulletin 83, 1–194.

Knaust, D., 2012. Trace fossil systematics. In: Knaust, D., Bromley, R.G. (Eds.), Trace Fossils as Indicators of Sedimentary Environments. Elsevier, Amsterdam (xx–xx).

Knaust, D., 2015. Siphonichnidae (new ichnofamily) attributed to the burrowing activity of bivalves: Ichnotaxonomy, behaviour and palaeoenvironmental implications. Earth Sci. Rev. 150, 497–519.

Knaust, D., Mikulas, R., Mangano, M.G., Netto, R., Buatois, L.A., 2022. The ichnospecies *Tambia gregaria* (Fritsch, 1908) comb. nov. from the Upper Ordovician of Czechia. Bull. Geosci. 97, 165–177.

Kotake, N., 2007. *Macaronichnus* isp. associated with *Piscichnus waitemata* in the Miocene of Yonaguni-jima Island, Southwest Japan. In: Miller, W.I. (Ed.), Trace Fossils:Concepts, Problems, Prospects. Elsevoer, Amsterdam, pp. 492–501.

Kotake, N., Nara, M., 2002. The ichnofossil *Piscichnus waitemata*: biogenic sedimentary structure produced by the foraging behavior using water jet. J. Geol. Soc. Jpn. 108, 1–2.

Krapovickas, V., Mancuso, A.C., Marsicano, C.A., Domnanovich, N.S., Schultz, C.L., 2013. Large tetrapod burrows from the Middle Triassic of Argentina: A behavioural adaptation to seasonal semi-arid climate? Lethaia 46, 154–169.

Krummeck, W.D., Bordy, E.M., 2018. *Reniformichnus katikatii* (new ichnogenus and ichnospecies): Continental vertebrate burrows from the Lower Triassic, main Karoo Basin, South Africa. Ichnos 25, 138–149.

Leonardi, G., Oporto, N.R., 1983. Biogenetic erosion structures (modern parrots' nests) on marine and fluvial cliffs in Southern Argentina. An. Acad. Bras. Cienc. 55, 293–295.

Lessa, E.P., Vassallo, A.I., Verzi, D.H., Mora, M.S., 2008. Evolution of morphological adaptations for digging in living and extinct ctenomyid and octodontid rodents. Biol. J. Linn. Soc. 95, 267–283.

Liu, J., Li, L., 2013. Large tetrapod burrows from the Permian Naobaogou Formation of the Daqingshan Area, Nei Mongol, China. Acta Geol. Sin. Engl. Ed. 87, 1501–1507.

Lockley, M., Meyer, C., 2010. Dinosaur Tracks and Other Fossil Footprints of Europe. Columbia University Press, New York, 323 pp.

Lockley, M.G., Helm, C.W., Cawthra, H.C., De Vynck, J.C., Perrin, M.R., 2021. Pleistocene golden mole and sand-swimming trace fossils from the Cape coast of South Africa. Quat. Res. 101, 169–186.

Locock, A.J., Bowerman, M., Li, L., Kraft, J.L., Gingras, M., 2020. Fossil, ichnofossil, or concretion: The enigma of "Dinocochlea ingens". Sediment. Geol. 399, 105619.

Lopes, R.P., Frank, H.T., Buchmann, F.S.d.C., Caron, F., 2017. *Megaichnus* igen. nov.: Giant paleoburrows attributed to extinct Cenozoic mammals from South America. Ichnos 24, 133–145.

Löwemark, L., 2015. Evidence for targeted elasmobranch predation on thalassinidean shrimp in the Miocene Taliao Formation, NE Taiwan. Lethaia 48, 227–234.

Lucas, S.G., 2015. *Thinopus* and a critical review of Devonian tetrapod footprints. Ichnos 22, 136–154.

Lucas, S.G., Gobetz, K.E., Odier, G.P., McCormick, T., Egan, C., 2006. Tetrapod burrows from the Lower Jurassic Navajo Sandstone, southeast Utah. New Mexico Museum of Natural History and Science Bulletin 37, pp. 147–154.

Lucas, S.G., Spielmann, J.A., Klein, H., Lerner, A., 2010. Ichnology of the Upper Triassic (Apachean) Redonda Formation, east-central New Mexico. New Mexico Museum of Natural History and Science Bulletin 47, pp. 1–75.

Lugn, A.L., 1941. The origin of *Daemonelix*. J. Geol. 49, 673–696.

Llanos, A.C., Crespo, J.A., 1952. Ecología de la vizcacha (*Lagostomus maximus maximus* Blainv.) en el nordeste de la provincia de Entre Ríos. Rev. Investig. Agric. 6, 289–378.

MacEachern, J.A., Bann, K.L., Bhattacharya, J.P., Howell Jr., C.D., 2005. Ichnology of Deltas: Organism Responses to the Dynamic Interplay of Rivers, Waves, Storms, and Tides. In: Giosan, L., Bhattacharya, J.P. (Eds.), River Deltas–Concepts, Models, and Examples. SEPM Society for Sedimentary Geology, pp. 49–85.

Machicote, M., Branch, L.C., Villarreal, D., 2004. Burrowing owls and burrowing mammals: are ecosystems engineers interchangeable as facilitators? Oikos 106, 527–535.

Marchetti, L., MacDougall, M.J., Buchwitz, M., Canoville, A., Herde, M., Kammerer, C.F., Fröbisch, J., 2024. Origin and early evolution of vertebrate burrowing behaviour. Earth Sci. Rev. 250, 104702.

Mares, M.A., Braun, J.K., Channell, R., 1997. Ecological observations on the octodontid rodent, *Tympanoctomys barrerae*, in Argentina. SW. Nat. 42, 488–493.

Marriott, S.B., Morrissey, L.B., Hillier, R.D., 2009. Trace fossil assemblages in Upper Silurian tuff beds: Evidence of biodiversity in the Old Red Sandstone of southwest Wales, UK. Palaeogeogr. Palaeoclimatol. Palaeoecol. 274, 160–172.

Martens, T., 2001a. Paläontologie, Biostratigraphie und Paläogeographie der Tambach-Formation (Oberrotliegend, Unterperm) im Thüringer Wald—ein Überblick. Beiträge zur Geologie von Thüringen 8, 181–212.

Martens, T., 2001b. Permokarbonische Tetrapodenfaunen in Pangaea—ein ungleicher Erforschungsgrad. Hallesches Jahrb. Geowiss. B 23, 45–68.

Martens, T., 2005. First burrow casts of tetrapod origin from the Lower Permian (Tambach Formation) in Germany. New Mexico Museum of Natural History & Science. Bulletin 30, 207.

Martin, L.D., Bennett, D.K., 1977. The burrows of the Miocene beaver *Palaeocastor*, western Nebraska, U.S.A. Palaeogeogr. Palaeoclimatol. Palaeoecol. 22, 173–193.

Martinell, J., de Gibert, J.M., Domènech, R., Ekdale, A.A., Steen, P.P., 2001. Cretaceous ray traces?: An alternative interpretation for the alleged dinosaur tracks of La Posa, Isona, NE Spain. Palaios 16, 409–416.

Masello, J.F., Pagnossin, M.L., Sommer, C., Quillfeldt, P., 2006. Population size, provisioning frequency, flock size and foraging range at the largest known colony of Psittaciformes: the Burrowing Parrots of the north-eastern Patagonian coastal cliffs. Emu Austr. Ornithol. 106, 69–79.

Massari, F., D'Alessandro, A., 2010. Icehouse, cool-water carbonate ramps: The case of the Upper Pliocene Capodarso Fm (Sicily): Role of trace fossils in the reconstruction of growth stages of prograding wedges. Facies 56, 47–58.

McClanahan, L.L., Shoemaker, V.H., Ruibal, R., 1976. Structure and function of the cocoon of a ceratophryd frog. Copeia 1976, 179–185.

McDonough, C.M., DeLaney, M.A., Le, P.Q., Blackmore, M.S., Loughry, W.J., 2000. Burrow characteristics and habitat associations of armadillos in Brazil and the United States of America. Rev. Biol. Trop. 48, 109–120.

McLoughlin, S., Mays, C., Vajda, V., Bocking, M., Frank, T.D., Fielding, C.R., 2020. Dwelling in the dead zone—Vertebrate burrows immediately succeeding the end-Permian extinction event in Australia. Palaios 35, 342–357.

Melchor, R., Perez, M., Villegas, P., Espinoza, N., Umazano, A., Cardonatto, M.C., 2023. Early Cretaceous lepidosaur (sphenodontian?) burrows. Sci. Rep. 13, 10209.

Melchor, R.N., Genise, J.F., Buatois, L.A., Umazano, A.M., 2012a. Fluvial Environments. In: Knaust, D., Bromley, R.G. (Eds.), Trace Fossils as Indicators of Sedimentary Environments, Developments in Sedimentology. Elsevier, pp. 329–378.

Melchor, R.N., Genise, J.F., Umazano, A.M., Superina, M., 2012b. Pink fairy armadillo meniscate burrows and ichnofabrics from Miocene and Holocene interdune deposits of Argentina: Palaeoenvironmental and palaeoecological significance. Palaeogeogr. Palaeoclimatol. Palaeoecol. 350–352, 149–170.

Melchor, R.N., Genise, J.F., Farina, J.L., Sánchez, M.V., Sarzetti, L., Visconti, G., 2010. Large striated burrows from fluvial deposits of the Neogene Vinchina Formation, La Rioja, Argentina: A crab origin suggested by neoichnology and sedimentology. Palaeogeogr. Palaeoclimatol. Palaeoecol. 291, 400–418.

Melchor, R.N., Loope, D.B., 2016. Communal therapsid burrows from interdune facies of the Lower Jurassic Navajo Sandstone, southern Utah, USA, Ichnia 2016. In: Fourth International Congress on Ichnology Indanha-A-Nova (Portugal), pp. 174–175.

Melchor, R.N., Umazano, A.M., Perez, M., Krause, J.M., 2016. Endemic bioturbation: distinctive Neogene eolian trace fossil assemblages dominated by large meniscate burrows from Patagonia, Argentina. In: Ichnia 2016, Fourth International Congress on Ichnology, Abstract Book, Indanha-A-Nova (Portugal), pp. 208–209.

Melnyk, S., Lazowski, C.N., Gingras, M.K., 2022. Sedimentological and ecological significance of a biodeformational structure associated with an unusual feeding behavior in gulls (*Larus* sp.). Ichnos 29, 84–92.

Melton, D.A., 1976. The biology of aardvark (Tubulidentata-Orycteropodidae). Mamm Rev. 6, 75–88.

Modesto, S.P., Botha-Brink, J., 2010. A burrow cast with *Lystrosaurus* skeletal remains from the Lower Triassic of South Africa. Palaios 25, 274–281.

Mueller, P., McCann, T., 2014. Ichnology of the Miocene Jaraba Formation, Almazán Basin, NE Spain—Morphological characteristics and paleoenvironmental implications of distinct continental trace fossils. Neues Jahrbuch Geol. Palaontol. Abhand. 274, 55–71.

Müller, A.H., 1969. Über ein neues Ichnogenus (*Tambia* n. g.) und andere Problematica aus dem Rotliegenden (Unterperm) von Thüringen. Mber. dt. Akad. Wiss. 11, 922–931.

Müller, J., Berman, D.S., Henrici, A.C., Martens, T., Sumida, S.S., 2006. The basal reptile *Thuringothyris mahlendorffae* (Amniota: Eureptilia) from the Lower Permian of Germany. J. Paleontol. 80, 726–739.

Muñiz, F., Belaústegui, Z., Cárcamo, C., Domènech, R., Martinell, J., 2015. *Cruziana*- and *Rusophycus*-like traces of recent Sparidae fish in the estuary of the Piedras River (Lepe, Huelva, SW Spain). Palaeogeogr. Palaeoclimatol. Palaeoecol. 439, 176–183.

Nascimento, D.L., Netto, R.G., Cenci, R., Villegas-Martín, J., 2024. Neoichnology of *Rhinella dorbignyi* (Bufonidae) burrows: improving the recognition and interpretation of toad burrows. Ichnos 31 (4), 317–333.

Nevo, E., 1979. Adaptive Convergence and Divergence of Subterranean Mammals. Annu. Rev. Ecol. Systemat. 10, 269–308.

Niedźwiedzki, G., Szrek, P., 2020. Non-tetrapod trace fossils from the Middle Devonian tetrapod tracksite at Zachełmie Quarry, Holy Cross Mountains, Poland. Palaeogeogr. Palaeoclimatol. Palaeoecol. 553, 109763.

Nielsen, J.K., Nara, M., Jacobsen, A.R., 2020. Bowl-shaped structures in a Pleistocene clastic carbonate wedge on the Island of Rhodes, Greece. Ichnos Int. J. Plant Anim. 27, 326–333.

Nyakatura, J.A., Allen, V.R., Lauströer, J., Andikfar, A., Danczak, M., Ullrich, H.-J., Hufenbach, W., Martens, T., Fischer, M.S., 2015. A Three-Dimensional Skeletal Reconstruction of the Stem Amniote *Orobates pabsti* (Diadectidae): Analyses of Body Mass, Centre of Mass Position, and Joint Mobility. PLoS One 10, e0137284.

Ojeda, R.A., Gonnet, J.M., Borghi, C.E., Giannoni, S., Campos, C.M., Diaz, G.B., 1996. Ecological observations of the red vizcacha rat, *Tympanoctomys barrerae* in desert habitats of Argentina. Mastozool. Neotrop. 3, 183–191.

Olsen, P.E., Et-Touhami, M., Whiteside, J.H., 2012. *Cynodontipus*: A procolophonid burrow—not a hairy cynodont track (Middle-Late Triassic: Europe, Morocco, eastern North America). In: Northeastern Section—47th Annual Meeting. GSA, p. 92.

Pearson, N.J., Gingras, M.K., Armitage, I.A., Pemberton, S.G., 2007. Significance of Atlantic sturgeon feeding excavations, Mary's Point, Bay of Fundy, New Brunswick, Canada. Palaios 22, 457–464.

Peterson, O.A., 1906. Suggestion regarding te probable origin of Daemonelix. Memoir. Carnegie Mus. 2, 185–191.

Platt, B.F., 2014. The foraging pits of the nine-banded armadillo, *Dasypus novemcinctus* (Mammalia: Xenarthra: Dasypodidae), and implications for interpreting conical trace fossils. Palaeontol. Electron. 17 (3.46A), 1–17.

Platt, B.F., Smith, J.J., 2023. Late Miocene paleoecology and paleoclimate in the central High Plains of North America reconstructed from paleopedological, ichnological, and stable isotope analyses of the Ogallala Formation in western Kansas, USA. Evolving Earth 1, 100019.

Ponomarenko, D., Ponomarenko, E., 2019. Describing krotovinas: A contribution to methodology and interpretation. Quat. Int. 502, 238–245.

Quintana, C.A., 1992. Estructura interna de una paleocueva, posiblemente de un Dasypodidae (Mammalia, Edentata), del Pleistoceno de Mar del Plata (Provincia de Buenos Aires, Argentina). Ameghiniana 29, 87–91.

Rafuse, D.J., Kaufmann, C.A., Gutiérrez, M.A., González, M.E., Scheifler, N.A., Álvarez, M.C., Massigoge, A., 2017. Taphonomy of modern communal burrow systems of the plains vizcacha (*Lagostomus maximus*, Chinchillidae) in the Pampas region of Argentina: implications for the fossil record. Hist. Biol. 1–17.

Raisanen, D.C.W., Hasiotis, S.T., 2018. New ichnotaxa of vertebrate burrows from the Salt Wash Member, Upper Jurassic Morrison Formation, south-eastern Utah (USA). Ann. Soc. Geol. Pol. 88, 181–202.

Reichman, O.J., Smith, S.C., 1990. Burrows and burrowing behavior by mammals. Curr. Mammal. 2, 197–244.

Reig, O., 1970. Ecological notes on the fossorial octodont rodent *Spalacopus cyanus* (Molina). J. Mammal. 51, 592–601.

Reolid, M., Cardenal, J., Rodríguez-Tovar, F.J., Alcalá, L., 2023. Environmental interpretation of giant trace fossils from the lower Kimmeridgian (Upper Jurassic) of the Iberian Chain (Spain) from lithofacies, microfossil and ichnofossil assemblages. Lethaia 56.

Ride, W.D.L., Cogger, H.G., Dupuis, C., Kraus, O., Minetti, A., Thompson, F.C., Tubbs, P.K., 2012. International Code of Zoological Nomenclature. https://www.iczn.org/the-code/the-code-online/.

Riese, D.J., Hasiotis, S.T., Odier, G.P., 2011. Synapsid burrows and associated trace fossils in the Lower Jurassic Navajo Sandstone, Southeastern Utah, U.S.A., indicates a diverse community living in a wet desert ecosystem. J. Sediment. Res. 81, 299–325.

Roig, F.A., Roig-Juñent, S., Corbalán, V., 2009. Biogeography of the Monte Desert. J. Arid Environ. 73, 164–172.
Romer, A.S., Olson, E.C., 1954. Aestivation in a Permian lungfish. Breviora 30, 1–8.
Rood, J.P., 1970. Ecology and social behavior of the desert cavy (*Microcavia australis*). Am. Midl. Nat. 83, 415–454.
Rood, J.P., 1972. Ecological and behavioural comparisons of three genera of Argentine cavies. Anim. Behav. Monogr. 5, 3–83.
Sanders, J., Davies, M., 1984. Burrowing behavior and associated hindlimb myology in some Australian hylid and leptodactylid frogs. Aust. Zool. 21, 123–142.
Scheibler, E., Liu, W., Weinandy, R., Gattermann, R., 2006. Burrow systems of the Mongolian gerbil (*Meriones unguiculatus* Milne Edwards, 1867). Mamm. Biol. 71, 178–182.
Schindler, T., Poschmann, M., Wuttke, M., 2005. Chondrichthyan feeding depressions in a subtidal coastal environment from the Mainz Basin (Oligocene, SW Germany). Neues Jahrbuch fur Geologie und Palaontologie Abhandlungen 237, 29–39.
Schmidt, B.K., Branch, W.R., 2005. Nests and eggs of the Black-headed Bee-eater (*Merops breweri*) in Gabon, with notes on other bee-eaters. Ostrich J. Afr. Ornithol. 76, 80–81.
Schultz, C.B., 1942. A review of the *Daimonelix* problem. University of Nebraska Studies—Studies in Science and Technology, 2, pp. 1–30.
Seilacher, A., 2007. Trace Fossil Analysis. Springer, Berlin, 226 pp.
Shimer, H.W., 1903. Adaptations to aquatic, arboreal, fossorial and cursorial habits in mammals. III. Fossorial adaptations. Am. Nat. 37, 819–825.
Sidor, C.A., Miller, M.F., Isbell, J.L., 2008. Tetrapod burrows from the Triassic of Antarctica. J. Vertebr. Paleontol. 28, 277–284.
Smith, J.J., Platt, B.F., Ludvigson, G.A., Thomasson, J.R., 2011. Ant-nest ichnofossils in honeycomb calcretes, Neogene Ogallala Formation, High Plains region of western Kansas, U.S.A. Palaeogeogr. Palaeoclimatol. Palaeoecol. 308, 383–394.
Smith, R.M.H., 1987. Helical burrow casts of therapsid origin from the Beaufort Group (Permian) of South Africa. Palaeogeogr. Palaeoclimatol. Palaeoecol. 60, 155–169.
Smith, R.M.H., Angielczyk, K.D., Benoit, J., Fernandez, V., 2021. Neonate aggregation in the Permian dicynodont *Diictodon* (Therapsida, Anomodontia): Evidence for a reproductive function for burrows? Palaeogeogr. Palaeoclimatol. Palaeoecol. 569, 110311.
Soibelzon, L.H., Pomi, L.H., Tonni, E.P., Rodriguez, S., Dondas, A., 2009. First report of a South American short-faced bears' den (*Arctotherium angustidens*): palaeobiological and palaeoecological implications. Alcheringa 33, 211–222.
Stimson, M., Hinds, S., King, O., McLean, D., Dafoe, L., MacRae, R.A., Park, A., 2022. Discovery of Visean to early Serpukhovian (Mississippian) aged tetrapod burrows from Midland and Lepreau Falls, New Brunswick, Canada: an ichnological, palynological and seismic investigation of the Millstream Subbasin Atlantic Geoscience, 58, pp. 32–33 (Abstracts).
Storm, L., Needle, M.D., Smith, C.J., Fillmore, D.L., Szajna, M., Simpson, E.L., Lucas, S.G., 2010. Large vertebrate burrow from the Upper Mississippian Mauch Chunk Formation, eastern Pennsylvania, USA. Palaeogeogr. Palaeoclimatol. Palaeoecol. 298, 341–347.
Sues, H.D., Olsen, P.E., 2015. Stratigraphic and temporal context and faunal diversity of Permian-Jurassic continental tetrapod assemblages from the Fundy rift basin, eastern Canada. Atl. Geol. 51, 139–205.
Sutton, M., Rahman, I., Garwood, R., 2014. Techniques for Virtual Palaeontology. John Wiley & Sons, Ltd, Chichester.
Tanner, L., Lucas, S., 2008. Tetrapod trace fossils from lowermost Jurassic strata of the Moenave Formation, northern Arizona, USA. Volumina Jurassica 6, 133–142.
Taylor, P.D., Sendino, C., 2011. A new hypothesis for the origin of the supposed giant snail *Dinocochlea* from the Wealden of Sussex, England. Proc. Geologists' Assoc. 122, 492–500.
Uchman, A., Torres, P., Johnson, M.E., Berning, B., Ramalho, R.S., Rebelo, A.C., Melo, C.S., Baptista, L., Madeira, P., Cordeiro, R., Ávila, S.P., 2018. Feeding traces of recent ray fish and occurrences of the trace fossil *Piscichnus waitemata* from the Pliocene of Santa Maria Island, Azores (Northeast Atlantic). Palaios 33, 361–375.
Varricchio, D.J., Martin, A.J., Katsura, Y., 2007. First trace and body fossil evidence of a burrowing, denning dinosaur. Proc. Biol. Sci. 274, 1361–1368.
Vassallo, A.I., Echeverría, A., 2009. Evolution of brain size in a highly diversifying lineage of subterranean rodent genus *Ctenomys* (Caviomorpha: Ctenomyidae). Brain Behav. Evol. 73, 138–149.

Verdolin, J.L., Lewis, K., Slobodchikoff, C.N., 2008. Morphology of burrow systems: A comparison of Gunnison's (*Cynomys gunnisoni*), white tailed (*C. leucurus*), black tailed (*C. ludovicianus*), and Utah (*C. parvidens*) praire dogs. SW. Nat. 53, 201–207.

Vizcaíno, S.F., Bargo, M.S., Cassini, G.H., Toledo, N., 2016. Forma y Función en Paleobiología de Vertebrados. Editorial de la Universidad Nacional de La Plata, La Plata.

Vizcaíno, S.F., Fariña, R.A., Mazzetta, G.V., 1999. Ulnar dimensions and fossoriality in armadillos. Acta Theriol. 44, 309–320.

Vizcaíno, S.F., Milne, N., 2002. Structure and function in armadillo limbs (Mammalia: Xenarthra: Dasypodidae). J. Zool. 257, 117–127.

Vizcaíno, S.F., Zárate, M.A., Bargo, M.S., Dondas, A., 2001. Pleistocene burrows in the Mar del Plata area (Argentina) and their probable builders. Acta Paleontologica Polonica 46, 289–301.

Voigt, E., Gittins, G., 1977. The "Lackfilm" method for collecting sedimentary peels: Archaeological applications. J. Field Archaeol. 4, 449–457.

Voigt, S., 2005. Die Tetrapodenichnofauna des kontinentalen Oberkarbon und Perm im Thüringer Wald—Ichnotaxonomie, Paläoökologie und Biostratigraphie. Cuvillier Verlag, Göttingen, 308 pp.

Voigt, S., Schneider, J.W., Saber, H., Hminna, A., Lagnaoui, A., Klein, H., Brosig, A., Fischer, J.A.N., 2011. Complex tetrapod burrows from the Middle Triassic red beds of the Argana Basin (Western High Atlas, Morocco). Palaios 26, 555–566.

Voorhies, M.R., 1975. Vertebrate burrows. In: Frey, R.W. (Ed.), The Study of Trace Fossils. Springer-Verlag, New York, pp. 325–350.

Walker, L.W., 1952. Underground with burrowing owls. Nat. Hist. 61, 78–81.

Ward, J.D., 1988. Eolian, fluvial and pan (playa) facies of the Tertiary Tsondab Sandstone Formation in the central Namib Desert, Namibia. Sediment. Geol. 55, 143–162.

Wells, K.D., 2007. The Ecology and Behavior of Amphibians. University of Chicago Press, Chicago and London.

White, C.R., 2005. The allometry of burrow geometry. J. Zool. 265, 395–403.

Whitehouse, F.W., 1934. A large spiral structure from the Cretaceous beds of Western Queensland. Memoir. Queensl. Mus. 10, 203–210.

Wood, H.E., Wood, A.E., 1933. *Daemonhelix* in the Pleistocene of Texas. J. Geol. 41, 824–833.

Woodruff, D.C., Varricchio, D.J., 2011. Experimental modeling of a possible *Oryctodromeus cubicularis* (Dinosauria) burrow. Palaios 26, 140–151.

Woods, R.W., 1970. The avian ecology of a tussock island in the Flakland Islands. Ibis 112, 15–24.

Woodward, B.B., 1922. On *Dinocochlea ingens*, n.gen.et sp., a gigantic Gastropod from the Wealden Beds near Hastings. Geol. Mag. 59, 242–247.

Wu, N.C., Alton, L.A., Clemente, C.J., Kearney, M.R., White, C.R., 2015. Morphology and burrowing energetics of semi-fossorial skinks (*Liopholis* spp.). J. Exp. Biol. 218, 2416–2426.

Yelinek, K., Chin, K., 2007. Probable dung beetle burrows associated with *Daemonelix*, beaver burrows in the Miocene Harrison Formation, Nebraska, U.S.A. In: Bromley, R.G., Buatois, L.A., Mángano, M.G., Genise, J.F., Melchor, R.N. (Eds.), Sediment-Organism Interactions: A Multifaceted Ichnology. SEPM Special Publications, pp. 343–350.

Yoshida, J., Carpenter, K., 2020. Burrows of small animals from the Lower Cretaceous Cedar Mountain Formation, Utah, USA. Fossils 108, 11–21.

Zachos, J., Pagani, M., Sloan, L., Thomas, E., Billups, K., 2001. Trends, Rhythms, and Aberrations in Global Climate 65 Ma to Present. Science 292, 686–693.

Zárate, M.A., Bargo, M.S., Vizcaíno, S.F., Dondas, A., Scaglia, O., 1998. Estructuras biogénicas en el Cenozoico tardío de Mar del Plata (Argentina) atribuibles a grandes mamíferos. Rev. Assoc. Argent. Sedimentol. 5, 95–103.

Zonneveld, J.P., Gingras, M.K., Beatty, T.W., 2010. Diverse ichnofossil assemblages following the P-T mass extinction, Lower Triassic, Alberta and British Columbia, Canada: Evidence for shallow marine refugia on the Northwestern Coast of Pangaea. Palaios 25, 368–392.

CHAPTER

13

The ichnology of reproduction in fossil tetrapods

Lida Xing[1,2,3,a], *Qiyan Chen*[2,a], *Donghao Wang*[2] and *Jens N. Lallensack*[4]

[1]Frontiers Science Center for Deep-time Digital Earth, China University of Geosciences (Beijing), Beijing, China; [2]School of the Earth Sciences and Resources, China University of Geosciences (Beijing), Beijing, China; [3]Dinosaur Odyssey Science Museum, Guangxi, China; [4]Departamento de Ciência da Computação, Universidade Federal de Minas Gerais, Belo Horizonte, Brazil

1. Introduction

1.1 Reproduction in extant tetrapods and the trace fossil record

There are two main parts to the reproduction of the animals: mating and parental care (Hsieh and Plotnick, 2020). Some typical physical features exhibited by extant tetrapods during reproduction include: sexual dimorphism (Pipoly et al., 2015; including the periodic temporary acquisition of plumage change in some birds), embryonic marsupial pups after birth (Tyndale-Biscoe and Renfree, 1987), beak ornamentation associated with nestling begging behaviors (Caro et al., 2016), and specialized organs for milk production in mammals (Oftedal, 2012) and others. The traces made during various reproductive processes mainly consist of those being produced during courtship display behaviors and nest ornamentation (Sanz and García-Navas, 2011; Rubalcaba et al., 2017), but also include a wide variety of nests and their infillings for heat insulation (De Zwaan and Martin, 2018; Mainwaring et al., 2023b), as well as hairs, feathers, and excrement left during the posthatching period in the nests (Guigueno and Sealy, 2012).

Analysis of the examples aforementioned and others seems to indicate that the fossil evidence related to reproduction is dominated by the organisms themselves rather than their

[a] These authors contributed equally to this work

Vertebrate Ichnology
https://doi.org/10.1016/B978-0-443-38351-9.00013-X

traces. Most of the former appear in stages during the life process or manifest themselves as soft tissues, which are less likely to be preserved as fossils compared with hard tissue such as bones and teeth (Briggs and McMahon, 2016). Consequently, our current knowledge of reproductive behavior of fossil taxa is mainly limited to body fossils (including egg fossils), such as special display structures related to sexual dimorphism (e.g., Chiappe et al., 1999), egg structures (e.g., Tanaka et al., 2015), and embryos and their relationship to incubation and brooding (Kundrát et al., 2008; Xing et al., 2022).

Traces have received less attention than body fossils in recent research, due to the odds of preservation of fossil evidence (e.g., unburied nests are susceptible to physical weathering) and the difficulty of identifying them (e.g., traces of display behaviors can only be easily identified when there is a large dataset in a population). In contrast to reproductive behaviors in the fossil record, feeding and habitat related behaviors have been disproportionately represented, while examples of other behavioral categories, especially social ones, are considerably rarer (Hsieh and Plotnick, 2020). Some of these behaviors have only begun to receive attention in the past few years, in part because trace fossils can provide information lacking in body fossils, especially skeletal remains. For example, proposed display traces can be linked to the courtship behavior of extant ground-nesting birds in the absence of fossil corroboration of similar sexual dimorphism (Lockley et al., 2016), and modern nests and patterns of egg clutches can be used to compare the hatching patterns of fossil taxa, thereby enabling understanding of the ancient incubation patterns of eggs and preincubation parental care (Mainwaring et al., 2023a). The traces of the latter, although rarer than fossil eggs, may even be considered to be more valuable than eggs containing embryos to the study of the ethology of extinct organisms to some extent, given the diversity of specializations adapted to their incubation conditions and nesting patterns within different groups of tetrapods. Considering the complexity of the whole process of reproduction, which involves not only the physiological processes of the individuals themselves (e.g., locomotion, feeding, egg-laying) but also the social behaviors of the animals, an interindividual interaction, which both body fossils and trace fossils can only serve as indirect evidence for the existence of the actions. Examples of indirect evidence include proposed courtship display traces discussed further. The study of related trace fossils in conjunction with the reproductive behaviors of extant tetrapods can expand the sources of behavioral information on fossil taxa, effectively filling the gaps in knowledge of the body fossil record, potentially limiting the multiplicity of interpretations of certain features of body fossils.

1.2 Terminology for nesting and courtship display

The terminology for nesting employed herein mainly follows that of Vila et al. (2010a,b), Sander (2008), Mueller-Töwe et al. (2002) and Carpenter (1999).

- Egg horizon: a stratigraphic layer containing in situ remains of eggshells, eggs, or clutches.
- Site: a specific location that contains one or more clusters of eggs.
- Egg cluster: an accumulation of eggs represented by an uncertain number of clutches.

- Clutch: an accumulation of eggs produced by a female during an oviposition event, however, in fossil nests "clutch" can refer to all eggs in the nest since parentage is unknown (Varricchio et al., 2008).
- Nest: a place prepared to receive and keep eggs (Carpenter, 1999, p. 153), partial clutches are often all that survives of fossil nests.
- Nest (structure): a preserved trace fossil showing a sediment mound, depression or other disruption of the original sedimentary structure caused by excavation by an adult tracemaker. It is generally conceded that plant material was at times incorporated into the nest structure even though it is seldom preserved, and the only exception in this article is a twig nest of a bird discussed in Grellet-Tinner et al. (2012).
- Cover: is the surrounding soil and vegetation that either partially or completely hides and insulates the egg(s), nest structure, and possibly as a nest sitter.
- Hatching window: a large opening on the upper side of an in situ egg lacking shell fragments within (Cousin et al., 1994 in Carpenter, 1999, p.214). Interior shell fragments may indicate predation or crushing rather than hatching.
- Nest site: the locality where parents choose to build their nest, shows the environment for nesting and incubation.
- Nest site selection: a preference of particular nesting environment (Nagy et al., 2019).
- Nest site fidelity: a behavior of returning to the same nest site or area during breeding seasons, usually effects survival of offspring of oviparous animals (Haas, 1998; Yang et al., 2017).
- Colonial nesting: a behavior in which a large number of animals of the same species nest at the same site, to avoid predators and enhance the oversight of food (Alexander, 1974; Krebs, 1974).
- Communal nesting: a behavior of producing eggs/pups as well as sharing parental responsibilities between multiple individuals in a nest/den. This behavior occurs in social animals including birds and mammals. Communal nesting is not in relation to communal nursing (Hayes, 2000; Bertram, 2014; Murray et al., 2020).
- Courtship display: a behavior aimed to facilitate attraction and mating with the opposite sex across animals, including attracting mates of single individuals and the approval between bounded pairs, and the performance of this behavior varies with distance (Mitoyen et al., 2019). The special nest scrape display in some shorebirds should be considered as both courtship display and nesting attempts (e.g., *Charadrius hiaticula*; Pienkowski, 1984).
- Lekking: a kind of courtship display behavior, which is primarily manifested as the aggregation of displaying males. Females visit leks solely for copulation and subsequently leave to raise their offspring without any further association. It is therefore evident that they cannot gain any other resources from males (Wiley, 1991).

While eggs are considered to be body fossils, nest structures and clutch patterns are considered to be trace fossils that represent egg-laying behaviors, including number of eggs, spatial arrangement, etc. (Martin, 2014; Hogan and Varricchio, 2023). With highly varied macroscopic and microscopic shell properties, fossil eggs are classified as ootaxa rather than ichnotaxa (Mikhailov et al., 1996; Carpenter, 1999; Martin, 2014).

Institutional abbreviations: Denver Museum of Nature and Science, Denver, Colorado, USA (DMNH)

2. History of study

Fossilized skeletons and eggs associated with reproduction were known to civilizations from both the Old World and the New World, either in their own right or through mythology (Carpenter, 1999; Mayor, 2013). The first scientifically documented fossils of reproduction were egg shell fragments found in 1859, but the they were not associated with dinosaurs until Philippe Matheron's study in 1869, which proposed to attribute them to *Hypselosaurus* found at the same site, or a gigantic bird (Carpenter et al., 1996).

The first recorded fossil nest (a clutch of elongated eggs) was excavated in 1923 in the Djadokhta Formation, Flaming Cliffs site, Mongolia by the Central Asiatic Expedition led by Roy Chapman Andrews for the American Museum of Natural History (Carpenter et al., 1996; Carpenter, 1999), and initially described by Brown and Schlaikjer (1940). Subsequently, in the mid-1970s at Egg Mountain in NW Montana, USA, *Maiasaura* nests containing the first known juveniles were reported (Horner and Makela, 1979), although a number of fossilized eggs from basal reptiles and dinosaurs had been found earlier in different areas (including Mongolia, India, North America, etc.) and some partial embryos have been documented (Romer and Price, 1939; Sternberg, 1956; Lehmann, 1971; Jain, 1989; Carpenter, 1999; Zelenitsky et al., 2000). Except for a morphological description by Brown and Schlaikjer in 1940 of the Mongolian nest, reports from the 1940 to the mid 1970s lacked descriptions of the morphology and functions of fossil nests.

Based on the distribution of nests at Egg Mountain in western Montana, evidence of nest site fidelity and possible colonial nesting behavior and site fidelity of *Maiasaura* or other hadrosaurs was first discussed in 1979 and 1982. The relative positions of *Maiasaura* hatchlings, nests, and adults may imply some degree of parental care (Horner and Makela, 1979; Horner, 1982). Subsequently, more studies began to focus on the functions of dinosaur nests or semi-complete clutches, such as nest temperature controls (Coombs, 1989), nest construction methods (Kerourio, 1981), the degree of parental involvement in brooding, and egg-laying strategy (Carpenter et al., 1996, part 2.1). Most reports of tetrapod nests and clutches focused on reporting or distinguishing the clutch patterns of different ootaxa (Carpenter, 1999), instead of characterizing trace fossils separately from eggs, or using them to reconstruct the incubation to nursery process.

The discovery of two theropod nests in the late 1900s greatly enriched researchers' understanding of the incubation of dinosaurs: one from the same area as the 1923 Mongolian nest was attributed to *Citipati* (Norell et al., 1995), and the other from the Egg Mountain locality attributed to *Troodon* (Varricchio et al., 1999). The latter has preserved one of the most complete nest structures, a shallow elongated depression enclosed by a low wall or "rim" (Varricchio et al., 1999). Both of these provide evidence of partially buried clutches, monoautochronic ovulation and nest attendence in *Troodon* (Hogan and Varricchio, 2023).

For the first decade of the new century, fossils and sites with high potentials for study continued to be discovered. Multiple nests of Sauropodomorpha (Titanosauria) with rims have enabled an understanding of how the nest structures of these large dinosaurs differed from those of megapodes with sediment-filled nests (Chiappe et al., 2004). Additionally, these trough nests provided new insights into a hindfoot scratch-digging (scraping) method during the nest-building processes of gigantic titanosaurs (Fortuny et al., 2007; Vila et al., 2010a,b). A burrow with an association of adult and young within a terminal chamber provides examples of different locations where dinosaurs raised their babies in extreme environments (Thescelosauridae, Ornithischia, Varricchio et al., 2007). During this time, the investigators also began to incorporate nest features into a phylogenetic perspective: the egg-filled clutch center of *Troodon* nests was used as one of the pieces of reproductive-related evidence in the study of dinosaur-bird evolutionary sequences (Grellet-Tinner et al., 2006).

In the last decade or so, new fossil localities have continued to provide fresh insights into various taxa, especially those for which the previous fossil record was generally sparse. For example, the discovery of large numbers of eggs buried in the same layer attested to the presence of colonial nesting behavior in both pterosaurs and fossil birds (Fernández et al., 2013; Wang et al., 2017); the discovery of in situ preservation of an adult crocodylian with eggs indicating likely preincubation parental care in extant crocodiles (Hastings and Hellmund, 2015); the retention of fossil avian twig-nest building material that attests to an evolutionary relationship between sister groups (Grellet-Tinner et al., 2012); and the possible existence of the earliest known parenting behavior hypothesized by the egg structure in dinosaurs among Jurassic sauropodomorph dinosaurs (Han et al., 2023).

Also, during this period, new methods of studies of extant nests (especially bird nest with large sample sizes) that have already shifted to evolutionary analyses with large datasets (Varricchio and Jackson, 2016; Fang et al., 2018; Nagy et al., 2019; Mainwaring et al., 2023a, 2023b; Deeming, 2023). These have examined the qualitative relationships between differences in nesting patterns and habitat selection preferences (Mayr, 2017), etc. Influenced by these studies, the research on fossil tetrapod reproduction traces, especially that focused on nesting behavior, also gradually shifted from discrete morphological and behavioral studies to a systematic evolutionary mindset, such as qualitative to semiquantitative relationships of nest structure and egg clutch patterns with hatching modes in fossil taxa, especially theropods for which the evidence is abundant (Murray et al., 2020; Hogan and Varricchio, 2023). This means it is necessary to reexamine previous discoveries, combining them with the detailed compilation and reclassification of fossilized evidence on animal behavior (Hsieh and Plotnick, 2020).

3. Ichnotaxa

As mentioned in the Terminology section, nest structures and clutch patterns are not assigned to ichnotaxa. Instead, they usually occur in association with named ootaxa. Although we categorized the nest-associated structures as traces in the Terminology section, the scarcity, strong correlation with egg fossils, and the dispersion of preserved characteristics among different nest-related specimens, in particular for the nest structure itself, render the separate naming of these structures less meaningful.

However, a standardized description of these traces is still required, which could further facilitate the integration of nesting traits with phylogenetic studies. We therefore provide a list of suggested descriptive features for references, which mainly refer to Nagy et al. (2019) (modern birds), Murray et al. (2020) (modern crocodiles), Grellet-Tinner et al. (2006) and Hogan and Varricchio (2023) (extant and extinct reptiles). For some traits that are difficult to obtain accurately from trace fossils, such as nest site preferences (Nagy et al., 2019; Murray et al., 2020), we made some adjustments that are more applicable to them. Some of the features have been previously discussed along with eggs, such as clutch size, here we propose to distinguish it from the former.

(1) Quantitative features:
- The minimum length and width of nest structures and complete clutches.
- The minimum height and absolute depth. If with the nest structure, the height of the rim, such as in nests of *Maiasaura*, *Troodon* and titanosaurs (Horner and Makela, 1979; Varricchio et al., 1999; Vila et al., 2010a,b) and primary depth of the depression should be included. The latter can also be obtained from buried or semiburied nests, like those produced by titanosaurs, by measuring the height from the very bottom of the clutch to the very top of the rim.
- Clutch size/clutch volume/number of eggs. Differs from the extant definition; clutch size only shows the total number of eggs in one nest, instead of separating them according to parentage.

(2) Qualitative features:
- Oviposition: En masse, monoautochronic or single oviduct. Usually distinguished by the way the eggs are arranged in groupings.
- Egg arrangements: directed or undirected; forming distinctive shapes or not (e.g., hollow ring-shaped clutches of oviraptorids).
- Level of egg burial/nest design: exposed, partial buried, or fully buried.
- Choice of nest sites: Ground (with or without low vegetation), cavities/holes (only ground cavities), water and others. A considerable number of arboreal or flying animals construct nests that are completely detached from the sediment. These nests are located in a variety of environments, including ledges, trees, tree cavities, and walls (Nagy et al., 2019). However, given that this is not reflected in the trace fossil record, we do not include these in our considerations. For those nests that utilize specialized materials, such as the use of termite mounds in modern alligatorids (Murray et al., 2020), they should be classified as "others" and discussed separately.

(3) The presence/absence of correlated behaviors:
- With or without signs of colonial nesting. This behavior is usually related to an in situ position, a high degree of synchronicity; a large number of clutches in the same site, and clutches should be proved to be laid in a single nesting season (Horner, 1982; Fernández et al., 2013; Tanaka et al., 2019).
- With or without signs of communal nesting. This behavior is typically observed in cases with a large clutch volume/adult body mass ratio, such as those observed in troodontids and oviraptorids (Varricchio et al., 2008; Yang et al. 2016, 2019).

- With or without signs of preincubation parental care: none, attending, or brooding. Few posthatching/birthing parental care reflected in the trace fossil, so it is not considered here.

Therefore, this section only addresses two ichnotaxa, the only two that have been proposed as trace fossils representing (failed) nest building attempts/possible courtship display. Although courtship display is quite commonly found in extant tetrapods (Porges, 1998; Woolley et al., 2004; Fusani, 2008), few direct examples have been found in the fossil records. Trace fossils that record the courtship behavior have rarely been found due to the difficulty in identification and interpretation (Lockley et al., 2016; Díaz-Martínez, 2023; Moklestad and Lucas, 2023), while skeletal remains could only indirectly attest to courtship display through sexual dimorphism (Chiappe et al., 1999). This section thus discusses only two types of traces belongs to theropod dinosaurs, one of which does not quality as an independent ichnogenus due to it being highly speculative.

3.1 *Ostendichnus* Lockley et al., 2016 (Figs. 13.1A–C and 13.2A–C)

3.1.1 *Type ichnospecies*

Ostendichnus bilobatus Lockley et al., 2016

3.1.2 *Valid ichnospecies*

Ostendichnus bilobatus Lockley et al., 2016

3.1.3 *Distribution*

Initially, *Ostendichnus* was described from the Dakota Group (Cenomanian), BLM National Conservation Areas, Colorado, United States. But it may also be present in the Gates Formation (Albian), Alberta, Canada, based on a single specimen.

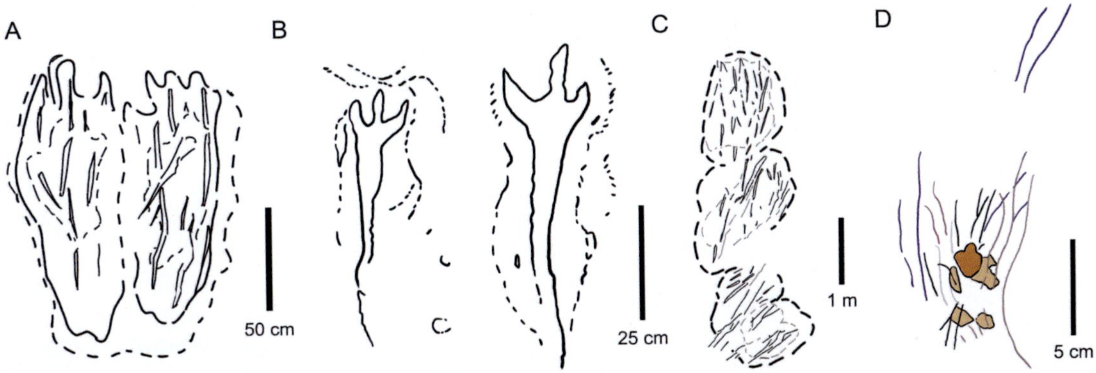

FIGURE 13.1 Interpreted outline drawings of *Ostendichnus* (A–C) and scratches of *Minisauripus* (D). The distal end points upwards. (A) Holotype from Roubideau creek site (RC) with trough-like structure. (B) Paratype from Roubideau creek site (RC) with trough-like structure. (C) *Ostendichnus* from Club Gulch site (CG) with bowl-like structure. (D) Scratches attributed to *Minisauripus*. *(A) Interpreted from Fig. 13.3B of Lockley et al. (2016). (B) Redrawn from Fig. 13.3C of Lockley et al. (2016), (C) Emended from Fig. 13.2A of Lockley et al. (2016), (D) Redrawn from Fig. 13.3A of Kim et al. (2016)*

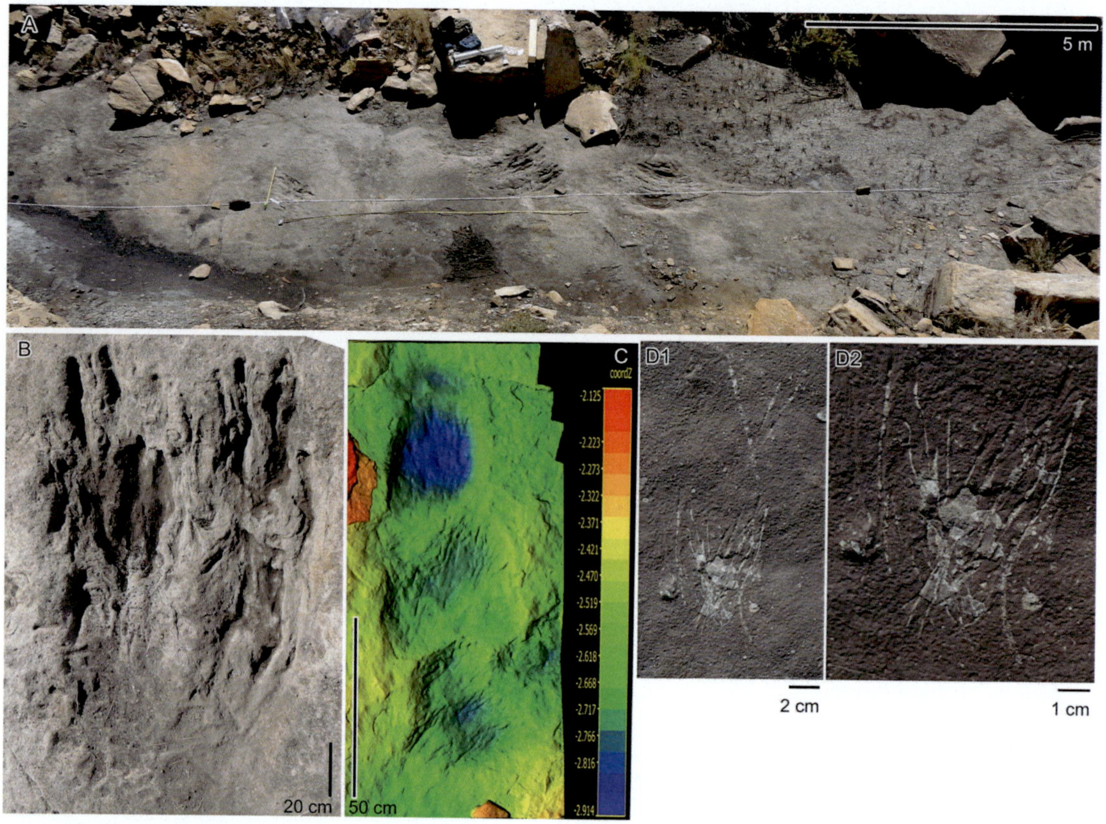

FIGURE 13.2 Photographs and depth color images of *Ostendichnus* (A–C) and scratches of *Minisauripus* (D). (A) Bedding plane of holotype in Roubideau Creek site (RC). (B) Holotype of *Ostendichnus*. (C) *Ostendichnus* from Club Gulch site (CG) with bowl-like structure. (D) scratches attributed to *Minisauripus*. *(A) Photo provided by the late Martin Lockley, (B) Photo provided by the late Martin G. Lockley, (C) Fig. 13.2A from Lockley et al. (2016). (D) Fig. 13.2 from Kim et al. (2016).*

3.1.4 Holotype

DMNH EPV.69,705, latex mold and fiberglass replica of large digging trace, within which a diagnostic theropod track occurs. (Paratypes: DMNH EPV.69,703, EPV.69,704, EPV.69,706, and EPV.69,707 latex molds and fiberglass replicas of large digging traces.)

3.1.5 Diagnosis (modified from Lockley et al., 2016)

Including a complex of parallel to subparallel scratch marks, which are bilaterally symmetrical, within large (up to 2 m in length), paired troughs (~60%) and single oval bowls (~40%) with the multiple well-defined digital scratch marks aligned with the long axis of the whole trace. Up to 10%–15% as deep as long. A few traces mostly with a single raised central ridge, separating left and right troughs, have a complete or partial tridactyl theropod tracks that is diagnostic.

3.1.6 *General description (modified form Lockley et al., 2016)*

Large, bilaterally symmetrical, scrapes-like paired troughs and single oval bowls; 0.75–2.00 m long and 0.50–1.25 m in width; depth various from 5 to 25 cm (Fig. 13.2B and C). Multiple well-defined digital scratch marks align with the whole trace, as does a raised medial ridge defining the long axis of the trace. Some scratch marks have sharp anterior terminations, indistinguishable from typical theropod digit traces. Additional features include sand crescents, where sediment was pushed back by posterior motion of digits into rearward peripheral piles, or thin apron kicked posteriorly and somewhat laterally to define the posterior end and the left and right margins of scrapes. Within some scrapes, a complete or partial theropod track is aligned with the scratch marks and the medial ridge.

3.1.7 *Discussion*

The ichnogenus *Ostendichnus* was first described by Lockley et al. (2016) with the ichnospecies *Ostendichnus bilobatus* for abundant large-scale scrapes from four separate sites within equivalent Cretaceous rock units in Colorado (South Platte and Naturita formations in the Dakota Group). the number of sites was brought to five with a total of more than 100 recognizable scrapes (Lockley et al., 2018). Nearby circular "bowls" comparable in length to the longest troughs but lacking claw marks were not considered *Ostendichnus* and therefore not described or counted (Fig. 13.1C, Fig. 13.2C). This kind of nest-scraping display traces may have been found before, but has not been independently pointed out or discussed. Whether these traces represent courtship display or failed nests (or both), they should be interpreted as a previously unrecognized type of calichnia trace (traces linked to nesting).

3.1.8 *Tracemaker*

Ostendichnus was attributed to theropods by Lockley et al. (2016) based on pes prints within a few configurations indicating bipedal tracemakers (Fig. 13.1). Narrow, acuminate claw traces within the troughs suggest scraping behavior. Two sizes of theropods were proposed for one site by Lockley et al. (2018) and confirmed for the other sites by Moklestad and Lucas (2023). Scaling functions from Thulborn (1989) were applied to the lengths of the footprints to derive the hip heights, which indicate ~1 m for the smaller theropod (about the size of the nearly contemporaneous *Deinonychus*) and ~2 m for the larger theropod (much larger than extant birds and about the size of the much older *Utahraptor*). As for the determination of the function of this type of trace, Lockley et al. (2016) eliminate or disregard three possibilities of these scraping marks, including: (1) nest sites or colonies; (2) digging for food, water, or shelter; and (3) territory-marking scrapes. Lockley et al. (2016, 2018) also assumed that courtship behavior occurs in all tetrapods (Isles, 2009; Lockley et al., 2016), including the nest scrape display behavior among shorebirds (Collias and Collias, 2014), and the lekking behavior in certain terrestrial birds such as kakapos (*Strigops habroptila*) (McDonald, 1989). Both deemed comparable with *Ostendichnus*, and the former has been considered to be a type of mating ritual instead of failed nest building attempt (Lockley et al., 2016).

Nevertheless, the evidence presented in Lockley et al. (2016) is insufficient to permit the differentiation between display and abandoned nesting traces when excluding the possibility of a failed nest site, which leads to a rediscussion of this possibility by Moklestad and Lucas (2023). This research primarily addresses four aspects that may contribute to the potential for failed nesting traces. Three of these factors were employed to identify them as nesting traces: their shape, riparian locality and orientation. The unsuitable substrate condition was used to elucidate the reasons why these nests were not completed, including encroachment of groundwater and the failures of tracemakers in loosening hard ground and removing roots (Moklestad and Lucas, 2023). The bowl-shaped structure in some of the traces is the most compelling evidence in support of the scrapes as nesting traces. This conclusion is based on the removal of claw marks and the median ridge between the troughs, which are signs of their further expansion into a hole or mound-like nest, and unlikely to occur during simplex courtship displays. The presence of both trough- and bowl-shaped scrapes at the same site precludes the possibility that the latter were formed as a result of weathering.

Further discussion is also required in detail of the hypothesis regarding the traces as failed nesting traces. The view of Moklestad and Lucas (2023) that a less open paleoenvironment can potentially help differentiate courting areas from nesting areas can, in fact, only be applied when it is predetermined that the sites are leks. As stated in Moklestad and Lucas (2023), it can be argued that multiple sets of scrapes at the same site could be formed consistently over multiple breeding seasons. At this juncture, it is conceivable that these sites may also be places utilized for displaying after pairing instead of leks, and cannot be distinguished from displaying ones, such as the "trail" or "display" traces from common ringed plovers even when they are not capable of producing eggs (Pienkowski, 1984). Analogous to the plovers referenced in Lockley et al. (2016) and the wandering albatross (*Diomedea exulans*) (Pickering and Berrow, 2001), some plover males continue to display after arriving at the nest site. Therefore, the environment cannot be used as a strong indicator to distinguish between displaying and nesting behavior. The explanation for the emergence of a similar orientation of *Ostendichnus* is also not very reliable when regarding them as nests. The example used by Moklestad and Lucas (2023) contains horned larks (*Eremophila alpestris*; Beason, 2020) and banded plovers (which should be piping plovers, *Charadrius melodus*; Elliott-Smith and Haig, 2020). It is notable that both nest orientations differ from *Ostendichnus*, which is not parallel to the direction of flow in the nearby water column as *Ostendichnus* (Cairns, 1982), so it is challenging to explain the apparent directionality by analogy. From the earlier discussion of the two hypotheses, it is evident that for modern birds, especially the ground-nesting shorebirds, which are repeatedly cited as analogs, the boundary between courtship displaying and nest-building is not distinct. Indeed, the two can even be seen as occurring simultaneously when nest scrape displays are made at nest sites (e.g., Pienkowski, 1984). We can confirm that *Ostendichnus* is a class of trace associated with reproduction. Given the occurrence of bowlshaped structures that are unrelated to displaying, we argue that *Ostendichnus* is more related to an unsuccessful attempt at nest building, although it does not completely rule out the presence of a display component in it.

3.2 *Minisauripus* Zhen et al., 1994 (Figs. 13.1D and 13.2D)

3.2.1 *Type ichnospecies*

Minisauripus chuanzhuensis Zhen et al., 1994

3.2.2 *Valid ichnospecies*

Minisauripus chuanzhuensis Zhen et al., 1994
Minisauripus zhenshuonani Lockley et al., 2008

3.2.3 *Distribution*

Only occurs in the Haman Formation (Albian), Korea, Sangchon-ri Tracksite (inferred), which is just on the south of the Gain-ri tracksite.

3.2.4 *General description (modified from Kim et al., 2016)*

Traces associated with *Minisauripus*, including a complex of subparallel scratch marks about 11 cm in length and 7 cm in width, with at least 25 separate very fine scratch marks. It is much smaller than *Ostendichnus*, which has the mean length of 101.3 cm and mean width of 73 cm, but with the same width/length proportion of about 0.7. All these traces are subparallel to the long axis of the complex and none is more than 1–2 mm in width. Two other separate scratches aligned with the long axis of the complex occur about 3 cm away. The complex has a roughly bilobed shape; the majority of the scrapes can be considered analogous to the straight claw mark in *Ostendichnus*, with the exception of the most lateral scratch on the right side, which exhibits a clear change of direction. The body of this complex only contains thread-like claw marks, without any bowl- or trough-shaped structures as in *Ostendichnus* (Fig. 1E) In contrast, the center of the complex contains shorter overlapping scratches and an indented area that may represent a track. Although it cannot be defined as a track definitively, the inferred track area is ~2.0 cm long and ~1.5 cm in width (l/w = 1.33). The width is also consistent with the width of the two sets of three scratch marks described below.

3.2.5 *Discussion of the tracemaker*

The ichnogenus *Minisauripus* was first described by Zhen et al. (1994) with the type ichnospecies *Minisauripus chuanzhuensis* defined later based on very small dinosaur tracks from the Emei tracksite, Xingfu Cliff, Sichuan (Jiaguan Formation, Barremian to Santonian, Hou et al., 2023), The ichnogenus was also reported by Lockley et al. (2008) and Kim et al. (2006, 2012) separately in Shandong (China), Sinsu Island and Changseon Island (South Korea), which add the feature of claw marks that would have been left on part of the substrate that further validates that the ichnogenus was created by small-bodied theropods. Though this presumed display trace is different from *Minisauripus*, it is comparable in size (Kim et al., 2012; Xing et al., 2016) with this kind of previously discovered trace, much smaller than the smallest bird footprint in the Gainri tracksite nearby (ichnogenus *Koreanornis*), and then all the known fossil bird traces found in the Korean Cretaceous (Kim et al., 2016). It is no larger than the largest *Minisauripus* (Kim et al., 2012).

In addition, *Minisauripus* typically is a narrow footprint with low digit divarication angles, whereas all known Cretaceous bird tracks from Korea have wide digit divarication angles. Also, most of the Korean bird trackways have inward rotation of the individual tracks, based on the orientation of digit III relative to the trackway mid line. By contrast, the scrapes described here indicate a slight outward rotation. The scratch or scrape marks described here best fit the size and shape of the feet of the trackmaker of *Minisauripus*. Support for this inference comes from the evidence that the *Minisauripus* trackmaker had very, sharp, narrow claws, which sometimes splayed outward. In some cases, the distal claw scrape marks described here show an outward curvature relative to the orientation of the digit trace. This is notable in the outer scrape of the right side tridactyl trace (Kim et al., 2016, Fig. 13.1E).

4. Principal localities

4.1 Display traces

Given that proposed fossil display traces have only been recognized in the last few years, the known occurrences are very limited.

The majority of the known sites are from the Dakota Group in Colorado, United States, although a similar trace has been found in an earlier layer in Alberta, Canada (Lockley et al., 2018).

There are five *Ostendichnus* sites located in Colorado: Dinosaur Ridge (multiple levels, in press), Duncan Road, Roubideau Creek, Club Gulch (two levels), and Club Gulch west (Lockley et al., 2016, 2018). Considering that the Club Gulch site has the most scrapes (~60), we consider this site as the most representative one among the five (Fig. 13.1A–D).

The Club Gulch scrapes occur in the Dakota Group, interpreted as crevasse splays and swampy delta plain deposits west of the Western Interior Seaway (Lockley et al., 2016: S1). This and other Dakota sites contain other traces created by ankylosaurs, avian and other nonavian theropods, ornithopods, pterosaurs, crocodilians, and turtles (Lockley et al., 2016; 2021).

4.2 Nesting traces

This section focuses on the archosaurs (mostly nonfossorial) (Oliveira et al., 2011; Hastings and Hellmund, 2015; Marsola et al., 2016; Murray et al., 2020), turtles (Jackson and Schmitt, 2008; Lawver and Jackson, 2014) and lepidosaurids (Martin et al., 2020). These have a clear tendency to nest, especially the nonavian dinosaurs that flourished in the late Mesozoic; their fossil eggs are relatively plentiful and relatively well preserved (Carpenter et al., 1994; Carpenter, 1999; Mainwaring et al., 2023a).

Nesting traces, which are often closely associated with fossilized eggs and/or shell fragments, are far more common than the display traces mentioned above. However, because of the typically incomplete preservation of nests including the decomposition of plant material and possible nest linings, or the scarcity of semicomplete clutches, the number of principal localities of such trace fossils is much lower than that of sites that preserve isolated eggs (Martin, 2014). Here, we limit ourselves to locations where large numbers of eggs in egg clutches occur together with nests structure. Such localities have only been reported for a limited number of tetrapod taxa, mainly within Archosauria.

4.2.1 *Nonavian dinosaurs (Figs. 13.3 and 13.4)*

Dinosaurs flourished in the Cretaceous, from which the greatest number of their eggs and nesting traces are preserved. The abundance of their nesting material far exceeds that of other tetrapod taxa. Significant discoveries are globally widespread on all of the continents except Antarctica (Carpenter et al., 1996; Carpenter, 1999).

4.2.1.1 Theropoda

The majority of known theropod nests are attributed to oviraptorosaurs and troodontids. As mentioned, one of the best-preserved dinosaur nests belongs to *Troodon formosus* and was found in the Two Medicine Formation (Campanian) at the Egg Mountain locality in Teton County, Montana, United States. An abundance of large, dispersituberculate eggshells represent a hatched or destroyed clutch have found before the near-complete nest (MOR 693) unearthed. This nest consists of a tight cluster of elongated, erected egg, and

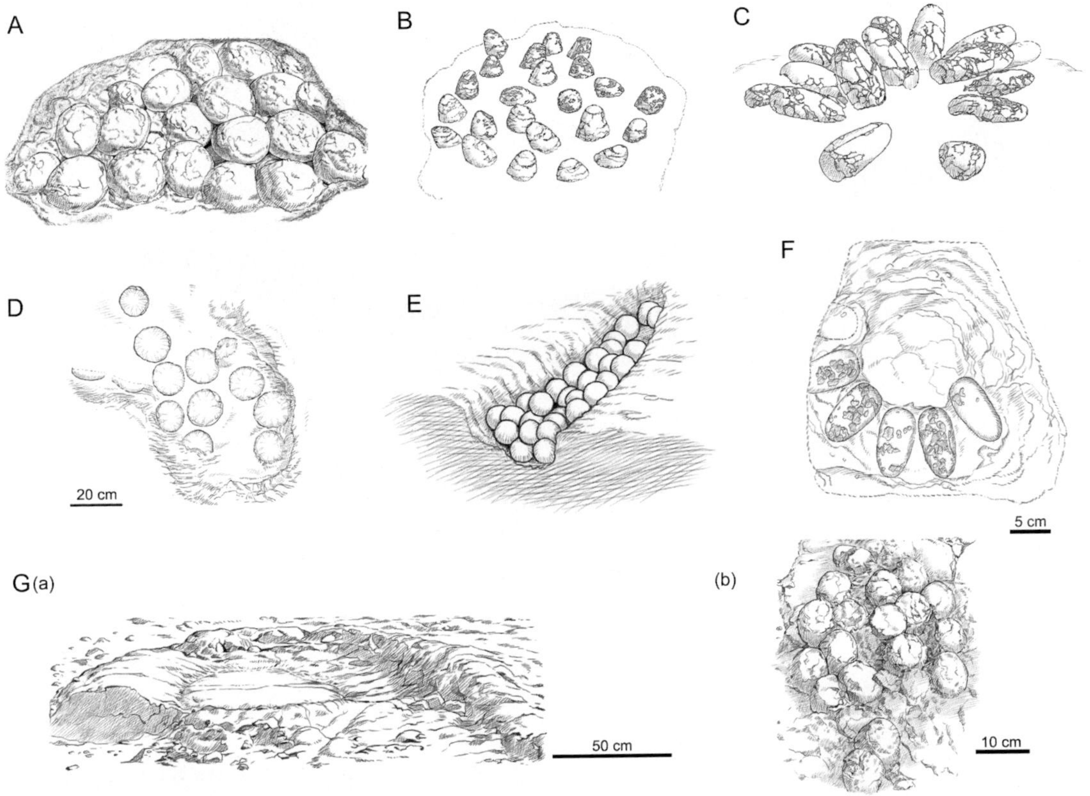

FIGURE 13.3 Typical nest structure and egg patterns among Dinosauria. Ornithischian: (A) Clutch pattern of *Lambeosaurus*. (B–C) Two clutch patterns of *Protoceratops*. Sauropods: (D–E) Typical clutch pattern (D) and hypothetical nest structure (E) of titanosaurs. Theropods: (F) Clutch pattern of *Montanoolithus* (Zelenitsky and Therrien, 2008). (G) Nest structure (a) and clutch pattern (b) of *Troodon*. *(A) Redrawn from Horner (1999), (B–C) redrawn from Thulborn (1992), (D-E) redrawn from Chiappe et al. (2004); Vila et al. (2010a,b), (G) from Varricchio et al. (1997)*

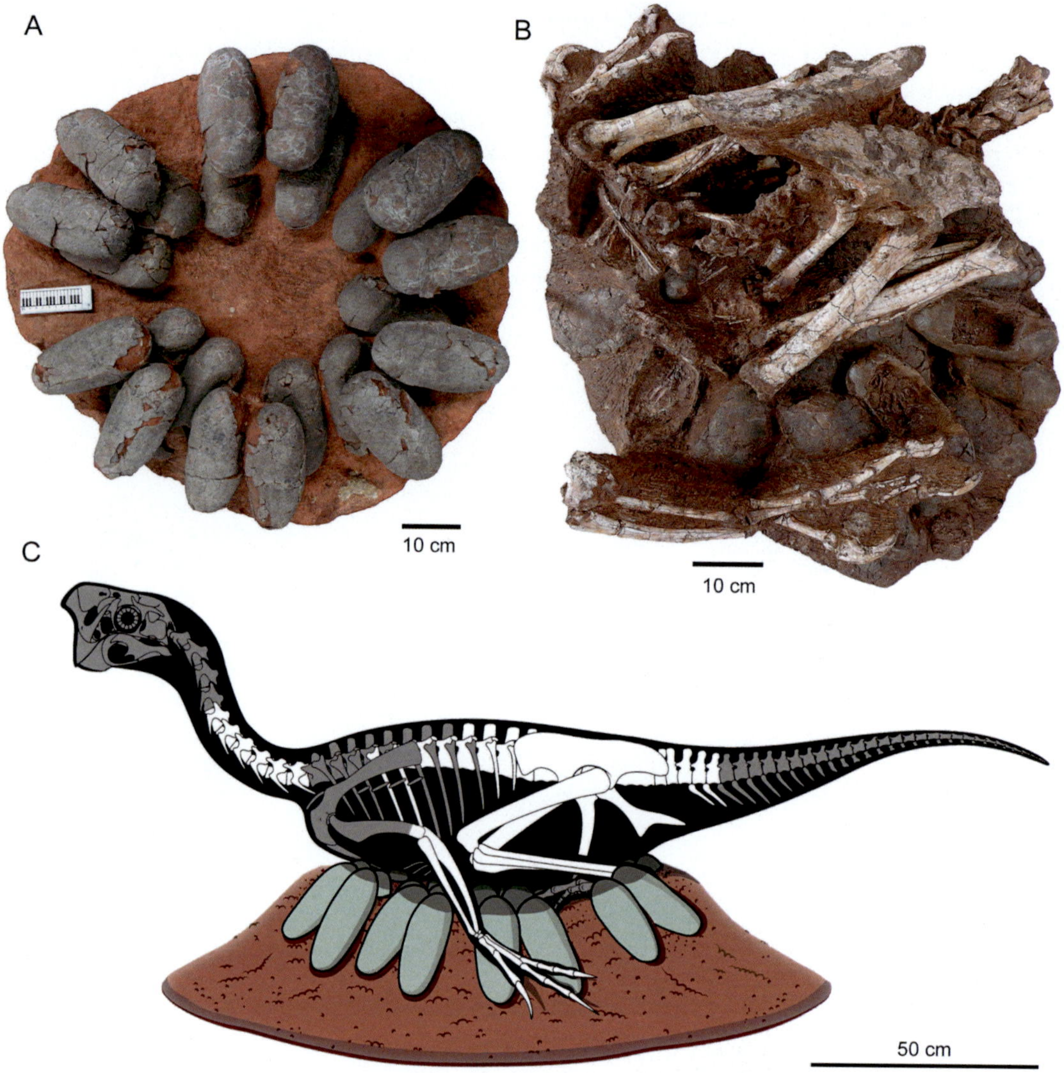

FIGURE 13.4 Typical clutches and possible incubating pattern of oviraptorosaurs. (A) Top view of an oviraptor ring-shaped clutch (PFMM 0,010,403,018, Fig. 13.1C of Wiemann et al., 2017). (B) An oviraptorid clutch with an adult skeleton preserved atop (Fig. 13.1A of Bi et al., 2021); (C) The restoration of possible incubating pattern of oviraptorosaurs (Fig. 13.1C of Bi et al., 2021).

near-complete nest structure including a shallow bowl-shaped depression with a distinct rim. MOR 693 is attributed to *Troodon* bon the basis of egg shape, surface texture, microstructure match with the identifiable embryonic remains and other eggs associated with an adult *Troodon*. The tight, erected placement fits the predescribed partial *Troodon* clutch (Varricchio et al., 1999, Fig. 13.3G). Egg Mountain preserves an isolated outcrop of stacked carbonate-rich sediments, stratigraphically surrounded by fine-grained, laminated to massive mudstone and siltstone, consisting of muddy micrites and calcareous mudstones representing the products of soil formation, bioturbation, and occasional overbank deposition (Lorenz, 1981; Horner et al., 1987). Aside from the nest built by *Troodon*, the Two Medicine Formation also yields various ankylosaurs, ceratopsians, ornithopods, tyrannosauroids and maniraptors, including the famous *Maiasaura* along with its nest and offsprings. Among these, the body fossils of dromaeosaurs, tyrannosaurs, *Troodon* itself, and the *Maiasaura* nest that can be seen at the same locality as the *Troodon* nest (Montellano, 1988; Dalman et al., 2018).

The nests of oviraptorosaurs have left behind the most abundant and informative fossil evidence of dinosaur reproductive biology (Yang et al., 2019; Figs. 13.3F and 13.4). The nests of *Citipati osmolskae* (Oviraptorosauria) are known for their lone adults preserved crouched in the center of each nest defined by concentric rings of eggs. This positioning and posture are unique in the fossil record globally, and both were excavated from the Djadokhta Formation (Campanian) at Ukhaa Tolgod, Mongolia (Norell et al., 1995, 2018). The location will be discussed with Ornithischia. Otherwise, another fragmentary oviraptorosaurian (*Oviraptor philoceratops*) specimen co-occurs with fossil eggs was found in Djadokhta Formation, yet at Bayan Mandahu, Inner Mongolia, China (Dong and Currie, 1996). Besides, the sole nests of oviraptorosaurs have also been found in East Asia: Small and medium-sized nests are primarily found in the Nanxiong Formation in the Nanxiong Basin (Guangdong, China) and Ganzhou (Jiangxi, China), while the large ones are primarily found in the Gaogou Formation at Yangcheng (Henan, China), Zhutian Formation at Dongyuan (Guangdong, China), and the Upper Cretaceous of Aphae-do (Jeollanam-do, Korea) (Tanaka et al., 2018; Yang et al., 2019). Both lithologic and geochemical characteristics of these Upper Cretaceous strata, mainly from central–southeastern China, indicate a long period of semiarid to arid paleoenvironments in the region (Yan et al., 2007; Wang et al., 2016). In addition to the typical dinosaurian oofauna in them and their adjacent strata, there are hadrosauroids, basal neoceratopsians, advanced sauropods, coelurosaurians, testudines, etc., which are also found in similar paleoenvironments in the Late Cretaceous (Xu et al., 2022; Fang et al., 2005), especially the abundant oviraptorosaurids in Jiangxi (Lü et al., 2015) and the caenagnathids in Henan (Pu et al., 2017).

4.2.1.2 Sauropodomorpha

The nests created by titanosaurs and basal sauropodomorphs yield important information on reproduction. The long known history is subdivided here with more information provided further.

Both of the two oldest known dinosaurian nests discovered in the Early Jurassic belong to basal sauropodomorphs, respectively attributed to *Massospondylus* and *Qianlong*, the latter with coburied skeletal fossils of adults and embryos (Reisz et al., 2012; Han et al., 2023). Only in the latter is there preservation of a large-scale nest site including eight nests and some shell fragments. The nest site of *Massospondylus* was found in the Upper Elliot Formation (Hettangian to Sinemurian; McPhee et al., 2017), Rooidraai, Golden Gate Highlands National Park, South Africa. This area has yielded multiple in situ clutches of eggs as well as fragmentary eggshells and bones, all from a 2 m thick interval of muddy siltstone (Reisz et al., 2012). Sedimentological data indicate that the climate in this region was at least seasonally arid at the time of deposition (Rainforth, 2003), with a sequence of relatively homogeneous muddy siltstones. Much of the fine grained silt in the upper Elliot has been interpreted as loess (Bordy et al., 2004), which indicates either an ephemeral lake margin in a flood basin or the vicinity of an oxbow lake was the setting for the dinosaurian nesting site (Reisz et al., 2012). The Elliot Formation crosses the Triassic/Jurassic boundary, and it is known for its abundance of early sauropodomorphs, some heterodontosaurids and other basal ornithischians, some crocodylomorphs, several theropods and other tetrapods, among them adult and embryonic bones of *Massospondylus* (Owen, 1854; Reisz et al., 2005).

The nest of *Qianlong* was found in the Zhenzhuchong Member of the Ziliujing Formation (possibly Sinemurian), Zhuanpo, Pingba District, Anshun City, Guizhou Province, southwestern China, where the paleoenvironment is similar to the *Massospondylus* nest site. The nest of *Qianlong* does not provide additional knowledge of nests, but it does demonstrate the earliest evidence of leathery eggs of dinosaurs (Han et al., 2023). We expect the discovery of abundant sauropodomorph fossils as in the Elliot Formation. Theropods, ornithischians and plesiosaurs have also found in the widely defined Ziliujing Formation (Dong, 1980; Dong et al., 1983). Other skeletal fossils are not present in the same basin as *Qianlong*, in Guizhou Province, and only tracks of sauropods and basal ornithischian occur (Xing et al., 2023).

The eggs of titanosaurs are numerous and widely distributed in the Late Cretaceous, especially in Argentina and India (Carpenter, 1999; Vila et al., 2010b). They represent a separate taxon named Megaloolithidae, but nests of this clade are typically inferred from the presence of egg clutches (Erben et al., 1979; Mohabey, 1996, Fig. 13.3D and E). Rarely preserved nest structures with rims were first found in 2000 (Loope et al., 2000) and thereafter redescribed and reinterpreted by Chiappe et al. (2004). This site is located in the Anacleto Formation (Campanian), Auca Mahuevo, Patagonia, Argentina. This area has yielded egg clutches in at least four stratigraphic layers in an 85 m thick sequence of sandstones, siltstones, and mudstones, in which four nests have been found with their subaerial rims (Chiappe et al., 2004). Most of the previously excavated clutches were stacked in multiple layers with no particular geometric arrangement of eggs. Silty, reddish brown, mottled mudstone with numerous slickensided surfaces indicative of paleosol development (Chiappe et al., 2000) and thin channel and crevasse splay sandstone lenses nearby (Chiappe et al., 2004), which represent fluvial, lacustrine, and floodplain environments. Aside from these nests of titanosaurs, this formation has diversity of body fossil on titanosaurs and theropods.

4.2.1.3 Ornithischia

The nests for which more information is available in this taxon were created by hadrosaurs and *Protoceratops*. Typical sites containing important information on reproduction are subdivided as follows (Fig. 13.3A–C). These clades provide less information on nest structure than those of theropods and sauropodomorphs, but retain more evidence of the posthatching stage by adult-multiple juveniles together in or just beside their nests (e.g., *Psittacosaurus*, Zhao et al., 2013; Hedrick et al., 2014). The wellknown *Maiasaura* nest was also discovered in the Two Medicine Formation (Campanian), in close proximity to the location where the *Troodon* nest was found, as discussed earlier (Horner and Makela, 1979).

The first nests of *Protoceratops* were discovered in the Djadochta Formation (Campanian), Flaming Cliffs site, Mongolia, in the early 19th century (Carpenter et al., 1996). In close proximity 2011, a nest containing a large number of juvenile skeletons of *Protoceratops andrewsi* was discovered in Tugrikin-shireh, Mongolia (Fastovsky et al., 2011). These two Djadochta localities in Gobi Desert, located in South Mongolia are represented by nonmarine, cemented reddish orange and pale orange to light gray, medium to fine grained sands and sandstones, which include minor deposits of calcareous concretions and orange brown silty clay (Dashzeveg et al., 2005), interpreted as arid and intermittent fluvial-lacustrine deposits. The clastic rocks are poorly cemented with a more homogeneous color in the Flaming Cliffs site while the Tugrikin shireh site appears as the opposite (Lefeld, 1971; Fastovsky et al., 1997). The Ukhaa Tolgod site also located in Gobi Desert and are found southwest of the Djadochta localities. *Citipatihas* been discovered at the Ukhaa Tolgod site that has particularly abundant cross-stratified and fine-structured sandstones with small amounts of conglomeratic lenses, which may indicate seasonal creeks with less volume but more hydraulical energy. Notable findings in these three Gobi localities include the first known fossils of *Oviraptor*, *Protoceratops*, *Saurornithoides*, and *Velociraptor*, the first confirmed dinosaur eggs (a partial nest of *Oviraptor*), as well as fossil mammals (Dashzeveg et al., 1995). The majority of these are found in unstructured sandstones, indicating burial in situ by high-energy wind and slope failure sand bearing events (Loope et al., 1999).

4.2.2 Other tetrapods

4.2.2.1 *Avialae* and *Pterosauria*

Both ground-nesting avialans and pterosaurs lack fossil evidence of nesting structures, despite inferences of their incubation behavior based on eggshell structure (Grellet-Tinner et al., 2007; Varricchio and Jackson, 2016; Mayr, 2017). However, they still have fossil evidence of probable colonial/communal nesting behavior (Fernández et al., 2013; Wang et al., 2017), so this section mainly describes this aspect.

The accumulation of possible *Neuquenornis* (a fossil bird 17 cm tall at hip; Molina-Pérez et al., 2019) eggs was found in the Bajo de la Carpa Formation (middle-upper Santonian) near the campus of the Universidad Nacional del Comahue, North of Neuquén city, Argentina. The eggs in this site can be correlated with basal ornithothoracines by the microstructure of egg shell and embryonic anatomy and are attributed to *Neuquenornis* from the upper strata by their egg volume (Schweitzer et al., 2002). Sixty-five eggs were scattered in this site within a 7 m by 6.5 m area, and their erect positions show that they were preserved in situ. Only two of them contained embryonic remains. Some are almost complete shells but lack bony remains; other eggs are partially preserved; and some are just large shell pieces (Fernández et al., 2013). This site was presumed a nesting colony mainly based on their in situ position with high degree of synchronicity (Kidwell and Behrensmeyer, 1993; Fernández et al., 2013), but none of these scattered eggs could consider as a clutch. Fernández et al. (2013) argued that the ellipsoidal shape of the eggs just as eggs laid in a clutch of three to eight in extant birds. This locality is located in the Neuquén Basin, and unlike some other regions, the Santonian rocks of this formation are a transgressive sequence that consisting of quartz-rich sandstone with light clay matrix and diagenetic carbonate cement, which indicates fluvial and aeolian systems (Sánchez et al., 2006) inferred to represent extensive flood plains (Garrido, 2010) with increasing abundance of aeolian sediments. This site not only contains these fossil eggs but also the Bajo de la Carpa crocodilian fauna dominated by crocodyliforms (Bonaparte, 1991).

More than 200 eggs and 16 embryos of *Hamipterus tianshanensis* (a pterosaur with 1.5–3.5 m wingspan; Wang et al., 2014) are preserved in sandstone of the Tugulu Group

(Early Cretaceous), Turpan-Hami Basin, Xinjiang, northwestern China. The eggs are elongated, mainly 5–6 cm in length and 2.5–3 cm in width. They were preserved with skeletal remains of *Hamipterus tianshanensis*, the only species in the bonebed. The fossils have been transported and disorganized, so the clutch size remains unknown (Wang et al., 2017). The potential for the preservation of pterosaur fossils is limited, particularly in the case of their eggs and embryos (Carpenter et al., 1994). A 2.2 m section has yielded as many as eight layers of pterosaur bones, which suggests that they may also exhibit site fidelity, similar to rookery-breeding seabirds (Wang et al., 2017). Additionally, the presence of some other extant nonavian dinosaurs, such as the ornithischian *Maiasaura* (Horner, 1982), has been documented. The possibility that they were buried after transportation from multiple locations cannot be ruled out. Wang et al. (2017) speculated that the large number of pterosaur fossils (~10–30 cm thick layers) could be the product of the same event. This could possibly suggest that *Hamipterus tianshanensis* nested colonially. The main locality where eggs have been collected is characterized by a succession of white to gray, medium to fine grained sandstones that were deposited in a fluvio-lacustrine environment. In addition to fossilized eggs, a large number of body fossils belonging to *Hamipterus*, which provides evidence of reproductive strategy, ontogeny, and behavior of pterosaurs, have been found in the strata of the same group at the same site (Wang et al., 2014, 2017).

4.2.2.2 Turtles

Although fossil turtle eggs have been found on all continents except Antarctica, and have a temporal record extending from the Middle Jurassic to the Pleistocene, the reproduction related traces are relatively scarce. Most descriptions of fossil turtle eggs focus on isolated specimens rather than in situ eggs and clutches, thus providing little or no information about site taphonomy (Lawver and Jackson, 2014). Nonetheless, the scarce fossil record of in situ clutches shows that turtle nesting was along the banks of water bodies (including typical fluvial and coastal environments). Where possible, periodic submerging of eggs was very analogous to the nesting forms of extant turtles living in similar habitats, especially those originating from the seashore (Fang et al., 2003; Zelenitsky et al., 2008; Bishop et al., 2011), some of which have shell structures that can be adapted to desiccating conditions. Recent studies report the world's first sea turtle nest depredation trace fossil in the Cretaceous (Bishop, 2021; Bishop et al., unpublished manuscript). The two nests mentioned in this paper are located in Colorado (USA) and Pula (Croatia). Based on the photographs provided by Bishop et al. (unpublished manuscript), both exhibit profile structures and covering pits that are relatively convincing in their morphological similarity to modern sea turtle nests. In combination with the location of the turtle eggs aforementioned, it is clear that there is considerable similarity in nesting behavior between Cretaceous sea turtles and modern sea turtles, indicating that the reproductive strategy of this taxon was conservative. However, the specific taxa of the tracemakers remains to be further investigated.

4.2.2.3 Lizards

Although fossils of lepidosaurid burrows are scarce, and the exact function of each burrow is mostly unknown, the only one with an identified tracemaker and fairly well-preserved features records clear nesting behavior. This nest is located in the Late Pleistocene of the Bahamas, though no eggs are preserved in it, the sign of active backfilling and the possible egg chamber can be distinguished by comparison with nest of extant iguanas (Martin et al., 2020). Combined with the nesting behavior of extant iguanas, the sex of the tracemaker can even be obtained (Hayes et al., 2004). Since all of this information seems to be based on analogies to the way extant taxa behave, no additional biological information of extinct taxa has been obtained from the fossil nest.

5. Nests and patterns of egg clutches

Nests of tetrapods vary across representative taxa, while few of them have fossil records of the nest structure and clutch pattern (see 4.2 Nesting traces): From their origin, mammalian burrows/dens have many functions including, but not limited to reproduction. Areas dedicated to reproduction can be partially discerned in the nests of extant species, but are difficult to find in the fossil record unless body fossils appear as indications (Lopes et al., 2017; Mainwaring et al., 2023b; Deeming, 2023). Sauropterygians and ichthyopterygians are the best studied Mesozoic marine reptiles. Skeletal evidence suggests that they were viviparous and did not lay eggs (Blackburn and Sidor, 2015). Pterosaurs theoretically nested, but only a large number of eggs representing partial clutches have been found (Grellet-Tinner et al., 2007; Wang et al., 2017). These have been utilized to speculate on colonial nesting, but do not provide further information on nest structure. Lepidosaurids (and other reptiles) as well as amphibians nesting traces are poorly studied. Lepidosaurians include very few preserved traces, most of which are burrows of doubtful tracemakers and functions (e.g., Doody et al., 2015; Hasiotis et al., 2004), only one of which has a clear reproductive purpose (Martin et al., 2020). For the amphibians, although parental egg protection is evident in extant taxa, only scattered tadpoles and eggs occur in fossil record and lack credible trace fossils (Gardner, 2016; Xing et al., 2021).

5.1 Burying versus open nesting

The degree of egg/nest burial during incubation can be difficult to determine given that all fossil nests are eventually buried. Fortunately, certain observations relate to burial such as the size of the parent, the ratio of parent to egg size, certain egg parameters such as size, shape, shell conductance, shell thickness, and shell composition, and the organic carbon content and/or plant material in the immediate overburden (Varricchio et al., 2008). Only theropod dinosaurs show a clear succession from fully buried nests to partially buried nests to open nests. There is no clear evidence of these kinds of changes with other taxa, such as crocodilians and turtles, where extant taxa also cover their eggs with sediment (Murray et al., 2020; Hogan and Varricchio, 2023; Bishop et al., unpublished manuscript).

The degree of egg burial correlates with the metabolic energy available to the egg from contact incubation (Hogan, 2024) and influences the presence or absence of egg-turning behavior in adults (Deeming, 1991), the latter of which is an important feature that occurs in birds. Partial egg burial was probably an intermediate state between subterranean and subaerial eggs (Hogan and Varricchio, 2023). Those practicing partial burial of their eggs include oviraptorosaurs and troodontids, which were located at important nodes in nonavian dinosaur–bird evolution (Varricchio et al., 1997; Tanaka et al., 2018). Apparently, partial egg burial played a significant role when transforming to Aves. However, the reason for the emergence of partial egg burial remains unclear (Hogan and Varricchio, 2023). It may be related to the strong adhesion and dense structure of the pennaceous feather found in the forelimb of pennaraptorans, which provides a stable thermal insulation environment (Yang et al., 2019; Hogan, 2024). However, further experiments are needed to support this hypothesis.

As temperature is the most crucial factor in the hatching success, Neornithes may have been the only clade to survive K-Pg boundary among Dinosauria partly by utilizing the properties of nests that are entirely unburied. This type of nest has weaker requirements for environmental conditions after hatching begins than buried or semiburied types (Martin et al., 2007; Varricchio and Jackson, 2016).

5.2 Arrangement of eggs

This section focuses on the oviraptorosaur and troodontid eggs mentioned in the previous section, the peculiar arrangement of which suggests a change in incubation patterns for these taxa. In other taxa that fully buried their eggs, such as the crocodilians, turtles, and sauropods, their eggs are discharged into the nest en masse, with an essentially random arrangement (Vail et al., 2010a; Lawver and Jackson, 2014; Murray et al., 2020; Hogan and Varricchio, 2023). The arrangement of eggs is also a feature that only occurs in the Maniraptora clade, but for which no evidence has been reported within basal taxa (i.e., therizinosauroid and alvarezsaurids, Kundrát et al., 2008; Agnolin et al., 2012).

Oviraptorosaurs have elongated and relatively larger eggs compared with the basal maniraptorans (e.g., therizinosauroid: Kundrát et al., 2008). Some of these features also appear in other advanced maniraptorans, such as the troodontids described earlier. The oviraptorosaur eggs are classified as Elongatoolithidae, and they are ordered in the nest in a rather distinctive, multilayered hollow-ringed clutch, in which the hollow portion is positively correlated in size with the body size of the spawner (Norell et al., 1995; Norell et al., 2018; Tanaka et al., 2018, Fig. 13.4A). The mode of incubation indicated by this particular arrangement is a matter of debate. It is generally believed that contact incubating improves the efficiency of energy utilization in the fully exposed and half exposed eggs (Varricchio and Jackson, 2016), but some studies have argued that this kind of arrangement hinders sufficient heat transfer to the inner (lower) ring(s) of eggs (Yang et al., 2019). But modern Charadriiformes and experimental evidence, especially the latter, may support the conclusion that contact incubation can be paired with this kind of egg arrangement (Hogan and Varricchio, 2023, Fig. 13.4B and C).

Troodontids possessed smaller, more avian-like asymmetrical eggs, but their morphology was still largely similar to that of the oviraptorids. These eggs were likely manipulated after laying, turned upright and held suberect by compacted sediment (Varricchio et al., 1999). Egg arrangements within these clutches would have permitted greater adult-egg contact for improved heat transfer and insulation (Varricchio and Jackson, 2016), which clearly indicates the brooding behaviors.

5.3 Clutch size and reproductive strategy

For both archosaurs and turtles, clutch size is positively correlated with body size in all extant taxa, with the degree of correlation within archosaurs broadly similar across different parental care patterns (Moll, 1979; Reisz et al., 2012; Lawver and Jackson, 2014). Reproductive strategies in extinct turtles seem generally similar to those of living species (Zelenitsky et al., 2008; Lawver and Jackson, 2014). Similarly, crocodilians exhibit significantly lower egg-laying numbers only in the small Mesozoic notosuchians (Marsola et al., 2016). In nonavian dinosaurs, although the number of eggs laid generally varies with body size, similarly sized maniraptora (~1 m high at the hip), those thought to have sat on their nests have varying clutch sizes: therizinosaurs (n = 7 in Kundrát et al., 2008; with the follow up material of n = 8), *Troodon formosus*, n = 24, Varricchio et al. (1999) and oviraptosaurs (*Citipati osmolskae*, n = 18 or 22 (estimated), Norell et al., 1995; Yang et al., 2019). The variability in clutch size may be a result of troodontids and oviraptorosaurs having communal nests, based on eggshell histology, elemental analyses of eggshells, and ratio of clutch volume to adult mass (Varricchio et al., 2008; Yang et al. 2016, 2019). Egg size is not clearly proportional to body size in large dinosaurs that are much larger than extant tetrapods (Ribeiro et al., 2014; Vila et al., 2010a). This may suggest that as body size increases, there is a greater tendency to select the R strategy during reproduction, while the likelihood of adults participating in incubation decreases at the same time.

There have been other variations in the reproductive strategies of dinosaurs, although the fossil evidence is also largely confined to the maniraptoran-bird clade. These variations are reflected in changes in oviposition patterns rather than clutch size (Hogan and Varricchio, 2023). The earliest possible evidence for this change occurs in alvarezsaurids, with the presence of skeletons preserved in association with two eggs corresponding to their body size (Agnolin et al., 2012). The two clades of Pennaraptora, oviraptorosaurs and troodontids, both show a clear monoautochronic character of oviposition of two eggs at a time (Varricchio et al., 1997; Clark et al., 1999), whereas birds have lost a functional oviduct and produce only one egg at a time This type of delayed incubation also likely results in synchronous hatching of the eggs (Varricchio and Jackson, 2016), which may have evolved as a general predator avoidance strategy.

5.4 Possible evidence for parental care

Parental care is present in many evolutionary clades of tetrapods and is universal within extant birds and mammals (Royle et al., 2012). It mainly consists of the attendance in the pre-incubation period of oviparous tetrapods and the care of the young in the posthatching/birthing period that includes the mammals. Both periods are difficult to document directly in fossils.

Attendance in incubation is widespread in the two extant branches of the archosaurs, but not found in squamates and turtles. Within the fossil record of archosaurs, this behavior is widely observed in all three evolutionary clades of dinosaurs, but lack of a definitive record for extinct crocodilians. It has been hypothesized that attendance in incubation may be plesiomorphic based on the relationship of clutch size to adult body mass in basal sauropods (Meng et al., 2004; Reisz et al., 2012; Hastings and Hellmund, 2015). However, as noted earlier, this trait is lost at least in the advanced sauropods. The majority of the above taxa were not directly involved in incubation, but rather in nest guarding. Only the pennaraptorans described above show direct incubation behavior. But considering the low number of eggs, it is possible that the clutches were incomplete, which caused the adult to stay in the nest (Yang et al., 2019). Posthatching/birthing parental care, however, is harder to demonstrate using body or trace fossil evidence. The cooccurrence of many juvenile individuals in a single nest does not necessarily indicate the behavior of offspring attendance or food provisioning by adults, but may be instead indicative of mutualistic behavior within the incubation prior to reaching adulthood (Zhao et al., 2013). The assumption of care of juveniles is not reliable when adults are present in close proximity, as in the case of *Maiasaura*.

6. Conclusions and perspectives

Courtship and nesting behavior are rarely recorded in body fossils, but can be preserved somewhat more often as trace fossils, which are valuable sources of data for studying the evolution of reproduction. These trace fossils have limitations. They are preserved in small quantities, the interpretations of them depend very much on comparisons with traces that can either be produced by specific extant organisms, or the study of general patterns, and it is analogous to that of footprints with orientations, repetition, even spacing, progression, parallelism, etc. Multiple interpretations of the same data are likely.

Reproductive behavior in tetrapods is represented in trace fossils by possible courtship scrapes and nesting traces, including clutch patterns. The former are currently identified only in a limited region from a single group of tetrapods, and the latter have started to be discussed independently from fossil eggs in the last ~50 years, and can serve as evidence to illustrate incubating behaviors and reproductive strategy. To some extent, the degree of parental involvement in the incubation–rearing process can be inferred. The study of the putative display scrapes (or possible failed nesting scrapes) in Colorado (USA) is still in its infancy, as it has only been discovered and interpreted in the last decade. Additional occurrences are confined to a single scrape pair in Alberta (Lockley et al., 2018). It is possible that occurrences in other locations have been overlooked in earlier studies. Further research, including multiple studies in progress, will provide more insights into both kinds of reproduction-related traces.

Acknowledgments

The authors thank reviewers Tom Moklestad, Spencer G. Lucas, and Adrian P. Hunt for their helpful considerations and constructive feedback, especially on grammar and the part of courtship behavior. This research was funded by the "Deep-time Digital Earth" Science and Technology Leading Talents Team Funds for the Central Universities for the Frontiers Science Center for Deep-time Digital Earth, China University of Geosciences (Beijing) (Fundamental Research Funds for the Central Universities; grant number: 2652023001), the 111 project (B20011), and Fundamental Research Funds for Central Universities (265QZ201903). The material is a contribution to the Deep-time Digital Earth (DDE) Big Science Program.

References

Agnolin, F.L., Powell, J.E., Novas, F.E., Kundrát, M., 2012. New alvarezsaurid (Dinosauria, Theropoda) from uppermost Cretaceous of north-western Patagonia with associated eggs. Cretac. Res. 35, 33–56.

Alexander, R.D., 1974. The evolution of social behavior. Annu. Rev. Ecol. Systemat. 5 (1), 325–383.

Barthel, K.W., Swinburne, N.H.M., Conway Morris, S., 1990. Solnhofen: A Study in Mesozoic Palaeontology.

Beason, R.C., 2020. Horned Lark (Eremophila alpestris), version 1.0. In: Birds of the World. Cornell Lab of Ornithology, Ithaca, NY, USA.

Bertram, B.C., 2014. The Ostrich Communal Nesting System, vol 47. Princeton University Press.

Bi, S., Amiot, R., de Fabrègues, C.P., Pittman, M., Lamanna, M.C., Yu, Y., Yu, C., Yang, T., Zhang, S., Zhao, Q., Xu, X., 2021. An oviraptorid preserved atop an embryo-bearing egg clutch sheds light on the reproductive biology of non-avialan theropod dinosaurs. Sci. Bull. 66 (9), 947–954.

Bishop, G., 2021. Ancient Sea Turtle Nests from Around the World: Academia Letters. Article 3082, 8 pp.

Bishop, G. A., Kržič, B., Meyer, B. K., et al., Modern and Ancient Sea Turtle Nest Depredation; and the First Protostegid Nest.

Bishop, G.A., Pirkle, F.L., Meyer, B.K., Pirkle, W.A., Rollins, H.B., Thomas, D.H., 2011. The foundation for sea turtle geoarchaeology and zooarchaeology: morphology of recent and ancient sea turtle nests, St. Catherines Island, Georgia, and Cretaceous Fox Hills Sandstone, Elbert County, Colorado. Geoarchaeology of St. Catherines Island, Georgia. Am. Mus. Natural History Anthropol. Papers 94, 247–269.

Blackburn, D.G., Sidor, C.A., 2015. Evolution of viviparous reproduction in Paleozoic and Mesozoic reptiles. Int. J. Dev. Biol. 58 (10–11-12), 935–948.

Bonaparte, J.F., 1991. Los vertebrados fósiles de la Formación Río Colorado, de la ciudad de Neuquén y cercanías, Cretácico superior, Argentina. Museo Argentino de Ciencias "Naturales" Bernardino "Rivadavia" e Instituto Nacional de Investigación de las Ciencias Naturales.

Bordy, E.M., Hancox, P.J., Rubidge, B.S., 2004. Basin development during the deposition of the Elliot Formation (Late triassic-Early jurassic), Karoo supergroup, South Africa. S. Afr. J. Geol. 107 (3), 397–412.

Briggs, D.E., McMahon, S., 2016. The role of experiments in investigating the taphonomy of exceptional preservation. Palaeontology 59 (1), 1–11.

Brown, B., Schlaikjer, E.M., 1940. The structure and relationships of Protoceratops. Ann. N. Y. Acad. Sci. 40 (3), 133–265.

Cairns, W.E., 1982. Biology and behavior of breeding piping plovers. Wilson Bull. 531–545.

Caro, S.M., Griffin, A.S., Hinde, C.A., West, S.A., 2016. Unpredictable environments lead to the evolution of parental neglect in birds. Nat. Commun. 7 (1), 10985.

Carpenter, K., 1999. Eggs, Nests, and Baby Dinosaurs: A Look at Dinosaur Reproduction. Indiana University Press.

Carpenter, K., Hirsch, K.F., Horner, J.R. (Eds.), 1994. *Dinosaur eggs and babies*. Cambridge University Press.

Carpenter, K., Hirsch, K.F., Horner, J.R. (Eds.), 1996. Dinosaur Eggs and Babies. Cambridge University Press.

Chiappe, L.M., Dingus, L., Jackson, F., Grellet-Tinner, G., Aspinall, R., Clarke, J., Coria, R., Garrido, A., Loope, D., 2000. Sauropod eggs and embryos from the Late Cretaceous of Patagonia. In: First International Symposium on Dinosaur Eggs and Babies Isona. Extended abstracts, Spain, pp. 23–29.

Chiappe, L.M., Schmitt, J.G., Jackson, F.D., Garrido, A., Dingus, L., Grellet-Tinner, G., 2004. Nest structure for sauropods: sedimentary criteria for recognition of dinosaur nesting traces. Palaios 19 (1), 89–95.

Chiappe, L.M., Shu an, J., Qiang, J., Norell, M.A., 1999. Anatomy and systematics of the Confuciusornithidae (Theropoda: Aves) from the late mesozoic of northeastern China. Bull. Am. Mus. Nat. Hist. 242, 1–89.

Clark, J.M., Norell, M., Chiappe, L.M., 1999. An oviraptorid skeleton from the Late Cretaceous of Ukhaa Tolgod, Mongolia, preserved in an avianlike brooding position over an oviraptorid nest. Am. Mus. Novit. 3265, 36.

Collias, N.E., Collias, E.C., 2014. Nest Building and Bird Behavior, vol 857. Princeton University Press.

Coombs, W.P., 1989. Modem Analogs for Dinosaur Nesting and Parental Behavior.

Cousin, R., Breton, G., Fournier, R., Watté, J.P., Carpenter, K., 1994. Dinosaur egglaying and nesting in France. In: Dinosaur Eggs and Babies. Cambridge University Press, New York, pp. 37–46.

Dalman, S.G., Lucas, S.G., Malinzak, D.E., 2018. Tyrannosaurid teeth from the upper cretaceous two Medicine Formation of Montana. New Mexico Mus. Natural History Sci. Bull. 79, 125–139.

Dashzeveg, D., Dingus, L., Loope, D.B., Swisher, C.C., Dulam, T., Sweeney, M.R., 2005. New stratigraphic subdivision, depositional environment, and age estimate for the upper cretaceous Djadokhta Formation, southern ulan nur basin, Mongolia. Am. Mus. Novit. 2005 (3498), 1–31.

Dashzeveg, D., Novacek, M.J., Norell, M.A., Clark, J.M., Chiappe, L.M., Davidson, A., McKenna, M.C., Dingus, L., Swisher, C., Altangerel, P., 1995. Extraordinary preservation in a new vertebrate assemblage from the Late Cretaceous of Mongolia. Nature 374 (6521), 446–449.

de Zwaan, D.R., Martin, K., 2018. Substrate and structure of ground nests have fitness consequences for an alpine songbird. Ibis 160 (4), 790–804.

Deeming, D.C., 1991. Reasons for the dichotomy in egg turning in birds and reptiles. In: Egg Incubation: Its Effects on Embryonic Development in Birds and Reptiles, pp. 307–323.

Deeming, D.C., 2023. Nest construction in mammals: a review of the patterns of construction and functional roles. Philosop. Transac. Royal Soc. B 378 (1884), 20220138.

Díaz-Martínez, I., Citton, P., Castanera, D., 2023. What do their footprints tell us? Many questions and some answers about the life of non-avian dinosaurs. J. Iber. Geol. 1–22.

Dong, Z.M., 1980. A new Plesiosauria from the lias of Sichuan basin. Vertebr. Palasiat. 18 (3), 191.

Dong, Z.M., Currie, P.J., 1996. On the discovery of an oviraptorid skeleton on a nest of eggs at Bayan Mandahu, Inner Mongolia, People's Republic of China. Can. J. Earth Sci. 33 (4), 631–636.

Dong, Z., Zhou, S., Zhang, H., 1983. Dinosaurs from the jurassic of sichuan. Palaeontologica Sinica, New Series C 162 (23), 1–136.

Doody, J.S., James, H., Colyvas, K., Mchenry, C.R., Clulow, S., 2015. Deep nesting in a lizard, déjà vu devil's corkscrews: first helical reptile burrow and deepest vertebrate nest. Biol. J. Linn. Soc. 116 (1), 13–26.

Elliott-Smith, E., Haig, S.M., 2020. Piping Plover (*Charadrius melodus*), version 1.0. In: Birds of the World. Cornell Lab of Ornithology, Ithaca, NY, USA.

Erben, H.K., Hoefs, J., Wedepohl, K.H., 1979. Paleobiological and isotopic studies of eggshells from a declining dinosaur species. Paleobiology 5 (4), 380–414.

Fang, X.S., Lu, L.W., Jiang, Y.G., Yang, L.F., 2003. Cretaceous fossil eggs from the Tiantai basin of Zhejiang, with a discussion on the extinction of dinosaurs. Geol. Bull. China 22 (7), 512–520.

Fang, X.S., Zhang, Z.Z., Zhang, X.Q., et al., 2005. Fossil eggs from the heyuan basin, guangdong, China. Geol. Bull. China 24 (7), 682–686.

Fang, Y.T., Tuanmu, M.N., Hung, C.M., 2018. Asynchronous evolution of interdependent nest characters across the avian phylogeny. Nat. Commun. 9 (1), 1863.

Fastovsky, D.E., Badamgarav, D., Ishimoto, H., Watabe, M., Weishampel, D.B., 1997. The paleoenvironments of tugrikin-shireh (Gobi Desert, Mongolia) and aspects of the taphonomy and paleoecology of Protoceratops (Dinosauria: ornithishichia). Palaios 59–70.

Fastovsky, D.E., Weishampel, D.B., Watabe, M., Barsbold, R., Tsogtbaatar, K.H., Narmandakh, P., 2011. A nest of *Protoceratops andrewsi* (Dinosauria, Ornithischia). J. Paleontol. 85 (6), 1035–1041.

Fernández, M.S., Garcia, R.A., Fiorelli, L., Scolaro, A., Salvador, R.B., Cotaro, C.N., Kaiser, G.W., Dyke, G.J., 2013. A large accumulation of avian eggs from the Late Cretaceous of Patagonia (Argentina) reveals a novel nesting strategy in Mesozoic birds. PLoS One 8 (4), e61030.

Fortuny, J., Vila, B., Galobart, A., 2007. Técnicas de documentación y representación tridimensionalde puestas de dinosaurio. In: Cantera Paleontológica. Diputación Provincial de Cuenca, pp. 181–192.

Fusani, L., 2008. Testosterone control of male courtship in birds. Horm. Behav. 54 (2), 227–233.

Gardner, J.D., 2016. The fossil record of tadpoles. Fossil Imprint 72.

Garrido, A.C., 2010. Estratigrafía del Grupo Neuquén, Cretácico Superior de la Cuenca Neuquina (Argentina): nueva propuesta de ordenamiento litoestratigráfico. Revista del Museo Argentino de Ciencias Naturales 12 (2), 121–177.

Grellet-Tinner, G., Chiappe, L., Norell, M., Bottjer, D., 2006. Dinosaur eggs and nesting behaviors: a paleobiological investigation. Palaeogeogr. Palaeoclimatol. Palaeoecol. 232 (2–4), 294–321.

Grellet-Tinner, G., Murelaga, X., Larrasoana, J.C., Silveira, L.F., Olivares, M., Ortega, L.A., Trimby, P.W., Pascual, A., 2012. The First Occurrence in the Fossil Record of an Aquatic Avian Twig-Nest With Phoenicopteriformes Eggs: Evolutionary Implications.

Grellet-Tinner, G., Wroe, S., Thompson, M.B., Ji, Q., 2007. A note on pterosaur nesting behavior. Hist. Biol. 19 (4), 273–277.

Guigueno, M.F., Sealy, S.G., 2012. Nest sanitation in passerine birds: implications for egg rejection in hosts of brood parasites. J. Ornithol. 153, 35–52.

Haas, C.A., 1998. Effects of prior nesting success on site fidelity and breeding dispersal: an experimental approach. The Auk 115 (4), 929–936.

Han, F., Yu, Y., Zhang, S., Zeng, R., Wang, X., Cai, H., Wu, T., Wen, Y., Cai, S., Li, C., Wu, R., 2023. Exceptional Early Jurassic fossils with leathery eggs shed light on dinosaur reproductive biology. Natl. Sci. Rev., nwad258

Hasiotis, S.T., Wellner, R.W., Martin, A.J., Demko, T.M., 2004. Vertebrate burrows from Triassic and Jurassic continental deposits of North America and Antarctica: their paleoenvironmental and paleoecological significance. Ichnos 11 (1–2), 103–124.

Hastings, A.K., Hellmund, M., 2015. Rare in situ preservation of adult crocodylian with eggs from the middle Eocene of Geiseltal, Germany. Palaios 30 (6), 446–461.

Hayes, L.D., 2000. To nest communally or not to nest communally: a review of rodent communal nesting and nursing. Anim. Behav. 59 (4), 677–688.

Hayes, W.K., Carter, R.L., Cyril Jr, S., Thornton, B., 2004. Conservation of an endangered Bahamian rock iguana. In: Iguanas: Biology and Conservation, pp. 232–257.

Hedrick, B.P., Chunling, G., Omar, G.I., Fengjiao, Z., Caizhi, S., Dodson, P., 2014. The osteology and taphonomy of a *Psittacosaurus* bonebed assemblage of the Yixian Formation (Lower Cretaceous), Liaoning, China. Cretac. Res. 51, 321–340.

Hogan, J.D., Varricchio, D.J., 2023. Chthonic severance: dinosaur eggs of the Mesozoic, the significance of partially buried eggs and contact incubation precursors. Philos. Trans. Royal Soc. B 378 (1884), 20220144.

Hogan, J.D., 2024. The egg-thief architect: experimental oviraptorosaur nesting physiology, the possibility of adult-mediated incubation, and the feasibility of indirect contact incubation. Paleobiology 1–15.

Horner, J.R., Makela, R., 1979. Nest of juveniles provides evidence of family structure among dinosaurs. Nature 282 (5736), 296–298.

Horner, J.R., 1982. Evidence of colonial nesting and 'site fidelity'among ornithischian dinosaurs. Nature 297 (5868), 675–676.

Horner, J.R., 1999. Egg clutches and embryos of two hadrosaurian dinosaurs. J. Vertebr. Paleontol. 19 (4), 607–611.

Horner, J.R., Czerkas, S.J., Olson, E.C., 1987. Ecologic and behavioral implications derived from a dinosaur nesting site. Dinosaurs Past and Present 2, 50–63.

Hou, L., Pei, J., Cai, Y., Zhou, Z., Tong, Y., Zhao, Y., 2023. Upper Cretaceous–Paleocene magnetostratigraphy of the Longmenshan foreland basin and its implications for the tectonic deformation of the Longmenshan fault zone. Tectonics 42 (2), e2022TC007660.

Hsieh, S., Plotnick, R.E., 2020. The representation of animal behaviour in the fossil record. Anim. Behav. 169, 65–80.

Isles, T.E., 2009. The socio-sexual behaviour of extant archosaurs: implications for understanding dinosaur behaviour. Hist. Biol. 21 (3–4), 139–214.

Jackson, F.D., Schmitt, J.G., 2008. Recognition of vertebrate egg abnormalities in the Upper Cretaceous fossil record. Cretac. Res. 29 (1), 27–39.

Jain, S.L., 1989. Recent dinosaur discoveries in India, including eggshells, nests and coprolites. Int. Symp. Dinosaur Tracks Traces 1, 99–108.

Kerourio, P., 1981. Nouvelles observation sur le de nidification et de ponte chez les dionsauriens du Cretace terminal du Medi de la France. Comptes rendus sommaire des seance de la Societe Geologique de France 1, 25–28.

Kidwell, S.M., Behrensmeyer, A.K., 1993. Taphonomic approaches to time resolution in fossil assemblages: introduction. Short Courses Paleontol. 6, 1–8.

Kim, J.Y., Kim, S.H., Kim, K.S., Lockley, M., 2006. The oldest record of webbed bird and pterosaur tracks from South Korea (Cretaceous Haman Formation, Changseon and Sinsu Islands): more evidence of high avian diversity in East Asia. Cretac. Res. 27 (1), 56–69.

Kim, J.Y., Lockley, M.G., Seo, S.J., Kim, K.S., Kim, S.H., Baek, K.S., 2012. A paradise of Mesozoic birds: the world's richest and most diverse Cretaceous bird track assemblage from the Early Cretaceous Haman Formation of the Gajin tracksite, Jinju, Korea. Ichnos 19 (1–2), 28–42.

Kim, K.S., Lockley, M.G., Lim, J.D., Buckley, L., Xing, L., 2016. Small scale scrapes suggest avian display behavior by diminutive Cretaceous theropods. Cretac. Res. 66, 1–5.

Krebs, J.R., 1974. Colonial nesting and social feeding as strategies for exploiting food resources in the great blue Heron (Ardea herodias). Behaviour 99–134.

Kundrát, M., Cruickshank, A.R., Manning, T.W., Nudds, J., 2008. Embryos of therizinosauroid theropods from the Upper Cretaceous of China: diagnosis and analysis of ossification patterns. Acta Zool. 89 (3), 231–251.

Lawver, D.R., Jackson, F.D., 2014. A review of the fossil record of turtle reproduction: eggs, embryos, nests and copulating pairs. Bull. - Peabody Mus. Nat. Hist. 55 (2), 215–236.

Lefeld, J., 1971. Geology of the Djadokhta Formation at Bayn Dzak (Mongolia). Palaeontol. Pol. 25, 101–127.

Lehmann, U., 1971. New aspects in ammonite biology. In: Proceedings of North America Paleontology Convocation I, pp. 1251–1269.

Lockley, M.G., Houck, K.J., Matthews, N., McCrea, R.T., Xing, L., Tsukui, K., Ramezani, J., Breithaupt, B., Cart, K., Martin, J., Buckley, L.G., 2018. New theropod display arena sites in the Cretaceous of North America: clues to distributions in space and time. Cretac. Res. 81, 9–25.

Lockley, M.G., Kim, J.Y., Kim, K.S., Kim, S.H., Matsukawa, M., Rihui, L., Jianjun, L., Yang, S.Y., 2008. *Minisauripus* the track of a diminutive dinosaur from the Cretaceous of China and South Korea: implications for stratigraphic correlation and theropod foot morphodynamics. Cretac. Res. 29 (1), 115–130.

Lockley, M.G., McCrea, R.T., Buckley, L.G., Deock Lim, J., Matthews, N.A., Breithaupt, B.H., Houck, K.J., Gierliński, G.D., Surmik, D., Soo Kim, K., Xing, L., 2016. Theropod courtship: large scale physical evidence of display arenas and avian-like scrape ceremony behaviour by Cretaceous dinosaurs. Sci. Rep. 6 (1), 18952.

Lockley, M.G., Kim, K.S., Lim, J.D., 2021. *Tetrapodosaurus* trackways from the Cretaceous of Colorado and Canada: observations on variability in ankylosaurian gaits. New Mexico Mus. Nat. History Sci. Bull. 7 (82), 227–235.

Loope, D., Schmitt, J., Jackson, F., 2000. Thunder platters: thin discontinuous limestones in Late Cretaceous continental mudstones of Argentina may be chemically infilled sauropod tracks. In: Geological Society of America, Abstracts with Programs A, vol 450.

Loope, D.B., Mason, J.A., Dingus, L., 1999. Lethal sandslides from eolian dunes. J. Geol. 107 (6), 707–713.

Lopes, R.P., Frank, H.T., Buchmann, F.S.D.C., Caron, F., 2017. *Megaichnus* igen. nov.: giant paleoburrows attributed to extinct Cenozoic mammals from South America. Ichnos 24 (2), 133–145.

Lorenz, J.C., 1981. Sedimentary and Tectonic History of the Two Medicine Formation, Late Cretaceous (Campanian), Northwestern Montana. Princeton University.

Lü, J., Pu, H., Kobayashi, Y., Xu, L., Chang, H., Shang, Y., Liu, D., Lee, Y.N., Kundrát, M., Shen, C., 2015. A new oviraptorid dinosaur (Dinosauria: Oviraptorosauria) from the Late Cretaceous of Southern China and its paleobiogeographical implications. Sci. Rep. 5 (1), 11490.

Mainwaring, M.C., Medina, I., Tobalske, B.W., Hartley, I.R., Varricchio, D.J., Hauber, M.E., 2023a. The evolution of nest site use and nest architecture in modern birds and their ancestors. Philosop. Trans. Royal Soc. B 378 (1884), 20220143.

Mainwaring, M.C., Stoddard, M.C., Barber, I., Deeming, D.C., Hauber, M.E., 2023b. The evolutionary ecology of nests: a cross-taxon approach. Philos. Trans. Royal Soc. B 378 (1884), 20220136.

Marsola, J.C.D.A., Batezelli, A., Montefeltro, F.C., Grellet-Tinner, G., Langer, M.C., 2016. Palaeoenvironmental characterization of a crocodilian nesting site from the Late Cretaceous of Brazil and the evolution of crocodyliform nesting strategies. Palaeogeogr. Palaeoclimatol. Palaeoecol. 457, 221–232.

Martin, A.J., 2014. Dinosaurs without Bones: Dinosaur Lives Revealed by Their Trace Fossils. Simon and Schuster.

Martin, A.J., Stearns, D., Whitten, M.J., Hage, M.M., Page, M., Basu, A., 2020. First known trace fossil of a nesting iguana (Pleistocene), the Bahamas. PLoS One 15 (12), e0242935.

Martin, T.E., Auer, S.K., Bassar, R.D., Niklison, A.M., Lloyd, P., 2007. Geographic variation in avian incubation periods and parental influences on embryonic temperature. Evolution 61 (11), 2558–2569.

Mayor, A., 2013. Fossil legends of the first Americans. In: Fossil Legends of the First Americans. Princeton University Press.

Mayr, G., 2017. Evolution of avian breeding strategies and its relation to the habitat preferences of Mesozoic birds. Evol. Ecol. 31 (1), 131–141.

McDonald, D.B., 1989. Correlates of male mating success in a lekking bird with male-male cooperation. Anim. Behav. 37, 1007–1022.

McPhee, B.W., Bordy, E.M., Sciscio, L., Choiniere, J.N., 2017. The sauropodomorph biostratigraphy of the Elliot Formation of southern Africa: tracking the evolution of Sauropodomorpha across the Triassic–Jurassic boundary. Acta Palaeontol. Pol. 62 (3), 441–465.

Meng, Q., Liu, J., Varricchio, D.J., Huang, T., Gao, C., 2004. Parental care in an ornithischian dinosaur. Nature 431 (7005), 145–146.

Mikhailov, K.E., Bray, E.S., Hirsch, K.F., 1996. Parataxonomy of fossil egg remains (Veterovata): principles and applications. J. Vertebr. Paleontol. 16 (4), 763–769.

Mitoyen, C., Quigley, C., Fusani, L., 2019. Evolution and function of multimodal courtship displays. Ethology 125 (8), 503–515.

Mohabey, D.M., 1996. A new oospecies, *Megaloolithus matleyi*, from the Lameta Formation (Upper Cretaceous) of Chandrapur district, Maharashtra, India, and general remarks on the palaeoenvironment and nesting behaviour of dinosaurs. Cretac. Res. 17 (2), 183–196.

Moklestad, T., Lucas, S.G., 2023. Elongated, non-avian theropod tracks (pes scrapes) in the Cretaceous of Colorado may represent failed attempts at nest building. In: New Mexico Museum of Natural History and Science: Bulletin 94, vol 94.

Molina-Pérez, R., Larramendi, A., Connolly, D., Cruz, G.Á.R., 2019. Dinosaur Facts and Figures: The Theropods and Other Dinosauriformes. Princeton University Press.

Moll, E.O., 1979. Reproductive cycles and adaptations. In: Harless, M., Morlock, H. (Eds.), Turtles: Perspectives and Research, pp. 305–332.

Montellano, M., 1988. *Alphadon halleyi* (Didelphidae, Marsupialia) from the Two Medicine Formation (Late Cretaceous, Judithian) of Montana. J. Vertebr. Paleontol. 8 (4), 378–382.

Mueller-Towe, I.J., Sander, P.M., Schuller, H., Thies, D., 2002. Hatching and infilling of dinosaur eggs as revealed by computed tomography. Palaeontographica Abteilung A-Stuttgart- 267 (1/6), 119–168.

Murray, C.M., Crother, B.I., Doody, J.S., 2020. The evolution of crocodilian nesting ecology and behavior. Ecol. Evol. 10 (1), 131–149.

Nagy, J., Hauber, M.E., Hartley, I.R., Mainwaring, M.C., 2019. Correlated evolution of nest and egg characteristics in birds. Anim. Behav. 158, 211–225.

Norell, M.A., Balanoff, A.M., Barta, D.E., Erickson, G.M., 2018. A second specimen of *Citipati osmolskae* associated with a nest of eggs from Ukhaa Tolgod, Omnogov Aimag, Mongolia. Am. Mus. Novit. 2018 (3899), 1–44.

Norell, M.A., Clark, J.M., Chiappe, L.M., Dashzeveg, D., 1995. A nesting dinosaur. Nature 378 (6559), 774–776.

Oftedal, O.T., 2012. The evolution of milk secretion and its ancient origins. Animal 6 (3), 355–368.

Oliveira, C.E., Santucci, R.M., Andrade, M.B., Fulfaro, V.J., Basilio, J.A., Benton, M.J., 2011. Crocodylomorph eggs and eggshells from the Adamantina Formation (Bauru Group), Upper Cretaceous of Brazil. Palaeontology 54 (2), 309–321.

Owen, R., 1854. Descriptive Catalogue of the Fossil Organic Remains of Reptilia and Pices Contained in the Museum of the Royal College of Surgeons of England.

Pickering, S.P.C., Berrow, S.D., 2001. Courtship behaviour of the wandering albatross *Diomedea exulans* at Bird Island, South Georgia. Mar. Ornithol. 29, 29–37.

Pienkowski, M.W., 1984. Breeding biology and population dynamics of Ringed plovers *Charadrius hiaticula* in Britain and Greenland: nest-predation as a possible factor limiting distribution and timing of breeding. J. Zool. 202 (1), 83–114.

Pipoly, I., Bókony, V., Kirkpatrick, M., Donald, P.F., Székely, T., Liker, A., 2015. The genetic sex-determination system predicts adult sex ratios in tetrapods. Nature 527 (7576), 91–94.

Porges, S.W., 1998. Love: an emergent property of the mammalian autonomic nervous system. Psychoneuroendocrinology 23 (8), 837–861.

Pu, H., Zelenitsky, D.K., Lü, J., Currie, P.J., Carpenter, K., Xu, L., Koppelhus, E.B., Jia, S., Xiao, L., Chuang, H., Li, T., 2017. Perinate and eggs of a giant caenagnathid dinosaur from the Late Cretaceous of central China. Nat. Commun. 8 (1), 14952.

Rainforth, E.C., 2003. Revision and re-evaluation of the Early Jurassic dinosaurian ichnogenus *Otozoum*. Palaeontology 46 (4), 803–838.

Rauhut, O.W., Foth, C., Tischlinger, H., 2018. The oldest *Archaeopteryx* (Theropoda: Avialiae): a new specimen from the Kimmeridgian/Tithonian boundary of Schamhaupten, Bavaria. PeerJ 6, e4191.

Reisz, R.R., Evans, D.C., Roberts, E.M., Sues, H.D., Yates, A.M., 2012. Oldest known dinosaurian nesting site and reproductive biology of the Early Jurassic sauropodomorph *Massospondylus*. Proc. Natl. Acad. Sci. USA 109 (7), 2428–2433.

Reisz, R.R., Scott, D., Sues, H.D., Evans, D.C., Raath, M.A., 2005. Embryos of an Early Jurassic prosauropod dinosaur and their evolutionary significance. Science 309 (5735), 761–764.

Ribeiro, V., Mateus, O., Holwerda, F., Araújo, R., Castanhinha, R., 2014. Two new theropod egg sites from the late Jurassic Lourinhã Formation, Portugal. Hist. Biol. 26 (2), 206–217.

Romer, A.S., Price, L.I., 1939. The oldest vertebrate egg. Am. J. Sci. 237 (11), 826–829.

Royle, N.J., Smiseth, P.T., Kölliker, M., 2012. The evolution of parental care. Oxford University Press, Oxford.

Rubalcaba, J.G., Fuentes, D., Veiga, J.P., Polo, V., 2017. Nest decoration as social signals by males and females: greenery and feathers in starling colonies. Behav. Ecol. 28 (5), 1369–1375.

Sánchez, M.L., Gómez, M.J., Heredia, S., 2006. Sedimentología y paleoambientes del Subgrupo Río Colorado (Cretácico Superior), Grupo Neuquén, en las bardas de la ciudad de Neuquén y alrededores. Rev. Asoc. Geol. Argent. 61 (2), 236–255.

Sander, P.M., 2008. Upper Cretaceous titanosaur nesting sites and their implications for sauropod dinosaur reproductive biology. Palaeontographica: Beiträge zur Naturgeschichte der Vorzeit. Abt. A, Palaeozoologie, Stratigraphie 284 (4), 69.

Sanz, J.J., García-Navas, V., 2011. Nest ornamentation in blue tits: is feather carrying ability a male status signal? Behav. Ecol. 22 (2), 240–247.

Schweitzer, M.H., Jackson, F.D., Chiappe, L.M., Schmitt, J.G., Calvo, J.O., Rubilar, D.E., 2002. Late Cretaceous avian eggs with embryos from Argentina. J. Vertebr. Paleontol. 22 (1), 191–195.

Sternberg, C.M., 1956. A juvenile hadrosaur from the oldman formation of Alberta. In: Edmond Cloutier. Queen's Printer.

Tanaka, K., Kobayashi, Y., Zelenitsky, D.K., Therrien, F., Lee, Y.N., Barsbold, R., Kubota, K., Lee, H.J., Chinzorig, T., Idersaikhan, D., 2019. Exceptional preservation of a Late Cretaceous dinosaur nesting site from Mongolia reveals colonial nesting behavior in a non-avian theropod. Geology 47 (9), 843–847.

Tanaka, K., Zelenitsky, D.K., Lü, J., DeBuhr, C.L., Yi, L., Jia, S., Ding, F., Xia, M., Liu, D., Shen, C., Chen, R., 2018. Incubation behaviours of oviraptorosaur dinosaurs in relation to body size. Biol. Lett. 14 (5), 20180135.

Tanaka, K., Zelenitsky, D.K., Therrien, F., 2015. Eggshell porosity provides insight on evolution of nesting in dinosaurs. PLoS One 10 (11), e0142829.

Thulborn, R.A., 1989. The gaits of dinosaurs. In: Gillette, D.D., Lockley, M.,J. (Eds.), Dinosaur Tracks and Traces. Cambridge University, Cambridge, pp. 40–50.

Thulborn, R.A., 1992. Nest of the dinosaur *Protoceratops*. Lethaia 25 (2), 145–149.

Tyndale-Biscoe, C.H., Renfree, M., 1987. Reproductive Physiology of Marsupials. Cambridge University Press.

Varricchio, D.J., Jackson, F.D., 2016. Reproduction in Mesozoic birds and evolution of the modern avian reproductive mode. Auk 133 (4), 654–684.

Varricchio, D.J., Jackson, F., Trueman, C.N., 1999. A nesting trace with eggs for the Cretaceous theropod dinosaur *Troodon formosus*. J. Vertebr. Paleontol. 19 (1), 91–100.

Varricchio, D.J., Jackson, F., Borkowski, J.J., Horner, J.R., 1997. Nest and egg clutches of the dinosaur *Troodon formosus* and the evolution of avian reproductive traits. Nature 385 (6613), 247–250.

Varricchio, D.J., Martin, A.J., Katsura, Y., 2007. First trace and body fossil evidence of a burrowing, denning dinosaur. Proc. Biol. Sci. 274 (1616), 1361–1368.

Varricchio, D.J., Moore, J.R., Erickson, G.M., Norell, M.A., Jackson, F.D., Borkowski, J.J., 2008. Avian paternal care had dinosaur origin. Science 322 (5909), 1826–1828.

Vila, B., Galobart, A., Oms, O., Poza, B., Bravo, A.M., 2010a. Assessing the nesting strategies of Late Cretaceous titanosaurs: 3-D clutch geometry from a new megaloolithid eggsite. Lethaia 43 (2), 197–208.

Vila, B., Jackson, F.D., Fortuny, J., Sellés, A.G., Galobart, À., 2010b. 3-D modelling of megaloolithid clutches: insights about nest construction and dinosaur behaviour. PLoS One 5 (5), e10362.

Wang, C.Z., Wang, J., Hu, B., Lu, X.H., 2016. Trace fossils and sedimentary environments of the Upper Cretaceous in the Xixia basin, southwestern Henan Province, China. Geodin. Acta 28 (1–2), 53–70.

Wang, X., Kellner, A.W., Jiang, S., Cheng, X., Wang, Q., Ma, Y., Paidoula, Y., Rodrigues, T., Chen, H., Sayão, J.M., Li, N., 2017. Egg accumulation with 3D embryos provides insight into the life history of a pterosaur. Science 358 (6367), 1197–1201.

Wang, X., Kellner, A.W., Jiang, S., Wang, Q., Ma, Y., Paidoula, Y., Cheng, X., Rodrigues, T., Meng, X., Zhang, J., Li, N., 2014. Sexually dimorphic tridimensionally preserved pterosaurs and their eggs from China. Curr. Biol. 24 (12), 1323–1330.

Wiemann, J., Yang, T.R., Sander, P.N., Schneider, M., Engeser, M., Kath-Schorr, S., Müller, C.E., Sander, P.M., 2017. Dinosaur origin of egg color: oviraptors laid blue-green eggs. PeerJ 5, e3706.

Wiley, R.H., 1991. Lekking in birds and mammals: behavioral and evolutionary issues. In: Advances in the Study of Behavior, vol 20. Academic Press, pp. 201–291.

Woolley, S.C., Sakata, J.T., Crews, D., 2004. Evolutionary insights into the regulation of courtship behavior in male amphibians and reptiles. Physiology & behavior 83 (2), 347–360.

Xing, L., Lockley, M.G., Li, C., Klein, H., Li, J., Wang, D., 2023. Two new dinosaur tracksites from the Jurassic Guizhou Province, China. Hist. Biol. 35 (1), 49–57.

Xing, L., Lockley, M.G., Yang, G., Cao, J., Benton, M., Xu, X., Zhang, J., Klein, H., Persons, W.S., Kim, J.Y., Peng, G., 2016. A new *Minisauripus* site from the Lower Cretaceous of China: tracks of small adults or juveniles? Palaeogeogr. Palaeoclimatol. Palaeoecol. 452, 28–39.

Xing, L., Niu, K., Ma, W., Zelenitsky, D.K., Yang, T.R., Brusatte, S.L., 2022. An exquisitely preserved in-ovo theropod dinosaur embryo sheds light on avian-like prehatching postures. iScience 25 (1).

Xing, L., Wang, D., Li, G., McKellar, R.C., Bai, M., Chen, H., Evans, S.E., 2021. Possible egg masses from amphibians, gastropods, and insects in mid-Cretaceous Burmese amber. Hist. Biol. 33 (7), 1043–1052.

Xu, L., Xie, J., Zhang, S., Choi, S., Kim, N.H., Gao, D., Jin, X., Jia, S., Gao, Y., 2022. Fossil turtle eggs from the Upper Cretaceous Gaogou Formation, Xiaguan-Gaoqiu basin, Neixiang County, Henan Province, China: interpretation of the transformation from aragonite to calcite in fossil turtle eggshell. Cretac. Res. 134, 105166.

Yan, Y., Xia, B., Lin, G., Cui, X., Hu, X., Yan, P., Zhang, F., 2007. Geochemistry of the sedimentary rocks from the Nanxiong Basin, South China and implications for provenance, paleoenvironment and paleoclimate at the K/T boundary. Sediment. Geol. 197 (1–2), 127–140.

Yang, H.Q., Zhao, L., Han, Q.H., Fang, S.G., 2017. Nest site preference and fidelity of Chinese alligator (*Alligator sinensis*). Asian Herpetol Res 8, 244–252.

Yang, T.R., van Heteren, A., Wiemann, J., Chen, C.J., Spiering, B., Farke, A., MacKenzie, A., Miller-Camp, J., 2016. Communal nesting behavior of dinosaurs revealed by statistical analyses of phosphorus distribution in, and external morphology of, eggshells. In: SVP 76th Annual Meeting, Program and Abstracts, vol 255.

Yang, T.R., Wiemann, J., Xu, L., Cheng, Y.N., Wu, X.C., Sander, P.M., 2019. Reconstruction of oviraptorid clutches illuminates their unique nesting biology. Acta Palaeontol. Pol. 64 (3).

Zelenitsky, D.K., Therrien, F., 2008. Unique maniraptoran egg clutch from the Upper Cretaceous Two Medicine Formation of Montana reveals theropod nesting behaviour. Palaeontology 51 (6), 1253–1259.

Zelenitsky, D.K., Carpenter, K., Currie, P.J., 2000. First record of elongatoolithid theropod eggshell from north America: the Asian oogenus *Macroelongatoolithus* from the Lower Cretaceous of Utah. J. Vertebr. Paleontol. 20 (1), 130–138.

Zelenitsky, D.K., Therrien, F., Joyce, W.G., B. Brinkman, D., 2008. First fossil gravid turtle provides insight into the evolution of reproductive traits in turtles. Biol. Lett. 4 (6), 715–718.

Zhao, Q., Benton, M.J., Xu, X., Sander, P.M., 2013. Juvenile-only clusters and behaviour of the Early Cretaceous dinosaur *Psittacosaurus*. Acta Palaeontol. Pol. 59 (4), 827–833.

Zhen, S., Li, J., Chen, W., Zhu, S., 1994. Dinosaur and bird footprints from the Lower Cretaceous of Emei County, Sichuan. Mem. Beijing Natural History 54, 105–120.

PART IV

The ichnology of geoconservation

CHAPTER

14

Vertebrate ichnosites as geological heritage: A global synthesis of the best practices in geoconservation

Carlos Neto de Carvalho[1,2], *Adrian P. Hunt*[3], *Luis Alcalá*[4], *Tobias Fischer*[5], *Charles Helm*[6], *Kevin Page*[7,8], *Andrea Baucon*[9], *Evelyn Kustatscher*[10], *Girolamo Lo Russo*[11], *Luca Arena*[9], *Karen Moreno*[12] *and Lida Xing*[13]

[1]Geology Office of the Municipality of Idanha-a-Nova, Naturtejo UNESCO Global Geopark, Idanha-a-Nova, Portugal; [2]Earth Surface Processes Group (RG2) – Instituto D. Luiz, Faculdade de Ciências da Universidade de Lisboa, Lisbon, Portugal; [3]Flying Heritage and Combat Armor Museum, Everett, WA, United States; [4]Parque de las Ciencias, Granada, Spain; [5]Natur- und UNESCO Geopark TERRA.vita, Osnabrück, Germany; [6]African Centre for Coastal Palaeoscience, Nelson Mandela University, Gqeberha, South Africa; [7]Geodiversity & Heritage, Thornedges, Longbarn, Sandford, Crediton, Devon, United Kingdom; [8]Camborne School of Mines (University of Exeter), Penryn Campus, Penryn, United Kingdom; [9]DISTAV, Università di Genova, Genova, Italy; [10]Museum of Nature South Tyrol, Bolzano, Italy; [11]Museo di Storia Naturale di Piacenza, Piacenza, Italy; [12]Universidad Austral de Chile, Facultad de Ciencias, Instituto de Ciencias de la Tierra, Valdivia, Chile; [13]Department of Natural History, Tyrolean State Museums, Hall in Tirol, Austria

1. Introduction

For the last 3 decades, an upsurge of interest among the scientific community and policy makers for the heritage significance of geology has been registered in several parts of the world, in relation to the development of geoconservation, but especially geotourism and

Vertebrate Ichnology
https://doi.org/10.1016/B978-0-443-38351-9.00014-1

geopark projects. Geodiversity can be defined as the natural diversity of geological (rocks, minerals, fossils), geomorphological (landforms, topography, physical processes), soil, and hydrological features, including their assemblages, structures, systems, and contributions to landscapes (Gray, 2013). Geoheritage (a blend of geology and heritage) is the geological aspect of natural and cultural heritage. "Geoheritage encompasses global, national, state-wide, and local features of geology, at all scales, that are intrinsically important sites or culturally important sites offering information or insights into the evolution of the Earth; or into the history of science, or that can be used for research, teaching, or reference" (Brocx and Semeniuk, 2007, p. 53). This means that only scientific and culturally meaningful areas or territories of geodiversity, or geosites, can be considered heritage, but they also depend on the scale of the demand for investment in research, protection, and use for cultural, educational, and tourist purposes. Crucially, many countries now have national inventories and legislative frameworks for the selection and protection of their key geological and paleontological sites.

At a global level, the importance of protecting Geoheritage was recognized by UNESCO in 1972 under the World Heritage Convention, which considers that "deterioration or disappearance of any item of the natural heritage constitutes a harmful impoverishment of the heritage of all the nations of the world," and in 1991 with the Declaration of the Rights of the Memory of the Earth. In 2004, the Council of Europe adopted the Recommendation Conservation of the Geoheritage and Areas of Special Geological Interest. The years 2007–09 were proclaimed the International Year of Planet Earth, and, in 2008, the resolution Conservation of Geodiversity and Geoheritage by the International Union for Conservation of Nature was approved. The establishment of the European Geoparks Network in 2000 and the Global Geoparks Network in 2004 led to another significant enhancement in geoconservation activities. Geoparks are sites of international geological significance, highlighted as part of an inventory of geosites, and are managed using a holistic concept of sustainable development. This approach is bottom-up, empowering local communities. Conservation of geological heritage and promotion of geotourism needs to take into account all natural and cultural heritages of the area and at the same time to improve the environmental, economic, and social conditions following a participative action plan. This action plan is evaluated every 4 years, in this way constantly pushing forward the sustainable conservation and use of geoheritage and the interests and demands of local communities.

With the 38th session of UNESCO's General Conference held in Paris on November 17, 2015, the Charter of the International Geoscience and Geoparks Program (IGGP) was adopted by the 195 member states of the United Nations, as a result of which the Global Geoparks Network became an integral part of UNESCO (such as World Heritage sites and Biosphere Reserves). At present, there are 229 UNESCO Global Geoparks in 50 countries (UNESCO, 2025). While Biosphere Reserves focus on the harmonized management of biological and cultural diversity, and World Heritage sites promote the conservation of natural and cultural sites of outstanding universal value (including geological ones such as the Messel Pit and the Dolomites as examples), UNESCO Global Geoparks give international recognition to sites that promote the importance and significance of protecting the Earth's geodiversity through actively engaging with local communities.

Beyond World Heritage sites and Global Geoparks, several other key initiatives and groups have been established to promote the recognition of geological and paleontological sites at a global level, including the NGO, ProGEO from 1993 (www.progeo.ngo), the Geoheritage Specialist Group (GSG) of the World Commission for Protected Areas (part of the International Union for the Conservation of Nature) (https://iucn.org/commissions/world-commission-protected-areas/our-work/geoheritage), and the International Union for Geological Sciences, International Commission on Geoheritage (https://iugs-geoheritage.org/). Crucially, ProGEO and IUCN's GSG have collaborated to establish a series of resolutions within IUCN to ensure the recognition of the importance of geological and paleontological heritage within the nature conservation system globally, leading to a proposal of "Key Geoheritage Areas" (KGAs) as global designations for such sites (https://portals.iucn.org/library/sites/library/files/resrecfiles/WCC_2020_RES_074_EN.pdf).

Geoconservation of in loco, open-air geosites is a key point of discussion in UNESCO Global Geoparks, as well as in natural protected areas. According to the European Association for Conservation of Geological Heritage (ProGEO, 2017), "Geoconservation is defined as a set of actions aimed at informing the management of geological sites, in- or ex-situ, comprising inventory and assessment, conservation, statutory protection, interpretation and monitoring of site." The academic assessment of geosites is usually based on criteria that take into account their scientific, educational, and tourism values, as well as fragility and vulnerability (Brilha, 2016). Frequently, geoheritage assessments take into account the scientific properties recognized by geoscientists. But, other factors, such as those related to the social role attributed to geological sites by local communities, should also be considered.

The integration of different points of view can contribute to the definition and qualification of heritage geosites within a sociocultural perspective (Endere and Prado, 2015). In practice, these assessments are adapted and modified by local authorities and attend also to different criteria, such as the historical and mythological relevance for local communities, proximity to urban areas and schools, conservation of green areas in towns and cities, definition of environmentally protected areas, location in public or private property and accessibilities, budget available for conservation, interpretation and promotion. Conservation of geological heritage became regulated in some countries under national and regional laws, and their management is under different organizations, from national and regional authorities, municipalities, NGOs, universities, or private. Regarding the integration of the geoparks concept in national legislation, good practices are reported for the European Union in Italy, Spain, and Portugal (Nikolova and Sinnyovsky, 2019).

Open-air paleontological sites are highly susceptible to weathering and anthropogenic activity even when they are correctly managed and protected. Geoconservation measures are increasingly applied to significant paleontological outcrops. Trace fossils represent an important component of Geoheritage, offering crucial insights into the diversity and evolution of life and environments through time. They can also be drivers for developing new tourism services with great innovation, especially in rural or remote areas (Laws and Scott, 2003; Cobos and Alcalá, 2017). Geotourism is a new and innovative segment of tourism that focuses on geoheritage. It involves the sustainable and socially responsible organization of tourism resources (their protection, conservation, and valorization). This segment also emphasizes the provision of products and services and the promotion of destinations with a focus on geoheritage (e.g., Hose, 1995; Dowling and Newsome,

2005; Neto de Carvalho et al., 2011; Farsani et al., 2011). Ichnosites (trace fossil relevant sites) are therefore a cultural and economic resource, but also fragile geoheritage that requires special geoconservation measures (Coster and Legal, 2021). However, the practice of protecting ichnosites in loco was almost nonexistent until the nineties of the 20th century, and with some remarkable exceptions, it all started with the impressive dinosaur tracksites.

Despite the symbolic meanings and interpretations that trace fossils have held for hunter-gatherers or local communities over thousands of years (e.g., Baucon et al., 2008; Troiano et al., 2024), it was only with the Scientific Revolution and the development of the concept of Heritage that the need to study and protect the fossils for future generations emerged. The first attempts to remove vertebrate tracks from their original stratigraphic context for preservation in museums began after the pioneering work of William Buckland and Edward Hitchcock in the early 1800s, who first understood these tracks to be biogenic in nature. For more than one century of the scientific development of paleontology, vertebrate tracks were treated mostly as curiosities without biostratigraphic application or evolutionary interest. In addition, their occurrences in very large exposures made it impossible to remove them to museums. This led to the sampling of footprints that depleted some of the classic tracksites.

However, the perspective changed dramatically and worldwide through the influential studies and energy of a single paleontologist, Martin Lockley (Lucas et al., 2024b). He started in the early 1980s with the study of dinosaur tracksites in Colorado, USA, with the aim of understanding dinosaur paleobiology and habitats. Soon Lockley would expand his research from the USA to Eastern Asia, Europe, South America, and Africa and included bird, pterosaur, synapsid, mammal, and hominin footprints (see references in Lucas et al., 2024a,b). The Denver Dinosaur Trackers Research Group, led by Martin Lockley, amassed thousands of fossil footprints and replicas in the University of Colorado at Denver Tracks Museum (1996–2012), which Lockley founded and was director of until 2 years before closing the doors. As a result of the many scientific outreach initiatives driven by the Research Group's studies, Martin Lockley became involved in several projects focused on the onsite protection and conservation of dinosaur tracksites in the USA, South Korea, Spain, Portugal, and Bolivia. It is clear from the number of papers referenced in this chapter with Martin Lockley as a coauthor that he made fundamental contributions to the research and protection of numerous and diverse tracksites. Therefore, this chapter is a tribute to Martin Lockley's work on the definition of vertebrate ichnosites as Geoheritage and their educational and geotourism value and use. It intends to provide a worldwide summary (Fig. 14.1) of best practices in geoconservation, providing brief scientific relevance of the ichnosites with a bibliography of references, but focusing on individual countries' policies and initiatives regarding legal protection, conservation strategies of vertebrate ichnosites and science outreach projects being implemented. Although the focus is more comprehensive with respect to onsite geoconservation of vertebrate ichnosites dating from the Carboniferous to the Quaternary, this chapter also introduces the main museum exhibitions devoted to vertebrate ichnology.

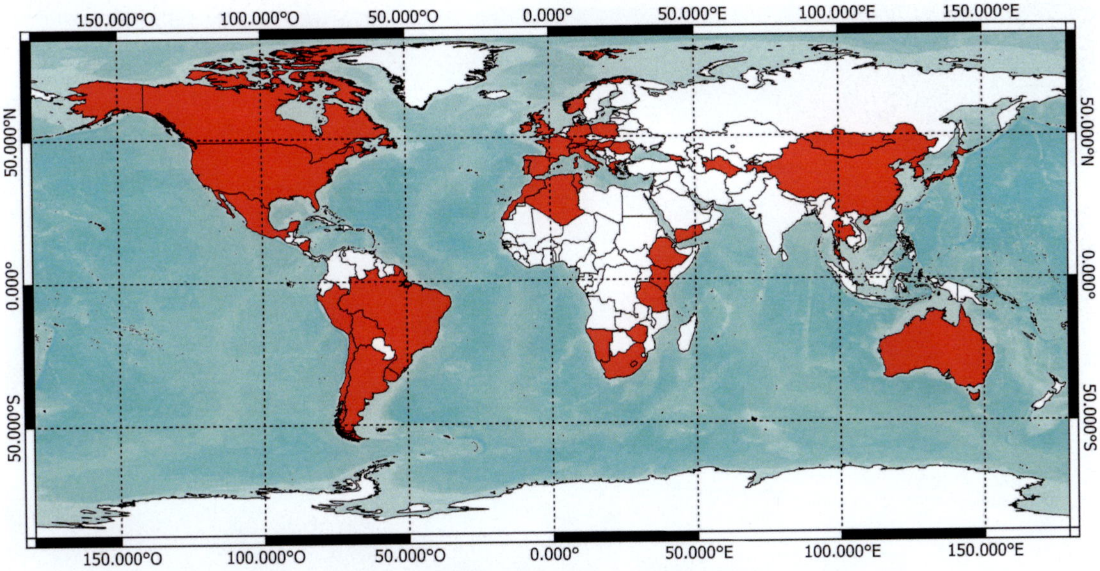

FIGURE 14.1 World map highlighting countries with protected vertebrate ichnological geoheritage reviewed in this chapter. **North America**: Canada, Mexico, Nicaragua, USA; **South America**: Argentina, Bolivia, Brazil, Chile, Peru, Uruguay; **Europe**: Croatia, France, Georgia, Germany, Greece, Hungary, Ireland, Italy, Norway, Poland, Portugal, Romania, Slovenia, Spain, Switzerland, UK. **Africa**: Algeria, Ethiopia, Kenya, Lesotho, Morocco, Namibia, South Africa, Tanzania, Tunisia, Zimbabwe. **Asia**: China, Japan, Mongolia, North Korea, Palestine, Republic of Korea, Republic of Yemen, Tajikistan, Thailand, Turkmenistan. **Oceania**: Australia. Map lines delineate study areas and do not necessarily depict accepted national boundaries.

2. Global overview of ichnological sites

2.1 North America

2.1.1 Canada

Tumbler Ridge is a remote coal-mining community in northeastern British Columbia, with a population of around 2500. It is situated in an area of Cretaceous track-bearing deposits, with over 80 identified tracksites. In 2000, two young local boys correctly identified the first dinosaur tracks in the region after they capsized while tubing down rapids on a creek close to town. At the time, the town's coal mines had both closed, and the community was predicted to become a "ghost town." The discovery recast the community in a positive light and led to the development of the Tumbler Ridge Museum. It is not an exaggeration to state that ichnology (and geoheritage) saved the town.

The presence of a museum dedicated largely to ichnology (Fig. 14.2A) was an important factor in the designation of the Tumbler Ridge UNESCO Global Geopark in 2014 (at the time, the second such designation in North America). This status, while highly important in raising the profile of regional geoheritage, does not automatically confer protection on the numerous ichnosites and body-fossil sites that have subsequently been identified. Furthermore, much of the topography is covered by boreal forest, and exposures are typically confined to canyon settings or are exposed by industrial activity. Most tracksites are therefore

FIGURE 14.2 Examples of geoconservation of vertebrate ichnosites in North America. (A) Dinosaur Discovery Gallery in the Tumbler Ridge Museum, Canada, with the focus on the regional ichnosites; (B) recovering a large block containing Cretaceous swim traces of giant crocodylians by crane from beside a highway, Tumbler Ridge UNESCO Global Geopark, Canada. (C) Blue Beach Fossil Museum, a private museum in Nova Scotia with a large collection of Carboniferous vertebrate tracks. (D) Digital model of the Acahualinca human footprints, which includes detailed bilingual exhibition on the excavations and different theories surrounding the footprints. (E) Clayton Lake State Park, New Mexico. *(D) From Lucas et al. (2024a), Fig. 5.*

ephemeral and are liable to be destroyed through floods or further industrial activity. Physical recovery and digital recording (through photogrammetry) thus often take precedence over site preservation and protection.

The result can be viewed as a "bottom-up" geoheritage success story. The UNESCO Global Geopark designation brings welcome positive attention to the region and its geoheritage, while the Museum accumulates and exhibits priceless specimens, including many holotypes. Private foundations, industry donors, and various levels of government have provided funding. The Tumbler Ridge municipal government and the Peace River Regional District have provided exemplary support. In both cases, referenda were held to gauge the level of public support, and the results were overwhelmingly in favor. An excellent collaborative relationship with industry has been developed, through which new track-bearing exposures are rapidly reported, and geoheritage projects are funded either directly or through in-kind donations. For example, one mining company built a rough, 200 m long road to access globally important crocodilian traces that had been identified on the mine property and then recovered and transported four large, trace-bearing rocks to the Museum (Lockley et al., 2021). In another case, a crane company donated funds, personnel, and equipment to recover giant crocodilian trace fossils from beside a highway (Fig. 14.2B), which had to be shut down for a few hours (Plint et al., 2022). A recovery by crane of a heavy slab containing tyrannosaurid tracks featured the town's mayor as the crane operator. Numerous recoveries of priceless specimens, including holotypes, have been accomplished by helicopter from remote locations, with funding from industry. Others have been recovered by canoe or on the backs of volunteers and safely accessioned in the Museum.

Two dinosaur tracksites close to Tumbler Ridge have been opened to tourists, under the supervision of museum guides. Occasional acts of vandalism have occurred at both sites, which reinforce the policy of not disclosing other ichnosite locations to the public.

The Gething Formation (Lower Cretaceous: Aptian) of the Peace River Canyon has been known as a significant source of tetrapod tracks since the 1920s (e.g., McLearn, 1923, 1931; Sternberg, 1931, 1932). Many tracks were flooded after the construction of the Peace Canyon Dam. The Twin Sisters Track Site, discovered in 2008, is in the vicinity of Carbon Creek and is distinct from most of the other Gething Formation sites in that it is not exposed on the bank of a creek or river but rather at the top of a large, flat hill (McCrea et al., 2014, 2017). The Carbon Creek (or "Twin Sisters") track site (British Columbia) was granted some protected status in the last few years through Section 14. status via the British Columbia Heritage Branch within the Ministry of Forests, Lands and National Resources Operations. But this did not stop it from being vandalized (although the perpetrators were caught and brought to justice with jail terms and fines). An area of 750 m^2 has been uncovered with 1200 tracks and trackways, including a theropod trackway with 23 tracks, a pair of sauropod trackways resembling *Parabrontopodus*, 12 vertebrate ichnotaxa including *Amblydactylus kortemeyeri*, *Magnoavipes* isp., large tetradactyl theropod tracks, and several tracks of a large ignotornid. Most of the tracks and trackways belonging to theropods are *Irenesauripus acutus*, but the site also includes the ornithopod tracks *Gypsichnites pascensis*. This tracksite is part of the Northern Dinosaur Trail that also includes the Grande Cache Dinosaur Track Sites, Philip J. Currie Dinosaur Museum, and the Tumbler Ridge Museum and Tumbler Ridge UNESCO Global Geopark.

An example of successful site protection and preservation comes from neighboring Alberta, the Canadian province immediately to the east of British Columbia. Large-scale tracksites in the Grande Cache area were designated as a provincial heritage resource in 2006. The majority of the tracks are found in three main areas within the former Smoky River Coal Mine where they were exposed after the removal of Coal Seam No. 4 of the Gates

Formation (Lower Cretaceous) (Spivak et al., 2006). One tracksite has over 10,000 documented tracks. The majority of tracksites include only *Tetrapodosaurus borealis*. Several track types are also known from the tracksites, including at least four types of theropod tracks, several types of bird tracks, and possible crocodilian swim traces (McCrea, 2000, 2003; McCrea et al., 2014). The area was granted Provincial Heritage Resource status by the government of Alberta in 2006 (Spivak et al., 2006). The sites contain compelling evidence of the gregarious behavior of ankylosaurs, as well as avian track holotypes (McCrea, 2003; McCrea et al., 2014; Spivak et al., 2006).

Canada currently has five UNESCO Global Geoparks. The Cliffs of Fundy UNESCO Global Geopark on Canada's east coast also contains dinosaur tracks. They, too, are housed in a regional museum (the Fundy Geological Museum), and their profile is similarly raised by the presence of the Geopark.

In Nova Scotia, a large and significant tracksite was found in 1994 within the intertidal zone of Northumberland Strait. The fossiliferous area, at lowest tide, is 1000 m^2, and much of this is often covered by sand (Van Allen et al., 2005). Collection at the site is controlled under legislation of the Province of Nova Scotia, and a Heritage Research Permit is required from the Curator of Special Places, Nova Scotia Museum. The tracks occur in the Gzhelian to possibly Asselian (late Carboniferous-early Permian) Cape John Formation of the Pictou Group. The occurrence of the footprints within the world's only known in situ walchian conifer forest provides exceptional insight into paleoecology (Van Allen et al., 2005). A *Batrachichnus*—*Amphisauropus*-dominated ichnoassemblage also includes *Dromopus*, *Varanopus*, and *Dimetropus* (Van Allen et al., 2005; Lucas et al., 2021).

The coastal exposure at the public Blue Beach is almost 5 km in length, with relatively undisturbed sedimentary strata forming wide, wave-cut benches. Blue Beach yields the most important earliest vertebrate tracks from the early Carboniferous (Mansky and Lucas, 2013). The Tournaisian Horton Bluff Formation is exposed along the western side of the Avon estuary. The vertebrate ichnofossils include an unnamed sarcopterygian crawling trail, tetrapod resting traces, *Undichna*, *Attenosaurus*, *Batrachichnus*, *Hylopus*, *Pseudobradypus*, and *Palaeosauropus* as well as diverse bromalites (Mansky et al., 2012; Mansky and Lucas, 2013).

The Blue Beach Fossil Museum in Hantsport (Fig. 14.2C) is a private museum that contains a very large collection of nearly 2200 footprint-bearing slabs and 7000 fossilized bone specimens from the Early Carboniferous of the Blue Beach area (https://bluebeachfossilmuseum.com/). The museum provides educational tours. Also, the Museum of Natural History in Halifax, a part of the Nova Scotia Museum, holds Paleozoic trackways from Nova Scotia (https://naturalhistory.novascotia.ca/). It has public displays and educational programs. The Royal Ontario Museum in Toronto (https://www.rom.on.ca/en) is a large government museum with significant collections of vertebrate tracks from western Canada. Finally, the Canadian Museum of Nature in Ottawa (https://nature.ca/en/) has significant collections of vertebrate ichnofossils, notably many tracks and replicas from the Gething Formation of the Peace River Canyon of British Columbia described by Sternberg. The museum has extensive displays, including of vertebrate trace fossils.

2.1.2 *Mexico*

Dugés (1894) first described Cenozoic tracks from Mexico, but it was not until the 1970s that extensive study began of vertebrate ichnofossils (Rodríguez-de la Rosa et al., 2004; Bravo-Cuevas and Rodríguez-de la Rosa, 2014; Pichardo, 2016). Very few tracksites are accessible to the public. Sonora Dinosaur Park is a geo-ecotourism project located in the northeastern part of the state. The park has yielded a large number of Late Cretaceous dinosaur footprints and bones. Tracks pertain to hadrosaurs and include a trackway with a protective wall. The park is a tourism project that includes a paleontology museum and involves education, culture, conservation, tourism, research, and sustainable development of the communities.

Several Mexican museums have vertebrate tracks in their collections, including the Museo Regional del Instituto Nacional de Antropología e Historia Guadalajara (Cenozoic mammals and birds), Museo Regional de La Laguna, and the Museo del Desierto (Saltillo, Coahuila) (human footprints).

2.1.3 *Nicaragua*

Acahualinca is one of the most well-known and well-preserved human footprint sites in the Americas. It is located on the shores of Lake Managua, among a chain of volcanoes that were the source of the volcaniclastic material into which the footprints are impressed (Lockley et al., 2007). Mapping and paleontological work on the footprints at the site accelerated with Martin Lockley's work in 2006 (Lucas et al., 2024a; Fig. 14.2D), who, with collaborators, published detailed descriptions of the prints, maps of the trackways, and defined the new ichnogenus and ichnospecies *Hominipes modernus* from a holotype at this site (Lockley et al., 2007, 2009; Kim et al., 2008; see Walk the World: the hominin trackways chapter in this book). In all, 322 human tracks were mapped and documented. Besides, other vertebrate tracks were described by Lockley et al. (2008) (see the Quaternary Vertebrate Tracksites chapter in this book). Two radiocarbon readings bracket the footprint layer, giving an estimated age of between 6.5 and 5.8 ka BP (see Lucas et al., 2024a). However, Schmincke et al. (2009, 2010) correlated the tephras at Acahualinca to the Masaya Triple Layer, which is dated at 2120 ± 120 years BP. Lucas et al. (2024a) rightly argue for the need for more direct evidence to constrain these dates.

A shelter was first built over the trackways in 1942 to protect the footprints. New buildings were constructed in the 1980s, but in 2015, the tracks were flooded, and the museum was closed to the public for planning and for reconstruction of the museum buildings (Lucas et al., 2024a). In addition to footprints, the museum features a small collection of pottery and other items of archeological interest from several sites in Nicaragua. The Acahualinca Tracks Museum was considered one of the most important hominin ichnosites in the world by Kim et al. (2008) because it combines accessibility, a large sample of well-preserved trackways and reliable geochronological data.

2.1.4 *USA*

There are numerous vertebrate tracksites in the United States, and many are on federally managed land in the western interior. Herein the more prominent localities that are preserved for public access are listed. According to Lockley and Schumacher (2021), 10 of the 12 most important USA sites occur in the western states of Colorado, New Mexico, Utah, and Wyoming. Properties managed by the National Park Service, notably National Parks and most National Monuments, yield abundant tracks and coprolites (e.g., Santucci et al., 1998, 2006, 2014; Hunt et al., 2012b), but only the most significant areas are highlighted (see also Quaternary Vertebrate tracksites chapter in this book).

2.1.4.1 Alabama

The Stephen C. Minkin Paleozoic Footprint Site is on private land that used to constitute the Union Chapel Mine (Buta and Kopaska-Merkel, 2016). The site preserves abundant vertebrate (and invertebrate) traces from the Early Pennsylvanian (Buta et al., 2005; Hunt et al., 2005). The Minkin site is not generally open to the public, but visits are easily arranged. The Alabama Paleontological Society conducts monthly collecting expeditions, and interested parties are welcome to participate. Specimens of scientific value must be donated to a museum.

The site was found in 1999. State law in Alabama requires surface mines to be reclaimed shortly after mining operations end. The preservation effort was spearheaded by local and national geologists and amateur paleontologist Steve Minkin, and the locality was named after him following his untimely death. The Minkin site is now preserved due to cooperation between the New Acton Coal Mining Company, the Alabama Department of Conservation and Natural Resources, and avocational and academic paleontologists.

2.1.4.2 Arizona

The Tuba City Tracksite/Moenave Dinosaur Tracks is just off Highway US-160 E at Mile Marker 316 near Tuba City on the lands of the Navajo Nation. This well-known site tracksite preserves Early Jurassic theropod tracks (principally *Kayentapus*). The tracks were first found by Sam Welles of the University of California, Berkeley, in the late 1930s (Lockley and Hunt, 1995a,b). Navajo tribal members are present at the site to give tours of the tracks.

Grand Canyon National Park was not established specifically to preserve fossils, but it contains outstanding vertebrate tracks from Paleozoic strata that were first described by Charles Gilmore of the National Museum of Natural History (Gilmore, 1927; see Marchetti et al., 2020, and references therein). Tracks are readily accessible to the general public, particularly on the Hermit Trail, and several visitor centers display tracks. Caves in the park preserve the largest known accumulations of Pleistocene vertebrate coprolites (Mead et al., 2020; Hunt and Lucas, 2022, and references therein).

2.1.4.3 California

Outstanding Cenozoic vertebrate ichnofossils are documented from four localities in Death Valley National Park (Scrivner, 1984; Scrivner and Bottjer, 1986; Nyborg and Santucci, 1999; Santucci et al., 2014). The tracks consist of mammalian and bird traces preserved within fine-grained fluvial-lacustrine sediments. The most abundant and diverse vertebrate ichnofossil locality in the park is within Copper Canyon, where lakeshore deposits preserve 12 bird ichnospecies, five cat ichnospecies, five artiodactyl ichnospecies, three horse ichnospecies, two undescribed tridactyl tracks, and one proboscidean (Santucci et al., 2011). The Copper Canyon tracks are readily accessible to hikers in the park.

Anza-Borrego Desert State Park is managed by the California Department of Parks and Recreation. Tetrapod ichnofauna are present in the Vallecito-Fish Creek Basin and the Borrego Badlands (Remeika, 1999, 2001, 2006; Bucheim et al., 1999). Track sites occur through a sequence of stratigraphic units including the Yuha Formation, Palm Spring Formation, and the Canebrake Conglomerate. Those track sites range in age from early Pliocene to Pleistocene (Remeika, 1999, 2001, 2006). The ichnofauna includes records of a small anisodactyl bird, an aquatic musteloid, a canid, felids, a perissodactyl, a monodactyl equoid, a tayassuid, a cervid, camelids, a large llamoid, and proboscideans (Remeika, 1999, 2001, 2006).

2.1.4.4 Colorado

Dinosaur Ridge is a segment of the Dakota Hogback west of Denver and was designated the Morrison Fossil Area National Natural Landmark in 1973. The nonprofit group Friends of Dinosaur Ridge, which was cofounded by paleontologist Martin G. Lockley, is Jefferson County Open Space's licensed tour operator of Dinosaur Ridge and the Triceratops Trail. The Dinosaur Ridge area includes a main visitor center, the Martin G. Lockley Discovery Center, an Early Cretaceous tracksite on the ridge, and a trail with Late Cretaceous tracks in the adjacent town of Golden. The Early Cretaceous tracks are dominantly ornithopod (*Caririchnium*) with some theropod (*Magnoavipes*) and other tracks (Lockley, 2001, 2003; Lockley et al., 2001; Lockley and Hunt, 1994a). The Late Cretaceous tracks mainly pertain to the ceratopsian track *Ceratopsipes* (Lockley and Hunt, 1995a,b). Dinosaur Ridge has a variety of tours and educational programs and attracts approximately 250,000 people per year.

Picket Wire Canyonlands are located in the Comanche National Grasslands of southeastern Colorado, managed by the US Forest Service. The biggest attraction is the largest dinosaur tracksite in North America. This includes about 2100 tracks over an 8200 m^2 area. More than 1900 prints in about 130 separate trackways extend across a quarter mile of bedrock along the banks of the Purgatoire River (Schumacher and Lockley, 2023). The tracks were dubbed "Dinosaur Lake" by Lockley et al. (1986) The majority of the tracks pertain to a sauropod (*Parabrontopodus*) with trackways up to 88 m long, some with over 100 successive pes prints. Parallel sauropod trackways indicate gregarious behavior, including a mixed herd of subadult and adult individuals. There are also more than 300 theropod (*Megalosauripus*) tracks (Schumacher and Lockley, 2023). Visiting the tracksite involves a round-trip hike of 18 km.

The US Forest Service announced in early 2024 that they have purchased three parcels of land in the southwestern part of Colorado at an elevation of more than 2800 m, which include the West Golden Hill tracksite, which will be developed for public access. The tracks were found in 2021 on private land and are preserved in an Upper Jurassic unit that may be the Junction Creek Member of the Bluff Sandstone (Goodell et al., 2021). The exposed surface is 3000 m^2. The site preserves one long circuitous sauropod trackway with 134 consecutive tracks that may represent *Parabrontopodus* or *Brontopodus* (Goodell et al., 2021). This is the second largest sauropod tracksite in the United States. The site is accessed by a steep 4-mile round trip trail that gains almost 500 m in elevation. Signage will be added at the tracksite.

2.1.4.5 Connecticut

In 1846, a stone quarry was opened in Middlefield, Connecticut, on property owned by Wesley Coe, a professor at Yale University and later a curator of zoology at that university's

Peabody Museum. Dinosaur tracks were discovered in the quarry soon after excavations began. After Coe's death, the footprint-bearing portion of the quarry was donated to Yale and was named the Peabody Museum Outdoor Exhibit. In 1976, ownership of the small site was transferred to the town of Middlefield, and the quarry was renamed Powder Hill Dinosaur Park, which is open free to the public. The tracks occur in the Lower Jurassic East Berlin Formation and represent theropods—*Eubrontes*, *Anchisauripus*, and *Grallator*.

In 1966, 2000 dinosaur tracks were found during construction of a state government building in the city of Rocky Hill. The Dinosaur State Park opened in 1968, and the trackways were designated a registered National Natural Landmark by the US Department of the Interior. The facility is managed by the Connecticut Department of Energy and Environmental Protection. More than 500 of the tracks are now enclosed within the Exhibit Center's geodesic dome, and the remaining 1500 were reburied for preservation. The tracks are from the Lower Jurassic East Berlin Formation and pertain to the theropod ichnotaxa *Eubrontes* and *Anchisauripus*.

2.1.4.6 Massachusetts

Dinosaur Footprints Wilderness Reservation was purchased for the public in 1935 and is currently managed with the assistance of the Massachusetts Department of Conservation and Recreation. There is no admission charge. Hitchcock (1836) first noted theropod trackways at this locality. Ostrom (1972) described tracks in a very influential paper in which he identified 22 trackways of *Eubrontes giganteus*, of which 19 were parallel. This paper was important in interpreting gregarious behavior in dinosaurs, although this has been questioned by Getty et al. (2012, 2015). The track-bearing beds are in the Lower Jurassic Portland Formation. Getty et al. (2015) identified a total of 805 dinosaur tracks, including examples of the dinosaur ichnotaxa *Eubrontes*, *Anchisauripus*, *Grallator*, and *Anomoepus*, as well as the crocodylomorph ichnogenus *Batrachopus*.

2.1.4.7 New Mexico

Vertebrate (and invertebrate) trackways from the Robledo Mountains Formation (Lower Permian) form the core of Prehistoric Trackways National Monument and are among the most extensive and most studied Paleozoic ichnofaunas (e.g., Hunt et al., 1993; MacDonald, 1994, 1995; Haubold et al., 1995; Lucas and Heckert, 1995; Lucas et al., 1998a,b, 2011; Voigt et al., 2013). The discovery and conservation of the tracks owe much to the efforts of avocational paleontologist Jerry MacDonald. The vertebrate ichnotaxa include *Batrachichnus salamandroides*, *B.* aff. *salamandroides*, *Dimetropus leisnerianus*, *D. nicolasi*, *Dromopus lacertoides*, cf. *Erpetopus willistoni*, *"Gilmoreichnus hermitanus"*, *Hyloidichnus bifurcatus*, *Limnopus vagus*, *Lunichnium rotterodium*, *Robledopus macdonaldi*, *Serpentichnus robledoensis*, *S. sigmoidalis*, *S.* cf. *robledoensis*, *S. robledoensis*, *Undichna bina*, *U. britannica*, *U. britannica*, and *U. unisulca* (Voigt et al., 2013).

Most National Monuments are administered by the National Park Service, which has a conservation focus. However, this National Monument is governed by the US Bureau of Land Management whose mandate is to promote multiple use, including at this location quarrying and off-road vehicle access. Thus, ironically, the viewing of fossil trackways fossils at the National Monument is very limited. There are no developed sites interpreting the fossils. To preserve them for ongoing and future scientific study, some trackway fossils have been removed and transported to the New Mexico Museum of Natural History and Science, making up the Jerry MacDonald Paleozoic Trackways Collection. The Las Cruces Museum of Nature and Science also has 9 and 4.5 m long trackway slabs on display.

A camper at Clayton Lake in 1982 first discovered the abundant dinosaur tracks (>400) at the dam spillway in the upper Mesa Rica Sandstone and the lower Pajarito Formation, which are the middle and upper part of the Dakota Group locally (upper Albian). The Clayton Lake State Park is now open to the public with interpretive signage and a raised walkway around the tracksite (Fig. 14.2E). Currently, the site is being extensively restudied by Spencer G. Lucas and collaborators. The ichnofossils were initially studied by Gillette and Thomas (1985, 1989), and subsequently, they were reevaluated by several other workers (e.g., Bennett, 1993; Lockley and Hunt, 1995a,b; Lockley et al., 1992, 2000; Hunt and Lucas, 1998, 2004; Lucas and Dalman, 2016). The vast majority of the dinosaur tracks at Clayton Lake are of ornithopods, many of which can be assigned to *Caririchnium* isp. Theropod tracks are of two kinds assigned to *Magnoavipes* isp. and cf. *Irenesauripus* isp. One quadrupedal ornithischian trackway is present that has characters of *Deltapodus* and thus may represent an ankylosaur trackway (Lucas and Dalman, 2016).

White Sands National Park (formerly National Monument) is managed by the National Park Service and is open to the public for a fee. Here, extensive Pleistocene lacustrine facies have long yielded vertebrate tracks (e.g., Lucas et al., 2007a,b). In the past few years, abundant new tracks have been discovered including multiple hominin footprints (Bennett et al., 2019, 2020, 2021; Bustos et al., 2018, 2024). These hominin footprints may date back to between ~23 and 21 ka and indicate the presence of humans in North America during the Last Glacial Maximum (Bennett et al., 2021). Nevertheless, there is ongoing debate about the age of the human tracks (e.g., Haynes, 2022; Rachal et al., 2022; Oviatt et al., 2023). Nonhuman trace fossils include canid, felid, camelid, proboscidean, and xenarthran tracks (Bustos et al., 2024). These tracks are rapidly lost to soil erosion after exposure, and they need to be ideally studied within the first few hours of exposure before they dry out and/or are covered by a fine layer of salt (Bustos et al., 2024).

2.1.4.8 Texas

In 1940, R.T. Bird of the American Museum of Natural History collected segments of a sauropod and a theropod trackway from a site in the bed of the Paluxy River (Glen Rose Formation; Lower Cretaceous), in what is now Dinosaur Valley State Park (Bird, 1939, 1944). Now the area is administered by Texas Parks and Wildlife. Jim Farlow and colleagues have conducted extensive studies of the tracks including detailed photomosaics of extant Paluxy tracksites, using GIS technology and LiDAR images in combination with study of historic maps and photographs (Farlow, 1987; Farlow et al., 1989, 2012, 2015). Five main tracksite areas have been mapped within Dinosaur Valley State Park. Each of these areas has named individual track sites. Tracksites occur in more than one track layer, but the largest and most spectacular footprints occur in the Main Track layer. There are two dinosaur footprint morphotypes in the Main Track layer: sauropod trackways (*Brontopodus*) and more numerous tridactyl footprints, most or all of which were made by large theropods (possible ornithopod prints occur in a track layer stratigraphically higher than the Main Track layer) (Farlow et al., 2012, 2015). The trackways collected by R.T. Bird suggest that at least one theropod was following a sauropod. The portion of the "chase sequence" that Bird collected for the Texas Science and Natural History Museum is part of the type trackway of *Brontopodus* (Farlow et al., 1989).

Dinosaur tracks are preserved in the riverbed and are not always visible. River erosion contributes to footprint destruction, but freezing of the rock surface during the winter is more damaging (Farlow et al., 2012, 2015). There is a rough balance between the rates at which tracks are destroyed by erosion, and "new" tracks are being exposed (Farlow et al., 2012, 2015).

2.1.4.9 Utah

The Moccasin Mountain Dinosaur is near Kanab and is administered by the US Bureau of Land Management. The tracksite is in the Early Jurassic Navajo Sandstone and preserves hundreds of specimens of *Kayentapus*, *Grallator*, *Otozoum*, *Batrachopus*, and *Brasilichnium* (Milner et al., 2011). The site is located in a sandstone wash. There is a parking area and interpretive displays. Access requires a four-wheel drive, high clearance vehicle. Brochures and children's activity handouts are available at the BLM Kanab Visitor Center in Kanab.

Warner Valley Dinosaur Track Site is located near Washington and is under the jurisdiction of the US Bureau of Land Management. Over 400 tracks are preserved in the Springdale Member of the Kayenta Formation (Lower Jurassic). All tracks represent theropods and are identified as *Eubrontes* and *Grallator* (Milner et al., 2011). Access requires driving 15 km down a dirt road to the signed trailhead. The site is popular because of its large number of tracks.

Mill Canyon Dinosaur Trailhead Interpretive Site is north of Moab and is administered by the US Bureau of Land Management. The site contains over 200 tracks and traces from the Early Cretaceous Ruby Ranch Member of the Cedar Mountain Formation. Tracks represent at least 10 different types of animals, including nonavian theropods, sauropods, ornithopods, ankylosaurs, birds, and crocodiles. Ichnotaxa include *Brontopodus*, *Irenesauripus*, cf. *Caririchnium*, *Dromaeosauripus*, and cf. *Tetrapodosaurus* (Lockley et al., 2014a,b). The interpretive trail is 400 m long and is composed of a series of marked paths and boardwalks that form a loop around the tracksite. This site is one of the most ichnologically diverse tracksites known from the Early Cretaceous in North America.

Copper Ridge Dinosaur Tracks Interpretive Site is north of Moab on land managed by the US Bureau of Land Management and was discovered by Linda Dale Jennings Lockley in 1989. There is a short walk of less than 200 m from a parking lot with a kiosk to the tracksite, which is in the Salt Wash Member of the Upper Jurassic Morrison Formation. The trackway includes traces of a turning sauropod and four theropods, including *Hispanosauropus* (Foster et al., 2006; Foster, 2015).

Dinosaur Stomping Grounds is north of Moab on lands administered by the US Bureau of Land Management. It is in the same Klondike Bluffs area as the Copper Ridge site. There is a 6.5 km round trip hike to view abundant theropod tracks. The tracks are preserved at the interformational contact between white, reworked dune sandstones from the aeolian Moab Member of the Curtis Formation and red, sabkha facies siltstones and sandstones of the upper tongue of the Summerville Formation (Lockley, 2022). They are Late Jurassic in age. At this site, there are over 2300 individual tracks exposed over an area of more than 8000 m^2. The tracks here represent *Megalosauripus*. This site represents part of the Moab Megatracksite that extends between Green River, Utah, and the Colorado–Utah State line (Lockley, 1991, 2022; Lockley and Hunt, 1995a,b).

Bull Canyon Overlook Interpretive Site is located north of Moab on lands under the jurisdiction of the US Forest Service. It is a dinosaur tracksite and scenic overlook of Bull Canyon and Fisher Mesa. There is a short gravel trail to Early Jurassic theropod dinosaur tracks with interpretive signage.

The Twenty-mile Wash Dinosaur Tracksite is in the Grand Staircase Escalante National Monument and is about 32 km from the town of Escalante. This area is managed by the US Bureau of Land Management. Several hundred tracks are present at the site in the Escalante Member of the Entrada Sandstone. Ichnotaxa include *Therangopodus*, *Megalosauripus*, and *Brontopodus* (Foster et al., 2000; Eves et al., 2012). There is a 2 km out-and-back trail and sparse signage.

Red Cliffs Dinosaur Track Site is preserved in the rocks of the Red Cliffs National Conservation Area managed by the US Bureau of Land Management. The site occurs in the transition zone between the Lower Jurassic Kayenta Formation and the Navajo Sandstone. The main site has 17 tracks of *Grallator* and *Kayentapus*. *Eubrontes* tracks can be seen near the end of the trail. Vehicle and trailer height and width are restricted. There is a day use fee of $5 per vehicle. The tracks and trackways of this site can be seen along the Silver Reef Trail.

The Red Fleet Reservoir Tracksite site occurs within Red Fleet State Park that is managed by Utah State Parks. The tracks are located down a 3.7 km long trail that ends at the lake edge. Three kiosks provide educational content. The tracks are preserved on a sloping rock layer. The tracks number over 350 and occur in the Early Jurassic Glen Canyon Sandstone (Navajo/Nugget Sandstone) and represent *Grallator* and *Eubrontes* (Hamblin and Bilbey, 1999; Hamblin et al., 2000). Even though the tracks are within the park, visitors do not have to pay the park entrance fee to visit them. Other tracks occur elsewhere in the park in several stratigraphic units (Hamblin and Bilbey, 1999; Hamblin et al., 2000).

St. George Dinosaur Discovery Site at Johnson Farm was discovered in 2000 by local optometrist Dr. Sheldon Johnson when leveling a hill on his property inside the city limits of St. George (Milner and Lockley, 2006; Milner et al., 2006a). Most of the site is preserved in a structure that opened to the public in 2005 with funds raised by the Johnsons and the DinosaurAH!torium Foundation. The track-bearing horizon at the site is in the lower portion of the Whitmore Point Member of the Moenave Formation (Early Jurassic). It yields a rich ichnofauna dominated by *Grallator cursorius* and *Eubrontes giganteus* and also has important specimens *of Batrachopus deweyi*. Other vertebrate ichnotaxa include *Anomoepus* isp., *Exocampe* isp., and *Undichna* isp. Abundant dinosaur swimming traces are assigned to *Characichnos* isp. (Hunt and Lucas, 2006; Milner et al., 2006b). The site is open to the public and free for schools within the Washington County School District.

Glen Canyon National Recreation Area is administered by the National Park Service, and it is located principally in Utah, but also partially in Arizona. Unlike many other properties managed by the National Park Service (National Parks, National Monuments), there is no fee for entrance to the area. Vertebrate tracks are common and visited by boaters and hikers. The tracks range in age from early Permian to Late Jurassic, but the majority are from the Lower Jurassic Glen Canyon Group (Lockley et al., 1998). Late Pleistocene coprolites are common in caves (Hunt and Lucas, 2022).

2.1.4.10 Wyoming

Red Gulch Dinosaur Tracksite in Big Horn County was discovered in 1997 by Erik P. Kvale, a geologist from the Indiana Geological Survey, and is managed by the US Bureau of Land Management. Over 1000 theropod tracks, arranged in at least 125 trackways, are preserved in the Middle Jurassic Canyon Springs Member of the Lower Sundance Formation (Breithaupt et al., 2001; Bates et al., 2009). In 2002, the site was renovated with a new boardwalk, interpretive signs, picnic tables, benches, trails, an upgraded access road, and graveled parking for up to 15 cars and 3 buses. Innovative investigative techniques have been applied to this site including integrated LiDAR and photogrammetric documentation, and the site is one of the most extensively documented dinosaur tracksites in the world (Breithaupt et al., 2001; Bates et al., 2009).

2.1.4.11 Museums in the United States with significant collections of vertebrate trace fossils

The following museums have significant collections of vertebrate trace fossils; the most important collections are noted. All these collections are accessible to researchers, most have displays open to the public, and they also have educational programming. There are many other museums with smaller but significant trace fossil collections in the United States such as the University of Michigan Museum of Paleontology (Triassic tracks), Yale Peabody Museum (Paleozoic tracks), and American Museum of Natural History (Cretaceous tracks).

2.1.4.11.1 Arizona Museum of Northern Arizona (Flagstaff, Arizona: https://musnaz.org/) includes a display facility that is open to the public, and it hosts many educational events. The vertebrate ichnofossils include important Permian, Triassic, and Jurassic tracks from northern Arizona and coprolites from the Triassic and Quaternary of Arizona and Utah. It is a regional repository for fossils from the National Park Service.

2.1.4.11.2 California Raymond M. Alf Museum of Paleontology (Claremont, California https://www.alfmuseum.org/visit/) is the only nationally accredited museum in the United States on a high school campus, The Webb Schools. The public exhibits and programs actively support educational programs for the area. The paleontological collections were started by Raymond Alf, who came to the school in 1929. The museum houses an outstanding collection of Cenozoic vertebrate tracks, principally from California.

University of California Museum of Paleontology (Berkeley, California: https://ucmp.berkeley.edu/) is a research institution but has active programs in education for area schools and community colleges. The most important vertebrate trace fossils are Triassic tracks from Arizona, the large collection of F.E. Peabody and S.P. Welles of material from the Moenkopi Formation (Peabody, 1948; Klein and Lucas, 2010).

2.1.4.11.3 Colorado Museum of Natural History, University of Colorado (Boulder, Colorado: https://www.colorado.edu/cumuseum/) includes dinosaur and other tracks and casts that were previously at the University of Colorado at Denver, and many were collected by Martin Lockley. Karen Chin has amassed a diverse collection of coprolites, principally from Mesozoic and Cenozoic localities in North America. The museum has displays that are open to the public and utilized by area schools.

2.1.4.11.4 District of Columbia National Museum of Natural History, United States National Museum (Washington, DC: https://www.si.edu/museums/natural-history-museum) is the "world's most popular natural history museum" with 4,400,000 visits in 2023 and diverse outreach and educational programs The most important vertebrate ichnofossils are Carboniferous and Permian tracks from Grand Canyon National Park. The museum also has a collection of coprolites from around the world.

2.1.4.11.5 Massachusetts Beneski Museum of Natural History, Amherst College (Amherst, Massachusetts: https://www.amherst.edu/museums/naturalhistory) has the world's largest collection of dinosaur, and related, tracks, among them many type specimens. The collection of tracks dates back to the beginning of vertebrate ichnology (e.g., Hitchcock, 1836) and is principally composed of Late Triassic and Early Jurassic tracks from the Newark Supergroup from the northeastern United States. The museum has public displays and provides educational field guides and lab materials to area schools.

2.1.4.11.6 New Mexico New Mexico Museum of Natural History and Science (Albuquerque, New Mexico: https://www.nmnaturalhistory.org/) is the most visited museum in the state, and school groups come from all around the region. The museum provides diverse educational resources including a field station in the nearby Sandia Mountains. The state-funded museum has diverse exhibits that include trace fossils. The museum has been at the forefront of vertebrate ichnology in the past 40 years, has sponsored conferences (e.g., Gillette and Lockley, 1989) and has published multiple scientific monographs on tracks and coprolites (e.g., Lucas and Heckert, 1995; Lucas et al., 2007a,b; Klein and Lucas, 2010; Lucas et al., 2010; Hunt et al., 2012a; Lockley and Lucas, 2014). The large track collection contains particularly important ichnofaunas from the Carboniferous, Permian, Triassic, and Cretaceous of western North America and coprolites from the Paleozoic, Mesozoic, and Cenozoic of western North America and elsewhere.

2.1.4.11.7 Texas Texas Science and Natural History Museum, University of Texas at Austin (Austin, Texas: https://sciencemuseum.utexas.edu/), formerly the Texas Memorial Museum, reopened in 2023. There are large public displays, and the museum provides educational resources for area schools. It has substantial fossil collections, including important Cenozoic tracks from West Texas.

2.1.4.11.8 Utah Prehistoric Museum, Utah State University Eastern (Price, Utah: https://eastern.usu.edu/prehistoric-museum/) is a medium-sized museum in the small town of Price. There are extensive public displays, notably Late Cretaceous dinosaur tracks from the coal mines near the town.

2.2 South America

2.2.1 Argentina

The paleontological heritage of Argentina is one of the richest and most diverse in South America and the best documented (Endere and Prado, 2015). Some of the dinosaur world's most spectacular finds are from Patagonia, and especially of Villa El Chocón, town of Neuquén province, SW Argentina, born from the construction of the hydroelectric dam of the same name. Studies in the town began in 1987, coordinated by the paleontologist Jorge O. Calvo and geologist César Gazzera from the Natural Sciences Museum of the National University of Comahue, which conducted a major ichnological study in the region (e.g., Calvo, 1991; Calvo and Lockley, 2001; Calvo and Mazzeta, 2004). In 1997, a law declared it a "National Historic Site" as the "Valley of the Dinosaurs" So far, a total of 10 ichnogenera of pterosaur, nonavian theropod, sauropod, and ornithopod tracks have been described (Krapovickas et al., 2020).

The importance of the dinosaurs for Villa El Chocón makes them the identity of this town located in the Patagonian Desert. Everything in this town, also called the "Land of Giants," is related to dinosaurs because the economic activity revolves around geotourism, from the "Footprints" camping site to the accommodation, the augmented reality viewpoint, and the life-size models of dinosaurs dispersed in the landscapes surrounding the town. The Ernesto Bachmann Museum was built after the discovery of *Giganotosaurus carolinii* in 1993 (Fig. 14.3A). Here, dinosaurs are shown as a pleasant, agile, and amusing concept. Young guides explain how the remains of the dinosaurs and other vertebrate fossils were found. Hiking trails give access to the tracksites surrounding the reservoir.

In 2006, a group of researchers from CRYCIT–CONICET in Mendoza discovered a dinosaur tracksite west of Malargüe in the upper part of the Anacleto Formation. They have determined that a group of titanosaurs moved south between Malargüe and the Las Leñas Valley during the middle Campanian. A new titanosaur sauropod ichnotaxon was named *Teratopodus malarguensis* (Tomaselli et al., 2021). The Malargüe City Council opened the Municipal Museum of Natural Sciences and is finishing the Municipal Paleontological Park—Dinosaur Footprints project, with the development of accessible sites and viewpoints, capacity training of the tour guides, and the technical intervention to seal the track bed.

Sanagasta in the Province of La Rioja shows evidence of a group of sauropods that nested repetitively and purposely at a Cretaceous hydrothermal site (Grellet-Tinner and Fiorelli, 2010). Several egg clutches (up to 35 eggs each) of sauropods were found at the site, counting both complete and fragmented eggs. All egg clutches were found in dig-out holes and were within 1–3 m of geothermal structures such as mounds, terraces, and geysers. The discovery of this colonial nesting locality shows not only nest fidelity over a long time, but a symbiotic relationship between egg clutches and a peculiar hydrothermal environment that favored their incubation (Grellet-Tinner and Fiorelli, 2010). The Dinosaur Park is part of a provincial geological reserve that covers 850 ha in Bolsón de Huaco, 4 km from the town of Sanagasta. Apart from a large entrance with beautiful wooden signs, there is an interpretive center with a small auditorium, and a well-labeled trail with attractive viewpoints (Fig. 14.3B) and paleoartists' dinosaur replicas.

The other attractive site is Estancia Laguna Manantiales in the Bosques Petrificados de Jaramillo National Park, in the province of Santa Cruz. There are Middle Jurassic footprints of small dinosaurs (<10 cm) in great abundance and diversity, including *Wildeichnus*, *Grallator*,

FIGURE 14.3 Geoconservation examples of ichnosites in South America. (A) Ernesto Bachmann Museum in El Chocón, SW Argentina. (B) The new infrastructure improves accessibility and safety in Sanagasta Dinosaur Park, Argentina. (C) The "Parque Cretacico" of Cal Orck'o. (D) Giant footprint exposed in El Vergel canyon, Torotoro National Park. (E) Deviation of the Peixe River to protect the tracksite in the Valley of Dinosaurs Park, Sousa, State of Paraíba. (F) New facilities for protection of the Cañada del Ombú tracksite, Uruguay. *(A) Photo by Juan Ignacio Canale. (B) Photo by Rafael di Robertis in wikicommons. (C) Photo by Marco Birchler. (D) Photo by Christian Meyer. (E) Photo by Bruno Marques in wikicommons. (F) Photo by Ney Araujo Ferreira.*

and the endemic *Delatorrichnus* and *Sarmientichnus*, and the mammalian *Ameghinichnus*, in a variety of types of progression (De Valais, 2011). The Park also has impressive, complete fossil trunks over a dozen meters long. This place is very extensive, and there are few guides, but for an enthusiastic individual, it will not be a problem to find information and see the tracks, which are very abundant.

The Provincial Natural Reserve Pehuén Co—Monte Hermoso is located in the districts of Coronel L. Rosales and Monte Hermoso, in the South of the Province of Buenos Aires. It covers an area of approximately 16.2 km^2 along the beach. Until now, 100 trackways, hundreds of isolated tracks and trampling areas, have been registered, all of which belong to about 24 different species of mammals and birds (see Quaternary Vertebrate Tracks and Trackways and Walk the World: chapters in this book, and numerous references therein). Human footprints dated from Late Pleistocene and Early Holocene are also present. Geological, Paleontological, and Archeological Provincial Reserve Pehuén Co—Monte Hermoso was included in the World Heritage tentative list of UNESCO in 2014. The "Sala de Interpretación Paleontológica Florentino Ameghino" belonging to the "Carlos Darwin" Municipal Museum in Pehuen Co provides interpretation of the ichnosites located in the vicinity.

2.2.2 *Bolivia*

The Cal Orck'o tracksite is located in a quarry of the approximately 4.5 km northwest of Sucre (Chuquisaca Department, Bolivia) in the Altiplano/Cordillera Oriental. The tracksite is located in a limestone bed belonging to the El Molino Formation and is dated to the middle Maastrichtian (Meyer et al., 2021b). The main track-bearing level is almost vertical with a surface area of ~65,000 m^2. The high-resolution mapping revealed a total of 12,092 individual dinosaur tracks in 465 trackways. Nine different morphotypes of dinosaur tracks have been documented (Meyer et al., 2021b). Among them are several trackways of theropods, ornithopods, ankylosaurs, and sauropods, with the latter group representing $^1/_4$ of the total number of footprints. This geosite is protected as Paleontological National Monument by the Decree 25211 from October 30, 1998, and was included in the World Heritage tentative list of UNESCO in 2003. The site was closed off to tourism from 2008 to 2016. The government of Bolivia attempted to have Cal Orck'o designated as a UNESCO World Heritage site in 2009, which would have provided USD $8-million in funding to help preserve the dinosaur tracks. However, the designation failed because the company owner of the quarry opposed the application.

The dinosaur-themed Cretaceous Park, less than 300 m away, faces the limestone wall and extends for 5000 m^2. It is one of South America's most attractive tourist parks, resulting from a municipal project of the city of Sucre with private sponsorship that was opened in 2006 (Fig. 14.3C). The fabulous 24 life-size replicas, true to their scientifically supported real-life appearance, are displayed in a natural setting representing the last period of the age of dinosaurs in South America. The park features one of the largest dinosaur sculptures in the world, a 36-meter-long replica of a *Titanosaurus*. Services offered by the dinosaur park include guided tours to the wall with the dinosaur footprints; an audiovisual room showing documentaries that help to understand the site's importance; a paleontological-themed museum; a gift shop; as well as a 46-meter-tall lookout with great views of Sucre and the surrounding countryside.

Torotoro National Park, an aspiring geopark, is located around 150 km from the city of Cochabamba. With the discovery of the first footprints in 1968, Torotoro has become one of the most important paleontological sites of the Bolivian Cretaceous. More than 3500 footprints have been recorded belonging to the Upper Cretaceous with probable producers such as titanosaurs, hadrosaurs, ankylosaurs, *Carnotaurus*, and *Buitreraptor*, among others.

Dinosaur footprints have been found all around the National Park, yielding a continuous record until the K/Pg boundary. Meyer et al. (2021b) mentioned several sites in Torotoro with an evaluation of the abundance, size of the track producers, and ichnotaxonomic assignation of sauropod trackways. The most representative and intact is Wayllas on which different species left footprints belonging to ankylosaurs, ceratopsids, hadrosaurs, and carnotaurs. Dinosaur footprints were also found in Kúchu Rodeo, Carreras Pampa, and on the banks of the rivers, such as the El Vergel Canyon (Fig. 14.3D). Guided visits to these tracksites are provided by the Park.

2.2.3 Brazil

The ichnofossil record from the Peixe River Basin, in the State of Paraíba, was declared a national geosite by the Brazilian Commission of Geological and Paleobiological Sites SIGEP 26 (Leonardi and Carvalho, 2002). Large and small theropod, in addition to ornithopod dinosaur, footprints can be found in Berriasian to lower Barremian formations, with a total of 535 trackways and tracks representing over 90 different dinosaurs and chelonians, distributed by 20 stratigraphic levels (Leonardi and Carvalho, 2021). The trackways in Passagem das Pedras were among the first finds of dinosaur remains in Brazil, dated from 1924. However, at the site of Serrote do Letreiro, it is unquestionable that prehistoric inhabitants of the area acknowledged the footprints and intentionally executed petroglyphs around them, establishing a symbolic connection between human graphic expression and the fossil record (Baucon et al., 2008; Troiano et al., 2024).

The Dinosaur Valley Natural Monument (Sousa) was established by the Decree n°14833, of December 20, 1992, as the first ichnosite with formal protection in Brazil. Parque Vale dos Dinossauros is a 40-hectare protected area established by the Paraíba State Government and Sousa Municipal City Hall (Leonardi and Carvalho, 2002). Due to frequent flooding, it had to include the modification of the river course that was deviated to avoid flooding of the tracksite (Carvalho and Da Rosa, 2008; Siqueira et al., 2011; Fig. 14.3E). The visitor center includes a museum and an auditorium, as well as parking for cars and buses. There are interpretive trails with indicative signs, and the site is fenced and has a gate to control the entry and exit of visitors. Guided visits are carried out throughout the year, by schools practicing Environmental Education, by tourists, and by researchers. It currently receives 2000 people per month.

The Predebon fossil site is one of the main geosites of the newest Quarta Colônia UNESCO Global Geopark, located in Rio Grande do Sul. The tracksite belongs to the Santa Maria Formation, Ladinian-Carnian in age. Nine morphotypes include *Rhynchosauroides retroversipes* and *Dicynodontipus protherioides*. They were produced by sphenodontid lacertoids, small advanced cynodonts, and possibly tritylodontids. Tracks of basal dinosaurs are also present and can possibly be attributed to *Staurikosaurus*, *Saturnalia*, and *Sacisaurus* (Silva et al., 2008a,b,c). Possible occasional bipedalism and climbing habit are demonstrated in *Rhynchosauroides retroversipes*; alternating gait and tail drag typical of basal amniotes is interpreted for nonmammalian cynodonts producing *Dicynodontipus protherioides*. Some of the oldest dinosaur tracks and also advanced cynodont tracks were found in this national geosite (Silva et al., 2008a,c). The site is nowadays a private paleontological park.

The Ouro Fossil Site, São Paulo, is best displayed on the sidewalks of the city of Araraquara; the density of slabs shows 5–10 isolated tracks and trackways, on an average route of 100 × 2.5 m. These slabs were quarried during the 19th century from the Late Jurassic Botucatu Formation. Over 200 slabs with footprints were deposited at the National Museum of Rio de Janeiro, Paleontological Collection at the National Department of Mineral Production (Rio de Janeiro), and Federal University of Paraná (Curitiba), but also can be found in collections at the Universidade Estadual Paulista (UNESP Rio Claro) and Department of Geology at the Federal University of Rio de Janeiro. Eight morphotypes of tridactyl dinosaurs, seven morphotypes of theromorphoids, and nine mammaloid morphotypes, including *Brasilichnium elusivum*, were described (Leonardi and Sarjeant, 1986). In 2008, Araraquara City Hall installed some slabs with trace fossils along the street of Rua dos Voluntários da Pátria with explanatory panels on them, creating an Open Air Museum. In the town of S. Carlos, the educational use of trace fossils on sidewalks has been developed (Francischini et al., 2020).

2.2.4 Chile

Since 1962, various sites with vertebrate footprints have been discovered in the Mesozoic deposits of Chile (e.g., Moreno et al., 2004), indicating the presence in the country of an interesting diversity of dinosaur species and other contemporary vertebrates. Quebrada Chacarilla contains several outcrops with vertebrate tracks from the Upper Jurassic–Lower Cretaceous. These were declared a Natural Monument in August 2004 (Moreno, 2008). The diversity of footprints observed in its inclined strata includes those of crocodiles, theropod dinosaurs (mainly theropods) from 7 cm to more than a meter long, sauropods, and ornithopods of large size (70–100 cm). These trackways indicate varied behaviors: walking, zigzagging, swimming, intra- and interspecific interaction, among others. Conservation projects for this site have been presented to the Illustrious Municipality of Pica since 2001 (P. Cañarte and K. Moreno). However, the efforts of the authorities have been focused on the assembly of sculptures in the town of Pica. Currently, visitors are mainly rally enthusiasts. The year of 2024 marked the 20th anniversary of the creation of the Chacarilla Dinosaur Footprints Natural Sanctuary, and there are conversations with the municipality of Pica to reactivate the idea of setting up an on-site interpretation center.

In the Region of Libertador Bernardo O'Higgins, 600 m east of the mountain town of Termas del Flaco, in an inclined layer of the Baños del Flaco Formation (Upper Jurassic, Tithonian), footprints of sauropod dinosaurs (ca. 70 cm), small ornithopods (ca. 15 cm), and small- to medium-sized theropods (12–35 cm) have been preserved. The scientific information available about this record is abundant in comparison with other places in the country (Casamiquela and Fasola, 1968; Moreno and Pino, 2002; Moreno and Benton, 2005). This reveals an interesting faunal association present in the shores of a subtropical lagoon, in the vicinity of a coral reef. The original description of the ichnospecies *Iguanodonichnus frenki* comes from here, although it has been reinterpreted as sauropod tracks (Moreno and Benton, 2005). The site is visited by hundreds of people during the spring months and summer, when ice and snow recede. Services include: tour guides, mainly self-taught training and horseback riding. A modest project was tendered in 2007 by the National Monuments Council with the aim of installing signs, providing information, and demarcating trails. However, due to the high frequency of visitors and the nature of the place (annual thaw, passage of animals), these efforts were ineffective. Patrimonia Consultores, a company awarded by the Council of National Monuments (CMN) and the National Heritage Service of the O'Higgins region, is currently working on the environmental and sociocultural baseline, paleontological diagnosis, zoning, and preparation of a proposal for the management of a historical and paleontological monument, Dinosaur Footprints, in Termas del Flaco.

2.2.5 *Peru*

The Early Cretaceous tracksite of Querulpa Chico, in the Province of Castilla, is dominated by 69 large theropod footprints. It is associated with coastal delta deposits of the Hualhuani Formation in the Arequipa Basin, subject to extensive arid or semiarid paleoclimate. Moreno et al. (2012) noted the high abundance of trackways belonging to distinct large theropods, possibly spinosaurids and/or carcharodontosaurids, and the exclusion of tracks of other animals, suggesting some degree of grouping behavior. Since 2019, the site has infrastructure and offers guided visits. The so-called "Parque Jurásico de Querulpa" (despite the trace fossils being dated from the Early Cretaceous) shows a cactus garden area with an observation platform and with models of different dinosaurs, providing misleading information on possible producers, with the exception of Giganotosaurus. It includes also a small, on-site museum, where findings made in the area are on display. The intervention was carried out as part of an interinstitutional agreement between the Provincial Municipality of Castilla and the Zafranal Project, with the aim of revaluing this attraction considered in the Tourism Development Plan "Between the Sea and Fire". This plan seeks to turn the Majes valley into one of the main tourist destinations by 2025. Numerous tour operators in Arequipa offer day tours to the Jurassic Park of Querulpa, often combined with a trip to the Petroglyphs of Toro Muerto and/or the Chancharay springs, located nearby.

2.2.6 *Uruguay*

The Cañada del Ombú tracksite in Tacuarembó, North of Uruguay, is protected as a Natural Historic Monument. The record of sauropod (the majority), theropod, and ornithopod dinosaur tracks and trackways from the Tacuarembo Formation (Late Jurassic-?Early Cretaceous) is described in detail by Mesa and Perea (2015), consisting of two trackways and four isolated tracks. The ichnofossils are preserved in sandstones of interdune facies of the Batovı (lower) Member. The project "Ruta de los Dinosaurios" is building an onsite museum in Cuchilla del Ombú. For several years, the only protection the tracksite had was a fence that prevented the passage of livestock and vehicles in the area. Recently, the construction of a roof and walkways allows visitors to better enjoy the footprints (Fig. 14.3F). Casts are used in the interpretation of the tracks, so visitors can touch and understand the morphological differences.

2.3 Europe

2.3.1 *Croatia*

The Ministry of Economy and Sustainable Development is currently responsible for protection of fossil and mineral sites in Croatia. According to the Nature Protection Act (Official Gazette 80/13, 15/18, 14/19, 127/19), there is the possibility to declare protected fossils as protected natural value. They are declared for their rarity, size, appearance, or educational and scientific significance (Irina Pleše, pers. comm.). Protected fossils are declared by the Minister with Decision. Only dinosaur tracksites have been protected as special paleontological reserves. From 23 ichnosites protected, 21 locations are located in the area of Istria, and the remaining two locations are in Hvar and Biokovo. For a synthesis of dinosaur tracksites in Croatia, see Dalla Vecchia (1997).

The localities of Ploče, Pogledalo/Barban, Plješivac/Kamik, Trstike/Debela Glava are managed by the Public Institution Brijuni National Park. The Bikovo locality is managed by Public Institution Biokovo Nature Park. The locality of Brbinšćica is managed by a public institution for management of the protected Parks in the area of Zadar county—Natura Jadera. The locality of Žukovo is managed by a public institution for management of the protected Parks in the area of Splitsko-Dalmatinska county. The Grakalovac and Fenoliga localities are managed by the local public institution Kamenjak Nature Reserve. Other localities are managed by a public institution for management of the protected Parks in the area of the Istrian County—Natura Histrica.

Considering interpretation of the ichnosites, the educational and recreational trail "Following the Dinosaur Path on Kamenjak" is located on the right side of the Kamenjak peninsula, only 300 m from the entrance. The total length of the path is about 600 m and includes models at original sizes of dinosaurs and other animals from the Cretaceous that lived in Kamenjak. Along the path there are panels that describe the geological curiosities of Kamenjak in Croatian, Italian, English, and German. The Grakalovac ichnosite is located within the Kamenjak Significant Landscape, and there is an educational panel on the site.

Fenoliga is a small island located northwest of Kamenjak. The island is approximately 150 m long and wide and is well known for the dinosaur trackways dated to the Cenomanian that were discovered by the Austrian paleontologist Adolf Bachofen-Echt in 1974. There are 146 theropod and sauropod footprints organized in six trackways (Mezga and Bajraktarević, 1999).

The Ploče ichnosite is located within the Brijuni National Park, and at the site there is an educational panel. A slab with five prints was recovered to the Tito Museum in Brijuni. Pogledalo/Barban is part of the dinopark in Brijuni. A dinosaur sculpture was placed on the site (Fig. 14.4A).

Finally, in Cape Vrbanj, more than 200 footprints on 4 track levels are dated as Early Cretaceous. At this place, a model of a theropod is also exhibited.

2.3.2 France

A law enacted in 2002 grants formal recognition to the geological heritage in France. In 2007, the French Ministry of Environment launched the inventory program for the national geological heritage, and the data are now being collected at a regional scale (De Wever et al., 2015). The ratified site data are stored and available for public use on the website (http://inpn.mnhn.fr) in a structure similar to natural data. A total of 25 tetrapod ichnosites were identified in this inventory, and we introduce some of the best practices examples.

Dinosaur footprints in the Kimmeridgian limestones of Villette are located in the Bourgogne-Franche-Comté region. Footprints of mainly sauropod dinosaurs are organized into around 10 trackways of several dozen steps each, some parallel to each other and others intersecting. Much smaller prints are attributed to juveniles (Le Loeuff et al., 2006). Developments for tourist and educational activities by the Commune of La Vallière provided simple fencing and informative panels in French only.

FIGURE 14.4 Examples of geoconservation projects in European ichnosites (part I). (A) Dinosaur model placed in the ichnosite at Pogledalo, Croatia. (B) Loulle ichnosite aerial view, France. (C) Saignon Oligocene tracksite after the cleaning and joint sealing project of Luberon UNESCO Global Geopark, France. (D) Dinosaur tracks and facility at Barkhausen, TERRA.vita UNESCO Global Geopark, Germany. (E) Bottrop tracks ice age Hall in Quadrat, Germany. (F) Ipolytarnóc tracksite complex as the gate to Novohrad-Nógrád UNESCO Global Geopark, in Hungary. *(A) Photo by Irina Pleše. (B) Photo by David Bottin. (E) Photo by GeoPark Ruhrgebiet. (F) Photo by Ipolytarnóc Nature Reserve.*

In the same region, the dinosaur footprint slab from Loulle was discovered in 2004 and is one of the most important French dinosaur tracksites. Around 20 trackways with nearly 1500 footprints made by sauropods and theropods are preserved in a beach limestone facies from the Upper Oxfordian (Cariou et al., 2014), on the floor of a disused 3800 m^2 quarry. Some prints are around 90 cm in diameter. Without protection, the fragility of the limestone slabs, winter frost, and trampling by visitors led to irreparable damage to this exceptional site. A large part of the site is now protected. A visitor walkway and panels help visitors to understand the environment in which the dinosaurs evolved in the Jurassic Period, 155 million years ago. Some of the main trackways were contrasted with colored sand (Fig. 14.4B). During the summer months, guided tours are organized once a week by prior arrangement.

The Saignon tracksite is located in the heart of the Luberon UNESCO Global Geopark (Provence) and is dated from the early Oligocene. It has been protected since 1987 by the Luberon Geologic National Nature Reserve, which is managed by the Luberon Regional Nature Park. The site consists of two main levels (Coster and Legal, 2021): the lower level is about 400 m^2 in area and yielded a 1000 mammal fossil footprints on the surface of a micritic limestone bed. Most of the tracks were made by small to large-sized artiodactyls and perissodactyls, some of which are organized into trackways (Fig. 14.4C). A large perissodactyl morphotype is *Rhinoceripeda voconcense* comb. nov., providing possible evidence of social behavior in early rhinocerotids (Belvedere et al., 2023). Based on fewer tracks a second, smaller perissodactyl ichnotaxon, *Lophiopus* isp., was identified by the same authors. Artiodactyl prints are attributed, when possible, to *Bifidipes velox*. This ichnocoenosis is characteristic of a post-Grande Coupure fauna with the presence of newcomer rhinocerotids that reached Europe from Asia and the persistence of endemic pre-Grande Coupure palaeotheriids (Belvedere et al., 2023). The surface yielding the footprints shows numerous natural fractures that were sealed in a recent geoconservation intervention. The upper level, stratigraphically 2 m above the previous one, is characterized by the presence of bird tracks and sedimentary structures, such as ripple marks and desiccation cracks (Coster and Legal, 2021). Currently, strategies for in-situ preservation by building a protective shelter are being assessed. The site management plan for permanent sheltering includes a visitor walkway and educational interpretation trails. A sustainable approach is being developed, comprising a green, energy-efficient building using a solar power system with a reduced environmental impact. The main benefits of building a protective cover include the conservation of the site, the creation of a place where visitors can see the behaviors and locomotion of past mammals as well as the production of solar energy (Coster and Legal, 2021).

The Regional Natural Park of the Monts d'Ardèche, located in south-eastern France, became the Monts d'Ardèche UNESCO Global Geopark in September 2014. This territory possesses numerous and rich Middle-Late Triassic vertebrate tracksites, including three-toed footprints of carnivorous dinosaurs.

Tridactyl *Grallator* footprints are the most frequent type of footprints at Le Sartre. Large *Otozoum* attributed to *Plateosaurus*, and *Brachychirotherium* tracks of four-legged pseudosuchian reptiles, also define the ichnosite (Szewczyk et al., 2020). The UNESCO Global Geopark label helped to formalize a long-standing partnership for the study of this ichnological patrimony between the RNP of the Monts d'Ardèche and the University of Burgundy. To date, 15 tracksites and nearly a 1000 individual tracks have been inventoried (Fara et al., 2021). The locality "Le Sartre" is the first equipped locality in France to show Late Triassic dinosaur tracks and to be freely accessible to the public.

The Pterosaur Beach in Crayssac, in the region of Occitanie, as the highlight of the Causses du Quercy Regional Natural Park and UNESCO Global Geopark, is a world reference site for tracks of pterosaurs and other reptiles from the end of the Jurassic (Mazin et al., 2003). This ichnosite is known for the multitude, diversity, and quality of the preserved traces, besides the evidence of a landing pterosaur. It is the only discovery in France of pterosaur footprints and shows the longest pterosaur trackway. The site is set up for visits, highlighted and protected by an impressive futuristic building of more than 2000 m^2 roofed by solar panels in a former quarry. The guided visits in French only are made in the darkness using low-angle lights to emphasize the exceptional quality of track preservation.

The Dinoplagne is an ichnosite of Tithonian age in the same region. A trackway of an exceptionally large titanosaurid sauropod defined as *Brontopodus plagnensis* enables analysis of the locomotory behavior and the variations in the consistency of the substrate (Mazin et al., 2017). This is the main trackway with more than 155 m in length. The 110 footprints reach an impressive 2 m in diameter (including pressure pads).

2.3.3 *Georgia*

The Sataplia Nature Reserve is located in the Tskaltubo district, about 260 km from Tbilisi. Established in 1935 to protect the Sataplia Cave and dinosaur footprints found in the area, this geotourism destination recently benefited from a refurbishment of the Sataplia Cave visitor center. One-hundred sixty-three dinosaur footprints were left in two limestone beds of the Lower Cretaceous. On the lower layer, there are 150 footprints left by up to 30 individuals. Based on these footprints, Leonid Gabunia in 1951 described the new ichnogenus *Satapliasaurus*, in which three ichnospecies were identified: *S. tschabukianii*, *S. dsocenidzei*, *S. kandelakii*; the footprints in the upper bed were attributed to camptosaurids (Gamkrelidze et al., 2021). A conservation center and museum have been constructed, together with glass walkways, viewing points, a cafeteria, and other visitor amenities.

2.3.4 *Germany*

In Germany, vertebrate track discoveries span from the Carboniferous to the Late Pleistocene. These include tracks of early reptiliomorph tetrapods from the Permian to the Triassic, dinosaurs from the Late Jurassic and Early Cretaceous, Late Pleistocene mammals, and early humans from the Paleolithic.

Germany's oldest tetrapod tracks were found in the Bochum district of Stiepel (Western Germany; (Voigt et al., 2021). Strollers discovered six footprints of an individual in a former quarry by the Ruhr River in 2012. The trackway is located on the underside of a sandstone layer of the approximately 316-million-year-old Finefrau Sandstone of the lower Witten Formation (Bashkirian). The tracks have been assigned to the ichnotaxon *Ichniotherium praesidentis* (Voigt et al., 2021; Schüppel and Wrede, 2022). They were produced by large Diadectomorpha, which anatomically lie between amphibians and reptiles. The Bochum Diadectomorpha tracks are thus 8 million years older than the oldest skeletal anatomical finds of the group (Voigt et al., 2021). The largest hind foot imprint measures about 20 cm in size, which is why the size of the animal is estimated to be one and a half to two meters long. To protect the tracks against weathering and vandalism, the track slab was recovered from the quarry in 2013. It is currently stored in the German Mining Museum in Bochum. The Geological Survey of North Rhine-Westphalia displays a cast of the track slab accessible to the public. The National GeoPark

Ruhrgebiet (Ruhr area) has placed an informational panel in the publicly accessible quarry (Schüppel and Wrede, 2022). Three additional, slightly younger footprints of the same track type were discovered between 1923 and 1957 in the Upper Carboniferous coal mines underground. The originals and their casts are now exhibited at the German Mining Museum in Bochum (Voigt et al., 2021). About 310-million-year-old *Ichniotherium praesidentis* tracks are also known from the underground Carboniferous coal districts of the Saarland.

From the Bromacker locality located between Georgenthal and Tambach-Dietharz near Gotha (Central Germany), several hundred slabs with tetrapod footprints have been recovered since 1887. Therefore, Bromacker was among the first evidence of Late Paleozoic tetrapod tracks and, due to ongoing scientific excavations, one of the most significant, productive, and best-documented terrestrial vertebrate faunas of the Early Permian (Voigt, 2002; Martens, 2018). All track finds originate from the Tambach Sandstone of the Tambach Formation and date back to the Cisuralian, around 290 million years ago. The former Tambach Basin, in which the sandstone was deposited, represents a rarely described, terrestrial highland area (Eberth et al., 2000). Furthermore, thanks to the large and well-preserved fossil record, a distinctive feature of the Bromacker locality is the possibility to correlate trackways with vertebrate skeletal finds. Trackways of *Ichniotherium cottae* and *Ichniotherium sphaerodactylum* have been associated with the diadectid reptiles *Diadectes absitus* and *Orobates pabsti* found at Bromacker (Voigt et al., 2007). *Orobates pabsti* was also digitally reconstructed and built as a robot (OROBot) to reconstruct the *Orobates* locomotion (Nyakatura et al., 2019). Moreover, several other connections between trackways and trackmakers have been carried out: the ichnotaxon *Dimetropus leisnerianus* was made by nontherapsid synapsids such as *Dimetrodon teutonis* (Romano et al., 2016) found only at the Bromacker locality and until now is the only described *Dimetrodon* outside North America. In addition to this, connections between the reptiliomorph amphibian *Seymouria sanjuanensis* and the trackways of *Amphisauropus kablikae* have been documented (Marchetti et al., 2017). Additional ichnoassociations include the ichnogenera *Varanopus* (bolosaurian parareptile), *Tambachinium* (varanopid), *Batrachichnus* and *Limnopus* (temnospondyls) as well as *Dromopus* (araeoscelid or varanopid; Marchetti, 2024).

Bromacker is now recognized as one of currently 147 National Geotopes by the Academy for Geosciences and Geotechnologies, the highest recognition for a geosite in Germany. The still-active excavation site can be visited during the summer season. Larger track slabs, information panels, as well as models of the trackmakers and the skeletons found have been set up on site by the UNESCO Global Geopark Thuringia Inselsberg—Drei Gleichen. The excavation site is also a station on the Saurian Discovery Trail of the UNESCO Global Geopark, featuring a total of 17 life-sized tetrapod sculptures. The "BROMACKER" project, supported by the German Federal Ministry of Education and Research and consisting of a cooperation among the Natural History Museum Berlin—Leibniz Institute for Evolution and Biodiversity Research, the Foundation Schloss Friedenstein Gotha, the Friedrich Schiller University Jena, and the UNESCO Global Geopark Thuringia Inselsberg—Drei Gleichen, includes the implementing of scientific work with new science communication methods such as social media communication and digital exhibitions. Moreover, the BROMACKER lab in Gotha allows the visitors to be informed about the newest results of the excavations and exhibits apart from fossil finds, and to see the scientific work on a fossil from recovery, through live-preparation and research, to presentation.

Another track-bearing sandstone, dated to the Wuchiapingian, is the aeolian Cornberg Sandstone in Hesse (Central Germany). Tetrapod trackways have been attributed to Anthracosauria, Captorhinida, and Pelycosauria (Fichter, 1994). Cornberger track slabs are on display in various museums, such as the Natural History Museum in Kassel, the Geoscience Center of the University of Göttingen, the Senckenberg Museum Frankfurt, and the Kreisheimatmuseum Rotenburg. A Sandstone Museum was established in the Cornberg Monastery, which focuses on sandstone as a building and sculpture material and features track slabs, reptile sculptures, and a Permian-era diorama of life. The former Cornberg Sandstone Quarry is now used as a near-natural and extracurricular place for learning and for cultural events.

In 1835, Johann Jakob Kaup described tetrapod trackways of the so-called "Handtier" ("hand animal") *Chirotherium barthii* from the Winzer quarry near Heßberg in Hildburghausen (Central Germany; Kaup, 1835). This work is the world's first scientific description of a fossil track and marks the beginning of the scientific exploration of fossil tetrapod tracks (Haubold, 2006). The tracks of five to six individuals were found on a bedding plane of the Thuringian *Chirotherium* Sandstone of the Solling Formation at the transition between the Olenekian and Anisian. The original area, likely about 200 m^2 in size, was already being dissected from the year of its first description, and the individual slabs were sold to museums across Europe. Today, about 100 track slabs are stored in 40 collections (Haubold, 2006). At the beginning of the 21st century, an attempt was made to reconstruct the original track slab using records such as the original track plan and the distributed finds and their casts. From the 50 m^2 reconstructed track surface, the *Chirotherium* monument was created in 2004 and placed on the marketplace of Hildburghausen. The monument consists of a 30 m^2 wall with assembled casts of 22 original slabs, as well as a life-like bronze sculpture of the trackmaker, a basal archosaur (Haubold, 2006).

In 2007, beneath the root ball of a fallen beech tree in Euerdorf-Saalrangen (Bavaria), *Chirotherium* footprints were discovered on a bedding plane of the Franconian *Chirotherium* Sandstone (Lagally et al., 2011). The tracks were uncovered, but notably, they were not divided into slabs but left in situ. The Euerdorf dinosaur trackway geosite was recognized as No. 98 of the "100 most beautiful geotopes in Bavaria" by the Bavarian State Ministry for Environment and Health (Lagally et al., 2011). On-site, the geosite is protected from being walked on by visitors through wooden boundaries. However, due to the lack of roofing, the geosite is highly susceptible to weathering. An informational panel from the State Ministry, in cooperation with the UNESCO Biosphere Reserve Rhön, provides details about the discovery, paleontology, and the prehistoric environment. Additionally, the geosite is part of the geological educational trail "Weg durch die Zeit" from Bad Kissingen-Euerdorf and serves as Geo-Point 3 as an external station of the Museum Terra Triassica in Euerdorf. The museum itself features life-sized models and dioramas depicting the Triassic world.

Tetrapod tracks from marginal-marine facies of carbonate deposits of the Anisian and Ladinian provide a rare insight into the terrestrial vertebrate faunas of Central Europe. From various geosites in Bavaria, Baden-Württemberg, Hesse, Lower Saxony, North Rhine-Westphalia, Thuringia, and Saxony-Anhalt, at least 19 different tetrapod ichnotaxa are known (Klein, 2020). During an excavation at a construction site for a sports facility in Borgholzhausen in the Teutoburg Forest (NW Germany, North Rhine-Westphalia) in 1996, 30 horizons with tetrapod trackways were uncovered (Albat, 1999; Grzegorczyk, 2005). Today, a 2.8 × 1.7-meter large track slab featuring tracks of *Rhynchosauroides peabodyi* and

Procolophonichnium haarmuehlensis is on display in the EarthAgeCenter of the Heimathaus Borgholzhausen (Diedrich, 2002; Fischer, 2022). Information panels from the UNESCO Global Geopark TERRA.vita in the exhibition and in the associated GeoGarden outdoors provide information about the life in the world at that time. Two life-sized models of the trackmakers, *Macrocnemus* and *Hescheleria*, offer visitors deeper insights into this period. The track slab has been protected as a movable paleontological monument (Grzegorczyk, 2005).

During an excavation supported by the UNESCO Global Geopark TERRA.vita in the Silbersee quarry near Hasbergen at the Teutoburg Forest south of Osnabrück (NW Germany, Lower Saxony), several smaller slabs and one large, approximately 2-m-long track slab containing *Rhynchosauroides* and *Procolophonichnium* tracks, were recovered, which are managed by the Museum Association Hüggel in Hasbergen. The excavation area is visited during tours into the Silbersee tunnels by the Geopark's TERRA.guides. Both, the Silbersee quarry and the tunnel, are protected as natural monuments. Information panels from the Geopark and life-sized models of the trackmakers recreate the living world (Fischer, 2022). Additionally, there is an excursion guide available (Diedrich, 2010).

The fortunate location of the region of today's Wiehen Hills (Northwestern Germany) at the northern edge of an island about 153 million years ago (Middle Kimmeridgian) enabled the preservation of tracks from migrating dinosaur herds. Thus, finds of sauropod tracks were reported in silty sandstones from a former quarry near Bad Essen-Barkhausen in 1921—the first sauropod tracks found in Europe and the first Late Jurassic sauropod tracks discovered worldwide (Lockley and Meyer, 2000; Fischer et al., 2021; Meyer et al., 2021a). Today, more than 130 dinosaur footprints of two different ichnospecies can be observed in the quarry (Fischer and Koch, 2022). The subcircular footprints were described as *Parabrontopodus barkhausensis* and attributed to small-sized sauropods. They are closely related to *Europasaurus holgeri*, whose skeletons were found in the geographically and stratigraphically close Langenberg quarry in the Harz Mountains (Fischer et al., 2021; Meyer et al., 2021a). Nine of the 10 trackways of *P. barkhausensis* run parallel to each other, indicating gregarious behavior of the trackmakers (Fischer et al., 2021; Meyer et al., 2021a; Fischer and Koch, 2022). The tridactyl tracks of three other trackways were described as *Jurabrontes teutonicus* and attributed to large megalosaurus (Meyer et al., 2021a; Fischer and Koch, 2022). The trackmakers crossed and overstepped the sauropod tracks. The dense network of joints prohibited the recovery of the tracks from the quarry. Therefore, the tracks were initially preserved as plaster casts. Regular in situ geoconservation measures did not begin until the 1960s. These included the installation of a concrete drainage bar, impregnations of the tracklayer surface, and the injection of cement slurry behind the layer surface to solidify the rock structure.

In 2000, the UNESCO Global Geopark TERRA.vita installed a glass roof that protects 80 m^2 of the approximately 200 m^2 large track surface (Fischer et al., 2021; Meyer et al., 2021a; Fig. 14.4D). Following additional geoconservation measures by the Geopark in the years 2020–2021, three more concrete drainages were installed, the eastern quarry edge was cemented, and fracture edges at layer transitions were sealed with lime jointing agent. After the track surface was cleared of plant growth, the tracks were taxonomically revised using 3D photogrammetry by Meyer et al. (2021a). Additionally, previously unreported footprints and tracks became visible (Meyer et al., 2021a; Fischer and Koch, 2022). Since 1936, the Dinosaur Tracksite Bad Essen-Barkhausen has been protected as a natural monument and was designated as a National Geotope in 2019.

The Dinosaur Tracksite Bad Essen-Barkhausen has been used for geotourism from an early stage. A field guidebook was created already in 1921, and since 1976, the geosite has had the status of an open-air museum. Today, the UNESCO Global Geopark TERRA.vita provides information about the dinosaur tracks, their makers, and the palaeoenvironment through two life-size models of the trackmakers and six informational panels (Fischer et al., 2021). The geosite is integrated into the hiking and cycling network of the Geopark and is the annual event location for the Day of the Geotope, a national day to celebrate geological heritage. The experiential education school camp Barkhausen continues to conduct education for a sustainable development program at the dinosaur tracks for 1800 students annually. Silicone casts are displayed in the permanent exhibitions at the Museum am Schölerberg in Osnabrück and in the Dinosaur Park Münchehagen. Latex molds and fiberglass replicas of the dinosaur tracks are stored in the collection of the University of Colorado Museum (Fischer et al., 2021).

The only site for Late Jurassic dinosaur tracks outside the Wiehen Hills in Germany is the Langenberg Quarry near Goslar in the UNESCO Global Geopark Harz · Braunschweiger Land · Ostfalen. The tracksite, which contained 20 footprints from probably two different large theropod ichnospecies, was destroyed by quarrying activities after its discovery in 2003. However, the five best-preserved tracks were recovered and are now stored in the Dinosaur Park Münchehagen (Lallensack et al., 2015).

Early Cretaceous dinosaur trackways have been found at 13 geosites in Northern Germany in the Obernkirchen Sandstone, a subunit of the Bückeberg Formation, also known as the "German Wealden" (Late Berriasian). Formed in the lagoonal and sandy barrier to back-barrier areas of a limnic-deltaic environment, the sandstone preserved dinosaur tracks from ankylosaurs, bipedal ornithischians, ornithopods, sauropods, and theropods, as well as other reptiles such as pterosaurs, turtles, and possibly crocodiles (Hornung et al., 2012a,b; Hornung and Reich, 2013). The first dinosaur track found in Germany was also described from the Obernkirchen Sandstone in 1879. Currently, most of the historical tracksites are no longer accessible or have been lost to quarry operations Hornung et al., 2012b. However, two geosites in Münchehagen and from Bückeberg Mountain stand out as exceptional tracksites.

In 1979, dinosaur tracks were discovered for the first time in Münchehagen in the Rehburg Mountains. Today, 256 footprints from eight large sauropods (*Rotundichnus muenchehagensis* nom. dub.), likely showing gregarious behavior, and 19 impressions from the trackway of a tridactyl biped are known (Fischer, 1998; Hornung et al., 2012b). The preservation of the tracks is not particularly good, as they were eroded prediagenetically and damaged by exposure before geoconservation measures were implemented. In 1987, the dinosaur tracks of Münchehagen were protected as a natural monument and recognized as a National Geotope in 2006 (Wings et al., 2012). Due to the rarity of long trackways and the likely gregarious behavior, the geosite is significant at a European level (Wings et al., 2012). The approximately 15,000 m^2 track-bearing surface was initially protected by a temporary structure and since 1992 a larger building covers the most important trackways (Fischer, 1998; Wings et al., 2012). Today, the track layer and the building are the main attractions of the 2.5-km loop trail through the Dinosaur Park Münchehagen, which operates in a private–public partnership. In this open-air museum, more than 300 life-sized dinosaur models and sculptures of other extinct animals from around the world are displayed.

In close proximity to the Dinosaur Park, dinosaur tracks from other quarries of Münchehagen are also known. More than 15 bipedal and quadrupedal ornithopod trackways were discovered in the F. Westling GmbH quarry in 2004 and 2005, from which a 20 m^2 large slab was recovered and is now on display at the Geoscience Centre of the University of Göttingen (Hornung et al., 2012b). A more than 15-meter-long theropod trackway was recovered from the quarry in 2010, too, and is on display in the Dinosaur Park Münchehagen. Casts from the site are stored in the Hanover State Museum. Track slabs from additional sites in Münchehagen are stored at the Geoscience Centre of the University of Göttingen (Hornung et al., 2012b).

Besides skeletal finds of dinosaurs and other reptiles, the Obernkirchen Sandstone from Bückeberg Mountain (Weserbergland) is known for its trackways from various saurians and other reptiles, including turtles and pterosaurs (Hornung et al., 2012b; Hornung and Reich, 2013). In the last still-active quarry, excavations by the Hanover State Museum have identified six to seven track levels since 2006, including one titled the "chicken yard", featuring several hundred tracks from tridactyl and didactyl theropods and bipedal ornithopods (Hübner, 2016; Richter and Böhme, 2016). It is accessible to visitors digitally through a 3D model. A larger bed surface in the quarry has been incorporated into the 4.6-kilometer-long Dinosaur Tracks Schaumburger Land Nature Trail. Starting at the Dinosaur Pavilion at the Youth Education and Recreation Center (JBF-Centrum), a former sandstone building of the quarry operation, nine interactive stations including special panels for children explain the dinosaurs of the Bückeberg Mountain, the mining history of the region, and the Bückeberg Mountain itself. The track-bearing surface is accessible through a ramp for reduced-mobility access. A wooden path and platforms guide visitors across the surface to protect the tracks. Regular guided tours are conducted by nature guides. Track slabs from the Bückeberg Mountain are displayed at the Hanover State Museum, the Geoscience Centre of the University of Göttingen, and the Dobergmuseum Bünde.

About 300,000 years ago, the current location of Schöningen in Lower Saxony (Northern Germany), located in the UNESCO Global Geopark Harz · Braunschweiger Land · Ostfalen, was next to a lake. On the muddy shore of this lake, *Homo heidelbergensis* left behind three footprints, the oldest human footprints found in Germany (Altamura et al., 2023; see Walk the World chapter in this book). Two of the three footprints were made by young individuals. In addition, tracks of elephants, identified as *Palaeoloxodon antiquus*, were found along with the footprint of a forest rhinoceros. The footprint of the latter belongs either to *Stephanorhinus kirchbergensis* or *Stephanorhinus hemitoechus*, and it represents the first discovery of such a footprint from the Pleistocene in Europe (Altamura et al., 2023). Excavations at the Paleolithic site of Schöningen have been ongoing since 1994. In 2013, the Forschungsmuseum Schöningen ("Research Museum") was built at the site. This futuristic and large museum displays finds from the excavations, including the world's oldest wooden hunting weapons, and introduces visitors to the field of archeological science. The museum is recognized as an extracurricular learning site and offers educational programs focusing on education for sustainable development. The museum also includes an outdoor area with an adventure trail.

A 150 m^2 area with 600 footprints in Bottrop-Welheim (Western Germany) provides a rare insight into a diverse, late Pleistocene Ice Age landscape populated by many large mammals (von Koenigswald et al., 1995, 1996). The animals left their tracks in the loamy sediment of the braided-river system of the Emscher River. In total, 30 trackways could be reconstructed, with more than half of the tracks produced by reindeer individuals of the species *Rangifer tarandus* (see Quaternary Vertebrate Ichnology chapter in this book).

Initially, in 1992, only isolated paw prints of the cave lion were found. Since the site was a construction pit for building a new sewage treatment plant, in situ preservation was impossible. Moreover, the salvage of the soft loam in which the tracks were imprinted was not feasible. Even before the entire area was exposed, researchers from the Institute of Paleontology at the University of Bonn and the Museum for Local and Early History in the Quadrat in Bottrop contemplated a very special and elaborate cast preservation method, which is explained in detail by von Koenigswald et al. (1995). Areas already exposed were first covered with plaster. After stabilizing the groundwater control of the construction pit, the entire track surface was uncovered and protected from winter weather using wooden boards and polyethylene or pond liners. The sand fillings in the footprints were moistened with water and removed with an industrial vacuum cleaner. Gas burners were used to dry the area. The track surface was then initially coated with a plastic layer using a specially developed epoxy resin system and fiberglass laminate. Finally, the entire layer surface was recovered in approximately 1-m-wide resin strips (von Koenigswald et al., 1995). Today, the track slab is displayed in the specially designed Ice Age Hall of the Museum for Local and Early History in the Quadrat in Bottrop (Fig. 14.4E).

2.3.5 *Greece*

A remarkable site was found at Trachilos beach, near the town of Kissamos in western Crete. In a marginal marine succession of Messinian age belonging to the Roka Formation, Gierliński et al. (2017) and Kirscher et al. (2021) dated and described more than 50 footprints, in two track-bearing levels, that the authors argued present hominin-like characteristics. The footprints were found under the protected area of an archeological site, an ancient coastal quarry. One month after the publication of this important find and all the excitement in national and international media, a local teacher removed nine footprints from the outcrop in different pieces and tried to sell them on the mainland. The teacher was arrested and seven footprints recovered, which are now in the local Archeological Museum of Kissamos. Meanwhile, the Natural History Museum of Crete working with the local municipality covered the tracksite with layers of geotextile and local beach sediment. The area was fenced and a panel interpreting the tracksite was placed to inform the visitors about the importance of the site and the need to cover it. In May 2024, one of us (CNC) had the opportunity to follow the geoconservation reassessment of the site made by Charalambos Fassoulas, from the University and Natural History Museum of Crete. Local inhabitants are aware of the site and any noticed movement inside the fenced area is reported to local authorities. The track-bearing levels with most of the footprints are well preserved since they have been kept covered since 2017. But the site is under the reach of coastal storms, and a research project including finding new geoconservation solutions is under way with the support of the municipality (Fassoulas, pers. comm.).

2.3.6 *Hungary*

The transnational Novohrad-Nógrád UNESCO Global Geopark situated in Northern Hungary and Southern Slovakia has several important Neogene fossil sites developed for geotourism. One of them is the lower Miocene paleontological locality complex at Ipolytarnóc, which has been well known since the middle 19th century. Ipolytarnóc Fossils Nature Conservation Area is the main geotouristic gateway to the Novohrad-Nógrád Geopark, where high-tech interpretation resources explain the geological background and fossil resources to visitors.

The site has been protected by the State since 1944 and managed by the Bükk National Park Directorate (Kordos et al., 2021). Over 7000 tracks attributed to 35 different ichnotaxa were reported, among which are mammal, bird, and crocodilian tracks (Kordos et al., 2021, and references therein). Permanent staff with guided tours along the established geological study trail in the early 1980s was the solution for effective geoconservation. Excavated areas, where tracks were exposed, were covered by conservation buildings, and new interpretation trails were opened in 1985 (Kordos et al., 2021; Fig. 14.4F). After 2005, further geotourism developments included 4D time travel movies and an augmented reality sandbox experiment, and these interactive exhibitions can provide half-day programs to visiting families. In the great conservation hall, holographic animations reconstruct the animals of the Miocene epoch and GUIDE@-HAND applications are available in three languages on the trail network. Because of the newest geotouristic developments, Ipolytarnóc has become an important nature tourism destination in the region.

2.3.7 *Ireland*

Valentia Island tetrapod trackway in Kerry County exhibits the oldest in-situ footprints of an amphibian venturing onto a land surface known anywhere in the world, dated from Middle Devonian (Stössel, 1995; Parkes and Morris, 1999, and references therein; see also Devonian tracks chapter in this book). This is open to the public and will soon be listed as a County Geological Site. A special case was made for the state to buy the site, and it is now protected and accessible to the general public, with local interpretation. It is effectively the Irish state's first geological national monument (Parkes, 2008).

2.3.8 *Italy*

Although Italy represented a predominantly marine area from the late Permian through to the Cretaceous, the country hosts a noteworthy vertebrate ichnoheritage, as evidenced by data collected since the end of the 19th century (Citton et al., 2015; Conti et al., 2020; Leonardi, 2008). The Italian ichnoheritage encompasses a wide temporal spectrum, with vertebrate ichnosites reported from the Carboniferous through the Pleistocene. It is worth noting that successful initiatives related to site protection and preservation have been implemented across the geologic timescale, with examples stemming from Palaeozoic, Mesozoic, and Cenozoic deposits. These initiatives underscore the value attributed to these important sites and the commitment to their long-term preservation. It should also be noted that the fossil sites further discussed are protected by the same legislative framework that protects the archeological heritage in Italy. Specifically, paleontological and archeological heritage is regulated under the Legislative Decree of January 22, 2004, n°42, and its subsequent updates. Accordingly, archeological artifacts and fossils are property of the State, which a priori recognizes their cultural value, prohibits their trade and arbitrary possession, and compels any private or public entity to report discovery and possession to the competent authorities.

The so far oldest tetrapod footprints come from the Carboniferous of Italy, respectively from the San Giorgio Formation of Sardinia (late Moscovian in age) and the Corona Formation of the Carnian Alps (early Gzhelian in age). They include exclusively anamniote tracks (*Batrachichnus*, *Limnopus*) and tetrapod tracks indet. resembling *Matthewichnus* (Marchetti et al., 2020 and references therein). Apart from being exhibited in temporary exhibitions (e.g., Dino & Co,

2011–12, Museum of Nature South Tyrol) and being featured in volumes for the public (Kustatscher et al., 2019), these fossils were, however, not used from a geoheritage point of view.

The Cisuralian tetrapod footprint record is much more abundant and diverse. The first discoveries of footprints in the Cisuralian mark also the beginning of ichnological studies in Italy (e.g., Curioni, 1870; for more details, see Leonardi, 2008, and references therein). The fossils of the Athesian Volcanic District have been exhibited in temporary and permanent exhibitions (Caldera, 2023–24, Museum of Nature South Tyrol), but no focused interpretation in loco or in museums has been carried out so far.

The Wuchiapingian ichnofossils of the Val Gardena Sandstone in the Bletterbach gorge include abundant and diverse tracks attributed to eureptiles (cf. *Dromopus*, *Rhynchosauroides*, *Paradoxichnium*, cf. *Protochirotherium*), abundant tracks attributed to parareptiles (*Procolophonichnium*, *Pachypes*), uncommon but diverse tracks attributed to therapsids (*Capitosauroides*, *Dicynodontipus*, *Dolomitipes*) and rare tracks attributed to anamniotes (cf. *Batrachichnus*) (e.g., Marchetti et al., 2020 and references therein). This is the most abundant, diverse, and best-preserved tetrapod ichnoassociation of the Lopingian worldwide. It is characterized by a strong affinity to Triassic ichnofaunas including the earliest ichnological evidence of the archosauriform radiation (e.g., Marchetti et al., 2020 and references therein).

Bletterbach is the perfect example of how the local community can start a bottom-up project of development of an important paleontological site. Locals helped international scientists (among which were famous names of Italian paleontology and geology such as Piero Leonardi, Umberto Nicosia, Giuseppe Leonardi, and Maria Alessandra Conti) during their fieldwork, both logistically and during the search for fossils. In the 1980s, the first guided tours began into the valley by the tourism association. In 1998, the first local museum was opened at Radein/Redagno in collaboration with the scientists of the Rome university. In 2000, the project Geoparc Bletterbach started that concluded in 2005 with the opening of the visitor center in Aldein/Aldino thanks to funds from the European Union (Ziel-2-Projekt). In 2001, the Bletterbach was protected as a natural monument for its geological and paleontological values. Fossils of the area have been featured in tens of local itinerary exhibitions since 2013 (e.g., DinoMITI, 2013; Museum of Nature South Tyrol & MUSE; Dino & Co, 2011–12, Museum of Nature South Tyrol), as well as in national and international documentaries. Today, Bletterbach is visited by approximately 50,000 visitors a year. In 2005, the Bletterbach site was officially included in a Geopark, under the management of the "Associazione pro Museo di Aldino/Museumsverein Aldein" Since 2005, the main entrance to this park has been located at the Visitor's Center, which is home to a small museum and a bookstore. This center is situated near Aldino/Aldein, to the right of Bletterbach Canyon. Correspondingly, on the canyon's left side, stands the Redagno/Radein GEOMuseum, which operates under the same organization (Fig. 14.5A).

However, the most impressive exhibit of the Bletterbach ichnoheritage is undoubtedly the Bletterbach gorge itself, colloquially known as Butterloch for the area near the waterfall. The Geopark employs trained guides who, during the tourist season (from May to October), lead groups of visitors through the valley. They showcase in situ fossil trackways and a wide range of geological attractions (Conti et al., 2020). In 2009, Bletterbach became one of the nine systems of the UNESCO World Heritage Dolomites. In 2022, the Saurian Trail was opened, a path through the forest that shows reconstructions of the tetrapods, plant, pollen, and plant–insect interactions carved in wood.

FIGURE 14.5 Examples of geoconservation projects in European ichnosites (part II). (A) Redagno/Radein GEOMuseum at Bletterbach natural monument, Italy. (B) Lower Jurassic Sołtyków Jets shedded tracksite in Poland; (C) The Avelino quarry natural monument in Portugal, with simple improvement of the access, fence and informative panel (during the ichnological training course of the Portuguese Association of Biology and Geology Teachers); (D) The former "Galinha quarry" with a large exposure of the longest Middle Jurassic sauropod trackways, which is now preserved as a natural monument, the first to be declared in Portugal in 1996; (E) "Balcón del Atlántico" interpretive mural in Matalascañas (Spain) devoted to the ecology of coastal Neanderthals based on the track record; (F) Upper Jurassic Ababuj tracksite in Teruel, asset of cultural interest (BIC) in Spain, which is accessible for people with reduced mobility or in wheelchairs and contains information in Braille.

Most Triassic ichnosites in Italy are located in the Southern Alps, but track-bearing localities are also described in the Western Alps, Northern Apennines, Maritime Alps, and Sardinia. The tetrapod footprints have been collected throughout the Triassic, from the Lower Triassic (Olenekian) Werfen Formation to the lower Middle Triassic (Anisian) where ichnoassociations are dominated by lizard-like footprints (e.g., *Rhynchosauroides*) with a gradual increase in footprints referable to archosauriforms (e.g., chirotheriids) (Mietto et al., 2020). Here the most important fossiliferous site from a geoheritage perspective is that of the Piz da Peres (Todesco et al., 2008; Mietto et al., 2020 and references therein). Field excavations (2004–08) carried out by the Museum of Nature South Tyrol and the MUSE-Trento allowed collection of a large ichnoassociation, plant fossils, amber droplets, and invertebrate tracks (Todesco et al., 2008) from the siliciclastic Richthofen Conglomerate of Illyrian (Upper Anisian) age. A partial skeleton has been found dispersed along the slope of the mountain (Renesto et al., 2020). Most of the footprints and trackways belong to *Rhynchosauroides tirolicus*. Less common but relatively diversified are the chirotheriid tracks that were referred to *Isochirotherium*, *Synaptichnium*, *Chirotherium*, *Brachychirotherium*, and *Sphingopus*, together with other tetrapod tracks such as *Rotodactylus* and *Procolophonichnium* (Avanzini and Wachtler, 2012). The site crops out in the Natural Park Fanes-Sennes-Prags, which is part of the System Northern Dolomites of the UNESCO World Heritage Site Dolomites. Fossils are exposed at the visitor center of the natural park as well as in the Museum of Nature South Tyrol and the MUSE-Trento. Due to the complexity of the exposure, nothing has been organized so far onsite.

A typical vertebrate ichnosite that is accessible to the public today is the boulder of Dolomia Principale/Hauptdolomit (Upper Triassic, late Carnian) with a path of dinosaurs at Mount Pelmetto (Pelmo Massif, Veneto Region). It was the first site discovered bearing dinosaur trackways in 1985. Thanks also to the help of the "Amici del Museo della Val Fiorentina" association, at least 10 blocks with footprints of various types and sizes have been identified (Mietto et al., 2020 and references therein). Most tracks and footprints are *Grallator*-like footprints formed by bipedal trackmakers attributed to small and primitive theropod dinosaurs. The museum of Selva di Cadore exposes a cast of the boulder, which can be reached with the help of two mountain trails.

The Dogna Valley, a narrow valley carved by the Dogna Creek and located in the Julian Alps, emerged as a significant paleontological site in 1994, when archosaurian trackways were unearthed from the Tuvalian (upper Carnian, Upper Triassic) of the Monticello Formation (Dalla Vecchia, 1996). These trackways are prominently exposed along Dogna Creek. Subsequent investigations in the same deposits have revealed the presence of nesting traces (Avanzini et al., 2007). The Dogna trackways are now displayed at the "Museum of the Territory of Dogna" The museum encompasses a paleontological section, a historical section, as well as an ethnographic section, allowing visitors to explore the rich history and cultural significance of the region. In proximity to Dogna, within the confines of the "Parco Naturale delle Dolomiti Friulane," tracks attributed to medium and large-sized ceratosaurs, prosauropods, and archosaurs—including aetosaurs and rauisuchians—are evident across a number of blocks from the Triassic Dolomia Principale Formation (Dalla Vecchia, 1998; Marzola and Dalla Vecchia, 2014). The taphonomy and the trackmakers at Malga Cjasevent (also known as Malga Casavento) are elucidated through a series of illustrative panels collectively titled "Triassic Park."

A Triassic ichnosite has also been documented in Lerici, near La Spezia in the northern Apennines, Italy. These tracks were found within the Triassic Montemarcello Formation (Nicosia and Loi, 2003). The track-bearing layers are Carnian in age and preserve two chirotherian- and three dinosaur-related forms. As such, these finds constitute some of the oldest known records of a genuine dinosaur-based ichnocoenosis (Nicosia and Loi, 2003). The popularization of the Lerici trackways was linked with the geopaleontological museum of Lerici Castle, which opened in 1998 and closed in 2015. It housed and displayed, in addition to a mineral collection, an extensive repertoire of fossil specimens from the surroundings of La Spezia. The museum was named after Ilario Sirigu, who, when 12 years old, discovered the fossil tracks during an excursion in the Lerici territory.

Situated just a few kilometers south of Rovereto in Trentino-Alto Adige, the Lavini di Marco tracksite is one of the most significant dinosaur tracksites in Europe (Avanzini et al., 2003a,b). The first tracks at this site were discovered in 1989 by Luciano Chemini, an amateur naturalist (Avanzini et al., 2003a,b). The discovery of the Lavini di Marco tracksite marked a critical turning point, triggering comprehensive searches for other dinosaur footprint-bearing outcrops across the Southern Alps (Avanzini et al., 2003a,b). Systematic field explorations carried out at the Lavini di Marco site since 1991 have led to the discovery of numerous additional tracks (Leonardi and Mietto, 2000). These trampled horizons at Lavini di Marco have been attributed to the middle-upper section of the Monte Zugna Formation (Jurassic: Hettangian–Sinemurian) (Avanzini et al., 2003a,b). They cover roughly 300,000 m^2 of monoclinal surfaces (Avanzini et al., 2003a,b), part of which can be observed by following a paleontological trail. The paleontological trail of Lavini di Marco starts from the parking lot near the Grotta Damiano Chiesa. Access to the trail is free. The loop path winds along a few kilometers, which can be completed in less than 2 h at a leisurely pace. Along the route, explanatory panels have been installed, engraved, and screen-printed on stone, to give everyone the chance to delve into the geological history of the area and its unique ichnological features. A wooden tower allows for a comprehensive view of the tracks surfacing along the "Colatoio Chemini" an outcrop named after the discoverer of the site. The Lavini di Marco paleontological site is fully protected under Provincial Law no. 37 of 1983.

The Altamura ichnosite, first discovered by geologists Sarti and Claps in 1999, is a Late Cretaceous site that boasts an impressive collection of 26,000 dinosaur tracks and 12 distinct trackways (Nicosia et al., 1999; Petti et al., 2022). Despite the initial discovery, the study and preservation of this palaeosurface were placed on hold until 2019. At this time, a thorough mapping survey was conducted to meticulously document its geological and paleontological features (Petti et al., 2022). The results of the survey have been compiled into a geothematic map, which purposedly aims to facilitate future conservation efforts, evaluation, and public dissemination of this significant paleontological heritage (Petti et al., 2022). Currently, the ichnosite is accessible upon request, i.e., a form has to be submitted to the local municipality. Explanatory panels made of concrete are available on-site to aid interpretation of the trackways. A gangway provides a top-down perspective of the ichnosite.

Since the discovery of the Altamura tracksite in 1999, the Mesozoic carbonate succession of the Apulia Carbonate Platform has emerged as a prolific source of dinosaur footprints (Antonelli et al., 2023). Numerous dinosaur tracksites have since been uncovered in the Murge Plateau and the Gargano promontory, including those at Mattinata, Borgo Celano, Bisceglie, and Lama Balice (Antonelli et al., 2023; Leonardi and Mietto, 2000; Nicosia et al., 1999; Petti et al., 2018, 2022). Specifically, three loose blocks bearing theropod footprints were discovered at a pier in Mattinata and subsequently assigned to the Upper Jurassic (Conti et al., 2005). Part of the track-bearing blocks are in the historical and archeological museum of Mattinata. In addition, some of the Mattinata blocks now form part of an exhibit in the nearby town of Borgo Celano, at the "Paleontological Museum of the Dinosaurs." It serves as a dinosaur-themed park and museum, featuring an array of dinosaur animatronics displayed outdoors. Alongside the Mattinata blocks, the park also showcases the Cretaceous dinosaur tracks discovered in the area of Borgo Celano (Petti et al., 2008). The dinosaur tracks of Bisceglie were discovered in the summer of 1996 in a quarry that is now abandoned (Sacchi et al., 2009). Ten blocks with dinosaur tracks were extracted and placed in the flower beds of the freely accessible municipal park of Sant'Andrea, locally known as the "Dinosaur Park" Some panels were installed with information about the trackways.

In September 2006, local hikers uncovered a significant collection of dinosaur footprints near Esperia, approximately 30 km south of Frosinone in Latium, Central Italy. This track-bearing stratum is part of a shallow-water limestone sequence, which is attributed to the Aptian stage and has yielded around 80 dinosaur footprints from both quadrupedal and bipedal dinosaurs (Petti et al., 2008). The dinosaur traces at this site are readily accessible to the public, with a wooden barrier in place for their protection. A cast of the trampled surface, capturing the Esperia dinosaur tracks, is on display in the Karst Museum of Esperia.

In the year 1994, Umberto Lenuzza and Ivano D'Aronco, inhabitants of Osoppo in the province of Udine, discovered an assemblage of trackways on Osoppo Hill. These were brought to light during restoration excavations of an old fort that had been damaged in the 1976 earthquake. Dalla Vecchia and Rustioni (1996) assigned the trackways to mammalian tracemakers, i.e., the equid *Hipparion*, a possible bovid, and a small rhinoceros. According to the aforementioned authors, the trackways, in general, and particularly those of *Hipparion*, enable the attribution of the ichnofossil-bearing layers to the Messinian. The fossil tracks are part of the visitor trail of the Osoppo fort; they are protected by railings, and there is an interpretive panel for the public.

Locally referred to as "devil's trails," human trackways have been discovered in association with other mammalian footprints in a Middle Pleistocene pyroclastic flow. This flow, originating from the Roccamonfina volcano in northwestern Campania, southern Italy, has been dated to approximately 385,000–325,000 years ago (Mietto et al., 2003; Panarello et al., 2020). This tracksite is located within the Roccamonfina Regional Park, where explanatory panels are provided to enhance visitor understanding of the site's geological and paleontological significance. Additionally, a metal gangway has been installed to offer a top-down perspective of the trampled surface, enabling visitors to fully appreciate the extent and arrangement of these trackways.

2.3.9 *Norway*

The Lower Cretaceous (Barremian) section at Isfjorden, Svalbard, consists of sandstones consistent with deposition on an alluvial flood plain. The tracks are situated on two different horizons stratigraphically below the original horizon found in 1960. Footprint evidence from Festningen suggests that during the Early Cretaceous, there was a diverse dinosaur fauna on Svalbard and that both theropods and ornithopods were present at the time (Hurum et al.,

2006). Festningen locality in Isfjorden, mid-Svalbard, is protected by law, but the tracksite is actively eroded by natural processes. The first 13 tracks found in 1960 were destroyed by sea erosion. There are guided tours focusing on dinosaur tracks.

2.3.10 *Poland*

The tracksite in the "Gagaty Sołtykowskie" (Sołtyków Jets) geological reserve in the village of Sołtyków is one of the main geological attractions in the Holy Cross Mountains (Babel and Leonowicz, 2015). Lower Jurassic strata, rich in numerous, well-preserved, dinosaur tracks and remains of Jurassic plants and insects are exposed in a former clay pit. The pit exposes continental fluvial-lacustrine sediments from the lower part of the Zagaje Formation. A wooden shed protects the best-preserved tracks (Fig. 14.5B). The three-digit footprints, known as *Kayentapus soltykovensis*, were made by dilophosaurs. The tracks run toward each other, which was interpreted as evidence of joint hunting (Pieńkowski, 2004b). The *Parabrontopodus* assemblage was made by two large sauropods, moving to the south, and four small, at most 2-year-old, sauropods, moving to the north (Gierliński and Pieńkowski, 1999). The tracksite is protected by law and remains in the area of Nature Reserve Gagaty Sołtykowskie.

Bałtów is a small village near Ostrowiec Świetokrz. Footprints were described as belonging to Lower Jurassic allosaurids and stegosaurids (Gierliński and Niedźwiecki, 2005). In 2004, Jura Park was established, featuring fossil footprints, life-size dinosaur models, and additional tourist infrastructure. Bałtów dinosaur theme park was the first large-scale dinosaur park in Poland.

The Museum of Nature and Technology in Starachowice, Holy Cross Mountains, has extensive collections of excellently preserved footprints from the Lower Triassic at Wióry (Klein and Niedźwiedzki, 2012).

2.3.11 *Portugal*

Regulation of natural heritage conservation in Portugal is made at a national level by the Decree n. 142/2008 that was revised by the Decree n. 242/2015, which, among other things, allowed the establishment of regional and local protected natural areas. There are several vertebrate tracksites in Portugal that are within protected natural areas and considered geoheritage (Santos et al., 2008, 2010), but several can hardly be recognized by the public, through the existence of informative materials, such as leaflets, websites, location on map apps, and regular offers of guided visits (e.g., Oxfordian theropod tracksite of Laje do Costado from Cabo Mondego Natural Monument (Carvalho et al., 2022), Middle Pleistocene proboscidean tracksite of Malhão beach in SW Alentejano and Costa Vicentina Natural Park (Neto de Carvalho, 2009), Serras do Bouro and Mangues Upper Jurassic dinosaur tracksites in Oeste UNESCO Global Geopark). For these sites, no conservation measures have been established.

Other ichnosites have informative panels and accessibility improvements, such as the Barremian iguanodontid tracksite from Salema Beach in Algarve, southern Portugal (Santos et al., 2013), or the two Kimmeridgian dinosaur tracksites of Serra da Pescaria and Salgado in Nazaré and Alcobaça, respectively, central Portugal. Some of these tracksites still require detailed scientific studies. The geological conditions of some of the vertebrate tracksites in

Portugal and the long exposure times put them at serious risk of disappearing, such as the Lower Cretaceous dinosaur tracksite of Praia Grande do Rodízio (Madeira and Dias, 1983), in Sintra-Cascais Natural Park near Lisbon, and the Middle Pleistocene mammal and bird tracksites in SW Alentejano and Costa Vicentina Natural Park (Neto de Carvalho et al., 2016), or from being eroded by exposure to rainwater and frequent visiting, as the case of the Bajocian lagerstätte of Cabeço da Ladeira, the "Jurassic Beach," in Serra de Aire e Candeeiros Natural Park. In this ichnosite, an echinoderm lagersttäte and some of the earliest evidences of ray feeding behaviors were found (Machado et al., 2021), and an urgent measure was taken to erect fencing and close the site in order to control visits.

The conservation of vertebrate tracks started in early times in Portugal, with the rescue in 1884 of a Late Jurassic theropod trackway by Jacinto Pedro Gomes (Gomes, 1916; Santos and Neto de Carvalho, 2016) that is now exhibited in the National Museum of Natural History and Science, in Lisbon. Other dinosaur footprints from the Cabo Mondego and S. Martinho do Porto localities can also be found at the Geological Museum of Lisbon. In the last 30 years, other regional institutions have rescued track slabs and isolated tracks from erosion in coastal areas, such as the Lourinhã Museum (Upper Jurassic dinosaur tracks and pterosaur trackway from the coastal area of the municipality; e.g., Mateus and Milàn, 2008), Natural History Society of Torres Vedras (numerous Upper Jurassic dinosaur tracks from the west coast, e.g., Castanera et al., 2020) and Portuguese Center for Geohistory and Prehistory (Lower Cretaceous dinosaur tracks from the coast of Cape Espichel; e.g., Figueiredo et al., 2017).

The last decade of the 20th century saw a surge of enthusiasm in Portugal about dinosaurs in general and protection of dinosaur tracksites, in particular. António Galopim de Carvalho, by then director of the National Natural History Museum, technically supported by Vanda Faria dos Santos and Martin Lockley, was responsible for the protection of some of the most important dinosaur ichnosites found in that period, as national Natural Monuments. Galopim de Carvalho, an active communicator of geology and geoconservationist in his 90s now, was a pioneer of the onsite conservation and protection of dinosaur ichnosites, such as Carenque and Serra de Aire.

In 1986, the Pego Longo-Carenque tracksite was discovered in an abandoned quarry at Quinta de Santa Luzia, near Lisbon. In a thin layer of limestone from the middle Cenomanian, 10–15 cm thick, there are around 200 footprints, including a 132 m long trackway, probably from an ornithopod. This ichnosite was at risk of being destroyed by the construction of the Estrada Circular Regional Exterior de Lisboa (CREL) highway, being protected thanks to the involvement of civil society in what became known by Galopim de Carvalho as the "Battle of Carenque" (Galopim de Carvalho, 1994). To avoid destroying the ichnosite, the solution found was to open tunnels through which the CREL passes. However, since the construction of the tunnels and the protection as a national Natural Monument in 1997, the tracksite was buried for protection, so the Carenque Natural Monument, with 6.09 ha, does not currently have any educational or tourist use. The sauropod figuratively represented at the entrance of the tunnels doesn't even have representation in the local track record.

Another Natural Monument is the Avelino ichnosite. In Zambujal (Sesimbra, south of Lisbon) and with 1.66 ha and good accessibility, the Avelino ichnosite includes Upper Jurassic *Parabrontopodus* trackways with different track sizes. The main slab is approximately 15 m long by 10 m wide, and dips 30 degrees, which facilitates observation and pedagogical use (Santos et al., 2016; Fig. 14.5C). Other protected tracksites located nearby include the famous

Pedra da Mua tracksite, showing gregarious behavior in Late Jurassic young sauropods and related to a medieval legend that was the reason for the construction of an important sanctuary in the 18th century, revealing in a mosaic panel one of the earliest pictures related to interpretation of vertebrate tracks (Antunes, 1976; Santos et al., 2016). On the north side of Lagosteiros bay, there is the Lagosteiros Natural Monument with Hauterivian theropod trackways (Antunes, 1976). All these natural monuments have signage, improved accessibilities and interpretation panels.

The Serra de Aire Dinosaur Tracks Natural Monument located near Fátima, in Serra de Aire e Candeeiros Natural Park, is the most famous dinosaur ichnosite in Portugal. At the place where the dinosaur footprints are situated today, there used to be a quarry, the so-called Pedreira do Galinha. The 20 existing trackways in a single bed are the largest and oldest sauropod tracks known (Bathonian), but also one some of the best preserved, consisting of more than 1000 footprints. The National Museum of Natural History created a working group that serves to demonstrate to official and scientific entities the paleontological importance of the find (e.g., Santos et al., 1994), which led to the deactivation of the large quarry and consequent creation of the monument, the first dinosaur ichnosite protected under law in Portugal, in 1996. The monument, with an area of 20 ha, was opened to the public on March 1, 1997, and in the same year, the first autonomous circuit for visiting and interpreting the tracksite was founded. In 2002, the Jurassic Garden was created with the aim of showing visitors current plants from characteristic botanical groups in the Mesozoic, and the "Aramosaurus," a sculpture representing the producer of the prints. The park also has an entertainment area, an Environmental Educational Center, a picnic area, a large mural illustrating the evolution of life on Earth throughout geological time, and several information panels along the approximately 1000 m of the trail that make the perimeter of the former quarry (Santos et al., 2008, 2010; of over 20,000 visitors per year, an intervention was made to protect and highlight the tracks and to create raised wooden walkways. This kind of walkways were also established in Vale de Meios Natural Monument, furthermore in a deactivated quarry located not far from the previous one, where over 700 large and well-preserved, mainly theropod footprints can be found on an 11,000 m^2 bedding surface, dated to the Bathonian (Razzolini et al., 2016).

2.3.12 *Romania*

Pârâul Bozului, in Vrancea County, Eastern Romania, is a Nature Reserve according to the Decree nr. 5/2000. The Pârâul Bozului paleontological reserve, dated from the Burdigalian, is of scientific importance due to the presence of trace fossils attributed to 35 species of vertebrates and five species of invertebrates, the most important of which are the footprints of aquatic birds (e.g., *Anatipeda., Ardeipeda., Gruipeda, Charadriipeda*) and mammals (*Bestiopeda, Felipeda, Canipeda, Hippipeda, Pericopeda* and especially *Proboscipeda enigmatica*), described among others in the classic paper of Panin and Avram (1962). A negative impact on the conservation of the tracks is the installation of vegetation in the crevices of the rocks, which accentuates their degradation, while, due to the configuration and the very steep slope angle of the beds, conservation measures are very difficult to apply.

Peştera Gheţarul de la Vârtop, or Vârtop Glacier Cave, in the region of Cluj, is a Natural Monument. This cave became famous because of the first description and dating of Neanderthal footprints (Onac et al., 2005; See Walk the World chapter in this book). Research in 1974 by a team led by Professor Iosif Viehmann from the "Emil Racoviță" Speleological Institute in Cluj led to the discovery of the Sălii Pașilor (Hall of Steps). It was named after three Neanderthal footprints, which were discovered in this chamber. One footprint was removed and studied at the Institute of Speleology in Cluj. The other two footprints were cut out and stolen soon after, and the perpetrators were never found. The cave is managed by the administration of the Apuseni Natural Park.

Urşilor Cave, also in the Apuseni mountains in Chişcău, was protected immediately after its discovery in 1975 and was later arranged for the tourism of one sector. Urşilor Cave is one of the most complex MIS 3 cave bear settlements of Europe: it incorporates a large number of cave bears' palaeoichnofabrics (bioglyphs: nests, footprints, scratch marks, fur imprints, etc.) (Diedrich, 2011; see Quaternary Ichnology chapter). The cave was opened to visitors on July 14, 1980. Every year the cave is visited by more than 200,000 tourists.

In 1965, Ciur-Izbuc Cave in the Carpathian Mountains of Romania was discovered to contain about 400 ancient human footprints (see Walk the World chapter in this book). The age of the prints (10–15 ka BP) was based partly on their association with cave bear (*Ursus spelaeus*) footprints and bones, and the assumption that cave bears became extinct near the end of the last ice age. Since their discovery, the human and bear evidence and the cave itself have attracted spelunkers and tourists, with the result that the ancient footprints are in danger of destruction by modern humans. In an effort to conserve the footprints and information about them and to reanalyze them with modern techniques, Ciur-Izbuc Cave was restudied in summer of 2012. Modern results are based on fewer than 25% of the originally described human footprints, the rest having been destroyed (Webb et al., 2014).

2.3.13 Slovenia

The Godovič tracksite is included in the Idrija UNESCO Global Geopark as one of the most important geosites. It includes six tetradactyl footprints dated to 220 million years ago (Late Triassic, Norian/Rhaetian) and produced by a biped dinosaur walking on soft ground along the shores of a tropical sea. The footprints were made by either a plateosaur (Plateosauridae)—one of the first dinosaurs—or a phytosaur (Phytosauria), a crocodile-like archosauriform (Dalla Vecchia, 1997). The site has informative panels and is included in the hiking trail system.

2.3.14 Spain

The declaration of a paleontological site as a legally protected asset is based, on the one hand, on a 1985 law on Spanish Historical Heritage whose maximum category of protection, applied to many dinosaur tracksites, is that of Asset of Cultural Interest (Bien de Interés Cultural, BIC), which equates them to the most valuable resources of Spanish heritage, such as the Roman Aqueduct of Segovia, the Cathedral of León, or the Royal Palace of Madrid. This law has been adapted in different ways in the legislation of some Spanish Autonomous Communities, with greater or lesser prominence of paleontology, but in all cases it is a form of protection related to assets of cultural interest. This is the case of the Catalan Late Cretaceous site of Fumanya, declared a Cultural Asset of National Interest (Bé Cultural d'Interès Nacional, BCIN).

In addition, a 2017 law on Natural Heritage and Biodiversity—and its predecessor—allows the establishment of protection for paleontological assets through another category, that of Natural Monument. This category of protection has been applied on fewer occasions, but this is the case for the dinosaur tracksite of Santisteban del Puerto in Jaén, and the tracksites of Asturias located in the municipalities of Colunga, Gijón, Ribadesella, and Villaviciosa, which were declared a Protected Natural Area under the category of Natural Monument. This area was also declared a Special Area of Conservation for Ichnite Sites (ES120004 7).

The Spanish Ministry of Culture and the corresponding Offices of the governments of six Spanish Autonomous Communities (Aragón -province of Teruel-, Castilla y León -Burgos-, Cataluña -Barcelona-, Comunidad Valenciana -Castellón and Valencia-, La Rioja and Principado de Asturias) jointly promoted a World Heritage Candidacy called Dinosaur Ichnites of the Iberian Peninsula (Icnitas de Dinosaurios de la Península Ibérica, IDPI). In the first phase of this nomination, 216 Spanish tracksites were included and, as a guarantee of presenting the maximum protection categories allowed by Spanish legislation, many of them were declared BIC. As a result, 65 dinosaur tracksites have been declared BIC in the category of Historic Site (40 in La Rioja, 14 in Aragón, 10 in the Comunidad Valenciana, and 1 in Castilla y León), according to the database of the Ministry of Culture:

https://www.cultura.gob.es/cultura/patrimonio/bienes-culturales-protegidos.html.

The candidacy was evaluated in 2005, and after decision 30COM 8B.26 of the World Heritage Committee (WHC), a new conceptual framework was proposed, and Portugal joined the candidacy. The new candidacy was presented in 2008, including just 11 tracksites (three in Portugal and eight in Spain). Selection of sites was established according to scientific criteria (FCPTD, 2009; Mampel et al., 2009), and the network of selected sites covering 100 Ma of the history of life represented the so-called "Dinosaurs Walking in a Drifting World, a terrestrial ichnological record in the core of the Pangea's fragmentation." After evaluation, decision 34COM 8B.7 of the WHC deferred the inscription in the WHL, asking the State Parties to justify the Outstanding Universal Value (OUV) of a property based on dinosaur tracks and considerations for a serial nomination with existing properties. In the end, the applicant States did not submit a new proposal, but the whole process benefited the heritage of the two countries thanks to the list of legally protected tracksites and the extensive documentation that was compiled for each of them, as well as the numerous research, conservation, and dissemination actions that were carried out.

The great diversity of types of dinosaur tracks found in Spain, their abundance, the breadth of geological time that they cover, their relation with many sites containing direct remains from the same period, and their comparison with other areas of similar characteristics (geological and paleontological), make the Spanish territory of the Iberian Peninsula one of the prime areas to study the behavior, evolution, and dispersion of dinosaurs. Indeed, sites with tracks of various types of vertebrates (crocodylomorphs, turtles, pterosaurs, mammals ...) have been described in Spain, but the largest inventory of vertebrate tracksites that have received attention for their protection and, as will be seen further, for their use as an educational and tourist resource, are those of dinosaurs.

Several hundred dinosaur tracksites cataloged in Spain (comprising tens of thousands of tracks) are an example of the extraordinary and diverse richness of the dinosaur ichnological record in Spain, bearing in mind that:

- they represent their faunal succession during some 90 million years (from the Late Jurassic to the Late Cretaceous) and shed much light on the changes that occurred in dinosaur faunas between the Jurassic and Cretaceous,
- the rich ichnological diversity and its close association with the direct remains found serve as proof of the evolution of dinosaurs and allow us to generate hypotheses about their dispersion and way of life in a very restricted area limited by a single tectonic plate, the Iberian Plate, which served as a bridge connecting the different faunas of the surrounding continental masses,
- the Spanish dinosaur tracksites show multiple processes associated with the formation of the sites, with numerous continental sedimentary facies and depositional environments (fluvial and intertidal/supratidal coastal areas, as well as palustrine and lacustrine environments),
- also record most types of dinosaur trackmakers: thyreophorans, ornithopods, sauropods (both wide and narrow-gauge trackways), and theropods (large, medium, and small carnivorous dinosaurs, including avian dinosaurs). As a consequence of the large number of investigations that have been carried out—as demonstrated by several doctoral theses (Moratalla, 1993; Castanera, 2013; García-Ortiz, 2016; Piñuela, 2016; Razzolini, 2016) and monographic works (e.g., Torcida Fernández-Baldor, 2006)—dozens of new ichnotaxa have been described from sites in Spain.

According to the Spanish dinosaur tracksites record, we can highlight three important areas that are representative of different epochs during which dinosaurs inhabited Laurasia (FCPTD, 2009; Alcalá, 2012):

- Dinosaur Coast featuring the Upper Jurassic sites of northern Spain formed during the Proto-Atlantic expansion. During this period, large sauropods (eusauropods, diplodocids and basal titanosauriforms), theropods of different sizes and types (carnosaurs, spinosaurus, ceratosaurs, and maniraptors), and thyreophorans (mainly stegosaurs) were abundant,
- A Changing World featuring sites emerging in the so-called Straight of Soria during the Jurassic-Cretaceous transition and the Early Cretaceous in Spain. This domain shows the faunal shift between typical Late Jurassic faunas (abundance of sauropods) and Early Cretaceous faunas (abundance of ornithopods). Thyreophorans, although present, were limited to very restricted areas,
- The End of an Era where sites cluster around the Pyrenees, formed in the Mesozoic basins present during the closure of the union between the Atlantic and the Tethys at the beginning of the Alpine Orogeny. The representation of sauropods is limited to the presence of titanosaurids and, among the ornithopods, to the abundance of hadrosaurids. Among the theropods, medium-sized and small theropods predominated. Ceratopsids either did not exist in this part of Laurasia or left no recognizable evidence.

It is well known that dinosaurs have a great popular impact, which has led to the development of research on them more than on other vertebrates. At first, dinosaur tracksites were the object of attention with various types of interventions, halfway between physical protection measures and geotourism use, and then they began to be timidly protected, also with certain legal protection measures. Suddenly, there was an explosion in the number of tracksites that became protected due to the submission of an application for inclusion of such sites on the UNESCO World Heritage List (WHL).

However, there are also other tracksites in Spain, less popular, but equally representative of Spanish extinct vertebrate associations. Thus, in Abiego (Huesca), there is a site with more than 1000 artiodactyl tracks, many of them isolated, from the Late Eocene-Oligocene. In Jumilla (Murcia), a site from the late Miocene shows tracks of equids, bovids, and camelids, as well as some carnivores; the site is protected by a shed and by the General Urban Development Plan (Plan General de Ordenación Urbana, PGOU). Another site in Jumilla shows tracks attributed to rhinoceroses, bovids, carnivores, a wading bird, and a reptile, which has also been provided with a cover. Finally, very recently, a Middle Pleistocene site with hundreds of mammal (including Neanderthals) and bird trackways and footprints named the Matalascañas Trampled Surface (MTS) was described in the Asperillo Cliff Natural Monument, in the Huelva Province of Andalucia (e.g., Neto de Carvalho et al., 2020, 2021, 2023). The site is only occasionally exposed by the coastal dynamic processes so the Municipality of Almonte set up a large-scale interpretative mural with the main findings and their significance close to the site (Fig. 14.5E).

The variability and wealth exhibited at dinosaur tracksites have led to a number of different research, conservation, and last but not least, awareness initiatives throughout the whole of the Spanish terrestrial Upper Jurassic and Cretaceous outcrops, turning tracksites into veritable museums in the field and generating resources for the different paleontological institutions that guarantee ongoing social use of the sites. Dissemination of tracksites has helped to revitalize areas that were socioeconomically depressed, serving as a clear example of the benefits that heightening the awareness of a natural resource have for all different sectors of the population at large (Cobos et al., 2020).

Indeed, dinosaur tracks are normally preserved in open-air in-situ sites, thus they need exceptional preservation measures as part of the natural environment (on the contrary, bones are usually well curated in city museums); and tracksites allow local development in rural areas. Many sites in the north-eastern sector of the peninsula, from Asturias to Murcia, have been made available for visitation as a geotourism resource for territorial development, even in some cases that have not necessarily been legally protected. But, as might be expected, actions of this type predominate in the communities/provinces with the most declared BICs, La Rioja, and Teruel. There are usually sheds covering the tracks and working as protection against runoff water or the possibility of rocky elements falling due to gravity. These measures of effective protection are always linked to make them available for visits, generally by means of information panels. Previously, excavations have usually been carried out to enlarge the surface of the site and reveal new tracks, as well as restoration and preventive conservation treatments: removing vegetation on the site and application of herbicide to prevent regrowth of plants; cleaning of the cracks and fissures; adhering the broken fragments (if any), and sealing fissures with mortar and, in some cases, with epoxy resin. In parallel to conservation actions, casts of some tracks are often made.

There are many known cases in Spain where ichnofossils have been used to promote geotourism. Usually, the tracksites are prepared, and often the actions are included in tourist routes, as is the case in La Rioja (Castanera et al., 2018), which include tracksites and museums/interpretation centers, such as those in Arnedo and Igea. In Teruel, there are also several examples of tracksites (one of them adapted for inclusive visits through information in the Braille system and wheelchair accessibility, Fig. 14.5F). A representative example is the creation of the DINOwalk (DINOpaseo), an urban paleontological route (Cobos and Alcalá, 2017), which serves as a starting point for a wider local walking trail, The Dinosaur Route (La Ruta del Dinosaurio) in El Castellar (Cobos et al., 2020). Tourists can enjoy dinosaur tracks, both in the streets of the town (replicas), and in the nearby natural sites where they were found (original ichnofossils). In turn, these resources of the small town of El Castellar are part of the great route The Dinosaur Track Road (La Carretera de las Huellas de Dinosaurio), a 70 km rural road that allows you to visit more than 20 tracksites. These actions are complementary to permanent facilities—such as the large network of Dinópolis museums—because they offer the possibility of learning about the geological characteristics and the specific places of origin of the fossils presented in them on site.

Numerous Upper Jurassic–Lower Cretaceous tracksites are found in the Maestrazgo UNESCO Global Geopark. Seven of these sites are classified as Assets of Cultural Interest by the Regional Government of Aragon, the highest level of protection established in Spanish legislation, and three of them are open to visitors to promote education and geotourism (Alcalá and Cobos, 2021). Las Cerradicas and Los Corrales del Pelejón are the only paleontological sites in the Geopark covered by a structure that protects them from weathering and other threats (such as the passage of livestock, visitors, etc.) and include interpretation. The Miravete 1 tracksite was partially prepared for visitors in 2013. A track is currently displayed in the main Dinópolis exhibition center at the city of Teruel, as a sample of what visitors can see if they choose to visit the Geopark.

In the Origens UNESCO Global Geopark, the Orcau 2 site is the second dinosaur tracksite described in the Late Cretaceous of the Pyrenees and the first in the Conca de Tremp basin. Iguanodontid tracks described as *Ornithopodichnites* magna and *Orcauichnites garumniensis* and large-size titanosaurid sauropod tracks compose the tracksite. Fences have been put in place to protect the outcrop, and the access was improved for the safety of people. An informative panel provides interpretation. Guided visits are available by the Conca Dellà Museum. The Tremp Learning Camp, official educational entity of the Catalan government, carries out visits and didactic activities in the Orcau-2 Geosite. It is visited by about 2000 students/year (www.geoparcorigens.cat). The paleontological site of the Sallent Valley is considered one of the most important nesting areas of dinosaurs found in Europe. In the Cretaceous viewpoint, there are interpretive panels and audio guides, both in four languages, describing the importance of the site. Guided visits are provided by the Dinosfera interpretation center (www.geoparcorigens.cat/en/whattodo/museums).

This model of tracksite routes and museums can be found in other areas of Spain, such as Asturias (various tracksites around the Jurassic Museum of Asturias, MUJA), Castilla y León (with the Dinosaur Museum of Salas de los Infantes and the Paleontological Museum of Villar del Río), Comunidad Valenciana (Paleontological Museum of Alpuente and Dinomania of Cinctorres), and Castilla La Mancha (Paleontological Museum of Castilla-La Mancha and associated interpretation centers), among others.

2.3.15 *Switzerland*

More than 20 localities with tracks from both sauropod and theropod dinosaurs at different stratigraphic levels have been found in Switzerland (Meyer and Thüring, 2003). In a quarry near the small village of Lommiswil, the first Late Jurassic dinosaur

(Kimmeridgian) tracks were reported from Switzerland. Individual tracks are up to a meter or more in diameter, and one trackway can be traced for about 90 m. Track morphology and trackway pattern clearly identify them as those of large sauropods. Four-hundred-fifty individual footprints can be seen, and more were exposed in recent years with ongoing quarry operations. The Natural History Museum of Solothurn is one of the most remarkable of its kind in Switzerland. The museum has existed in its present form since the end of the 1970s, but its roots go back almost 200 years. The museum program includes guided visits to the dinosaur tracks near Lommiswil. The viewing platform there on the quarry offers a good view of the dinosaur tracksite. Replicas of the footprints are found in the Museum.

The footprints discovered in 1976 at Vieux Emosson are regarded as Ladinian or Carnian in age. The footprint assemblage of Vieux Emosson, located at 2400 m altitude in the Swiss Alps, includes a short trackway with shallow but well-preserved tracks referred to the ichnogenus *Isochirotherium* (Demathieu and Weidmann, 1982; Avanzini and Cavin, 2009; Klein et al., 2016). Preliminary observations in situ and of the footprint casts made in 1979 and deposited in the Natural History Museum of Geneva made by these authors suggest that the Vieux Emosson footprints are mainly "track preservation variations" of chirotherian tracks. The area was declared a protected site by the government of the Valais canton on November 9, 1983. An educational trail was created in 1998 by the Municipality of Finhaut with the support of Espace Mont-Blanc. This trail was completely modernized in 2015 in collaboration with the Natural History Museum of Geneva and with the support of the Canton of Valais and the Swiss Confederation. It is one of three "Finhaut Trails" an educational project that aims to showcase the commune's history and heritage. The proposed route is fully signposted and is shown on local hiking maps. The footprints are usually visible from mid-July (before then, they are covered by snow).

In 2002, a new dinosaur tracksite was discovered on public land on the future course of the "Transjurane" highway in Courtedoux, Canton Jura, northwestern Switzerland. The uncovered area with 1500 m^2 exhibits over 650 sauropod and theropod tracks, of which about 400 constitute 17 sauropod trackways referable to *Parabrontopodus* and two trackways of theropod dinosaurs. The sauropod trackway pattern reveals evidence of local, shoreline-controlled travel by several groups of sauropods during the Late Jurassic (Marty et al., 2003). Since its discovery, the protection and the conservation of the site in situ was promoted by the "Section de paléontologie" (Marty et al., 2004). According to Swiss law, the site is a natural object of high scientific value and has to be protected from destruction, as, in Switzerland, heritage sites of national value are rated with the same level of significance as federal infrastructure projects, such as highways. Due to the favorable location of the site at one end of a viaduct, the site could be protected by the construction of an additional small highway bridge. Such a building could provide ideal conditions for the conservation, scientific research, and public viewing of the site. The "In the Footsteps of Giants" educational trail at Courtedoux enables exploration of the tracksite under the guidance of the Jurassica Museum (https://www.jurassica.ch/en/Satellites/Dinotec.html).

The recent extension and renovation of the Technical Division (DIVTEC) of the Centre Jurassien d'Enseignement et de Formation (CEJEF) allowed the discovery of dinosaur footprints in Porrentruy. Visitors can see sauropod and theropod trackways in the building's inner courtyard along with life-size artist's impressions of dinosaurs. Two clocks also allow visitors to better understand geological time (millions of years) by comparing it to a more precise measurement of time (milliseconds).

An exhibition with a good overview of the Swiss tracksites can be found in the Aathal Dinosaur Museum (https://sauriermuseum.ch/).

2.3.16 United Kingdom

Although the geological record of the British and Irish islands is remarkably complete—a consequence of long being "on the edge" of a continent—this also means that a dominance of marine facies through significant time intervals has limited the potential for the preservation of terrestrial vertebrate trackways. Nevertheless, although vertebrate trackways through much of the preserved geological record are few—except in the Permian-Triassic and Mid Jurassic to Early Cretaceous intervals—many of those recorded are of significant scientific importance and hence incorporated within national conservation sites. As a consequence, there is a significant conservation "case-history" associated with many of these sites.

As part of a vision to rebuild Britain after the Second World War (1939–45), a specialist working group was established to produce a preliminary list of geological and geomorphological sites for legal conservation, the later framework coming as the National Parks and Access to the Countryside Act in 1949. The Act established parallel concepts of "geological monuments" and "physiographic features" (i.e., geomorphological sites) that received legal protection within "Sites of Scientific Interest"—later Sites of Special Scientific Interest, or SSSIs (Page and Wimbledon, 2009).

Initial site selection, however, was not systematic and tended to highlight only well-known, "classic" localities, and it took the establishment of the Geological Conservation Review (GCR) in 1977, to begin the process of developing a true inventory of nationally important geological and geomorphological sites (Wimbledon et al., 1995; Ellis et al., 1996). The GCR process selected sites for subsequent legal designation as SSSIs within thematic "blocks," each subject area led by an appropriate specialist, ultimately leading to an initial selection of over 3000 sites across England, Scotland and Wales. The north of Ireland, however, was not covered by the GCR process, although a parallel process of site selection was initiated—the "Earth Science Conservation Review" or "ESCR"—based on similar principles, with sites being legally notified and protected within "Areas of Scientific Interest" or "ASIs."

Although vertebrate trackways would be expected to have been selected within one of the nine themed "blocks" for vertebrate paleontology (see Ellis et al., 1996, Fig. 56), only three sites overall were originally explicitly listed for trackways, both in the Permian-Triassic Reptilia block—Masonshaugh Qy (Cummingston, Morayshire, Scotland) and Hayes Point to Bendrick Rocks (South Glamorgan, Wales)—see further (Benton and Spencer, 1995). A subsequent addition, as discussed further, was Ardley Trackways within the Jurassic-Cretaceous Reptilia block.

Other sites from which footprints and trackways have been recorded are generally included within "Stratigraphy" (i.e., Earth History) themes, for instance, as part of a sedimentary system or selected for an invertebrate paleontology interest, although some are included in Paleontology sites where bone material is the primary GCR interest. These themes include: Non-Marine Devonian (Barclay et al., 2005), Permian-Triassic (Benton et al., 2002), Permian-Triassic Reptilia (Benton and Spencer, 1995), Aalenian-Bajocian, Bathonian (both Jurassic; Cox and Sumbler, 2002), Jurassic-Cretaceous Reptilia (Benton and Spencer, 1995), Portlandian-Berriasian (Jurassic-Cretaceous boundary), and Wealden (nonmarine Lower Cretaceous; Radley and Allen, 2012).

The process of formally designating these GCR sites as legally protected interest features as, or within, SSSIs began in earnest after the passing of the 1981 Wildlife and Countryside Act, which very significantly strengthened the provisions of the 1949 Act. Crucially, the 1981 Act introduced the concept of "notifiable activities" as a list of actions that might require prevention or management within sites notified for different interest features, including both ecological and geological. Activities could be selected from this list to produce a "tailored" selection of "Potentially Damaging Operations"—also known as "Operations Liable to Damage" or "OLDs"—appropriate to the interest features of the SSSI being legally notified. In the context of geological sites, such operations might include the collection of geological specimens. The legislation, therefore, puts the onus on the legal owner or occupier of the site to notify the applicable conservation agency (see further) if they intended to carry out any of the restricted activities and seek advice—potentially even including a site visit from a specialist from the agency. If the agency subsequently decided that the activity could take place in some form, subject to certain conditions, they would issue a "Consent." Where such an operation would result in damage to the interest features of the site, however, the proposals could be refused.

The "PDO" process generally worked well if the landowner or manager (i.e., "occupier") was seeking consent for an operation they intended to carry out, or, if a third party, such as a geoscience researcher, was applying to sample from a protected site—in which case the consent would still go to the landowner. However, if a third party carried out any activity on the site, including the removal of geological materials such as fossils and minerals, without contacting the landowner and, through them, applying for a formal consent, the legislation could still hold the landowner responsible for the damage (i.e., for failing to adequately protect the site, themselves), and hence, liable for prosecution. The landowner would, therefore, be expected to take the necessary legal action, themselves, against the perpetrators, something that they might not always be in a position to do, especially if the identity of the former could not be established. The absence of a specific legal statute in the UK to protect geological heritage, in particular paleontological, is probably the largest single factor in the frequent failure of conservation systems across the UK to adequately protect geological sites, especially paleontological, from irresponsible or illegal damage and theft.

This lack of provision of a specific legal category for paleontological heritage in particular—as well as the ever-present risk of third party instigated loss and damage—has meant that there has been an all-to-common reliance on nonstatutory "Codes of Conduct" in an attempt to influence the actions of site users. Only in Scotland, however, is such a "Code" backed by legal statute, in this case the Nature Conservation (Scotland) Act 2004, and was updated in 2023 (https://www.nature.scot/doc/scottish-fossil-code-and-guidance). Crucially, as the Scottish Fossil Collecting Code attempts to balance the needs of science, and hence conservation, against the aims and objectives of a wide range of actual or potential site users, including educational groups and specimen collectors, it represents a leading example of good practice globally (Page and Wimbledon, 2009; Scottish Natural Heritage, 1996; recently updated and available).

Although the 1981 Wildlife and Countryside Act remains the primary structure through which natural heritage conservation sites are notified and legally protected in England, Wales, and Scotland, its enforcement has been modified through subsequent legislation such as the Countryside and Rights of Way Act (2002) in England and Wales and the Nature Conservation Act 2004 in Scotland. In the north of Ireland, however, legislative frameworks parallel but similar to those elsewhere in the UK including the Nature Conservation and Amenity Lands (Northern Ireland) Order 1985 and the Environment (Northern Ireland) Order. To add further complexity to the legislative differences across the UK, are the differences in philosophies and approaches by the agencies responsible for enforcement across the Country (Page and Wimbledon, 2009).

Beyond the national protected site network of SSSIs, a second network of "nonstatutory" geological sites exist across most of the UK, primarily selected by local or regional voluntary groups using local or regional criteria. Known collectively as "Regionally Important Geological/Geomorphological sites" (RIGS) or "Local Geodiversity Sites" (LGS), these sites have a limited level of protection through national Town and Country (i.e., "Spatial") planning legislation. Unlike GCR sites and SSSIs, however, there is no single national listing of RIGS sites available for the UK, so it is difficult to assess their contribution to the conservation of vertebrate trackways. In addition, although there is a common perspective that RIGS form a "second tier" of geological conservation sites, of lower scientific importance than nationally selected GCR/SSSIs, this could not be further from the truth, as many are of at least equal scientific importance at a national level and some are of irrefutable international scientific importance (Page, 2017).

Although fossil vertebrate trackways are typically viewed as large—and hence robust—features, many across the UK are far from that—and beyond the ever-present issue of irresponsible or illegal fossil collection, both climate and strongly eroding coastlines are major damaging factors. In coastal areas, high tidal ranges and storms will rapidly erode and hence destroy most in-situ features, and an ancient right-of-access to all intertidal areas means that both in-situ and external features can remain vulnerable to collection or damage—irrespective of whether the site is legally protected or not. Inland, however, natural exposures at appropriate levels are few, and, in the wet, temperate climate, tend to rapidly become overgrown with natural vegetation or weather and gradually disintegrate. In addition, however, the threat of illicit collection—or damaging attempts to collect remains—exists unless appropriate security and/or access controls can be provided. Crucially, therefore, the combination of natural threats such as erosion and weathering and the ever-present risk of removal by collectors means that in-situ conservation of vertebrate trackways is rarely a realistic option across the UK. Nevertheless, various approaches to the conservation of such trackways have been applied.

Although there are a few sites with some evidence of Devonian trackways, Clashach Quarry in Morayshire, Scotland, has some of the oldest "managed" trackways in the UK of Permian age (Permian - Permian-Triassic Reptilia and Permian-Triassic (reds beds) GCR sites): The site includes exposures on both foreshore and in a quarry complex, the latter showing large bedding plane surfaces with trackways made by two or three species of dicynodont, mammal-like reptiles (Benton and Spencer, 1995; Benton et al., 2002; Ogilvie et al., 2000). The key exposures with the trackways are associated with a ridge of unworked sandstone between the quarry complex and the sea, but as quarrying continues in adjacent areas, there is a possibility that new finds might be made, but also trackways could be destroyed. NatureScot, formerly Scottish Natural Heritage, has produced conservation guidance for the quarry and monitors the area (e.g., SNH, 1999).

Ironically, reopening of a key area of the quarry in the late 1990s to supply stone to face the new Museum of Scotland led to the discovery of new trackways but also put them at risk. Fortunately, however, some of the footprints were saved and installed, originally with interpretation, close to the coastal footpath adjacent to the site—a unique facility in Scotland for trackways outside of a museum context. However, in the meantime part of a trackway on the adjacent shore was illegally removed using a rocksaw, forcing the National Museums of Scotland to intervene to remove what was left of the trackway before it was also vandalized—and in the process removing a resource used for education for the local community. As elsewhere across the UK, this scenario demonstrates that leaving trackways in-situ places them at risk, and the best conservation strategy is to record them, for instance as a scanned 3D model, or selective removal to a secure location.

Natural degradation due to weathering of blocks in the interpretative area—and in-situ prints on the beach—will always be an ongoing problem, however, which may necessitate further intervention, for instance, removal to a covered and hence climate-protected location. In this context, it is also crucial to manage any interpretative provision and unfortunately at Clashach it has not been, although community involvement has removed some of the vegetation that was beginning to obscure the blocks.

One of the best-known vertebrate footprint sites in the UK is the Late Triassic, Hayes Point to Bendrick Rock GCR Site (Permian-Triassic (red beds) site network). The site shows large bedding plane surfaces of sandstone in basin-margin facies in the upper part of the Mercia Mudstone Group of Norian age. At least five ichnogenera have been recorded, including *Grallator*, *Anchisauripus*, *Chirotherium* or *Eosauropus*, *Evazoum*, and *Otozoum*. The first two form genera were made by *Coelophysis*-like theropods, and the remainder probably by prosauropods and possibly archosaurs (Benton and Spencer, 1995; Benton et al., 2002). The site lies on the northern margin of the marine Bristol Channel, but, despite an exceptional tidal range, the area is relatively sheltered and not subject to the full force of Atlantic storms.

The footprints have been known since at least the 1970s, and key areas were excavated by the National Museum of Wales (NMW) in Cardiff, where they are displayed. Almost inevitably, however, with the massive rise of largely uncontrolled commercial collecting activity across the UK from the later 1970s to the present (Page and Wimbledon, 2009; Page, 2018), the remaining trackways remained at risk, and in 2005, around 25 m^3 were removed by commercially motivated collectors. Bizarrely, rather than concealing the theft, parts of the trackway were openly put on sale, including via internet sites and a fossil shop in Dorset. Fortunately, this "mistake" allowed the conservation authority in Wales, CCW, in cooperation with the Welsh Police, to identify the location of various parts of the stolen slab and recover them in a series of Police "raids" (https://www.geolsoc.org.uk/GeositesDinosaurBendrick; https://geoconservationlive.org/sites/bendrick-rock/).

Note that this police action was carried out on the basis of damage to a site protected under national legislation. If the footprints had been removed from a "nonstatutory" site, such as an RIGS, however, only action by the landowner, for instance, in the context of theft, would be possible to aid recovery, as palaeontological heritage does not have a special status within UK law.

The problem remains, however, that the Bendrick Rock site is well known, including through scientific publications that are often freely available to all, including commercial and other fossil collectors, but also ironically through periodic, even conservation-focussed publicity, and hence, there have been further cases of damage (https://cardiffjournalism.co.uk/thecardiffian/2021/02/02/dinosaur-footprint-site-near-barry-damaged-by-looters/; Page, 2024, in press). Only where 24 h security is available can such risks be considered adequately addressed, although as the Bendrick Rock site is adjacent to an industrial complex, some collaboration to at least monitor visitors to the site might still be possible, which could aid follow-up action if another incident of large-scale collection occurs.

However, the site is now owned and managed by a national Geoconservation group, the British Institution for Geological Conservation (BIGC; www.geoconservationlive.org), so that innovative approaches to conserving and promoting the site are being explored. These include a high-resolution drone survey of the entire site to assist the identification and mapping of key footprint-bearing bedding surfaces, as well as to monitor change, including that due to coastal erosion—as well the inevitable gradual, but entirely natural, attrition of in-situ footprints. There are many challenges, however, not least the marginal urban/industrial context of the site in an area of low income and related socio-economic issues. As well as being a key part of a global ichnological heritage, however, the Bendrick Rock site provides an open-air laboratory through which to explore various conservation and educational measures that can be of great relevance to sites elsewhere. Somewhat ironically, perhaps, as the illegally collected slab has now largely been recovered and is now safely installed in a national museum—along with the previous, scientifically collected material—it ensures that whatever may or may not take place at the site in the future, at least part of the interest feature is now safely conserved off-site as geological heritage in a cultural context, and hence safe from theft and natural degradation.

Although the presence of a horizon with dinosaur footprints within a marine-marginal marine Middle Jurassic sequence (as described by Cox and Sumbler, 2002) at Ardley Quarry in Oxfordshire has been known for some time, it was not until 1997 that it was "officially" reported. The site was very extensive, and over 40 theropod and sauropod trackways have been recognized, some of which were over 180 m in length and hence the longest ever recorded within the UK and some of the longest in Europe (Mossman et al., 2003; Day et al., 2004). As the site was an active quarry and waste disposal site, conservation UK legislation is insufficiently strong to halt the operation, as compensation might be payable to the operator. Instead, it was agreed that the trackways could be buried by landfill after detailed recording by a scientific team, one argument being that they would be protected from natural weathering—although completely inaccessible, potentially in perpetuity. Only a protective membrane, as required by law to prevent contaminating leachate reaching local groundwater systems would separate the footprint surface from the refuse. As one of the recording team is quoted as saying: "It is difficult economically to stop the operation. Covering them up is a good way of preserving them. It's a shame but it would be too costly to preserve them any other way." (https://www.independent.co.uk/news/science/dinosaur-tracks-are-longest-in-britain-633902.html).

A 60 m section of a trackway, however, was cast and inset within a lawn in front of the Oxford University Museum to form part of a "Dinosaur Garden" (https://oumnh.ox.ac.uk/learn-oxfordshire-dinosaurs), potentially the only opportunity for the general public to "view" any of the discoveries, as access to the working site is restricted. A buried section of the trackway site—out with the boundary of the working quarry and landfill site—was designated as an SSSI in 2010, however, as part of the Jurassic-Cretaceous Reptilia GCR block. This unusual approach of designating an area with no current exposure is an unavoidable compromise by a national conservation agency when confronted with the limitations of UK conservation legislation in the face of economic factors but that still place a duty on the organization to designate and conserve features of nature conservation importance.

The first finds of dinosaur footprints in Scotland on the Isle of Skye generated an inevitable media interest after their discovery in 1982. Several trackways, including of theropods, ornithopods, and sauropods, are now known within the non-marine Middle Jurassic succession of Staffin Bay area and on the Trotternish Peninsula (Clark et al., 2004; Brusatte and Clark, 2015—the general geology of the area is described by Cox and Sumbler, 2002 in the context of a Bathonian GCR site, also see Morton and Hudson, 1995). The preservation and position of the footprints at An Corran in Staffin Bay led the conservation authority, then Scottish Natural Heritage, and the Hunterian Museum in Glasgow, to agree to molds for casting being made of part of the trackway in order to safeguard the trackway in the latter institute. Crucially, the local community has effectively taken on the role of site monitoring, to help ensure that remaining in-situ footprints are not illegally removed or damaged, as well as preparing an interpretive panel. One such case in 2016 involved an attempt to make casts of the footprints using plaster, which could have seriously damaged them, and a subsequent police action to try and apprehend the culprit (https://eartharchives.org/articles/the-skye-dinosaur-footprint-vandal-lessons-for-assault-on-us-national-monuments/index.html).

Unlike for English localities, under Scottish Law, removal of anything from any site requires the landowner's permission, even though access to most land is unrestricted, unlike in England (Page and Wimbledon, 2009). This context strengthens the context for any Police action as any removal of materials represents theft, and with the Scottish Fossil Collecting Code in place, the legal responsibilities of any would be collector are made very clear. In the context of the An Corran locality in particular, a Nature Conservation Order was placed on the site and adjacent areas as the strongest legal mechanism available to protect the site (https://www.nature.scot/landforms-and-geology/protecting-our-geodiversity/codes-researchers-and-collectors/helping-safeguard-skyes-fossil-heritage-faqs).

The North Yorkshire coast (Middle Jurassic) and Isle of Wight, England (Lower Cretaceous) show similarities in that they include actively eroding coastlines but also that trackways are often recorded as individual isolated footprints recorded on the foreshore after erosion from the cliffs (Radley and Allen, 2012; Lockwood et al., 2014; Whyte et al., 2010). Under such circumstances, virtually only removal to a safe, ideally institutional location is a viable conservation option, not least as once on the beach, footprints will be rapidly abraded or become buried in beach sand. Although the ruggedness of the coastline and related access have tended to deter the removal of footprints from the North Yorkshire coast, on the Isle of Wight, due to a more open coastline, it has become a significant conservation issue and landowners, including the National Trust, have taken measures to control illicit removal. The Yorkshire footprints and occasional trackways are mainly developed in the Bajocian, Ravenscar Group and include sauropods, ornithopods, and a stegosaurid. Volunteer participation in recording several key trackways of swimming tracks has been important (Whyte et al., 2010). Although some specimens have been recovered, the large size of many and the remote locations mean that regular, detailed recording—to take advantage of new rock falls—is generally the best conservation tool in the area. The key localities that yield the footprints are included within both the Aalenian-Bajocian and Bathonian GCR blocks.

Footprints and trackways from the Isle of Wight come from the nonmarine "Wealden facies" of the Lower Cretaceous, which ranges from Valanginian to Barremian age and includes Jurassic-Cretaceous Reptilia and Wealden GCR sites. Ornithopods and theropods are typical with records going back to at least the 1850s, and significant skeletal finds, in particular, have contributed to the area sometimes being called 'Dinosaur Island'. Fossil collecting on the island is promoted on many websites, including by both the National Trust—a national nongovernmental heritage management organization (https://www.nationaltrust.org.uk/visit/isle-of-wight/compton-bay-and-downs/fossil-hunting-and-geology-at-compton-bay-and-downsand) and the Natural History Museum in London (https://www.nhm.ac.uk/discover/isle-of-wight-dinosaur-island.html), as well as by many others.

Conservation messages, however, are not easy to find, although the former do state as part of their guidance. Nevertheless, fossil collecting has become very intensive on the island and in several areas, and action has had to be taken in several cases to try and reduce loss and damage, including of footprints—despite many of the prints forming large blocks on the beach.

The presence of dinosaur trackways in the quasi to nonmarine Purbeck Group close to the Jurassic-Cretaceous boundary of the "Isle of Purbeck" and the Isle of Portland of east and south Dorset, respectively, is well known. As the trackways are typically developed in thin-bedded limestones, removal of individual footprints or sections is relatively straightforward—although very few appear to be conserved in any institution (but see Ensom and Delair, 2007). As many of the finds have been made in working quarries, inevitably many trackways are likely to have been destroyed. A section of one example on Portland, however, was removed by the quarry company and placed in a "storage area" as separated blocks (https://wessexcoastgeology.soton.ac.uk/portdino.htm#:~:text=The%20dinosaur%20footprints%20are%20present,extent%20protected%20and%20under%20observation). Due to lack of further intervention, the blocks are now partially overgrown and significantly deteriorating as weathering takes place.

In Purbeck, however, an important attempt to conserve and make available to visitors an in-situ section of such a trackway can be visited at Spyway Quarry; indeed, the site is considered to be "... the largest in situ easily accessible UK dinosaur tracksite." (Butler, 2023). The site shows around 100 large sauropod footprints and is owned and managed by the National Trust, although unstaffed and with open access. The nature of the trackway, however, on a continuous slab, means that systematic attempts to remove sections would be very difficult—although the footprints still remain vulnerable to damage by more "improvized" attempts to remove them. The biggest threats to the site, however, are due to the weathering of the limestone surface—from 2013—combined with abrasion of the surface by the estimated 10,000 visitors a year from 2016. The effects of these and other factors are currently being monitored, including using 3D imaging, in order to provide the National Trust with guidance for future management of the area (Butler, 2023).

Although the abrasion issue could be addressed with more visitor direction or artificial surfacing within the site—or even provision of suitable protective "overshoes" (cf. Matthews and McIlroy, 2019)—natural weathering will mean that the locality has a very real lifespan before it is so deteriorated as to be of limited visitor interest. As elsewhere across the UK, the only effective solution to this problem would be the installation of some sort of canopy across the area, the costs of which, however, are likely to be prohibitive.

Although there are several records of vertebrate and even human footprints in Quaternary and Holocene sediments across the UK (see Walk the World chapter in this book), both in mineral workings and in foreshore locations, the conservation prescription is quite clear, as only removal or casting from the soft sediment substrate is a conservation option—although many have largely only been recorded rather than recovered. Some of the best known of these tracksites are of Mesolithic and Neolithic age from the Formby area of the Sefton Coast in NW England as preserved in Flandrian coastal sediments (Roberts, 2009, 2014; Burns, 2021; Wiseman et al., 2022). The rich assemblages include human, birds, deer, bovids, boar, beaver and lynx. Many of the footprints were recorded photographically with later studies using more photogrammetry and hence facilitating 3D modeling. This information has been used to compile a comprehensive database. In general, workers on the site consider that those trackways, as preserved in soft, muddy sediment, are ephemeral and likely to be removed very soon—even "overnight"—after exposure by erosion (Wiseman et al., 2022), and there appears to have been little or no attempt to excavate and preserve any tracks offsite. Although the area is not within a GCR site, ownership and management by the National Trust help ensure that use of the area respects the sensitivity of these extremely delicate deposits.

2.4 Africa

2.4.1 Algeria

Several sites have been discovered in Lower Cretaceous strata in the El Bayadh area (western Saharan Atlas). Around the town of El Bayadh, three dinosaur track sites have been found, and, not far away, three more tracksites (Chabou et al., 2015). The two closest sites to the town (sites 1 and 2) are located 2 km northeast of El Bayadh, and a third (site 3) is 1 km west of the town. At site 1, two track levels were recognized in bioclastic limestone beds of Valanginian age, over an area of about 1 km^2. About 100 tridactyl footprints with clawed digits of theropod dinosaurs forming 12 separate trackways were discovered. About 20 half-moon-shaped small footprints related to sauropods were also found at this site. Site 2 is located only 500 m east of the site 1 and shows two micritic limestone beds, also dated from the Valanginian. Fifty isolated footprints, including about 30 tridactyl tracks of small size, and about 40 half-moon-shaped footprints, were found. Site 3 is located 1 km west of the locality of El Bayadh, on the left bank of Oued Merires. About 10 tridactyl footprints of theropod dinosaurs were found on the surface of a steeply dipping limestone bed of the same age as the remaining sites.

Another significant site is the El Kohol Daiet Sidi el Arbi tracksite situated south of El Bayadh on sandstones. The tracks are located on a bedding surface of 50 m^2, and comprise oval-shaped sauropod and large theropod footprints (Bessedik et al., 2008). The oval-shaped footprints are very large, with an average length of 120 cm and a width of 110 cm; those that are best preserved show claw impressions. The discovery of new tracksites over the past 20 years in the region of El Bayadh, and the media coverage given to these finds, has revived the scientific study of dinosaur footprints in Algeria and raised awareness of the need for their protection (Chabou et al., 2015). The local authorities issued a decree for the protection of these sites (Mahboubi et al., 2007). An information panel was erected near site 1, indicating that the site is protected against any act of vandalism. Since then, no other protective measures have been taken, and the site has been damaged at several locations. For example, site 2 disappeared following the floods that occurred in Autumn 2011 (Chabou et al., 2015).

2.4.2 *Ethiopia*

The Gombore II-2 butchering site (Altamura et al., 2018) in the Upper Awash Valley of Ethiopia, and those found in the test-pits excavated next to the Gombore II Open air museum site (Altamura et al., 2020), provided tracksites with hominin and other vertebrate tracks (see Quaternary Vertebrate tracksites and Walk the World chapters in this book). The two localities are only a few dozen meters apart. All of the Gombore Gully is included in a protected and fenced nature and archeological area under the control of the Ethiopian Federal Authority for Research and Conservation of Cultural Heritage (ARCCH) and of the Oromia region (with a local director). In addition to Gombore Gully, which is interdicted to outsiders and fenced off, the entire Melka Kunture area (ca. 100 km^2) is also protected as an archeological area, although there are numerous farms and settlements. Within the fenced area of Gombore Gully, both Gombore II-2 butchering site and Gombore II Open air museum are displayed and available to the public, with a tourist path connected to a nearby museum. The first site shows casts of archeological materials unearthed in the 1990s, and the latter displays a portion of an archeological deposit covered by a local-style roof. Unfortunately, the ichnological evidence found next to the two museum sites was imprinted in soft sediments and could not be left on display exposed to open air weather conditions. After excavation, they were covered with tarps and soil to protect them. However, casts were made, and some human and hippo tracks were removed in sediment blocks and taken to the ARCCH deposits. In the future it is also planned to make and place casts or 3D prints obtained from laser scans and 3D models on site (Altamura, pers. comm.). The whole Melka Kunture site is currently a candidate as a UNESCO site, and there is hope that in the coming years it can therefore be even more protected and enhanced.

2.4.3 *Lesotho*

The Lower Moyeni site in Lesotho, in the upper Elliot Formation, has been covered by a government-funded building, the "Dinosaur Footprints Visitors" Centre, creating an enclosed space designed to minimize erosion and preserve the many Early Jurassic tracks and trackways (Fig. 14.6A). A bridge allows visitors to examine the trackways from above without walking on the surface. Unfortunately, as an example of unintended consequences, the ceiling has provided birds with a roosting site, leading to defecation on the tracks, which can no longer be enjoyed with the benefit of angled sunlight. The site was identified by the legendary Paul Ellenberger, who reported on it in detail (Ellenberger, 1970, 1974). More recent reappraisals of the site have been performed by Smith et al. (2009), Wilson et al. (2009) and Marsicano et al. (2014). The surface contains a diverse assemblage of footprints of carnivorous and herbivorous dinosaurs, crocodilians, and temnospondyl amphibians.

FIGURE 14.6 Examples of geoconservation of vertebrate tracksites in Africa. (A) Lower Moyeni site in Lesotho showing the building that has been placed over the surface. (B) Helicopter recovery of a purported ammoglyph with a geometric pattern on South Africa's Cape south coast. (C) Laetoli footprint cast and diorama at Olduvai gorge Museum, Ngorongoro Lengai UNESCO Global Geopark; (D) the signaled "Géosite Beni Ghedir" in a hiking trail in Tunisia. *(A) Photo credit Charles Helm. (B) Photo credit Richard Webb. (C) Photo credit Utalii Plus Tanzania. (D) Photo credit wildyness.com.*

2.4.4 Kenya

The Ileret Footprint site was initially discovered by a team of local and international researchers over the course of 2004–2007, in Marsabit County, on the eastern shore of Lake Turkana. Numerous hominin fossils have been found near Ileret, including *Homo erectus* footprints and other animal tracks dating back to about 1.5 million years ago (Hatala et al., 2016), making them the second oldest hominin footprints ever found after those at Laetoli, Tanzania (see Walk the World: the Hominin track ways chapter in this book). Five distinct excavated sites have 97 hominin tracks, and intensive use of lake margin habitats by *Homo erectus* groups was inferred (Roach et al., 2016). The tracksites are a finding of a longer excavation project of the northern Kenyan region known as the "cradle of mankind." In collaboration with the National Museums of Kenya and with financial contributions from the British Council Cultural Protection Fund, the International Center for the Study of the Preservation and Restoration of Cultural Property is engaged in a community-centered project focused on the long-term preservation of fossil footprints (ICCROM, 2023). Two major factors pose risks to this heritage resource. Climate change has caused erratic rainfall patterns, resulting in significant soil erosion over the past decade. In addition, the region has witnessed unprecedented floodwaters and prolonged droughts, exacerbating erosion. The area is also threatened by urban sprawl. The neighboring Dassanech community plays a vital role in the preservation of this heritage resource, having actively participated in its exploration, discovery, and preservation. The project aims to not only preserve the fossil but also to engage local communities in heritage preservation and boost their livelihoods through sustainable tourism. Until 2025, detailed research and documentation and in situ preservation will be undertaken. At the end of the preservation phase, the project partners plan to work with the local community to develop a cultural center and exhibition on site.

2.4.5 Morocco

The M'Goun UNESCO Global Geopark, in the Béni Mellal-Khénifra region, was designated in 2014. The M'Goun UNESCO Global Geopark is located some 100 km from Marrakech, in the middle of the chain of the central High Atlas. It includes famous and spectacular footprints of sauropod and theropod dinosaurs and many deposits of bones. Six sites with dinosaur footprints include the list of geosites of M'Goun Geopark, four of them located in Iouaridène, besides Ait Blal and Ibaqualliwn. They are composed of sauropod (*Breviparopus taghbaloutensis*) and theropod (*Megalosauripus*) trackways and isolated tracks dated to Early and Middle Jurassic (Amine et al., 2018, and references therein). Since June 2023, the newly opened Museum of the M'Goun Geopark in Azilal, Béni-Mellal province, provides information regarding these geosites.

Discovered in 2013, the Anza tracksite is one of several sites in the Western High Atlas of Morocco that contain well-preserved and diverse dinosaur and pterosaur trackways. The occurrence of rare tracks at Anza combined with the stratigraphic position (Late Cretaceous, Coniacian–Santonian age), coastal location, and easy accessibility, have inspired the scientific community, nongovernmental organizations, and local community, to increase public awareness of the scientific, geoeducational, and geotourist relevance of the tracksite, as well as to develop strategies for site conservation and protection (Lkebir et al., 2020). The 300 m^2 tracksite reveals more than 323 pterosaur and theropod (*Grallator*) footprints, isolated or organized in 50 trackways (Masrour et al., 2017). To address the issue of degradation of the tracksite due to natural processes, including global sea level rise, authorities have begun a new project of casting. The replica will be displayed in the Earth Science Museum at the Faculty of Sciences in Agadir. The local NGOs, along with local volunteers, set up informative panels, created a parking lot and accessibility to the site, and provide guided visits. Moreover, they have enhanced the area by creating zones for green areas, and including a model of a theropod. A number of groups have undertaken many efforts to increase public awareness of the importance of the site, including development of informative leaflets and workshops (Lkebir et al., 2020).

In the Imilchil area of the Meknès-Tafilalet region, central Morocco, Middle Jurassic dinosaur tracksites have been described at 14 sites (Gierliński et al., 2009; Oukassou et al., 2019; Klein et al., 2023; Masrour et al., 2023). The Bathonian-Late Jurassic (Isli Formation) track assemblage consists of theropod, sauropod, and ornithopod footprints, as well as crocodylomorph and isolated pterosaur footprints, whereas the Bajocian deposits (Imilchil Formation) yielded a theropod-sauropod-thyreophoran track assemblage. There are two tracksites with some protection measures, one protected by the local population (because it is inside their property) and the second one protected by the local authority.

Chouaib Doukkali University in the coastal Atlantic city of El Jadida houses the largest Late Paleozoic and Mesozoic tetrapod footprints collection, with more than 2000 specimens, although not open to the public (Lagnaoui, pers. comm.).

2.4.6 Namibia

The Otjihaenamparero dinosaur tracks are a set of different fossil tracks located at the Otjihaenamparero farmstead, 23 km east of the small town of Kalkfeld in the Otjozondjupa Region in central Namibia. These Middle Jurassic tracks were first reported as dinosaur imprints in 1925, and the site was in 1951 declared a national monument (Wagensommer et al., 2016). All imprints were produced by tridactyl bipedal dinosaurs. There are two crossing trackways, counting more than 30 individual imprints of 12–15 cm in length and 80 cm apart. An additional trackway consists of several imprints of 7 cm length that are 28–33 cm apart. The longest trackways in the Etjo Sandstone are about 32 m (105 ft) in length. They belong to at least two different forms of the ichnogenus *Saurichnium*, namely *Saurichnium damarense* and *Saurichnium tetractis*. The dinosaur footprints, which were declared a national monument in 1951, are open to visits that are run by the farm owner with paid admission.

2.4.7 South Africa

The Karoo Basin of South Africa and Lesotho is renowned for both its body fossils and trace fossils, from the Carboniferous to the Jurassic. Bordy et al. (2023) presented six selected geoheritage sites of global geoscientific importance that exemplify that heritage. Two of these sites, Gansfontein and Moyeni (see Lesotho section), feature vertebrate trace fossils, and both enjoy a degree of protection. A plaque at the 1000 m^2 Gansfontein paleosurface, near Fraserburg in South Africa's Northern Cape province, proclaims that the Permian site is recognized as a geoheritage locality by the Geological Society of South Africa (GSSA), which also fenced off the surface as a means of protecting it. The site was exposed in 1968 following a flood event. The crevasse splay deposit records abundant and diverse trace fossils registered after successive flooding events, including tetrapod trackways (some of which have been attributed to dinocephalians and pareiasaurs) and fish trails. The site has been formally described (Smith, 1993; Ronchi et al., 2023), laser scanned, and the tracks have been assigned to ichnogenera following photogrammetric modeling (Marchetti et al., 2019). Bordy et al. (2023) described the site as "of international importance as the best-preserved and most-diverse assemblage of middle Permian therapsid trackways." The condition of the paleosurface is slowly deteriorating as a result of natural weathering. A tour operator (who also keeps the site clean) guides visitors through 10 stops on a route along the paleosurface. A fossil display in nearby Fraserburg provides further information.

The Asante Sana site, in a private nature reserve in the Eastern Cape province of South Africa, featured a 34 m^2 Permian bed containing the trackways of six large tetrapods, attributed to a group of dicynodonts, as well as a smaller dicynodont trackway (De Klerk, 2002). Heroic efforts were made to preserve the site, including digging trenches, impregnating the surface with resin, installing a roofed structure over the bed, and leaving some areas unexcavated.

Despite these measures, the fractured, friable siltstone surface has progressively become substantially eroded.

Pleistocene vertebrate ichnosites on the Cape coast present a different problem. Fortunately, many of them are in protected areas (national parks, provincial nature reserves, and private nature reserves), but this nominally protected status does not afford them protection against the ravages of erosion, cliff-collapse events, and storm surges. Furthermore, the aeolianite surfaces on which the tracks occur form attractive canvases for graffiti artists (Helm et al., 2021). Hominin ichnosites are of obvious paleo-anthropological importance (see Walk the World: the hominin track ways chapter in this book), either as tracksites or as paleoart—the term "ammoglyph" was coined for patterns made by hominins in sand, that are now recognizable in rock (Helm et al., 2019). One tracksite, dated to ~153 ka, contains the oldest track thus far attributed to *Homo sapiens* (Helm, 2023).

Because of the ephemeral nature of the sites, recovery and digital documentation and replication have become a priority. Although South Africa contains several UNESCO World Heritage Sites, it has not yet embraced the concept of UNESCO Global Geoparks. Examples of successful initiatives include the recovery of what came to be known as "Eve's footprints" (a hominin trackway) from the west coast by helicopter (Roberts 2008). Although the site was in the West Coast National Park, graffiti had almost impinged on the tracks. The track-bearing slab was accessioned in the Iziko South Africa Museum, and a replica and exhibits were installed in an interpretive center in the national park.

Another helicopter recovery took place on a remote stretch of coastline (Fig. 14.6B), recovering an ammoglyph and transporting it to the Blombos Museum of Archeology in Still Bay, where it is on exhibit. Had this not happened, this important specimen would almost certainly have been destroyed in a severe 2023 storm surge (Helm et al., 2024). Further ammoglyphs (Helm et al., 2019) were recovered from the Garden Route National Park in collaboration with a team of SANParks officials using 4WD vehicles. Given the absence of a suitable local repository, they were also transported to the Blombos Museum of Archeology, where they are on exhibit. Within a year of this successful recovery, a landslide buried the site, underlining the importance of rapid recovery efforts. Finally, Pleistocene elephant tracks have been identified in the coastal section of the Addo Elephant National Park on the southeast coast. This creates the potential for an exhibit on ichnology in the Interpretive Center in the Addo Main Rest Camp, and for the recovery and exhibition of suitable specimens. These initiatives are actively being pursued with SANParks.

The southern African experience with protecting and preserving ichnological geoheritage sites is thus mixed. The benefits of full documentation and replication, whether through casting, laser scanning, or photogrammetry, are clearly evident for any sites for which formal protection is considered.

2.4.8 *Tanzania*

Ngorongoro Lengai UNESCO Geopark and World Heritage in Arusha, northern Tanzania, managed by the Ngorongoro Conservation Area Authority (www.ncaa.go.tz/geopark/), includes two protected sites of the most important hominin tracksites in the world, Laetoli and Engare Sero. Besides, the Geopark is a popular stopover for visitors to other well-known national parks such as Lake Manyara and the Serengeti. Due to its offering a wide variety of tourism opportunities, it is considered a "must-visit" destination by most travelers, especially those interested in exploring geological wonders such as the Ngorongoro Crater, Lake Natron, and Lake Eyasi.

The South Tourist Route explores the world-famous Laetoli Footprints. The Laetoli Footprint site, located 45 km south from the Olduvai Gorge, is among the world's most famous archeological sites, providing indisputable evidence that bipedalism in human ancestors was achieved 3.7 million years ago (see Walk the World chapter in this book). Laetoli has been excavated since 1938; however, it was in the 1970s when Mary Leakey turned her attention to the area. Shortly after, Mary and her team discovered numerous mammal fossil remains, including a well-preserved specimen of archaic *Homo sapiens* and a magnificent holotype (*Australopithecus afarensis*) a.k.a "Lucy". In 1976, two members of her team, including Andrew Hill, fortuitously discovered the 27-meter-long trackway composed of 70 hominin footprints of Site G, that have been mostly attributed to *Australopithecus afarensis* (Leakey and Hay, 1979; Leakey and Harris, 1987; Meldrum et al., 2011). In total, there are 18 localities in the area with mammal and bird footprints. In 1979, scientists tried protecting site G by reburying the tracksite. But the backfill contained acacia seeds that took root and grew through some of the prints; meanwhile termite nests appeared in the area, forcing researchers to reexcavate the site using mini powered tools and surgical instruments. The Tanzania Field School has conducted a pilot conservation project at site 7 for the past several years, in conjunction with the Tanzanian Antiquities Department, and with the consultation and permission of local authorities and the Maasai villages in the vicinity (Musiba et al., 2008). Tanzanian teachers (both high school and elementary school) and local people participated in all aspects of the conservation project. In 1993, measures were taken to prevent degradation of Site G (Feibel et al., 1995). The original trackway was remolded and new casts were made; a part of it can be found in many museums of the world. As the trackway is very fragile, the new replica cast was used to guide reexcavation in the field. A team of specialists reexcavated half of the trackway to record its condition, stabilize the surface, extract dead roots and rebury it with synthetic geotextile materials. This allows the trackway surface to breathe, and protects it against root growth. In 2015, two new sets of footprints were unearthed approximately 150 m south of the old footprints site, at Site S (Masao et al., 2016). The two individuals were walking on the same surface and in the same direction as the three individuals at the old site, thus raising the possibility that more footprints may be discovered. Most researchers agree that a hominin species made the footprints at Laetoli. The site is in the registered National Historic Sites of Tanzania.

The Olduvai Gorge Museum exhibits numerous fossils, lithics, and skeletons of many extinct animals excavated in the region. Mary Leakey founded the museum in 1970. In 2017, the Ngorongoro Conservation Area Authority replaced the original museum structure by constructing a new museum and visitor center based on the traditional boma architecture. It is a museum dedicated to the recognition and understanding of the Olduvai Gorge and Laetoli fossil sites. The exhibits at the Olduvai Gorge Museum are centered on the research and artifacts that have come from the surrounding area. There is one hall dedicated to the Leakey family and their scientific contributions. Many of the artifacts are original, but some are casts, such as the hominin fossils. The adjacent hall is dedicated to the Laetoli footprints site. A cast that was made of part of the footprint trackway in 1996 by the J. Paul Getty Museum is on display, together with a large illustration depicting three *Australopithecus afarensis* walking through the area ca. 3.7 million years ago (Fig. 14.6C). Accompanying this are several panels illustrating the process of formation of the

Laetoli footprints. The Olduvai Gorge Museum receives roughly 3000 visitors each day during peak season from all around the world.

Not far from the famous Laetoli site, and covering an area of 300 m^2, Engare Sero is the largest human fossil footprint site to have ever been discovered in the African continent. The site was discovered in 2006 through a research project that led to the discovery of over 400 hominin and animal tracks, including those of zebras and bovids (see Quaternary Vertebrate Tracksites chapter in this book). Footprints from this site represent one of the country's best-preserved records of anatomically modern *Homo sapiens* (see Walk the World chapter in this book). The footprints date between 19–6 ka BP and represent the distinct paths of at least 20 different individuals (Liutkus-Pierce et al., 2016; Hatala et al., 2020). The footprints were preserved by volcanic mudflows from the nearby Ol Doinyo Lengai, which is still active. Since 2009, the Engare Sero research project team has been excavating and analyzing the Engare Sero footprints, led by Appalachian State University professor Cynthia Liutkus-Pierce and including Human Origins Program research scientist Briana Pobiner. According to their finds, the footprints were left by a group of mostly adult females traveling together (Liutkus-Pierce et al., 2016). The prints lie in an ephemeral river channel that is sometimes flooded. Winds often whip across the arid site, dragging grit across the footprints and producing erosion. Across the site, a given footprint wears down by a tenth to a sixth of a millimeter each year (Zimmer et al., 2018).

It is still possible to see human footprints in the mudflat shores of Lake Natron. Between 2013 and 2015, local officials built a small rock wall around the prints to divert runoff. They also installed a fence around the footprints to prevent people from walking or driving across the site. The area is visited by hiring Maasai guides from nearby villages.

2.4.9 *Tunisia*

The official name of the site is "Site paléontologique à gisements fossilifères du Jurassique de Beni Ghedir" and it is signposted as "Géosite Beni Ghedir" (Fig. 14.6D) at the foot of the hill near the town of Ghomrassen, southern Tunisia. Approximately 130 tridactyl footprints distributed over an area of 200 m^2, preserved on Middle Jurassic (Callovian) beds exposed at the Beni Ghedir site, represent the oldest evidence of a dinosaur fauna in Tunisia (Contessi and Fanti, 2012). The site shows some models in real size of the dinosaur producers. The experience starts at Tataouine's Earth Memory Museum, a small museum with a collection of fossils, including dinosaurs and models of different reptiles.

2.4.10 *Zimbabwe*

Ntumbe River Dinosaur Footprints Sites is located in the Chewore Safari Area, 11 km SW of the Mkanga Bridge National Park Station, close to the northern border of Zimbabwe with Zambia, in the remote southern part of the World Heritage Site that comprises Mana Pools National Park and the Sapi and Chewore Safari Areas. Chewore South Safari Area is situated on the Angwa River in the lower Zambezi Valley. In 1984, the first of several trackways with 14 large footprints of large dinosaurs in mudstones was discovered by Mike Aldersey, an Australian hunter at the Ntumbe River area near the Chewore Safari Area. Since then, several other sets of tracks, at a total of four sites, have been located by Tim Broderick of the Geological Department of the Natural History Museum with Professor Theagarten Lingham-Soliar and Zimbabwean geologist Ali Ait-Kaci. In 1990, an expedition sponsored by Operation

Raleigh and the Geologist's Association of London went in search of further dinosaur footprints, and a further 31 were discovered. They include some of the best-preserved dinosaur trackways in Africa and date from Middle to Late Jurassic. Site 1 shows an *Allosaurus*-like theropod trackway of 48 m composed of 45 footprints. Site 2 includes the first giant sauropod footprints found in Sub-Saharan Africa (Ait-Kaci et al., 2004). From their dimensions of 96 cm and absence of characteristic crescent-shaped imprints of the front legs, these are thought to be from the rear legs of *Brachiosaurus*. Sadly, at least one of the five footprints here has been destroyed by elephants or floods in the stream bed in the last few years. Site 3 lies in a shallow tributary of the Ntumbe River, and at four close locations has at least 88 of the distinctive footprints attributed to *Allosaurus*. Site 4 is within the red sandstones with the first trackway having 10 small footprints (<5 cm) attributed to an immature dinosaur plus a second trackway with slightly larger footprints. The area is in a very remote hunting country, and permission to visit is required from the Zimbabwean Parks Authority. These dinosaur footprint sites are seldom visited except by small parties of visitors in the company of an experienced and armed professional guide and ZPWA armed game scouts. The Natural History Museum of Zimbabwe in Bulawayo preserves casts of 14 footprints of a large bipedal dinosaur found in the Chewore Safari Area. Other casts are found in the Geological Survey of Zimbabwe in Harare.

2.5 Asia

2.5.1 China

Between 1911 and 1930, China's modern legal framework for preserving cultural relics gradually evolved. As early as 1928, regional authorities became aware of foreign archeological missions excavating fossilized skeletons in north-western China, Shaanxi-Gansu, and Mongolia (Huang, 2012). The Legalization of Antiques Preservation was enacted in 1930, marking China's inaugural official law on cultural relics. In 1982, China introduced the Law of the People's Republic of China on the Protection of Cultural Relics. Chapter I, Article 2, specifically includes that "Fossils of paleovertebrates and paleoanthropoids of scientific value shall be protected by the State in the same way as cultural relics." However, divergent interpretations exist regarding whether the mentioned paleovertebrate fossils encompass all such tetrapod fossils or solely those associated with human activities (Quaternary). Currently, those caught smuggling, theft, damage, reselling, resurrecting, or illegal transfer of vertebrate fossils with scientific value, including tetrapod skeletons and tracks, will be convicted and sentenced in accordance with the Criminal Law of the People's Republic of China (Article 328, crime of stealing and digging up vertebrate fossils) and the relevant provisions.

Various detailed fossil protection policies concerning palaeoichnologic fossils have been instituted and continue to be implemented. These policies encompass the Measures for the Management of Paleontological Fossils from 2002 to 2019, the Regulations on the Protection of Fossils established in 2011, and the Measures for the Implementation of the Regulations on the Protection of Fossils enacted in 2013. The Regulations on the Protection of Fossils, Article 2, indicate that "the term "fossil" as mentioned in this Regulation refers to the body fossils and trace fossils of animals and plants formed in the past periods of geological history and preserved in the Earth's crust". Article 7 "designates paleontological fossils with significant scientific research value or limited occurrence as key protection fossils. This classification extends to large or widely distributed fossilized traces, such as paleovertebrate tracks."

The grading system of vertebrate track fossils was included in the National Grading Standards for Fossils in 2012. Vertebrate track fossils are divided into three levels, with Level 1 being larger or concentrated tracks with clear features indicative of locomotion; Level 2 being a more concentrated group of well-defined tracks; and Level 3 being a single well-defined track with clear features, and a small number of less-defined tracks belonging to a same trackway.

Since August 2000, China has been establishing tremendous national geoparks, boasting 47 UNESCO Global Geoparks (data from spring 2024 at www.unesco.org/en/iggp/geoparks#full-list-of-unesco-global-geoparks) and 281 national geoparks (including qualified ones) as of 2021 (Wang and Zhu, 2023). In June 2019, the Chinese government issued the "Guidelines on the Establishment of a System of Nature Protected Areas with National Parks as the Main Body". This categorizes protected areas into national parks, nature reserves, and nature parks based on ecological value and protection levels, where geoparks fall under the nature parks category. National geoparks step on the "red line of ecological protection," requiring mandatory and stringent protection while hindering economic development. Consequently, there's reduced initiatives among local governments to declare new national geoparks. However, traditional perspectives imply that establishing a UNESCO Global Geopark necessitates a preexisting national geopark as a foundational requirement. Consequently, the Guidance's implementation has resulted in reduced applications for both National Geoparks and Global Geoparks. Therefore, after 2019, the development of China's national geoparks has entered a phase of sluggish growth.

We have counted 19 geoparks in China that have preserved tetrapod trace fossils (Table 14.1). Among them, there are three UNESCO Global Geoparks, covering an area of 3772 km^2, 12 National Geoparks, covering an area of 3624 km^2, and the others are provincial level, covering an area of 345 km^2. Eight geoparks were established between 2000 and 2009, and 11 geoparks were established between 2011 and 2020. All the geoparks are currently in operation.

These geoparks are distributed in various regions of China, with the highest concentration in Shandong Province in eastern China (five geoparks), followed by the Sichuan Basin (three geoparks). This distribution corresponds to the distribution and density of tetrapod track fossils in China. There is only one geopark recorded for Triassic tracks, four for Jurassic tracks, and the others for Cretaceous records. Except for the Triassic nothosaur swim traces, the other types are all dinosaur tracks, and the most obvious type are saurischian tracks, which account for 79% of the geoparks. Ornithopod tracks are also not uncommon, accounting for 53% of the geoparks. Tracks of pterosaurs, ceratopsians, and ankylosaurs are rare (Table 14.1).

These geoparks exhibit varying degrees of protection for tetrapod track fossils. UNESCO Global Geoparks, benefiting from greater financial resources and opportunities, demonstrate enhanced capabilities in protecting tracks. For instance, the Yanqing Global Geopark in China collaborates with the Lesvos Island UNESCO Global Geopark in Greece to employ protective agents on the Jurassic dinosaur track layers to extend the life of the tracks (Zhang, 2021). During winter, a comprehensive approach is adopted, utilizing felt cloth to cover and shield the tracks situated in the mountains, mitigating the adverse effects of severe, cold climates (Fig. 14.7A). In contrast, the level of track protection in National Geoparks is diverse. While

TABLE 14.1 List of Geoparks (divided in UNESCO, National, and Provincial) in China with protected vertebrate tracksites (ichnofauna present indicated).

Name of the Geopark	Location	Level	Area (km²)	Strata	Type of tracks									Opened	Supporting museum
					The.	Sau.	Orn.	Thy.	Cer.	Pte.	Not.	Tes.	Ave.		
Yanqing Global Geopark of China	Yanqing, Beijing	Global	620	J-K boundary Tuchengzi Fm.	■	■	■							2013	Qianjidian dinosaur tracks Showroom, Yanqing Geological Museum
Zigong Unesco Global Geopark	Zigong, Sichuan	Global	1630	J3 Shaximiao Fm.	■	■	■							2008	Zigong Dinosaur Museum
Jiayin Dinosaur National Geopark	Jiayin, Heilongjiang	National	38	K2 Yuliangzi Fm.			■							2001	Jiayin Shenzhou Dinosaur Museum
Qiyun Mountains National Geopark	Xiuning, Anhui	National	110	K2 Xiaoyan Fm.	■									2002	
Liujiaxia Dinosaur National Geopark	Yongjing, Gansu	National	15	K2 Hekou Fm.	■	■	■			■			■	2005	Liujiaxia Dinosaur Museum
Lufeng Dinosaur National Geopark	Lufeng, Yunnan	National	260	J1 Lufeng Fm. J2 Chuanjie Fm.	■	■	■	■						2004	Lufeng Dinosaur Museum
Zhucheng Dinosaur National Geopark	Zhucheng, Shandong	National	9	K1 Qingshan Gr., K1 Laiyang Gr.	■	■	■					■		2014	Zhucheng Chinese Tyrannosaur Museum, Zhucheng Dinosaur Museum
Chishui Danxia National Geopark	Zunyi, Guizhou	National	1200	K2 Guanshan Fm.	■									2012	
Qijiang National Geopark of Chongqing	Qijiang, Chongqing	National	108	K2 Guanshan Fm.	■	■	■			■			■	2009	Qijiang Geopark Museum
Ordos National Geopark	Ordos, Neimenggu	National	179	K1 Zhidan Gr.	■	■							■	2011	Otoke Geological Museum, Otoke Field Geological Relics Museum
Danxia Landform National Geopark in Chengde	Chengde, Hebei	National	48	K1 Zhangjiakou Fm.			■							2011	
Xinjiang Hami Pterodactyl-Yardang National Geopark	Hami, Xinjiang	National	1564	K1 Tugulu Gr.	■									2020	
Luoping Biota National Geopark	Luoping, Yunnan	National	79	T2 Guanling Fm.							■			2011	Luoping Geopark Museum
Laiyang Cretaceous National Geopark	Laiyang, Shandong	National	15	K1 Laiyang Gr.	■									2011	Laiyang Paleontological Fossil Museum, Tsintaosaurus Site Museum, No. 2 Dinosaur Site Museum
Chicheng Jurassic Provincial Geopark	Chicheng, Hebei	Provincial	115	J-K boundary Tuchengzi Fm.	■	■	■							2013	
Malingshan Provincial Geopark in Tancheng	Tancheng, Shandong	Provincial	25	K1 Dasheng Gr.	■	■								2007	
Junan Dinosaur Tracks Provincial Geopark	Linyi, Shandong	Provincial	60	K1 Dasheng Gr.	■									2005	
Jishan Provincial Geopark in Linshu	Linyi, Shandong	Provincial	59	K1 Dasheng Gr.	■	■			■				■	2012	
Nanxiong Dinosaur Provincial Geopark	Nanxiong, Guangdong	Provincial	85	K2 Zhutian Fm.			■							2013	

most tracks are safeguarded by structures, certain tracks undergo 3D scanning, and others remain exposed outdoors. Preservation approaches in provincial geoparks are relatively uniform, primarily involving outdoor exposure, with limited utilization of digital data collection methods.

A total of 28 museums with tracks were identified (Fig. 14.7B as example), primarily comprising geological museums and museums of natural history scattered across different provinces in China (Table 14.2). While some non-state-owned or private museums were included in the count, those affiliated with geoparks were excluded (Fig. 14.7C as an example). The majority of these museums showcase local tracks, although there are noteworthy exceptions. Notably, museums in Beijing and Nanjing boast comprehensive displays.

FIGURE 14.7 Some examples of strategies in geoconservation and science communication in Asia. (A) Qianjiadian Site in Yanqing UNESCO Global Geopark, Beijing. Visitors can see the tracks from the J3-K1 Tuchengzi Fm. at close range from the plank road. There are baffles on the track surface to prevent falling rocks. (B) Yanguoxia Site in Yongjing, Gansu; the museum at the foot of the mountain where site one is located. (C) Xixia Dinosaur relics Park, in the town of Danshui, Funiushan UNESCO Global Geopark. The area is famous for the egg clutches and nests of sauropod, theropod and ornithopod dinosaurs, besides turtles found in the late Cretaceous Gaogou formation. The impressive park includes 500 m of tunnels and a 9-meter-deep pit, aimed to intersect several egg nests in cross section and from underneath. (D) Shar Tsav site of Manlai soum (Mongolia) visiting facilities. (E) Jeo-ri tracksite (South Korea) in a road cut and its impressive protective structure. (F) Goseong Dinosaur Museum in Sangjogam Park. *(A) Photo credit Chen Haiying. (B) Photo credit Shenna Wang. (C) Photo credit Carlos Neto de Carvalho. (D) Photo credit Tour Mongolia. (E) Photo credit LifeinKorea.com. (F) Photo credit VisitKorea.*

TABLE 14.2 Museums in China with important collections of dinosaur tracks.

Name of the Museum	Location	Opened	Type of tracks											State
			The.	Sau.	Om.	Thy.	Pte.	Cro.	Tes.	Ave.	Mam.	Fis.	Lim.	
National Natural History Museum of China	Beijing	1959	■					■						Open
Geological Museum of China	Beijing	1916	■											Open
Nanjing Museum of Natural	Nanjing, Jiangsu	2004	■		■				■	■				Open
Paleozoological Museum of China	Beijing	1995	■		■									Open
Anhui Geological Museum	Hefei, Anhui	2012	■							■				Open
Hunan Geological Museum	Changsha, Hunan	2012	■											Open
Nanjing Geological Museum	Nanjing, Jiangsu	1916			■			■						Open
Zhejiang Museum of Natural History	Hangzhou, Zhejiang Anji, Zhejiang	1929 2018	■	■	■	■	■			■				Open
Chongqing Museum of Natural History	Chongqing	2015	■							■	■			Open
Paleontological Museum of Liaoning	Shenyang, Liaoning	2011	■				■			■				Open
Inner Mongolia Museum of Natural History	Huhehaote, Neimenggu	2018	■	■						■				Open
Geological Museum of Guizhou	Guiyang, Guizhou	2020	■											Open
Ningxia Geological Museum	Yinchuan, Ningxia	2011	■	■										Open
Henan Natural Historty Museum	Zhenzhou, Henan	2008	■											Open
Gansu Geological Museum	Lanzhou, Gansu	2010	■	■	■									Open
Geological Museum of Heilongjiang Province	Haerbin, Heilongjiang	2007			■									Open
Chengdu Museum of Natural History	Chengdu, Sichuan	2022	■	■	■									Open
Museum of China University of Geosciences	Beijing	1952	■	■	■									Open
Yifu Museum of China University Of Geosciences	Wuhan, Hubei	1982	■									■	■	Open
Yanji Dinosaur Museum	Yanji, Jilin	2022		■	■									Open
Nanxiong Museum	Nanxiong, Guangdong	1982			■									Open
Zhucheng Dinosaur Museum	Zhucheng, Shandong	1997	■	■	■									Open
Zhucheng Museum	Zhucheng, Shandong	1958		■	■									Open
Dongyang Museum	Dongyang, Zhejiang	2014					■							Open
Hezheng Museum of Ancient Animal Fossil	Hezheng, Gansu	2003	■											Open
Shenmu Museum	Shenmu, Shanxi	2012	■		■									Open
Yingliang Stone Natural History Museum	Quanzhou, Fujian	2015	■			■				■				Open
Dinosaur Odyssey Museum	Nanning, Guangxi	2024	■	■	■	■	■		■	■				To be opened

It is worth noting that there are few public museums exclusively focused on track displays, primarily due to the limited diversity and quantity of available track specimens. The latter is likely due to the engineering difficulty of collection or a greater emphasis on in situ conservation. Among the non-state-owned museums, the Yingliang Stone Natural History Museum specializes in exhibiting tetrapod tracks from the Wuerhe region in Xinjiang, while the Dinosaur Odyssey Museum features tetrapod tracks from various regions across China. In general, most track collections consist of isolated individuals or slabs showcased within display cases.

To our current knowledge, the only ichnosite from the Quaternary of China is Dingdingmen in the city of Luoyang, Henan Province (see Walk the World: the hominin track ways chapter in this book). The Dingding Gate was the south-central, and the main gate of the outer city wall of Luoyang, the eastern capital city during the Sui and Tang dynasties. According to historical sources, the gate was built in 605 AD and was destroyed near the end of the Northern Song Dynasty (918–960 AD). During the excavations, significant evidence relating to the Silk Road was found in front of the gate, including rutted wheel tracks, human footprints, and camel hoof prints (Su et al., 2020). After 3D scanning the whole surface, it was reburied in order to protect the trampled surface. A cast was made that was then placed on top of the original surface. The artificial soil used for casting has now settled and is the same in appearance as the original (Su et al., 2020). The heritage value of the Dingding Gate site rests on the claim that it was the imperial gate of the Sui and Tang dynasties with close material connections to the Silk Road. For this reason, the reconstitution of the Dingding Gate was made in 2009 (Museum of the Site of the Dingding Gate) to protect the exposed ruins and to mark the importance of the site. Together with historical records, the Dingding Gate was recognized as the eastern origin of the Silk Road from the seventh to tenth centuries (Su et al., 2020). Luoyang Silk Road Dingding Gate was subsequently included as one of the sites of the 2014 UNESCO World Heritage Silk Roads: The Routes Network of Chang'an-Tian Shan Corridor.

2.5.2 *Japan*

Japan shows a large number of sites with proboscidean footprints, mostly from the Pliocene-Pleistocene (Matsukawa and Shibata, 2015; see Quaternary Vertebrate Ichnology chapter). Despite the large number their protection status is quite unclear, and apparently none of them were included in a geoconservation strategy and open to the public. Casts were made and tracks were removed and are stored or exhibited among collections of local museums.

The number of dinosaur tracksites is comparatively much less impressive (Matsukawa et al., 2005), and only one with purported footprints is protected. Along route 299 in the town of Kanna, Gunma Prefecture, stands a tracksite in the mountain cliff known as the Sazanami-iwa ("ripple rock"), and dated from the Aptian. It was discovered in 1953 during the construction of the road. Often referred to as the "ripple marks of Sebayashi" after the former village, it is located in, the rock was thoroughly researched by Gunma University upon its discovery. Later, in 1960, the "holes" were identified as dinosaur footprints, the first fossil purported footprint site to be found in Japan, despite the fact that they are taxonomically indeterminate (Matsukawa et al., 2005). Following this, Kanna started to brand itself as a "dinosaur kingdom," opening a museum called the Kanna Dinosaur Center in 1987, exhibiting local fossils and Mongolian dinosaurs. Kanna Dinosaur Center includes dinosaur skeleton displays, a fossil digging experience, an exciting live theater, and a workshop of fossil replication. Though the museum's collection is rather impressive, the "dinosaur kingdom" of Kanna remains an obscure destination due to its difficult accessibility. The Sazanami-iwa was protected by the local Gunma prefecture as a natural monument.

2.5.3 Mongolia

The Mongolian Gobi Desert is perhaps the largest dinosaur fossil reservoir in the world. The region is especially important as regards dinosaur fossils from the Late Cretaceous. Among several sites with dinosaur tracks known (e.g., Ishigaki et al., 2009) Shar Tsav stands out and is situated 80 km northeast from Khanbogd soum and 108 km from Manlai soum in the south, 600 km southeast of the capital Ulaanbaatar. Tracks of dinosaurs at the Shar Tsav site of Manlai soum, Umnugobi have been taken under protection (Fig. 14.7D). The "Gobi Oyu" development supporting fund built a protective facility, as a first stage action for developing tourism at the site. Authorities of Umnugobi aimag, the Paleontological Institute, and the Tourism Union also took part in the construction of the facility.

The site from the Maastrichtian of the Nemegt Formation was discovered in 1995 by Mongolian and Japanese researchers and a joint expedition conducted active field research in 1996, revealing over 2800 prints of dinosaurs. Later detailed research studies in 2001 and 2010 detected over 18,000 footprints and trackways of four to five types of herbivorous and carnivorous dinosaurs (Ishigaki et al., 2009). Therefore, it is a unique paleontological heritage site also proving that dinosaurs lived in groups. A protective facility and footbridges have been built over the main findings to protect the footprints from weathering agents, as well as from human improper actions. The project is planned to be implemented for 10 years with the aim to make the area a tourist region. Shar Tsav is part of the "Cretaceous Dinosaur Fossil Sites in the Mongolian Gobi" included in the Tentative List of UNESCO in 2014 by the Mongolian National Commission for UNESCO.

2.5.4 North Korea

Located on the east side of the Pyongyang-Kaesong Highway between the county-seats of Pyongsan and Kumchon, Dinosaur Footprint and Fossils site is accessible when heading north. Housed inside a white plaster building, the main set of footprints were discovered in January 1990 during road maintenance. The original discovered trackway was about 100 m long, 15 m wide, and at a 20-degree slant, emerging from a rock face. The tracks are thought to have been left behind by two dinosaurs, an adult and a baby, walking together. Only part of the trackway is on display. At the entrance of the small hall, there is a foreign dinosaur documentary dubbed in Korean about the end of the dinosaur age that can be watched before heading to the back of the building where the footprints are on display. There are classic science and dinosaur paintings such as the hand-painted timeline of the universe, starting with the Big Bang, and one of some happily swimming dinosaurs. Outside of the hall, there are some more footprints on display, supposedly from a pterosaur, as well as some fossil plants. The footprints are thought to date back to 80–100 million years ago. The scientific study of this tracksite is unknown to us.

2.5.5 Palestine

Dinosaur tracks at Beit Zeit, West of Al-Quds (Jerusalem) were discovered in a *moshav* in 1962. The area of 400 m^2 was uncovered, and numerous additional tracks were found, now including more than 200 tracks. Several types of tracks can be distinguished, which might belong to three producers. There are several 10–15 m trackways that are made up of

15–20 footprints, and other shorter trackways, only 3–4 m long, consisting of no more than four to six tracks (Khalaf-von Jaffa, 2014). An old open-sided shelter covers the area, but rain and problematic conditions have been eroding the invaluable footprints. Some simple interpretation and dinosaur models are provided at the site.

2.5.6 *Republic of Yemen*

Madar is a village about 50 km north of the Yemeni capital of Sanaa. In 2003, a local journalist noted the presence of dinosaur tracks in a limestone bed and brought them to the attention of geologists at the University of Sanaa. The main site is approximately 3 km west off the main road and has been signposted and fenced by the Yemen Geological Survey. Additional tracks have been recognized nearby, within the villages of Arhab and Bait al Washr. At the main site tracks of 11 sauropods and one ornithopod dating from 150 million years ago were found (Schulp et al., 2008). Footprints of both juveniles and adults were described by the authors. The discovery was the first of dinosaur footprints in the Arabian Peninsula. The discovery has since been signposted for tourists, and efforts are underway to list the site with UNESCO. The Arhab area possesses all the requirements to become a geopark (Al-Wosabi and Al-Aydrus, 2015), despite the dramatic civil war situation that has lasted for 10 years; these include a large area, the occurrence of geological and paleontological features of regional and international importance, several archeological sites, willingness of the local community to be involved in a geopark project, a society in urgent need of socioeconomic development, elements of scientific and cultural tourism, and ease of access to the region. Establishing a geopark in the region will safeguard and preserve the geological, paleontological, and archeological heritage and improve the quality of life for the local communities.

2.5.7 *South Korea*

Sites with dinosaurs found throughout the southern coast of Korea are in the Tentative List of UNESCO and include the largest sites with eggs and tracks of dinosaurs from the Cretaceous in the world. The fossil site in Haenam Uhangri is protected under law as a natural monument. It is located in the Uhangri Formation, dated from the Campanian. In fact, this is a composite site, comprising many fossil sites, including three that have enclosed protective structures. A total of 263 ornithopod footprints have been unearthed at the fossil sites under the No. 1 protective structure. In addition, footprints of bipedal and quadrupedal dinosaurs have been found. Pterosaur footprints have been found at the fossil sites under the No. 2 protective structure. When the tracks were excavated, they were considered to be the longest pterosaur trackways in the world, and were thus important for interpreting the ecosystems of pterosaurs. The fossil site under the No. 3 protective structure is large (45 m long and 9 m wide). Deep and distinct footprints occur at this site, interpreted as being undertracks made by bipedal sauropods (Song, 2010).

Enclosed protective structures with air conditioning were built at Haenam Uhangri, following the discovery of a dinosaur footprint within the region in 1992. The Haenam Uhangri Dinosaur Museum was founded to promote the area's significance as a fossil site. The museum provides materials and exhibitions related to dinosaurs as well as Ulhang-ri as a natural heritage site. However, although it is possible to control the internal microclimate therein, condensation occurs, frequently as a result of microclimatic differences in the interior structure, and thus, additional measures are required (Yoo and Lee, 2015).

The Dinosaur Egg Site in Bibong-ri at Deokryang Bay is famous for the dinosaur egg nests of 1.5 m in diameter with perfectly preserved dinosaur's eggs. Dinosaur (sauropod and ornithopod) egg clutches of the Upper Cretaceous Seonso Conglomerate are found in at least six horizons consisting of sandy mudstones to mudstones (Paik et al., 2004). About 10 hatcheries and 100 fossils of dinosaur eggs were discovered on the rock. The Dinosaur Egg Site in Bibong-ri of Boseong is very important for studying dinosaurs' habitat and life. Therefore, it is designated and protected as a Natural Monument. Boseong Bibong Dinosaur Park shows full-sized dinosaur skeletons and animatronics, both indoors in a modern facility and outdoors.

The Mudeungsan UNESCO Global Geopark is located in the southern part of the Korean peninsula near the city of Gwangju and is centered around Mount Mudeung. The fossil site in Hwasun-gun is famous for the tracks of carnivorous dinosaurs arranged in 60 very long trackways (longest 52 m) and over 1500 footprints from the Cretaceous Neungju Group in a quarry near Seoyu-ri (Huh et al., 2006). Tracks found in the area belong mostly to theropods, especially small-sized theropods. On the basis of morphology and size, the theropod tracks have been classified into three types, including *Magnoavipes*, *Ornithomimipus*, or *Xiangxipus*, and tracks belonging to large theropods (Huh et al., 2006). In addition to the carnivorous dinosaurs, tracks were found of 12 herbivorous dinosaurs, as well as fossilized plants. The Hwasun site, previously noted for its abundant theropod trackways, also reveals one horizon (L2) with at least six parallel trackways of small, blunt-toed ornithopods with wide tracks assigned to the ichnogenus *Ornithopodichnus* (Lockley et al., 2012). The site also reveals a second horizon (L4) that yields a single clear trackway of a much larger, blunt-toed ornithopod. Recently, small pterosaur footprints assigned to *Pteraichnus* isp. were also described from the tracksite (Jung et al., 2022). Hwasun is a local geotourist attraction of the geopark, with the site fenced and presenting bilingual interpretation panels.

The Jeo-ri dinosaur tracksite from the Lower Cretaceous Sagok Formation in the Uiseong County area in Gyeongsangbuk-do (North Gyeongsang Province) was the first dinosaur tracksite designated as a Korean Natural Monument. The fossil site at Uiseong Jeori was discovered in 1988 during road construction and is located in mudstone and sandstone of the Sagok Formation. More than 380 footprints were attributed to theropods (1 trackway), sauropods (12 trackways), and ornithopods (10 trackways) (Lee et al., 2023). The trackways show a preferred orientation to the north and northwest with some similarly sized ornithopod trackways, suggesting gregarious behavior. The 1600 m^2 area of the Jeo-ri tracksite is located in a road cut and protected from erosion by an impressive protective structure (Fig. 14.7E), which delivers adequate ventilation but is vulnerable to frost weathering in winter (Yoo and Lee, 2015). Therefore, it is considered necessary to extend the drain at the back of the protective structure for the site. Conservation treatment was conducted in 2003, painting the footprints and using cement to distinguish between fossils and matrix. However, this work received fierce criticism because the original shape of the tracks was severely destroyed, and cleaning treatment of the paint was therefore carried out in 2012 (Yoo and Lee, 2015). Interpretation is basic but provided in both Korean and English.

Korea has the world' s most dense concentration of footprints of dinosaurs and birds along its south coast in Goseong County. This fossil site in the city of Yeosu is distributed in beds found in five islets in the Samcheonpo bay, including Sado, Choodo, Nangdo, Jeokgeumdo, and Mokgeumdo. Not only is it famous for its quantity of tracks, but includes also an 84-meter-long trackway. The fossil site at Goseong-gun is famous for the quantity and variety of fossils, since more than 4000 dinosaur footprints and about 420 trackways were discovered. The Jindong Formation is one of the few localities where sauropod, ornithopod, and bird tracks all occur in abundance, probably due to latitudinal/climatic controls (Lockley et al., 2006). The sauropod tracks, which include wide-gauge forms allied to *Brontopodus*, are characterized by a high proportion of small individuals. The most abundant dinosaur trackways are those of iguanodontids (cf. *Caririchnium* or *Iguanodontipus*) that often traveled in herds. Bird tracks, including the large ichnospecies *Jindongornipes kimi*, an intermediate-sized form, *Goseongornipes markjonesi*, and a small ichnospecies *Koreanaornis hamanensis*, occur at several stratigraphic levels (Lockley et al., 2006).

First opened in November of 2004, the Goseong Dinosaur Museum, the main dinosaur museum in Korea, is located in the Sangjogam Park, where the first dinosaur footprints were discovered in South Korea. While the museum is a tourist attraction that showcases dozens of breathtaking murals, enhanced recreations, and exhibition halls, the scenery beyond the museum's doors is equally incredible (Fig. 14.7F). With four stories, five halls, a film room, a gift shop, and more, visitors can explore five main exhibits, including "Capital of Dinosaur," "Dinosaur's Footprint of Goseong," "Dinosaurs of the Cretaceous Period," "Dino Land," and "Old Trace." Visitors can see genuine egg fossils of *Oviraptor* and *Protoceratops*, as well as dinosaur models. The museum is connected to the Sangjogam Beach, where dinosaur footprints are preserved. The pathways are accessible until reaching several series of steep stairs, which lead to the water.

2.5.8 *Tajikistan*

Close to Dushanbe, the Shirkent Valley was established as a Nature Reserve in 1991, followed by the discovery of three sites of dinosaur tracks from different ages containing more than 400 trackways. The most important site is located on the right bank of the Shirkent River, slightly above a village with the same name. Two trackways are seen here on the surface of a limestone bed. The length of the tracks is 50 cm on the average, the width up to 30 cm, and the step length about 75 cm. The unusual *Macropodosaurus gravis* tracks from the Early Cretaceous (Albian) locality of Shirkent-1 show that the trackmaker was bipedal and plantigrade. Through a comparison of *Macropodosaurus* tracks with those of other dinosaurs and by a juxtaposition of track morphology and pes skeletal morphology, it is argued that they belong to an aberrant group of theropods, the segnosaurs (Sennikov, 2021).

2.5.9 *Thailand*

Khon Kaen National Geopark is located on the Mesozoic sandstone syncline of the Khorat Plateau, northeastern Thailand. The geopark is included in the Phu Wiang National Park, covering the Wiang Kao and Phu Wiang districts, with an area of approximately 1038 km^2. The territory is well known for dinosaur remains, including one dinosaur tracksite found in the Phra Wihan Formation and dated from the Early Cretaceous (Le Loeuff et al., 2009). This formation is the main track-bearing unit of Thailand with five footprint sites: Hin Lat Pa Chad (Khon Kaen Province), Phu Faek (Kalasin Province), Khao Yai National Park (Prachin Buri Province), Phu Kao (Nong Bua Lam Phu Province), and Phu

Hin Rong Kla National Park (Loei Province). More than 60 different dinosaur tracks are organized in 10 trackways. Most of them are footprints of small theropods and ornithischians. In addition to the footprints of ornithomimosaurs, sizable reptilian and small animal footprints have also been found that may indicate ancient reptiles. The paw prints of a small family of dinosaurs are imprinted on fine white sandstone (Singtuen et al., 2023). The geopark has a dinosaur museum and research center, which also acts as a tourist attraction with the cooperation of the Khon Kaen Provincial Administrative Organization, Department of Mineral Resources, Tourism Authority of Thailand, and academic organizations (i.e., Khon Kaen University, Mahasarakham University, and Phu Wiang Wittayayon School) (Singtuen et al., 2022).

Thailand's Department of Mineral Resources has just discovered a new set of dinosaur footprints in the Phetchabun-Khon Kaen border area. The footprints are estimated to be over 225 million years old, making them the oldest dinosaur footprints ever found in Thailand and Asia. The footprints were discovered by a team of paleontologists from the Department of Mineral Resources during a survey of the area in January 2024. They come from a layer of mudstone, are well-preserved, and show anatomical details of the dinosaurs' feet, including toes, claws, and soles. The footprints are from a variety of dinosaurs, including sauropods, theropods, and ornithopods. The discovery of the footprints is significant because it provides new insights into the diversity and ecology of dinosaurs in Thailand during the Late Triassic period. The footprints also help to fill in the gaps in the understanding of the evolution of dinosaurs in Asia.

The Dinosaur Footprint Park in the Highway 212 Phanom Subdistrict, Tha Uthen at Nakhon Phanom province features footprints imprinted into the region's reddish-brown sandstone approximately 100 million years ago. This site, a former sandstone quarry where large blocks were extracted and used to delimit the banks of the Mekong River, was acquired in 2007 by the Department of Mineral Resources. Further excavations and the building of a roof were realized in 2008 (Le Loeuff et al., 2009). It revealed an impressive assemblage of more than 200 footprints of small theropods, ornithopods, and crocodiles. Most of the footprints are theropod footprints, consistently smaller than *Asianopodus*. It seems that the parallel theropod trackways were not left by a single group but by several small packs of theropods, as evidenced by the varying calculated speeds (Le Loeuff et al., 2009). The site is a tourist attraction and shows models in real size of the theropod producers, besides old-style bilingual interpretive panels.

2.5.10 Turkmenistan

Dinosaur Plateau (Turkmen: *Dinozawrlar platosy*) is a large limestone slab lying on the slope of the Köýtendag mountains in the Lebap velayat of Turkmenistan. The Hojapil-Ata site extends over an area of approximately 28,500 m^2 on the south flank of the Koitendag valley. A landslide exposed the trampled layer, although no data are available to date this event. Historically, tracks exposed at the Hojapil-Ata site have been known by the locals for centuries. The name of the locality, which translates into "ancient elephants" or "elephant feet," reflects the idea that these tracks were left by the elephants used in the army of Alexander the Great, which conquered eastern Turkmenistan in the fourth century B.C. Twenty-four trackways with a total of 913 tracks are preserved; the longest trackways measure 271 m (HA-1, 108 tracks), 216 m (HA-3, 139 tracks), 188 m (HA-4, 122 tracks), and 127 m (HA-2, 91 tracks). The *Therangospodus/Megalosauripus* track association found in the Upper Jurassic peritidal deposits shows several similarities with coeval ichnoassemblages from Europe, suggesting the presence of different medium- and large-sized theropod dinosaurs in the two regions. The large morphotype is attributed to *Megalosauripus uzbekistanus* (Fanti et al., 2013). The large number of visitors that visit the main track-bearing surface combined with natural weathering has led to the disappearance of several tracks in the last 2 decades, with many others have been significantly damaged (estimated erosion rate 1 mm/10 years in the higher section, up to 6 mm/10 years in the lowermost areas of the site; Fanti et al., 2013). Notwithstanding, the government of Turkmenistan is working to include the Dinosaur Plateau in the UNESCO World Heritage List.

2.6 Oceania

2.6.1 Australia

Australia's rich and diverse paleontological geoheritage includes many unique and world-famous localities (Percival, 2014). Though some sites are well managed and adequately protected, others, due to their remoteness and difficulties related to surveillance, have suffered from vandalism or theft. Percival (2014) provides a list of significant Australian fossil sites, as compiled by the Australian Heritage Council. Only some of the protected sites under law received proper geoconservation measures and are open to geotourism.

Dinosaur Stampede National Monument in Queensland shows 3300 dinosaur footprints dated between 104 and 92 million years ago in Lark Quarry Conservation Park. The Dinosaur Stampede National Monument was included in the Australian National Heritage List on 2004, for values of rarity and research. It is considered to be the site of the world's only known record of a dinosaur stampede (Thulborn and Wade, 2007), with fossilized footprints interpreted as a predator stalking and causing a stampede of around 180 two-legged dinosaurs (see different interpretations of this site in Romilio et al., 2013). *Skartopus* and *Wintonopus* trackmakers were thought to have stampeded past the cf. *Tyrannosauripus* trackmaker, walking in the opposite direction. The large tracks attributed to *Tyrannosauripus* were later interpreted as being produced by a large herbivore similar to *Muttaburrasaurus*, rather than by a predatory theropod, and were suggested to belong to the ichnogenus *Amblydactylus* instead (Romilio and Salisbury, 2011).

In 2015, the Winton Shire Council invited the Australian Age of Dinosaurs Museum of Natural History (AAOD) to take over the operation of public guided tours at Dinosaur Stampede National Monument (www.australianageofdinosaurs.com). Part of the trackway that could be destroyed by floods was relocated to the AAOD Museum in Winton. The Lark Quarry buildings and infrastructure (Fig. 14.8A) have been designed to protect the tracksite using ecologically sustainable design (ESD) principles, for maximum water, waste, and energy efficiency. This building protects the trackway from damage by stabilizing temperature and humidity fluctuations, stops water running over the footprints, and keeps people and wildlife off the footprints, themselves. A movie presentation in the Trackways Theater shows animated recreations of the dinosaur stampede—as it may have happened circa. 95 million years ago. The Trackways Shop has a range of custom merchandize depicting the stampede, including dinosaur foot casts.

FIGURE 14.8 Examples of protection of dinosaur tracksites in Australia. (A) Facilities of the Dinosaur Stampede national monument (AAOD). (B) An indoor view of the March of the titanosaurs exhibition at the AAOD Museum in Winton, Queensland (Steven Lippis/AAOD).

A 70-meter-long trackway made by a large titanosaur, besides other ornithopod and coelurosaur footprints, was found in 2016 in a creek bed at Karoola Station, central Queensland. One of the sauropod footprints included a single, three-toed footprint preserved within. According to Poropat et al. (2021), the footprint was made by a medium-sized theropod dinosaur, similar to Winton's own *Australovenator wintonensis*, which crossed the tracksite before the sauropod when the silt was still fairly wet at depth. Twenty paleontologists and volunteers excavated the tracksite over the course of more than 2 years and moved its entirety to a purpose-built facility at the AAOD Museum, where it is now open to the public (Fig. 14.8B). The trackway relocation was one of the biggest projects of its kind in the world, comparable with the Paluxy River trackway relocation in Texas, USA.

The Dinosaur Coast near Broome (Western Australia) reveals thousands of dinosaur tracks along beaches from Roebuck Bay near Broome, north to Minarrinay (Coulomb Point). They are found in intertidal exposures of the Broome Sandstone along the Dampier Peninsula 100 km coastline. The dinosaur tracks formed part of the West Kimberley National Heritage Area (Place ID 106063) since 2011. Theropod, sauropod, ornithopod, and thyreophoran tracks and trackways in 21 morphotypes dated to 140−127 Ma have been described (Salisbury et al., 2016). The Dinosaur Coast Management Group (DCMG) is a nonprofit organization, formed in 2015 to protect and promote the dinosaur tracks of the Dampier Peninsula and to educate the public about their local cultural and wide scientific importance. For thousands of years, dinosaur tracks have been part of the cultural heritage of the people of the Dampier Peninsula and greater West Kimberley. There is a strong link between dinosaur tracks and Aboriginal creation stories. The Goolarabooloo Law Bosses hold the responsibility for the continuity of the Songline and its capacity to sustain life in its fullest. This includes keeping the coastal land clear of any development that would interfere with their law and culture, or damage their song cycle, places and sites. Paddy Roe, with the support of his Goolarabooloo community, established 31 years ago the Lurujarri Heritage Trail, a 9-day cultural walking journey along the song cycle that involves leading visitors throughout this 80 km stretch of coastal terrain, traveling as a group and sharing their heritage. Dinosaur Coast Display at the Broom Historical Museum provides basic information about the paleontological heritage. The "Dinosaur Track Guide App" and "Information Brochure" have general location information and tips on how to identify a track in the coastal area.

In 2003, human footprints were discovered preserved in silt in Mungo Lake, part of Willandra Lakes World Heritage in New South Wales. Excavations revealed about 450 human footprints dated between 23–19 ka by optically stimulated luminescence (Webb et al., 2006; Westaway et al., 2013; see Walk the World: the hominin track ways chapter in this book). The undisturbed stratigraphic context of Willandra Lakes World Heritage provides outstanding evidence for the economic life of *Homo sapiens sapiens* to be reconstructed. Archeological remains such as hearths, stone tools, and shell middens show a remarkable adaptation to local resources. The evidence of occupation deposits establishes that humans had dispersed as far as Australia by 42 ka BP. Sites also illustrate human burials that are of great antiquity, such as a cremation dating to around 40 ka BP, the oldest ritual cremation site in the world (Bowler et al., 1970), and traces of complex plant-food gathering systems that date back before 18 ka BP associated with grindstones to produce flour from wild grass seeds, at much the same time as their use in the Middle East (reviewed in Allen and Holdaway, 2009). Since inscription as World Heritage in 1981, the discovery of the human fossil trackways has added to the understanding of how early humans interacted with their environment. Conservation strategies for the human tracksite were formulated after discussion between the First Nations custodians, archeologists, and government agencies (Macgregor, 2022). The excavated site was reburied on the recommendation of a detailed management plan. The decision to rebury the footprints was the only practical option for preservation due to the harsh local conditions and the remote location.

3. Conclusions

The worldwide burst of scientific research, especially on dinosaur tracksites, in the last 30 years has been followed by the need to protect ichnosites under regional or national laws as geoheritage, to prevent degradation and theft, and ultimately to give educational and economic use. The general public and national, regional, and local authorities are realizing that ichnosites are part of the cultural, public heritage that must be protected, and which may be important to the local economic reactivation and/or development through geotourism, therefore contributing to an increase in population self-esteem, as well as in helping to rescue regional cultural values and identity (Carvalho and Da Rosa, 2008). We detailed in this chapter how several countries are dealing with the tracksites and the paleontological heritage in general by legislation for more effective protection. There have been efforts by ichnologists working with national authorities in Europe (Spain and Portugal), Asia (South Korea), and South America (Bolivia and Argentina) to nominate fossil footprints sites for World Heritage inscription, highlighting outstanding values that go beyond scientific relevance (Alcalá et al., 2016).

The development of Global Geoparks in the last 20 years fostered the geoconservation of numerous vertebrate tracksites, in countries such as Brazil, China, France, South Korea, Spain, or Tanzania, under a logic of bottom-up sustainable development through innovative geotourism. The exhaustive examples found worldwide in this chapter reveal that dinosaur tracksites with long and impressive trackways, or well-preserved footprints and evidence of remarkable behaviors and ecology, are by far the largest number of vertebrate ichnosites protected and valued. Undoubtedly, dinosaurs have a formidable appeal for science education and geotourism, and the latter is especially important if the fossils are exhibited in the same place where they were discovered, through in situ displays and facilities. This is almost obligatory for dinosaur tracksites of a certain size, as it is not easy, nor advisable, to remove the layer containing the tracks in order to transport them to a museum. This also happened in the past, usually leaving part of the tracksite in the field exposed to erosion and theft, as is happening still nowadays mostly with Cenozoic tracksites, including hominin tracksites. With notable exceptions for Cenozoic mammal footprints, in Ipolytarnóc (Hungary) or the hominin footprints of Laetoli (Tanzania), it is uncommon to find non-dinosaur tracksites that have been subjected to geoconservation measures. One of the reasons is that Cenozoic sites are more challenging regarding onsite exhibition, especially Quaternary tracksites that can be eroded at the speed of river flooding or changing tides. Nevertheless, massive limestone and sandstone track beds from the Paleozoic and Mesozoic are also susceptible to intensive weathering and erosion, at the same rate as they were exposed by erosional agents. Therefore, although there are outstanding examples of canopies covering ichnosites, made of steel, wood or concrete, and even sometimes covered with solar panels, most of the examples presented are just basic geoconservation projects, integrating the tracksites in hiking trails or providing simple (panels, leaflets, apps), or more comprehensive interpretation in local museums. Photogrammetry and 3D printing came to revolutionize the record of tracksites since these methods allow with high accuracy the reproduction of footprints from the field, in materials that are easier to archive or exhibit in museums, and can be shared among institutions. They are the solution to prevent the complete loss of reliable scientific information, but they cannot replace efforts to preserve original tracksites that can be a source of social (educational) and economic (geotourism) revenues for local communities.

In several countries, even in those like the United Kingdom, where geoconservation took pioneering steps and legislation as we detailed in this chapter, removal of tracks to museums is the preferable methodology to avoid ichnosite theft and degradation, but also shows a clear lack of investment/interest in geoheritage. The least comprehensive examples of tetrapod track collections in museums, also summarized in this chapter, reveal that most of them are not exhibited under proper conditions. Notwithstanding, excellent examples of onsite museums and interpretive centers presented here are clearly the way to foster long-term, continuous research, educational activities and large numbers of visitors contributing decisively to develop tourism destinations impacting the socioeconomic wealth of local communities. In fact, local tourism can be predicated on local paleontological finds; however, it is deemed considerably more efficient if such attractions are backed with an extensive infrastructure of recreational theme parks, as defended by Antczak (2020). The incredible national or public–private partnerships (and less commonly private) investments made to build museum complexes and parks, with outstanding examples in China, Germany, or South Korea, not only enclose the ichnosite in large buildings, preventing for good their destruction.

They also disclose a proud sense of heritage for local communities and a major attraction for visitors, especially when the scientific research supports their outstanding universal value, and keeps providing new scientific findings.

Acknowledgments

We would like to acknowledge many contributors and commentators to this paper, namely N. Abbassi (IASBS Zanjan), M. Alivernini (Thüringen Inselsberg—Drei Gleichen UNESCO Global Geopark), F. Altamura (Italo-Spanish Archaeological Mission at Melka Kunturé and Balchit), M. Avanzini (MUSE - Museo delle Scienze, Trento), M. Belvedere (University of Florence), M.J. Benton (BIGC/University of Bristol), E. Bordy (University of Cape Town), I. Bouchemla (Mouloud Mammeri University of Tizi-Ouzou), A. Camens (Flinders University), I.S. Carvalho (Universidade Federal do Rio de Janeiro), S. Dasgupta (Indian Institute of Technology of Bombay), E. Doyle (Burren and Cliffs of Moher UNESCO Global Geopark), B. Evans (BIGC/University of Bristol), C. Fassoulas (University/Natural History Museum of Crete), G. Iliopoulos (University of Patras), P.W. Jackson (Geological Museum in Trinity College), S. Jensen (Univ. Extremadura), M. Kavčič (Idrija UNESCO Global Geopark), B. Kear (The Evolution Museum), A. Lagnaoui (Université Hassan 1er), C.J. Macfadyen (NatureScot), D. Marty (Naturhistorisches Museum Basel), O. Medina (Facultad de Arquitectura y Ciencias del Hábitat de Sucre), V. Mesa (Facultad de Ciencias, UdelaR), C.A. Meyer (University of Basel), J. Milan (Geomuseum Faxe), F.M. Petti (MUSE, Museo delle Scienze Trento), I.Ž. Pleše (Ministry of Economy and Sustainable Development of Croatia), G. Puras Castells and Xavi Mir (Origens UNESCO Global Geopark), N. Schumacher (Ruhrgebiet National GeoPark), M.J. Simms (National Museums Northern Ireland), T. Solbakk (Geological Survey of Norway), W.A. Wimbledon (BIGC/University of Bristol), and Jianping Zhang (China University of Geosciences - Beijing). A final appreciation to the outstanding patience and review work of the editors Spencer Lucas (New Mexico Museum of Natural History) and Hendrik Klein (Saurierwelt Paleontological Museum).

References

Ait-Kaci, A.A., Lingham-Soliar, T., Broderich, T., 2004. Giant sauropod tracks from the Middle-Late Jurassic of Zimbabwe in close association with theropod tracks. Lethaia 37 (4), 467–470.

Albat, F., 1999. Tetrapodenfährten im Unteren Muschelkalk von Borgholzhausen – ein Grabungsbericht. Geol. Paläontol. Westfalen 52, 19–39.

Alcalá, L., 2012. The IDPI (Icnitas de Dinosaurios de la Península Ibéríca) WHL candidacy in perspective. In: Huh, M., Kim, H.J., Park, J.Y. (Eds.), The 11th Symposium on Mesozoic Terrestrial Ecosystems Gwangju, South Korea, pp. 255–258.

Alcalá, L., Cobos, A., 2021. Dinosaur tracksites from the Maestrazgo UNESCO Global Geopark (Teruel, Spain). Geoconserv. Res. 4 (2), 413–426.

Alcalá, L., Lockley, M.G., Cobos, A., Mampel, L., Royo-Torres, R., 2016. Evaluating the dinosaur track record: an integrative approach to understanding the regional and global distribution, scientific importance, preservation, and management of tracksites. In: Falkingham, P.L., Marty, D., Richter, A. (Eds.), Dinosaur Tracks: The Next Steps. Indiana University Press, Bloomington, Indianapolis, pp. 101–111.

Allen, H., Holdaway, S.J., 2009. The archaeology of Mungo and the Willandra Lakes: Looking back, looking forward. Archaeol. Ocean. 44 (2), 96–106. https://doi.org/10.1002/j.1834-4453.2009.tb00052.x.

Altamura, F., Bennett, M.R., D'Août, K., Gaudzinski-Windheuser, S., Melis, R.T., Reynolds, S.C., Mussi, M., 2018. Archaeology and ichnology at Gombore II-2, Melka Kunture, Ethiopia: everyday life of a mixed-age hominin group 700,000 years ago. Sci. Rep. 8, 2815.

Altamura, F., Bennett, M.R., Marchetti, L., Melis, R.T., Reynolds, S.C., Mussi, M., 2020. Ichnological and archaeological evidence from Gombore II OAM, Melka Kunture, Ethiopia: an integrated approach to reconstruct local environments and biological presences between 1.2 and 0.85 Ma. Quat. Sci. Rev. 244, 106506. https://doi.org/10.1016/j.quascirev.2020.106506.

Altamura, F., Lehmann, J., Rodríguez-Álvarez, B., Urban, B., van Kolfschoten, T., Verheijen, I., Conard, N.J., Serangeli, J., 2023. Fossil footprints at the late Lower Paleolithic site of Schöningen (Germany): a new line of research to reconstruct animal and hominin paleoecology. Quat. Sci. Rev. 310, 108094. https://doi.org/10.1016/j.quascirev.2023.108094.

Al-Wosabi, M., Al-Aydrus, A.A., 2015. Dinosaur footprint sites in Arhab area: an aspiring geopark in Yemen. In: Errami, E., Brocx, M., Semeniuk, V. (Eds.), From Geoheritage to Geoparks. Geoheritage, Geoparks and Geotourism. Springer, Cham, pp. 167–182. https://doi.org/10.1007/978-3-319-10708-0_12.

Amine, F., Berrahma, M.'H., Aarab, A., 2018. Inventory and management of dinosaurs traces: a case study of Azilal region, Morocco. Int. J. Glob. Sci. Res. 5 (2), 793–814.

Antczak, M., 2020. Are fossils enough? Palaeontological tourism based on local dinosaur discoveries. Geogr. Tour. 8 (2), 15–27.

Antonelli, M., Petti, F.M., Conti, J., Sacco, E., Petruzzelli, M., Spalluto, L., Wagensommer, A., 2023. Lower Cretaceous dinosaur footprints from the Molfetta tracksite (Apulia, southern Italy). Cretac. Res. 142, 105388. https://doi.org/10.1016/j.cretres.2022.105388.

Antunes, M.T., 1976. Dinossáurios eocretácicos de Lagosteiros. Ciências da Terra 1, 1–35.

Avanzini, M., Cavin, L., 2009. A new *Isochirotherium* trackway from the Triassic of Vieux Emosson, SW Switzerland: stratigraphic implications. Swiss J. Geosci. 102 (2), 353–361.

Avanzini, M., Dalla Vecchia, F.M., Mietto, P., Piubelli, D., Preto, N., Rigo, M., Roghi, G., 2007. A vertebrate nesting site in Northeastern Italy reveals unexpectedly complex behavior for Late Carnian reptiles. Palaios 22, 465–475. https://doi.org/10.2110/palo.2005.p05-137r.

Avanzini, M., Leonardi, G., Mietto, P., 2003a. *Lavinipes cheminii* ichnogen., ichnosp. nov., A possible Sauropodomorph track from the Lower Jurassic of the Italian Alps. Ichnos 10, 179–193. https://doi.org/10.1080/10420940390256195.

Avanzini, M., Leonardi, G., Mietto, P., Piubelli, D., 2003b. The dinosaur ichnsite at the Lavini di Marco (Calcari Grigi Group, Lower Jurassic). Acta Geol. 80, 31–36.

Avanzini, M., Wachtler, M., 2012. *Sphingopus ladinicus* isp. nov. from the Anisian of the Braies Dolomites (Southern Alps, Italy). Boll. Soc. Paleontol. Ital. 51, 63–70.

Babel, M., Leonowicz, P., 2015. Sołtyków – dinosaur tracks and something for jewellery specialists. In: Barski, M., et al. (Eds.), The Holy Cross Mountains– 25 Journeys through Earth History. University of Warsaw, Faculty of Geology, Warszawa, p. 190.

Barclay, W.J., Browne, M.A.E., McMillan, A.A., Pickett, E.A., Stone, P., Wilby, P.R., 2005. The Old Red Sandstone of Great Britain, Geological Conservation Review Series, No. 31. Joint Nature Conservation Committee, Peterborough, p. 393.

Bates, K.T., Breithaupt, B.H., Falkingham, P.L., Matthews, N.A., Hodgetts, D., Manning, P.L., Foss, S.E., Cavin, J.L., Brown, T., Kirklan, J.I., Santucci, V.L., 2009. Integrated LiDAR & photogrammetric documentation of the Red Gulch dinosaur tracksite (Wyoming, USA). In: Foss, S.E., Cavin, J.L., Brown, T., Kirkland, J.I., Santucci, V.L. (Eds.), Proceedings of the Eighth Conference on Fossil Resources, Utah, pp. 101–103.

Baucon, A., Privitera, S., Bonacossi, D.M., Canci, A., Neto de Carvalho, C., Kyriazi, E., Laborel, J., Laborel-Deguen, F., Morhange, C., Marriner, N., 2008. Principles of Ichnoarchaeology: new frontiers for studying past times. Acta Geol. 83, 43–74.

Belvedere, M., Fabre, E., Pandolfi, L., Legal, S., Coster, P., 2023. Stepping into Oligocene. A reassessment of the early Oligocene mammal tracks from Saignon (SE France). Hist. Biol. 1–17. https://doi.org/10.1080/08912963.2023.2286275.

Bennett, S.C., 1993. Reinterpretation of problematical tracks at Clayton Lake State Park, New Mexico: Not one pterosaur, but several crocodiles. Ichnos 2, 37–42.

Bennett, M.R., Bustos, D., Belvedere, M., Martinez, P., Reynolds, S., Urban, T., 2019. Soft sediment deformation below mammoth tracks at White Sands National Monument (New Mexico) with implications for biomechanical inferences from tracks. Palaeogeogr. Palaeoclimatol. Palaeoecol. 527, 25–38.

Bennett, M.R., Bustos, D., Odess, D., Urban, T.M., Lallensack, J.N., Budka, M., Santucci, V.L., Martinez, P., Wiseman, A.L., Reynolds, S.C., 2020. Walking in mud: remarkable Pleistocene human trackways from White Sands National Park (New Mexico). Quat. Sci. Rev. 249, 106610.

Bennett, M.R., Bustos, D., Pigati, J.S., Springer, K.B., Urban, T.M., Holliday, V.T., Reynolds, S.C., Budka, M., Honke, J.S., Hudson, A.M., Fenerty, B., 2021. Evidence of humans in North America during the last glacial maximum. Science 373 (6562), 1528–1531.

Benton, M.J., Cook, E., Turner, P., 2002. Permian and Triassic Red Beds and the Penarth Group of Great Britain. Geological Conservation Review Series, No. 24. Joint Nature Conservation Committee, Peterborough, p. 337.

Benton, M.J., Spencer, P.S., 1995. Fossil reptiles of Great Britain. Geol. Conserv. Rev. Ser. 10, 386.

Bessedik, M., Mammeri, C., Belkebir, L., Mahboubi, M., Adaci, M., Hebib, H., Bensalah, M., Mansour, B., Mansouri, M.E.H., 2008. Nouvelles données sur les ichnites de dinosaures de la région d'El Bayadh (Crétacé inférieur, Algérie). Palaeovertebrata 36 (1–4), 7–35.

Bird, R.T., 1939. Thunder in his footsteps. Nat. Hist. 43, 254–261.

Bird, R.T., 1944. Did Brontosaurus ever walk on land? Nat. Hist. 53, 60–67.

Bordy, E.M., Smith, R.M., Choiniere, J.N., Rubidge, B.S., 2023. Selected Karoo geoheritage sites of palaeontological significance in South Africa and Lesotho. Geol. Soc. London Spec. Publ. 543. https://doi.org/10.1144/SP543-2022-20.

Bowler, J.M., Jones, R., Allen, H., Thorne, A.G., 1970. Pleistocene human remains from Australia: a living site and human cremation from Lake Mungo, Western New South Wales. World Archaeol. 2 (1), 39–60.

Bravo-Cuevas, V., Rodríguez-de la Rosa, R., 2014. A summary of the Mesozoic vertebrate tracks of Mexico. In: Rivera-Sylva, H., Carpenter, K., Frey, E. (Eds.), Dinosaurs and Other Reptiles from the Mesozoic of Mexico. Indiana University Press, pp. 181–192.

Breithaupt, B.H., Southwell, E.H., Adams, T., Matthews, N.A., 2001. Innovative documentation methodologies in the study of the most extensive dinosaur tracksite in Wyoming. In: Santucci, V.L., McClelland, L. (Eds.), Proceedings of the 6th Fossil Research Conference Proceedings Volume. Geologic Resources Division Technical Report NPS/NRGRD/GRDTR-01/01, pp. 113–122.

Brilha, J., 2016. Inventory and quantitative assessment of geosites and geodiversity sites: a review. Geoheritage 8, 119–134.

Brocx, M., Semeniuk, V., 2007. Geoheritage and geoconservation – history, definition, scope and scale. J. Roy. Soc. West Aust. 90, 53–87.

Brusatte, S.L., Clark, N.D.L., 2015. Theropod dinosaurs from the middle Jurassic (Bajocian–Bathonian) of Skye, Scotland. Scot. J. Geol. 51, 157–164.

Bucheim, H.P., Leggitt, V.L., Biaggi, R.E., Loewen, M.A., 1999. Palaeoenvironments of avian trackway sites within the "Olla member," Palm springs formation, Anza-Borrego Desert State Park. San Bernardino County Museum Assoc. Q. 46 (2), 47–52, 57.

Burns, A., 2021. The Mesolithic footprints retained in one bed of the former saltmarshes at Formby point, Sefton Coast, North West England. In: Pastoors, A., Lenssen-Erz, T. (Eds.), Reading Prehistoric Tracks. Springer, Cham, pp. 295–316.

Bustos, D., Jakeway, J., Urban, T.M., Holliday, V.T., Fenerty, B., Raichlen, D.A., Budka, M., Reynolds, S.C., Allen, B.D., Love, D.W., Santucci, V.L., 2018. Footprints preserve terminal Pleistocene hunt? Human-sloth interactions in North America. Sci. Adv. 4 (4), eaar7621.

Bustos, D., Santucci, V.L., Odess, D., Martinez, P.J., Connelly, C.J., 2024. Incredible discoveries and devastation of paleontological resources in a changing world preserved at White Sands National Park. Parks Stewardship Forum 40 (1), 69–70.

Buta, R.J., Kopaska-Merkel, D.C., 2016. Footprints in Stone, Fossil Traces of Coal-Age Tetrapods. University of Alabama Press Tuscaloosa, USA.

Buta, R.J., Rindsberg, A.K., Kopaska-Merkel, D.C. (Eds.), 2005. Pennsylvanian Footprints in the Black Warrior Basin of Alabama. Alabama Paleontological Society Monograph, vol 1.

Butler, R., 2023. Condition, use and future of palaeontological heritage at Spyway and Horn Park quarries, Dorset. In: Matthews, J. (Ed.), XIth International ProGEO Symposium, 9–11th October 2023, Charnwood Forest, UK (Abstracts), p. 29.

Calvo, J.O., 1991. Huellas de Dinosaurios en la Formación Río Limay (Albiano-Cenomaniano), Picun Leufu, provincia de Neuquén, República Argentina. (Ornithischia-Saurischia: Sauropoda-Theropoda). Ameghiniana 28 (3–4), 241–258.

Calvo, J.O., Lockley, M.G., 2001. The first pterosaur tracks from Gondwana. Cretac. Res. 22, 585–590.

Calvo, J.O., Mazzeta, G.V., 2004. Nuevos hallazgos de huellas de dinosaurios en la Formación Candeleros (Albiano-Cenomaniano), Picún Leufú, Neuquén, Argentina. Ameghiniana 41 (4), 545–554.

Cariou, E., Olivier, N., Pittet, B., Mazin, J.M., Hantzpergue, P., 2014. Dinosaur track record on a shallow carbonate-dominated ramp (Loulle section, Late Jurassic, French Jura). Facies 60 (1), 229–253. https://doi.org/10.1007/s10347-013-0368-y.

Carvalho, I.S., Cunha, P.P., Figueiredo, S., 2022. Dinoturbation in upper Jurassic siliciclastic levels at Cabo Mondego (Lusitanian Basin, Portugal): evidences in a fluvial-dominated deltaic succession. Palaeoworld 31 (3), 455–477. https://doi.org/10.1016/j.palwor.2021.09.001.

Carvalho, I.S., Da Rosa, A.A.S., 2008. Paleontological tourism in Brasil: examples and discussion. Arq. Mus. Nac. 66 (1), 271–283.

Casamiquela, R.M., Fasola, A., 1968. Sobre pisadas de dinosaurios del Cretácico Inferior de Colchagua (Chile). Departamento de Geología 30, 1–24. Universidad de Chile.

Castanera, D., 2013. Aspectos paleoecológicos a partir del registro icnológico de tetrápodos en el intervalo Jurásico-Cretácico de la Cordillera Ibérica (Cameros oriental y Maestrazgo). In: Tesis Doctoral Inédita. Universidad de Zaragoza.

Castanera, D., Pascual, C., Canudo, J.I., Barco, J.L., 2018. Bringing together research, geoconservation and reaching a broad public in the form of a geotourism project: the ichnite route of Soria (Spain). Geoheritage 10, 393–403. https://doi.org/10.1007/s12371-017-0249-9.

Castanera, D., Silva, B.C., Santos, V.F., Malafaia, E., Belvedere, M., 2020. Tracking Late Jurassic ornithopods in the Lusitanian Basin of Portugal: ichnotaxonomic implications. Acta Palaeontol. Pol. 65 (2), 399–412.

Chabou, M.C., Laghouag, M.Y., Bendaoud, A., 2015. Dinosaur track sites in Algeria: a significant national geological heritage in danger. In: Errami, E., Brocx, M., Semeniuk, V. (Eds.), From Geoheritage to Geoparks, first ed. Springer, Cham, pp. 157–166.

Citton, P., Nicosia, U., Sacchi, E., 2015. Updating and reinterpreting the dinosaur track record of Italy. Palaeogeogr. Palaeoclimatol. Palaeoecol. 439, 117–125. https://doi.org/10.1016/j.palaeo.2015.01.018.

Clark, N.D.L., Booth, P., Booth, C., Ross, D.A., 2004. Dinosaur footprints from the Duntulm formation (Bathonian, Jurassic) of the Isle of Skye. Scot. J. Geol. 40, 13–21.

Cobos, A., Alcalá, L., 2017. Palaeontological heritage as a resource for promoting geotourism in the rural setting: el Castellar (Teruel, Spain). Geoheritage 10, 405–414.

Cobos, A., Alcalá, L., Royo-Torres, R., 2020. The dinosaur route in El Castellar (Teruel, Spain): Palaeontology as a factor of territorial development and scientific education in a sparsely inhabited area. Geoheritage 12, 52. https://doi.org/10.1007/s12371-020-00474-3.

Contessi, M., Fanti, F., 2012. Vertebrate tracksites in the Middle Jurassic-Upper Cretaceous of South Tunisia. Ichnos 19 (4), 211–227.

Conti, M.A., Leonardi, G., Mietto, P., 2020. The saga of tetrapod tracks, trackmakers and trackers in Italy. J. Mediterr. Earth Sci. 12, 13–28.

Conti, M.A., Morsilli, M., Nicosia, U., Sacchi, E., Savino, V., Wagensommer, A., Di Maggio, L., Gianolla, P., 2005. Jurassic dinosaur footprints from Southern Italy: footprints as indicators of constraints in paleogeographic interpretation. Palaios 20, 534–550. https://doi.org/10.2110/palo.2003.p03-99.

Coster, P., Legal, S., 2021. Oligocene trace fossil geoconservation and geoheritage: on the mammals tracks of the Luberon UNESCO Global Geopark. Geoconserv. Res. 4 (2), 613–620.

Cox, B.M., Sumbler, M.G., 2002. British Middle Jurassic stratigraphy. Geol. Conserv. Rev. Ser. 26, 508.

Curioni, G., 1870. Osservazioni geologiche sulla Val Trompia. Rend. Ist. Lomb. Sci. Lett. Arti Mem. 3, 1–60.

Dalla Vecchia, F.M., 1996. Archosaurian Trackways in the Upper Carnian of Dogna Valley. Natura Nascosta, pp. 5–17.

Dalla Vecchia, F., 1997. Terrestrial tetrapod evidence on the Norian (Late Triassic) and Cretaceous carbonate platforms of northern Adriatic region (Italy, Slovenia and Croatia). In: Proceedings of the International Symposium "Mesozoic Vertebrate Faunas of Central Europe". Sargetia, Acta Musei Devensis, Series Scientia Naturae, vol 17, pp. 177–201.

Dalla Vecchia, F.M., 1998. Impronte di rettili terrestri nella Dolomia Principale (Triassico superiore) delle Prealpi Carniche (Pordenone, Friuli). Atti Tic. Sci. Terra 7, 87–107.

Dalla Vecchia, F.M., Rustioni, M., 1996. Mammalian trackways in the Conglomerato di Osoppo (Udine, NE Italy) and their contribution to its age determination. Mem. Sci. Geol. Padova 48, 221–232.

Day, J., Norman, D.B., Gale, A., Powell, H., 2004. A Middle Jurassic dinosaur trackway site from Oxfordshire, UK. Palaeontology 47 (2), 319–348. https://doi.org/10.1111/j.0031-0239.2004.00366.x.

Demathieu, G., Weidmann, M., 1982. Les empreintes de pas de reptiles dans le Trias du Vieux Émosson (Finhaut, Valais, Suisse). Eclogae Geol. Helv. 75, 721–757.

De Klerk, W.J., 2002. A dicynodont trackway from the *Cistecephalus* assemblage zone in the Karoo, east of Graaff-Reinet, South Africa. Palaeontol. Afr. 38, 73–91.

De Valais, S., 2011. Revision of dinosaur ichnotaxa from the La Matilde formation (Middle Jurassic), Santa Cruz province, Argentina. Ameghiniana 48 (1), 28–42.

De Wever, P., Alterio, I., Egoroff, G., Cornée, A., Bobrowsky, P., Collin, G., et al., 2015. Geoheritage, a national inventory in France. Geoheritage 7 (3), 205–247. https://doi.org/10.1007/s12371-015-0151-2.

Diedrich, C., 2002. Die Wirbeltierfundstelle Borgholzhausen (Teutoburger Wald, NW-Deutschland) aus der Oolith-Zone (Unterer Muschelkalk, Mitteltrias). Paläontol. Z. 76 (1), 35–56.

Diedrich, C.G., 2010. Silbersee im Hüggel. Eine Wattwanderung im Mittelgebirge. Natur- und Geopark TERRA.vita - Nördlicher Teutoburger Wald, Wiehengebirge, Osnabrücker Land e. V., Osnabrück, p. 22.

Diedrich, C.G., 2011. An overview of the ichnological and ethological studies in the cave bear den in Urşilor Cave (Western Carpathians, Romania). Ichnos 18 (1), 9–26.

Díaz-Martínez, I., 2013. Icnitas de dinosaurios bípedos de La Rioja (Cuenca de Cameros, Cretácico Inferior): Icnotaxonomía y aplicación paleobiológica. In: Tesis doctoral inédita, Universidad de La Rioja.

DinoMITI, 2013. The exhibition of fossil reptiles of the UNESCO Dolomites.

Dowling, R., Newsome, D. (Eds.), 2005. Geotourism. Elsevier, Oxford, p. 290.

Dugés, A., 1894. Felis fósil de San Juan de los Lagos. La Naturaleza 2, 412–423.

Eberth, D.A., Berman, D.S., Sumida, S.S., Hopf, H., 2000. Lower Permian terrestrial paleoenvironments and vertebrate paleoecology of the Tambach Basin (Thuringia, central Germany): the upland holy grail. Palaios 15, 293–313.

Ellenberger, P., 1970. Les niveaus paleontologiques de premiere apparition des Mammiferes primordiaux en Afrique du Sud et leur ichnologie. Etablissement de zones stratigraphiques detaillees dans le Stormberg du Lesotho (Afrique du Sud) (Trias superieur a Jurassique). In: Haughton, S.H. (Ed.), I.U.G.S. 2nd Symposium on Gondwana Stratigraphy and Palaeontology. Pretoria, Council for Scientific and Industrial Research, Pretoria, South Africa, pp. 343–370.

Ellenberger, P., 1974. Contribution a la classification des pistes de vertebres du Trias: Les types du Stormberg d'Afrique du Sud (II): Le Stormberg Superior. In: Palaeovertebrata Memoir Extraordinaire. Laboratoire de Paleontologie des Vertebres, Montpellier.

Ellis, N.V., Bowen, D.Q., Campbell, S., Knill, J.L., McKirdy, A.P., Prosser, C.D., Vincent, M.A., Wilson, R.C.L., 1996. An introduction to the geological conservation review. Geol. Conserv. Rev. Ser. 1, 131.

Endere, M.L., Prado, J.L., 2015. Characterization and valuation of paleontological heritage: a perspective from Argentina. Geoheritage 7, 137–145. https://doi.org/10.1007/s12371-014-0124-x.

Ensom, P.C., Delair, J.B., 2007. Dinosaur tracks from the lower Purbeck strata of Portland, Dorset, southern England. Geosci. South-West Engl. 11, 309–325.

Eves, R.L., Pollock, G.L., Davis, L.E., 2012. Dinosaur tracks and trackways in the Escalante member of the Entrada formation (Middle Jurassic), Twentymile Wash, Grand Staircase-Escalante national monument, Utah, the compass. Earth Sci. 84 (3), 4. https://digitalcommons.csbsju.edu/compass/vol84/iss3/4.

Fanti, F., Contessi, C., Nigarov, A., Esenov, P., 2013. New data on two large dinosaur tracksites from the Upper Jurassic of Eastern Turkmenistan (Central Asia). Ichnos 20 (2), 54–71.

Fara, E., Moreau, J.-D., Szewczyk, L., Klee, N., Gand, G., Vennin, E., 2021. Inventory and conservation of Triassic vertebrate tracks in the Monts d'Ardèche UNESCO Global Geopark, France. Geoconserv. Res. 4 (2), 317–328.

Farlow, J.O., 1987. Lower Cretaceous Dinosaur Tracks, Paluxy River Valley, Texas. South- Central Section. Geological Society of America, Baylor University, pp. 1–50.

Farlow, J.O., Pittman, J.G., Hawthorne, J.M., 1989. *Brontopodus birdi*, Lower Cretaceous sauropod footprints from the U.S. Gulf coastal plain. In: Gillette, D.D., Lockley, M.G. (Eds.), Dinosaur Tracks and Traces. Cambridge University Press, Cambridge, pp. 371–394.

Farlow, J.O., Bates, K.T., Bonem, R.M., Dattilo, B.F., Falkingham, P.L., Gildner, R., Jacene, J., 2015. Dinosaur Footprints from the Glen Rose Formation (Paluxy River, Dinosaur Valley State Park, Somervell County, Texas). Early-and Mid-Cretaceous Archosaur Localities of North-Central Texas. Field Trip Guidebook, pp. 14–37.

Farlow, J.O., O'Brien, M., Kuban, G.J., Dattilo, B.F., Bates, K.T., Falkingham, P.L., Piñuela, L., 2012. Dinosaur tracksites of the Paluxy River valley (Glen Rose Formation, Lower Cretaceous), Dinosaur Valley State Park, Somervell county, Texas. In: Proceedings of the V International Symposium about Dinosaur Palaeontology and Their Environment, Burgos, pp. 41–69.

Farsani, N.T., Coelho, C., Costa, C., Neto de Carvalho, C., 2011. Geoparks and geotourism: concepts, theories and paradigms. In: Farsani, N.T., Coelho, C., Costa, C., Neto de Carvalho, C. (Eds.), Geoparks & Geotourism – New Approaches to Sustainability for the 21st Century. Brown Walker Press, Florida, pp. 5–60.

FCPTD (Fundación Conjunto Paleontológico de Teruel-Dinópolis), coord, 2009. Dinosaur ichnites of the Iberian Peninsula. In: World Heritage Candidacy. Ministerio de Cultura de España and Ministerio do Ambiente, do Ordenamento do Territorio e do Desenvolvimiento Regional, Portugal, p. 460.

Feibel, C.S., Agnew, N., Latimer, B., Demas, M., Marshall, F., Waane, S.A.C., Schmid, P., 1995. The Laetoli hominid footprints—a preliminary report on the conservation and scientific restudy. Evol. Anthropol. Issues News Rev. 4 (5), 149–154. https://doi.org/10.1002/evan.1360040502.

Fichter, J., 1994. Permische Saurierfährten. Ein Diskussionsbeitrag zu der Bearbeitungsproblematik der Tetrapodenfährten des Cornberger Sandsteins (Perm, Deutschland) und des Coconino Sandsteins (Perm, USA). Philippia 7 (1), 61–82.

Figueiredo, S., Dinis, P., Belo, J., Rosina, P., Strantzali, I.B., 2017. A new record of a possible ornithopod footprint from the Lower Cretaceous of Cabo Espichel (Sesimbra, Portugal). Boll. Soc. Paleontol. Ital. 56 (2), 217–231.

Fischer, R., 1998. Die Saurierfährten im Naturdenkmal Münchehagen. Mitt. Geol. Inst. Univ. Hannover 37, 3–59.

Fischer, T., 2022. Valuable Landscapes and Landscape Elements in the Nature and UNESCO Global Geopark TERRA.vita. Natur- und Geopark TERRA.vita - Nördlicher Teutoburger Wald, Wiehengebirge, Osnabrücker Land e. V., Osnabrück, p. 196.

Fischer, T., Chellouche, P., Lockley, M.G., Meyer, C.A., 2021. The dinosaur tracks of Bad Essen-Barkhausen in the UNESCO Global Geopark TERRA.vita (NW Germany): 100 years of development from an industrial quarry to an open-air museum. Geoconserv. Res. 4 (2), 378–388.

Fischer, T., Koch, R., 2022. Exkursion durch das Osnabrücker Bergland (UNESCO Global Geopark TERRA.vita, NW Deutschland). In: ARGE NDG, LBEG, Geologischer Dienst NRW (Eds.), 82. Tagung der Arbeitsgemeinschaft Norddeutscher Geologen vom 7. bis 10. Juni 2022 in Osnabrück. Vortragskurzfassungen, Posterpräsentationen, Exkursionsführer. Geologischer Dienst NRW, Krefeld, pp. 116–144.

Foster, J.R., 2015. Theropod dinosaur ichnogenus *Hispanosauropus* identified from the Morrison Formation (Upper Jurassic), western North America. Ichnos 22 (3–4), 183–191.

Foster, J.R., Hamblin, A.H., Lockley, M.G., 2000. The oldest evidence of a sauropod dinosaur in the western United States and other important vertebrate trackways from Grand Staircase-Escalante National Monument, Utah. Ichnos 7 (3), 169–181.

Foster, J.R., Lockley, M.G., Lucas, S.G., 2006. The vertebrate ichnological record of the Morrison Formation (Upper Jurassic, north America). New Mexico Museum Nat. History Sci. Bull. 36, 203–216.

Francischini, H., Fernandes, M.A., Kunzler, J., Rodrigues, R., Leonardi, G., Carvalho, I.S., 2020. The ichnological record of Araraquara sidewalks: history, conservation, and perspectives from this urban paleontological heritage of southeastern Brazil. Geoheritage 12, 50. https://doi.org/10.1007/s12371-020-00472-5.

Galopim de Carvalho, A.M., 1994. Dinossáurios e a Batalha de Carenque. In: Editorial Notícias, Lisboa, p. 292.

Gamkrelidze, I., Okrostsvaridze, A., Koiava, K., Maisadze, F., 2021. Geotourism Potential of Georgia, the Caucasus. Springer.

García-Ortiz, E., 2016. Análisis de los yacimientos de icnitas de La Rioja (N de España) como recurso patrimonial y aplicación de nuevas tecnologías a su estudio. Tesis doctoral inédita. Universidad de León.

Getty, P.R., Judge, A., Csonka, J., Bush, A.M., 2012. Were Early Jurassic dinosaurs gregarious? Reexamining the evidence from dinosaur footprint reservation in Holyoke, Massachusetts. In: Thomas, M.A. (Ed.), Guidebook for Fieldtrips in Connecticut and Massachussets, Geological Society of America Northeast Section, 47th Annual Meeting, Hartford, Connecticut, State Geological and Natural History of Connecticut Guidebook, vol. 9, pp. A1–A18.

Getty, P.R., Hardy, L., Bush, A.M., 2015. Was the *Eubrontes* track maker gregarious? Testing the herding hypothesis at powder hill dinosaur park, Middlefield, Connecticut. Bull. Peabody Mus. Nat. Hist. 56, 95–106.

Gierliński, G.D., Menducki, P., Janiszewska, K., Wicik, I., Boczarowski, A., 2009. A preliminary report on dinosaur track assemblages from the Middle Jurassic of the Imilchil area, Morocco. Geol. Q. 53 (4), 477–482.

Gierliński, G., Niedźwiedzki, G., 2005. New saurischian dinosaur footprints from the Lower Jurassic of Poland. Geol. Q. 49, 99–104.

Gierliński, G.D., Niedźwiedzki, G., Lockley, M.G., Athanassiou, A., Fassoulas, C., Dubicka, Z., Boczarowski, A., Bennett, M.R., Ahlberg, P.E., 2017. Possible hominin footprints from the late Miocene (c. 5.7 Ma) of Crete? PGA (Proc. Geol. Assoc.) 128 (5–6), 697–710.

Gierliński, G., Pieńkowski, G., 1999. Dinosaur track assemblages from the Hettangian of Poland. Geol. Q. 43, 329–346.

Gillette, D.D., Thomas, D.A., 1985. Dinosaur tracks in the Dakota Formation (Aptian-Albian) at Clayton Lake State Park, Union County, New Mexico. New Mexico Geol. Soc. Guidebook 36, 283–288.

Gillette, D.D., Thomas, D.A., 1989. Problematic tracks and traces of late Albian (early Cretaceous) age, Clayton Lake State dinosaur tracks in the Dakota Formation (Aptian-Albian) at Clayton Lake State Park, Union County, New Mexico, USA. In: Gillette, D.D., Lockley, M.G. (Eds.), Dinosaur Tracks and Traces. Cambridge University Press, Cambridge, pp. 337–342.

Gillette, D.D., Lockley, M.G. (Eds.), 1989. Dinosaur Tracks and Traces. Cambridge University Press, Cambridge, U.K., pp. 1–454

Gilmore, C.W., 1927. Fossil footprints from the Grand Canyon. Smithson. Misc. Collect. 77 (9), 1–39.

Gomes, J.P., 1916. Descoberta de rastos de saurios gigantescos no Jurássico do Cabo Mondego. Comun. dos Serv. Geol. Port. 11, 132–134.

Goodell, Z.G., Lockley, M.G., Lucas, S.G., Schumacher, B.A., Smith, J.A., Trujillo, R., Xing, L., 2021. A high-altitude sauropod trackway site in the Jurassic of Colorado: the longest known consecutive footprint sequence reveals evidence of sharp turning behavior. In: Lucas, S.G., Hunt, A.P., Lichtig, A.J. (Eds.), Fossil Record 7. New Mexico Museum of Natural History and Science Bulletin 84.

Gray, M., 2013. Geodiversity: Valuing and Conserving Abiotic Nature, second ed. Wiley Blackwell, Chichester, UK.

Grellet-Tinner, G., Fiorelli, L., 2010. A new Argentinean nesting site showing neosauropod dinosaur reproduction in a Cretaceous hydrothermal environment. Nat. Commun. 1, 32. https://doi.org/10.1038/ncomms1031.

Grzegorczyk, D., 2005. In die Denkmallisten eingetragene paläontologische Bodendenkmäler in Westfalen-Lippe (Stand 2004). Geol. Paläontol. Westfalen 63, 51–77.

Hamblin, A.H., Bilbey, S.A., 1999. A dinosaur track site in the Navajo-Nugget sandstone, red Fleet reservoir, Uintah County, Utah. In: Gillette, D.D. (Ed.), Vertebrate Paleontology in Utah, vols. 99–1. Utah Geological Survey Miscellaneous Publication, pp. 51–57.

Hamblin, A.H., Bilbey, S.A., Hall, J.E., 2000. Prehistoric animal tracks at Red Fleet State Park, Northeastern Utah. In: Sprinkel, D.A., Chidsey Jr, T.C., Anderson, P.B. (Eds.), Geology of Utah's Parks and Monuments, vol 28. Utah Geological Association Publication, pp. 1–10.

Hatala, K.G., Harcourt-Smith, W.E., Gordon, A.D., Zimmer, B.W., Richmond, B.G., Pobiner, B.L., Green, D.J., Metallo, A., Rossi, V., Liutkus-Pierce, C.M., 2020. Snapshots of human anatomy, locomotion, and behavior from Late Pleistocene footprints at Engare Sero, Tanzania. Sci. Rep. 10, 7740. https://doi.org/10.1038/s41598-020-64095-0.

Hatala, K., Roach, N., Ostrofsky, K., Wunderlich, R.E., Dingwall, H.L., Villmoare, B., Green, D.J., Harris, J.W.K., Braun, D.R., Richmond, B.G., 2016. Footprints reveal direct evidence of group behavior and locomotion in *Homo erectus*. Sci. Rep. 6, 28766. https://doi.org/10.1038/srep28766.

Haubold, H., 2006. Die Saurierfährten *Chirotherium barthii* KAUP, 1835 – das Typusmaterial aus dem Buntsandstein bei Hildburghausen/Thüringen und das "Chirotherium-Monument". Veröffentlichungen Naturhistorisches Museum Schleusingen 21, 3–31.

Haubold, H., Hunt, A.P., Lucas, S.G., Lockley, M.G., 1995. Wolfcampian (Early Permian) vertebrate tracks from Arizona and New Mexico. NMMNH&S 6, 135–165.

Haynes, C.V., 2022. Evidence for humans at White Sands National Park during the last glacial maximum could actually be for Clovis people ~13,000 years ago. PaleoAmerica 8 (2), 95–98.

Helm, C.W., 2023. Pleistocene Vertebrate Trace Fossils from the Cape South Coast of South Africa: Inferences and Implications. Ph.D. thesis. Nelson Mandela University, p. 400. http://hdl.handle.net/10948/60589.

Helm, C.W., Cawthra, H.C., De Vynck, J.C., Helm, C.J.Z., Rust, R., Stear, W., 2019. Patterns in the sand: a Pleistocene hominin signature along the South African coastline? PGA (Proc. Geol. Assoc.) 130 (6), 719–740. https://doi.org/10.1016/j.pgeola.201b9.08.004.

Helm, C.W., Cawthra, H.C., Cowling, R.M., De Vynck, J.C., Lockley, M.G., Marean, C.W., Dixon, M.G., Helm, C.J.Z., Stear, W., Thesen, G.H.H., Venter, J.A., 2021. Protecting and preserving South African aeolianite surfaces from graffiti. Koedoe 63 (1), a1656. https://doi.org/10.4102/koedoe.v63i1.1656.

Helm, C.W., Cawthra, H.C., De Vynck, J.C., Helm, C.J.Z., Rust, R., W. Stear, W., 2024. Rescuing ancient palaeoart in coastal South Africa. In: Kumar, G. (Ed.), Study of Palaeoart of the World: A Quest for Understanding the Evolution of Human Constructs of Reality. Pathak Publisher and Distributors, New Delhi, pp. 359–366.
Hitchcock, E., 1836. Ornithichnology—description of the foot marks of birds, (Ornithichnites) on New Red Sandstone in Massachusetts. Am. J. Sci. 29, 307–340.
Hornung, J.J., Böhme, A., Reich, M., 2012a. Berriasian–Valanginian (Early Cretaceous) vertebrate faunas of northwestern Germany. Centenary Meeting of the Paläontologische Gesellschaft. Programme, Abstracts and Field Guides – Terra Nostra 82–83.
Hornung, J.J., Böhme, A., van der Lubbe, T., Reich, M., Richter, A., 2012b. Vertebrate tracksites in the Obernkirchen Sandstone (late Berriasian, Early Cretaceous) of northwest Germany — their stratigraphical, palaeogeographical, palaeoecological, and historical context. Paläontol. Z. 86 (3), 231–267.
Hornung, J.J., Reich, M., 2013. The first record of the pterosaur ichnogenus *Purbeckopus* in the late Berriasian (early Cretaceous) of northwest Germany. Ichnos 20, 164–172.
Hose, T.A., 1995. Selling the story of Britain's stone. Environ. Interpretation 10 (2), 16–17.
Huang, H.Y., 2012. Background analysis of the birth of legalization of antiques preservation (1911–1930). J. Nat Hist. Acad. 1–44.
Hübner, T., 2016. Elusive Ornithischian tracks in the famous Berriasian (Lower Cretaceous) Chicken Yard tracksite of Northern Germany: quantitative differentiation between small tridactyl trackmakers. In: Falkingham, P.L., Marty, D., Richter, A. (Eds.), Dinosaur Tracks. The Next Steps. Indiana University Press, Indiana, pp. 311–330.
Huh, M., Paik, I.S., Lockley, M.G., Hwang, K.G., Kim, B.S., Kwak, S.K., 2006. Well-preserved theropod tracks from the Upper Cretaceous of Hwasun County, southwestern South Korea, and their paleobiological implications. Cretac. Res. 27, 123–138.
Hunt, A.P., Lockley, M.G., Lucas, S.G., MacDonald, J.P., Hotton III, N., Kramer, J., 1993. Early Permian tracksites in the Robledo mountains, south central New Mexico. In: Lucas, S.G., Zidek, J. (Eds.), Vertebrate Paleontology in New Mexico: New Mexico Museum of Natural History and Science Bulletin, vol 2, pp. 23–31.
Hunt, A.P., Lucas, S.G., 1998. Tetrapod ichnofaunas from the Lower Cretaceous of northeastern New Mexico. NMMNH&S 14, 163–167.
Hunt, A.P., Lucas, S.G., 2004. Multiple parallel dinosaur tail drags from the Early Cretaceous of New Mexico. J. Vertebr. Paleontol. 24 (Suppl. to no. 3), 73A.
Hunt, A.P., Lucas, S.G., 2006. The significance of the St. George dinosaur discovery site at Johnson farm: a Jurassic Konzentrat-Ichnolagerstätte, Utah, USA. In: Harris, J.D., Lucas, S.G., Spielmann, J.A., Lockley, M.G., Milner, A.R.C., Kirkland, J.I. (Eds.), The Triassic-Jurassic Terrestrial Transition, vol 37. New Mexico Museum of Natural History and Science Bulletin, pp. 282–288.
Hunt, A.P., Lucas, 2022. Coprolites in caves: Late Pleistocene coprofaunas of the American Southwest and their significance. NMMNH&S 88, 343–359.
Hunt, A.P., Milàn, J., Lucas, S.G., Spielmann, J.A. (Eds.), 2012a. Vertebrate Coprolites: New Mexico Museum of Natural History and Science Bulletin, vol 57, p. 387.
Hunt, A.P., Lucas, S.G., Pyenson, N.D., 2005. The significance of the Union Chapel mine site: a lower Pennsylvanian (Westphalean A) ichnological Konzentrat-lagerstätte, Alabama, USA. In: Buta, R.J., Rindsberg, A.K., Kopaska-Merkel, D.C. (Eds.), Pennsylvanian Footprints in the Black Warrior Basin of Alabama: Alabama Paleontological Society Monograph, vol. 1. Alabama Paleontological Society, pp. 3–14.
Hunt, A.P., Santucci, V.L., Tweet, J.S., Lucas, S.G., 2012b. Vertebrate coprolites and other bromalites in National Park Service areas. In: Hunt, A.P., Milàn, J., Lucas, S.G., Spielmann, J.A. (Eds.), Vertebrate Coprolites: New Mexico Museum of Natural History and Science Bulletin, vol 57, pp. 343–353.
Hurum, J.H., Milàn, J., Hammer, Ø., Midtkandal, I., Amundsen, H., Sæther, B., 2006. Tracking polar dinosaurs - new finds from the Lower Cretaceous of Svalbard. Norw. J. Geol. 86, 397–402.
ICCROM, 2023. New Project to Preserve 1.5-Million-Year-Old Human Footprint Fossil in Kenya. www.iccrom.org.
Ishigaki, S., Watabe, M., Tsogtbaatar, K., Saneyoshi, M., 2009. Dinosaur footprints from the Upper Cretaceous of Mongolia. Geol. Q. 53 (4), 449–460.
Jung, J., Huh, M., Unwin, D.M., Smyth, R.S.H., Hwang, K.-G., Kim, H.-J., Choi, B.D., Xing, L., 2022. Evidence for a mixed-age group in a pterosaur footprint assemblage from the early Upper Cretaceous of Korea. Sci. Rep. 12, 10707. https://doi.org/10.1038/s41598-022-14966-5.

Kaup, J.J., 1835. Über Thierfährten bei Hildburghausen. Neues Jahrbuch für Mineralogie, Geogonosie, Geologie und Petrefaktenkunde 1835, 227–228.

Khalaf-von Jaffa, N.A.B., 2014. The ceratosaur dinosaur *Elaphrosaurus bambergi* Janensch, 1920 tracks from Beit Zeit, West of Al- Quds (Jerusalem), occupied Palestine. Gazelle 117, 1–33.

Kim, J.Y., Kim, K.S., Lockley, M.G., Matthews, N., 2008. Hominid ichnotaxonomy: an exploration of a neglected discipline. Ichnos 15 (3–4), 126–139. https://doi.org/10.1080/10420940802467868.

Kirscher, U., Atfy, H., Gärtner, A., Dallanave, E., et al., 2021. Age constraints for the Trachilos footprints from Crete. Sci. Rep. 11, 19427.

Klein, H., 2020. Tetrapodenfährten des Muschelkalks. In: Deutsche Stratigraphische Kommission (Ed.), Stratigraphie von Deutschland XIII. Muschelkalk. Teil 1. Schriftenreihe der Deutschen Gesellschaft für Geowissenschaften, 91 (1), pp. 226–235.

Klein, H., Gierliński, G.D., Oukassou, M., Saber, H., Lallensack, J.N., Lagnaoui, A., Hmmina, A., Charrière, A., 2023. Theropod and ornithischian dinosaur track assemblages from Middle to? Late jurassic deposits of the central high Atlas, Morocco. Hist. Biol. 35 (3), 320–346.

Klein, H., Lucas, S.G., 2010. Review of the tetrapod ichnofauna of the Moenkopi Formation/Group (Early-Middle Triassic) of the American southwest. NMMNH&S 50, 67.

Klein, H., Niedźwiedzki, G., 2012. Revision of the Lower Triassic tetrapod ichnofauna from Wióry, Holy Cross Mountains, Poland. NMMNH&S 56, 1–62.

Klein, H., Wizevich, M.C., Thüring, B., Marty, D., Thüring, S., Falkingham, P., Meyer, C.A., 2016. Triassic chirotheriid footprints from the Swiss Alps: ichnotaxonomy and depositional environment (cantons Wallis & Glarus). Swiss J. Palaeontol. 135, 295–314.

Kordos, L., Meszaros, I., Szarvas, I., 2021. Tracking a "prehistoric Pompeii" , Rhinoland and Crocodilia: new discoveries and interpretations of Ipolytarnóc (N Hungary) lower Miocene track site. Geoconserv. Res. 4 (2), 621–634.

Krapovickas, V., Garrido, A.C., Canale, J.I., 2020. Dinosaur tracks at Cañadón de las Campanas and Villa El Chocón area (Late Cretaceous, Candeleros Formation), Patagonia, Argentina. Rev. Asoc. Geol. Argent. 77 (3), 447–462.

Kustatscher, E., Nowak, H., Corradini, C., 2019. Tra terra e mare – le Alpi Carniche 300 milioni di anni fa / Zwischen Land und Meer – die Karnischen Alpen vor 300 Millionen Jahren. – Naturmuseum Südtirol. Bozen, p. 80.

Lagally, U., Rohrmüller, J., Glaser, S., Loth, G., Pürner, T., 2011. In: Hundert Meisterwerke – die schönsten Geotope Bayerns. Bayerisches Landesamt für Umwelt, Augsburg, p. 288.

Lallensack, J.N., Sander, P.M., Knötschke, N., Wings, O., 2015. Dinosaur tracks from the Langenberg Quarry (Late Jurassic, Germany) reconstructed with historical photogrammetry: evidence for large theropods soon after insular dwarfism. Palaeontol. Electron. 18 (2.31A), 1–34.

Laws, E., Scott, N., 2003. Developing new tourism services: dinosaurs, a new drive tourism resource for remote regions? J. Vacat. Mark. 9, 368–380.

Leakey, M.D., Harris, J.M. (Eds.), 1987. Laetoli: A Pliocene Site in Northern Tanzania. Clarendon Press, Oxford., p. 561

Leakey, M.D., Hay, R.L., 1979. Pliocene footprints in the Laetoli beds at Laetoli, northern Tanzania. Nature 278, 317–323.

Lee, J.H., Lockley, M.G., Jung, S., Ha, Y., Cho, K.R., Lim, J.D., Kim, K.S., 2023. Dinosaur tracksites from the Cretaceous Sagok Formation in the Uiseong Regional Geopark, South Korea: historical, palaeobiological perspectives. Hist. Biol. 35 (9), 1656–1664. https://doi.org/10.1080/08912963.2022.2106563.

Le Loeuff, J., Gourrat, C., Landry, P., Hautier, L., Liard, R., Souillat, C., Buffetaut, E., Enay, R.A., 2006. Late jurassic sauropod track site from southern Jura (France). C. R. Palevol 5, 705–709.

Le Loeuff, J., Saenyamoon, T., Souillat, C., Suteethorn, V., Buffetaut, E., 2009. Mesozoic vertebrate footprints of Thailand and Laos. Geol. Soc. Spec. Publ. 315, 245–254. https://doi.org/10.1144/SP315.17.

Leonardi, G., 2008. Vertebrate ichnology in Italy. Acta Geol. 83, 213–221.

Leonardi, G., Carvalho, I.S., 2002. Icnofósseis da Bacia do Rio do Peixe, PB - O mais marcante registro de pegadas de dinossauros do Brasil. In: Sítios Geológicos e Paleontológicos do Brasil. Ministério das Minas e Energia - Governo Federal do Brasil, Brasília, vol 1, pp. 101–111.

Leonardi, G., Carvalho, I.S., 2021. Dinosaur Tracks from Brazil: A Lost World of Gondwana. Indiana University Press, p. 462.

Leonardi, G., Mietto, P., 2000. Dinosauri in Italia, le piste giurassiche dei Lavini di Marco (Trentino) e gli altri resti fossili italiani. Accademia Editoriale, Pisa-Roma.

Leonardi, G., Sarjeant, W.A.S., 1986. Footprints representing a new Mesozoic vertebrate fauna from Brazil. Mod. Geol. 10, 73–84.

Liutkus-Pierce, C.M., Zimmer, B.W., Carmichael, S.K., McIntosh, W., Deino, A., Hewitt, S.M., McGinnis, K.J., Hartney, T., Brett, J., Mana, S., Deocampo, D., 2016. Radioisotopic age, formation, and preservation of Late Pleistocene human footprints at Engare Sero, Tanzania. Palaeogeogr. Palaeoclimatol. Palaeoecol. 463, 68–82. https://doi.org/10.1016/j.palaeo.2016.09.019.

Lkebir, N., Masrour, M., Pascual-Arribas, C., de Ducla, M., Hernández-Medrano, N., Pérez-Lorente, F., 2020. Short report: the Anza tracksite: ichnological heritage and geoeducational significance of dinosaur and pterosaur tracks in coastal Morocco. J. Afr. Earth Sci. 171, 103949.

Lockley, M.G., 1991. The Moab Megatracksite: a preliminary description and discussion of millions of Middle Jurassic tracks in Eastern Utah. In: Guidebook for Dinosaur Quarries and Tracksites Tour: Western Colorado and Eastern Utah, 1991 Grand Junction Geological Society, pp. 59–65.

Lockley, M.G., 2001. A field guide to Dinosaur Ridge. In: Friends of Dinosaur Ridge and University of Colorado at Denver Dinosaur Trackers Research Group, Fully revised third ed., p. 34.

Lockley, M.G., 2003. Fossil footprints of the dinosaur Ridge and fossil trace areas. In: Friends of Dinosaur Ridge, Morrison Colorado, p. 66.

Lockley, M.G., 2022. The distribution of theropod-dominated ichnofaunas in the Moab Megatracksite area, Utah: implications for Late Jurassic palaeobiology along an arid coast. Hist. Biol. 34 (9), 1717–1751.

Lockley, M.G., Garcia-Vasquez, R., Espinoza, E., Lucas, S.G., 2007. Notes on a famous but 'forgotten' human footprint site from the Holocene of Nicaragua. NMMNH&S 42, 97–102.

Lockley, M.G., Garcia-Vasquez, R., Espinoza, E., Lucas, S.G., 2009. America's most famous hominid footprints: history and context of tracks from Nicaragua. Ichnos 16, 55–69.

Lockley, M.G., Gierlinski, G.D., Dubicka, Z., Breithaupt, B.H., Matthews, N.A., 2014a. A preliminary report on a new dinosaur tracksite in the Cedar Mountain Formation (Cretaceous) of eastern Utah. NMMNH&S 62, 279–285.

Lockley, M.G., Gierlinski, G.D., Houck, K., Lim, J.D., Kim, K.S., Kim, D.Y., Kim, T.K., Kang, S.H., Hunt-Foster, R., Li, R., Chesser, C., 2014b. New excavations at the Mill canyon dinosaur track site (Cedar Mountain formation, lower Cretaceous) of eastern Utah. NMMNH&S 62, 287–300.

Lockley, M.G., Houck, K.J., Prince, N.K., 1986. North America's largest dinosaur trackway site: implications for Morrison Formation paleoecology. Geol. Soc. Am. Bull. 97 (10), 1163–1176.

Lockley, M.G., Houck, K., Yang, S.Y., Matsukawa, M., Lim, S.-K., 2006. Dinosaur-dominated footprint assemblages from the Cretaceous Jindong Formation, Hallyo Haesang national park area, Goseong county, South Korea: evidence and implications. Cretac. Res. 27 (1), 70–101.

Lockley, M.G., Huh, M., Kim, S., 2012. *Ornithopodichnus* and pes-only sauropod trackways from the Hwasun tracksite, Cretaceous of Korea. Hist. Biol. 19 (1), 93–100.

Lockley, M.G., Hunt, A.P., 1994. Fossil Footprints of the Dinosaur Ridge Area. Friends of Dinosaur Ridge, University of Colorado at Denver Dinosaur Trackers Research Group and Morrison Museum of Natural History, p. 53.

Lockley, M.G., Hunt, A.P., 1995a. Dinosaur Tracks and Other Fossil Footprints of the Western United States. Columbia University Press, New York, p. 338.

Lockley, M.G., Hunt, A.P., 1995b. The first ceratopsid tracks and associated ichnofauna from the Laramie Formation (Upper Cretaceous: Maastrichtian) of Colorado. J. Vertebr. Paleontol. 15, 592–614.

Lockley, M., Hunt, A.P., Holbrook, J., Matsukawa, M., Meyer, C., 1992. The dinosaur freeway: a preliminary report on the Cretaceous megatracksite, Dakota Group, Rocky Mountain front range, and high plains, Colorado, Oklahoma and New Mexico. In: Flores, R.M. (Ed.), Field Guidebook, Mesozoic of the Western Interior: SEPM 1992 Theme Meeting: Fort Collins, Rocky Mountain Section. SEPM (Society for Sedimentary Geology), pp. 39–54.

Lockley, M.G., Hunt, A.P., Meyer, C., Rainsforth, E.C., Schultz, R.J., 1998. A survey of fossil footprint sites at Glen Canyon National Recreation Area (Western USA): a case study in documentation of trace fossil resources at a National Reserve. Ichnos 5, 177–211.

Lockley, M.G., Lucas, S.G. (Eds.), 2014. Fossil Footprints of Western North America. New Mexico Museum of Natural History and Science Bulletin, vol 62, p. 508.

Lockley, M.G., Lucas, S.G., Hunt, A.P., 2000. Dinosaur tracksites in New Mexico: a review. NMMNH&S 17, 9–16.

Lockley, M.G., Lucas, S.G., Hunt, A.P., Garcia Vasquez, R., Espinoza, E., 2008. Non-hominid vertebrate tracks from the Holocene of Nicaragua. NMMNH&S 44, 393–397.

Lockley, M., Meyer, C., 2000. Dinosaur Tracks and Other Fossil Footprints of Europe. Columbia University Press, New York, p. 323.
Lockley, M.G., Plint, A.G., Helm, C.W., Sharman, K.J., Vannelli, K.M., 2021. Crocodylian swim tracks from the Gates Formation (Albian), British Columbia, Canada: comparisons with Cretaceous crocodilian ichnology in western USA. Cretac. Res. 128, 104967. https://doi.org/10.1016/j.cretres.2021.104967.
Lockley, M.G., Schumacher, B.A., 2021. The process of ranking the geo-heritage 'values' of dinosaur tracksites in the USA and globally. NMMNH&S 82, 177–184.
Lockley, M.G., Wright, J.L., Matsukawa, M., 2001. A new look at *Magnoavipes* and so-called big bird tracks from Dinosaur Ridge (Cretaceous, Colorado). Mt. Geol. 38 (3), 137–146.
Lockwood, J.A.F., Lockley, M.G., Pond, S., 2014. A review of footprints from the Wessex formation (Wealden group, lower Cretaceous) at Hanover point, the Isle of Wight, southern England. Biol. J. Linn. Soc. 113, 707–720.
Lucas, S.G., Allen, B.D., Morgan, G.S., Myers, R.G., Love, D.W., Bustos, D., 2007a. Mammoth footprints from the upper Pleistocene of the Tularosa basin, Doña Ana county, New Mexico. Cenozoic VTT&T 42, 149–154.
Lucas, S.G., Dalman, S.G., 2016. The early Cretaceous Clayton Lake dinosaur tracksite, northeastern New Mexico. NMMNH&S 74, 127–140.
Lucas, S.G., Espinoza, E., Alvarado, G.E., 2024a. Acahualinca, Nicaragua, a scientifically significant site of ancient human footprints in the New World. In: Clary, R.M., Pyle, E.J., Andrews, W.M. (Eds.), Geology's Significant Sites and Their Contributions to Geoheritage. Geological Society, London, Special Publications, p. 543. https://doi.org/10.1144/SP543-2022-255.
Lucas, S.G., Estep, J.W., Hoffer, J.M. (Eds.), 1998a. Permian Stratigraphy and Paleontology of the Robledo Mountains, New Mexico. New Mexico Museum of Natural History and Science Bulletin, vol 12, p. 98.
Lucas, S.G., Heckert, A.B. (Eds.), 1995. Early Permian Footprints and Facies. New Mexico Museum of Natural History and Science Bulletin, vol 6, p. 301.
Lucas, S.G., Heckert, A.B., Estep, J.W., Hunt, A.P., Anderson, O.J., 1998b. Stratigraphy, paleontology and depositional environments of the Lower Permian Robledo mountains Formation of the Hueco group, Robledo mountains, New Mexico. NMMNH&S 12, 29–41.
Lucas, S.G., Houck, K.J., Hunt, A.P., Lamontagne, J.J., Minney, G.K., Taylor, L.H., 2024b. Martin G. Lockley (1950–2023): prime mover of vertebrate footprint studies. Ichnos. https://doi.org/10.1080/10420940.2024.2309493.
Lucas, S.G., Spielmann, J.A., Klein, H., Lerner, A.J., 2010. Ichnology of the Redonda formation. NMMNH&S 47, 75.
Lucas, S.G., Spielmann, J.A., Lockley, M.G. (Eds.), 2007b. Cenozoic Vertebrate Tracks and Traces. New Mexico Museum of Natural History and Science Bulletin, vol 42, p. 330.
Lucas, S.G., Stimson, M.R., King, O.A., Calder, J.H., Mansky, C.F., Hebert, B.L., Hunt, A.P., 2021. Carboniferous tetrapod footprint biostratigraphy, biochronology and evolutionary events. In: Lucas, S.G., Schneider, J.W., Wang, X., Nikolaeva, S. (Eds.), The Carboniferous Timescale. Geological Society, London, Special Publications, p. 512. https://doi.org/10.1144/SP512-2020-235.
Lucas, S.G., Voigt, S., Lerner, A.J., MacDonald, J.P., Spielmann, J.A., Celeskey, M.D., 2011. The Prehistoric Trackways National Monument, Permian of southern New Mexico, U.S.A. Ichnol. Newsl. 28, 10–14.
MacDonald, J.P., 1994. Earth's First Steps: Tracking Life before the Dinosaurs. Johnson Books, Boulder, p. 290.
MacDonald, J.P., 1995. History of the discovery of fossil footprints in southern New Mexico. NMMNH&S 6, 1–12.
Macfadyen, C.J., 2006a. Scottish fossil collecting Code. Palaeontol. Associ. Newsl. 62, 100.
Macfadyen, C.J., 2006b. Missing Birk Knowles fossils: SNH calls it a day. Earth Heritage 27, 8–9.
Macgregor, C., 2022. Preserving the ancient human trackways site in the Willandra lakes world heritage area. Stud. Conserv. 67 (Suppl. 1), 150–155. https://doi.org/10.1080/00393630.2022.2050616.
Machado, S., Mergulhão, L., Pereira, B.C., Pereira, P., Carvalho, J., Anacleto, J.A., Neto de Carvalho, C., Belo, J., Paredes, R., Baucon, A., 2021. Geoconservation in the Cabeço da Ladeira Paleontological Site (Serras de Aire e Candeeiros Nature Park, Portugal): Exquisite Preservation of Animals and Their Behavioral Activities in a Middle Jurassic Carbonate Tidal Flat. Geosciences 11, 366. https://doi.org/10.3390/geosciences11090366, 2021.
Madeira, J., Dias, R., 1983. Novas pistas de dinossáurios no Cretácico Inferior. Comun.. Geol. 69 (1), 147–158.
Mahboubi, M., Bessedik, M., Belkebir, L., Adaci, M., Hebib, H., Bensalah, M., Mammeri, A., Mansour, B., Mansouri, M.E.H., 2007. Premières découvertes d'empreintes de pas de dinosaures dans le Crétacé inférieur de la région d'El Bayadh (Algérie). BSGF 18 (2), 127–139.

Mampel, L., Cobos, A., Alcalá, L., Luque, L., Royo-Torres, R., 2009. An integrated system of heritage management applied to dinosaur sites in Teruel (Aragón, Spain). Geoheritage 1, 53–73. https://doi.org/10.1007/s12371-009-0005-x.

Mansky, C.F., Lucas, S.G., 2013. Romer's gap revisited: continental assemblages and ichno-assemblages from the basal Carboniferous of Blue beach, Nova Scotia, Canada. NMMNH&S 60, 244–273.

Mansky, C.F., Lucas, S.G., Spielmann, J.A., Hunt, A.P., 2012. Mississippian bromalites from Blue beach, Nova Scotia, Canada. NMMNH&S 57, 161–170.

Marchetti, L., 2024. Stratigraphy of the Early Permian Bromacker locality (Tambach Formation, Sakmarian, Germany). Permophiles 76, 33–35.

Marchetti, L., Francischini, H., Lucas, S.G., Voigt, S., Hunt, A.P., Santucci, V.L., 2020. Paleozoic vertebrate ichnology of Grand Canyon National Park. In: Santucci, V.L., Tweet, J.S. (Eds.), Grand Canyon National Park Centennial Paleontological Resource Inventory. Natural Resource Report NPS/GRCA/NRR—2020/2103, pp. 333–379.

Marchetti, L., Klein, H., Buchwitz, M., Ronchi, A., Smith, R.M.H., De Klerk, W.J., Sciscio, L., Groenewald, G.H., 2019. Permian-Triassic vertebrate footprints from South Africa: ichnotaxonomy, producers and biostratigraphy through two major faunal crises. Gondwana Res. 72, 139–168. https://doi.org/10.1016/j.gr.2019.03.009.

Marchetti, L., Mujal, E., Bernardi, M., 2017. An unusual *Amphisauropus* trackway and its implication for understanding seymouriamorph locomotion. Lethaia 50 (1), 162–174.

Marsicano, C.A., Wilson, J.A., Smith, R.M.H., 2014. A Temnospondyl trackway from the Early Mesozoic of western Gondwana and its implications for basal tetrapod locomotion. PLoS One 9 (8), e103255. https://doi.org/10.1371/journal.pone.0103255.

Martens, T., 2018. Scientific Importance of the Fossillagerstätte Bromacker (Germany, Tambach Formation, Lower Permian) – vertebrate fossils. Wissenschaftliche Bedeutung der Fossillagerstätte Bromacker (Deutschland, Tambach-Formation, Unteres Perm) – Wirbeltierfossilien. Cuvillier, Göttingen, p. 52.

Marty, D., Cavin, L., Hug, W.A., Jordan, P., Lockley, M.G., Meyer, C.A., 2004. The protection, conservation and sustainable use of the Courtedoux dinosaur tracksite, Canton Jura, Switzerland. Rev. Paléobiol. 9, 39–49.

Marty, D., Cavin, L., Hug, W.A., Meyer, C.A., Lockley, M.G., Iberg, B., 2003. Preliminary report of the new Courtedoux dinosaur tracksite from the Upper Kimmeridgian of Switzerland. Ichnos 10, 209–219.

Marzola, M., Dalla Vecchia, F.M., 2014. New dinosaur tracks from the Dolomia Principale (Upper Triassic) of the carnic Prealps (Friuli-Venezia Giulia, NE Italy). BSPI 53, 1–18.

Masao, F.T., Ichumbaki, E.B., Cherin, M., Barili, A., Boschian, G., Iurino, D.A., Menconero, S., Moggi-Cecchi, J., Manzi, G., 2016. New footprints from Laetoli (Tanzania) provide evidence for marked body size variation in early hominins. Elife 5, e19568. https://doi.org/10.7554/eLife.19568.

Masrour, M., Boutakiout, M., Ceniceros, J.M., Rolland, T., Pérez-Lorente, F., 2023. Morphometric and biomorphic variability of theropod footprint characters from Imilchil tracksites (mid-? Late jurassic, central high Atlas, Morocco). J. Afr. Earth Sci. 207, 105048.

Masrour, M., Pascual-Arribas, C., De Ducla, M., Nieves, H.M., Pérez-Lorente, F., 2017. Anza palaeoichnological site. Late Cretaceous. Morocco. Part I. The first African pterosaur trackway (manus only). J. Afr. Earth Sci. 134, 766–775.

Mateus, O., Milàn, J., 2008. Ichnological evidence for giant ornithopod dinosaurs in the Upper Jurassic Lourinhã Formation, Portugal. Oryctos 8, 47–52.

Matsukawa, M., Shibata, K., 2015. Review of Japanese Cenozoic (Miocene-Modern) vertebrate tracks. Ichnos 22, 261–290. https//doi.org/10.1080/10420940.2015.1064407.

Matsukawa, M., Shibata, K., Kukihara, R., Koarai, K., Lockley, M.G., 2005. Review of Japanese dinosaur track localities: implications for ichnotaxonomy, paleogeography and stratigraphic correlation. Ichnos 12 (3), 201–222.

Matthews, J., McIlroy, D., 2019. On the adhesion of sediment to footwear and the implications for geoconservation. Geoheritage 11 (4), 1749–1756.

Mazin, J.-M., Billon-Bruyat, J.-P., Hantzpergue, P., Lafaurie, J.-P., 2003. Ichnological evidence for quadrupedal locomotion in pterodactyloid pterosaurs: trackways from the Late Jurassic of Crayssac (south western France). In: Buffetaut, E., Mazin, J.-M. (Eds.), Evolution and Paleobiology of Pterosaurs. Geological Society of London, London, pp. 283–296.

Mazin, J.-M., Hantzpergue, P., Olivier, N., 2017. The dinosaur tracksite of Plagne (early Tithonian, Late Jurassic; Jura Mountains, France): the longest known sauropod trackway. Geobios 50 (4), 279–301.

McCrea, R.T., 2000. Vertebrate Palaeoichnology of the Lower Cretaceous (lower Albian) Gates Formation of Alberta [M.S. Thesis]: Saskatoon, Saskatchewan. University of Saskatchewan, p. 184.
McCrea, R.T., 2003. Fossil tracks from Tumbler Ridge: a brief history of collaboration between amateurs and academics. In: Seventh Annual Symposium of Alberta Palaeontological Society, pp. 41–48.
McCrea, R.T., Buckley, L.G., Lockley, M.G., Xing, L., Matthews, N., Pigeon, T., Amos, L., 2017. The "six peaks dinosaur track site": a diverse fossil vertebrate track site from the Gaylard member of the Gething Formation (Early Cretaceous: Aptian) from the Carbon Creek Basin, northeastern British Columbia. CSVP 2017, 31–35.
McCrea, R.T., Buckley, L.G., Plint, A.G., Currie, P.J., Haggart, J.W., Helm, C.W., Pemberton, S.G., 2014. A review of vertebrate track–bearing formations from the Mesozoic and earliest Cenozoic of western Canada with a description of a new theropod ichnospecies and reassignment of an avian ichnogenus. In: Lockley, M.G., Lucas, S.G. (Eds.), Fossil Footprints of Western North America, New Mexico Museum of Natural History and Sciences Bulletin, vol 62, pp. 5–93.
McLearn, F.H., 1923. Peace River Canyon coal area, BC. In: Geological Survey of Canada Summary Report, Part B, pp. 1–46.
McLearn, F.H., 1931. Environment of dinosaur tracks in the Peace River Canyon. Bull. Geol. Soc. Am. 42, 362.
Mead, J.I., Tweet, J.S., Santucci, V.L., Tobin, B., Chambers, C.L., Thomas, S.C., Carpenter, M.C., 2020. Pleistocene/Holocene cave fossils from Grand Canyon National Park: ice age (Pleistocene) flora, fauna, environments, and climate of the Grand Canyon, Arizona. In: Santucci, V.L., Tweet, J.S. (Eds.), Grand Canyon National Park: Centennial Paleontological Resource Inventory (Non-sensitive Version). Fort Collins, National Park Service, Natural Resource Report NPS/GRCA/NRR—2020/2103, pp. 403–463.
Meldrum, D.J., Lockley, M.G., Lucas, S.G., Musiba, C., 2011. Ichnotaxonomy of the Laetoli trackways: the earliest hominin footprints. J. Afr. Earth Sci. 60, 1–12.
Mesa, V., Perea, D., 2015. First record of theropod and ornithopod tracks and detailed description of sauropod trackways from the Tacuarembó Formation (Late Jurassic-?Early Cretaceous) of Uruguay. Ichnos 22, 109–121.
Meyer, C.A., Belvedere, M., Englich, B., Lockley, M.G., 2021a. A reevaluation of the Late Jurassic dinosaur tracksite Barkhausen (Wiehengebirge, Northern Germany). Paläontol. Z. 95 (3), 537–558.
Meyer, C.A., Marty, D., Thüring, B., Thüring, S., Belvedere, M., 2021b. The Late Cretaceous dinosaur track record of Bolivia – review and perspective. J. S. Am. Earth Sci. 106, 102992.
Meyer, C.A., Thüring, B., 2003. Dinosaurs of Switzerland. Comptes Rendus Palevol 2 (1), 103–117.
Mezga, A., Bajraktarević, Z., 1999. Cenomanian dinosaur tracks on the islet of Fenoliga in southern Istria, Croatia. Cretac. Res. 20 (6), 735–746.
Mietto, P., Avanzini, M., Belvedere, M., Bernardi, M., Dalla Vecchia, F.M., Porchetti, S., Gianolla, P., Petti, F.M., 2020. Triassic tetrapod ichnofossils from Italy: the state of the art. In: Romano, M., Citton, P. (Eds.), Tetrapod Ichnology in Italy: The State of the Art, J. Mediterr. Earth Sci. 12, pp. 83–136.
Mietto, P., Avanzini, M., Rolandi, G., 2003. Human footprints in Pleistocene volcanic ash. Nature 422. https://doi.org/10.1038/422133a, 133–133.
Milner, A.R.C., Birthisel, T.A., Kirkland, J.A., Breithaupt, B.H., Matthews, N.A., Lockley, M.G., Santucci, V.L., et al., 2011. Tracking Early Jurassic dinosaurs across southwestern Utah and the Triassic–Jurassic transition. In: Field Trip Guide Book, 71st Annual Meeting of the Society of Vertebrate Paleontology, Paris Las Vegas, Las Vegas, Nevada, pp. 1–107.
Milner, A.R.C., Lockley, M.G., 2006. History, geology and paleontology: St. George Dinosaur Discovery Site at Johnson farm, Utah. In: Reynolds, R.E. (Ed.), Making Tracks across the Southwest: Zzyzx, California State University, Desert Studies Consortium and. LSA Associates Inc., pp. 35–48
Milner, A.R.C., Lockley, M.G., Johnson, S.B., 2006a. The story of the St. George Dinosaur Discovery Site at Johnson farms: an important new Lower Jurassic dinosaur tracksite from the Moenave Formation of southwestern Utah. NMMNH&S 37, 329–345.
Milner, A.R.C., Lockley, M.G., Kirkland, J.I., 2006b. A large collection of well-preserved theropod dinosaur swim tracks from the Lower Jurassic Moenave Formation, St. George, Utah. NMMNH&S 37, 315–328.
Moratalla, J.J., 1993. Restos indirectos de dinosaurios del registro español: Paleoicnología de la Cuenca de Cameros (Jurásico Superior–Cretácico Inferior) y Paleoología del Cretácico Superior. PhD. Thesis (not published). Universidad Autónoma de Madrid.
Moreno, K., 2008. Valoración y estado de conservación de huellas de vertebrados mesozoicos en Chile. In: I Simposio – Paleontología en Chile. Libro de Actas, pp. 13–17.

Moreno, K., Benton, M., 2005. Occurrence of sauropod dinosaur tracks in the Upper Jurassic of Chile (redescription of *Iguanodonichnus frenki*). J. S. Am. Earth Sci. 20, 253–257.

Moreno, K., Blanco, N., Tomlinson, A., 2004. New dinosaur footprints from the Upper Jurassic of northern Chile. Ameghiniana 41, 535–544.

Moreno, K., De Valais, S., Blanco, N., Tomlinson, A.J., Jacay, J., Calvo, J.O., 2012. Large theropod dinosaur footprint associations in western Gondwana: Behavioural and palaeogeographic implications. Acta Palaeontol. Pol. 57 (1), 73–83.

Moreno, K., Pino, M., 2002. Huellas de dinosaurios (Theropoda-Ornitopoda-Sauropoda) de la Formación Baños del Flaco, VI Región, Chile: paleoambiente y paleoetología. Rev. Geol. Chile 29, 191–206.

Morton, N., Hudson, J.D., 1995. Field guide to the Jurassic of the Isles of Raasay and Skye, Inner Hebrides, NW Scotland. In: Taylor, P.D. (Ed.), Field Geology of the British Jurassic. Geological Society, London, pp. 209–280.

Mossman, D.J., Bruning, R., Powell, H.P., 2003. Anatomy of a Jurassic theropod trackway from Ardley, Oxfordshire, U.K. Ichnos 10, 2–4. https://doi.org/10.1080/10420940390257941.

Musiba, C.H., Mabula, A., Selvaggio, M., Magori, C.C., 2008. Pliocene animal trackways at Laetoli: research and conservation potential. Ichnos 15, 166–178.

Nicosia, U., Loi, M., 2003. Triassic footprints from Lerici (La Spezia, northern Italy). Ichnos 10, 127–140. https://doi.org/10.1080/10420940390256203.

Nicosia, U., Marino, M., Mariotti, N., Muraro, C., Panigutti, S., Petti, F.M., Sacchi, E., 1999. The Late Cretaceous dinosaur tracksite near Altamura (Bari, southern Italy). Geol. Rom. 7, 237–247.

Nyakatura, J.A., Melo, K., Horvat, T., Karakasiliotis, K., Allen, V.R., Andikfar, A., Andrada, E., Arnold, P., Lauströer, J., Hutchinson, J.R., Fischer, M.S., Ijspeert, A.J., 2019. Reverse-engineering the locomotion of a stem amniote. Nature 565, 351–355.

Neto de Carvalho, C., 2009. Vertebrate tracksites from the Mid-Late Pleistocene eolianites of Portugal: the first record of elephant tracks in Europe. Geol. Q. 53 (4), 407–414.

Neto de Carvalho, C., Belaústegui, Z., Toscano, A., Muñiz, F., Belo, J., Galán, J.M., Gómez, P., Cáceres, L.M., Rodríguez-Vidal, J., Cunha, P.P., Cachão, M., Ruiz, F., Ramirez-Cruzado, S., Giles-Guzmán, F., Finlayson, G., Finlayson, S., Finlayson, C., 2021. First tracks of newborn straight-tusked elephants (*Palaeoloxodon antiquus*). Sci. Rep. 11, 17311. https://www.nature.com/articles/s41598-021-96754-1.

Neto de Carvalho, C., Figueiredo, S., Belo, J., 2016. Vertebrate tracks and trackways from the Pleistocene eolianites of SW Portugal. Comun. Geol. 103 (especial I), 101–116.

Neto de Carvalho, C., Muñiz, F., Cáceres, L.M., Rodríguez-Vidal, J., Medialdea, A., Del Val, M., Cunha, P.P., García, J.M., Giles-Guzmán, F., Carrión, J.S., Belaústegui, Z., Toscano, A., Gómez, P., Galán, J.M., Belo, J., Cachão, M., Ruiz, F., Ramirez-Cruzado, S., Finlayson, G., Finlayson, S., Finlayson, C., 2023. Neanderthal footprints in the "Matalascañas trampled surface" (SW Spain): new OSL dating and Mousterian lithic industry. Quat. Sci. Rev. 313, 108200.

Neto de Carvalho, C., Muñiz, F., Galán, J.M., Rodríguez-Vidal, J., Gómez, P., Toscano, A., Ruiz, F., Belo, J., Belaústegui, Z., Cáceres, L.M., Finlayson, G., Finlayson, S., Giles-Guzmán, F., Finlayson, C., 2020. First vertebrate tracks and palaeoenvironment in a MIS-5 context in the Doñana national park (Huelva, SW Spain). Quat. Sci. Rev. 243. https://doi.org/10.1016/j.quascirev.2020.106508.

Neto de Carvalho, C., Rodrigues, J.C., Canilho, S., Amado, S., 2011. Geopark Naturtejo, bajo los auspicios de la UNESCO: la construcción participativa de un destino geoturístico em Portugal. Tierra Tecnol. 40, 52–56.

Nikolova, V., Sinnyovsky, D., 2019. Geoparks in the legal framework of the EU countries. Tourism Manag. Perspect. 29, 141–147.

Nyborg, T.G., Santucci, V.L., 1999. The Death Valley National Park Paleontological Survey. National Park Service, Geological Resources Technical Report, p. 66. NPS/NRGRD/GRDTR-99/01.

Ogilvie, S., Glennie, K., Hopkins, C., 2000. Excursion guide 13: the Permo-Triassic sandstones of Morayshire, Scotland. Geol. Today 16, 185–190.

Onac, B.P., Viehmann, I., Lundberg, J., Lauritzen, S.-E., Stringer, C., Popiţ, V., 2005. U/Th ages constraining the Neanderthal footprint at Vartop Cave, Romania. Quat. Sci. Rev. 24 (10e11). https://doi.org/10.1016/j.quascirev.2004.12.001, 1151e1157.

Ostrom, J.H., 1972. Were some dinosaurs gregarious? Palaeogeogr. Palaeoclimatol. Palaeoecol. 11, 287–301.

Oukassou, M., Klein, H., Lagnaoui, A., Charrière, A., Saber, H., Gierliński, G.D., Lallensack, J.N., Hminna, A., Boumaalif, A., Oussou, A., Ouarhache, D., 2019. *Polyonyx*-like tracks from Middle-? Upper Jurassic red beds of

Morocco: implications for sauropod communities on southern margins of Tethys. Palaeogeogr. Palaeoclimatol. Palaeoecol. 536, 109394.

Oviatt, C.G., Madsen, D.B., Rhode, D., Davis, L.G., 2023. A critical assessment of claims that human footprints in the Lake Otero Basin, New Mexico, date to the last glacial maximum. Quat. Res. 111, 138–147. https://doi.org/10.1017/qua.2022.38.

Page, K.N., 2017. Integrating international, national and regional inventories – the reality: a case study from Devon, SW England, UK. In: De Wever, P., et al. (Eds.), Geoheritage Inventories: Challenges, Achievements and Perspectives, MNHN Paris.

Page, K.N., 2018. Fossils, Heritage and Conservation: managing demands on a precious resource. In: Reynard, E., Brilha, J. (Eds.), Geoheritage, first ed. Elsevier, Chennai, pp. 107–128.

Page, K.N., Wimbledon, W.A.P., 2009. The conservation of Jurassic heritage in the UK – a critical review of current practice and effectiveness. Vol Jurass. 6, 163–173.

Paik, I.S., Huh, M., Kim, H.J., 2004. Dinosaur egg-bearing deposits (Upper Cretaceous) of Boseong, Korea: occurrence, palaeoenvironments, taphonomy, and preservation. Palaeogeogr. Palaeoclimatol. Palaeoecol. 205, 155–168.

Panarello, A., Palombo, M.R., Biddittu, I., Di Vito, M.A., Farinaro, G., Mietto, P., 2020. On the Devil's tracks: unexpected news from the Foresta ichnosite (Roccamonfina volcano, central Italy). J. Quat. Sci. 35, 444–456. https://doi.org/10.1002/jqs.3186.

Panin, N., Avram, E., 1962. Nouvelles empreintes de vertebres dans le Miocene de la zone subcarpathique roumaine. Stud. Cercet. Geol. 7, 455–484.

Parkes, M.A., Morris, J.H., 1999. The Valentia Island tetrapod trackway. In: Barettino, D., Vallejo, M., Gallego, E. (Eds.), Towards the Balanced Management and Conservation of the Geological Heritage in the New Millennium. Sociedad Geológica de España, Madrid, pp. 65–68.

Parkes, M.A., 2008. A history of geoconservation in the Republic of Ireland. In: Burek, C.V., Prosser, C.D. (Eds.), The History of Geoconservation, vol 300. Geological Society, London, Special Publications, pp. 237–248. https://doi.org/10.1144/SP300.18.

Peabody, F.E., 1948. Reptile and Amphibian Trackways from the Moenkopi Formation of Arizona and Utah, vol 27. University of California Publications, Bulletin of the Department of Geological Sciences, pp. 295–468.

Percival, I.G., 2014. Protection and preservation of Australia's palaeontological heritage. Geoheritage 6, 205–216. https://doi.org/10.1007/s12371-014-0106-z.

Petti, F., Petruzzelli, M., Conti, J., Spalluto, L., Wagensommer, A., Lamendola, M., Francioso, R., Montrone, G., Sabato, L., Tropeano, M., 2018. The use of aerial and close-range photogrammetry in the study of dinosaur tracksites: Lower Cretaceous (upper Aptian/lower Albian) Molfetta ichnosite (Apulia, southern Italy). Palaeontol. Electron. https://doi.org/10.26879/845.

Petti, F.M., Antonelli, M., Sacco, E., Conti, J., Petruzzelli, M., Spalluto, L., Cardia, S., Festa, V., La Perna, R., Marino, M., Marsico, A., Sabato, L., Tropeano, M., Barracane, G., Montrone, G., Piscitelli, A., Francescangeli, R., 2022. Geothematic map of the Altamura dinosaur tracksite (early Campanian, Apulia, southern Italy). Geol. Field Trips 14, 1–19. https://doi.org/10.3301/GFT.2022.01.

Petti, F.M., D'Orazi Porchetti, S., Conti, M.A., Nicosia, U., Perugini, G., Sacchi, E., 2008. Theropod and sauropod footprints in the Early Cretaceous (Aptian) Apenninic carbonate platform (Esperia, Lazio, Central Italy): a further constraint on the palaeogeography of the central-Mediterranean are. Studi trent. Acta Geol. 83, 323–334.

Pichardo, R.S., 2016. Las huellas de dinosaurios en México y su estudio. Rev. Digit. Univ. 16 (11). http://www.revista.unam.mx/vol.16/num11/art87/.

Pieńkowski, G., 2004b. Sołtyków – unikalny zapis paleoekologiczny wczesnojurajskich środowisk kontynentalnych. Tomy Jurajskie 2, 1–16.

Piñuela, L., 2016. Huellas de dinosaurios y de otros reptiles del Jurásico Superior de Asturias. In: Tesis Doctoral Inédita. Universidad de Oviedo.

Plint, A.G., Charles, W., Helm, C.W., Lockley, M.G., 2022. Crocodylian and Dinosaur Trace Fossil Assemblages from Crevasse Splay/levee and Floodplain Lake Environments: Middle Cenomanian Dunvegan Formation, Northeast British Columbia, Canada. Historical Biology. https://doi.org/10.1080/08912963.2022.2043294.

Poropat, S.F., White, M.A., Ziegler, T., Pentland, A.H., Rigby, S.L., Duncan, R.J., Sloan, T., Elliott, D.A., 2021. A diverse Late Cretaceous vertebrate tracksite from the Winton Formation of Queensland, Australia. PeerJ 9, e11544. https://doi.org/10.7717/peerj.11544.

ProGeo, 2017. Geodiversity, Geoheritage & Geoconservation, the ProGEO Simple Guide, 2017. https://www.iucn.org/sites/dev/files/progeo_leaflet_en_2017.pdf.

Rachal, D.M., Mead, J.I., Dello-Russo, R., Cuba, M.T., 2022. Deep-water delivery model of *Ruppia* seeds to a near-shore/terrestrial setting and its chronological implications for late Pleistocene footprints, Tularosa Basin, New Mexico. Geoarchaeology 37 (6), 923–933. https://doi.org/10.1002/gea.21937.

Radley, J.D., Allen, P., 2012. The non-marine Lower Cretaceous Wealden strata of southern England. Proc. Geol. Assoc. 123 (2), 235–244.

Razzolini, R., 2016. Morphological variation and ichnotaxonomy of dinosaur tracks. Linking footprint shapes to substrate and trackmaker's anatomy and locomotion. In: Tesis Doctoral Inédita. Universitat Autònoma de Barcelona.

Razzolini, N.L., Oms, O., Castanera, D., Vila, B., Santos, V.F., Galobart, A., 2016. Ichnological evidence of megalosaurid dinosaurs crossing Middle Jurassic tidal flats. Sci. Rep. 6, 31494. https://doi.org/10.1038/srep31494.

Remeika, P., 1999. Identification, Stratigraphy, and Age of Neogene Vertebrate Footprints from the Vallecito-Fish Creek Basin, Anza-Borrego Desert State Park, California, vol 46. San Bernadino County Museum Association Quarterly, pp. 37–45.

Remeika, P., 2001. The Fish Creek Canyon ichnofauna: a Pliocene (Blancan) vertebrate footprint assemblage from Anza-Borrego Desert State park, California. In: Santucci, V.L., McClelland, L. (Eds.), Proceedings of the 6th Fossil Resource Conference. Geological Resources Division, Technical Report, NPS/NRGRD/GRDTR-01/01, pp. 55–75.

Remeika, P., 2006. Fossil footprints of Anza-Borrego. In: Jefferson, G.T., Lindsay, L. (Eds.), Fossil Treasures of the Anza-Borrego Desert: The Last Seven Million Years. Sunbelt Publications, El Cajon, pp. 311–327.

Renesto, S., Kustatscher, E., Gianolla, P., 2020. A putative juvenile specimen of *Eusaurosphargis dalsassoi* from the Anisian (Middle Triassic) of Piz Da Peres (Dolomites, Northern Italy). Riv. Ital. Paleontol. Stratigr. 126, 277–288.

Richter, A., Böhme, A., 2016. Too many tracks: preliminary description and interpretation of the diverse and heavily dinoturbated Early Cretaceous "chicken yard" ichnoassemblage (Obernkirchen tracksite, northern Germany). In: Falkingham, P.L., Marty, D., Richter, A. (Eds.), Dinosaur Tracks. The Next Steps. Indiana University Press, Indiana, pp. 334–357.

Roach, N.T., Hatala, K.G., Ostrofsky, K.R., Villmoare, B., Reeves, J.S., Du, A., Braun, D.R., Harris, J.W., Behrensmeyer, A.K., Richmond, B.G., 2016. Pleistocene footprints show intensive use of lake margin habitats by *Homo erectus* groups. Sci. Rep. 6, 26374.

Roberts, D.L., 2008. Last Interglacial hominid and associated vertebrate fossil trackways in coastal eolianites, South Africa. Ichnos 15 (3), 190–207. https://doi.org/10.1080/10420940802470482.

Roberts, G., 2009. Ephemeral, subfossil mammalian, avian and hominid footprints within Flandrian sediment exposures at Formby point, Sefton coast, north west England. Ichnos 16, 33–48. https://doi.org/10.1080/10420940802470730.

Roberts, G., 2014. The Lost World of Formby Point. The Alt Press, Liverpool.

Rodríguez-de la Rosa, R.A., Aguillón-Martínez, M.C., López-Espinoza, J., Eberth, D.A., 2004. The fossil record of vertebrate tracks in Mexico. Ichnos 11 (1–2), 27–37.

Romano, M., Citton, P., Nicosia, U., 2016. Corroborating trackmaker identification through footprint functional analysis: the case study of *Ichniotherium* and *Dimetropus*. Lehtaia 49 (1), 102–116.

Romilio, A., Salisbury, S.W., 2011. A reassessment of large theropod dinosaur tracks from the mid-Cretaceous (late Albian–Cenomanian) Winton Formation of Lark Quarry, central-western Queensland, Australia: a case for mistaken identity. Cretac. Res. 32 (2), 135–142.

Romilio, A., Tucker, R.T., Salisbury, S.W., 2013. Reevaluation of the Lark quarry dinosaur tracksite (late Albian–Cenomanian Winton formation, central-western Queensland, Australia): no longer a stampede? J. Vertebr. Paleontol. 33 (1), 102–120.

Ronchi, A., Marchetti, L., Klein, H., Groenewald, G.H., 2023. A Middle Permian oasis for vertebrate and invertebrate life in a high-energy fluvial palaeoecosystem of Southern Gondwana (Karoo, Republic of South Africa). Geosciences 13 (11), 325. https://doi.org/10.3390/geosciences13110325.

Sacchi, E., Conti, M.A., D'Orazi Porchetti, S., Logoluso, A., Nicosia, U., Perugini, G., Petti, F.M., 2009. Aptian dinosaur footprints from the Apulian platform (Bisceglie, Southern Italy) in the framework of periadriatic ichnosites. Palaeogeogr. Palaeoclimatol. Palaeoecol. 271, 104–116. https://doi.org/10.1016/j.palaeo.2008.09.018.

Salisbury, S.W., Romilio, A., Herne, M.C., Tucker, R.T., Nair, J.P., 2016. The Dinosaurian ichnofauna of the Lower Cretaceous (Valanginian–Barremian) Broome sandstone of the Walmadany area (James Price point), Dampier peninsula, western Australia. J. Vertebr. Paleontol. 36 (1), 1–152. https://doi.org/10.1080/02724634.2016.1269539.

Santos, V.F., Barriga, F., Galopim de Carvalho, A.M., Mergulhão, L.M., Monteiro, C., Silveira, S.C., Fernandes, M., Alho, J.M., Coelho, R., 2010. Dinosaur track sites from Portugal and its significance as natural heritage. Schrift. Dtsch. Geol. Ges. 66, 124–125.

Santos, V.F., Caetano, P., Pólvora, A., 2016. Sesimbra Geocircuit and dinosaur tracksites. Comun. Geol. 103 (Especial I), 153–158.

Santos, V.F., Callapez, P.M., Rodrigues, N.P.C., 2013. Dinosaur footprints from the Lower Cretaceous of the Algarve basin (Portugal): new data on the ornithopod palaeoecology and palaeobiogeography of the Iberian peninsula. Cretac. Res. 40, 158–169. https://doi.org/10.1016/j.cretres.2012.07.001.

Santos, V.F., da Silva, C.M., Rodrigues, L.A., 2008. Dinosaur track sites from Portugal: scientific and cultural significance. Oryctos 8, 77–88.

Santos, V.F., Lockley, M.G., Meyer, C.A., Carvalho, J., Galopim de Carvalho, A.M., Moratalla, J.J., 1994. A new sauropod tracksite from the Middle Jurassic of Portugal. Gaia 10, 5–13.

Santos, V.F., Neto de Carvalho, C., 2016. Gomes (1915-1916) to Ichnia 2016 – one hundred years of vertebrate ichnology in Portugal. In: Baucon, A., Neto de Carvalho, C., Rodrigues, J. (Eds.), Ichnia 2016: Abstract Book. UNESCO Geopark Naturtejo/International Ichnological Association, Idanha-A-Nova, pp. 284–285.

Santucci, V.L., Hunt, A.P., Lockley, M.G., 1998. Fossil vertebrate tracks in National Park Service areas. Dakoterra 5, 107–114.

Santucci, V.L., Hunt, A.P., Nyborg, T., Kenworthy, J.P., 2006. Additional fossil vertebrate tracks in National Park Service areas. America's antiquities—100 years of managing fossils on federal lands. NMMNH&S 34, 152–158.

Santucci, V.L., Nyborg, T.G., Buchheim, P.H., Nick, K.E., 2011. Vertebrate Paleontological Resources of Death Valley National Park, California: Field Trip Guide Book – 71st Annual Meeting of the Society of Vertebrate Paleontology, pp. 127–155. Nevada State Museum – Paleontological Papers Number 1.

Santucci, V.L., Tweet, J., Bustos, D., Nyborg, T., Hunt, A.P., 2014. An inventory of Cenozoic fossil vertebrate tracks and burrows in National Park Service areas. In: Lockley, M.G., Lucas, S.G. (Eds.), Fossil Footprints of Western North America. New Mexico Museum of Natural History and Science Bulletin, vol 62, pp. 469–488.

Schmincke, H.U., Kutterolf, S., Perez, W., Rausch, J., Freundt, A., Strauch, W., 2009. Walking through volcanic mud: the 2100 year old Acahualinca footprints (Nicaragua) I: stratigraphy, lithology, volcanology and age of the Acahualinca section. Bull. Volcanol. 71, 479–493.

Schmincke, H., Kutterolf, S., Wendy Perez, W., Rausch, J., Freundt, A., Strauch, W., 2010. Walking through volcanic mud: the 2,100-year-old Acahualinca footprints (Nicaragua) II: The Acahualinca people, environmental conditions and motivation. Int. J. Earth Sci. 99, 279–292.

Schulp, A.S., Al-Wosabi, M., Stevens 3, N.J., 2008. First dinosaur tracks from the Arabian peninsula. PLoS One 3 (5), e2243. https://doi.org/10.1371/journal.pone.0002243.

Schumacher, B.A., Lockley, M.G., 2023. North America's largest assemblage of Late Jurassic dinosaur trackways, a new look at the Purgatoire Valley dinosaur tracksite (Dinosaur Lake), Morrison Formation, southeastern Colorado, United States. Hist. Biol. 35 (7), 1146–1161.

Schüppel, K., Wrede, V., 2022. Nationaler GeoPark Ruhrgebiet. Springer, Berlin, p. 215.

Scottish Natural Heritage, 1996. Scotland's hills and mountains: a concordat on access. SHN, Inverness.

Scrivner, P.J., 1984. Stratigraphy, Sedimentology and Vertebrate Ichnology of the Copper Canyon Formation (Neogene), Death Valley National Monument [MS Thesis]. University of Southern California, Los Angeles, p. 334.

Scrivner, P.J., Bottjer, D.J., 1986. Neogene avian and mammalian tracks from Death Valley National Monument, California: their context, classification and preservation. Palaeogeogr. Palaeoclimatol. Palaeoecol. 57, 285–331.

Sennikov, A.G., 2021. The plantigrade Segnosaurians: sloth Dinosaurs or bear Dinosaurs? Paleontol. J. 55 (10), 1158–1185.

Silva, R.C., Carvalho, I.S., Fernandes, A.C.S., 2008a. Pegadas de dinossauros do Triássico (Formação Santa Maria) do Brasil. Ameghiniana 45 (4), 783–790.

Silva, R.C., Ferigolo, F., Carvalho, I.S., Fernandes, A.C.S., 2008b. Lacertoid footprints from the upper Triassic (Santa Maria Formation) of southern Brazil. Palaeogeogr. Palaeoclimatol. Palaeoecol. 262 (3–4), 140–156.

Silva, R.C., Carvalho, I.S., Fernandes, A.C.S., Ferigolo, J., 2008c. Pegadas teromorfóides do Triássico Superior (Formação Santa Maria) do Sul do Brasil. Rev. Bras. Geociencias 38 (1), 100–115.

Singtuen, V., Anumat, A., Chartramthitikun, N., et al., 2023. Evaluation and geoconservation of dinosaur footprint paleontological heritage at the Khon Kaen National Geopark in northeastern Thailand. Geoheritage 15, 21. https://doi.org/10.1007/s12371-023-00794-0.

Singtuen, V., Vivitkul, N., Junjuer, T., 2022. Geoeducational assessments in Khon Kaen National Geopark, Thailand: implication for geoconservation and geotourism development. Heliyon 8 (12), e12464. https://doi.org/10.1016/j.heliyon.2022.e12464.

Siqueira, L.M.P., Polck, M.A.R., Hauch, A.C.G., Silva, C.A., Chaves, F.B., Yamamoto, I.T., Araujo, J.P., Andrade, J.A.F.G., Filgueira, B.T.M., Trindade, M.H.P.A., Machado, R.R., Santucci, R.M., 2011. Paleontological sites at Rio do Peixe Basins (Northeastern Brazil): Georeferencing, vulnerability diagnosis and protection measures. Anuário do Instituto de Geociências – UFRJ 34 (1), 9–21.

Smith, R.M.H., 1993. Sedimentology and ichnology of floodplain paleosurfaces in the Beaufort group (Late Permian), Karoo sequence, South Africa. Palaios 8, 339–357. https://doi.org/10.2307/3515265.

Smith, R.M.H., Marsicano, C.A., Wilson, J.A., 2009. Sedimentology and paleoecology of a diverse Early Jurassic tetrapod tracksite in Lesotho, southern Africa. Palaios 24, 672–684. https://doi.org/10.2110/palo.2008.p08-115r.

Song, J.Y., 2010. Reinterpretation of unusual Uhangri dinosaur tracks from the view of functional morphology. J. Paleontol. Soc. Korea 26 (2), 95–105.

Spivak, D.N., Wetzel, D.E., Caillaiu, J., 2006. Historical resources act designation of the Grande Cache dinosaur tracksite (lower Cretaceous, Albian), Grande Cache, Alberta. NMMNH&S 34, 36–37.

Sternberg, C.M., 1931. Lower Cretaceous dinosaur tracks in Peace River Canyon, British Columbia. Bull. Geol. Soc. Am. 42, 362.

Sternberg, C.M., 1932. Dinosaur tracks from Peace River, British Columbia. Nat. Museum Canada Annu. Rep. 1930, 59–85.

Stössel, I., 1995. The discovery of a new Devonian tetrapod trackway in SW Ireland. J. Geol. Soc. London 152, 407–413.

Su, X., Sigley, G.G., Song, C., 2020. Relational authenticity and reconstructed heritage space: a balance of heritage preservation, tourism, and urban renewal in Luoyang Silk Road Dingding Gate. Sustainability 12, 5830. https://doi.org/10.3390/su12145830.

Szewczyk, L., Vennin, E., Moreau, J.-D., Gand, G., Verolet, M., et al., 2020. Tracking dinosaurs in coarse-grained sediments from the Upper Triassic of Ardèche (southeastern France). Palaios 35 (11), 447–460. https://doi.org/10.2110/palo.2020.025ff.

Thulborn, T., Wade, M., 2007. Dinosaur stampede in the Cretaceous of Queensland. Lethaia 12 (3), 275–279.

Todesco, R., Wachtler, M., Dellantonio, E., Avanzini, M., 2008. First report on a new late Anisian (Illyrian) vertebrate tracksite from the Dolomites (Northern Italy). Studi Trent. Sci. Nat., Acta Geol. 83, 247–252.

Tomaselli, M.B., David, L.D.O., Riga, B.J.G., Coria, J.P., Mercado, C.R., Guerra, M., Tiviroli, G.S., 2021. New titanosaurian sauropod tracks with exceptionally well-preserved claw impressions from the Upper Cretaceous of Argentina. Cretac. Res. 129, 104990. https://doi.org/10.1016/j.cretres.2021.104990.

Torcida Fernández-Baldor, F., coord, 2006. Huellas que perduran. Icnitas de Dinosaurios: Patrimonio y Recurso. Fundación del Patrimonio Histórico de Castilla y León, Valladolid, p. 457.

Troiano, L.P., Santos, L.B., Aureliano, T., Ghilardi, A.M., 2024. A remarkable assemblage of petroglyphs and dinosaur footprints in Northeast Brazil. Sci. Rep. 14, 6528. https://doi.org/10.1038/s41598-024-56479-3.

UNESCO, 2025. UNESCO Global Geoparks. https://www.unesco.org/en/iggp/geoparks/about.

Van Allen, H.E.K., Calder, J.H., Hunt, A.P., 2005. The trackway record of a tetrapod community in a walchian conifer forest from the Permo-Carboniferous of Nova Scotia. In: Lucas, S.G., Zeigler, K.E. (Eds.), The Nonmarine Permian. New Mexico Museum of Natural History and Science Bulletin, vol 30, pp. 322–332.

Voigt, S., 2002. Zur Geschichte der Tetrapodenfährtenfunde in den Sandsteinbrüchen bei Tambach-Dietharz (1887-1908). Abh. Ber. Mus. Nat. Gotha 22, 47–58.

Voigt, S., Berman, D.S., Henrici, A.C., 2007. First well-established track-trackmaker association of Paleozoic tetrapods based on *Ichniotherium* trackways and diadectid skeletons from the Lower Permian of Germany. J. Vertebr. Paleontol. 27 (3), 553–570.

Voigt, S., Ganzelewski, M., Zewe, G., 2021. Deutschlands älteste Vierfüßerfährte – Neues zum Bochumer Sensationsfund. GeoPark Ruhrgebiet Geoparknews 1/2021, 4–7.

Voigt, S., Lucas, S.G., Krainer, K., 2013. Coastal-plain origin of trace-fossil bearing red beds in the Early Permian of southern New Mexico, U.S.A. Palaeogeogr. Palaeoclimatol. Palaeoecol. 369, 323–334.

von Koenigswald, W., Walders, M., Sander, M., 1995. Jungpleistozäne Tierfährten aus der Emscher-Niederterrasse von Bottrop-Welheim. Munchner Geowissenschaftliche Abdhandlungen (A)27, 5–50.

von Koenigswald, W., Sander, P.M., Walders, M., 1996. The upper Pleistocene tracksite Bottrop-Welheim (Germany). Acta Zool. Cracov. 39 (1), 235–244.

Wagensommer, A., Latiano, M., Mocke, H.B., Porchetti, S., 2016. Dinosaur diversity in an Early Jurassic African desert: the significance of the Etjo Sandstone ichnofauna at the Otjihaenamaparero locality (Namibia). N. Jb. Geol. Paläont. Abh. 281 (2), 155–182.

Wang, L.L., Zhu, Q.F., 2023. Characteristics and geological significance of geoheritage in the Arxan Global Geopark, Inner Mongolia. Geoscience 37 (4), 1075–1088.

Webb, D., Robu, M., Moldovan, O., Constantin, S., Tomus, B., Neag, I., 2014. Ancient human footprints in Ciur-Izbuc Cave, Romania. Am. J. Biol. Anthropol. 155 (1), 128–135.

Way Act, 2002. Countryside and Rights of Way Act 2002. Croner-i.

Webb, S., Cupper, M.L., Robins, R., 2006. Pleistocene human footprints from the Willandra Lakes, southeastern Australia. J. Hum. Evol. 50 (4), 405–413.

Westaway, M.C., Cupper, M.L., Johnston, H., Graham, I., 2013. The Willandra Fossil Trackway: Assessment of ground penetrating radar survey results and additional OSL dating at a unique Australian site. Aust. Archaeol. 76, 84–89.

Whyte, M.A., Romano, M., Watts, W., 2010. Yorkshire dinosaurs: a history in two parts. In: Moody, R.T.J., Buffetaut, E., Naish, D., Martill, D.M. (Eds.), Dinosaurs and Other Extinct Saurians: A Historical Perspective. Geological Society, London, Special Publications, vol 343, pp. 189–207. https://doi.org/10.1144/SP343.11.

Wilson, J.A., Marsicano, C.A., Smith, R.M.H., 2009. Dynamic locomotor capabilities revealed by early dinosaur trackmakers from southern Africa. PLoS One 4 (10), e7331. https://doi.org/10.1371/journal.pone.0007331.

Wings, O., Falk, D., Knötschke, N., Richter, A., 2012. Excursion Guide B1: The Early Cretaceous Dinosaur Trackways in Münchehagen (Lower Saxony, Germany) – The Natural Monument 'Saurierfährten Münchehagen' and the adjacent Wesling Quarry. In: Richter, A., Reich, M. (Eds.), Dinosaur Tracks 2011. An International Symposium, Obernkirchen, April 14-17, 2011. Abstract Volume and Field Guide to Excursions. Universitätsdrucke Göttingen, Göttingen, pp. 113–142.

Wimbledon, W.A.P., Benton, M.J., Bevins, R.E., Black, G.P., Bridgland, D.R., Cleal, C.J., Cooper, R.G., May, V.J., 1995. The development of a methodology for the selection of British sites for conservation. Part 1. Mod. Geol. 20, 159–202.

Wiseman, A.L.A., Vicari, D., Belvedere, M., de Groote, I., 2022. Neolithic track sites from Formby Point, England: new data and insights. J. Archaeol. Sci. 44, 103546.

Yoo, J.H., Lee, C.H., 2015. Materials and microclimatic characteristics of dinosaur footprint sites in Uiseong and Haenam Area, Korea. Indian J. Sci. Technol. 8 (1), 1–7. https://doi.org/10.17485/ijst/2015/v8iS1/60774.

Zhang, J., 2021. Comparative analysis of Chinese national geopark and UNESCO Global Geopark. Geol. Rev. 67 (2), 459–466.

Zimmer, B., Liutkus-Pierce, C., Marshall, S.T., Hatala, K.G., Metallo, A., Rossi, V., 2018. Using differential structure-from-motion photogrammetry to quantify erosion at the Engare Sero footprint site, Tanzania. Quat. Sci. Rev. 198, 226–241.

Index

'*Note:* Page numbers followed by "f" indicate figures and "t" indicate tables.'

C

D

G

H